The Healthcare Library of Northern Ireland
and
Queen's University Biomedical Library

Tel: 028 9097 2710
E-mail: BiomedicalLibrary@qub.ac.uk

For due dates and renewals:
QUB borrowers see 'MY ACCOUNT' at
www.qub.ac.uk/lib

HSC borrowers see 'MY ACCOUNT' at
www.healthcarelibrary.qub.ac.uk

This book must be returned not later
than its due date but may be recalled
earlier if in demand.

Fines are imposed on overdue books.

Fines are imposed on overdue books

Vogel · Motulsky

HUMAN GENETICS

Problems and Approaches

Third Edition

Springer

Berlin
Heidelberg
New York
Barcelona
Budapest
Hong Kong
London
Milan
Paris
Santa Clara
Singapore
Tokyo

F. Vogel   A.G. Motulsky

HUMAN GENETICS

Problems and Approaches

Third, Completely Revised Edition

With 439 Figures and 205 Tables

 Springer

Emeritus Professor Dr. Dr. h.c. Friedrich Vogel
Institut für Humangenetik und Anthropologie
Im Neuenheimer Feld 328, 69120 Heidelberg, Germany

Arno G. Motulsky, M.D.
Emeritus Professor of Medicine and Genetics
University of Washington
Division of Medical Genetics, School of Medicine
Seattle, WA 98195, USA

Library of Congress Cataloging-in-Publication Data
Vogel, Friedrich, 1925 – . Human genetics: problems and approaches / F. Vogel, A. G. Motulsky. – 3rd completely rev. ed. p. cm. Includes bibliographical references and index.
ISBN 3-540-60290-9 (hc)
1. Human genetics. I. Motulsky, Arno G., 1923–. II. Title.   QH431.V59   1996   573.2'1–dc20
95-52392 CIP

ISBN 3-540-60290-9 Springer-Verlag Berlin Heidelberg New York

ISBN 3-540-16411-1 2. Auflage Springer-Verlag Berlin Heidelberg New York Tokyo
ISBN 0-387-16411-1 2nd edition Springer-Verlag New York Heidelberg Berlin Tokyo

ISBN 3-540-09459-8 1. Auflage Springer-Verlag Berlin Heidelberg New York
ISBN 0-387-09459-8 1st edition Springer-Verlag New York Heidelberg Berlin

Book production: PRO EDIT GmbH, 69216 Heidelberg, Germany
Coverdesign: Struve & Partner, 69126 Heidelberg, Germany
Typesetting: Appl, 86650 Wemding, Germany
Printing and Binding: Stürtz AG, 97080 Würzburg, Germany

SPIN: 10080248   27/3136 – 5 4 3 2 1 0 – Printed on acid-free paper

To our wives and children

Preface to the Third Edition

The first two editions of this book, published in 1979 and in 1986, were well received by the scientific community. Translations into Italian, Japanese, and Russian suggest that this book was regarded useful in many parts of the world. Meanwhile, human genetics has seen dramatic developments, and the "molecular revolution" has attracted thousands of scientists, including many molecular biologists, to this field. About 3700 human genes have already been mapped to chromosomal sites. Many such genes have been cloned, and the various mutations causing disease have been identified. Novel mutational mechanisms such as expanded trinucleotide repeats have been discovered in conditions such as Huntington's disease and the fragile X syndrome of mental retardation. Gene action now can often be elucidated by studying the pathway from gene to phenotype following positional cloning rather than working in the opposite direction, as was customarily done before the tools of "new genetics" were available. In an increasing number of genetic diseases, the pathogenic mechanisms have been elucidated with positive consequences for prevention and treatment. It therefore became necessary to rewrite almost completely major portions of this book.

These developments are now making genetics arguably the leading basic science for medicine, as well as a recognized medical speciality. But all these changes do not mean that the entire framework of human genetics had to be reconstructed. What we have witnessed during the two recent decades has not been what T. Kuhn would have called a "paradigm clash." On the contrary – the existing genetic theory has proved to be so flexible that it could integrate all the new results, including novel mechanisms of nonmendelian inheritance (such as genomic imprinting), without running into major problems. We, therefore, were able to combine old and new concepts and results so as to present a comprehensive picture of modern human genetics.

The two of us have seen – and have actively participated in – these fascinating scientific developments over more than 40 years. This experience has been an unusual intellectual privilege and has helped us to stress continuity where younger colleagues might have seen a completely new beginning. The attempt by two authors to deal with the entire field of human genetics might be considered audacious or even foolhardy. Yet, we have undertaken this arduous task to articulate a cohesive exposition of our field.

Many colleagues helped by discussing parts of this book and making useful suggestions. In the following, we shall mention some names but we are sure that these are not all who have helped us by their criticism and suggestions: W. Buselmaier, M. and T. Cremer, C. Epstein, Ch. Fischer, U. Francke, S. Gartler, W. Hager, A. Jauch, J. Krüger, P. Propping, G. Rappold, B. Royer-Pokora, T. M. Schroeder-Kurth, O. Steinlein, P. Vogt. We owe a particular note of thanks to Dr. Peter Newell (University of Washington) for his meticulous reading of the 2nd edition of this book and his many suggestions, which helped us to clarify various statements and concepts.

After discussion with the publisher – mainly R. Lange – we also changed some aspects of the layout and organization. Subdivision of the material into many more chapters and citing the literature immediately after each chapter (and not at the end of the book) should facilitate reading. The new figures were drawn by Edda

Schalt und Christine Schreiber. T. Barton did an excellent job of copy editing. F. Vogel thanks the Center for Advanced Study in the Behavioral Sciences, Stanford, for extending a second invitation to him for the academic year 1993/94. His stay was supported by a grant from the Volkswagen Foundation. Both authors wish to thank the Rector of the Institute for Advanced Studies (Wissenschaftskolleg) in Berlin for making it possible to spend some time together there for planning and working on this edition.

The authors again would like to thank Springer-Verlag – especially D. Götze, B. Lewerich, and R. Lange – for publishing this volume.

Heidelberg, Seattle, Spring 1996   Friedrich Vogel
                                                                Arno G. Motulsky

Preface to the First Edition

Human genetics provides a theoretical framework for understanding the biology of the human species. It is a rapidly growing branch of science. New insights into the biochemical basis of heredity and the development of human cytogenetics in the 1950s heightened interest in this field. The number of research workers and clinicians who define themselves as full-time or part-time human and medical geneticists has increased sharply, and detailed well-founded knowledge has augmented exponentially. Many scientists and physicians are confronted with genetic problems and use concepts and methodology of human genetics in research and diagnosis. Methods developed in many different fields of the biologic, chemical, medical, and statistical sciences are being utilized toward the solution of genetic problems. The increasing number and sophistication of well-defined and elegantly solved problems helps to refine an extensive framework of genetic theory. These new conceptual insights in their turn lead to solutions of new questions. To mention only one example, the structure of hemoglobin genes has been elucidated using methods derived from protein chemistry and DNA technology. It is an exciting experience to participate in these developments!

Moreover, scientific progress in genetics has practical implications for human well-being. Improved knowledge of the genetic cause of an increasing number of human diseases helps to refine diagnosis, to find new therapeutic approaches, and above all, to prevent genetic diseases. So far, human genetics has had less of an impact on the behavioral and social sciences. It is possible that genetic differences involved in shaping personality structure, cognitive faculties, and possibly human social behavior may be at least as important as genetic variation affecting health and disease. The data, however, are less clear and more controversial. These problems are discussed in detail in the text. The rapid progress of human genetics in recent decades has attracted – and is still attracting – an increasing number of students and scientists from other fields. Various elementary textbooks, more advanced monographs of various branches of the field, and the original journal literature are the usual sources of introduction to human genetics. What seems to be lacking, however, is a fairly thorough and up-to-date treatise on the conceptual basis of the entire field of human genetics and its practical applications. Often, the absence of a broadly based background in the field leads to misunderstanding of its scope, unclear goals for research, improper selection of methods, and imbalanced theoretical discussions. Human genetics is based on a powerful theory, but this implicit conceptual foundation should be made explicit. This goal is the purpose of this book. It certainly is a formidable and possibly even too audacious task for two sole authors. However, both of us have been active in the field for more than 25 years. We have worked on various problems and with a variety of methods. Since the early years of our careers, we have met occasionally, followed each other's writings, and were often surprised by the similarity of our opinions and judgments despite quite different early medical and scientific backgrounds. Moreover, our knowledge of the literature turned out to be in part overlapping and in part complementary. Since we are working in different continents, AGM had a better knowledge of concepts and results in the USA, while FV knew more of the continental European literature. Moreover, both of us have extensive experience as editors of journals in human genetics and one (FV) published a fairly comprehensive textbook in Germany some time ago (*Lehrbuch der allgemeinen Humangenetik,* Springer 1961), parts of which were still useful for the new book. We finally decided to take the risk, and, by writing an "advanced" text, to expose our deficiencies of knowledge, shortcomings of understanding, and biases of judgment.

A text endeavoring to expose the conceptual framework of human genetics cannot be dogmatic and has to be critical. Moreover, we could not confine ourselves to hard facts and well-proved statements. The cloud of conjectures and hypotheses surrounding a rapidly growing science had to be depicted. By doing so, we face the risk of being disproved by further results.

A number of colleagues helped by reading parts of the manuscript on which they had expert knowledge and by making useful suggestions: W. Buselmaier, U. Ehling, G. Flatz, W. Fuhrmann, S. Gartler, Eloise Giblett, P. Propping, Laureen Resnick, and Traute M. Schroeder. They should not be held responsible for possible errors. J. Krüger was of supreme help in the statistical parts. Our secretaries, Mrs. Adelheid Fengler and Mrs. Gabriele Bauer in Heidelberg, Mrs. Sylvia Waggoner in Seattle, and Mrs. Helena Smith in Stanford gave invaluable aid. The figures were drawn by Edda Schalt and Marianne Lebküchner. Miriam Gallaher and Susan Peters did an expert job of copy editing. The authors are especially grateful to Dr. Heinz Götze and Dr. Konrad F. Springer, of Springer Publishing Company, for the excellent production. The work could not have been achieved had the two authors not been invited to stay at the Center for Advanced Study in the Behavioral Sciences at Stanford (California) for the academic year of 1976/1977. The grant for AGM was kindly provided by the Kaiser Family Foundation, while the Spencer Foundation donated the grant for FV.

The cover of this book shows the mythical first human couple, Adam and Eve, as imagined by Albrecht Dürer (1504). They present themselves in the full beauty of their bodies, ennobled by the genius and skill of a great artist. The drawing should remind us of the uniqueness and dignity of the human individual. Human genetics can help us to understand humanity better and to make human life happier. This science is a cardinal example of Alexander Pope's statement. "The proper study of mankind is man."

Spring 1979

Friedrich Vogel, Heidelberg
Arno G. Motulsky, Seattle

Contents Overview

Contents

Introduction

No theory solves all the puzzles with which it is confronted at a given time,
nor are the solutions already achieved often perfect. On the contrary,
it is just the incompleteness and imperfection of the existing data-theory fit that, at any time,
define many of the puzzles that characterize normal science.

T. S. Kuhn, *The Structure of Scientific Revolutions*, 1962

Human Genetics as Fundamental and Applied Science. Human genetics is both a fundamental and an applied science. As a fundamental science, it is part of genetics – the branch of science that examines the laws of storage, transmission, and realization of information for development and function of living organisms. Within this framework, human genetics concerns itself with the most interesting organism – the human being. This concern with our own species makes us scrutinize scientific results in human genetics not only for their theoretical significance but also for their practical value for human welfare. Thus, human genetics is also an applied science. Its value for human welfare is bound to have repercussions for theoretical research as well, since it influences the selection of problems by human geneticists, their training, and the financing of their research. Because of its continued theoretical and practical interest, human genetics offers fascination and human fulfillment unparalleled by work in fields that are either primarily theoretical or entirely practical in subject matter.

Science of Genetics. Genetics has become a mature science. It is based on a powerful and penetrating theory. The profundity of a theory depends on the depth of the problems that it sets out to solve and can be characterized by three attributes: the occurrence of high-level constructs, the presence of a mechanism, and high explanatory power [2]. In genetics, the high-level "construct" is the gene as a unit of storage, transmission, and realization of information. Since the rediscovery of Mendel's laws in 1900, genetic mechanisms have been worked out step by step to the molecular level – deciphering of the genetic code, analysis of transcription and translation, the function of gene-determined proteins, the fine structure of the genetic material, and DNA sequences outside of genes. The problems of regulation of gene activity in the development and function of organisms are currently a principal goal of fundamental research. So far, the explanatory power of the theory has not nearly been exhausted.

How Does a Science Develop? Kuhn (1962) [9] described the historical development of a science as follows: In the early, protoscientific stage, there is substantial competition among various attempts at theoretical foundation and empirical verification. Basic observations suggest a set of problems that, however, is not yet visualized clearly. Then, one "paradigm" unifies a group within the scientific community in the pursuit of a common goal, at the same time bringing into sharper focus one or a few aspects of the problem field, and suggesting a way for their solution. If the paradigm turns out to be successful, it is accepted by an increasing part of the scientific community, which now works under its guidance, exploring its possibilities, extending its range of application, and developing it into a scientific theory.
This concept of a paradigm has three main connotations:

1. It points to a piece of scientific work that serves as an "exemplar," suggesting ways in which a certain problem should be approached.
2. It delimits a group of scientists who try to explore this approach, expand its applicability, deepen its theoretical basis by exploration of basic mechanisms, and enhance its explanatory power.
3. Finally, while an elaborate theory must not – and, in most cases, does not – exist when a paradigm is initiated, its germ is already there, and a successful paradigm culminates in the elaboration of this theory.

This process of developing a science within the framework of a paradigm has been described by Kuhn as "normal science." The basic theory is taken increasingly for granted. It would be sterile at this stage to doubt and reexamine its very cornerstones; instead, it is applied to a variety of problems, expanded in a way that is comparable to puzzle solving. From time to time, however, results occur that, at first glance, defy explanation. First, this leads to attempts at accommodating such results within the theoretical framework by additional ad hoc hypotheses. These attempts are often successful; sometimes, however, they fail. If in such a situation an alternative paradigm is brought forward that explains most of the phenomena accounted for by the old theory as well as the new, hitherto unexplained phenomena, a scientific "revolution" may occur. The new paradigm gains support from an increasing majority of the scientific community, it soon develops into a new – more explanatory – theory, and the process of normal science begins anew.

This portrayal of scientific development has been criticized by some philosophers of science [10]. The concept of "normal" science as outlined above does not appeal to some theorists. Working within the framework of a given set of concepts has been denounced as dull, boring, and in any case not as science should be. According to these philosophers, scientists ought to live in a state of permanent revolution, constantly questioning the basic foundations of their field, always eager to put them to critical tests and, if possible, to refute them [14–17]. Many scientists actively involved in research, on the other hand, have readily accepted Kuhn's view; he has apparently helped them to recognize some important aspects in the development of their own fields.

Central Theory of Genetics Looked at as a Paradigm. While Kuhn's concepts were developed on the basis of the history of the physical sciences, his description well fits the development of genetics. Up to the second half of the nineteenth century, the phenomena of heredity eluded analysis. Obviously, children were sometimes – but by no means always – similar to their parents; some diseases were shown to run in families; it was possible to improve crops and domestic animals by selective breeding. Even low-level laws were discovered, for example Nasse's law that hemophilia affects only boys but is transmitted by their mothers and sisters (Sect. 4.1.4). However, a convincing overall theory was missing, and attempts at developing such at theory were unsuccessful. In this situation, Mendel, in his work *Versuche über Pflanzenhybriden* (1865) [11] first improved a procedure; he complemented the breeding experiment by counting the offspring. He then interpreted the results in terms of the random combination of basic units; by assuming these basic units, he founded the gene concept – the nuclear concept underlying genetic theory (Sect. 1.4).

Since the rediscovery of his work in 1900, Mendel's insight has served as a paradigm in all three connotations: it provided an exemplar as to how breeding experiments should be designed and evaluated, it resulted in the establishment of a scientific community of geneticists, and it led to the development of a deep and fertile scientific theory. A special problem that has not been answered satisfactorily, in our opinion, concerns the question of why acceptance of Mendel's paradigm had to wait for as long as 35 years after these experiments were published. It would be too simplistic to blame academic arrogance and short-sightedness of contemporary biologists who did not

want to accept the work of a "nonacademic" outsider, even if this factor may indeed have been one of the components for this neglect. We believe rather that the many new biological discoveries in the 35 years following Mendel's discovery were of such a revolutionary nature as to qualify as a scientific crisis in the Kuhnian sense and therefore required a completely new approach.

Soon after the rediscovery of Mendel's laws in 1900, however, an initially small, but quickly growing group of scientists gathered who developed genetics in an interplay between theory and experiment and launched the major scientific revolution of the twentieth century in the field of biology.

Human Genetics and the Genetic Revolution. Meanwhile, the biological revolution of the nineteenth century – evolutionary theory – had been accepted by the scientific community. One major consequence was the realization that human beings had evolved from other, more "primitive" primates, that humans are part of the animal kingdom, and that the laws of heredity which had been found to apply for all other living beings are also valid for our species. Hence, Mendel's laws were soon applied to traits that were found in human pedigrees – primarily hereditary anomalies and diseases. Analyzing the mode of inheritance of alcaptonuria – a recessive disease – Garrod (1902) [6] clearly recognized the cardinal principle of gene action: genetic factors specify chemical reactions (Sect. 1.5). This insight also required 30 years before being incorporated into the body of "normal" science.

Elucidation of inheritance in humans did not begin with Mendel's paradigm. Many relevant observations had been reported before, especially on various diseases. Moreover, another paradigm had been founded by F. Galton in his work on *Hereditary Talent and Character* (1865) [5] and in later works: to derive conclusions as to inheritance of certain traits such as high performance, intelligence, and stature, one should measure these traits as accurately as possible and then compare the measurements between individuals of known degree of relationship (for example, parents and children, sibs, or twins) using statistical methods. This approach did not contain the potential for elucidating the mechanisms of heredity. On the other hand, it seemed to be much more generally applicable to human characteristics than Mendelian analysis; predigree analysis in terms of Mendel's laws was hampered by the fact that most human traits simply could not be classified as alternate characteristics, as could round and shrunken peas. Human characteristics are usually graded and show no alternative distribution in the population. Moreover, the phenotypes are obviously determined not only by the genetic constitution but by external, environmental influences as well – the result of an interaction between "nature and nurture" (Galton). Therefore, naive attempts at applying Mendel's laws to such traits were doomed to failure. For traits that are regarded as important, such as intelligence and personality, but also for many diseases and mental retardation, there was only the choice between research along the lines suggested by Galton or no research at all. Investigations on genetic mechanisms would have to await elucidations of the genetics of other, more accessible organisms. Under these circumstances, scientists chose to follow Galton. This choice had not only theoretical reasons; it was strongly influenced by the desire to help individuals and families by calculating risk figures for certain diseases, thereby creating a sound basis for genetic counseling. More important, however, was the concern of some scientists about the biological future of the human species, which they saw threatened by deterioration due to relaxation of natural selection. The motives for their research were largely eugenic: it seemed to provide a rational foundation for measures to curb reproduction of certain groups who were at high risk of being diseased or otherwise unfit.

History of Human Genetics: A Contest Between Two Paradigms. The two paradigms – Mendel's gene concept and Galton's biometric approach – have developed side by side from 1900 up to the present; many present-day controversies, especially in the field of behavior genetics but also those concerning strategies in the genetic elucidation of common diseases, are immediately understandable when the history of human genetics is conceived as a contest between these two paradigms. This does not mean that the two paradigms are mutually exclusive; in fact, correlations between relatives as demonstrated by biometric analysis were interpreted in terms of gene action by Fisher in 1918 [4]. Some human geneticists have worked during some part of their career within the framework of the one paradigm, and during another within the framework of the other paradigm. By and large, however, the two streams of research have few interconnections and may even become further polarized because of highly specialized training for each group, epitomized by the biochemical and molecular genetic laboratories for the one and the computer for the other group.

In the first decades of this century the biometric paradigm of Galton appeared to be very successful. Genetic variability within the human population was believed to be established for normal traits such as stature or intelligence as well as for a wide variety of pathologic conditions such as mental deficiency and psychosis, epilepsy, and common diseases such as diabetes, allergies, and even tuberculosis. Mendelian analysis, on the other hand, seemed to be confined to rare hereditary diseases; the ever repeated attempts at expanding Mendelian explanation into the fields of normal, physical characteristics and common diseases were usually undertaken without critical assessment of the inescapable limitations of Mendelian analysis. The first major breakthrough of Mendelian genetics was the establishment of the three-allele hypothesis for the AB0 blood groups by Bernstein in the 1920 s [1] (Sect. 4.2.2). Further progress, however, had to await the development of genetic theory by work on other organisms such as *Drosophila,* bacteria, and viruses, especially bacteriophages.

The advent of molecular biology in the late 1940 s and 1950 s had a strong influence on human genetics and, indeed, brought the final breakthrough of Mendel's paradigm. A major landmark was the discovery by Pauling et al. in 1949 [13] that sickle cell anemia is caused by an abnormal hemoglobin molecule.

The foundation of human chromosome research in the late 1950 s and early 1960 s (Sect. 2.1) came as a second, important step. At present, most investigations in human genetics have become a part of mainstream research within the framework of genetic theory. The human species, regarded by most early experimental geneticists as a poor tool for genetic research, is now displaying definite advantages for attacking basic problems. Some of these advantages are the large size of available populations, the great number and variety of known mutants and chromosome anomalies, and the unparalleled detailed knowledge of human physiology and biochemistry in health and disease. The elucidation of human genome structure in the near future will further facilitate both basic and applied research in human genetics.

One would expect that such breakthroughs have led to the establishment of Mendel's paradigm as the only leading paradigm in human genetics. This, however, is not the case. In spite of the fact that genetic theory is now pervading many fields that seemed to be closed to it only a short time ago, the paradigm of Galton – biometric analysis – has attained an unsurpassed level of formal sophistication over the past 10–25 years. The availability of computers has greatly facilitated the development and application of biometric techniques. Moreover, in some fields, such as behavior genetics, the application of genetic theory – Mendel's paradigm – is still hampered by severe difficulties (Chap. 15), and here biometric methods have dominated for a long time. In the same field, however, they are most severely criticized.

Progress in Human Genetics and Practical Application. The achievements of molecular biology and chromosome research have not only altered human genetics as a pure science, but have also brought marked progress in its application for human welfare. At the beginning, this progress did not appear very conspicuous; the diagnosis of hereditary diseases was improved, and many, hitherto unexplained malformations were accounted for by chromosome aberrations. The first practical success came in the early 1950s when the knowledge of enzyme defects in phenylketonuria (Sect. 7.2.2.9) and galactosemia led to succesful preventive therapy by a specific diet. However, a breakthrough on a much larger scale was achieved when the methods of prenatal diagnosis for chromosome aberrations and for some metabolic defects were introduced in the late 1960s and early 1970s (Sect. 18.2). Suddenly, genetic counseling could now be based not only on probability statements but, in an increasing number of cases, on certainty of individual diagnoses. This scientific development coincided with a growing awareness in large parts of the human population that unlimited human reproduction must not be accepted as a natural law but can – and should – be regulated in a rational way. Introduction of oral contraceptive agents signaled this awareness. The chance to avoid the births of severely handicapped children is now accepted by a rapidly increasing proportion of the population. At the same time, better knowledge of pathophysiological pathways is improving the chances for individual therapy of hereditary diseases, including the promise somatic gene therapy by introduction of genes into cells of functional tissues (see Sect. 19.1). Applications of human genetics as a practical tool to prevent suffering and disease have found wide resonance and have now one of the most rewarding approaches in preventive medicine. In many countries, the politically responsible bodies have already created, or are now creating the institutions for widespread application of the new tools.

Effects of Practical Applications on Research. These practical applications have led to a marked increase in the number of research workers and the amount of work within the past 20–30 years. From the beginning of the twentieth century up to the early 1950s, human genetics had been the interest of a mere handful of scientists for most of whom it was not even a full-time occupation. Many of the pioneers were trained and worked much of their lifetime as physicians in special fields of medicine, such as Waardenburg and Franceschetti in ophthalmology, and Siemens in dermatology. Others were interested in theoretical problems of population genetics and evolution and chose problems in human genetics as the field of application for their theoretical concepts, most notably J. B. S. Haldane and R. A. Fisher. Still others had their point of departure in physical anthropology.

This heterogenous group of scientists did not form a coherent scientific community. For a long time, there was almost no formal infrastructure for the development of a scientific specialty. There were almost no special departments, journals, and international conferences. This lack of focus resulted in a marked heterogeneity in quality and content of scientific contributions.

All this has changed. There are now departments and units of human and medical genetics in many countries; universities and medical schools have introduced special curricula, many journals and other publications exist, and numerous congresses and conferences are being held. The overall impression is that of an active and vigorous field which is growing exponentially.

Dangers of Widespread Practical Application for Scientific Development. This development, however, satisfactory as it is, has also a number of potentially undesirable consequences:

a) Research is promoted primarily in the fields of immediate practical usefulness related to hereditary diseases; fields of less immediate practical importance may be neglected.

b) In the past, the contact with fundamental research in molecular genetics and cell biology was not intensive enough. This may have led to a slowdown in the transfer of scientific concepts and experimental approaches from these fields. Fortunately, this has changed recently with the advent of recombinant DNA techniques. It is remarkable today how fast some results of basic research are being transferred into practical application.

c) Areas of great importance for our understanding of human evolution (and possibly of human history), as well as for the functioning of human society and its institutions, are neglected if mainstream research in human genetics is directed exclusively to medical problems. Population, evolutionary genetics, on the one hand, and behavior genetics, on the other, are the two branches that suffer most. If these fields were to be excluded from mainstream research in human genetics, they will all too easily lose meaningful contact with human biology.

d) Much medical research applies established methods to answer straightforward questions. Many studies collect data with new techniques. Individual results are often not of great import, but the ensemble of such data are the essential building blocks for the future progress of normal science. Much of such work is being carried out in human and medical genetics and is quite essential for many medical and anthropological applications. However, there is continued need in human genetics to develop testable hypotheses and try to test their consequences from all viewpoints.

Human geneticists must not neglect the further development of genetic theory. Basic research is needed in fields in which the immediate practical application of results is not possible but might in the long run be at least as important for the future of the human species as current applications in diagnostic and preventive medicine.

Advantages of Practical Application for Research. The needs of medical diagnosis and counseling have also given strong incentives to basic research. Many phenomena that basic research tries to explain would simply be unknown had they not been uncovered by study of diseases. We would be ignorant regarding the role of sex chromosomes in sex determination had there not been patients with sex chromosomal anomalies. Phenomena such as spontaneously enhanced chromosome instability in Fanconi's anemia or Bloom's syndrome with all its consequences for somatic mutation and cancer formation (Chap. 10) were discovered accidentally in the process of examining certain patients for diagnostic reasons. Genetic analysis of the "supergene" determining the major histocompatibility complex in humans contributes much to our fundamental understanding of how the genetic material above the level of a single gene locus is structured, and how the high genetic variability within the human population can be maintained (Sect. 5.2.5). However, research in this field would certainly be much less active had there not been the incentive of improving the chances of organ transplantation.

Whether we like it or not, society pays increasing amounts of money for research in human genetics because we want to have practical benefits. Hence, to promote basic research, we must promote widespread practical applications. To guarantee progress in practical application for the future as well – and not only in the field of medicine – basic research needs to be supported. This is also the only way to attract good research workers and to maintain – or even improve – scientific standards. This paradox creates priority problems for all those concerned with research planning.

Human Genetics and the Sociology of Science. The discussion above should have demonstrated that human genetics – as all other sciences – has not developed in a sociological vacuum, following only the inherent logical laws of growth of theory and experimental testing. Human genetics is the work of social groups of human beings who are subject to the laws of group psychology and are influenced by the society at large in their attitudes toward research and their selection of problems. Unfortunately, sociological investigations of group formation and structure in human genetics have not been carried out. Another group active in the foundation of molecular biology, that which introduced the bacteriophages of *Escherichia coli* into the analysis of genetic information, has been studied extensively [3]. We know from this and from other examples that, during a phase in which a new paradigm is being founded, the group that shares this paradigm establishes close within-group contacts. The normal channels of information exchange such as scientific journals and congresses are superseded by more informal information transfer through telephone calls, e-mail communications, preprints, and personal visits. Within the group, influential personalities serve as intellectual and/or organizational leaders. Outside contacts, on the other hand, are often loose. When the acute phase of the scientific revolution is over, the bonds within the group are loosened, and information is again exchanged largely by normal channels of publication.

Similar developments can be observed in the field of human genetics. In Sect. 2.1 we shall sketch the group structure of the British chromosome workers in the late 1950 s when the first human chromosome aberrations were discovered and clinical cytogenetics was founded. Other examples are the groups active in the elucidation of the major histocompatibility complex (Sect. 5.2.5) and in the assignment of gene loci to chromosome segments. (Sect. 5.1).

Of similar influence on population genetics has been the first "big science" research project in human genetics – the Atomic Bomb Casualty Commission (ABCC, now the Radiation Effects Research Foundation, RERF) project that was launched in the late 1940 s in Japan by American and Japanese research workers to examine the genetic consequences of the atomic bombs in Hiroshima and Nagasaki (Sect. 11.1.4). In later years, this project led, for example, to comprehensive studies of the genetic effects of parental consanguinity. The second endeavor of this type is the "Human Genome Project" – the attempt at analyzing and sequencing the entire human genome by coordinated international cooperation (see Sect. 18.4).

Many, if not most of the more interesting developments in the field were not initiated by investigators who would declare themselves human geneticists, or who worked in human genetics departments. They were launched by research workers from other fields such as general cytogenetics, cell biology, molecular biology, biochemistry, and immunology, but also from clinical specialities such as pediatrics, hematology, ophthalmology, and psychiatry. A common theme running through many recent developments has been the application of nongenetic techniques from many different fields such as biochemistry and immunology to genetic concepts. On the other hand, techniques originally developed for solving genetic problems, especially for molecular studies of DNA, are being introduced at a rapidly increasing rate into other fields of research, for example in both medical research and practical medicine. In fact, most recent progress in human genetics comes from such interdisciplinary approaches. The number of research workers in the field has increased rapidly. Most did not start as human geneticists but as molecular biologists, medical specialists, biochemists, statisticians, general cytogeneticists, etc. They were drawn into human genetics in the course of their research. This very variety of backgrounds makes discussions among human geneticists stimulating and is one of the intellectual assets of the present state of our field. However, such diversity is also a liability as it may lead to an overrating of one's small specialty at the expense of a loss of an overview of the whole field [7a]. With increasing complexity of research methods, specialization within human genetics has become inevitable. However, this brings with it the danger that the outlook of the scientist narrows, whole fields are neglected, and promising research opportunities remain unexploited.

Human Genetics in Relation to Other Fields of Science and Medicine. The rapid development of human genetics during recent decades has created many interactions with other fields of science and medicine. Apart from general and molecular genetics and cytogenetics, these interactions are especially close with cell biology, biochemistry, immunology, and – with many clinical specialties. Until recently, on the other hand, there have been few if any connections with physiology. One reason for this failure to establish fruitful interactions may be a difference in the basic approach: genetic analysis attempts to trace the causes of a trait to its most elementary components. Geneticists know in principle that the phenotype is produced by a complex net of interactions between various genes, but they are interested more in the components than in the exact mechanism of such interactions. At present, genetic analysis has reached the level of gene structure and the genetic code; a final goal would be to explain the properties of this code in terms of quantum physics. A malevolent observer might compare the geneticist with a man who, to understand a book, burns it and analyzes the ashes chemically.

The physiologist, on the other hand, tries to read the book. However, he often presupposes that every copy of the book should be exactly identical; variation is regarded as a nuisance. To put it differently, physiology is concerned not with the elements themselves but with their mode of interaction in complicated functional systems (see Mohr [12]). Physiologists are more concerned with the integration of interacting systems than with the analysis of their components. The analysis of regulation of gene activities by feedback mechanisms, for example, the Jacob-Monod model in bacteria, and some approaches in developmental genetics of higher organisms have taught geneticists the usefulness of thinking in terms of systems. On the other hand, methods for molecular analysis of DNA have been introduced into physiology at an increasing scale. Genes for receptors and their components, for example for neurotransmitters, and genes for channel proteins are being localized in the genome and analyzed at the molecular level. Hence, the gulf between physiology and genetics is now being bridged. With the increasing interest of human geneticists in the genetic basis of common diseases and individual genetic variation in response to influences such as nutrition and stress, genetic concepts are increasingly influencing the many branches of medicine that, in the past, have profited relatively little from genetic theory. Molecular biology is developing increasingly into a common basis for many branches of science, and most biomedical scientists are nowadays becoming better acquainted with the principles of genetics. A field of molecular medicine is emerging.

Future of Human Genetics. Research methods in science are becoming increasingly complicated and expensive, and human genetics is no exception. As a necessary consequence, mastering of these methods increasingly requires specialization in a narrow field. Purchase of big instruments creates financial difficulties. Hence, selection of research problems is often directed not by the intrinsic scientific interest of the problems or the conviction that they could, in principle, be solved, but by the availability of research methods, skilled coworkers, and instruments. The tendency toward specialization will inevitably continue, and it is possible that, in this process, important parts of human genetics will be resolved into fields that are mainly defined by research methods, such as biochemistry, chromosome research, immunology, or molecular biology (see 7a). Already now, prenatal diagnosis – including cell culture and chromosome determination – has occasionally become the domain of the obstetrician; hereditary metabolic diseases are often studied and treated by pediatricians with little genetic training.

Survival of an established field of science has no value in itself. If a field dies because its concepts and accomplishments have been accepted and are being successfully integrated into other fields, little is lost.

In human genetics, however, this state has not been reached. Many concepts of molecular biology, often in combination with "classical" methods such as linkage analysis, are now being applied to humans. Since psychiatric genetics and genetic analysis of variation in function of sensory organs are already being studied in this manner, we predict that ultimately, when our understanding of behavioral genetics reaches the level of gene action in the neurosciences, we may even expect an impact of genetics on the social sciences.

Fields of Human and Medical Genetics. The field of human genetics is large, and its borders are indistinct. The development of different techniques and methods has led to the development of many fields of subspecialization. Many of these overlap and are not mutually exclusive. The field of *human molecular genetics* has its emphasis in the identification and analysis of genes at the DNA level. Methods such as DNA digestion by restriction endonucleases, Southern blotting, and the polymerase chain reaction (PCR) are being applied. *Human biochemical genetics* deals with the biochemistry of nucleic acids, proteins, and enzymes in normal and mutant individuals. Laboratory methods of the biochemist are being used (chromatography; enzyme assays). *Human cytogenetics* deals with the study of human chromosomes in health and disease. *Immunogenetics* concerns itself largely with the genetics of blood groups, tissue antigens such as the HLA types, and other components of the immune system. *Formal genetics* studies segregation and linkage relationships of Mendelian genes and investigates more complex types of inheritance by statistical techniques.

Clinical genetics deals with diagnosis, prognosis, and to some extent treatment of various genetic diseases. Diagnosis requires knowledge of etiological heterogeneity and acquaintance with many disease syndromes. *Genetic counseling* is an important area of clinical genetics and requires skills in diagnosis, risk assessment, and interpersonal communication. *Population genetics* deals with the behavior of genes in large groups and concerns the evolutionary forces of drift, migration, mutation, and selection in human populations. The structure and gene pool of human populations are studied by considering gene frequencies of marker genes. In recent years population geneticists have become interested in the epidemiology of complex genetic disease that require biometric techniques for their studies. *Behavioral genetics* is a science that studies the hereditary factors underlying behavior in health and disease. Behavior geneticists attempt to work out the genetic factors determining personality and cognitive skills in human beings. The genetics of mental retardation and various psy-

chiatric diseases are also considered. The field of sociobiology tries to explain social behavior by using biological and evolutionary concepts.

Somatic cell genetics is the branch of human genetics that studies the transmission of genes at the cellular level. Cell hybridization between different species has become an important tool for the cartography of human genes. *Developmental genetics* studies genetic mechanisms of normal and abnormal development. The field has started to expand recently under the influence of molecular methods, and with strong emphasis on animal experimentation. *Reproductive genetics* is the branch of genetics that studies details of gamete and early embryo formation by genetic techniques. This area is closely related to reproductive physiology and is developing rapidly. *Pharmacogenetics* deals with genetic factors governing the disposal and kinetics of drugs in the organism. Special interest in human pharmacogenetics relates to adverse drug reactions. *Ecogenetics* is an extension of pharmacogenetics and deals with the role of genetic variability affecting the response to environmental agents.

Clinical genetics has grown very rapidly in recent years because of the many practical applications of diagnosis and counseling, intrauterine diagnosis, and screening for genetic disease. Most research in human genetics is currently carried out in clinical genetics, cytogenetics, molecular and biochemical genetics, somatic cell genetics, and immunogenetics under medical auspices. Research in formal and population genetics has benefited enormously from the general availability of the computer.

Unsolved and Intriguing Problems. With the rapid increase in knowledge over recent years new and often unexpected problems have arisen. At a time when hereditary traits were defined by their modes of inheritance, the relationship between genotype and phenotype appeared relatively simple. This straightforward relationship seemed correct when some hereditary diseases were shown to be caused by enzyme defects, and when hemoglobin variants turned out to be due to amino acid replacements caused by base substitutions. With increasing knowledge of the human genome, however, many hereditary traits with phenotypes that had been considered identical turned out to be heterogeneous. These were caused either by mutations in different genes or by different mutations within the same genes. However, even mutations that are identical by the strictest molecular criteria sometimes have striking phenotypic differences. Analysis of such genotype-phenotype relationships by the study of genetic and environmental modifiers poses intriguing future problems in human genetics.

"Simple" Mendelism suggested that autosomal genes from the paternal and maternal sides contributed identically to the child's phenotype. In most instances this expectation was correct; exceptions were overlooked or set aside. More recently, however, we have learned that the contributions of the paternal and maternal genomes to the developing embryo may vary and can lead to different disease phenotypes depending on which parent transmitted the mutant gene or chromosome segment. The term "genomic imprinting" was coined to describe this phenomenon. Problems of "phenogenetics" were much discussed by geneticists in the first decades of the twentieth century [7] but were set aside later when mainstream research moved in other directions. These problems are now returning with full force. They are not only of theoretical interest since their solution has practical consequences for the diagnosis treatment, and prognosis of some genetic diseases.

Possible Function of a Textbook. In his book on "*The Structure of Scientific Revolutions,*" Kuhn in 1962 [9] described the function of textbooks not very flatteringly: they are "pedagogic vehicles for the perpetuation of normal science" that create the impressions as if science would grow in a simple, cumulative manner. They tend to distort the true history of the field by only mentioning those contributions in the past that can be visualized as direct forerunners of present-day achievements. "They inevitably disguise not only the role but the very existence of . . . revolutions . . ."

Below we shall proceed in the same way: we shall describe present-day problems in human genetics as we see them. The result is a largely affirmative picture of normal science in a phase of rapid growth and success. Anomalies and discrepancies may exist, but we often do not identify them because we share the "blind spots" with most other members of our paradigm group. The "anticipation" phenomenon in diseases such as myotonic dystrophy is one example (Sect. 4.1.7). This disease tends to manifest more severely and earlier in life with each generation. Obviously, this observation did not appear not compatible with simple mendelism. Therefore, it was explained away by sophisticated statistical arguments which we cited in earlier editions of this book. In the meantime, however, anticipation has been shown to be a real phenomenon, caused by a novel molecular mechanism. What we can do is to alert the reader that human genetics, as all other branches of science, is by no way a completed and closed complex of theory and results that only needs to be supplemented in a straightforward way and without major changes in conceptualization. Our field has not developed – and

will not develop in the future – as a self-contained system. Rather, human genetics, as all other sciences, is an undertaking of human beings – social groups and single outsiders – who are motivated by a mixture of goals such as search for truth, ambition, desire to be acknowledged by one's peer group, the urge to convince the society at large to allocate resources in their field – but also the wish to help people and to do something useful for human society.

Therefore, we shall emphasize the history and development of problems and approaches. Occasionally, we shall ask the reader to step back, reflecting with us as to why a certain development occurred at the time it did, why another development did not occur earlier, or why a certain branch of human genetics did not take the direction that one would have expected logically. Inevitably, this implies much more criticism than is usually found in textbooks. Such criticism will – at least partially – be subjective, reflecting the personal stance of the authors. Our goal is to convince the reader that a critical attitude improves one's grasp of the problems and their possible solutions – it is not our intention to convince him that we are always right.

We would have liked to give more information on the ways in which sociological conditions within the field and – still more important – the developments in the society at large have influenced the development of human genetics, and the ways in which thinking on these problems has in turn influenced the societies. The eugenics movement in the United States and the *Rassenhygiene* ideology in Germany have had a strong – and sometimes devastating – influence on human beings as well as on the social structure of society at large. Too little systematic research has been carried out, however, to justify a more extended discussion than that presented in Sect. 1.8 [8]. Much more historical research along these lines is all the more urgent, as many of the ethical problems – inherent, for example, in the sterilization laws of many countries during the first decades of the twentieth century – are now recurring with full force in connection with prenatal diagnosis, selective abortion and the possibility of germinal gene therapy (Sects. 18.2, 19.2). Scientists and physicians working in human genetics were actively involved in and sanctioned ethically abhorrent measures in the past such as killing severely malformed newborns and mentally defectives in Nazi Germany – and how will future generations judge our own activities? These are intriguing questions. They show the Janus face of human genetics: it is a fundamental science – guided by a fertile theory and full of fascinating problems. It is also an applied science, and its applications are bound to have a strong impact on society, leading to novel and difficult philosophical, social, and ethical problems.

References

1. Bernstein F (1924) Ergebnisse einer biostatistischen zusammenfassenden Betrachtung über die erblichen Blutstrukturen des Menschen. Klin Wochenschr 3 : 1495–1497
2. Bunge M (1967) Scientific research, vols 1, 2. Springer, Berlin Heidelberg New York
3. Cairns J, Stent GS, Watson JD (1966) Phage and the origin of molecular biology. Cold Spring Harbor Laboratory, New York
4. Fisher RA (1918) The correlation between relatives on the supposition of Mendelian inheritance. Trans R Soc Edinb 52 : 399–433
5. Galton F (1865) Hereditary talent and character. Macmillans Magazine 12 : 157
6. Garrod AE (1902) The incidence of alcaptonuria: a study in chemical individuality. Lancet 2 : 1616–1620
7. Harwood J (1993) Styles of scientific thought. The German genetics community 1900–1933. University of Chicago Press, Chicago
7a. Hirschhorn K (1996) Human genetics: a discipline at risk for fragmentation. Am J Hum Genet 58 : 1–6 (William Allan Award Address 1995)
8. Kevles DJ (1985) In the name of eugenics. Genetics and the uses of human heredity. Knopf, New York
9. Kuhn TS (1962) The structure of scientific revolutions. University of Chicago Press, Chicago
10. Lakatos I, Musgrave A (1970) Criticism and the growth of knowledge. Cambridge University Press, New York
11. Mendel GJ (1865) Versuche über Pflanzenhybriden. Verhandlungen des Naturforschenden Vereins, Brünn
12. Mohr H (1977) The structure and significance of science. Springer, Berlin Heidelberg New York
13. Pauling L, Itano HA, Singer SJ, Wells IC (1949) Sickle cell anemia: a molecular disease. Science 110 : 543
14. Popper KR (1963) Conjectures and refutations. Rutledge and Kegan Paul, London
15. Popper KR (1970) Normal science and its dangers. In. Lakatos I, Musgrave A (eds) Criticism and the growth of knowledge. Cambridge University Press, New York, pp 51–58
16. Popper KR (1971) Logik der Forschung, 4th edn. Mohr, Tübingen (Original 1934)
17. Watkins JWN (1970) Against "normal science." In: Lakatos I, Musgrave A (eds) Criticism and the growth of knowledge. Cambridge University Press, New York, pp 25–38

1 History of Human Genetics

"Man is a history-making creature who can neither repeat his past nor leave it behind."

W. H. Auden (From: D. H. Lawrence)

The history of human genetics is particularly interesting since, unlike in many other natural sciences, concepts of human genetics have often influenced social and political events. At the same time, the development of human genetics as a science has been influenced by various political forces. Human genetics because of its concern with the causes of human variability has found it difficult to either remain a pure science or one of strictly medical application. Current concerns regarding the heritability of IQ and the existence of inherited patterns of behavior again have brought the field into public view. A consideration of the history of human genetics with some attention to the interaction of the field with societal forces is therefore of interest. We will concentrate our attention on historical events of particular interest for human genetics and refer to landmarks in general genetics only insofar as they are essential for the understanding of the evolution of human genetics.

1.1 The Greeks

Prescientific knowledge regarding inherited differences between humans has probably existed since ancient times. Early Greek physicians and philosophers not only reported such observations but also developed some theoretical concepts and even proposed "eugenic" measures.

In the texts that are commonly ascribed to Hippocrates, the following sentence can be found:

Of the semen, however, I assert that it is secreted by the whole body – by the solid as well as by the smooth parts, and by the entire humid matters of the body ... The semen is produced by the whole body, healthy by healthy parts, sick by sick parts. Hence, when as a rule, baldheaded beget baldheaded, blue-eyed beget blue-eyed, and squinting, squinting; and when for other maladies, the same law prevails, what should hinder that longheaded are begotten by longheaded?

This remarkable sentence not only contains observations on the inheritance of normal and pathological traits but also a theory that explains inheritance on the assumption that the information carrier, the se-

men, is produced by all parts of the body, healthy and sick. This theory became known later as the "pangenesis" theory. Anaxagoras, the Athenian philosopher (500–428 B. C.), had similar views: ". . . in the same semen are contained hairs, nails, veins, arteries, tendons, and their bones, albeit invisible as their particles are so small. While growing, they gradually separate from each other." Because, he said, "how could hair come out of non-hair, and flesh out of non-flesh?" (Fragment 10; see Capelle [9]). In his opinion, men produced the seed, women the breeding ground (Freeman, p. 272).

A comprehensive theory of inheritance was developed by Aristotle (see [3]). He also believed in a qualitatively different contribution by the male and the female principles to procreation. The male gives the impulse to movement whereas the female contributes the matter, as the carpenter who constructs a bed out of wood. When the male impact is stronger, a son is born who, at the same time, is more like his father, when the female, a daughter, resembling the mother. This is the reason why sons are usually similar to their fathers and daughters are similar to their mothers. Barthelmess (1952; our translation) [3] writes: "Reading the texts from this culture, one gets the overall impression that the Greeks in their most mature minds came closer to the theoretical problems than to the phenomena of heredity." Aristotle's assertion even provides an early example of how observation can be misled by a preconceived theoretical concept. Sons are not more similar to their fathers, nor daughters to their mothers.

Plato, in the *Statesman (Politikos),* explained in detail the task of carefully selecting spouses to produce children who will develop into bodily and ethically eminent personalities. He wrote (Sect. 310; translation by Skemp 1952):

They do not act on any sound or self-consistent principle. See how they pursue the immediate satisfaction of their desire by hailing with delight those who are like themselves and by disliking those who are different. Thus they assign far too great an importance to their own likes and dislikes.
The moderate natures look for a partner like themselves, and so far as they can, they choose their wives from women of this quiet type. When they have daughters to bestow in mar-

riage, once again they look for this type of character in the prospective husband. The courageous class does just the same thing and looks for others of the same type. All this goes on, though both types should be doing exactly the opposite . . .

Because if a courageous character is reproduced for many generations without any admixture of the moderate type, the natural course of development is that at first it becomes superlatively powerful but in the end it breaks out into sheer fury and madness . . .

But the character which is too full of modest reticence and untinged by valor and audacity, if reproduced after its kind for many generations, becomes too dull to respond to the challenges of life and in the end becomes quite incapable of acting at all.

In the *Republic,* Plato not only requires for the "guards" (one of the highest categories in the social hierarchy of his utopia) that women should be common property; children, should be educated publicly but the "best" of both sexes should beget children who are to be educated with care. The children of the "inferior," on the other hand, are to be abandoned (Politeia, 459 a ff.). Democritus, on the other hand, writes: "More people become able by exercise than by their natural predisposition." Here (as in other places), the nature-nurture problem appears already.

1.2 Scientists Before Mendel and Galton

The literature of the Middle Ages contains few allusions to heredity. The new attitude of looking at natural phenomena from an empirical point of view created modern science and distinguishes modern humans from those in earlier periods. This approach succeeded first in investigation of the inorganic world and only later in biology. In the work *De Morbis Hereditariis* by the Spanish physician Mercado (1605), the influence of Aristotle is still overwhelming, but there are some hints of a beginning emancipation of reasoning. One example is his contention that both parents, not only the father, contribute a seed to the future child. Malpighi (1628–1694) proposed the hypothesis of "preformation," which implies that in the ovum the whole organism is preformed in complete shape, only to grow later. Even after the discovery of sperm (Leeuwenhoek, van Ham and Hartsoeker, 1677), the preformation hypothesis was not abandoned altogether, but it was believed by some that the individual is preformed in the sperm, only being nurtured by the mother. The long struggle between the "ovists" and the "spermatists" was brought to an end only when C. F. Wolff (1759) attacked both sides and stressed the necessity of further empirical research. Shortly thereafter experi-

mental research on heredity in plants was carried out by Gaertner (1772–1850) and Koelreuter (1733–1806). Their work prepared the ground for Mendel's experiments [3].

The medical literature of the eighteenth and early nineteenth centuries contains reports showing that those capable of clear observation were able to recognize correctly some phenomena relating to the inheritance of diseases. Maupertuis, for example, published in 1752 an account of a family with polydactyly in four generations and demonstrated that the trait could be equally transmitted by father or by mother. He further showed, by probability calculation, that chance alone could not account for the familial concentration of the trait. Probably the most remarkable example, however, was Joseph Adams (1756–1818) (see [32]), a British doctor who, in 1814, published a book with the title *A Treatise on the Supposed Hereditary Properties of Diseases,* which sought to provide a basis for genetic counseling. The following findings are remarkable:

a) Adams differentiated clearly between congenital "familial" (i.e. recessive) and "hereditary" (i.e. dominant) conditions.

b) He knew that in familial diseases the parents are frequently near relatives.

c) Hereditary diseases need not be present at birth; they may manifest themselves at various ages.

d) Some disease predispositions lead to a manifest disease only under the additional influence of environmental factors. The progeny, however, is endangered even when the predisposed do not become ill themselves.

e) Intrafamilial correlations as to age of onset of a disease can be used in genetic counseling.

f) Clinically identical diseases may have different genetic bases.

g) A higher frequency of familial diseases in isolated populations may be caused by inbreeding.

h) Reproduction among persons with hereditary diseases is reduced. Hence, these diseases would disappear in the course of time, if they did not appear from time to time among children of healthy parents (i.e., new mutations!).

Adams' attitude toward "negative" eugenic measures was critical. He proposed the establishment of registries for families with inherited diseases.

C. F. Nasse, a German professor of medicine, correctly recognized in 1820 one of the most important formal characteristics of the X-linked recessive mode of inheritance in hemophilia and presented a typical comprehensive pedigree. He wrote (our translation):

All reports on families, in which a hereditary tendency towards bleeding was found, are in agreement that the bleeders are persons of male sex only in every case. All are explicit on this point. The women from those families transmit this tendency from their fathers to their children, even when they are married to husbands from other families who are not afflicted with this tendency. This tendency never manifests itself in these women. . . .

Nasse also observed that some of the sons of these women remain completely free of the bleeding tendency.

The medical literature of the nineteenth century shows many more examples of observations and attempts to generalize and to find rules for the influence of heredity on disease can be found. The once very influential concept of "degeneration" should be mentioned. Some features that older authors described as "signs of degeneration" in the external appearance of mentally deficient patients are now known to be characteristic of autosomal chromosomal aberrations or various types of X-linked mental retardation.

In the work of most of the nineteenth century authors, true facts and wrong concepts were inextricably mixed, and there were few if any criteria for getting at the truth. This state of affairs was typical for the plight of a science in its prescientific state. Human genetics had no dominant paradigm. The field as a science was to start with two paradigms in 1865: biometry, which was introduced by Galton, and Mendelism, introduced by Mendel with his pea experiments. The biometric paradigm was influential in the early decades of the twentieth century, and many examples and explanations in this book utilize its framework. With the advent of molecular biology and insight into gene action, the pure biometric approach in genetics is on the decline. Nevertheless, many new applications in behavioral or social genetics, where gene action cannot yet be studied, rely on this paradigm and its modern elaborations. The laws that Mendel derived from his experiments, on the other hand, have been of almost unlimited fruitfulness and analytic power. The gene concept emerging from these experiments has become the central concept of all of genetics, including human genetics. Its possibilities have not been exhausted.

1.3 Galton's Work

In 1865, F. Galton published two short papers with the title "Hereditary Talent and Character". He wrote [15]:

The power of man over animal life, in producing whatever varieties of form he pleases, is enormously great. It would seem as though the physical structure of future generations was almost as plastic as clay, under the control of the bree-

der's will. It is my desire to show, more pointedly than – so far as I am aware – has been attempted before, that mental qualities are equally under control.

A remarkable misapprehension appears to be current as to the fact of the transmission of talent by inheritance. It is commonly asserted that the children of eminent men are stupid; that, where great power of intellect seems to have been inherited, it has descended through the mother's side; and that one son commonly runs away with the talent of the whole family.

He then stresses how little we know about the laws of heredity in man and mentions some reasons, such as long generation time, that make this study very difficult. However, he considers the conclusion to be justified that physical features of humans are transmissible because resemblances between parents and offspring are obvious. Breeding experiments with animals, however, had not been carried out at that time, and direct proof of hereditary transmission was therefore lacking even in animals. In humans, "we have . . . good reason to believe that every special talent or character depends on a variety of obscure conditions, the analysis of which has never yet been seriously attempted." For these reasons, he concluded that single observations must be misleading, and only a statistical approach can be adequate.

Galton evaluated collections of biographies of outstanding men as to how frequently persons included in these works were related to each other. The figures were much higher than would be expected on the basis of random distribution.

Galton himself was fully aware of the obvious sources of error of such biological conclusions. He stressed that "when a parent has achieved great eminence, his son will be placed in a more favorable position for advancement, than if he had been the son of an ordinary person. Social position is an especially important aid to success in statesmanship and generalship . . . ."

"In order to test the value of hereditary influence with greater precision, we should therefore extract from our biographical list the names of those that have achieved distinction in the more open fields of science and literature." Here and in the law, which in his opinion was "the most open to fair competition," he found an equally high percentage of close relatives reaching eminence. This was especially obvious with Lord Chancellors, the most distinguished lawyers of Great Britain.

Galton concluded that high talent and eminent achievement are strongly influenced by heredity. Having stressed the social obstacles that inhibit marriage and reproduction of the talented and successful, he proceeded to describe a utopic society,

In which a system of competitive examination for girls, as well as for youths, had been so developed as to embrace every important quality of mind and body, and where a considerable sum was yearly allotted. . . . to the endowment of such marriages as promised to yield children who would grow into eminent servants of the State. We may picture to ourselves an annual ceremony in that Utopia or Laputa, in which the Senior Trustee of the Endowment Fund would address ten deeply-blushing young men, all of twenty-five years old, in the following terms . . . .

In short, they were informed that the commission of the endowment fund had found them to be the best, had selected for each of them a suitable mate, would give them a substantial dowry, and promised to pay for the education of their children.

This short communication already shows human genetics as both a pure and an applied science: on the one hand, the introduction of statistical methods subjects general impressions to scientific scrutiny, thereby creating a new paradigm and turning pre-science into science. Later, Galton and his student K. Pearson proceeded along these lines and founded biometric genetics. On the other hand, however, the philosophical motive of scientific work in this field is clearly shown: the object of research is an important aspect of human behavior. The prime motive is the age-old "γνῶθι σεαυτόν" ("know yourself," inscription on the Apollo temple at Delphi).

Hence, with Galton, research in human genetics began with strong eugenic intentions. Later, with increasing methodological precision and increasing analytic success, such investigations were increasingly removed from this prime philosophical motive. This motive helps to understand the second aspect of Galton's work: the utopian idea to improve the quality of the human species by conscious breeding. During the Nazi era in Germany (1933–1945) we saw how cruel the perverted consequences of such an idea may become (Sect.1.8). Even these experiences, however, are sometimes forgotten, as testified by some aspects of recent discussions in genetic manipulation and engineering (Chap.19). Nevertheless, the question first posed by Galton remains, even more than ever, of pressing importance: What will be the biological future of mankind?

1.4 Mendel's Work

The other leading paradigm was provided by Mendel in his work *Experiments in Plant Hybridization,* which was presented on 8 February and 8 March, 1865 before the *Naturforschender Verein* (Natural Science Association) in Brünn (now Brno, Czech Republic) and subsequently published in its proceedings [31]. It has frequently been told how this work went largely unnoticed for 35 years and was rediscovered

independently by Correns, Tschermak, and de Vries in 1900 (see [3]). From then on, Mendel's insights triggered the development of modern genetics, including human genetics.

Mendel was stimulated to carry out his experiments by observations on ornamental plants, in which he had tried to breed new color variants by artificial insemination. Here he had been struck by certain regularities. He selected the pea for further experimentation. He crossed varieties with differences in single characters such as color (yellow or green) or form of seed (round or angular wrinkled) and counted all alternate types in the offspring of the first generation crosses and of crosses in later generations. Based on combinatorial reasoning, he gave a theoretical interpretation: the results pointed to free combination of specific sorts of egg and pollen cells. In fact, this concept may have occurred to Mendel before he carried out his studies. He may have verified and illustrated his findings by his "best" results, since agreement between the published figures and their expectation from the theoretical segregation ratios is too perfect from a statistical point of view. The interpretation of this discrepancy remains controversial. In any case, there is no question that Mendel's findings were correct.

Mendel discovered three laws: the law of uniformity, which states that after crossing of two homozygotes of different alleles the progeny of the first filial generation (F_1) are all identical and heterozygous; the law of segregation, which postulated 1 : 2 : 1 segregation in intercrosses of heterozygotes and 1 : 1 segregation in backcrosses of heterozygotes with homozygotes; and the law of independence, which states that different segregating traits are transmitted independently.

What is so extraordinary in Mendel's contribution that sets it apart from numerous other attempts in the nineteenth century to solve the problem of heredity? Three points are most important:

1. He simplified the experimental approach by selecting characters with clear alternative distributions, examining them one by one, and proceeding only then to more complicated combinations.
2. Evaluating his results, he did not content himself with qualitative statements but counted the different types. This led him to the statistical law governing these phenomena.
3. He suggested the correct biological interpretation for this statistical law: The germ cells represent the constant forms that can be deduced from these experiments.

With this conclusion Mendel founded the concept of the gene, which has proved so fertile ever since. The history of genetics since 1900 is dominated by analysis of the gene. What had first been a formal concept derived from statistical evidence has emerged as the base pair sequence of DNA, which contains the information for protein synthesis and for life in all its forms [14].

1.5 Application to Humans: Garrod's Inborn Errors of Metabolism

The first step of this development is described in this historical introduction: A. Garrod's (1902) [16] paper on "The Incidence of Alkaptonuria: A Study in Chemical Individuality." There are two reasons for giving special attention to this paper. For the first time, Mendel's gene concept was applied to a human character, and Mendel's paradigm was introduced into research on humans. Additionally, this work contains many new ideas set out in a most lucid way. Garrod was a physician and in later life became the successor of Osler in the most prestigious chair of medicine at Oxford [193]. His seminal contribution to human genetics remained unappreciated during his lifetime. Biologists paid little attention to the work of a physician. Their interest was concentrated more on the formal aspects of genetics rather than on gene action. The medical world did not understand the importance of his observations for medicine. Garrod first mentioned the isolation of homogentisic acid from the urine of patients with alkaptonuria by Walkow and Baumann (1891). Then, he stated as the most important result of the investigations carried out so far:

As far as our knowledge goes, an individual is either frankly alkaptonuric or conforms to the normal type, that is to say, excretes several grammes of homogentisic acid per diem or none at all. Its appearance in traces, or in gradually increasing or diminishing quantities, has never yet been observed . . . .

As a second important feature "the peculiarity is in the great majority of instances congenital . . . ." Thirdly: "The abnormality is apt to make its appearance in two or more brothers and sisters whose parents are normal and among whose forefathers there is no record of its having occurred." Fourthly, in six of ten reported families the parents were first cousins, whereas the incidence of first-cousin marriages in contemporary England was estimated to be not higher than 3%. On the other hand, however, children with alkaptonuria are observed in a very small fraction only of all first-cousin marriages.

There is no reason to suppose that mere consanguinity of parents can originate such a condition as alkaptonuria in their offspring, and we must rather seek an explanation in some peculiarity of the parents, which may remain latent for generations, but which has the best chance of asserting itself in the offspring of the union of two members of a family in which it is transmitted.

Then, Garrod mentioned the law of heredity discovered by Mendel, which "offers a reasonable account of such phenomena" that are compatible with a recessive mode of inheritance. He cited another remark of Bateson and Saunders (Report to the Evolution Committee of the Royal Society, 1902) with whom he had discussed his data:

We note that the mating of first cousins gives exactly the conditions most likely to enable a rare, and usually recessive, character to show itself. If the bearer of such a gamete mates with individuals not bearing it the character will hardly ever be seen; but first cousins will frequently be the bearers of similar gametes, which may in such unions meet each other and thus lead to the manifestation of the peculiar recessive characters in the zygote.

After having cited critically some opinions on the possible causes of alkaptonuria, Garrod proceeded:

The view that alkaptonuria is a "sport" or an alternative mode of metabolism will obviously gain considerably in weight if it can be shown that it is not an isolated example of such a chemical abnormality, but that there are other conditions which may reasonably be placed in the same category.

Having mentioned albinism and cystinuria as possible examples, he went on: "May it not well be that there are other such chemical abnormalities which are attended by no obvious peculiarities [as the three mentioned above] and which could only be revealed by chemical analysis?" And further:

If it be, indeed, the case that in alkaptonuria and the other conditions mentioned we are dealing with individualities of metabolism and not with the results of morbid processes the thought naturally presents itself that these are merely extreme examples of variations of chemical behaviour which are probably everywhere present in minor degrees and that just as no two individuals of a species are absolutely identical in bodily structure neither are their chemical processes carried out on exactly the same lines.

He suggested that differential responses toward drugs and infective agents could be the result of such chemical individualities. The paper presents the following new insights:

a) Whether a person has alkaptonuria or not is a matter of a clear alternative – there are no transitory forms. This is indeed a condition for straightforward recognition of simple modes of inheritance.
   The condition is observed in some sibs and not in parents.
   The unaffected parents are frequently first cousins.

This is explained by the hypothesis of a recessive mode of inheritance according to Mendel. The significance of first-cousin marriages is stressed especially for rare conditions; this may be a precursor to population genetics.

b) Apart from alkaptonuria several other similar "sports" such as albinism and cystinuria may exist. This makes alkaptonuria the paradigm for the "inborn errors of metabolism." In 1908 Garrod published his classic monograph on this topic [17].

c) These sports may be extreme and therefore conspicuous examples of a principle with *much more widespread applicability*. Lesser chemical differences between human beings are so frequent that no human being is identical chemically to anyone else.

From these concepts Garrod drew more far-reaching conclusions, which are often overlooked. In a book published in 1931 and recently reprinted with a lengthy introduction by Scriver and Childs, Garrod suggested that hereditary susceptibilities or diatheses are a predisposing factor for most common diseases and not merely for the rare inborn errors of metabolism. These concepts were precursors of current work to delineate the specific genes involved in the etiology of common disease [46].

Throughout this book the principle of a genetically determined biochemical individuality will govern our discussions. Garrod's contribution may be contrasted with that of Adams. Apart from the "familial" occurrence of some hereditary diseases, Adams observed a number of phenomena that were not noted by Garrod, such as the late onset of some diseases, the intrafamilial correlation of age of onset, and the genetic predisposition leading to manifest illness only under certain environmental conditions. However, Adams did not have Mendel's paradigm. Therefore, his efforts could not lead to the development of an explanatory theory and coherent field of science. Garrod did have this paradigm and used it, creating a new area of research: human biochemical genetics.

1.6 Visible Transmitters of Genetic Information: Early Work on Chromosomes

Galton's biometric analysis and Mendel's hybridization experiments both started with visible phenotypic differences between individuals. The gene concept was derived from the phenotypic outcome of certain crossings. At the time when Mendel carried out his experiments nothing was known about a possible substantial bearer of the genetic information in the germ cells. During the decades to follow, however, up to the end of the nineteenth century, the chromosomes were identified, and mitosis and meiosis were analyzed. These processes were found to be highly regular and so obviously suited for orderly distribution of genetic information that in 1900 the parallelism of Mendelian segregation and chromosomal distribution during meiosis was realized, and chromosomes were identified as bearers of the genetic information. [11]

Many research workers contributed to the development of cytogenetics [2, 3]. O. Hertwig (1875) first observed animal fertilization and established the continuity of cell nuclei: *omnis nucleus e nucleo*. Flemming (1880–1882) discovered the separation of sister chromatids in mitosis; van Beneden (1883) established the equal and regular distribution of chromosomes to the daughter nuclei. Boveri (1888) found evidence for the individuality of each pair of chromosomes. Waldeyer (1888) (see [11]) coined the term "chromosome."

Meanwhile, Naegeli (1885) had developed the concept of "idioplasma," which contains – to use a modern term – the "information" for the development of the next generation. He did not identify the idioplasma with any specific part of the cell. W. Roux seems to have been the first to set out by logical deduction which properties a carrier of genetic information was expected to have. He also concluded that the behavior of cell nuclei during division would perfectly fulfill these requirements. The most important specific property of meiotic divisions, the ordered reduction of genetic material, was first recognized by Weismann.

These results and speculations set the stage for the identification of chromosomes as carriers of the genetic information, which followed shortly after the rediscovery of Mendel's laws and apparently independently by different authors (Boveri; Sutton; Correns, 1902; de Vries, 1903).

Chromosome studies and genetic analysis have remained intimately connected in cytogenetics ever since. Most basic facts were discovered and concepts developed using plants and insects as the principal experimental tools.

The development of human cytogenetics was delayed until 1956 when the correct number of human chromosomes was established as 46 by use of rather simple methods. It should be stressed that this delay could not be explained by the introduction of new cytological methods at that time. In fact, this discovery could have been made many years earlier. The delay was probably related to the lack of interest in human genetics by most laboratory-oriented medical scientists. Human genetics did not exist as a scientific discipline in medical schools since the field was not felt to be a basic science fundamental to medicine. Hereditary diseases were considered as oddities that could not be studied by the methodology of medical science as exemplified by the techniques of anatomy, biochemistry, physiology, microbiology, pathology, and pharmacology. Thus, most geneticists worked in biology departments of universities, colleges, or in agricultural stations. They were usually not attuned to problems of human biology and pathology, and there was little interest to study the human chromosomes. The discovery of trisomy 21 as the cause of Down syndrome and the realization that many problems of sex differentiation owe their origin to sex chromosomal abnormalities established the central role of cytogenetics in medicine. Further details in the development of cytogenetics are described in Chap. 2.

1.7 Early Achievements in Human Genetics

1.7.1 AB0 and Rh Blood Groups

The discovery of the AB0 blood group system by Landsteiner in 1900 [25] and the proof that these blood types are inherited (von Dungern and Hirschfeld, 1911 [47]) was an outstanding example of Mendelian inheritance applied to a human character. Bernstein in 1924 [7] demonstrated that A, B, and 0 blood group characters are due to multiple alleles at one locus. The combined efforts of Wiener, Levine, and Landsteiner 25–30 years later led to discovery of the Rh factor and established that hemolytic disease of the newborn owes its origin to immunological maternal-fetal incompatibility [26–28]. The stage was set for the demonstration in the 1960s that Rh hemolytic disease of the newborn can be prevented by administration of anti-Rh antibodies to mothers at risk [43; 56].

1.7.2 Hardy-Weinberg Law

Hardy [19], a British mathematician, and Weinberg [52], a German physician, at about the same time (1908) set out the fundamental theorem of population genetics, which explains why a dominant gene does not increase in frequency from generation to generation. Hardy published his contribution in the United States in *Science*. He felt that this work would be considered as too trivial by his mathematics colleagues. Weinberg was a practicing physician who made many contributions to formal genetics. He developed a variety of methods in twin research [51] and first elaborated methods to correct for biased ascertainment in recessive inheritance [53].

1.7.3 Developments Between 1910 and 1930

The years between 1910 and 1930 saw no major new paradigmatic discoveries in human genetics. Most of the data in formal genetics (such as linkage, nondisjunction, mutation rate) as well as the mapping of chromosomes were achieved by study of the fruit fly, largely in the United States, but also in many other parts of the world. Many scientists tried to apply the burgeoning insights of genetics to humans. British scientists exemplified by Haldane excelled in the elaboration of a variety of statistical techniques required to deal with biased human data. The same period saw the development of the basic principles of population genetics by Haldane and Fisher in England and by Wright in the United States. This body of knowledge became the foundation of population genetics and is still used by workers in that field. In 1918, Fisher was able to resolve the bitter controversies in England between the Mendelians, on the one hand, and followers of Galton (such as Pearson) on the other, by pointing out that correlations between relatives in metric traits can be explained by the combined action of many individual genes. Major steps in the development of medical genetics during this period were the establishment of empirical risk figures for mental and affective disorders by the Munich school of psychiatric genetics and the introduction of sound criteria for such studies.

1.8 Human Genetics, the Eugenics Movement, and Politics

1.8.1 United Kingdom and United States

The first decade of the century saw the development of eugenics in Europe and in the United States [1, 12, 22, 29, 41, 45]. Many biological scientists were impressed by their interpretation of an apparently all-pervasive influence of genetic factors on most normal physical and mental traits as well as on mental retardation, mental disease, alcoholism, criminality, and various other sociopathies. They became convinced that the human species should be concerned with encouragement of breeding between persons with desirable traits (positive eugenics) and discourage the sick, mentally retarded, and disabled from procreation (negative eugenics). Galton became the principal early proponent of such ideas. Various eugenic study units were established in the United States (Eugenics Record Office, Cold Spring Harbor) and the United Kingdom. Much of the scientific work published by these institutions was of poor quality. Particularly, many different kinds of human traits such as "violent temper" and "wandering trait" were forced into Mendelian straightjackets. Most serious geneticists became disenchanted and privately disassociated themselves from this work. For various reasons, including those of friendship and collegiality with the eugenicists, the scientific geneticists did not register their disagreement in public. Thus, the propagandists of eugenics continued their work with enthusiasm, and the field acquired a much better reputation among some of the public than it deserved. Thus, many college courses on eugenics were introduced in the United States.

These trends had several important political consequences. Eugenics sterilization laws were passed in many states in the United States, which made it possible to sterilize a variety of persons for traits such as criminality for which no good scientific basis of inheritance existed. The attitude that led to the introduction of these laws is epitomized by United States Supreme Court Justice Holmes' statement that "three generations of imbeciles are enough."

Eugenic influences also played an important role in the passing of restrictive immigration laws in the United States. Using a variety of arguments the proponents of eugenics claimed to show that Americans of northwestern European origin were more useful citizens than those of southern European origin or those from Asia. Since such differences were claimed to be genetic in origin, immigration from southern and eastern European countries and from Asia was sharply curtailed. Similar trends were also operative in the United Kingdom. While solid work in human genetics was carried out by a few statistical geneticists, there was also much eugenic propaganda, including that by the distinguished statistician Pearson, the successor to Galton's academic chair in London.

Kevles [23] has published a wide-ranging and insightful history of eugenics and human genetics in the Anglo-Saxon countries. His book is the most carefully researched and exhaustive study of the uses and abuses of eugenic concepts.

1.8.2 Germany

In Germany [5, 6, 18, 54, 55] eugenics took the name of *Rassenhygiene* from a book of that title published in 1895 by Ploetz [42]. The *Rassenhygiene* movement became associated with mystical concepts of race, Nordic superiority, and the fear of degeneration of the human race in general and that of the German *Volk* in particular by alcoholism, syphilis, and increased reproduction of the feebleminded or persons from the lower social strata. Some representatives of

this movement became associated with a dangerous type of sociopolitical prejudice: antisemitism. They warned the public against contamination of German blood by foreign, especially Jewish, influences. Most followers of the racial hygiene concept were nationalistic and opposed to the development of an open society that allows individual freedom and democratic participation. They shared this attitude with a significant segment of the educated classes in Germany. General eugenic ideas divorced from racism and other nationalist notions were often espoused by intellectuals who were concerned about the biological future of mankind. Thus, socialists publicized such views in Germany [18]. In 1931, two years before Hitler's coming into power, the German Society of Racial Hygiene added eugenics to its name. However, all efforts in this area soon became identified with the Nazi ideology.

Prominent German human geneticists identified themselves with the use of human genetics in the service of the Nazi state. Recognized scientists, such as Fischer, F. Lenz, Rüdin, and von Verschuer, accepted Nazi leadership and Nazi philosophy. While most of the propaganda for the new racial hygiene was not formulated by scientists but by representatives of the Nazi party, men such as Fischer and von Verschuer participated in spreading Nazi race ideology. Jews were declared foreign genetic material to be removed from the German *Volk* [48]. A eugenic sterilization law was already passed in 1933 that made forced sterilization obligatory for a variety of illnesses thought to be genetic in origin. Heredity courts were established to deal with appropriate interpretation of the sterilization law [39]. This law was hailed by some eugenicists in the United States even at the end of the 1930s [24]. Sterilization laws for eugenic indications were also passed in some Scandinavian countries around the same time but allowed voluntary (in contrast to forced) sterilization [39].

The exact role of the German human geneticists in the increasing radicalization and excesses of the application of Nazi philosophy has begun to be assessed by archival study [38]; von Verschuer's role in sponsoring twin and other genetic research by his former assistant Mengele in the Auschwitz concentration and extermination camp is clear. We have no record that any voices were raised in public by these men in protest against "mercy killings" of the mentally retarded and newborn children with severe congenital defects nor against the mass killings of Jews. The new historical evidence suggests that von Verschuer must have had some idea of such events, since he had continued contact with Mengele when the mass killings at Auschwitz were at their height. The "final solution" to the "Jewish problem" resulted in the murder of about 6 million

Jews in the early 1940s [44]. While we have no record that human geneticists favored this type of "solution," their provision of so-called scientific evidence for a justification of Nazi antisemitism helped to create a climate in which these mass murders became possible. This episode is one of the most macabre and tragic chapters in the history of man's inhumanity to man in the name of pseudoscientific nationalism.

1.8.3 Soviet Union

Eugenics was initiated in the Soviet Union [12, 18] in the 1920s by the establishment of eugenics departments, a eugenic society, and a eugenics journal. Eugenic ideals soon clashed with the official doctrine of Marxism-Leninism, however, and these efforts were abandoned by the late 1920s. Scientists who had become identified with eugenics left the field to work with plants and animals.

Interest in the medical application of human genetics nevertheless persisted. A large institute of medical genetics was established in the Soviet Union during the 1920s. Its director, L. E. Levit, disappeared in the 1930s, and human genetics was officially declared a Nazi science. The later ascendance of Lysenko stifled all work in genetics, including that of human genetics, and no work whatever was carried out in this field until the early 1960s, after Lysenko's domination had ceased [214]. The reintroduction of human genetics into the Soviet Union occurred by way of medical genetics. A textbook of medical genetics was published by Efroimson in 1964. A new institute of medical genetics was established in 1969 under the directorship of the cytogeneticist Bochkov. Work in many areas of medical genetics is now carried out in that institution and in other places.

1.8.4 Human Behavior Genetics

Vehement discussion continues regarding the role of genetic determinants in behavior, IQ, and personality. Some scientists entirely deny genetic influences on normal behavior or social characteristics such as personality and intellect. This attitude toward genetics is shared by some psychologists and social scientists and even a few geneticists who are concerned about the possible future political and social misuse of studies in human behavioral genetics and sociobiology that claim to show strong genetic determinants of intelligence and social behavior.

We do not agree with those who deny any genetic influence on behavior or social traits in humans. However, we also caution against a too ready acceptance of results from biometric comparison of twins and other relatives, which claim high heritabilities for many of these traits. Genetic data and pseudodata may be seriously misused by political bodies. However, as biologists and physicians impressed by biological variation under genetic control, we would be surprised if the brain did not also show significant variation in structure and function. Such variation is expected to affect intellect, personality, behavior, and social interactions. The extent to which genetic variation contributes to such traits, and especially the biological nature of such variation, will have to await further studies (see Chaps. 15–17).

1.9 Development of Medical Genetics (1950–the Present)

1.9.1 Genetic Epidemiology

In the 1940s and 1950s a number of institutions pioneered in research on epidemiology of genetic diseases. T. Kemp's institute in Copenhagen, J. V. Neel's department in Ann Arbor, Michigan, and A. C. Stevenson's in Northern Ireland and later in Oxford contributed much to our knowledge on prevalence, modes of inheritance, heterogeneity, and mutation rates of various hereditary diseases. Recent years have seen a renaissance in this area, with special attention to complex analysis of common diseases. Utilization of new laboratory methods including DNA techniques together with more powerful methods of biometric analysis [34, 37] and linkage studies provide powerful new approaches in this area.

1.9.2 Biochemical Methods

The years after World War II brought a rapid expansion in the field of human genetics by the development of both biochemical and cytological methods. Human genetics, which had been the concern largely of statistically oriented scientists, now entered the mainstream of medical research. The demonstration by Pauling et al. [40] that sickle cell anemia is a molecular disease was a key event in this area. The hemoglobins allowed detailed study of the consequences of mutation. The genetic code was found to be valid for organisms as far apart as viruses and humans. Many detectable mutations were found to be single amino acid substitutions, but deletions of various sorts and frameshift mutations similar to those discovered in micro-organisms were discovered. The nucleotide sequences of the hemoglobin genes were worked out using techniques developed in biochemistry and molecular genetics. Many inborn errors of metabolism were shown to originate in various enzyme deficiencies, each caused by a genetic mutation that changes enzyme structure. Methemoglobinemia due to diaphorase deficiency and glycogen storage disease were the first enzyme defects to be demonstrated (Sect. 7.1).

1.9.3 Genetic and Biochemical Individuality

Work on variants of the enzyme glucose-6-phosphate-dehydrogenase helped to establish the concept of extensive mutational variation. Biochemical individuality explained some drug reactions and led to the development of the field of pharmacogenetics [33]. Marked biochemical heterogeneity of human enzymes and proteins was first shown [20]. The uniqueness of humans, which is apparent by the physiognomic singularity of each human being was shown to apply at the biochemical and immunological level as well. Here, as in several other fields (such as the hemoglobin variants and the mechanism of sex determination), studies in humans led the way to generally valid biological rules. The significance of polymorphism for the structure of populations including that of man is being widely studied by population geneticists. The hypothesis that some expressed polymorphisms are the genetic substrate against which the environment acts to determine susceptibility and resistance to common disease led to the development of the field of ecogenetics [34]. The histocompatibility gene complex has become an important paradigm for the understanding of why several genes with related function occur in closely linked clusters. This locus may be of great importance to understand susceptibility to autoimmune and other diseases. More recently an enormous amount of apparently unexpressed genetic variation has been demonstrated at the DNA level.

1.9.4 Cytogenetics, Somatic Cell Genetics, Prenatal Diagnosis

After cytogenetic techniques became available, they were applied to detect many types of birth defects and intersex states. A specific type of malignancy, chronic myelogenous leukemia, was shown to be caused by a unique chromosomal aberration. Banding techniques developed by Caspersson in 1969 made it possible to visualize each human chromosome and gave cytogenetic methods added powers of resolution.

Soon, biochemical and cytogenetic techniques were combined in somatic cell genetics. Specific enzyme defects were identified in single cells grown in tissue cultures. The development of methods to hybridize human with mouse cells by Henry Harris and Watkins [21] and Ephrussi and Weiss [13] soon allowed the assignment of many genes to specific chromosomes and the construction of a human linkage map.

The developments in somatic cell genetics led to the introduction of prenatal diagnosis in the late 1960s, when amniocentesis at the beginning of the second trimester of pregnancy was developed. This allowed tissue cultures of amniotic cells of fetal origin, permitting both cytogenetic and biochemical characterization of fetal genotypes, assignment of sex, and the diagnosis of a variety of disorders in utero. In the early 1980s chorion villus biopsy – a procedure done during the first trimester of pregnancy – was introduced, and is being widely used. The discovery that neural tube defects are associated with increases in α-fetoprotein of the amniotic fluid permits intrauterine diagnosis of an important group of birth defects [8]. Ultrasound methods to visualize the placenta and to diagnose fetal abnormalities added to the diagnostic armamentarium. This noninvasive method allows phenotypic diagnosis of a variety of fetal defects more and more frequently.

Clinical Genetics. The field of clinical genetics is growing rapidly [30]. Many hospitals are establishing special clinics in which genetic diseases can be diagnosed and genetic counseling provided. The heterogeneity of genetic disease has been increasingly recognized. Genetic counseling is now intensified to provide patients and their families with information on natural history, recurrence risks and reproductive options. Screening programs of the entire newborn populations for diseases such as phenylketonuria are being introduced in many countries, and other screening programs such as those to detect carriers of Tay-Sachs disease have undergone extensive trials [10] (Chap. 18).

Accompanying these clinical developments was a period in which the evolution of new scientific concepts in human genetics per se became somewhat less prominent since many human geneticists were devoting their work to clinical problems to which they could make important practical contributions. With the advent of the new DNA techniques (Chap. 3) this has changed rapidly. Basic work in human genetics is now performed increasingly by a variety of scientists such as cell biologists, molecular biologists, biochemists, and others, who do not necessarily have training in human genetics [36]. However, human genetics is identified with medical genetics in many of its activities. The scientific developments of the past decades are being widely applied in practical medicine.

1.9.5 DNA Technology in Medical Genetics

Advances in molecular genetics and DNA technology are being applied rapidly to practical problems of medical genetics [49]. Since understanding of the hemoglobin genes was more advanced than that of other genetic systems, the initial applications related to the diagnosis of hemoglobinopathies. Several methods are now being utilized. Inherited variation in DNA sequence that is phenotypically silent was found to be common, supplying a vast number of DNA polymorphisms for study. Just as everyone's physiognomy is unique, each person (except for identical twins) has a unique DNA pattern. DNA variants are being used in family studies as genetic markers to detect the presence of closely linked genes causing monogenic diseases. Direct detection of genetic disease has been achieved by utilizing nucleotide probes that are homologous to the mutations that are searched for. The polymerase chain reaction, together with rapidly increasing knowledge on human DNA sequences, has opened up new opportunities for direct diagnosis at the DNA level. Occasionally, a specific restriction enzyme may detect the mutational lesion. Different DNA mutations at the same locus usually cause an identical phenotypic disease. This finding makes direct DNA diagnosis without family study difficult unless the specific mutation that causes the disease is known.

Efforts to construct a map of the human genome are nearing completion. Several hundred DNA markers that are evenly spaced over all chromosomes provide the necessary landmarks for detection of the genes for all monogenic diseases and aid in unraveling the contribution of specific genes to common diseases.

The use of DNA to treat genetic diseases is also under study. Current efforts are concerned with the insertion of the DNA of normal genes into somatic cells (somatic gene therapy). In vitro and animal experiments are under way using, for example, retroviruses as vectors for these genes. Human experiments have been initiated but no definitive cures have get been reported (See Chaps. 7; 19). Germinal gene therapy, i.e., insertion of normal genes into defective germ cells, or fertilized eggs for treatment of human genetic disease has never been carried out. Such an approach is highly controversial, and is even prohibited by law in some countries.

1.9.6 Unsolved Problems

Human genetics had been most successful by being able to guide work that was made possible by the development of techniques from various areas of biology using Mendelian concepts. Important basic frontiers that are still being extended concern problems of gene regulation, especially during embryonic development, control of the immune system and of brain function. Human genetics is likely to contribute to these problems by imaginative use of the study of genetic variation and disease applying newer techniques. In medical genetics, the problem of common diseases including many birth defects requires study of the specific genes involved in such diseases and many additional insights into the mechanisms of gene action during embryonic development.

At first glance, the history of human genetics over the past 40 years reads like a succession of victories. The reader could conclude that human geneticists of the last generation pursued noble science to the benefit of mankind. However, how will posterity judge current efforts to make use of our science for the benefit of mankind as we understand it? Will the ethical distinction between selective abortion of a fetus with Down syndrome and infanticide of severely malformed newborns be recognized by our descendants? Are we again moving down the "slippery slope?"

Conclusions

Theories and studies in human genetics have a long history. Observations on the inheritance of physical traits in humans can even be found in ancient Greek literature. In the 18th and 19th centuries observations were published on the inheritance of numerous diseases, including empirical rules on modes of inheritance. The history of human genetics as a theory-based science began in 1865, when Mendel published his *Experiments on Plant Hybrids* and Galton his studies on *Hereditary Talent and Character*. A very important step in the development of human genetics and its application to medicine came with Garrod's demonstration of a Mendelian mode of inheritance in alcaptonuria and other inborn errors of metabolism (1902). Further milestones were Pauling's elucidation of sickle cell anemia as a "molecular disease" (1949), the discovery of genetic enzyme defects as the causes of metabolic disease (1950s, 1960s), the determination that there are 46 chromosomes in humans (1956), the development of prenatal diagnosis by amniocentesis (1968–1969), and the large-scale introduction of molecular methods (1970s, 1980s, and 1990s).

Concepts appropriated from human genetics have often influenced social attitudes. Abuses have occurred, such as with legally mandated sterilization, initially in the United States and later more extensively in Nazi Germany, where the killing of mentally impaired patients was followed by the genocide of Jews and Gypsies. More recently, controversies have attended a number of activities, such as the Human Genome Project and some of the applications of genetic knowledge in the diagnosis, prevention, and therapy of an ever wider range of disorders.

References

1. Allen GE (1975) Genetics, eugenics and class struggle. Genetics 79 : 29–45
2. Baltzer F (1962) Theodor Boveri – Leben und Werk. Wissenschaftliche Verlagsgesellschaft, Stuttgart
3. Barthelmess A (1952) Vererbungswissenschaft. Alber, Freiburg
4. Bearn AG (1993) Archibald Garrod and the individuality of man. Clarendon, Oxford
5. Becker PE (1988) Zur Geschichte der Rassenhygiene. Thieme, Stuttgart (Wege ins Dritte Reich, vol 1)
6. Becker PE (1990) Sozialdarwinismus, Rassismus, Antisemitismus und völkischer Gedanke. Thieme, Stuttgart (Wege ins Dritte Reich, vol 2)
7. Bernstein F (1924) Ergebnisse einer biostatistischen zusammenfassenden Betrachtung über die erblichen Blutstrukturen des Menschen. Klin Wochenschr 3 : 1495–1497
8. Brock DJH (1977) Biochemical and cytological methods in the diagnosis of neural tube defects. Prog Med Genet 2 : 1–40
9. Capelle W (1953) Die Vorsokratiker. Kröner, Stuttgart
10. Committee for the Study of Inborn Errors of Metabolism, Assembly of Life Sciences NRC (1975) Genetic screening. Programs, principles, and research. National Academy of Sciences, Washington
11. Cremer T (1986) Von der Zellenlehre zur Chromosomentheorie. Springer, Berlin Heidelberg New York
12. Dunn LC (1962) Cross currents in the history of human genetics. Am J Hum Genet 14 : 1–13
13. Ephrussi B, Weiss MC (1965) Interspecific hybridization of somatic cells. Proc Natl Acad Sci USA 53 : 1040
14. Falk R (1984) The gene in search of an identity. Hum Genet 68 : 195–204
15. Galton F (1865) Hereditary talent and character. Macmillans Magazine 12 : 157
16. Garrod AE (1902) The incidence of alcaptonuria: a study in chemical individuality. Lancet 2 : 1616–1620
17. Garrod AE (1923) Inborn errors of metabolism. Frowde, London (reprinted by Oxford University Press, London 1963)
18. Graham LR (1977) Political ideology and genetic theory: Russia and Germany in the 1920's. Hastings Cent Rep 7 : 30–39
19. Hardy GH (1908) Mendelian proportions in a mixed population. Science 28 : 49–50
20. Harris H (1969) Enzyme and protein polymorphism in human populations. Br Med Bull 25 : 5
21. Harris H, Watkins JF (1965) Hybrid cells from mouse and man: artificial heterokyryons of mammalian cells from different species. Nature 205 : 640
22. Joravsky D (1970) The Lysenko affair. Harvard University Press, Boston
23. Kevles DJ (1985) In the name of eugenics. Genetics and the uses of human heredity. Knopf, New York
24. Kühl S (1994) The Nazi connection. Eugenics, American racism, and German national socialism. Oxford University Press, Oxford
25. Landsteiner K (1900) Zur Kenntnis der antifermentativen, lytischen und agglutinierenden Wirkungen des Blutserums und der Lymphe. Zentralbl Bakteriol 27 : 357–362
26. Landsteiner K, Wiener AS (1940) An agglutinable factor in human blood recognized by immune sera for rhesus blood. Proc Soc Exp Biol 43 : 223
27. Levine P, Stetson RE (1939) An unusual case of intragroup agglutination. JAMA 113 : 126–127
28. Levine P, Burnham L, Katzin EM, Vogel P (1941) The role of isoimmunization in the pathogenesis of erythroblastosis fetalis. Am J Obstet Gynecol 42 : 925–937
29. Ludmerer K (1972) Genetics and American society. Johns Hopkins University Press, Baltimore
30. McKusick VA (1975) The growth and development of human genetics as a clinical discipline. Am J Hum Genet 27 : 261–273
31. Mendel GJ (1865) Versuche über Pflanzenhybriden. Verhandlungen des Naturforschenden Vereins, Brünn
32. Motulsky AG (1959) Joseph Adams (1756–1818). Arch Intern Med 104 : 490–496
33. Motulsky AG (1972) History and current status of pharmacogenetics. In: Human Genetics. Proceedings of the 4th International Congress of Human Genetics. Paris, September 1971. Excerpta Medica, Amsterdam, pp 381–390
34. Motulsky AG (1977) Ecogenetics: genetic variation in susceptibility to environmental agents. In: Human Genetics. Proceedings of the 5th International Congress of Human Genetics. Mexico City, 10–15 October 1976. Excerpta Medica, Amsterdam, pp 375–385
35. Motulsky AG (1978) The genetics of common diseases. In: Morton NE, Chung CS (eds) Genetic epidemiology. Academic, New York, pp 541–548
36. Motulsky AG (1978) Presidential address: Medical and human genetics 1977: trends and directions. Am J Hum Genet 30 : 123–131
37. Motulsky AG (1984) Genetic epidemiology. Genet Epidemiol 1 : 143–144
38. Müller-Hill B (1984) Tödliche Wissenschaft. Rowohlt, Hamburg
39. Nachtsheim H (1952) Für und wider die Sterilisierung aus eugenischer Indikation. Thieme, Stuttgart
40. Pauling L, Itano HA, Singer SJ, Wells IC (1949) Sickle cell anemia: a molecular disease. Science 110 : 543
41. Penrose LS (1967) Presidential address – the influence of the English tradition in human genetics. In: Crow JF, Neel JV (eds) Proceedings of the 3rd International Congress of Human Genetics. Johns Hopkins University Press, Baltimore, pp 13–25
42. Ploetz A (1895) Die Tüchtigkeit unserer Rasse und der Schutz der Schwachen: ein Versuch über Rassenhygiene

und ihr Verhältnis zu den humanen Idealen, besonders zum Sozialismus. Fischer, Berlin

43. Pollack W, Gorman JG, Freda VJ (1969) Prevention of Rh hemolytic disease. Prog Hematol 6 : 121–147

44. Reitlinger G (1961) The final solution. Barnes, New York

45. Rosenberg CE (1976) No other gods. On science and American social thought. Johns Hopkins University Press, Baltimore

46. Scriver CR, Childs B (1989) Garrod's inborn factors in disease. Oxford University Press, Oxford

47. Von Dungern E, Hirszfeld L (1911) On the group-specific structures of the blood. III. Z Immunitatsforsch 8 : 526–562

48. Von Verschuer O (1937) Was kann der Historiker, der Genealoge und der Statistiker zur Erforschung des biologischen Problems der Judenfrage beitragen? Forschungen zur Judenfrage 2 : 216–222

49. Weatherall DJ (1985) The new genetics and clinical practice. Oxford University Press, Oxford

50. Weiling F (1994) Johann Gregor Mendel. Forscher in der Kontroverse. V. Med Genet 6 : 35–50

51. Weinberg W (1901) Beiträge zur Physiologie und Pathologie der Mehrlingsgeburten beim Menschen. Arch Gesamte Physiol 88 : 346–430

52. Weinberg W (1908) Über den Nachweis der Vererbung beim Menschen. Jahreshefte des Vereins für vaterländische Naturkunde in Württemberg 64 : 368–382

53. Weinberg W (1912) Weitere Beiträge zur Theorie der Vererbung. IV. Über Methode und Fehlerquellen der Untersuchung auf Mendelsche Zahlen beim Menschen. Arch Rass Ges Biol 9 : 165–174

54. Weindling PF (1989) Health, race and German politics between national unification and nazism 1870–1945. Cambridge University Press, Cambridge

55. Weingart P, Kroll J, Bayertz K (1988) Rasse, Blut und Gene. Suhrkamp, Frankfurt

56. Zimmerman D (1973) Rh. The intimate history of a disease and its conquest. Macmillan, New York, p 371

2 The Human Genome: Chromosomes

*Before a renewed, careful control has been made of the chromosome number
in spermatogonial mitoses of man we do not wish to generalize our present findings
into a statement that the chromosome number of man is 2n = 46, but it is hard to avoid
the conclusion that this would be the most natural explanation of our observations.*

H.J. Tjio and A. Levan, Hereditas 42: 1–6, 1956

2.1 Human Cytogenetics: A Successful Late Arrival

The chromosome theory of Mendelian inheritance was launched in 1902 by Sutton and Boveri. In the same year Garrod, establishing the autosomal-recessive mode of inheritance for alkaptonuria and commenting on metabolic individuality in general, created the paradigm of "inborn errors of metabolism." Simple modes of inheritance were soon established for many other human disorders. A few years later Bridges (1916) [10] examined in *Drosophila* the first case of a disturbance in chromosome distribution during meiosis and named it "nondisjunction." Cytogenetics of animals and plants flourished during the first half of the century, and almost all important phenomena in the field of cytogenetics were discovered during this period. Moreover, cytogenetic methods helped to elucidate many basic laws of mutation.

One might have expected that these results and concepts of general cytogenetics would soon flow into human genetics, helping to explain phenomena that are genetic in origin but did not fit the expectations derived from Mendel's laws. This transfer, however, was not to occur until 50 years later. The age of human cytogenetics began only in the 1950 s when Tjio and Levan (1956) [131] and Ford and Hamerton (1956) [31] established the diploid human chromosome number of 46. Lejeune et al. (1959) [69] discovered trisomy 21 in Down syndrome, and Ford et al. (1959) [32] and Jacobs and Strong (1959) [56] established that Turner and Klinefelter syndromes are caused by X chromosomal anomalies.

The late arrival of human cytogenetics is usually ascribed to shortcomings in the preparation methods of chromosomes. Indeed, the jumbled masses of chromosomes in old illustrations demonstrate the difficulties encountered by the pioneers who tried to count human chromosomes. Still, it is hard to conceive that development of more adequate methods would have been delayed so long had the cytogeneticists realized there are human anomalies awaiting explanation. Some human geneticists did consider the possibility that certain anomalies are due to chromosomal aberrations. For example. Waardenburg in 1932 [135] remarked (our translation):

The stereotyped recurrence of a whole group of symptoms among mongoloids offers an especially fascinating problem. I would suggest to the cytologists that they examine whether it is possible that we are dealing with a human example of a certain chromosome aberration. Why should it not occur occasionally in humans, and why would it not be possible that – unless it is lethal – it would cause a radical anomaly of constitution? Someone should examine in mongolism whether possibly a "chromosomal deficiency" or "nondisjunction" – or the opposite, "chromosomal duplication" – is involved. ... My hypothesis has at least the advantage of being testable. It would also explain the possible influence of maternal age.

He then remarked that the very rare familial occurrence of Down syndrome and the concordance of monozygotic twins are compatible with this hypothesis. Waardenburg, a practicing ophthalmologist who in his spare time became one of the foremost specialists for inherited eye disease, had no opportunity to put his suggestion into practice. The cytogeneticists of his time, however, lacking appropriate methods, did not carry through. The spark was there, but nothing caught fire.

2.1.1 History and Development of Human Cytogenetics

First Observations on Human Mitotic Chromosomes [130]. Research on human cytogenetics could be said to have begun with the work of Arnold (1879) [3] and Flemming (1882) [33], who for the first time examined human mitotic chromosomes. In the following years a number of reports appeared with various estimates of the number of human chromosomes. Outstanding among these early contributions was the work of von Winiwarter (1912) [141]. He examined the testicular histology from four men aged 21, 23, 25, and 41 years, fixed and cut into sections only 7.5 μm in diameter, which impeded chromosome counting. Thirty-two spermatogonial mitoses could be evaluated; for 29 he counted 47 elements, 46 for two oth-

Fig. 2.1. An early picture of a spermatogonial mitosis. (From von Winiwarter 1912 [134])

Fig. 2.2. The sex bivalent in a first meiotic anaphase. (From Painter 1923 [91])

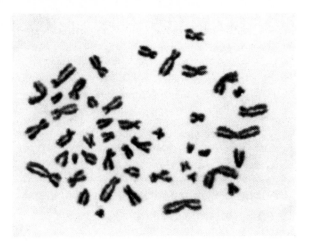

Fig. 2.3. A metaphase of a human embryonic lung fibroblast grown in vitro. From the first report in which the human chromosome number was established as 46: (From Tjio and Levan 1956 [131])

ers, and 49 for one (Fig. 2.1). Sixty diplotene plates were evaluated; 57 showed 24 elements, two seemed to have 25, and one 23. In diplotene, he even observed the sex chromosomes but explained them as one chromosome that was transported to one pole. He concluded that human males have 47 and females 48 chromosomes. His evidence for females was slender, as he found only three clear oogonial mitoses in a 4-month-old fetus. The results were compatible with there being 48 chromosomes.

The report with the strongest impact was that of Painter in the 1920 s [91]. He examined the testicles of three individuals from the Texas State Insane Asylum. In all three cases the cause for the removal of the testes was "excessive self-abuse coupled with certain phases of insanity." The results were based mainly on examinations from two of the three individuals. In a preliminary report (1921) he described the chromosome number as 46 or 48, but in the definitive report (1923) he had decided in favor of 48 chromosomes. In first meiotic divisions he was able to demonstrate the sex bivalent, consisting of the X and Y chromosomes, which at anaphase migrated to opposite poles (Fig. 2.2).

In the following years, a chromosome number of 48 in humans was supported in a number of reports [351]. However, two technical difficulties impeded further progress:

1. Sectioning by the usual histological techniques often disturbed mitoses.
2. The chromosomes tended to lie on top of each other and even to clump together.

These difficulties were ultimately overcome by:

a) Use of suspensions of intact cells that might be squashed or simply air-dried rather than of histological sections.
b) Subjecting cells to a brief treatment with a hypotonic solution, causing them to swell and burst, thus spreading out the chromosomes for better definition.

The hypotonic shock technique paved the way for easy chromosome counting [52, 53]. (Interestingly enough, even 30 years later Painter's estimate of 48 was so strongly imprinted on investigators' minds that in the first study on human chromosomes using the new technique the human chromosome number was reported as 48 [52]).

An Old Error Is Corrected and a New Era Begins [32]. In the summer of 1955 Levan, a Swedish cytogeneticist, visited Hsu in New York and learned the technique of squash preparation using hypotonic shock. He and Tjio then improved the technique by shortening the hypotonic treatment and adding colchicine, a chemical that arrests cells in metaphase to increase the number of countable cells. They examined lung fibroblasts of four human embryos. To their surprise they found a chromosome number of 46 in most of 261 metaphases. Figure 2.3 shows one example. In discussing their findings they mentioned three Swedish investigators who had studied mitoses in liver cells of aborted human embryos a year earlier. This study was discontinued because they were unable to find 48 chromosomes; in all cells, they found only 46.

This evidence was soon supplemented by Ford and Hamerton (1956) [31]. They examined testicular tissue from three men of relatively advanced age. In the great majority of all metaphase I cells, 23 bivalents were found, confirming the results of Tjio and Levan. Spermatogonial mitoses were difficult to find, but a few clear counts confirmed the chromosome number to be 46.

With these results, the stage was set for the development of clinical cytogenetics. Still it was to be almost another three years, however, before the first abnormal karyotypes in humans were reported.

Solution of an Old Riddle: Down Syndrome (Mongolism) Is Due to Trisomy 21. In the spring of 1959 Waardenburg's suggestion was finally followed. Lejeune et al. [69] reported chromosome studies from fibroblast cultures in nine children with Down syndrome. Fifty-seven diploid cells were regarded as technically perfect. In all of them the chromosomes numbered 47. The supernumerary chromosome was described as small and "telocentric." Meiotic nondisjunction was suggested as the most likely explanation for the additional chromosome.

First Reports on Trisomies and Monosomies of Sex Chromosomes. Barr and Bertram [4] discovered the "X chromatin," an intranuclear body 0.8–1.1 µm in size, which is commonly located at the periphery of the interphase nuclei of females and is not present in males. The discovery was accidental since it originated in an investigation of the effects of fatigue on the central nervous system in cats. What first seemed to be a sex difference only in the neurons of cats turned out to be a normal finding characteristic of the nuclear inheritance of female mammals including human females. Corresponding structures, the drumsticks, were found by Davidson and Smith in 1954 [19] in polymorphonuclear neutrophil leukocytes. The obvious next step was the examination of X chromatin in cells of patients with disturbances in sexual development. Here, most male patients with the Klinefelter syndrome (Sect. 2.2.3.2) turned out to be X chromatin-positive in spite of their predominantly male phenotype, whereas most female patients with the Turner syndrome (Sect. 2.2.3.3) were X chromatin-negative – again in contrast to their female phenotype. If the X chromatin was directly related to the X chromosomes, this finding pointed to X chromosome anomalies in these two syndromes.

This suspicion was strengthened when the frequency of X-linked color vision defects in patients with Klinefelter syndrome was found to be lower than in normal males but much higher than would be expected in XX females [85, 98].

This situation found its explanation when Jacobs and Strong (1959) [56], examining the chromosomes from bone marrow mitoses in Klinefelter patients, found 47 chromosomes, whereas both parents had a normal karyotype. The supernumerary chromosome belonged to the group of chromosomes including X chromosomes; the karyotype was tentatively identified as XXY.

Soon after this first report, the XXY karyotype in the Klinefelter syndrome was confirmed in many more cases and is now known as the standard karyotype in this condition. At the same time, the result in the Klinefelter syndrome was complemented by chromosome examinations in another syndrome in which a discrepancy between phenotypic and nuclear sex seemed to exist: the Turner syndrome. Ford et al. (1959) [32] showed that the karyotype had only 45 chromosomes, obviously with one X and no Y chromosome. A third anomaly, with 47 chromosomes and three X chromosomes, was soon described in a slightly retarded woman with dysfunction of the sexual organs (Jacobs et al. 1959) [57]. The analytic possibilities afforded by human sex chromosome anomalies for sex determination in humans are discussed in Sect. 2.2.3.

Birth of Human Cytogenetics 1956–1959: A Scientific Revolution. Kuhn [63] pointed to the difference between the progress of "normal science" and the occasional occurrence of "scientific revolutions" (see "Introduction"). From the standpoint of human genetics, the development of cytogenetics between 1956 and 1960 resembled such a "revolution." Based on a new method rather than on a new concept, the whole field attained a new dimension. Since only methods and not concepts changed, this advance did not render most earlier work outdated but supplemented it in many directions. Any discussion of gene regulation, linkage, the structure of genetic material, spontaneous and induced mutations, population genetics, human evolution, and the practical use of genetic knowledge in the prevention of genetic disease were now obsolete without due regard to human cytogenetic data and concepts.

From the viewpoint of experimental geneticists, human cytogenetics appeared much humbler. The many advances were viewed as the belated application to humans of concepts that had been known for many years, sometimes even half a century or more. More recently, human cytogenetics has reached the stage at which the unique advantage of human material for the solution of more general biological questions such as localization of gene action during interphase is now being explored.

What triggered this revolution? As is often the case, it was a technical improvement: the hypotonic treatment for spreading chromosomes, accompanied by the examination of single isolated nuclei rather than tissue sections.

Paradigm Group in Early Human Cytogenetics. Tjio and Levan [131] discovered the correct number of chromosomes but – apparently too far removed from medicine – did not see the potential application to human pathology. This step, taken by a group of British scientists, was one in which concepts of basic cytogenetics underwent a most fortuitous combination with experience in medicine. The same step, however, was taken by a French scientist outside the academic structure of medicine, J. Lejeune [69]. During a scientific revolution, the group of scientists working on the new paradigm usually creates its own network of scientific interaction. The early phases of human cytogenetics offer an interesting case of research in the history of science. A participant, D. G. Harnden, gave us the following interesting information.

The leading figures in the British group were C. Ford in Harwell and W. M. Court Brown in Edinburgh, both working in units sponsored by the Medical Research Council. Ford's interest in human chromosomes grew out of his work on mouse tumors and meiotic cells; Court Brown decided to work with human chromosomes because as an epidemiologist he felt it was necessary to combine epidemiological with basic biological studies. The two groups soon established relationships; for example, Patricia Jacobs, a nonmedical cytogeneticist, was sent by Court Brown to Ford where she adapted the bone marrow culture technique developed by Lajtha for the examination of human chromosomes. Harnden developed the technique of growing fibroblasts from skin biopsies, which he felt were more readily available than bone marrow.

The Edinburgh group was located in a hospital and had easy access to clinical material. Here, a physician, J. Strong examined the case of Klinefelter syndrome. Harwell, where Ford worked, was an atomic energy biological station and had no direct hospital connections; however, a relationship with Guy's Hospital in London soon developed, and P. Polani suggested looking at Turner syndrome. Cooperation between the two groups was intensive; there was a great deal of interaction by letter, telephone, visits, and exchange of material. Human geneticists, such as P. Polani, L. S. Penrose, and J. Edwards, sent material to Harwell and advised the clinically inexperienced laboratory workers on medical matters.

The idea of examining Down syndrome offered itself to the British workers as the obvious next choice after the search for aberrations in Klinefelter and Turner syndromes led to success. The idea seems to have originated independently in Harwell and Edinburgh, where the workers were well advanced with their study before they knew of Lejeune's work.

The success of the two British groups was made possible by a lucky combination of persons with different but complementary backgrounds. Close cooperation developed for a few years during which the "paradigm" shared by the group revealed its explanatory power. Later, the cooperation slowly abated. At the same time, however, two other investigators had independently recognized the possibilities of the new methods. One was Lejeune in France; the other was the team of Fraccaro and Lindsten in Sweden, who started work on the Turner syndrome without knowledge of the investigations at Harwell.

Steps in the Development of Human Cytogenetics. The most important steps in the development of human cytogenetics were as follows:

1956	Tjio and Levan and Ford and Hamerton established the number of chromosomes in the diploid human cell [131; 31].
1959	Lejeune discovered trisomy 21 in Down's syndrome; Ford et al. and Jacobs and Strong found the XXY karyotype for the Klinefelter syndrome and the XO karyotype for the Turner syndrome.
1960	Moorhead et al. [80] published the method for chromosome preparation from short-term lymphocyte cultures. Two autosomal trisomies, later identified as trisomies 13 and 18, were described by Patau et al. [94] and Edwards et al. [27]. Nowell and Hungerford [86] described the Philadelphia chromosome in chronic myeloid leukemia.
1963	The first deletion syndrome, the cri du chat syndrome, was observed by Lejeune et al. [70].
1964/65	Schroeder et al. (1964) [116] and German et al. (1965) [41] discovered a genetically determined increased chromosome instability in Fanconi anemia and Bloom syndrome, respectively.
1968/70	Chromosome banding techniques were introduced. This permitted unequivocal identification of all human chromosomes [13].
1975	Yunis [137] introduced high-resolution banding methods.
Late 1980	Nonradioactive in situ hybridization methods: "chromosome painting."

Clinical Cytogenetics: The Most Popular Speciality of Human Genetics. Between 1960 and the late 1980ies, human – and especially clinical – cytogenetics developed into the most popular branch of human genetics. One reason was that the causes of many previously unexplained malformation syndromes became apparent. Another likely reason is that after relatively simple manipulations a "real" particulate appearance could actually be seen under the microscope. Visual images appeal to the medically trained and to many nonmedical biologists. In striking contrast, the more abstract concepts of formal genetics and population genetics do not attract the majority of physicians and biologists.

The great increase in the popularity of clinical cytogenetics is all the more remarkable since during the first decade almost no practical significance of these results was seen for medical therapy or prevention apart from diagnosis and genetic counseling. This changed dramatically when antenatal diagnosis became possible. In the last decade of the 20th century, molecular genetics has become the most popular field of human genetics.

2.1.2 Normal Human Karyotype in Mitosis and Meiosis

2.1.2.1 Mitosis

Cell Cycle. Figure 2.4 diagrams the cell cycle of a dividing mammalian cell. After mitosis the cell is in the G_1 phase; all chromosomes are present in nonduplicated form. If the cell does not divide any more, for example, as a mature neuron, the chromosomes remain in this state (G_0). A dividing cell now enters the next mitotic cycle. RNA and proteins are synthesized, and the cell becomes ready for DNA replication, which takes place in the S phase. Various parts of the chromosomes replicate asynchronously, as evidenced when [^{3}H]thymidine is added at a certain time during the S phase. Then, only chromosomes that have not finished replication take up the labeled compound and can be identified by autoradiography.

A certain amount of "unscheduled" or repair replication goes on during the G_2 phase, in which the cell prepares for mitosis (M). During the G_1 phase, the material of every chromosome of the diploid set (2 n) is present once. In the G_2 phase, on the other hand, every chromosome has doubled into two identical elements that are called sister chromatids. The material of every chromosome is now present twice (2×2 n = 4 n). During or after replication, the two sister chromatids exchange segments repeatedly so that the two chromatid arms of a mitotic chromosome have parts of both chromatids (Sister Chromatid Exchange, SCE). This can be made visible by using a specific staining technique after treatment with bromodeoyuridine – a thymine analogue (Fig. 2.5) [65].

Mitosis. The phases of mitosis are set out in Fig. 2.6. Mitosis starts by condensation of chromatin (Fig. 2.6 a; early prophase). At the end of prophase the chromosomes are clearly visible; the two sister chromatids lie side by side. Meanwhile, the nuclear membrane has dissolved, the nucleolus has disappeared, and the spindle is formed. The spindle consists of microtubules formed by a protein called tubulin and are visible under the microscope as spindle fibers. They connect the centromeric regions of the chromosomes with the centrioles. With dissolution of the nuclear membrane, prophase is finished, and the cell is now in metaphase. The centromeres are located in the equatorial plane between the two spindle poles. Now the two chromatids of each chromosome begin to separate, until they are connected only at the centromeric region. Finally, the centromeres also separate, to form half-chromosomes that are drawn to the opposite poles by the spindle fibers. The function of spindle microtubules can be demonstrated by colchicine treatment, which inhibits aggregation of tubulin and dissolves microtubules. This disturbs chromosome arrangement in the equatorial plane and inhibits their anaphase movement. Separation of chromatids oc-

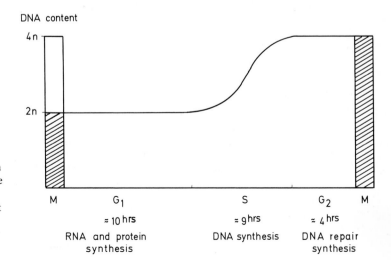

Fig. 2.4. Cell cycle of a dividing mammalian cell. In the G_1 phase the diploid chromosome set (2 n) is present once. After DNA synthesis (S phase) the diploid chromosome set is present in duplicate (4 n). *M*, Mitosis; ▨, DNA content during mitosis. See text for details

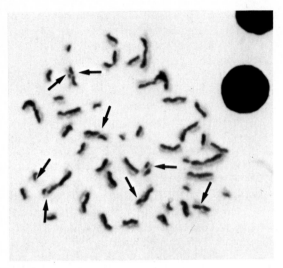

Fig. 2.5. Sister chromatid exchanges in a normal human metaphase. *Arrows,* locations of sister strand exchanges. (Courtesy of Dr. T. M. Schroeder-Kurth)

curs even in the presence of colchicine. In the last phase of mitosis, telophase, chromosomes are decondensed, spindle fibers disintegrate, the tubulin is stored in the cell, a new nuclear membrane is formed, and cell division begins. Chromosomes can most easily be examined in metaphase.

2.1.2.2 Preparation and Staining
of Mitotic Metaphase Chromosomes

Preparation. [48, 62, 119] In principle, chromosome preparations can be made from all tissues and all suspensions that contain mitoses. In humans, direct preparations from bone

marrow and preparations from short-term blood cultures or from long-term fibroblast cultures or other growing cells can be used. The most convenient technique is blood culture, as blood from patients is easily available, whereas bone marrow puncture or skin biopsy for fibroblast cultivation is more invasive and unpleasant. Bone marrow preparations have the advantage that in vivo mitoses can be examined directly without delay.

The blood of healthy individuals contains no dividing cells. Therefore cell divisions must be stimulated artificially. This is possible, for example, by phytohemagglutinin (PHA). One hour after PHA incubation of a blood sample, the small (T) lymphocytes show RNA synthesis, and about 24 h later DNA synthesis follows. The leukocyte suspension is grown in a culture medium for 72 h, and then chromosome preparations are made. To arrest as many cells as possible in the prometaphase or metaphase, spindle formation is prevented by a drug with colchicine-like effect, preferably colcemid. Under special conditions, culture time may be reduced to 48 h.

To obtain preparations in which the chromosomes are spread out in one plane, the cells are treated for a short time (10–30 min) with a hypotonic solution. The cells are then fixed with ethanol and acetic acid; a drop of the cell suspension is placed on the slide, air-dried, and stained.

Bone marrow preparations require sternal or iliac crest puncture. Cells are cultivated for only about 2 h with colcemid. Preparation differs in some details from that described above. Fibroblast cultures are prepared from skin biopsies, and the skin is minced into very small pieces and grown in culture medium in such a way that the tissue particles stick to the surfaces. After about 10 days cells start growing on these surfaces and after about 21 days, are brought in suspension, prepared, stained, and examined.

Staining. The simplest staining methods use Giemsa solution, 2% acetic orcein, or 2% karmin solution. These dyes stain the entire chromosome uniformly and intensively. To obtain a

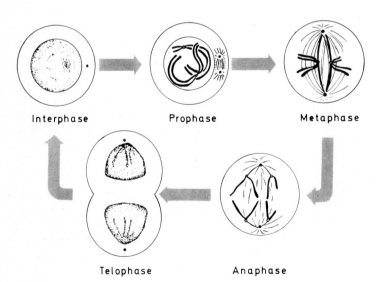

Interphase   Prophase   Metaphase

Telophase   Anaphase

Fig. 2.6. Mitosis. Only 2 of the 46 chromosomes are drawn. See text for details. (Courtesy of Dr. W. Buselmaier)

more detailed picture of chromosome structure and to identify single chromosomes or chromosome segments, banding
methods are used.

Banding Methods. Caspersson et al. [13] discovered that fluorescence of chromosomes after quinacrine mustard staining
shows a distinctive sequence of bands for each chromosome.
Each human chromosome could be identified by this staining
method. Soon afterward, it was shown that very similar
banding patterns could be elicited by Giemsa staining with
certain additional techniques.

At the Paris conference in 1971 on standardization and nomenclature of chromosomes [92], all the data available at
that time were compared. It turned out that all methods revealed the same structures, although some techniques exposed some chromosome segments more clearly, while others
worked better with other segments.

Available Methods [25]. The various types of bands were
named after the techniques by which they are revealed most
clearly.

a) Q bands (quinacrine) are the fluorescing bands visible after
   staining with quinacrine mustard or similar compounds.
b) G bands (Giemsa) are revealed by Giemsa staining with
   various additional techniques, which insure that only the
   most readily staining chromosome segments take up the
   dye. Q and G bands are identical.
c) R bands (reverse) are stained after controlled denaturation by heat. They are located between the Q or G bands,
   behaving as a photographic negative in relation to its positive image.
d) C bands (constitutive heterochromatin) are localized in
   the pericentromeric regions.
e) T bands (telomeric) mark the telomeric regions of chromosomes.

For references regarding techniques, see [127].

Differences Revealed by the Banding Methods. The
chemical differences revealed by the banding methods
are still under investigation. Two main hypotheses are
usually discussed: the DNA hypothesis and the protein
hypothesis. The DNA hypothesis is based on the observation that various parts of human chromosomes
differ in their content of A-T (adenine-thymine) and
G-C (guanine-cytosine) base pairs. Quinacrine is attached mainly to A-T rich segments [25, 90]. The protein hypothesis, on the other hand, is based on the observation that proteolytic treatment induces the appearance of G bands. Different kinds of DNA are
linked in the chromosome to different protein species.
The banding pattern probably depends on properties
of the combined DNA-protein complex.

A number of functional differences have been worked
out between Q bright (G dark) and Q dark (G bright)
bands (from [127]) (Table 2.1).

Silver Staining of Nucleolus Organizer Regions [42, 114, 118]. A
silver staining method is specific for the nucleolus organizer
regions. They appear as black dots on the yellow-brownish

Table 2.1. Characterization of human G dark and G bright bands

G bright	G dark
Lower AT/GC ratio	Higher AT/GC ratio
Rich in SINE repeats and Alu sequences (see Sect. 3.1.1.1)	Rich in LINE repeats (see Sect. 3.1.1.1)
Early replicating	Late replicating
Correspond to pachytene interchromomeres	Correspond to pachytene chromomeres
Contain "housekeeping" genes	Genes tend to be tissue specific
Rich in transcribed genes	Sparse genes, simple sequence DNA

The bands have been maintained amazingly well during evolution (Sect. 14.2.1), and chromosome breaks occur mainly
near the borders of bands. The order within chromosomes,
which is represented in part by the bands, appears to be
much more far-reaching than shown in Table 2.1. We return
to this problem in later chapters (see Saitoh and Laemmli
1994 [108]).

background of the chromosomes (Fig. 2.7). Only those nucleolus organizers are stained that were functionally active
during the preceding interphase [118]. Methods of "chromosome painting (FISH) are described in Sect. 3.1.3.3.

Chromosomes from Human Spermatozoa. A method has been
described for making chromosome preparations directly
from human spermatozoa through their
incubation with zona pellucida-free golden hamster oocytes
[106] which has been used to study human spermatozoa
[76]. Since no fertilization takes place, the mixing of human
sperm and hamster eggs is ethically entirely acceptable.

2.1.2.3 Normal Human Karyotype in Mitotic Metaphase Chromosomes

Banding Techniques. The human karyotype stained
with a number of banding techniques is shown in
Fig. 2.8. Every chromosome can be identified. Figure 2.10 gives a schematic representation of G or
Q bands, together with the number assigned to every
band. The single chromosomes, together with their
most frequently observed "normal" variants, may be
described as follows [54]:

Individual Characterization of Human Chromosomes.
Group A (nos. 1–3). Large, metacentric and submetacentric chromosomes; no. 1 is the largest metacentric
chromosome. The centromere is in the middle, the
centromere index (length of short arm divided by total chromosome length × 100) being 48 or 49. Close

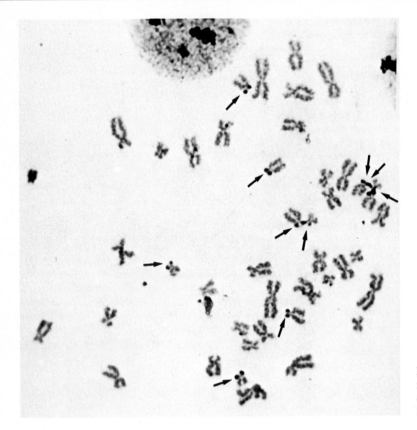

Fig. 2.7. Silver staining *(arrows)* of the nucleolus organizer regions of acrocentric chromosomes. (Courtesy of Dr. T. M. Schroeder-Kurth)

to the centromere, in the proximal part of the long arm, a "secondary constriction" is found fairly frequently. This constriction causes the occasional elongation of the long arm (Fig. 2.9). The extended segment may be very thin compared to the rest of the chromosome, suggesting "uncoiling" of the chromatid supercoil characterizing the metaphase chromosomes. This "uncoiler" phenomenon, as all individual variants of chromosome morphology, is transmitted to all cells, including about one-half of the germ cells, thus fulfilling the formal requirements of a simple dominant mode of inheritance. This "uncoiler-1 locus" was utilized to map the Duffy locus to chromosome 1 (Sect. 5.1.2). The secondary constriction shows little fluorescence with Q banding, but it does show a dark G band.

The largest submetacentric chromosome is no. 2.

Chromosome 3 is about 20% shorter than no. 1 and can therefore be distinguished easily. With Q banding, the proximal part of the long arm shows a brightly fluorescent band. The intensity of fluorescence varies strikingly between different individual chromosomes but is constant for the same chromosome.

Group B (nos. 4, 5). Large, submetacentric chromosomes. R and G banding shows striking differences between these two chromosomes.

Group C (nos. 6–12). Medium-sized, submetacentric chromosomes. Chromosomes 6, 7, 8, 11, and 12 are relatively submetacentric; no. 9 frequently shows a secondary constriction in the proximal part of its long arm. All chromosomes can definitely be identified by Q or G banding. The secondary constriction in no. 9 stains neither with quinacrine nor with Giemsa stain. Chromosomes 11 and 12 show very similar patterns.

In contrast to the other chromosomes of this group, the X chromosome varies considerably in length. In general, it is similar to the longer C chromosomes. In female cells, one of the two X chromosomes still replicates in the late S phase, whereas replication of the other C chromosomes is complete except for short segments.

Group D (nos. 13–15). These acrocentric chromosomes look quite different from other human chromosomes. All three pairs may have satellites; their short-arm region shows strong interchromosomal variability. The proximal short arms are of varying length, satellites may be lacking or especially large; they may or may not show fluorescence; in some cases, double (tandem) satellites are observed. The long arms of all three D chromosomes are clearly distinguishable by Q and G banding. The following criteria are used for definition of variants in the D–G groups: The length of a short arm is compared

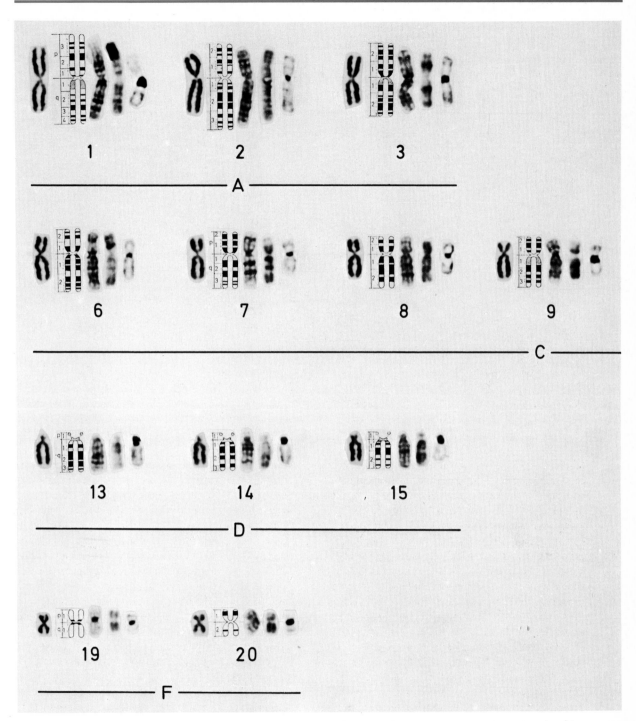

Fig. 2.8. Karyotype of a human male stained conventionally and using different banding techniques. *From left to right,* conventional staining; schematic representation of banding patterns; *G* banding; *R* banding; *C* banding. (Courtesy of Dr. T. M. Schroeder-Kurth)

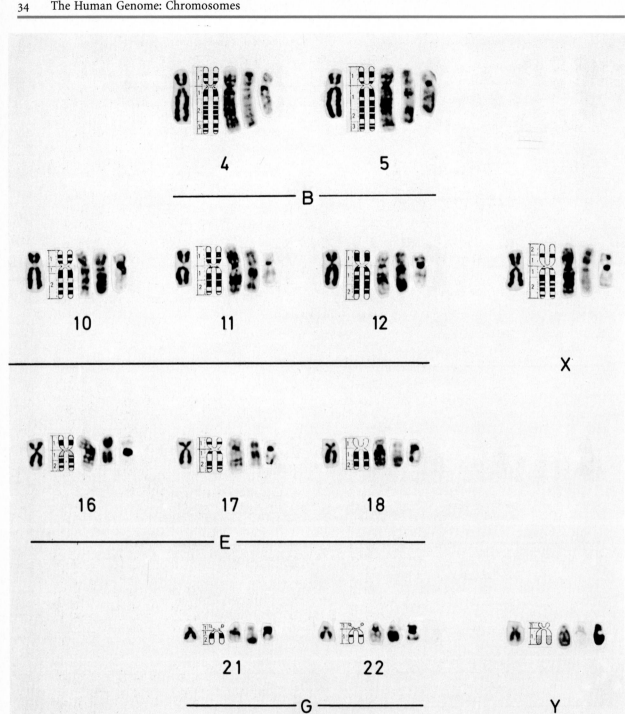

Fig. 2.8. *(continued)*

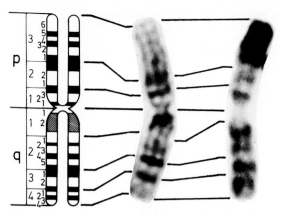

Fig. 2.9. Chromosome 1: comparison of G and R banding with schematic representation. (Courtesy of Dr. T. M. Schroeder-Kurth)

with that of the short arm of no. 18 from the same cell. Normally it is shorter. It is called long (ph +), if it is as long as the short arm of no. 18, and very long, if it is longer. Large satellites are called (ps +), double satellites (pss), shortened short arms with or without satellites (ph –). The frequency of D group heteromorphism (See next section) has been given as 3.7% (8 of 216) with banding techniques and 2.3% (411 of 24 440) without banding techniques [422] (Fig. 2.12).

Group E (nos. 16–18). Relatively short, metacentric or submetacentric chromosomes. In general, its total length is normally somewhat more than one-third of no. 1 but shows striking variations. The long arm shows a secondary constriction in about 10% of all cases. The length of a proximal G band varies with this constriction. Chromosome 18 is about 5%–10% shorter than no. 17 and has shorter long arms (cen-

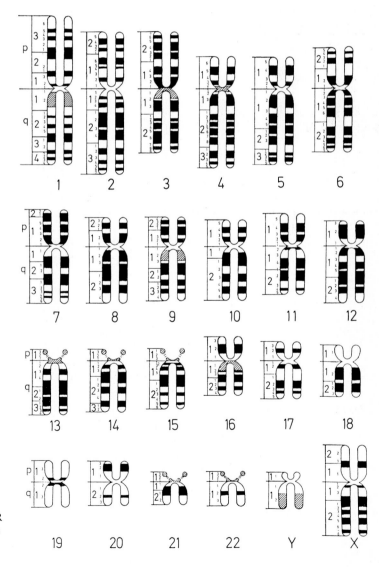

Fig. 2.10. Banding patterns according to the Paris nomenclature (G, Q, and R banding). *Black,* positive G and Q bands and negative R bands; *hatched,* variable regions. (From Paris Conference 1971 [92]; see, however Fig. 2.14)

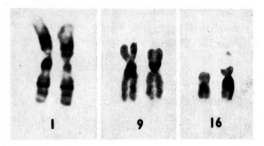

Fig. 2.11. Heteromorphism of constitutive heterochromatin at the secondary constrictions of chromosomes *1, 9,* and *16,* C banding. (From Koske-Westphal and Passarge 1974 [62])

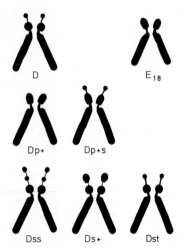

Fig. 2.12. Heteromorphism of acrocentric marker chromosomes of group D or G. *First row,* normal D and no. 18 chromosomes for reference. *Second row,* Dp + : short arms at least as long as the short arms of a no. 18 chromosome; p + s: normal-sized satellites on elongated short arms, *third row, D,* chromosomes with structural variations of the satellite region. *ss,* double satellites; *s +,* enlarged satellites; *st,* enlarged satellite stalks. (From Zankl and Zang 1974 [138])

tromeric index 31 in no. 17 as compared with 26 in no. 18); no. 17 replicates early, no. 18 late.

Group F (nos. 19, 20). Banding patterns of these small chromosomes are distinctively different.

Group G (nos. 21, 22). These small acrocentric chromosomes can easily be distinguished by their banding patterns. The variability of their short-arm region is quite as large as that of D chromosomes. The same variants as in D chromosomes are usually distinguished (Fig. 2.12). Satellites and short arms may show weak, moderate, or strong fluorescence and G banding. A small percentage have an elongated short arm. Other variants – such as giant satellites, elongated and shortened short arms – are much rarer. The short arms of D and G chromosomes contain the nucleolus organizer regions that are stained specifically by the silver method.

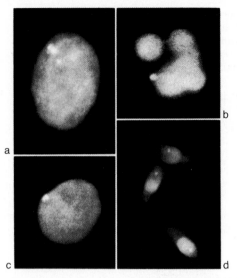

Fig. 2.13 a–d. Quinacrine mustard stainings of different cell nuclei of chromosomally normal men. **a** Buccal smear. The Y chromatin appears as a double structure. **b** Granulocyte from a blood smear. The Y chromatin is protruding as a small appendage. **c** Large lymphocyte from a blood smear. **d** Sperms. The Y chromatin is found near the border of the strongly fluorescent part of the sperm head (× 2400). (From Schwarzacher and Wolf 1974 [119])

The Y chromosome is usually but not always larger than the G group chromosomes. The long arm shows individual and transmissible variations in length. It shows distal brilliant fluorescence after quinacrine staining. In most cases two strongly fluorescing bands – in some rare cases even three – can be distinguished. In a population study, striking variants in length were found in 5.6% of 2444 newborns. In most cases the Y chromosome was elongated; in 5%, it was longer than an F chromosome, and in 0.33% longer than no. 18; 0.25% of the sample had very small Y chromosomes.

The distally brilliant fluorescent part of the long arm of the Y chromosome can be identified in the interphase nucleus as a bright dot about 0.3–1.0 μm in diameter. Figure 2.13 shows this "Y chromatin" in an epithelial cell, in a granulocyte, in a large lymphocyte, and in sperms [119].

Measurements of mitotic chromosomes meet with some difficulties, as the centromere position cannot always be determined accurately. A set of rules was laid down at the Paris conference in 1971 [92] Table 2.2 contains some typical measurements.

Chromosome Heteromorphisms. As mentioned in the description of single chromosomes, these do not always appear completely identical in all individual

Table 2.2. Measurements of relative length (percentage of total haploid autosome length) and centromere index (length of short arm divided by total chromosome length $\times$ 100): chromosomes stained with orcein or the Giemsa 9 method and preidentified by Q band patterns (From Paris Conference 1971) [449]

Chromosome no.	Relative length	Centromere index
1	8.44 ± 0.433	48.36 ± 1.166
2	8.02 ± 0.397	39.23 ± 1.824
3	6.83 ± 0.315	46.95 ± 1.557
4	6.30 ± 0.284	29.07 ± 1.867
5	6.08 ± 0.305	29.25 ± 1.739
6	5.90 ± 0.264	39.05 ± 1.665
7	5.36 ± 0.271	39.05 ± 1.771
X	5.12 ± 0.261	40.12 ± 2.117
8	4.93 ± 0.261	34.08 ± 1.975
9	4.80 ± 0.244	35.43 ± 2.559
10	4.59 ± 0.221	33.95 ± 2.243
11	4.61 ± 0.227	40.14 ± 2.328
12	4.66 ± 0.212	30.16 ± 2.339
13	3.74 ± 0.236	17.08 ± 3.227
14	3.56 ± 0.229	18.74 ± 3.596
15	3.46 ± 0.214	20.30 ± 3.702
16	3.36 ± 0.183	41.33 ± 2.74
17	3.25 ± 0.189	33.86 ± 2.771
18	2.93 ± 0.164	30.93 ± 3.044
19	2.67 ± 0.174	46.54 ± 2.299
20	2.56 ± 0.165	45.45 ± 2.526
21	1.90 ± 0.170	30.89 ± 5.002
22	2.04 ± 0.182	30.48 ± 4.932
Y	2.15 ± 0.137	27.17 ± 3.182

Data from 95 cells from 11 normal subjects (six to ten cells per person). Average total length of chromosomes per cell: 176 µm. Standard deviations are an average of the standard deviations found in each of 11 subjects (six to ten cells per subject).

members of a population. Chromosome "heteromorphisms" are observed especially in the satellite regions of acrocentric chromosomes, in the length of the Y heterochromatic part, and in the "secondary constrictions" of chromosomes 1 and 9. However, they also occur in heterochromatic segments of other chromosomes (for heterochromatin, see Sect. 3.1.1.2). In aneuploidies they have been used for identification of the parental origin of a chromosome (Sect. 9.2.3). In a number of chromosomes, fragile sites have been discovered, i.e., chromosome sites showing an increased risk of chromosome or chromatid breaks. Such breaks are especially easily induced by folic acid depletion of the culture medium [54]. A fragile site on the tip of the long arm of the X is of particular interest, being associated with a characteristic form of mental retardation (Sect. 15.2.1.2).

High-Resolution Banding [49]. Chromosomes in prophase and prometaphase are less tightly condensed than metaphase chromosomes. By suitable treatment of lymphocyte cultures with methotrexate for partial synchronization of cell cycles, a stage in which a relatively high number of cells are in prophase or prometaphase can be selected for preparation. Shortening of Colcemid treatment helps to decrease the degree of condensation. In such a preparation, single bands as revealed by standard methods can be resolved into subbands. The degree of resolution depends on the stage at which the cell has been picked. Some authors have described up to 2000 bands [137]; about 800–1200 bands can normally be seen in late prophase [36] (Fig. 2.14). The method cannot replace standard methods in routine diagnosis; it is, however, useful for more precise identification of breakpoints and small aberrations, such as in families with balanced and unbalanced translocations, or particularly in tumor cytogenetics.

Electron-Microscopic Images of Human Chromosomes [107, 117]. A number of methods in electron microscopy have been used to gain insight into the overall structure of human chromosomes. Present models of the organization of genetic material in eukaryotes are discussed in Sect. 3.1.1.5. The evidence from electron microscopy does not contradict models assuming a chromatin thread that is supercoiled in several orders. Three types of fibrils have been found: one has a diameter of approx. 250 Å, a second measures about 100 Å, and a third only 30–50 Å. There seems to be good evidence that the latter fiber is the genetically active chromatin. A pure DNA double helix has a diameter of approx. 20 Å; hence 30–50 Å corresponds to a DNA fiber together with proteins (histone and nonhistone). The 100 Å fiber seems to be a secondary coil of the 30–50 Å fiber, and the 250 Å fiber may be a tertiary coil. Probably about nine of the 250 Å fibers are somehow bundled together, and two of these bundles seem to again form a coiled structure that can be discerned in appropriate electron-microscopic pictures and seems to be characteristic for each chromosome [107]. In some preparations, relics of a membrane, presumably the nuclear membrane, are seen. Some investigators regard these as evidence that the interphase chromosomes are fixed to the membrane at various points. In view of the numerous steps of preparation for electron microscopy, it is difficult to decide whether these – or other – details reflect in vivo structure or are merely preparative artifacts.

2.1.2.4 Meiosis

Biological Function of Meiosis. While in the usual type of cell division, or mitosis, the number of divisions in daughter cells remains constant, the meiotic process is designed to reduce the number of chromosomes from the diploid number (46 in humans) to one-half this number (23 in humans). Fertilization of two germ cells, each with the haploid number, recon-

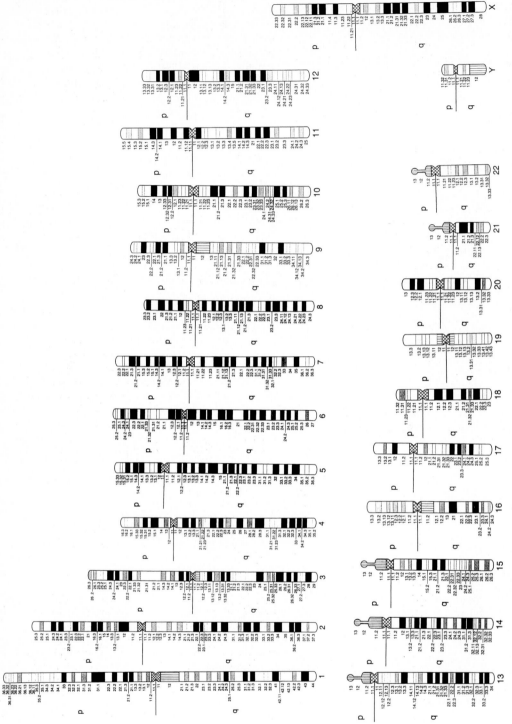

Fig. 2.14. Human chromosomes with ~ 850 bands. Relative lengths of chromosomes and bands are based on exact measurements; intensity of bands indicated by variable shading is based on quantitative density measurements. (From Francke 1994 [36])

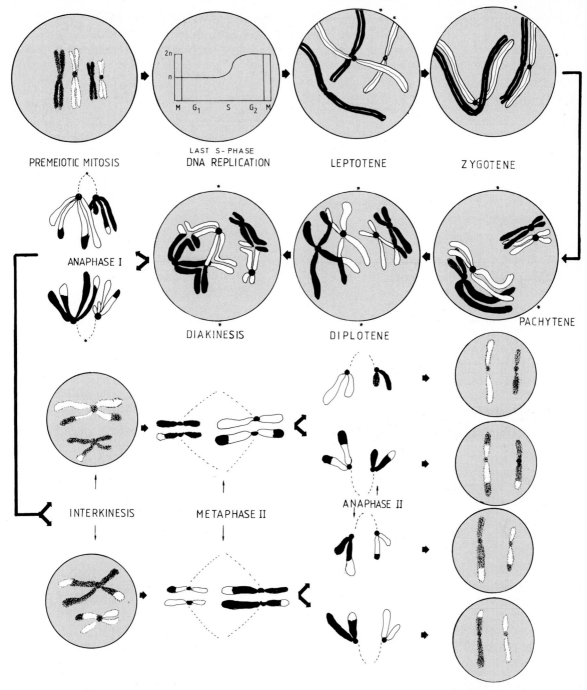

Fig. 2.15. The stages of meiosis. *Black*, paternal; *white*, maternal chromosomes. The figure depicts male meiosis; in female meiosis, polar body formation occurs

stitutes the diploid number of 46 in the zygote and in all of its descendant cells. Chance alone determines which of two homologous chromosomes ends up in a given germ cell. Genetic variability is thus enhanced. The somatic cell is diploid, containing both members of a pair of homologous chromosomes (2 n), whereas the germ cell is haploid, containing only one of each pair (n) The last regular DNA synthesis occurs during the interphase before the first meiotic division and precedes the meiotic phases shown in Fig. 2.15.

Meiotic Division I. Prophase I: Long chromosome threads become visible (leptotene) followed by pairing of homologous chromosomes, frequently from the chromosome ends (zygotene). The exact molecular mechanism of chromosome pairing is not yet known. The paired homologous chromosomes are connected by the so-called synaptonemal complex, a characteristic double structure. After completion of pairing, the chromosomes become shorter through contraction (pachytene). A longitudinal cleft in each pair of chromosomes then becomes visible; four chromatids of each kind are seen side by side (diplotene). Non-sister chromatids are separated, while sister chromatids remain paired. In this phase, chromatin crossings – "chiasmata" – between non-sister chromatids become visible.

Metaphase I: The chromosomes are ordered in the metaphase plane as the centromeres are drawn to the poles. The homologous chromosomes are drawn somewhat apart but are still kept together by the chiasmata, frequently at the ends.

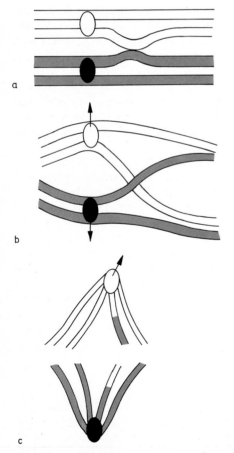

Fig. 2.16 a–c. Crossing over and chiasma formation. **a** Homologous chromatids are attached to each other. **b** Crossing over with chiasma formation occurs. **c** Chromatid separation occurs

Anaphase I: The chiasmata are first "terminalized," i.e., they seem to migrate to the chromosome ends and are then resolved. The paired chromosomes separate and migrate to the opposite poles. The daughter nuclei are formed (interkinesis).

Meiotic Division II. This is in principle a mitotic division of the replicated haploid set of chromosomes. As noted above, meiosis begins after replication. The genetic material, which during division I becomes fourfold (2×2 homologous chromosomes) is, at the completion of division II, ordinarily distributed to four cells. A second, important aspect of meiosis is the random distribution of nonhomologous chromosomes, which leads to a very large number of possible combinations of possible germ cells. In humans with 23 chromosome pairs, the number of possible combinations in one germ cell is $2^{23} = 8\,388\,608$. The number of possible combinations of chromosomes in an offspring of a given pair of parents is $2^{23} \times 2^{23}$ and is further enhanced by crossing over during pairing of homologous chromosomes. The morphological counterpart of crossing over is chiasma formation. Every chiasma corresponds to one crossing over event involving two non-sister chromatids (Fig. 2.16). For some time, it was disputed whether crossing over occurs during regular DNA synthesis – by "copy choice" – or after regular DNA synthesis – by breakage of non-sister chromatids at homologous sites and subsequent crosswise reunion (Fig. 2.17). The controversy now appears to be resolved in favor of the exchange hypothesis. Prophase I shows no regular, but much unscheduled DNA synthesis, which could easily indicate the reunion phase of crossing over.

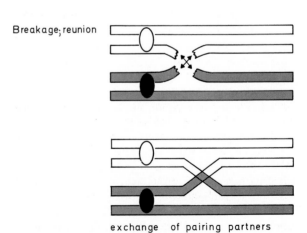

Fig. 2.17. Breakage and reunion of nonsister chromatids in crossing over

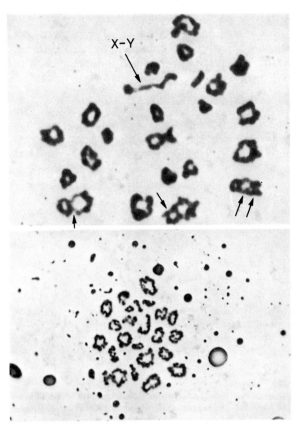

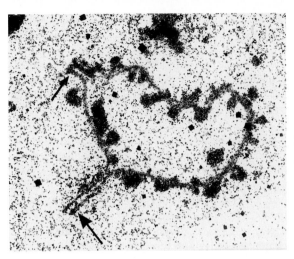

Fig. 2.19. Pairing of the human X and Y chromosomes in early meiosis: pairing of short arms *(lower left)* and tips of long arms *(upper left)*. *Left,* Y chromosome. (Courtesy of A. C. Chandley)

Fig. 2.18. Meiotic metaphases I (diakinesis) of a male with clearly visible XY bivalent. *Arrows,* chiasmata (From Kjessler 1966)

Meiosis in the Human Male. From the beginning of puberty human spermatocytes continuously undergo meiosis. After the second meiotic division DNA and mitochondria are densely packed during sperm development, and the sperm acquires the ability to move actively.

Male diakinesis chromosomes are seen in Fig. 2.18. The homologous chromosomes are still close together at the ends, whereas the centromeric regions have already started on their way to the poles. The sex bivalent is clearly distinguished from all the others by end-to-end association of the X and Y chromosomes. No chiasma formation can be seen. During pachytene and prophase the sex bivalent is already prematurely condensed and contained in a "sex vesicle." Part of the short – arm region of the X chromosome and the short arm of the Y are paired (Fig. 2.19), and hybridization studies with DNA probes have shown these regions to be structure-homologous. Male meiotic chromosomes can now be studied in great detail by cytogenetic methods [14]. Female meiotic chromosomes have been studied extensively by the FISH technique (see Sect. 3.1.3.3) [15].

Genes located on a common segment of the X and Y chromosome would be indistinguishable from autosomal genes if free recombination occurred between them. Such "pseudoautosomal" DNA sequences with X and Y linked alleles have been identified [12, 105]. Haldane in 1936 [46] suggested the existence of incomplete sex linkage caused by occasional crossovers between X and Y for some human genes postulated to reside on a common X and Y chromosomal segment. However, no plausible evidence for such partial sex linkage in humans exists. In recent years the pseudoautosomal segment of X and Y chromosomes has been studied extensively, and a number of genes have been identified [101]. We return to these data below.

The range of variation and the average number of chiasmata per cell are given in Table 2.3. Some bivalents may contain several, up to five or even six, chiasmata. From the number of chiasmata the genetic map length (Sect. 5.1.2) of the human genome has been estimated to be approx. 25.8 M in the male; it is longer but cannot be estimated in the female [82] because suitable chromosomal preparations are not available. In the house mouse, the only other mam-

Table 2.3. Number of chiasmata in male meiosis (1st division)

No. indi- viduals	Age range	No. of cells	Chiasmata/cell		Chiasmata/ bivalent, mean
			Range	Mean	
48	15–79	817	39–64	54.4	2.36

Data from various authors (see Hamerton 1971 [48]).

mal for which an estimate is available, map length is estimated to be between 16.2. and 19.2 M [48]. (One morgan, M, is the unit of map distance between linked genes; this unit measures recombination frequencies. One centimorgan, cM, means 1% recombination and equals the crossover frequencies between genes.)

Meiosis in the Human Female. In all mammals, oogenesis differs substantially from spermatogenesis; the overall timing is described in Fig. 9.13, 9.14; Sect. 9.3.3. Figures 2.20 and 2.21 show the cell processes (see also [64]). The oocytes are already formed in the late embryonic stage; after the diplotene, the cell enters a stage in which the chromosomes have a lampbrushlike appearance (dictyotene). Meiosis re-

mains arrested at this stage for many years. After birth, most oocytes degenerate. After puberty, some oocytes start growing, finish the first meiotic division, and enter prophase II and metaphase II. At the same time, the egg is ovulated. Meiosis is finished only after fertilization. Nuclear membranes are formed around the haploid male and female chromosome sets in the fertilized egg, and the zygote now contains two "pronuclei." At this stage it is especially susceptible to disturbances, for example, by mutagenic agents (Sect. 11.1.3). Several hours later the two pronuclei fuse to form a diploid nucleus, and the zygote starts dividing.

Study of female meiotic chromosomes is difficult, and few satisfying pictures have been published. Linkage analysis has shown crossing over in females

Fig. 2.20. Mitosis and meiosis in the human female. Until the 3rd month only mitotic divisions are to be seen (*A,* interphase; *B,* metaphase; *C,* anaphase). Then the first meioses become visible (*D,* leptotene; *E,* zygotene). Until the 7th month new oocytes enter meiosis. The first pachytene *(F)* and diplotene *(G)* stages are observed in the 7th month. Meiosis does not proceed any further. Instead, the tetrads are again stretching, a nuclear membrane and a nucleolus are formed, and the cell enters a "resting phase," the dictyotene *(H).* The function of the cells *(I)* around the oocytes is nutritional; they will later form follicles in which oocytes are embedded. (From Ohno et al. 1962; see also Bresch and Hausmann 1972)

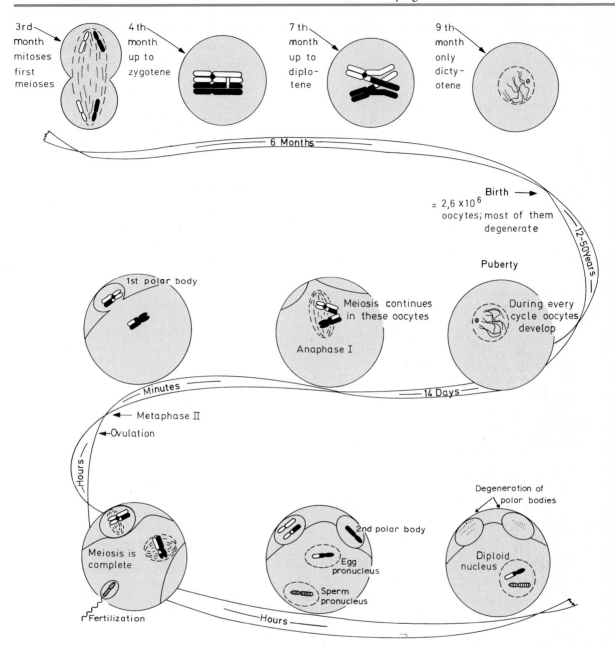

Fig. 2.21. Meiosis in the human female. Meiosis starts after 3 months of development. During childhood the cytoplasm of oocytes increases in volume, but the nucleus remains unchanged. About 90% of all oocytes degenerate at the onset of puberty. During the first half of every month the luteinizing hormone *(LH)* of the pituitary stimulates meiosis which is now almost completed (end of the prophase that began during embryonic age; metaphase I, anaphase I, telophase I and – within a few minutes – prophase II and metaphase II). Then meiosis stops again. A few hours after metaphase I is reached ovulation is induced by *LH*. Fertilization occurs in the fallopian tube. Then the second meiotic division is completed. Nuclear membranes are formed around the maternal and paternal chromosomes. After some hours the two "pronuclei" fuse, and the first cleavage division begins. (From Bresch and Hausmann 1972)

to be more frequent than in males (Sect. 5.1.2). Hence, more chiasmata should be expected.

In females, only one of the four meiotic products develops into an oocyte, the others becoming polar bodies that under normal conditions are not ferti-

lized. It is usually assumed that the risk of a chromosome being discarded in a polar body is unrelated to its genetic properties. This assumption is correct for most gene mutations, as shown by their undisturbed segregation ratios (50 : 50, 25 : 75, etc.). In structurally abnormal chromosomes disturbances of their segregation into germ cells may be explained by nonrandom expulsion of normal and abnormal chromosomes into the polar body.

Sex Difference in Meiosis. There are two principal aspects by which meiosis differs in males and females:

1. In males all four division products develop into mature germ cells, whereas in females only one of them becomes a mature oocyte, while the others are lost.
2. In males, meiosis immediately follows a long series of mitotic divisions; it is completed when spermatids start developing into mature sperms. In females, meiosis begins at a very early stage of development, immediately after a much smaller series of mitotic divisions. It is then arrested for many years and is only finished after fertilization.

These sex differences are important in human genetics. The fact that only one of the four division products develops into a mature oocyte, and the three polar bodies contain little or no cytoplasm enables this oocyte to transmit to the new zygote a full set of cytoplasmic constituents such as mitochondria and messenger RNA (Chap. 8). These differences in cell kinetics are probably responsible for sex differences in mutations rates for trisomies, on the one hand, and point mutations, on the other (Chap. 9; Sect. 10.1).

2.2 Human Chromosome Pathology

2.2.1 Syndromes Due to Numeric Anomalies of Autosomes

Mechanisms Creating Anomalies in Chromosome Numbers (Numerical Chromosome Mutations) (Fig. 2.22). Anomalies in chromosome numbers may be caused by various mechanisms:

a) The most important mechanism is nondisjunction. Chromosomes that should normally be separated during cell division stick together and are transported in anaphase to one pole. This may occur at mitotic division but is observed more frequently during meiosis. The exact reasons are unknown, but in humans the acrocentric chromosomes run a higher risk of being involved (Sect. 9.2). Meiotic nondisjunction was discovered by Bridges in 1916 in *Drosophila* [10]. For every gamete with one additional chromosome, another one is formed with one less chromosome. After fertilization with a normal gamete the zygote is ei-

ther trisomic or monosomic. Somatic nondisjunction in mitotic cell division during early development may lead to mosaics with normal, trisomic, and monosomic cells.

b) A second mechanism leading to numerical abnormalities is loss of single chromosomes, presumably due to "anaphase lagging" when one chromosome may lag behind the others. Chromosome loss leads to mosaics with one euploid and one monosomic cell population. In the mouse the pronucleus stage, i. e., the time between impregnation and fusion of the two haploid parental nuclei, is especially susceptible to loss of the paternal X chromosome. This phase, and possibly the first cleavage stages, presumably are equally vulnerable in humans, since many mosaics are formed during these stages (Sect. 9.2.3).

c) A third mechanism is polyploidization. Here all chromosomes are present more than twice in every cell. In humans, except for tumor cells, only triploidy is observed. The chromosome number is 3 n = 69.

An abnormal number of chromosomes in a cell (aneuploidy) increases the risk of further irregularities, such as chromosome loss due to anaphase lagging in subsequent cell divisions. For many mosaics with two cell populations of equal proportions, one trisomic and one euploid, this is the most plausible explanation (Sect. 9.2.3). The partnerless chromosome seems to interfere in these cases with normal chromosome pairing.

Down Syndrome. With an incidence at birth of 1–2/1000, Down syndrome is the most frequent chromosome aberration syndrome in humans, and a common condition encountered in genetic counseling services. Figure 2.23 demonstrates how physical differences between the three main racial groups are overshadowed by the similarity due to this syndrome. Figure 2.24 represents schematically the most frequent clinical symptoms. The following observations on Down syndrome are important:

a) The condition is a well-defined syndrome. In spite of appreciable variability of signs, the clinical diagnosis is rarely in doubt for experienced clinicians.
b) Its frequency increases with the age of the mother.
c) In most cases, affected individuals are the only ones with the condition in an otherwise healthy family; in a small minority of families more than one case is observed.
d) Monozygotic (MZ) twins are usually concordant, while the great majority of dizygotic twins are discordant. This rule, however, has exceptions: Discordant MZ pairs occasionally occur [22]. This is probably caused by chromosome loss in the cells forming one of the twins.
e) Males with Down syndrome have no children. However, at least 17 women with this syndrome have reproduced. Among their 19 children, including one pair of MZ twins, seven had Down syndrome, eight were normal, two were

Nondisjunction in 1st meiotic division   or   in 2nd meiotic division

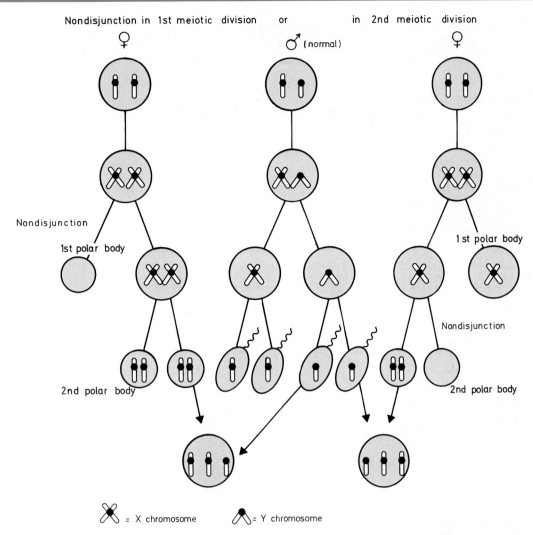

Fig. 2.22. Nondisjunction of the X chromosome in the first *(left)* and second *(right)* meiotic division in a woman. Fertilization by a normal Y sperm. An XXY individual can result from either first or second meiotic division nondisjunction

retarded without Down syndrome, one was stillborn without signs of Down syndrome, and two stillborn MZ twins with normal karyotypes were counted as one individual. All mothers and affected children in whom chromosome examinations were carried out showed the standard karyotype 47,21 + ; one of the retarded children without Down syndrome had a normal karyotype 46,XY.

f) Life expectancy of the patients is reduced [96]. In an Australian sample published in 1963 [17], 31.1% had died by the end of their 1st year and 46% by the end of the 3rd. Life expectancy is reduced in later life as well. All these patients develop amyloid plaques in the brain in middle age, consistent with Alzheimer's disease and exhibit the resultant dementia. Moreover, infections appear more commonly, suggesting defects in immune defense. Congenital cardiac anomalies are also increased. With antibiotic therapy and heart surgery, the patients are now surviving much longer (Fig. 2.25).

g) Expressivity of phenotypic features is variable. Congenital heart disease, for example, is present in some but not all patients, and the same is true for many other clinical signs (Fig. 2.24). This increased variability of phenotypic manifestation is characteristic for all chromosomal aberration syndromes in humans.

h) There is a 20-fold increased risk of dying from acute leukemia. The reasons are unknown. Three different hypotheses come to mind: a higher risk of aneuploidy due to mitotic disturbances in blood stem cells, a lower resistance against infection with a leukemogenic virus, a lower efficiency in the function of repair enzymes for which experimental evidence is available (Sect. 10.3).

Discussion of other aspects of Down syndrome:

- Sect. 8.4.3:   Gene action
- Sect. 18.1:   Genetic counseling, prenatal aspects
- Sect. 9.2.2:   Mutation, maternal age
- Sects. 15.2.2.1:   Psychophysiological aspects

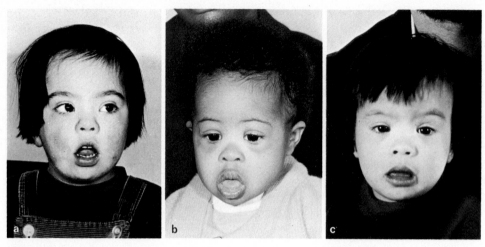

Fig. 2.23 a–c. Children with Down syndrome. **a** European. **b** African-American. **c** Oriental-European. The common features of Down syndrome are more impressive than the racial differences. (Courtesy of Dr. T. M. Schroeder-Kurth)

Growth failure

Mental retardation

Flat occiput

Dysplastic ears

Many "loops" on the finger tips

Simian crease

Medial axial triradius

Unilateral or bilateral absence of one rib

Intestinal stenosis

Umbilical hernia

Dysplastic pelvis

Hypotonic muscles

Big toes widely spaced

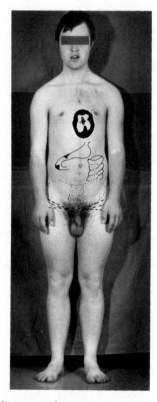

Broad flat face

Slanting eyes

Epicanthus

Short nose

Small and arched palate

Big wrinkled tongue

Dental anomalies

Short and broad hands

Clinodactyly

Congenital heart disease

Megacolon

Fig. 2.24. The main clinical findings of Down syndrome

Standard Karyotype in Down Syndrome. The G chromosomes of a patient with Down syndrome are seen in Fig. 2.26 (G and Q banding). The banding patterns of nos. 21 and 22 can be distinguished easily, no. 21 being the stronger and broader fluorescent and having one to two dark G bands. Chromosome 22 has a dark G band at the proximal long arm and a fainter band more distally.

It is now generally acknowledged that every patient with this syndrome has the supernumerary chromosome, either as free trisomy 21 or as a translocation chromosome formed from a chromosome 21 and another chromosome such as no. 21, 22, 13, 14, or 15. Observations on some rare cases with reciprocal translocations suggest that the distal part of the long arm of chromosome 21, and especially the band 21q22, is responsible for this phenotype [45]. For example, a girl with the tandem duplication of one chromosome 21 except for band 21q22 (Fig. 2.27) in addition to another free no. 21 was moderately mentally retarded but did not show most of the features of Down syndrome. Trisomy only of 21q22 led to mild manifestation of this syndrome [44]. More recently numerous cases have been analyzed by cytogenetic and molecular methods in which short duplications have resulted in partial Down's syndrome. The responsible area can now be identified within about 400 kb of DNA [127]. DNA sequences apparently responsible for parts of the syndrome are seen in Fig. 2.28, which also contains data on gene loci located on chromosome 21.

Down syndrome was known as a clinical entity long before trisomy 21 was discovered. The other autosomal chromosome syndromes were buried in the great number of multiple malformations and could be singled out only after the abnormal chromosome complement had been discovered. In retrospect, some of these syndromes are so singular that they probably could have been delineated on purely clinical grounds.

Other Autosomal Trisomies. Patau et al. in 1960 [94] described the first case of an autosomal trisomy other than trisomy 21. This discovery was the result of a deliberate search, which was guided by a hypothesis specified by the authors as follows:

On genetic grounds it was not to be expected that the addition of an autosome to the normal complement would have a similarly restricted effect (as in X trisomy). Only one type of autosomal trisomic has been reported to date, and although the extra chromosome is one of the two smallest autosomes . . ., its presence in triplicate results in mongolism. . . . It was to be expected that other autosomal trisomics, if they should be at all viable, would also display multiple congenital disturbances.

Their systematic search along these lines produced three cases in which trisomies were found: two involving trisomy 18 and one with trisomy 13. At the same time, trisomy 18 (first incorrectly labeled as tri-

Fig. 2.25. Man with Down syndrome at the age of 64 (Courtesy of Dr. G. Tariverdian)

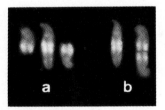

Fig. 2.27. a Tandem duplication of one chromosome 21 except for band 21q22 *(middle chromosome)* in a mildly mentally retarded child without most signs of Down syndrome. (From [44]) **b** The abnormal chromosome *(left)* and, for comparison, two normal no. 21 placed closely opposite each other *(right)*. Here, in distinction to the duplication chromosome, the band 21q22 is visible twice

Fig. 2.26. D and G chromosomes of a patient with Down syndrome. Q and G staining. Note the broad G band in the proximal region of 21p, which distinguishes no. 21 from no. 22. (Courtesy of Dr. T. M. Schroeder-Kurth)

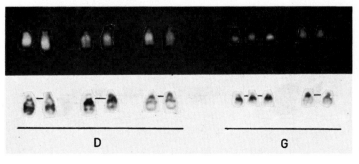

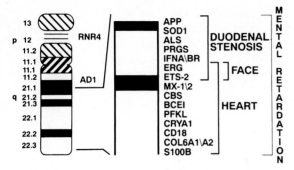

Fig. 2.28. Mapping of Down syndrome phenotypes to regions of chromosome 21. The maps show the phenotypic effects produced by triplication of various segments of the chromosome. Location of various genes, for example: *APP*, amyloid beta precursor protein; *SOD*, superoxide dismutase 1; *CBS*, cystathionine beta synthase; *COLGA1/A2*, collagen type VI, alpha 1 and alpha 2. (From Therman and Sussman 1993 [127])

somy 17) was also discovered by Edwards et al. [27]. The main signs and symptoms of both trisomies are seen in Figs. 2.29 and 2.30.

The method of nonradioactive in situ hybridization (chromosome painting, Sect. 3.1.3.3) makes it possible

to diagnose these (and other) trisomies not only in metaphases but also in interphase cells, as well as shown for trisomy 18 (Fig. 3.13).

Discussion of other aspects of trisomies 13 and 18:

- Sect. 18.2:   Genetic counseling, prenatal aspects
- Sect. 9.2:   Mutation, maternal age

In subsequent years all attempts to discover new autosomal trisomy syndromes among newborns failed, and these were assumed to be invariably lethal, especially as chromosome studies on spontaneous abortions revealed a variety of other trisomies. Discovery of three new syndromes, trisomies 8, 9 and 22, had to await development of the banding techniques [1, 58, 61]. Here, again, the children had severe and complex malformations. Trisomies 8 and 9 occur only as mosaics with a normal cell line in live newborns. These three trisomy syndromes are very rare [111].

Triploidy. Apart from a doubtful mosaic [7] the first discovered cases of triploidy were two aborted fetuses [23, 95]. Later, examinations of spontaneous abor-

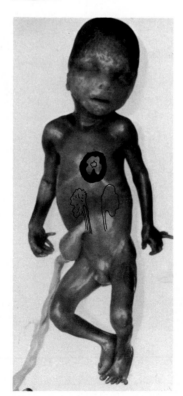

Coloboma- microphthalmus

Mental retardation

Growth failure

Low set and deformed ears

Deafness

Simian crease

Distal axial triradius

Atrial septal defect

Ventricular septal defect

Dextrocardia

S -shaped fibular radial arch

Increased segmentation of poly- morphonuclear granulocytes Higher frequency of drumsticks and C- appendages

Microcephaly

Arrhinencephaly

Hypertelorism

Cleft lip and palate

Polydactyly, flexion deformities of fingers Deformed finger nails

Kidney cysts

Double ureter

Hydronephrosis

Hydroureter

Umbilical hernia

Developmental uterine abnormalities

Cryptorchidism

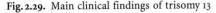

Fig. 2.29. Main clinical findings of trisomy 13

Growth failure

Mental retardation

Dolichocephaly with
protruding occiput

Retroflexion of head

Arches on three or more
finger tips

Absence of skin creases above
distal joints

Simian crease

Short sternum

Horse shoe kidney

Abduction deformity of hips

Muscular hypertonus

Pes equinovarus

Prominent heel

Dorsal flexion of big toes

Open skull sutures and wide
fontanelles at birth

Hypertelorism

High arched eyebrows

Low set and deformed ears

Micrognathia

Flexion deformities of fingers

Persistent ductus arteriosus

Ventricular septal defect

Meckel's diverticulum

Absence of labia majora

Prominent external genitalia

Hydramnios

Small placenta

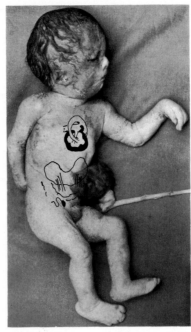

Fig. 2.30. Main clinical findings of trisomy 18

tions confirmed that triploidy is not rare, but a small number of cases were also observed among children born alive [84]. In 1974, more or less detailed information was available from 275 cases of triploid abortuses with a gestation time of less than 20 weeks. Twenty-two fetuses survived past the fetal age of 28 weeks; five died in utero, and the others survived for only a few hours or days after birth. All children surviving for longer than a few days, were triploid-diploid mosaics.

Triploids born alive have low birth weight, large posterior fontanelles with underdevelopment of the occipital and parietal skull bones and other nonspecific malformations also found in other autosomal aberrations. Those with 69,XXY karyotypes have grossly malformed genitals with a small penis and/or hypospadias, bifid scrotum, and nondescended testicles. Some of the mosaics have survived. Their clinical features are not very distinct, but the diagnosis may be suspected in mentally retarded children with abnormal placentas, syndactyly, abnormal genitals, and asymmetry.

Various errors in germ cell formation may lead to triploidy (Fig. 2.31). Some cause differences in the ratios of XXX, XXY, and XYY individuals among triploids. Evidence available so far indicates a high frequency of double fertilization or failure of the first meiotic division in the oocyte [56, 68]. Among triploid fetuses two different phenotypes have been identified by study of 19 such fetuses [78]. In type I (two cases) the fetus is relatively well grown; the head may be

normal or slightly microcephalic. The placenta is large and cystic (hydatidiform mole). The additional chromosome complement comes from the father. Type II, on the other hand, shows intrauterine growth retardation, relative macrocephaly, and a small, noncystic placenta. In these instances the additional chromosome complement comes from the mother. These results – as many others – point to different contributions of parental genotypes to embryonic development (genomic imprinting, see Sects. 4.1.7; 8.2). Tetraploidies and tetraploid moles have occasionally been observed. These are much rarer than triploidies and are not compatible with extrauterine life [127]. Since identification of individual chromosomes with molecular methods has become possible, another mechanisms of abnormal chromosomal transmission from parents to children has come to attention. Instances of uniparental disomy, i.e., origin of both chromosomes of a pair from one parent in a diploid individual, have been discovered (see Sect. 8.2).

Mosaics. Individuals with two or more genetically different cell populations are referred to as mosaics. They are found relatively often in numerical chromosome aberrations of the sex chromosomes but also in autosomal aberrations. A mosaic may be formed either by mitotic nondisjunction or by loss of single chromosomes due to anaphase lagging (Fig. 2.32). Frequencies of such mitotic errors have been determined for Down syndrome. The risk in trisomic zy-

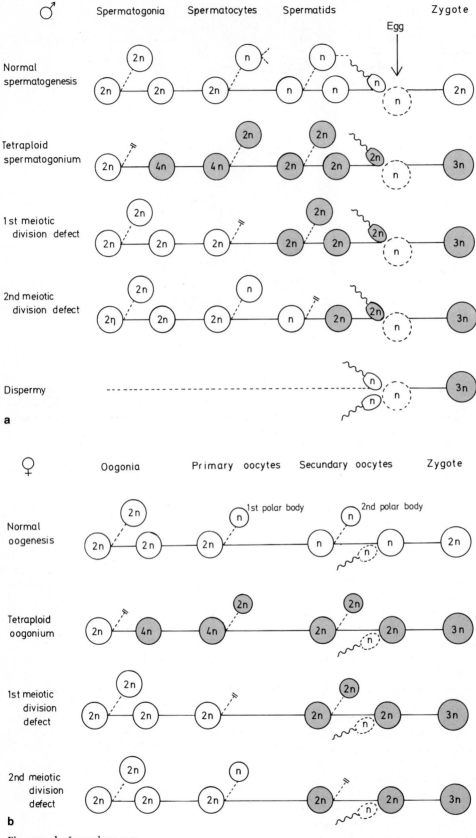

Fig. 2.31a, b. Legend see p. 51

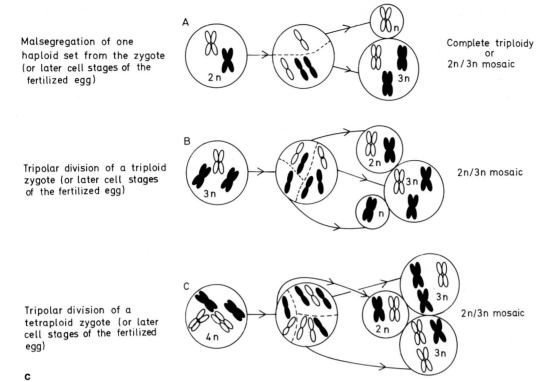

Malsegregation of one haploid set from the zygote (or later cell stages of the fertilized egg)

Complete triploidy or 2n/3n mosaic

Tripolar division of a triploid zygote (or later cell stages of the fertilized egg)

2n/3n mosaic

Tripolar division of a tetraploid zygote (or later cell stages of the fertilized egg)

2n/3n mosaic

c

◁ **Fig. 2.31 a–c.** Anomalies in oogenesis, spermatogenesis, or fertilization that may lead to triploidy. **a** In the male, triploidy may result from a tetraploid spermatogonium, from a disturbance in the first or second meiotic division (spermatocytes or spermatids), or from fertilization by two sperms. **b** In the female the same mechanism may occur as in the male germ cells. **c** Abnormal division of the zygote or later cell stages of the fertilized egg giving rise to complete triploidy or mosaicism. (From Niebuhr 1974 [84])

gotes is about 400 times higher than in euploid zygotes for anaphase lagging, and about 70 times higher for mitotic nondisjunction. These calculations were based on relative incidences of different types of mosaicism (Sect. 10.1) and on analysis of the maternal age effect.

Mosaics that result from meiotic nondisjunction with subsequent loss of the supernumerary chromosome due to anaphase lagging are expected to show the same increase with maternal age as do standard trisomies. Mosaics that result from mitotic nondisjunction, on the other hand, are not expected to show any increase of maternal age. Hence, the proportion of mosaics due to anaphase lagging may be roughly estimated in a large series from comparison of the maternal age effect between mosaic and standard trisomies. An exact estimate is made difficult by the fact that some mosaics escape diagnosis, unless a very large number of cells is counted. Moreover, cases with a small number of aberrant cells, and correspondingly with few or no phenotypic abnormalities, are detected only occasionally – mainly if there are trisomic cells in their germinal tissue, and hence cases with trisomies have occurred among their pro-

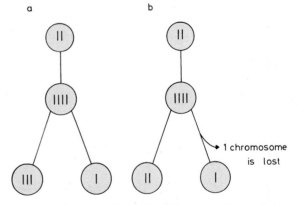

Fig. 2.32. Mitotic nondisjunction (**a**) and anaphase lagging (**b**). **a** After the homologous chromosomes have doubled to form four chromatids, three of them come into one division product in the next anaphase; the other division product only gets one chromosome. **b** One chromosome is lost during anaphase movement

geny. Still, among published mosaics 17%–30% have been estimated to be due to mitotic nondisjunction. As expected, the maternal age is especially low in cases with 34% or fewer trisomic cells among those examined [103]. The overall frequency of mosaics among all cases with clinical symptoms of Down syndrome is around 2%. Milder cases are more likely to be mosaics. Hook has published useful tables to assess the number of cells that need to be examined to determine the extent of mosaicism.

2.2.2 Syndromes Due to Structural Anomalies of Autosomes

2.2.2.1 Karyotypes and Clinical Syndromes

First Observations of Down Syndrome. Once trisomy 21 had been identified as the cause of Down syndrome, the obvious question was whether this trisomy is present in all cases. If not, the karyotypes of the exceptions would be of great interest. As the risk for meiotic nondisjunction was known to increase with the age of the mother, and as a single event could lead to only one affected offspring at a time, exceptions had to be sought primarily among affected children of younger mothers and in exceptional families with more than one affected patient.

Polani et al. in 1960 [99] examined three such children with Down syndrome. In one girl, the first child of a 21-year-old mother and 23-year-old father, they found 46 chromosomes. There were four normal chromosomes of the G (21, 22) group. One chromosome of the D (13–15) group had an elongated short arm. The authors suspected that one chromosome 21 was translocated to the short arm of the D chromosome. A short time afterward this suspicion was confirmed in familial cases. Two healthy mothers of the three patients and the common grandmother had only 45 chromosomes and only three free G chromosomes. However, one of the D chromosomes – the investigators suspected it would be no.15 – had an elongated short arm. If this arm included the missing chromosome 21, the karyotype of these women was

"balanced," and the genetic material was completely present. In some of their offspring, on the other hand, the translocation chromosome included most of the chromosome 21 material occurring in their cells together with two normal chromosomes no.21; these children were effectively trisomic for 21 and developed Down syndrome in spite of their normal number of chromosomes. This karyotype was unbalanced.
At about the same time the first G/G translocation was reported [33]. Shortly afterward, studies of the meiotic division I in a balanced heterozygote revealed a trivalent, i.e., a figure consisting of three chromosomes, giving conclusive evidence that the unusual chromosomes found in these families are, indeed, translocation chromosomes [47].

Frequency of Translocation Down Syndrome. Translocation Down syndrome explains a number of familial cases, but not all. Standard trisomies 21 may also occur repeatedly in the same families, pointing to constitutional factors or a mosaic status in the parents (germ cell precursors, Sects. 9.2; 10.1). Table 2.4 shows the frequencies of translocation cases, inherited and noninherited, among affected children of younger as compared with older mothers. Most translocations are of the types 14/21 or 21/21. There are, however, a few reciprocal translocations in which, apart from chromosome 21, other, nonacrocentric chromosomes are involved.

Gaps and Breaks. Chromosomes must first be broken to form any kind of rearrangements. Under the light microscope it may be difficult to distinguish chromosome breaks from achromatic regions called gaps. Such gaps may be true breaks, but they may also be local despiralizations. Chromosome breakage is frequently analyzed in mutation research. Therefore there must be some agreement as to which aberrations are to be considered as breaks and which as gaps. One of the proposed agreements is illustrated in Fig. 2.33. The distinctions listed are very conservative and probably underestimate the number of breaks. Breaks and gaps may occur during interphase before and after replication. If occurring before replication, the lesion is visible in the following meta-

Table 2.4. Incidence of translocations among children with Down syndrome (from Mikkelsen 1971)

Maternal age under 30				Maternal age over 30			
Total no. of patients	No. of translocations			Total no. of patients	No. of translocations		
	Sporadic	Inherited	Parents not examined		Sporadic	Inherited	Parents not examined
1431	69	32	14	1058	7	5	4
Total	115 = 8.04%			Total	16 = 1.51%		

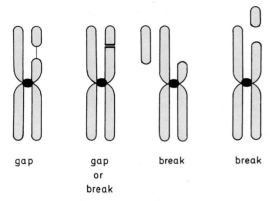

gap   gap   break   break
          or
         break

Fig. 2.33. Definition of chromosome gaps and breaks. *From left to right,* in a gap, the separate segment is not dislocated; there might be a tiny connection. If this connection is lacking, it cannot be decided whether a gap or a break has occurred. The two *right-hand* figures clearly show breaks with different locations of the broken segment. (Courtesy of Dr. T. M. Schroeder-Kurth)

phase in both daughter chromatids (isochromatid break). After replication only one chromatid is affected (chromatid break). These various types of breaks and gaps are shown in Fig. 2.34.

A break not affecting the centromere produces a shorter chromosome with a centromere and an acentric fragment. This fragment may or may not form a small ring but, lacking a centromere, runs a high risk

of being lost during the subsequent mitosis. Hence, chromosome breakage often leaves behind a cell deficient in a chromosome segment. In some cases, however, chromosomes broken at two breaking points rejoin under the influence of repair enzymes. If the broken ends do rejoin the chromosomes – and with them the cell – is again intact. In fact, experience with repair-deficient human diseases (Sect. 10.3) suggests that this may happen over and over again in many human tissues. In other cases, broken ends join with break points from other chromosomes, homologous or nonhomologous, one condition being that the two breaks have occurred within a reasonably short time period and reasonably close to each other. This leads to the various types of chromosome rearrangements.

Intrachromosomal Rearrangements (Intrachanges). A single chromosome may break at two different sites, and the intermediate part may rejoin upside down. This rearrangement does not lead to disturbances in mitosis, especially if the breaks have occurred in the G_1 phase. Inversions can be diagnosed by use of banding methods and/or in situ hybridization with multiple mapped markers. Inversions can be diagnosed by use of banding methods and/or in situ hybridization with multiple mapped markers when the centromere is not included (paracentric inversions). A shift in centromeric position readily identifies

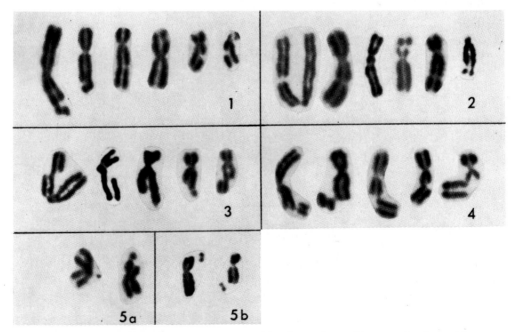

Fig. 2.34. *1–5.* Various types of chromosome gaps and breaks. *1,* Chromatid gaps; *2,* isochromatid gaps; *3,* chromatid breaks; *4,* isochromatid breaks; *5,* minutes and fragments: single *(a);* double minute *(b).* (From Gebhart [39])

pericentric inversions. Inversion heterozygotes are not particularly rare in human populations. There may be difficulties in chromosome pairing at meiosis, which may lead to partial elimination of certain types of germ cells in inversion heterozygotes (Fig. 2.35). These difficulties do not occur in homozygotes (for a review see [62].) Inversions – and especially pericentric inversions – have played a major role in the phylogeny of higher primates (Sect. 14.2.1).

Another type of intrachange is the ring chromosome (Fig. 2.36). Here two telomeres are usually lost as fragments, and the open ends rejoin. A ring chromosome may or may not be able to undergo mitosis, depending on whether the two chromatids joined in a crossed manner. If there is no sister strand exchange between the breakage points during DNA replication, the ring may replicate, forming two separate rings with one centromere each. They can pass the next mitosis without difficulty. One sister strand exchange forms one big ring with two centromeres. This structure is normally destroyed in the next mitosis. Two sister strand exchanges may form two rings that are entwined with each other. Various possibilities are detailed in Fig. 2.36, some of which are demonstrated with one case report (p. 55). Occasionally chromatid breakage and ring formation occurs in the G_2 phase, and patterns

such as that shown in Fig. 2.37 are observed in single cells.

Interchromosomal Rearrangements (Interchanges). In many cases, joining occurs between different chromosomes, homologous or nonhomologous. If breakage occurs in the G_1 phase, joining follows in the G_1 (or early S) phase before DNA replication. If each of the resulting chromosomes happens to have one centromere, the translocation chromosomes may pass through the next mitosis without difficulties. If one of the resulting chromosomes happens to get two centromeres, a dicentric chromosome is formed. Depending on the exact mode of replication, it may be able to pass the next mitosis, under the following conditions: (a) the centromeres migrate to the same pole and, (b) replication and sister chromatid exchange between the two centromeres has not led to interwining of the two chromatids (Fig. 2.38). If breakage and rejoining occur after DNA replication, only one sister chromatid of each chromosome is affected. The rejoined sister chromatids are still paired with their unaffected partners. This leads to the interchange patterns shown in Fig. 2.39 in the first mitotic division after reunion. The mitotic anaphase proceeds without further difficulties if the two centromeres happen to be located on different elements (Fig. 2.39; classes I, III, and V). If the centromeres

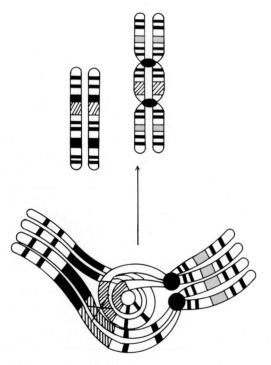
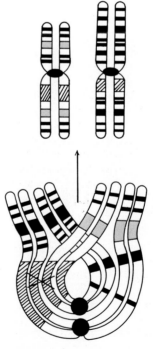

Fig. 2.35. Chromosome mispairing during meiosis in heterozygotes of pericentric *(left)* and paracentric *(right)* inversions. In the two figures, crossing over is assumed in the segments that are marked by an *X*. As a consequence, abnormal chromosomes are found that lead to aneuploidy of zygotes in the next generation

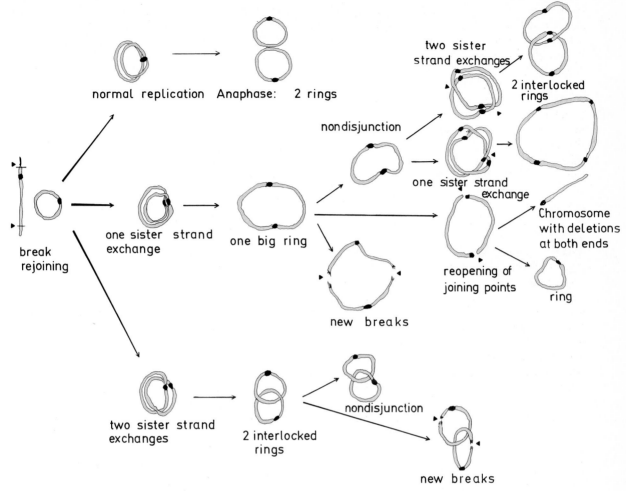

Fig. 2.36. Ring chromosome formation in the G$_1$ phase and fate of the ring chromosome in the following mitotic divisions. See text for details

are located on the same configuration, the resulting daughter cell is in any case aneuploid. Either the centromeres migrate to different poles, in which case an "anaphase bridge" is formed, and the chromosome finally breaks again, or the two centromeres migrate to the same pole, which can only happen with nonhomologous reunions (Fig. 2.39, classes V, VI, and VII). In the latter case the problem is postponed to the next mitosis, in which the chromosome appears as dicentric. It may or may not survive this mitosis. In any case, under the conditions mentioned above interchanges cause a great deal of cell loss due to aneuploidy or mitotic disturbance.

In human somatic tissue many of these mitotic disturbances are visible even in conventionally stained cells, without special chromosome preparations. Figure 2.40 shows an anaphase bridge and a so-called micronucleus in human bone marrow cells. Micronuclei are formed from chromosomes or chromosome fragments not taking part in normal mitosis at the same time. This leads to the phenomenon of premature chromosome condensation. The main nucleus is observed at metaphase with normally contracted chromatids, while the chromosomes of the micronucleus show a prophaselike condensation. These cytological anomalies have become important for quick evaluation of mutagenic agents (Sect. 11.2). Premature chromosome condensation can also be elicited in vitro by fusion of a cell in interphase with another cell that is just preparing for mitosis [122].

The method of chromosome painting described below (Sect. 3.1.3.3) has made it possible not only to show chromosome rearrangements in mitoses but above all to analyze them in great detail in interphase cells (See Fig. 3.11 for an example).

In meiosis, translocations may lead to disturbances as homologous chromosome segments tend to pair with each other. In metaphase I, they form so-called

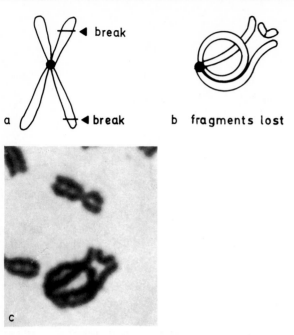

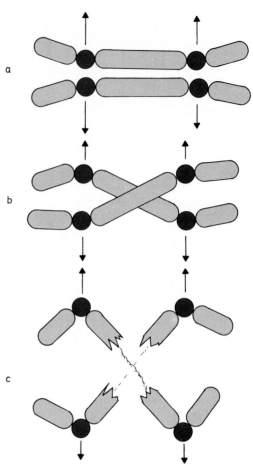

Fig. 2.37 a–c. Formation of a ring chromosome in the G_2 phase. **a** Two breaks in one of two sister chromatids. **b** Rejoining of the broken ends; pairing of fragments with the homologous chromatid segments. **c** The same ring chromosome in a human metaphase. (Courtesy of Dr. T. M. Schroeder-Kurth)

Fig. 2.38 a–c. Mitotic anaphase of a dicentric chromosome. **a** Both centromeres migrate to the same pole; chromosomes remain intact. **b** The centromeres migrate to opposite poles. Anaphase bridges are formed. **c** The chromosomes are broken

chains-of-three if three chromosomes are involved, for example, in a balanced translocation carrier. If four chromosomes are involved, a chain-of-four may be formed. This event may or may not lead to further aneuploidy, depending on the anaphase movements of the four centromeres. If the two centromeres of one element happen to move to one pole – and if the chromatids happen not to be intertwined between the centromeres – normal anaphase may take place. Very often, however, additional chromosome breakage occurs. This is one of the reasons why meiosis is such a good filter for removal of chromosome rearrangements.

Chromosome breakage together with its consequences may be observed in somatic cells and in germ cells. Breakage in somatic cells has become important for mutation research and is discussed in Chap. 11. Breakage in germ cells may or may not be transmissible to the next generation. If transmitted, it often causes zygote death during the embryonic stage. In a certain number of cases, however, the aberration is compatible with postnatal life, leading to a chromosome aberration syndrome. Before some of these syndromes are described, the universally accepted nomenclature of human karyotype description should be explained. This nomenclature was devised by a group of cytogeneticists and was last brought up to

date [79a, 92]. The following is a list of the abbreviations used:

t	Translocation
h	Changes in length of secondary constrictions
cen	Centromer
del	Deletion
der	Derivative chromosome
dup	Duplication
ins	Insertion
inv ins	Inverted insertion
rcp	Reciprocal translocation*
rec	Recombinant chromosome*
rob	Robertsonian translocation* ("centric fusion")
tan	Tandem translocation*
ter	Terminal or end (pter, end of short arm; qter, end of long arm)
:	Break (no reunion, as in a terminal deletion)
::	Break and join
→	From – to

(*: optional, where it is desired to be more precise than merely using t.)

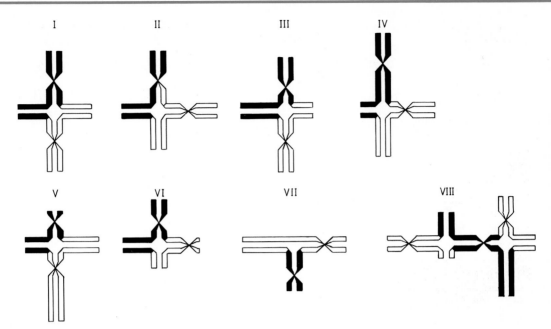

Fig. 2.39 I–VIII. Classes of interchanges found after translocation during the G$_2$ phase. Involvement of two homologous chromosomes. **I** Alternate position of the centromeres; exchange of fragments of equal length. **II** Adjacent position of the centromeres; exchange of fragments of equal length. **III** Alternate position of the centromeres; exchange of fragments of different length. **IV** Adjacent position of the centromeres; exchange of fragments of unequal length. Involvement of two nonhomologous chromosomes. **V** Alternate position of the centromeres. **VI** Adjacent positions of the centromeres. **VII** Triradial configuration (loss of fragments required). Complex of nonhomologous chromosomes (more than two). **VIII** One example of a figure with three chromosomes involved

Description of Human Karyotypes. The total number of chromosomes is given, followed by the sex chromosome complement. Then it is indicated which chromosomes are too many, or too few, or structurally altered. Some examples are the following:

46,XX	Normal female karyotype.
46,XY	Normal male karyotype.
47,XY, + 21	Male karyotype with 47 chromosomes; the additional chromosome has been identified als no. 21.
46,XY,1q +	Male karyotype with 46 chromosomes; the long arm (q) of one chromosome 1 is longer than normal.
47,XY, + 14p +	Male karyotype with 47 chromosomes, including an additional chromosome no. 14, which has en elongated short arm (p)*.
45,XX,−14,−21, + t(14q21q)	Female karyotype with a balanced Robertsonian translocation between a D and G group chromosome.
46,XY,−5,−12,t(5p12p), t(5q12q)	Male karyotype with two translocations involving interchange of both whole arms of chromosomes 5 and 12. The breaks have occurred at or very near the centromere, and no information is available as to which centromere is included in either product.

* (Petit, p = small or short.)

Changes in length of secondary constrictions, or negatively staining regions, should be distinguished from increases or decreases in arm length due to other structural alterations by placing the symbol h between the symbol of the arm and the + or − sign. For example:

46,XY,16qh +	Male karyotype with 46 chromosomes, showing an increase in length of the secondary constriction on the long arm of chromosome 16.

All symbols for rearrangements are to be placed before the designation of the chromosome or chromosomes involved, and the rearrangement chromosome or chromosomes should always be placed in parentheses:

46,XX,r(18)	Female karyotype with 46 chromosomes, including a ring (r) chromosome 18.

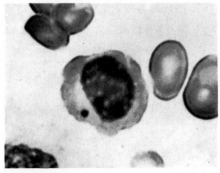

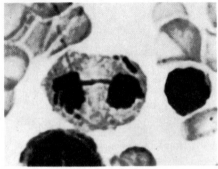

Fig. 2.40. a Micronucleus formation due to chromosome aberration in a bone marrow cell of a patient with Fanconi anemia. **b** Anaphase bridge due to a dicentric chromosome; same patient. (From Schroeder 1970 [169])

46,X,i(Xq)	Female karyotype with 46 chromosomes, including one normal X and one isochromosome (i) for the long arm of X.

The intensity of Q banding patterns is described as follows:

Negative	No or almost no fluorescence
Pale	As on distal 1p
Medium	As the two broad bands on 9q
Intense	As the distal half of 13q
Brilliant	As on distal Yq

Chromosome Band Nomenclature. Each chromosome is considered to consist of a continuous series of bands, with no unbanded areas. The bands are allocated to various regions along the chromosome arms and delimited by specific chromosome landmarks. The bands and the regions to which they belong are identified by numbers, with the centromere serving as the point of reference for the numbering scheme. In designating a particular band, four items are required: the chromosome number, the arm symbol, the region number, and the band number within that region. For example, 1p33 indicates chromosome 1, short arm, region 3, band 3. The region and band numbers may be taken from Fig. 2.11. The following examples illustrate the principle of description:

Isochromosomes (abbreviated and full description):

46,X,i(Xq) 46,X,i(X)(qter → cen → pter)	Break points are at or close to the centromere and cannot be specified. The designation indicates that both complete long arms of the X chromosome are present and are separated by the centromere.

Terminal deletion:

46,XX,del(1)(q21) 46,XX,del(1)(pter → q21)	This indicates a break at band 1q21 and deletion distal of the long-arm segment. The remaining chromosome consists of the entire short arm and that part of the long arm lying between the centromere and band 1q21.

Reciprocal translocations:

46,XY,t(2;5)(q21;q31) 46,XY,t(2;5)(2pter → 2q21:: 5q31 → 4qter; 5pter → 5q31:: 2q21 → 2qter)	Breakage and reunion have occurred at bands 2q21 and 5q31 in the long arms of chromosomes 2 and 5, respectively. The segments distal to these bands have been exchanged between the two chromosomes. Note that the derivative chromosome with the lowest number (i.e., no. 2) is designated first.

These examples should aid in understanding the symbols used in the following text and in cytogenetic publications. High-resolution banding requires logical extension of this nomenclature (see Fig. 2.14). In chromosome 21, for example, the long arm is subdivided into q11, q21, and q22; q22 can be subdivided into q22.1, q22.2, and q22.3.

Deletion Syndromes. An individual who is heterozygous for a deletion is monosomic for a part of the chromosome. De Grouchy et al. in 1963 were apparently the first to publish a case with del 18p–. The first deletion syndrome was established by Lejeune et al. [70], also in 1963. They described three children with a deletion of the short arm of chromosome 5 (del 5 p–). In addition to the usual signs of autosomal chromosome aberration such as developmental retardation and low birth weight, the children showed a moonlike face with hypertelorism. Their appearance was not extraordinarily peculiar, but they had a striking cry that resembled that of a cat (cri du chat = cat cry; Fig. 2.41).

There are various mechanisms by which a deletion may be formed: (a) true terminal deletion, (b) interstitial deletion, and (c) translocation. A number of reports have pointed to a translocation in the cri du chat syndrome.

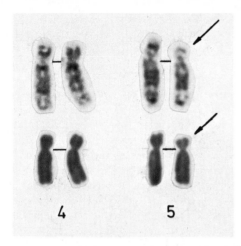

Fig. 2.41. Partial karyotype of a case with cri du chat syndrome – deletion 5p

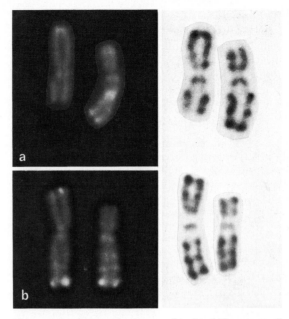

Fig. 2.42 a, b. *Aneusomie de recombinaison?* Karyotype of a malformed proband and his normal mother (Q and G banding). For explanation, see text. **a** Partial karyotype of no. 10 of the mother. **b** Partial karyotype of no. 10 of the son. (From Dutrillaux et al. 1973 [340])

Intrachanges: Paracentric and Pericentric Inversions. Paracentric inversions (i.e., those not including the centromere) in humans are difficult to diagnose. These are discussed in the context of chromosome evolution (Sect. 14.2.1). Since the early 1960s numerous reports on presumed pericentric inversions (i.e., with inclusion of the centromere) have been published. Some of the probands had various anomalies such as mental retardation or malformation. Others were phenotypically normal, but they or their wives had repeated abortions. Again, others showed no abnormality at all. When only the conventional staining methods were available, pericentric inversions were thought to be rare.

After banding methods had been introduced, higher frequencies were reported. Chromosome 9 seemed to be affected frequently, and a relatively high prevalence was reported from Finland [31]. These inversions do not influence chromosome segregation during meiosis and do not bring about an increased risk of prenatal death for heterozygotes, as evidenced by the normal (corrected) segregation ratios. Observations of this type may yield clues for mechanisms of chromosome evolution (Sect. 14.2.1).

The probands of this Finnish study had been examined for diagnostic purposes. Therefore it is not surprising that they showed a variety of anomalies. These anomalies, however, did not show a common pattern; moreover, relatives with the inversion were clinically normal. Hence these pericentric inversions very probably did not influence the phenotypes of their carriers nor their fecundity or the number of miscarriages.

Small inversions may be fairly frequent in particular populations. They may not influence health or fecundity at all. If the inversions are larger, impairment

of normal meiosis is more likely. However, inversion carriers are euploid, and phenotypic abnormalities are therefore not to be expected, except for possible position effects.

Aneusomie de Recombinaison. Occasionally families have been observed in which one parent seemed to have the same aberration as the child – for example, a pericentric inversion or a translocation. Yet, the parent was phenotypically normal, whereas the child showed a severe malformation syndrome. In some of these cases, chance coincidence of a harmless chromosome variant with a malformation syndrome of different origin is the most likely explanation. However, crossing over between the abnormal chromosome and its normal counterpart in the displaced region might lead to unbalanced germ cells. This mechanism was suggested by Lejeune and Berger in 1965 [68], but its confirmation had to await introduction of the banding techniques.

The first case in which this mechanism could be demonstrated was a boy with multiple malformations [26]. Figure 2.42 shows the no. 10 chromosomes of the proband and his mother. Apparently the mother had a large pericentric inversion. Crossing over within this inversion led to an abnormal chromosome, making the child trisomic for the segment q4. More such cases can be discovered by high-resolution banding.

Ring Chromosomes. The situation is different in ring chromosomes, as ring formation implies loss of usually telomeric chromosome segments. Individuals with ring chromosomes resemble the corresponding deletion carriers. They may, for example, have the cri du chat syndrome if 5p is affected [83]. In other cases the symptoms are striking, depending on the size of the deleted segments.

Figure 2.35 shows the fate of a ring chromosome in mitosis; in most cases the ring replicates and passes mitosis normally. Sometimes, one sister strand exchange occurs, and a double ring with two centromeres is formed; in others, double sister strand exchange leads to two interlocked rings. In the next interphase, the double ring may again have one or two (or more) sister strand exchanges, leading either to interlocked double rings or to a fourfold ring. Thus, an indefinite number of combinations is possible.

Fragments. Chromosome fragments are usually lost in mitosis or meiosis, unless retaining a centromere (or a part of it), and they may segregate as supernumerary marker chromosomes. Such markers were not rare in a Danish random sample of newborns (Sect. 9.1.2.1); in some cases phenotypic abnormalities were reported. Their chromosomal origin can now often be identified by chromosome painting (Sect. 3.1.3.3).

Isochromosomes. Chromosomes are occasionally found that consist of two identical arms. Such chromosomes are known as isochromosomes and presumably originate by abnormal division of metaphase chromosomes, as shown in Fig. 2.43. If the chromosome involved has arms of unequal length, isochromosomes for the short arm or isochromosomes for the long arm may result.

Isochromosomes are observed relatively frequently for the X chromosome; an isochromosome of the long arm of the X, i (Xq), leads to the Turner syndrome, since this chromosome is always inactivated and only the normal X is active (Sect. 2.2.3.3).

Interchanges: Centric Fusions (Robertsonian Translocations). Centric fusion is the most frequent type of chromosome rearrangement in human populations. The first reported cases of translocation Down syndrome were due to centric fusion between the long arm of chromosome 21 and one chromosome of the group 13–15 or 21–22 (D or G group). Similar cases have since been observed repeatedly. Centric fusion accounts for only a small percentage among all cases of Down syndrome, and many of these are due to new mutation. Only the five acrocentric pairs undergo centric fusion. In the interphase nucleus the short arms and centromeric regions of these chromosomes are located close to the nucleolus, the short arms containing the nucleolus organizers, which carry genes for rRNA. Participation of single acrocentric pairs in centric fusions is nonrandom. While the data in Table 2.5 are biased due to ascertainment from couples seeking prenatal diagnosis, unselected data from newborns show the

Table 2.5. Different types of Centric fusions in couples referred for prenatal diagnosis (from the European Collaborative Study, Boué and Gallano 1984 [9])

Chromosomes involved	14	15	21	22
13	210	13	24	3
14		6	156	4
15			12	4
21				16

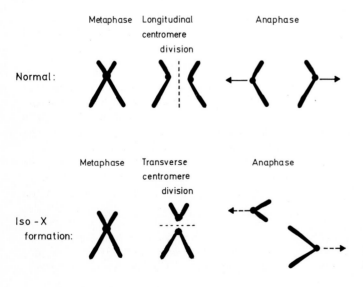

Fig. 2.43. Formation of an isochromosome by abnormal cleavage of the centromere

same pattern [109]. Moreover, a detailed study of available observations has shown convincingly that the differences between the various types are real. Preferences of certain unions could be caused by structural homologies and, consequently, higher exchange rates between their respective centromeric regions [129].

Centric fusion indicates that the short arms of the two participating chromosomes and possibly one of the centromeres are lost (Fig. 2.44). This means that ribosomal RNA genes are lost as well; indeed, the average number of rRNA genes as estimated by DNA-RNA hybridization studies is lower in so-called balanced carriers of centric fusions than in the normal population [11; 24]. Functionally this makes little difference as, to the best of our knowledge, these carriers are perfectly healthy.

Figure 2.45 shows the possible combinations of chromosomes in germ cells of a carrier for a 14/21 translocation, and for a 21/21 translocation. After fertilization with a normal sperm there are six different possibilities. However, the first two, monosomy 14 and trisomy 14, are never observed, and monosomy 21 is – at least in the great majority of all cases – lethal. Each of the remaining three – trisomy 21, balanced, and normal – might be expected to have a probability of one-third. This expectation, however, is not confirmed by experience: if the mother is the carrier, the probability is around 15%, and if the father is the carrier, it is no more than 5%. However, the risk of a balanced zygote is the expected segregation rate of approx. 50%. The problem is discussed in greater detail in Sect. 2.2.2.

In a 21/21 translocation – or a 21/21 isochromosome – prospects are much more gloomy: either the child is trisomic and affected with Down syndrome, or the aneuploidy is lethal.

Interchanges: Reciprocal Translocations. Unlike centric fusions, reciprocal translocations do not necessarily entail loss of material. The broken parts are joined with other chromosomes. Therefore, the balanced zygote has 46 and not 45 chromosomes. Among the daughter cells, the types seen in Fig. 2.47 can be expected. Most often only the partially trisomic or the partially monosomic are found, the others presumably being lethal.

A typical pedigree is seen in Fig. 2.46 [112]. Two sibs, 11 and 9 years old, showed various, severe clinical signs such as mental retardation, anomalies of skull and face, hypoplastic and hypotonic skeletal muscles, and club feet. Figure 2.49 shows the karyotype of the mother who had a balanced reciprocal translocation; a part of the short arm of chromosome 10 is translocated to the short arm of chromosome 7. Hence, the children had a partial trisomy 10. Interestingly, in addition to the concordant clinical signs mentioned above, they also showed some discordances; some clinical signs occurred only in one of the sibs, for example, epileptic seizures, cleft lip, and palate. Such phenotypic differences between car-

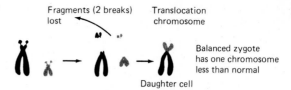

Fig. 2.44. Principle of a centric fusion (Robertsonian translocation). Two acrocentric chromosomes lose their short arms, and the long arms fuse. The translocation chromosome may have one or two centromeres; in the latter case one centromere must be suppressed. In any case the balanced zygote has one chromosome less than normal (as compared with the reciprocal translocation, which leads to a balanced zygote with normal chromosome number)

riers of identical chromosomal aberrations are common (see below).

Phenotypes in Autosomal Chromosome Aberrations. The most conspicuous phenotypic feature in autosomal chromosome aberration syndromes is the considerable overlap of signs and symptoms. The main findings are:

General
    Low birth weight (small for date)
    Failure to thrive
    Mental retardation (usually severe)
    Short stature
Head and face
    Microcephaly
    Incomplete ossification
    Micrognathia
    Anomalous positioning of eyes
    "Dysmorphic facies"
    Low-set, deformed ears
Various anomalies of hands and feet
    Anomalous dermatoglyphic patterns
Internal organs
    Congenital defects of heart and/or great vessels
    Cerebral malformations
    Malformations of the genitourinary system

The following symptoms generally do not point to an autosomal chromosome anomaly, but exceptions occur: mental retardation without additional malformation; malformations associated with normal mental capacity; isolated, single malformations.

Apart from these common malformations many, but not all, autosomal aberrations show other more or less specific clinical patterns. Moreover, the common signs may be more or less severely expressed. The signs caused by a specific aberration usually form a pattern that is characteristic for this aberration and makes a preliminary diagnostic impression possible on clinical grounds. These symptom complexes

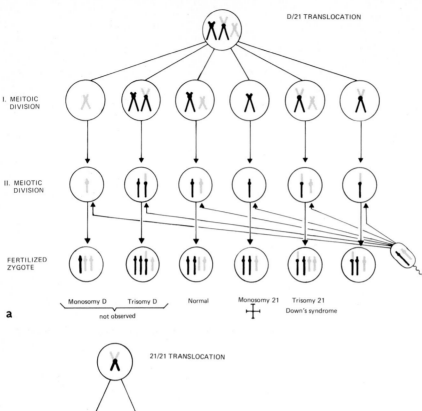

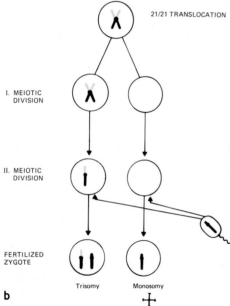

Fig. 2.45. a Diagram of germ cell formation if the mother is the carrier of a 14/21 translocation in a balanced state: the one chromosome 14 has acquired the translocated long arm of chromosome 21. As a result there is only one free chromosome 21. Since this free chromosome 21 and the two 14-chromosomes combine according to the laws of chance, six different germ cells can theoretically be formed, and it should be possible to find six different types of zygotes after fertilization with a normal sperm. However, three of the possible six have not been observed. The remainder are either normal, balanced, or trisomic in a ratio that can only be established empirically. **b** Formation of germ cells with a 21/21 translocation and a 21-iso-chromosome. The possibilities are that either the translocation chromosome forms the germ cell – a resulting zygote would be functionally trisomic and the child would manifest Down syndrome – or the translocation chromosome does not get into the germ cell, in which case the zygote is missing a chromosome 21 and dies

help in deciding when chromosome study is indicated.

Among patients showing a common aberration the variability of many of the findings tends to be great. Some persons with Down syndrome, for example, may be only mildly retarded, whereas most are severely mentally retarded; cardiac malformations are found in many, and intestinal atresia in only a few. This variability is presumably due in part to the fact that the abnormal chromosome is superimposed on very different genotypes, i.e., the genetic background

differs. Still, variability in development seems to be abnormally enhanced. Increased lability of embryonic development by yet unknown mechanisms has been postulated.

It is remarkable that trisomies cause any anomaly at all. Their carriers have a full set of genetic material, and no single gene function is altered or lost. In addition, we know from heterozygotes of autosomal-recessive diseases (Sect. 7.2.2.8) that for most enzymes one-half of the normal production is sufficient to maintain normal function. It is not readily

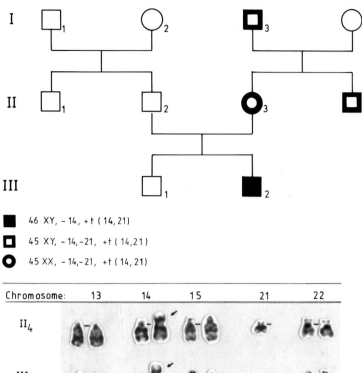

Fig. 2.46. Pedigree of a male patient with Down syndrome and a 14/21 Robertsonian translocation *(III,2)*. Mother *(II,3)*, her brother *(II,4)*, and maternal grandfather *(I,3)* have a balanced karyotype. The cytogenetic findings in I,3 and II,3 were identical to those in II,4. (Courtesy of Dr. T. M. Schroeder-Kurth)

■ 46 XY, – 14, + t (14, 21)

◻ 45 XY, – 14, –21, + t (14, 21)

● 45 XX, – 14, –21, + t (14, 21)

Fig. 2.47. Reciprocal translocation. Two breaks in different chromosomes create two chromosomal fragments that are reciprocally translocated or exchanged to the broken chromosome from which they did not originate

apparent why the presence of three rather than two gene products as present in trisomies should make so much difference. These problems are discussed in Sect. 8.4.3.

Moreover, the common symptoms of all the autosomal syndromes are independent of the single chromosomes involved. One might note that the organ systems mainly involved are known to have a long and complex embryonic development, and that therefore many different genes are needed for this development to proceed normally. However, this explanation is also very general, and, besides, the necessary genes are all there.

What is the nature of the genetic disturbance leading to these chromosome syndromes? They may be due either to surplus activity or a defect of single genes, or to failure in the regulation of genes during embryonic development [29]. Analysis of autosomal

chromosome aberrations should therefore be suitable to teach us much regarding the mechanisms of gene action during development and gene regulation in humans. In Down syndrome, for example, a region of no more than 400 kb appears to be responsible for the clinical signs (see Sect. 2.2.1). This problem is taken up in Sect. 8.4.3. For a special problem – the development of sexual characteristics – studies of patients with numerical and structural chromosome aberrations have already been highly instructive. Before sex chromosomal aberrations are discussed, however, some remarks are necessary regarding segregation and prenatal selection of unbalanced, and possible clinical signs in "balanced" translocation carriers. In addition to their theoretical interest, these questions have an important practical relevance for risk assessment and genetic counseling.

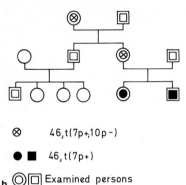

⊗   46,t(7p+,10p-)

●■   46,t(7p+)

b ◎⊡ Examined persons

Fig. 2.48. Pedigree of two siblings (mother: balanced translocation; children: partial trisomy 10p + ; Fig. 2.49)

2.2.2.2 Segregation and Prenatal Selection of Translocations: Methodological Problems

The problem of segregation and prenatal selection of translocations has often been studied, but there were many contradictions, and the studies have failed to give a clearcut picture. Recently, however, many of these problems have been solved by Schäfer [109]. In principle we follow his analysis.
Translocations are relatively rare. No single research group is therefore able to collect sufficient data for definite conclusions. Hence, data published in the literature must be analyzed. Such data, however, are subject to many biases. Some of these are discussed below (Sect. 4.3.4 and Appendix 2); for example, sibships in which abnormal chromosomes segregate are generally ascertained only if at least one sib is affected. It also makes a difference whether all sibships with at least one

affected sib in the population are ascertained or only a small fraction. These problems can be handled relatively simply if hereditary diseases are involved, since in this case families are always ascertained through at least one "proband" who suffers from the disease in question. With translocations, however, families might be ascertained, for example, because of multiple abortions or through an unbalanced proband at birth or at prenatal diagnosis. Balanced carriers are, occasionally, discovered in population surveys, and a family study is initiated. It is impossible to correct completely for all biases – especially as necessary information is lacking in many published reports. The correction by Schäfer, however, appears to be the optimum attainable with the cases presently available. The analysis is presented in greater detail in Appendix 2.

The study was based on 1050 families with segregating translocations, with altogether 2109 pairs of parents and 4745 progeny. Moreover, 556 reported instances of pathological effects in carriers of balanced translocations were collected, as were results from 814 prenatal diagnoses, and about 130 000 individuals examined in various screening programs were evaluated. This statistical study led to detailed risk estimates for various kinds of pregnancy outcome, and to results on phenotypic effects in unbalanced and balanced translocation carriers. For an understanding of these results it is necessary to visualize the consequences of translocations during meiosis.

Segregation of Translocations in the First Meiotic Division. In the first meiotic division homologous chromosomes pair. This rule also applies to translocated chromosome segments: they pair with their *original* partners. This leads to complexes of *four* chromosomes in reciprocal translocations and

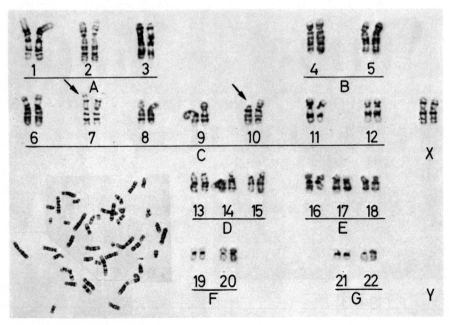

Fig. 2.49. G-banded karyotype of the mother of two children of Fig. 2.48 with a reciprocal translocation involving chromosome 7 and 10 *(arrows)*. The two children (Fig. 2.48) showed the elongated chromosome 7 (with the no. 10) short arm but two normal chromosomes no. 10. (From Schleiermacher et al. 1974 [112])

of *three* chromosomes in Robertsonian translocations. As in normal meiosis, spindle fibers fix at the centromeres, and homologous chromosomes are moved to opposite poles. This regular anaphase movement may lead to four different division products with equal probabilities (Fig. 2.50):

Fig. 2.50a: Let the two normal chromosomes be A_1, B_1; they come into one division product (haploid cell), and the two translocation chromosomes, A_2, B_2, come into the other (= alternate disjunction).

Fig. 2.50b: One normal and one translocated chromosome come into one haploid cell (= adjacent-1 disjunction). Here, two possibilities exist: A_1B_2 or A_2B_1. Each of the four combinations has the probability 0.25.

A_1B_1 is karyotypically normal and A_2B_2 is balanced, since the two chromosomes A_2 and B_2 have exchanged segments. A_1B_2 and A_2B_1, however, are unbalanced. In addition, other, abnormal types of segregation may occur as a result of the chromosomal aberration. For example, homologous centromeres might occasionally come into the *same* division product (adjacent-2, Fig. 2.51), or three centromeres may come into one, one centromere into the second division product (3:1 disjunction; Fig. 2.52).

Expectations for Unbalanced Zygotes. Using corrections for ascertainment biases, the average risk for abnormal, unbalanced offspring at birth is estimated to be 7% for carrier mothers and 3% for carrier fathers. These risk figures are estimated for *all* balanced carriers. For carriers in families in which unbalanced zygote offspring have already been observed the risk figures for sons and daughter together are higher (14% of all births for mothers and 8% for fathers); the risk for unbalanced zygote sons of carrier fathers is especially low (5%). When the mothers are carriers, 66% of all unbalanced zygote offspring are found to be of the adjacent-1 type, 3% adjacent-2, and 31% 3:1. With carrier fathers, 90% were adjacent-1, 3% adjacent-2, and 8% 3:1 (see above).

For translocation carriers ascertained through prenatal diagnosis the estimated risk figures are: 11.7% unbalanced offspring for mothers and 12.1% for fathers.

The low *overall* estimate (7% for carrier mothers; 3% for fathers) is due to the fact that only about one-half of all translocations can give rise to malformation syndromes at birth; the others invariably lead to fetal death.

For *Robertsonian* translocations involving chromosome 21, a 13% risk of unbalanced offspring has been estimated when

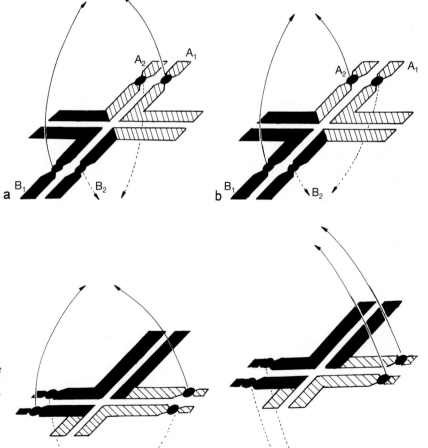

Fig. 2.50. Schematic representation of a translocation quadrivalent predisposed either to alternate (**a**) or to adjacent-1 (**b**) disjunction. The disjunction in **a** leads to normal and balanced zygotes; that in **b** leads to two types of unbalanced zygotes

Fig. 2.51 a, b. Both chromosomes involved in a translocation are acrocentric. One of the paired arms that does *not* carry the centromere is much longer than the arms carrying the centromere. Here it may (rarely) occur that homologous centromeres are transported into the same daughter cell (adjacent-2 disjunction). This type of disjunction is also possible if one of the two chromosomes is a chromosome no. 9

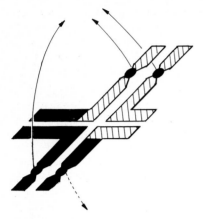

Fig. 2.52. A 3:1 segregation with formation of tertiary trisomy (and monosomy). The length of the paired segments between the centromeres makes chiasma formation likely; the chiasma on the right cannot be terminalized properly

the mother is a carrier; the risk is 3% with a carrier father. DqDq carriers, on the other hand, run practically no risk of having unbalanced offspring. The same seems to apply for Dq22q translocations. For genetic counseling the figures for Robertsonian translocations can be used with confidence, but those for reciprocal translocations are too crude since they are derived from many types of translocations involving a great number of different chromosomes; a refinement is, however, possible. For example, the risk is higher if the carrier who asks for advice has been ascertained through an unbalanced rather than a balanced proband. Moreover, special parameters, such as the length of the chromosomes involved in the translocation and, especially, the size of the trisomic (or monosomic) chromosome segments, should be considered: as a rule of thumb, the larger these segments, the stronger is the prenatal selection against the aneuploid germ cell or zygote.

Increasing data also suggest that the risks are higher when the total length of the two chromosomal segments involved in the imbalance is less than about 2% of total genome length. With longer chromosomal structures affected there is usually very early loss of the unbalanced zygote. Partial monosomies are more damaging than partial trisomies. Ideally such a risk assessment should be based on empirical

data about precisely the same translocation that is observed in the family in question [327]. This, however, is impractical in most cases, as too many different translocations exist.

Boué and Gallano [276] reported results from a collaborative study of 71 European prenatal diagnosis centers, with 2356 observations on the karyotypes of fetal cells in diagnoses performed in couples in which one parent had a balanced structural rearrangement. The results (Table 2.6) demonstrate that the risk of future children showing the unbalanced karyotype is much larger if the parent with a balanced karyotype has been ascertained through an unbalanced child than if he or she has been ascertained through a spontaneous abortion. The authors also subdivided their data by considering the various autosomes involved; there were no conspicuous differences. However, the smaller the involved chromosome fragments, the larger was the risk of unbalanced offspring. Larger imbalances produce more phenotypic damage and are therefore followed more often by early and often unrecognized loss of the zygote. There were 232 prenatal diagnoses in instances in which one of the parents had a balanced Robertsonian translocation (excluding those involving at least one chromosome 21). Fifteen chromosomally normal, 133 balanced, and 28 (12%) unbalanced fetuses with malformations were observed. This confirms the deviation from the expected 1:1 segregation between normal and balanced zygotes. Moreover, earlier results were confirmed that there were no unbalanced offspring from carriers of Robertsonian translocations including chromosomes 13, 14, 15, and 22.

Phenotypic Deviations in Balanced Translocation Carriers. Balanced translocation carriers have a complete set of genetic material and should therefore be phenotypically normal. As a rule, this expectation is borne out by experience. However, many reports have claimed a somewhat higher incidence of malformations, mental retardation, and minor birth defects. Analysis of the available evidence has shown that multiple malformations and mental deficiency are rare among balanced translocation carriers but more common than among karyotypically normal subjects. As a rule, such clinical findings are encountered in *sporadic* or in familial 14q21q translocations:

Table 2.6. Frequencies of fetuses with unbalanced anomalies and birth defects with different modes of ascertainment. (From Boué and Gallano 1984 [9])

Chromosomal Defect	Ascertainment					
	Infants with unbalanced chromosomal anomalies		Spontaneous abortions		Other ascertainment	
	n	%	*n*	%	*n*	%
Reciprocal translocations:	54/260	20.8	7/205	3.4	10/144	6.9
Inversions:	6/8	75	0/25	0	1/85	<1

Numbers refer to the frequency of unbalanced aberrations among the total number of fetuses studied.

In this familial group undetected mosaics might be suspected. In sporadic translocations, breaks within single genes or disconnection of genes from their functionally important location ("position effects") might explain the phenotype. Most carriers of familial translocations are clinically unaffected. Selection against unbalanced zygotes leads to an increase in the abortion rate; translocation carriers are overrepresented among couples with multiple miscarriages. Infertility has been observed among male carriers; in females it occurs only with X-autosomal translocations. However, these findings should not obscure the fact that the great majority of translocation carriers, familial and sporadic, have a normal phenotype.

These results on human translocations have implications for the mutation rate, as well as for fitness, selection, and the theory of evolution. These aspects are discussed in Sects. 9.1.2, 12.2.1, and 14.2.

Premature Sister-Chromatid Separation (Heterochromatin Repulsion). There is a rare, autosomal-recessive complex malformation syndrome with abnormally short limbs, abnormal maxilla, and other, severe malformations: Roberts syndrome (268300). Pro- and metaphase chromosomes show an abnormally premature separation of centromeres, the short arm of acrocentrics, and the Yqh heterochromatin [40, 133].

2.2.2.3 Small Deletions, Structural Rearrangements, and Monogenic Disorders: Contiguous Gene Syndromes [113, 132]

Deletions and other unbalanced structural rearrangements leading to loss of genetic material generally cause such fundamental disturbances of embryonic development that the resulting zygote does not survive to birth. Sometimes, however, the DNA sites affected by the structural defect do not lead to embryonic or fetal lethality but may be associated with the clinical signs of a known monogenic disorder. Occasionally the clinical findings of more than one hereditary disease may be present. If the resulting condition does not prevent reproduction, the resulting phenotype segregates. One of us (F. V.) predicted such events already in his 1961 textbook, referring to them as a "broad mutation." If the deletion is located at a site where the gene for a monogenic disease is located, deletion of this gene may cause the clinical findings that usually accompany this Mendelian disorder. In fact, 5%–10% of monogenic diseases are associated with microscopically undetectable gene deletions. However, a deletion may involve additional contiguous genes; therefore

the resulting phenotype may show features of several monogenic disease. Hence, a deletion or a breakpoint of a structural rearrangement may point to the chromosomal site where previously unmapped genes are located (deletion mapping; see Sect. 5.1.2). For example, a patient showed signs not only of Duchenne muscular dystrophy but also of X-linked chronic granulomatous disease, retinitis pigmentosa, and of McLeod syndrome, i.e., absence of the blood group Kell precursor substance [37]. A small deletion was found on the short arm of the X chromosome (Xp21) and therefore helped in the localization of four different genes on the short arm of the X chromosome and in the subsequent cloning of these genes. Translocation breakpoints of X-autosomal translocations had pointed before to the same region as the site of the gene for Duchenne muscular dystrophy [43] (see also [132]), but there had been no clues to the location of the three other genes. Their mapping to this area was confirmed by later studies.

Another example is retinoblastoma (180200), the malignant eye tumor. In cases with bilateral tumors, visible constitutional chromosomal anomalies that are generally deletions involving the region 13q14 are found in over 10% of cases using high-resolution banding methods (see Sect. 10.4.3). Similar findings were made in another childhood malignancy, Wilms tumor (194080), a tumor of the kidney that may be transmitted with autosomal dominant inheritance or, more often, is nonhereditary. In some instances Wilms tumor is associated with aniridia (106210), genitourinary malformation, and mental retardation (WAGR syndrome). In such cases a deletion at 11p13 is usually found. Aniridia without Wilms tumor is a well-known autosomal-dominant disease; aniridia occurs in 1%–2% of Wilms tumor patients. Conversely, more than 10% of aniridia patients, particularly the sporadic cases due to fresh mutations, develop Wilms tumor. Often a deletion or chromosomal rearrangement leading to loss of genetic material is found in aniridia. Other syndromes caused by small deletions include are the Miller-Dieker (247200), Langer-Giedion (150230), di George (188400), Beckwith-Wiedemann (130650), Alagille (118450), Prader-Willi, and Angelman syndromes (176270; 234400; Table 2.7). The latter two syndromes are remarkable because they exhibit the phenomenon of genomic imprinting: the same segment of chromosome 15 (15q11.2-q12) is deleted, but in the Prader-Willi syndrome the chromosome with the deletion comes from the father while in Angelman syndrome the maternal chromosome is deleted. In Beckwith-Wiedemann syndrome the defective chromosome is always transmitted from the mother (genomic imprinting is discussed in Sect. 8.2).

Table 2.7. Contiguous genes syndromes (excluding those on the X chromosomes). Note that the specific deleted genes have not been identified in most instances

Disease	Localization	Clinical signs	Special remarks
Retinoblastoma (180 200)	13q14	Childhood eye tumors (children with visible deletions may show microcephaly, slight anomalies of face and hands, and mental retardation)	About 5%–10% of all cases have chromosomal aberrations, but mostly rearrangements. The others only have the tumor. This is a point mutation, inheritance dominant.
Wilms tumor, WAGR syndrome (194 070, 109 210)	11q13	Kidney tumors, bilateral aniridia in combination with genitourinary malformations and mental retardation possible	A combination of any of these anomalies is possible. About 10% of aniridia patients also develop Wilms' tumor. Inheritance dominant.
Miller-Dieker syndrome (24 700)	17p13	Lissencephaly; facial anomalies; mental retardation	Cases with and without visible deletions occur.
Beckwith-Wiedemann syndrome (130 650)	11p15	Enlarged tongue; gigantism; exomphalos; hypoglycemia; often adrenal carcinoma or nephroblastoma	Mode of inheritance: irregularly dominant. All known 11p15 arrangements are inherited from the mother. In many instances, no chromosomal aberration
Langer-Giedion syndrome (150 230)	8q24.1	Sparse hair, lax skin, pear-shaped nose, multiple cartilaginous exostoses, microcephaly, mental retardation	This is trichorhino-phalangeal syndrome type II. Type I has no mental retardation and lax skin. Autosomal dominant.
di George and Sprintzen syndromes (188 400)	22q11	Cellular immune deficiency, hypoparathyroidism, other anomalies; region of 3rd and 4th pharyngeal pouch	Cases with deletions and with chromosomal rearrangement involving 22q and some with submicroscopic deletions are observed. Autosomal dominant.
Alagille syndrome (118 450)	20p11	Cholestasis; pulmonal valvular stenosis; arterial stenosis; abnormal vertebrae; neurological signs and learning difficulties	Clinical variability, deletions and structural anomalies; autosomal dominant with variable expressivity.
Prader-Willi and Angelman syndromes (176 270, 234 400)	15q11	Prader-Willi syndrome: hypotonia, small hands and feet, hypogonadism, obesity, small stature, mental retardation Angelman syndrome: "happy puppet" face, mental deficiency; paroxysmal laughing; seizures; lack of speech	In most instances, structural alterations, mostly deletions of 15q11.2. Genomic imprinting: Angelman syndrome if the abnormal chromosome comes from the mother; Prader-Willi syndrome if it comes from the father

In X-linked disorders, chromosome rearrangements involving one or even more than one monogenic disease have been described. This may well be caused by ascertainment bias; such translocations usually originate in the male germ line [132]. Therefore X-autosomal translocations almost always come to attention in females, since males inherit the Y and not the X chromosome from their fathers. However, rare girls with a phenotype usually found only in boys tend to be examined especially thoroughly.

Contiguous gene syndromes have led to an attempt at subdividing the genome into deletion-viable and deletion-nonviable segments. If a gene is located in a deletion-viable region, somewhat larger deletions causing additional clinical signs may be observed in patients with a well-known hereditary disease (generally due to a new mutation). This is not the case if the gene is located in a nonviable region. Here large deletions kill the zygote in an early state of development. These syndromes have also been studied at the DNA level since they offer a good change for discovery and analysis of interesting chromosomal segments carrying disease genes. Moreover, with the development of new cytogenetic methods – such as in situ hybridization (FISH) – more such syndromes will probably be discovered. This will help in distinguishing deletion-viable from deletion-nonviable chromosomal regions – an important piece of information for human developmental genetics (Chap. 8).

2.2.3 Sex Chromosomes

2.2.3.1 First Observations

Nondisjunction of Sex Chromosomes and Sex Determination in Drosophila. Meiotic nondisjunction was discovered by Bridges in 1916 [10] in the sex chromosome of *Drosophila melanogaster*. Morgan in 1910 [81] had earlier described the X-linked mode of inheritance, and at the same time had elucidated the X–Y mechanism of sex determination in *Drosophila*. In his experiments a few exceptions had occurred that did not conform to the predictions of X-linkage. Bridges explained them by an anomaly in the mechanism of meiosis. *Drosophila* has four chromosome pairs, three pairs of autosomes, and two sex chromosomes. Just as in humans, the males have the complement XY, the females XX. Hence, each normal male germ cell has either one X or one Y chromosome; all female germ cells have an X. In crosses between an affected homozygote for the X-linked recessive trait white and a wild-type or normal male, all male offspring would be expected to have white eyes as their mothers. All daughters should be heterozygous and have normal red eyes. As a rule, this expectation was fulfilled. In exceptional cases, however, male offspring had normal red eyes, and some females were white-eyed. This was shown by Bridges to be due to nondisjunction of the maternal X chromosome leading to an oocyte with either two or no X chromosomes. Fertilization with sperm from a wild-type male was expected to lead to four different types of zygotes: XXX, XXY, XO, and YO. YO was not observed; apparently, zygotes without an X chromosome cannot survive. The other three types were observed and gave evidence regarding the mechanism of sex determination:

a) XXX ⎤
b) XXY ⎦ Female phenotype

c) XO   Male phenotype: sterile

Hence, the phenotypic sex in this fruit fly depends on the number of X chromosomes. One X chromosome makes a male, more than one X chromosome makes a female. The Y is also involved in sex determination, as XO males are sterile.

XO Type in the Mouse. The X-linked mutation scurfy (sf) appeared first by spontaneous mutation. The animals have scurfy skin. The hemizygous males are sterile; therefore, the strain can be maintained only by crossing heterozygotes (X^{sf}/X^{+}) with normal males (X^{+}/Y). From this mating, scurfy and normal males are expected in a segregation ratio of 1 : 1; all females should be normal. From time to time, however, an exceptional sf female is observed. As with male hemizygotes, they are sterile. However, their ovaries can be transplanted to normal females, which have been mated with wild-type males. The sons are all sf; the daughters are all normal but fall into two groups, those transmitting sf and those not transmitting it. Further analysis showed that these daughters have two different karyotypes, X^{+}/O and X^{+}/X^{sf}; the first group does not transmit sf, the second does. This experiment showed that, contrary to the findings in *Drosophila*, XO is a fertile female in the mouse. Hence, in this animal, the Y and not the X chromosome is decisive for the phenotypic sex. Subsequently, the XO types of the mouse have been found to be fairly frequent. In most cases the condition is caused not by meiotic nondisjunction but by chromosome loss after fertilization. In mutation research, this chromosome loss has become an important tool for assessing mutagenic activity (Sect. 11.1.3). Not long after the XO type, the XXY type was also discovered in the mouse. It is a sterile male unlike *Drosophila* where the XXY type was female.

First X Chromosomal Aneuploidies in Humans: XXY, XO, XXX. Jacobs and Strong in 1959 [56] studied a 42-year-old man with the typical features of Klinefelter syndrome (Fig. 2.53), including gynecomastia, small testicles, and hyalinized testicular tissue (Fig. 2.54). X chromatin in cells of buccal smears and drumsticks in granulocytes were found. Chromosome examination from bone marrow revealed an additional, submetacentric chromosome "in the medium size range." The authors felt that the patient very probably had the constitution XXY. However, "The possibility can not be excluded ... that the additional chromosome is an autosome carrying feminizing genes." The patient's parents both had normal karyotypes with 46 chromosome; hence, nondisjunction had occurred in one of their germ cells. Shortly afterward, the XXY status for Klinefelter syndrome was confirmed in many other cases.

At the same time, the XO type was discovered by Ford et al. [32]. Their patient, a 14-year-old girl, presented clinically as having Turner syndrome (Fig. 2.55) and was X chromatin-negative. The modal number of chromosomes in bone marrow cells was 45; there were only 15 "medium length metacentric

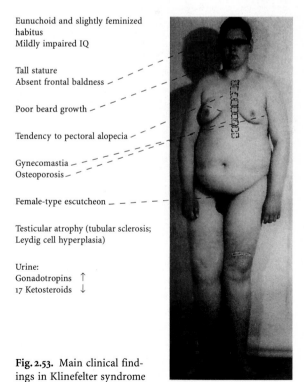

Eunuchoid and slightly feminized habitus
Mildly impaired IQ

Tall stature
Absent frontal baldness

Poor beard growth

Tendency to pectoral alopecia

Gynecomastia
Osteoporosis

Female-type escutcheon

Testicular atrophy (tubular sclerosis; Leydig cell hyperplasia)

Urine:
Gonadotropins ↑
17 Ketosteroids ↓

Fig. 2.53. Main clinical findings in Klinefelter syndrome

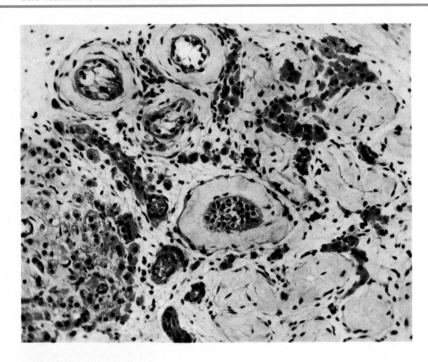

Fig. 2.54. Hyalinized testicular tissue in Klinefelter syndrome. The normal tubules are lacking and are replaced by hyalinized tissue

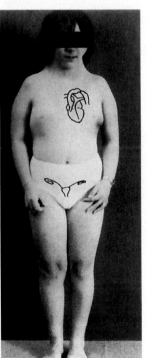

Small stature

"Sphinx" face
"Carp" mouth

Low set nuchal hair

Pterygium colli

Coarctation of the aorta

Shield-shaped thorax
Widely spaced nipples

Poor breast development

Cubitus valgus

Rudimentary ovaries
Gonadal streak
Primary amenorrhea

Shortened metacarpal IV
Finger nail hypoplasia

Dorsal metacarpal and metatarsal lymphedema (at birth)

Multiple pigmented naevi

Urine:
Gonodotropins   ↑
17 Ketosteroids ↓
Estrogens   ↓

Fig. 2.55. Main clinical findings in Turner syndrome

dividual with female phenotype. Noting the XXX state of *Drosophila*, they remarked that this anomaly was still unknown in humans.

This gap was soon closed by the report of a 35-year-old woman with poor development of external sexual characteristics and secondary amenorrhea, who showed 47 chromosomes with an additional X : 47,XX, + X. In this case, two tissues – bone marrow and fibroblasts – were examined, and both showed the same aneuploidy. In many of her buccal smear cells, and in some granulocytes, the patient had two X chromatin bodies. The following generalizations emerged:

1. Unlike *Drosophila*, the phenotypic sex in humans is determined by the presence or absence of the Y chromosome, not by the number of X chromosomes. In this respect humans are similar to mice, but the XO type in mice is a fertile female; in humans, it is a female with a nonfunctioning ovary.
2. The number of X chromatin bodies is one fewer than the number of X chromosomes.

These two observations were soon to become the cornerstones of our knowledge and hypotheses on sex determination and on the genetic activity of X chromosomes.

Discussion of other aspects of Klinefelter and Turner syndromes:

– Sect. 2.2.3.3:   Molecular basis of sex determination
– Sect. 9.1:   Chromosomal mutations, maternal age
– Sect. 15.2.2:   Brain development, behavioral aspects

chromosomes" as in normal males. The evidence strongly suggested a chromosome constitution XO. The authors, comparing this result with that known from *Drosophila*, concluded that, contrary to the fly, the XO type in humans may lead to an "agonadal" in-

2.2.3.2 X Chromosomal Aneuploidies in Humans: Current Knowledge

Difference Between X Chromosomal and Autosomal Aneuploidies. Soon after these first discoveries a great number of other aneuploidies of sex chromosomes were described. As a group they show some remarkable differences from the autosomal aneuploidies discussed before.

a) Mean intelligence is often below the norm, but the extent of mental retardation is not nearly as pronounced as in the autosomal conditions; many probands have normal intelligence, and in a few it is even above average (Sect. 15.2.2).

b) The phenotypic disturbances most severely affect development of the sexual organs and sexual hormone dependent growth. Other malformations do occur – mainly in Turner syndrome, but except for the small stature of Turner patients, they usually are less frequent and less severe.

In brief, X chromosomal aneuploidy does not disturb embryonic development nearly as much as does autosomal aneuploidy. The reason is that normal women have two, normal men only one X chromosome. This difference led to the development in evolution of a powerful mechanism of gene dosage compensation that happened to benefit carriers of X aneuploidies (see below).

Clinical Classification of X Chromosomal Aneuploidies; Mosaics. The most important numerical and structural anomalies of the X chromosome are found in Table 2.8. In general, the number of additional X chromosomes enhances the severity of mental retardation. The number of X chromatin bodies is one less than the number of X chromosomes. Table 2.8 also lists the most frequent mosaics. It does not include interchanges involving the X chromosome. The same rules apply here as for reciprocal translocations of autosomes. Great variability in symptoms has been observed.

Some of these interchanges became important for our theoretical concepts of X inactivation. The most severe, somatic malformations are found in Turner syndrome where more cytogenetic heterogeneity exists than in the other X-chromosomal aneuploidies. Some subcategorizations have been proposed on clinical grounds [48, 97], the most important being the distinction between simple gonadal agenesis without additional symptoms and Turner syndrome with the findings shown in Fig. 2.55. But the cytogenetic data show little if any correlation with these categorizations.

Theoretically, XO zygotes should be somewhat more frequent than any other types, since they can be pro-

Table 2.8. Numeric and structural X chromosomal aneuploidies in humans

Karyotype	Phenotype	Approx. incidence
XXY	Klinefelter syndrome	1/700 ♂
XXXY	Klinefelter variant	≈ 1/2500 ♂
XXXXY	Low grade mental deficiency; severe sexual under-development; radioulnar synostosis	Very rare
XXX	Sometimes mild mental retardation; occasionally disturbances of gonadal function	1/1000 ♀
XXXX XXXXX	Physically normal; severe mental retardation	Rare
XXY/XY and XXY/XX mosaics	Klinefelter-like, sometimes with milder symptoms	≈ 5%–15% of all Klinefelter-like patients
XXX/XX mosaics	Like XXX	Rare
XO	Turner syndrome	≈ 1/2500 ♀ at birth
XO/XX and XO/XXX mosaics	Turner syndrome; varying different degrees of manifestation	Not uncommon
Various structural anomalies of X chromosomes	Variable	Not uncommon
XYY	Increased stature; occasional behavioral abnormalities	1/800 ♂
XXYY	Increased stature; otherwise resembling Klinefelter syndrome	Rare

duced by nondisjunction in both sexes and both meiotic divisions. This expectation does not fit the observed data, as all the karyotypes together that lead to Turner syndrome are much rarer than XXX or XXY. This finding points to strong selection against germ cells without the X chromosome and/or to strong intrauterine selection against XO zygotes. The latter expectation is corroborated by observations on abortions, among which the XO type is, indeed, frequent. Another line of evidence points in the same direction: the risk of nondisjunction in general increases with the age of the mother (Sect. 9.2.2). For XXY and XXX karyotypes this increase can be clearly demonstrated; but not for the

XO karyotypes. Hence, it is assumed that surviving XO zygotes are the result not of meiotic but of mitotic nondisjunction or of early chromosome loss. The relatively greater proportion of mosaics in this group compared with XXX and XXY fits this hypothesis.

XYY zygotes, on the other hand, can be formed only by nondisjunction during the second meiotic division in males. Nevertheless, they are about as frequent as XXY zygotes. Therefore, the probability for nondisjunction of Y chromosomes appears to be much higher than combined probabilities for X chromosome nondisjunction. Mosaics have been observed for all types. The mechanisms for mosaic formation are discussed in Sect. 10.1.

Intersexes (see also [29a]). From clinical observation, three types of intersexes are distinguished:

1. True hermaphroditism: germ cells of both sexes are present.
2. Male pseudohermaphroditism: only testicles are observed.
3. Female pseudohermaphroditism: only ovaries can be found.

Unfortunately, this simple categorization is not supported by the cytogenetic evidence. Many different karyotypes can be found, even 46,XX males. Many intersexes are mosaics for cells with different sex chromosome complements in various combinations. The phenotype of 45,XX/46,XY mosaics, for example, may be ovarian dysgenesis, gonadal dysgenesis with male pseudohermaphroditism, or "mixed gonadal dysgenesis" – one gonad being a streak, the other a dysplastic testicle. Of the true hermaphrodites, some have a 46,XX karyotype. Others are 46,XX/46,XY mosaics, or XY, or 46,XX mosaics. The XX/XY state may originate from any of nine different mechanisms, such as fertilization of the oocyte by two different sperms, fusion of two fertilized eggs, mitotic errors during cleavage, or exchange of blood stem cells between dizygotic twins of different sex during embryonic life (Sect. 6.3.3).

The primary function of the sex-determining factors is induction of gonads. Gonadal function in turn determines development of the other sexual organs and the secondary sex characters. Disturbances of gonad induction may be caused either by an abnormal sex chromosome complement or by other interfering factors not directly involving the sex chromosomes. In the latter case, the intersex may have a normal XX or XY constitution. Balanced structural changes involving the X often lead to infertility in both sexes. For details on male-determining factors of the Y chromosome see Sect. 8.5.

2.2.3.3 Dosage Compensation for Mammalian X Chromosomes

Nature of the X Chromatin. After the X chromatin had been discovered by Barr and Bertram in 1949 [4], there were speculations regarding its nature. In analogy to *Drosophila* it was first thought to consist of heterochromatic parts of the two X chromosomes. Demonstration of its bipartite nature seemed to corroborate this conclusion. However, Ohno et al. in 1959 [88, 89] showed it to represent one single X chromosome. In diploid prophase preparations of regenerating rat liver cells, the X chromatin body of the preceding interphase was not resolved as heterochromatic regions of two chromosomes. Instead, a rather large chromosome, heavily condensed along its entire length, was regularly observed. In sharp contrast, no such condensed chromosome was seen in male cells. It was concluded that each X chromatin body represents a single X chromosome. This conclusion was confirmed in other mammals, and Taylor in 1960 [125] demonstrated by labeling of the late S phase with [³H]thymidine that the female heterochromatic X shows DNA replication only near the end of the S phase in somatic cells of the Chinese hamster. Taylor's finding was confirmed in many other mammalian cells. The heterochromatization of the X occurs in an early embryonic stage. Cleaving mammalian zygotes have no X chromatin. The time of its first appearance in various species ranges from the blastocyst to early primitive streak stages, with a cell number of about 50 in the pig to probably thousands in humans, and sometimes before and sometimes after implantation. In the human trophoblast X chromatin appears on the 12th day of development and in the embryo proper on the 16th day. It is formed rather suddenly in the entire embryo.

Evidence from aneuploid human individuals with more than two X chromosomes shows that only one X chromosome remains in the euchromatic stage, whereas all others are heterochromatic.

In monozygous female twins the process of X inactivation may be distinctly nonrandom [59a], leading to activity of one X chromosome in the first twin and of the other in the second twin. In one such pair one twin developed Duchenne muscular dystrophy, whereas the other one was normal [102].

X Inactivation as the Mechanism of Gene Dosage Compensation: Lyon's Hypothesis. In 1961 Lyon [72] (see also [75]) made the step from morphological evidence to function, concluding that the heteropyknotic X chromosome may be either paternal or maternal in origin and is functionally inactive. With this, she formulated one of the most fertile hypotheses in mammalian genetics:

The evidence had two parts. First, the normal phenotype of XO females in the mouse shows that only one active X chromosome is necessary for normal development, including sexual development. The second piece of evidence concerns the mosaic phenotype of female mice heterozygous for some sex-linked mutants. All sex-linked mutants so far known affecting coat color cause a "mottled" or "dappled" phenotype, with patches of normal and mutant color, in females heterozygous for them.

It is here suggested that this mosaic phenotype is due to the inactivation of one or the other X chromosome in embryonic development ... This hypothesis predicts that for all sex-localized gene action the heterozygote will have a mosaic appearance, and that there will be a similar effect when autosomal genes are translocated to the X chromosome. When the phenotype is not due to localized gene action, various types of result are possible. Unless the gene action is restricted to the descendants of a very small number of cells at the time of inactivation, these original cells will ... include both types. Therefore, the phenotype will be intermediate between the normal and hemizygous types, or the presence of any normal cell may be enough to ensure a normal phenotype, or the observed expression may vary as the proportion of normal and mutant cells varies, leading to incomplete penetrance in heterozygotes.

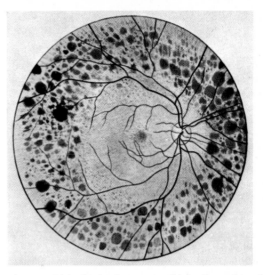

Fig. 2.56. Right fundus in a 6-year-old daughter of a male patient with X-linked ocular albinism. The distribution of pigment in this heterozygote is clearly patchy. (From Franceschetti and Klein 1964, [35])

In the same year, Lyon tentatively explained an observation on a human X-linked disease in the same way: In X-linked ocular albinism, male hemizygotes lack retinal epithelial pigment and have a pale eye fundus. Heterozygous females have irregular retinal pigmentation, with patches of pigment and patches lacking pigment, so that the fundus has a stippled appearance. Figure 2.56 shows this condition. Lyon also predicted that mosaicism should be demonstrable in other X-linked genes, among them the variants of the enzyme glucose-6-phosphate dehydrogenase (G6PD).

Fur color in mice as affected by X-linked mutations or a stippled appearance of the eye fundus as shown in X-linked ocular albinism in man are phenotypic characteristics separated from primary gene action by the process of differentiation. Hence, interpretation of the origin of such phenotypes can always be disputed. These findings served to suggest the hypothesis of X-inactivation but were not sufficient to prove it. A critical test of such a hypothesis should utilize simpler and less ambiguous situations. X-linked gene products whose presence can be detected at the protein level provided the experimental material. The first X-linked gene for which such an analysis became possible was the human G6PD locus. Indeed, Beutler [5], without knowledge of Lyon's hypothesis, had independently developed the concept of X-inactivation by observations on human G6PD variants. In spite of the fact that females have two and males only one copy of the G6PD gene, the average level of G6PD enzyme activity was found to be identical in both sexes as well as in in-

dividuals possessing additional numbers of X chromosomes (XXX, XXY). Hence a mechanism of dosage compensation must have been at work. If a female is heterozygous for an electrophoretic G6PD variant, the hypothesis of random inactivation predicts that in some cells the X chromosome with the normal allele, in others that with the variant allele will be active. Therefore a given single cell will be capable of determining only one of the two enzyme variants. Such mosaicism was, indeed, first characterized by Beutler et al. [6] in red cells by ingenious but indirect methods and later confirmed by a number of authors with various techniques [38, 74]. One approach utilized cloning of fibroblasts in tissue culture. In the Black population, the G6PD gene is polymorphic, two frequent alleles, Gd A and Gd B, being present. Cloned cells of fibroblast cultures from black women heterozygous for these alleles showed either the Gd A or the Gd B variant (Fig. 2.57) but not both, as found in their normal tissues. When women heterozygous for one of the G6PD deficiency variants were examined, the same phenomenon was observed: some cell clones had normal acitivity, and in others there was little activity.

Other evidence came from leiomyomata of the uterus in women heterozygous for the A and B G6PD variants [71a]. Tumor tissues invariably showed only one of the two mutant types while normal uterine tissues showed both types. This finding was possible only under three conditions:

1. Only one allele is active.
2. The whole tumor originated from a single cell, i.e., it represents a single cell clone.

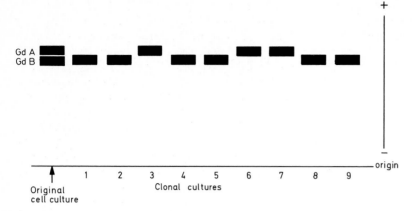

Fig. 2.57. Electrophoretic separation of G6PD components in ultrasound-treated cells from a tissue culture of a heterozygous woman with genotype Gd AB. The original cell culture shows the two G6PD components GdA and GdB. The clones derived from single cells show either GdA or GdB but not both. (From Harris 1980 [50])

3. The individual X chromosomes remain either active or inactive over the whole period of tumor growth.

Hence this experiment not only corroborated the Lyon hypothesis but also helped to establish the principle that tumors usually originate from a single cell.

Experiments with single cells were also carried out with another X-linked enzyme anomaly: the hypoxanthine-guanine phosphoribosyl transferase (HPRT) defect, and random inactivation was confirmed. This enzyme defect has been utilized for examining many problems of human gene action. Its discussion is therefore deferred to Sect. 7.2.2.6.

Other Examples of X Inactivation in Humans. By a variety of methods X inactivation has now been shown for a number of X-linked conditions in man. Especially interesting is the demonstration of retinal mosaicism in red-green color blindness [8]. By shining a very narrow beam of red or green light into the retinas of women heterozygous for color vision defects, patches of defective color perception were found, as would be expected if the retina were a mosaic consisting of normal and color-defective clones.

Anhidrotic ectodermal dysplasia is a rare X-linked condition. Affected males show absence of teeth, hypotrichosis, and absence of sweat glands. Patches with and without sweat glands can be recognized in many heterozygous females [93].

In chronic granulomatous disease with leukocyte malfunction (306 400), the bactericidal activity of granulocytes is very much reduced; they ingest staphylococci normally but are defective in their ability to digest them. Heterozygous females for the X-linked variety of this disease have two populations of leukocytes, the normal and the abnormal [100, 136]. In many other X-linked diseases observations have

been made that are compatible with the predictions of the Lyon hypothesis. In some diseases, such as X-linked agammaglobulinemia (Bruton), there is selection against cells in which the X carrying the mutation is active.

Cells in Which the Second X Is Not Inactivated [73, 74]. The X chromatin body becomes visible around the 16 th day, in the blastocyst stage. Functional inactivation is unlikely to occur much earlier. If one X were to be *always* inactive, the difference in phenotype between the normal male (XY) and Klinefelter syndrome (XXY) as well as between normal females (XX) and Turner syndrome patients (XO) would require explanations other than the possibility of full gene action of the X chromosomes prior to inactivation. There is good evidence that no inactivation occurs in oocytes as well as in male germ cells. In the mouse, the enzyme LDH is specified by an autosomal gene while G6PD, as in humans, by an X-linked gene. In fertilized XO oocytes G6PD was found to be half as active as in XX oocytes, whereas LDH activity was the same in both [28]. This is a gene dosage effect as predicted if both X chromosomes were active.

One human blood group system, the Xg system, has an X-linked mode of inheritance. The Lyon hypothesis would predict that heterozygous females have two distinct types of red blood cells, each carrying only the antigen determined by the X chromosome active in its precursor cell. However, it became clear early in the study of X inactivation that this prediction was wrong: two different erythrocyte populations could not be detected. One possibility was that the Xg antigen is not produced by the red cells but is taken up from their environment, for example the serum. This, however, was disproved by observation of a blood chimera (Sect. 6.3.3) – a woman who had received blood stem cells from her dizygotic twin during embryonic life in addition to her "own" red cells. Some of her red blood cells were O and Xg^a+, others were AB and Xg^a negative. Had the Xg come from the serum, all cells would have had the same Xg type – the genetically "own" type of the proband.

This riddle was solved when the Xg locus was shown to be located close to the tip of the short arm of the X chromosome and when at least one other locus that is closely linked with the Xg locus – that of the enzyme steroid sulfatase – was also shown to escape inactivation; the distal part of the short arm of the human X is *not* inactivated [104] since, unlike the rest of the inactivated X, this region replicates early.

In recent years a number of additional, noninactivated genes have been discovered which are located at Xp21.2–p22.1 and Xp11, respectively, but two such genes, RPS4X and XIST, have been localized on the long arm. The latter is active only in the inactivated X and could be involved in the inactivation process itself. Hence, inactivation does not depend on position of genes but on some property of individual genes [21]. Genes that are not inactivated are located all along the X chromosome. On the other hand, one gene, ADP/ATP translocase, is not inactivated, and is located in the pseudoautosomal region. An almost identical homologue on Xq13–q26, however, is inactivated [110]. Here, position appears to be important.

There may also be genetic factors influencing inactivation.

X Inactivation and Abnormal X Chromosomes [126].
When the first abnormal X chromosomes in humans – such as isochromosomes of the long arm, ring chromosomes, or deletions of parts of the long arm – were observed, the rules of inactivation appeared simple: the abnormal X was always inactivated, leaving the cell with one normal active X. Two hypotheses have been proposed to explain this specific inactivation pattern. According to the selection hypothesis, the normal and the abnormal X were originally inactivated at random in the same way as two normal X. Cells with inactivation of the abnormal X chromosome would be genetically grossly unbalanced and for this reason would have a lower division rate than the effectively normal cells with inactivation of the normal X chromosome. The second hypothesis regarded inactivation as an inherent property of the abnormal X [87].

In the meantime, a number of translocations with involvement of the X chromosome have been discovered which show that neither of these two hypotheses can be entirely correct. There are three such groups of translocations: those with 46 chromosomes and a balanced reciprocal translocation (practically all of them the X-autosome type), those with 46 chromosomes and an unbalanced X-autosomal or X/X translocation; and those with 45 chromosomes and an unbalanced X-autosomal translocation. Only the first group is considered here. Observations of the second and third basically confirm the conclusions reached from the first.

In most cases of such translocations the normal X chromosome is inactivated; the phenotype is variable, depending on gene disruptions. Some families have been observed where in one member, the normal, and in another member, the abnormal X chromosome was inactivated. For example, in one family the mother had a balanced X/21 translocation (Fig. 2.58). One translocation chromosome was formed from

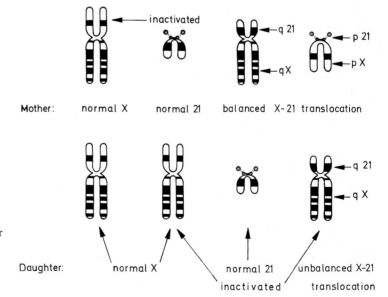

Fig. 2.58. X dosage compensation in a mother and a daughter with different sets of X chromosomes. The mother has two X/21 translocation chromosomes and one normal X which is inactivated; the daughter has two normal X chromosomes, one of which is inactivated, and one translocation chromosome, which is also inactivated. (From Summit et al. 1973 [124])

the long arms, the other from the short arms of one X and one no. 21, the break point being close to the centromere. In this woman the normal X chromosome was inactivated, as evidenced by its late replication. Her cells showed one X chromatin body. Her daughter had the large translocation chromosome, but not the smaller one, and two normal X chromosomes. One of the latter was inactivated but – unlike in the mother's case – the translocation chromosome was also inactivated. Hence, dosage compensation was achieved in mother and daughter alike. In the daughter, however, inactivation extended beyond the boundaries of the X into the translocated long arm of no. 21, creating additional clinical findings similar to those occasionally described in monosomy 21.

This case shows that the hypothesis according to which inactivation is determined by the structure of the abnormal X cannot generally be true. In this family, however, and in others in which only one inactivation pattern was found, the pattern was one that left the cell genetically relatively balanced.

The abnormal phenotypes of most of these balanced translocation carriers involving the X chromosome are remarkable, as in all balanced translocations between autosomes the carriers are normal. Two explanations are possible, between which the observations do not discriminate. Either the continuity of a certain region of the long arm is needed for full female differentiation – in which case the defective phenotype would be due to a position effect – or the inactivation of the same X in all cells is responsible for the abnormal phenotype, possibly due to functional hemizygosity of a recessive gene. Conceivably, one or the other of the mechanisms is at work in different cases.

Another observation is relevant for the mechanism of X inactivation. Whereas many cases of isochromosomes i(Xq) have been observed, only few cases of i(Xp) have become known despite the fact that such cases should be produced just as frequently by abnormal centromere division [120]. On the other hand, Xq deletions are known to occur. All these observations led to the hypothesis that an inactivation center exists on the proximal part of the long arm of the X chromosome. If this center is present in the abnormal chromosome, it can be inactivated. If two centers are present – as in some unbalanced X translocations – two X chromatin bodies may be formed. If no center is left – as in most i(Xp) chromosomes – inactivation cannot occur, and the zygote, being functionally trisomic for the short arm of the X chromosome, will as a rule be grossly unbalanced and incapable of normal development. In a study of three cases with a partial Xq deletion the inactivation center was localized near the border between the proximal Q-dark and Q-bright region (~q13) [128].

Some observations even seem to indicate that the inactivation impulse created by this center may extend beyond the limits of the X chromosome in the direction of the short arm but not the long arm. Figure 2.59 shows the various types of abnormal X chromosomes, their inactivation patterns, and phenotypes.

Much more work on X-autosome translocation has been reported in the mouse than in humans. Some data in this species also support the hypothesis that an inactivation center exists. It might even be possible to enhance or diminish the activity of this center, as measured by the degree of inactivation of an autosomal segment translocated to an X by selection experiments.

There are many hypotheses and some studies on the molecular mechanism of X inactivation. So far, however, no definite conclusions can be drawn.

2.2.4 Chromosome Aberrations and Spontaneous Miscarriage [66]

Incidence of Prenatal Zygote Loss in Humans. About 15% of all pregnancies in humans end by recognizable spontaneous abortion, defined as pregnancy termination before the 22nd week (body weight of the embryo: 500 g or less). However, there is good evidence in humans as well as in other mammals that many more zygotes are lost at an earlier stage of development; they are often severely malformed [66]. It appears that almost 50% of all conceptuses may be lost within the first 2 weeks of development, before the pregnancies are recognized [121]. In humans this early zygote loss usually goes unnoticed.

Incidence of Chromosome Aberrations. In 1961 two abortuses with triploidy were described [23, 95], and in 1963 the first two series of cytogenetic studies on abortions [16] showed a surprisingly high proportion of chromosomally abnormal abortuses. In the years thereafter a great number of series were published. One survey mentions 3714 specimens from fairly unbiased series, 1499 of which (40.4%) had chromosomal aberrations [66]. There was considerable variation among these series in the fraction of abnormal karyotypes, probably due to selective factors such as maternal age, failures of tissue culture, or gestational age which to be the most important parameter. In Fig. 2.60, the available data are broken down according to gestational age. The highest frequency of miscarriages was found between the 8th and the 15th weeks of gestation, while only about 5% miscarriages occur during the last weeks of pregnancy. The relatively low frequency in the early weeks is explained by longer retention of aberrant embryos in the uterus and by the fact that

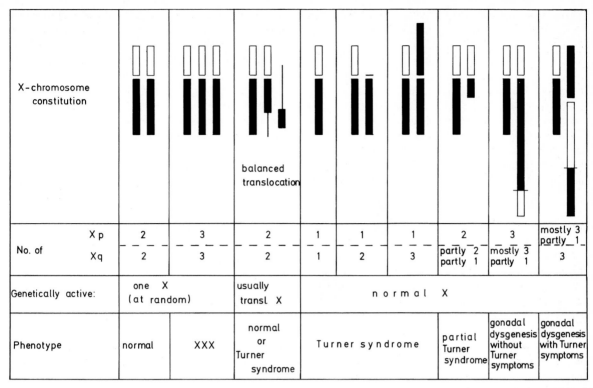

Fig. 2.59. Monosomy, disomy, and trisomy for different parts of the human X chromosome, their inactivation patterns, and phenotypic effects. *Black,* long arms; *white,* short arms. This figure is based on evaluation of numerous reported cases. For a more detailed report see [130]. (From Therman and Pätau 1974 [126])

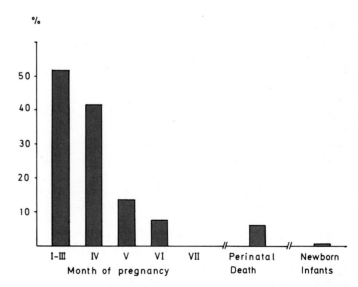

Fig. 2.60. Frequency of chromosome anomalies among 1641 spontaneous abortuses in relation to duration of pregnancy. The frequency among 675 infants dying during the perinatal period and among 59 749 newborns is also shown. (Data of various authors; from Lauritsen 1976 [72])

such early pregnancies are often unrecognized. Taking 15% as the incidence of perceptible spontaneous abortions among all recognized pregnancies, antenatal zygote loss due to chromosome aberrations can be estimated at about 5%–6%. This is about ten times more than the incidence of chromosome aberrations among living newborns (about 0.5%–0.6%; Sect. 9.1.2.1). Moreover, this figure does not include loss of zygotes before implantation in the uterus. There is now good evidence from various sources, including mutation experiments (Sect. 11.2), that preimplantation loss may be even higher. Obviously, spontaneous abortion is a powerful tool for early elimination of defective zygotes.

Types of Chromosome Aberrations in Aborted Fetuses.
With the beginning of abortion studies it became obvious that the distribution of types of chromosome anomalies observed in abortions differs from that in newborns. Some aberrations, such as the XO type, are present in newborns as well as in abortions; others, for example triploidies, lead almost always to miscarriage and are compatible with birth of a living child only in exceptional cases (Sect. 2.2.1); others such as trisomy 16 are observed exclusively in aborted fetuses. Creasy et al. [18] have published comprehensive data:

A total of 2607 presumptive spontaneous miscarriages were studied. Due to various technical and other problems, chromosomes could be studied in less than half of them. This might have introduced certain biases.

Of the 941 singleton abortuses 287 (30.5%) were chromosomally abnormal. Table 2.9 shows frequencies of the main groups of trisomies. One-half were primary autosomal trisomies, nearly one-quarter were X monosomies, and one-eighth were polyploid. The remainder were mostly monosomies or translocations. A total of 183 autosomal trisomies were discovered (Table 2.9). Additional chromosomes were observed most frequently for chromosome 16, and for chromosomes 21, 15, and 22. No examples of an extra chromosome 1, 5, 6, 7, 11, 12, 17, or 19 were detected. Among the 36 twin pregnancies, karyotyping was possible in 26 cases in at least one twin. No chromosome anomalies were found.

Table 2.9. Frequency of different autosomal trisomies in aborted fetuses (percentage in 183 cases; from Creasy et al. 1976 [18])

1	–
2	4.48
3	1.12
4	1.90
5	–
6	0.53
7	1.60
8	3.72
9	3.72
10	2.13
11	–
12	–
13	2.36
14	6.50
15	10.04
16	32.11
17	–
18	5.58
19	–
20	1.90
21	12.54
22	9.76
Total	99.99

Phenotypes of Abortuses. There were significant phenotypical differences between conceptuses with the various chromosome complements. The presence of an extra chromosome 2 or 3, for example, may be incompatible with the formation of an embryo and leads to the production of an empty sac.

Trisomy 9 seems to result in limited and abnormal embryonic development; this is in accordance with the occasional observation of living but severely malformed newborns (Sect. 2.2.1). A fair amount of embryonic development, albeit with malformations, seems to be compatible with all types of trisomy D. Trisomy 16, on the other hand, leads to severe and early developmental disturbance; empty sacs and severely disorganized embryos are observed in most cases. In contrast, trisomy 18 causes much less disturbance, again compatible with the relatively more frequent survival into postnatal age. Of the two types of trisomy G, trisomy 21 is compatible with better development than trisomy 22. However, the authors estimated from their own as well as from literature data that more than 60% of all zygotes with trisomy 21 are aborted!

The widest variability in phenotypic manifestation has been encountered in this study – as well as in many others – among XO zygotes, which constitute the most frequent single karyotype among all observed abortions. From apparently normal embryos to empty sacs, a wide range of phenotypes was observed. Characteristic are hygromata, i.e., edematous thickening of tissue, also observed in living XO newborns (Sect. 2.2.3).

The 12 triploids were mostly embryos and fetuses with various malformations (Sect. 2.2.1). In contrast to this, tetraploids were nearly all intact empty sacs; two of them with abnormal amniotic cavities.

In a collection of 3714 abortions from recent studies [121], somewhat more than half of abnormal karyotypes were trisomies, ≈ 20% were monosomies, ≈ 18% polyploidies, ≈ 3% structural anomalies, and the rest others. All types of trisomies, excepting trisomy 1, were observed, albeit with greatly varying frequencies. These frequencies led to a number of calculations [121] on the overall frequency of numerical aberrations (trisomies and monosomies together). If monosomies and trisomies of all autosomes are assumed to occur in identical frequencies, and if early elimination accounts for unequal abortions frequencies, about 10%–30% of all human zygotes would be abnormal chromosomally at conception. To some extent such speculations are corroborated by results from chromosome studies in human sperm [76]. Among 6821 sperm chromosome complements, an overall frequency of 3.9% for aneuploidies was found –3.3% hypohaploid and 0.7% hyperhaploid. The hypohaploids may be explained in part by artifacts; however, this explanation can hardly apply to the hyperhaploids. For hyperhaploids the relative frequencies of involvement of single autosomes did not deviate significantly from expectations, assuming

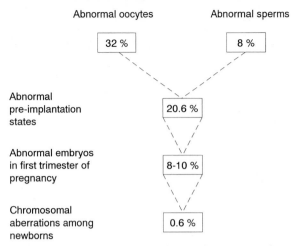

Fig. 2.61. Approximate frequencies of germ cells and embryos with chromosomal aberrations in various states of development up to birth

the same probability of nondisjunction for all autosomes. X and Y hyerploids were much more common than expected [76]. For aneuploids produced during oogenesis, no such data are available. It is well known, however, that nondisjunction during oogenesis is much more common (or leads to a fertilized zygote much more often) than nondisjunction during spermatogenesis (Sect. 9.2.1). On the other hand, it is doubtful whether observations on human oocytes artificially fertilized in vitro (two-thirds with chromosomal aberrations [2]) can be regarded as representative of the normal situation.

Some Conclusions. A number of conclusions can be derived from studies on chromosomes in abortions. Various chromosomes participate in very unequal proportions in the recognizable overall zygote loss. This nonrandom pattern becomes especially obvious when the numbers and percentages of autosomal trisomies are compared. These findings do not indicate differences in nondisjunction rates during meiosis or during early cleavage divisions. A higher nondisjunction risk, however, is likely for the five acrocentric pairs of the D and G groups as well as for the X and Y chromosomes. The apparent differences in frequency of trisomies for the remaining autosomes could easily be caused by the different times of zygote death. For example, if trisomy 1 leads to zygote death before or during morula formation, all no. 1 trisomies will go unnoticed. Phenotypic variability may be wide even among groups of cytogenetically uniform zygotes. It is especially striking between zygotes with different karyotypes. Some – such as trisomy 21 – are "near misses." Others, for example, trisomy 16, are not compatible even with early stages of embryonic development and are therefore invariably lethal. Figure 2.61 presents a rough estimate of the proportion of zygotes eliminated up to birth by chromoso-

mal aberrations. Comparison of aneuploid abortuses – together with tissue cultures of surviving carriers of aneuploidies using biochemical and morphological analysis – may therefore become an important tool for elucidation of genetic regulation during embryonic development. This topic is taken up in detail in Sect. 8.4.3.

Conclusions

The normal diploid human chromosome complement has 23 chromosome pairs, 22 pairs of autosomes, and one pair of sex chromosomes (XX in females, XY in males). Each chromosome can be identified individually with appropriate methods. An increasing number of developmental anomalies in humans have been explained by deviations in the number or structure of chromosomes. The great majority of chromosomal aberrations of the autosomes leads to spontaneous miscarriage. Trisomy 21 (Down syndrome) is the only autosomal aberration with survival into adulthood. Its characteristic phenotype includes severe mental retardation. Most sex chromosome abnormalities have mild phenotypic features and are relatively common in the human population.

References

1. Aller V, Albisqueta JA, Perez A, Martin MA, Goday C, Del Mazo J (1975) A case of trisomy 8 mossaicism 47, XY, + 8/46, XX. Clin Genet 7 : 232–237
2. Angell RR, Hitken RJ, van Look PFA, Lumsden MA, Templeton AA (1983) Chromosome abnormalities in human embryos after in vitro fertilization. Nature 303 : 336–338
3. Arnold J (1879) Beobachtungen über Kernteilungen in den Zellen der Geschwülste. Virchows Arch Pathol Anat 78 : 279
4. Barr ML, Bertram LF (1949) A morphological distinction between neurones of the male and the female and the behavior of the nucleolar satellite during accelerated nucleoprotein synthesis. Nature 163 : 676–677
5. Beutler E (1963) Autosomal inactivation. Lancet 1 : 1242
6. Beutler E, Yeh M, Fairbanks VF (1962) The normal human female as a mosaic of X chromosome activity: studies using the gene for G-6-PD as a marker. Proc Natl Acad Sci USA 48 : 9
7. Böök JA, Santesson B (1960) Malformation syndrome in man associated with triploidy (69 chromosomes). Lancet 1 : 858–859
8. Born G, Grützner P, Hemminger HJ (1976) Evidenz für eine Mosaikstruktur der Netzhaut bei Konduktorinnen für Dichromasie. Hum Genet 32 : 189–196
9. Boué A, Gallano P (1984) A collaborative study of the segregation of inherited chromosome structural rear-

rangements in 1356 prenatal diagnoses. Prenat Diagn 4 : 45–67

10. Bridges CB (1916) Non-disjunction as proof on the chromosome theory of heredity. Genetics 1 : 1–52, 107–163

11. Bross K, Dittes H, Krone W, Schmid M, Vogel W (1973) Biochemical and cytogenetic studies on the nucleolus organizing regions (NOR) of man. I. Comparison of trisomy 21 with balanced translocations t(DqGq). Hum Genet 20 : 223–229

12. Burgoyne PS (1986) Mammalian X and Y crossover. Nature 319 : 258–259

13. Caspersson T, de la Chapelle S, Foley GE, Kudynowski J, Modest EJ, Simonsson E, Wagh V, Zech L (1968) Chemical differentiation along metaphase chromosomes. Exp Cell Res 49 : 219

14. Chandley AC (1988) Meiosis in man. Trends Genet 4 : 79–83

15. Chang EY, Gartler SM (1994) A fluorescent in situ hybridization analysis of X chromosome pairing in early human female meiosis. Hum Genet 94 : 389–394

16. Clendenin TM, Benirschke K (1963) Chromosome studies on spontaneous abortions. Lab Invest 12 : 1281–1292

17. Collman RD, Stoller A (1963) A life table for mongols in Victoria, Australia. J Ment Defic Res 7 : 53

18. Creasy MR, Crolla JA, Alberman ED (1976) A cytogenetic study of human spontaneous abortions using banding techniques. Hum Genet 31 : 177–196

19. Davidson WM, Smith DR (1954) The nuclear sex of leucocytes. In: Overzier (ed) Intersexuality. Academic, New York, pp 72–85

20. Davies KE (1981) The applications of DNA recombinant technology to the analysis of the human genome and genetic disease. Hum Genet 58 : 351–357

21. Davies KE (1991) The essence of inactivity. Nature 349 : 15–16

22. De Wolff E, Schärer K, Lejeune J (1962) Contribution à l'étude des jumeaux mongoliens. Un cas de monozygotisme hétérocaryote. Helv Paediatr Acta 17 : 301–328

23. Delhanty JA, Ellis JR, Rowley PT (1961) Triploid cells in a human embryo. Lancet 1 : 1286

24. Dittes H, Krone W, Bross K, Schmid M, Vogel W (1975) Biochemical and cytogenetic studies on the nucleolus organizing regions (NOR) of man. II. A family with the 15/21 translocation. Hum Genet 26 : 47–59

25. Dutrillaux B, Lejeune J (1975) New techniques in the study of human chromosomes: methods and applications. Adv Hum Genet 5 : 119–156

26. Dutrillaux B, Laurent C, Robert JM, Lejeune J (1973) Inversion péricentrique, inv (10), chez la mère et aneusomie de recombinaison, inv (10), rec (10), chez son fils. Cytogenet Cell Genet 12 : 245–253

27. Edwards JH, Harnden DG, Cameron AH, Crosse VM, Wolff OH (1960) A new trisomic syndrome. Lancet 1 : 787

28. Epstein CJ (1969) Mammalian oocytes: X chromosome activity. Science 163 : 1078

29. Epstein CJ (1986) The consequences of chromosome imbalance: principles, mechanisms and models. Cambridge University Press, New York

29 a. Ferguson-Smith MA, Goodfellow PN (1995) SRY and primary sex-reversal syndromes. In: Scriver CR, Beaudet AL, Sly WS, Valle D (eds) The metabolic and molecular bases of inherited disease. McGraw-Hill, New York, pp 739–748

30. Flemming W (1882) Beiträge zur Kenntnis der Zelle und ihrer Lebenserscheinungen. III. Arch Mikrosk Anat 20 : 1

31. Ford CE, Hamerton JL (1956) The chromosomes of man. Nature 178 : 1020–1023

32. Ford CE, Miller OJ, Polani PE, de Almeida JC, Briggs JH (1959) A sex-chromosome anomaly in a case of gonadal dysgenesis (Turner's syndrome). Lancet 1 : 711–713

33. Fraccaro M, Kaijser K, Lindsten J (1960) Chromosomal abnormalities in father and mongoloid child. Lancet 1 : 724–727

34. Fraccaro M, Lindsten J, Ford CE, Iselius L et al (1980) The 11q; 22q translocation: a European collaborative analysis of 43 classes. Hum Genet 56 : 21–51

35. Franceschetti, Klein (1964) In: Becker PE (ed) Humangentik. Thieme, Stuttgart

36. Francke U (1994) Digitized and differentially shaded human chromosome ideograms for genomic applications. Cytogenet Cell Genet 65 : 206–219

37. Francke U, Ochs HD, de Martinville B et al (1985) Minor Xp21 chromosome deletion in a male associated with expression of Duchenne muscular dystrophy, chronic granulomatous disease, retinitis pigmentosa and McLeod syndrome. Am J Hum Genet 37 : 250–267

38. Gartler SM, Sparkes RS (1963) The Lyon-Beutler hypothesis and isochromosome X patients with the Turner syndrome. Lancet 2 : 411

39. Gebhard (1970) In: Vogel F, Röhrborn G (eds) Chemical mutagenesis in mammals and man. Springer, Berlin Heidelberg New York

40. German J (1979) Roberts' syndrome. I. Cytological evidence for a disturbance in chromatid pairing. Clin Genet 16 : 441–447

41. German J, Archibald R, Bloom D (1965) Chromosomal breakage in a rare and probably genetically determined syndrome of man. Science 148 : 506

42. Goodpasture C, Bloom SE (1975) Visualization of nucleolar organizer regions in mammalian chromosomes using silver staining. Chromosoma 53 : 37–50

43. Greenstein RM, Reardon MP, Chan TS (1977) An X-autosome translocation in a girl with Duchenne muscular dystrophy (DMD): evidence for DMD gene location. Pediatr Res 11 : 457

44. Habedank M, Rodewald A (1982) Moderate Down's syndrome in three siblings having partial trisomy 21q22.2 and therefore no SOD-1 excess. Hum Genet 60 : 74–77

45. Hagemeijer A, Smit EME (1977) Partial trisomy 21. Further evidence that trisomy of band 21q22 is essential for Down's phenotype. Hum Genet 38 : 15–23

46. Haldane JBS (1936) A search for incomplete sex linkage in man. Ann Eugen 7 : 28–57

47. Hamerton JL (1968) Robertsonian translocation in man. Evidence for prezygotic selection. Cytogenetics 7 : 260–276

48. Hamerton JL (1971) Human cytogenetics, vol 1/2. Academic, New York

49. Harnden DG, Lindsten JE, Buckton K, Klinger HP (1981) An international system for human cytogenetic nomenclature. High resolution banding. Birth Defects 17(5)

50. Harris H (1980) The principles of human biochemical genetics, 4th edn. North-Holland, Amsterdam

51. Heberer G (1940) Die Chromosomenverhältnisse des Menschen. In: Just G (ed) Handbuch der Erbbiologie des Menschen, vol 1. Springer, Berlin, pp 2–30

52. Hsu TC (1952) Mammalian chromosomes in vitro. I. The karyotype of man. J Hered 43 : 167

53. Hsu TC, Pomerat CM (1953) Mammalian chromosomes in vitro. II. A method for spreading the chromosomes of cells in tissue culture. J Hered 44 : 23–29

54. Jacobs PA (1977) Human chromosome heteromorphisms (variants). Prog Med Genet [New Ser] 2 : 251–274

55. Jacobs PA, Morton NE (1977) Origin of human trisomy: a study of heteromorphism and satellite asssociations. Ann Hum Genet 45 : 357–365

56. Jacobs PA, Strong JA (1959) A case of human intersexuality having a possible XXY sex-determining mechanism. Nature 183 : 302–303

57. Jacobs PA, Baikie AG, MacGregor TN, Harnden DG (1959) Evidence for the existence of the human "superfemale." Lancet 2 : 423–425

58. Jacobsen P, Mikkelsen M, Rosleff F (1974) The trisomy 8 syndrome: report of two further cases. Ann Genet (Paris) 17 : 87–94

59. Johannisson R, Gropp A, Winking H, Coerdt W, Rehder H, Schwinger E (1983) Down's syndrome in the male. Reproductive pathology and meiotic studies. Hum Genet 63 : 132–138

59a. Jorgensen AL, Philip J, Raskind WH, Matsushita M, Christensen B, Dreyer V, Motulsky AG (1992) Different patterns of x-inactivation in monozygotic twins discordant for red-green color vision deficiency. Am J Hum Genet 51 : 291–298

60. Kaiser P (1988) Pericentric inversions: their problems and clinical significance. In: Daniel (ed) The cytogenetics of mammalian autosomal rearrangements. Wiley, New York, pp 163–247

61. Kakati S, Nihill M, Sinah A (1973) An attempt to establish trisomy 8 syndrome. Hum Genet 19 : 293–300

62. Koske-Westphal T, Passarge E (1974) Die Chromosomen des Menschen und ihre Untersuchung in somatischen Zellen. In. Vogel F (ed) Erbgefüge. Springer, Berlin Heidelberg New York, pp 261–323 (Handbuch der allgemeinen Pathologie, vol 9)

63. Kuhn TS (1962) The structure of scientific revolutions. University of Chicago Press, Chicago

64. Kurilo LF (1981) Oogenesis in antenatal development in man. Hum Genet 57 : 86–92

65. Latt SA (1973) Microfluorometric detection of deoxyribonucleic acid replication in human metaphase chromosomes. Proc Natl Acad Sci USA 70 : 3395–3399

66. Lauritsen JG (1977) Genetic aspects of spontaneous abortion. Laegeforeninges, University of Aarhus

67. Lauritsen JG, Bolund L, Friedrich U, Therkelsen AL (1979) Origin of triploidy in spontaneous abortuses. Ann Hum Genet 43 : 1–6

68. Lejeune J, Berger R (1965) Sur deux observations familiales de translocations complexes. Ann Genet (Paris) 8 : 21–30

69. Lejeune J, Gautier M, Turpin MR (1959) Étude des chromosomes somatiques de neuf enfants mongoliens. C R Acad Sci (Paris) 248 : 1721–1722

70. Lejeune J, Lafourcade J, Berger R, Vialatte J, Roeswillwald M, Seringe P, Turpin R (1963) Trois cas de délétion partielle du bras court d'un chromosome 5. C R Acad Sci (Paris) 257 : 3098–3102

71. Lifschytz E, Lindsley DL (1972) The role of X-chromosome inactivation during spermatogenesis. Proc Natl Acad Sci USA 69 : 182–186

71a. Linder A, Gartler SM (1965) Glucose-6-phosphate dehydrogenase mosaicism: utilization as a cell marker in the study of leiomas. Science 150 : 67–69

72. Lyon MF (1961) Gene action in the X-chromosome of the mouse. Nature 190 : 372–373

73. Lyon MF (1961) Gene action in the X-chromosome of mammals including man. In Gredda CL (ed) Proceedings of the 2nd International Conference of Human Genetics. Rome, August 1961, pp 1228–1229

74. Lyon MF (1968) Chromosomal and subchromosomal inactivation. Annu Rev Genet 2 : 31–52

75. Lyon MF (1988) X-chromosome inactivation and the location and expression of X-linked genes. Am J Hum Genet 42 : 8–16

76. Martin RH, Rademaker A (1990) The frequency of aneuploidy among individual chromosomes in 6821 human sperm chromosome complements. Cytogenet Cell Genet 53 : 103–107

77. McFadden DE, Kalousek DK (1991) Two different phenotypes of fetuses with chromosomal triploidy: correlation with paternal origin of the extra haploid set. Am J Med Genet 38 : 535–538

78. McKusick VA (1995) Mendelian inheritance in man, 11th edn. Johns Hopkins University Press, Baltimore

79. Mikelsaar A-V, Schmid M, Krone W, Schwarzacher HG, Schnedl W (1977) Frequency of agstained nucleolus organizer regions in the acrocentric chromosomes of man. Hum Genet 31 : 13–17

79a. Mitelman F (ed) (1995) An international system for human cytogenetic nomenclature. Recommendations of the International Standing Committee on Human Cytogenetic Nomenclature. Karger, Basel, p 114

80. Moorhead PS, Nowell PC, Mellman WJ, Battips DM, Hungerford DA (1960) Chromosome preparations of leukocytes cultured from human peripheral blood. Exp Cell Res 20 : 613–616

81. Morgan TH (1910) Sex-limited inheritance in drosophila. Science 32 : 120–122

82. Morton NE, Lindsten J, Iselius L, Yee S (1982) Data and theory for a revised chiasma map of man. Hum Genet 62 : 266–270

83. Nagakome Y, Limura K, Tangiuchi K (1973) Points of exchange in a human no. 5 ring chromosome. Cytogenet Cell Genet 12 : 35–39

84. Niebuhr E (1974) Triploidy in man. Hum Genet 21 : 103–125

85. Nowakowski H, Lenz W, Parada J (1959) Diskrepanz zwischen Chromatinbefund und genetischem Geschlecht beim Klinefelter-Syndrom. Acta Endocrinol (Copenh) 30 : 296–320

86. Nowell PC, Hungerford DA (1960) A minute chromosome in human chronic granulocytic leukemia. Science 132 : 1497

87. Ohno S (1967) Sex chromosomes and sex-linked genes. Springer, Berlin Heidelberg New York

88. Ohno S, Makino S (1961) The single-X nature of sex chronatin in man. Lancet 1 : 78–79

89. Ohno S, Kaplan WD, Kinosita R (1959) Formation of the sex chromatin by a single X chromosome in liver cells of Rattus norvegicus. Exp Cell Res 18 : 415–418

90. Pachmann U, Rigler R (1972) Quantum yield of acridines interacting with DNA of defined base sequence. Exp Cell Res 72 : 602

91. Painter TS (1923) Studies in mammalian spermatogenesis. II. The spermatogenesis of man. J Exp Zool 37 : 291–321

92. Paris Conference (1972) Standardization in human cytogenetics (1971). Cytogenetics 11 : 313–362

93. Passarge E, Fries E (1973) X-chromosome inactivation in X-linked hypohidrotic ectodermal dysplasia. Nature [New Biol] 245 : 58–59

94. Patau K, Smith DW, Therman E, Inhorn SL, Wagner HP (1960) Multiple congenital anomaly caused by an extra chromosome. Lancet 1 : 790–793

95. Penrose LS, Delhanty JDA (1961) Triploid cell cultures from a macerated foetus. Lancet 1 : 1261

96. Penrose LS, Smith GF (1966) Down's anomaly. Churchill, London

97. Polani PE (1962) Sex chromosome abnomalies in man. In: Hamerton JL (ed) Chromosomes in medicine. Heinemann, London, pp 73–139

98. Polani PE, Bishop PMF, Lennox B, Ferguson-Smith MA, Stewart JSS, Prader A (1958) Color vision studies in the X-chromosome constitution of patients with Klinefelter's syndrome. Nature 182 : 1092–1093

99. Polani BE, Briggs JH, Ford CE, Clarke CM, Berg JM (1960) A mongol child with 46 chromosomes. Lancet 1 : 721–724

100. Quie PG, White JG, Holmes B, Good RA (1967) In vivo bactericidal capacity of human polymorphonuclear leucocytes: diminished activity in chronic granulomatous disease of childhood. J Clin Invest 46 : 668–679

101. Rappold GA (1993) The pseudoautosomal regions of the human sex chromosomes. Hum Genet 92 : 315–324

102. Richards CS, Watkins SC, Hoffman EP et al (1990) Skewed X in activation in a female MZ twin results in Duchenne muscular dystrophy. Am J Hum Genet 46 : 672–681

103. Richards BW (1969) Mosaic mongolism. J Ment Defic Res 13 : 66–83

104. Ropers HH, Migl B, Zimmer J, Fraccaro M, Maraschio PP, Westerveld A (1981) Activity of steroid sulfatase in fibroblasts with numerical and structural X chromosome aberrations. Hum Genet 57 : 345–356

105. Rouyer F, Simmler MC, Johnsson C, Vergnaud G, Cooke HJ, Weissenbach J (1986) A gradient of sex linkage in the pseudoautosomal region of the human sex chromosomes. Nature 319 : 291–295

106. Rudok E, Jacobs PA, Yanaginachi R (1978) Direct analysis of the chromosome constitution of human spermatozoa. Nature 274 : 911–913

107. Ruzicka F (1973) Über die Ultrastruktur menschlicher Metaphase-Chromosomen. Hum Genet 17 : 137–144

108. Saitoh Y, Laemmli UK (1994) Bands arise from a differential folding path of the highly AT-rich scaffold. Cell 76 : 609–622

109. Schaefer MSD (1983) Segregation and Pathologie autosomaler familiärer Translokationen beim Menschen. Medical dissertation, University of Kaiserslautern

110. Schiebel K, Weiss B, Woehrle D, Rappold G (1993) A human pseudoautosomal gene, ADP/ATP translocase, escapes X-inactivation whereas a homologue on Xq is subject to X-inactivation. Nat Genet 3 : 82–87

111. Schinzel A (1984) Catalogue of unbalanced chromosome aberrations in man. De Gruyter, Berlin

112. Schleiermacher E, Schliebitz U, Steffens C (1974) Brother and sister with trisomy 10p: a new syndrome. Hum Genet 23 : 163–172

113. Schmickel RD (1986) Contiguous gene syndromes: a component of recognizable syndromes. J Pediatr 109 : 231–241

114. Schnedl W (1978) Structure and variability of human chromosomes analyzed by recent techniques. Hum Genet 41 : 1–10

115. Schroeder TM (1970) In: Vogel F, Röhrborn G (eds) Chemical mutagenesis in mammals and man. Springer, Berlin Heidelberg New York

116. Schroeder TM, Anschütz F, Knopp A (1964) Spontane Chromosomenaberrationen bei familiärer Panmyelopathie. Hum Genet 1 : 194–196

117. Schwarzacher HG (1970) Die Ergebnisse elektronenmikroskopischer Untersuchungen an somatischen Chromosomen des Menschen. Hum Genet 10 : 195–208

118. Schwarzacher HG, Wachtler F (1983) Nucleolus organizer regions and nucleoli. Hum Genet 63 : 89–99

119. Schwarzacher HG, Wolf U (1974) Methods in human cytogenetics. Springer, Berlin Heidelberg New York

120. Simpson JL (1976) Disorders of sexual differentiation. Etiology and clinical delineation. Academic, New York

121. Sperling K (1984) Frequency and origin of chromosome abnormalities in man. In: Obe B (ed) Mutation in man. Springer, Berlin Heidelberg New York, pp 128–146

122. Sperling K, Rao PN (1974) The phenomenon of premature chromosome condensation: its relevance to basic and applied research. Humangenetik 23 : 235–258

123. Stern C (1959) The chromosomes of man. J Med Educ 34 : 301–314

124. Summitt RL, Martens PR, Wilroy RS (1973) X-autosome translocation in normal mother and effectively 21-monosomic daughter. J Pediatr 84 : 539–546

125. Taylor JH (1960) Asynchronous duplication of chromosomes in cultured cells of Chinese hamster. J Biophys Biochem Cytol 7 : 455–464

126. Therman E, Patau K (1974) Abnormal X chromosomes in man. Origin, behavior and effects. Hum Genet 25 : 1–16

127. Therman E, Susman M (1993) Human chromosomes. Structure, behavior, effects, 3rd edn. Springer Verlag New York Heidelberg Berlin

128. Therman E, Sarto GE, Palmer CG, Kallio H, Denniston C (1979) Position of the human X inactivation center on Xp. Hum Genet 50 : 59–64

129. Therman E, Susman B, Denniston C (1989) The nonrandom participation of human acrocentric chromosomes in Robertsonian translocations. Ann Hum Genet 53 : 49–65

130. Therman E, Laxova R, Susman B (1990) The critical region of the human Xq. Hum Genet 85 : 455–461

131. Tjio HJ, Levan A (1956) The chromosome numbers of man. Hereditas 42 : 1–6

132. Tommerup N (1993) Mendelian cytogenetics. Chromosome rearrangements associated with mendelian disorders. J Med Genet 30 : 713–727

133. Van den Berg DJ, Francke U (1993) Roberts syndrome: a review of 100 cases and a new rating system for severity. Am J Med Genet 47 : 1104–1123

134. Von Winiwarter H (1912) Études sur la spermatogenèse humaine. I. Cellule de Sertoli. II. Hétérochromosome et mitoses de l'épithelium seminal. Arch Biol (Liege) 27 : 91–189

135. Waardenburg PJ (1932) Das menschliche Auge und seine Erbanlagen. Bibliogr Genet 7

136. Windhorst DB, Holmes B, Good RA (1967) A newly defined X-linked trait in man with demonstraion of the Lyon effect in carrier females. Lancet 1 : 737–739

137. Yunis JJ (1981) Mid-prophase human chromosomes. The attainment of 2000 bands. Hum Genet 56 : 293–298

138. Zankl H, Zang KD (1974) Quantitative studies on the arrangement of human metaphase chromosomes. II. The association frequency of human acrocentric marker chromosomes. Hum Genet 23 : 259–265

3 The Human Genome: Genes and DNA

In nature's infinite book of secrecy,
a little I can read.

W. Shakespeare, Anthony and Cleopatra

3.1 Organization of Genetic Material in Human Chromosomes

In the first two decades of modern research on human chromosomes many aspects of the organization of genetic material in chromosomes were analyzed; however, there was little concrete information on how this knowledge could be integrated with information from molecular biology into a molecular chromosome model. More recently, however, especially since the "new genetics" began to develop in the 1970s, new information has accumulated rapidly. At present, answers are emerging to many questions thought to be unanswerable only a few years ago. In this section we present an overview. The new genetics has had an impact on many aspects of human genetics; we return to these results in many of the later sections.

3.1.1 Chromatin Structure

3.1.1.1 Single-Copy and Repetitive DNA

Too Much DNA in a Human Genome? Shortly after the genetic code was deciphered in the early 1960s, scientists were impressed by the abundance of DNA in eukaryotic cells. According to various studies, the DNA content of a diploid human cell is of the order of 7.3×10^{-12} g (range 6.6–8.0). On the basis of molecular weights it can be calculated that a nucleotide pair of adenine and thymine (A = T) has a weight of 1.025×10^{-21} g, whereas a nucleotide pair with guanine and cytosine (G = C) weighs 1.027×10^{-21} g. Hence, the total diploid set has about 7.1×10^9 nucleotide pairs:

$$\frac{7.3 \times 10^{-12}}{1.026 \times 10^{-21}} = 7.1 \times 10^9$$

If this DNA consisted of structural genes coding for proteins, and if the average protein – as the hemoglobin genes – were comprised of about 150 amino acids, 6–7 million genes could be accommodated in the human genome [113, 114]. This figure is known to be too high by about two orders of magnitude; "informative" DNA sequences are intercalated with stretches that are *not* translated into an amino acid sequence (see below). Some have a specific function; for others no such a function has been detected so far, and it remains a matter of speculation.

Repetitive DNA [89, 90]. One major development was the discovery that the DNA of higher organisms contains a large proportion of repetitive DNA sequences. When DNA is isolated and cut into fragments of about equal length, the double-stranded structure can be separated into single strands by heating in the presence of salt solutions. They move freely and meet one another at random. When the temperature is lowered, single strands meeting a complementary partner connect to form DNA double strands. This procedure offers a simple method of identifying complementary DNA strands.

When bacterial DNA is heat-denatured in this way, and the fraction of newly reannealed double-strand DNA is registered in relation to concentration of molecules (C_o) and reaction time (t) the relationship is linear; logarithmic plotting gives a sigmoid curve, the $C_o t$ curve (Fig. 3.1). A similar experiment with human DNA fragments (approximately 600 bases in length) leads to an entirely different curve. Immediately after the experiment has started, a small percentage of the DNA is double-stranded. The steep slope of the curve immediately afterwards shows that a further DNA fraction reanneals about 50 000 times faster than bacterial DNA; still another fraction reanneals 10–1000 times as fast as bacterial DNA. The remaining more than 50 % of DNA shows kinetics similar to that found in bacteria. These data can be explained as follows: A small percentage of human DNA consists of regions whose complementary sequences are located on the same strand but in reverse (= palindromic) order. This DNA can reanneal very rapidly by simply folding together. Another fraction contains repetitive sequences that reanneal to form DNA double strands; here, the speed of reannealing depends on the number of identical (or near-identical) repeats. Finally, there is also single-copy DNA with reaction kinetics similar to that found in bacteria (Fig. 3.1).

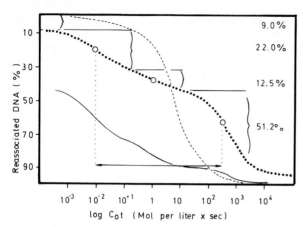

Fig. 3.1. Reannealing kinetics of bacterial DNA and human DNA of various fragment lengths. Percentage of double-stranded, reannealed DNA is plotted against the product of DNA concentration (C_0) and time *(t)*. Dotted, *sigmoid curve* (---) corresponds to bacterial DNA and is characteristic for single-copy DNA. The *curve in the middle* (....) shows the reassociation profile of human DNA fragments 600 bases in length. It can be subdivided into four classes: 9% have an un-measurably fast $C_0t_{1/2}$ value; 22% a value of 10^{-2}; 12.5% a value of 1.0 and 51.2% a value of 495. $C_0t_{1/2} = 10^{-2}$ means that reannealing is about 50 000 times faster than $C_0t_{1/2} = 495$. *Lower curve* (—), reaction kinetics with a fragment length of 1.3 kb. These reanneal much faster. This means that most segments contain repetitive sequences. Only about 10% of the DNA behave as single-copy DNA. (Data from Schmid and Deininger 1975 [89]; figure from Sperling 1984 [96])

How Are Single-Copy and Repetitive DNA Located Relative to Each Other? Various studies have shown that somewhat more than 50% of human DNA consists of single-copy stretches about 2 kb long. These are interspersed with – mainly intermediary – repetitive sequences that can be subdivided roughly into long interspersed repeat sequences (LINE) and short interspersed sequences (SINE). The basic type of LINE has approximately 6000 nucleotides, and the basic type of SINE has some 500 nucleotides [51]. Alu sequences (Sect. 3.2) are one example; shorter sequences are discussed below. As mentioned above (Sect. 3.1.1.1), LINE sequences are found mainly in the G dark chromosome bands and SINE sequences in the G light bands. In addition, highly repetitive DNA sequences formed by a millionfold repetition of oligonucleotides have been found in specific areas, such as the centromeric regions (see Sect. 2.1.2.2) or the long arm of the Y chromosome. They often show conspicuous individual quantitative and qualitative differences with no apparent effects on the phenotype. The single-copy DNA comprises the structural genes, but these genes occupy only a minor portion of this DNA. The sequence pattern described is very widespread and occurs in species as widely separated

from each other as mammals, amphibia, gastropods, and even flagellates (for details see [91]). This widespread occurrence of a relatively stable pattern suggests an important function, which, however, has not been identified so far (but see Sect. 3.1.1.5; 3.3). In some species, such as *Drosophila melanogaster* and *Chironomus tentans*, no such interspersion with short-period DNA sequence is found.

This information, together with knowledge on the structure of human genes (see below), allows a rough estimate of the number of genes in the human genome [115]. Let us assume that half the haploid genome, i.e., 1.75×10^9 nucleotide pairs, are single-copy sequences containing genes. If the number of genes is taken to be approximately 100 000, this leads to 1.75×10^4 or slightly fewer than 20 000 base pairs per gene. If we assume some 50 000 genes, this figure is doubled. Let us compare these figures with the gene lengths actually found: The longest gene found so far, the gene for dystrophin (whose defects lead to Duchenne and Becker muscular dystrophies) comprises $\sim 2 \times 10^6$ base pairs (see Table 3.4). Some other, "famous" genes, for example those for clotting factor VIII, cystic fibrosis, and phenylalanine hydroxylase, are of the order of magnitude of $1–2 \times 10^5$ base pairs. The hemoglobin β-chain gene, on the other hand, is much smaller (about 10^3 nucleotide pairs). Hence, the range is enormous. It is probable that some of the "big" genes have been discovered early merely because they are big, thereby offering more mutational sites, and the corresponding hereditary diseases therefore relatively common. So far we do not know the average gene length, but many more than 100 000 genes can hardly be accommodated in the genome. On the other hand, the number of mRNA species in nerve cells of the rat has been estimated to be in the order of 30 000 [99]. Even considering an increase due to relatively common alternative splicing (i.e., that the same gene is often responsible for more than one mRNA), a large fraction of all genes appears to be active in neuronal tissue. Similarly, a large number of cDNAs constructed from naturally occurring human brain mRNA have been detected. Whether such cDNAs, with uncertain but probable function, could be patented has given rise to heated discussions [116]. Various approaches to estimating the number of human genes have led to a figure of approximately 60 000–70 000 [52].

Repetitive DNA Sequences with Specific Functions. Some intermediate repetitive sequences contain genes necessary in all cells and in each phase of individual development (ribosomal RNA, histone, transfer RNA). In general, the genes for ribosomal RNA (rRNA) are part of the nucleolus organizer region; the nucleolus contains an rRNA pool. In humans, the nucleolus organizer regions comprise a part of the short arms of acrocentric chromosomes (nos. 13–15, 21, 22). In vitro RNA-DNA hybridization techniques have been used to estimate the number of rRNA genes in humans [10, 11]. By comparison with the fraction of human DNA that hybridizes with rRNA and the total amount of DNA in a human cell, the average total number of ribosomal genes has been estimated to be of the order of 416–443.

The multigene family formed by the numerous genes for variable sequences of immunoglobulins (Sect. 7.4.1) has so many similar copies that the corresponding DNA sequences must be expected to be intermediate repetitive. Other multigene families, some of which may contribute to the repetitive fraction, are discussed in Sect. 3.1.3.10.

Satellite DNA. Much of the DNA, especially in the highly repetitive sequences, has been characterized as *satellite DNA.* When fragmented DNA is centrifuged in a cesium chloride density gradient, a main band or peak is noted. In addition, however, some smaller peaks – or shoulders of the main peak – are often visible. The DNA of these smaller peaks is called satellite DNA; the number and location of satellite DNA peaks is characteristic for the species (Fig. 3.2). Their location in the cesium chloride gradient is determined by their base composition. A separate peak can be seen only if the base composition deviates from that of the main DNA fraction. Within the chromosomes, satellite DNA is usually confined to the constitutive heterochromatin; in humans it has also been found outside the centromeric areas in the Y chromosome, and in chromosomes 1, 9, and 16. It consists of short but highly repetitive DNA sequences that might be present in several million copies. (Satellite DNA must not be confused with the satellite regions of acrocentric chromosomes. The use of the same term is an unfortunate coincidence of nomenclature.)

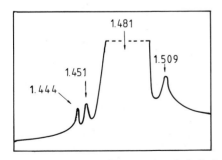

Fig. 3.2. Human satellite DNA. Analytical ultracentrifugation of total placental DNA in a cesium sulfate gradient in the presence of silver ions demonstrating the presence of satellites I (1444), II (1451), and III (1509). (From Miklos and John, Amer. J. Hum. Genet. 31, p. 266, 1979)

The function of satellite DNA is unknown and has therefore been the subject of much speculation. Since its discovery, cytogeneticists have been struck by its location within parts of the chromatin known for many decades from microscopic analysis as heterochromatin.

The Structure of Telomeres. Telomeres are the physical ends of eukaryotic chromosomes [8]. It is known from early work by Muller and McClintock (see [123]) that intact telomeres do not undergo fusion with other chromosome segments. Apparently the telomere provides a protective "cap" for the end of the chromosome. Telomeric DNA sequences and structures are similar among otherwise widely divergent eukaryotes. They consist of a very simple, tandemly repeated DNA sequence. In humans it is $(AGGGTT)_n$ [72]. The G-rich sequence runs 5′ to 3′ (for explanation see Sect. 3.1.3) to the end; it protrudes approximately 12–16 nucleotides beyond the complementary C-rich strand. The end is protected against chemical and enzymatic influences by specific, telomere-binding proteins.

3.1.1.2 Heterochromatin

Definitions and Properties. The name "heterochromatin" was coined by Heitz in 1928 (see [78]), who wrote (our translation): "In *P(ellia) epiphylla* [a moss] some parts of five of the nine chromosomes behave differently. In telophase, they do not become invisible as do the remaining parts and the other four chromosomes but can be observed as such in young interphase nuclei and also in nuclei of adult cells." Maintenance of the condensed state in interphase has remained the main characteristic of heterochromatin [12]. Later, other peculiarities were discovered. Most DNA replication during S phase, for example, occurs somewhat later in heterochromatin than in euchromatic chromosome segments. Two subclasses are usually distinguished: constitutive and facultative heterochromatin. In humans the facultative fraction is represented by the inactivated X chromosome in females and in males having more than one X (Sect. 2.2.3.3).

Heteromorphisms: Function and Relationship to Satellite DNA [55]. There is a large amount of interindividual variability in heterochromatin (Sect. 2.1.2.3), more than in the euchromatic parts of the genome. Such variants are called "heteromorphisms." In addition to the regions mentioned above (secondary constrictions of nos. 1, 9, 16), heteromorphisms are found mainly in the centromeric and satellite regions of acrocentric chromosomes. It has been known for many years that no classical genes can be assigned to constitutive heterochromatin, but most research workers are reluctant to assign no function at all to it. Many functions have been suspected. Examples include stabilization of chromatin structure and a "bodyguard" function for protection of the more valuable euchromatic DNA sequences against external impacts [41].

These considerations suggest that the phenomena observed by classic cytogeneticists which led to the concept of heterochromatin are closely related to more recent data on highly repetitive DNA and satellite DNA, which are derived from entirely different experimental approaches. Satellite DNA, highly repetitive DNA, and heterochromatin are located mainly close to the centromeres, but can also be found in other regions of some chromosomes (1, 9, 16, Y).

3.1.1.3 The Nucleosome Structure of Chromatin
[53, 63]

Chemical Composition of Chromatin. In addition to DNA, chromosomes contain a number of proteins. Together with the double helix of DNA these proteins form chromatin. Most abundant are the histones, positively charged alkaline proteins with a molecular weight of about 10 000–20 000. These can be subdivided into five classes (H1, H2A, H2B, H3, and H4). Other, so-called nonhistone proteins are present in varying but generally smaller amounts. The nonhistone fraction is heterogeneous and includes, for example, a number of enzymes.

Nucleosomes [53]. The chromatin thread consists of repeat units made up of a set of histone molecules in association with about 200 DNA base pairs. The set of histones consists of two each of the four types H2A, H2B, H3, and H4. They are folded in a globular fashion, forming a cyclinder. The DNA component of a nucleosome has two parts: a "core" of 140 base pairs, and a "linker" which varies in length from about 15 to about 100 base pairs, depending on the cell type. Such linkers apparently connect the nucleosomes with one other. Histone H1, which is about twice as long as the other histones, keeps the entire structure close together. When it is removed (which is not difficult experimentally) the chain becomes much looser (Figs. 3.3, 3.4). There is only one H1 molecule per nucleosome. The DNA is wrapped around the set of eight histones, forming a roughly spherical particle with a diameter of about 100 Å. Such particles lie close to each other along the length of a chromatin fiber. The exact way in which DNA is associated with histones is unknown; however, the double-helix structure is apparently undisturbed. Studies

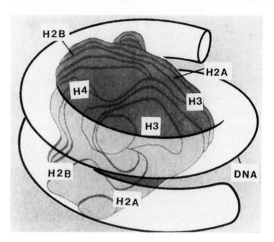

Fig. 3.3. Nucleosome. – A DNA sequence, 146 bp long, is wound one and three-fourths times around the histone octamer. This octamer consists of a central $(H3)_2(H4)_2$ tetramer and two H2A-H2B dimers. Approximate size: 11 nm diameter, 4–5 nm height. (From Knippers 1990 [51])

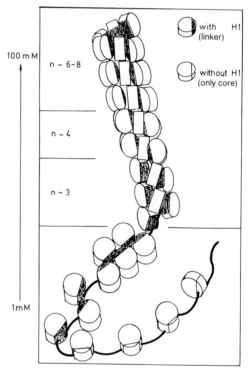

Fig. 3.4. Schematic representation of the nucleosome structure of chromatin. In in vivo experiments the precise structure depends on the salt concentration. At 100 mmol NaCl, 6–8 nucleosomes are combined in one turn of the chromatin thread *(above)*. At lower salt concentrations, only 3–4 nucleosomes/turn can be seen *(middle)*. In the absence of salt, nucleosomes have little contact with each other. (From Knippers 3rd ed. 1982 [51])

with DNA-RNA hybridization techniques (see Sect. 3.1.3.3) indicate that a wide variety of functionally different DNA stretches occur in the nucleosomes, from single-copy to repeated-sequence DNA, and actively transcribed sequences as well as those belonging to constitutive heterochromatin. Apparently the packaging in nucleosomes does not interfere with transcription [54]. Virtually all chromosomal DNA in eukaryotic cells is probably packaged in nucleosomes. The evidence for the nucleosome structure comes mainly from three lines of work: chains of particles are observed in electron micrographs of chromatin, X-ray diffraction studies suggest a repeat unit of chromatin, and enzymatic digestion with micrococcus nucleases allow the isolation of single nucleosomes.

3.1.1.4 Integration of the Chromatin Thread in Chromosome Structure

Interphase. The chromosome in interphase can be visualized as an elementary fibril that consists of a sequence of nucleosomes connected by linkers. This fibril does not extend through the entire nucleus but occupies certain domains [61, 62]. Chromatin is highly spiralized. Some aspects of the precise order of this spiralization are still being disputed; the fibers observed by various methods, and representing increasing orders of spiralization in inter- and metaphase are described in Table 3.1.

Mitotic and Meiotic Chromosomes. As seen in Table 3.1, chromosomes in mitosis and meiosis show a much higher degree of coiling than in interphase.

Table 3.1. Fibrils in inter- and metaphase: degree of shortening compared to the DNA double helix, diameter in Å.

Fibril	Degree of shortening		Diameter
	To the next unit	Compared with DNA	
DNA	1	1	10 Å
Nucleosome	7	7	100 Å
Nucleoprotein fibre (spheroid, superbead, elementary fibril)	6	42	200– 300 Å
Chromatid in interphase (chromonema)	40	1600	1000–2000 Å
Chromatid metaphase	5	8000	5000–6000 Å

Their banding patterns are discussed in Sect. 2.1.2.3. The number of subbands into which a band can be resolved varies with the degree of condensation (e. g., mitotic prophase or metaphase) and the quality of the staining method. An upper limit of so-called chromomers appears to be approximately 30 000–100 000 base pairs in length (see below [93]). Considering the number of base pairs/haploid genome ($\approx 3.5 \times 10^9$) and the number of bands seen even in the best preparations (up to $\approx$ 2000; Sect. 2.1.2.3), no one has yet even come close to achieving this level of resolution.

Studies on replication patterns of mitotic chromosomes have shown that dark G bands (= light R bands and, as a rule, bright fluorescent Q bands) usually replicate in the second phase of DNA synthesis (S phase). The single band in prometaphase chromosomes appears to be the unit of replication (which consists of a number of replicons, i. e., replication appears to start at several points within this unit at about the same time). There has been speculation that the organization into units of replication is of some functional significance. These units contain many highly repetitive and nontranscribed DNA sequences. The number of visible bands depends on the degree of chromosome condensation, as outlined in Fig. 3.5. As mentioned, G light areas (= R bands) contain more genes, especially "house-keeping" genes; G dark (= R light) bands contain fewer genes, especially those for thissue-specific functions.

In the human genome, regions with prominent R-bands are found especially in chromosome regions 3p, 6p, 11q, 12q, 17q, and 19 (p or q). Linkage studies (Sect. 5.1.2) have indeed localized more genes in these areas than would have been expected with a random distribution of genes. Furthermore, the number of recognized abortions that are trisomic for these regions is lower than expected, indicating very early and therefore undetectable lethality [38].

Many studies suggest a fine structure of the metaphase chromosome with certain "areas of constraint" (probably identical with dark G areas) alternating with other areas in which loops may be formed under certain conditions [108].

3.1.1.5 Integrated Model of Chromosome Structure

These data, together with results from molecular biology (see below), suggest an integrated model of the human chromosome. It consists basically of a single DNA double helix combined with histones in nucleosomes over its whole length. In some regions this double helix is composed mainly of repetitive sequences; highly repetitive satellite DNA stretches may be interspersed. Areas in which such repetitive se-

Fig. 3.5. Production of chromosome banding (G-banding) patterns by coiling of the chromatic strand, which includes lightly stained as well as dark areas! Note that the number of microscopically visible G bands decreases with increasing density of coiling. (From Schwarzacher 1976 [93])

quences are abundant, primarily the centromeric regions and secondary constrictions, are characteristic of (constitutive) heterochromatin. Elsewhere, the double helix consists mainly of unique sequences which are interspersed with low or intermediate repetitive regions. Classical cytogenetic methods show these segments to have the properties of euchromatin. Under certain conditions they may show more or less extended loops.

In especially well-suited cells (e. g., the relatively large oocytes of amphibia) chromosome structure can be studied in detail, and even transcription can be observed [70, 108].

3.1.2 The Genetic Code

One of the principal achievements in the 1960s that made the new genetics possible was the deciphering of the genetic code. Using synthetic trinucleotides, it was shown that a given triplet of bases specifies the ribosomal "translation" of a given amino acid. Soon the codons (three nucleotides coding for an amino

acid) for all amino acids had been established (Table 3.2). All amino acids except tryptophan and methionine have more than one codon, i. e., the code is degenerate. The base in the third position of a codon has reduced specificity, since four codons differing only in the third base are synonymous and represent the same amino acid. This feature and the tendency for similar amino acids (polar, etc.) to be specified by related codons ensure that the random mutational alteration of a nucleotide has minimal effects. Three codons specify termination signals; wherever these three triplets appear, translation is stopped. The AUG codon for methionine specifies the initiation of translation by N-formyl methionine at the beginning of a polypeptide chain.

The genetic code is universal and is used by organisms as far apart as viruses and humans – an impressive demonstration of the unity of life and its common origin on the planet earth. In recent years some minor exceptions to codon usage have been noticed in mitochondria, in that UGA is a tryptophan rather than a termination codon.

3.1.3 Fine Structure of Human Genes

Around 1970 molecular biology seemed to have reached a certain degree of completeness. The structure of DNA [120], the mechanisms of DNA replication, the "central dogma" of gene action – transcription and translation – and some major aspects of gene regulation were well established. Since the basic structures and processes had been analyzed mainly in micro-organisms, special features in eukaryotes (including humans) presented a number of additional problems; however, entirely new results were not expected at that time. In the early 1970s, however, a completely new development was triggered mainly by development of a new research tool: *restriction endonucleases*. The recombinant DNA technology that has developed since that time has opened the way to large-scale industrial production of gene products such as biologically important proteins, and to genetic manipulation of various organisms by artificial gene transfer. Our understanding of structure and function of the genetic material – especially in eukaryotes, including humans – has deepened far beyond our keenest hopes. Completely unexpected facts have been discovered, with implications for both theoretical and practical fields such as gene action, population genetics, evolution, and genetic counseling including prenatal diagnosis (Chap. 7; 14; 18.). This enormous progress has also given rise to widespread public concern over possible novel consequences of genetic engineering, either by inadvertent production of dangerous germs or even by manipulation of hu-

Table 3.2. Genetic code

First base		Second base					Third base
DNA		A	G	T	C		DNA
	mRNA	U	C	A	G	mRNA	
A	U	UUU UUC } Phe UUA UUG } Leu	UCU UCC UCA UCG } Ser	UAU UAC } Tyr UAA UAG } TERM	UGU UGC } Cys UGA TERM UGG Trp	U C A G	A G T C
G	C	CUU CUC CUA CUG } Leu	CCU CCC CCA CCG } Pro	CAU CAC } His CAA CAG } Gln	CGU CGC CGA CGG } Arg	U C A G	A G T C
T	A	AUU AUC AUA } Ile AUG Met	ACU ACC ACA ACG } Thr	AAU AAC } Asn AAA AAG } Lys	AGU AGC } Ser AGA AGG } Arg	U C A G	A G T C
C	G	GUU GUC GUA GUG } Val	GCU GCC GCA GCG } Ala	GAU GAC } Asp GAA GAG } Glu	GGU GGC GGA GGG } Gly	U C A G	A G T C

TERM, Terminator (stop) codon.

man embryos. Many of these ethical concerns were first articulated by the scientists actively involved in this work. At present, most scientists regard concerns about safety as largely unfounded; nevertheless, many ethical problems remain, and new ones continue to arise.

Whereas in earlier decades human and medical genetics developed as a relatively separate branch of science, large parts of the field have now been incorporated into mainstream research in molecular genetics. This development makes it more difficult to delineate the field. A textbook of human genetics cannot describe in detail all techniques of molecular biology that have led to such enormous scientific progress in human genetics. More specific sources should be used [51, 121, 122]. However, the principles of the new approaches need to be understood by all human and medical geneticists, as well as by students and research workers interested, for example, in evolution or behavior genetics.

3.1.3.1 Analysis of Human Genes

Much of the work in human molecular genetics has dealt with the discovery and analysis of human genes. Such analysis is described here with some characteristic examples. First, some instruments used in the analyses are described.

3.1.3.2 Restriction Endonucleases

Germinal Observations. In the course of his work on infectivity of the λ phage of various strains of *E. coli,* Arber [2] discovered that the DNA of this phage is fragmented – and infectivity influenced – by passage of the phage through the bacterium; classical recombination processes or mutations are not involved. Moreover, this is not specific to phage DNA; any foreign DNA is cut by these bacteria in the same way. This cutting can be regarded as a defense mechanism of the cell against foreign DNA, and is performed, as further studies have shown, by enzymes referred to as restriction endonucleases. This immediately raised the question of why these enzymes do not cut the DNA of their own cells. The answer found by Arber was that these enzymes acted only at certain specific recognition sites of the DNA, and that these sites are protected by methylation. The restriction endonucleases that were discovered first did not cut DNA at their specific recognition sites, but at other random sites. The first restriction enzyme that cleaved DNA at a sequence specific site, the *Hind* enzyme, was discovered by Smith at the end of the 1960s (see [2]). It was used first by Nathans [74] to construct a cleavage or restriction map of the genetic material of an organism – the SV40 virus. Berg [6] recognized the special advantages of DNA double strands in which the two strands are cut so that "sticky ends" are produced. One of the two strands is several bases longer than the other; these bases are now free to pair with other bases, for example, from another piece of DNA with sticky ends [268]. By this means DNA from various sources and various species can be joined to produce recombinant DNA.

Principles of DNA Recombination Technology [22]. A great number of such restriction enzymes (> 150) that cleave DNA at specific sites have now been discovered [51]. The enzyme *Ri,* for example, cuts DNA double strands in such a way that two adhesive ends are produced:

$$\downarrow$$
$$G-A-A-T-T-C$$
$$||| \; || \; || \; || \; || \; |||$$
$$C-T-T-A-A-G$$
$$\uparrow$$

Adhesive ends of different DNA molecules split by this enzyme connect by fourfold A = T pairing. A ligase is required for closing the gap. The various restriction endonucleases [51] differ in their sequence recognition sites. They can be used for a variety of purposes. A common application is the production of proteins in large quantities by introducing the relevant coding DNA into the micro-organisms or human cell lines. This use is important practically. Proteins otherwise available in only minute amounts can be produced in large quantities. The following principle is used. Apart from their single ring-shaped chromosome, bacteria often carry additional small, double-stranded DNA rings known as plasmids. Plasmids replicate autonomously and include genes for resistance against antibiotics and for substances that kill other bacteria, the colicins (Fig. 3.6). Such plasmids can be selected so that they are split by a restriction enzyme at one site only. Double-stranded DNA that has been cut by the same restriction enzyme may be introduced and replicated together with the rest of the plasmid within the bacterium

(Fig. 3.7). The source of the exogenous DNA is immaterial. It could come, for example, from human cells.

Besides bacterial plasmids, λ phages (the objects of Arber's studies) are also being used as DNA vectors. Part of the λ genome is not essential for lytic growth of the phage; instead, the phage is able to take up foreign DNA and to propagate it together with its own genome after infection of a bacterium. Certain tricks make it possible to improve the cloning capacity of plasmids by combining them with parts of the λ phage, especially the λ gene *cos.* In this way so-called cosmids are formed – especially well-suited vectors for human DNA in which up to 45 kb can be cloned [122]. Extremely useful vectors are yeast artificial chromosomes (abbreviated YACs [14, 51, 122]) in which DNA sequences up to 1000 kb can be cloned – in distinction to the vectors mentioned above (Fig. 3.8).

Once recombinant DNA has been replicated and amplified together with plasmid or phage DNA in transformed or transfected cells, two questions arise:

1. How can clones containing hybrid DNA be recognized among the progeny of transformed cells or viable bacteriophages?
2. How are specific DNA segments identified among the many cloned fragments?

Bacterial cells can be selected, for example, if the plasmids contain a resistance factor against an antibiotic, and if the culture is grown in the presence of this antibiotic. A variety of selection methods have been developed in recent years.

For genetic engineering of proteins, it is not only necessary to select and amplify certain DNA regions but also to induce them to express genetic activity and produce the required protein. This requires that the desired DNA sequences be combined with the machinery that promotes DNA transcription and translation, and that secondary processing at the transcriptional and translational levels be carried out correctly.

Identification and Analysis of Genes: Southern Blotting (Figs. 7.53, 12.5). The technique often used for the analysis of genes was described by Southern in 1975 [95]. The total DNA of human cells is cut by a restriction enzyme into thousands of fragments with a size range of 10^2 to 10^5 bp. The fragments are separated according to molecular weight by gel electrophoresis in agarose, and the double helices are separated by alkali treatment to produce single strands. The fragments are then blotted onto a nitrocellulose or nylon filter and fixed at 80 °C. The resultant pattern represents a replica of the electrophoretically separated DNA from the original agarose gel. The DNA frag-

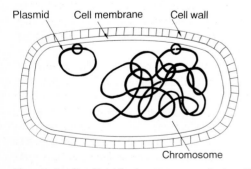

Fig. 3.6. *E. coli* cell with chromosome and plasmid. (From Klingmüller 1976 [50]

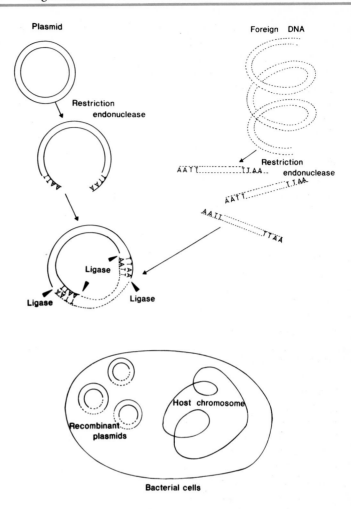

Fig. 3.7. Principle of introduction of foreign DNA into a bacterial plasmid using the RI endonuclease. (Adapted from Vosberg 1977 [118])

ments can be identified by hybridization with radioactive DNA probes that are specific for genes or chromosomal regions. Any fragment that contains part or all of the probed gene sequence is seen as a dark band after autoradiography.

Probes and Gene Libraries. An essential condition for such an analysis is availability of a gene-specific DNA probe that can be labeled and used for hybridization. In cases where messenger RNA is available, a specific probe can be produced by the enzyme reverse transcriptase. This enzyme catalyzes transcription of the mRNA nucleotide sequence into the complementary DNA sequence, so-called cDNA. mRNA is not used directly for hybridization, since it is too unstable or is not available in sufficient quantities. mRNA from a given tissue can be used to construct a cDNA gene library. Such cDNA libraries contain mainly single-sequence DNA specific for transcribing structural genes or their parts, and DNA sequences from their immediate neighborhood. They are used mainly for finding and characterizing the genes active in a given

tissue (see also [71]). Genomic libraries are available. They are produced by cutting DNA into segments using restriction enzymes and amplifying the resulting DNA pieces in a vector. They are used, for example, for finding complementary sequences in the genome. This permits investigation of the distribution of homologous sequences in the genome. Often a restriction *polymorphism* is discovered inside a certain DNA sequence; i.e., variation in restriction sites between different individuals. In these cases, such probes may be used for classical linkage studies in human families by methods described in Sect. 5.1.2.

Working with such genomic libraries is often unsatisfactory because of the sheer size of the human genome and the large number of fragments from which the "interesting" ones must be selected. For many problems it is preferable to have *chromosome-specific* banks. Construction of such banks, however, requires that specific chromosomes can be isolated. Such preparations are now possible by cytofluorimetric chromosome sorting [16a].

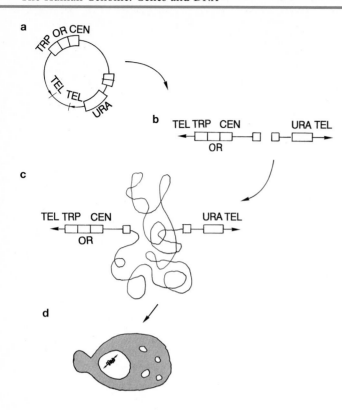

Fig. 3.8 a–d. Cloning large DNA fragments as YAC's. The cloning vector in circular configuration is first produced in bacteria and then cleaved to provide the two "arms" containing the telomeres *(TEL)*, a replication origin *(OR)*, a centromere *(CEN)*, and the genes *(URA, TRP)* used for selecting the transformed yeast clones. When these arms are ligated to a very large human DNA fragment (100–1000 kb) extracted from cultured cells, the DNA fragment is transformed into a YAC. After introduction into yeast it is able to replicate (thanks to the OR sequence), to stabilize its ends (thanks to the TEL sequences), and to share its copies among the daughter cells during division (thanks to the CEN sequence). (From Jourdan 1993)

Artificial oligonucleotides are another powerful tool for the construction of cDNA [102], especially in cases where no mRNA is available. Using our knowledge of the genetic code, they may be constructed, for example, for structural genes of protein with known amino acid sequences. Machines for construction of any desired nucleotide sequences are available.

When a gene has been identified – especially when mRNA is available – its fine structure can be analyzed by combining various methods. The ultimate goal of such studies is elucidation of the complete nucleotide sequence of the entire genetic region. Once the nucleotide sequence has been established, the mRNA can be searched in various tissues using hybridization methods (see below). The protein sequence can be deduced using the genetic code, and its function can be elucidated. This approach has become one of the most powerful for elucidating physiological mechanisms as well as their abnormalities in disease (for an example, see Sect. 3.1.3.9).

3.1.3.3 Nucleic Acid Hybridization

Principle. Under suitable conditions, single DNA strands tend to reanneal with complementary strands to form double strands. This property is used to identify electrophoretically separated DNA fragments by Southern blotting (Sect. 3.1.3.2). Most naturally occurring DNA forms double helices. In a DNA double helix the pyrimidine base cytosine (C) pairs with the purine base guanine (G), whereas the pyrimidine base thymine (T) pairs with the purine base adenine (A). These complementary base pairs (C ≡ G; T = A) are joined by hydrogen bonds, which can be resolved relatively easily. They have a strong tendency to reanneal under suitable conditions (temperature, salt concentration), forming double helices again when single-stranded DNA chains are mixed. The origin of single-stranded DNA is not important for reannealing; it does not even require complete complementarity of the single strands and even works when a certain fraction of the bases in each strand do not match (Fig. 3.9). Single-stranded DNA pairs hybridize even with RNA, provided only that complementarity of bases is maintained.

"Gene Walking." The hybridization technique may be used, for example, for analysis of a large gene when only part of its sequence is available. The probe is hybridized with sequences from a DNA library. A hybridizing library sequence is generally longer than the probe; its ends overlap with another library sequence and hybridize with it in part. Its free end hybridizes with the next, until a long portion, a "contig", for example an entire structural gene, has been assembled. In this way, the structural gene for the human blood clotting factor VIII, an extraordinarily long gene with 180 000 bases, was reconstructed starting with an oligonucleotide probe

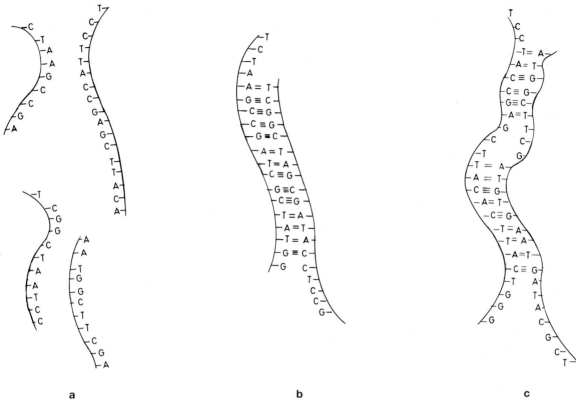

Fig. 3.9 a–c. Principle of DNA (or RNA) hybridization. **a** Nucleotide single strands in solution. **b** The single strands hybridize according to the pairing rules: thymine pairs with adenine, cytosine pairs with guanine. **c** Hybridization may also occur if matching of the two chains is not perfect, provided that the differences are not too large

only 36 bases in length. The above-mentioned method of first identifying a specific mRNA and then constructing a cDNA by reverse transcription was not feasible – mainly because of low mRNA concentration. The DNA sequence of the oligonucleotide probe was therefore inferred from the amino acid sequence of a factor VIII protein fragment; the precision achieved was sufficient for hybridization. For the entire analysis of the factor VIII gene, see Sect. 3.1.3.7.

In Situ Hybridization with Radioactive Probes. A technique especially well-suited for analysis of eukaryotic genomes is in situ hybridization [24, 42, 77]. Here, metaphase chromosome preparations are treated in situ with a radioactive DNA probe under conditions permitting hybridization with the chromosomal DNA. In this way, the gene of interest can be localized on a specific chromosomal segment. As an example, the essential steps are now discussed with reference to myosin heavy-chain gene [82].
A cDNA probe for the myosin heavy-chain gene was available from the rabbit. Since homologous structural genes from various mammals are generally similar, DNA hybridization between DNA of mammalian species is not impaired. This probe was amplified in a

plasmid and labeled with ^{3}H by *nick translation.* In this procedure, DNA is incubated with a small amount of DNAase I; this enzyme produces several single-strand breaks. Radioactive nucleotide building blocks are then added, together with a polymerase; these blocks are introduced into the DNA structure.
Mitotic chromosome preparations were obtained from lymphocyte cultures. After DNA denaturation, the ^{3}H-labeled DNA probe was incubated in the metaphase spreads for 16–18 h at 40 °C. After removal of unbound or nonspecifically associated probe DNA, autoradiography was carried out for 3 weeks. Following staining with quinacrine mustard (Q staining) the preparations were photographed. Figure 3.10 shows distribution of the label among human chromosome bands. The bulk of the label was localized to the short arm of chromosome 17, band 17 pl 2 → pter. Hence it was concluded that this myosin heavy-chain gene is located on this chromosome. Since this experiment was performed with a cDNA probe, it is theoretically uncertain that the myosin gene identified in this way is indeed an active gene; it could also be a "pseudogene," i.e., a DNA se-

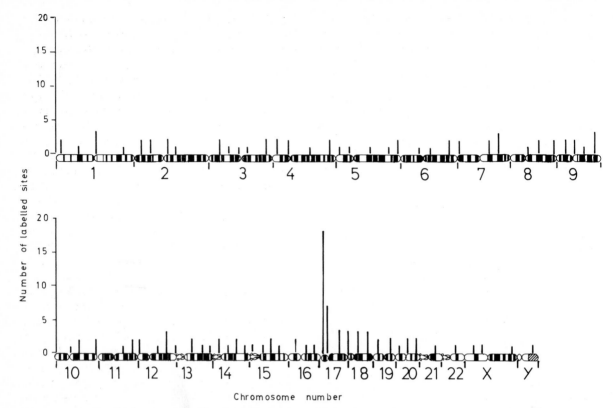

Fig. 3.10. Grain distribution in 36 metaphase spreads after in situ hybridization with a myosin heavy-chain cDNA probe. This histogram results from an analysis based on the division of the human haploid karyotype into 110 equal segments (chromosomes drawn in one quasi-continuous sequence). The number of labeled sites was plotted for each segment. Clustering of grains was found on the short arm of chromosome 17. (From Rappold and Vosberg 1984 [82])

quence that is structurally homologous to the active myosin gene but does not code for myosin, having lost important flanking sequences outside the transcribed portion. Such pseudogenes have been discovered, for example, within the α- and β-globin regions (Sect. 7.3.2). An increasing number of human genes have been localized using this technique (Sect. 5.1.2).

This method requires knowledge of the DNA structure and availability of a gene or cDNA fragment or sequences of the gene to be studied. If the protein sequence is known, artifical oligonucleotides can be constructed as inferred from the genetic code. However, since the code is degenerate, i.e., several different triplets of nucleotides specify the same amino acid, the actual codon used in the specific gene under study is not known so that the codon used most frequently to specify a given amino acid for oligonucleotide construction is chosen.

In recent years this method has often been replaced by nonisotopic in situ hybridization. The introduction of these methods has opened an entirely new chapter of molecular cytogenetics; therefore they are described in some detail.

Fluorescence In Situ Hybridization. The method of fluorescence in situ hybridization (FISH) [401, 476], also known as chromosome painting, is now being applied for the solution of an increasing number of problems as well as for diagnosis of chromosomal abnormalities; it represents a rewarding "marriage" between cytogenetics and molecular genetics. The principle of FISH is the hybridization of a known DNA or chromosomal sequence tagged with a fluorescent dye to its homologous counterpart in a conventional chromosome preparation followed by visualization of the target sequence by fluorescent techniques, often in different colors (Fig. 3.11).

FISH is a multistep procedure. The following steps are necessary: (a) preparation and labeling of DNA probes, (b) preparation of metaphase chromosomes, cell suspensions, or tissue sections, (c) denaturation of double-stranded DNA of the probe and of the specimen to be studied to obtain single-stranded DNA, (d) in situ hybridization, and (e) fluorescent probe detection, generally by fluorescence microscopy.

(a) During the labeling reaction modified nucleotide analogs are incorporated by hybridization. These are linked to haptens, for example, biotin or digoxigenin. Alternatively, nucleotide analogs are conjugated directly to fluorochromes.

(b) Preparation of metaphase and prophase chromosomes follows standard cytogenetic protocols. The method has the significant advantage that chromosomes can also be identified and analyzed in interphase; this has turned out to be especially useful for chromosome identification in amniotic fluid cells or in nuclei of tissue sections [18].

(c) The probe molecules and target DNA are denatured by heat. If complex DNA probes are used, an additional preannealing step with an excess of unlabeled DNA prior to hybridization with the labeled probe is required to prehybridize dispersed repetitive sequences which are present not only in the target DNA but are distributed over the entire genome; otherwise they would obscure hybridization by the labeled probe (chromosome in situ suppression, CISS, hybridization [62]). This pretreatment allows specific delineation of target sequences without the need to prepare DNA probes consisting entirely of target specific sequences.

(d) Hybridization is carried out for about 16 h, and is much faster than detection with radioactive probes.

(e) Fluorochromes linked to avidin are used to detect the target sequence. Avidin has a high affinity to biotin or to autoantibodies specifically recognizing a hapten used for probe labeling. Numerous fluorochromes are available which emit blue, green, or red colors. Hybridization is visualized by fluorescence microscopy with various filters. Digital imaging devices such as charge-coupled device (CCD) cameras add to the sensitivity of the method and allow quantification of fluorescent images. Confocal laser scanning microscopy [16] is used to study three-dimensional targets in cell nuclei.

Generally, FISH has high sensitivity and resolving power. Depending on the complexity of the DNA probes used, entire chromosomes can be visualized using composite probes established from sorted human chromosomes. On the other hand, even targets as small as 500 bp are recognizable on metaphase chromosomes. This equals the sensitivity of the isotopic detection method. Spatial resolution, however, is much higher: on metaphase chromosomes, two targets separated by about 5 Mbp can be resolved. Interphase chromatin is less condensed; therefore, resolution is increased to about 100 kb [56, 107]. In histone-depleted interphase nuclei spatial resolution has been increased to a few kilobases [104].

The possibility of visualizing several chromosomal targets simultaneously has broadened the spectrum of FISH applications. However, the amount of suitable haptens for probe labeling and of fluorochromes with clearly different emission spectra limits the number of differently colored targets. Therefore single probes may be labeled by a combination of fluorochromes in various ratios [74]. For example, the combination of three fluorochromes allows detection of seven probes – three with single colors, three with combinations of two, and one with a combination of all three colors [57, 83] (Fig. 3.11 c,d). Resolution can be increased by using fluorochromes in different ratios [20, 21] in combination with digital imaging devices.

The Fish method has been used for solving the following problems: (a) The centromeric regions of almost all human chromosomes can be visualized specifically. This is useful for diagnosis of numerical chromosomal aberrations in interphase cells [17]. (b) The method can be applied in combination with G banding. This increases its diagnostic usefulness. (c) Entire chromosomes can be painted by composite probes prepared from sorted chromosome preparations [18, 62, 79]. Using this method, it was shown that chromosomes in interphase nuclei are not distributed at random but occupy certain domains (Fig. 3.11 a). Distribution of these domains relative to each other apparently does not follow strict rules: in most instances, homologous chromosomes are not located close to each other pointing to a possible influence on gene action. (d) In addition to DNA probes which are available in increasing numbers, probes for any desired chromosome or band can be prepared from microdissected chromosomes [61].

The FISH method is now being applied in many field of human genetics and cell biology. For example, genes and DNA sequences have been mapped within the genome (Sect. 5.1.2). Other applications such as evolution are mentioned elsewhere in this volume (Sect. 14.2.1). In medical genetics, the method has become an important tool for solution of problems that cannot be solved by classical cytogenetic methods, such as identification of the origin of supernumerary chromosomes and various translocations [49, 103]. Chromosome damage due to mutagenic agents can be studied more easily [80]. Another clinical application is in Duchenne muscular dystrophy, where 70% of the mutations are small deletions that cannot be detected by conventional cytogenetic techniques [83] (Fig. 3.11 b). The most important advance allows study of chromosomes and their aberrations in interphase nuclei. This will become more important for the diagnosis of malignant tumors, especially since cells in solid tumors and from paraffin-embedded tumor sections can be studied [22, 88]. However, FISH analysis will probably not replace classical cytogenetic methods, even if the banding method (Fig. 3.11 d) [57] is improved by the creation of "chromosome bar codes." Both approaches have their advantages and disadvantages. Banding analysis requires the preparation of prometaphase or metaphase spreads, which is often difficult in solid tumors, and the resolving power is limited to several megabases even after high-resolution banding. In contrast, FISH can be used to study a chromosome region, its numerical representation, and its involvement in translocations at all stages of the cell cycle. Here the investigator needs to select the chromosomal regions to be studied. Classical cytogenetic methods, on the other hand, allow analysis of the en-

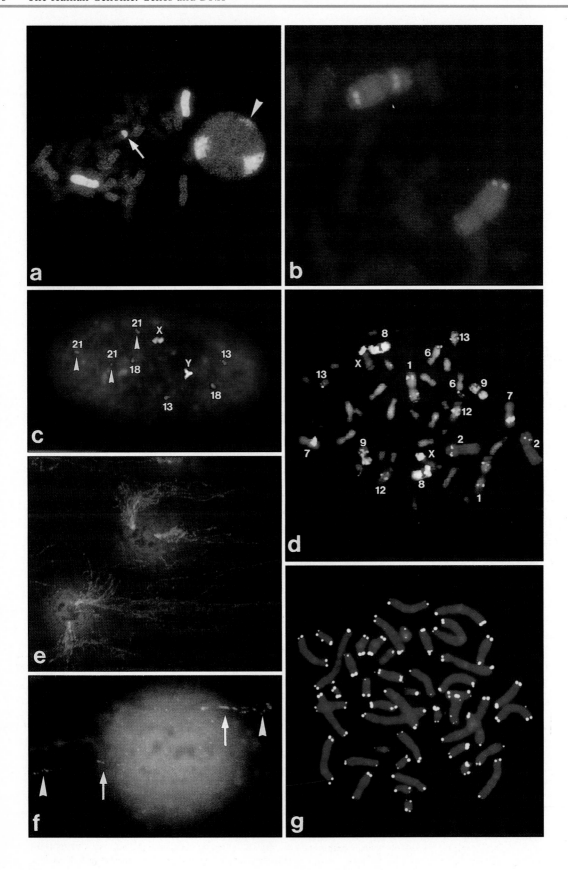

tire karyotype in metaphases prepared by one procedure. Different problems require different diagnostic
strategies.

Comparative Genomic Hybridization. The molecular cytogenetic technique known as comparative genomic hybridization
(CGH) is based on quantitative two-color in situ hybridization
[21, 45]. CGH allows the detection of genetic imbalances in
DNA prepared from tumors or other cells of interest, for example, gains and losses of chromosomes and detection of sequences larger than a few Mbp in size. Figure 3.12 f demonstrates the
principle. Total genomic DNA from the cells to be studied and
from a normal control individual is isolated. If the cells to be
studied are tumor cells, normal tissue from the same individual
may be used as control. The two genomic DNAs are labeled differently and are mixed in a 1:1 ratio. They are then hybridized to
normal human metaphase spreads which may come from any
individual. Since the two DNAs have been painted with different colors, fluorescence intensities along the chromosomes of
the reference metaphase represent the relative concentrations
of DNA sequences. If chromosomes or chromosome subregions

are present in identical copy numbers in both the tumor and the
reference DNA, the observed fluorescence is a blend with equal
contribution of the two colors (for example, yellow as the mixture of red and green). If certain chromosomes or chromosome
subregions are lost in most of the cells to be tested, the resulting
color is shifted to that of the normal chromosomes, for example, red. A gain of DNA in the test material, on the other hand,
leads to a more intense green staining at the site of this gain;
for example, *locally amplified DNA.* In many instances gross
chromosomal aberrations are visible directly in the fluorescence microscope. Precise quantitative analysis, however, requires measurement of fluorescence intensity using a special
CCD camera. The usefulness of the method in tumor diagnosis
has been demonstrated [46], for example, in gliomas [92].

3.1.3.4 *Sequencing of DNA* [39, 51, 58]

Nucleotide Sequence and the Genetic Code. Methods
for determining the sequence of amino acids in
polypeptide chains have been known since the

◁ **Fig. 3.11 a–g. a** Partial metaphase spread from a patient with
the karyotype 46,XX,8, + t(6;8) (6 pter ⇔ p 22 :: 8 p 23 ⇔ 8 q-
ter) after painting with a chromosome 6 specific library
probe. Two normal chromosomes 6 and the translocated fragment in the der(8) *(arrow)* are delineated on the metaphase
spread. An adjacent interphase nucleus shows two normally
sized chromosome domains and a small domain indicating
the interphase domain of the translocation fragment *(arrowhead)* (From Popp et al. 1993) **b** Carrier detection of Duchenne muscular dystrophy is performed using FISH to metaphase chromosomes from a female relative of a patient with
a known deletion of exon 45 of the dystrophin gene. Cohybridization of a chromosome specific library for the X chromosome (rhodamine, *red fluorescence*), a reference clone specific for the long arm of the X chromosome and a cosmid clone
specific for exon 45 of the dystrophin gene (FITC, *yellow
fluorescence*). Only one of the X chromosomes reveals signals
on the short arm, indicating a deletion, thus allowing the diagnosis "carrier" for Duchenne's muscular dystrophy. (Courtesy of A. Jauch) **c** Example of multicolor FISH with probes
for chromosomes 13 (pseudocolored, *red*), 18 (pseudocolored, *violet*), 21 (pseudocolored, *green*), X (pseudocolored,
yellow), and Y (pseudocolored, *white*) to cultured amniotic
fluid cells. Images were taken with a black-and-white CCD
camera coupled to an epifluorescence microscope. The multi-
colored composite image was simultaneously merged using
the software package Gene Join [83]. Chromosome 13 and 18
specific probes reveal two signals each. Both sex chromosomes showed one signal. However, three signals were observed for the chromosome 21 specific probe, indicating trisomy for chromosome 21 *(arrowheads).* Note that in the cultured cells, the signals sometimes appear as doublets, indicating DNA replication at the hybridization site. **d** Simultaneous
combinatorial FISH of directly fluorochromated whole chromosome painting libraries allowing multicolor painting of
chromosomes 1, 2, 7, 8, 9, 12, and X. Ten YAC clones were selected to provide additional green or red bars on these painted metaphase chromosomes, resulting in distinct fluores-

cence pattern for each chromosome: *1,* light blue, two green,
one red; *2,* red, one green; *7,* brown, one green; *8,* light purple,
one green, one red; *9,* green, one red; *12,* dark purple, one
green, one red; *X,* dark blue, one green, one red. Two clones
provided additional signals on nonpainted chromosomes 6
and 13. (From Lengauer et al. 1993 [57]) **e** An increased mapping resolution for FISH analysis can be obtained using interphase Halo preparations after high salt histone extraction.
Two products of this extraction procedure can be discerned:
(a) The DNA Halo fraction that consists of histone-depleted,
"naked" DNA strings. This fraction is sensitive to digestion
with DNase I. In situ the DNA loops extend circular around
a remaining fraction, the nuclear matrix. (b) The nuclear
matrix fraction, or nuclear scaffold, located in the center of
the Halo structures in situ is resistant to DNase I digestion.
More recent evidence suggests that certain DNA sequences
such as ribosomal genes or telomere consensus sequences co-
habite with the matrix fraction, indicating their contribution
to the interphase chromatin architecture. This demonstrates
the hybridization pattern of the paracentromeric heterochromatic DNA probe pUC 1.77 on Halo preparations. (Courtesy
of T. Ried and E. Schröck). **f** An example of Halo structures
in situ after two-color FISH using single cosmid clones from
a contig specific for the telomere of 7q. The clone adjacent
to the telomere was labeled with biotin (detected with FITC,
green fluorescence, arrow), whereas six other subtelomeric
clones were labeled with digoxigenin (detected with TRITC,
red fluorescence, arrowhead). On interphase Halos the cosmid
next to the telomere was mostly observed in the matrix fraction whereas the cosmids expanding in the subtelomeric region labeled the surrounding DNA loops. (Courtesy of
T. Ried and E. Schröck). **g** The telomere consensus repeat sequence (TTAGGG)n is distributed in varying copy numbers
on all human telomeres and is conserved during evolution.
FISH with a biotinylated oligonucleotide containing the repeat sequence labels all telomeres on human metaphase chromosomes. Note that the signal intensity varies due to differences in target number sequences. (Courtesy of E. Schröck)

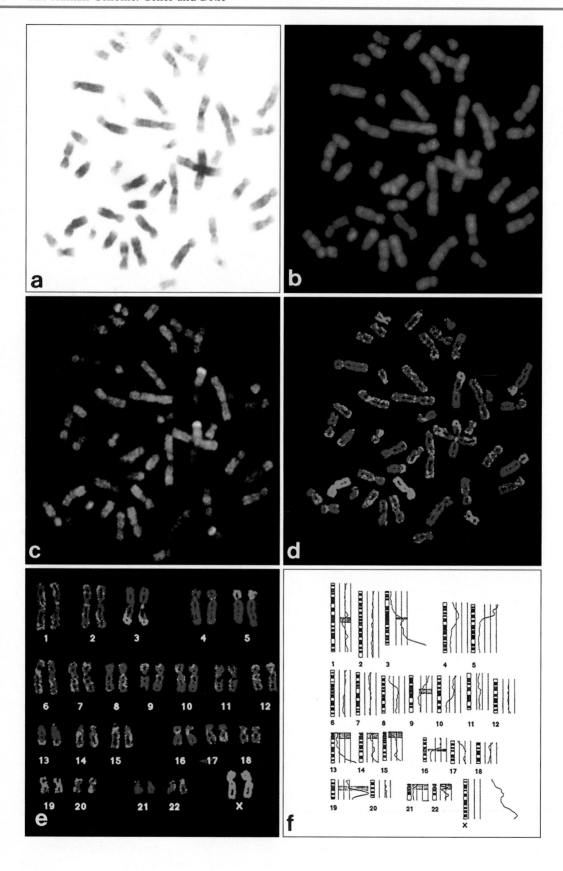

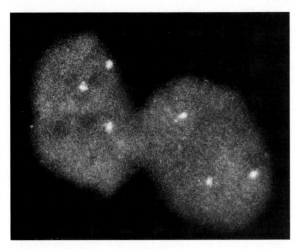

Fig. 3.13. Amniotic fluid cell nuclei of a fetus with trisomy 18 after CISS hybridization with the biotinylated Alu-PCR amplified YAC clone HTY 3045 (mapped to 18 q 23) detected with avidin-FITC. Nuclei were counterstained with propidium iodide

1950s. The problem was relatively easy, since the 20 amino acids found in naturally occurring proteins all have different properties. The nucleotide sequence of DNA, on the other hand, is relatively monotonous, consisting only of the four bases guanine, cytosine, adenine, and thymine. However, appropriate methods for DNA sequencing are now available [51] (Fig. 3.14).

A long nucleic acid molecule is cut into smaller segments by agents cutting it at specific sites. Then, the nucleotide sequence for each segment is determined separately. The sequence of segments in the entire chain is ascertained by the principle of overlapping ends; identical chains are cut a second time by an agent with different specificity, and the sequences of fragments produced by the two cuttings are compared. In this way the entire sequence can be combined like a puzzle. Nucleotide sequences within these fragments can be identified by the methods of Maxam and Gilbert (see [122]) or – in many cases

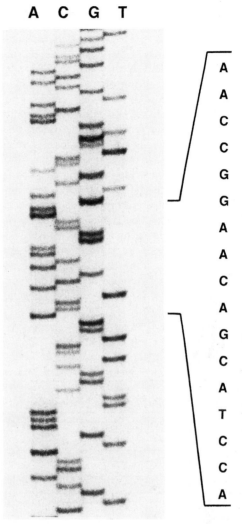

Fig. 3.14. DNA sequencing. Autoradiogram of a chain-terminator sequencing gel using denatured double-stranded plasmid as template. The base-specific ends in each lane are produced by incorporating different chain-terminating dideoxynucleotides. ^{32}P was incorporated during the reaction, and the reaction products were separated by denaturing polyacrylamide gel electrophoresis. Part of the sequence is indicated to the right of the gel. (Courtesy of O. Steinlein)

◁ **Fig. 3.12 a–f.** Comparative genomic hydridization (CGH) of DNA extracted from a small cell lung carcinoma (From Ried et al. 1994) **a** Staining of normal metaphase chromosomes with 4′-6′-diamindino-2-phenylindole. The G-like banding pattern was used for chromosome identification. **b** The same metaphase spread as in **a** after visualization of red fluorescence (rhodamine) specific for the digoxigenin-labeled reference DNA. Note that all chromosomes (except the X chromosomes) are labeled uniformly. **c** The same metaphase spread as in **a, b** after visualization of the green fluorescence (FITC) specific for the biotinylated tumor DNA. Note that the staining intensity varies due to DNA gains and losses in the tumor genome. **d** Display of the fluorescence ratio image of the same metaphase as in **a–c**. The FITC/rhodamine fluorescence intensity quotient is visualized. *Blue,* balance between the FITC and rhodamine fluorescence values; *green,* gain DNA sequences in the tumor genome; *red,* loss of DNA sequences in the tumor gene. **e** Karyotypelike arrangement of the chromosomes of the metaphase spread shown in **d.** Note that the CGH reveals identical patterns on the homologous chromosomes. **f** Mean of the ratio profile calculation of five metaphase spreads of the same small cell lung cancer case as presented in **a–e.** *Three vertical lines* (to the right of chromosome idiograms), different values of the fluorescence ratio between the tumor and the reference DNA: *Left line,* underrepresentation; *central line,* state; *right line,* overrepresentation of DNA sequences in the tumor genome. The ratio profile *(curve)* was computed as a mean value of five metaphase spreads

more conveniently – by that of Sanger et al., in which the base sequence can be read immediately from the electrophoresis gel (Fig. 3.14). Whereas the sequencing of nucleic acids was formerly a formidable job, it has now become relatively easy and very fast. Sequencing machines are being offered commercially. In many instances, two steps are necessary after a sequence has been established: (a) The sequence must be compared with other sequences determined previously to ascertain whether the same or a similar sequence has already been found anywhere in nature. Such homologies may provide important hints as to biological relationships, evolution, and function. Computerized data bases are now available (Appendix 4). (b) The sequence must be examined as to whether it contains a gene that can be transcribed. This is possible only if an open reading frame (ORF) is present, i.e., if no stop codons (see table 3.2) are present. Such an ORF, however, does not verify the presence of a gene; it could, for example, be a pseudogene.

3.1.3.5 Polymerase Chain Reaction: A Method that Has Revolutionized Molecular Biology

All the methods described so far, such as Southern blotting and DNA sequencing, require a certain minimum amount of DNA (approx. 10^5–10^6 molecules [119]. However, frequently the DNA (or any other nucleic acid) to be studied is available in only much smaller quantities. This difficulty in was overcome the past by cloning in a vector and multiplying this vector. An alternative, very fast, and elegant method was introduced a few years ago and has found wide acceptance: the polymerase chain reaction (PCR) [3, 73]. The significant advantage of the PCR method is the tiny starting amount of DNA that allows amplification.

The logic of the reaction is simple. A pair of primers complementary to both strands of a DNA molecule and flanking a target region of interest are used to direct DNA synthesis in opposite and overlapping directions. In each cycle, each of the two strands acts as a template for the generation of two new duplex mole-

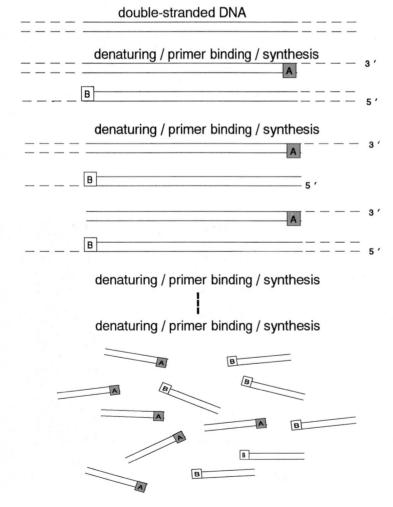

Fig. 3.15. The principle of PCR (polymerase chain reaction). The method requires several steps: DNA denaturing (by increase of temperature), primer binding, DNA replication (by addition of *Taq* polymerase and reducing of temperature), repeated until a very large number of copies of the original DNA have been obtained. *A, B,* The primers on either side of the DNA sequence to be multiplied; *interrupted lines* indicate that DNA replication sometimes continues beyond the DNA sequence to be studied

cules. Theoretically, this leads to the doubling of target sequences in each round of DNA synthesis and to an exponential increase in the amount of DNA. A limitation of this method is that the design of primers requires knowledge of base sequences of short DNA stretches on both sides of the sequence to be studied. Each cycle is initiated by melting double-stranded DNA (opening the H bridges) at $91°–95°C$ for about 1 min to obtain single-stranded templates. The primer oligonucleotides are then annealed to the two strands followed by a short pulse of DNA synthesis. This reaction requires a temperature of $58°–72°C$ and a polymerase for DNA synthesis from constituents in the medium. Usually *Taq* polymerase is used, which comes from the bacterium thermus aquaticus. This bacterium lives in hot springs; therefore its polymerase is heat-adapted and is not destroyed at melting temperature. Hence, it is not necessary to add new polymerase at each cycle. This has made it possible to automate the entire procedure (Fig. 3.16).

PCR machines are on the market and are being used widely. PCR is efficient, specific, and very sensitive. The theoretical number of product molecules is 2^n for *n* cycles, meaning about one million molecules after only 20 cycles. In practice, this number is slightly smaller for technical reasons. Some replication errors might occur, and these must be taken care of. Sensitivity is so high that even one single-copy DNA molecule such as DNA from one human sperm can be amplified and studied [60]. Care needs to be taken not to amplify DNA that originated from the operator or the environment.

The method can also be applied for RNA amplification after reverse transcription into cDNA (RT-PCR). It has found wide application both in research and in genetic diagnostics. Here, assays have been developed by which many DNA sequences can be amplified at the same time. Figure 3.16 shows such an assay developed for the diagnosis of deletions in the dystrophin gene in Duchenne and Becker muscular dystrophies (310 200). In 1992 the Hoffmann-LaRoche Company acquired all patents related to the PCR reaction and requires royalty payments in all instances in which the method is used for commercial purposes. Diagnostic use in medical genetics, microbiology, or in any other field falls under this commercial use category since it does not involve research but is used for clinical service. Such an interpretation becomes "sticky" when large numbers of patients need to be studied in pilot studies validating genetic diagnosis in medical genetics. At what point does research end and medical practice begin? (See also [116]).

The following describes analytical approaches to the study of three genes: one in which the analysis was relatively easy, another in which it was more complicated, and a third in which analysis turned out to be very time consuming. This discussion provides the opportunity to describe briefly additional research concepts and methods. Examples selected are (a) the β-globin gene; (b) the gene for clotting factor VIII, mutations of which cause hemophilia A; and (c) the gene for Huntington disease (HD).

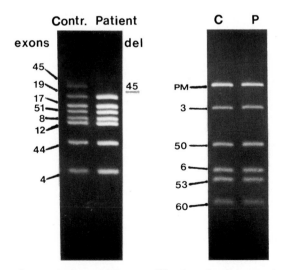

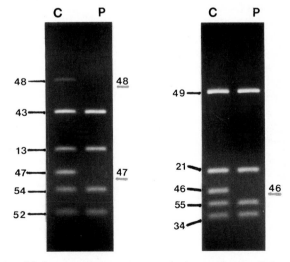

Fig. 3.16. PCR multiplex amplification (in four runs) of 24 exons and the promoter region of the dystrophin gene. *Patient (P)* shows deletion of exons 45–49; *C,* normal control.

Amplified exon sequences are arranged according to their length, not according to their sequence within the gene. (Courtesy of M. Cremer)

Discussion of other aspects of Duchenne and Becker muscular dystrophies:

- Sect. 3.1.3:   List of human mutations analyzed so far
- Sect. 5.1.2:   Linkage and localization
- Sect. 9.3:   Types of mutation and mutation rates
- Sect. 18.1 + App. 8:   Genetic counseling, prenatal diagnosis

3.1.3.6 Analysis of the β-Globin Gene

The Paradigmatic Role of the β-Globin Gene. The first step requires identification of this gene in the large amount of human DNA. This has been done by isolating DNA from human cells and treating it with a specific restriction enzyme. The many resultant DNA fragments are then separated according to their length by agarose gel electrophoresis. These are transferred to a nitrocellulose filter by Southern blotting and denatured. The DNA is now single-stranded and fixed to the filter. The next step is identification of DNA fragments containing the β-globin gene. This requires a radioactive probe made from β-globin mRNA using the enzyme reverse transcriptase (mRNA → cDNA). In this special case it was a big advantage that the mRNA was available. The cDNA probe could then be used for hybridization with the DNA on the filter. Autoradiography shows the positions of the fragments containing globin genes. This technique was also used for localizing the β-gene on a human chromosome to the short arm of chromosome 11 (11 p) by in situ hybridization.

Further characterization of other members of the β-globin gene family requires enrichment of the DNA specifying these genes by cloning in a vector such as a bacterial plasmid (see Sect. 3.1.3.2). The Hb β gene family was analyzed by comparing the DNA sequence of the transcribing region and the cDNA by electron microscopy, by sequencing DNA regions inside and outside transcribed sequences, and by identifying regulatory sequences. One of the first and most conspicuous results of such studies was that hybrid molecules between the β-globin genomic DNA and cDNA showed peculiar loops in electron microscopy [102]. These were caused by DNA regions not represented in the cDNA and obviously not transcribed, since cDNA is a true copy of mRNA. In the Hb β gene two such intervening sequences (introns) were discovered that separated three distinct units (exons) at the DNA level. Meanwhile, studies on many other eukaryotic genes have shown such introns to be the rule rather than the exception, quite in contradistinction to bacteria and viruses, where genes are continuously transcribed.

Exons often represent functional subunits of the gene; they may have developed from separate genes during evolution (Sect. 14.2.3). Hence, analysis of the β-hemoglobin gene and other hemoglobin genes had a result of general significance: demonstration of the exon-intron structure of eukaryotic genes (Fig. 3.18).

These and later studies confirmed inferences from family studies on abnormal hemoglobins (Sect. 4.3.2) that there is only one functional Hb β gene whereas, for example, the genes for the α and γ chains are present in duplicate. In addition pseudogenes have been discovered: DNA regions very similar in DNA sequence to functional genes which are not transcribed because of inactivating mutations in the coding or flanking areas.

Figure 7.33 shows the Hb β region. In addition to the gene itself and a pseudogene, there are two γ genes, one δ gene (for the Hb δ chain found in Ab A₂), and genes for an early embryonic hemoglobin. The molecular analysis confirmed inferences regarding the structure of this gene region from formal genetics and protein biochemistry (Sect. 7.3.2) but provided entirely new information about gene structure and organization.

Additional studies on this and other genes have also shed some light on how transcription occurs, and on how mature mRNA is produced (Figs. 3.17, 3.18). First, the entire gene is transcribed, including introns and flanking sequences proximal and distal to the exon intron gene complex. Then transcripts of introns are removed stepwise, the transcripts of exons are spliced, and a "cap" is added at the 5′ end and a poly-A sequence at the 3′ end. Finally, the processed mRNA leaves the nucleus to proceed to the ribosomes and acts as template for protein biosynthesis. The DNA sequences for the various globin genes are known, and many general problems of gene organization and gene action have been examined using these genes as models.

Discussion of other aspects of hemoglobin genes:

- Sect. 7.3:   Gene action
- Sect. 9.4:   Mechanisms of spontaneous mutation
- Sect. 12.2.1.6:   Population genetics
- Sect. 14.2.3:   Human evolution
- Chap. 18:   Genetic counseling, screening, prenatal diagnosis

3.1.3.7 Structure of the Factor VIII (Antihemophilic Factor) Gene

Antihemophilic Factor (Factor VIII). Hemophilia A is a "classical" hereditary disease with an X-linked recessive mode of inheritance (Sect. 4.1.4). The gene is located at

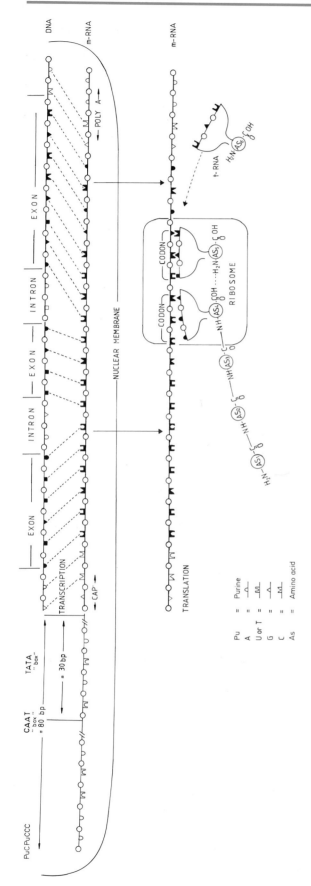

Xq 28. Analysis of the blood clotting process led in the 1950s to identification of a plasma protein – the antihemophilic factor (factor VIII) – as the gene-determined protein lacking in this disease. Factor VIII is necessary for the first step of blood coagulation, thromboplastin formation. Substitution with factor VIII concentrates is now the established therapy of hemophilia A and nowadays enables hemophilia patients to lead a near-normal life. The protein molecule, however, turned out to be large and complex, and its synthetic production appeared impossible. Now this situation has changed dramatically. The structure of the factor VIII gene has been elucidated, and expression of this gene was achieved in tissue culture; therapy with recombinant Factor VIII is therefore possible (see below) [31, 32, 105, 112].

Research Strategy in Elucidating the Factor VIII Gene. This was achieved independently, but with similar strategies, by two private gene technology companies: Genentech in San Francisco and Genetics Institute in Boston. The Genentech group proceeded as follows:

A DNA oligonucleotide probe only 36 bases long was synthesized based on the amino acid sequence of a tryptic peptide from human factor VIII. This very short DNA probe was used to screen a genomic DNA library in bacteriophage λ that had been derived from an individual with karyotype 49, XXXXY. Hence, a higher concentration of X-specific probes was achieved not by chromosome sorting but by using a naturally occurring abnormality. Clones identified in this way by DNA hybridization (Sect. 3.1.3.3) had overlapping ends, allowing initial identification of part of the DNA sequence (Fig. 3.19). This part was then used to place factor VIII mRNA in a T-cell hybridoma cell line. This mRNA was used to produce cDNA for the entire coding part of the gene. This portion (~ 9 kb) was then sequenced. The boundaries of exons were established by comparison of cDNA with genomic DNA. The complete gene consists of 186 000 base pairs; it has 26 exons ranging in length from 69 to 3106 base pairs; introns may be as long as

Fig. 3.17. The DNA sequence which is depicted here as single nucleotide chain is characterized by a specific sequence of bases. At the 5′ side, where transcription begins, two characteristic sequences, CAATT and TATA, were described 80 and 30 base pairs (bp) apart. By analogy with the bacterial genome, it was concluded that the CAATT sequence serves as recognition site for RNA polymerase whereas the TATA sequence serves as promoter region for polymerase-induced transcription. DNA is transcribed into the complementary mRNA sequence – first including the introns. Then, the mRNA is processed stepwise; the introns are removed; a "cap" is added at the 5′ end, and a poly-A sequence at the 3′ end. The processed mRNA then passes through the nuclear membrane and attaches to the ribosomes, where the genetic information is translated into a protein sequence

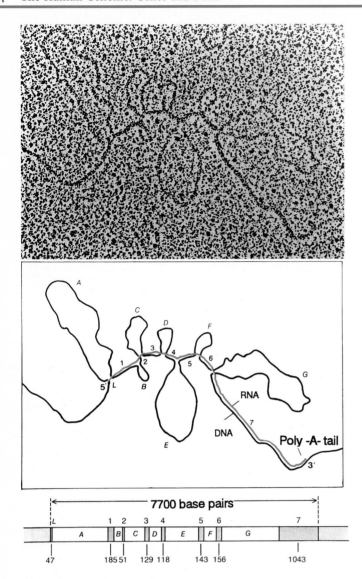

Fig. 3.18. Electron microscopic study of the structure of the ovalbumin gene, a characteristic mammalian gene: DNA containing the ovalbumin gene is hybridized with mRNA. The eight exons of the gene *(1–8)* hybridize with the complementary mRNA regions; its seven introns *(A–G)* form loops. *Above*, electron microscopic image; *middle*, interpretation; *below*, schematic drawing of the gene. (From Chambon 1981)

32.4 kb. The factor VIII protein consists of 2351 amino acids (Fig. 3.20).

To achieve expression of the gene in mammalian tissue, the complete 7-kb protein-coding sequence was assembled from portions of overlapping cDNA and inserted into a plasmid; in this plasmid, factor VIII sequences were placed between promoter sequences and a polyadenylation sequence of viral origin. This plasmid was then introduced into a hamster cell line by a chemical method (calcium-phosphate coprecipitation), and gene expression was measured using monoclonal antibodies against part of the factor VIII protein. This indicated a 300-fold increase over the control cells and comprised as much as 7% of the normal plasma activity, as shown by additional biochemical studies.

The other group (Genetics Institute), working along similar lines, achieved comparable results.

Significance of These Studies. A variety of methods from molecular biology were combined in an ingenious way to analyze an unusually complex gene. This result is significant for several reasons:

1. This was the first gene of such length and complexity to be analyzed in humans and, indeed, in any eukaryote.
2. This structure permitted new insights regarding evolution of this gene [60], including an unexpected homology of approximately 35% of the amino acid sequence with the copper-binding protein ceruloplasmin (see Sect. 7.2.3).
3. The result offers prospects for an improved therapy of hemophilia A (see also Sect. 7.2.2.9). Substitution treatment with factor VIII preparations is one of the success stories of therapy in hereditary diseases: Life expectation of hemophilics has increased

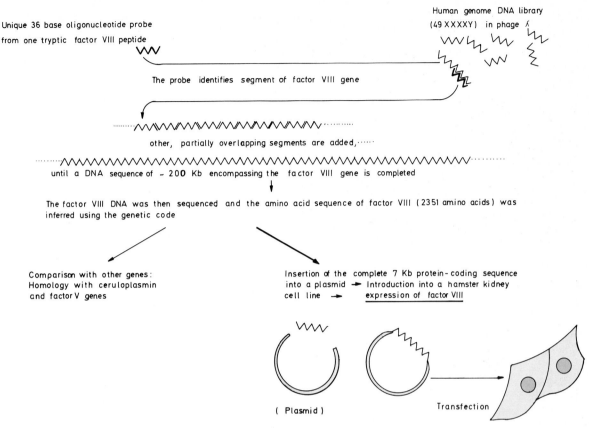

Unique 36 base oligonucleotide probe
from one tryptic factor VIII peptide

Human genome DNA library
(49 XXXXY) in phage λ

The probe identifies segment of factor VIII gene

other, partially overlapping segments are added, ...

until a DNA sequence of ~ 200 Kb encompassing the factor VIII gene is completed

The factor VIII DNA was then sequenced and the amino acid sequence of factor VIII (2351 amino acids) was inferred using the genetic code

Comparison with other genes:
Homology with ceruloplasmin
and factor V genes

Insertion of the complete 7 Kb protein-coding sequence
into a plasmid → Introduction into a hamster kidney
cell line → expression of factor VIII

(Plasmid)

Transfection

Fig. 3.19. Analysis of the human factor VIII gene, starting from a 36-base oligonucleotide probe of a tryptic peptide from factor VIII

dramatically, and many of them are now leading near-normal lives. This therapy, however, was not without some serious flaws. About two-thirds of hemophilia patients treated with factor VIII (usually obtained from very large donor pools) became infected with the HIV virus in the early 1980s. This danger could be circumvented when HIV testing became available in 1984 and HIV-positive blood was excluded. Factor VIII preparations produced by recombinant DNA techniques are now on the market and can entirely avoid such problems.

Discussion of other aspects of hemophilia A:

– Sects. 4.1.4, 5.1.2:   Formal genetics, linkage
– Sect. 7.2:   Gene action
– Sect. 9.4.2:   Molecular types of spontaneous mutation

An Exercise in the Sociology of Science. This analysis was achieved by two very large groups, as shown by the large number of authors of each paper. This effort approaches "big science" and is more customary in certain branches of physics such as high-energy physics. "In the application of molecular biology to practical affairs, it is for the time being clear that bigger battalions move faster" [66]. And these battalions have been set up not by universities or research institutes in basic research, but by private companies with the ultimate purpose of producing marketable products. This link between research and commerce is new to human geneticists, but has become widespread over recent years in molecular biology. Should this development cause serious concern? Is there a danger that commercial interests will influence the development of science too heavily, diverting manpower and resources from scientifically important problems to those promising profits in the near future?

The danger may not be too great. After all, close relationships between basic research and industrial use have been customary not only in applied fields, such as engineering, but also in chemistry for a long time. To the best of our knowledge, it has not impaired the quality of basic research. For biologists, however, the situation is new. It requires careful observation.

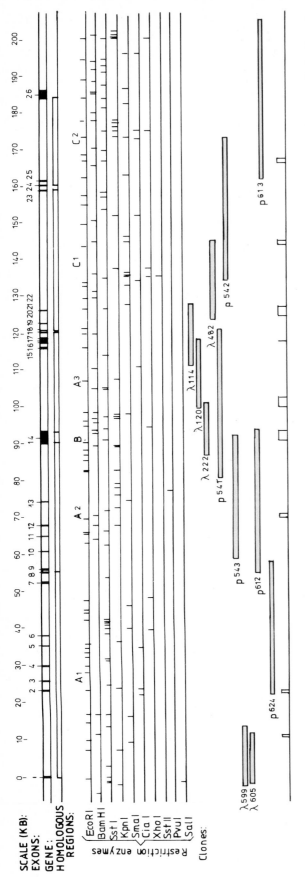

3.1.3.8 The Gene for Huntington Disease (HD)

Characteristics of Huntington disease include onset during adult life, involuntary choreiform movements, and progressive dementia (for more details see Sect. 4.1.2). There is no effective treatment.

Analysis. Of the three genes discussed here, the elucidation of the gene for HD turned out to be most difficult and took a decade's work by many laboratories. The gene was localized as early as 1983 to the short arm of chromosome 4 by linkage with a DNA marker. Most observers, including the investigators, hoped that the specific genetic defect would be known soon thereafter. However, 10 years elapsed before the crucial discovery [65]. Many difficulties contributed to this delay. The major obstacle was that nothing was known about the gene product. For the β-globin gene mRNA from which a cDNA probe could be constructed was available. For clotting factor VIII, at least a partial amino acid sequence was known and could be used to construct an oligonucleotide that served as a starting point of gene cloning. In HD, however, there were no clues whatever.

Analysis was performed using a series of steps also known as "positional cloning" [64] (Sect.3.1.3.9). First, the gene on chromosome 4p was localized more precisely. This required more genetic markers which were generated by a variety of techniques. Using these markers, genetic and physical maps (Chap.5) were constructed. The 4p telomere (on telomeres, see Sect. 3.1.1.1) which contained the HD gene was cloned in a YAC, and YACs as well as cosmid contigs of these regions were established.
This led to the discovery of a number of genes identified by open reading frames (ORFs). Analysis of recombination events in HD kindreds identified a candidate region of 2.2 Mbp located between the markers D4S10 and D4S98 at 4p16.3. (For nomenclature of genes and genetic markers see Appendix 10.)
This saturation with markers permitted to trace the origin of HD mutations by studying haplotypes and linkage disequilibrium (Chap.5). It turned out that multiple mutations had occurred to cause the disease, because the HD mutation was observed with widely varying sets of closely linked markers (haplotypes). This is not surprising, since kindreds included in these studies came from many different contries. However, about one-third of all these mutations may have had a common origin; they were found in the same haplotypes [64]. At the same time, the marker studies, and especially the haplotypes shared by these mutations, indicated

Fig.3.20. The factor VIII gene. *Open bar,* the gene; *filled bars,* the 26 exons. *Lower series of lines,* location of the recognition sites of the ten restriction endonucleases used for identification. *Gray boxes,* the extent of human DNA contained in each λ phage (λ) and cosmid (p) clone. (From Gitschier et al. 1984 [31])

that a 500-kb segment between markers D4S180 and D4S182 was the most likely site of the genetic defect. Hence, the candidate region was reduced by this step to less than one-fourth of the region identified earlier (Fig. 3.21).

To identify all possible gene transcripts in this area the technique of exon amplification was used – an application of the PCR technique (Sect. 3.1.3.5) [13]. By this technique, three genes were discovered: the α-adducin gene (ADDA), a putative transporter gene (IT10C3) in the distal portion of this segment, and in the central portion a gene for a G protein coupled receptor kinase. These genes were sought in HD patients, but no abnormalities were discovered.

The same method was applied to the proximal portion of the 500-kb segment. Here the search proved successful: a large gene spanning some 210 kb was discovered. It was slightly longer than the factor VIII gene described above and encodes a previously unknown protein which was named IT15. The reading frame contains a polymorphic (CAG) trinucleotide repeat; at least 17 alleles that varied from 11 to 34 CAG copies were discovered in the normal population. On HD chromosomes the length of this repeat was found to be increased substantially, with a range from 39 to 66 copies. Moreover, there appeared to be some correlation with the age at onset in that the longest repeats were detected in patients with juvenile onset of HD.

This result led to the conclusion that an increase in repeat length of CAG repeats is the mutation causing the phenotype of HD, and that IT15 is the long-sear-

ched for HD gene. Figures 3.22 and 3.23 show a northern blot analysis of the IT15 mRNA and Fig. 3.24 the cDNA clones defining this transcript. An open reading frame of 9432 bases appears to begin with a potential initiator methionine codon. This led to the prediction of a 348-kDa protein with 3144 amino acids. It is possible that translation starts at a later methionine position, and that the protein is therefore somewhat shorter. The gene product was termed huntingtin.

Table 3.3. Repeat length in HD patients compared with normal controls. (From MacDonald et al. 1993 [65])

Range of allele sizes (number of repeats)	Normal chromosomes (*n*)	HD chromosomes (*n*)
48 or more	0	44
42–47	0	30
30–41	2	0
25–30	2	0
24	169	0
Total	173	74

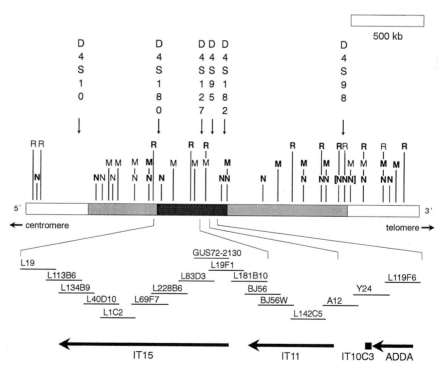

Fig. 3.21. Long-range restriction map of the HD candidate region. The HD candidate region determined by recombination events *(hatched bars between D4S10 and D4S182)* is expanded to show the cosmid contig (averaging 40 kb per contig). *Arrows,* transcriptional orientation (5′ → 3′). Locus names *(above)* denote selected polymorphic markers used in HD families. Restriction sites are given for Not1 *(N)*, MLUL *(M)*, and NRUL *(R)*. *Brackets around the N symbols,* presence of additional clustered Not1 sites. (From MacDonald et al. 1993 [65])

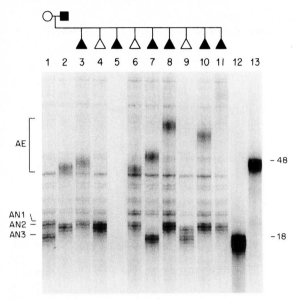

Fig. 3.22. PCR analysis of (CAG) repeats with some offspring displaying juvenile onset of HD. *Triangles,* progeny; birth order has been changed for confidentiality; *AN1, AN2, AN3,* positions of the allelic products from normal chromosomes; *AE,* range of PCR products from the HD chromosomes. The intensity of background constant bands, which represent a useful reference for comparison with the PCR products, varies with slight differences of PCR conditions. The PCR products from cosmids L191 F1 and GUS72-2130 *(lanes 12, 13)* have 18 and 48 CAG repeats, respectively. (From MacDonald et al. 1993 [65])

Fig. 3.23. Northern blot analysis of the IT 15 transcript. Results of the hybridization of IT 15 A to a "northern blot" of RNA from normal *(lane 1)* and HD homozygous *(lanes 2, 3;* lane 3 with more (CAG) copies than lane 2) lymphoblasts. A single RNA of approx. 11 kb was detected in all three samples, with slight variations due to unequal RNA concentrations (From MacDonald et al. 1993. [65])

Near its 5′ end, i.e., a short distance after translation begins, the sequence of the normal gene analyzed in this experiment contained 21 CAG repeats, encoding glutamine. All HD patients showed a much higher number of repeats. Similar gene amplification mutants have been observed in a few other diseases (Sect. 9.4). However, in the fra(X) mutation, for example, the amplified region is not transcribed.

The study had still another interesting result: the $(CAG)_n$ bands were not identical in members of the same families (Fig. 3.22). This means that the repeat number is unstable. Above a certain limit (which must still be defined in greater detail) changes from one generation to the next occur; the repeat number may increase, but a decrease has been observed, as well.

Prevalence and Mutation Rate. It has often been discussed how often new mutations occur in relation to the overall number of patients (see Sect. 9.3). Most investigators agree that the mutation rate is low. The disease has become relatively common (1:10 000) because in populations without birth control and genetic counseling, HD patients have no selective disad-

vantage since they have their offspring before onset of clinical symptoms; – possibly even a slight advantage, the nature of which is unknown. Occasionally, however, experienced medical geneticists have noted HD patients who seem to be the first ones in their families, and where a new mutation could not be excluded. The authors of the paper discussed here have studied this problem in two pedigrees in which a new mutation would have been possible. In both instances transmission was observed; the patients were not new mutants.

On the other hand, several new mutants have been observed as a result of a systematic search. An "intermediate" allele with 30–38 CAG repeats in phenotypically normal individuals appears to enhance the risk factor for such a new mutation, which has been observed so far only in the male germ line; this risk increases with advancing paternal age [33]. (For paternal age effects in new mutations, see Sect. 9.3.3).

Practical Questions. The data on the HD gene raise a number of practical questions regarding genetic counseling and prenatal diagnosis and their ethical consequences. These problems are discussed in Chap. 18.

Sociological observations. There are similarities and differences in the efforts that led to analysis of this gene compared with those of the factor VIII gene (Sect. 3.1.3.7) from the point of view of the sociology of science. Similarly to the factor VIII gene, the gene

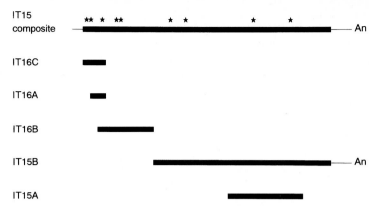

Fig.3.24. Schematic presentation of cDNA clones defining the IT 15 transcript. Five cDNAs are represented under a schema of the composite IT 15 sequence. *Thin line,* untranslated regions; *thick line,* coding sequence, *stars,* the positions of the exon clones 5′ to 3′. From 22 bases 3′ to the putative initiator Met ATG, the sequence was compiled from the cDNA clones and exons shown. There are nine bases of sequence intervening between the 3′ end of IT 16 B and the 5′ end of IT 15 B. These were identified by PCR amplification of first-strand cDNA and sequencing of the PCR product. At the 5′ end of the composite sequence the cDNA clone IT 16 C terminates 27 bases upstream of the $(CAG)_n$. However, when IT 16 C was identified, the authors had already generated genomic sequence surrounding the $(CAG)_n$. This sequence matched the IT 16 C sequence and extended it 337 bases upstram [65].

for HD was discovered as a result of the common effort of numerous groups. The publication reporting discovery of the HD gene has 60 authors, even many more than listed on the titles of the key reports on the hemophilia gene. Moreover, other groups not included in the HD Collaborative Research Group contributed to the outcome in one way or the other. However, in contrast to the factor VIII group, the HD team included scientists from universities and other nonprofit research institutions; it was not an activity by profit-oriented private companies. Probably one of the reasons is that these studies did not hold the promise of a marketable product such as factor VIII. It is comforting that scientists working in academic institutions, who are normally regarded as highly individualistic, are able to organize extensive collaboration when such cooperation promises better and faster results.

Comparison of the Three Analyses. Fig.3.25 depicts the steps necessary for the analysis of the three genes described. Much more work was necessary to identify the HD gene than the two others. The decisive difference was presence or absence of previous knowledge regarding gene action at the level of mRNA, or at least at the level of the protein from which the mRNA sequence could be inferred. Where such information is not available, analysis is much more cumbersome and time-consuming but – as a side product – allows the elucidation of many more genes whose products are unknown.

The key to a deeper analysis was provided by positional cloning or, to use an older expression, by reverse genetics. If a gene of unknown mode of action can be localized, identified, and its DNA sequenced, the mRNA and protein sequence can be deduced, as shown for the HD gene. Once this information is known, the defect that alters the normal gene sequence to produce a hereditary disease can be determined, and gene function can be studied. Until positional cloning became possible, scientists proceeded from the phenotype to detect biochemical differences between normal and mutant individuals, and were able to detect the mutant protein in some cases. Inferring gene structure from the protein defect was therefore sometimes possible. In positional cloning, however, the path is reversed. First the gene is identified at the DNA level; this knowledge is the first step for an analysis "from the bottom up." This approach is demonstrated using an example in which this methodology was followed for the first time: chronic granulomatous disease (CGD; 306 400) [19, 86]. The basic principles of positional cloning can be widely applied so that an increasing number of basic processes in normal physiology can be approached in this manner. The great power of this and related methods and concepts explains why molecular genetics has become a key science for better understanding of medicine.

3.1.3.9 Positional Cloning

Chronic Granulomatous Disease. Patients with CGD suffer from repeated and severe bacterial and fungal infections. Often the infective organisms persist in phagosomal vacuoles of neutrophil granulocytes and macrophages. These foci are encapsulated and form

a. Hbβ gene

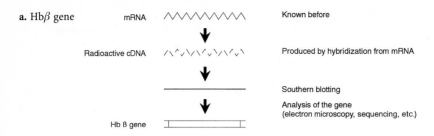

mRNA		Known before
Radioactive cDNA		Produced by hybridization from mRNA
		Southern blotting
		Analysis of the gene (electron microscopy, sequencing, etc.)
Hb ß gene		

b. Factor VIII gene

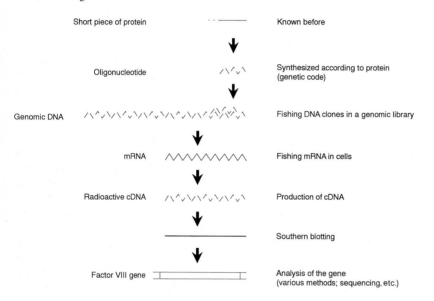

Short piece of protein		Known before
Oligonucleotide		Synthesized according to protein (genetic code)
Genomic DNA		Fishing DNA clones in a genomic library
mRNA		Fishing mRNA in cells
Radioactive cDNA		Production of cDNA
		Southern blotting
Factor VIII gene		Analysis of the gene (various methods; sequencing, etc.)

c. Huntington disease gene

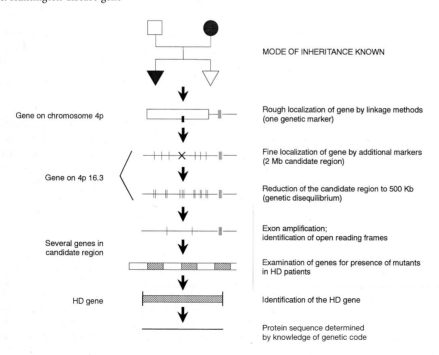

		MODE OF INHERITANCE KNOWN
Gene on chromosome 4p		Rough localization of gene by linkage methods (one genetic marker)
Gene on 4p 16.3		Fine localization of gene by additional markers (2 Mb candidate region)
		Reduction of the candidate region to 500 Kb (genetic disequilibrium)
Several genes in candidate region		Exon amplification; identification of open reading frames
		Examination of genes for presence of mutants in HD patients
HD gene		Identification of the HD gene
		Protein sequence determined by knowledge of genetic code

granulomas. Treatment is difficult, requiring continuous therapy with antibiotics. There is genetic heterogeneity; the mode of inheritance may be X-linked recessive, but several autosomal recessive types are known. Normally, after infecting a person bacteria are incorporated into neutrophil leukocytes. This stimulates these cells to produce O_2 radicals (O_2^- and H_2O_2) which contribute to bacterial destruction. Such an oxidative burst is initiated by activation of the enzyme NADPH. The following reaction is triggered by an oxidase which activates O_2 to form O_2^-:

$$NADPH + 2\,O_2 \rightarrow NADPH^+ + 2\,O_2^- + H^+.$$

In a next step, H_2O_2 is produced. The oxidase consists of many components, only some of which had been analyzed biochemically. In patients with CDG the "oxidative burst" is lacking, but the reasons for the defects were unknown.

The difficulties were overcome by a genetic strategy – the isolation and analysis of the X-linked gene for CGD. In a patient who suffered from CGD and, in addition, from Duchenne's muscular dystrophy and the so-called MacLeod phenotype, a small deletion in Xp21 that comprises these genes was identified [26]. The deletion detected in this patient as well as similar deletions in other CGD patients localized the X chromosomal site of the CGD gene. In comparisons of mRNA between normal and affected cells, normal cells stimulated to produce an oxidative burst were expected to produce a specific mRNA which was lacking in CGD cells. Such an mRNA was found and was used to construct a cDNA. The transcript was sequenced, and an open reading frame was identified. The DNA aberrations in CGD patients, including an intragenic deletion, confirmed the identity of this gene. The mother of this patient showed the expected heterozygote pattern: one normal and one abnormal CGD gene.

With knowledge of the gene and its mRNA, biochemical analysis of gene action became possible; this is the decisive step in "reverse genetics." An important component of the oxidase which triggers the oxidating burst is a cytochrome $^b 558$. It consists of two subunits – a gp91 glycoprotein and a p22 protein (Fig. 3.26). The gp91 glycoprotein is determined by the X-linked gene identified in this study and is deficient in CGD patients who, in most cases, have point mutations in this gene. The p22 and, later, the p47 and p67 proteins were protein later shown to be deficient in autosomal recessive types of this disease.

In recent years, the same principle – analysis of a biochemical pathway "bottom-up" – has often been followed. Elucidation of the gene for cystic fibrosis (CF), the most common autosomal-recessive disease in northwestern European populations, offers a key example. Here the specific mechanism of gene action had eluded all attempts at elucidation until the gene was identified. Knowledge of the CF gene and its mutations clarified the nature of the normal gene affecting membrane function and is setting the stage for the study of the intermediate steps between genotype and phenotype to explain the detailed pathophysiology of the disease. In HD, the function of the huntingtin gene product, and how expanded CAG repeats cause the characteristic symptoms, will hopefully be known soon.

A Classical Example of Positional Cloning: The Cystic Fibrosis Gene. One of the best examples for positional cloning – and at the same time, potentially one of the most useful discoveries medically – is elucidation of the gene that is mutant in CF (219700) [48, 84, 85]. This is a severe, autosomal recessive disease characterized mainly by increased stickiness of secretions from exocrine glands and an increase in their NaCl content. This may lead to meconium ileus immediately after birth, and during infancy and childhood to loss of function of the exocrine pancreas as well as multiple infections of bronchi and lungs. The pancreatic dysfunction can be treated with the missing enzyme, but the lung infections require careful pulmonary toilet and frequent antibiotics. These infections are difficult to treat and often lead to death during the first decades of life. However, improved attention to the details of therapy now gives a mean life expectancy of 28 years in the United States.

No prior information, either on mRNA or on protein structure, was available. Linkage that localized the gene to the long arm of chromosome 7 and the region was further narrowed by DNA markers. The search for the gene used "chromosome walking" (Sect. 3.1.3.2) and another technique called "chromosome jumping." The gene turned out to be about

---

◁ **Fig. 3.25 a–c.** Analysis of three characteristic human genes. **a** The gene of the Hb β chain. The mRNA was known before, and therefore localization and analysis of the gene (at 11p) required only a few steps. **b** The gene determining blood clotting factor VIII. A short amino acid sequence of the gene-determined protein was known. Knowledge of the genetic code permitted synthesis of a specific oligonucleotide which helped

in precise localization and analysis of the factor VIII gene on the X chromosome (in Xq28). **c** Huntington disease gene. Only the autosomal-dominant mode of inheritance was known. Linkage studies using DNA markers have led to localization of this gene to 4p. However, almost 10 years and many analytical steps were required to identify the gene, its structure, and the gene-determined protein by positional cloning

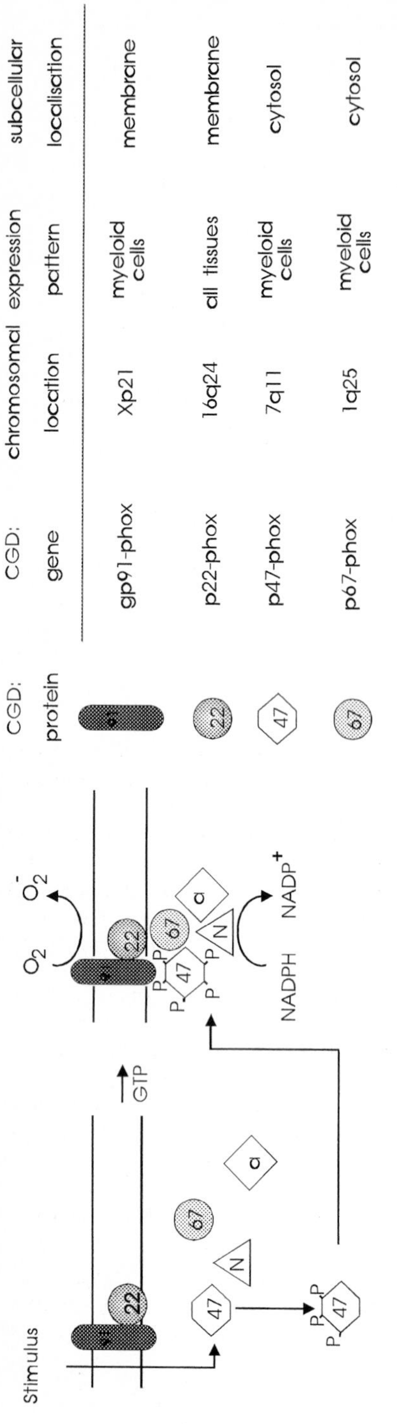

CGD: gene	chromosomal location	expression pattern	subcellular localisation
gp91-phox	Xp21	myeloid cells	membrane
p22-phox	16q24	all tissues	membrane
p47-phox	7q11	myeloid cells	cytosol
p67-phox	1q25	myeloid cells	cytosol

CGD: protein

◁ **Fig. 3.26.** Components and activation of phagocyte NADPH oxidase and genes encoding these proteins. Granulocytes from patients with CGD lack functional NADPH oxidase and therefore the superoxide producing system. In unstimulated cells, the gene products of gp 91-phox (91) and p22-phox (22) are present in the cell membrane, and form the cytochrome b_{558}. Present in the cytosol as free components are: p 47-phox (47), p 67-phox (67), the NADPH binding component N and a cytosolic p 21^{rac}. Stimulation of the cells causes the cytosolic components to associate with the membrane components, possibly by phosphorylation (P) of p 47-phox, indicated by ^{32}-P labeled p 47. The interaction leads to formation of the catalytically active enzyme complex. Also listed are the genes encoding these components and their chromosomal localization, cellular expression pattern and subcellular localization. Mutation of these genes produce the different forms of CGD (One X linked and three different autosomal types). (Modified from Smith and Curnutte 1991)

tion as to its function [84]. There were similarities with several, membrane-associated proteins involved in the transport of ions through membranes. This information together with knowledge of the amino acid sequence permitted researchers to construct a model that shows the way in which different domains of this protein are arranged in relation to the membrane (Fig. 3.27). This model agreed excellently with what was already known about the genetic defect in this disease: chloride channels are defective, leading to a higher NaCl concentration in the sweat of patients. Gene and protein were termed "cystic fibrosis transmembrane conductance regulator (CFTR)." Further studies led to the discovery of many additional mutations [111] that are mostly single nucleotide substitutions. About 300 CF mutations are known, among which the Δ 508 mutation is found in 70 % of central or northwest Europeans.

This opened a rich field for exploring the precise functional consequences of these mutants and relating them to the well-known variation in severity and course of the disease. To mention only one example, a recent study has shown that in the protein altered by the Δ 508 mutation chloride channel activity is not altered, but the protein appears unable to reach its predetermined place in the membrane. Hence, a possible therapeutic approach may be the restoration of normal membrane conductance by "escorting the mutant protein to the cell surface" [59].

250 kb long, having 27 exons. The cDNA had 4560 nucleotide pairs coding for 1480 amino acids. About 70 % of CF mutations show a deletion of three nucleotides coding for one phenylalanine at codon position 508 (Δ 508 mutation).

Comparison of the inferred amino acid sequence of the CF protein with known sequences gave informa-

Discussion of other aspects of cystic fibrosis:	
– Sect. 12.1.3:	Types of mutation
– Sect. 12.1.3:	Population genetics of hereditary diseases
– Sect. 14.3:	Human evolution
– Sect. 18.1, 19.2:	Genetic counseling, prenatal diagnosis, somatic gene therapy

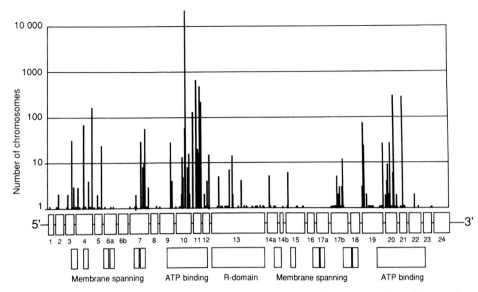

Fig. 3.27. Exon-intron structure, functional compartmentalization, and number of mutants leading in the homozygous state (mostly in compound heterozygotes) to CF, according to their relative frequencies. (From Tsui et al. 1992 [126])

3.1.3.10 Gene Families

Examples of Gene Families. The globin regions, α- and β-globin genes together, provide a good example of a gene family, i.e., a group of functionally related genes having similar structure and presumably a common origin in evolution (see Sect. 7.3.2). Other gene families include the genes for ribosomal RNA and the myosin and actin genes. Probably the largest gene family in humans and other mammals is that comprising genes for various cell recognition molecules as well as immunoglobulins – genes for the major histocompatibility complex (MHC) and for the T-cell receptor and others (Sect. 5.2.5; 7.4). There appears to be no general rule for the location of gene families on chromosomes. Some are clustered in the same chromosome regions, showing close linkage with or without linkage disequilibrium (Sect. 5.2.4; 5.2.5). The hemoglobin family forms two clusters, the Hb α cluster on chromosome 16 and the Hb β cluster on chromosome 11. Other gene families such as the muscle protein genes are scattered over many different chromosomes.

Genes for Actin and Myosin. It is the biological function of muscle to perform mechanical work by contracting. This process, which involves transforming chemical energy into mechanical energy, has been made possible by nature by creating extremely long, multinucleated cells, a large portion of which are occupied by contractile elements – the myofibrils – arranged as parallel bundles in the axis of contraction. Mechanical labor is performed by interaction of two kinds of protein molecules – myosins and actins. In addition to muscle contracture, actins are involved in many other cell func-tions, such as maintenance of cytoskeletal structure, cell motion, and mitosis.

The genes determining both types of proteins – actin and myosin – have been examined [36, 97]. The number of actin genes has been estimated at about 20. They have been preserved well through evolution. In addition to mammals and *Drosophila*, they have also been found, for example, in yeast and slime molds. α-Actins from humans, rabbits, and rat skeletal muscle have been found to be identical; untranslated regions are only partially identical. The split between skeletal muscle and heart muscle actin genes must have occurred a long time before the split between these various species of mammals [37].

Similarly to actins, myosins exist in humans as multiple isozymes. These isozymes appear in a distinctive order during individual development. Studies on DNA have again shown a multigene family which consists of many – probably more than ten – genes scattered over the genome.

A New Principle in Genetic Analysis. These observations on muscle protein gene families introduce a new principle into genetic analysis. Until recently, genetic analysis always started with genetic variability. Such variability might be identified at the phenotypic level, for example by the presence of a hereditary disease, or at some intermediary levels by lack of a functional protein, an electrophoretic protein variant, or a different antigenic site at the cell surface. This phenotypic variability was then traced by classical Mendelian segregation analysis to a corresponding variability at the gene level. Basic mechanisms of gene action at all levels could often be elucidated using genetic variants as tools for research. For the actin and myosin gene families, however, no normal or pathological genetic variants were known. Genetic analysis started at the level of the proteins and the genes themselves, irrespective of any interindividual variation. This was possible because mRNA was available in relatively large amounts. Meanwhile a gene-determined anomaly of myosin has been shown to cause a form of dominantly inherited cardiomyopathy (Sect. 7.6.7). Examination of the human gene map shows an increasing number of genes that have been identified and localized in the absence of any known genetic variability within the human species. Table 3.4 lists a number of characteristic human genes.

3.1.3.11 Genetic Variability Outside Coding Genes

In a classical analysis in 1978 Kan and Dozy [47] detected a DNA polymorphism closely linked to the β-globin gene which permitted prenatal diagnosis of sickle cell anemia in many cases following family studies. Since then a great many DNA polymorphisms have been detected (see Sect. 12.1.2).
Genetic variability at the DNA level – especially outside transcribed DNA regions – is much more common than anticipated on the basis of protein data (Sect. 12.1.2). Therefore its analysis contributes to population history. It is also important for the genetic theory of evolution, for example, with respect to the much-discussed question of the degree to which genetic differences between species and between population groups within a species are due to natural selection or random genetic drift (neutral hypothesis; see Sect. 14.2.3). Moreover, analysis of restriction polymorphisms casts new light on the molecular mechanisms of mutation (Sect. 9.4). Data indicating that

DNA polymorphisms are rarer in the X chromosome than in autosomes [15] extend Ohno's [76] observation that the X chromosome has been more strictly preserved during evolution. It is possible that functional constraints regarding the structure of the X apply not only to coding genes but to the entire genetic material.

Most important, however, is the contribution of such data for a detailed analysis of the arrangement and order of genes in the human genome. This is discussed in Chap. 5.

3.2 The Dynamic Genome

The research tools of new genetics have increased our knowledge of the structure of the genetic material. The simplistic model of a chromosome as a string of pearls – with the genes as pearls – fits the real data less than ever before. One aspect of our increased knowledge is the realization that the genetic material appears much less static than it did in the past. Some brief remarks on the dynamics of the genome are therefore necessary.

Movable Elements and Transposons. Maize (Indian corn) is distinguished by beautifully mottled corn cobs (Fig. 3.28). The genetics of this phenomenon was examined by McClintock [68]. She showed that "controlling elements" occur which can be transposed in the genome from one location to the other, and may cause increased instability of genes leading to somatic mutations in the tissue of this plant. The beautifully

Table 3.4. Some human genes, mutations of which may cause hereditary diseases, ordered by size

Gene or gene product	Length (kb)	No. of exons	Located at	Length of cDNA (kb)	Hereditary disease
Dystrophin	~ 2400	79	Xp	~ 16	Duchenne's and Becker's muscular dystrophies
Thyroglobulin	> 300	36	8q	~ 8.7	Dominantly inherited goiter
NF 1	~ 300	51	17q	~ 8.5	Neurofibromatosis type 1
CFTR	~ 250	27	7q	~ 6	Cystic fibrosis
Huntingtin	~ 210		4p	~ 10	Huntington disease
Rb 1	~ 190	27	13q	~ 4	Retinoblastoma
Factor VIII	~ 186	26	Xq	~ 9	Hemophilia A
vWF gene	~ 178	52	12p	~ 8.5	von Willebrand disease
Phenylalanine hydroxylase	~ 90	13	12q		Phenylketonuria
LDL receptor	~ 45	18	19p	~ 5.5	Familial hypercholesterolemia
HGPRT	~ 44	9	Xq		Lesch-Nyhan syndrome
Factor IX	~ 34	7	Xq	~ 2.8	Hemophilia B
Glucokinase	> 25	12	7p		Maturity-onset diabetes of the young
β-Myosin heavy chain	~ 23	38	14q	~ 6	Hypertrophic cardiomyopathy
A-1-antitrypsin	~ 10.2	5	14q		Chronic obstructive lung disease
Hemoglobin β-chain	~ 1.5	3	11p	~ 0.5	Hemoglobinopathies, thalassemias

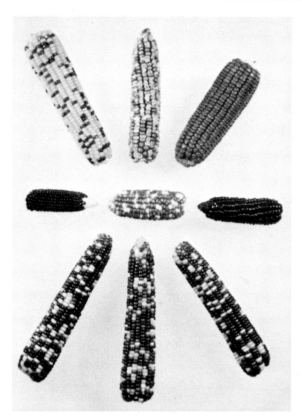

Fig. 3.28. Hybrid maize, different varieties. Note the mottled appearance of the cobs. (From Singleton, Elementary Genetics Princeton etc: Van' Nostrand 1962)

variegated pattern of maize is caused by such somatic mutations. Their specific properties have been analyzed in an extended series of elegant studies. For a long time, these controlling elements seemed to be a unique exception, until Taylor in 1963 [100] described "phage-induced mutations in *E. coli*"; this phage is now called Mu (mutator). Shortly afterwards Starlinger and Saedler (see [98]) described the Is elements (insertion sequences) in bacteria.

Movable Elements in Bacteria. Transposable elements are now defined as specific DNA segments than can repeatedly insert into a few or many sites in a genome [28, 98]. Three classes of such elements are distinguished in prokaryotes:

1. Is elements (simple insertion sequences) contain no known genes unrelated to insertion function. They are generally shorter than 2 kb.
2. Tn elements (transposons) behave formally as Is elements but are generally larger than 2 kb and contain additional genes unrelated to insertion function. They often contain two copies of an Is element.
3. Episomes are complex, self-replicating elements, often containing Is and Tn elements.

Such movable elements have been shown by DNA sequencing and other techniques to share the following properties:

Is sequences terminate in perfect or nearly perfect inverted repeats of 20–40 bp; most Tn elements terminate in long (800–1500 bp) Is-like sequences.

Movable elements when integrated into the host genome are flanked by short duplications (4–12 bp) of host DNA. Elements generating host repeats of the same size share more or less pronounced homologies, which indicate evolutionary relationships.

Movable elements can be inserted at multiple sites of the host genome by nonhomologous recombination; sometimes they insert at specific sites. This insertion has often been shown to involve a copy of the element, whereas the original element stayed at the donor site (Fig. 3.29).

If a transposable element inserts into a structural gene, gene mutations result as a consequence of gene splitting. Moreover, chromosome aberrations such as deletions, duplications, inversions, and translocations may be induced by inserted transposable elements.

Transposable Elements in Eukaryotes. The controlling elements in maize were the first transposable elements detected. Later, such elements became known in other eukaryotes, such as in *Drosophila* [34]. In this organism, insertion mutants are produced at high frequency at specific, preferred loci. Three sources of such insertions exist: (a) They may come from gene sequences dispersed over the genome. Elements of these gene sequences are known as transposons. (b) As a second possible source of transposing DNA, the repetitive sequences of constitutive, centromeric heterochromatin have been discussed. (c) RNA viruses known to be harbored by *Drosophila* have been implicated as possible sources; the enzyme reverse transcriptase may transcribe viral RNA into DNA that is then inserted. These elements are called retroposons.

In addition to important aspects of their structure, such as presence of inverse repeat units at their ends, transposable elements in *Drosophila* share some properties with those described in bacteria, mainly the ability to induce gene mutations at an unusually high rate, the instability and high reversion rate of these mutations, the independence of normal DNA replication, and the frequent induction of chromosomal aberrations. Transposable elements have also been described in yeast [28].

Significance of Movable Elements in Evolution? There have been speculations that gene transfer by movable elements may have played a role in evolution. Possibilities for genetic change would indeed be increased enormously if, in addition to the classical means of

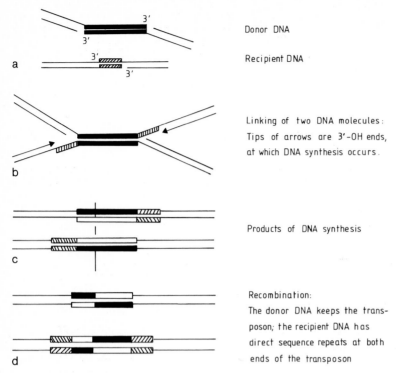

Donor DNA

Recipient DNA

Linking of two DNA molecules:
Tips of arrows are 3'-OH ends,
at which DNA synthesis occurs.

Products of DNA synthesis

Recombination:
The donor DNA keeps the transposon; the recipient DNA has
direct sequence repeats at both
ends of the transposon

Fig. 3.29 a–d. A model in which transposition is explained using known principles of molecular genetics. **a** At the two ends of the transposon (or the Is elements) of the upper DNA molecule one of the two DNA strands is cut by a restriction enzyme. **b** In the same way the recipient DNA strands are opened at opposite ends. **c** Then DNA replication leads to doubling of the transposon as well as the flanking sequences of the recipient. **d** Finally, recombination takes place. The donor DNA retains the transposon; the recipient DNA has the transposon as well as the directly repeated flanking sequences. (From Shapiro 1979 Proc. Natl. Acad. Sc. U.S.A. 76, 1933)

transmission of genetic material from parent to offspring, "transverse" transmission often occurred – possibly even between quite different species. It should be remembered that, for example, transduction of genes from one bacterium to the other by phages has been known for many years and is now being used in eukaryotes, including mammalian cells, for gene transfer and genetic engineering (Sect. 19.2). It is likely that such processes also occur in nature. Moreover, a sequence homologous to the human hemoglobin gene has been discovered in leguminous plants [9]. Its function may be "to ensure an adequate oxygen supply for bacteroid respiration in module tissue." The presence of this gene could possibly be explained by transfer from an insect or a mammal.

Movable Elements in the Human Genome? So far only very few movable elements have been shown convincingly to occur in the human genome. However, as with the genome of *Drosophila*, the human genome also contains interspersed and centromeric repeated DNA segments (Sect. 3.1.1.1) – sometimes even with palindromic sequences – that could be good candi-dates for transposing elements by analogy. For example, the oncogenes have structural homologies with cellular RNA viruses (retroviruses; Sect. 10.4.2); retroviruslike repetitive elements have been identified in human DNA [67]; and DNA viruses have been shown to be mutagenic in mammalian cells [29]. Two de novo insertions into the coagulation factor VIII gene apparently have been the mutations causing hemophilia A. A special group of dispersed repetitive sequences has been discovered in the human genome – the Alu sequences [91].

It has been mentioned before that much of the human DNA is organized according to the *Xenopus* pattern, i.e., single-copy sequences approx. 1–2 kb in length, interspersed with repeated sequences about 0.1–0.3 kb long. We also mentioned that some of these sequences are palindromic, containing repeated sequences in inverse order (Sect. 3.1.1.1). Whereas in the *Xenopus* genome these repeat sequences are part of many different sequence families, much of this material shows strong homologies in mammals such as in rodents or primates [91]. In humans, approx. 3%–6% of the total DNA consists of such 300-bp repetitive sequences, and some 60% of this material has

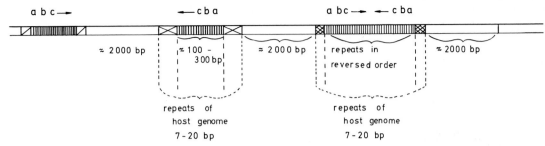

Fig. 3.30. Short-period interspersion pattern with Alu elements. Repeated sequences of about 100–300 bp in length alternate with single-copy sequences of approx. 2000 bp. Repeats may occur in two directions (abc → ; ← cba), or two repeated sequences in reversed order may immediately follow each other. Alu elements are flanked by short direct repeats of the "host" genome (7–20 bp) that are different from one "insertion" site to the next. About one-third of all 100–300 bp repeats are members of the Alu family

been shown by restriction enzyme analysis to be homologous. The number of such Alu sequences is now estimated at about 500 000 for the haploid genome. They can occur in straight, in reversed, or in palindromic order (Fig. 3.30). On both sides they are flanked by direct repeats ranging in length from 7 to 20 bp. Unlike the Alu sequences themselves, these repeats are unique to each different Alu sequence. Such repeats have also been found to flank bacterial transposons, as well as movable elements in eukaryotic DNA. Therefore it was concluded that Alu sequences originated as movable elements, and that the flanking repeats may have resulted from duplication of the DNA sequence at the target site of Alu insertion.

Alu sequences are often found within primary RNA transcripts; they are often but not always removed by RNA processing (Sect. 3.1.3.6). It has been speculated that they might be distributed over the genome by their relatively short mRNA transcript, which might then be transcribed to DNA by reverse transcriptase, and inserted at various places. Since they have been conserved during mammalian evolution, as evidenced by their partial homology between primates (including man) and rodents, they may have an important function. By analogy with similar elements in other eukaryotes, such as *Zea mays* and *Drosophila,* they may be involved in gene expression, mutation (germ line and somatic), or recombination in germ and somatic cells.

A first step to transposition of certain DNA sequences could be the formation of extrachromosomal, circular DNA repeats, which has been observed for sequences normally localized between Alu repeat clusters in aging human fibroblasts in vitro ([94] see also [43]). As noted above, the Alu sequences are part of the group of SINE sequences which are found mainly in the G light bands of chromosomes. The SINE category comprises, in addition, RNA pseudogenes and other components. The LINE sequences, on the other hand, are found mainly in the G dark bands. They

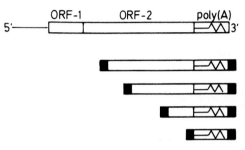

Fig. 3.31. LINE sequences in the mammalian genome. *Above,* the prototype of a LINE sequence, which is about 6000 bp long and has two open reading frames *(ORF1, ORF2)*. The 3′ end has a poly(A) tail, typical for a nonviral retrotransposon. However, the great majority of LINE sequences deviate from the prototype; some sequences at the 5′ end are lacking. The poly(A) tail, as well as sequence repeats at each end *(black blocks)* are always present. (From Weiner et al. 1986 [91])

are much longer, up to 6000 bp per unit or more. Their number is estimated to be 20 000–50 000. They are believed to have been acquired relatively recently during evolution (Fig. 3.31). The function of all these elements is unknown.

Gene Conversion. Another phenomenon observed in experimental genetics is *gene conversion* [58]. Various data from the hemoglobin gene suggest that this event also occurs fairly frequently in the human genome (Sect. 7.3.5; see also Fig. 3.32).

Gene conversion is the modification of one of two alleles by the other, which would alter, for example, a heterozygote Aa into a homozygote AA. Winkler, who over 50 years ago was the first to discuss this concept, proposed a "physiological interaction" between alleles. Studies in yeast have shown that atypical recombination is involved. The process is shown in Fig. 3.32. Crossing over always involves destruction of DNA sequences around the crossover site. Normally, the destroyed sequences are repaired using the se-

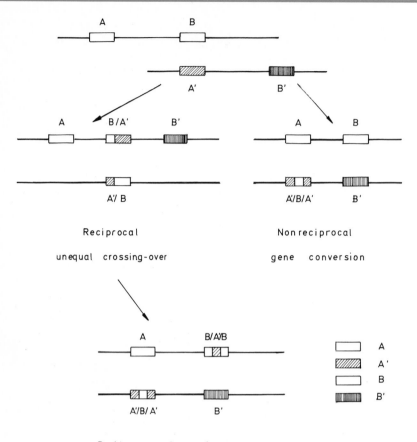

Fig. 3.32. Gene conversion vs. double unequal crossing over. A pair of homologous genes A, A^1 as well as B, B^1 are arranged in tandem. Because of homology, an "illegitimate" alignment between B and A^1 has occurred. Recombination within the gene produces a triplicated gene (A, B/A^1, B) on one strand and a deletion product (A^1/B) on the other strand. In a future generation a person who is a double heterozygote for the triplicated and single gene (as shown) might undergo an additional unequal crossover creating the gene products as shown. In gene conversion there is direct copying of part of the gene B into the middle of gene A^1 residing on the other DNA strand. The event is nonreciprocal, and the strand carrying A and B remains unchanged. The haplotype created by conversion is identical to one produced by a double crossover as shown. Since neither recombination product can be ascertained in humans, gene conversion would not be distinguished from such a double crossover. Statistically, however, a single conversion event is much more likely to occur than an event requiring two crossovers.

quences of the sister chromatid as template. In this way the original double helix is restored. Sometimes, however, the double-strand structure is repaired using the strand of the homologous chromosome as template. In this case, an abnormal segregation ratio is observed.

Gene conversion has also been observed in somatic tissue, especially in plants. Similar, deviant recombination processes are probably involved. This is not surprising since somatic pairing of homologous chromosomes and somatic crossing over has been observed in many species (see, for example [35]). There is some evidence that gene conversion occurred in the HLA region, in the hemoglobin variants genes, and in the color vision gene complex (Sect. 7.3.5).

Does the Genome Fluctuate? How Constant Is the Genetic Information and Its Transmission? When we observe inheritance of a monogenic disease or variants of a genetic polymorphism, such as the ABO or MN blood groups (Sect. 4.1.1), we are impressed by the regularity of genetic transmission indicating constancy of the genetic material over many generations. This regularity has been confirmed over and over again; we are justified in concluding that the few apparent exceptions are accounted for by other than biological explanations, such as false paternity. Our confidence in the all-pervading dependability of heredity is somewhat attenuated by the fact that new mutations are observed occasionally (Chap. 9), but mutation rates are usually very low. Moreover, mutations, once they have occurred, obey the rules of genetic transmission.

Recent new findings in molecular biology, however, raise problems: minisatellites may have enormous mutation rates, genes are cut into pieces, jump around in the genome, and convert their alleles to their own structure. They may even be introduced into our genome by a vagrant virus and not by the time-honored and pleasant method practiced by our parents and forebears. Must we forget our elementary genetics and be prepared to distrust all rules? Fortunately, the old rules can still be trusted in most instances. As Arber said in his Nobel lecture on genetic exchange: "In spite of possessing a multitude of natural mechanisms to promote exchange between genetic materials of unrelated origin, *E. coli* and other high organisms have succeeded in achieving a relatively high overall stability in their genetic makeup." The new results deepen our understanding of how the genetic material is structured, and how it works. Conceivably, they may even help to prevent genetic disease. But the old rules are still valid in the overwhelming majority of situations.

3.3 Attempts at Understanding Additional Aspects of Chromatin Function

We have tried to describe important elements necessary for carrying and transmitting genetic information: genes that are integrated into chromosomes and other elements of chromosomes that are necessary for their function, as well as repetitive DNA structures of unknown function, if there is any function for them at all. We should remember that evolution does not always find the optimum solution to a problem. Evolution is the result of an interplay among random processes. Often this leads to solutions that appear amazingly "clever." But this is not necessarily the case. A certain solution only needs to work somehow under prevailing conditions (see also Sect. 14.2.3). Hence, not all repetitive or apparently otherwise unused sequences in the human genome necessarily have a function; some may simply be the remnants of earlier processes or "scars" from battles with other living organisms such as viruses.

The human mind, however, strives for clarity. It is not surprising therefore that some authors have proposed hypotheses to explain the function of these sequences by viewing the chromosome as a whole. Trifonov [109, 110], for example, postulated, in addition to the well-known genetic code (Table 3.2) a "chromatin code," which was specified as a "chromatin folding code" by Vogt [117]: the DNA thread is folded in several orders, the lowest of which is the nucleosome structure (Sect. 3.1.1.3). Metaphase chromosomes show three higher orders of coiling which

may be partially preserved in interphase (Table 3.1). Moreover, the interphase chromatin is connected with a scaffold inside the nucleus. Base sequences, on the other hand, are not entirely without significance for tertiary structure; for example, periodically spaced blocks of at least two $(A)_n$ repeats induce a bending of the DNA double helix. Moreover, tracts of A = T pairs have difficulties assembling a nucleosome structure. These and similar observations have led to the hypothesis "that tandem repeated simple sequence units, interspersed in the genome, are predestined to establish a locus-specific organization of their DNA structure and the corresponding chromatin domain" [117]. This tertiary structure may offer specific opportunities for interaction with proteins, for example, enzymes, that control transcription.

This hypothesis suggests that differences in these structures occur in differentiated cells, i.e., in cells with specialized functions. Indeed, there are results indicating such differences in structure of interphase chromatin among specialized cells. Figure 3.33 presents a piece of an uncoiled chromosome arm as it might be imagined at present.

Within the nuclear genomes of warm-blooded vertebrates a compositional compartmentalization has been demonstrated; these compartments (isochores) consist of at least some 300 kb and are characterized by different GC levels; they comprise coding as well as noncoding sequences [7] and may be important for function.

3.4 The Genome of Mitochondria

Structure and Function of Mitochondria. Mitochondria are organelles found in the cytoplasm. Their number and form vary depending on the function of the cell; a mammalian liver cell, for example, contains approximately 1000–1500 mitochondria. All of them have certain structural aspects in common (Fig. 3.34): a matrix, an internal and an external membrane. The internal membrane forms characteristic folds, sometimes "cristae," in other cases "tubuli." Important biochemical functions, such as aerobic oxidation (oxidative phosphorylation, OXPHOS), take place in the mitochondria, which are often called the power plants of the body. Energy is stored as ATP. Of the three energy sources in our food, amino acids and fats can be degraded only by aerobic oxidation; this oxidation takes place in mitochondria. Moreover, they harbor the citric acid cycle. Mitochondria form an ordered multienzyme system; distribution of enzymes in a functionally meaningful order guarantees ordered sequences of biochemical reactions.

Mitochondria multiply, as living organisms, by division; de novo synthesis is impossible. They also have ribosomes, which are, however, smaller than those found in the cytoplasm. This and other evidence suggest that mitochondria originated as external micro-organisms that underwent a

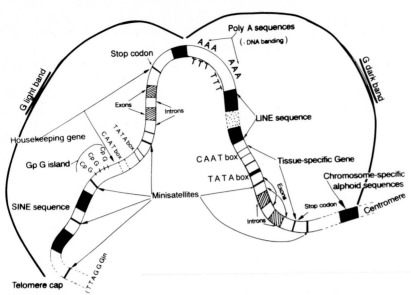

Fig. 3.33. Structure and expression of an idealized gene, including various promoter elements, an enhancer, and the transcribed region of the gene. The gene shown has two introns, each beginning with GT and ending with AG; they are spliced out to generate the mature mRNA. Not only promoter elements but also enhancers can be tissue specific. For the various elements see the various sections of Chap. 3.

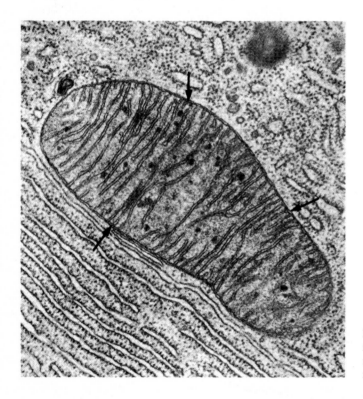

Fig. 3.34. Electron-microscopic photograph of a mitochondrium, × 53 000, *Arrows,* (outer and inner) membrane. (From Nielsen et al. 1970, Fundamental concepts of Biology, New York: Wiley 1970)

symbiotic relationship with the cell some time early in evolution, have since been integrated into the cell, but still maintain some of their specific properties.

The Genome of Mitochondria. It has been known for some time that mitochondria contain DNA of their own, which also codes for genes, for example, for tRNA. Genes necessary for mitochondrial enzymes, on the other hand, are often (not always) located on the chromosomes of nuclear DNA [4].

Some time ago, a well-organized effort by research groups at the MRC Laboratory of Molecular Biology in Cambridge led to full elucidation of the DNA sequence and gene organization of the human mitochondrial genome (mtDNA [1]; Fig. 3.35).

This genome has 16 568 bp, which are arranged in a circular fashion (see also [81]). It contains genes for 12 S and 16 S rRNA, 22 different tRNAs, cytochrome *c* oxidase subunits I, II, and III, ATPase subunit 6, cytochrome *b*, and eight other protein-coding genes. In contrast to the nuclear genome in chromosomes, the sequence of the mitochondrial genome shows extreme economy: The genes have no or only a few noncoding bases among them. In contrast to nuclear DNA, both DNA strands are transcribed and translated; moreover, there is a third, short, genetically active strand. In many cases the termination codons are not coded in the DNA but are created posttran-

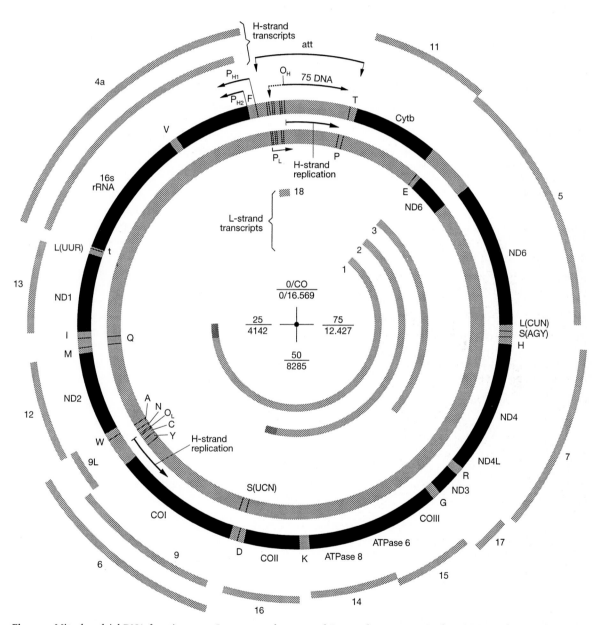

Fig. 3.35. Mitochondrial DNA function map, *Inner, outer double circles,* C-rich light *(L)* and the G-rich heavy *(H)* strands, respectively; dark gray regions, rRNA and peptide genes; light blocks between larger genes, tRNA genes; *inner, outer light arcs,* stable, processed, L- and H-strand transcripts; *OH, OL,* origins of H- and L-strand replications; *PH,* PL, H- and L-strand promoters. In the 7 S DNA D-loop region, I–III are curved sequence blocks (CSBs); *heavy arrow between II and III,* L-strand transcript processing site for generating H-strand replication primers. *t* (within MTTL 1), bidirectional transcription termination sequence. (From Wallace 1989)

scriptionally. Moreover, the genetic code of human mtDNA differs from the universal code in some important aspects: UGA (expressed, as usual, in terms of the mRNA code) codes for tryptophan and not for termination, AUA codes for methionine, not isoleucine; AGA and AGG are termination rather than arginine codons. In addition the third codon positions – which are the main source of codon degeneracy – are more often A or C (and less often G or T) than in the nuclear genome.

DNA Polymorphism and the Question of Hereditary Diseases Due to Mitochondrial Mutations. After the nucleotide sequence of human mitochondria became known, several DNA cleavage site polymorphisms were discovered. These have become important for our understanding of human evolution and are therefore discussed in Sect. 14.3.

Mitochondria are abundant in oocytes, but in sperm only four mitochondria (produced by fusion of a greater number) are found at the neck of the sperm head; they do not enter the oocyte at fertilization. Therefore the entire set of mitochondria in all cells of any individual comes from the mother [40]. Would a mutation at a suitable site cause a hereditary disease? Such a disease would be transmitted by the mother only, and to *all* her children (examples are given in Sect. 4.1.9).

One could argue on theoretical grounds that this mode of inheritance would be very unlikely since every oocyte contains multiple mitochondria. Even if a mutation were to occur in one of them, the overwhelming majority would still have the nonmutant DNA site (or gene); therefore, there could be no phenotypic effect. However, the same argument also holds for DNA polymorphisms of mtDNA, and these polymorphisms are nevertheless transmitted to all children by maternal inheritance. What is the reason for this peculiar behavior? Do all mitochondria in the oocyte derive from one stem mitochondrium [40]. This problem is discussed in Sect. 4.1.9 together with mitochondrial diseases which indeed do occur.

3.5 New Genetics and the Gene Concept

Intriguing Problems. Concepts and methods from molecular biology have enhanced our knowledge of the human genome far beyond expectations within only a few years. This has opened up new prospects for genetic diagnosis, prevention of diseases, and probably also for therapy. All these achievements should therefore be offered to as many persons as possible. However, up-to-date services in medical ge-

netics have drastically increased the requirements for equipment, personnel, and technical skill. In many countries these developments create difficult problems. Are human societies able and prepared to pay for such services on a large scale? Can a responsible consultant even recommend introduction of such expensive services in developing countries with many urgent medical and economic problems? Completely new problems can arise. There are still cultural traditions in some populations that regard a son as more valuable than a daughter. Will the availability of modern techniques of sex choice lead to complex and unforeseen consequences? (See also Chap. 19)

What Is a Gene? In classical, formal genetics a gene was the common unit of mutation, recombination, and action. Genes were initially regarded as arranged in a linear fashion on chromosomes, as pearls on a string (Fig. 3.36). Detailed genetic analysis, however, proved this concept too crude. For example, there are closely linked genes in *Drosophila* where two mutations on the same chromosome side by side (in *cis* position) have a smaller phenotypic effect than the same two mutations on homologous chromosomes, opposite to each other (in *trans* position). Often there is no phenotypic effect at all. Genes showing such *cis-trans* effects were called pseudoalleles (see also Chap. 5). When biochemical analysis became possible, such *cis-trans* effects were shown to occur when the two mutations involve two different sites within the same functional gene, i.e., the codons determining one protein (e.g., an enzyme). When such mutations occur in *cis* position a functionally intact protein can be formed by the homologous gene; the two normal sites can complement each other. Mutations in *trans* position, on the other hand, do not show this complementation, since no intact protein is formed. In micro-organisms as well as in human cell cultures, (Sect. 7.2.2.3) such complementation groups can easily be analyzed.

Immediately after the advent of molecular genetics in the 1950s a new conceptualization became necessary. Benzer in 1957 [5] proposed partition of the gene concept into three aspects: the units of recombination (recon), mutation (muton), and action (cistron; from the *cis-trans* effect). In subsequent years the recon and the muton were found to be as small as a single nucleotide, the smallest unit in the genetic material; the cistron was identified with that segment of DNA which codes for one polypeptide chain. Of these three terms only the cistron became popular among geneticists; it was synonymous with the functional gene.

With the increasing awareness of the complexities of the genetic material – introns, promoter sequences, pseudogenes, etc. – the delimitation of "genes" became more and more blurred. At present it is not at

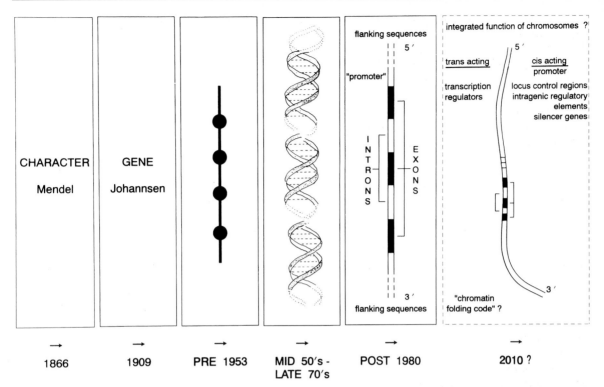

Fig. 3.36. Historical development of the gene concept. Genes were first postulated by Johannsen to account for the "characters" shown by Mendel to be responsible for hereditary transmission of traits. No material basis for the gene could be postulated until the demonstration of linkage of genes on chromosomes led to the "bead-on-a-string" model. The beads were considered to be the genes, and the string reflected the chromosome. The material basis of the gene remained undefined. The demonstration of DNA as the genetic material led to the definition of a gene as a specified sequence of DNA that codes for a polypeptide sequence (three nucleotide bases specify one amino acid). It was soon realized that the unit of polypeptide specification ("cistron") differs from the unit of recombination ("recon"), which is different from the unit of mutation ("muton"). The "muton" could be as small as a single DNA base. It was then shown that vast stretches of DNA do not code for proteins, that some DNA sequences are regulatory, and that structural genes are interrupted by noncoding intervening sequences (introns). Coding sequences of the structural genes are known as exons. The borders of the flanking and "regulatory" sequences upstream (5′) and downstream (3′) remain poorly defined. It is therefore impossible at present to define the boundaries of a gene precisely. The coding and intervening sequences of a gene can be accurately specified (see also [24]). Evidence is now accumulating that DNA sequences outside of coding genes and their control regions may have additional functions, for example, in a spatial configuration of chromosomes (see text).

all clear which of the long DNA sequences interspersed between the coding parts, including repetitive sequences such as LINE and SINE elements, are necessary for gene function. The concepts of "chromatin code" [109, 110] and "chromatin folding code" [117] may help in understanding the functional significance of a complex chromatin structure in which genes are embedded.

New Results on Structure of Genes and Formal Genetics. The discussions in Chap. 3 on gene structure could create the impression that most results of classical genetic analysis are now outdated. This, however, is not the case. The principles of formal genetics remain applicable and necessary for analyzing modes of inheritance in families, the presence or absence of linkage between nonallelic genes, and the basic properties of genes in populations. The situation might be compared to that encountered in physics: Quantum mechanics has helped us in understanding the nature of light much better than ever. However, classical physics such as geometrical optics is not only as correct as it always was, but also necessary for many practical applications, such as constructing spectacles or microscopes. It is therefore a necessary part of every physics textbook.

Conclusions

The information-carrying structure of chromosomes is deoxyribonucleic acid (DNA). The diploid human genome consists of approx. $6-7 \times 10^9$ nucleotides located within chromosomes where they are packaged with proteins. The base sequence of DNA specifies the information for each of the 20 amino acids, and their sequence determines the uniqueness of a protein. DNA also regulates transcription, which may vary depending on the state of embryonic development and differentiation of cells. Increasingly detailed knowledge of the human genome at the DNA level forms the basis for our understanding of genetic transmission and gene action. Extensive interindividual variation at the DNA level has been discovered using methods such as nucleic acid hybridization, Southern blotting, DNA sequencing, and the polymerase chain reaction. Such DNA variation has been useful in a variety of practical and theoretical studies including those on linkage, forensic and paternity identification, and human evolution.

Molecular techniques have allowed researchers to define and analyze numerous human genes and their corresponding genetic diseases. These include genes for hemoglobins (hemoglobinopathies), muscle protein dystrophin (X-linked muscular dystrophy), blood coagulation factor VIII (hemophilia A), the huntingtin gene (Huntington disease), the transmembrane conductance regulator gene (CFTR; cystic fibrosis), and many others. The base sequence of the mitochondrial genome and several diseases caused by mitochondrial mutations have also been defined.

References

1. Anderson S, Bankier AT, Barrel BG, de Bruyn MHL, Coulson AR, Dromin J, Eperon IC, Nierlich DP, Rue BA, Sanger F, Schreier PH, Smith AJH, Staden R, Young IG (1981) Science and organization of the human mitochondrial genome. Nature 290 : 457–465

2. Arber W (1979) Promotion and limitation of genetic exchange. Science 205 : 361–365

3. Asiki RK, Scharf S, Faloona F et al (1985) Enzymatic amplification of -globin genomic sequences and restriction site analysis for diagnosis of sickle cell anemia. Science 230 : 1350–1354

4. Attardi G, Schatz G (1988) Biogenesis of mitochondria. Annu Rev Cell Biol 4 : 289–333

5. Benzer S (1957) The elementary units of heredity. In: McElroy WD, Glass B (eds) The chemical basis of heredity. John Hopkins University Press, Baltimore, pp 70–93

6. Berg P (1981) Dissections and reconstructions of genes and chromosomes. Science 213 : 296–303

7. Bernardi G, Bernardi G (1986) Compositional constraints and genome evolution. J Mol Evol 24 : 1–11

8. Blackburn EH (1991) Structure and function of telomeres. Nature 350 : 569–573

9. Brisson N, Verma DPS (1982) Soybean leghemoglobin gene family: normal, pseudo, and truncated genes. Proc Natl Acad Sci USA 79 : 4055–4059

10. Bross K, Krone W (1972) On the number of ribosomal RNA genes in mna. Hum Genet 14 : 137

11. Bross K, Krone W (1973) Ribosomal cistrons and acrocentric chromosomes in man. Hum Genet 18 : 71–75

12. Brown SW (1966) Heterochromatin. Science 151 : 417–435

13. Buckler AJ, Chang DD, Graw SL et al (1991) Exon amplification: a strategy to isolate mammalian genes based on RNA splicing. Proc Natl Acad Sci USA 88 : 4005–4009

14. Burke DT, Carle GF, Olsen MV (1987) Cloning of large segments of exogenous DNA into yeast by means of artificial chromosome vectors. Science 236 : 806–812

15. Cooper DN, Schmidtke J (1984) DNA restriction fragment length polymorphism and heterozygosity in the human genome. Hum Genet 66 : 1–16

16. Cremer C, Cremer T (1978) Considerations on a laser-scanning microscope with high resolution and depth of field. Microsc Acta 81 : 31–44

16a. Cremer C, Gray JW, Ropers HH (1982) Flow cytometric characterization of a Chinese hamster x man hybrid cell line retaining the human Y chromosome. Hum Genet 60: 262–266

17. Cremer T, Landegent J, Brueckner A et al (1986) Detection of chromosome aberrations in the human interphase nucleus by visualization of specific target DNAs with radioactive and non-radioactive in situ hybridization techniques: diagnosis of trisomy 18 with probe L1.84. Hum Genet 74 : 346–352

18. Cremer T, Lichter P, Borden J et al (1988) Detection of chromosome aberrations in metaphase and interphase tumor cells by in situ hybridization using chromosome-specific library probes. Hum Genet 80 : 235–246

19. Curmutte JT, Babior BM (1987) Chronic granulomatous disease. Adv Hum Genet 16 : 229–297

20. Dauwerse JG, Wiegant J, Raap AK et al (1992) Multiple colors by fluorescence in situ hybridization using ratio-labelled DNA probes create a molecular karyotype. Hum Mol Genet 1 : 593–598

21. Du Manoir S, Speicher MR, Joos S et al (1993) Detection of complete and partial chromosome gains and losses by comparative genomic in situ hybridization. Hum Genet 90 : 590–610

22. Emery AEH (1984) An introduction to recombinant DNA. Wiley, Chichester

23. Emmerich P, Jauch A, Hofmann M-C et al (1989) Methods in laboratory investigation: interphase cytogenetics in paraffin embedded sections from human testicular germ cell tumor xenographs and in corresponding cultured cells. Lab Invest 61 : 235–242

24. Falk R (1984) The gene in search of an identity. Hum Genet 68 : 195–204

25. Fraccaro M, Kaijser K, Lindsten J (1960) Chromosomal abnormalities in father and mongoloid child. Lancet 1 : 724–727

26. Francke U, Ochs HD, de Martinville B et al (1985) Minor Xp21 chromosome deletion in a male associated with expression of Duchenne muscular dystrophy, chronic granulomatous disease, retinitis pigmentosa and McLeod syndrome. Am J Hum Genet 37 : 250–267

27. Gall JR, Pardue ML (1969) Formation and detection of RNA-DNA hybrid molecules in cytological preparations. Proc Natl Acad Sci USA 63 : 378–383

28. Gartler SM, Sparkes RS (1963) The Lyon-Beutler hypothesis and isochromosome X patients with the Turner syndrome. Lancet 2 : 411

29. Geissler E, Theile M (1983) Virus-induced gene mutations of eukaryotic cells. Hum Genet 63 : 1–12

30. Giles RE, Blanc H, Cann HM, Wallace DL (1980) Maternal inheritance of human mitochondrial DNA. Proc Natl Acad Sci USA 77 : 6715–6719

31. Gitschier J, Wood WI, Goralka TM, Wion KL, Chen EY, Eaton DH, Vehar GA, Capon DJ, Lawn RM (1984) Characterization of the human factor VIII gene. Nature 312 : 326–330

32. Gitschier J, Wood WI, Tuddenham EGD, Shuman MA, Goralka TM, Chen EY, Lawn RM (1985) Detection and sequence of mutations in the factor VIII gene of haemophiliacs. Nature 315 : 427–430

33. Goldberg YP, Kremer B, Andrew SE et al (1993) Molecular analysis of new mutations for Huntington's disease: intermediate alleles and sex of origin effects. Nature Genet 5 : 174–179

34. Green MM (1980) Transposable elements in Drosophila and other diptera. Annu Rev Genet 14 : 109–120

35. Grüneberg H (1966) The case for somatic crossing over in the mouse. Genet Res 7 : 58–75

36. Habedank M, Rodewald A (1982) Moderate Down's syndrome in three siblings having partial trisomy 21 q22.2 and therefore no SOD-1 excess. Hum Genet 60 : 74–77

37. Hanauer A, Levin M, Heilig R, Daegelen D, Kahn A, Mandel JL (1983) Isolation and characterization of DNA clones for human skeletal muscle alpha-actin. Nucleic Acids Res 11 : 3503–3516

38. Hassold TJ, Jacobs PA (1984) Trisomy in man. Annu Rev Genet 18 : 69–77

39. Hindley J (1983) DNA sequencing. In: Work TS, Burden RH (eds) Laboratory techniques in biochemistry and molecular biology. Elsevier, Amsterdam

40. Howell N, Halvorson S, Kubacka I et al (1992) Mitochondrial gene segregation in mammals: is the bottleneck narrow? Hum Genet 90 : 117–120

41. Hsu TC (1975) A possible function of constituitive heterochromatin: the bodyguard hypothesis. Genetics 79 [Suppl]:137–150

42. John H, Birnstiel ML, Jones KW (1969) RNA-DNA hybrids at cytological levels. Nature 223 : 582–587

43. Jones RS, Potter SS (1985) L1 sequences in HeLa extrachromosomal circular DNA: evidence for circularization by homologous recombination. Proc Natl Acad Sci USA 82 : 1989–1993

44. Jourdan B (1993) Traveling around the human genome. Inserm/CNRS, Marseille

45. Kallioniemi A, Kallioniemi O-P, Sudar D et al (1992) Comparative genomic hybridization for molecular cytogenetic analysis of solid tumors. Science 258 : 818–821

46. Kallioniemi O-P, Kallioniemi A, Sudar D et al (1993) Comparative genomic hybridization: a rapid new method for detecting and mapping DNA amplification in tumors. Semin Cancer Biol 4 : 41–46

47. Kan YW, Dozy AM (1978) Polymorphism of DNA sequence adjacent to the human β globin structural gene. Its relation to the sickle mutation. Proc Natl Acad Sci USA 75 : 5631–5635

48. Kerem BS, Rommens JM, Buchanan JA et al (1989) Identification of the cystic fibrosis gene: genetic analysis. Science 245. 1073–1080

49. Klever M, Grond-Ginsbach CJ, Hager H-D et al (1992) Chorionic villus metaphase chromosomes and interphase nuclei analyzed by chromosome in situ suppression (CISS) hybridization. Prenat Diagn 12 : 53–59

50. Klingmüller W (1976) Genmanipulation und Gentherapie. Springer, Berlin Heidelberg New York

51. Knippers R, Philippsen P, Schafer KP, Fanning E (1990) Molekulare Genetik, 5th edn. Thieme, Stuttgart

52. Kornberg A (1992) DNA replication, 2nd edn. Freeman, New York

53. Kornberg RD (1977) Structure of chromatin. Annu Rev Biochem 46 : 931–945

54. Kornberg RD, Lorch Y (1991) Irresistible force meets immovable object: transcription and the nucleosome. Cell 67 : 833–836

55. Kurnit DM (1979) Satellite DNA and heterochromatin variants: the case for unequal mitotic crossing over. Hum Genet 47 : 169–186

56. Lengauer C, Eckelt A, Weith A et al (1991) Selective staining of defined chromosomal subregions by in situ suppression hybridization of libraries from laser-microdissected chromosomes. Cytogenet Cell Genet 56 : 27–30

57. Lengauer C, Speicher MR, Popp S et al (1993) Chromosomal bar codes produced by multicolor fluorescence in situ hybridization with multiple YAC clones and whole chromosome painting probes. Hum Mol Genet 2 : 505–512

58. Lewin B (1983) Genes. Wiley, New York

59. Li C, Ramjeesingh M, Reyes E et al (1993) The cystic fibrosis mutation (F 508) does not influence the chloride channel activity of CFTR. Nature Genet 3 : 311–316

60. Li H, Gyllensten UB, Cui X et al (1988) Amplification and analysis of DNA sequences in single human sperm and diploid cells. Nature 335 : 414–417

61. Lichter P, Cremer T (1992) Human cytogenetics: a practical approach, 2nd edn. IRL/Oxford University Press, Oxford

62. Lichter P, Cremer T, Borden J et al (1988) Delineation of individual human chromosomes in metaphase and interphase cellls by in situ suppression hybridization using recombinant DNA libraries. Hum Genet 80 : 224–234

63. Lilley DMJ, Pardon JF (1979) Structure and function of chromatin. Annu Rev Genet 13 : 197–233

64. MacDonald ME, Novelletto A, Lin C et al (1992) The Huntington's disease candidate region exhibits many different haplotypes. Nature Genet 1 : 99–103

65. MacDonald ME, Ambrose CM, Duyao MP et al, the Huntington's Disease Collaborative Research Group (1993) A novel gene containing a trinucleotide repeat that is expanded and unstable on Huntington's disease chromosomes. Cell 72 : 971–983

66. Maddox J (1984) Who will clone a chromosome? Nature 312 : 306

67. Mager DL, Henthorn PS (1984) Identification of a retroviruslike repetitive element in human DNA. Proc Natl Acad Sci USA 81 : 7510–7514

68. McClintock B (1956) Controlling elements and the gene. Cold Spring Harbor Symp Quant Biol 21 : 197–216

69. McGinnies W, Krumlaut R (1992) Homeobox genes and axial patterning. Cell 68 : 283–302

70. Miller OJ, Beatty BR (1969) Visualization of molecular genes. Science 164 : 955–957

71. Monaco AP (1994) Isolation of genes from cloned DNA. Curr Opin Genet Dev 4 : 360–365

72. Moyzis RK, Buckingham JM, Cram LS et al (1990) A highly conserved DNA sequence, (TTAGGG)n, present at the telomers of human chromosomes. Proc Natl Acad Sci USA 85 : 6622–6626

73. Mullis KB, Faloona FA (1987) Specific synthesis of DNA in vitro via a polymerase-catalyzed chain reaction. Methods Enzymol 155 : 335–350

74. Nathans D (1980) Restriction endonucleases, Simian virus 40, and the new genetics. Science 206 : 903–909

75. Nedelof P, van der Flier S, Wiegant J et al (1990) Multiple fluorescence in situ hybridization. Cytometry 11 : 126–131

76. Ohno S (1967) Sex chromosomes and sex-linked genes. Springer, Berlin Heidelberg New York

77. Pardue ML, Gall JG (1969) Molecular hybridization of radioactive DNA to the DNA of cytological preparations. Proc Natl Acad Sci USA 64 : 600–607

78. Passarge E (1979) Emil Heitz and the concept of heterochromatin: longitudinal chromosome differentiation was recognized fifty years ago. Am J Hum Genet 31 : 106–115

79. Pinkel D, Landegent J, Collins C et al (1988) Fluorescence in situ hybridization with human chromosome specific libraries: detection of trisomy 21 and translocation of chromosome 4. Proc Natl Acad Sci USA 85 : 9138–9142

80. Popp S, Cremer T (1992) A biological dosimeter based on translocation scoring after multicolor CISS hybridization of chromosomal subsets. J Radiat Res 33 : 61–70

81. Prezant TR, Agapian JV, Fischel-Ghodsian N (1994) Corrections to the human mitochondrial RNA sequences. Hum Genet 93 : 87–88

82. Rappold GA, Vosberg H-P (1984) Chromosomal localization of a human myosin heavy-chain gene by in situ hybridization. Hum Genet 65 : 195–197

83. Ried T, Baldini A, Rand T, Ward DC (1992) Simultaneous visualization of seven different DNA probes by in situ hybridization using combinatorial fluorescence and digital imaging microscopy. Proc Natl Acad Sci USA 89 : 1388–1392

84. Riordan JR, Rommens JM, Kerem B-S et al (1989) Identification of the cystic fibrosis gene: cloning and characterization of complementary DNA. Science 245 : 1066–1073

85. Rommens JM, Ianuzzi MC, Kerem B-S et al (1989) Identification of the cystic fibrosis gene: chromosome walking and jumping. Science 245 : 1059–1065

86. Royer-Pokora B, Kunkel LM, Monaco AP et al (1986) Cloning the gene for an inherited human disorder – chronic granulomatous disease – on the basis of its chromosomal location. Nature 322 : 32–38

87. Rudkin GT, Stollar BD (1977) High resolution detection of DNA. RNA hybrids in situ by indirect immunofluorescence. Nature 265 : 472–473

88. Scherthan H, Cremer T (1994) Methodology of non-isotopic in situ hybridization in paraffin embedded tissue sections. In: Adolph KW (ed) Methods in molecular genetics, vol 2. Academic, New York

89. Schmid CW, Deininger PL (1975) Sequence organization of the human genome. Cell 6 : 345–358

90. Schmid CW, Jelinek WR (1982) The Alu family of dispersed repetitive sequences. Science 216 : 1065–1070

91. Schmidtke J, Epplen JT (1980) Sequence organization of animal nuclear DNA. Hum Genet 55 : 1–18

92. Schröck E, Thiel G, Lozanova T et al (1994) Comparative genomic hybridization on human malignant gliomas reveals multiple amplification sites and non-random chromosomal gains and losses. Am J Pathol 144 : 1203–1218

93. Schwarzacher HG (1976) Chromosomes. Springer, Berlin Heidelberg New York (Handbuch der mikroskopischen Anatomie des Menschen, vol 1/3)

94. Shmookler Reis RJ, Lumpkin CK, McGill JR, Riabowol KT, Goldstein S (1983) Extrachromosomal circular copies of an "inter-Alu" unstable sequence in human DNA are amplified during in vitro and in vivo ageing. Nature 301 : 394–398

95. Southern EM (1975) Detection of specific sequences among DNA fragments separated by gel electrophoresis. J Mol Biol 98 : 503–517

96. Sperling K (1984) Genetische Sektion – Anatomie der menschlichen Gene. In: Passarge E (ed) Verlag Chemie, Darmstadt, pp 73–100

97. Sperling K, Rao PN (1974) The phenomenon of premature chromosome condensation: its relevance to basic and applied research. Humangenetik 23 : 235–258

98. Starlinger P (1980) IS elements and transposons. Plasmid 3 : 241–259

99. Suthcliffe JG, Kiel M, Bloom FE, Milner RJ (1987) Genetic activity in the CNS, a tool for understanding brain function and dysfunction. In: Vogel F, Sperling K (eds) Human genetics. Proceedings of the 7th International Congress, Berlin 1986. Springer, Berlin Heidelberg New York, pp 474–483

100. Taylor AL (1963) Bacteriophage-induced mutation in Escherichia coli. Proc Natl Acad Sci USA 50 : 1043–1051

101. Teplitz RL (1986) The use of synthetic oligonucleotides in and prenatal diagnosis of genetic disease. In: Teplitz RL et al (eds) First International Symposium on the Role of Recombinant DNA in Genetics, Crete

102. Tilghman SM, Tiemeyer DC, Seirman JG, Peterlin BM, Sullivan M, Maizel JV, Leder P (1978) Intervening sequence of DNA identified in the structural portion of a mouse β-globin gene. Proc Natl Acad Sci USA 75 : 725–729

103. Tkachuk DC, Pinkel D, Kuo W-L et al (1991) Clinical applications of fluorescence in situ hybridization. Genet Anal Techn Appl 8 : 67–74

104. Tocharoentanaphol C, Cremer M, Schröck E et al (1994) Multicolor fluorescence in situ hybridization on metaphase chromosomes and interphase Halo-preparations using cosmid and YAC clones for the simultaneous high resolution mapping of deletions in the dystrophin gene. Hum Genet 93 : 229–235

105. Tode JJ, Knopf JL, Wozney JM, Sutzman LA, Bueker JL, Pittman DD, Kaufman RJ, Brown E, Shoemaker C, Orr EC, Amphlett GW, Foster WB, Coe ML, Knutson GJ, Fass DN, Hewick RM (1984) Molecular cloning of a cDNA encoding human antihaemophilic factor. Nature 312 : 342–347

106. Tommerup N (1993) Mendelian cytogenetics. Chromosome rearrangements associated with mendelian disorders. J Med Genet 30 : 713–727

107. Trask BJ, Massa H, Kenwick S, Gitschier J (1991) Mapping of human chromosome Xq28 by two-color fluorescence in situ hybridization of DNA sequences to interphase nuclei. Am J Hum Genet 48 : 1–5

108. Trendelenburg MF (1983) Progress in visualization of eukaryotic gene transcription. Hum Genet 63 : 197–215

109. Trifonov EN (1985) Curved DNA. CRC Crit Rev Biochem 19 : 89–106
110. Trifonov EN (1989) The multiple codes of DNA sequences. Bull Math Biol 51 : 417–432
111. Tsui L-C (1992) The spectrum of cystic fibrosis mutations. Trends Genet 8 : 392–398
112. Vehar GA, Keyt B, Eaton D, Rodriguez H, O'Brian DP, Rotblat F, Oppermann H, Keck R, Wood WI, Harkins RN, Tuddenham EGD, Lawn RM, Capon DJ (1984) Structure of human factor VIII. Nature 312 : 337–342
113. Vogel F (1964) Eine vorläufige Abschätzung der Anzahl menschlicher Gene. Z Menschl Vererbungs Konstitutionslehre 37 : 291–299
114. Vogel F (1964) Preliminary estimate of the number of human genes. Nature 201 : 847
115. Vogel F (1989) Humangenetik in der Welt von heute. Springer, Berlin Heidelberg New York (12th Salzburger Vorlesungen)
116. Vogel F, Grunwald R (eds) (1994) Patenting of human genes and living organisms. Springer, Berlin Heidelberg New York (Sitzungsberichte der Heidelberger Akademie der Wissenschaften. Mathematisch-naturwissenschaftliche Klasse, Suppl 1993/1992)
117. Vogt P (1990) Potential genetic functions of tandem repeated DNA sequence blocks in the human genome are based on a highly conserved "chromatin folding code". Hum Genet 84 : 301–336
118. Vosberg HP (1977) Molecular cloning of DNA. An introduction into techniques and problems. Hum Genet 40 : 1–72
119. Vosberg HP (1989) The polymerase chain reaction: an improved method for the analysis of nucleic acids. Hum Genet 83 : 1–15
120. Watson JD, Crick FHC (1953) The structure of DNA. Cold Spring Harbor Symp Quant Biol 18 : 123–132
121. Watson JD, Hopkins NH, Roberts JW et al (1987) Molecular biology of the gene, 4th edn. Benjamin/Cummings, Menlo Park
122. Watson JD, Gilman M, Witkowski J, Zoller M (1992) Recombinant DNA, 2nd edn. Freeman, New York
123. Zakion VA (1989) Structure and function of telomeres. Annu Rev Genet 23 : 579–604

4 Formal Genetics of Humans: Modes of Inheritance

The law of combination of the differing traits, according to which the hybrids develop,
finds its foundation and explanation in the proven statement that the hybrids produce germ
and pollen cells ... which originate from the combination of the traits by fertilization.

G. Mendel, Versuche über Pflanzenhybriden, 1865

4.1 Mendel's Modes of Inheritance and Their Application to Humans

Mendel's fundamental discoveries are usually summarized in three laws:

1. Crosses between organisms homozygous for two different alleles at one gene locus lead to genetically identical offspring (F_1 generation), heterozygous for this allele. It is unimportant which of the two homozygotes is male and which is female (law of uniformity and reciprocity). Such reciprocity applies only for genes not located on sex chromosomes.

2. When these F_1 heterozygotes are crossed with each other (intercross), various genotypes segregate: one-half are heterozygous again, and one-quarter are homozygous for each of the parental types. This segregation 1:2:1 is repeated after crossing of heterozygotes in the following generations, whereas the two types of homozygotes breed pure. As noted above (Sect. 1.4), Mendel interpreted this result correctly, assuming formation of two types of germ cells with a 1:1 ratio in heterozygotes (law of segregation and law of purity of gametes).

3. When organisms differing in more than one gene pair are crossed, every single gene pair segregates independently, and the resulting segregation ratios follow the statistical law of independent segregation (law of free combination of genes).

This third law applies only when there is no linkage (Chap. 5). Human diploid cells have 46 chromosomes: the two sex chromosomes and 44 autosomes forming 22 pairs of two homologues each. The pairs of homologues are separated during meiosis, forming haploid germ cells or gametes. After impregnation, paternal and maternal germ cells unite to form the zygote, which is diploid again. Sex is determined genotypically; women normally have two X chromosomes, men have one X and one Y chromosome (Sect. 2.1.2).

For an understanding of the statistical character of segregation ratios in humans it is important to realize that the number of germ cells formed (Sect. 9.3.3) is very large, particularly among males. Only a very small sample comes to fertilization. Regarding single gene loci this sampling process can generally be regarded as random.

Two alleles may be termed A and A′. The set of combinations described in Fig. 4.1 are possible. As noted above, these theoretical segregation ratios are probabilities; segregation ratios found empirically should be tested by statistical methods to determine whether they are compatible with the theoretical ratios implied by the genetic hypothesis.

The mating type of identical homozygotes (AA × AA or A′A′ × A′A′) is uninteresting except where it permits conclusions regarding genetic heterogeneity of a recessive condition (Sect. 4.3.5). Mating between the two different homozygous types (AA × A′A′) is usually rare and is therefore of little practical importance. Matings between homozygotes and heterozygotes (AA′ × AA) and between two heterozygotes (A′A × A′A) are most important practically as explained below.

Mendel found that a genotype does not always determine one distinct phenotype. Frequently heterozygotes resemble (more or less) one of the homozygotes. Mendel called the allele that determines the phenotype of the heterozygote dominant, the other recessive. With more penetrating analysis, some human geneticists have concluded that these terms may be misleading and should be abandoned. In fact, at the level of gene action, genes are not dominant or recessive. At the phenotypic level, however, the distinction is important and useful. Biochemical mechanisms of dominant hereditary diseases (Sect. 7.6) usually differ from those of recessive conditions (Sect. 7.2). Hence the mode of inheritance gives a hint regarding the biochemical mechanism likely to be involved.

In recent years, with the introduction of methods permitting analysis at a level closer to gene action, an increasing number of instances have become known in which each of two alleles in a heterozygous state has a distinct phenotypic expression. If both are inherited and phenotypically expressed, this mode of inheritance is sometimes called codominant.

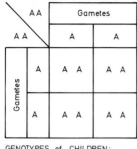

Fig. 4.1. Mating types with two alleles

4.1.1 Codominant Mode of Inheritance

The first examples of codominance in man were found in the genetics of blood groups; the MN blood types (111300; numbers refer to identifying numbers of diseases listed in [41]) may serve as an example (Table 4.1). When methods for genetic analysis at the protein level became available, many more examples were soon discovered (Sect. 12.1). The example in Table 4.1 clearly points to a genetic model with two alleles, M and N, the phenotypes M and N being the two homozygotes and MN the heterozygote. This example is used below for a statistical comparison between expected and observed segregation ratios. The "aberrant" cases in parentheses, which at first glance seem to contradict the genetic hypothesis, were the result of false paternity – a frequent finding in most such investigations.

4.1.2 Autosomal Dominant Mode of Inheritance

The first description of a pedigree showing autosomal dominant inheritance of a human anomaly was Farabee's [15] paper in 1905 on "Inheritance of Digital Malformations in Man." Textbooks usually refer to the condition as brachydactyly (short digits), but from the original paper it is clear that not only were the pha-

Table 4.1. Family studies of the genetics of MN blood types (from Wiener et al. 1953 [77])

Mating type	Number of families	Types of children			Total children
		M	N	MN	
M × M	153	326	0	(1)	327
M × N	179	(1)	0	376	377
N × N	57	0	106	0	106
MN × M	463	499	(1)	473	973
MN × N	351	(3)	382	411	796
MN × MN	377	199	196	405	800
	1580	1028	685	1666	3379

Parentheses, false paternity.

langes of hands and feet shortened, but the number of phalanges was also reduced (Fig. 4.2). In addition, stature was low (average of 159 cm in three males), apparently due to shortness of legs and inferentially also of arms. In every other aspect, Farabee wrote,

The people appear perfectly normal ... and seem to suffer very little inconvenience on account of their malformation. The ladies complain of but one disadvantage in short fingers, and that is in playing the piano; they cannot reach the full octave and hence are not good players.

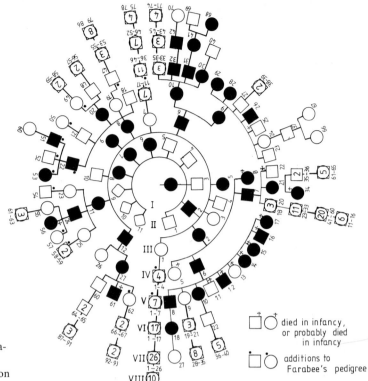

Fig. 4.2. The brachyphalangy pedigree of Farabee [15]. *Black symbols*, affected females (●) and males (■); *numbers*, point to their position in the pedigree

Figure 4.3 shows the pedigree. There are 36 affected in generations II–V, 13 of which are male and 23 female. Among the unaffected 18 are male and 15 female. The trait is transmitted from one of the parents to about half the children; transmission is independent of sex. Unfortunately, Farabee did not consider the children of the unaffected. Had he done so, he would have found them free from the anomaly. Many other pedigrees have shown absence of the trait among offspring of parents who do not carry the dominant gene. More recently the family has been reexamined [32]. The children of the unaffected family members and some affected family members were added, and X-ray examination confirmed that not only hands and feet were affected but the distal limb bones as well. The basic defect is thought to affect the epiphyseal cartilage. It is now called brachydactyly, type A1 (112500).

Affected patients are heterozygous for an autosomal allele leading to a clearcut and regular abnormality in the heterozygote. Therefore the trait is, by definition, dominant. The family shows two other characteristics that have since been found to be widespread:

1. The anomalies were described as being almost identical in all family members, and in each person appearing in all four extremities. This is a frequent finding in malformations with a regular mode of inheritance. The reason for the symmetry is evident considering that the same genes act on all four extremities.

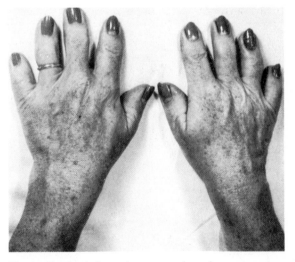

Fig. 4.3. Brachyphalangy in one member of a younger generation of Farabee's pedigree. (From Haws and McKusick 1963 [32])

2. The anomaly affected the well-being of its bearers only very little. This lack of health impairment is typical for such extended pedigrees. Reproduction is normal. Otherwise the trait would not be transmitted and would soon disappear. This is why, especially in the more serious domi-

nant conditions, extended pedigrees are the exception rather than the rule. Most diseases caused by mutations observed in the present generation have originated rather recently, often even in the germ cell of one of the parents (see Sect. 9.3).

Late Manifestation, Incomplete Penetrance, and Variable Expressivity. Sometimes a severe dominant condition manifests only during or after the age of reproduction. Here extended pedigrees are usually observed in spite of the severity of the condition. The classic example is Huntington disease (143100), a degenerative disease of the nerve cells in the basal ganglia (caudate nucleus and putamen) leading to involuntary extrapyramidal movements, personality changes, and a slow deterioration of mental abilities.

Wendt and Drohm [73] carried out a comprehensive study of all cases of Huntington disease in the former West Germany. The distribution of ages at onset is presented in Fig. 4.4. The great majority of their patients were married when they developed clinical symptoms. Even among thousands of patients the authors were not able to locate a single case that could be ascribed with confidence to a new mutation. Similar results were obtained in another study in Michigan [52] (see also [28, 29]). Analysis of the gene is described in Sect. 3.1.3.8.

Another phenomenon occasionally encountered in dominant traits is incomplete penetrance [62]). Penetrance is a statistical concept and refers to the fraction of cases carrying a given gene that manifests a specified phenotype. The transmission seems occasionally to skip one generation, leaving out a person who judging from the pedigree must be heterozygous, or the fraction of those affected among sibs (after appropriate corrections, Sect. 4.3.4) turns out to be lower than the expected segregation ratio. An example is retinoblastoma (180200), a malignant eye tumor of children. Bilateral cases (and cases with more than one primary tumor) are always dominantly inherited, whereas most unilateral, single tumors are nonhereditary, probably being caused by somatic mutation (Chap. 10). Even in pedigrees otherwise showing regular dominant inheritance, however, apparent skipping of a generation is observed occasionally (Fig. 4.5). Calculation of the segregation ratio in a large sample showed that about 45% of sibs were affected instead of the 50% expected in regular dominant inheritance. The penetrance of all cases (unilateral and bilateral) is therefore about 90%. Penetrance in families with bilateral cases is higher than in those with unilateral cases.

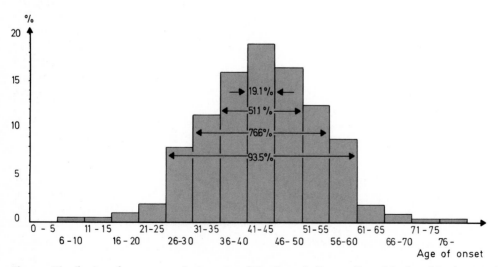

Fig. 4.4. Distribution of ages at onset in 802 cases of Huntington's disease. (From Wendt and Drohm 1972 [73])

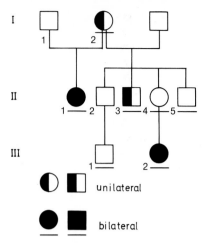

Fig. 4.5. Incomplete penetrance in retinoblastoma. The unaffected woman II,4 must be heterozygous, her mother I,2 and her daughter III,2 being affected; ☐ personally observed, (Personal observation, F. V.)

Discussion of other aspects of retinoblastoma:

- Table 3.4:   The gene
- Sect. 7.6:   Mechanisms of dominant inheritance
- Sect. 9.3:   Mutation rate
- Sect. 10.4.3:   Theory of cancer formation
- Sect. 12.2.1.1: Relaxation of natural selection
- Appendix 7:   Genetic counseling

Discussion of other aspects of neurofibromatosis:

- Table 3.4:   The gene
- Sect. 9.3:   Mutation rate
               Mechanisms of cancer formation
- Sect. 18.1: Genetic counseling

In many cases, penetrance is a function of the methods used for examination; higher penetrance is observed with detection methods (clinical or laboratory) that are closer to gene action.

In many dominant conditions the gene may manifest in all heterozygotes, but the *degree of manifestation* may be different. An example is neurofibromatosis (162 200). Some cases may show the full-blown picture with many tumors of the skin, café-au-lait spots, and systemic involvement, whereas other cases – even in the same families – may show only a few café-au-lait spots. The term used to describe this phenomenon is "variable expressivity" [62]. While such terms as "incomplete penetrance" and "variable expressivity" are often needed to convey quick understanding about certain phenomena, they may become dangerous if we forget

that they do not explain a biological mechanism but rather are labels for our ignorance.

It is indeed somewhat surprising that so many dominant conditions show such a large interindividual variability in age at onset and severity of manifestation. It would be more understandable if such variability were observed only between different families. Our knowledge of molecular biology (Sect. 9.4) suggests that the mutational events leading to these conditions are almost always slightly different between families. Indeed, there is usually an intrafamilial correlation between age at onset and severity of manifestation. For Huntington disease, for example, Wendt and Drohm [73] calculated a correlation coefficient of +0.57 for age at onset for affected family members. But there usually remains appreciable variability within families, in which the abnormal genes are identical by descent. It is again no more than a label for our ignorance when we invoke the "genetic background" or the action of all other genes for help. In Huntington disease, molecular analysis of the gene has provided at least a partial explanation: the number of repeats in the DNA triplet GAT is higher in patients with onset at a very young age. Unfortunately, there is no correlation between the number of repeats and age at onset in most patients who develop their clinical disease in the fourth to sixth decades of life.

Effect of Homozygosity on Manifestation of Abnormal Dominant Genes. An abnormal gene is called dominant when the heterozygote clearly deviates from the normal. Indeed, almost all bearers of dominant conditions in the human population are heterozygotes. From time to time, however, two bearers of the same anomaly do marry and have children. One quarter of these are then homozygous. This has been observed in several instances, especially when the spouses were relatives. The first example was probably that described by Mohr and Wriedt in 1919 [43]. In a consanguineous marriage between two bearers of a moderate brachydactyly (11 260) a child was born who not only lacked fingers and toes but also showed multiple malformations of the skeleton and died at the age of 1 year. A sister, however, had only the moderate anomaly, as did her parents [43].

Further examples of homozygosity of dominant anomalies are known. In one family, two parents with hereditary hemorrhagic teleangiectasia had a child showing multiple, severe internal and external telangiectasias who died at the age of 2.5 months [57]. Similarly, a very severe form of epidermolysis bullosa was observed in two of eight children of a couple, both of whom were afflicted with a mild type of this disease.

Another couple, both having a myopathy affecting the distal limb muscles, had 16 children, three of whom showed atypical and especially severe symptoms: the long flexors and the proximal hip muscles were also afflicted, and onset was earlier in life [72].

Epithelioma adenoides cysticum (132 700) is a dominant skin disease characterized by multiple nodular tumors. One female patient, whose parents were both affected, had especially severe symptoms, and her eight children all showed this anomaly (Fig. 4.6) [21]. Further examples include achondroplasia (100 800), Ehlers-Danlos syndrome (130 000), and others [822]. All these cases indicate that homozygotes of dominant anomalies are more severely affected than heterozygotes. It is therefore of interest that there appears to be no clinical difference between heterozygotes and homozygotes for Huntington disease, which is therefore a truly dominant disease as defined by Mendel. Clearly a different mechanism must apply to the pathogenesis of such a condition as compared with most other autosomal-dominant diseases, where dose effects are observed [74].

Given what we know about gene action, this is not surprising. In familial hypercholesterolemia (143 890) for example, the mechanism of action of a dominant gene is known. A decreased number of receptors for a regulatory substance (low-density lipoprotein) showed the expected differences between heterozygotes and affected homozygotes: 50 % decrease and complete absence or very much reduced activity of receptors, respectively (Sect. 7.6.4). Affected homozygotes show massive hypercholesteremia and usually die of myocardial infarction before the age of 30 years.

As noted above, Mendel called a gene dominant when the phenotype of the heterozygote resembled that of one homozygote. The examples of more severe manifestation of dominant genes in the homozygous than in the heterozygous state show that this strict definition is not maintained in human genetics. Here, all conditions are called dominant in which the heterozygote deviates consistently and perceptibly from the normal homozygote – irrespective of the phenotype of the anomalous homozygote. In Mendel's strict definition, most or even all dominant conditions in humans would be "intermediate." However, the more lenient connotation of "dominance" is now in general use.

4.1.3 Autosomal-Recessive Mode of Inheritance

The mode of inheritance is called recessive when the heterozygote does not differ phenotypically from the normal homozygote. In many cases special methods uncover slight detectable differences (Sect. 7.2.2.8). Contrary to dominant inheritance, in which almost all crosses are between heterozygotes and homozygous normals (Sect. 4.1.2), the great majority of matings observed in recessive anomalies involve heterozygous and phenotypically normal individuals. Since the three genotypes AA, Aa, and aa occur in the ratio 1:2:1 among the offspring, the probability of a child's being affected is 25 %. At the turn of the century when Garrod wrote his paper on alkaptonuria (Sect. 1.5) the "familial" character of recessive diseases was evident, as family size was large. Today, however, two-children families are generally predominant in industrialized societies. This means that the patient with a recessive disease is very often the only one affected in an otherwise healthy family. However, once an affected child has been born, the genetic risk for any further child of the same parents is 25 %. This is important for genetic counseling.

Xeroderma pigmentosum is an autosomal recessive disease (278 700). After exposure to ultraviolet light erythema develops, especially in the face, followed by atrophy and telangiectases. Finally, skin cancers develop that, if untreated, lead to death. Figure 4.7 b shows a typical pedigree; here the parents are first cousins. The rate of consanguinity among parents of patients with rare recessive diseases is well above the population average. Usually these parents have inherited this gene from a common ancestor (Sect. 13.2). In Garrod's days this was a powerful tool for recognizing rare recessive diseases; among ten families of alkaptonurics for which this information was available, the parents were first cousins in six cases (Sect. 1.5). Today, however, the consanguinity rate has decreased in most industrialized societies. Hence, even if the rate of consanguinity in families with affected children is substantially increased above the population average, this does not necessarily lead to the appearance of consanguineous mating when a limited num-

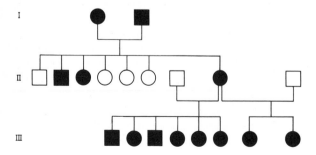

Fig. 4.6. Woman homozygous for epithelioma adenoides cysticum and her progeny in two marriages. (From Gaul 1953 [21]). The pedigree was complemented in 1958 by Ollendorff-Curth [45]

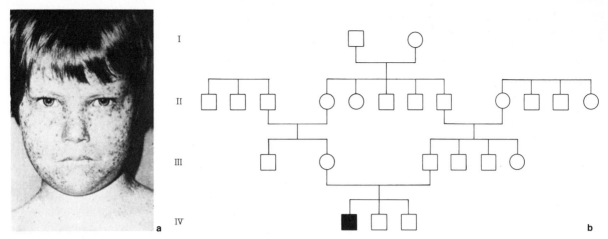

Fig. 4.7 a, b. Xeroderma pigmentosum. **a** Girl with this condition (Courtesy of Dr. U. W. Schnyder) **b** Pedigree of single case with first-cousin marriage. (From Dorn 1959 [14]).

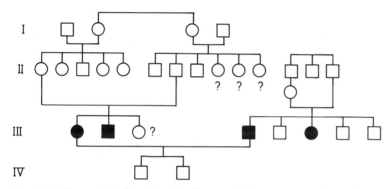

Fig. 4.8. Pedigree of deaf-mutism showing genetic heterogeneity. Both parents are affected with a hereditary type of deaf-mutism; they have affected sibs and come from consanguineous marriages; however, the two sons are not deaf. They are compound heterozygotes for different deaf mutism genes. (From Mühlmann 1930 [44]).

ber of families are studied particularly if the abnormal gene is not too rare. This phenomenon together with the small average family size makes it increasingly difficult to recognize an autosomal-recessive mode of inheritance with certainty. Fortunately, however, we no longer need to depend solely on formal genetics. When a rare disease, especially in a child, shows signs of being an inborn error of metabolism, and especially when an enzyme defect can be demonstrated, a recessive mode of inheritance can be inferred in the absence of evidence to the contrary. For purposes of genetic counseling, it must be assumed.

As a rule the vast majority of patients with autosomal-recessive diseases are children of two heterozygotes. Especially decisive for recessive inheritance are the rare matings of two homozygotes with the same anomaly. If both parents are homozygous for the same recessive gene, their mating should exclusively produce affected children. A number of such

examples are reported in albinism (203100, 203200). Some marriages between albinos, however, have produced normally pigmented children [63]. Unless these children are all illegitimate, this proves that the parents must be homozygous for different albino mutations, i.e., more than one albino locus must exist in man. This is the kind of proof that formal genetics can provide to indicate genetic heterogeneity of diseases demonstrating an autosomal recessive mode of inheritance and the same (or a very similar) phenotype. Genetic heterogeneity has now been shown in albinism by biochemical methods [54].

Another condition for which genetic heterogeneity has been proven in this way is deaf-mutism (Fig. 4.8). Since environmental causes can also cause deafness, it is remarkable that in the pedigree shown here both spouses have an affected sibling, and both parents are consanguineous. The hypothesis of genetic heterogeneity has since been confirmed for this condition with a variety of methods.

Pseudodominance in Autosomal Recessive Inheritance.
Occasionally matings between an unaffected heterozygote and an affected homozygote are observed.
One parent is affected, and the expected segregation
ratio among children is 1:1. Since this segregation
pattern mimics that found with dominant inheritance, this situation is aptly named "pseudodominance." Fortunately for genetic analysis, such matings are very rare.

Garrod's alkaptonuria (203500; Sect.1.5) provides an
example. In all families described since Garrod the
autosomal-recessive mode of inheritance had been
confirmed until 1956 when a family with a phenotypically similar but apparently dominant form was reported (Fig. 4.9) – a surprising finding. Some years
later the authors had to disavow their conclusions:
further family investigations had shown typical, recessive alkaptonuria. A number of marriages between
relatives (homozygotes × heterozygotes) had led to
pseudodominance. If an individual suffering from a
recessive disease mates with a normal homozygote,
all children are heterozygotes and hence phenotypically normal. As soon as we learn to treat recessive
diseases successfully, marriages of affected but treated homozygotes will increase.

Expressivity is generally more uniform within the
same family in recessive than in dominant disorders.
Incomplete penetrance seems to be rare. Variability
between families, however, may be appreciable.

Compound Heterozygotes. When a more penetrating
biochemical analysis becomes possible, alleles of different origin frequently have slightly different properties. In an increasing number of instances when the
gene is analyzed, and the mutations can be identified,
such differences can be explained by the properties of
the gene-determined proteins and the impairment of
their specific functions. The genes of hemoglobin α
and β chains offer an extreme example. Homozygosity
of a mutation within the Hbβ gene, for example, may
lead to sickle cell anemia or thalassemia major, depending on the precise place of the base substitution.
If there are different substitutions within the two alleles, the resulting phenotype might differ from any
one of the two true homozygotes. The phenotype of
the compound heterozygote who has the sickle cell
mutation in one allele and the HbC mutation in the
other is different from that of either homozygote (SS
or CC). It depends on the population structure how often homozygous patients with a recessive disease are
true homozygotes carrying precisely the same mutation twice, and how often they are compound heterozygotes who carry in their two chromosomes different
mutations of homologous genes (Fig. 4.10).

We can be reasonably sure that an affected homozygote carries two copies of the same mutation if both

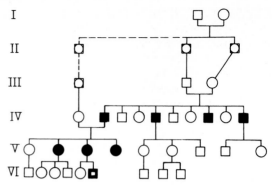

Fig. 4.9. Pedigree of pseudodominance of alkaptonuria, an
autosomal-recessive condition. ▣, Suspected alcaptonuric;
◯, sex unknown. (From Milch [42])

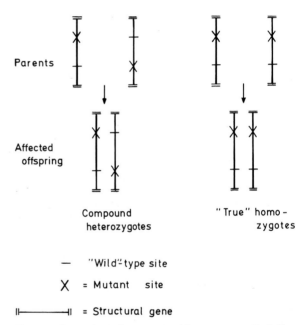

— = "Wild"-type site

✗ = Mutant site

‖———‖ = Structural gene

Fig. 4.10. Formation of a compound heterozygote. Each line
represents the mutant locus on one chromosome in a parent.
Among the many possibilities for mutation, two are shown. If
parents are heterozygous for mutations that are at identical
sites, the affected child is a "true" homozygote; otherwise,
he or she is a compound heterozygote

copies have a common origin; for example, if his parents are first cousins and if the condition is very rare.
Another source of identity by descent are cases from
an isolate in which a single mutation – which has
been introduced by one individual – became frequent, such as the skin disease called Mal de Meleda
on the Croatian island of Mljet (Sect.13.3.2). Even in
a larger and genetically heterogeneous population
group, however, the majority of homozygotes may
carry the same gene twice. This happens especially
when the gene had a selective advantage some time

in the past. The *CFTR* (cystic fibrosis) gene is one example: about 60%–70% of all abnormal alleles in northwestern European populations are of the type Δ 508, meaning that about 40%–50% of patients are indeed homozygous for this mutation ($0.7 \times 0.7 = 0.49$) (Sect. 12.1.3). In other diseases the great majority of "homozygous" individuals are in fact compound heterozygotes. With the progress of DNA studies of human genes this question will be answered directly in an increasing number of instances.

4.1.4 X-Linked Modes of Inheritance

In humans, every mating is a Mendelian backcross with respect to the X and Y chromosomes:

		Paternal gametes	
		X	Y
Maternal gametes	X	$^1/_4$ XX	$^1/_4$ XY
	X	$^1/_4$ XX	$^1/_4$ XY
Total		$^1/_2$ XX♀	$+$ $^1/_2$ XY♂

This implies that on average female and male zygotes are formed at a 1:1 ratio. This, however, is not quite true. The sex ratio at birth (known as the secondary sex ratio in contrast to the primary sex ratio at conception) is slightly shifted in favor of boys (102–106 boys/100 girls). The primary sex ratio is not known exactly, but there are hints that it is also somewhat variable. The formal characteristics of X-linked modes of inheritance can easily be derived from the mode of sex determination. [Many studies on the (primary and secondary) sex ratio have been published. Chromosome studies on abortions should reflect the primary sex ratio and point to a value not too far from 100 (boys and girls in a ratio of 1:1). However, the primary and secondary sex ratio also depend on the interval between sexual intercourse and ovulation, frequency of intercourse, general cultural conditions, and even war and peace. After artificial insemination, the fraction of male offspring appears to be appreciably increased.

X-Linked Recessive Mode of Inheritance. If we use A for the dominant, normal wild-type and a for the recessive alleles, the following matings are possible:

a) AA♀ × A♂. All children have the phenotype A. Neither this nor the analogous mating aa × a is useful for genetic analysis.

b) AA♀ × a♂. All sons have one of the mother's normal alleles. They are healthy. All daughters are heterozygous Aa. They are phenotypically healthy, but carriers of the abnormal allele. In the analogous, very rare mating aa♀ + A♂ all sons are affected (a), and all daughters are heterozygous (Aa).

c) Aa♀ + A♂. This type is most important. All daughters are phenotypically normal; half are heterozygous carriers. Half of their sons are hemizygous a and affected. The analogous mating Aa♀ × a♂ is extremely rare. There is a 1:1 ratio of affected and heterozygotes among female children and an 1:1 ratio of affected and normals among males.

The principal formal characteristics of X-linked recessive inheritance can be summarized as follows: Males are predominantly – and in rare X-linked conditions almost exclusively – affected. All their phenotypically healthy but heterozygous daughters are carriers. If no new mutation has occurred, and the mother of the affected male is heterozygous, half of his sisters are heterozygous carriers. Among sons of heterozygous women, there is a 1:1 ratio between affected and unaffected.

Strictly speaking, transmission from affected grandfathers via healthy mothers to affected grandsons is helpful, but not altogether decisive for locating the gene on the X chromosome. An autosomal gene with manifestation limited to the male sex could show the same pattern. The fact that all sons of affected men are unaffected, however, is decisive unless the wife is a heterozygous carrier which may not be unusual for common X-linked traits. This criterion can create difficulties in interpretation when a disease is so severe that the patients do not reproduce.

The two most famous and, from a practical standpoint, very important examples are hemophilia A and B (306700, 306900). Due to its alarming manifestations, hemophilia has been known to doctors for a long time and has given rise to the formulation of Nasse's rule (Sect. 1.2). Figure 4.11 shows the famous pedigree of Queen Victoria's descendants in the European royal houses. One of the hemophilics was the Czarevich Alexei of Russia, and in this case genetic disease influenced politics. Rasputin's power over the imperial couple was based at least partially on his ability to comfort the Czarevich when he was frightened by bleedings. Much larger pedigrees have been described, probably the most extensive being that of hemophilia B in Tenna, Switzerland. As a rule, however, the pedigrees observed in practice are much smaller. Frequently there is only one sibship with affected brothers, or the patient is even the only one affected in an otherwise healthy family. Again, as in dominant conditions (Sect. 4.1.2), this is caused by the reduced reproductive capacity of the patients, which leads to the elimination of most severe hemophilia genes within one or a few generations after they have been produced by new mutation. As expected, almost all hemophilia patients are males. However, there are a few exceptions. Figure 4.12 shows a pedigree from former Czechoslovakia in which a hemophilic had married a heterozygote (who was his double first cousin because in their parents' generation two brothers had married two sisters). The homozygous sisters both had moderately severe hemophilia similar to their affected male relatives.

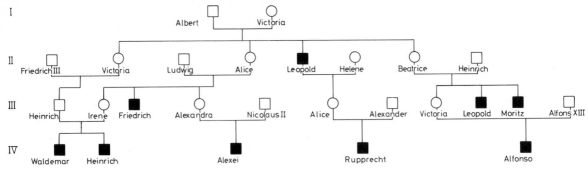

Fig. 4.11. Pedigree of X-linked recessive hemophilia A in the European royal houses. Queen Victoria (I,2) was heterozy-gous; she transmitted the mutant gene to one hemophilic son and to three daughters

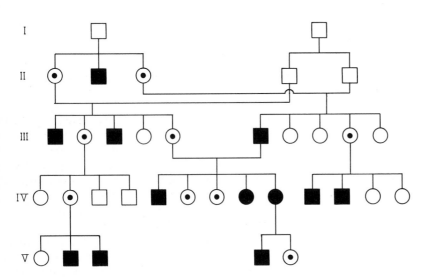

Fig. 4.12. Pedigree of two female homozygotes for X-linked hemo-philia. The parents are double first cousins. ⊙, Obligatory hetero-zygotes. (From Pola and Svojitka 1957 [50])

Some X-linked conditions have reached consider-able frequencies. The most widespread are red-green color vision defects (Sect. 15.2.1.5) and var-iants of the enzyme glucose-6-phosphate dehydro-genase (Sect. 7.2.2.2), but various types of X-linked mental retardation (Sect. 15.2.1.2) are also common.

X-Linked Dominant Mode of Inheritance. An X-linked dominant condition manifests itself in hemizygous men and heterozygous women. However, all sons of affected males are free of the trait unless their moth-ers are also affected, and the sons' children are also unaffected. On the other hand, all daughters of affec-ted males are affected. Among children of affected women the segregation ratio is 1:1 regardless of the child's sex, just as in autosomal-dominant inheri-tance. If affected individuals have a normal rate of re-production, about twice as many affected females as males are found in the population.

Since only children of affected males provide infor-mation in discriminating X-linked dominant from autosomal-dominant inheritance, it is difficult or

even impossible to distinguish between these modes of inheritance when the available data are scarce.

The first clearcut example was described by Siemens in 1925 [55] in a skin disease that he named "keratosis follicularis spinulosa decalvans cum ophiasi" (308 800). The disease manifests follicular hyperkeratosis leading to partial or total loss of eyelashes, eyebrows, and head hair. Severe manifesta-tions were, however, confined to the male members of this fa-mily.

Since then it has been confirmed for all traits with an X-linked dominant mode of inheritance that males are on average more severely affected than females. This finding is no surprise since heterozygous wo-men have a normal allele for compensation, but a sa-tisfactory explanation became possible only when random inactivation of one of the X chromosomes in females was discovered (Sect. 2.2.3.3).

Another example of X-linked dominant inheritance is vita-min D resistant rickets with hypophosphatemia (307 800) [78]. In the pedigree shown in Fig. 4.13, all 11 daughters of

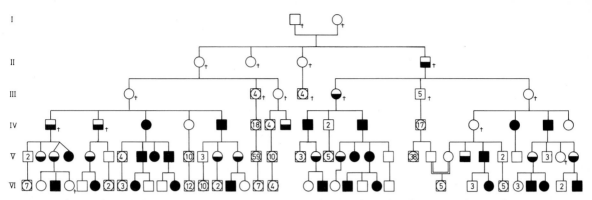

Fig. 4.13. Pedigree of X-linked dominant vitamin D resistant rickets and hypophosphatemia. ■, Hypophosphatemia and rickets; ▨, hypophosphatemia without rickets. (From Winters et al. 1957 [78])

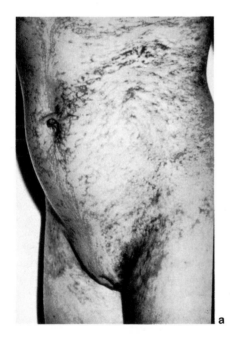

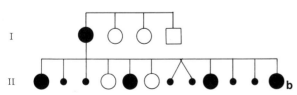

Fig. 4.14. a Incontinentia pigmenti (Bloch-Sulzberger; courtesy of Dr. W. Fuhrmann). Note the marble cake appearance of skin. **b** Pedigree of incontinentia pigmenti. •, Spontaneous abortion; ●, incontinentia pigmenti. (From Lenz 1961 [38])

the affected men suffered from rickets or had hypophosphatemia; all 10 of their sons, however, were healthy. The affected women have both affected and healthy sons and daughters. The probability for the mode of inheritance to be autosomal-dominant and for the affected males to have only affect-

ed daughters and only healthy sons is less than 1:10 000. Moreover, in this family male members also tended to be more severely affected than females.

X-Linked Dominant Inheritance with Lethality of the Male Hemizygotes [75]. Females with X-chromosomal diseases tend to have milder symptoms than males, as noted above. In some cases the male zygotes may be so severely affected that they die before birth, and only the females survive. This would result in pedigrees containing only affected females, and among their children affected daughters, normal daughters, and normal sons would be found in the ratio of 1:1:1. Among the male hemizygotes who did not die in very early pregnancy, spontaneous abortions (or male stillbirths) would be expected. W. Lenz in 1961 [38] was the first to show that this mode of inheritance exists in humans in the condition known as incontinentia pigmenti (Bloch-Sulzberger; 308300).

Around the time of birth the girls affected with this disease develop inflammatory erythematous and vesicular skin disorders. Later, marblecakelike pigmentations appear (Fig. 4.14a). The syndrome additionally comprises tooth anomalies. Figure 4.14b shows a typical pedigree. The alternative hypothesis would be that of an autosomal-dominant mode of inheritance with manifestation limited to the female sex. The two hypotheses would have the following consequences:

a) With autosomal-dominant sex-limited inheritance, and after proper correction (Sect. 4.3.4), there would be a 1:1 ratio of affected to unaffected among sisters of propositae. All brothers would be healthy. If the population sex ratio is assumed to be 1:1, a sex ratio of 2♂:1♀ would be expected among healthy sibs. With X-linked inheritance, on the other hand, the expected number of healthy brothers is much lower, because one-half of the male zygotes are expected to die before birth (possibly leading to an increased rate of spontaneous miscarriages). Among healthy sibs a 1♂:1♀ ratio would be expected.

b) With autosomal-dominant inheritance the abnormal gene may come from the father or from the mother. Therefore more remotely related affected relatives are to be expected among paternal as well as among maternal relatives. With X-linked inheritance, on the other hand, the gene must come from the mother. Considering the rarity of the condition, additional cases would not occur in the father's family.

c) With autosomal-dominant inheritance the loss of mutant genes per generation would be relatively small compared to the total number of these mutations in the population, since the male carriers, being free of symptoms, would reproduce normally. Therefore, assuming genetic equilibrium (Sect. 9.3.1), the number of new mutations would be small compared to the overall number of cases in the population. With X-linked inheritance, on the other hand, the loss of zygotes is high due to death of the hemizygote. Hence many of the cases in the population are caused by recent mutation, and extensive pedigrees are rare [4].

The available statistical evidence has consistently supported the hypothesis of an X-linked dominant mode of inheritance with lethality of the male hemizygote. According to Carney et al. [10], 593 female and 16 male cases have been reported. Among the female patients 55% had a positive family history. How can the sporadic males be explained? Of course, the phenomenon of *Durchbrenners* (Hadorn [24] used the term "escapers" – the occasional survival of individuals affected with a lethal genotype) is well known, but Lenz [39] suggested a more specific explanation, assuming, on the basis of a suggestion by Gartler and Francke [20], that a mutation occurs in only one halfstrand of the DNA double helix of either the sperm or the oocyte.

Meanwhile, a few other conditions have been added to this category. One is the orofaciodigital syndrome type I (311 200) which consists of a number of malformations of mouth and tongue, a median cleft lip, and various malformations of digits [19]. Other examples may include focal dermal hypoplasia, X-linked chondrodysplasia punctata, ornithine transcarbamylase deficiency (311 250; lethality in the *neonatal* hemizygous male), and partial lipodystrophy with lipotrophic diabetes. Another example is a rare disease of agenesis of the corpus callosum with flexor spasms, epileptic seizures, and chorioretinal abnormalities (Aicardi syndrome [1]). All cases except one patient with Klinefelter syndrome [41] were girls. They died in infancy. Hence, they must have been sporadic, new mutants. One observation in sisters was explained by germ cell mosaicism in one of the parents. Moreover, single pedigrees suggesting this mode of inheritance have been published for some other conditions, for example, a special variant of limb girdle muscular dystrophy [3, 75].

Genes on the Y Chromosome. Until the 1950s most geneticists were convinced that the human Y chromosome contained genes that occasionally mutate, giving rise to a Y-linked (or holandric) mode of inheritance with male-to-male transmission and males solely being affected. Stern in 1957 [58] reviewed the evidence with the result that the time-honored text-book example of Y-linked inheritance of the porcupine man (severe ichthyosis) could no longer be maintained as valid. The only characteristics for which Y-linked inheritance can still be discussed are hairy pinnae, i.e., hair on the outer rim of the ear. A number of extensive pedigrees have been published that show male-to-male transmission. However, the late onset, usually in the third decade of life, and the extremely variable expressivity and high prevalence in some populations (up to 30%), makes distinction from a multifactorial mode of inheritance with sex limitation very difficult. Y-linkage can therefore not be fully accepted for this trait.

The Y chromosome contains genes for male differentiation as well as for spermatogenesis. The evidence is discussed in Sect. 8.5.

In experimental animals, segregation ratios deviating from those expected from Mendelian expectations were occasionally reported, one example being the T locus of the mouse [6].

Other cases for which abnormal segregation has been asserted are less well-documented. Since families with many children have become the exception in most industrial societies, the prospect for tracking down and verifying abnormal segregation of pathological genes is becoming more difficult.

4.1.5 "Lethal" Factors [24]

Animal Models. Mutations showing a simple mode of inheritance often lead to more or less severe impairment of their bearer's health. There is even evidence (Sect. 4.1.4) that some X-linked conditions prevent the male hemizygote from surviving to birth. It can be assumed that mutations exist which interfere with embryonic development of their carriers so severely as to cause prenatal death.

The first reported case of a lethal mutation in mammalian genetics was the so-called yellow mouse. L. Cuénot [12] reported an apparent deviation from Mendel's law in 1905. A mutant mouse with yellow fur color did not breed true. When yellow animals were crossed with each other, normal gray mice always segregated out. All yellow mice were heterozygous. They all had the same genetic constitution A^Y/A^+; A^Y is a dominant allele of the agouti series, the wild allele of which is termed A^+. When A^Y/A^+ heterozygotes were mated with A^+/A^+ homozygotes, the expected 1:1 ratio between yellow and gray mice was observed. In 1910 it was found that A^Y/A^Y homozygotes are formed but die in utero. Abnormal embryos were later discovered in the expected frequency of 25%.

In this case the allele that is lethal in the homozygous state can be recognized in the heterozygotes by the yellow fur color.

Cases of this sort are exceptional. Generally heterozygotes of lethals are not readily recognizable; therefore lethals occurring spontaneously are difficult to ascertain even in experimental animals and much more so in man.

Usually a lethal mutation kills the embryo in a characteristic phase of its development ("effective lethal phase" [24]. This can easily be explained by the assumption that the action of the mutant gene would be required for further development in this phase.

Lethals in Humans. In humans many different types of lethals must occur since many metabolic pathways and their enzymes are essential for survival. It is likely that many still undetected enzyme defects do indeed occur but are not compatible with zygote survival. Moreover, many types of defects of inducer substances needed during embryonic development, and enzymes involved in nucleic acid and protein synthesis, may occur and add to the high incidence of zygote death, which has so far been unexplainable genetically. This problem is discussed from a different standpoint in the context of population genetics (Chap. 8).

According to current estimates, about 15%–20% of all recognized human pregnancies end in spontaneous miscarriage. Studies on other mammals suggest that an appreciable number of additional zygote losses go unnoticed, as death occurs during migration through the fallopian tubes. How much of this zygote wastage is due to genetic factors is unknown. A high proportion is caused by numerical or structural chromosome aberrations (Sect. 2.2.4). However, there are certainly other maternal causes for abortion as well. While it seemed hopeless to try to relate any proportion of antenatal (or even postnatal) zygote loss to autosomal-dominant or recessive lethals, it appeared more reasonable to speculate about X-linked lethals, as these could influence the sex ratio.

4.1.6 Modifying Genes

So far we have considered phenotypic traits depending on one gene only. However, the phenotypic expression of one gene is usually influenced by other genes. Experiments with animals, especially mammals, show the importance of this "genetic background." One way to overcome analytic difficulties caused by such variation is the use of inbred strains where all animals are genetically alike.

The genetic background is a fairly diffuse concept, but in a number of cases it has been possible to show that penetrance or expressivity of a certain gene can be influenced by another, which is called a "modifier gene" when expressivity is influenced. When penetrance is suppressed altogether, the term "epistasis" (and "hypostasis" of the suppressed gene)

is used. In experimental animals cases have been analyzed in which the interaction of two mutations at different loci leads to a completely new phenotype. The classic example is the cross of chickens with "rose" combs and "pea" combs, which leads to the "walnut" comb in homozygotes for both of these mutations. To the best of our knowledge, a similar situation has not been described in man. Modifier genes and epistasis, however, have been demonstrated.

Modifying Genes in the AB0 Blood Group System. The best analyzed examples of modifying genes are offered by the AB0 blood group systems. Occurrence of the ABH antigens in saliva (and other secretions) depends on the secretor gene Se. Homozygotes se/se are nonsecretors; heterozygotes Se/se and homozygotes Se/Se are secretors. Hence, se is a recessive suppressor gene. Other rare suppressor genes even prevent the expression of ABH antigens on the surface of erythrocytes.

Bhende et al. [2] discovered a phenotype in 1952 which they called "Bombay" (211100). The erythrocytes were not agglutinated either by anti-A, anti-B or anti-H. The serum contained all three of these agglutinins. Later another family was discovered showing that the bearers of this unusual phenotype did have normal AB0 alleles, but that their manifestation was suppressed (Fig. 4.15; a woman, II, 6, has a Bombay phenotype but transmitted the B allele to one of her daughters). It was further shown that A can also be suppressed, and the available family data suggested an autosomal-recessive mode of inheritance. In the family shown in Fig. 4.15, the parents of the proposita are first cousins.

The locus is not linked to the AB0 locus. The gene pair was named H, h, the Bombay phenotype representing the homozygote, h/h. The gene has been cloned (see [41]). Depending on the nature of the suppressed allele, the phenotype is designated O_hA_1, O_hA_2, or O_hB. The phenotype has a frequency of

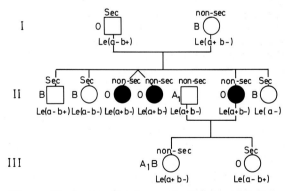

Fig. 4.15. The Bombay blood type. Manifestation of the B antigen is suppressed by a recessive gene x. Note that an O mother (II,6) has an A_1B child. (From Bhende et al. 1952 [8])

about 1 in 13000 among Maharati-speaking Indians in and around Bombay. A variant with reduced activity is common in the population isolate on Reunion Island [22]. It is caused by the defect of an enzyme that converts a precursor substance into the H antigen, which in turn is a precursor of the A and B antigens [31, 46, 51]. A second gene pair Yy, the rare homozygous conditions of which partially suppresses the A antigen, has been postulated, and subsequently a number of additional families with this condition have been reported.

Sex-Limiting Modifying Genes. In other, less directly accessible traits the action of modifying genes has been analyzed with statistical methods.

Haldane [25] tried in 1941 to identify such genes in Huntington disease, using the family data assembled by Bell in 1934 [5]. Harris in 1948 [30] examined the problem in a condition called diaphyseal aclasis (133700), which is characterized by multiple exostoses near the cartilaginous epiphyses.

The mode of inheritance is dominant; however, the condition is about twice as common in males as in females. It may be transmitted in some families through unaffected females but not through unaffected males. Statistical analysis of the comprehensive pedigree data collected by Stocks and Barrington [60] suggests in part of the families independent segregation of a factor leading to incomplete penetrance only in females: a sex-limiting modifying gene.

Modification by the Other Allele. Phenotypic expression of a gene may be modified not only by genes at other loci but also by the "normal" allele. One example comes from the genetics of the Rh factor (Sect. 5.2.4). Occasional blood specimens, when tested with an anti-Rh D serum, give neither a strong positive nor a negative reaction but an attenuated positive reaction. These are called D^u. In most cases a special allele is responsible for this effect, but there are exceptions. In several families the D^u reaction was observed only in family members having Cde as the homologous allele (Fig. 4.16).

Modification by variation in related genes. Sickle cell anemia caused by homozygosity for HbS (see section 7.3.2) becomes clinically less severe in the presence of several genetic conditions that increase the amount of fetal hemoglobin in the affected red cells. Similarly, the presence of the common alpha thalassemia gene (see section 7.3.2), makes for a milder disease manifestation.

Modification by a DNA Polymorphism Within the Same Gene. Analysis at the molecular level is revealing new and unsuspected phenomena, including those regarding modification of gene action. Prions are especially interesting proteins. Mutations within the prion gene (176640) may cause hereditary diseases such as Creutzfeldt-Jakob disease (CJD), Gerstmann-Straussler disease (GSD), or familial fatal insomnia (FFI). The same mutation (Asp $\rightarrow$ Asn-178) may lead either to CJD or to FFI, depending on a normal polymorphism within the same gene but at a different site: the allele Val 129 segregated in CJD and the allele Met 129 segregated in FFI [23].

Study of various modifying genes and their mechanism is promising to be an important feature for our understanding of the variability of genetic diseases. The causes of clinical variability in monogenic diseases are:

- Genetic heterogeneity
  - Intra-allelic: different mutations at same locus
  - Inter-allelic: different mutations at other loci
- Modifying genes
  - Additional polymorphisms altering protein conformation
  - Other, as yet unknown mechanisms
- Exposure to various environmental factors required for clinical end result
- Random additional somatic mutations of allele at same locus (e.g., tumors)
- Imprinting (parental origin of mutation)

4.1.7 Genomic Imprinting and Anticipation

A time-honored concept popular among physicians in the nineteenth and early twentieth centuries was anticipation. They observed that some hereditary diseases begin earlier in life and follow a more severe course as they progress through generations: the grandfather appeared to be mildly affected; the father was definitely ill, and in the son the disease manifests itself with full force. Anticipation was closely associated with another concept called "degeneration": in some families general, mental, and physical qualities were thought to deteriorate through the generations. These ideas be-

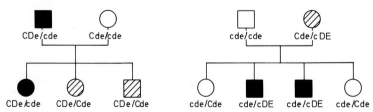

Fig. 4.16. Modification by the homologous allele in the Rh system. ●, D^+ blood with normal reaction; ◑, weak reaction (D^u variant); ○, D^- blood. The haplotype Cde reduces expression of the D factor to D^u. (From Ceppellini et al. 1953 [11])

came popular not only among physicians but also among the general public, and were expressed in literary works such as Thomas Mann's novel *Die Buddenbrocks*. In two diseases that tend to manifest during adult life, anticipation seemed to be obvious: Huntington disease and myotonic dystrophy (160 900) [17]. In the latter, myotonia is associated with relatively mild muscular dystrophy, cataracts, and sometimes mental retardation, or dementia. This disease shows an unusual degree of variability in age at onset, and earlier onset as well as a more severe course in some patients of the most recent generation.

When Mendel's laws were rediscovered, anticipation did not fit the new, and otherwise so successful, theory. Therefore scientists interested in genetic problems tried to explain these phenomena away with sophisticated arguments (which we also used in the first two editions of this book). Weinberg [71] pointed out, for example, that anticipation can easily be simulated if families were ascertained directly by patients of the youngest generation who were affected early in life. Their parents and grandparents, on the other hand, who were ascertained through these young probands, could be recognized only if onset of the disease was so late that they had a chance to have children.

Penrose [49], one of the best human geneticists of his time, explained in great detail that anticipation could be feigned if ascertainment through the youngest generation combined with dissimilarity of age at onset between parents and children, but similarity between sibs. This would be expected if, in a dominant condition, the normal allele influenced the degree of manifestation of the mutant allele (allelic modification; Fig. 4.17). There can be little doubt that explanations given by Weinberg and Penrose are correct in some instances. However, in Huntington disease and myotonic dystrophy – the explanation turned out to be different; in both cases anticipation is real. Two lines of evidence have led to this conclusion: (a) recent statistical analysis, suggested in part by observations in the mouse, has led to new and more specific insights, and (b) molecular analysis detected a novel type of mutation whose effects increased with passage through succeeding generations.

In Huntington disease, patients with early onset are more likely to have inherited their mutant genes from the father, whereas late onset is more common when the gene comes from the mother. In myotonic dystrophy, on the other hand, cases of very early onset are less rare; the babies have signs of the disease even at birth. This occurs (almost?) exclusively when the mothers are affected.

Such differences have also been observed in some other monogenic diseases (Table 4.2). The phenomenon is called "genomic imprinting." The influence of maternal and paternal genomes on development

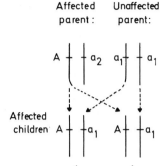

Fig. 4.17. Allelic modification. If manifestation of a dominant, abnormal gene A is modified by the normal allele, and if the allele a_1 causes severe and a_2 milder manifestation of A, there is a correlation in the degree of manifestation between affected sibs but not between affected parent and child. An affected child cannot receive the modifying a_2 allele

of the embryo differs from one another. Imprinting has also been shown for certain chromosomal aberrations such as the Prader-Willi and Angelman syndromes (Sect. 2.2.2.3). This is discussed in greater detail in Sect. 8.2.

The second insight that has contributed to explaining anticipation in Huntington disease and myotonic dystrophy is the molecular analysis of mutants. In Sect. 3.1.3.8 the IT 15 gene and the mutations leading to Huntington disease were described: amplification of a $(CAG)_n$ repeat beyond approx. 42 copies causes the disease. Moreover, these amplification products are unstable; the predominant tendency appears to

Table 4.2. Dominant diseases in which parental origin influences the disease (modified from Reik [53])

Disorder	Chromosome	Observations
Huntington disease	4	Early onset frequently associated with paternal transmission
Spinocerebellar ataxia	6	Early onset with paternal transmission
Myotonic dystrophy	19	Congenital form almost exclusively with maternal transmission
Neurofibromatosis I	17	Increased severity with maternal transmission
Neurofibromatosis II	22	Earlier onset with maternal transmission
Wilms tumor	11	Loss of maternal alleles in sporadic tumor
Osteosarcoma	13	Loss of maternal alleles in sporadic tumors

be toward an increase in copy numbers by further rounds of amplification; a reduction in copy numbers may occur but is apparently rarer. Higher copy number, on the other side, correlates with earlier onset: a convincing explanation for anticipation.

In myotonic dystrophy an analogous explanation has been found [9, 18, 27]. Here an unstable, amplified sequence was found in the 3′ untranslated region of a gene whose product was predicted to be a member of a protein kinase gene family [9]. It is a $(CTG)_n$ repeat. In normal individuals between 5 and 27 units are found, but affected patients may have between 50 and some 2000 repeats or even more. The repeat number tends to increase over the generations [27]; it is correlated with age at onset and severity of the disease, explaining anticipation. So far, however, no one knows how the phenomena of genomic imprinting are related to this type of mutation.

The anticipation story is interesting from the point of view of science development. In the "Introduction" we mentioned Kuhn's analysis [36]. Once a new paradigm has been established in the scientific community, scientists try to extend existing knowledge by using the theoretical concept on which the paradigm is founded. If data are found that apparently do not conform to theoretical prediction, this often does not induce them to abandon or even to modify their theory. Rather, they try every possibility to explain why and in what way the deviant data are wrong or misinterpreted. This is precisely what happened in myotonic dystrophy and Huntington disease. Facts were acknowledged only after (in the course of quite different studies) a modification of the basic theory had become necessary.

Interestingly, some physicians were not fooled by this conformity of the profession. Fleischer [17], a well-known ophthalmologist interested in hereditary eye diseases proposed anticipation in a myotonic dystrophy kindred in 1918. Klein [34] in 1958 regarded anticipation in this disease as real. We all should learn from this experience: we should keep our eyes open, take our observations seriously, and try not to argue them away when they seem to contradict our theoretical preconceptions.

4.1.8 Total Number of Conditions with Simple Modes of Inheritance Known So Far in Humans

For many years McKusick [41] has undertaken the task of collecting and documenting known conditions with simple modes of inheritance in man. Table 4.3 is based on the 11th edition. Since the 3rd addition was published (in 1971), the number of autosomal-dominant traits (confirmed and unconfirmed) has increased from 943 to 4458, that of autosomal-re-

Table 4.3. Number of known traits with simple modes of inheritance in man (from McKusick 1994 [41]); OMIM update Febr. 1996

Mode of inheritance	Number of traits Total
Autosomal-dominant	4669
Autosomal-recessive	4669
X-linked	443
Total	7862

cessive traits from 783 to 1730, and that of X-linked from 150 to 412.

While genetic polymorphisms are included (Sect. 12.1.2), most conditions listed in this register are rare. Many are rare hereditary diseases. At first glance the list is impressive. However, more detailed scrutiny of the conditions shows that our knowledge of these rare diseases is not nearly as good as it should and could be. There are several reasons:

a) Most hereditary diseases have become known by occasional observation of affected patients and their families. With rare diseases there is no other way to assess whether they do or do not have a genetic basis.

b) Some recessive diseases have become known because they happened to be frequent in special populations, primarily in isolates. Isolate studies permit examination of the manifestation of recessive diseases caused by a single mutation. One problem with this approach is that chance determines which genes are studied.

c) Most human and medical geneticists are working in relatively few industrialized countries. However, genes for rare diseases show a very unequal distribution in different populations. This is particularly true for recessives but has also been shown for dominants with normal or only slightly lowered biological fitness, i.e., when the incidence is not determined by the mutation rate. Hence the developing countries can be expected to abound with hereditary anomalies and diseases that are unclassified to date. Any medical geneticist who has ever walked through, say, an Indian village knows that this suggestion is not merely a theoretical speculation.

d) Genetic defects with simple modes of inheritance have a good chance of being detected when they show a clearcut phenotype that is readily recognizable. This is why the inherited conditions of the skin and eye are relatively well known. Other defects, however, may cause anomalies or diseases in some families that are precipitated by environmental factors. Most of such hidden defects are unknown at present.

e) The real significance of hereditary disease and its total impact can be established only by studies in large populations, using epidemiological methods. Such studies offer the opportunity to detect heterogeneity in etiology and to aid in distinguishing genetic and nongenetic causes. They afford the only basis on which genetic parameters such as mutation rates, biological fitness, and the relative incidence of mild and severe mutations of the same gene can be established. They also help in predicting the long-term and public health effects of medical therapy and of genetic counseling for future generations. Many epidemiological studies in human genetics were carried out in the 1940s and 1950s. A few institutes played a leading role – most notably Kemp's Institute in Copenhagen. Here a genetic register of the Danish population was established, and on this basis studies for several hereditary conditions were performed, such as achondroplasia (100 800), polycystic kidneys (173 900), limb malformations, and others.

Other active groups operated in Northern Ireland and in Michigan in the United States. Studies on single groups of diseases were also undertaken in the United Kingdom, the United States, Sweden, Finland, Switzerland, and Germany. Judged by present standards, many studies were necessarily imperfect. However, much that we know about incidence, different genetic types, mutations rates, and biological fitness we owe to these studies.

Difference in the Relative Frequencies of Dominant and Recessive Conditions in Humans and Animals? At first glance, there appears to be a difference between humans and experimental animals in the relative frequencies of dominant and recessive conditions. Of the better known mutants of *Drosophila melanogaster* 200 are recessive and only 13 (6.1%) dominant. In the chicken, 40 recessive and 28 dominant mutations have been reported. In the mouse only 17 of 74 mutants are dominant (23%) and the rest recessive. In the rabbit 32 recessive and 6 dominant mutations have been found. (Instances of multiple allelism are counted as one gene locus.) In humans, on the other hand, more dominant than recessive conditions are known. This discrepancy, however, is likely to be caused by diagnostic bias. Our species observes itself most carefully; therefore, defects are detectable that would probably escape observation when present in experimental animals. It would be difficult, for example, to detect brachy-

dactyly in the mouse. This condition, however, leads to a much more severe defect when homozygous (Sect. 4.1.2). Hence such a defect, dominant in man, would be counted as recessive in the mouse. Another reason might be that the population of industrialized countries is not in equilibrium for recessive genes. The frequency of consanguineous matings has dropped sharply, and therefore the chance of a recessive gene meeting another mutation in the same gene and becoming homozygous is reduced. A new equilibrium will be reached only in the very distant future when recessive genes could become sufficiently frequent again (Sect. 13.1.1). In our opinion, there is no significant reason to assume that humans are unique in regard to the ratio of dominant and recessive mutations.

4.1.9 Diseases Due to Mutations in the Mitochondrial Genome

As shown in Sect. 3.4, the mitochondrial genome, mtDNA, consists of a ring-shaped chromosome with more than 16 000 bp. It encodes a small (12 S) and a large (16 S) rRNA, 22 tRNAs, and 13 polypeptides. All these polypeptides are subunits of the mitochondrial energy-generating pathway, oxidative phosphorylation (OXPHOS). OXPHOS encompasses five multiunit enzyme complexes, arrayed within the mitochondrial inner membrane; most of the peptides necessary for building these enzyme complexes are encoded in nuclear genes.

At fertilization the oocyte contains about 200 000 mtDNAs. Once fertilized, the nuclear DNA replicates and the oocyte cleaves, but the mtDNA does not replicate until after the blastocyst is formed. Since the blastocyst cells that are destined to become the embryo proper constitute only a small fraction of all blastocyst cells, and only a fraction of these cells enter the female germ line, few of the oocyte's mtDNA molecules are found in the primordial germ cells. However, it is questionable whether this mechanism is sufficient for creating an mtDNA "population" in human cells that is as homogeneous, as is normally found, especially if we consider the fact that a single mitochondrium contains 5–10 mtDNA molecules.

Most proteins necessary for development of the mitochondria themselves are produced by nuclear genes. Therefore some of the diseases due to malfunction of mitochondria are caused by defects of such genes; they follow classical Mendelian modes of inheritance [67, 69]. On the other hand, diseases due to defects of genes in the mitochondrial genome are transmitted as the mitochondria themselves, i.e. from the mother to all children, irrespective of sex. However, considering the great number of mitochondria that a

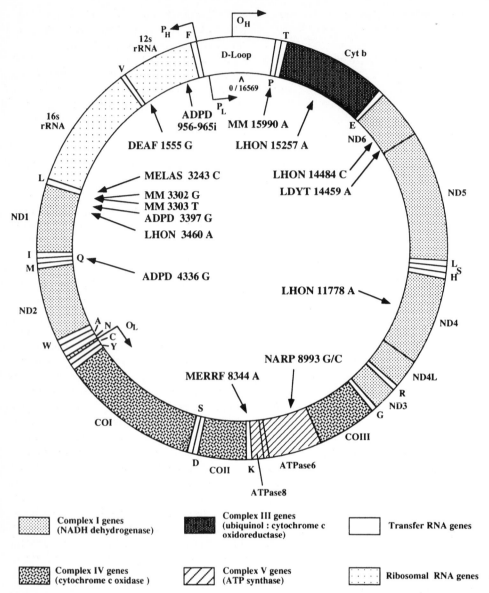

Fig. 4.18. Human tDNA map showing locations of genes and mutations Definitions of gene symbols and mutations, example: MTTK*MERRF8344A.MTTK is the altered mtDNA (MT) gene for tRNA (Lys) (TK); Myoclonic epilepsy and Ragged Red Fiber disease (MERRF) is the most characteristic clinical presentation, 8344 is the altered nucleotide, and A is the pathogenic base. (From Wallace 1994 [97])

oocyte contains, and the number of genomes per mitochondrium, it is not surprising that a child may inherit from its mother more than one type of mitochondrial genome; cells containing variable proportions of affected mitochondria are "heteroplasmic." During further development, one genome may become more abundant; different cell lineages may even become "homoplasmic" for different mitochondrial genomes. This may explain in part the enormous phenotypic variation between individuals with the same mitochondrial disease. A heteroplasmic mtDNA mutation may reduce the function of the gene-determined peptide. In most instances this is unimportant, but in a few cells the fraction of mitochondria containing the mutant increases to the extent that OXPHOS enzyme activity decreases until it falls below the cellular or tissue energetic threshold, i.e., the minimum activity necessary to sustain oxidative phosphorylation.

Four categories of diseases due to mutations in the mitochondrial genome may be distinguished (Fig. 4.18) [69]. In the first we find missense mutations with relatively mild phenotypic effects. These are transmitted maternally and appear to be homoplasmic. The second category comprises deleterious point mutations. Of course they can be transmitted maternally only if they are heteroplasmic. The third category, deletion mutants, occur by new mutations during early development, and these are therefore heteroplasmic. In the fourth category of diseases, certain mutations may be present that diminish OXPHOS activity somewhat but at onset not sufficiently to cause functional damage. During life time, however, additional random mutations accumulate in somatic cells, reducing their OXPHOS capacity until the threshold is reached. Then a degenerative disease of advanced age such as Alzheimer or Parkinson disease might ensue.

Leber Optical Atrophy. An example of the first category is Leber optical atrophy, (308 900) [67, 68]. In this disease, rapid vision loss occurs during young adult age; cardiac dysrhythmia is common. Variation in severity of the disease is strong; males are more often and on average more severely affected than females; the proportion of transmitting females in the family is much larger than expected if the mutation were X-linked. Transmission, however, occurs exclusively through females [65]. Molecular analysis revealed a $G \rightarrow A$ transition leading to an $Arg \rightarrow His$ replacement in the gene for the NADH subunit 4. The Arg residue must be important for function since it has been conserved in evolution from flagellates and fungi to humans. The mutation is homoplasmic; hence the clinical variability as well as the sex difference must have other causes that are still unknown. Other mitochondrial mutations in closely related genes have occasionally been described [33].

Two diseases apparently belong to the second category – deleterious but heteroplasmic point mutations. In a large kindred, Leber's disease was found to be associated with infantile bilateral striatal necrosis. In this family four phenotypes were found: normal, Leber disease, striatal necrosis, and the combination of the two diseases. All members were related through the female line. Since careful analysis has shown no deletion, the disease appears to be due to a deleterious but heteroplasmic point mutation. Depending on the preponderance of the aberrant mtDNA, the clinical signs vary [67]. The second disease of this class is one combining myoclonic epilepsy and mitochondrial myopathy – both conditions with huge interindividual variation [13].

Deletions. The third category is that of sporadic and heteroplasmic deletions. These occur as somatic mutations; since all deletions in one individual are identical, they must have arisen by clonal expansion of a single molecular event. Therefore a selective advan-

tage of mutant cells has been suggested [67]. Figure 4.19 shows such deletions. Clinical manifestations again depend on the distribution of mutant mitochondria. A family has been described [79] in which multiple deletions of mtDNA behaved as one autosomal-dominant trait. The affected individuals suffered from progressive external ophthalmoplegia, progressive proximal weakness, bilateral cateract, and precocious death.

Diseases of Advanced Age. The fourth category comprises diseases of advanced age that have not found satisfactory explanations so far. In both Alzheimer and Parkinson diseases, for example, pedigrees have been observed in which relatively early onset in middle age is combined with an autosomal-dominant mode of inheritance. In the majority of these cases, however, an accumulation of affected individuals within families is found but no combination of clear-cut Mendelian mode of inheritance with onset at more advanced age. Here, mildly to moderately deleterious germ line mutations established in the distant past, and present in a certain proportion of the population in combination with somatic mutations occurring during lifetime of the individual, may lead to such degenerative diseases. For example, a homoplasmic mutation among whites at nucleotide base pair 4336 leading to a tRNA mutant has been observed in 5 % of Alzheimer and Parkinson disease mutations, but appears to be much rarer in the general population. It may contribute to the multifactorial origin of these diseases [61].

In general, mutations within the mitochondrial genome affect mainly organ systems that depend on intact oxidation – central nervous system and muscles. Probably the number of known diseases due to mutations in the mitochondrial genome will increase in future (Table 4.4).

4.2 Hardy-Weinberg Law and Its Applications

4.2.1 Formal Basis

So far the application of Mendel's laws in man has been considered from the standpoint of the single family. What, however, are the consequences for the genetic composition of the population? The field of research that considers this problem is called population genetics (Chap. 12). Some basic concepts are introduced here.

These concepts revolve around the so-called Hardy-Weinberg law, discovered by these two authors independently in 1908 [26, 70]. In 1904 Pearson [48] – in the process of reconciling the consequen-

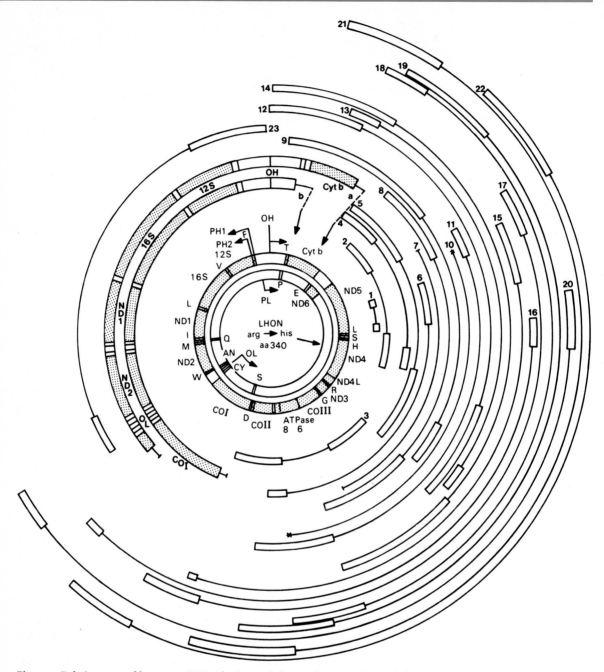

Fig. 4.19. Deletion map of human mtDNA. The inner circles show localization of genes and mutations. (See also Fig. 4.18). The arcs no. 1–23 show the mtDNA regions that were lost in various deletions. The open bars at the end of the arcs show regions of uncertainty. Deletion 1 was found in a patient with Myoclonic Epilepsy and Ragged Red Fibre Disease (MERRF) together with stroke-like symptoms. Deletions 2–23 were found in ocular myopathy patients with symptoms of varying severity. Deletion 10 was found in about one third of all ocular myopathy patients. The * at the ends of deletion 10 represents the associated 13 base pairs direct repeat. The two partial mtDNA maps labeled "a" and "b" to the left of the function map indicate the regions that were tandemly duplicated in patients with ocular myopathy associated with diabetes mellitus. The insertion sites around MTCYB (cytb) are indicated by arrows. (From Wallace 1989 [96])

Table 4.4. The Mitochondrial Chromosome (from McKusick [41])

Location (nt)	Symbol	Title	MIM #	Disorder*
577–647	MTTF	tRNA phenylalanine	590070	
648–1601	MTRNR1	12S rRNA	561000	Deafness, aminoglycoside-induced, 580000
1602–1670	MTTV	tRNA valine	590105	
1671–3229	MTRNR2	16S rRNA	561010	Cloramphenicol resistance, 515000
3230–3304	MTTL1	tRNA leucine 1 (UUA/G)	590050	MELAS syndrome, 540000; MERRF syndrome, 545000; Cardiomyopathy; Diabetes-deafness syndrome, 520000
3307–4262	MTND1	NADH dehydrogenase 1	516000	Leber optic atrophy, 535000
4263–4331	MTTI	tRNA isoleucine	590045	Cardiomyopathy
4400–4329†	MTTQ	tRNA glutamine	590030	Cardiomyopathy
4402–4469	MTTM	tRNA methionine	590065	
4470–5511	MTND2	NADH dehydrogenase 2	516001	Leber optic atrophy, 535000
5512–5576	MTTW	tRNA tryptophan	590095	
5655–5587†	MTTA	tRNA alanine	590000	
5729–5657†	MTTN	tRNA asparagine	590010	Ophthalmoplegia, isolated
5826–5761†	MTTC	tRNA cysteine	590020	Ophthalmoplegia, isolated
5891–5826†	MTTY	tRNA tyrosine	590100	
5904–7444	MTCO1	cytochrome c oxidase I	516030	
7516–7445†	MTTS1	tRNA serine 1 (UCN)	590080	
7518–7585	MTTD	tRNA aspartic acid	590015	
7586–8262	MTCO2	cytochrome c oxidase II	516040	
8295–8364	MTTK	tRNA lysine	590060	MERRF syndrome, 545000
8366–8572	MTATP8	ATP synthase 8	516070	
8527–9207	MTATP6	ATP synthase 6	516060	Leigh syndrome; NARP syndrome, 551500
9207–9990	MTCO3	cytochrome c oxidase III	516050	Leber optic atrophy 535000
9991–10058	MTTG	tRNA glycine	590035	
10059–10404	MTND3	NADH dehydrogenase 3	516002	Leber optic atrophy, 535000
10405–10469	MTTR	tRNA arginine	590005	Leber optic atrophy, 535000
10470–10766	MTND4L	NADH dehydrogenase 4L	516004	
10760–12137	MTND4	NADH dehydrogenase 4	516003	Leber optic atrophy, 535000
12138–12206	MTTH	tRNA histidine	590040	
12207–12265	MTTS2	tRNA serine 2 (AGU/C)	590085	
12266–12336	MTTL2	tRNA leucine 2 (CUN)	590055	
12337–14148	MTND5	NADH dehydrogenase 5	516005	
14673–14149†	MTND6	NADH dehydrogenase 6	516006	Leber optic atrophy, 535000
14742–14674†	MTTE	tRNA glutamic acid	590025	
14747–15887	MTCYB	cytochrome b	516020	
15888–15953	MTTT	tRNA threonine	590090	
16023–15955†	MTTP	tRNA proline	590075	

* In addition to the disorders caused by point mutations in individual genes, deletions involving more than one mitochondrial gene have been identified in Pearson syndrome (557000), early-onset chronic diarrhea with villus atrophy (520100), and Kearns-Sayre syndrome (530000), among others.

† Transcribed from light chain (L) in opposite direction from all the other genes which are transcribed from the heavy chain (H).

ces of Mendel's laws for the population with biometric results – had already derived this law for the special case of equal gene frequencies of two alleles.

The law in its more general form may be formulated as follows: Let the gene frequencies of two alleles in a certain population be p for the allele A and q for the allele B; $(p + q = 1)$. Let mating and reproduction be random with respect to this gene locus. The gene frequencies then remain the same, and the genotypes

AA, AB, and BB in the F_1 generation occur in the relative frequencies p^2, $2pq$, and q^2, the terms of the binomial expression $(p + q)^2$. In autosomal genes, and in the absence of disturbing influences, this proportion is maintained through all subsequent generations.

Derivations from the Hardy-Weinberg Law. We assume that at the beginning the proportions of genotypes AA, AB, and BB in the population of both males and females are D, $2H$, and R, respectively. Symbolically, the distribution of genotypes in both sexes may be written as:

$$D \times AA + 2H \times AB + R \times BB \qquad (4.1)$$

From this the distribution of mating types for random mating is obtained by formal squaring:

$$(D \times AA + 2H \times AB + R \times BB)^2 = D^2 \times AA \times AA$$
$$+ 4DH \times AA \times AB + 2DR \times AA \times BB + 4H^2 \times$$
$$AB \times AB$$
$$+ 4HR \times AB \times BB + R^2 \times BB \times BB$$

The distribution of genotypes in the offspring of the different mating types is:

AA $\times$ AA	AA
AA $\times$ AB	$\frac{1}{2}$AA + $\frac{1}{2}$AB
AA $\times$ BB	AB
AB $\times$ AB	$\frac{1}{2}$AA + $\frac{1}{2}$AB + $\frac{1}{4}$BB
AB $\times$ BB	$\frac{1}{2}$AB + $\frac{1}{2}$BB
BB $\times$ BB	BB

Inserting these distributions for the mating types into Eq. 4.1 yields the distribution of genotypes in the F_1 generation:

$$(D^2 + 2DH + H^2)AA + (2DH + 2DR + 2H^2 +$$
$$2HR)AB + (H^2 + 2HR + R^2)BB = p^2AA + 2pqAB$$
$$+ q^2BB$$

where $p = D + H$, $q = H + R$ are the frequencies of the alleles A and B, respectively, in the parental generation. Thus, the distribution of genotypes in the offspring generation is uniquely determined by the gene frequencies in the parental population:

$$D' = p^2, \quad 2H' = 2pq, \quad R' = q^2.$$

As:

$$p' = D' + H' = p^2 + pq = p,$$
$$q' = H' + R' = pq + q^2 = q,$$

the gene frequencies in the F_1 generation are equal to those in the parental generation. Thus, the genotype distribution in the next generation (F_2) is also the same as in the F_1 generation, and this holds true for all following generations.

This means that in autosomal inheritance these proportions are expected in the first generation and are maintained in the following generations. For X-linked genes the situation is slightly more complicated (Sect. 12.1.1). At the same time, the concept of gene frequencies $p + q = 1$ was created.

The Hardy-Weinberg law can also be rephrased, indicating that random mating is equivalent to drawing random samples of size 2 from a pool of genes containing the two alleles A and a with relative frequencies p and q. One of the advantages of this law is that frequencies of genetic traits in different popula-

tions can be expressed and compared in terms of gene frequencies.

Apart from making it possible to simplify population descriptions, the Hardy-Weinberg law can also help to elucidate modes of inheritance in cases where the straightforward approach through family studies would be too difficult. The classic examples are the AB0 blood types.

4.2.2 Hardy-Weinberg Expectations Establish the Genetic Basis of AB0 Blood Group Alleles

Multiple Allelisms. So far, only two different alleles for each locus have been considered. Frequently, however, more than two different states for one gene locus, i.e., more than two alleles, are possible. Examples of such "multiple allelism" in humans and experimental animals abound. Two of the classics are the white series in *Drosophila melanogaster* and the albino series in rabbits.

The formal characteristics can easily be derived:

a) In any one individual a maximum of only two alleles can be present (unless there are more than two homologous chromosomes, as in trisomics).
b) Between these alleles, crossing over can be disregarded as they are located at homologous loci. The qualifications of this second condition are dealt with in Sect. 3.5, in connection with modern concepts of the gene. Here the simplest formal model is described, using the AB0 blood groups as an example.

Genetics of the AB0 Blood Groups. The AB0 blood groups were discovered by Landsteiner in 1900 [37]. Compared to other blood group systems their most important property is the presence of isoantibodies that have led to frequent transfusion accidents. These accidents helped in the discovery of blood groups. The first relevant genetic theory was developed by von Dungern and Hirszfeld in 1911 [64]. To explain the four phenotypes A, B, 0, and AB they assumed two independent pairs of alleles (A, 0; B, 0), with dominance of A and B. In 1925 Bernstein [7] tested this hypothesis using the Hardy-Weinberg expectations for the first time. He found their concept to be wrong and replaced it by the correct explanation – three alleles with six genotypes, leading to the four phenotypes due to the dominance of A and B over 0.

The most obvious method to discriminate between these two hypotheses is by family investigation. However, differences between them are to be expected only in matings in which at least one parent carries group AB (Table 4.5). The two-locus hypothesis allows for 0 children while the three-allele hypothesis does not. Although AB is the rarest group, the early literature contained some reports of supposedly 0 children with AB parents; these children were either misclassified or illegitimate. Bernstein, however, was not misled by these ob-

Table 4.5. Comparison of the two theories for inheritance of AB0 blood groups (adapted from Wiener 1943 [76])

Parents	Children expected from the hypothesis of	
	Two gene pairs	Multiple alleles
0×0	0	0
$0 \times A$	0, A	0, A
$0 \times B$	0, B	0, B
$A \times A$	0, A	0, A
$A \times B$	0, A, B, AB	0, A, B, AB
$B \times B$	0, B	0, B
$0 \times AB$	0, A, B, AB	A, B
$A \times AB$	0, A, B, AB	A, B, AB
$B \times AB$	0, A, B, AB	A, B, AB
$AB \times AB$	0, A, B, AB	A, B, AB

Table 4.6. Expectations from multiple allele hypothesis for the AB0 system (from Bernstein 1925 [7])

Pheno-type	Geno-type	Frequency	
0	aabb	$(1-p)^2(1-q)^2 = p'^2q'^2$	
B	aaBB	$(1-p)^2q^2$	$\left.\begin{array}{c} \\ \\ \end{array}\right\} = p'^2(1-q'^2)$
	aaBb	$2(1-p)^2q(1-q)$	
A	AAbb	$p^2(1-q)^2$	$\left.\begin{array}{c} \\ \\ \end{array}\right\} = (1-p'^2)q'^2$
	Aabb	$2\,p(1-p)(1-q)^2$	
AB	AABB	p^2q^2	
	AaBB	$2\,p(1-p)q^2$	$= (1-p'^2)(1-q'^2)$
	AABb	$2\,p^2q(1-q)$	
	AaBb	$2\,p(1-p)2\,q(1-q)$	

servations. His argument goes as follows. It may be assumed that the two-gene pair theory is correct; p may be the gene frequency of A, $1 - p = p'$ of a; q the frequency of B, $1 - q = q'$ of b. The frequencies to be expected in the population are presented in Table 4.6.

This leads to the following relationships ($\overline{A}$, $\overline{B}$: frequencies of phenotypes):

$$\overline{0} \times \overline{AB} = \overline{A} \times \overline{B}$$

and

$$\overline{A} + \overline{AB} = 1 - p'^2; \quad \overline{B} + \overline{AB} = 1 - q'^2$$

Thus, it follows:

$$(\overline{A} + \overline{AB}) \times (\overline{B} + \overline{AB}) = \overline{AB}$$

These identities can be tested. It turned out – and has turned out ever since – that $(\overline{A} + \overline{AB}) \times (\overline{B} + \overline{AB}) > \overline{AB}$, and $\overline{0} \times \overline{AB} < \overline{A} + \overline{B}$. The differences are so large – and so consistent – that an explanation by chance deviations is inadequate. The first alternative possibility considered by Bernstein was heterogeneity within the examined population. This explanation, however, proved insufficient. On the other

hand, it could be shown that the distributions in all populations for which data were available are in perfect agreement with expectations derived from the multiple-allele hypothesis.

To understand Bernstein's argument a fresh look at the Hardy-Weinberg law is necessary. Up to now it has been derived here for the special case of two alleles only. However, it can also be shown to apply for more than two alleles. Assuming n alleles $p_1, p_2, \ldots, p_n$, the relative frequencies of genotypes are given by the terms of the expansion of $(p_1 + p_2 + \ldots p_n)^2$. It follows for the special case of A, B, and 0 with the frequencies p, q, and r that the distribution of genotypes is:

$$p^2(AA) + 2\,pq(AB) + 2\,pr(A0) + q^2(BB) + 2\,pr(B0) + r^2(B0).$$

Now, we follow Bernstein again (our translation): "for the classes" (phenotypes):

$$\overline{0} = 00 \quad \overline{B} = B0 + BB \quad \overline{A} = A0 + AA \quad \overline{AB} = AB$$

the following probabilities can be derived:

$$r^2 \quad 2\,qr + q^2 \quad 2\,pr + p^2 \quad 2\,pq$$

It follows:

$$\overline{0} + \overline{A} = (r + p)^2$$
$$\overline{0} + \overline{B} = (r + q)^2$$

and therefore:

$$q = 1 - \sqrt{\overline{0} + \overline{A}}$$
$$q = 1 - \sqrt{\overline{0} + \overline{B}}$$
$$q = 1 - \sqrt{\overline{0}}$$

and the relation:

$$1 = p + q + r = 1 - \sqrt{\overline{0} + \overline{B}}$$
$$+ 1 - \sqrt{\overline{0} + \overline{A}} + \sqrt{\overline{0}}$$

This can be tested using the AB0 phenotype distributions in various populations of the world. The criterion is that the gene frequencies calculated with this formula must add to 1. In addition, expected genotype frequencies can be calculated from these gene frequencies and can be compared with observed frequencies. Apart from the correctness of the genetic hypothesis, however, this result requires still another condition. There must be random mating with regard to this characteristic.

In the data analyzed by Bernstein the agreement already was excellent, and this has proven to hold true for the huge amount of data collected ever since. One example may help in understanding the principle of calculation. The following phenotype frequencies were reported from the city of Berlin ($n = 21104$): 43.23% A ($n = 9123$), 14.15% B ($n = 2987$), 36.60% 0 ($n = 7725$), and 6.01% AB ($n = 1269$).

Using Bernstein's formula, the gene frequencies are:

$$p = 1 - \sqrt{(0.3660 + 0.1415)} = 0.2876$$
$$q = 1 - \sqrt{(0.3600 + 0.4323)} = 0.1065$$
$$r = \sqrt{0.3660} \qquad\qquad = \frac{0.6050}{0.9991}$$

Thus:

$$p + q + r = 0.9991$$

At first glance, this result agrees well with the expectation, i.e., 1. As a statistical test for examining whether the deviation is significant, the χ^2 method can be applied [59]:

$$\chi_1^2 = 2\,n\left(1 + \frac{r}{pq}\right) D^2$$

$$D = 1 - (p + q + r)$$

In our example, the result is:

$$\chi_1^2 = 0.88$$

This confirms that the values found are in good agreement with the genetic hypothesis and with the assumptions of random mating for the AB0 system.

In a later paper Bernstein showed how the difference D may be utilized to correct the calculated gene frequencies. The uncorrected gene frequencies may be named p', q', and r', and the following formulas may be used:

$$p = p'(1 + D/2)$$
$$q = q'(1 + D/2)$$
$$r = (r' + D/2)(1 + D/2)$$

and for the example:

$$p = 0.2876(1 + 0.00045) = 0.2877$$
$$q = 0.1065(1 + 0.00045) = 0.1065$$
$$r = (0.6050 - 0.00045)(1 + 0.00045) = 0.6057$$

In the process of testing the two genetic hypotheses for the AB0 system Bernstein developed a method for calculating gene frequencies. Such methods have become important practically and are therefore treated in a separate section (Appendix 1).

Meaning of a Hardy-Weinberg Equilibrium. Populations showing agreement of the observed genotype proportions with the expectations of the Hardy-Weinberg Law are said to be "in Hardy-Weinberg equilibrium." This equilibrium must be distinguished from that between alleles, which is discussed in the contexts of selection (Sect. 12.2.1) and of mutation (Sect. 9.3.1). The Hardy-Weinberg equilibrium is an equilibrium of the distribution of genes in the population ("gene pool") among the various genotypes. Under random mating this equilibrium is reestablished after one generation, possibly with changed gene frequencies if it is disturbed by opposing forces.

It follows from our discussion, however, that the Hardy-Weinberg law can be expected to be valid only when the following prerequisites are not violated:

a) The matings must be random with respect to the genotype in question. This can safely be assumed for such traits as blood groups or enzyme polymorphisms. It cannot be assumed for visible characteristics such as stature, and still less for behavioral characteristics such as intelligence. This should be kept in mind when measures used in quantitative genetics, (for example, corre-lations between relatives), are interpreted in genetic terms.

b) A deviation from random mating is caused by consanguineous matings. If the consanguinity rate in a population is high, an increase in the number of homozygotes must be expected (Sect. 13.1.1). It is even possible to estimate the frequency of consanguinity in a population by means of the deviations from the Hardy-Weinberg proportions.

c) Recent migrations might disturb the Hardy-Weinberg proportions.

d) Occasionally selection is mentioned as a factor leading to deviations. This may be true but need not necessarily apply. As a rule, selection tends to cause changes in gene frequencies; selection before reproductive age, for example, in the prenatal period, or during childhood and youth, does not influence the Hardy-Weinberg proportions in the next generation at all. If genotypes are tested among adults in a situation in which a certain genotype had been selected against in children, this genotype is found to decrease in frequency. Even assuming appreciable selection in a suitable age group, ascertainment of statistically significant deviations from Hardy-Weinberg proportions requires large sample sizes – larger than are usually available. Sometimes the absence of significant selection is inferred from the observation that Hardy-Weinberg proportions are preserved in a population. This conclusion, however, unless carefully qualified may easily be wrong. Considering all the theoretical possibilities for disturbance, it is indeed amazing how frequently the Hardy-Weinberg proportions are found to be preserved in the human population.

e) Formally, a deviation from the Hardy-Weinberg law may be observed if the population is a mixture of subpopulations that do not completely interbreed (random mating only within subpopulations), and consequently the gene frequencies in these subpopulations differ. This was first described by Wahlund in 1928 [66], who gave a formula for calculating the coefficient F of the apparent inbreeding from the variance of the gene frequencies between the subpopulations.

f) Another cause of deviation may be the existence of a hitherto undetected ("silent") allele, a heterozygous carrier of which cannot be distinguished from a homozygous carrier of the usual allele. C.A.B.Smith (1970) [56], however, has pointed out that a silent allele causes a significant deviation from the Hardy-Weinberg law only when it occurs at a sufficiently high frequency for the homozygote to be detected.

4.2.3 Gene Frequencies

One Gene Pair: Only Two Phenotypes Known. In rare autosomal-recessive diseases only one gene pair is present, and only two phenotypes are usually known when the heterozygotes cannot be identified, or, as is usually the case, when direct data on population frequencies of heterozygotes are not available. This also applies for blood group systems for which only one type of antiserum is available. Here the frequency of homozygotes aa being q^2, the gene frequency is simply $\sqrt{aa}$. There is no way to test the assumption of random mating.

Table 4.7 [97] is slightly oversimplified; some of the frequencies given vary in different populations (Sect. 12.1). However, the data point out how much more frequent the heterozygotes are, especially for rare conditions. This is important for genetic counseling, and for the much-discussed problem of the number of lethal or detrimental genes for which the average human being might be heterozygous (Sect. 13.1.2). Methods for the calculation of gene frequencies are given in Appendix 1.

4.3 Statistical Methods in Formal Genetics: Analysis of Segregation Ratios

4.3.1 Segregation Ratios as Probabilities

During meiosis – and in the absence of disturbances – germ cells are formed in exactly the relative frequencies expected from Mendel's laws. A diploid spermatocyte heterozygous for alleles A and a produces two haploid sperms with A, and two with a. If all the sperms of a given male come to fertilization, and none of the zygotes die before birth, the segregation ratio among his offspring would be exactly 1:1. There would be no place for any statistics.

Organisms in which such an analysis is indeed possible are yeast and the bread mould *Neurospora crassa,* which has become important in biochemical genetics. In the development of such an organism, there is a phase in which the diploid state has just been reduced to the haploid, and all four meiotic products lie in a regular sequence. They can be removed separately, grown, and examined ("tetrad analysis"). Expected segregation ratios are found with precision.

In higher plants and animals, including humans, only a minute sample of all germ cells comes to fertilization. In the human female about 6.8×10^6 oogonia are formed; the number of spermatogonial stem cells in the male is estimated at about 1.2×10^9 (Sect. 9.3.3); the actual number of sperm is a multiple of this figure. Hence any given germ cell has a very small probability of coming to fertilization. In addition, the sampling process is usually random with respect to a given gene pair A,a. This means that for the distribution of genotypes among germ cells coming to fertilization the rules of probability theory apply, and empirically found segregation ratios may show deviations from their statistical expectations.

Modern humans are fairly accustomed to thinking in statistical terms when solving daily problems. These experiences help us to understand simple applications of probability theory. Everyone, for example,

Table 4.7. Differing homozygote and heterozygote frequencies for different gene frequencies (with examples of recessive conditions; adapted from Lenz 1983 [40])

Homozygote frequency q^2	Gene frequency q	Heterozygote frequency $2pq$	Approximate homozygote frequencies in European populations
0.64	0.8	0.32	Lp(a-) lipoprotein variant
0.49	0.7	0.42	Acetyl transferase, "slow" variant (Sect. 7.5.1)
0.36	0.6	0.48	Blood group 0
0.25	0.5	0.50	Nonsecretor (se/se)
0.16	0.4	0.48	Rh negative (dd)
0.09	0.3	0.42	Lactose restriction (northwestern Germany)
0.04	0.2	0.32	Le(a–b–) negative
0.01	0.1	0.18	β-Thalessemia (Cyprus)
1:2500	1:50	1:25	Pseudocholinesterase (dibucaine-resistant variant), cystic fibrosis; α_1-antitrypsin deficiency
1:4900	1:70	1:35	Adrenogenital syndrome (Canton Zurich)
1:10000	1:100	1:50	Phenylketonuria (Switzerland; USA)
1:22500	1:150	1:75	Albinism; adrenogenital syndrome with loss of NaCl
1:40000	1:200	1:100	Cystinosis
1:90000	1:300	1:150	Mucopolysaccharidosis type 1
1:1000000	1:1000	1:500	Afibrinogenemia

readily recognizes that the following rationale is wrong.

A young mother had always wished to have four children. After the third, however, there was a long pause. The grandmother asked her daughter whether she had now decided differently. Answered the daughter: "Yes, in principle, I would still like four children. But I read in the newspaper that every fourth child born is Chinese. And a Chinese child . . . there I am reluctant."

In another example, the mistake is less obvious. The parents of two albino children visit a physician for genetic counseling. They wish to know the risk of a third child also being albino. The physician knows that albinism is an autosomal-recessive condition, with an expected segregation ratio of 1:3 among children of heterozygous parents. He also knows that sibships in which all sibs are affected are very rare. Hence, he informs the parents: "As you already have two affected children, the chance that the third child will also be affected is very small. The next child should be healthy." The actual risk, of course, remains 25% (Sect. 4.3.2).

At textbook on human genetics cannot teach probability theory and basic statistics. Therefore, it is assumed that the reader has some knowledge of the basic concepts of probability theory, that he knows the most important distributions (binomial, normal, and Poisson distribution), and has some idea of standard statistical methods. The following presents some applications to problems in human genetics. We are aware of the danger that this section may be used as a "cookbook," without understanding of the basic principles and recommend that the reader become familiar with these principles, for example, in the opening chapters of Feller's *Probability Theory and Its Applications* [16].

4.3.2 Simple Probability Problems in Human Genetics

Independent Sampling and Prediction in Genetic Counseling. The physician who gave the wrong genetic counsel to the couple with two albino children, did not take into account that the fertilization events leading to the three children are independent of each other, and that each child has the probability of $^1/_4$ to be affected, regardless of the genotypes of any other children. The probabilities for each child must be multiplied. He was right when he said that illness of all three children is rare in a recessive condition: The probability is $(^1/_4)^3 = ^1/_{64}$ for all three children to be affected; the family to be counseled however, already had two such children and the probability of this occurring was only $(^1/_4)^2 = ^1/_{16}$. It takes only one event with the probability $^1/_4$ to complete the three-child family with albinism, $^1/_{16} \times ^1/_4 = ^1/_{64}$. It is also intuitively

obvious that there is no way for a given zygote to influence the sampling of gametes of the same parents many years later. Chance has no memory!

All possible combinations of affected and unaffected siblings in three-child families can be enumerated as follows (A = affected; U = unaffected):

UUU, AUU, UAU, AAU,   UUA, AUA, UAA, AAA

In recessive inheritance, the event U has the probability $^3/_4$. Thus, the first of the eight combinations *(UUU)* has the probability $(^3/_4)^3 = ^{27}/_{64}$. This means that of all heterozygous couples having three children $^{27}/_{64}$, or fewer than 50% have only healthy children. On the other hand, all three children are affected in $(^1/_4)^3 = ^1/_{64}$ of all such families. There remain the intermediate groups. Three-child families with one affected child and two healthy ones in that order obviously have the probability $^1/_4 \times ^3/_4 \times ^3/_4 = ^9/_{64}$. However, we are not particularly interested in the sequence of healthy and affected children. Therefore the three cases of such families, *UUA, UAU,* and *AUU,* can be treated as equivalent, giving $3 \times ^9/_{64} = ^{27}/_{64}$. The group with two affected can be treated accordingly, giving $3 \times ^1/_4 \times ^1/_4 \times ^3/_4 = ^9/_{64}$. As a control, let us consider whether the various probabilities add up to 1:

$$\frac{27 + 27 + 9 + 1}{64} = 1$$

This is a special case of the binomial distribution. There are two consequences for Mendelian segregation ratios one theoretical, the other extremely practical. First, it follows that among all families for which a certain segregation ratio must be expected, an appreciable percentage – 27 of 64 in a three-child family with recessive inheritance – cannot be observed because chance has favored them by not producing any affected homozygotes. Hence, the segregation ratio in the remainder is systematically distorted. Special methods have been devised to correct for this "ascertainment bias" (Sect. 4.3.4). Secondly, and this is a most practical conclusion, with limitation of the number of children to two or three, most parents both of whom are heterozygous for a recessive disease will not have more than one affected child. Since the probability of affected children occurring in another branch of the family is very low – and the rate of consanguinity in current populations of industrialized countries has likewise decreased – almost all affected children represent sporadic cases in an otherwise healthy family; there is no distinct sign of recessive inheritance. Any subsequent child, however, again runs the risk of $^1/_4$. The layman usually does not know that the condition is inherited. Therefore, genetic counseling must be actively offered to these families.

Differentiation Between Different Modes of Inheritance. In Sect. 4.1.4, an X-linked dominant pedigree is shown (Fig. 4.13) for vitamin D resistant rickets and hypophosphatemia. What is the probability of such pedigree structure if the gene is in fact located on one autosome? Only the children of affected males are informative because among children of affected women a 1:1 segregation irrespective of sex must be expected. The seven affected fathers have 11 daughters, all of whom are affected. The probability of this outcome with autosomal inheritance is $(^1/_2)^{11}$. The same fathers have 10 sons who are all healthy, giving a probability of $(^1/_2)^{10}$. Hence, the combined probability of 11 affected daughters and 10 healthy sons is:

$$(^1/_2)^{21} = \frac{1}{2\,097\,152}$$

This probability is so tiny that the alternative hypothesis of an autosomal-dominant mode of inheritance is convincingly rejected. The only reasonable alternative is the X-linked dominant mode. This hypothesis is corroborated independently by the observation (Sec. 4.1.4.4) that on average male patients are more severely affected than female.

This is different for a rare skin disease (Brauer keratoma dissipatum). For this condition a Y-chromosomal mode of inheritance has been considered – and indeed all nine sons of affected fathers in a published pedigree show the trait, whereas five daughters in both generations are unaffected. This gives:

$$(^1/_2)^9 \times (^1/_2)^5 = (^1/_2)^{14} = \frac{1}{16\,384}$$

Hence, the probability of this pedigree having occurred by chance as an autosomal-dominant trait is very low indeed. There is an important difference, however, from the example of vitamin D resistant rickets. Other pedigrees showing autosomal-dominant inheritance are unknown for this type of rickets, and all observations confirm the location of this gene on the X chromosome. For Brauer keratoma dissipatum, on the other hand, some families have been observed exhibiting very similar phenotypes that show clearcut autosomal-dominant inheritance. It is therefore likely that the described pedigree has been selected from an unknown number of observations because of its peculiar transmission. The calculation is misleading as the "universe" from which this sample of observations was drawn (all pedigrees with the same phenotype) is much larger (and ill-defined), and the sample (the pedigree) is biased. The trait seems to be autosomal-dominant.

Another, more obvious example of an error in the definition of the sample space is the mother, above, who did not want a Chinese baby.

4.3.3 Testing for Segregation Ratios Without Ascertainment Bias: Codominant Inheritance

Apart from these limiting cases, calculation of exact probabilities for certain families or groups of families is usually impracticable. Therefore statistical methods are used that are either based on the parameters of the "normal" distribution, which is a good approximation of the binomial distribution (parametric tests), or derive directly from probabilistic reasoning (nonparametric tests). One method that is especially well suited for genetic comparisons is the χ^2 test. This enables us to compare frequencies of observations in two or more discrete classes with their expectations. The most usual form is:

$$\chi^2 = \Sigma \frac{(E - O)^2}{E}$$

(E = expected number; O = observed number). In Farabee's pedigree with dominant inheritance (Sect. 4.1.2), there are 36 affected and 33 unaffected children of affected parents. With dominant inheritance, E is $^1/_2$ of all children, i.e., 34.5:

$$\chi_1^2 = \frac{(36 - 34.5)^2}{34.5} = \frac{(33 - 34.5)^2}{34.5} = 0.13$$

The probability p for an equal or greater deviation from expectation can be taken from a χ^2 table for 1 degree of freedom. The number of degrees of freedom indicates in how many different ways the frequencies in the different classes can be changed without altering the total number of observations. In this case the content of class 2, unaffected, is unequivocally fixed by the content of class 1. Therefore, the number of degrees of freedom is 1. In general the number of degrees of freedom is equal to the number of classes less 1.

A second example is taken from the codominant mode of inheritance (Sect. 4.1.2). Table 4.1 summarizes Wiener's family data for the MN blood types. Are the resultant segregation ratios compatible with the genetic hypothesis? For this problem, matings MM × MM, MM × NN, and NN × NN give no information. Expectations in the matings MM × MN and NN × MN are 1:1, in the mating MN × MN 1:2:1. This leads to Table 4.8 for the χ^2 test: For 4 degrees of freedom we find in the χ^2 table: $p = 0.75$. This is very good agreement with expectation.

Dominance. The situation becomes slightly more complicated when one allele is dominant and the other recessive. This is the case, for example, in the AB0 blood group system. Here, the phenotype $\overline{A}$ consists of the genotypes AA and A0. The expected segregation ratios among their offsprings differ. Some of the heterozygous parents A0 can be recognized, for example, in matings with $\overline{0}$ partners by the finding of 0 children. Others have only $\overline{A}$ children just by chance. Special statistical methods are necessary to calculate correct expectations and to compare empirical observations with these expectations [55a].

4.3.4 Testing for Segregation Ratios: Rare Traits

Principal Biases. If the condition under examination is rare, families are usually not ascertained at random; one starts with a "proband" or "propositus," i.e., a person showing the condition. This leads to

Table 4.8. Comparison between expected and observed segregation figures in the MN data of Wiener et al. (1953) Table 4.1 [77]

Mating type	MM	MN	NN	χ^2	Degrees of freedom
MM $\times$ MN	$\dfrac{(499-486)^2}{486}$	$\dfrac{(473-486)^2}{486}$	–	0.6955	1
MN $\times$ MN	$\dfrac{(199-200)^2}{200}$	$\dfrac{(405-400)^2}{400}$	$\dfrac{(196-200)^2}{200}$	0.1475	2
MN $\times$ NN	–	$\dfrac{(411-396.5)^2}{396.5}$	$\dfrac{(382-396.5)^2}{396.5}$	1.0605	1

an *ascertainment bias,* which must be corrected. The bias can be of different kinds, depending on the way in which the patients have been ascertained.

a) Family or truncate selection. All individuals suffering from a specific disease in a certain population at a certain time (or within certain time limits) are ascertained. The individual patients are ascertained independently of each other, i.e., the second case in a sibship would always have been found. Such truncate ascertainment is possible, for example, if the condition always leads to medical treatment, and all physicians report every case to a certain registry – as when an institute carries out an epidemiological study. As a rule, case collections approaching completeness are possible only in ad hoc studies of research workers specializing in a condition or group of conditions. Here, the ascertainment bias is caused exclusively by the fact that only those sibships are ascertained that contain at least one patient. As noted above, however, (Sect. 4.3.3), this leaves out all sibships in which no affected individual has occurred just by chance. Their expected number is:

$$\sum_s q^s n_s \tag{4.2}$$

(s = number of siblings/sibship; p = segregation ratio; $q = 1 - p$; n_s = number of sibships of size s). In recessive disorders, $p = 0.25$. The smaller the average sibship size, the stronger is the deviation from the 3:1 ratio in the ascertained families.

b) Incomplete multiple (proband) selection; single selection as limiting case. It is rare that all individuals in a population are ascertained; frequently a study starts, for example, with all patients in a hospital population who have a certain condition. Here an additional bias must be considered: the more affected members a sibship has, the higher is its chance to be represented in the sample. This bias causes a systematic excess of affected persons, which is added to the excess caused by truncate selection as explained above.

Koller (1940) [35] gave a simple example that demonstrates the nature of this excess. Let us assume that the probands are ascertained during examination of only a single year's group of conscripts. The population comprises a number of families with three children, at least one of whom has the disease, and one of whom is a member of the current year's group. Ascertainment of the family depends on the presence of an affected child in the one year group examined. Thus, all families with three affected siblings but only two-thirds of the families with two affected and one-third of those with only one affected are ascertained.

The methods of correction described below are reliable only if the probability for ascertainment of consecutive siblings is independent of the ascertainment of the first one. In an examination of conscripts, as described above, this may be the case. Most studies, however, begin with a hospital population or some other group of medically treated persons. Here, according to general experience, subsequently affected children are much more frequently brought to a hospital when another child has been treated successfully. The opposite trend, however, is also possible. Becker (1953) [2], for example, collected all cases of X-linked recessive Duchenne's muscular dystrophy in a restricted area of southwestern Germany. He had good reason to think that ascertainment was complete for this area. Nevertheless, brothers developing muscular dystrophy as the second or later cases in their sibships were generally not ascertained as probands (i.e., through hospitals and physicians) but through the first proband in the family. In his interviews with the parents Becker found the reason. In the case of the first patient in the sibship the parents usually consulted a physician. Then, however, they discovered that in spite of examinations and therapeutic attempts, the course of the disease could not be influenced. Hence they refrained from presenting a second child to the hospital or the physician.

c) Apart from these biases, which can be statistically corrected to a certain degree, there are other biases that cannot be corrected. Frequently, for example, a genetic hypothesis is discussed on the basis of families sampled from the literature. Experience shows that such sampling usually leads to reasonable results in autosomal-dominant and X-linked recessive disorders. Autosomal-recessive diseases, however, are more difficult to handle. Families with an impressive accumulation of affected sibs have a higher chance of being reported than those with only one or two affected members. This selection for "interesting" cases was more important early in the twentieth century because families generally had more children. Furthermore, recessive conditions discovered today are usually interesting from a clinical and biochemical point of view as well.

These biases can be avoided only by publishing all cases and by critical interpretation of data from the literature. A statistically sound correction is impossible, as such bias has no simple and reproducible direction.

To summarize, the method of segregation analysis depends on the way in which families are ascertained. It follows that the method of ascertainment should always be described carefully. Above all, the probands should always be fully indicated. It is also of interest whether the author during his case collection has become aware of any ascertainment biases.

These considerations show that complete (truncate) ascertainment of cases in a population, and within defined time limits, is the optimal method of data collection.

Methods for Correcting Bias. Two different types of correction are possible: test methods and estimation methods.

In a test method the observed values are compared with the expected values, which have been corrected for ascertainment bias. The first such test method was published by Bernstein in 1929, cf. [35]; it corrected for truncate selection. The expected number of affected E_r is:

$$E_r = sn_s \frac{p}{1 - q^s} \tag{4.3}$$

in all sibships of size n (definition of symbols as in Eq. 4.2). A similar test method can also be used for proband selection.

Test methods answer a specific question: do the observed proportions fit the expected values according to a certain genetic hypothesis?

In many if not in most actual cases, the question is more general: What is the unbiased segregation ratio

in the observed sibships? This is an estimation problem. The earliest method was published in 1912 by Weinberg [71] and was called the sib method. Starting from every affected sib in the sibship, the number of affected and unaffected among the sibs is determined. This method is adequate for "truncate selection," i.e., when each affected person is, at the same time, a proband. The sib method is the limiting case of the "proband method" used when the families are ascertained by incomplete multiple proband selection. The number of affected and unaffected siblings is counted, starting from every proband. A limiting case is single selection. Here each sibship has only one proband, and the counting is done once among the sibs.

These estimates converge with increasing sample size to the parameter p, the true segregation ratio; they are *consistent*. It was realized early, however, that they are not fully *efficient*, except for the limiting case of single selection, i.e., they do not make optimal use of all available information. Therefore improvements have been devised by a number of authors. Today such simple methods are no longer used. Moreover, the problems to be solved by segregation analysis are usually more complex. For example, the families to be analyzed may be a mixture of genetic types with various modes of inheritance; there may be admixture of "sporadic" cases, due either to new mutation or to environmental factors; penetrance may be incomplete, or the simple model of a monogenic mode of inheritance may be inadequate for explaining familial aggregation, and a multifactorial genetic model must be used (for the conceptual basis of such multifactorial models, see Chap. 6). Computer programs are now available for carrying out such analyses; they are available either from their authors' institutions or through an international network of program packages (Internet). Some of these also offer programs for comparing predictions from various genetic models. For details see Appendix 3.

4.3.5 Discrimination of Genetic Entities: Genetic Heterogeneity

It is a common experience in clinical genetics that similar or identical phenotypes are caused by a variety of genotypes. The splitting of a group of patients with a given disease into smaller but genetically more uniform subgroups has been a major topic of research in medical genetics over recent decades. Frequently such heterogeneity analysis is another aspect of the application of Mendel's paradigm (Sect. 6.1.1.6) and its consequences: carrying genetic analysis through different levels ever closer to gene action.

It appears at first glance that with modern biological methods discrimination of genetic entities on descriptive grounds, i.e., on the level of the clinical phenotype, would no longer hold interest. In our opinion, however, knowledge of the phenotypic variability of genetic disease in humans is needed for many reasons:

a) Such knowledge provides heuristic hypotheses for systematic application of the more penetrating methods from biochemistry, molecular biology, immunology, micromorphology, and other fields.
b) Treatment will often depend upon manipulation of gene disordered biochemistry and pathophysiology of a given disease.
c) We require insight into the genetic burden of the human population.
d) Better data are needed for many of our attempts to understand the problems of spontaneous and induced mutation.

Genetic Analysis of Muscular Dystrophy as an Example. One group of diseases in which analysis using the clinical phenotype together with the mode of inheritance proved to be successful are the muscular dystrophies. These conditions have in common a tendency to slow muscular degeneration, incapacitating affected patients who often ultimately die from respiratory failure. There are major differences in age at onset, location of the first signs of muscular weakness, progression of clinical symptoms, and mode of inheritance. These criteria were used by medical geneticists to arrive at the following classification of muscular dystrophies:

I. X-linked muscular dystrophies
  1. Severe type (Duchenne) (310200)
  2. Juvenile or benign type (Becker; 310100)
  3. Benign type with early contracture
    (Cestan-Lejonne and Emery-Dreifuss; 310300)
  4. Hemizygous lethal type
    (Henson-Muller-de Myer; 309950)
II. Autosomal-dominant dystrophy
  Facio-scapulo-humeral type
    (Erb-Landouzy-Déjérine; 158900)
III. Autosomal-recessive muscular dystrophies
  1. Infantile type
  2. Juvenile type
  3. Adult type
  4. Shoulder girdle type

This classification is based on many reports from various populations and, for the rarer variants, on reports of pedigrees. It does not include pedigrees in which affected members showed involvement only of restricted parts of the muscular system, such as distal and ocular types. Congenital myopathies were also

excluded. The main criteria for discrimination are obvious from the descriptive terms used in the tabulation; for details, see Becker [3]. At present, various mutations of the X-linked dystrophin gene are known at the molecular level which lead to the Duchenne and Becker types (Sect. 9.4). The gene for Emery-Dreyfus disease has been localized to distal Xq28; this gene and the others are still awaiting molecular characterization.

Multivariate Statistics. The critical human mind is an excellent discriminator. However, statistical methods for identifying subgroups within a population on the basis of multiple characteristics are now available (multivariate statistics). Such methods can also be applied to the problem of making discrimination of genetic entities more objective.

4.3.6 Conditions Without Simple Modes of Inheritance

The methods discussed so far are used mainly for genetic analysis of conditions thought to follow a simple mode of inheritance. In many diseases, however, especially in some that are both serious and frequent, there are problems:

a) Diagnosis of the condition may be difficult. There are borderline cases. Expressed more formally: the distribution of affected and unaffected in the population is not an outright alternative (examples: schizophrenia; hypertension; diabetes).
b) It is known from various investigations, including twin studies, that the condition is not entirely genetic but that certain environmental factors influence manifestation (example: decline of diabetes in European countries during and after World War II).
c) The condition is so frequent that clustering of affected patients in some families must be expected simply by chance (examples: some types of cancer).
d) It can be concluded from our knowledge of pathogenic mechanisms that the condition is not a single disease but a complex of symptoms common to a number of different causes (example: epilepsy). In fact, it is becoming apparent that diagnoses such as hypertension and diabetes subsume groups of heterogeneous disease entities.

In no such case can a genetic analysis that starts from the phenotype be expected to lead to simple modes of inheritance (for more complete discussion, see Chap. 6). However, for many such conditions, two questions of practical importance arise:

1. What is the risk of relatives of various degrees being affected? Is it higher than the population average?
2. What is the contribution of genetic factors to the disease? Under what conditions does the disease manifest itself?

Familial aggregation can be assessed by calculation of empirical risk figures. Twin studies and comparisons of incidence among relatives of probands with those in the general population are required to answer the questions discussed in Sect. 6.3. Here, we discuss risk figures.

Empirical Risk Figures. The expression "empirical risk" is used in contrast to "theoretic risks" as expected by Mendelian rules in conditions with simple modes of inheritance. The early methods were developed largely by the Munich school of psychiatric genetics in the 1920s with the goal of obtaining risk figures for psychiatric diseases.

The basic concept is to examine a sufficiently large sample of affected patients and their relatives. From this material, unbiased risk figures for defined classes of relatives are calculated. These figures are used to predict the risk for relatives in future cases. This approach makes the implicit assumption that risk figures are generally constant "in space and time", i.e., among various populations and under changing conditions within the same population. Considering the environmental changes influencing the occurrence of many diseases such as diabetes, this assumption is not necessarily true but is useful as a first approximation.

The approach can be extended to include the question of whether two conditions A and B have a common genetic component, leading to increased occurrence of patients with disease A among close relatives of patients with disease B.

Selecting and Examining Probands and Their Families. In conditions that have simple modes of inheritance, the selection of probands is usually straightforward. The modes of ascertainment are discussed in Sect. 4.3.4. For empirical risk studies the same rules apply. In fairly frequent conditions, complete ascertainment of cases in a population is rarely if ever feasible and is also unnecessary in these investigations. In most situations, a defined sample of probands, such as all cases coming to a certain hospital for the first time during a predefined time period can be used. The mode of ascertainment is single selection, or very close to it. This approach simplifies correction of the ascertainment bias among sibs of probands. The empirical risk figures can be calculated by counting affected and unaffected among the sibs,

excluding the proband. Risk figures among children ascertained through the parental generation are unbiased and need no correction.

Frequently, the diagnostic categories are not clearcut. In these cases, criteria for accepting a person as a proband must be defined unambiguously beforehand, and all possible biases of selection should be considered. Are more severe cases normally admitted to the hospital selected for study? Are patients selected from a particular social or ethnic group? Are there any other biases that might influence the comparability of the results? Genuinely unbiased samples are hardly if ever available, but the biases should be known. Most importantly, such biases should be independent of the problem to be analyzed. For example, it would be a mistake to consider only patients who have similarly affected relatives.

The goal of the examinations is to obtain maximal and precise information about the probands and their families as far as possible. Methods for achieving this goal, however, vary. Clinical experience and the study of publications on similar surveys are helpful.

Once the proband and his family are ascertained, the relatives should be noted as completely as possible, and information on their health status must be collected. Here, personal examination by the investigator and historical information provided by the patients and their relatives are indispensible. Such data should be backed by hospital records and various laboratory and radiological studies. Even results of clinical examinations should be regarded with scepticism since not all physicians are equally knowledgeable and careful, and official documents, such as death certificates, are often unreliable regarding diagnostic criteria.

In most cases, the determination of genetic risk figures answers the question of whether the risk is higher than in the average population. Sometimes adequate incidence and/or prevalence data from a complete population in which the study is carried out or a very similar one are available. More often than not, however, a control series must be examined with the same criteria as used for the "test" populations. If possible, examination on normal controls and their relatives should be performed in a "blind" way; i.e., the examiners should be unaware of whether the persons studied come from the patient or the control series. It is a good idea to use matched controls, i.e., to examine for every patient a control person matched in all criteria but not related to the condition to be investigated (such as age, sex, ethnic origin, etc.).

Statistical Evaluation, Age Correction. In conditions that manifest at birth, such as congenital malformations affecting the visible parts of the body, further calculations are straightforward: the empirical risk for children is given by the proportion of affected in the sample. In many cases, however, onset occurs during later life, and the period at risk may be extended. Here the question asked is: What is the risk of a person's becoming affected with the condition, provided he or she lives beyond the manifestation period? The appropriate methods of age correction have been discussed extensively in the earlier literature [35]; one much-used is Weinberg's "shortened method." First, the period of manifestation is defined on the basis of a sufficiently large sample (usually larger than the sample of the study itself). Then all relatives who dropped out of the study before the age of manifestation are discarded. The dropping out may be for any of a variety of reasons: death, loss of contact due to change of residence, or termination of the study. All persons dropping out during the age of manifestation are counted as one-half, and all who have survived the upper limit of manifestation age are counted full.

Example. Among children of schizophrenics, 50 were affected and 200 unaffected. Of these, 100 have reached the age of 45 and 100 are between the age of 15 and 45 (i. e., the age of manifestation for the great majority of schizophrenic cases). Thus, the corrected number of unaffected is: $200 - ^1/_2 \times 100 = 150$; the empirical risk is:

$$\frac{50}{150 + 50} = 25\%$$

Chapter 16 deals in detail with practical problems, taking schizophrenia and affective disorders as examples.

Theoretical Risk Figures Derived from Heritability Estimates? There are suggestions that empirical risk figures should be replaced by theoretical risk figures computed from heritability estimates for the multifactorial model (Sect. 6.1.1.4), after data are found to agree with expectations from such a model. This could be done when the data compared with a simple diallelic model. Such heritability estimates can be achieved by comparing the incidence of the condition in the general population with that in certain categories of relatives, for example, sibs or, with caution, from twin data. In theory the method permits inclusion of environmental, for example, maternal, effects. Its disadvantage, however, is that it depends critically on the assumption that the genetic model fits the actual situation sufficiently well. Since the genetic model chosen may not apply to the data at hand, there is danger that the sophisticated statistical approach suggests a spuriously high degree of precision of the results.

Conclusions

The transmission of traits determined by single genes, including hereditary diseases, follows Mendel's laws. Autosomal-dominant, autosomal-recessive, and X-linked modes of inheritance can be identified on the basis of the location of mutant genes on autosomes or on the X chromosome, and noting the phenotypic distinction between homozygotes and heterozygotes. Mutations in mitochondrial DNA are transmitted from the mother to all children. Deviations from the classical Mendelian transmission scheme may occur as a consequence of "genomic imprinting," where the parental origin of the mutation determines the phenotype. "Anticipation," with earlier age of onset in succeeding generations, may owe its origin to unstable mutations. Genotype frequencies in populations follow the Hardy-Weinberg Law, which can be used to estimate gene frequencies. In rare traits, such as those in most hereditary diseases, pedigrees are often ascertained via affected individuals and their sibships; when such pedigrees are used to calculate Mendelian segregation ratios, the resulting "ascertainment bias" in favor of affected persons must be corrected by appropriate statistical methods.

References

1. Aicardi J, Chevriee JJ, Rousselie F (1969) Le syndrome spasmes en flexion, agénésie calleuse, anomalies chorio-rétiniennes. Arch Fr Pediatr 26 : 1103–1120
2. Becker PE (1953) Dystrophia musculorum progressiva. Thieme, Stuttgart
3. Becker PE (1972) Neues zur Genetik und Klassifikation der Muskeldystrophien. Hum Genet 17 : 1–22
4. Becker PE (ed) (1964–1976) Humangenetik. Ein kurzes Handbuch in fünf Bänden. Thieme, Stuttgart
5. Bell J (1934) Huntington's chorea. Treasury of human inheritance 4. Galton Laboratory
6. Bennett T (1975) The T-locus of the mouse. Cell 6 : 441–454
7. Bernstein F (1925) Zusammenfassende Betrachtungen über die erblichen Blutstrukturen des Menschen. Z Induktive Abstammungs Vererbungslehre 37 : 237
8. Bhende YM, Deshpande CK, Bhata HM, Sanger R, Race RR, Morgan WTJ, Watkins WM (1952) A "new" blood-group character related to the ABO system. Lancet 1 : 903–904
9. Brook JD, McCurrach ME, Harley HG et al (1992) Molecular basis of myotonic dystrophy. Cell 68 : 799–808
10. Carney G, Seedburgh D, Thompson B, Campbell DM, MacGillivray I, Timlin D (1979) Maternal height and twinning. Ann Hum Genet 43 : 55–59
11. Ceppellini R, Dunn LC, Turri M (1955) An interaction between alleles at the Rh locus in man which weakens the reactivity of the Rho factor (Du). Proc Natl Acad Sci USA 41 : 283–288
12. Cuénot L (1905) Les races pures et leurs combinaisons chez les souris. Arch Zool Exp Genet 3 : 123–132
13. De Vries DD, de Wijs IJ, Wolff G et al (1993) X-linked myoclonus epilepsy explained as a maternally inherited mitochondrial disorder. Hum Genet 91 : 51–54

14. Dorn H (1959) Xeroderma pigmentosum. Acta Genet Med Gemellol (Roma) 8 : 395–408

15. Farabee (1905) Inheritance of digital malformations in man. In: Papers of the Peabody Museum for American Archeology and Ethnology, vol 3. Harvard University Press, Cambridge/MA, p 69

16. Feller W (1970/71) An introduction to probability theory and its applications, 2nd edn. Wiley, New York

17. Fleischer B (1918) Über myotonische Dystrophie mit Katarakt. Grafes Arch Ophthalmol 96 : 91–133

18. Fu Y-H, Pizzuti A, Fenwick RG et al (1992) An unstable triplet repeat in a gene related to myotonic dystrophy. Science 255 : 1256–1258

19. Fuhrmann W, Stahl A, Schroeder TM (1966) Das oro-facio-digitale Syndrom. Humangenetik 2 : 133–164

20. Gartler SM, Francke U (1975) Half chromatid mutations: Transmission in humans? Am J Hum Genet 27 : 218–223

21. Gaul LE (1953) Heredity of multiple benign cystic epithelioma. Arch Dermatol Syph 68 : 517

22. Gerard G, Vitrac D, Le Pendu J, Muller A, Oriol R (1982) H-deficient blood groups (Bombay) of Reunion island. Am J Hum Genet 34 : 937–947

23. Goldfarb LG, Petersen RB, Tabatan M et al (1992) Fatal familial insomnia and familial Creutzfeldt-Jakob disease: disease phenotype determined by a DNA polymorphism. Science 258 : 806–808

24. Hadorn E (1955) Developmental genetics and lethal factors. Wiley, London

25. Haldane JBS (1941) The relative importance of principal and modifying genes in determining some human diseases. J Genet 41 : 149–157

26. Hardy GH (1908) Mendelian proportions in a mixed population. Science 28 : 49–50

27. Harley HG, Rundle SA, Reardon W et al (1992) Unstable DNA sequence in myotonic dystrophy. Lancet 339 : 1125–1128

28. Harper PS (ed) (1991) Huntington's disease. Saunders, London

29. Harper PS (1992) The epidemiology of Huntington's disease. Hum Genet 89 : 365–376

30. Harris H (1948) A sex-limiting modifying gene in diaphyseal aclasis (multiple exostoses). Ann Eugen 14 : 165–170

31. Harris H (1980) The principles of human biochemical genetics, 4th edn. North-Holland, Amsterdam

32. Haws DV, McKusick VA (1963) Farabee's brachydactylous kindred revisited. Johns Hopkins Med J 113 : 20–30

33. Huoponen K, Lamminen T, Juvonen V et al (1993) The spectrum of mitochondrial DNA mutations in families with Leber hereditary optical neuroretinopathy. Hum Genet 92 : 379–384

34. Klein D (1958) La dystrophie myotonique (Steinert) et la myotonie congénitale (Thomsen) en Suisse. J Genet Hum 7 [Suppl]

35. Koller S (1940) Methodik der menschlichen Erbforschung. II. Die Erbstatistik in der Familie. In: Just G, Bauer KH, Hanhart E, Lange J (eds) Methodik, Genetik der Gesamtperson. Springer, Berlin, pp 261–284 (Handbuch der Erbbiologie des Menschen, vol 2)

36. Kuhn TS (1962) The structure of scientific revolutions. University of Chicago Press, Chicago

37. Landsteiner K (1900) Zur Kenntnis der antifermentativen, lytischen und agglutinierenden Wirkungen des Blutserums und der Lymphe. Zentralbl Bakteriol 27 : 357–362

38. Lenz W (1961) Zur Genetik der Incontinentia pigmenti. Ann Paediatr (Basel) 196 : 141

39. Lenz W (1975) Half chromatid mutations may explain incontinentia pigmenti in males. Am J Hum Genet 27 : 690–691

40. Lenz W (1983) Medizinische Genetik, 6th edn. Thieme, Stuttgart

41. McKusick VA (1995) Mendelian inheritance in man, 11th edn. Johns Hopkins University Press, Baltimore

42. Milch RA (1959) A preliminary note of 47 cases of alcaptonuria occurring in 7 interrelated Dominical families, with an additional comment on two previously reported pedigrees. Acta Genet (Basel) 9 : 123–126

43. Mohr OL, Wriedt C (1919) A new type of hereditary brachyphalangy in man. Carnegie Inst Publ 295 : 1–64

44. Mühlmann WE (1930) Ein ungewöhnlicher Stammbaum über Taubstummheit. Arch Rassenbiol 22 : 181–183

45. Ollendorff-Curth (1958) Arch Derm Syph 77 : 342

46. Ott J (1977) Counting methods (EM algorithm) in human pedigree analysis. Linkage and segregation analysis. Ann Hum Genet 40 : 443–454

47. Pauli RM (1983) Editorial comment: dominance and homozygosity in man. Am J Med Genet 16 : 455–458

48. Pearson K (1904) On the generalized theory of alternative inheritance with special references to Mendel's law. Philos Trans R Soc [A] 203 : 53–86

49. Penrose LS (1947/49) The problem of anticipation in pedigrees of dystrophia myotonica. Ann Eugen 14 : 125–132

50. Pola V, Svojitka J (1957) Klassische Hämophilie bei Frauen. Folia Haematol (Leipz) 75 : 43–51

51. Race RR, Sanger R (1975) Blood groups in man, 6th edn. Blackwell, Oxford

52. Reed TE, Neel JV (1959) Huntington's chorea in Michigan. II. Selection and mutation. Am J Hum Genet 11 : 107

53. Reik W (1989) Genomic imprinting and genetic disorders in man. Trends Genet 5 : 331–336

54. Scriver CR, Beaudet AL, Sly WS, Valle D (eds) (1989) The metabolic basis of inherited disease, 6th edn. McGraw-Hill, New York

55. Siemens HW (1925) Über einen, in der menschlichen Pathologie noch nicht beobachteten Vererbungsmodus: dominant geschlechtsgebundene Vererbung. Arch Rassenbiol 17 : 47–61

55a. Smith CAB (1956/7) Counting methods in genetical statistics. Ann Hum Genet 21: 254–276

56. Smith CAB (1970) A note on testing the Hardy-Weinberg law. Ann Hum Genet 33 : 377

57. Snyder LF, Doan CA (1944) Is the homozygous form of multiple teleangiectasia lethal? J Lab Clin Med 29 : 1211–1216

58. Stern C (1957) The problem of complete Y-linkage in man. Am J Hum Genet 9 : 147–165

59. Stevens WL (1950) Statistical analysis of the ABo blood groups. Hum Biol 22 : 191–217

60. Stocks P, Barrington A (1925) Hereditary disorders of bone development. Treasury of human inheritance 3, part 1

61. Stoffel M, Froguel P, Takeda J et al (1992) Human glucokinase gene: isolation, characterization and identification of two missense mutations linked to early-onset non-insulin dependent (type 2) diabetes melitus. Proc Natl Acad Sci USA 89 : 7698–7702

62. Timoféef-Ressovsky NW (1931) Gerichtetes Variieren in der phänotypischen Manifestierung einiger Generationen

von Drosophila funebris. Naturwissenschaften 19 : 493–497

63. Trevor-Roper PD (1952) Marriage of two complete albinos with normally pigmented offspring. Br J Ophthalmol 36 : 107

64. Von Dungern E, Hirszfeld L (1911) Über gruppenspezifische Strukturen des Blutes. III. Z Immunitatsforsch 8 : 526–562

65. Waardenburg PJ, Franceschetti A, Klein D (1961/1963) Genetics and ophthalmology, vols 1,2. Blackwell, Oxford

66. Wahlund S (1928) Zusammensetzung von Populationen und Korrelationserscheinungen vom Standpunkt der Vererbungslehre aus betrachtet. Hereditas 11 : 65–105

67. Wallace DC (1989) Report of the Committee on Human Mitochondrial DNA. Cytogenet Cell Genet 51 : 612–621

68. Wallace DC (1989) Mitochondrial DNA mutations and neuromuscular disease. Trends Genet 5 : 9–13

69. Wallace DC (1994) Mitochondrial DNA sequence variation in human evolution and disease. Proc Natl Acad Sci USA 91 : 8739–8746

70. Weinberg W (1908) Über den Nachweis der Vererbung beim Menschen. Jahreshefte des Vereins für vaterländische Naturkunde in Württemberg 64 : 368–382

71. Weinberg W (1912) Methoden und Fehlerquellen der Untersuchung auf Mendelsche Zahlen beim Menschen. Arch Rassenbiol 9 : 165–174

72. Welander L (1957) Homozygous appearance of distal myopathy. Acta Genet (Basel) 7 : 321–325

73. Wendt GG, Drohm D (1972) Die Huntingtonsche Chorea. Thieme, Stuttgart (Fortschritte der Allgemeinen und Klinischen Humangenetik, vol 4)

74. Wexler NS, Young AB, Tanzi RE et al (1987) Homozygotes for Huntington's disease. Nature 326 : 194–197

75. Wettke-Schöfer R, Kantner G (1983) X-linked dominant inherited diseases with lethality in hemizygous males. Hum Genet 64 : 1–23

76. Wiener AS (1943) Additional variants of the Rh type demonstrable with a special human anti-Rh-serum. J Immunol 47 : 461–465

77. Wiener AS, di Diego N, Sokol S (1953) Studies on the heredity of the human blood groups. I. The MN types. Acta Genet Med Gemellol (Rome) 2 : 391–398

78. Winters RW, Graham JB, Williams TF, McFalls VC, Burnett CH (1957) A genetic study of familial hypophosphatemia and viatmin D-resistant rickets. Trans Assoc Am Physicians 70 : 234–242

79. Zevani M, Serudei S, Gellera C et al (1989) An autosomal dominant disorder with multiple deletions of mitochondrial DNA starting at the D-loop region. Nature 339 : 309–311

5 Formal Genetics of Humans: Linkage Analysis and Gene Clusters

If "to take a possible example, an equally close linkage" (as between the genes for hemophilia and color blindness) "were found between the genes for blood group" and that "determining Huntington's chorea, we should be able, in many cases, to predict which children of an affected person would develop this disease and to advise on the desirability or otherwise of their marriage."

J. B. S. Haldane and J. Bell (1937) The linkage between the genes for colour-blindness and haemophilia in man. Proc. Roy. Soc. B 123, 119.

5.1 Linkage: Localization of Genes on Chromosomes

Genes are located in a linear fashion on the chromosomes. This has the logical consequence that genes located on the same chromosome are transmitted together, i.e., that their segregation is not independent. On the other hand, it is know from cytogenetics that chiasmata are formed during the first meiotic division, and that certain chromosome segments are exchanged between homologous chromosomes (crossing over; see Sect. 2.1.2.4). Hence, even genes located on the same chromosome are not always transmitted together; the probability of transmission of two linked genes depends on the distance between them and on the frequency with which they are separated by crossing over. If located on a fairly long chromosome, and if the distance is large enough that numerous crossing over events occur between them, genes located on the same chromosome may even seem to segregate independently. Such genes are syntenic but not linked. It was the great achievement of Morgan and his school in the first two decades of the twentieth century to exploit linkage for localizing genes relative to each other on chromosomes and developing gene maps in the fruit fly *Drosophila melanogaster*.

Studies on linkage and gene mapping in humans have lagged behind this development for many decades. Sophisticated statistical techniques were developed to get around the difficulty that directed breeding experiments are impossible in humans, and information from naturally occurring families must be used. The application of such techniques, however, was only sparsely rewarded by detection of linkage. A breakthrough occurred only when the new techniques of somatic cell genetics and especially cell fusion were introduced. These techniques permitted the assignment of genes to specific chromosomes and even chromosome segments. Later, methods taken from molecular biology, especially the ubiquity of common DNA variants, brought further progress [7]. Today's human gene map contains many genes, the number of successful assignments is growing quickly, and new insights into the organization of the genetic material are forthcoming.

In the following we describe first the principle of the classic approach to gene localization as introduced by Morgan and his followers. This provides an opportunity to introduce some general concepts. We then discuss statistical methods for detecting and measuring linkage in humans; examples are given in Appendix 7. The various groups of DNA markers are described next, followed by the principle of cell fusion and its use in localizing genes on chromosomes, as well as the application of radioactive and nonisotopic in situ hybridization for this purpose. Genetic maps are compared to physical maps, and the use of linkage studies as analytical tools in genetic analysis of common diseases with complex etiology and pathogenesis is assessed.

5.1.1 Classic Approaches in Experimental Genetics: Breeding Experiments and Giant Chromosomes

According to Mendel's third law, segregation of two different pairs of alleles is independent; all possible zygotes of two pairs of alleles are formed by free recombination. Mating between the double heterozygote AaBb and the double homozygote aabb leads to:

Paternal gametes		AB	Ab	aB	ab
Maternal gametes	ab	$1/4$ AaBb	$1/4$ Aabb	$1/4$ aaBb	$1/4$ aabb

The four genotypes are formed in equal proportions.

Soon after Mendel's laws were rediscovered Bateson et al. in 1908 [2] found an exception from this rule in the vetch, *Lathyrus odoratus*. Certain combinations were observed more frequently and others less frequently than expected. In some cases, the two parental combinations – in our example AaBb (father) and aabb (mother) – were increased among the progeny; in other cases the two other types Aabb or aaBb were more frequent.

Paternal gametes		AB	Ab	aB	ab
Maternal gametes	ab	AaBb	Aabb	aaBb	aabb
First case (coupling)		$\frac{1}{2}-\Theta$	Θ	Θ	$\frac{1}{2}-\Theta$
Second case (repulsion)		Θ	$\frac{1}{2}-\Theta$	$\frac{1}{2}-\Theta$	Θ

Θ = Recombination fraction Θ.

The alleles of the parental combination seemed either to attract one another or to repel one another. Bateson et al. [2] coined the terms "coupling" for the former phase and "repulsion" for the latter. Morgan in 1910 [45] recognized that coupling and repulsion are two aspects of the same phenomenon (i.e., location of two genes on the same or homologous chromosomes). He coined the term "linkage." Coupling occurs when the genes A and B are localized in the doubly heterozygous parent on the same chromosome $\frac{AB}{ab}$, and repulsion occurs when they are localized on homologous chromosomes $\frac{Ab}{aB}$. The terms *cis* and *trans* are more frequently used to refer to genes in coupling or repulsion, respectively. If linkage is complete, only two types of progeny can occur. More frequently, however, all four types are found, albeit two types in smaller numbers. Morgan explained this finding by exchange of chromosome pieces between homologous chromosomes during meiotic crossing over. He also recognized that the frequency of crossing over depends on the distance between two gene loci in one chromosome. Using recombination analysis as an analytic tool, he and his coworkers succeeded in locating a great number of gene loci in *Drosophila* and in establishing chromosome maps. Their results were confirmed in the early 1930s when Heitz, Bauer, and Painter discovered the giant chromosomes of some *Dipterae*. With this experimental tool many gene localizations known from indirect evidence could be confirmed by direct inspection when they were accompanied by small structural chromosomal variation. In the meantime linkage analyses have been carried out in a great number of species.

Linkage and Association. It is sometimes assumed that genes which are linked should always show a certain association in the population, i.e., that the chromosomal combinations AB or ab (coupling) occur more frequently than the combinations Ab or aB (repulsion). However, this is not the case in a randomly mating population. Even if the linkage is fairly close, repeated crossing over in many generations causes all four combinations, AB, ab, Ab, and aB, to be randomly distributed in the long run. As a rule, association of genetic traits does not point to linkage. This rule, however, has exceptions. Some combinations of closely linked genes do indeed occur more often than expected with random distribution. Such "link-

age disequilibrium" was first postulated in humans for the rhesus blood types (Sect. 5.2.4) and has also been proven for the major histocompatibility complex (MHC), especially the HLA system (Sect. 5.2.5) and for many DNA polymorphisms. It has now been shown for many mutations and the neighboring DNA polymorphisms (Sect. 12.1.3). Linkage disequilibrium may have three reasons:

1. The population under examination originated from a mixture of two populations with different frequencies of the alleles A,a and B,b, and the time elapsed since the mixing of the populations was not sufficient for complete randomization.
2. Two mutants, for example, DNA markers, are located so closely together that an insufficient number of generations has elapsed to separate them by recombination since the two mutations occurred in one chromosome.
3. Certain combinations of alleles at linked gene loci are maintained in high frequency by natural selection.

These problems are discussed in greater detail in connection with the MHC system (Sect. 5.2.5) and in the discussion on association between HLA and disease (Sect. 6.2.3).

5.1.2 Linkage Analysis in Humans

Direct Observation of Pedigrees. Linkage analysis by classic methods in humans is difficult since no directed breeding occurs. However, in some cases pedigree inspection can provide information. Linkage is excluded, for example, if one of the genes under scrutiny can be localized to the X chromosome while the other is on an autosome. By the same token, there is a high probability of demonstrating formal linkage if both genes are X-linked. Even in this case, however, formal linkage may not be demonstrable since the loci may be so far from each other that crossing over separates them. Similar considerations hold for genes located on a given autosomal chromosome. The term synteny refers to two or more genes being situated on the same chromosome, regardless of whether formal linkage can be demonstrated. Either a large pedigree or a number of smaller pedigrees must be screened to assess the extent of crossing over. Figure 5.1a shows a pedigree with red-green color blindness (30380, 30390) and hemophilia. The males in the sibships at risk either have both conditions or are normal. The genes are in the coupling (or *cis*) state. The pedigree in Fig. 5.1b shows the opposite; here these genes are in the repulsion (or *trans*) phase.

In some cases linkage between gene loci localized on an autosome can also be established by simple inspection of an extensive pedigree. Figure 5.2 shows a large pedigree in which Huntington disease segregates together with a DNA polymorphism of the G8 probe of *Hin*dIII – [20]. Four allelic variants of this probe are observed in this pedigree, A, B, C, and D. The Huntington gene invariably segregates together with allele C. One individual, VI, 5 (arrow), so far has been unaffected by Huntington disease, but she will be affected later, provided that her father, (who has not been tested), does not happen to have transmitted another chromosome that carries a C allele not linked to the Huntington gene. The pedigree points to close linkage between the locus for this DNA polymorphism and

the Huntington gene. Some cross-overs in other such pedigrees have been detected, and the recombination fraction is 4% or less. The example of this pedigree shows that the chromosomal phase of alleles at two loci (*cis* or *trans*) can often be ascertained with great precision even in one large pedigree, and that recombinants can be identified if (at least) three generations are available for analysis, and if there are many sibs.

Statistical Analysis. In most cases linkage analysis is more difficult. Extensive pedigrees such as that in Fig. 5.2 are exceptional; most available families consist of two parents and their children. Here the

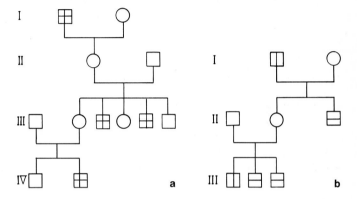

Fig. 5.1 a, b. Pedigrees with red-green color blindness (⊟), hemophilia (⊡), or both conditions (⊞). **a** Both abnormal genes in coupling. (From Madlener 1928) **b** In repulsion. (From Stern 1973 [77])

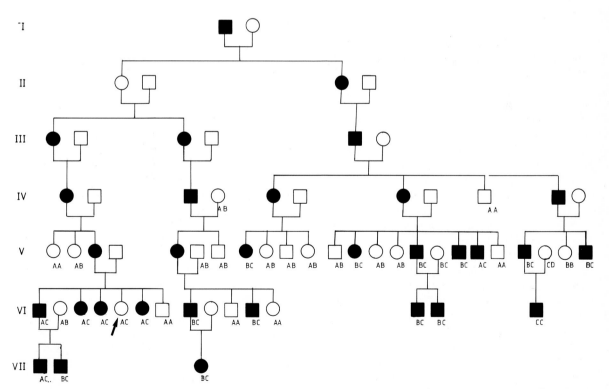

Fig. 5.2. Large pedigree from Venezuela with Huntington disease. *A, B, C,* Three different "alleles" of a DNA polymorphism. The Huntington gene is transmitted together with "al-lele" C. One individual, VI,5 (*arrow*) has so far been unaffected. She will most likely be affected later (See text). (From Gusella et al. 1983 [19])

problem is that the chromosomal phase is usually unknown: a double heterozygote may be AB/ab (cis) or Ab/aB (trans). When the alleles are randomly distributed in the population (linkage equilibrium), the two types are expected in about equal frequencies: an AB/ab person forms germ cells in the ratio:

AB	Ab	aB	ab
$\dfrac{1-\Theta}{2}$	$\dfrac{\Theta}{2}$	$\dfrac{\Theta}{2}$	$\dfrac{1-\Theta}{2}$

whereas a heterozygote Ab/aB forms germ cells in the ratio:

AB	Ab	aB	ab
$\dfrac{\Theta}{2}$	$\dfrac{1-\Theta}{2}$	$\dfrac{1-\Theta}{2}$	$\dfrac{\Theta}{2}$

Expectations for germ cells are then in any case:

AB	Ab	aB	ab
$\dfrac{1-\Theta}{2}$	$\dfrac{\Theta}{2}$	$\dfrac{\Theta}{2}$	$\dfrac{1-\Theta}{2}$

or

$\dfrac{\Theta}{2}$	$\dfrac{1-\Theta}{2}$	$\dfrac{1-\Theta}{2}$	$\dfrac{\Theta}{2}$

which adds up to:

$\frac{1}{4}$	$\frac{1}{4}$	$\frac{1}{4}$	$\frac{1}{4}$

irrespective of Θ. It even remains true if $\Theta = 0$ (very close linkage).

All four types of germ cells occur with the same frequencies, regardless of the probability of recombination Θ. Linkage does not lead to any association of alleles A,B or a,b in the population (Exception: linkage disequilibrium; Sect. 5.2). Another criterion for linkage must be found, one that is independent of the phase of the double heterozygote.

Such a criterion would be the *distribution* of children within sibships. In mating of AB/ab persons (*cis* phase) most children show the allele combinations of their parents; in matings of Ab/aB (*trans* phase) most children show these alleles in a new combination. How can these deviations from random distribution within sibships be measured and used for establishing linkage and determining the probability of recombination? Bernstein in 1931 [4] was the first to develop such a method. It has now been replaced by the method of "logarithm of differences" (lod) scores as developed by Haldane and Smith (1947) [22] and Morton [43–46] and is generally used to assess linkage. Its principle can be described as follows:

The probability P_2 that the observed family data conform to the behavior of two loci under full recombination without any linkage is calculated and similarly, the probability P_1 that the identical family data are the result of two linked loci under a specified recom-

bination fraction (Θ) is estimated for various families. The ratio of these two probabilities is the likelihood ratio and expresses the odds for and against linkage. This ratio $\dfrac{P_1(F/\Theta)}{P_2(F/(^1/_2))}$ must be calculated for each family F.

A man may be doubly heterozygous for the gene pairs A,a and B,b. His wife may be homozygous for the two recessive alleles aa, bb. Assume that his two sons, as the father, are doubly heterozygous, i. e., they inherited the dominant alleles A and B from the father. This probability is $^1/_2 \times ^1/_2 = ^1/_4$ for each son if the genes segregate independently. If the gene loci are closely linked without crossing over, the probability for occurrence of this pedigree may be calculated as follows. Either the genes occur in coupling state: AB/ab, then the possibility for common transmission to each of the two sons is $^1/_2$ (transmission of the combination ab would also have a probability of $^1/_2$), or the genes occur in repulsion state Ab/aB, where transmission of both dominant alleles to the same son requires crossing over. With close linkage and in absence of crossing over the probability here of common transmission = 0. Hence, the total probability for transmission of the combination aB to either son is $^1/_2$ and the likelihood ratio is $P_1/P_2 = (^1/_2)/(^1/_4) = 2$ in favor of close linkage. Likelihood ratios for the various degree of loose linkage can be calculated in the same way.

For convenience the logarithm of the ratio is used, and a lod score z (meaning "log odds" or "log probability ratio") is used:

$$z = \log_{10} \frac{P(F|\Theta)}{P(F|(^1/_2))} \tag{5.1}$$

Here, $P(F|\Theta)$ denotes the probability of occurrence for a family F when the recombination fraction is Θ. Using the logarithms instead of the probabilities themselves has the advantage that the score of any newly found family can be added, giving a combined score $z = \Sigma z_i$ for all families examined.

Equation 5.1 implies an identical recombination fraction for both sexes. Since sex differences in recombination rates have been described [65] (see below), the z score in actual data should be computed separately for the sexes:

$$z = \log_{10} \frac{P(F/\Theta, \Theta')}{P(F/(^1/_2, ^1/_2))} \tag{5.2}$$

where Θ is the recombination fraction in females and Θ' in males.

It follows from the definition of the likelihood ratio that the higher its numerator, the stronger is the deviation in the direction of linkage. In terms of logarithms the higher the z score, the better is the evidence for linkage. A lod score of 3 or higher is generally considered as proof of linkage. Minor corrections for dominance and for ascertainment of pedigrees with rare traits but are not dealt with here [71].

The score $z(\Theta, \Theta')$ for the entire set of data is the sum of the scores of the separate families. For a first approach $\Theta = \Theta'$ is assumed to simplify the calculations. After linkage has been established, a possible sex difference can be looked for.

Computer programs for detection and estimation of linkage are now available (App. 7). They also allow for testing whether a part only of the observed families show linkage (= linkage heterogeneity). These programs permit to make optimal use of linkage information even in large and sometimes complicated pedigrees. For a detailed account of reasoning on linkage as well as methods of analysis, see Ott [57].

The use of lod Scores. The ideal mating for linkage studies involves a double heterozygote, i.e., a person heterozygous for two different traits, with a person homozygous for the two genes. The following types of families do not contribute information regarding linkage:

a) Families in which neither parent is doubly heterozygous
b) Families in which there cannot be any observable segregation
c) Families in which the phases of the parents are unknown and there is only one examined child

Most linkage studies involve analysis of two common markers or of a common gene with a gene for a rare genetic disease. Opportunities to study linkage between two rare genes hardly ever exist. The ideal family for linkage studies is a kindred with at least three generations, many matings, and a large number of offspring. Such families are becoming rare in Western societies. An alternative approach involves testing of many small families. This may even have an advantage if more than one gene locus causes a special phenotype. In these instances the study of a single, large pedigree with linkage may create the impression that this gene locus is the only one whose mutations cause the phenotype in question, whereas analysis of many, smaller pedigrees may point to other loci as well, and hence to genetic heterogeneity.

As discussed in Sect. 2.1.2.4, the human male genome has a map length of approx. 25.8 M. With about 3.5×10^9 bp/haploid genome (Sect. 3.1.1.1), this means that 1 cM corresponds to approx. 1.356×10^6 bp (or 1356 kb). However, crossing-over sites are not distributed equally, as is discussed below. Morton and Collins [51] collected estimates from five different series and reported a figure of 24.28–30.02 M for autosomes.

When linkage has been established and a maximum likelihood estimate of Θ achieved, the question of heterogeneity should be examined. If, for example,

linkage between the locus for a genetic polymorphism and a rare dominant condition has been established, linkage analysis can help to prove genetic heterogeneity if only part of the family data shows linkage. This occurs very often [41]; the statistical problem posed by such a situation is tricky. Ott [55] has proposed using the χ^2 statistic to compare hypotheses: linkage without heterogeneity vs. nonlinkage, linkage with heterogeneity vs. linkage without heterogeneity, and linkage with heterogeneity vs. nonlinkage. It is also possible to estimate the proportion of families showing linkage in the data set studied.

The human genome has now been so saturated with genetic markers that it is often advisable to estimate linkage not only for two loci but for several markers at once (multipoint linkage). Computer programs for linkage analysis are available; the LINKAGE package, for example, has proven to be very useful. Appendix 7 (Table A7) contains necessary information; two calculated examples are also given, the first showing lod scores for one large pedigree and the second demonstrating analysis of several small pedigrees of different structure, with genetic heterogeneity.

Recombination Probabilities and Map Distances. Once a number of linkages have been established, the next step is to estimate map distances between these loci. These distances are expressed in morgans and centimorgans, 1 cM (map unit) meaning 1% recombination ($\Theta = 0.01$) for small map distances. For larger distances this value must be corrected for double crossing over. Various methods have been proposed. Given a recombination frequency Θ, the map distance can be read directly from Fig. 5.3.

The Sib Pair Method. The use of lod scores is the ideal method if the mode of inheritance of the two traits to be tested for linkage has been established. Examples include testing of linkage for two genetic markers or for a marker and a clearcut monogenic disease. At least two generations should be available. The analysis becomes more difficult if penetrance of the mutant gene is incomplete, and a definitive phenotype cannot be assigned. While inclusion of a penetrance term in the analysis may be possible, introduction of this and other adjustments can be hazardous since it may lead to false claims, particularly if the data are manipulated in various ways until a "positive" linkage result is obtained.

In general, if the mode of inheritance cannot be established, or when data from only one generation are available, it is preferable to use the sib method first suggested by Penrose in the 1930s [60] (see also [5, 65–67]; Fig. 5.4). Its rediscovery has been called by Ott [56] "the cutting of the Gordian knot." (Alexander the Great was challenged to disentangle this

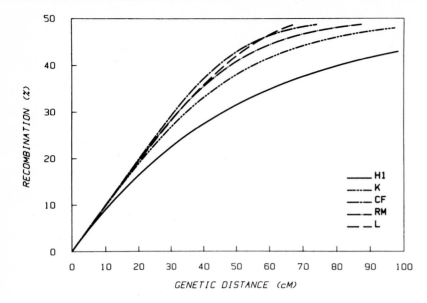

Fig. 5.3. Genetic distance (in centimorgans) in relation to the recombination fraction (in percentage), according to estimates from various authors. *H1,* Haldane function with no interference; *K,* Kosambi; *CF,* Carter and Falconer; *RM,* Rao and Morton; *L,* Ludwig. (From White and Lalouel 1987 [86])

knot, which no one had been able to do previously; he cut through it with his sword.) This is because the detection of linkage with this method does not depend on correct assignment of the mode of inheritance but only on the influence of a gene that contributes nonnegligibly to the trait and on linkage of this gene with a marker. Penrose explains, "The method is based on the principle that, when pairs of sibs are taken at random, certain types of sibling pairs will be more frequent if there is linkage than if there is free assortment of the characters studied." The method is used as follows: Codominant genetic markers with several alleles are studied in a series of sib pairs (or other pairs of relatives) both of whom are affected with a disease whose linkage relationship to the marker is to be investigated. If there is *no* linkage to the marker, 25 % of affected sib pairs share both maternal and paternal alleles of the marker, 50 % share one of the marker alleles, and 25 % differ for both marker alleles. If the marker is linked to a gene that contributes to causing the disease, the proportion of sib pairs with the disease who share one or two marker alleles is increased over the expected 25 % or 50 %, respectively.

The problem is straightforward when the marker status of both parents is known, and the identity by descent of the marker can be established. However, the method can also be used, albeit with more difficulty, if only marker information on the affected sib pairs is available, and the parents are not investigated (see also [10]). Problems arise here if the affected parent is homozygous for the marker, or if the unaffected parent contributes a marker allele to the child that is identical to the marker that cosegregates with the disease gene [50]. This problem can be avoided in

other of two ways: (a) *Unaffected* sib pairs in the same families can be studied. Such pairs are expected to be more similar in the alternative marker alleles. However, the disease must have 100 % penetrance so that there is certainty that an unaffected sib really does not carry the disease gene. (b) An approach that is becoming increasingly feasible is to study haplotypes of several closely linked markers or of multiallelic variable-number tandem repeats (VNTRs) rather than of only a single restriction fragment length polymorphism (RFLP) marker. Under these conditions each parent often has a unique haplotype (or VNTR) at the site under study with four haplotypes (or VNTR types) between the two parents. A child inherits only two haplotypes (or VNTRs), one from each parent. With no linkage 25 % of sib pairs share two haplotypes or VNTRs, 50 % are identical for one, and 25 % for none. With linkage, statistically significant increases over the 25 % and 50 % proportions of shared haplotypes (or VNTRs) are obtained.

This method has been adapted for pairs of relatives other than sibs as well; computer programs for testing linkage and estimating map distances are available (see also Appendix 7).

Haplotype analysis is especially useful if general circumstances favor the view that all patients suffer from a genetic disease which can be traced back to one single mutation. (See the discussion of such "founder effects" in Sect. 13.3.2). Here, the time since this mutation occurred may not have been sufficient to randomize marker loci around this mutation by repeated crossing over; it may still exist within the same haplotype in most instances. If the mode of inheritance is autosomal-recessive, patients may be

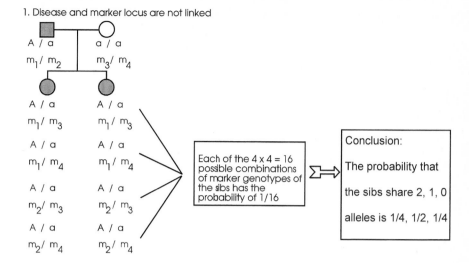

1. Disease and marker locus are not linked

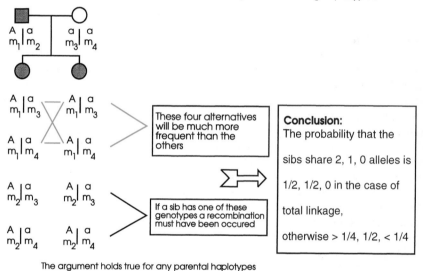

2. Disease and marker locus are linked, for example the following haplotypes:

Fig. 5.4. Principle of the sib pair method for finding linkage. Assume a mating between a parent heterozygous for a dominant disease allele A and a parent homozygous for a normal allele a; each of the four parental chromosomes have different marker alleles at a marker locus. The possible alternatives are presented. Blue bars connect the more common sib pairs

The argument holds true for any parental haplotypes

homozygous for this haplotype. In this way, the autosomal-recessive gene for benign recurrent intrahepatic cholestasis (BRIC), an autosomal-recessive condition occurring in the Tyrolian Alps, was mapped to chromosome 18. If its main precondition – evidence for a founder effect – is met, the method is very efficient statistically [24a].

Results for Autosomal Linkage, Sex Difference, and Parental Age. The first autosomal linkage in man was found by Mohr [44] between the Lutheran blood groups and the AB0 secretor locus. Some years later linkage between the Rh loci and elliptocytosis (166 900) was established and used to detect genetic heterogeneity of elliptocytosis, since not all families with elliptocytosis showed this linkage. A short time later, linkage between the AB0 blood group locus

and the dominant nail-patella syndrome (161 200) locus was found. This linkage established for the first time a sex difference of recombination probabilities in humans: map distance between these loci was 8 cM for males and 14 cM for females. A great many linkage have since been examined for sex differences; in the majority of instances a higher recombination fraction has been observed in females than in males. The same sex difference had been known for a while in the mouse [63]. It conforms to Haldane's rule [21] that crossing over is generally more frequent in the homogametic than in the heterogametic sex. In humans, however, this rule has exceptions. In the distal portion of 11 p, for example, the recombination frequency appears to be higher in males [87]. It appears that such a higher male recombination rate may be characteristic for chromosome parts close to telomeres. There is also good evidence that the absolute exchange rate is higher in chromosome parts close to the telomere [86]. However, the overall recombination rate is definitely higher in females [15]. In *Drosophila,* there is no crossing over at all in the male.

There has been considerable discussion as to whether recombination frequency is also influenced by parental age. In the mouse, the data are consistent with decreasing recombination rates with aging in females and increasing rates in males. Weitkamp [85] found a significantly increased incidence of recombinants with increasing birth order in humans for eight closely linked pairs of loci, indicating a parental age effect. There was no difference between males and females for this effect. A similar parental age effect was found for the Lutheran/secretor and Lutheran/myotonic dystrophy (160 900) pairs but not for AB0/nail-patella or Rh/PGD pairs.

In a survey of cytogenetically determined chiasma frequencies from 204 males reported in the literature, little or no linear trend with age was found [36]. No cytogenetic data are available for females. The discrepancy between formal recombination data and chiasma frequencies is unexplained [86].

Information from Chromosome Morphology. Pairs or clusters of autosomal loci found to be linked (linkage groups) could not be assigned to specific chromosomes by a formal methodology of family study. The first chromosomal localization was accomplished as follows [14, 64].

The long arms of chromosome 1 frequently show a secondary constriction close to the centromere. In about 0.5 % of the population, this constriction appears much thinner and longer than normal. The variant is dominantly inherited. An uncoiler locus (Un-l) appears to be localized on chromosome 1. Linkage studies show close linkage between the blood group Duffy locus and the Un-l trait; $\Theta = 0.05$. Linkage between Duffy and congenital zonular cataract (116 200) had been discovered earlier. Hence, a linkage group with three loci, cataract, Duffy, Un-l could be assigned [14].

Another possibility to localize genes on specific chromosomes was afforded by deletions. If a gene locus whose mutation has a dominant effect is lost by deletion, the absence of that gene may occasionally have a phenotype similar to the dominant mutation. More extensive symptoms may also be present, since more genetic material than a single gene would be expected to be lost. In 1963 a retarded child with bilateral retinoblastoma was found to have a deletion of the long arm of one D chromosome [37]. This chromosome was later identified as no. 13, and this 13q14 deletion has been found in a number of other cases with retinoblastoma and additional anomalies. Patients with

retinoblastoma without additional symptoms usually have no deletion. The localization of this gene (Rb1) has since been confirmed by DNA marker studies (see Sect. 10.4.3).

Another approach, thought to be more generally useful, is the quantitative examination of enzyme activities in cases with chromosome anomalies. Most enzymes show a clearcut gene dose effect in heterozygotes, i.e., heterozygotes for an enzyme deficiency have approx. 50 % of enzyme activity. Therefore a similar gene dose effect might be expected when a gene locus is localized on a chromosome segment that has been lost by deletion.

The results of many early studies of this sort proved disappointing. Later, however, an increasing number of such gene dosage effects have been described in vitro, on trisomic and monosomic cells [33] (Sect. 8.4.3). To mention only one example, the activity of the enzyme phosphoribosylglycinamide synthetase was studied in several cases of partial monosomy and full and partial trisomy 21, as earlier studies had suggested a gene dosage effect for this enzyme. In regular trisomy 21 an excess was found with a ratio of trisomy 21 to normal of 1.55. A ratio of 0.99 was found in 21q21 → 21pter monosomy; 0.54 in 21q22 → 21qter monosomy; 0.88 in 21q21 → 21pter trisomy; and 1.46 in 21q22.1 trisomy. Therefore the phosphoribosylglycinamide synthetase gene locus could be localized in subband 21q22.1 [11]. Utilization of variants in chromosome morphology (heteromorphisms), such as the secondary constriction on chromosome 1 mentioned above, along with gene dosage studies, slowly opened the way to linkage and gene localization. A new method, using cell fusion, has led to much more rapid progress.

5.1.3 Linkage Analysis in Humans: Cell Hybridization and DNA Techniques

First Observations on Cell Fusion. The history of cell fusion is related by Harris [24]. Binucleate cells were observed in 1838 by J. Mueller in tumors, and afterwards by Robin in bone marrow, by Rokitansky in tuberculous tissue, and by Virchow both in normal tissues and in inflammatory and neoplastic lesions. The view that some of these cells were produced by fusion of mononucleate cells derived from the work of de Bary in 1859, who observed that the life cycle of certain myxomycetes involves the fusion of single cells to form multinucleated plasmodia. The earliest reports of multinucleated cells in lesions that can be identified with certainty as being of viral origin appear to be those of Luginbuehl (1873) and Weigert (1874), who described such cells at the periphery of smallpox pustules.

Following the introduction of tissue culture methods, numerous observations were made on cell fusion in cultures of animal tissue (see [23]). Enders and Peebles (1954) found that the measles virus induces cells in tissue culture to fuse to form multinucleated syncytia. Okada (1958) showed that animal tumor cells in suspension can be fused rapidly to form multinucleated giant cells using high concentrations of hemagglutinating parainfluenza virus (Sendai virus).

In 1960 Barski, identified cells generated by spontaneous fusion in a mixed culture of two different but related mouse tumor cell lines. These cells contained the chromosome complements of both parent cells within a single nucleus. This phenomenon was then examined by Ephrussi et al., who concluded that not only closely related mouse cells could be hybridized; even larger genetic differences did not exclude spontaneous cell fusion. However, it soon became obvious that the frequency of spontaneous cell fusion is very low, and that many cell types never fuse spontaneously. Fusion frequency must be increased in some manner. Furthermore, isolation of hybrid cells was possible only when culture conditions gave these cells a selective advantage.

Both problems were soon solved. Littlefield (1964) isolated the rare products of spontaneous fusion in mixed cultures by a technique adopted from microbial genetics. Fusion of two cells deficient in two different enzymes resulted in hybrids that recovered the complete enzyme set by complementation. Only these cells survived selection against the deficient cells.

Harris and Watkins (1965) [213] enhanced the fusion rate of various cells by treatment with UV-inactivated Sendai virus. Along with introduction of this method, they showed that fusion can be induced between cells from widely different species, and that the fused cells are viable. With this work, widespread use of the cell fusion method in various branches of cell biology began.

First Observation of Chromosome Loss in Human-Mouse Cell Hybrids and First Assignment of a Gene Locus. Weiss and Green (1967) [85] fused a stable, aneuploid mouse cell line, a subline of mouse L cells, with a diploid strain of human embryonic fibroblasts. The mouse cell line was deficient in the thymidine kinase (TK) locus and did not grow in hypoxanthine-aminopterin-thymidine (HAT) medium, a culture medium selective for cells containing the human TK locus (188300).

Cultures were initiated by mixing the two types of cells and growing them on standard medium. After 4 days cultures were placed in the selective HAT medium. This led to degeneration of the mouse cells, leaving a single layer of human cells. After 14–

21 days hybrid colonies could be detected growing on the human cell monolayer. A number of these colonies were then isolated, grown for a longer time period, and examined. They turned out to maintain the mouse chromosome complement, but 75%–95% of the human chromosomes were lost. One human chromosome, however, was present in almost all cells growing in the HAT medium. This suggested that the locus for thymidine kinase is localized on this chromosome. Therefore control experiments were carried out with a bromodeoxyuridine-containing medium. Bromodeoxyuridine, a base analogue for thymine, is accepted by TK in place of thymine and selects against cells containing this enzyme. A special chromosome described as "having a distinctive appearance" was present in almost all HAT cultures but in none of the bromodeoxyuridine cultures. It was concluded that the TK locus is indeed localized to this chromosome. Shortly thereafter the chromosome bearing the TK locus was found to be no. 17 [43] (Fig. 5.5).

This work led to two principles which were later decisive for the use of cell hybridization in linkage work:

1. Hybrids between mouse and human cells tend to lose many human chromosomes. It was later shown that this loss is random, and therefore examining a great number of hybrids one can expect to find a cell that has kept any one specified chromosome.
2. By using an appropriate selective system it is possible to select cells with a certain enzyme activity and to localize the gene loci specifying this enzyme to a specific chromosome.

Whereas genetics has historically been the science of genetic variability within a species, the hybridization method permits the localization of genes that do not show genetic variability in humans, provided only that the gene products of the human and nonhuman cells can be identified. One means of identification is the use of a selective system.

Since 1967 selective systems have been developed for several enzymes. One example uses the hypoxanthine phosphoribosyltransferase locus on the X chromosome (Sect. 7.2.2.6). This system can be used for selection not only of other X-linked loci but also of autosomal loci if a part of the autosome has undergone translocation with the X chromosome. It is also possible to assign loci for which no selective system exists, provided that enzymes produced by the two species have recognizable differences such as electrophoretic variation.

In hybrid cultures chromosome breakage and rearrangement are relatively frequent events. This chromosomal behavior made possible the suitable selection of hybrid clones containing identifiable parts of

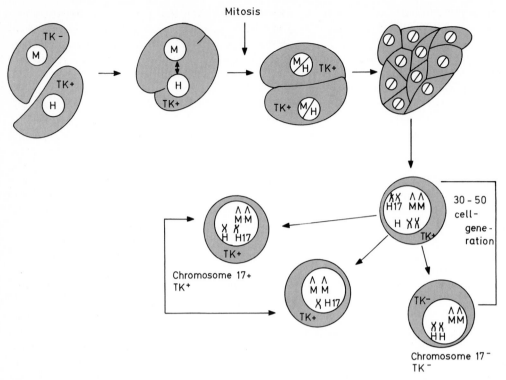

Fig. 5.5. The principle of gene localization on an autosome. Thymidine kinase deficient mouse cells (*M, TK⁻*) are grown in mixed cell culture with normal human cells (*H, TK⁺*). The cells are fused spontaneously, chemically, or by Sendai virus. After 30–50 cell generations the cells have lost part of their human chromosomes. Only cells having kept chromosome 17 show thymidine kinase activity (two cells at *left*). Cells without chromosome 17 show no *TK* activity (cell at *bottom right*)

chromosomes, thereby combining the advantages of deletion mapping and hybridization.

Other Sources of Information for Gene Localization. Another method utilized for gene localization is the in situ DNA-RNA hybridization technique (see Sect. 3.1.3.3). This method brings labeled cDNA of the gene to be localized (or DNA produced from RNA by the enzyme reverse transcriptase) into contact with chromosomal DNA, which is made single-stranded by appropriate techniques. The cDNA sequences that match their DNA counterpart on the chromosome hybridizes at a specific chromosomal locus. In recent years this method has been complemented by nonisotopic in situ hybridization, which requires less time and also offers improved resolution power.

DNA Polymorphisms and Gene Assignment. In recent years the detection of an increasing number of DNA restriction site polymorphisms and other DNA markers has opened up an additional approach to mapping of the human genome. In addition to these – now classical – RFLPs, other markers have been detected, among them the minisatellites [25], short DNA sequences distributed in great number over the human genome and occurring with variable numbers of repeats; this number is different in almost every individual. Therefore the information content for linkage studies is very high. Another such system, that of so-called microsatellites, consists of $(CA)_n$ repeats that occur in great numbers in the genome; the number of repeats per unit is extremely variable, as well. Localization of individual $(CA)_n$ probes in the genome has been achieved by identifying and using specific DNA sequences on both sides of these markers that allows amplification by the polymerase chain reaction (PCR; Sect. 3.1.3.5). Table 12.3 shows the types of available DNA markers. Genes for many important hereditary diseases could be localized on specific chromosomal sites using such markers. Model calculations have been performed [7, 35, 75] showing that only a few hundred such markers distributed at random along the human genome are necessary for the entire human genome to be mapped and for at least one marker closely linked to the gene locus of a given hereditary disease to be found that can be used in genetic counseling and prenatal diagnosis (Sect. 18.1; 18.2). Immortalized lymphoblastoid cell lines from large three-generation families with known genotypes for many marker loci have become available for the study of new markers [87]. In addition to this set of families in Salt Lake City, Utah, another set has become available in France – the CEPH families [13, 15]. Both sets consist of nuclear families (both parents and a great number of children) which have been typed with many genetic markers. DNA from such families is available to the scientific community for further mapping. With increasing numbers of available markers, analysis of linkage relationships not only between two gene loci (e.g., one disease gene and one marker) but also between a disease gene and a set of markers – a haplotype – is becoming more and more important. Such haplotypes are being used increasingly in many studies as well as in genetic counseling and prenatal diagnosis. The proportion of kindreds in which the combination of genotypes is informative for linkage – and hence for prenatal diagnosis – can be enhanced appreciably by using such sets of markers.

In recent years progress in establishing linkage – or synteny – and in assigning loci to certain chromosomes has been rapid. Published conferences are held every 2 years, and newsletters are circulated. In this rapidly evolving field of science centered around a special set of methods, scientists have created their own system of scientific interaction independent of "official" channels such as scientific journals to keep pace with the rapid progress.

Gene Symbols To Be Used. In cooperative scientific activities such as mapping of the human genome, certain terminological conventions are necessary. Such conventions have been established by a committee on standardized human gene nomenclature [41] and by another committee consisting of an international group of linkage specialists [28] for gene symbols and for human linkage maps in general. The rules for gene symbols include: only uppercase letters, no hyphens, no more than four or five letters or numbers, etc. For details see Appendix 8.

Present Status of Gene Localization and Assignment to Autosomes. All the information noted above has been brought together in a documentation that is being publishing continuously, and is available in a database (see Appendix 7). A selection of important genes is contained in Fig. A7.5.

Linkage of X-Linked Gene Loci. Assignment of loci to the X chromosome is no problem when the pedigrees show the typical pattern of X-linked inheritance. Assignment of genes to special segments of the X requires the techniques described above, often supported by family studies.

The X chromosome (after chromosome 1) is still the second most completely saturated human chromosome (Fig. 5.8). Well-known genes for human diseases have been localized. But even with these localizations many X-linked genes need to be mapped for example, it is estimated that up to 80 (or even more) different types of X-linked mental retardation may exist. For only few of these have genes been localized so far.

Genetic and Physical Map of the Homologous Segment of X and Y Chromosomes. When a genetic map of a chromosome or chromosome segment becomes, known, it ultimately needs to be complemented by a physical map. The final goal is to identify the DNA sequence of all genes within this area. The following presents an introduction using the pseudoautosomal region of Xp and Yp as examples.

This region is located in the Giemsa light band Xp22.3 and in Yp11.32 (Fig. 2.19). Various authors have constructed partial physical maps; the latest estimate runs at about 2560 kb for its entire length [62]. There is a certain interindividual length variation. A number of families from the CEPH family pool [13] were studied with 11 exactly localized DNA markers; this established very high recombination frequencies in males and lower but clearly elevated frequencies in females. In males, the genetic map is 55 cM long and in females 8–9 cM. Hence the sex difference in this case is opposite to Haldane's rule. Moreover, it is about 20–25 times higher than the average of all chromosomes in the male and 6 times higher in the female. Differences in recombination rates between chromosome regions have also been demonstrated for other chromosomes.

The first step of physical mapping has been restriction of this area with "rare cutter enzymes" which cut DNA into regions with many CpG islands. A striking accumulation of such rare cutter sites was found within the most distal 500 kb of the pseudoautosomal region, which also shows most individual DNA variants (Fig. 5.6; 5.7). As a rule, CpG islands indicate the presence of genes. So far it has not been clarified whether their accumulation in this area indicates an extremely high density of genes or a stretch of noncoding, CpG-rich DNA with some other function. Closer scrutiny of this terminal area by chromosome jumping has revealed at least five regions in which the CpG islands are concentrated. Further analysis, for example, constructing of a contig of YAC clones, is in progress (Figs. 5.6, 5.7).

Some genes are also located in the pseudoautosomal region; there is, for example, a gene for the enzyme ADP/ATP translocase (3 001 500) [68] Figure 5.7 shows the map of the pseudoautosomal region.

The Y Chromosome. It is the Y chromosome [83, 88] which determines male sex. Recent analysis has suc-

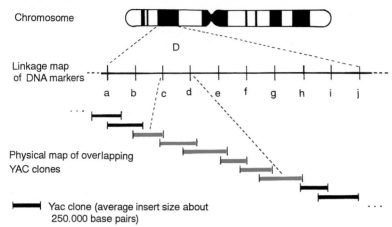

Fig. 5.6. Principle of integrating a linkage map with the physical map, using YAC clones. The overlapping clones cover the entire area of the D gene, as well as adjacent areas on both sides, permitting determination of the entire DNA base sequence. *Blue bars,* disease gene D lies on one of the YAC clones

ceeded in localizing specific factors involved in sex determination to certain segments of this chromosome. As in many other instances, analysis of pathological conditions has contributed to understanding of the normal state, such as the study of men with two X but apparently no Y chromosome. As early as 1966, Ferguson-Smith [17] postulated XY translocations which were expected to transfer to the X a small – but for male development decisive – part of the Y chromosome. This expectation has been confirmed by many observations [88]. Since meiotic pairing of X and Y chromosomes occurs in the pseudoautosomal (Fig. 2.19) and in adjacent nonhomologous regions, and since pairing errors provide a plausible mechanism for such translocations, the search for the testis-determining factor (TDF) soon concentrated on the short arm. Here the SRY (sex reversal gene on Y) gene was finally identified [70]. This gene gives rise to a mRNA of about 1 kb; it apparently codes for a protein controlling transcription. This fits well with the prediction that the TDF is a "master regulatory gene" [88].

This factor alone, however, is not sufficient for male differentiation [43]. Other factors appear to have important supportive functions, such as the ZFY factor, which is also located close to the boundary of the pseudoautosomal region [58].

A factor, or group of factors, has recently attracted interest because analysis of abnormalities has thrown new light on certain defects of spermatogenesis. A region within the euchromatic segment of the long arm appears to be important for normal spermatogenesis, since deletions within this region lead to arrested sperm formation either in an early stage, i.e., not even functional spermatogonia are formed, or in postmeiotic stages [83]. The first discovered deletions

were so large that they could be recognized by cytogenetic methods [80]. Small deletions were later identified by molecular techniques [83], and their recognition has become important for differential diagnosis of male infertility.

In addition to genes involved in testis development and spermatogenesis, the Y chromosome may also harbor genes to promote growth and stature.

Linkage Analysis with Genetically Ill-Defined, Quantitative Traits? Earlier linkage studies sometimes included quantitative traits with multifactorial inheritance. The idea behind these studies was that linkage would detect major genes influencing these phenotypes. Theoretically this approach is certainly correct. If a measurable characteristic shows fairly close linkage with a genetic marker, this indeed points to linkage of a major gene with this marker locus. If linkage can be shown for two measurable characters, both may be influenced by two major genes. The same argument applies to diseases with multiple, complex causes, including many common diseases. Practically, however, great care is needed. The reason is threefold:

1. If many quantitative characters are included in such an analysis, chance alone yields some statistically significant values that suggest linkage.
2. Linkage predicts correlations in families but not in the population. The prediction, however, depends on random mating. Some measurable, graded characteristics are sometimes associated with assortative mating.
3. If diseases without clearcut Mendelian mode of inheritance are studied, attribution of phenotypes to genotypes is often dubious.

Earlier studies often yielded disappointing results. More recently, however, the method for studying affected pairs of close relatives has been introduced (or, rather, Penrose's sib pair method was rediscovered; Sect. 5.1.2). The logistic difficulties with this method, – especially the large number of pairs of relatives needed for a significant result – are being

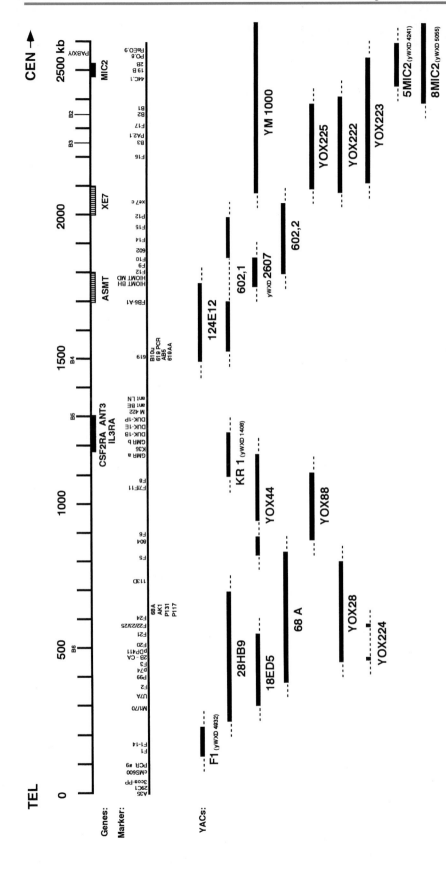

Fig. 5.7: Schematic representation of the 2600 kb pseudoautosomal region (PAR1) on the short arm of the human sex chromosomes on Xp22.3 and Yp11.32. A YAC map encompassing 19 YACs is shown. The YACs are drawn proportional to their size (*horizontal black bars*) with the YAC identification number *below*. Their position can vary by ± 50 kbp (*dotted lines*). *Dotted segments between black bars*, internal deletions within YACs. Hybridisation of probes against Southern blots of YAC DNA are used to identify overlaps. Three small gaps (areas not covered by a YAC) are shown on the map. Several of the YACs are chimeric, and only the portion within the PAR is drawn. Six genes which are known within the pseudoautosomal region have been mapped on this contig. The mapping position of the the colony-stimulating factor receptor subunit α (*CSF2RA*), interleukin receptor subunit α (*IL3RA*), adenine nucleotide translocase 3 (*ANT3*), and a cell surface antigen (*MIC2*) is indicated as *black bar* below the scale. *Vertical dashed line*, the assumed interval for the acetylserotonin methyltransferase (*ASMT*) and a gene with unknown function (*XE7*) (Courtesy Dr. G. Rappold)

overcome by appropriate large-scale international cooperation. In some diseases, for example the most important mental diseases such as schizophrenia (Sect. 16.2), an approach of this sort together with painstakingly precise clinical classification may ultimately lead to a more incisive analysis of gene action and specific pathogenetic aberrations. However, there are numerous difficulties. For example, if genetic heterogeneity exists, and if a disease is caused in part by different major genes in various population groups, collection of numerous sib pairs from these populations may conceal linkage present in one of these populations. We return to these problems in Chap. 16.

DNA Variants in Linkage. The large number of DNA polymorphisms provide many new markers, and most linkage work is now beeing carried out with DNA variants (Sects. 12.1, 12.2).

Linkage disequilibrium (i. e., failure to demonstrate free assortment; see Sect. 5.2) has frequently been found between the sites of various markers at a given locus. Since these sites are physically very close, crossovers between them are rare, and many generations must pass before linkage equilibrium is reached. Furthermore, current data suggest that recombination rates at closely linked markers may vary considerably between different chromosomal locations. Thus, both "hot" and "cold" spots of recombination appear to exist [18, 52].

Practical Application of Results from Linkage Studies. In the past the main interest of linkage studies was theoretical. Practical applications, however, are emerging increasingly. If, for example, gene A causes a rare hereditary disease manifesting itself later in life, and B is a genetic marker closely linked to A and segregating in the same family, the disease can be predicted in a young individual, and this prediction may be used in genetic counseling. Often prenatal diagnosis is also possible (Sect. 18.2). While in an increasing number of instances genetic diagnosis can be based on direct study of the mutant gene itself, indirect diagnosis using linkage with genetic markers will remain important if the gene has not yet been cloned.

5.2 Gene Loci Located Close to Each Other and Having Related Functions

5.2.1 Some Phenomena Observed in Experimental Genetics

Closely Linked Loci May Show a Cis-Trans *Effect.* When series of multiple alleles were analyzed in *Drosophila*, crossing over within these series was ob-

served occasionally, indicating that what had been considered as one "gene" can be subdivided by genetic recombination. Such alleles were termed "pseudoalleles" by McClintock in 1944 [40]. In some a so-called *cis-trans* effect was shown. When two mutations were localated side by side on the same chromosome (*cis* position), the animal was phenotypically normal, but when they were localized on homologous chromosomes (*trans* position), a phenotypic anomaly was seen [39].

Explanation in Terms of Molecular Biology. In fungi, bacteria, and phages, genetic recombination is normally observed within functional genes, i. e., DNA regions carrying information for one polypeptide chain. A *cis-trans* effect is now considered to be typical for two mutations that are not able to complement each other functionally, i. e., that are located within the same structural gene. Complementation between two mutations, by the same token, is regarded as an indication that these mutations are located in different functional genes. A gene has many mutational sites and may be subdivided by recombination. Complementation tests are often used to test genetic, biochemically characterized conditions for heterogeneity.

A Number of Genes May Be Closely Linked. Close linkage has frequently been described between mutations affecting closely related functions, which are perfectly able to complement each other functionally and show no *cis-trans* effect. In bacteria such as *E. coli,* gene loci for enzymes acting in one sequence have been found to be closely linked and arranged in the sequence of their metabolic pathway. Their activity is subject to a regulating mechanism by a common operator and promoter [31].

5.2.2 Some Observations in the Human Linkage Map

Types of Gene Clusters that Have Been Observed. The first impression when examining the human linkage map is that most loci are distributed fairly at random. However, there are exceptions:

a) The loci for human hemoglobins γ, δ, and β are closely linked.
b) The immune globulin region comprises a number of loci responsible for synthesis of immunoglobulin chains. The same is true for genes of the T cell receptor (chromosome 14q11).
   Also exceptional is a cluster of genes, which are obviously related functionally, being involved in he immune response. This is the major histocom-

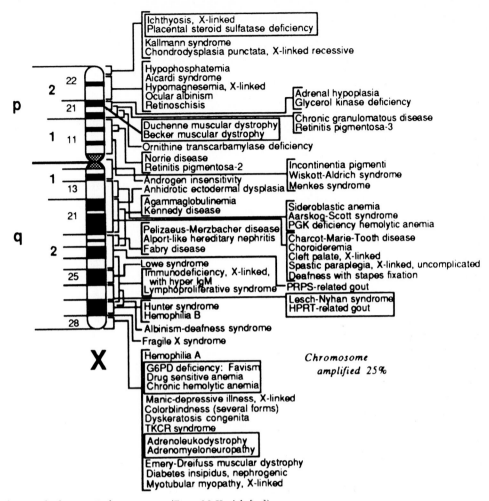

Fig. 5.8. Disease gene loci on the human X chromosome. (From McKusick [41])

patibility complex (MHC) cluster including various components of complement on chromosome 6.

In many cases, incisive genetic analysis has been possible by establishing the amino acid sequence of proteins, and the differences between the closely linked genes have become known even at the level of base sequence within the transcribing strand of the DNA – sometimes without any knowledge of the gene product.

c) No less than four gene loci involved in the glycolytic pathway are located on chromosome 1.

d) A number of genes determining closely related enzymes seem to be closely linked, for example, pancreatic and salivary amylase on chromosome 1, and guanylate kinase 1 and 2 on the same chromosome.

e) The protan and deutan loci for red-green color blindness are located in the same cluster on the X chromosome.

Clusters Not Observed So Far. As mentioned above, functionally related genes in bacteria are frequently closely linked; they are subject to common control within an operon. One would expect that, in humans, such operons would also occur. Linkage data known so far, however, provide no positive evidence. Two genes linked in the same operon in bacteria are those for galactose-1-phosphate uridyltransferase and galactokinase. In humans these genes are localated on chromosomes 3 and 17, respectively. Similarly, the gene for G6PD is located on the X chromosome, and that for 6-PGD, the following enzyme in the shunt pathway, is situated on autosome 1. Genes belonging to one gene family are sometimes but by no means always located close together. For genes involved in the immune system, including those for immunoglobulin synthesis, T cell receptors, and genes in the MHC system, this location has functional significance (Sect. 7.4.1).

5.2.3 Why Do Gene Clusters Exist?

They Are Traces of Evolutionary History. In some cases clustering is simply left over from the evolutionary history of these genes. Early in evolution there was one locus for a given gene. Then gene duplication occurred and offered the opportunity of functional diversification. The first duplication paved the way for further duplications due to unequal crossing over (Sect. 5.2.8) and hence for further functional specialization.

With no further chromosomal rearrangements the gene clusters remain closely linked. It is unknown whether in these cases close linkage is necessary for orderly function. While it may be so in some cases, this explanation is not needed to explain clustering. Evolutionary explanations are sufficient.

For example, the red and green color vision genes appear to have arisen by gene duplication. These genes are discussed in Sect. 15.2.1.5.

Duplication and Clustering May Be Used for Improvement of Function. The clustering of genes is without obvious functional significance. It would be surprising, however, if evolution were never to take advantage of this situation, combining products of such gene clusters to form higher functional units. This may be the case for the hemoglobin molecule since in the β cluster the ε, γ, β, and δ genes are arranged in the sequence of their successive activation during individual development (Sect. 7.3.2). In the immunoglobulins and T cell receptors close linkage of a number of genes, possibly a great many, has become important functionally (Sect. 7.4), as their gene products combine to form various classes of functional molecules.

5.2.4 Blood Groups: Rh Complex (111 700), Linkage Disequilibrium

The history of the rhesus blood types provides a fascinating illustration of how science develops. First, a new phenomenon was discovered. Scientists soon realized that it eludes explanation by conventional concepts. Then a long-lasting scientific controversy arose as to the most appropriate extension of these concepts. In this controversy, a new explanatory principle was created that survived the controversy in this special case, and that could be applied to an increasing number of other observations. Finally, the problem was solved, and the controversy ended – by new methods.

History. In 1939 Levine and Stetson [38] discovered a novel antibody in the serum of a woman who had just delivered a macerated stillborn child and had received blood transfusions from her AB0-compatible husband. Of 101 type 0 bloods only 21 showed a negative reaction with this antibody. There was no association with AB0, MN, or P systems.

The following year Landsteiner and Wiener [34], immunizing rabbits with the blood of rhesus monkeys, obtained an immune serum that gave positive reactions with the erythrocytes of 39 of 45 individuals. Later the antibodies were compared with those of Levine and Stetson and thought to give reactions with the same antigens. This was subsequently found to be not quite true, and now the antigen uncovered by the true anti-rhesus antibody is called LW –, in honor of Landsteiner and Wiener. Rh typing in humans is always carried out with sera of human origin, i. e., according to Levine and Stetson's observation. The following discussion relates only to reactions with these human sera.

The great practical importance of the rhesus system became apparent when transfusion accidents were traced to this antibody, and especially when erythroblastosis fetalis, a common hemolytic disease of the newborn, was explained by Rh-induced incompatibilities between mother and fetus. The red blood cells of about 85 % of all whites give positive reactions; family examinations showed that Rh-positive individuals are homozygous Rh/Rh or heterozygous Rh/rh, whereas the rh-negative individuals are homozygous rh/rh.

In 1941 Wiener discovered a different antibody that reacted with the cells of 70 % of all individuals and was independent of the basic Rh factor (Rh', according to Wiener). A third related factor was discovered in 1943. These three factors were found in all possible combinations with one another, and the combinations were inherited together. Wiener proposed the hypothesis that these serological "factors" are properties of "agglutinogens," and that these agglutinogens are determined by one allele each of a series of multiple alleles. The agglutinogens were thus thought to determine the factors in different combinations. This descriptive hypothesis is so general that it indeed explains all the complexities discovered later.

Fisher's Hypothesis of Two Closely Linked Loci. R. A. Fisher developed a more specific hypothesis. At that time another antibody had been detected, anti-Hr. In 1943 Fisher (see [61]) examined a tabulation prepared by Race, containing the data accumulated so far. He recognized that Rh' and Hr were complementary. All humans have either Rh', Hr, or both. Individuals with both antigens never transmit them together to the same child, and a child always receives one of the two. Fisher explained these findings by proposing one pair of alleles for the two antigens. The pair was named C/c. In analogy, an additional pair of alleles D/d was postulated for the original antigens Rh$^+$ and rh$^-$, and a third pair of alleles for the third factor that had been discovered. To explain the genetic data close linkage between these three loci was assumed.

Fisher's hypothesis predicted discovery of the two missing complementary antigens d and e. This prediction was later fulfilled for e but not for d. In developing this hypothesis Fisher went one step further. In the British population, there are three classes of frequency of the Rh gene complexes (Figs. 5.9, 5.10). Fisher explained this finding by suggesting that the rare combinations could have originated from the more frequent ones by occasional crossing over. All four combinations of the less common class may have originated from occasional crossing over between the most frequent combinations; not, however, CdE. This complex needs inclusion of a second-order chromosome. Therefore the hypothesis explains why CdE is so rare. Still another prediction is possible. In every crossing over leading, on the one hand, to

Cde, CDE, or cdE the complex cDe must also be produced. It follows that the frequency of the three former combinations together should equal the frequency of cDe. Frequencies actually found were: cDe, 0.0257; Cde + cdE + CDE, 0.0241 (in blacks, however, cDe has a high frequency).

Furthermore, Fisher believed the sequence of the three loci to be D-C-E, since cdE – which must have originated by crossing over between D and E from the genotype cDE/cde – is much more frequent in comparison with this genotype than Cde in relation to genotype CDe/cde (crossing over between C and D) and CDE in relation to CDe/cDE (crossing over between C and E).

Confirmation and Tentative Interpretation of the Sequential Order. In the half century since Fisher's hypothesis many new observations have been made, the most important for the question of sequence being the combined antigens, for example, ce. These compound antigens were all compatible with the sequence D-C-E, whereas no such antigen suggesting close linkage between D/d and E/e has emerged. Fisher's hypothesis leads to two questions:

1. If the rare types have been formed by occasional crossing over from the more frequent ones, cases of crossing over should occasionally turn up in family studies. One such family has indeed been reported [76]: the father was CDe/cde, the mother cde/cde, four children were cde/cde, and three others CDe/cde, all in concordance with genetic theory. The sixth-born child, however, was Cde/cde. As the discrepancy involved father and child, it

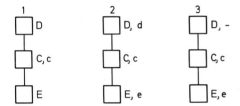

Fig. 5.9. A hypothetical structure of the Rh complex. *1*, On the basis of the evidence known in 1941; *2*, antigens predicted by Fisher and Race; *3*, antigens discovered; antigen *d* was not found

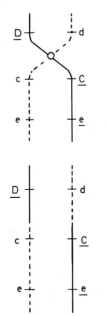

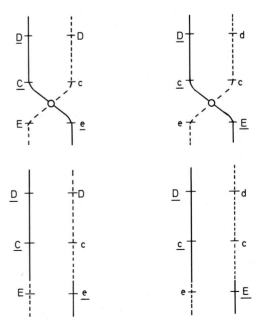

Fig. 5.10. Postulated production of three rare Rh haplotypes from the more common ones by crossing over. Each diagram refers to a different crossing over event. (From Race and Sanger 1975 [61])

could be argued that the child was illegitimate. This, however, was made unlikely by blood and serum groups and by the fact that the family belonged to a religious sect with especially strict rules against adultery.

2. How should we envision the structure of the Rh gene(s) in the light of evidence from molecular genetics? There are two possibilities in principle:
   a) The Rh complex is one gene with many mutational sites. Mutational changes are expressed as antigenic differences.
   b) The Rh complex is composed of a number of closely linked genes, possibly three, and the main antigens reflect genetic variability at these genes. One important criterion is the *cis-trans* effect found in mutations affecting the same functional gene (Sect. 4.8). As the ce compound antigen can be found in *cis*-phase CE/ce but not in *trans*-phase Ce/cE, Race and Sanger (1969) [60] tentatively concluded that C/c and E/e may be located within the same functional gene.

Molecular Basis. The Rh specificities have now been identified as membrane polypeptides. Molecular studies of the gene(s) have shown that in all D^+ individuals, two closely related Rh genes in each haploid genome appear to be present. One of these genes is missing in D^- individuals [12]. The authors concluded that one of the two genes controls the D polypeptide whereas the C/c and E/e specificities are coded by the second gene. These observations explain at the molecular level why no anti-d serum has been found. They also confirm the sequence D-C-E postulated by Fisher, as well as the above-mentioned conclusion of Race and Sanger [60] that C/c and E/e appear to be located in the same gene product. However, the case has not been definitely closed. It is still possible that there are indeed three loci [26]. If there are two loci, both original hypotheses were partially correct: the specificities C,c,E,e are located within the same gene-determined protein, as postulated by Wiener (which does not exclude occasional intragenic crossing over); the D specificity, on the other hand, is located in a second, closely linked gene, as postulated by Fisher. Moreover, attempts at understanding genetics of the Rh system led to the development of a new concept by Fisher that has found widespread application in many fields of human genetics: linkage disequilibrium.

Linkage Disequilibrium. Linkage normally does not lead to association between certain traits in the population (Sect. 5.1.1). Even if initially there is a nonrandom distribution of linkage phases, repeated crossing over randomizes the linkage groups, and in the end

the coupling and repulsion phases for two linked loci are equally frequent. There is linkage equilibrium. However, when the population begins with a deviation from this equilibrium, for example, because two populations with different gene frequencies have merged, or because a new mutation has occurred on one chromosome, the time required to reach equilibrium depends on the closeness of linkage: the closer the linkage, the longer the time until equilibrium is reached. It is never reached if certain types have a selective disadvantage.

A selective disadvantage for certain Rh complexes that could lead to their becoming less frequent has not been demonstrated so far; selection works against heterozygotes (Sect. 12.2.1.4), but this does not mean that a general disadvantage has never existed; neither has a conclusive explanation in terms of population history been postulated. Fisher's hypothesis, by answering some questions, has posed a number of others. However, the concept of linkage disequilibrium proved to be still more important in the genetic analysis of DNA polymorphisms (Sect. 12.1.2) and the Major Histocompatibility (MHC) complex:

5.2.5 Major Histocompatibility Complex [73, 79]

History. It had long been known that skin grafts from one individual to another (allotransplants) are usually rejected after a short time. In 1927 K.H.Bauer [3] observed that rejection does not occur when skin is transplanted from one monozygotic twin to the other (isotransplant). Such a transplant is accepted just as a transplant in the same individual (autotransplant). This showed the rejection reaction to be genetically determined. In the following years skin, and later kidney, transplantations between monozygotic twins were occasionally reported. Research on histocompatibility antigens in humans began only when leukocytes were shown to be useful as test cells.

Dausset observed in 1954 that some sera of polytransfused patients contain agglutinins against leukocytes. He later showed that sera from seven such patients agglutinated leukocytes from about 60% of the French population, but not the leukocytes of the patients themselves. Twin and family investigations soon established that these isoantigens are genetically determined. Other isoantigens (now part of the HLA-B) were discovered by van Rood. Another important achievement was the microlymphocyte toxicity test introduced by Terasaki and McClelland in 1964, which is now the most frequently used method (Figs. 5.11, 5.12). Subsequently the number of detected leukocyte antigens increased rapidly, and in 1965 it was suggested that most of

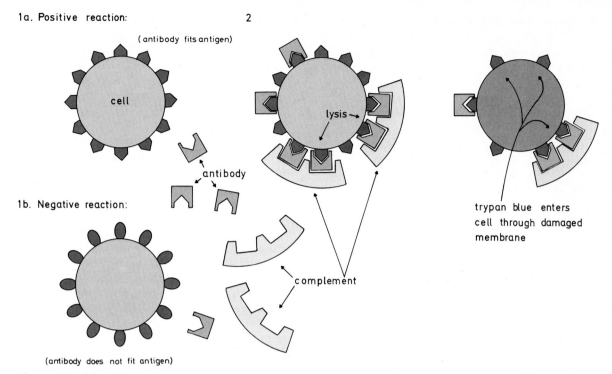

1a. Positive reaction: 2

(antibody fits antigen)

cell

lysis →

antibody

1b. Negative reaction:

trypan blue enters
cell through damaged
membrane

complement

(antibody does not fit antigen)

Fig. 5.11. Principle of the lymphocytotoxicity test: A cell having an appropriate antigen reacts with a specific antibody and complement. As a result, trypan blue enters the cell through the damaged membrane and indicates that the cell surface antigen has been recognized by a specific antibody

these antigens were components of the same genetic system. At the histocompatibility workshop in 1967, 16 different teams typed identical samples from Italian families. Here the basic relationships among the different antigens were established. Finally, Kissmeyer-Nielsen [30] proposed the hypothesis of two closely linked loci (now A and B) controlling two series of alleles.

More recently, especially since the PCR technique became available (Sect. 3.1.3.5), scientists have begun to study MHC genes directly at the DNA level. This has led to a splitting up of serologically defined gene loci, especially among the class II antigens (HLA D-DR; see Fig. 5.13).

Social Phenomenon: Formation of a "Paradigm Group." The same sociological phenomenon noted for linkage studies occurred in histocompatibility research as well. A group that maintained close contacts was formed. Special international meetings were organized, and direct exchange of information was intensified. The third histocompatibility workshop organized by Ceppellini in 1967 played a major role in this process. Contacts among research workers had to be especially intensive, as HLA typing depends vitally on exchange of antisera. A study on this "paradigm" group in the late 1960s and early

1970s and the role played by research workers such as Bodmer, Dausset, Ceppellini, Kissmeyer-Nielsen, van Rood, and Terasaki in the development of its activities would be of much interest for the history and sociology of modern biological research. Rapid progress in this field was fueled not only by the intrinsic scientific interest but also by the hope that success rates in organ transplantations could be improved.

Main Components of the MHC on Chromosome 6. The linkage group of the MHC is presented in Fig. 5.13. There are now three classes of MHC antigens. As revealed by studies using mainly molecular methods, each class can be subdivided into a great number of subclasses which are not described here (for details see [81]). Class I comprises the HLA-B, HLA-C, and HLA-A loci (in this order). In class II the HLA-D) loci are found together with some other related, transcribed areas. Between these two classes a heterogeneous group of genes is located which have been named MHC class III despite the fact that at least some of these genes, such as those for 21-hydroxylase appear to have no functional relationship to the MHC system.

The function of this system has been elucidated; it plays an important role in the immune response and

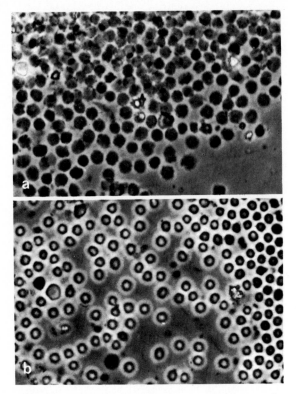

Fig. 5.12 a, b. Lymphocytotoxicity test. **a** Positive reaction. **b** Negative reaction. Positive reaction is indicated by staining of the cells. (Courtesy of Dr. J. Greiner)

is described in Sect. 7.4. Here only some genetic aspects are considered.

Table 5.1 presents a survey, together with allele frequencies (See also Figs. 5.14; 5.15). The concept of four series of alleles is based on the following lines of evidence:

a) No individual possesses more than two antigens from any of the series.

b) Recombination between these series has been observed, for example, between the loci for A and B series, 40 crossovers among 4614 meiotic divisions were described up to 1975, giving a combined ($\varphi + \vartheta$) recombination frequency of 40/4614 = 0.0087 = 0.87 cM. Ten A-B recombinants informative for the C series have been reported. In eight of these the C antigen followed B, and in two it followed A. Therefore C is located between A and B, closer to B.

c) When two antigens from the same series are present together in a parent, he or she always transmits one of them – never both or none – to the child. The segregation ratio is 0.5, corresponding to a simple codominant mode of inheritance.

d) Hardy-Weinberg proportions have been demonstrated for each of the allele series separately in large population samples.

e) Serological cross reactions occur almost exclusively within the series, not between them. This points to a close biochemical relationship of the antigens within a given series. Figure 5.14 shows transmission en bloc to four of the five children examined and crossing over between A and C in a fifth child.

Complement Components. Complement consists of a series of at least ten different factors present in fresh serum. The factors are called C_1, C_2, C_3, etc., and C_1 is activated by antibodies that react to their corresponding antigens. Then C_1 activates C_4, this activates C_2, and so on. The end result of this "complement cascade" is damage to the cell membrane carrying the antigen and often lysis of the cell. Moreover, activated complement components have a number of other biological properties, such as chemotaxis or histamine release. They are important immune mediators in the body's defense against microbial infection.

The complement system can be activated not only via C_1 (the classic pathway) but also via C_3 through an alternative pathway involving the "properdin factors." The factor B(BF) acts as "proactivator" for C_3.

For some of the complement factors hereditary deficiencies have been described, and polymorphisms are known. C_2, C_3, and C_4, are polymorphic. The loci of C_2 and C_4 A and B are in class III, together with the properdin factor B with the main alleles BF^F and BF^S. The locus for C_3, on the other hand, is located on chromosome 19.

Significance of HLA in Transplantation. One of the main motives for rapid development of our knowledge of HLA antigens has been the hope of improving the survival rate of transplanted organs, primarily kidneys. Indeed, kidneys from HLA-identical and AB0-compatible siblings have a survival rate in the recipient almost equaling that of monozygotic twins. The survival rate is worse in unrelated recipients even if HLA matching is as perfect as possible, and AB0 compatibility is secured. This shows that, apart from the major histocompatibility system – the HLA system – there must be other systems of importance for graft survival. This is not surprising. A great number of such systems are known in the mouse. These systems lead to host-versus-graft reactions in almost all transplantations (Fig. 5.16). These reactions, however, can frequently be managed by immunosuppressive therapy. The chances for survival, and the survival times, of transplanted kidneys have increased substantially. The same is true for transplantation of other organs, such as heart, liver, and pancreas.

Region of MHC class I gene loci (HLA-A, B, C, E, F, G)   **a**

TM

Region of MHC class II gene loci (HLA-DR, DQ, DO, DN, DP)   **b**

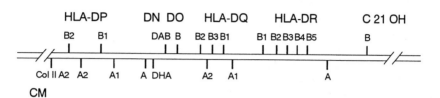

CM

Region of MHC class III gene loci   **c**

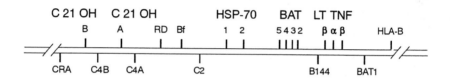

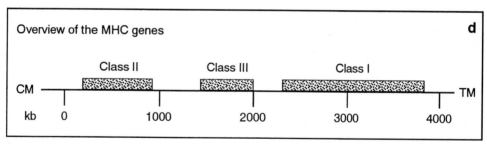

Fig. 5.13 a–d. Sequence MHC class I, II, and III genes on the short arm of chromosome 6. **a–c** Detailed view of the three regions, together with their subregions. **d** Overview of approx. 4000 kb. *TM,* Telomere; *CM,* centromere. *Letters, numbers, and their combinations,* the genes and their subregions. Genes of class III (HSP-70, DAT, C2/OH, CT) have no direct functional relationship with the immune response. (From Albert 1993 [1])

Considering the high degree of polymorphism and the low gene frequencies of HLA alleles, successful matching of potential recipients with donor kidneys from others than sibs requires large-scale international organizations. Such organizations have now been founded. Once kidneys – or other transplantable organs – become available due to the accidental death of an individual, a center is notified in which persons in need of such an organ are registered, together with their HLA status. The donor is typed, and the recipient whose HLA status best fits receives the organ.

Linkage Disequilibrium. One of the most conspicuous properties of the HLA system is that some HLA alleles tend to occur more frequently together than expected by chance. Table 5.2 shows some examples. The A1,B8 haplotype, for example, occurs about five times as often as expected.

Table 5.1. Gene frequencies (percentages) of HLA-A, -B, -C, -DR, and DQ traits in various populations (from Albert 1994 [1])

HLA	Europeans	Orientals	Africans	HLA	Europeans	Orientals	Africans
A1	14.2	1.0	8.1	B56	1.1	1.5	0.3
A2	28.9	28.1	17.5	B57	2.9	0.7	2.9
A3	13.2	1.5	6.7	B58	0	1.2	0
A11	6.3	11.7	1.9	B60	3.8	6.5	2.3
A23	1.4	0.1	8.0	B61	2.1	11.7	1.5
A24	10.3	31.4	4.8	B62	6.1	9.6	2.6
A25	2.4	0	0	B63	0.7	0	1.9
A26	3.2	7.2	4.5	B64	1.1	0	1.3
A28	4.7	2.1	9.9	B65	2.6	0.2	1.6
A29	2.9	0.4	4.9	B67	0	0.1	0
A30	3.5	2.3	11.0	B71	0.1	0.4	0.8
A31	2.9	5.2	1.6	B72	0.3	0.5	7.1
A32	3.9	0.4	2.3	B73	0.1	0.2	0
A33	1.4	6.0	3.9	BX	0.4	1.6	1.3
A34	0.1	0.3	5.1				
A36	0.1	0.1	3.2	Cw1	3.3	16.3	1.0
A43	0	0	1.3	Cw2	4.1	1.0	11.9
A66	0.2	0.5	0.3	Cw3	12.6	27.3	8.3
AX	0.4	1.7	5.0	Cw4	11.6	5.3	14.0
				Cw5	6.9	0.6	3.0
B7	11.5	4.7	12.1	Cw6	8.6	3.8	12.9
B8	9.6	0.2	5.5	Cw7	24.3	12.1	24.1
B13	2.9	3.8	1.6	Cw8	3.7	0.3	3.5
B18	5.5	0.3	4.2	CX	24.9	33.3	21.3
B27	3.4	1.6	1.9				
B35	10.5	10.2	7.1	DR1	9.5	5.0	5.1
B37	1.6	0.6	1.3	DR2	15.8	15.1	15.1
B38	2.5	0.7	1.6	DR3	12.0	1.8	14.9
B39	2.0	0.4	0	DR4	12.7	21.8	7.6
B41	0.9	0.1	2.3	DR7	12.0	2.9	13.2
B42	0.2	0.5	5.8	DR8	3.0	7.3	0.8
B44	12.3	6.0	7.7	DR9	0.8	11.5	1.5
B45	0.4	0.1	2.3	DR10	0.8	0.5	2.3
B46	0.1	3.6	0	DR11	12.3	4.0	16.5
B47	0.2	0.4	0	DR12	2.0	7.2	3.4
B48	0	1.6	0	DR13	5.4	2.9	3.8
B49	1.8	0.3	2.3	DR14	5.8	6.8	10.7
B50	1.1	0.3	0.6	DRX	7.9	13.2	5.3
B51	6.2	7.8	1.9				
B52	2.0	7.3	0.6	DQ1	32.3	30.2	40.1
B53	0.5	0.3	6.7	DQ2	18.1	5.0	23.1
B54	0.1	6.7	0	DQ3	23.3	32.7	24.6
B55	1.6	2.1	0	DQX	26.3	32.1	12.2

Consider two alleles at two linked loci, with frequencies p_1 and p_2. With free recombination between them their combined frequency, i.e., the haplotype frequency h, should be $p_1 \times p_2$. If such a result is obtained, the two loci are said to be in linkage equilibrium. If the haplotype frequency h is higher than expected with free recombination, there is linkage disequilibrium (Δ, deviation from linkage equilibrium), which is often symbolized as $\Delta = h - p_1 p_2$. Haplotype and gene frequencies can be estimated from family and population data. In families the hap-

lotypes of parents can in most cases be derived from those of their children. In the family shown in Fig. 5.14, for example, one of the maternal haplotypes should be 3, 1, 22, as she has transmitted it to three of her children. Frequencies of the single alleles can be deduced from the same sample of unrelated parents, and the extent of linkage disequilibrium Δ can be calculated. In a sample from a random mating population, deviation from linkage equilibrium can be tested in a 2×2 table by the χ^2 test as shown in Table 5.3 for a Danish sample [79].

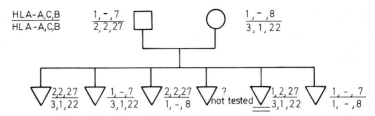

Fig. 5.14. Pedigree with crossing over between the HLA-A and HLA-C genes within the MHC. Crossing over between A and C must have occurred in the father's germ line, making for the haplotype 1, 2, 27 in the fifth child. (From Svejgaard et al. 1975 [79])

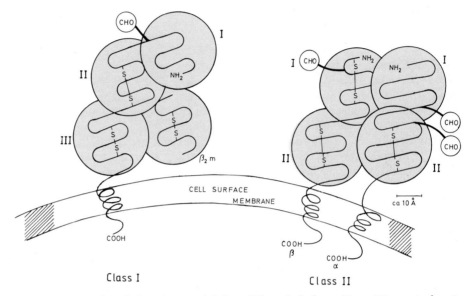

Fig. 5.15. Arrangement of class I (HLA-A, B, C) and class II antigen (HLA-D/DR) domains which may yield similar overall structures. Roman numerals denote domains. $\beta_2 m$, β-Microglobulin; *CHO*, carbohydrate. (From Kämpe et al. 1983 [27])

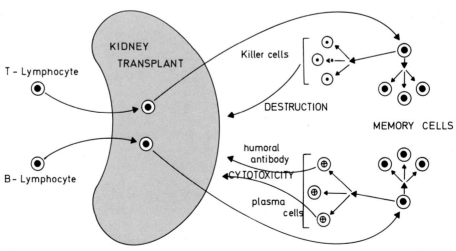

Fig. 5.16. Simplified diagram of the activation of the immune system by a kidney allotransplant. The transplant is recognized as foreign to the host organism by its T and B lymphocytes. This leads to activation of cellular and tumor immune response. (From Svejgaard et al. 1975 [79])

Table 5.2. Linkage disequilibrium (gametic association; from Svejgaard, 1975 [889])

Haplotype			Frequency (%)	
A	B	D	Observed	Expected
A1	B8		9.8	2.1
A3	B7		5.4	2.1
	B8	Dw3	8.6	1.4
	B7	Dw2	3.9	1.8

The expected haplotype frequencies were calculated under the assumption of no association.

In the HLA system the deviations from linkage equilibrium are indeed striking. The situation is similar to that encountered with the Rh system (Sect. 5.2.4), but there is one important difference: in the Rh system only one case of recombination has been discovered, whereas many cases are known for the HLA system. Hence, genetic data point to much closer linkage in the Rh system than among the MHC genes. This conclusion has been corroborated by molecular studies in both systems (see above).

The observation of linkage disequilibrium – together with identification of immune response (Ir) genes in the mouse – initiated the investigations of HLA associations with diseases (Sect. 6.2.3).

Linkage disequilibrium may have either of two main causes:

1. Two populations homozygous for different haplotypes mixed a relatively short time ago, and repeated crossing over at a low rate has so far not been sufficient to lead to random distribution of alleles.
2. Certain combinations of alleles on closely linked gene loci caused a selective advantage for their bearers and have therefore been preserved.

To be able to decide between these two possibilities Bodmer (1972) [6] calculated how long a linkage disequilibrium would need to disappear in a random mating population.

For these calculations he used the work of Jennings (1917), according to which Δ decreases to zero at a rate of $1-\Theta$ per generation, where Θ is the recombination fraction between the two loci. Between the HLA-A and HLA-B loci Θ was found to be of the order of magnitude of 0.008. Taking linkage disequilibrium between HLA-A1 and B8 as an example, Δ values of about 0.06–0.1 have been found in European populations. On the other hand, Δ values between 0.01–0.02 are not statistically significant with reasonable sample sizes. Therefore it is meaningful to examine how many generations are needed to reduce Δ from 0.1 by a factor of 5 to 0.02.

Table 5.3. Association of HLA-A1 and B8 in unrelated Danes (2 × 2 table; from Svejgaard et al. 1975 [79])

	Number of individuals		Total
	B8+	B8-	
A1+	376	235	611
A1-	91	1265	1356
Total	467	1500	1967

The table is often given as follows:

First antigen	Second antigen	+/+ a	+/- b	-/+ c	-/- d	Total n
A1	B8	376	235	91	1265	1967

where, for example, +/- means number of individuals possessing the first character (A1) and lacking the second (B8). The χ^2 is:

$$\chi_1^2 = \frac{(ad - bc)^2 N}{(a+b)(c+d)(a+c)(b+d)} = 699.4$$

corresponding to the correlation coefficient:

$$r = \sqrt{\chi^2/n} = \sqrt{699.4/1967} = 0.60$$

Gene frequencies for A1 and B8 can be calculated by Bernstein's formula:

$$p = 1 - \sqrt{1 - \alpha}$$

(where α is the antigen frequency) as 0.170 and 0.127, respectively.

The Δ value can be calculated by the formula

$$\Delta = \sqrt{\frac{d}{n}} - \sqrt{\frac{(b+d)(c+d)}{n^2}} = 0.077$$

Thus, the frequency of the HLA-A1, B8 haplotype is

$$h_{A1, B8} = p_{A1}\, p_{B8} + \Delta_{A1, B8} = 0.170 \times 0.127 + 0.077 = 0.099.$$

Using the above principle of Jennings, we obtain:

$$(1 - \Theta)^n = (1 - 0.008)^n = 1/5; \quad n \approx 200$$

This means that Δ would be reduced to an insignificant value within about 200 generations of random mating, i.e. 5000 years, taking a generation as around 25 years.

This period is approximately the length of time since agriculture first came to parts of northern Europe and is certainly a very short time considering the evolutionary life span of the human species. The fact that such a significant Δ could be eroded in so short a time in the absence of selection suggests at least that this particular combination of HLA-A1, B8 is being maintained at its comparatively high frequency by some sort of interactive selection [584]. We consider it likely that selection will also be found to explain some of the other common cases of linkage disequilibrium and that the effect of recent population mixture will be shown to be of minor importance. Certain haplotypes seem to have a selective advantage that keeps them more frequent than others. This selective advantage, on the other hand, cannot be directly related to the diseases for which associations have been shown so far, as they are too rare. Besides, the onset of most of them is usually delayed until after the age of reproduction. Infectious diseases have probably been the most important selective forces for maintaining the MHC polymorphism as well as linkage disequilibrium. This topic is discussed in Sects. 6.2.3; 12.1.2).

The Normal Function of the System. The HLA determinants are localized at the surface of the cell and are strong antigens. They exhibit the most pronounced polymorphism of expressed genes known so far in humans, with abundant linkage disequilibrium. Disease associations have been shown between HLA antigens and diseases for which an autoimmune mechanism had previously been suspected. Furthermore, similar systems are known in all other mammals examined so far (see [1]). Finally, there is close linkage with other loci concerned with the immune response. All this evidence together is very suggestive of a system that regulates the contact of cells with their environment. In recent years, this function has been elucidated in detail (see Sect. 7.4). These genes are important mediators of the immune reaction. Such cell recognition mechanisms may be important in embryonic development and differentiation, especially when they are present on only certain cell types. However, such hypothesis would not explain the selective advantage of the high degree of polymorphism in this system.

Another possible function is protection against viral or bacterial infection. Antigenic material of human origin may be incorporated in the outer membrane of the virus, which is thereby made less recognizable to another human host. However, if the virus contains MHC material from a genetically different individual, it is more readily inactivated by the immune system. Such a mechanism would also explain why the extreme polymorphism of the MHC system has a selective advantage. Further elucidation of the MHC will teach us a great deal about how the organism handles its interaction with the environment. This knowledge is important to our understanding of how natural selection has shaped our genetic constitution in the past, and how recent changes in our environment may influence it in the future.

To broaden the empirical basis for such understanding, however, it may be useful to ask whether there are other examples in nature of such gene clusters with related functions? Can their analysis provide us with hints for a better understanding of the MHC cluster? There is indeed one such example that has already been analyzed very carefully – mimicry in butterflies [857]. It cannot described here for lack of space, since it has no direct relationship with human genetics. But for the reader interested in more general, philosophical aspects of science, it is highly interesting showing how certain general principles may be used by nature in quite different contexts (See also earlier editions of this book).

5.2.6 Genes with Related Functions on the Human X Chromosome?

Unlike the autosomes, the X chromosome has remained relatively unchanged during mammalian evolution. There is good evidence pointing to homology among different species [53]. The X chromosome has the unique feature that it is present in one copy in the male and in two copies in the female – a difference that is not entirely compensated for by X inactivation (Sect. 2.2.3.3). Is there any hint that X-linked genes are a nonrandom sample of all human genes? Do they show any peculiarities? This question has been examined by classifying all known X-linked and autosomal mutations into four categories [82]:

1. Those affecting the sense organs (eye, inner ear), skin, and teeth.
2. Those affecting the brain and nervous system.
3. Structural anomalies of the skeleton, muscles and connective tissue, of inner organs systems such as heart and digestive tract; cell surface antigens (blood and histocompatibility antigens), and tumors.
4. Genes affecting metabolism; blood clotting and other hematological diseases; enzyme and serum and protein polymorphisms; endocrinological disorders; an increasing number of these genes have been discovered by molecular methods, without the help of known mutations.

Figure 5.17 shows the result, which is based on more than 2000 gene loci: categories 1 and 2 are much more frequent in X-linked mutations. In addition, category 4 contains a number of X-linked mutations influencing endocrine functions of the neuropituitary that with equal justification could be included in category 2. Hence, the "higher" functions of the nervous system and sense organs seem to be overrepresented on the human X chromosome, whereas genetic

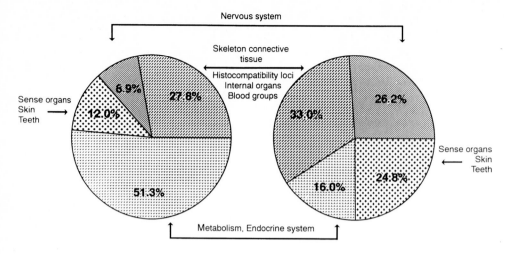

Autosomal mutations   **X-linked mutations**

Fig. 5.17. Possible clustering of mutations affecting sense organs, skin, teeth, and the nervous system on the human X chromosome. Four groups of phenotypes in 1029 autosomal and 92 X-linked mutations. Only confirmed mutations. (From [41])

specification of metabolic processes, but to some degree also of body structure, is underrepresented.

Ascertainment and classification of human mutations have never been very systematic. The conditions for recognizing X-linked genes differ from those for autosomal ones. This bias could simulate a difference between the X chromosome and the autosomes. However, a paucity of X-linked genes is very likely for metabolic processes. In *Drosophila*, where ascertainment can be better controlled, significant nonrandomness of mutations affecting various organ systems has been described [16]. Should the difference between X chromosome and autosomes in humans prove to be real, the following questions arise: has this difference anything to do with the special properties of X-linked genes in the regulation of gene action? Do these genes run a lower risk of becoming recessive lethals by mutation? Has this been an important selective advantage? Or is this clustering merely the trace of the evolutionary history of these genes?

5.2.7 Unequal Crossing Over

Discovery of Unequal Crossing Over. In the early years of work with *Drosophila* some authors observed that the bar mutation, an X-linked dominant character, occasionally reverts to normal, whereas in other cases homozygotes for the allele produce offspring with a new and more extreme allele, later called "double bar." Sturtevant (1925) [78] showed that this peculiar behavior is not due to mutations but to unequal crossing over, producing, on the one hand, a chromosome with two bar loci (double bar) and, on the other, a chromosome with no bar locus at all. When the giant salivary chromosomes of *Drosophila* permitted visual testing of genetic hypotheses, Bridges (1936) [9] showed that the simple, dominant bar mutation is caused by a duplication of some chromosomal bands. The reversion corresponds to the unduplicated state, whereas double bar is caused by a triplication of that band. Both reversion and triplication can be produced by a single event of unequal crossing over. Bridges did not yet formulate clearly the obvious reason for this event: the mispairing of "structure-homologous" but not "position homologous" chromosome sites (Fig. 5.18).

Unequal Crossing Over in Human Genetics. Haptoglobin [8], a transport protein for hemoglobin, is found in the blood serum and shows a polymorphism, the most common alleles being HP^{1F}, HP^{1S} and HP^2. Smithies et al. (1962) [72] discovered that the allele HP^2 is almost twice the length of each of the two alleles HP^{1F} and HP^{1S}, as evidenced by the composition of its polypeptide chain. In the HP^2 chain the amino acid sequence of the HP^1 chain is repeated almost completely. They concluded that the HP^2 allele might have been produced by gene duplication. Moreover, they predicted that unequal crossing over might again occur with a relatively high probability between HP^2 alleles, producing, on the one hand, an allele similar to HP^1 and, on the other, an allele comprising the genetic information almost in triplicate. Repeated occurrence of this event might lead to still longer alleles and hence to a polymorphism of allele lengths

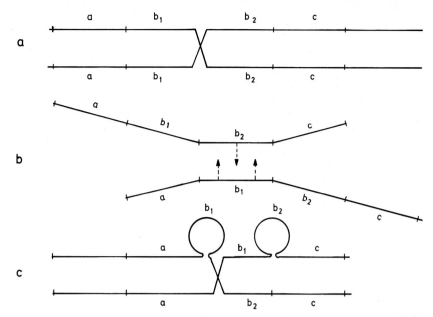

Fig. 5.18 a–c. The principle of unequal crossing over. **a** Normal pairing and crossing over. The two genes b_1 and b_2 are assumed to have very similar DNA sequences. **b** Genes b_1 and b_2 are pairing. This leads to a shift of the two homologous chromosomes relative to each other. **c** Such pairing requires formation of two loops in the upper chromosome

in the population. Indeed, such alleles have occasionally been observed and are known as Johnson-type alleles [74].

There is an essential difference between the first unique event that produces the almost double-sized gene (for example, HP^2) from a single gene HP^1, and the unequal but homologous crossing over that becomes possible as soon as the first duplicated allele is present in the population [32].

First Event. Given a pair of homologous chromosomes, both partner chromosomes consist of largely identical sequences of nucleotides. Normally these partner chromosomes pair at meiosis, and there can be no unequal crossing over. To allow mispairing and thus unequal crossing over, an initial duplication is necessary. Mechanisms for such a duplication are known in cytogenetics, the simplest being two breaks at slightly different sites in adjacent homologous chromatids during meiosis and subsequent crosswise reunion. Another mechanism would be mispairing due to homology of short base sequences in nonhomologous positions. Our present knowledge of the structure of DNA sequences suggests ample opportunities for such a mispairing (slippage).

If the sites of breakage are separated only by the length of one structural gene, this event results in two gametes that do not contain this gene at all, together with two others containing it in duplicate (Fig. 5.19). The gametes containing a relatively large

deletion have a high risk of not being transmitted because of lethality of the ensuing embryo. On the other hand, a gamete with the duplication is likely to develop into a diploid individual, providing for the first time a chance for mispairing of homologous sequences and therefore for unequal crossing over.

Consequences of Unequal Crossing Over. The consequences are seen in Fig. 5.19. As long as the duplication remains heterozygous, all gametes contain either one or two copies of the duplicated gene. When the duplication becomes homozygous, however, larger allele sequences may be formed. Unequal crossing over may lead, on the one hand, to gametes with only one copy and, on the other, to gametes containing three, and in subsequent generations, more than three copies (Figs. 5.19, 5.20).

If the probability of unequal crossing over is not too low, high variability is soon found in the number of homologous chromosomal segments that resemble each other in structure but not in position. If selection favors a certain number of such chromosomal segments, which may be as small as a single gene, this number soon becomes the most common one. Selection relaxation leads to an increase in variability in both directions: the proportion of individuals with a very high number of such genes as well as those with a low gene number gradually increases [32]. Another genetic mechanism resembling unequal crossing over in some aspects is gene conversion where nonreciprocal products result.

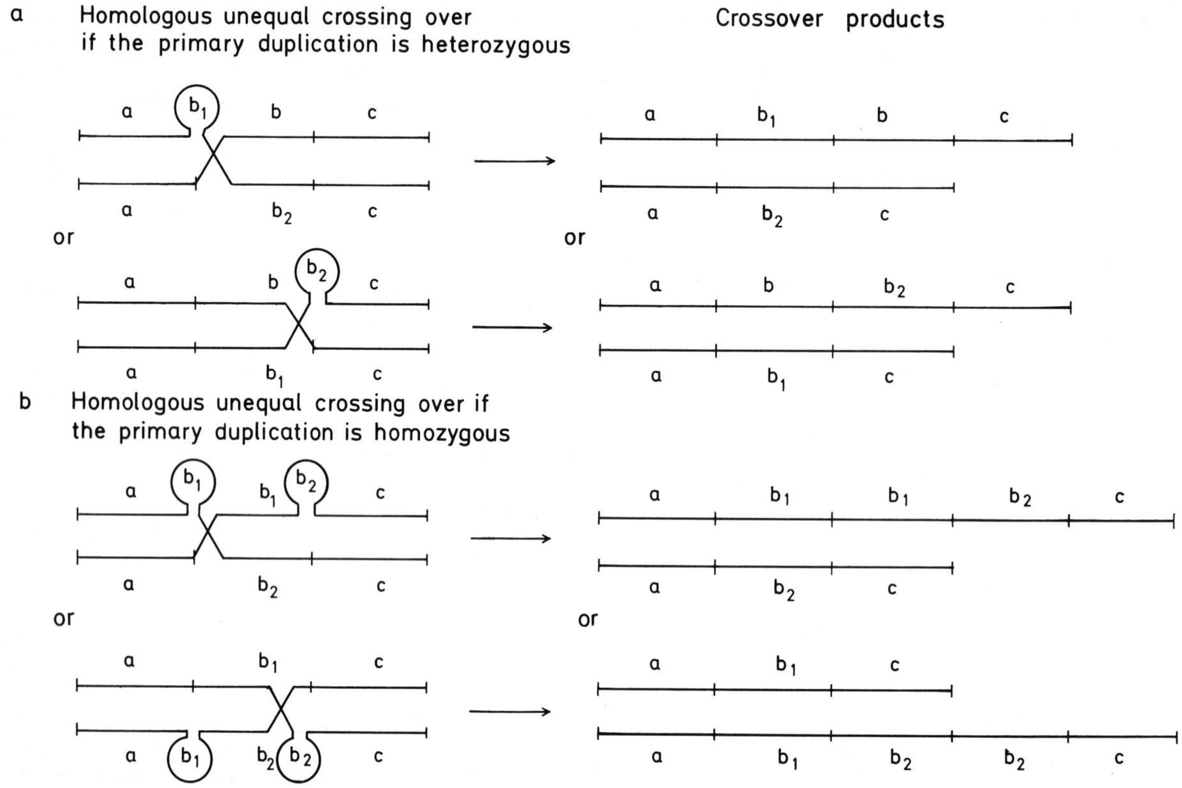

a Homologous unequal crossing over
if the primary duplication is heterozygous

Crossover products

b Homologous unequal crossing over if
the primary duplication is homozygous

or

or

Fig. 5.19 a, b. Unequal crossing over between structure-homologous but not position-homologous genes. **a** Unequal crossing over always leads to one crossover product with two genes b (b_1b or bb_2) and to another with only one gene

(b). Formation of larger allele sequences becomes possible if the primary duplication is homozygous. In this case a chromosome with three alleles b (b_1 b_1 b_2 or b_1 b_2 b_2) may be formed (From Krüger and Vogel 1975 [32])

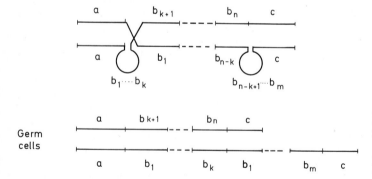

Germ
cells

Fig. 5.20. The consequence of unequal crossing over. In subsequent generations chromosomes with (theoretically) unlimited numbers of alleles may be formed. Unequal crossing over between any of them may lead to still larger (or still shorter) haplotypes. $b_1 \ldots b_k \ldots b_n$ refer to homologous genes

Other examples besides the haptoglobin genes are the closely linked hemoglobin β- and δ-genes and the color vision pigment locus (Sect. 15.2.1.5). Here the Lepore-type mutants and the X-linked color vision genes may be caused by unequal crossing over (Fig. 7.49). Moreover, there are many examples for moderately or highly repetitive DNA sequences within which unequal crossing over should be possible. The presence of short repetitive DNA sequences such as minisatellites (Sect. 12.1.2) provides ample op-

portunities for pairing 'slippage,' leading to unequal crossing over. The high mutation rate within such areas (sometimes even a few percent per meiosis (Sect. 12.1.2) a well as the resulting huge interindividual variability show that this is not merely a theoretical speculation. Other repeated DNA sequences are those coding for the immunoglobulins (Sect. 7.4). Increasing knowledge of the functional significance of repeated DNA sequences will bring a better understanding of the significance of unequal crossing over.

Intrachromosomal Unequal Crossing Over. With structure-homologous but not position-homologous genes, such as those found in multigene families (Sect. 3.1.3.10), unequal crossing over becomes possible not only between homologous chromosomes but also between sister chromatids (intrachromosomal unequal crossing over). Theoretical considerations have shown that this process could have played a role in molecular evolution [29].

Conclusions

A few years after the rediscovery of Mendel's laws early in the twentieth century the first exception to Mendel's third law (independent segregation) was discovered: genes located sufficiently close to one other on the same chromosomes often segregate together – they are linked. The frequency of recombination increases with increasing distance between these genes. Genes on the same chromosomes but located far apart from each other, however, may even segregate independently if the distance between them is greater – these are syntenic, but not linked. A great number of genetic markers are available for localizing human genes, and statistical methods for assessing linkage in the human genome and determining the distance between gene loci have been developed. Methods from cell, biochemical, and molecular genetics have helped in localizing genes to specific chromosomes and chromosome segments. Such techniques make it possible to localize genes for both normal and abnormal traits and to define the nature of such genes by positional cloning. The identification of genes involved in susceptibilities to common diseases with complex causes by linkage studies remains a major challenge.

While genes involved in the same biochemical pathways are seldom located close together, some clusters of closely linked genes exist that have related functions; the genes of the major histocompatibility complex, for example, have been analyzed particularly thoroughly.

References

1. Albert E (1993) Immungenetik. In: Gemsa, Resch (eds) Lehrbuch der Immunologie. Thieme, Stuttgart
2. Bateson W, Saunders, Punnett RG (1908) Confirmations and extensions of Mendel's principles in other animals and plants. Report to the Evolution Committee of the Royal Society, London
3. Bauer KH (1927) Homoiotransplantation von Epidermis bei eineiigen Zwillingen. Beitr Klin Chir 141 : 442–447
4. Bernstein F (1931) Zur Grundlegung der Chromosomentheorie der Vererbung beim Menschen. Z Induktive Abstammungs Vererbungslehre 57 : 113–138
5. Bishop MJ (1994) Guide to human genome computing. Academic, London
6. Bodmer WF (1972) Population genetics of the HL-A system: retrospect and prospect. In: Dausset J, Colombani J (eds) Histocompatibility testing. Munksgaard, Copenhagen, pp 611–617
7. Botstein D, White RL, Skolnick M, Davis RW (1980) Construction of a genetic linkage map in man using restriction fragment length polymorphisms. Am J Hum Genet 32 : 314–331
8. Bowman BH, Kurosky A (1982) Haptoglobin: the evolutionary product of duplication, unequal crossing over, and point mutation. Adv Hum Genet 12 : 189–261
9. Bridges CB (1936) The bar "gene," a duplication. Science 83 : 210
10. Cantor RM, Rotter JI (1992) Analysis of genetic data: methods and interpretation. In: King A, Rotter JI, Motulsky AG (eds) The genetic basis of common diseases. Oxford University Press, New York, pp 49–70
11. Chadefaux B, Allord D, Rethoré MO, Raoul O, Poissonier M, Gilgenkrantz S, Cheruy C, Jérôme H (1984) Assignment of human phosphoribosylglycinamide synthetase locus to region 21 q 22.1. Hum Genet 66 : 190–192
12. Colin Y, Cherif-Zahar B, LeVankim C et al (1991) Genetic basis of the RhD-positive and the RhD-negative blood group polymorphism as determined by Southern analysis. Blood 78 : 2747–2752
13. Dausset J, Cann H, Cohen D et al (1990) Centre d'études du polymorphisme humain (CEPH): collaborative genetic mapping of the human genome. Genomics 6 : 575–577
14. Donahue RP, Bias WB, Renwick JH, McKusick VA (1968) Probable assignment of the Duffy blood-group locus to chromosome 1 in man. Proc Natl Acad Sci USA 61 : 949
15. Donis-Keller H, Green P, Helms C et al (1987) A genetic linkage map of the human genome. Cell 51 : 319–337
16. Elston RC, Glassman E (1967) An approach to the problem of whether clustering of functionally related genes occurs in higher organisms. Genet Res 9 : 141–147
17. Ferguson-Smith MA (1966) X-Y chromosomal interchange in the aetiology of true hermaphroditism and of XX Klinefelter's syndrome. Lancet 2 : 475–476
18. Gerhard DS, Kidd KK, Kidd JR, Egeland JA (1984) Identification of a recent recombination event within the human β-globin gene cluster. Proc Natl Acad Sci USA 81 : 7875–7879
19. Gusella JF, Wexler NS, Conneally PM, Naylor SL, Anderson MA, Tanzi RE, Watkins PC, Ottina C, Wallace MR, Sakaguchi AY, Young AB, Shoulson I, Bonilla E, Martin JB (1983) A polymorphic DNA marker genetically linked to Huntington's disease. Nature 306 : 234–238
20. Gusella JF, Tanzi RE, Anderson MA, Hobbs W, Gibbons K, Raschtchian R, Gilliam TC, Wallace MR, Wexler NS, Conneally PM (1984) DNA markers for nervous system diseases. Science 225 : 1320–1326
21. Haldane JBS (1922) Sex ratio and unisexual sterility in hybrid animals. J Genet 12 : 101–109
22. Haldane JBS, Smith CAB (1947) A new estimate of the linkage between the genes for colour blindness and haemophilia in man. Ann Eugen 14 : 10–31
23. Harris H (1970) Cell fusion. Clarendon, Oxford

24. Harris H, Watkins JF (1965) Hybrid cells from mouse and man: artificial heterokyryons of mammalian cells from different species. Nature 205 : 640

24a. Houwen RHJ, Baharloo S, Blankenship K et al. (1994) Genome screening by searching for shared segments: mapping a gene for benign recurrent intrahepatic cholestasis. Nature [Genet] 8 : 380–386

25. Jeffreys AJ, Wilson V, Thein SL (1985) Hypervariable "minisatellite" regions in human DNA. Nature 314 : 67–73

26. Kajii E, Umenishi F, Iwamoto S, Ikemoto S (1993) Isolation of a new cDNA clone encoding an Rh polypeptide associated with the Rh blood group system. Hum Genet 91 : 157–162

27. Kämpe O, Larhammer D, Wiwan k, Scheuning L, Claesson L, Gustafsson K, Pääbo S, Hyldig-Nielsen JJ, Rask L, Peterson PA (1983) Molecular analysis of MHC antigens. In: Möller E, Möller G (eds) Genetics of the immune response. Plenum, New York, pp 61–79

28. Kang S-S, Wong PWK, Susmano A et al (1991) Thermolabile methylenetetrahydrofolate reductase: an inherited risk factor for coronary artery disease. Am J Hum Genet 48 : 536–545

29. Kimura M (1983) The neutral theory of molecular evolution. Cambridge University Press, Cambridge

30. Kissmeyer-Nielsen F (ed) (1975) Histocompatibility testing 1975. Munksgaard, Copenhagen

31. Knippers R, Philippsen P, Schafer KP, Fanning E (1990) Molekulare Genetik, 5th edn. Thieme, Stuttgart

32. Krüger J, Vogel F (1975) Population genetics of unequal crossing over. J Mol Evol 4 : 201–247

33. Kurnit DM (1979) Down syndrome: gene dosage at the transcriptional level in skin fibroblasts. Proc Natl Acad Sci USA 76 : 2372–2375

34. Landsteiner K, Wiener AS (1940) An agglutinable factor in human blood recognized by immune sera for rhesus blood. Proc Soc Exp Biol Med 43 : 223

35. Lange K, Boehnke M (1982) How many polymorphic genes will it take to span the human genome? Am J Hum Genet 34 : 842–845

36. Lange K, Page BM, Elston RC (1975) Age trends in human chiasma frequencies and recombination fractions. I. Chiasma frequencies. Am J Hum Genet 27 : 410–418

37. Lele KP, Penrose LS, Stallard HB (1963) Chromosome deletion in a case of retinoblastoma. Ann Hum Genet 27 : 171

38. Levine P, Stetson RE (1939) An unusual case of intragroup agglutination. JAMA 113 : 126–127

39. Lewis EB (1951) Pseudoallelism and gene evolution. Cold Spring Harbor Symp Quant Biol 16 : 159–174

40. McClintock B (1944) The relation of homozygous deficiencies to mutations and allelic series in maize. Genetics 29 : 478–502

41. McKusick VA (1995) Mendelian inheritance in man, 11th edn. Johns Hopkins University Press, Baltimore

42. Migeon BR, Miller SC (1968) Human-mouse somatic cell hybrids with single human chromosome (group E): link with thymidine kinase activity. Science 162 : 1005–1006

43. Mittwoch U (1992) Sex determination and sex reversal: genotype, phenotype, dogma and semantics. Hum Genet 89 : 467–479

44. Mohr J (1954) A study of linkage in man. Munksgaard, Copenhagen (Opera ex domo biologiae hereditariae humanae universitatis hafniensis 33)

45. Morgan TH (1910) Sex-limited inheritance in drosophila. Science 32 : 120–122

46. Morton NE (1955) Sequential tests for the detection of linkage. Am J Hum Genet 7 : 277–318

47. Morton NE (1956) The detection and estimation of linkage between the genes for elliptocytosis and the Rh blood type. am J Hum Genet 8 : 80–96

48. Morton NE (1957) Further scoring types in sequential tests, with a critical review of autosomal and partial sex linkage in man. Am J Hum Genet 9 : 55–75

49. Morton NE (1976) Genetic markers in atherosclerosis: a review. J Med Genet 13 : 81–90

50. Morton NE (1993) Genetic epidemiology. Annu Rev Genet 27 : 521–538

51. Morton NE, Collins A (1990) Standard maps of chromosome 10. Ann Hum Genet 54 : 235–251

52. Murray JC, Mills KA, Demopulos CM, Hornung S, Motulsky AG (1984) Linkage disequilibrium and evolutionary relationships of DNA variants (RFLPs) at the serum albumin locus. Proc Natl Acad Sci USA 81 : 3486–3490

53. Ohno S (1967) Sex chromosomes and sex-linked genes. Springer, Berlin Heidelberg New York

54. Ohta T (1993) Diversifying selection, gene conversion, and random drift: interactive effects on polymorphism at MHC loci.

55. Ott J (1983) Linkage analysis and family classification under heterogeneity. Ann Hum Genet 47 : 311–320

56. Ott J (1990) Cutting a Gordian knot in the linkage analysis of complex human traits. Am J Hum Genet 46 : 219–221

57. Ott J (1991) Analysis of human genetics linkage. Johns Hopkins University Press, Baltimore

58. Page DC, Mosher R, Simpson EM et al (1987) The sex-determining region of the human Y chromosome encodes a finger protein. Cell 51 : 1091–1104

59. Penrose LS (1934/35) The detection of autosomal linkage in data which consist of pairs of brothers and sisters of unspecified parentage. Ann Eugen 6 : 133–138

60. Race RR, Sanger R (1969) Xg and sex chromosome abnormalities. Br Med Bull 25 : 99–103

61. Race RR, Sanger R (1975) Blood groups in man, 6th edn. Blackwell, Oxford

62. Rappold GA (1993) The pseudoautosomal regions of the human sex chromosomes. Hum Genet 92 : 315–324

63. Reid DH, Parsons PH (1963) Sex of parents and variation of recombination with age in the mouse. Heredity 18 : 107

64. Renwick JH (1969) Progress in mapping human autosomes. Br Med Bull 25 : 65

65. Risch N (1990) Linkage strategies for genetically complex traits. I. Multilocus models. Am J Hum Genet 46 : 222–238

66. Risch N (1990) Linkage strategies for genetically complex traits. II. The power of affected relative pairs. Am J Hum Genet 46 : 229–241

67. Risch N (1990) Linkage strategies for genetically complex traits. III. The effect of marker polymorphism on analysis of affected relative pairs. Am J Hum Genet 46 : 242–253

68. Schiebel K, Weiss B, Woehrle D, Rappold G (1993) A human pseudoautosomal gene, ADP/ATP translocase, escapes X-inactivation whereas a homologue on Xq is subject to X-inactivation. Nature Genet 3 : 82–87

69. Sheppard PM (1975) Natural selection and heredity, 4th edn. Hutchinson, London

70. Sinclair AH, Berta P, Palmer MS et al (1990) A gene from the human sex-determining region encodes a protein

with homology to a conserved DNA-binding motif. Nature 346 : 240–244

71. Smith SM, Penrose LS, Smith CAB (1961) Mathematical tables for research workers in human genetics. Churchill, London

72. Smithies O, Connell GE, Dixon GH (1962) Chromosomal rearrangements and the evolution of haptoglobin genes. Nature 196 : 232

73. Snell GD, Dausset J, Nathenson S (1977) Histocompatibility. Academic, New York

74. Sørensen H, Dissing J (1975) Association between the C_3^F gene and atherosclerotic vascuolar diseases. Hum Hered 25 : 279–283

75. Southern EM (1982) Application of DNA analysis to mapping the human genome. Cytogenet Cell Genet 32 : 52–57

76. Steinberg AG (1965) Evidence for a mutation or crossing over at the Rh-locus. Vox Sang 10 : 721

77. Stern C (1973) Principles of human genetics, 3rd edn. Freeman, San Francisco

78. Sturtevant AH (1925) The effects of unequal crossing over at the bar locus in drosophila. Genetics 10 : 117

79. Svejgaard A, Hauge M, Jersild C, Platz P, Ryder LP, Staub Nielsen L, Thomsen M (1979) The HLA system. An introductory survey, 2nd edn. Karger, Basel

80. Tiepolo l, Zuffardi O (1976) Localization of factors controlling spermatogenesis in the nonfluorescent portion of the human Y chromosome. Hum Genet 34 : 119–124

81. Trowsdale J, Ragoussis J, Campbell JD (1991) Map of the human MHC. Immunol Today 12 : 429–467

82. Vogel F (1969) Does the human X chromosome show evidence for clustering of genes with related functions? J Genet Hum 17 : 475–477

83. Vogt P, Keil R, Kirsch S (1993) The "AZF" function of the human Y chromosome during spermatogenesis. In: Sumner AT, Chandley AC (eds) Chromosomes today, vol 2. Chapham and Hall, London, pp 227–239

84. Weiss MC, Green H (1967) Human-mouse hybrid cell lines containing partial complements of human chromosomes and functioning human genes. Proc Natl Acad Sci USA 58 : 1104–1111

85. Weitkamp LR (1972) Human autosomal linkage groups. Proceedings of the 4th International Congress of Human Genetics, Paris 1971. Excerpta Medica, Amsterdam, pp 445–460

86. White R, Lalouel J-M (1987) Investigation of genetic linkage in human families. Adv Hum Genet 16 : 121–228

87. White R, Leppert M, Bishop DT, Barker D, Berkowitz J, Brown C, Callahan P, Holm T, Jerominski L (1985) Construction of linkage maps with DNA markers for human chromosomes. Nature 313 : 101–105

88. Wolf U, Schempp W, Scherer G (1992) Molecular biology of the human Y chromosome. Rev Physiol Biochem Pharmacol 121 : 148–213

6 Formal Genetics of Humans: Multifactorial Inheritance and Common Diseases

The criterion of the scientific status of a theory is its falsifiability, or refutability or testability.

(K. R. Popper, in "Conjectures and Refutations," 1963, p. 37)

6.1 Levels of Genetic Analysis

The paradigm that resulted from Mendel's work with pea crosses (Sect. 1.4) has since developed in several steps to the level of the gene in its definition as an information-carrying segment of the DNA double helix. The principal goal of genetic analysis is to pursue this path for the character under investigation. In the early part of this volume, characteristics were selected as examples of basic principles in which the genotype-phenotype relationship was fairly straightforward: in the "normal" range, blood groups; in the "abnormal" range, rare hereditary diseases.

However, there are many normal characteristics for which genetic variability obviously exists but no simple mode of inheritance can be found. These include stature and body proportion, physiognomic features ("this child is the absolute image of its father"), skin color, and blood pressure. Many diseases may have a complex of various causes, but often the liability may differ between individuals and may be genetic in origin. In earlier years, a Mendelian framework was often superimposed naively on such data, without testing the formal requirements for simple modes of inheritance. More recently, there has been a tendency to describe observations on complex traits in terms of biometric models that were often based on oversimplifying assumptions. Thus, the biological significance of the resulting conclusions may occasionally be questionable.

We first set out the logical background for a discussion of genetic hypotheses and distinguish a number of levels at which genetic analysis has become possible.

6.1.1 Findings at the Gene – DNA Level

The DNA sequence contains the information for a sequence of amino acids in a polypeptide chain, and therefore it is our goal to trace genetic differences to the DNA level. This first became possible for the hemoglobin variants (Sect. 7.3) and later for some other proteins, initially inferring alterations in DNA from the observed changes in amino acids and knowledge of the genetic code. Later, these inferences were confirmed in many cases where one amino acid is replaced by another, but other changes such as deletions, frameshifts, and amplification mutants also could be detected by direct DNA studies (Sects. 7.3 and 9.4). In these instances, a genetic variant has been pursued down to its primary cause – a specific change at the DNA level: the "information carrier."

The increasing availability of methods for DNA analysis made it possible to elucidate novel types of mutations, such as those interfering with the regulation of transcription and others producing defective splicing during processing of gene transcripts. Most of the initial work of this type was carried out with the human β-hemoglobin system and the resulting phenotypic changes presented as so-called β-thalassemias, which are characterized by absent (β^0) or deficient (β^+) β-chain production. As more DNA probes for different genes became available, the elucidation of mutations at the DNA level began to accelerate. Different mutations analogous to those observed in the hemoglobinopathies, as well as other types, occur in all hereditary variants and diseases analyzed so far (see Sect. 9.4) [25]. It can be expected that in the future most studies on mutational alterations in humans will be carried out directly on DNA rather than at the gene product level.

6.1.1.1 Analysis at the Gene Product – Biochemical Level

Here the identification of the mutant site within the gene is not possible, but the individual gene in which the mutation has occurred can be inferred. There are several possibilities:

a) Specific proteins can be characterized by biochemical methods. Genetic variability reflects differences in proteins or enzymes. When a protein consists of more than one polypeptide chain, identification of the individual polypeptide chains involved may be possible. Examples include the numerous serum protein and enzyme polymorphisms.

b) Many proteins serve as enzymes, catalyzing specific metabolic steps. Therefore, when a specific genetic block has been demonstrated, the enzyme defect identified, and all other biochemical explanations excluded, a mutation in a specific gene specifying the involved enzyme can be inferred. The next step is characterization of the enzyme protein (see above).

c) Another subgroup within this category comprises examples in which an antigenic profile of the cell surface can be identified using specific antibodies. Examples include the blood groups and the HLA types (Sect. 5.2.5). This method permits identification not only of specific gene loci but also – within certain limits – of structural differences within these gene loci.

Until recently, genetic analysis had to be carried out as shown for the hemoglobin variants: from the analysis of the protein to the DNA. Very often this approach met various obstacles since the affected protein was unknown. Identification and analysis of genes at the molecular level opened an alternative method: the gene is identified first at the DNA level, its sequence and sequence alterations due to mutation are then identified. The amino acid sequence in the protein is inferred from the genetic code. Finally, and as the most important step of this analysis, the function of the protein can be ascertained. This approach has been termed "positional cloning," (or formerly "reverse genetics,") indicating the change of strategy in comparison with the classical model of gene identification. Positional cloning is now becoming the preferred means of analysis. Chronic granulomatous disease (306400) and cystic fibrosis (219700) are described as examples in Sect. 3.1.3.9. Only the results of an analysis at the gene product level are strictly comparable to results from thoroughly examined species in experimental genetics, such as *Drosophila melanogaster,* the mouse, maize, silkworm, and others. In many of these mutations no specific protein changes, enzyme defects, or aberrant antigen profiles could be demonstrated, but breeding experiments and recombination analysis offered an efficient alternative path for identifying individual genes. In the meantime the direct DNA approach is usually possible in experimental genetics as well. It might be interesting to examine how this purely formal gene concept, which long dominated the thinking of experimental geneticists has influenced thinking and conceptualization in fundamental genetics. Human biochemical genetics in its early phases was further advanced than biochemical genetics dealing with other species. The synthesis of insights from biochemistry and genetics led to rapid progress in microbial and fungus genetics. Biochemical human genetics remains more advanced than that of other mammalian species. However, in spite of recent progress in DNA techniques, biochemical genetics, and linkage analysis, many genetic analyses in humans still need to be carried out at a less sophisticated level.

6.1.1.2 Analysis at the Qualitative Phenotypic Level: Simple Modes of Inheritance

Conclusions often must be based on phenotypic differences far removed from the primary gene action. Still, the relationship between genotype and phenotype is so straightforward that simple Mendelian modes of inheritance can often be inferred with certainty. However, the involvement of a specific gene cannot be identified definitely: The same phenotype, with the same mode of inheritance, may be caused by mutations at several different gene loci.

Rare Conditions Qualitatively Different from the Normal. This category comprises most inherited diseases. An individual either is normally pigmented or lacks skin pigment (albinism). When measurements are possible, they show two different classes: affected and unaffected. When measurements of metabolites in blood or urine can be done, the values are distributed in two modes. Such findings many point to an enzyme defect and to identification of the involved gene. Examples are the increased excretion of homogentisic acid in the urine of alkaptonurics – Garrod's paradigm for inborn errors of metabolism (Sect. 1.5) – and the distribution of phenylalanine in the blood plasma of phenylketonurics in comparison to normals (Fig. 6.1). Conditions of this sort are usually rare. Whenever a more discriminating analysis is performed, even diseases with similar phenotypes and identical modes of inheritance often prove be genetically heterogeneous. Criteria for such heterogeneity include:

a) Children from matings between two homozygotes for a recessive condition are phenotypically normal, indicating that each homozygote parent carries a different recessive gene.

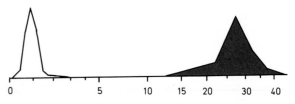

Fig. 6.1. Phenylalanine level in the blood plasma of healthy individuals and phenylketonurics *(shaded)* expressed in mg%. (Adapted from Penrose 1951)

b) Linkage analysis shows close linkage with a marker gene in a number of the analyzed families; in others, the two genes segregate independently. The first analyzed example was linkage of one of the loci for dominant elliptocytosis with the Rh locus on chromosome 1 [784].

c) Different protein or enzyme defects are demonstrated in various families by biochemical analysis. Conditions initially considered homogeneous turn out to have a different genetic basis (examples: hemophilia A and B, glycogen storage diseases, hereditary hemolytic anemias; Sect. 7.2.2.2).

Frequent Variants; Bimodal Distribution. In another group of traits analyzed at the phenotypic level, the phenotypes do not show a clearcut distribution into two classes. Not every individual can be attributed to a specific class; there is overlap. Yet, as long as the area of overlap is relatively small, the overall distribution remains bimodal.

An example from pharmacogenetics illustrates this phenomenon. After administration of a single dose of the tuberculostatic drug isomiazid the plasma level in various individuals was different, and the distribution showed two modes (Fig. 6.2). This suggested a simple mode of inheritance of drug biotransformation. This hypothesis was confirmed by family studies. Homozygotes Ac^s/Ac^s (Ac^s = allele for slow inactivation) showed high drug levels, whereas heterozygotes Ac^s/Ac^r (Ac^r = allele for rapid inactivation) and homozygotes Ac^r/Ac^r had low drug levels. This difference was caused by a variant of the enzyme *N*-acetyltransferase. These findings confirmed the genetic hypothesis put forward when the bimodal distribution of isoniazid concentrations was discovered.

In the absence of other, more conclusive criteria, a simple mode of inheritance may be inferred if the variable under investigation is measurable and shows a bimodal distribution. Exceptions occur in either of two directions:

1. A bimodal (or multimodal) distribution may be simulated by other mechanisms. This is especially likely when some threshold effect is involved. However, even when a threshold effect seems to be absent, a bimodal distribution may arise as a secondary effect when the variable under investigation shows a tendency to self-enhancement once it has reached a certain level. Blood pressure is an example where kidney damage may lead to a further increase in blood pressure. Bimodality may also exist if the variable depends, in addition to genetic factors, on environmental influences. A population study in a tropical country, for example, found a bimodal distribution of IgE blood levels. A simple mode of inheritance was suspected. Upon closer scrutiny, the population studied was seen to consist of two subpopulations: hospital personnel and inmates of a prison, who suffered from many more intestinal worms and therefore had higher IgE levels. Bimodality often suggests but does not prove a monogenic mode of inheritance. It may have other reasons. A monogenic mode of inheritance can be proven only by family studies. Certain methods of ascertainment may also simulate bimodality.

2. More frequently a bimodal or trimodal distribution is hidden because of overlap between the genotypic classes. The means of the two distributions may be so similar that bimodality is obscured. Harris and Smith [59] examined the conditions under which a combination of two normal distributions can lead to a bimodal distribution:

a) Two normal distributions with identical variances combine to a bimodal distribution only if the differ-

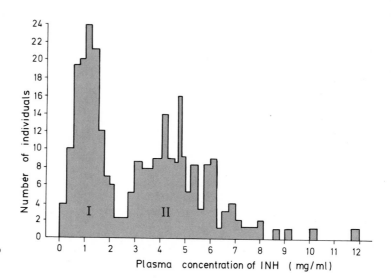

Fig. 6.2. Plasma concentration of isoniazid (*INH*) in 267 members of 53 families; bimodal distribution. The antimode is between 2 and 3 mg%. (Adapted from Evans et al. 1960 [38])

b) When the variances differ, the difference of the means must be at least equal to a certain multiple of the smaller standard deviation, which varies from 2 (when the variances are equal) to about 2.6 (when the variances are extremely different).

c) When the means are closer together, so that no bimodal distribution results, and the number of individuals in the two distributions are not too different, a "bitangential" distribution indicates that two different distributions may be involved (Fig. 6.3).

In practice, such bitangential distributions are difficult to evaluate. A skewed distribution is seen frequently for biological variables and cannot be reliably distinguished from bitangentiality. Variables occurring in nature only rarely show an ideal normal distribution, and chance deviations must be taken into account. Comparison of Figs. 6.2 and 6.3 gives an impression of how chance deviations may distort the distribution of empirically observed values with moderate sample size.

To summarize: *A unimodal distribution of a quantitative trait may be compatible with a monogenic mode of inheritance. However, without additional data, discrimination from multifactorial models is impossible.*

Figure 6.4 shows the opposite example: the enzyme activity distributions of electrophoretically identified genetic variants of glutamate-pyruvate transaminase (GPT) are clearly different. However, a distribution of GPT enzyme activity in the population that disregards the various allele genotypes is almost normal and might be interpreted as multifactorial in origin. Yet, the total distribution in the population owes its origin to only two alleles *(GPT1, GPT2)* and their corresponding phenotypes (GPT1, GPT2-1, GPT2). Many examples of this kind have been found when analyzing the variability of enzyme levels in various enzyme polymorphisms (Sect. 12.1.2).

Recognition of bimodality may be especially difficult if the two classes have very different frequencies, for example, if the number of individuals in one type is 10% or less than in the other (Fig. 6.5). Here it may be doubtful whether the smaller mode is genuine, representing a genetically different group. This result could as well be due to:

a) Chance deviations.
b) A threshold: this possibility should be seriously considered where the values constituting the second mode are close to zero.

Chance deviations can be excluded by examining more individuals. Under such circumstances a threshold effect cannot be ruled out; however, family studies are usually necessary, especially if the mode of inheritance is autosomal-dominant or X-linked. In such families the two genotypes are expected in about equal frequencies, with the consequence that bimodality is more clearcut than in a population sample. When families of probands from the lesser mode are examined, the following criteria confirm a simple mode of inheritance, if present, in the majority of families:

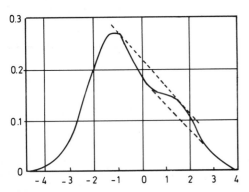

Fig. 6.3. Bitangential distribution. (From Harris and Smith 1951)

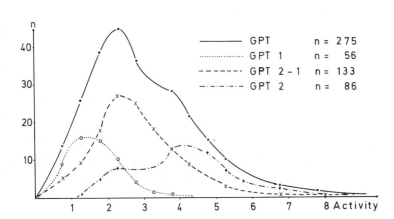

Fig. 6.4. Distribution of enzyme activities for three GPT genotypes, almost combining to a somewhat skewed normal distribution. (Data from Becker 1976 [5])

a) Clearcut bimodal distribution of the trait within sibships; the sibs who can be assigned to the mode reflecting normals are distributed in a manner similar to the general population.

b) Parents' values are also bimodal, with the additional condition that at least one of the two parents falls into the second mode.

c) An approximate 1:1 ratio of unaffected to affected should be found in these sibships. An approximate 3:1 ratio should be found in sibships where both parents fall into the second mode, i.e., when both parents are heterozygous.

With multifactorial inheritance spurious bimodality may result due to a threshold at zero. As a result the distribution of sibs whose values can be assigned to the first mode shows a lower mean than that of the population. Moreover, parents' values can more frequently be assigned to the first mode, with a lower mean value than that of the population (Fig. 6.6).

One trait for which an autosomal-dominant mode of inheritance has been inferred using these criteria is the low voltage electroencephalogram (EEG) [4, 153]. The human brain constantly produces certain voltage oscillations which after passing through suitable amplifiers can be drawn as curves on paper. As a rule, several leads (8–16) from different points of the head are taken simultaneously. The proband relaxes and rests with closed eyes but does not sleep. Sleeping patterns are different and are used for special diagnostic purposes.

The resting EEG of the healthy adult consists of a few wave types, the α waves being especially noticeable. In addition, β waves (>13/s) and a few ϑ waves (4–8/s) may occur (Fig. 6.7). These few elements, however, may be formed, distributed, and combined in so many different ways that a comparison with handwriting is inviting. Almost every human being has his own characteristic EEG, which remains constant over many years in the absence of diseases such as epilepsy and brain tumors and excepting such transitory physiological states as severe fatigue and intoxication. During childhood and youth the EEG develops from irregular forms with relatively slow waves to the final pattern which is reached by about age 19 at the latest and changes only very slowly with advancing age. Individual differences in speed of development are striking, leading to a high variability during child-

hood. Twin studies have shown the normal, individual EEG pattern to be almost exclusively genetically determined.

In about 4% of the adult population an EEG type is found with the following characteristics:

a) The occipital α waves are completely absent or can be seen for only a short time and with very low amplitude.

b) The EEG may therefore look absolutely flat or show an irregular pattern with β or ϑ waves of low amplitude.

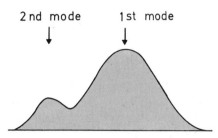

Fig. 6.5. Bimodal distribution of a quantitative trait in the human population. One of the two types is much more common than the other

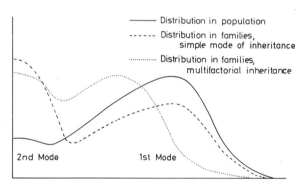

Fig. 6.6. Bimodal distribution when the second mode is close to 0. Discrimination between a simple diallelic mode of inheritance and a multifactorial model. Note that with multifactorial inheritance, the first mode is shifted to the left, whereas with diallelic inheritance, the first mode is identical in the population and in the families in which segregation occurs in the two modes

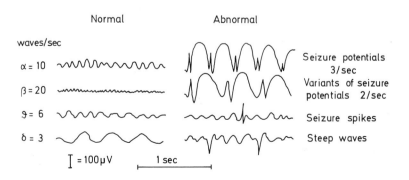

Fig. 6.7. Wave types of the human electroencephalogram

c) Contrary to the normal α EEG, there is no reaction when the eyes are opened. After the eyes are closed, a very few α waves may but do not necessarily appear.

The first requirement for genetic analysis is a measurement to quantify the extent of occipital α formation. One such measurement is the so-called α index, defined as follows:

$$I = \frac{\text{number of } \alpha \text{ waves}/10\,\text{s}}{\alpha \text{ frequency} \times 10\,\text{s}}$$

Figure 6.8 shows the distribution of this index in 61 sibships ascertained through parents or siblings with low-voltage EEG. This distribution has two maxima, one at about 70–80 and the other at 0, the latter corresponding to the low-voltage type. At first glance this distribution seems to favor the hypothesis of a simple mode of inheritance. Since the second maximum is 0, however, a fictitious bimodal distribution could be due to a threshold at 0.

Are there additional arguments in favor of a monogenic inheritance? The distribution among the parents is, again, bimodal; the distribution around the first mode corresponds well with the distribution in the general population. Still more important, in all 19 families ascertained by a child, at least one parent had a low-voltage EEG (Fig. 6.9). Analysis of the segregation ratio gave an estimate close to 75% for all families with two affected parents. For families with only one affected parent, the estimated overall segregation ratio proved to be somewhat below its expectation (≈50%). (The expectations are somewhat higher than 0.5 and 0.75, respectively since (because of the frequency of the characteristics) some homozygotes are expected among the parents.)

In this case, an autosomal-dominant mode of inheritance can be inferred by analysis of the distribution of a quantitative measure (the α index), together with family studies. Meanwhile, a gene whose mutations cause the low-voltage EEG has been assigned to the distal part of 20q by linkage studies with DNA markers [137, 138]. However, there is genetic heterogeneity; only about one-third of all examined families show this linkage.

In principle, similar criteria can be used in studying X-linked recessive inheritance. Here, however, analysis of the distribution may be more difficult in females since a trimodal distribution is expected: two homozygotes and one heterozygote. An example is the glucose-6-phosphate dehydrogenase level in G6PD deficiency. Since there is much overlap between normals and heterozygotes as well as between heterozygotes and hemizygotes; no clear trimodality can be discerned (see Sect. 7.2.2.2).

6.1.1.3 Genetic Analysis at the Quantitative Phenotypic-Biometric Level

Additive Model. In many cases phenotypic variability is so complex that the action of single mutations can no longer be identified. Here genetic conclusions deduced from similarities among relatives are necessarily of a more general kind. Still the "multifactorial" genetic models applied in such examples have certain characteristics in common – and more importantly some predictions derived from these models are fulfilled when tested on the observed data.

The simplest possible model assumes cooperation of a number of gene pairs. It is assumed that an upper-case allele (A or B but not a or b) contributes to the trait ("positive" allele) while the lowercase allele (a or b but not A or B) is silent and has no effect on the trait ("negative" allele). The phenotypic character varies gradually, depending only on the relative number of positive and negative alleles, whose contribution is assumed to be equal and additive in this model. These gene pairs may be named A, a; B, b; C, c; D, d; etc. Thus it makes no difference for the phenotype whether the genotype is AABBccdd ..., AaBbCcDc ..., or aabbCCDD ... (additive polygeny). This model is used in explanations for a variety of concepts. It should be made clear that the model represents an abstraction and is oversimplified. In reality, the contributions of genes acting in a multifactorial system almost always differ in quantity and quality. Some are more important than others.

Let us assume n gene pairs with gene frequencies $p = q = 0.5$ for the positive and the negative alleles. The distribution of the phenotypic classes on an arbitrary quantitative scale is given by the binomial formula $(p + q)^{2n}$ (Fig. 6.10). The higher the number of gene pairs, the more individuals are found in the central (i.e., more average values) as distinguished from the peripheral part of the distribution (i.e., more extreme values). At first glance the distinctions appear

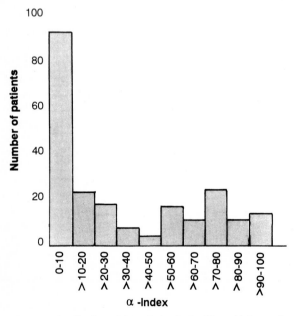

Fig. 6.8. Distribution of the α-index in families with low-voltage EEG. (From Anokhin et al. 1992 [4])

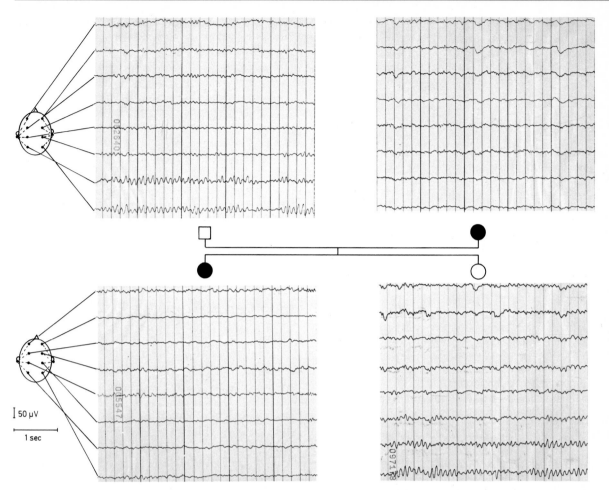

Fig. 6.9. Family observation: mother and first daughter have a low-voltage EEG; father and second daughter have an α EEG. Unipolar leads [153]

to provide a criterion for the number of gene pairs contributing to a characteristic when the empirical distribution is compared to a number of theoretical ones. However, such an inference holds only if each gene contributes just as much to the variable at the extremes of the distribution as it does in the center. This assumption can be challenged by the general – and often biologically plausible – hypothesis that at the extremes further deviation in the same direction is more difficult to achieve. For instance, for measurements of biologically active substances and enzymes values of less than zero do not exist.

The argument has in fact been used to estimate the number of gene pairs involved in the genetic variability of skin pigmentation [139]. In our opinion, it is likely that the number of skin pigmentation genes cannot be very high since segregation of very light individuals, on the one hand, and very dark ones, on the other, is not infrequent in matings of hybrid persons.

The distributions in Fig. 6.10 have only one mode; they are unimodal. Furthermore, they are similar in shape to the "normal" distribution. This similarity becomes closer with increasing number of gene pairs (n): The normal distribution is the limiting case for the binomial distribution with increasing n. It can be shown that this approximation becomes just as good when gene frequencies of positive and negative alleles are not equal. Higher values of n are required to reach the same degree of approximation as in the symmetric case. In general, a unimodal distribution of a variable – with the shape of the distribution curve more or less approximating the normal distribution – is typical for this genetic model of additive polygeny. However, neither a unimodal distribution, the shape of the distribution, nor the shape of the curve depend on the specific features of this model – equal and additive contributions of genes. They are good indicators for multifactorial inheritance in a more general sense.

On the other hand, as shown in Sect. 6.1.1.2, these properties do not exclude the possibility of a "major gene" with a simple mode of inheritance. In fact it is

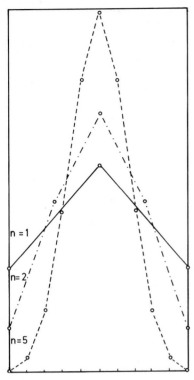

Fig. 6.10. Distribution of genotypes according to the binomial distribution $(p + q)^{2n}$ with $p = q = 0.5$ for 1, 2, and 5 gene pairs ($n = 1$, 2, 5) *Abscissa,* quantitative variation of measurements; vertical axis: No. of individuals

Table 6.1. Low-voltage EEG: segregation analysis for 61 sibships with children above the age of 19 years: expected and observed segregation ratio

	Mating types	
	$+ \times +$	$+ \times -$
Expected	0.759	0.509
Observed	0.625	0.420

The expectations are not precisely 0.75 and 0.5 because a certain small fraction of homozygotes must be expected among + individuals. The difference between expected and observed ratios for mating type $+ \times -$ is marginally significant ($p = 0.03$; see also [4]).

Table 6.2. Genotypes and phenotypes in additive polygenic inheritance

Phenotype	Genotype	Frequency	$p_1 = p_2 = q_1 = q_2 = 0.5$	
+4	AA BB	$p_1^2 p_2^2$	0.0625	
+2	AA Bb	$p_1^2 2p_2 q_2$	0.125	0.25
	Aa BB	$2p_1 q_1 p_2^2$	0.125	
0	AA bb	$p_1^2 q_2^2$	0.0625	
	aa BB	$q_1^2 p_2^2$	0.0625	0.375
	Aa Bb	$2p_1 q_1 2p_2 q_2$	0.25	
−2	Aa bb	$2p_1 q_1 q_2^2$	0.125	0.25
	aa Bb	$q_1^2 2p_2 q_2$	0.125	
−4	aa bb	$q_1^2 q_2^2$	0.0625	
			1.000	

biologically likely that only a few major genes are the principal genetic factors in several diseases acting against a background of many genes of less pathophysiological significance.

The first condition for establishing a unimodal and nearly normal distribution for the character is that the character in question can be measured, and that a quantitative scale is used. It is possible, for example, to distribute all adult men into two alternative classes: those taller than 1.67 m and those shorter than 1.67 m. With such limited information, family investigations could easily lead to the conclusion that variability in human stature depends on a dominant gene with incomplete penetrance. The example seems trivial and the point that it makes is self-evident, but the older literature is full of examples of this type of error.

A genetic hypothesis cannot be based exclusively on the population distribution of the variable. Family data are also needed. What type of family data does the model predict? This is examined using the simplest possible case: two gene pairs, A, a and B, b, acting additively and equally. The gene frequencies can be expressed as p_1, p_2 and q_1, q_2, respectively. There are nine possible genotypes with five different phenotypes, occurring with the frequencies given in Table 6.2 (Fig. 6.12). The frequencies of the possible mating types and the distribution of the children's genotypes for each mating type can be calculated. For the special case of the last column (all gene frequencies equal and 0.5), the calculation is given in Table 6.3. From these distributions of genotypes, the corresponding phenotypic distributions of the children can be deduced (Table 6.4).

The properties of the examined model are:

a) All resulting distributions have essentially the same form: they are symmetric and unimodal.
b) If the parents' phenotypes are identical, the mean of the children equals the parents' phenotype. If the parents' phenotypes are different, the childrens' mean is exactly the mean of the parents (midparent value).
c) Greater heterozygosity of the parents is accompanied by an increase in the expected variance

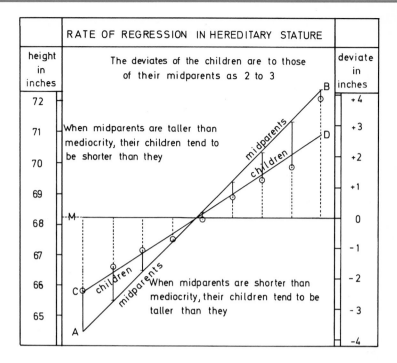

Fig. 6.11. Midparent-child correlation for stature; regression to the mean. (Original drawing and wording by F. Galton)

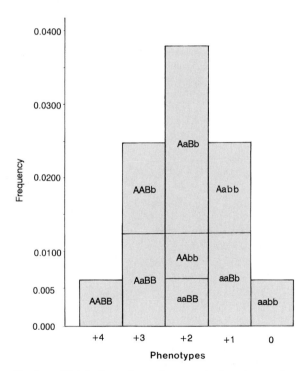

Fig. 6.12. Distribution of genotypes and phenotypes of a quantitative trait with two additively acting gene pairs A,a and B,b. Random mating; all gene frequencies are assumed to be 0.5. The presence of a gene designated by an uppercase letter (A or B) is assumed to contribute +1 to the measured phenotype. Genes designated with lower case letters (a or b) make no quantitative contributions to the phenotype. Gene effects are additive (e.g. AABB = 4, aabb = 0)

among the children. It is highest with mating types 0×0 and zero in mating types $+4 \times +4$, -4×-4, and $+4 \times -4$.

d) The mean of the children of all persons with the same phenotype (for example, children of all persons with type +4) deviates half as much from the population mean as does the phenotype of these parents (for example, the mean phenotype of children of +4 parents is +2).

This model is highly particularized and simple; however the analysis proves fairly cumbersome. For examining the general case (n gene pairs; gene frequencies $p_1 \ldots p_n$; $q_1 \ldots q_n$), the method should be changed. First we assume a gene pair of heterozygotes with the phenotypic effect α, whereas one homozygote (AA) has the effect 2α and the other (aa) has the effect 0. Thus we again assume that heterozygotes hold exactly the middle between the two homozygotes. The mean and variance of the corresponding characteristic, x, can now derived as follows:

$$E_x = \frac{p^2(2\alpha) + 2pq(\alpha)}{p^2 + 2pq + q^2} = \frac{2p\alpha(p + q)}{(p + q)^2} = 2p\alpha, \qquad (6.1)$$

$$V_x = E(x^2) - (Ex)^2 = p^2(4\alpha^2) + 2pq(\alpha^2) - 4p^2\alpha^2 = 2pq\alpha^2 \quad (6.2)$$

(using $p + q = 1$). α may be visualized as the contribution of the allele A to the character x. V_x is the genetic variance in the population.

The more general case of n gene pairs with gene frequencies $p_i = p_1, p_2 \ldots p_n$ for the genes $A_1, A_2 \ldots A_n$ and $q_i = q_1, q_2 \ldots q_n$ for the genes $a_1, a_2, \ldots, a_n$ can now be examined:

$$E_x = 2\alpha \sum_{i=1}^{n} p_i \qquad V_x = 2\alpha^2 \sum_{i=1}^{n} p_i q_i$$

Table 6.3. Mating types, their frequencies, and segregation ratios among children with two additive gene pairs and $p_1 = p_2 = q_1 = q_2 = 0.5$

Mating	Freq.	AABB	AABb	AaBB	AAbb	aaBB	AaBb	Aabb	aaBb	aabb
AABB × AABB	0.003906	1	–	–	–	–	–	–	–	–
× AABb	0.015625	$1/2$	$1/2$	–	–	–	–	–	–	–
× AaBB	0.015625	$1/2$	–	$1/2$	–	–	–	–	–	–
× AAbb	0.007813	–	1	–	–	–	–	–	–	–
× aaBB	0.007813	–	–	1	–	–	–	–	–	–
× AaBb	0.031250	$1/4$	$1/4$	$1/4$	–	–	$1/4$	–	–	–
× Aabb	0.015625	–	$1/2$	–	–	–	$1/2$	–	–	–
× aaBb	0.015625	–	–	$1/2$	–	–	$1/2$	–	–	–
× aabb	0.007813	–	–	–	–	–	1	–	–	–
AABb × AABb	0.015625	$1/4$	$1/2$	–	$1/4$	–	–	–	–	–
× AaBB	0.031250	$1/4$	$1/4$	$1/4$	–	–	$1/4$	–	–	–
× AAbb	0.015625	–	$1/2$	–	$1/2$	–	–	–	–	–
× aaBB	0.015625	–	–	$1/2$	–	–	$1/2$	–	–	–
× AaBb	0.062500	$1/8$	$1/4$	$1/8$	$1/8$	–	$1/4$	$1/8$	–	–
× Aabb	0.031250	–	$1/4$	–	$1/4$	–	$1/4$	$1/4$	–	–
× aaBb	0.031250	–	–	$1/4$	–	–	$1/2$	$1/4$	–	–
× aabb	0.015625	–	–	–	–	–	$1/2$	$1/2$	–	–
AaBB × AaBB	0.015625	$1/4$	–	$1/2$	–	$1/4$	–	–	–	–
× AAbb	0.015625	–	$1/2$	–	–	–	$1/2$	–	–	–
× aaBB	0.015625	–	–	$1/2$	–	$1/2$	–	–	–	–
× AaBb	0.062500	$1/8$	$1/8$	$1/4$	–	$1/8$	$1/4$	–	$1/8$	–
× Aabb	0.031250	–	$1/4$	–	–	–	$1/2$	–	$1/4$	–
× aaBb	0.031250	–	–	$1/4$	–	$1/4$	$1/4$	–	$1/4$	–
× aabb	0.015625	–	–	–	–	–	$1/2$	–	$1/2$	–
AAbb × AAbb	0.003906	–	–	–	1	–	–	–	–	–
× aaBB	0.007813	–	–	–	–	–	1	–	–	–
× AaBb	0.031250	–	$1/4$	–	$1/4$	–	$1/4$	$1/4$	–	–
× Aabb	0.015625	–	–	–	$1/2$	–	–	$1/2$	–	–
× aaBb	0.015625	–	–	–	–	–	$1/2$	$1/2$	–	–
× aabb	0.007813	–	–	–	–	–	–	1	–	–
aaBB × aaBB	0.003906	–	–	–	–	1	–	–	–	–
× AaBb	0.031250	–	–	$1/4$	–	$1/4$	$1/4$	–	$1/4$	–
× Aabb	0.015625	–	–	–	–	–	$1/2$	–	$1/2$	–
× aaBb	0.015625	–	–	–	–	$1/2$	–	–	$1/2$	–
× aabb	0.007813	–	–	–	–	–	–	–	1	–
AaBb × AaBb	0.062500	$1/16$	$1/8$	$1/8$	$1/16$	$1/16$	$1/4$	$1/8$	$1/8$	$1/16$
× Aabb	0.062500	–	$1/8$	–	$1/8$	–	$1/4$	$1/4$	$1/8$	$1/8$
× aaBb	0.062500	–	–	$1/8$	–	$1/8$	$1/4$	$1/8$	$1/4$	$1/8$
× aabb	0.031250	–	–	–	–	–	$1/4$	$1/4$	$1/4$	$1/4$
Aabb × Aabb	0.015625	–	–	–	$1/4$	–	–	$1/2$	–	$1/4$
× aaBb	0.031250	–	–	–	–	–	$1/4$	$1/4$	$1/4$	$1/4$
× aabb	0.015625	–	–	–	–	–	–	$1/2$	–	$1/2$
aaBb × aaBb	0.015625	–	–	–	–	$1/4$	–	–	$1/2$	$1/4$
× aabb	0.015625	–	–	–	–	–	–	–	$1/2$	$1/2$
aabb × aabb	0.003906	–	–	–	–	–	–	–	–	1

The following considerations, which for the sake of simplicity are given for one gene pair only, are valid for n gene pairs as well.

We now consider the relationships between parents and children and that among siblings. To simplify the calculation, α is assumed to be 1, making the phenotypic value of homozygotes AA = 2, that of heterozygotes Aa = 1, and that of homo-zygotes aa = 0. Table 6.5 shows the frequencies of all possible parent-child combinations. They can be explained as follows. The frequency of AA mothers among all mothers is p^2. Each of their children gets one A gene. The probability that this gene meets another A gene in the zygote is p. This gives the overall frequency $p^2 \times p = p^3$. For the other maternal geno-types an analogous calculation can be made. The overall dis-

Table 6.4. Distribution of children in additive polygenic inheritance (the lacking 5 categories, 0×-2, 0×-4, -2×-2, -2×-4, -4×-4, can be calculated according to the same rule)

Parents' genotypes	+4 AABB	+2 AABb; AaBB	0 AAbb; aaBB; AaBb	−2 Aabb; aaBb	−4 aabb
$+4 \times +4$	1				
$+4 \times +2$	0.5	0.5			
$+4 \times 0$	0.1666	0.6667	0.1666		
$+4 \times -2$		0.5	0.5		
$+4 \times -4$			1		
$+2 \times +2$	0.25	0.5	0.25		
$+2 \times 0$	0.08333	0.41667	0.41667	0.08333	
$+2 \times -2$		0.25	0.5	0.25	
$+2 \times -4$			0.5	0.5	
0×0	0.02778	0.22222	0.50000	0.22222	0.02778

Table 6.5. Frequency of parent-child (father- or mother-child combinations) in the human population in a randomly mating population (see text)

Parent	Children		
	AA	Aa	aa
AA	p^3	$p^2 q$	–	p^2
Aa	$p^2 q$	pq	pq^2	$2pq$
aa	–	pq^2	q^3	q^2

tribution for the whole population (parents as well as children) is of course $p^2 + 2pq + q^2$ (marginal sums in Table 6.5).

The variable now under investigation may be referred to as x_1 in the parents and x_2 in the child. Equations 6.1 and 6.2 derived above then lead to:

$$\bar{x}_1 = \bar{x}_2 = 2p \tag{6.3}$$

$$V_{x1} = V_{x2} = 2pq \tag{6.4}$$

In general, the covariance of the two variables, x_1 and x_2, is defined as:

$$\text{Cov}(x_1, x_2) = E(x_1 x_2) - \bar{x}_1 \bar{x}_2$$

Here $E(x_1 x_2)$ is defined as: $\sum x_{1i}\, x_{2i} p(x_{1i}, x_{2i})$
The values $x_{1,i}$ and $x_{2,i}$ represent the phenotypic expression of the trait, in our example 2, 1 and 0. The $p(x_{1i}, x_{2i})$ are the corresponding entries in Table 6.5; $p(2, 2)$, for example, has the value q^3. It follows:

$$\text{Cov}(x_1, x_2) = 4p^3 + 4p^2 q + pq - (2p)^2 = pq$$

and for the correlation coefficient r_{PC} between parent and child:

$$r_{PC} = \frac{\text{Cov}(x_1, x_2)}{\sigma x_1 \sigma x_2} = \frac{pq}{2pq} = 0.5$$

This important result was derived by Fisher in 1918 [44]. In a random mating population and with additive gene action the correlation coefficient between a parent and child is 0.5. It

can be shown in a similar way that under the same conditions the correlation coefficient between full sibs is also 0.5. These correlation coefficients are independent of the gene frequencies p_1 and q_1 and state in a statistical way that parents and children as well as sibs have 50% of their genes in common. The correlation coefficient for relatives is a follows (these values also refer to the average proportion of genes shared by the various types of relatives): monozygotic twins, 1.0; dizygotic twins, 0.5; sibs, parent-child (first-degree relatives), 0.5; aunts, uncles, etc. (second-degree relatives), 0.25; first cousins (third-degree relatives), 0.125.

The situation becomes more complicated when A is more or less dominant over a. In this case, the correlation coefficients are influenced by the gene frequencies. The parent-child correlation is no longer equal to the sib-sib correlation, but – apart from the case $q = 1$ – is lower.

6.1.1.4 Heritability Concept

The concept of heritability is widely used in quantitative genetics. The graded characteristic under examination, expressed in metric units, may be called the "value." The value measured in a given individual is its phenotypic value. This phenotypic value for most biological characteristics owes its origin to both genetic and environmental factors. Environment is considered in a broad sense, i.e., comprising all non-genetic circumstances that influence the phenotypic value. The two components are usually called the genotypic value and the environmental value (Falconer [40] uses the term environmental deviation):

$$P = G + E$$

where P = phenotypic value, G = genotypic value, and E = environmental value.

The phenotypic values of all individuals in a population have a mean and a variance around this mean. The variance is distinguished from other measures of variability by one mathematical property: different variances can be added to give a common variance, and, conversely, a common phenotypic variance V_P can be broken down into its components, such as the genotypic variance V_G and the environmental variance V_E:

$$V_P = V_G + V_E$$

However, the addition rule for variances applies only if genotypic and environmental values are independent of each other, i.e., when they are not correlated. If there is a correlation between the two, the covariance of G and E must be added:

$$V_P = V_G + V_E + 2 \, \text{Cov}_{GE}$$

Let us take an example from the area of genetics that first introduced these concepts – agricultural studies [40]. It is normal practice in dairy husbandry to feed cows according to their milk yield. Cows that produce more milk are given more food. Human society often behaves in a similar way toward its own members, as is discussed in the section on behavior genetics.

Another assumption is that specific differences in environments have the same effect on the various genotypes. When this is not so, there is an interaction between genotype and environment, giving an additional component to the variance V_I. Even in experimental animals this component can be measured only under special conditions.

The genotypic value V_G can be subdivided into several components: an additive component (V_A) and a component (V_D) measuring the deviation due to dominance and epistasis from the expectation derived from the additive model. The dominance variance is contributed by heterozygotes (Aa) that are not exactly intermediate in value between the corresponding homozygotes (aa and AA). The variance contributed by epistasis refers to the action of genes that affect the expression of other genes. Hence the concept of additive variance does not imply the assumption of purely additive action of the genes involved. Even the action of genes showing dominance or epistasis tends to have an additive component. The whole genotypic variance can be written:

	Genetic Variance		Environmental variance		Measurement variance	

$$V_P = \overbrace{V_A + V_D} \; + \; \overbrace{V_E + V_I} \; + \; \overbrace{\text{Cov}_{GE} + V_M}$$

This introduces a new component (V_M) that relates to the variability in measurement of the same character at different times. The value may represent truly different values, such as different test results on different days, or measurement errors, such as differences in test results of the same blood specimen and differences on repeated testing of the same individual. If all

these variables are known, they can be incorporated into the calculations. In the following, covariance between heredity and environment (Cov_{GE}) and interaction variance (V_I) is assumed to be 0 and neglected. Measurement variance (V_M) is also neglected but is considered in the discussion of twin methods (Appendix 6).

For convenience it is useful to introduce a new concept: heritability, defined as:

$$h^2 = \frac{V_A}{V_P}$$

the value of this variable ranges from 0 to 1 (0%–100%), expressing the contribution of additive genetic elements to the phenotype under study. In other words, heritability is a population statistical parameter that expresses the (additive) genetic contribution to the trait under study in a single percentage value. A low value implies few contributions of additive genes to the trait, while a high value suggests a larger contribution. The concept was developed for purposes of selection in plant and animal breeding of economically useful traits such as milk production in cows and egg laying in chickens. The additive part of genetic variability is most important for these purposes. Any other genetic component, such as dominance, tends to reduce the accuracy of prediction. In humans, however, we are more interested in the total genetic variability, whether it is additive or not.

In human genetics, therefore, heritability as defined above $\left(\dfrac{V_A}{V_P} \right)$ is often called "heritability in the narrow sense" and is expressed by another definition:

$$h^2 = \frac{V_G}{V_P}$$

where V_G and V_P refer to the total genotypic and phenotypic variance, respectively. This formulation is known as the heritability in the broad sense, or degree of genetic determination.

There is a relationship between heritability in the narrow sense (h_N^2) and the theoretical correlation coefficients between relatives as given above. For the most important degrees of relationship the following formulas apply:

Monozygotic twins:	$h^2 = r$
Sib-sib or dizygotic twins:	$h^2 = 2r$
One parent – one offspring:	$h^2 = 2r$
Midparent-offspring:	$h^2 = r/\sqrt{1/2} = r/0.7071$
First cousins:	$h^2 = 8r$
Uncle-nephew:	$h^2 = 4r$

Properties of h². In considering the biological significance of heritability measurements, its properties need careful scrutiny:

a) Heritability is a ratio. A ratio changes when either the numerator or the denominator changes. h^2 increases when the numerator (V_G, genotypic, or V_A, additive variance) increases, or the denominator (V_E, environmental variance) decreases. Otherwise stated, a more similar environment will raise heritability!

b) The estimation of heritability is based on theoretical correlations between relatives. These correlations are valid only for random mating. Assortative mating leads to other correlations and unless taken into consideration produces systematic errors in the estimation of h^2. The correlations resulting from assortative mating were calculated first by Fisher in 1918 [44] (see also [21]; for a more complete treatment see [169, 170]). These correlations can be used for adjustment of h^2.

c) An estimation of h^2 is strictly valid only when the assumption is made that covariance and interaction between genotypic and environmental values are 0.

Falconer tried to escape this dilemma for covariance by proposing the following convention. If the genetic constitution of an individual creates environmental conditions that improve or worsen his phenotype, this phenomenon can be included as part of the genotypic value. Formally, this is correct, even if it tends to obscure the problems involved in genotype-phenotype relationships. For animal breeding this convention may be useful. Applied to humans, however, it leads to difficulties.

More difficulties arise when these concepts are applied to interpreting heritability values from twin data (Appendix 5). The interaction term poses another difficulty in interpretation, one for which no satisfactory solution has yet been proposed. Correlations between relatives do not prove genetic variability; they may also be caused by common environmental influences within families. In animal breeding, where the environment can be controlled, this factor might either be neglected, or quantified. In humans, this is almost impossible. (Attempts at quantification have been made using the method of path coefficients but readily quantifiable data are rarely available.) The problem will be mentioned repeatedly in this and other chapters of this book.

6.1.1.5 An Example: Stature

An example of a biometric study in which heritability can be estimated is Galton's classic work on inheritance of stature (data from [68]). He measured 204 parental couples along with their 928 adult children. There was a methodological difficulty due to the lower average stature of women. Galton overcame this by multiplying all measurements of females by 1.08, thus adjusting them to the male measurements; on average the stature of males in his sample was 1.08 times the stature of females. Having made this correction, he determined the midparent value for each couple: 1/2(male + female). The results of the study are presented in the correlation Table 6.6. A correlation is apparent on inspection of the table. The correlation coefficient is:

$$r_{pc} = 0.59 \quad p < 0.01$$

where r_{pc} is midparent-child correlation.

This value can be used to calculate h^2. The midparent-offspring correlation is:

$$h^2 = \frac{r}{\sqrt{1/2}}$$

with random mating. This gives:

$$h^2 = \frac{0.59}{0.7071} = 0.834$$

Obviously, stature is predominantly determined genetically, but there is a component of $0.166 = 1 - 0.834$ not accounted for by additive genetic variance. This may be due mainly to "environmental factors." Do these data offer any hints of environmental influences?

The same data may be arranged in a different way (Tables 6.6, 6.7). Here, another divergence from expectation is obvious. With additive gene action, the children's mean is expected to be exactly one-half of parents' values, i.e., should be identical to the midparent value. This, however, is not the case. Rather, the data show that if the midparent value is

Table 6.6. Stature (inches) of parents and adult children (Galton, from Johannsen 1926 [38])

Stature of midparent	Stature of children							
	60.7	62.7	64.7	66.7	68.7	70.7	72.7	74.7
64	2	7	10	14	4	–	–	–
66	1	15	19	56	41	11	1	–
68	1	15	56	130	148	69	11	–
70	1	2	21	48	83	66	22	8
72	–	–	1	7	11	17	20	6
74	–	–	–	–	–	–	4	–
	5	39	107	255	287	163	58	14

Table 6.7. Stature (inches) of mid-parent and mean stature of children (from Johannsen 1926 [38])

Stature of mid-parent	64.5	65.5	66.5	67.5	68.5	69.5	70.5	71.5	72.5
Mean stature of children		65.8	66.7	67.2	67.6	68.3	68.9	69.5	69.9	72.2

higher than the population mean, the children's mean is lower than that of their parents. On the other hand, if the midparent value is lower than the population mean, the children's mean is higher. Therefore the children's mean tends to deviate from the parents' mean in the direction of the population mean.

This phenomenon was observed by Galton and termed "regression to the mean." It can also be shown in other, similarly continuously distributed characteristics (Fig. 6.11).

What is the reason for this divergence from genetic expectations? Individuals who can be ranged at the extremes of a distribution curve presumably obtain not only the genetic factors that make for the extreme phenotype but probably have benefited in addition from unusual environmental circumstances. Furthermore, specific gene-gene and gene-environmental interactions may have been operative in causing their extreme phenotypes. Their children on average are less likely to benefit from the special environmental influences and gene-environmental interaction that placed the parent in the extreme categories. Their phenotypic values are therefore likely to be closer to the mean of the population – a regression to the mean.

6.1.1.6 Quantitative Genetics and the Paradigms of Mendel and Galton

How do the two paradigms on which human genetics was founded relate to each other? The gene concept developed from Mendel's experiments (Sect. 1.4), while Galton's paradigm was based on concepts of correlation between human relatives and regression analysis. The two concepts can be linked theoretically to each other. Results of correlation studies in relatives can be interpreted in terms of action of individual genes, as was first shown extensively by Fisher (1918) [44]. Such correlation studies using biometric methods can complement genetic analysis.

Paradigms of Mendel and Galton: Explanatory Power. As mentioned in the "Introduction," a paradigm comprises three main aspects: an exemplary approach, a group of scientists who follow this approach, and at least the germ of a scientific theory. The long-term success of a paradigm depends mainly on the depth and explanatory power of this theory. Therefore it may be useful to compare the two paradigms regarding values of their underlying theories, using criteria developed in the philosophy of science [18]. According to Bunge,

The basic desiderata of scientific theory construction are the following: (1) To *systematize knowledge* by establishing logical relations among previously disconnected items; in parti-

cular, to explain empirical generalizations by deriving them from higher-level hypotheses. (2) To *explain facts* by means of systems of hypotheses entailing the propositions that express the facts concerned. (3) To *increase knowledge* by deriving new propositions (e.g., predictions) from the premises in conjunction with relevant information. (4) To *enhance the testability* of the hypotheses, by subjecting each of them to the control of the other hypotheses of the system ...

A few scientific theories comply not only with the basic desiderata (1)–(4) but also with the following additional goals: (5) To *guide* research either (a) by posing or reformulating fruitful problems, or (b) by suggesting the gathering of new data which would be unthinkable without the theory, or (c) by suggesting entire new lines of investigation. (6) *To offer a map of a chunk of reality*, i.e., a representation ... of real objects and not just a summary of actual data and a device for producing new data.

Bunge used Darwin's theory of evolution as an example of a theory that fulfills all the above criteria. In general, the ability of a theory to fulfill its task depends on its depth. Criteria for the depth of a theory are: "The occurrence *of high-level constructs;* the presence of a *mechanism,* and a *high explanatory power.* The three propositions are intimately linked: it is only by introducing high-brow (transempirical) concepts that unobservable 'mechanisms' can be hypothesized, and only what is hypothesized to occur in the depths can explain what is observed at the surface."

Less deep theories are closer to the phenomena; these are therefore called "phenomenological"; in distinction to these are theories hypothesizing definite "mechanisms," which are often called "representational" or "mechanistic." Often such deep mechanistic theories reward the scientists with an unexpected bonus: their explanatory power extends beyond the range of phenomena for the explanation of which they had been created.

When the theories that developed within the paradigms of Galton and Mendel are compared using these criteria, Galton's approach appears as a phenomenological theory. K. Pearson, Galton's master student, pointed out as early as 1904 that quantitative comparison of phenotypes between relatives with biometric methods leads to "a purely descriptive statistical theory." To some degree, it may systematize knowledge, but it offers only relatively low-level and nonspecific hypotheses, i.e., similarity between relatives can be explained by heredity – or, more specifically, by additive gene action either with or without a contribution of dominance, or of environmental contributions. Such propositions are of a very general nature, and there is only occasional enhancement by additional hypotheses. For an example, see the Carter effect described in Sect. 6.1.2.3; here the higher incidence of a birth defect in relatives of female probands is predicted and explained by the additional hypothesis of identical distribution of liability genes in both sexes, despite the unequal sex distribution among the probands. Bunge's conditions 5 and 6 are not fulfilled at all: Problems are not reformulated in a fruitful way, nor does the theory suggest gathering of new data. It suggests only the obvious: comparison of relatives.

Compare with this result the theory founded by Mendel's paradigm. Soon after its discovery a high-order construct – the unit of transmission, recombination, and function now called a *"gene"* – was introduced. This opened the way for investigating the *mechanisms* of gene replication, transmission, recombination, and action. Stepwise elucidation of these mechanisms constitutes the history and present situation of genetics. The *explanatory power* of this theory has been proven far beyond the range of phenomena for which it was originally developed. At present the theory explains not only transmission of traits from parent to offspring but also between different cell generations of an organism, for example, cancer formation.

Its explanatory power has not yet been exhausted. Returning to our classification of genetic analysis (i.e., at the DNA-gene level, at the gene product-biochemical level, at the qualitative phenotypic level, and at the quantitative phenotypic-biometric level), the Galtonian biometric paradigm provides answers at the level most remote from gene action. Research with biometric genetic methods can be said to be guided by "black box" theory. Two external observable variables – the measurements of parents and children or other sets of relatives – are compared with each other, but the mediating biological variables are unknown and remain in a black box (Fig. 6.13).

All of human development, structure, and function is ultimately controlled by genes. Differences among human beings can be demonstrated by unique physiological, biochemical, and immunological features of each individual. Genetic determination of these features can be shown by family resemblance. The degree of resemblance depends upon the closeness of the relationship. Monozygous twins share all their genes and are more similar than any other relatives. Sibs share 50 % of their genes, while more remote relatives share only a small fraction.

A comparison of relatives using biometric techniques for the measurement of any human phenotype is therefore likely to show genetic factors underlying that trait. A "heritability" or the proportion of total variability attributed to genetic causation is expected to be higher than 0. Since the principal biological basis of *human behavior* lies in the brain, and the brain is as likely to show genetic variation as any other organ, some genetic factors are likely to determine behavior on a priori grounds. The teasing apart of shared genes and a shared environment in the family setting becomes particularly difficult in behavioral traits and causes problems in interpretation. Polemics with potentially explosive sociopolitical ramifications have resulted (see Chap. 17).

Analysis of any human trait and particularly of human behavior yields more meaningful information if the phenotype under investigation can be studied by Mendelian techniques at the level of gene action. The black box is thus opened, and the unknown mediating variable is replaced by a known biological mechanism.

Considering the marked differences in scientific value of both theories, one might ask why some work in human genetics is still being performed along Galtonian lines. The principal explanation is that analysis guided by advanced genetic theory is often impossible. Most human phenotypes – such as behavioral phenotypes and disease liabilities – simply cannot be studied directly according to Mendelian principles. The underlying variables to be analyzed must first be isolated using additional, sometimes sophisticated, biological techniques of all types. Finding such variables requires specialized knowledge in normal and abnormal human biology using the methodology of the various biomedical sciences. Counting and measurement of more simple and obvious phenotypes, on the other hand, can often be carried out. This is why Galtonian techniques have often been the first approach to further analysis and can often bring practically useful results despite their theoretical weakness.

Here the new methods from molecular genetics may bring a significant change. Linkage studies with DNA markers help in localizing and analyzing genes without prior knowledge of mechanisms of their action. Once the gene is known, its mechanism of action and its contribution to a complex phenotype can be identified. At the moment, however, this ap-

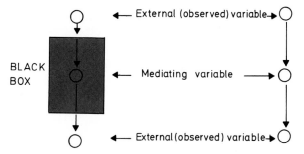

Fig. 6.13. Difference between a "black box" hypothesis and a hypothesis involving a mechanism. In a black box hypothesis *(left)* the mediating variable by which one observable variable influences the other is unknown. In an explanatory hypothesis *(right)* the mediating variable can be deduced from scientific theory; a mechanism for the influence of one observable variable to the second can be found

proach – despite its theoretical appeal – has not met with practical success. Still, the Galtonian approach continues to be important for formulation of hypotheses, for selection of traits to be studied with more incisive methods, and for devising research strategies. Human traits that are under control of a large number of genes, each contributing to the total variability, are difficult to study by the Mendelian approach. For some of these traits, however, the conventional biometric paradigm, which presupposes a large number of genes of small effect, may not be appropriate. One or several genes with major effect that are detectable individually with biological techniques may be at work, with the remaining genes providing the "genetic background."

Consequently we should use Galtonian techniques as long as no better alternatives are available. However, we should refrain from making them a goal in themselves; the challenge that *in principle* a better alternative is possible should always be kept in mind. It is all too easy for groups of geneticists trained mainly in statistical methods and the use of computers to develop highly sophisticated statistical methods for computing heritabilities, developing path analyses for contribution of various factors from heredity, family environment, economic status in a phenotype, or comparing genetic models with and without contributions for major genes to a certain phenotype. However, the final result is often disappointing to biologically oriented scientists, who demand more incisive data. Statistical methods are of great importance for analysis in human genetics. They should be used, for example, in testing biologically well-founded hypotheses that have been proposed under the guidance of a powerful biological theory. With increasing insight into the genetic and environmental determinants of complex phenotypes, such as birth defects, diseases, and behavioral characteristics, statistical methods for their analysis are also becoming more sophisticated. Interactions between various genes, as well as between genes and environmental determinants must be considered. The geneticist – who is necessarily specialized in only one of the many fields requiring expert knowledge – should be careful not to loose sight of the factual basis of this analysis. It is necessary to maintain a clear conceptual distinction of the various levels at which basic data are available and of the level at which genetic analysis is performed.

This should be kept in mind when the following sections on more complex models of inheritance are studied.

6.1.2 Multifactorial Inheritance in Combination with a Threshold Effect

6.1.2.1 Description of the Model

The previous section described genetic analysis at the quantitative phenotypic-biometric level for normal traits with a unimodal and nearly normal distribution in the population. The simple model of additive polygenic inheritance was shown to account for these properties so that parent-offspring and sib-sib correlations can be used to estimate heritability h^2.

In many diseases and malformations, however, clear qualitative distributions apply: the individual either suffers from a given disease or is free from it. However, neither family investigations nor chromosome studies have been able to uncover a simple mode of inheritance or a visible chromosome abnormality. Family studies show an increased empirical risk for near relatives to be affected with the same condition (familial aggregation). In many cases pathophysiological considerations suggest a complicated etiology. Various biological influences are often obvious, and environmental factors such as malnutrition, infective organisms, and unknown agents are additionally implicated. When all these genetic and environmental influences together exceed a certain threshold, the organism's capacity to cope breaks down and the individual becomes ill or dies.

The terms "threshold" and "liability" are often used in discussion of multifactorial inheritance. A threshold implies a sharp qualitative difference beyond which individuals are affected. While the concept of a threshold is useful for models of multifactorial inheritance, it is seldom likely to exist. The concept of a liability implies a graded continuum of increasing susceptibility to the disease. It is more difficult to deal with analytically, but biologically this mechanism is more likely to apply in most situations.

In hereditary diseases with simple mode of inheritance, malfunction caused by a mutation at a single gene prevents normal function. In other conditions the mutation only leads to difficulties in special circumstances, such as in the monogenically determined drug reaction (Sect. 7.5.1). Most conditions, however, are so complex that direct analysis of all contributing factors becomes impossible. Obviously many different genes are involved. We are left with the black box situation – genetic analysis can be carried out more readily by statistical than biological methods.

To make genetic predictions at this complex level, several assumptions are necessary:

a) The genetic liability to disease is more or less normally distributed, and the distribution shows one mode.

b) The liability is caused by a great number of genes acting additively, each contributing equally to the liability.

c) The individual becomes sick or malformed when liability exceeds a certain threshold. This threshold may be sharply defined; in most cases, however, there is a threshold area within which additional environmental circumstances determine whether the individual becomes ill (Fig. 6.14).

Obviously this model oversimplifies the actual situation, but it may be useful as the first step in understanding certain common diseases and malformations.

Animal Experiments. Certain observations in experimental genetics of mammals have been explained by threshold characters such as polydactyly in the guinea pig [173]. Two strains were crossed, one with the normal three toes on the hind feet, the other with four toes. Among the F_1 animals only a few had four toes, whereas in the second generation of $F_1 \times F_1$ crosses about one-fourth of all animals showed four toes. Genetic analysis suggested that the two strains differed in additive alleles at four gene loci: any animal could have a maximum of eight and a minimum of zero plus alleles. In matings of two homozygous animals (8×0) (Fig. 6.15) the F_1 generation being heterozygous should have four plus alleles. This genotype leads to four toes only in exceptional cases. In the F_2 generation $(F_1 \times F_1)$ all combinations of plus alleles occur, giving a continuous distribution. In this case it was shown in principle that additive gene action may indeed be associated with a threshold character (Fig. 6.15). In another example not only the discontinuous but also the continuous phase was demonstrated, i.e., a phenotypic effect of the quantitatively varying liability. Grueneberg [55, 56] analyzed such a system in the mouse. In mice of the inbred CBA strain a third molar tooth is frequently lacking. In 133 of 744 CBA animals at least one of the four third molars was missing. In the C57 black strain, however, this molar is almost always present. Crossing between the two strains (CBA × C57) showed that the mode of inheritance is not simple, in spite of the fact that the character (tooth present or absent) is a clearcut alternative trait. Even in animals of the CBA strain with the extra tooth it was on the average much smaller than in the C57 black strain (Fig. 6.16). Therefore in the CBA strain tooth size varies continuously down to a certain minimum size. Below this threshold the tooth is not formed at all. Grueneberg called this phenomenon "quasi-continuous variation." The threshold is not absolutely sharp, and there seems instead to be a threshold area. The multifactorial genetic basis was revealed principally by the strong difference between the two strains and by the interstrain crosses. Within the genetically uniform CBA strain, variability was caused by environmental influences.

Demonstration of both continuously distributed liability and discontinuous thresholds has been tried repeatedly in humans (see, for example [28]), but in

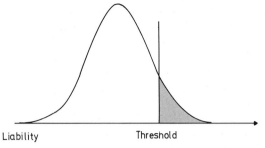

Fig. 6.14. Multifactorial inheritance in combination with a threshold effect – the simplest situation. The disease liability (for definition see text) shows a normal distribution; individuals to the *right* of the threshold are affected with the disease

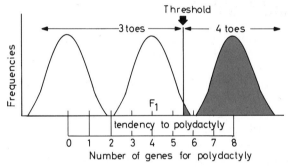

Fig. 6.15. Multifactorial inheritance in combination with a threshold: presence of an extra toe in guinea pigs. Two parental stocks, one showing three toes and the other four toes. A fraction of the F_1 hybrids shows four toes. Genetic analysis showed eight genes to be responsible for this trait. The number of animals showing this extra toe depends on the number of "plus" genes. (From Wright 1931 [173])

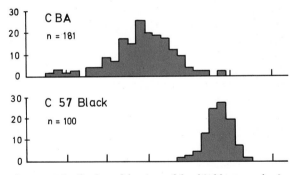

Fig. 6.16. Distribution of the sizes of the third lower molar in two inbred mouse strains, CBA *(above)* and C57 black *(below)*. (From Grüneberg 1952 [56])

most cases only the discontinuous phase is observed; the individual is either affected or unaffected. To ascertain the type of familial pattern that can be expected with threshold characteristics, a theoretical model is examined below.

6.1.2.2 Simple Theoretical Model

Note the model set out in Sect. 6.1.1.3: two gene pairs A, a; B, b with equal and additive contributions to the phenotype and gene frequencies $p_1 = p_2 = q_1 = q_2 = 0.5$. This genotype is assumed to determine a liability, leading to a manifest disease when three or four plus alleles (A or B) are present (Fig. 6.17). The relative numbers of affected and unaffected children from the mating types, plus × plus, plus × minus, and minus × minus are seen in Fig. 6.18.

These expectations are remarkably similar to those of a simple autosomal-dominant mode of inheritance: expectations for mating plus × plus are almost identical when a certain number of homozygotes among the plus parents are assumed. For matings plus × minus the expectation is somewhat but certainly not very much lower with the additive model. Still, regular dominance with full penetrance in the heterozygote is always clearly distinguishable, especially when more than two generations of a family can be investigated. With incomplete penetrance, however, the problem of discrimination from multifactorial inheritance with a threshold becomes very difficult; some sibships must now be expected in which both parents are unaffected, and the segregation ratio is lower than 0.5. Here, however, a comparison between sibships from plus × minus and minus × minus matings may help. In the multifactorial model a lower ratio of affected is expected when both parents are unaffected than when one parent is affected, whereas with simple autosomal-dominance and incomplete penetrance segregation ratios should be identical under such circumstances. This argument could be challenged by the assertion that penetrance may be influenced by the genetic background. The problem then becomes largely semantic; it is obvious from the outset that the assumption of equal contributions of all genes to the phenotypes is an oversimplification. However, if their contribution were unequal, at what degree of contribution of one locus to phenotypic variability should we start talking about a "major gene"? However, the stronger the contribution of single genes, the better are the prospects of identifying such genes by linkage analysis.

The case examined above is a very special one. The following criteria for multifactorial inheritance and against the simple diallelic mode of inheritance follow intuitively from the simple, special model set out above but could be derived in a more stringent way [738].

6.1.2.3 How Should the Model Be Used for Analysis of Data? [157]

These theoretical results must be used with caution in the analysis of actual data. As mentioned repeatedly, the multifactorial model is an abstraction and presents an oversimplified picture of the way in which multiple gene loci cooperate to create a liability. In addition, the data normally available are limited and therefore tend to have high sampling variances.

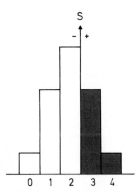

Fig. 6.17. Multifactorial inheritance of two gene pairs A, a, B, b in combination with a threshold: phenotypes – (□) and + (■) depending on the number of genes A, B in a random mating population. Gene frequencies A = B = a = b = 0.5. Five phenotypes (0, 1, 2, 3, 4) are possible (also see Fig. 6.12)

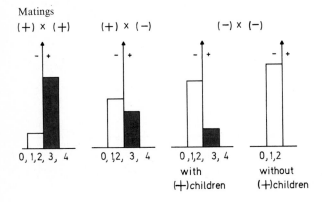

Fig. 6.18. Relative frequency of children + and − from four different mating types in the genetic model described in Fig. 6.17

Qualitative (or Semiquantitative) Criteria for Multifactorial Inheritance. Four such criteria can be derived:

1. The twin criterion: if concordance in MZ twins is more than four times higher than in DZ twins, a multifactorial model is more adequate than a simple diallelic model (Table 6.8). The opposite does not apply: a concordance ratio of less than 4 does not necessarily rule out the multifactorial hypothesis.
2. Segregation ratio of affected and unaffected sibs in matings plus × minus and minus × minus: if affected sibs are more than 2.5 times more frequent in matings with one affected parent than in matings with two unaffected parents, the multifactorial model should be preferred. Here, again, a ratio of less than 2.5 does not exclude multifactorial inheritance.
3. Sex ratio of affected persons: many anomalies for which multifactorial inheritance should be considered show a sex difference in incidence. In most cases only a small part of this sex difference is related directly to the sex chromosomes; most is due to physiological differences between the sexes. It is therefore reasonable to assume that the genotypic liability shows the same distribution in both sexes, but that the threshold is different. Consequently an affected person belonging to the sex with lower incidence has on average a higher personal liability than affected individuals of the other sex. This higher liability manifest itself as a higher frequency of affected relatives. The sex with the lower incidence should have a higher proportion of affected relatives, when the same degrees of relationship are compared. This argument was first put forward by Carter [20] and is sometimes called the Carter effect. Carter demonstrated it in pyloric stenosis, an anomaly in the newborn in which the thickening of the pyloric muscle prevents release of stomach contents into the duodenum. While this defect is much more frequent in infant boys than in girls, there is a higher incidence among relatives of affected girls than among relatives of affected boys (Table 6.9 and Fig. 6.19).

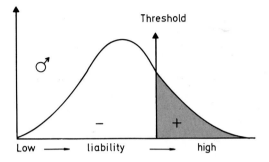

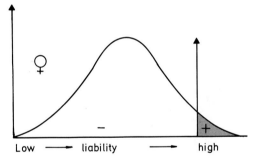

Fig. 6.19. A multifactorial condition may be more common in one sex than in the other. Pyloric stenosis, for example, is more common in males than in females. The genetic liability, on the other hand, can be assumed to be identical in both sexes. The position of the threshold differs. As a consequence the average affected male expresses the trait with a lower genetic liability than the average affected female. Therefore the incidence of this condition among relatives of male probands is expected to be lower than among relatives of affected female probands who carry more of the predisposing genes than affected males. This phenenomen is sometimes called the "Carter effect" [20]

Table 6.8. Twin concordance with various modes of inheritance

	MZ twins	DZ twins
Autosomal-dominant inheritance[a]	100%	50%
Autosomal-recessive inheritance[a]	100%	25%
Multifactorial inheritance with environmental influences[b]	~40%–60%	~4%–8%

[a] Theoretical expectations,
[b] Empirical findings.

Table 6.9. Pyloric stenosis: frequency among the close relatives of male and female probands (from Fuhrmann and Vogel 1983 [46], adapted from [20])

	Brother	Sister	Son	Daughter	Nephew	Niece	Male cousin	Female cousin
Males (n = 281)	5/230 2.17%	5/242 2.07%	19/296 6.42%	7/274 2.55%	5/231 2.16%	1/213 0.47%	6/1061 0.57%	3/1043 0.29%
Females (n = 149)	11/101 10.89%	9/101 8.91%	14/61 22.95%	7/62 11.48%	4/60 6.67%	1/78 1.28%	6/745 0.81%	2/694 0.29%

4. Consanguinity: the models examined above assume random mating. With consanguinity, however, the distribution of the liability in the population has a higher variance:

$$V_F = V_O \times (1 + F) \qquad (6.5)$$

Here, F = the inbreeding coefficient, V_F = variance among all progeny from matings with inbreeding coefficient F, V_O = variance with random mating (Fig. 6.20). Figure 6.21 shows the increased incidence among children from first-cousin matings (F = 1/16) compared with the random mating population. The much larger increase observed with monogenic, autosomal-recessive inheritance is given for comparison. In most cases, however, autoso-mal-dominant inheritance with reduced penetrance rather than autosomal-recessive inheritance is the obvious alternative to the multifactorial model. Therefore a moderate increase of the condition with inbreeding is an additional argument favoring a multifactorial model over an autosomal-dominant model, provided that admixture of families with a rare, autosomal-recessive type can be excluded.

Quantitative Criteria. It is not entirely satisfactory to use only semiquantitative criteria for a genetic model; methods for quantitative comparison are thus needed. Such methods have been developed.
The incidences of the character in question are determined among the different degrees of relatives of the

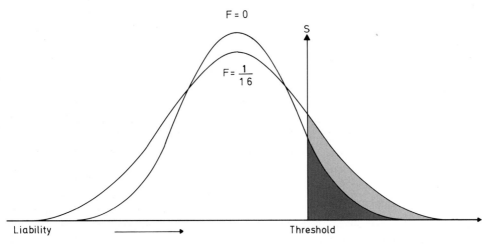

Fig. 6.20. Distribution of genetic liability with random mating and with $F = \frac{1}{16}$ (first-cousin marriage). The areas at the *right* of the threshold indicate the increase in frequency of a threshold character. *S*, Threshold

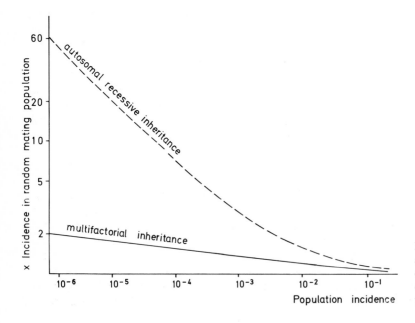

Fig. 6.21. Increased incidence of autosomal-recessive and multifactorial characters among children from first-cousin matings compared with population incidence

probands. Then, the joint probability of all these incidence figures together is compared with its theoretical expectation from the multifactorial model, on the one hand, and the diallelic model, on the other. Within the diallelic model hypotheses can be specified, such as autosomal-dominant inheritance with various degrees of penetrance or autosomal-recessive inheritance. Another alternative often considered is a polygenic background in combination with a major gene; i.e., a "mixed model." Appendix 3 provides many more details, a table comprising a number of available computer programs, and information on how to obtain access to these programs.

Comparison of Genetic Models. The principle of such a comparison can be described as follows: If the incidence of a trait in the population is known, each genetic model leads to a certain distribution of affected individuals among various categories of relatives of the probands. If the mode of inheritance is autosomal-dominant, for example, about the same proportion of affected persons is often found among parents as among sibs. With autosomal recessive inheritance, on the other hand, affected persons occur mainly among sibs but much more rarely among parents or children. With polygenic inheritance and a threshold, the proportion of affected sibs is often higher if one parent is affected than if both parents are unaffected; this difference does not exist, or is much smaller, if a dominant gene with incomplete penetrance is involved. Expectations for various degrees of relationship with the proband can be calculated for various genetic hypotheses and can then be compared with empirical observations. The functions derived for the various genetic hypotheses are called likelihood functions; for the polygenic model they permit estimation of the two important parameters: incidence of the trait in the population and heritability (h^2; see above). From these estimates, expected frequencies of the trait among various types of relatives can be derived, and compared with the actually observed frequencies.

When these likelihood functions for various genetic models have been established, they can be compared to ascertain which of them best fits the data set. In practice the ratio of likelihood functions of two models, or its logarithm, is often used; this principle is explained in greater detail in the context of lod scores used for linkage analysis in Sect. 5.1.2.

Let us assume that two models have been compared: polygenic inheritance in combination with a threshold and an autosomal dominant mode of inheritance with incomplete penetrance. Four outcomes are possible:

1. There is no difference between the empirical data and either model; no model can be excluded.

2. Only the model of dominant inheritance of a single gene can be excluded.
3. Only the polygenic model can be excluded.
4. Both models can be excluded.

6.1.2.4 If the Statistical Analysis Gives No Clear Answer, How Should We Decide?

The above discussion indicates that compatibility of a set of data with a genetic model does not mean that this model is correct. A completely different model could fit the same data equally well. As shown above and in Appendix 3, there is considerable overlap in expectations between the special models chosen here as examples for a diallelic single locus and a polygenic model, especially if the condition studied is common. It is a general rule that scientific hypotheses can be refuted when the observations do not fit but cannot be accepted until all other possible and plausible hypotheses have been excluded. However, the human geneticist working with anomalies that have simple modes of inheritance tends to forget this rule, as the relationship between observation and genetic hypothesis is quite straightforward in the usual cases of simple modes of inheritance.

How should one proceed when the statistical data do not permit a decision in favor of either of these hypotheses? The most obvious answer would be to leave the problem open. However, the tendency of family data to evade clear interpretation is a challenge; furthermore, the diseases involved are frequent and practically important so that more thorough research into their causes is needed. Hence some guidance for further studies and genetic counseling may be desirable.

The hypothesis of a major gene has many advantages for research strategy. In earlier times, however, when only the analytical way from phenotype to protein to gene was available, search for a major biological cause for the disease often led to disappointment. In such situations, identification of genes by a linkage approach, followed by identification of gene action and its deficiencies by positional cloning, aroused new hope. This stretegy is pursued in psychiatric genetics, for example, schizophrenia research (Chap. 16). It must be admitted, however, that the results have not been too impressive so far.

The genetic hypothesis of multifactorial inheritance is more cautious and conservative; adopting it as a preliminary description of the data, we remain aware that it represents analysis at the level most remote from the gene action: the black box must still be opened up. In thinking about strategies, we are not guided in one direction by an overpowering genetic hypothesis but remain open-minded toward various possibilities. If pursuing one of them should indeed lead to discovery of major gene action, we would be overjoyed, as this would bring our analysis down to a more genetic or biochemical plane. However, if the attempt does not succeed, we are still open to considering how a smaller deviation in a physiological parameter – which may be present even in only a fraction of our probands – could interact with other small deviations to cause a truly multifactorial disease liability.

Therefore, if one cannot be reasonably certain that single gene action applies by a clearcut genetic or biochemical criterion or both, acceptance of the more general multifactorial model is the wiser decision. However, in many cases a major gene has not really been excluded. This may have consequences for our attitude toward genetic risks due to mutagenic agents (Sect. 11.1): adopting a multifactorial model without reservation may lead us to underestimate genetic threats. To avoid this error, some experiences from genetic research in experimental mammals deserve consideration.

6.1.2.5 Radiation-Induced Dominant Skeleton Mutations in the Mouse: Major Gene Mutations that Would Not Be Discovered in Humans

Experimental work with mammals has shown how major gene action may be hidden in the phenotypic variability of the organism. Such major genes may be exposed by suitable breeding studies or by phenotypic analysis of induced mutations. One example is discussed here which is also important for risk assessment of mutation induction in humans [128].

Genetic damage due to dominant mutations induced by a mutagenic agent, for example, radiation, can be assessed by comparing first-generation descendants from treated and untreated animals, but for many characteristics it is difficult to distinguish between the effects of newly occurring genetic damage and the variation existing within a strain. Such studies have been performed mainly for cataracts or skeletal anomalies. These defects are especially interesting for the question of multifactorial versus single-gene inheritance. Many of the induced mutations would not have been identified from their phenotypic effects as determined by single-gene mutations or small deletions had they not been the results of a mutation experiment; for example, they showed low penetrance, very variable expressivity, and in the next generation segregation ratios which were much lower than those expected with a simple dominant mode of inheritance. These experiments indicate that even in

situations in which the phenotype appears to point to multifactorial inheritance the influence of major genes may still be involved.

However, these results should be generalized to human mutations only with caution: radiation-induced mutations are often deletions, while "spontaneous" mutations are predominantly single base substitutions that may cause less impressive clinical signs.

6.1.2.6 Isolation of Specific Genetic Types with Simple Diallelic Modes of Inheritance Using Additional, Phenotypic Criteria

It is often possible to define specific Mendelian subtypes of diseases from a large heterogeneous group of patients. The combination of careful clinical analysis and laboratory studies with genetic analysis may often be successful in isolating genetic from nongenetic entities. Early success was obtained in cases of mental retardation [112], deafness, and blindness. With the development of improved nosology coupled with careful clinical observation, many patients with mental retardation previously thought to be unclassifiable can now be categorized into specific entities. The various types of X-linked mental retardation, especially the common type with fragile-X, are a good example for a very common condition [150].

Discussion of other aspects of X-linked mental retardation with Fra X chromosome:

- Sect. 9.3:   Mutation rate
- Sect. 15.2.1.2:   Types of mental retardation
- Sect. 18.2; 18.1: Genetic counseling, prenatal diagnosis

Studies on blind and deaf children in residential institutions have also been successful. In deafness and blindness about 50 % of all residential cases have been shown to be genetic in origin. Practically all cases were of Mendelian rather than multifactorial origin. Many different types were found.

It is almost a general rule that among the multifactorial diseases, rare (and not so rare) Mendelian variants can often be distinguished. Thus, X-linked deficiency of hypoxanthine guanine phosphoribosyl transferase causes 1 % of cases of gout. Some cases of hypertension are caused by the rare inherited pheochromocytoma. Cancer of the esophagus may be rarely caused by a gene producing keratomas of the palms and soles at the same time (Fig. 6.22).

There are a number of syndromes in which cancer occurs (see Chap. 10) as part of a more comprehensive pleiotropic pattern. Occasionally families show dominant inheritance of a more or

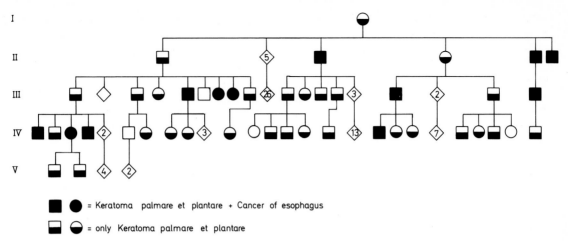

Fig. 6.22. Cancer of the esophagus as an additional symptom in patients with a special, autosomal-dominant type of kera-

toma palmare et plantare. (From Howell-Evans et al. 1958 [64])

less common cancer. Here early onset and multiple occurrence of cancers helps to delimit these major gene manifestations from the usual cancer type. Cancer of the breast, with an early age at onset, often occurs as a Mendelian dominant trait and is found in about 0.5 % of the female population (Chap. 10). In the pedigree in Fig. 6.22 the ages of the cancer patients are 34, 37, 38, 43, 44, 45, 46, 52, and 63 years, all but the last being very unusual for cancer of the esophagus. In dermatology, many benign and malignant tumors are observed as isolated cases as well as in families. Here, as in other conditions, single tumors in one patient favor a nongenetic origin whereas multiple tumors tend to be inherited and frequently show an autosomal-dominant mode of inheritance (see [53]).

6.1.2.7 How Can an Apparently Multifactorial Condition Be Analyzed Further, When Special Types with Simple Modes of Inheritance Cannot Be Isolated?

A Complex Functional Defect Is Caused by a Combination of Small Aberrations. The additive model used for a somewhat more quantitative understanding of multifactorial inheritance is an oversimplistic abstraction. In reality the variability is not unidimensional, and a variety of various genetically determined physiological influences may cooperate to induce a certain condition. It should be possible to isolate some of these influences.

In two series of children with strabismus examined by one investigator [121], the figures in Table 6.23 were found for parents and for siblings resulting from different mating types. Of 12 monozygotic twin pairs, 11 were concordant, whereas only 7 of 27 dizygotic pairs showed concordance. The data point strongly toward multifactorial inheritance. Incomplete dominance cannot be excluded but would require additional influence by the genetic background.

Table 6.10. Frequency of manifest strabismus (+) among sibs of children with strabismus. (From Richter 1967 [121]; Vogel and Krüger 1967 [157])

Mating type of parents	Number of propositi	Number of sibs	Manifest strabismus in sibs
Series of 697 patients (4–7 years old)			
+×+	24	33	11 (33.3 %)
+×−	288	301	95 (30.6 %)
−×−	385	478	98 (20.5 %)
Series of 136 school children (12 years old)			
+×+	6	6	3 (50.0 %)
+×−	61	120	52 (43.3 %)
−×−	69	82	2 (29.3 %)
Population frequency of strabismus: 3 %–4 %			

Richter's twin series on strabismus

	Concordant	Discordant	Total
MZ twins	11	1	12
DZ twins	7	20	27
Total	18	21	39

It is known that strabismus is the end result of a number of minor physiological aberrations. Each of these alone can be overcome to achieve normal eye muscle coordination. When several of these aberrations are combined, the regulatory capacity of the visual system decompensates, and squinting results. Such aberrations occur more frequently among close relatives of the proband. In the pedigree in Fig. 6.23 three patients squint; two parents show isolated heterophoria (slight motor weakness). One parent had an isolated anomaly of refraction, another shows heterophoria. The eyes of one parent were completely normal. The conclusions from this study –

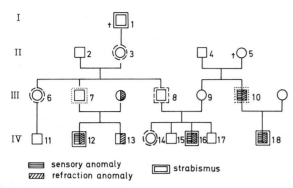

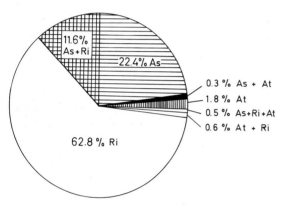

Fig. 6.23. Strabismus in three members of a family. Other relatives show different, minor anomalies. *Dotted and stippled lines,* various borderline results. Sensory anomalies observed in such pedigrees include, for example, amblyopia and imperfect binocular vision. (From Richter 1967 [121])

Fig. 6.24. Relative frequencies of probands manifesting one, two or even three atopic diseases. *As,* Asthma; *Ri,* rhinitis; *At,* atopic dermatitis. (Data from Zurich, Switzerland; Schnyder 1960)

that strabismus is a multifactorial condition, and that some physiological factors can be isolated – have been confirmed later and were extended in a different population [61].

An attempt to analyze the genetic susceptibility to congenital dislocation of the hip was successful in demonstrating presumably polygenic factors affecting the concavity of the acetabulum together with a possibly monogenic factor affecting joint laxity [28].

Family investigations that consist of careful examination of family members for related and associated abnormalities may help in understanding the relative importance of elements which, when combined with one another, lead to a complex functional defect. This is possible even if single gene action cannot be identified.

A Multifactorial System Comprises a General Disposition that May Lead to a Group of Related Diseases; Specific Dispositions Influencing the Clinical Manifes-

tation Pattern. The group of "atopic diseases" comprises atopic dermatitis, bronchial asthma, and hay fever. Figure 6.24 shows the relative frequencies of probands manifesting one, two, or three atopies in the population of Zurich [127].

Family investigations are most compatible with multifactorial inheritance. A further question can be posed: is the influence of the genes on the liability to atopic diseases only one-dimensional and quantitative, or are there other genes influencing the organ specificity of disease manifestation?

If the liability shows a one-dimensional distribution, skin atopies (dermatitis) and respiratory atopies (asthma, hay fever) should occur in the same ratio among relatives of probands with either skin atopies or respiratory atopies. On the other hand, if organ-specific factors are involved, a certain accumulation of similar atopies among the probands' relatives should appear.

Figure 6.25 illustrates this comparison; among first-degree relatives of asthmatics, respiratory atopies are much more frequent, whereas among relatives of dermatitis patients, atopic dermatitis prevails. Thus, within the multifactorial genetic system determining the genetic liability to atopic diseases there are factors increasing the liability in general that act side by side with others influencing special organ manifestations.

Such an analysis is more satisfying than mere attempts to fit overall incidence figures with expectations derived from genetic models that are oversimplified from the outset. Despite this improvement, genetic analysis remains at the biometric level, remote from gene action. "Breaking open the black box" is now on the way. It has been shown, for example, that clinical ragweed pollinosis (hay fever) is influenced by interaction between two gene loci, one of which regulates basic IgE production while the other acts on IgE production in reaction to the allergen. The latter is identical with or closely linked to the HLA-A2 allele. Genetic influences at other levels of the immune response are very well possible [117]. A probable selective advantage of atopic genotypes under more primitive living conditions is discussed in Sect. 12.2.1.8. For many other genetic aspects see [94].

6.2 Genetic Polymorphism and Disease

6.2.1 New Research Strategy

Various strategies have been tried for achieving deeper insight into the pathophysiological mechanisms of multifactorial diseases. The comparison of disease

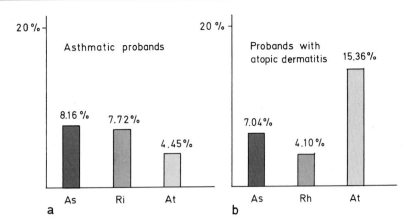

Fig. 6.25 a, b. Frequency of atopic dermatitis and respiratory atopies in relatives of probands with asthma (**a**), with atopic dermatitis (**b**). *As,* Asthma; *Ri,* rhinitis; *At,* atopic dermatitis. (Data from Dorn and Schwarz, in [4a]; Vol. IV, 1964)

phenotypes in families and analysis of the results using biometrical methods has led to disappointment because it is difficult to find criteria for distinguishing the various models. A closer look at phenotypes in probands and their relatives has met with limited but encouraging success in some instances, as shown above using atopic diseases and strabismus as examples. In principle, linkage studies using DNA markers – identification of "major genes" – and analyzing their mechanism of action "bottom-up" is a more generally applicable strategy (Sect. 3.1.3.9). However, the results have not been very encouraging so far. Therefore it may be a promising alternative strategy to try a similar "reverse" approach but not to start at the gene-DNA level but at the gene product–biochemical level. The multifactorial model assumes the cooperation of many genes; some of these are expected to be frequent. Often a disease phenotype makes it difficult to suspect which aspects might be caused by special genes. Sometimes, however, pathophysiological or epidemiological evidence suggests analysis at an intermediary level. A high blood cholesterol concentration, for example, increases the risk of coronary heart disease. Therefore it is a good strategy to study genetic determination of blood cholesterol; it is closer to the level of gene action and may therefore give clearer results (see Sect. 6.4.2.2). One group of genetically determined traits that have been shown to influence genetic disease susceptibilities at intermediary levels are genetic polymorphisms. Work on such polymorphisms has succeeded in uncovering and analyzing genetic variability at many gene loci which determine cell surface antigen structures, enzyme proteins, and serum proteins with many different – and in many cases unknown – functions. Hence it is not farfetched to examine whether some of these polymorphisms are parts of the multifactorial system influencing disease liabilities.

Harris and Hopkinson [58] showed that at least one-third of structural genes determining blood enzymes

exist as polymorphisms. Thus not every "normal" person has the same gene product, and variant proteins and enzymes are commonly found. It is not very meaningful to pick polymorphisms at random and study their associations with certain diseases. Chances for serendipitous findings are too low. Rather, those polymorphisms should be selected for which a pathophysiological relationship with the disease in question may be inferred. However, initial associations between genetic polymorphisms and diseases were not discovered in this "rational" way.

6.2.2 Disease Association of the Blood Groups

6.2.2.1 AB0 Blood Groups

Shortly after the AB0 blood groups were discovered associations with certain diseases were suspected. During the 1920s the first phase in examination of these associations reached its culmination. At that time almost every common disease was thought by some authors to be associated with the blood groups. Most of these studies, however, were carried out with insufficient numbers and inadequate methods. Results proved largely contradictory. Therefore, the reaction of most scientists during the following years was understandable: In their – basically justified – criticism, they threw out the "baby with the bath water," and for a long time the blood groups were regarded as not being associated with disease.

Wrong Hypothesis Leads to an Important Discovery. The first hint of blood group disease associations was the discovery of serological mother-child incompatibility in the Rh system. A short time thereafter associations with common diseases were discovered.

In 1953 Aird et al. [1] described an association between group A and cancer of the stomach. Stocks had shown in 1950 that mortality in stomach cancer is higher on the average in northern English than in southern English cities.

Aird et al. proposed that the difference is genetically determined. At that time, mouse strains with high and low cancer frequencies had been described. In their search for a possible genetic parameter they came across an analogy with AB0 distribution. In northern England blood group 0 is more frequent while A is more common in the south. Their working hypothesis was that group 0 is associated with stomach cancer, causing its higher incidence in the north. To examine this hypothesis they collected cancer cases in various English and Scottish cities and compared their AB0 distribution with that of carefully selected control samples, generally patients treated in the same hospitals for a number of different complaints.

Table 6.11 shows the result. Contrary to the working hypothesis, a significant association with group A and not 0 was found. Hence the higher cancer mortality in the north could not be caused by the higher frequency of group 0 in this area. This study triggered a flood of investigations on blood group associations of disease.

Statistical Standard Method. [172] Before the most important results are presented, the standard statistical method used for their analysis must be explained. Incidence of two characters (or groups of characters, for example, A versus 0 or A + B + AB versus 0) are compared in two samples, the patient sample and the control sample. The ratio:

$$x = \frac{A(\text{Pat}) \times 0(\text{Con})}{0(\text{Pat}) \times A(\text{Con})}; \quad y = \ln x \quad (6.6)$$

has the expected value of 1 when the ratio A/0 is identical in the two samples, i.e., if there is no association [A(Pat) is the absolute number of A individuals in the patients' sample and A(Con) the absolute number of A individuals among controls]. Otherwise the ratio x is higher or lower than 1. This ratio x is commonly called the "relative incidence." In our example its meaning is that the incidence of stomach cancer in persons of group A is x times higher than that in persons of group 0. The deviation of x from 1 can be tested for significance as follows:

$$V = \frac{1}{w} = \frac{1}{A(\text{Pat})} + \frac{1}{0(\text{Pat})} + \frac{1}{A(\text{Con})} + \frac{1}{0(\text{Con})}$$

χ^2 of deviation $= y^2 w$ (1 df).

A number of estimates of x can be combined to a common estimate as follows:

$$Y = \frac{\sum wy}{\sum w} \quad Y = \ln X;$$

χ^2 of deviation: $Y^2 \sum w$ (1 df)

χ^2 of heterogeneity: $\sum w y^2 - Y^2 \sum w$

(df = number of single comparisons − 1).

Standard deviation of Y: $\sigma = \dfrac{1}{\sqrt{\sum w}}$

A Flood of Investigations and Their Results [102, 154]. Within a period of about 15 years a number of associations for common diseases were detected (Table 6.12). Apart from stomach cancer, which was examined in at least 101 different samples, several

Table 6.11. Differences in relative frequencies of groups A and 0 in patients with cancer of the stomach and controls in the United Kingdom (from Aird et al. 1953 [1])

Cancer cases		Controls		Percentages		Difference	χ^2_1
A	0	A	0	A(Pat)	A(Con)		
1442	1424	1269	1581	50.31	44.53	+5.78	19.2 ($p = 10^{-6}$)

The samples of patients and controls consisted of subsamples from seven places; all showed the same deviation. There was no statistical heterogeneity.

other malignant neoplasias showed that the risk of being affected was somewhat higher for patients with blood group A. A tendency in this direction was also found in a number of nonmalignant diseases such as rheumatic diseases, pernicious anemia, and – with a stronger effect – in thrombotic and thromboembolic diseases. An association with group 0, on the other hand, was shown for gastric and duodenal ulcers. These data indicate that if blood group A has a small but significant disadvantage in predisposing its carriers to some diseases comprising some of the most frequent causes of death in our society, a higher frequency of blood group 0 in healthy, elderly persons than in the general population might be expected. This expectation was confirmed in one study of persons aged over 75, who at the time of examination were still in reasonable good health [69].

Possible Biases. Large-scale statistical investigations of this type are subject to certain biases:

a) Selection of appropriate controls. Human populations are not uniform in their blood group distributions. In spite of conformity of the data to Hardy-Weinberg proportions there may be hidden stratification of subgroups differing in gene frequencies at the gene loci under examination. If the controls happen to be taken consistently from a subgroup other than the patients, a spurious association might result. For example, if blood group 0 confers especially good health on its bearers, the incidence of 0 in samples of blood donors who are more likely to be a particularly healthy subgroup of the population might be high.

b) Publication only of positive results. It is an understandable wish of research workers to have their work rewarded by "positive" results, i.e., in this case, by discovering an association. It is therefore possible that only those who find a "significant" association (possibly by chance) publish their data. Others, who have been less "lucky,"

Table 6.12. Significant associations between blood groups and noninfectious disease

Diagnosis	No. of series	No. of Patients	Controls	Comparison	Relative incidence Mean	χ^2	(df = 1)	Heterogeneity χ^2	df	p
Neoplasias of the intestinal tract										
Cancer, stomach	101	55434	1852288	A : 0	1.22	386	0.0027	2178	100	0.0027
Cancer of colon and rectum	17	7435	183286	A : 0	1.11	14	0.0027	10	16	N.S.
Malignant tumors of salivary glands	2	285	12968	A : 0	1.64	13	0.0027	15	1	0.0027
Cancer, pancreas	13	817	108408	A : 0	1.24	8	0.01	15	12	N.S.
Cancer, mouth and pharynx	2	757	41098	A : 0	1.25	8	0.01	1	1	N.S.
Other neoplasias										
Cancer, cervix	19	11927	197577	A : 0	1.13	31	0.0027	29	18	0.05
Cancer, corpus uteri	14	2598	160602	A : 0	1.15	10	0.0027	18	13	N.S.
Cancer, ovary	17	2326	243914	A : 0	1.28	27	0.0027	19	15	N.S.
Cancer, breast	24	9503	355281	A : 0	1.08	11	0.0027	31	23	N.S.
Multiple primary cancers	2	433	7823	A : 0	1.43	10	0.0027	1	1	N.S.
Nonmalignant tumors										
Nonmalignant salivary tumors	2	581	12968	A : 0	2.02	55	0.0027	23	1	0.01
Other internal diseases										
Duodenal ulcers	44	26039	407518	0 : A	1.35	395	0.0027	81	43	0.01
				0 : A + B + AB	1.33	447	0.0027	84	43	0.01
Gastric ulcers	41	22052	448354	0 : A	1.17	96	0.0027	79	40	0.01
				0 : A + B + AB	1.18	125	0.0027	63	40	0.05
Duodenal and gastric ulcers	6	957	120544	0 : A	1.53	27	0.0027	19	5	0.01
				0 : A + B + AB	1.36	19	0.0027	24	5	0.01
Ulcer, without differentiation between stomach and duodenum	11	4199	88239	0 : A	1.15	15	0.0027	9	10	N.S.
				0 : A + B + AB	1.18	25	0.0027	17	10	N.S.
Bleeding ulcers (gastric and duodenal)	2	1869	28325	0 : A	1.46	53	0.0027	0	1	N.S.
				0 : A + B + AB	1.51	73	0.0027	1	1	N.S.
Rheumatic diseases	17	6589	179385	A : 0	1.23	50	0.0027	29	16	0.01
				A + B + AB : 0	1.23	57	0.0027	33	16	N.S.
Pernicious anemia	13	2077	119989	A : 0	1.25	20	0.0027	12	12	N.S.
Diabetes mellitus	20	15778	612819	A : 0	1.07	14	0.0027	38	19	0.01
				A + B + AB : 0	1.07	16	0.0027	42	19	0.01
Ischemic heart disease	12	2763	218727	A : 0	1.18	14	0.0027	23	11	0.05
				A + B + AB : 0	1.17	15	0.0027	29	11	0.01
Cholecystitis and cholelithiasis	10	5950	112928	A : 0	1.17	26	0.0027	10	9	N.S.
Eosinophilia	3	730	1096	A : 0	2.38	46	0.0027	1	2	N.S.
				A + B + AB : 0	2.13	49	0.0027	1	2	N.S.
Thromboembolic disease	5	1026	287246	A : 0	1.61	46	0.0027	23	4	0.0027
				A + B + AB : 0	1.60	49	0.0027	23	4	0.0027

leave their data unpublished. The accumulation of positive publications leads to a spurious association.

It has been shown that these biases cannot be responsible for the associations found [154]. It is comforting that data collected for many other diseases yield consistently negative results despite the fact that patient and control samples are collected identically and evaluated so that such biases should be at work. Congenital malformations are an example. The entire group that includes congenital heart disease, harelip and cleft palate, malformation of kidney and urinary tract, hydrocephalus, and others showed no blood group association even though 4762 patients and 156 716 controls were examined [154].

Failure to Find a Mechanism. In the early years of studies of blood groups and diseases many authors speculated on biological reasons for these associations. The role of blood group substances, for example, in the secretions of stomach and duodenum, was thought to be responsible. Indeed, tumors of organs containing much of this substance, for example, salivary glands or ovaries, showed especially strong associations. A more general hypothesis related these associations to stronger immune responses in group 0 carriers than in group A persons. This hypothesis led to studies in population genetics (Sect. 12.2.1.8) but was not pursued experimentally for the common diseases such as duodenal ulcer and cancer of the stomach. For example, it is now known that the bacterium *Helicobacter pylori* is almost always found in the pyloric region of ulcer patients and probably plays a major role in the causation of the disease. New experimental evidence seems to corroborate this conclusion (Sect. 12.2.1.8). Better understanding will possibly have to await more thorough immunological knowledge about the role of the cell surface, specifically its glycoproteins, in interaction with other cells and with environmental influences. The fact that attempts to demonstrate a convincing mechanism for the associations seemed to have failed may have contributed to a disappointment among research workers. In recent years the flood of work on blood group associations has dried up almost completely.

It has also become clear that the total contribution of the AB0 genes to the genetic etiology of these diseases is probably small, as shown in an analysis of the contribution of blood group 0 to peptic ulcer [643]. These studies therefore, while statistically clearcut in the case of peptic ulcer, cancer of the stomach, and some other conditions, have not aided immediately in further understanding of the genetic and environmental etiology of these diseases.

6.2.2.2 The Kell System

Kell System Mutations, Acanthocytosis, and Chronic Granulomatous Disease. Apart from disease associations of common blood groups, some examples of hereditary anomalies due to rare or modifying genes of "blood group" genes are known. Section 4.1.6 describes modifying genes in the AB0 system; to our knowledge, they have not been studied for possible significance for the health of their bearers. However, the Kell blood group system provides examples of direct associations between rare blood group alleles and disease. They are especially interesting because the Kell substance is known to be involved in the structure of cell membranes.

Various alleles at the autosomal Kell locus exist in populations of European origin; there are two alleles K, k, the rarer allele K has a frequency of 0.05.

An allele of the Kell system (Js) is found in 14%–20% of African-Americans but is extremely rare in other populations and therefore constitutes an excellent marker gene for African origin. Hemolytic disease of the newborn rarely is caused by anti-Kell antibodies; however, if it occurs, the basic mechanism is similar to that of Rh hemolytic disease.

In addition to the autosomal locus for the Kell antigen, an X-linked locus that codes for a precursor substance of Kell known as Kx has been identified. All normal persons have Kx antigenic activity on both red and white cells. Some individuals are homozygous for a "silent" allele (Ko) of the autosomal Kell locus. In such cases none of the usual Kell antigens but a strong Kx reaction can be detected. This finding is compatible with the interpretation that unconverted Kx material specified by the X-linked locus is the only Kell-related antigen in homozygous carriers of the Ko or silent allele. Such persons are clinically and hematologically normal. Mutations at the Kx locus have been identified and may lead to phenotypic expression in red or white cells or in both types of cells.

The McLeod phenotype of red cells [168] is caused by an X-linked point mutation or deletion causing absence of the Kx substance. Absence of the Kx antigen from the red cells causes a membrane abnormality associated with acanthocytosis ("spiny" red cells) and increased red cell destruction. The severity of hemolysis may range from compensated blood destruction to severe hemolytic anemia. Abetalipoproteinemia [15] – the usual cause of acanthocytosis (145 950) – is not present. The red cell anomalies clearly are caused by absence of Kx since red cells lacking all Kell antigens except Kx (Ko) are morphologically normal. As expected, the fully expressed McLeod phenotype is seen only in males.

Mothers of males with the McLeod or mutant Kx phenotype are expected to be heterozygotes for both the normal Kx and the mutant Kx allele. The principle of X inactivation (Sect. 2.2.3.3) postulates that such women would have a mosaic population comprising cells expressing both the normal Kx and the mutant Kx allele. In fact, red cell populations consisting of normal and abnormal cells have been observed. Such mosaicism has been demonstrated by both immunological and morphological techniques since the Kx-negative cells were acanthocytes. The abnormal cells were outnumbered by the normal cells due to the shortened red cell survival of the Kx-negative cells. The X-linked Kx antigen presumably codes for a membrane protein. Mutations at that locus produce pathological membrane alterations leading to the morphological red cell abnormalities and hemolysis.

The separate genes for Kx and the X-linked type of chronic granulomatous disease are very closely linked on Xp21. Several cases of contiguous deletions have been reported where both genes were absent, with the resulting phenotypic pattern of both conditions.

6.2.3 The HLA System and Disease [142, 149]

As explained above (Sect. 5.2.5), the major histocompatibility complex (MHC) on chromosome 6 is homologous with the H2 complex of the mouse [2]. Immunization of inbred mouse strains with a variety of apparently unrelated antigens (synthetic polypeptides, serum proteins, cell surface antigens) induces high levels of antibodies in some strains and low levels (or no response) in others. The quantity of antibodies induced is controlled by immune response loci, which are part of the H2 complex. Linkage with the H2 complex has since also been demonstrated in mice for susceptibility to virus-induced cancer and infection by the lymphocytic choriomeningitis virus and for genetic factors predisposing to autoimmune thyroiditis [124].

In thyroiditis the connection has been established between a particular transplantation antigen type, a specific antithyroglobulin antibody response, and severity of the disease. This was an important step toward elucidation of the mechanism of the association. (It is interesting that in humans an association between autoimmune thyroiditis and the antigen HLA-B8 has been described.)

Results suggest the hypothesis that in humans immune response genes may also be closely linked with the HLA genes. As linkage disequilibrium had been demonstrated for the well-defined genes of the human HLA complex, it might also be assumed for these as yet hypothetical immune response genes. Therefore disease associations with HLA types were anticipated.

The first anomaly to be examined in man was Hodgkin disease, a malignant disorder of the lymphatic system. A collaborative study on 523 patients showed a significant association with HLA 1 [3]. Examinations of other malignant diseases such as acute lymphatic and myelogenous leukemia produced conflicting results. Much more striking associations, however, were found for a number of nonmalignant diseases, such as ankylosing spondylitis, gluten-sensitive enteropathy (sprue), Reiter disease, multiple sclerosis, and psoriasis (Table 6.13). In some cases the extent of the associations is enormous. In ankylosing spondylitis a relative incidence (Sect. 6.2.2.1) of 87 was found, i.e., the disease is 87 times more likely in bearers of the HLA type B27 than in the rest of the population.

While almost all patients with ankylosing spondylitis have the HLA-B27 type, most carriers of HLA-B27 do not have ankylosing spondylitis. The frequency of HLA-B27 in the white population of the United States is about 5%, while the frequency of ankylosing spondylitis is about 1/2000. However, careful clinical and radiological investigations have shown minor symptoms, and X-ray findings suggestive of mild ankylosing spondylitis were found in about 20% of B27 carriers. Narcolepsy is the other disease for which an unusually high association has been found – in this case with DR2 [82]. However, narcolepsy is a rare disease, and the DR2 type is common.

Several differences stand out when HLA associations are compared with AB0 associations. Most HLA associations are much stronger. For AB0 blood groups and disease, most relative incidences were less than twice the incidence of controls, whereas for the HLA associations the frequencies were usually much higher. The data suggest that the contribution of HLA types to the multifactorial systems causing these diseases is more substantial than that of AB0 genes to diseases found to be associated with them. Therefore attempts to elucidate mechanisms have a better chance of success.

Probable Mechanisms of HLA–Disease Associations. The function of the HLA system in the immune response is that of recognition of antigens and presenting them (Sect. 7.4) to T (class I) or B (class II) lymphocytes. The diseases for which HLA associations have been described are therefore most likely caused by specific differences in recognition or presentation of certain HLA antigens.

Serological differentiation at the HLA locus has been complemented by analysis of fine structure using molecular methods (see Fig. 5.13). This led to further subdivision of gene loci and to better specification of the sites responsible for disease associations. In insulin-dependent (type 1) diabetes, where DR3 and DR4 associations are common, for example, certain subgroups of DR3 and DR4 specifications show no association with diabetes at all, whereas about 95% of the DR4$^+$ diabetics have the subtype DQβ3.2 [6]. In lupus erythematosus DQα2 and DQβ2 appear to be important determinants [32].

A disease association of this sort may have two main causes:

1. It may be caused by a biological function of the allele (or haplotype) itself.
2. It may be due to another gene that is closely linked with the allele or haplotype under study.

When studies on HLA–disease associations began, many scientists believed the latter alternative. They postulated that HLA genes "hitchhike" with closely

Table 6.13. Associations between HLA and some diseases (see Svejgaard et al. 1983 [143]; Albert 1993 [2]; de Vries and van Rood 1992 [30]) with additions

Condition	HLA	Relative risk	Remarks and specified associations
Hodgkin disease	A1	1.4	Amiel [3], the first association,
Idiopathic hemochromatosis	A3	8.2	now somewhat doubtful
	B14	4.7	
Behçet disease	B5	6.3	
Congenital adrenal hyperplasia	B47	15.4	
Ankylosing spondylitis	B27	87.4	
Reiter disease	B27	37.0	
Acute anterior uveitis	B27	10.4	
Subacute thyroiditis	B35	13.7	
Psoriasis vulgaris	Cw6	13.3	
Dermatitis herpetiformis	DR3	15.4	DQA1*501, DQB1*201
Celiac disease	DR3	10.8	
	DR7		
Sicca syndrome	DR3	9.7	
Idiopathic Addison disease	DR3	6.3	
Graves disease	DR3	3.7	
Insulin-dependent diabetes	DR3	3.3	
	DR4	6.4	DQβ
	DR2	0.2	
Myasthenia gravis	DR3	2.5	
	B8	2.7	
Systemic lupus erythematosus (SLE)	DR3	5.8	DQα2, DQβ2a
Idiopathic membranous nephropathy	DR3	12.0	
Multiple sclerosis	DR2	4.1	
Optic neuritis	DR2	2.4	
Goodpasture syndrome	DR2	15.9	
Rheumatoid arthritis	DR4	4.2	
Pemphigus (Jews)	DR4	14.4	
IgA nephropathy	DR4	4.0	
Hydralazine-induced SLE	DR4	5.6	
Hashimoto thyroiditis	DR5	3.2	
Pernicious anemia	DR5	5.4	
Juvenile rheumatoid arthritis	DR5	5.2	DR8, A2, DQA1*04*05
		3.6	DPB1*201
Narcolepsy	DR2	49	
IgA deficiency	DR3		
Scleroderma	DR5		

linked immune response genes or even unrelated genes. For example, the gene for hemachromatosis or iron storage disease is closely linked but not part of the HLA-A gene. However, it is unlikely that the increase in iron absorption characteristic of this disease is biologically related to immune function. Therefore this gene is presumably located by chance in this area. In other instances associations of a disease with HLA-A or HLA-B alleles were discovered which turned out to be due to DR alleles in linkage disequilibrium with HLA-A or HLA-B alleles. At present it is believed that the majority of HLA associations are caused by the HLA specificities themselves or their molecular subtypes. It can be generalized that most

such associations are found in disorders characterized as autoimmune diseases where the organism produces immune factors directed against self.

The fact that a disease shows association with HLA may give hints as to its pathogenesis. In multiple sclerosis, for example, immunological investigations guided by the HLA associations revealed a specifically decreased cellular immunity to measles and other paramyxoviruses [42, 54].

Multiple sclerosis and narcolepsy both are associated with the DR2 allele. Could narcolepsy be caused by the failure of successful interaction with the same or similar "slow" virus that has been suggested as the cause of multiple sclerosis? In diabetes, the observation that insulin-dependent diabetes shows associations with HLA antigens, but mature onset diabetes does not, indicates different basic causes and suggests that the etiology is different and that in insulin-dependent diabetes an autoimmune or virus etiology is involved. Much lower concordance for identical twins with this type of diabetes than with mature-onset diabetes as well as higher personal risk for family members of patients with mature-onset diabetes lead to the same conclusion [145] (see sect. 6.4.2.1).

Some evidence even points to two types of juvenile diabetes mellitus: one type, which is associated with HLA-D3 (and -B8) seems to be caused by an autoimmune mechanism, and another one, associated with D4 in which the patients are often insulin antibody responders [125, 126]. Hence, analysis of HLA associations may help to refine nosological classification of a group of diseases and to detect genetic heterogeneity.

Linkage and Association. The two phenomena of linkage and association should be carefully distinguished. Linkage refers to two genes being located on the same chromosome within detectable distance of each other. The term association is often used when a higher frequency of a given gene is found in a certain disease or trait. Association does *not* imply that the gene involved in the disease and the marker gene are located on the same chromosome. Confusion regarding these concepts can arise in studies of HLA gene frequencies in disease [98]. The HLA complex is located on chromosome 6. The gene for 21 hydroxylase deficiency in the homozygous state produces congenital adrenal hyperplasia (209 100) and is closely linked to the HLA complex. Similarly, the gene of one type of dominant spinocerebellar ataxia (164 400) is linked to the HLA complex [73]. Data on the iron storage disease hemochromatosis (141 600) (inherited as an autosomal-recessive trait with occasional heterozygote manifestations) show that the gene for this disease also is linked to the HLA complex. These diseases are monogenic conditions, and their respective loci are situated within measurable distance of the HLA complex on the 6th chromosome. There is no reason to believe that the genes for these diseases and HLA genes are physiologically related. The diseases with HLA associations (Table 6.13) are not simple monogenic conditions but have usually been shown to be of multifactorial origin. In one group of HLA-related diseases (chronic hepatitis, myasthenia gravis, rheumatoid arthritis, Addison disease, thyrotoxicosis, juvenile diabetes, celiac disease, and multiple sclerosis), associations with DR specificities of the HLA system have been demonstrated. The common factor in these conditions is the presence of autoantibodies, and they have been

classified as autoimmune diseases or at least as immunologically associated diseases. Familial aggregation without clear Mendelian pattern has been detected when appropriate studies were carried out.

Organ-specific autoantibodies lead to the manifestations of the various autoimmune diseases. Whether additional genes on other chromosomes are implicated is unknown. The production of autoantibodies probably requires appropriate environmental stimuli, which often are viral in origin, as suggested for diabetes, hepatitis, and multiple sclerosis. Persons with certain DR specificities are presumably more susceptible to the formation of such autoantibodies than those who lack such genes; this explains at least part of the genetic susceptibility to autoimmune diseases. DR genes of the HLA complex and autoimmune diseases show association but no demonstrated linkage.

Studies on HLA types have also been performed in some infectious diseases such as tuberculosis and leprosy. Unfortunately, this problem has not been studied extensively, considering the great importance of such diseases for human evolution in the past (see Sect. 12.2.1.6). This is probably due to professional specialization. Immunologists and other medical scientists conducting HLA research seldom have interests in problems of natural selection and evolution. Nevertheless, some investigations on leprosy have been carried out [30]. While comparisons between leprosy patients and controls have failed to provide consistent results, family studies point to an association with certain HLA-DR types in this disease.

In view of the important role of T cells in HIV infections, HLA association with AIDS-related infections is of great interest. The few available studies reveal no consistent pattern. One study, however, produced an interesting result [136]: 32 hemophilia patients were treated with a single batch of factor VIII contaminated with HIV. Eighteen became antibody positive and showed a continuous decline in circulating T cells over 4 years. While only 26% of hemophilia patients had the HLA haplotype A1 B8 DR3, 8 of 11 patients with this haplotype became HIV positive. Of the 18 seropositive hemophilia patients 11 had clinical signs of AIDS – among them all the 8 with the A1 B8 DR3 haplotype. These data suggest an association of this haplotype with a relatively rapid course of AIDS.

Since some of the diseases which show HLA associations lead to death, or contribute to shortening of life span, it should be expected that their absence has a positive effect on life expectancy. Indeed, a study on 102 persons over the age of 90 years in Okinawa, Japan, revealed a strong association with HLA-DRw9 and HLA-Dr1, with relative incidences of 5.2 and 13.3, respectively [144]. In Japanese, HLA-DRw9 is associated with autoimmune disease, whereas HLA-DR1 may provide relative protection against infections. Studies on ABO groups in elderly persons have shown a similar protection effect of group O (see above Sect. 6.2.2.1).

6.2.4 α_1-Antitrypsin Polymorphism and Disease [39]

α_1-Antitrypsin Polymorphism. The ABO blood groups show a weak association with a great number of diseases, but no convincing explanation for these associations existed up to recently. HLA antigens show a much stronger association with a great number of diseases. The pathogenic mechanism has not been worked out in detail, but there are reasonable and experimentally testable hypotheses. The α_1-antitrypsin polymorphism is associated in adults mainly with one disease, chronic obstructive pulmonary disease. Its mechanism has been elucidated to some extent.

The antiproteolytic activity of human serum was detected in 1897 by Camus and Gley and by Hahn. Landsteiner in 1900 showed this activity to be located in the albumin fraction. Of the six antiproteases identified in the human serum α_1-antitrypsin and α_2-macroglobulin have the highest concentrations. Both are able to inhibit a great number of proteases, including

thrombin. Antiproteolytic activity is measured by hydrolysis of artificial substrates by trypsin in the presence of the serum to be tested. There is a close correlation between immunologically measured concentration and activity. The concentration increases quickly, for example, with bacterial infection, after injection of typhoid vaccine, and during pregnancy. Synthesis occurs in the liver. Interindividual differences were first observed in 1963 [83]. A simple recessive mode of inheritance was proposed for low α_1-antitrypsin levels. Electrophoretic techniques and isoelectric focusing demonstrated many phenotypes (Fig. 6.26). The genetic basis of this heterogeneity is a series of multiple alleles. The locus was named PI (protein inhibitor), the various alleles PI^M, PI^Z, etc., and the phenotypes M/M, M/Z, etc. In all populations examined so far the PI^M alleles are the most frequent, with a common gene frequency of 0.9 or higher. Other, rarer alleles are designated by letters. The position of these letters in the alphabet gives an approximation of electrophoretic mobility. Nucleotide sequence analysis of cloned cDNA has demonstrated, for example, that the Z variant is caused by a single nucleotide substitution. The gene is located on 14q31-32. It has 10.2 kb, and five exons. Two variants, Z and S, are especially important because the α_1-antitrypsin level is appreciably reduced. In another, very rare allele, PI^-, no activity of the enzyme at all is found in the homozygote (null allele). The heterozygote PI^M/PI^- has an M phenotype with only 50% of the normal concentration. Intravenous injection of typhoid vaccine and diethylstilbestrol leads to a 100% increase in activity of subjects with the MM type. Heterozygotes of the MZ type show a moderate increase, whereas in homozygotes of the ZZ type hardly any increase is seen.

Association with Chronic Obstructive Pulmonary Disease. Eriksson 1965 [37] reported 33 homozygotes of the ZZ type; at least 23 had definite symptoms of chronic obstructive pulmonary disease (COPD). On the basis of his family data Eriksson estimated ob-

Table 6.14. α_1-Antitrypsin concentration of different α_1-antitrypsin phenotypes (adapted from Kueppers 1975; see Kueppers [79])

Phenotype	Percentage of normal (MM = 100%)
M/M	100
S/S	50–60
Z/Z	10–15
M/S	70–90
M/Z	55–65

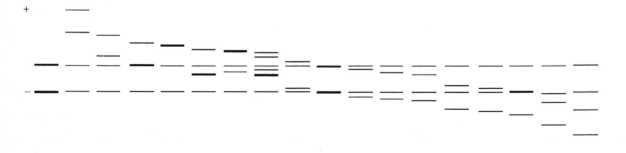

M1   BM1   CM1   DM1   EM1   GM1   IM1   FM1   LM1   M1M3 M1M4 M1M2 M1N   M1P   M1V   M1S   M1X   M1Z

Fig. 6.26. Schematic representation of isoelectric focusing patterns of the more common α-antitrypsin types (gradient between pH 3.5 and 5). (From Kueppers 1992 [79])

structive pulmonary disease to be at least 15 times more frequent among these homozygotes than in the general population. This observation has been confirmed by many investigators in a great number of patients. In one group of 295 patients with this diagnosis 20 PI^Z homozygotes were detected, whereas not even one would have been expected on the basis of the gene frequency. Usually the first symptoms are recognized during the 3rd or 4th decade of life; patients with the usual variety of COPD usually become affected during their 50s or 60s. Loss of lung tissue and blood vessels in the lower lobes is characteristic. Interestingly, even before this defect was discovered, a special group of patients with these symptoms had been delineated from the majority of patients with COPD. The question whether heterozygotes also have a higher frequency of COPD has been much discussed. It has been claimed that heterozygotes have an approximately threefold risk of developing COPD compared with M homozygotes [39]. Pulmonary function tests in heterozygotes show more frequent abnormalities. Their symptoms tend to appear later in life.

Among homozygotes only 70%–80% develop obstructive emphysema, and in heterozygotes the frequency is much lower [79]. Environmental factors appear to influence manifestation: α_1-antitrypsin also inhibits proteolytic enzymes released by granulocytes or macrophages. It is therefore likely that these enzymes, which are normally released during inflammatory process, are insufficiently inactivated.

When a patient is exposed to recurrent bronchial irritation, such as that caused by smoking or frequent infections, these enzymes cause digestive damage of the lungs. Tobacco smoking enhances the danger of bronchial infections and hastens the progress of the disease [39] (Fig. 6.27). Therefore, Z/Z homozygotes and heterozygotes should be advised to refrain from smoking and to avoid jobs leading to bronchial irritation. Bronchitis should be treated early and intensively. "If we are able to change the smoking habits of a PI^Z individual, this may add 15 years to his life" [39].

Another disease associated with low α_1-antitrypsin values in homozygotes is childhood cirrhosis of the liver. This association is firmly established but is seen less frequently than chronic pulmonary disease. Cryptogenic cirrhosis in adults also appears to be more common in ZZ homozygotes.

Significance of the New Research Strategy. The α_1-antitrypsin polymorphism is remarkable since the mechanism of lung damage can be explained. This result is in striking contrast to the associations described in the AB0 system and even in the HLA system. The situation is simpler than in these systems. One genotype is involved not in all but in a high fraction of cases, and the disease is rather specific and was therefore identified even before the biochemical cause became known. The Z/Z state can be regarded as a recessive disease with "incomplete penetrance." There are probably many other such recessive dis-

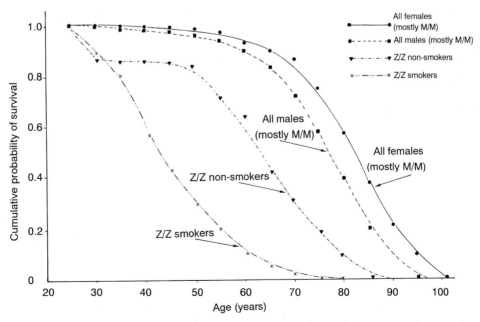

Fig. 6.27. Mortality and survival of homozygotes for the Z variant of α_1-antitrypsin in smokers and nonsmokers. The strong influence of smoking is obvious. (From Kueppers 1992 [79])

eases still buried within the large groups of frequent multifactorial conditions, either because the phenotype is difficult to define, or because all parameters for delineation have not yet been considered.

The contribution of the PI^Z allele to the disease liability of heterozygotes seems to be quite similar to the ABo and HLA situations; COPD occurs only in a minority of cases; it resembles the more usual type of obstructive pulmonary disease, and the influence of additional environmental and genetic factors appears to be strong.

The example of this disease association shows the more recently developed indirect research strategy at its best. A genetic polymorphism is analyzed at the genetic level. The product of a single gene is identified using biochemical methods. The consequences of the polymorphism for gene action are sought. In this case the quantity of protein is defective, and enzyme activity is low so that the organism does not adequately respond to environmental challenges such as infections. This specific functional weakness leads to disease, especially with frequent exposure to certain environmental influences such as bronchial irritation.

It is likely that such a research strategy will become useful for genetic analysis in other areas, where genotype-phenotype relationships are so complex that a clear analysis from the phenotype down to the genes by Mendelian methods is barred, and methods of quantitative genetics must be relied upon. Or, to put it in a slightly different way: This strategy will help in applying Mendelian approaches to situations in which Galtonian techniques now are the only applicable methods.

Disease Associations of Other Polymorphisms [102]. Apart from the three systems described above, disease associations have been examined (and in some cases made likely) for a number of other polymorphisms, for example, blood group systems [154]; haptoglobins, and tasting of phenylthiourea. Some of these are described in the context of population genetics (Chap. 12). Especially interesting are the associations that have been reported between the apolipoprotein E polymorphism and atherosclerosis as well as Alzheimer's disease [152, 153]. In variants of the third complement component and some diseases, the C3^F allele seems to be associated with rheumatoid arthritis [14, 41], and hepatitis [41]. The C6 complement deficiency has been found in about half the patients with meningococcal meningitis. If confirmed, these associations are of considerable interest because plausible hypotheses about biological mechanisms and genetic consequences can be considered.

Since the goal of such studies is stepwise elucidation of the influence of potentially many genes on a disease liability, and since we have every reason to assume that genes do not simply act in an additive way but interact with each other, it is meaningful to look specifically for such interactions. For example, HLA types and immunoglobulins interact in the normal immune response. Therefore it has been an obvious step to look for interactions of HLA types with immunoglobulin allotypes (GM; see Sect. 12.1.6). In such a study it has been shown that the haplotype HLA-B*,DR3 entails an about 15 times increased risk for chronic hepatitis. However, additional presence of the type GM (a+X+) raises this risk to 40 times [23].

6.3 Nature-Nurture Concept: Twin Method

In discussing methods of quantitative genetics the use of twin data for quantitative assessment of the degree of genetic determination was mentioned repeatedly. Indeed, twin investigations have played a major role in the history of human genetics. At one time, the twin method was even regarded as the "royal road" to genetic analysis in humans. In one field of major importance, behavior genetics, many of our conclusions are still based on twin data. Therefore critical assessment of the twin method, its advantages and limitations, is important.

6.3.1 Historical Remarks

Introduction of the twin method is generally accredited to Galton (1876) [47], who also adopted the alternative terms "nature" and "nurture". Here Galton was – either consciously or unconsciously – following the terminological lead of Shakespeare, whose character Prospero in *The Tempest* says of Caliban: "A devil, a born devil, on whose nature nurture can never stick." It is doubtful, however, whether Galton recognized the essence of the problem. Very probably he did not know that there are two types of twins, monozygotic (MZ) and dizygotic (DZ). This distinction was discovered only a short time previously by C. Dareste, who reported it to the Société d'Anthropologie in 1874 (see [162]). Most likely, Galton intuitively had the right idea but no clear concept about the usefulness of the twin method. Such intuitive insights are frequently observed in the history of science when a new paradigm, emerges and do not detract from the usefulness of the paradigm.

Poll (1914) [113] was the next to use the twin method in assessing genetic determination. However, he failed since methods for distinguishing MZ from DZ twins were lacking. After Poll single twin pairs continued to be described, but zygosity diagnosis remained ambiguous.

The method was given a sound foundation with the work of Siemens (1924) [131]. Siemens' achievement was threefold:

1. He showed that twin series of a size suitable for meaningful statistical evaluation can easily be found when schools are asked for help. This made the investigation of normal variability possible.
2. He developed a reliable method for zygosity diagnosis. Up to that time, research workers had tried in vain to achieve such a diagnosis using a single criterion. Siemens, however, showed that reliable separation of these groups is possible when a large number of criteria are examined. Each of these criteria may on average show only somewhat more similarity in MZ twins, but all of them together can separate the two groups with a high degree of reliability.
3. Siemens proposed to examine not only MZ but also DZ twins. DZ twins are no more similar genetically than are other siblings, having on the average 50 % of their genes in common by descent. However, as MZ twins, they are born at the same time and are exposed to similar environmental conditions.

6.3.2 Basic Concept

The twin method is founded on the biological peculiarity that MZ twins originate from division of one zygote. Therefore they must as a rule be identical genetically. A group of genetically identical individuals is called a clone. It follows that any phenotypic differences between MZ twins must be largely caused by environmental influences. Here environment is defined in the widest possible sense: anything that is not fixed genetically. However, MZ twins are alike only for characteristics under control of germ line genes. Somatic mutations do occur, particularly affecting the globulins of the antibody system. Thus, differences in gene products resulting from somatic mutations would be expected in MZ twins. However, we are not aware of a study of this phenomenon.

To recognize whether and to what degree a characteristic is determined genetically and to what degree its variability is modified by environmental influences, the degree of similarity between MZ twins must be measured. As DZ twins are thought to be influenced by the same environmental differences but to have only one-half of their genes in common by descent, they are used as suitable controls.

The following sections shown how this concept can be quantified and discuss its limitations.

6.3.3 Biology of Twinning

Dizygotic Twins. In most mammals (rodents, carnivores, some ungulates) every birth is multiple. In every ovulation the ovary discharges several ova, which may be fertilized by one sperm each. The marmoset monkey regularly gives birth to DZ twins. In higher ungulates, such as horses and cattle, and in the higher primates including humans, generally only one ovum is discharged at ovulation. There are occasional exceptions. If two oocytes are discharged at the same time and fertilized by different sperms, DZ twins result. In the same way, polyovulation occasionally leads to trizygotic triplets and quadrizygotic quadruplets. However, not all triplets, quadruplets, and quintuplets arise in this manner.

It follows that DZ twins do not necessarily have the same father. The two oocytes may be fertilized by sperm from different men with whom the mother has had intercourse around the time of ovulation. One such possible case from Nazi Austria is of interest [49].

The fraternal twins were 25 at the time of examination. The legal father was a Jew, the mother was not Jewish. At that time Austria was incorporated into Nazi Germany, and racial laws discriminating against Jews were introduced. Since there was interest in exonerating the children from the "blemish" of being half Jewish, the mother reported an extramarital relationship at the time the twins were conceived. The sexual partner was still available for testing. The AB0 and MN blood group tests (the only systems routinely available at that time) gave the following results:

Legal father	B, M
Alleged father	A, MN
Mother	O, M
Twin brother	B, M
Twin sister	A, MN

If one accepts these tests as accurate, the results are:

1. Exclusion of the legal father for the girl – first, as she had to have inherited the A allele from one of her parents, and, secondly, as her N allele was not accounted for.
2. Exclusion of the alleged father for the boy, as neither he nor the mother had the allele B.

Theoretically, a third man other than the alleged sexual partner could have been the father of both twins, but examination with additional anthropological methods gave suggestive evidence that the boy indeed resembled more the legal father and the daughter the alleged father. More recent DZ twin pairs with two fathers have been reported and were discovered in the course of a disputed paternity case. In one case one of the two fathers was black and the other white.

Anastomosis of blood vessels, which is quite normal between MZ twins during fetal development, may occur in exceptional cases in DZ twins as well. This may lead to mutual transfusion of blood stem cells since early embryos are immunologically tolerant to each other. The result are twins who are blood chimeras with two populations of genetically different blood cells [33, 108]. In cattle a vascular connection between DZ twins is the rule and leads to partial sexual transformation in the female partner of fraternal twin pairs. Such a masculinized calf is known as a freemartin.

Monozygotic Twins. Much more interesting is the formation of MZ twins. In a certain sense they may be said to be the most extreme state of duplication. Less extreme duplications, such as Siamese twins, or double-headed infants, are observed occasionally in humans. Many of these duplications are fatal.

Some unusual types of twins, however, have survived and become famous, for example, the "Siamese twins" Chang and Eng (Fig. 6.28), who were born in Thailand in 1811. At the age of 18 the twins went to the United States and made a living by displaying themselves in curiosity shows. Later, they married two sisters. Eng had 12, and Chang had 10 children. They settled in the Carolinas and grew tobacco. At age 61 Chang had a stroke and died 2 years later from bronchitis. Eng who had been healthy up to that moment survived his brother by only 2 h. Chang and Eng were connected by a tissue bridge about 10 cm wide reaching from the lower end of the sternum almost down to the navel. At postmortem examination this bridge was discovered to contain liver tissue connecting the two livers. Hence any attempt to separate the brothers surgically would hardly have been successful in 1872. Today even more extensive connections between such twins have been severed.

The factors inducing division at an early cleavage stage and giving rise to MZ development in humans are unknown. MZ twins, however, have been produced in experimental embryology – many decades ago in amphibians and more recently in mammals. Mirror-image similarity between human MZ twins has been discussed repeatedly. Since strong asymmetry can be experimentally produced in animals, true asymmetry in some MZ pairs would not be surprising.

Very rarely, twin pregnancies occur in which an oocyte and a polar body (see Sect. 2.1.2.4) are fertilized by different sperm. One such event led, in addition to birth of a normal child, to an acardiac monster. The abnormal twin had been produced by fertilization of the first polar body (diploid) by a separate sperm, as shown by chromosome heteromorphisms and HLA haplotypes [11].

Frequency of Twinning [17]. Table 6.15 shows incidence figures in various populations for MZ and DZ twin births. The proportion of MZ twins was calculated by Weinberg's difference method, which is based on the fact that MZ twins are always of the same sex.

Fig. 6.28. The Siamese twins Chang and Eng. (From Lotze, R.: Zwillinge. Oehringen: Ferd. Rau 1937)

Table 6.15. Incidence of DZ and MZ twin births per 10000 births (from Propping and Krüger 1976 [115])

Country	Time period	DZ	MZ
Spain	1951–1953	59	32
Portugal	1955–1956	56	36
France	1946–1951	71	37
Austria	1952–1956	75	34
Switzerland	1943–1948	81	36
West Germany	1950–1955	82	33
Sweden	1946–1955	86	32
Italy	1949–1955	86	37
England, Wales	1946–1955	89	36
USA whites	?	67	39
USA blacks (California)	1905–1959	110	39
USA Chinese	?	22	48
USA Japanese	?	21	46
Japan	1955–1962	24	40

Among DZ twins, on the other hand, one-half are of the same sex, the other half are oppositely sexed. Therefore:

Frequency of DZ twins = 2 × DZ of opposite sex
Frequency of MZ twins = Frequency of all twins − DZ twins

This method gives only approximate results since somewhat more boys are born than girls. Moreover, there is some inconclusive evidence that DZ twins of the same sex are more frequent than expected. This may be caused by the fact that the time elapsed between ovulation and cohabitation influences the sex ratio at fertilization. It is certain that these deviations are small, and Table 6.16 therefore presents a fairly good approximation of the truth.

The frequency of MZ births shows little variability among different populations. Frequencies of DZ births, on the other hand, differ, the highest DZ frequency being found among African blacks, with variability among tribes. The Yoruba in Nigeria have a twinning frequency of 4.5%; 93.3% of such twins are dizygous. Among African-Americas DZ twins are born more frequently than among American whites. In Europe the rate of dizygosity is about 8/1000 births. However, here higher as well as lower frequencies have been observed in some populations. On the Åland Islands the twinning rate was 15.2/1000 between 1900 and 1949. The lowest rates are found in Mongoloid populations, especially among Japanese.

The differences in DZ twinning rates among the main racial groups are maintained when the data are corrected for maternal age and birth order.

Factors Influencing Frequency of Twin Births: Maternal Age and Birth Order. The probability of a twin birth increases with maternal age. This increase affects exclusively DZ births, as recognized by Weinberg in 1901. Subsequent work confirmed a maternal age effect, showing that the DZ rate increases from 10% at puberty at a rate of 0.7%–0.8% a year up to the age of 35–39 years and drops afterward [78, 91, 115]. The reason for the maternal age effect is probably an increase in the gonadotropin level (FSH). This hormone has been shown to increase with maternal age, and it could easily cause an increased tendency to polyovulation. In the Yoruba, for example, women having had two twin births show the highest FSH levels and mothers of single-birth children the lowest FSH levels. The hypothesis is corroborated by the fact that women treated with gonadotropic hormones for sterility due to anovulatory cycles frequently have multiple births. Discontinuation of birth control pills does not influence the twinning rate [93]. The reduced DZ twinning rates during the last years of reproductive age may be due to the ovaries being unable to discharge more ova in spite of high FSH levels. The DZ twinning rate increases not only with maternal age but also with birth order.

Genetic Factors. At the beginning of this century Weinberg [164, 165] recognized that twin births show clustering in families. This familial aggregation is true for DZ twins only. After appropriate corrections

Table 6.16. Incidence (percentage) of congenital malformations in twins and singletons per 1000 births (from Propping and Vogel 1976 [116])

Source	Approximate sample size	Single-tons	Twins
Hendricks	~35 000	3.3	10.6
Stevenson et al.	421 000	12.7	14.4
Hay and Wehrung	10 200 000	5.8	6.2
Onyskowová et al.	240 000	13.2	26.4
Emanuel et al.	25 000	13.2	23.2

for maternal age, the probability of future DZ twin births for the mother is about four times the frequency of DZ twins in the population. The chance for her female relatives is also increased; for her sisters it equals her own chance. For male DZ twins and for fathers of DZ twins, on the other hand, the chance is not increased. The mode of inheritance seems to be multifactorial; FSH levels may well be the major genetically determined cause.

For MZ twins there are no indications of any genetically influenced variability. The recurrence probability for MZ mothers does not exceed the population average. Mothers of DZ twins are on average about 1–2 cm taller than mothers of either MZ twins or singletons [19].

Decrease in Twin Births in Industrialized Countries. In almost all industrialized countries there has been a decrease in the frequency of twinning since World War II, and the old rule of one twin birth to 80 single births no longer holds. In Germany during the 1970 s, for example, there was less than one twin birth for 100 single births. Figure 6.29 shows the decrease for Germany. This decline was already apparent in the period after World War I and, following a short peak in the late 1930 s, became very pronounced after 1945. The decrease has commonly been explained by a maternal age effect. The average number of pregnancies has decreased during this period, and most pregnancies have occurred at an age at which the chance for twins is lower. However, this explanation is insufficient and explains only a small part of the decrease. Considering known physiological and genetic data, the following hypothesis seems to be of interest [115].

Polyovulation is correlated with fecundity, i.e., the probability per cohabitation of conceiving a child. One common factor applying to both polyovulation and fecundity is the FSH level. In earlier years highly fecund women contributed more than their average share to the birth rate, thereby enhancing the number of DZ births. The number of children today is widely

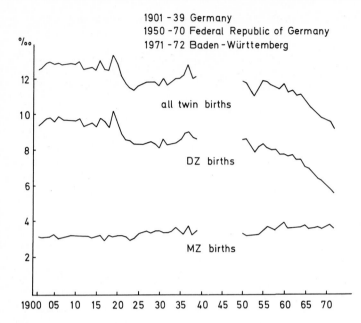

Fig. 6.29. Decrease in twin births in Germany during recent years. The decrease is due entirely to the decline in DZ twins. (From Propping and Krüger 1976 [115])

limited by parents using birth control, and the significance of biological fecundity for actual reproduction has decreased. Hence the number of DZ births has decreased as well. This hypothesis is supported by statistical data. In 1946, for example, twin frequency increased steeply in the United States. One year earlier many soldiers had come home, and birth control was presumably less extensively practiced. As Fig. 6.29 shows, Germany experienced an increase in twin births in the late 1930s. This was at the time of Nazi propaganda in favor of large families, which led to an appreciable increase in the birth rate. Considered from this standpoint, the racial gradient for twin frequencies (black–white–Mongoloid) could possibly be due to natural selection for fecundity. In Africa high infant mortality made it necessary to exploit the reproductive capacity of women to the extreme, whereas in Japan birth control had been practiced for centuries, probably diminishing the selective advantage of high fecundity.

In recent years the frequency of DZ twins (as well as other multiple births) seems to have increased again due to hormonal treatment. The number of twin pregnancies in the early phases of pregnancy appears higher than in later phases and at birth since an appreciable number die very early as shown by ultrasound studies [81].

Frequencies of Multiple Births of More than Two Children. Hellin's rule – i.e. that the frequency of twin births = t, triple births = t², etc. – holds true only very approximately. All combinations of mono-, di-, and trizygosity, etc. are possible. The famous Dionne quintuplets in Canada, for example, were monozygous.

6.3.4 Limitations of the Twin Method

Systematic Differences Between Twins and Nontwins. The purpose of twin studies to assess the role of genetic factors in human traits and diseases is to obtain results that apply not only to twins but to the whole population. Any twin study must pose the following question: do twins differ from nontwins in the trait under study? Any differences might impair the general validity of any conclusions drawn from a twin sample.

Comparison of several biologic characters has shown differences between twins and nontwins during embryonic development. Twins suffer from a higher frequency of abnormalities during pregnancy and at birth. Their lower birth weight can be attributed only partly to the shorter duration of gestation. The stillbirth rate and infant mortality in early life are considerably higher in multiple births than in single ones; in later years, twins run a higher risk than nontwins of becoming mentally retarded, presumably at least partly due to complications during pregnancy and at birth (see below). Even the mean IQ of both MZ and DZ twins is slightly lower than that of control populations.

Do nongenetic factors act differently upon MZ and DZ twins? Could this difference alter the probability of manifestation of the condition under investigation? This is important because the basic concept (Sect. 6.3.2) of the twin method assumes that MZ and DZ twins are exposed to identical prenatal and postnatal environmental factors. Birth weight may serve as a simple and measurable indicator. In an extensive survey Carney et al. (see [116]) compared the average birth weights of 572 individuals from twin pairs of identical sex, classified by sex, placentation, and zygosity. The mean birth weights calculated are presented in Table 6.17.

MZ twins of both sexes weigh less than DZ twins. The type of chorion and placenta has no effect on the mean birth weight of the surviving individuals. Therefore it appears likely that zygosity rather than placentation is responsible for the difference in birth weight.

Some observations on X inactivation in females indicate that the division of the zygote occurs after X inactivation (and is a fairly disruptive process). It therefore may happen that all cells in which a certain X-linked gene has been inactivated end up in one twin while the cells with active X chromosomes are found in the cotwin. This phenomenon leads to clinical expression of X-linked traits (such as Duchenne muscular dystrophy or color blindness) in only one member of a female twin pair who are heterozygotes for the X-linked trait. Two of the MZ Dionne quintuplets were colorblind! Moreover, the normal mosaic spots may be larger in twins than in singletons [106].

Comparison of individuals within monochorionic pairs (who are always monozygotic) demonstrates differences in birth weight of more than 1000 g. Such differences may be the result of arteriovenous anastomoses leading to a "transfusion syndrome," consisting of chronic malnutrition with reduction in cytoplasmic mass of parenchymal organs and markedly reduced hemoglobin and serum protein in the donor twin. Since more than 20 % of all MZ twins have only one chorion, the transfusion syndrome could account for considerable intrapair differences in birth weight that are not observed in DZ twins [7].

It follows directly that birth weight is not one of the characters for which meaningful use of twin data can be made to estimate, for example, heritability. Intrauterine development also influences other characteristics. Table 6.18 shows the overall frequencies of congenital malformations in twins and in single children as reported by several authors. Although frequencies of congenital malformation vary greatly in the five series, (presumably due to different definitions) on the whole these anomalies are more frequent in twins in every analysis published. This tendency is more obvious when particular types of malformation are considered. The risk among twins is enhanced at least for congenital heart disease, anencephalus, hydrocephalus, and cleft lip and palate. In all four instances the risk is higher in same-sexed twins than in those of opposite sex. This indicates that MZ may be affected more often than DZ twins. The "transfusion syndrome" could easily account for this difference. If this explanation is correct, malfor-

Table 6.17. Birth weights of twins. (From Carney et al. 1972, see [116])

Males 2659 g ($n = 304$)		Females 2547 g ($n = 268$)	
Dichorionic	Mono-chorionic	Dichorionic	Mono-chorionic
$n = 196$ 2703 g	$n = 108$ 2579 g	$n = 162$ 2577 g	$n = 106$ 2500 g
DZ $n = 160$ 2728 g MZ $n = 36$ 2595 g	MZ $n = 108$ 2579 g	DZ $n = 144$ 2601 g MZ $n = 18$ 2385 g	MZ $n = 106$ 2500 g

Table 6.18. Incidence of selected congenital malformations in twins and singletons per 1000 births (see [120])

Type of malformation	Source of data	Single-tons	Twins Total	Same sex	Different sex
Congenital heart disease	a	0.74	1.65	1.82	1.27
	b	2.8	6.3	–	–
	c	0.59	0.71	0.81	0.49
Anencephaly	a	0.92	1.24	1.52	0.64
	b	1.3	1.2	–	–
	c	0.23	0.37	0.45	0.22
Hydro-cephalus	a	0.61	0.72	0.91	0.32
	b	1.0	3.1	–	–
	c	0.30	0.40	0.45	0.31
Cleft lip and/or cleft palate	a	1.21	0.34	1.68	0.64
	b	0.8	0.4	–	–
	c	1.11	1.07	1.10	1.01

Sources: a, Stevenson et al. (1966); b, Edwards (1968); c, Hay and Wehrung (1970) (see [116]).

mations should be expected in only one of the twins. This expectation is in fact fulfilled. Sirenomely is a rare condition in which the lower body is underdeveloped, and the two legs are not separated. On the basis of data collected by Lenz (1973) [86], the incidence of this malformation can be estimated at about 1:1000 in MZ births and 1:60 000 in the general population. Figure 6.30 shows a MZ pair which is extremely discordant – very probably due to an anomaly of embryonic development in one twin.

Twin studies on congenital malformations using unselected case series show relatively low concordance rates in MZ twins (see Table 6.18). However, the twin method can produce only ambiguous results in regard to these malformations [116]. As in regard to all other conditions, the possible influence of intrauterine factors in twin pregnancies should be considered before a twin study is undertaken.

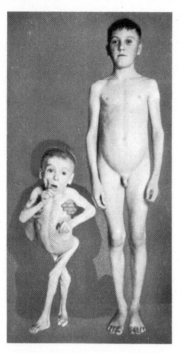

Fig. 6.30. MZ twins at the age of 10 years. Discordant dwarfism. The second born, dwarfed twin never learned to walk or talk. He showed an undiagnosed developmental disturbance of the skeleton and died a short time after examination. The twin brother showed normal stature but a bilateral coloboma of the iris, which was absent in the malformed twin. (From Grebe 1959 [50])

Peculiarities of the Twin Situation in Postnatal Life. Can twins be regarded as "normal" children? Can the results of measurements be extrapolated to the nontwin population? The following considerations need comment.

IQ results are lower in twins than in singletons especially in younger age groups, as determined in 1947 for all Scottish school children of age 11, including 794 single twin individuals, and for 95 237 French school children between the ages of 6 and 13, including 808 twin individuals. Zazzo (1960) recalculated these test results and found a mean IQ of 93 in the twins, compared to the population mean of 100. This difference in IQ scores can even be found at the end of the 2nd decade, as demonstrated in the study by Husén [66] on all male military conscripts (1948–1952) in Sweden. The IQ variability was also greater in twins, and the frequency of mental retardation was twice as high as in the general population.

The reasons for a lower IQ seem to be manifold. Premature birth with a higher risk of minor brain damage may be one factor; the larger burden on the family that must care for two young children at the same time could be another.

Twins form a social group. They depend less on exchange of information with the outside world because they have each other [158, 159]. Studies in central Europe have shown that they frequently develop a "private language." These findings probably explain why twins learn to speak later than other children. This process is more intensive in MZ than in DZ twins. MZ twins generally spend more time together; they frequently tend consciously toward uniformity, whereas DZ twins rather tend to stress the difference. The tendency toward uniformity, however, is stronger in female than in male MZ twins. *Protesting against identity* especially among male MZ twins occurs and may even lead to "twin hostility" [160]. One twin may more frequently follow the father and the other the mother, and the identifications may lead to conspicuous behavioral differences between the twins. Another phenomenon that is frequently observed is role differentiation: one is the spokesperson to the outside world and usually answers when both are addressed; the other may be the decision-maker for problems involving both twins. Or one may dominate and the other may be more submissive. This role differentiation occurs in both MZ and DZ twins but seems more frequent in MZ twins. This may lead to spurious discordance in MZ twins for behavioral traits.

Obviously these peculiarities of the twin situation must be considered, especially when personality traits such as "extraversion" or "neuroticism" are investigated. Still more important, these attitudes are influenced by changing cultural patterns. In earlier years the identification of twins with each other was usually encouraged, for example, by identical clothing or by sending them to the same school. Today, many educators recommend emphasizing differences. These special conditions of twin life influence mainly personality characteristics, making the twin method especially controversial in behavior genetics (Chap. 15, 16). In theory these difficulties could be prevented by examining twins who were separated early in life and reared apart. In practice, however, such twins are rare, and the very fact that an ideal study from a scientific viewpoint would require adoption into homes with quite different environments makes them exceptional.

These postnatal peculiarities of the twin situation may not be quite as biased for somatic diseases as for behavioral genetics. For a chronic infectious disease, such as tuberculosis or leprosy, for example, it may be important whether a twin has more contact with one of the parents if this parent is infectious for this disease. Fortunately, less biased results can be expected for multifactorially determined common diseases of adults.

6.3.5 Diagnosis of Zygosity

Every twin study requires a reliable method for zygosity diagnosis. Since Siemens [132] established the principle of polysymptomatic similarity diagnosis in 1924, this problem has largely been solved. Study of genetic markers such as blood polymorphism and DNA tests has made twin diagnosis independent of the personal judgment and experience of the investigator. Details are given in Appendix 5.

6.3.6 Application of the Twin Method to Alternatively Distributed Characteristics

The twin method serves three purposes in this area:

1. The difference in concordance between MZ and DZ twins can be used to determine whether genetic variability plays a role in a given disease.
2. Penetrance (i.e., the probability of manifestation of the disease) can be estimated.
3. The conditions of manifestation can be examined.

In earlier phases of twin research most work was centered around the first two aspects. Recently, however, the third has often been emphasized. Four approaches have been used to achieve these purposes [89]:

1. *Case Reports.* Descriptions of cases of concordant or discordant twin pairs, especially MZ pairs, are being published, often for the sake of their curiosity value. The scientific value of this approach lies in the possibility of a thorough analysis of discordant MZ pairs. In rare disorders case reports may provide the only available information on whether the conditions have any genetic basis. A single carefully analyzed discordant MZ pair for any disease shows that the condition cannot be exclusively genetically determined; it may point to epigenetic effects that modify gene expression.
2. *Accumulated Case Reports.* Even when a number of case reports are accumulated, the same limitations apply as for single case reports. Such a compilation of twin pairs is not representative, but systematic analysis of MZ twins discordant for a certain disease may provide valuable information on the environmental and other epigenetic conditions affecting manifestation. A relatively small series of discordant twins could produce much better information on risk factors influencing the manifestation of disease than would surveys of large populations in which the problem of adequate control samples might be insuperable.
3. *"Limited Representative" Sampling.* This method is the most frequent approach for obtaining large, unbiased series of twin pairs. Luxenburger [89] con-

sidered this approach to be of "limited representative value" because the sample is taken not from a defined region and time period but from a patient series. All twin individuals are ascertained within a population of affected patients of the disease under investigation. The co-twins are examined to determine whether they are affected as well. It is important, however, that *all* twins within the patient population be ascertained; otherwise concordant pairs usually have a higher chance of being included in the sample than do discordant ones. Successful ascertainment of all twins is achieved when the frequency of detected twins in the series equals that of the frequency of twins in the general population. The proportion of like-sexed and different-sexed twins and – after zygosity diagnosis – the proportion of MZ and DZ should agree with that in the general population. Practical application of this method would be much easier if the routine questionnaire for hospital patients were supplemented by the question: is the patient a twin?

4. *Unlimited Representative Sample.* Here every twin in a population is ascertained and examined regardless of whether he or she suffers from the condition under investigation. Usually complete birth registers over a span of several years must be screened. The number of individuals who must be examined is enormous in comparison to "limited representative sampling." If the condition has an incidence, say, of 0.5 % and the number of twin individuals is 1:50, with the "limited representative" approach 10 000 persons must be screened to find 200 twin individuals. For unlimited representative sampling, a population of 2 million must be screened. On the other hand, for some traits, especially mental illnesses, this approach – which has been undertaken in Denmark, Norway, Sweden, and Finland [24, 43, 76, 147] – has produced results showing interesting variations from those of investigations using limited representative sampling (Chap. 15, 16). In Budapest, Hungary, all twins born since 1970 are being registered [29].

A recent large twin sample is a modified unlimited representative sample. All male twins who were registered in the United States Armed Services during World War II are being sampled by a twin registry maintained by the National Research Council in Washington, D.C. A variety of investigations have been carried out with these data or are in progress.

6.3.7 An Example: Leprosy in India

A twin study on leprosy in India may serve as an example for application of the twin method [22]. Leprosy is caused by *Mycobacterium leprae* (Hansen's bacillus); however, not ev-

Table 6.19. Concordance in 102 twin pairs with leprosy, 62 MZ and 40 DZ twins (from Chakravartti and Vogel 1973 [22])

Sex	MZ pairs	DZ pairs
	Concordant	Concordant
Males	24, 60.0%	5, 22.7%
Females	13, 59.1%	1, 16.7%
Males + females	–	2, 16.7%
Total	37, 59.7%	8, 20.0%

$\chi^2 = 15.53; p \sim 0.003$.

Table 6.20. Concordance and discordance for type of leprosy in MZ and DZ leprous twins (from Chakravartti and Vogel 1973 [22])

	Concordant for type	Discordant for type	Sum total
MZ twins	32	5	37
DZ twins	6	2	8
Total	38	7	45

Only twin pairs concordant for leprosy are included.

eryone exposed to the bacillus becomes infected, and not everyone who becomes infected develops clinical symptoms. Furthermore, the infection produces varying manifestations depending on the immunological state of the organism. One patient may show only depigmented and anesthetic macules (tuberculoid leprosy) while another have diffuse infiltrations (lepromatous leprosy).

The apparent differences in susceptibility may have many causes, but available information makes genetic influence probable. Two such causes are the clustering of the same type of leprosy among near relatives and racial differences in relative frequencies of different leprosy types. In whites and blacks the tuberculoid type is more frequent while in Orientals the lepromatous leprosy prevails. Other studies with smaller patient series, although not entirely satisfactory on methodological grounds, have suggested genetic influences in leprosy as well [22].

The twin study discussed here was carried out in endemic leprosy areas in West Bengal and Andhra Pradesh, India, where at least 2%–4% of the total population is known to be affected. A determined effort was made to ascertain all twins suffering from leprosy within these districts. First, all those in permanent and mobile leprosy clinics were asked (a) are you a twin, and (b) do you have any twin pairs in your family or in your village? The investigation was then extended to village surveys. A total of 102 twin pairs with at least one twin affected by leprosy were examined.

Table 6.19 shows that the concordance rate in MZ is significantly higher than in DZ. In addition, in many of the affected

MZ pairs both the course of the disease and the extent of the lesions show striking similarity. The intrapair differences at the age of onset of all concordant (MZ and DZ) twin pairs tends to be smaller in MZ than in DZ twins.

As leprosy may show different clinical manifestations, this disease allows analysis of concordance regarding the particular type of leprosy present. Table 6.20 shows that 5 of the 37 MZ pairs concordant for leprosy were discordant as to leprosy type, with one being tuberculoid and the other lepromatous. These pairs offered the opportunity for an additional finding. It has been suggested in the past that a simple mode of inheritance is responsible for lepromatous leprosy. A possible candidate is impaired function of T lymphocytes. However, the discovery of five MZ twins concordant for leprosy but discordant for the type renders this possibility unlikely. Thus twin studies, apart from providing evidence on how genetic variability influences susceptibility, may help to illuminate more specific hypotheses on pathogenesis.

A possible bias must be considered. A determined effort was made to ascertain all twin individuals with leprosy in the regions under investigation. However, as the relative frequencies show, ascertainment of MZ twins was much more complete than that of DZ twins. (In the Indian population the ratio MZ/DZ is very similar to that in Europe.) The reason for this bias is found in certain living conditions in this part of India. Most of the individuals examined lived in rural areas with a high illiteracy rate; most of the local inhabitants did not even know their exact age. Therefore a twin pair was usually recognized as such only when the similarity could not be overlooked. Under these circumstances DZ twins frequently would not even be noticed. Sometimes the sibs themselves did not realize they were twins.

How might incomplete ascertainment of DZ twins have influenced the result? As ascertainment of concordant pairs is generally favored, the differences between MZ and DZ pairs may be underestimated. More important, however, is the question: were the MZ pairs completely ascertained? Probably not. Some pairs may have successfully hidden their disease; many of the pairs lived in beggars' colonies and received no treatment; twins from the upper castes may have escaped the survey due to treatment by private physicians; some patients may have purposely given wrong answers to avoid the social stigma of leprosy for their afflicted twins or relatives. As most of these factors apply to both MZ and DZ twins, regardless of concordance or discordance, a stronger influence on concordance ratio in MZ due to proband selection is unlikely. Nevertheless, the concordance figure may still be too high.

As to environmental risk factors, analysis of discordant MZ pairs confirmed that continuous and intensive contact with infectious cases is most important. Therefore concordance figures of the same high order of magnitude can be expected only in those areas in which leprosy is highly endemic. Infection is almost ubiquitous, and contracting the disease depends mainly on inherited susceptibility. In populations with lower incidence of leprosy, infections depend more on chance. Therefore, low concordance rates between MZ twins may be expected.

Surveys on tuberculosis show similar results [161]. Earlier studies gave concordance rates in the order of magnitude encountered in the leprosy study above. Patients included in these studies grew up at a time when almost everyone in industrialized areas such as western Europe or the United States was exposed to the infection as evidenced by positive tuberculin tests. A more recent study found concordance to be lower [133]. In recent years, the risk of infection had been reduced appreciably.

Even for a purely somatic condition such as leprosy, observed concordance rates and hence conclusions as to the degree of genetic determination or liability are valid only for the environmental conditions prevalent in the population in which the twins are living. Generalization to other populations should be subject to careful consideration of living conditions. In western Europe, for example, leprosy disappeared during the seventeenth and eighteenth centuries without any

therapy, only due to improvement in living conditions. There was probably little or no influence of genetic changes.

6.3.8 Twin Studies in Other Common Diseases

Table 6.21 lists some diseases in which the twin method has helped to establish the significance of genetic factors in suceptibility. The first three entries are malformations, and therefore the transfusion syndrome may have influenced the concordance rate. In all diseases higher concordance in MZ twins than in DZ twins was noted. The data fulfill the "twin criterion" (Sect. 6.1.2.3) for multifactorial inheritance.

Table 6.22 gives figures for five infectious diseases. High concordance is not in itself sufficient to establish the existence of genetic factors in susceptibility; the difference between MZ and DZ must be considerable. For example, before immunization was introduced almost every child contracted measles. Therefore concordance in both MZ and DZ twins was naturally high, indicating that genetic factors are of no particular importance. The data in Table 6.22 were collected at a time when these diseases were very common.

Analysis of discordance can shed some light on genetic versus environmental factors in disease. Studies by Lemser as early as 1938 [84] showed, for example, that pregnancies – and especially multiple pregnancies – can lead to manifestation of diabetes in predisposed mothers, as in a number of cases one co-twin became diabetic after several pregnancies, whereas the other one, who became pregnant less frequently, remained healthy (Fig. 6.31). However, the information supplied by twin studies on genetic aspects of disease susceptibility tends to be general and nonspecific. It therefore is not surprising that twin studies for internal diseases have lost popularity in recent years. The feeling seems to prevail that, compared with the input of time and resources, the return in terms of new and specific knowledge is too meager.

Table 6.21. Twin results in multifactorial diseases (excluding mental diseases; data from von Verschuer 1959 [161]; Jörgensen 1974 [69]; Berg 1983 [8])

Condition	Twin pairs	n	Concordant n	Concordant %	MZ more frequently concordant than DZ
Clubfoot	MZ	35	8	22.9	10.0
	DZ	135	3	2.3	
Cong. dislocation of hip	MZ	29	12	41.4	14.8
	DZ	109	3	2.8	
Cleft lip and palate	MZ	125	37	29.6	6.4
	DZ	236	11	4.7	
Cancer	MZ	196	34	17.4	1.6
	DZ	546	59	10.8	
Coronary heart disease	MZ	50	23	46.0	4
	DZ	59	7	11.8	
Diabetes mellitus	MZ	181	101	55.8	4.9
	DZ	394	45	11.4	
Hyperthyroidism	MZ	49	23	47.0	15.1
	DZ	62	4	6.5	4.1

Table 6.22. Twin concordances in four infectious diseases (from Jörgensen 1974 [69]; Chakravartti and Vogel [22])

Disease	Twins n	MZ n	Concord.	Conc. (%)	DZ n	Concord.	Conc. (%)	MZ higher than DZ
Measles	3645	1629	1586	97.4	2016	1901	94.3	1.03
Scarlet fever	702	321	175	54.6	381	179	47.1	1.16
Pneumonia	800	328	106	32.3	412	86	18.2	1.77
Tuberculosis	1316	386	204	52.8	930	192	20.6	2.56
Leprosy	102	62	37	59.1	40	8	10.0	2.90

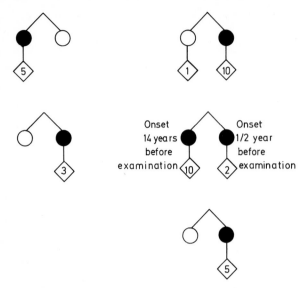

Fig. 6.31. Five MZ adult pairs discordant for diabetes mellitus. The twin sister with the higher number of pregnancies developed diabetes mellitus, whereas the sister with only a few or no pregnancies remained healthy or developed diabetes much later. (Data from Lemser 1938 [84])

However, recent twin studies have helped to establish that genetic factors play a role in the biotransformation of all drugs examined so far (Sect. 7.5.1). Some valuable information may still be gathered by this method, if it is used correctly and is supplemented by other methods (for examples see Chap. 15, 16).

6.3.9 Twin Method in Investigating Continuously Distributed Characteristics

To what degree is the variability of a characteristic in a population genetically determined? One condition for a meaningful investigation is that the characteristic be measurable. This would seem to be self-evident, but in behavior genetics (Chap. 15) the problem of appropriate instruments for measurement is in fact a major problem. Once this problem is resolved, the next problem would be: how do we obtain a twin sample? Schools, colleges, and military conscription usually provide good possibilities for locating an "unlimited representative sample" (Sect. 6.3.6). However, it is not easy to find a really unbiased sample. At least one bias that is always present is that MZ twins are more likely to volunteer than DZ twins. It is plausible that cooperation is correlated with personality variables; this source of selection may therefore introduce a bias into some studies in behavior genetics.

When measuring the characteristic in question, an important but much neglected aspect are errors in measurement.

Heritability Estimates from Twin Data. The concept of heritability has been introduced above (Sect. 6.1.1.5). In a continuously distributed characteristic, such as stature, heritability was estimated from comparison between parents and children. Twin data offer an alternative means to obtain heritability estimates. The method is discussed in Appendix 5. It is suggested that three alternative estimates of h^2 should be calculated:

h_1^2 from comparison of MZ with DZ pairs
h_2^2 from comparison of MZ pairs with age-matched, unrelated control pairs of the same series
h_3^2 from the intraclass correlation coefficients within the whole series of MZ, on the one hand, and DZ, on the other

These three estimates are subject to various biases, i.e., an unbiased estimate of h^2 from twin data is impossible. This is especially true for the estimate from intraclass correlation coefficients h_3^2, which has been used most in the literature.

Most heritability estimates from twin studies are based on unrealistic assumptions. For example, twins are assumed to represent an unbiased sample of the population, and the twins examined are taken as an unbiased sample of all twins. It is assumed that the environment of twins is identical to that of the general population, and therefore that identical environmental influences act on MZ and DZ twins. This assumption may well be the most troublesome, since MZ twins often search out and create more similar environments. However, interaction between heredity and environment as well as covariance between heredity and environment are usually neglected. Effects of dominance cannot be separated from additive genetic variance. The limitation of the heritability concept precludes any hints as to the number of genes operative and any clues as to the genetic mechanism involved.

All these considerations should warn us not to take heritability estimates from twin data too literally. They are crude measures that may serve as a first orientation in estimating a genetic component in the phenotypic variability of a certain character. They pose questions rather than answering them.

6.3.10 Meaning of Heritability Estimates: Evidence from Stature

High heritability has been found for stature. This means that environmental variations found in the population from which twins come have little influence on the phenotypic value. One might conclude that stature is a genetically stable character that cannot be altered by any changes in the environment, except possibly in extreme situations such as by severe malnutrition. This conclusion has been shown to be wrong.

Increase in Stature During the Past Century [85]. Over the past one and a half centuries an appreciable increase in stature has been observed in many countries. Examination of skeletons from the Neolithic to the middle of the nineteenth century shows mean body height to have remained constant over this period (Table 6.23). The recent increase in stature is most notable in population subgroups of lower socioeconomic status (Table 6.24). Moreover, it is found even at birth and, particularly, after the first year of life. At the same time, a substantial decrease in the mean age at menarche has been observed. Statistical analysis of the data on stature excludes genetic influences. The reduction of inbreeding in recent times compared with more intensive inbreeding in earlier periods is not sufficient to explain the observed increase in body height (Chap. 13). Natural selection would have been expected to lead to decreased stature since during the same time period population subgroups with lower socioeconomic status who were shorter tended to have more children. There is convincing evidence that this increase in stature is caused by environmental factors – better nutrition during childhood, especially during the first years of life, and reduction in infectious diseases, mainly infections of the gastrointestinal tract leading to diarrhea.

Lesson to Be Learned from This Example. High heritability for a given trait as measured under certain environmental conditions does not prevent this trait from being influenced strongly by secular trends in environmental conditions to which the whole population or appreciable parts of it may be subjected.

This phenomenon is especially true for characteristics developing over a prolonged period of time, during which there may be exposures to various changing external influences. However, not every environmental change influences such a trait. Some – and even those that plausibly would have an effect – might still be without influence. A prediction in general terms is impossible; every situation may be different. This topic is taken up again in the section on behavior genetics.

6.3.11 Twin-Family Method [71, 88]

It usually is burdensome to amass twin or family material for a proposed study. Therefore if one plans to investigate a certain condition by means of both approaches, examination of the families of twins is an obvious possibility for reducing the effort involved.

The two designs may be linked in a more special way. The fact that some MZ twins are concordant and others discordant for a given disorder may be caused by one or both of two reasons: (a) manifestation of the disorder is influenced by nongenetic factors, and (b)

Table 6.23. Average stature (cm) of adult males (from Lundmann; see Lenz 1959 [85])

	Sweden	Norway	Denmark
Stone Age	169.5	164	170.0
Bronze Age	166.5	–	166.5
Iron Age	167.0	167.0	168.0
Middle Ages	167.5	167.0	–
1855	167.5	168.0	165.5
1939	174.5	174.5	171.5

Table 6.24. Stature (cm) of Swiss conscripts (from Lenz 1959 [85])

Kanton Luzern	1897–1902	1927–1932	Increase
Merchants and students	166.6	171.2	4.4
Factory laborers	161.8	167.0	5.2
Farmers	163.1	166.1	3.0
Kanton Schwyz	**1887**	**1935**	
Intellectuals	167.0	170.6	3.6
Heavy physical labor laborers	164.0	169.4	5.5
Light physical labor laborers	163.2	168.0	4.8
Farmers	162.9	168.7	5.8
Factory laborers	155.9	169.6	13.7
City of Zürich	**1910**	**1930**	
Merchants and students	169.6	172.7	3.1
Tailors	166.5	169.5	3.0
Factory workers	166.4	170.5	4.1
Farmers	165.8	168.4	2.6
Blacksmiths	165.7	168.8	3.1

there are two different types – an inherited and a noninherited form. The two possible explanations can be distinguished by comparing empirical risk figures for the disorder in close relatives of concordant as compared with discordant MZ pairs. If there is heterogeneity, so that one type of the disease is nongenetic, the risk of relatives of discordant MZ pairs is no higher than that in the general population, while discordance caused by nongenetic factors leads to similar risks among relatives of concordant or discordant twins.

To our knowledge, Luxenburger [89] was the first to use the twin-family method. He showed that in schizophrenia risks are about equal in the groups of relatives of concordant and discordant MZ twin pairs, suggesting that sporadic nonhereditary cases are rare or nonexistent.

6.3.12 Co-Twin Control Method [48]

Since MZ twins are very similar or identical in a number of characters, one can examine whether and to what degree various environmental influences are capable of changing a given characteristic. Often a characteristic changes spontaneously over time; a disease may remit spontaneously, and this may be attributed erroneously to intervention or external influences.

MZ twins offer a good investigative opportunity by exposing one twin to the influence under study and the other not. This method allows the most perfect experimental device by complete control of possible host variability. The method can be compared to using inbred animal strains in experimental medicine.

Although the method was developed for investigating educational influences on human behavioral characteristics, it can also be used in a more general way, for example, to test whether certain therapeutic measures are useful.

One study [75] examined 22 MZ and 28 DZ twins by psychological test procedures to determine whether certain aspects of intelligence can be improved by "psychological exercise." The tests were first carried out without any prior training. Then the twin with the poorer performance received training once every week for 5 weeks. At the end of the 5 weeks the twins were examined again, and an increase in performance was found in the trained twins but not in their co-twins.

A Swedish study [103] compared two methods of teaching reading and writing in ten MZ and eight DZ twins of the same sex. The advantages commonly claimed for the analytic method, which starts with the reading of whole words rather than single letters, were not confirmed. Certain advantages of the more traditional method in which individual letters are taught to be combined into words were, however, detected. Due to the different structure of languages, this result cannot immediately be generalized. Replication of the study in an English-speaking population would be interesting, as the sound of letters depends much more on their context within words than it does in other European languages.

6.4 Contribution of Human Genetics to Concepts and a Theory of Disease [156]

6.4.1 General Principles

The Concepts of Disease and Diagnosis [156]. Who is ill? Almost everyone suffers illness for some period during life. The distinction between a slight discomfort and a disease requiring help from a physician is not sharp. It depends on the present state of medicine, but still more on general living conditions, societal attitudes toward health and disease, and the organization of medical care. This is epitomized by the huge differences in the numbers of physicians (and other care givers) per capita in affluent industrial societies, on the one hand, and in many developing countries, on the other. Even in countries with a high standard of living, however, not every episode of being unwell can – and should – be treated. Epidemiological studies in an urban population of Sweden, for example, have shown that more than one-half of adults within the past year had suffered from symptoms that, in the opinion of the examiners, would have justified psychiatric or psychotherapeutic treatment [57].

When someone is ill, physicians use their skills to reach a diagnosis [167]. This means that an attempt is made to try to classify the patient's complaints into a specific disease category. The systematics of disease classification is called nosology. Unlike zoological or botanical systematics, there is no natural system of diseases. Doctors work with a heterogeneous classification that has grown historically with the development of medicine. Disease categories often have blurred borders and may overlap. This is not necessarily wrong. Medicine is only partially a science and does not strive for a perfect classification system. Rather, the assignment of a diagnosis aids in predicting the natural history of the disease and may help in devising appropriate treatments. Diagnostic classification helps physicians to make use of the past experiences of their profession.

To repeat: medicine is not a science. It is an art and an applied science or *techne* (Plato). It uses science to work out guidelines for diagnosis, prognosis, therapy, and prevention. Here, the science of genetics has contributed to many useful concepts and methods in recent decades.

Diseases with Simple Causes. Scientific understanding requires theoretical foundations. Requirements for a good scientific theory are discussed in Sect. 6.1.1.6. A theory underlying the explanation of all diseases does not exist and probably never will. However, theories are possible for certain aspects of disease. For example, the concept of a *disease* that is determined by one single *cause* proved to have a high explanatory value. For example, the multiple and varied signs of tuberculosis occur only following infections with *Mycobacterium tuberculosis*. The specific development of the tubercular infection and the natural history of tuberculosis in an individual depend on many additional circumstances, including genetic factors. A theory of disease that is centered around the concept of a disease unit produced by a single cause is specific, requires an elucidation of mechanisms, and therefore has a high explanatory value. As a first choice, such a theory is preferable to a concept based on mere description of disease signs, such as coughing or hemoptysis, or on the low-order constructs on which the organ pathology of the nineteenth century was based, such as "chronic productive inflammation of the lungs." The goal of scientific research on disease is the replacement of descriptive pathology by more explanatory diagnostic concepts.

Obviously this is easier when a distinct disease has a single cause. A century ago this concept was successfully applied to various infectious diseases. Understandably, this success encouraged scientists to apply the concept to conditions in which a single cause does not exist, and where diagnostic criteria were fuzzy. A probable example is schizophrenia. Here the search for the one major biological or psychological cause has been unsuccessful, although it is conceivable that a major cause has eluded research [114].

Hereditary diseases with simple, monogenic inheritance are excellent examples for the successful application of the monocausal disease concept. Using mutations of the hemoglobin genes as examples, it can be shown how genetic analysis based on Mendel's paradigm and its extension into molecular biology not only permitted identification of the *causes* of disease but also paved the way for elucidation of the *mechanisms* by which well-defined mutations cause impairment of function, i.e., disease (Sect. 7.3). It is noteworthy, however, that interaction with other genes and possibly with the environment determines the severity of monogenic diseases. Sickle cell anemia is a well-studied example. Higher HbF levels entail milder clinical manifestations in sickle cell anemia, and various well-defined mutations that cause elevated levels of fetal hemoglobin (hereditary persistence of fetal hemoglobin) ameliorate the clinical picture. However, even more subtle alterations in the chromosomal environment around the HbS mutation (as defined by DNA variant haplotypes) apparently affect critical HbF regulatory sites [163]. Thus, the "Senegalese" type of sickle cell anemia is associated with more HbF, a preponderance of the HbG γ chains, and a lower proportion of irreversibly sickled cells compared with the mutationally identical "Benin" type of sickle cell anemia which differs in DNA haplotype [104, 105]. The simultaneous presence of α thalassemia is another modifying factor associated with a less severe clinical pattern. Our increasing ability to define specific genetic determinants that affect clinical severity in sickle cell anemia provide an excellent model for elucidating the pathogenesis and clinical severity of other genetic diseases by analysis of interacting genes.

The hemoglobin variants demonstrate another phenomenon. Mutations within the same gene may lead to quite different phenotypes. Methemoglobinemia, for example, is a different disease than sickle cell anemia get both are caused by different mutations affecting the Hbβ gene. Other mutations at different sites of the same gene, with different effects, have been observed. Conversely, genetic heterogeneity, i.e., causation of similar or even identical phenotypes by mutations at different gene loci, is also quite common, so that a variety of causes may lead to the same end effect.

In chromosomal aberrations the *causes* of many birth defects have been identified. Chromosomal aberration syndromes are defined unequivocally by their abnormal chromosomal constitution. But the *mechanisms* by which these aberrations lead to abnormal phenotypes, i.e. the pathways from genotype to phenotype, remain poorly understood (Chap. 8; see also [36]).

The situation is *different* in many genetically *influenced* diseases and anomalies. In these – as in schizophrenia – a single cause cannot be identified, and in many cases may not exist. The same pathogenetic process causing the disease might be triggered by a variety of causes – either alone or in combination. Some of these causative factors may be genetic, while others are "environmental" including somatic (e.g., allergens), behavioral (e.g., feeding and drinking habits), social (e.g., influences of parents, school, occupation), and other factors. Often a preliminary description in terms of "multifactorial inheritance with or without threshold effect" (Sect. 6.1.2) allows some preliminary conclusions. However, the identification and analysis of the specific genetic and environmental components contributing to a disease risk is the *next* goal. A genetic liability may have different causes and components between one individual and family and the next, as shown in the discussion of hyperlipidemias and coronary heart disease (Sect. 6.4.2.2). The same is true for a disease that has been called "the nightmare of the medical geneticist" (Neel, [107]) – diabetes mellitus.

Genetics of Diabetes Mellitus [26, 107, 110, 126]. The developments in our progressive understanding of diabetes mellitus illustrate how a common disease is gradually becoming better understood. Very early in medical history the disease was diagnosed when there was thirst, polyuria, weight loss, weakness, coma, and death associated with sweet-tasting urine. Today, a quantitative fasting blood glucose level with a cutoff point of 140 mg per 100 ml is used for diagnosis. This cutoff point is arbitrary, however, and therefore causes difficulties in classification and problems for genetic analysis. This is a typical example of the general situations described above: a slightly increased blood glucose level in itself is no disease; the individual does not even feel it. But it points to an increased risk for potentially serious complications. Diabetes is highly heterogeneous, i.e., different genetic and possibly nongenetic causes produce a clinical condition that is diagnosed as diabetes. Both common and uncommon varieties of diabetes exist. The two most common are known as type I and II diabetes and can be differentiated by many numerous criteria (Table 6.25). Their etiology is different, and they run true to form, i.e., there is familial aggregation limited to the type of diabetes observed in the index case. Although familial aggregation is less striking in the more severe type I diabetes, the pathophysiology is better understood [126]. There is increasing evidence that this disease is caused by a viral

insult to the islets of Langerhans in the pancreas, followed by the production of anti-islet autoantibodies which gradually destroy the islets. This process leads to insulin deficiency and characteristic clinical findings. However, not every person develops the disease.

Susceptibility is strongly increased by the presence of two alleles at the HLA-DR locus, DR3 and DR4. Compound heterozygotes DR3/DR4 run a still higher risk if the risks were simply additive. Susceptibility appears to be more or less confined to suballeles that can be identified with molecular methods (Sect. 6.2.3). Moreover, association with DR3 or DR4 permits subdivision of type I diabetes into two subtypes: the DR3-associated type shows primarily autoantibodies, whereas in the DR4-associated type a virus etiology appears more likely (Fig. 6.32). Thus this is another example of categorization of an apparently uniform disease into two subtypes.

However, even among identical twins there is only 50% concordance, suggesting that various other factors, such as lack of equal exposure, random events, and poorly understood environmental factors also play an important role in the etiology.

Type II diabetes is common in the middle-aged and elderly but is usually mild. Genetic factors play an important role, as evidenced by the high identical twin concordance rate. The nature of the genetic factors and their mode of transmission have not yet been elucidated.

Further heterogeneity (based on a variety of autoimmune phenomena in type I diabetes and by various criteria such as obesity in type II diabetes) is likely but has not yet found general acceptance.

In contrast to type I and type II diabetes, which do not follow Mendelian inheritance, a rare form of the disease with early onset and a mild course without complications is transmitted by autosomal-dominant inheritance. This condition is known as maturity-onset diabetes of the young (MODY). In some families a functional defect of the enzyme glucokinase has been observed [140], but there is genetic heterogeneity at the DNA level even for the MODY group.

A variety of very rare types of diabetes have been differentiated. These involve insulins with amino acid substitutions that make for lessened activity of the insulin molecule [129]. Defective conversion of proinsulin to insulin owing to amino acid sustitution at critical sites has also been described as an autosomal-dominant trait [122]. Most forms of diabetes, however, are not associated with structurally abnormal insulin.

Insulin receptor action has been studied extensively [125]. Various receptor abnormalities manifesting as a decreased number of insulin receptors or decreased insulin binding capacity have been found in a rare

Table 6.25 Types I and II diabetes (from Olefsky 1985 [110])

	Type I	Type II
Prevalence	0.2%–0.3%	2%–4%
Proportion of all diabetes[a]	7%–10%	90%–93%
Onset	<30 years	>40 years
Body fat	Lean	Obese (~80%)
Ketoacidosis	Common	Rare
Insulin deficiency	Absolute	Rare
Therapy	Insulin	Diet
Complications	Vasculopathy, neuropathy, nephropathy	Infrequent and late
MZ twin concordance	40%–50%	100%
First-degree relatives affected	5%–10%	10%–15%
HLA D3/D4 association	Yes	No
Circulating pancreatic islet cell autoantibodies	Yes	No
Other autoimmune phenomenon	Occasional	No
Insulin secretion	Severe deficiency	Variable
Insulin resistance	Occasional; insulin antibodies	Usual: postreceptor defects?

[a] All other "diabetes" forms are quite rare: <1%.

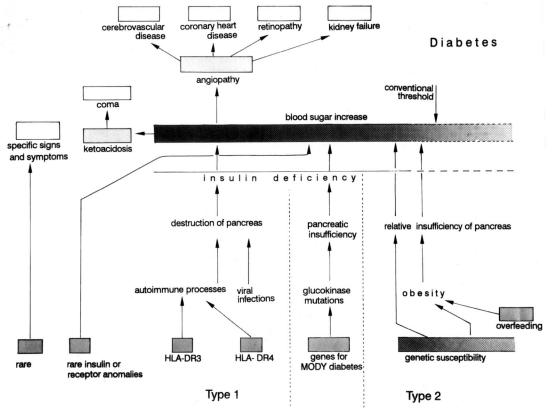

Fig. 6.32. Genetic determination and pathogenetic pathways leading to diabetes mellitus. Numerous genes – often in combination with environmental factors – may lead to an increase in blood sugar which may cause, in addition to clinical signs of diabetes (fatigue, thirst, etc.), complications such as ketoacidosis leading to coma, and angiopathy leading to cer- ebrovascular disease, coronary heart disease, retinopathy, and kidney failure. Note difference between type 1 and type 2 diabetes. MODY stands for mature onset diabetes of the young, is inherited as an autosomal dominant trait, and may be caused by glucokinase mutations. (See text)

type of insulin-resistant diabetes and in some rare genetic conditions such as leprechaunism, lipodystrophy, ataxia telangiectasia, and acanthosis nigricans. A list of some 60 rare conditions, in fact, has been collected in which, in addition to other clinical signs, hyperglycemia is observed [126].

Discussion of other aspects of diabetes mellitus:

- Sect. 3.13.10: Glucokinase deficiency in MODY diseases
- Sect. 6.2.3:   Association with HLA alleles
- Sect. 6.3.8:   Studies on twins
- Sect. 9.2.4:   Nondisjunction of chromosomes
- Sect. 14.3:   Population genetics, natural selection

Disease Concepts and Diagnosis [166, 167]. When a diagnosis is made, a certain cluster of clinical signs and laboratory data are subsumed as a disease unit. This attitude is justified when a single major cause can be pinpointed, such as in infectious and monogenic genetic diseases. However, the great majority of diseases are defined phenomenologically, or in more recent times by some metric characteristics, such as blood sugar in diabetes or blood pressure for hypertension.

In medical practice, this procedure is often successful since medical diagnosis is meant to serve as a guide to therapy. Therapy may not require the careful heterogeneity analysis necessary for genetic investigations. Therefore it may be reasonable and advisable to stop diagnostic procedures at a point at which no further benefit for the patient can be expected. If the management remains identical, it is no longer important for practical purposes to differentiate between subtle diagnostic categories.

However, there may be pitfalls in that a given category may be too superficial to provide appropriate therapeutic guidance for all patients encompassed by that diagnosis. For example, a diagnosis of fever 100 years ago comprised many different diseases which today can be subclassified and demand different therapies. Similarly, a diagnosis of anemia 75 years ago was all that could be specified in pale patients with too little blood. Today we know of different types of hereditary and acquired anemia, often requiring specific treatments. Blood transfusion would be an inappropriate therapy for *all* anemias, since iron deficiency anemias can be treated specifically with iron, pernicious anemia with vitamin B_{12}, and hereditary spherocytosis with splenectomy. Another example: although we treat hypertension today empirically with many different drugs, it is likely that better understanding of the heterogeneous mechanisms of hypertension will in future lead to a more specific treatment appropriate for certain subgroups of patients. We know already that blacks with hypertension respond better to diuretics than to beta blockers (in contrast to whites), but we do not yet know the reason for this variable response.

The medical geneticist always requires a highly specific diagnosis with special attention to heterogeneity in order to give appropriate genetic advice regarding recurrence risks and prenatal diagnosis (Sect. 18.1; 18.2). Since diseases with similar manifestations may be attributable to different genetic mechanisms or may not be genetic at all.

In practice, diagnosis is often a stepwise process. First, a crude classification is made, for example, diabetes. Then a variety of methods are used for a subclassification, for example type I or type II diabetes. A number of questions are asked: at what age did the disease begin? How did it begin? Is the patient slim or obese? And so on. At the same time, the physician attempts to determine whether one of the many rare diseases is present in which diabetes is only one clinical sign within a more complex syndrome; diagnosis of such a syndrome might be important for the therapeutic strategy.

Information regarding the health status in family members is necessary for more precise classification. All these data are used to select an optimum course of action: precise diagnosis → prognosis → specific therapy → prevention of complications and prevention in the family.

Normal Variation and Disease. The distinction between a disease and the upper limits of normal variation is important. For example, hypertension is not a disease (although often considered as such), since it represents the designation for a certain percentage of the population whose blood pressure is higher than an arbitrary cutoff value. The risk of complications of hypertension increases with higher blood pressure levels, but there is no threshold value at which the risk disappears. The diagnosis of "hypertension" in a sense is inappropriate. Hypertension per se is a "risk factor" for coronary heart disease, stroke, and renal failure, rather than an illness.

As we learn more about various genetic risk factors conferring susceptibility to certain diseases, similar problems arise. Many persons carry the HLA DR3 or DR4 determinant, but only a small percentage develop type I diabetes. The relative risk is about five times that of a person not carrying these HLA types. The absolute risk of HLA D3 and D4 carriers for developing diabetes remains quite small. Homozygotes for the Pi Z phenotype often develop COPD, but not all gene carriers become ill. Such Pi Z homozygotes are not diseased, but may become so in the future.

One of the goals of medical genetics is the elaboration of "marker profiles" to aid in identifying subgroups at high risk of developing certain diseases – particularly if measures can be taken to prevent, defer, or ameliorate the deleterious effects of the genetic predisposition by environmental manipulation. This approach is particularly promising since the development of disease frequently requires interaction between genetic susceptibilities and environmental factors. Such preventive medicine of the future will be "tailor made" to the unique genotypes of an individual rather than directed at the entire population.

Creating the scientific basis for recommending such rules will be the specific goal of ecogenetics, discussed in Sect. 7.5.2. The theory of disease develops into a theory of health preservation. Explaining some of the aspects hinted at in this section in greater detail is a major pupose of the chapters to follow – especially our considerations on gene action (Chap. 7) and mutation (Chap. 9).

6.4.2 Current Status of the Genetics of Common Diseases [97, 99]

Genetic diseases caused by chromosomal aberrations and by Mendelian mutations affecting single genes are relatively well understood. Their mechanisms can be approached by studies of individual gene action (see Chap. 7) or by considering how gross chromosomal defects cause developmental damage (see Chap. 2). Various data show that familial aggregation is frequent in many other diseases. Appropriate studies (see Chap. 6) ideally need to be carried out to confirm that familial aggregation is caused by common genes rather than by a common family environment.

A variety of experimental designs to discriminate between the role of environment and heredity have been established. Such designs include study of identical twins reared apart in different environments and the comparison of disease frequency in adopted children with that in their biological and adopted relatives (see Chap. 15). When identical twins even in different environments show greater concordance than

DZ twins in similar environments, genetic rather than environmental factors are suggested. Similarly, when adopted children resemble their biological parents rather than their adoptive parents, genetic factors appear certain. The frequency of a trait or disease is also often studied in spouses who share the same environment in comparison with biological relatives who share both heredity and environment. Absence of correlation between spouses after many years of living together when relatives have an increased disease frequency helps to argue for genetic factors.

Based on various investigations of this type it has been concluded that genetic factors are operative in the following groups of diseases:

a) Common birth defects (i.e., neural tube defects, cleft lip and palate, club foot, congenital heart disease, and others)
b) Common psychoses (schizophrenia and affective disorders)
c) Common diseases of middle life (diabetes, hypertension, coronary heart disease)

Family studies carried out in these diseases, with rare exceptions, were unable to demonstrate Mendelian inheritance. Based on models of polygenic gene action it has been inferred that many unspecified genes acting together with environmental factors are operative in the etiology of these disorders. The biological action of the involved genes remains largely unknown and is considered a "black box." It is often assumed that the number of operative genes is relatively large, and that the contribution of each of the postulated individual genes to the pathogenesis of the disease is relatively small. When the disease occurs as a qualitative phenomenon with dichotomized classes of "healthy" and "ill," such as in congenital malformations, a threshold is assumed (Sect. 6.1.2).

Our concept of multifactorial inheritance is summarized in Fig. 6.33. We would like to emphasize the potential role of one or a few major genes in many supposedly multifactorial traits. A relatively small number of potentially identifiable major genes may contribute to the genetic etiology and explain most of the genetic variation. Such genes do not act in a vacuum. The ensemble of all other genes against which such major genes act constitute the "genetic background." It is well known that the genetic background may modify and influence expression of major genes (Sect. 4.1.7).

Particularly in birth defects one finds that a role is played by random factors [161] that are not determined by either genetic or environmental agents but act stochastically. One can visualize how some cardiac defects may occur because the complex dynamic sequence of twisting and turning to form the normal heart may fail to synchronize by chance only.

Further research on the genetics of common diseases is most likely to be fruitful by attention to the action of individual genes to be studied by combined genetic, biochemical, immunological, clinical, and statistical methods. Biometric approaches alone are unlikely to provide new insights.

6.4.2.1 Biological and Pathophysiological Approaches to the Genetic Etiology of Common Diseases

Heterogeneity Analysis: Differentiation of Monogenic Subtypes from the Common Varieties. Often rare subvarieties of clear Mendelian inheritance such as X-linked hypoxanthine phosphoribosyl transferase deficiency in gout [74, 130] and familial hypercholesterolemia in coronary heart disease [57] need to be differentiated from the multifactorial common diseases by appropriate clinical, laboratory, and genetic methods. Similar considerations apply to most other common diseases.

Clinical Population Genetics. Clinical, laboratory, and family studies of unselected cases of heterogeneous disease groupings such as mental retardation [111], deafness, blindness, and coronary heart disease [52] can distinguish familial from sporadic cases. Appropriate statistical and biochemical analyses of the sporadic cases can establish a certain number as being caused by monogenic and mostly autosomal-recessive inheritance. Family study may establish genetic heterogeneity as manifested by different modes of monogenic inheritance in some cases and multifactorial transmission in others. Linkage studies may help to identify major genes.

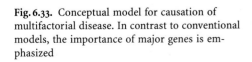
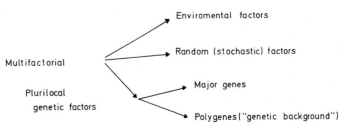

Fig. 6.33. Conceptual model for causation of multifactorial disease. In contrast to conventional models, the importance of major genes is emphasized

Polymorphism and Disease. Some of the extensive polymorphisms of human genes may represent part of the genetic basis for differential susceptibility to common diseases. Associations of HLA and disease are often of particular interest and may be related to differential responses to autogenous antigens. Various disease associations with AB0 types are less clear from a pathophysiological viewpoint [154]. In general the approach of relating polymorphic marker genes is most successful when markers that are pathophysiologically related to the disease can be studied. Random genetic markers investigated in random diseases are less likely to produce meaningful data.

Heterozygotes for Rare Diseases May Be More Susceptible to a Functionally Related Common Disease. Even with rare autosomal-recessive disease there are many heterozygotes in the population (Table 4.6). Such heterozygotes may be at higher risk for common diseases that are pathophysiologically related to the enzyme defect [155]. For example, heterozygotes for the rare autosomal-recessive methemoglobinemia do not develop methemoglobinemia unless they ingest methemoglobin-forming drugs. Normal homozygotes have sufficient methemoglobin reductase to reduce the methemoglobin formed by the drug while heterozygotes have sufficient amounts (50% of normal) under usual conditions but not enough when methemoglobin formation is excessive following drug administration.

Other examples are discussed in Sect. 7.5.1.

6.4.2.2 Genetics of Coronary Heart Disease
[92, 100, 119, 135]

The prevalence of coronary heart disease varies widely throughout the world. In general the highest frequency is found in western countries and in subpopulations leading western life-styles; the frequency in developing populations is usually quite low. Temporal trends in the United States show the strong effects of environmental factors; there was increasing mortality of coronary heart disease in midcentury and significant lowering of the coronary death rate since that time [87]. The important role of environment is also shown by the increasing rate of coronary heart disease when migrants from low-frequency countries (e.g., Japan) move to high-frequency areas (e.g., United States) [70].

Genetically oriented studies of coronary artery disease and atherosclerosis in general aim to (a) detect genetic differences between individuals which predispose to atherosclerosis; (b) differentiate genetic from environmental determinants; (c) understand the pathophysiology of the condition; and (d) identify sub-

populations at risk for preventive measures. With the development of appropriate genetic and other "markers" an increasing number of persons can be identified whose genetically determined constitution makes them more susceptible to certain environmental factors (including diet) that cause coronary atherosclerosis. Twin studies have shown a significantly higher frequency of coronary heart disease among MZ twins than among DZ twins. However, such studies are difficult in view of problems with diagnostic endpoints. An ideal twin study for coronary heart disease would require angiography (or a noninvasive technique outlining coronary vessels) that delineates the extent of coronary atherosclerosis.

There is general agreement about familial aggregation of coronary atherosclerosis. The frequency of coronary heart disease is about two to six times higher in patients' families than in control families (for references see [8, 123, 141]). The following facts regarding familial aggregation are noteworthy:

1. Familial aggregation increases with decreasing age of affected patients, i.e., in premature coronary heart disease. Genetic factors are less important in coronary heart disease of older patients.
2. While women have a lower frequency of coronary heart disease than men, affected women have a stronger familial aggregation than men. The less frequently affected sex has a higher extent of genetic "loading" (Sect. 6.1.2.3).
3. A family history of premature coronary heart disease (<55 years) emerges as the strongest risk factor for coronary heart disease and appears to be stronger than all other risk factors (see below [109]).
4. Hyperlipidemia, hypertension, and diabetes represent risk factors for coronary heart disease that have strong genetic determinants. However, various studies suggest that familial aggregation may not be entirely accounted for by these three well-known genetic risk factors [109]. Additional familial factors appear to contribute to familial aggregation.
5. Familial aggregation does not necessarily mean genetic determination. Families share similar environments that may include agents causing a higher frequency of coronary heart disease in family members. There is even evidence of assortative mating for coronary risk factors. Spouses' families show the same extent of familial aggregation of coronary heart disease as the families of their affected husbands. Spouses of affected patients also have a higher frequency of coronary heart disease than the control population [146]. Assortative mating presumably occurs for social class, life-style, smoking, and dietary habits. A significant compo-

nent of familial aggregation is therefore mediated by environmental factors shared by families. Furthermore, there are likely to be complex genetic-environmental interaction such that individuals with different genotypes react differently to various environmental influences such as diet.

6. A family history of premature coronary heart disease regardless of the nature of the various genetic and environmental factors serves to identify high-risk individuals and families.

Risk Factors. Extensive epidemiological work has been carried out to identify a variety of risk factors for coronary atherosclerosis. Advancing age, male sex, hypertension, hypercholesterolemia, low levels of high-density lipoprotein (HDL), and diabetes carry a particularly strong risk.

Other risk factors that have been implicated include hypertriglyceridemia, high levels of apolipoprotein B and lipoprotein (a), low levels of HDL and apolipoprotein (apo) A$_1$, sedentary life, and obesity. Inborn variation in thickness of the vascular intima and of musculoelastic layers has also been suggested.

These findings are compatible with a multifactorial etiology of coronary heart disease. Genetic factors can be elucidated by studying the extent of genetic contribution to the various risk factors.

Hyperlipidemias. Current data suggest that hypercholesterolemia and low HDL levels are strong risk factors [67] (Table 6.26). Hypertriglyceridemia may be a mild risk factor but is not generally accepted as an independent risk factor [100]. Among the inherited hyperlipidemias several conditions need to be differentiated.

Familial Hypercholesterolemia [45, 51, 100]. The best understood condition is autosomal-dominant familial hypercholesterolemia (Sect. 7.6.4), with a heterozygote frequency of about 1/500 in the United States. Heterozygotes have elevated levels of cholesterol and low-density lipoprotein (LDL) cholesterol, often in the upper 5% and usually in the upper 1% of the population range. Fifty percent of affected males have some manifestation of coronary heart disease by the age of 50. Tendon xanthomas and early corneal arcus may occur. Clinical manifestation in females occurs 10–15 years later. Other risk factors, such as hypertension, cigarette smoking, and low HDL levels, interact with the gene for familial hypercholesterolemia in hastening the onset of atherosclerotic symptoms. The homozygote state is extremely rare (1 in one million), and coronary heart disease occurs in adolescence or even earlier. Familial hypercholesterolemia occurs in many countries and populations. Laboratory detection is difficult since there is no generally available laboratory test for unequivocal diagnosis of the characteristic LDL receptor defect. Tests for LDL receptor function can be performed only on a research basis (Sect. 7.6.4). Molecular diagnosis is possible once the specific LDL receptor defect is known, but the existence of several dozen different LDL receptor mutations [63] makes screening impractical.

Familial hypercholesterolemia needs to be distinguished by clinical, laboratory, and genetic criteria from various hypercholesterolemias caused by other

Table 6.26. Common hyperlipidemias associated with coronary heart disease

Name	Prevalence	Physiological abnormality	Defect	Genetics	Frequency in unselected myocardial infarct survivors	
					<60 years	Average age[a]
Familial hyper-cholesterolemia	1/500	Diminished LDL breakdown	Abnormal LDL receptor	Autosomal dominant	3%– 6%	46 years
Polygenic hyper-cholesterolemia	5%	Several		"Polygenic"	Increased	58 years
Familial combined hyperlipidemia	0.3–1%	Increased apo B synthesis	Unknown; heterogeneous	Autosomal dominant[b]	11%–20%	52 years
Familial hyper-triglyceridemia	1%	Increased VLDL synthesis	Unknown	Autosomal dominant[b]	4%– 5%	57 years
Type III hyper-lipidemia (remnant removal disease)	1/10000	Diminished remnant catabolism	Abnormal apo E binding and additional factors	Homozygote for apo E$_2$	1%– 2%	50s

[a] Average age of male patients with myocardial infarct.
[b] May be multifactorial in some families.

acquired and genetic entities. Only about 1 in 25 persons with a cholesterol level in the upper 5 percentile carries the gene for familial hypercholesterolemia. About 3%–8% of unselected male patients aged 60 years or below with myocardial infarct are heterozygous for familial hypercholesterolemia (Table 6.26). With decreasing age at the first myocardial infarction the frequency of this conditions increases. The linked restriction enzyme fragment length polymorphism at the LDL locus can be used for preclinical diagnosis in informative families where the phenotypic diagnosis of cholesterol elevation is not quite clear, but where at least one definitely diagnosed patient is available.

Familial Combined Hyperlipidemia. Studies of patients with hyperlipidemia have produced the definition of a familial disorder characterized by elevation of both cholesterol and triglyceride (type IIb), of cholesterol alone (type II), or of triglyceride alone (type IV). This disorder has also been termed "multiple lipoprotein type hyperlipidemia" or familial combined hyperlipidemia and is probably common in the general population (1/100–1/300) [12]. The transmission pattern in families often suggests an autosomal-dominant mode of inheritance. Formal complex segregation analysis of families in Seattle, Washington, could not detect a single gene [168]; however, a similar method suggested a single gene in the original Seattle families as well as in 55 British Families [27]. Full penetrance is not reached until the late 20 s. The diagnosis of this condition cannot be made in a single individual although mixed hyperlipidemia (i.e., elevation of both cholesterol and triglyceride levels) suggests its presence. Detection therefore requires extensive family studies. Several large pedigrees with the disorder have been documented. The condition is seen in approximately 10 % of unselected myocardial infarct survivors under the age of 60 years. Among probands with familial combined hyperlipidemia without myocardial infarction the familial aggregation of premature coronary heart disease is striking [16]. It has been suggested that apo B elevation may be a better marker than other lipid parameters and will allow more ready detection of familial combined hyperlipidemia [62].
However, an elevated apo B level is not specific or pathognomic for familial combined hyperlipidemia.
Linkage studies so far have failed to localize the gene or genes for this condition. However, a localization at the apo A_1-CIII locus has been suggested in a subset of families with familial combined hyperlipidemia [171].

Familial Hypertriglyceridemia. A common autosomal-dominant condition associated with segregation of high triglyceride levels alone has been postulated.

Children may not present with this entity. Research into hyperglyceridemia is rendered difficult by the considerable lability of triglyceride levels, which are influenced by a variety of dietary and other factors including alcohol. Isolated hypertriglyceridemia of any sort as a risk factor for coronary heart disease has been questioned. No basic defect for the claimed autosomal-dominant variety of hypertriglyceridemia has been demonstrated, but increased very low density lipoprotein (VLDL) triglyceride synthesis has been suggested.

Broad Beta Disease of Type III Hyperlipoproteinemia (Remnant Removal Disease). Type III or dysbetalipoproteinemia occurs only in homozygotes for an apo E variant (E_2-E_2) that is found in about 1% of European populations. However, additional lipid-raising factors, such as familial combined hyperlipidemia and various secondary forms of hyperlipoproteinemia, are required to produce this disease. This disorder therefore is the result of the interaction of two genetic disorders or of a genetic and an acquired disease. While the frequency of the underlying polymorphism is high (1%), the actual condition is rare (1/10 000).

High-Density Lipoprotein. Most studies have shown that low HDL levels predispose to coronary artery disease in populations that eat high-fat diets, such as in the United States. The higher HDL levels in women (compared with men) may account for their delayed onset of coronary artery disease. The role of HDL in preventing coronary artery disease has been attributed to its function of removing cholesterol from atherosclerotic plaques. HDL levels tend to be inversely correlated with triglyceride levels. Since apo A_1 is a constituent of HDL, its levels are affected by the apo A_1 concentration. Common genetic variants of apo A_1 have not yet been detected. Nevertheless, family and twin studies (but not spouse-pair comparisons) show correlations in HDL level with a calculated heritability level of about 40 % [13].

6.4.2.3 Lipid-Related Polymorphisms

Apolipoprotein E Polymorphism. A common polymorphism of apo E exists and can easily be detected by molecular techniques. Table 6.27 shows the amino acid sequence difference between the three apo E types and their population frequencies in the white population. The apo E_2 allele reduces cholesterol, LDL cholesterol, and apo B levels while the apo E_4 allele raises these levels as compared with the common standard type apo E_3. The population frequency of the resultant genotypes in a North American white

Table 6.27. Apo E polymorphisms (from Motulsky and Brunzell 1992 [100])

Allele	Gene product	Composition	Typical frequency	Average allelic effect on total cholesterol (mg/dl)
$\varepsilon2$	E2	pos. 112:cys pos. 158:cys	0.109	−14
$\varepsilon3$	E3	pos. 223:cys pos. 158:arg	0.760	−0.16
$\varepsilon4$	E4	pos. 112:arg pos. 158:arg	0.131	+7

Table 6.28. Effect of apo E genotypes on plasma cholesterol levels in Minnesota (from Motulsky and Brunzell 1992 [100])

Genotype	Population frequency (%)	Average cholesterol levels (mg/dl) in normocholesterolemic men
$\varepsilon2\ \varepsilon2$	0.46	133
$\varepsilon2\ \varepsilon4$	3.15	183
$\varepsilon3\ \varepsilon3$	59.5	192
$\varepsilon3\ \varepsilon2$	12.7	182
$\varepsilon3\ \varepsilon4$	23.9	193
$\varepsilon4\ \varepsilon4$	0.9	207
All genotypes	100.0	191

population is shown on Table 6.28. The differences in cholesterol levels between the relatively rare homozygotes for E_2E_2 and E_4E_4 is considerable: about 70 mg/dl. Clearly detectable increases in coronary heart disease ascribable to this polymorphism would be detectable only among the rare (less than 1%) E_4-E_4 homozygotes.

It is of great interest that a significantly greater frequency of E_4 has been detected fortuitously in many studies of European and Japanese patients with Alzheimer's disease, particularly of the late-onset familial variety. The relative risk of Alzheimer's disease for heterozygotes carrying the E_4 gene is three to four times that of the control population while that for the rarer E_4E_4 homozygotes is higher still. The mechanisms remain under study. While the E_4 polymorphism is useful for our understanding of Alzheimer's disease, it should *not* be used for predictive identification of high risk since many patients with E_4 will not develop Alzheimer's disease, and, conversely, a large number of patients who have Alzheimer's disease do not have the E_4 type.

Lipoprotein (a) Polymorphism. Lp(a) is a unique glycoprotein attached to apo B. The gene and its protein sequence are highly homologous to plasminogen on chromosome 6 where both genes have been mapped. LP(a) levels are strongly associated with coronary artery disease. Lp(a) is deposited in coronary arteries where it may promote thrombosis by interfering with fibrinolysis or thrombogenesis or both. In one study, about 30% of patients with premature heart coronary artery disease had levels above the 95th percentile of the normal population. In men younger than 60 years the excess or attributable risk of an Lp(a) level in the upper quartile was 28% as compared to 13% in those aged between 60–70 years [100]. The Lp(a)-related risk appears to be independent of the risk for cholesterol, HDL,

apolipoproteins 1 and 2 and triglyceride levels but is probably additive to other lipid-related risks. A significant proportion of the familial aggregation of coronary artery disease in the absence of monogenic hyperlipidemia has been related to increased Lp(a) levels [34].

There is marked genetic heterogeneity at the Lp(a) locus, with more than 20 common alleles [72, 80], and therefore most persons are compound heterozygotes. Each allele determines a specific number of multiple tandem repeats of a unique coding sequence that determines the so-called kringle-4 protein structure. (A "kringle" is the shape of a Danish roll.) The size of the Lp(a) protein is inversely related to Lp(a) levels, but not all variations in level are related to the kringle polymorphism. The Lp(a) levels in the population differ 1000-fold. Because of its resemblance to plasminogen, elevated Lp(a) levels may be a link between atherogenesis and thrombosis.

6.4.2.4 Associations of Coronary Heart Disease with Genetic Markers [8, 10, 96, 101, 134]

Protein Markers. Genetic polymorphisms that are biochemically and pathophysiologically related to a disease may represent the "genetic background" which makes certain individuals more likely to be affected. An analysis of such polymorphisms may assemble a group of markers that in the aggregate contribute significantly to disease susceptibility. A variety of "blind" markers were first studied in coronary atherosclerosis simply because they were available. Most of them contribute only little to the total etiology. Individuals with blood group A of the AB0 system have a higher chance of thrombotic coronary heart disease and have a higher cholesterol level as well. Minor effects that raise cholesterol levels are also exerted by the nonsecretor gene, the haptoglobin[2] genes, and the Gm[a] genes.

DNA Markers in Population Association Studies. The association of various hyperlipidemias with abnormalities of apolipoproteins and their receptors has led to searches for an altered frequency of one or another DNA variant of lipid-related genes in patients with hyperlipidemia or coronary heart disease. Since such population association studies can be executed more readily than family studies, much work has been carried out with this approach.

The rationale is to identify mutations of an apo gene that predispose to hyperlipidemia and coronary heart disease by demonstrating an altered frequency of a closely linked DNA variant (such as restriction fragment length polymorphism, variable-number tandem repeat, or a DNA haplotype) that is cosegregating with the lipid mutation predisposing to coronary heart disease. It is assumed that a yet undefined mutation of an apo gene occurred and then expanded in frequency over the generations with cosegregation of a closely linked DNA marker. The method therefore requires strong linkage disequilibrium (Sect. 5.2.4) between the original mutation and the marker and, most importantly, identity of the mutation in various patients who share the ancestral mutation. Coronary artery disease, with its heterogeneous genetic etiology as a diagnostic endpoint, is less probable to give meaningful results in such studies than a specific category of hyperlipidemia, which is more likely to be mutationally homogeneous. Close ethnic matching is essential, since different populations (even among whites) have different DNA marker frequencies which may cause spurious results in comparisons between patients and controls of somewhat different ethnic origin. Furthermore, test results are usually compared on relatively small samples, and even then a statistically "significant" result may have no biological significance unless appropriate corrections are made for the number of tests carried out and, ideally, after the observations have been replicated in a second sample. Various investigations with such DNA marker studies have been summarized [60, 65], but most results have proven to be spurious when further studies were performed. This approach has therefore been less useful than linkage and segregation studies in searching for the critical mutations predisposing to coronary heart disease. A few associations, however, have been confirmed such as higher levels of cholesterol, LDL cholesterol, and apo B associated with the common variant of an apo B *Xba*1 polymorphism that does not alter the amino acid sequence of apo B [9, 31]. The underlying mutation presumably affects regulation of the apo B gene, but it has not been identified. However, the effects of this polymorphism are relatively small: an increase of about 3 mg/dl cholesterol for the apo B enhancing allele and a decrease of 3 mg/dl for

the apo B diminishing variant. The difference between the respective homozygotes, however, is about 19 mg/dl.

Homocysteine and Arterosclerosis [12 a, 99 a]. Elevated homocysteine levels are frequently observed in patients with arteriosclerotic vascular disease affecting the coronary, cerebrovascular and peripheral circulation. Homocysteine is a graded risk factor for coronary artery disease independent of lipid and other risk factors. About 10 % of coronary mortality may be attributed to elevated homocysteine levels. High homocysteine levels can be diminished by increased folic acid intake, suggesting a therapy to reduce the frequency of vascular disease. A very common polymorphism of the enzyme methylene tetrahydrofolate reductase (MTHF) is seen in 5–15 % of the population in the *homozygote* state, and is associated with increased homocysteine levels under conditions of suboptimal folic acid nutrition. Folic acid will also reduce the increased homocysteine levels in such persons. The MTHF variant is observed at significantly higher frequencies in patients with premature vascular disease, including those with coronary artery disease, and can be readily detected by molecular techniques. The MTHF polymorphism is an example of a common genetic variant interacting with a nutritional factor, i. e., folic acid, in the complex pathogenesis of arteriosclerosis.

Other Genetic Factors. The familial aggregation of coronary heart disease in the absence of lipid elevation suggests the operation of other genetic and environmental factors that do not affect lipids. Much work still needs to be done here. The response of blood vessels to atherogenic stimuli and the elucidation of the genes involved in hypertension, which is a risk factor for coronary heart disease, are only a few of the possible areas of future research.

Implications. Can premature coronary heart disease be prevented? The declining frequency in mortality from these diseases in the past 25 years in the United States suggests that various environmental changes can affect these conditions. A high priority must therefore be assigned to the identification of high-risk groups.

Hypertension is a strong risk factor for both coronary and cerebrovascular disease. The condition can be readily identified, and population screening is clearly justifiable. Since hypertension is familial, identification of a hypertensive person should lead to blood pressure screening of first-degree family members, and initiation of antihypertensive treatment, if required.

Population screening for hyperlipidemias raises logistical and operational problems but has gained increasing acceptance in many quarters. Large-scale testing for cholesterol has been recommended. Cholesterol levels can be reduced by diet and in more severe hypercholesterolemia by a variety of lipid-lowering drugs. Since elevated cholesterol levels are correlated with mortality from coronary artery disease, and cholesterol reduction decreases the frequency of myocardial infarcts, cholesterol testing followed by appropriate diets and/or drug treatment has been advocated. There is increasing evidence that drug therapy can retard the progression of arteriosclerotic lesions (assessed by angiography) as well as reduce mortality. However, for every coronary artery event that is prevented, many individuals must be treated. The fact that most coronary artery disease occurs at a rather modest level of cholesterol elevation is another argument advanced by critics of population screening. They feel that a healthy prudent diet and healthy living be followed by everyone in the population. Such an approach, together with the mass media advocating the cessation of smoking, exercise, and blood pressure monitoring, has been validated in North Karelia, Finland [118], and in the Stanford, California, area. It is difficult, however, to maintain momentum for long periods. Furthermore, as compliance with these measures does not prevent coronary artery disease in those at highest risk, a good argument can be made to implement both the population approach and a scheme in which those at highest risk are identified by screening. No comprehensive test battery to predict coronary artery disease has yet been validated for disease prediction beyound the recommendations of testing for cholesterol (LDL) and HDL and possibly for Lp(a).

Targeted screening of individuals with a family history of coronary heart disease is a possibility. For instance, school children could be given questionnaires asking their parents about familial coronary heart diseases with the aim of identifying high-risk families for appropriate intervention. The logistical problems would be surmountable, but this approach would raise problems of confidentiality and privacy.

Triggered screening following a clinical diagnosis of premature coronary heart disease is already feasible and strongly recommended. Following such a diagnosis family members would be studied for lipids and hypertension, and those found to be affected could be treated. This mode of ascertainment is "retrospective," but it can be instituted relatively simply by physician education. Since many patients are seen in hospitals, hospital medical staff should be sensitized to the desirability of initiating such procedures.

Although many hyperlipidemias are influenced by genetic factors and may sometimes be entirely genetically determined, attempts to define the specific genetic background of the hyperlipidemias by family studies are rarely pursued since effective antihyperlipidemic treatment can be carried out successfully without knowing the specific genetics of a given hyperlipidemia.

Considerably more investigative work on the complex genetic basis of hyperlipidemia needs to be carried out [134]. This work is likely to lead to better understanding of the underlying pathophysiology and to identification of those at highest risk. We already know more about genetic risk factors in coronary artery disease than in any other common, genetically influenced disease. Nevertheless, gene-gene interaction between the various lipid risk factors is likely and needs to be clarified. Similarly, gene-environmental interaction probably also occurs, with the result that different individuals react in variable ways to the same environmental or dietary factors. We already know of genetic strains of animals that absorb cholesterol variably, and it is therefore likely that individuals also do not respond equally to high fat diets.

Furthermore, variable responses to antihyperlipidemic agents may occur due to different pathophysiology of the underlying lipid disorder and to possible differences in drug metabolism. No comprehensive test battery to predict coronary artery disease has yet been validated for disease prediction beyond the recommendations of testing for cholesterol (LDL) and HDL and possibly for Lp(a).

Conclusions

Genetic analysis may be performed at various levels. An increasing number of genes can now be identified at the gene-DNA level, allowing definitive inference regarding the corresponding gene product at the biochemical level through knowledge of the genetic code. In earlier times biochemical analysis preceded DNA studies. However, when DNA analysis is not yet possible, analysis at the *qualitative* phenotypic level often makes it possible to identify single-gene or monogenic inheritance. Analysis at the *quantitative* phenotypic-biometric level by calculating correlations between relatives and similar methods to provide heritability estimates may suggest genetic influences for a given trait, but heritability estimates provide no information regarding the number or nature of the genes involved. Developing a genetic model of multifactorial inheritance in association with a threshold was the first step in explaining the genetic basis of some common diseases with complex etiology. Analysis of the specific individual genes involved and their interaction with environmental factors is the next step in understanding the genetic basis of complex disease. Comparison of concordance rates in monozygotic and dizygotic twins permits inferences regarding genetic influences on disease susceptibility. Well-controlled studies of associations with genetic traits such a AB0 blood groups, HLA types, and DNA variants occasionally provide clues toward a better understanding of pathogenetic mechanisms. Genetic and environmental analysis of susceptibility to coronary heart disease offers a good example of ongoing genetic analysis at various levels.

References

1. Aird I, Bentall HH, Roberts JAF (1953) A relationship between cancer of stomach and the AB0 groups. BMJ 1 : 799–801
2. Albert E (1991) Immunogenetik. In: Lehrbuch der Immunologie. Thieme, Stuttgart
3. Amiel JL (1967) Study of the leucocyte phenotypes in Hodgkins's disease. In: Curtoni ES, Mattuiz PC, Tosi RM (eds) Histocompatibility testing 1967. Munksgaard, Copenhagen, pp 79–81
4. Anokhin A, Steinlein O, Fischer C et al (1992) A genetic study of the human low-voltage electroencephalogram. Hum Genet 90 : 99–112
4 a. Baraitser M (1982) The genetics of neurological disorders. Oxford University Press, Oxford
5. Becker PE (ed) (1964–1976) Humangenetik, ein kurzes Handbuch in fünf Bänden. Thieme, Stuttgart
6. Bell JI, Todd JA, McDevitt HO (1989) The molecular basis of HLA-disease association. Adv Hum Genet 18 : 1–41
7. Benirschke K, Kim CK (1973) Multiple pregnancy. N Engl J Med 288 : 1276–1284, 1329–1336
8. Berg K (1983) Genetics of coronary heart disease. Prog Med Genet [New Ser] 5 : 35–90
9. Berg K (1986) DNA polymorphism at the apolipoprotein level. Clin Genet 301 : 515–520
10. Berg K (1991) Atherosclerosis and coronary artery disease. In: Nora JJ, Berg K, Nora AH (eds) Cardivascular diseases: genetics epidemiology and prevention. Oxford University Press, New York, pp 3–40
11. Bieber FR, Nance WE, Morton CC, Brown JA, Redwine FO, Jordan RL, Mohanakumar T (1981) Genetic studies of an acardiac monster: evidence of polar body twinning in man. Science 213 : 775–777
12. Boman H, Ott J, Hazzard WR, Albers JJ, Cooper MN, Motulsky AG (1978) Familial hyperlipidemia in 95 randomly ascertained hyperlipidemic men. Clin Genet 13 : 108
12 a. Boushey CJ, Beresford SAA, Omenn GS, Motulsky AG (1995) A quantitative assessment of plasma homocysteine as a risk factor for vascular diseases, probable benefits of increasing folic acid intakes. JAMA 274 : 1049–1957
13. Breslow JL (1989) Familial disorders of high density lipoprotein metabolism. In: Scriver CR, Beaudet AL, Sly WS, Valle D (eds) The metabolic basis of inherited disease, 6th edn. McGraw-Hill, New York, pp 1251–1266
14. Brönnestam R (1973) Studies on the C3 phenotype and rheumatoid arthritis. Hum Hered 23 : 206–213
15. Brown MS, Goldstein JL (1979) Abetalipoproteinemia. In: Goodman RM, Motulsky AG (eds) Genetic diseases among Ashkenazi Jews. Raven, New York
16. Brunzell JD, Schrott HG, Motulsky AG, Bierman EL (1976) Myocardial infarction in the familial forms of hypertriglyceridemia. Metabolism 25 : 313–320
17. Bulmer MG (1970) The biology of twinning in man. Clarendon, Oxford
18. Bunge M (1967) Scientific research, vols 1, 2. Springer, Berlin Heidelberg New York
19. Carney G, Seedburgh D, Thompson B, Campbell DM, MacGillivray I, Timlin D (1979) Maternal height and twinning. Ann Hum Genet 43 : 55–59
20. Carter CO (1961) The inheritance of congenital pyloric stenosis. Br Med Bull 17 : 251–254
21. Cavalli-Sforza LL, Bodmer WF (1971) The genetics of human populations. Freeman, San Francisco
22. Chakravartti MR, Vogel F (1973) A twin study on leprosy. Thieme, Stuttgart (Topics in human genetics, vol 1)
23. Childs B, Moxon ER, Winkelstein JA (1992) Genetics and infectious diseases. In: King RA, Rotter JI, Motulsky AG (eds) The genetic basis of common diseases. Oxford University Press, New York, pp 71–91
24. Christian JC (1978) The use of twin registers in the study of birth defects. Birth Defects 14 : 167–178
63. Cooper DN, Krawczak M (1993) Human gene mutation. Bioscience Scientific, Oxford
26. Cudworth AG, Wolf E (1982) The genetic susceptibility to type I (insulin-dependent) diabetes mellitus. Clin Endocrinol Metab 11 : 389–408
27. Cullen P, Farren B, Scott J, Farrall M (1994) Complex segregation analysis provides evidence for a major gene acting on serum triglyceride levels in 55 British families with familial combined hyperlipidemia. Arteriosclerosis Thromb 14 : 1233–1249
28. Czeizel A, Tusnády G, Vaczó G, Vizkelety T (1975) The mechanism of genetic predisposition in congenital dislocation of the hip. J Med Genet 12 : 121–124
30. De Vries RRP, van Rood JJ (1992) Immunogenetics and disease. In: King RA, Rotter JI, Motulsky AG (eds) The genetic basis of common disease. Oxford University Press, New York, pp 92–104
31. Deeb SS, Failor RA, Brown BG, Brunzell JD, Albers JJ, Wjsman E, Motulsky AG (1992) Association of apolipoprotein B gene variants with plasma apoB and LDL cholesterol levels. Hum Genet 88 : 463
32. DeJuan D, Martin-Villa JM, Gomez-Reino JJ et al (1993) Differential contribution of C4 and HLA-DQ genes to systemic lupus erythematosus susceptibility. Hum Genet
33. Dunsford I, Bowley CC, Hutchinson AM, Thompson JS, Sanger R, Race RR (1953) A human blood-group chimera. BMJ 2 : 81
34. Durrington PN, Hunt L, Ishola M et al (1988) Apolipoproteins (a), AI, and B and parental history in men with early onset ischaemic heart disease. Lancet 1 : 1070–1073
35. Edwards JH (1965) The meaning of the associations between blood groups and disease. Am J Hum Genet 29 : 77

36. Epstein CL (1986) The consequences of chromosome imbalance: principles, mechanisms and models. Cambridge University Press, New York

37. Eriksson S (1965) Studies in alpha 1-antitrypsin deficiency. Acta Med Scand 177 : 175

38. Evans DAP, Manley K, McKusick VA (1960) Genetic control of isoniazid metabolism in man. BMJ 2 : 485

39. Fagerhol MK, Cox DW (1981) The Pi polymorphism: genetic, biochemical and clinical aspects of human alpha1-antitrypsin. Adv Hum Genet 11 : 1–62

40. Falconer DS (1981) Introduction to quantitative genetics, 2nd edn. Oliver and Boyd, Edinburgh

41. Farhud DB, Ananthakrishnan R, Walter H (1972) Association between the C3 phenotypes and various diseases. Hum Genet 17 : 57–60

42. Finkelstein S, Walford RL, Myers LW, Ellison GW (1974) HL-A antigens and hypersensitivity to brain tissue in multiple sclerosis. Lancet 1 : 736

43. Fischer M, Harvald B, Hauge M (1969) A Danish twin study of schizophrenia. Br J Psychiatry 115 : 981–990

44. Fisher RA (1918) The correlation between relatives on the supposition of Mendelian inheritance. Trans R Soc Edinb 52 : 399–433

45. Fredrickson DS, Goldstein JL, Brown MS (1978) The familial hyperlipoproteinemias. In: Stanbury JB, Wyngaarden JB, Fredrickson DS (eds) The metabolic basis of inherited disease, 4th edn. McGraw-Hill, New York, pp 604–655

46. Fuhrmann W, Vogel F (1983) Genetic counseling, 3rd edn. Springer, Berlin Heidelberg New York (Heidelberg science library 10)

47. Galton F (1876) The history of twins as a criterium of the relative powers of nature and nurture. J Anthropol Inst

48. Gesell A, Thompson H (1929) Learning and growth in identical infant twins: an experimental study by the method of co-twin control. Genet Psychol Monogr 6 : 5–124

49. Geyer E (1940) Zwillingspärchen mit zwei Vätern. Arch Rassenbiol 34 : 226–236

50. Grebe H (1959) Erblicher Zwergwuchs. Ergeb Inn Med Kinderheilkd 12 : 343–427

51. Goldstein JL, Brown MS (1979) LDL receptor defect in familial hypercholesterolemia. Med Clin North Am 66 : 335–362

52. Goldstein JL, Hazzard WR, Schrott HG, Bierman EL, Motulsky AG (1973) Hyperlipidemia in coronary heart disease. II. Genetic analysis of lipid levels in 176 families and delineation of a new inherited disorder. J Clin Invest 54 : 1544–1568

53. Gottron HA, Schnyder UW (eds) (1966) Vererbung von Hautkrankheiten. Springer, Berlin Heidelberg New York (Handbuch der Haut- und Geschlechtskrankheiten, suppl 7)

54. Grosse-Wilde H, Bertrams J, Schuppien W, Netzel B, Ruppelt W, Kuwert EK (1977) HLA-D typing in 111 multiple sclerosis patients: distribution of four HLA-D alleles. Immunogenetics 4 : 481–488

55. Grüneberg H (1952) Quasi-continuous variations in the mouse. Symp Genet 3 : 215–227

56. Grüneberg H (1952) Genetical studies in the skeleton of the mouse. IV. Quasi-continuous variations. J Genet 51 : 95–114

57. Halldin J (1984) Prevalence of mental disorder in an urban population in central Sweden. Acta Psychiatr Scand 69 : 503–518

58. Harris H, Hopkinson DA (1972) Average heterozygosity per locus in man: an estimate based on the incidence of enzyme polymorphisms. Ann Hum Genet 36 : 9–20

59. Harris H, Smith CAB (1948) The sib-sib age of onset correlation among individuals suffering from a hereditary syndrome produced by more than one gene. Ann Eugen 14 : 309–318

60. Hegele RA, Breslow JL (1987) Apolipoprotein genetic variation in the assessment of atherosclerosis susceptibility. Genet Epidemiol 4 : 163–184

61. Hegeman JP, Mash AJ, Spivey BE (1974) Genetic analysis of human visual parameters in populations with varying incidences of strabism. Am J Hum Genet 26 : 549–562

62. Hershon K, Brunzell J, Albers JJ, Haas L, Motulsky A (1981) Hyper-apo-B-lipoproteinemia with variable lipid phenotype (familial combined hyperlipidemia). Arteriosclerosis 1 : 380 a

63. Hobbs HH, Russell DW, Brown MS, Goldstein JL (1990) The LDL receptor locus in familial hypercholesterolemia: mutational analysis of a membrane protein. Annu Rev Genet 24 : 133–170

64. Howell-Evans W, McConnell RB, Clarke CA, Sheppard PM (1958) Carcinoma of the oesophagus with keratosis palmaris and plantaris (tylosis). Q J Med [New Ser] 27 : 413–429

65. Humphries SE (1988) DNA polymorphisms of the apolipoprotein genes – their use in the investigation of the genetic component of hyperlipidaemia and atherosclerosis. Atherosclerosis 72 : 89–108

66. Husén T (1959) Psychological twin research. Almquist and Wiksell, Stockholm

67. Inkeles S, Eisenberg D (1981) Hyperlipidemia and coronary atherosclerosis: a review. Medicine (Baltimore) 60 : 110–123

68. Johannsen W (1926) Elemente der exakten Erblichkeitslehre, 3rd edn. Fischer, Jena (1st edn 1909)

69. Jörgensen G (1974) Erbfaktoren bei häufigen Krankheiten. In: Vogel F (ed) Erbgefüge. Springer, Berlin Heidelberg New York, pp 581–665 (Handbuch der allgemeinen Pathologie, vol 9)

70. Kagan A, Rhoads GC, Zeegan PD, Nichaman MZ (1971) Coronary heart disease among men of Japanese ancestry in Hawaii: the Honolulu heart study. Isr J Med Sci 7 : 1573

71. Kallmann FJ (1946) The genetic theory of schizophrenia. An analysis of 691 schizophrenie twin index families. Am J Psychiatry 103 : 3

72. Kamboh MI, Ferrell RE, Kottke BA (1991) Expressed hypervariable polymorphism of apolipoprotein (a). Am J Hum Genet 49 : 1063–1074

73. Kämpe O, Larhammar D, Wiwan k, Scheuning L, Claesson L, Gustafsson K, Pääbo S, Hyldig-Nielsen JJ, Rask L, Peterson PA (1983) Molecular analysis of MHC antigens. In: Möller E, Möller G (eds) Genetics of the immune response. Plenum, New York, pp 61–79

74. Kelley WN, Greene ML, Rosenbloom FM, Henderson JF, Seegmiller JE (1969) Hypoxanthine-guanine phosphoribosyltransferase deficiency in gout. Ann Intern Med 70 : 155

75. Kloimwieder R (1942) Die Intelligenz in ihren Beziehungen zur Vererbung, Umwelt und Übung. Z Menschl Vererbungs Konstitutionslehre 25 : 582–617

76. Kringlen E (1967) Heredity and environment in the functional psychoses. Heinemann, London

77. Krüger J (1973) Zur Unterscheidung zwischen multifaktoriellem Erbgang mit Schwellenwerteffekt und einfachem diallelem Erbgang. Hum Genet 17 : 181–252

78. Krüger J, Propping P (1976) Rückgang der Zwillingsgeburten in Deutschland. Dtsch Med Wochenschr 101 : 475–480

79. Kueppers FA (1992) Chronic obstructive pulmonar disease. In: King RA, Rotter JI, Motulsky AG (eds) The genetic basis of common diseases. Oxford University Press, New York, pp 222–239

80. Lackner C, Boerwinkle E, Leffert CC et al (1991) Molecular basis of apolipoprotein (a) isoform size heterogeneity as revealed by pulsed-field gel electrophoresis. J Clin Invest 87 : 2077–2086

81. Landy HL, Weiner S, Corson SL, Batzer FR, Bolognese RJ (1986) The "vanishing twin." Ultrasonographic assessment of disappearance in the first trimester. Am J Obstet Gynecol 155 : 14–19

82. Langdon M, van Dam M, Welsh KL et al (1984) Genetic markers in narcolepsy. Lancet 2 : 1178–1180

83. Laurell C-B, Eriksson S (1963) The electrophoretic a1-globulin pattern of serum a 1-antitrypsin deficiency. Scand J Clin Lab Invest 15 : 132

84. Lemser H (1938) Zur Erb-und Rassenpathologie des Diabetes mellitus. Arch Rassenbiol 32 : 481

85. Lenz W (1959) Ursachen des gesteigerten Wachstums der heutigen Jugend. In: Akzeleration und Ernährung. (Fettlösliche Wirkstoffe, vol 4)

86. Lenz W (1973) Vererbung und Umwelt bei der Entstehung von Mißbildungen. Humanbiologie 124 : 132–145

87. Levy RI (1981) Declining mortality in coronary heart disease. Arteriosclerosis 1 : 312–325

88. Luxenburger H (1935) Untersuchungen an schizophrenen Zwillingen und ihren Geschwistern zur Prüfung der Realität von Manifestationsschwankungen. Z Gesamte Neurol Psychiatr 154 : 351–394

89. Luxenburger H (1940) Zwillingsforschung als Methode der Erbforschung beim Menschen. In: Just G, Bauer KH, Hanhart E, Lange J (eds) Methodik, Genetik der Gesamtperson. Springer, Berlin, pp 213–248 (Handbuch der Erbbiologie des Menschen, vol 2)

90. Malinow MR (1990) Hyperhomocystinemia: a common and easily reversible risk factor for occlusive atherosclerosis. Circulation 81 : 2004–2006

91. McArthur N (1953) Statistics in twin birth in Italy. 1949 and 1950. Ann Eugen 17 : 249

92. McGill HC Jr (1968) Fatty streaks in the coronary arteries and aorta. Lab Invest 18 : 560–564

93. Métneki J, Czeizel A (1980) Contraceptive pills and twins. Acta Genet Med Gemellol (Rome) 29 : 233–236

94. Meyers DA, Marsh DG (1992) Allergy and asthma. In: King RA, Rotter JI, Motulsky AG (eds) The genetic basis of common disease. Oxford University Press, New York, pp 130–149

95. Morton NE (1956) The detection and estimation of linkage between the genes for elliptocytosis and the Rh blood type. Am J Hum Genet 8 : 80–96

96. Morton NE (1976) Genetic markers in atherosclerosis: a review. J Med Genet 13 : 81–90

97. Motulsky AG (1978) The genetics of common diseases. In: Morton NE, Chung CS (eds) Genetic epidemiology. Academic, New York, pp 541–548

98. Motulsky AG (1979) The HLA complex and disease. Some interpretations and new data in cardiomyopathy. N Engl J Med 300 : 918–919

99. Motulsky AG (1982) Genetic approaches to common diseases. In: Bonné-Tamir B (ed) Medical aspects. Liss, New York, pp 89–95 (Human genetics, part B)

99 a. Motulsky AG (1996) Nutritional ecogenetics: homocysteine-related arterioclerotic vascular disease, neural tube defects, and folic acid (Invited editorial). Am J Hum Genet 58 : 17–20

100. Motulsky AG, Brunzell JD (1992) The genetics of coronary arteriosclerosis. In: King RA, Rotter JI, Motulsky AG (eds) The genetic basis of common disease. Oxford University Press, Oxford, pp 150–169

101. Mourant AE, Kopec AC, Domaniewska-Sobczak K (1978) Blood groups and diseases. Oxford University Press, London

102. Mourant AE, Kopec AC, Domaniewska-Sobczak K (1978) Blood groups and diseases. Oxford University Press, London

103. Naeslund J (1956) Metodiken rid den första läsundervisningen. En översikt och experimentella bidrag. Svenska, Uppsala

104. Nagel RL, Labie D (1985) The consequences and implications of the multicentric origin of the Hb S gene. In: Stamatoyannopoulos G, Nienhuis A (eds) Experimental approaches for the study of hemoglobin switching. Liss, New York, pp 93–103

105. Nagel RL, Fabry ME, Pagnier J, Zouhoun I, Wajcman H, Baudin V, Labie D (1985) Hematologically and genetically distinct forms of sickle cell anemia in Africa. The Senegal type and the Benin type. N Engl J Med 312 : 880–884

106. Nance W (1990) Do twin Lyons have larger spots? Am J Hum Genet 46 : 646–648

107. Neel JV, Fajons SS, Conn JW, Davidson RT (1965) Diabetes mellitus. In: Neel JV, Shaw MW, Schull WJ (eds) Genetics and the epidemiology of chronic disease. GPO, Washington, pp 105–132

108. Nicholas JW, Jenkins WJ, Marsh WL (1957) Human blood chimeras. A study of surviving twins. BMJ 1 : 1458

109. Nora JJ, Lortscher RH, Spangler RD, Nora AH, Kimberling WJ (1980) Genetic-epidemiologic study of early-onset ischemic heart disease. Circulation 62 : 503–508

110. Olefsky JM (1985) Diabetes mellitus. In: Wyngaarden JB, Smith LH Jr (eds) Cecil textbook of medicine, 17 th edn. Saunders, Philadelphia, pp 1320–1341

111. Penrose LS (1938) (Colchester survey) A clinical and genetic study of 1280 cases of mental defect. HMSO, London (Special report series of the Medical Research Council, vol 229)

112. Penrose LS (1962) The biology of mental defect, 3 rd edn. Grune and Stratton, New York

113. Poll H (1914) Über Zwillingsforschung als Hilfsmittel menschlicher Erbkunde. Z Ethnol 46 : 87–108

114. Propping P (1983) Genetic disorders presenting as "schizophrenia". Karl Bonhoeffer's early view of the psychoses in the light of medical genetics. Hum Genet 65 : 1–10

115. Propping P, Krüger J (1976) Über die Häufigkeit von Zwillingsgeburten. Dtsch Med Wochenschr 101 : 506–512

116. Propping P, Vogel F (1976) Twin studies in medical genetics. Acta Genet Med Gemellol (Rome) 25 : 249–258

117. Propping P, Voigtländer V (1983) Was ist gesichert in der Genetik der Atopien? Allergologie 6 : 160–168

118. Puska P, Tuomilehto J, Salonen J et al (1979) Changes in coronary risk factors during a comprehensive five-year community programme to control cardiovascular diseases (North Karelia project) BMJ 2 : 1173–1178

119. Rao CD, Elston RC, Kuller LH, Feinleib M, Carter C, Havlik R (eds) (1984) Genetic epidemiology of coronary heart disease. Past, present and future. Liss, New York

120. Renwick JH (1969) Progress in mapping human autosomes. Br Med Bull 25 : 65

121. Richter S (1967) Zur Heredität des Strabismus concomitans. Humangenetik 3 : 235–243

122. Robbins DC, Blix PM, Rubenstein AH, Kanazawa Y, Kosaka K, Tager HS (1981) A human proinsulin variant at arginine 65. Nature 291 : 679–681

123. Robertson FW (1981) The genetic component in coronary heart disease – review. Genet Res (Cambr) 37 : 1–16

124. Rose NR, Vladutio AO, David CS, Shreffler DC (1973) Autoimmune murine thyroiditis. V. Genetic influence on the disease in BSVS and BRVR mice. Clin Exp Immunol 15 : 281–287

125. Rotter JI, Rimoin DL (1981) The genetics of the glucose intolerance disorders. Am J Med 79 : 116–126

126. Rotter JI, Vadheim CM, Rimoin DL (1992) Diabetes mellitus. In: King RA, Rotter JD, Motulsky AG (eds) The genetic basis of common diseases. Oxford University Press, New York, pp 413–481

127. Schnyder UW (1955) Neurodermitis und Allergie des Respirationstraktes. Dermatologica 110 : 289

128. Selby PB, Selby PR (1978) Gamma-ray-induced dominant mutations that cause skeletal abnormalities in mice. II. Description of proved mutations. Mutat Res 51 : 199–236

129. Shoelson S, Haneda M, Blix P, Nanjo A, Sanke T, Inouye K, Steiner D, Rubenstein A, Tager H (1983) Three mutant insulins in man. Nature 302 : 540–543

130. Short EM (1992) Hyperuricemia and gout. In: King RA, Rotter JI, Motulsky AG (eds) The genetic basis of common disease. Oxford University Press, New York, pp 482–506

131. Siemens HW (1924) Die Zwillingspathologie. Springer, Berlin

132. Siemens HW (1924) Die Leistungsfähigkeit der zwillingspathologischen Arbeitsmethode. Z Induktive Abstammungs Vererbungslehre 33 : 348

133. Simonds B (1963) Tuberculosis in twins. Putnam, London

134. Sing CR, Orr JD (1976) Analysis of genetic and environmental sources of variation in serum cholesterol in Techumseh, Michigan. III. Identification of genetic effects using 12 polymorphic genetic marker systems. Am J Hum Genet 28 : 453–464

135. Smith C (1971) Recurrence risks for multifactorial inheritance. Am J Hum Genet 23 : 578–588

136. Steel CM, Beatson D, Cuthbert RJG et al (1987) HLA haplotype A, B8DR3 as a risk factor for HIV related disease. Lancet 1 : 1185–1188

137. Steinlein O, Anokhin A, Mao Y-P et al (1992) Localization of a gene for the human low-voltage EEG on 20 q and genetic heterogeneity. Genomics 12 : 69–73

138. Steinlein O, Smigrodzki R, Lindstrom J, Anand R, Köhler M, Tocharoentanophol C, Vogel F (1994) Refinement of the localization of the gene for neutonal nicotinic acetylcholine receptor a4 subunit (CHRNA4) to human chromosome 20 q13.2-q13.3. Genomics 22 : 493–495

139. Stern C (1953) Model estimates of the frequency of white and near-white segregants in the American negro. Acta Genet (Basel) 4 : 281–298

140. Stoffel M, Froguel P, Takeda J et al (1992) Human glucokinase gene: isolation, characterization and identification of two missense mutations linked to early-onset non-insulin dependent (type 2) diabetes melitus. Proc Natl Acad Sci USA 89 : 7698–7702

141. Stone NJ (1979) Genetic hyperlipidemia and atherosclerosis. Artery 5 : 377–397

142. Svejgaard A, Jersild C, Staub Nielsen L, Bodmer WF (1974) HL-A antigens and disease. Statistical and genetical considerations. Tissue Antigens 4 : 95–105

143. Svejgaard A, Platz P, Ryder LP (1983) HLA and disease 1982. Immunol Rev 70 : 193–218

144. Takata H, Suzuki M, Ishii T et al (1987) Influence of major histocompatibility complex region genes on human longevity among Okinawan Japanese centenarians and nonagenarians. Lancet 2 : 824–826

145. Tattersall RB, Pyke DA (1972) Diabetes in identical twins. Lancet 2 : 1120–1125

146. Ten Kate LP, Boman H, Daiger SP, Motulsky AG (1984) Increased frequency of coronary heart disease in relatives of wives of myocardial infarct survivors: assortative mating for lifestyle and risk factors. Am J Cardiol 53 : 399–403

147. Tienari P (1963) Psychiatric illnesses in identical twins. Acta Psychiatr Scand [Suppl]:171

148. Timoféef-Ressovsky NW (1931) Gerichtetes Variieren in der phänotypischen Manifestierung einiger Generationen von Drosophila funebris. Naturwissenschaften 19 : 493–497

149. Tiwari JL, Terasaki PI (1985) HLA and disease associations. Springer, New York

150. Turner G, Jacobs PA (1984) Mental retardation and the fragile X. Adv Hum Genet 13:n

151. Utermann G, Hardewig A, Zimmer F (1984) Apolipoprotein E phenotypes in patients with myocardial infarction. Hum Genet 65 : 237–241

152. Utermann G, Kindermann I, Kaffarnik H, Steinmetz A (1984) Apolipoprotein E phenotypes and hyperlipidemia. Hum Genet 65 : 232–236

153. Vogel F (1970) The genetic basis of the normal human electroencephalogram (EEG). Hum Genet 10 : 91–114

154. Vogel F (ed) (1974) Erbgefüge. Springer, Berlin Heidelberg New York (Handbuch der allgemeinen Pathologie, vol 9)

155. Vogel F (1984) Relevant deviations in heterozygotes of autosomal-recessive diseases. Clin Genet 25 : 381–415

156. Vogel F (1990) Humangenetik und Konzepte der Krankheit. In: Sitzungsberichte der Heidelberger Akademie der Wissenschaften, Mathematisch-naturwissenschaftliche Klasse, 6. Abhandlung. Springer, Berlin Heidelberg New York, pp 335–353

157. Vogel F, Krüger J (1967) Multifactorial determination of genetic affections. Proceedings of the 3rd International Congress of Human Genetics. Johns Hopkins University Press, Baltimore, pp 437–445

158. Von Bracken H (1934) Mutual intimacy in twins: types of structure in pairs of identical and fraternal twins. Character Personality 2 : 293–309

159. Von Bracken H (1969) Humangenetische Psychologie. In: Becker PE (ed) Humangenetik, ein kurzes Handbuch, vol 1/2. Thieme, Stuttgart, pp 409–562

160. Von Verschuer O (1954) Wirksame Faktoren im Leben des Menschen. Steiner, Wiesbaden

161. Von Verschuer O (1958) Die Zwillingsforschung im Dienste der inneren Medizin. Verh Dtsch Ges Inn Med 64 : 262–273

162. Waardenburg PJ (1957) The twin study method in wider perspective. Acta Genet (Basel) 7 : 10–20

163. Wainscoat JS, Thein SL, Higgs DR, Bell JI, Weatherall DJ, Al-Awamy BH, Serjeant GR (1985) A genetic marker for elevated levels of haemoglobin F in homozygous sickle cell disease. Br J Haematol 60 : 261–268

164. Weinberg W (1902) Beiträge zur Physiologie und Pathologie der Mehrlingsgeburten beim Menschen und Probleme der Mehrlingsgeburtenstatistik. Z Geburtsh Gynakol 47 : 12

165. Weinberg W (1909) Der Einfluß von Alter und Geburtenzahl der Mutter auf die Häufigkeit der ein- und zweieiigen Zwillingsgeburten. Z Geburtsh Gynakol 65 : 318–324

166. Wieland W (1975) Diagnose. Überlegungen zur Medizintheorie. De Gruyter, Berlin

167. Wieland W (1983) Systematische Bemerkungen zum Diagnosebegriff. Münstersche Beiträge zur Geschichte und Theorie der Medizin 20 : 17–34

168. Williams WR, Lalouel JM (1982) Complex segregation analysis of hyperlipidemia in a Seattle sample. Hum Hered 32 : 24–26

169. Wilson SR (1973) The correlation between relatives under the multifactorial model with assortative mating. I. The multifactorial model with assortative mating. Ann Hum Genet 37 : 189–204

170. Wilson SR (1973) The correlation between relatives under the multifactorial model with assortative mating. II. The correlation between relatives in the equilibrium position. Ann Hum Genet 37 : 205–215

171. Wojchechowski AP, Farral M, Cullen P, Wilson TME, Bayliss JD, Farren B, Griffin BA, Caslake MJ, Packard CJ, Shepherd J, Thakker R, Scott J (1991) Familial combined hyperlipidemia linked to the apolipoprotein AI-CIII-AIV gene cluster on chromosome 11 q23-q24. Nature 349 : 161–164

172. Woolf B (1955) On estimating the relation between blood group and disease. Ann Hum Genet 19 : 251–253

173. Wright S (1931) Evolution in mendelian populations. Genetics 16 : 97–159

7 Gene Action: Genetic Diseases

This investigation reveals … a clear case of a change produced in a protein molecule by an allelic change in a single gene involved in synthesis.

L. Pauling, "Sickle cell anemia, a molecular disease," Science, 1949

7.1 Aspects of the Problem

Gene Action and Genetic Strategies. How do genes manage to determine the development and function of the organism? This is the basic problem of genetic biology that should be solved. The topics discussed earlier such as the structure of the genetic material and Mendelian segregation are interesting scientifically because they help in solving this basic question. In Chap. 6 Mendelism is described as the central paradigm of genetics. Thus, a Mendelian mode of inheritance for a certain phenotype points to a specific change within the information-carrying DNA, and the study of linkage with DNA variants as markers leads to precise localization of the gene causing this phenotype. After the nature of the genetic alteration has been defined, the path of gene action between genotype and phenotype can be elucidated. For a long time linkage studies were known in principle to be the "King's road" leading to gene localization. Linkage study in humans is more difficult than in experimental organisms because of the lack of directed matings. Although statistical techniques needed for human gene mapping have been available for some time, practical implementation was impeded by the lack of suitable genetic markers that could be used as landmarks for gene localization. This has fundamentally changed with discovery of the very frequent DNA polymorphisms. (Sect. 12.1.2).

Before these methods were available, only phenotypes (i.e., end results) could be studied. Therefore "classic" genetic analysis had to follow the the path from the phenotype as a starting point in several steps down to the "bottom": the gene and its mutation. As biochemical genetics developed, this approach was often remarkably successful. It led to clearcut results at the gene product level such as in enzyme defects and hemoglobin variants, and allowed inferences of the DNA mutation from knowledge of the genetic code that related an amino acid alteration to the corresponding nucleotide change in the DNA. However, research workers soon became aware of the limits of this approach. Attempts to dig deeper into the biological causes were often doomed to failure. Biochemical data did not point unequivocally to a certain defect, or more often the phenotype could not be related to a gene-determined defect in an enzyme or protein.

Here the more recent "bottom-up" approach often helped. After mapping and identifying the gene, its normal and abnormal protein product could be defined, allowing the elucidation of its normal function and of the abnormal phenotype. As shown in Chap. 3.1.3.9, such "positional cloning" has been increasingly successful, such as in Huntington disease and cystic fibrosis. In these diseases a definite phenotype and the mode of inheritance were apparent, but neither the location of their respective genes nor the structure or function was known. Gene localization had to be carried out "blindly," requiring a search over the entire genome since there were no clues as to location of these genes. After mapping was accomplished, elucidation of the mutations at the DNA level in each case resulted in definition of previously unknown proteins (huntingtin and CFTR) that were mutant and caused the disease. However, the exact mechanism of how a specific mutation causes the disease remains under study. While there are excellent clues to cell membrane dysfunction in cystic fibrosis, the pathway from the altered genotype to phenotype in Huntington disease remains completely unknown.

In other cases, such as in some autosomal dominant cardiomyopathies, localization of the disease phenotype to chromosome 14q led to a search for a biologically plausible candidate gene that mapped to this site. A muscle protein myosin had earlier been localized to this position and was found to be mutant in the affected cardiomyopathy kindreds. In this example, the normal myosin gene had been analyzed before the disease was mapped. With the broad application of molecular methods in many fields of biology and medicine such analyses of "normal" genes without assistance by known monogenic traits or diseases are becoming increasingly common. Modern genetic analysis no longer requires mutations for gene identification!

The next logical step would be analysis of traits in which the mode of inheritance is more complex, and the gene or genes are unknown. Here linkage analysis is technically much more difficult but may point to a "major gene" (Sect. 6.1.2). The action of this gene may then be analyzed "bottom-up," as described.

Hereditary Diseases as Analytical Tools for the Elucidation of Gene Action. The examples above deal with hereditary diseases. This emphasizes an important principle of genetic analysis. Classical genetic analysis starts with distinguishable abnormal phenotypes that exhibit a Mendelian mode of inheritance. Such traits permit a thorough analysis of gene action. "Normal" traits, on the other hand, often are distributed continuously; correlations among relatives suggest genetic determination which cannot be readily interpreted in terms of gene action and biochemical mechanisms. Therefore "classical" genetics is largely a genetic analysis of anomalies. However, from this analysis, conclusions can be inferred as to normal function. In humans, this trend toward analysis of genetic diseases is reinforced by the setting in which genetic studies are performed. Patients with certain diseases are more readily available for genetic studies than healthy individuals. Moreover, the motivation of research workers is stronger if there is hope of practical and useful results.

For all of these reasons, we mainly use hereditary diseases as models for exploring genetic concepts that apply to normal development and function.

The Sequence of Problems to be Discussed. In the following sections we begin with the conceptually relatively simple and well-analyzed cases: enzyme defects. The hemoglobin variants follow as examples for successful analysis of gene action and its variation. Some special topics are then discussed, such as genetic variation leading to disease only when specific environmental conditions are present (pharmacogenetics; ecogenetics). A survey of genetic mechanisms in dominant conditions reveals a great variety of disturbances of different normal functions. In the next step from the simple to the more complex we discuss briefly the intricate immune system.

Immunogenetics has developed over recent years into a field of its own, but many of its aspects are also important for other fields of human genetics. This leads to the most complex, and least understood field of gene action: embryonic development and its disturbances (Chap. 8). In this context new concepts are described which apparently contradict predictions from Mendelism, such as genomic imprinting.

7.2 Genes and Enzymes

7.2.1 One-Gene/One-Enzyme Hypothesis

Early Forerunners. Garrod in 1902 [75] (Sect. 1.5) related the gene defect in alkaptonuria to a specific inability of the organism to degrade homogentisic acid. The obvious next problem was to determine the specific mechanism responsible for this inability. Metabolic steps are catalyzed by enzymes. Therefore alterations of enzymes offered a plausible explanation. This had already been discussed by Driesch in 1896 and proposed by Haldane in 1920 (see [109]) and by Garrod in 1923 [96]. Important early analytic steps in biochemical genetics were the analyses of eye color mutants in the flour moth *Ephestia kühniella* by Kühn [147] and Butenandt [44] and in *Drosophila* by Beadle and Ephrussi (1936) [15]. These first attempts chose mutants in insects that had been analyzed by genetic methods to elucidate the mechanisms of gene action. This approach, however, had only limited success as the problem proved to be too complex for a direct attack. A more successful approach required two conditions:

1. A simpler test organism had to be found that provided better opportunities for experimentation.
2. The problem needed to be examined by looking for genetic explanations of the biochemical phenotypes rather than by providing biochemical explanations for genetically defined traits.

Both requirements were met by the work of Beadle and Tatum in 1941 [16] and that of Beadle in 1945 [14].

Beadle's and Tatum's Simple Organism and Method of Attack. The paper by these two investigators begins as follows (our italics):

From this standpoint of physiological genetics the development and functioning of an organism consists essentially of an integrated system of chemical reactions controlled in some manner by genes. It is entirely tenable to suppose that these genes ... control or regulate specific reactions in the system either by acting directly as enzymes or by determining the specificities of enzymes. Since the components of such a system are likely to be interrelated in complex ways, and since the synthesis of the parts of individual genes are presumably dependent on the functioning of other genes, it would appear that there must exist orders of directness of gene control ranging from simple one-to-one relations to relations of great complexity. In investigating the roles of genes, the physiological geneticist usually attempts the physiological and biochemical bases of already known hereditary traits. This approach ... has established that many biochemical reactions are in fact controlled in specific ways by specific genes. Furthermore, investigations of this type tend to support the assumptions that gene and enzyme specificities are of the same order. There are, however, a number of limitations inherent in this approach. Perhaps the most serious of these is that the investigator must in general confine himself to a study of nonlethal heritable characters. Such characters are likely to involve more or less non-essential so-called "terminal" reactions . . .. A second difficulty . . . is that the standard approach to the problem implies the use of characters with visible manifestations. Many such characters involve morphological variations, and these are likely to be based on systems of biochemical reactions so complex as to make analysis exceedingly difficult.

Considerations such as those just outlined, have led us to investigate the general problem of the genetic control of developmental and metabolic reactions by *reversing the ordinary procedure* and, instead of attempting to work out the chemical basis of known genetic characters, to set out to determine *if and how genes control known biochemical reactions*. The ascomycete neurospora offers many advantages for such an approach and is well suited for genetic studies. Accordingly, our program has been built around this organism. The procedure is based on the assumption that X-ray treatment will induce mutations in genes concerned with the control of known specific chemical reactions. If the organism must be able to carry out a certain chemical reaction to survive on a given medium, a mutant unable to do this will obviously be lethal on this medium. Such a mutant can be maintained

and studied, however, if it will grow on a medium to which has been added the essential product of the genetically blocked reaction.

Beadle and Tatum [16] then described their experimental design (Fig. 7.1). The complete medium contained agar, inorganic salts, malt extract, yeast extract, and glucose. The minimal medium, on the other hand, contained only agar, salts, biotin, a disaccharide, and fat or other carbon source. Mutants that grow on the complete but not on the minimal medium were tested systematically by the gradual adding of the complete medium components to ascertain which component the mutants were unable to synthesize.

In this manner, mutants were isolated that are unable to synthesize growth factors, such as pyridoxin, thiamine, and p-aminobenzoic acid. These defects were shown to be caused by mutations at specific gene loci. This work inaugurated an abundance of investigations on neurospora, bacteria, and yeast, in which such mutants were analyzed, and "genetic blocks" in single metabolic steps were related to specific enzyme defects. Soon this approach became an important tool for better assessment of single steps within metabolic pathways.

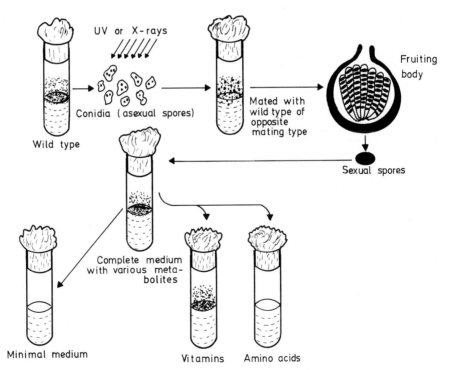

Fig. 7.1. The experimental design for discovering biochemical mutants in neurospora. The X-ray or UV-induced mutation does not impair fungus growth on complete medium. On minimal medium, however, the fungus cannot grow. Addition of vitamins restores growth capacity shown as "stippling." Addition of amino acids causes no growth. The figure sug-

gests that the mutation has affected a gene, which influences vitamin metabolism. The next step in the experiment would be to determine which vitamin is able to restore normal function. The genetic block is found in the metabolism of this vitamin. (Modified from Sinnott et al. 1958 [222])

The one-gene/one-enzyme hypothesis was established and now had a solid experimental foundation. This hypothesis proved to be highly fertile during the decades that followed. Analysis of enzyme defects and variants soon also provided evidence for genetic blocks in which only the function of the enzyme was impaired while an enzyme protein was still present that had kept its antigenic properties (cross-reacting material; CRM). In other cases the enzyme had an altered temperature optimum for its action. Some variants could be explained by a mutation causing altered activity of a series of enzymes by affecting a common control unit. From such studies there emerged the concept of regulation of bacterial gene action, which included the operon concept.

First Enzyme Defects in Humans. The first genetic disease in humans for which an enzyme defect could be shown was a recessively inherited type of methemoglobinemia (Gibson and Harrison in 1947 [100] and Gibson in 1948 [99]; 250 800). The enzyme deficient in these cases is the NADH-dependent methemoglobin reductase. The first systematic attempt to elucidate a group of human metabolic diseases was made in 1951 by the Cori's in glycogen storage disease [64].

The Cori's showed first that the structure of liver glycogen in ten cases of what was then called von Gierke disease (232 200) was within the normal range of variation in eight and definitely abnormal in two cases. It was also obvious that liver glycogen, which accumulates in excessive amounts, is not readily available for blood sugar formation, as the patients show a tendency to hypoglycemia. Many enzymes are required for the conversion of glycogen to glucose in the liver. Two of them, amylo-1,6-glucosidase and glucose-6-

phosphatase, were selected as possible candidates to be the deficient enzyme. Liberation of phosphate from glucose-6-phosphate was measured in liver homogenates at various pH levels. Figure 7.2 shows the results. In a normal liver, a high level of activity is noted, with a maximum at pH 6–7. Severe liver damage due to cirrhosis leads only to a moderate decrease. In a fatal case of von Gierke disease, on the other hand, no activity at all could be detected; this was confirmed in a second case. Two patients with milder disease symptoms showed markedly reduced activity.

It was concluded that there is an enzyme defect of glucose-6-phosphatase in these fatal cases of von Gierke disease. At the same time, however, in most of the milder cases the activity of this enzyme was not reduced below the level found in liver cirrhosis; only the two patients shown in Fig. 7.2 had moderately reduced values. The Cori's offered no explanation for the latter results. They also noted that abnormal muscle storage of glycogen cannot be explained by a lack of glucose-6-phosphatase, since this enzyme is normally absent from muscle. For cases with muscular glycogenosis they suggested a defect of amylo-1,6-glucocidase at the possible explanation. This prediction was soon confirmed, as Forbes [87] discovered this defect in a clinical case of glycogen storage disease involving both heart and skeletal muscles.

Currently, many different enzyme defects have been discovered in glycogen storage disease. For details see [121].

While the patterns of manifestation differ somewhat among the various types, there is much overlap in clinical manifestations. The mode of inheritance is, with only one exception, autosomal-recessive. Had the enzyme defects not been discovered, glycogen

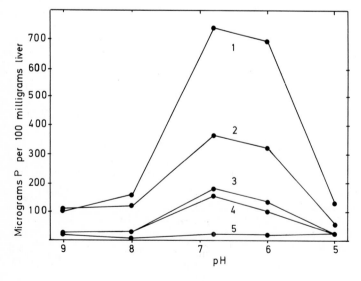

Fig. 7.2. Liberation of phosphate from glucose-6-phosphate in liver homogenates from various patients as a measure of glucose-6-phosphatase activity. *1,* High phosphate liberation in a patient with normal liver function; *2,* moderate reduction in cirrhosis of the liver; *3, 4,* marked reduction in two patients with the milder form of glycogen storage disease; *5,* complete absence of enzyme activity in a patient with severe von Gierke disease. (From Cori and Cori 1952 [64])

storage disease would be regarded as a single genetic disease with intrafamilial correlations as to severity, details of symptoms, and time of death. Hence, we have here an example of the way in which genetic heterogeneity, which can only be suspected at the phenotypic level (Sect. 4.3.5), is confirmed by analysis at the biochemical level: identification of specific genes by analysis of enzyme activities.

In the following years, enzyme defects were discovered with increasing momentum, and in the 1994 edition of his *Mendelian Inheritance in Man* McKusick [164] was able to list about 250 autosomal-recessive conditions in which a specific enzyme defect had been located. Many achievements in methodology have contributed to this progress. Most are beyond the scope of this volume, but some played a key role in the development of concepts and methods in biochemical and molecular genetics.

Steps in Understanding Human Enzyme Defects. A number of steps have been especially important for continuing developments:

1934 Phenylketonuria was detected by Følling [86].
1941 Beadle and Tatum [16] introduced the one-gene/one-enzyme hypothesis.
1948 Gibson [99] described the first enzyme defect in a human disease (recessive methemoglobinemia).
1952 Glucose-6-phosphatase deficiency in von Gierke disease was discovered by the Cori's [64].
1953 Jervis [127] demonstrated the lack of phenylalanine hydroxylase in phenylketonuria. Bickel [28] reported the first attempt to alleviate an enzyme defect by a low-phenylalanine diet.
1955 Smithies [225, 226] introduced starch gel electrophoresis.
1956 Carson et al. [53] discovered the defect of glucose- 6-phosphate dehydrogenase (G6PD) in drug-induced hemolytic anemia.
1957 The enzyme defect in galactosemia (P-gal-transferase deficiency) was described by Kalckar et al. [131], showing that an identical enzyme deficiency exists in humans and bacteria.
1961 Krooth and Weinberg [148] demonstrated the enzyme defect of galactosemia in fibroblasts cultured in vitro.
1967 Seegmiller et al. [218] discovered the defect of hypoxanthine-guanine-phosphoribosyl transferase (HPRT) in Lesch-Nyhan syndrome.
1968 Deficiency of excision repair in xeroderma pigmentosum was described by Cleaver [59].
1970 Elucidation of the enzyme defects in the mucopolysaccharidoses by Neufeld [186], allowing full understanding of the catabolic pathways of mucopolysaccharide metabolism.

1974 Demonstration by Brown and Goldstein [38] that genetically determined overproduction of an enzyme (HMG CoA reductase) in familial hypercholesterolemia is caused by a membrane receptor defect (low-density lipoprotein receptor), which modulates intracellular HMG CoA reductase activity.
1977 Demonstration by Sly et al. [224] that the mannose 6-phosphate components of lysosomal enzymes are recognized by fibroblast receptors. Genetic processing defect prevents binding of lysosomal enzymes with failure to enter cells and secretion into plasma (I cell disease).
1980 Defect of receptor cyclase coupling protein in pseudohypoparathyroidism.
1983 Cloning of the gene for phenylalanine hydroxylase by Woo et al. [269].
1989 Identification of the gene for cystic fibrosis (cystic fibrosis transmembrane regulator, CFTR) and identification of the basic defect [205].
Early 1990s Successful treatment of Gaucher's disease by enzyme therapy.

7.2.2 Genes and Enzymes in Humans: Present State of Knowledge

Scope and Limitations of this Review. Each enzyme defect presents special problems in methodology and interpretation. Limitations of space enforce a short and highly selective discussion of these problems, two groups of which are selected.

1. Those of significance for an understanding of general principles of genetic determination and control in man.
2. Those of significance for diagnosis of enzyme defects and their contribution to our understanding of disease.

For other diseases, we refer the reader to more specialized monographs [215] and to the many reviews on single diseases or groups of diseases.

7.2.2.1 Discovery and Analysis of Enzyme Defects

Difference in Research Strategy Between Humans and Neurospora. The progress in analysis of enzyme defects in neurospora and bacteria was achieved by a novel research strategy. Instead of searching for biochemical explanations of known mutants, mutants were induced and screened as to whether they affected known metabolic steps. Such an approach can work only if mutations indeed cause genetic blocks due to enzyme defect. Moreover, recovery of mutants is limited to those actually leading to enzyme defects irrespective of whether they are a large or a small proportion of all occurring mutations. In practice this limitation proved useful since it helped to make the problem analytically accessible, allowing for concepts of genetic control mechanisms of enzyme activity to emerge.

In humans, the approach via known metabolic pathways is barred, as we can neither induce mutations artificially nor screen them in a system comparable to the selective systems for identification of auxotrophic mutants in neurospora. We must begin with the phenotype and try to find and analyze the underlying enzyme defect. The obvious disadvantages of this approach are its dependence on chance observations of individuals with rare diseases, and limitation of this approach to instances in which the phenotype gives some hints as to where the biochemical anomaly should be sought. There are also advantages, however. In no experimental animal are so many individuals constantly being examined for their state of health as in humans; moreover, only in our species does the variety of methods available for analysis range from refined clinical descriptions down to characterizations of enzyme proteins. As a consequence, a rich spectrum of phenotypes is offered for observation. The "bottom-up" strategy of positional cloning provides an alternative approach for identification of biochemical pathways and enzyme defects, since it is very much easier to determine the DNA sequence of a gene and infer from it the protein sequence than to work at the protein level (Sect. 3.1.3.9).

Clinical Symptoms Leading to the Detection of Enzyme Defects. How are enzyme defects discovered? Many ways exist. The defect of glucose-6-phosphatase in von Gierke disease offers the simplest example. The disorder had been known for a long time, and the clinical symptoms suggest an anomaly in a specific metabolic pathway. As soon as this pathway is sufficiently well-known, and enzyme assays are available, the research worker must ascertain which enzyme is defective. However, difficulties may occur. These may be primarily technical. For example, many enzyme defects in humans are due not to complete lack of the enzyme but to mutationally altered enzyme properties causing anomalies, such as reduced substrate affinity. Most in vitro assays use high substrate concentration and thus may enable even an altered enzyme to exhibit normal enzyme activity [1186]. In vitro assays therefore may not reflect in vivo activity. Sometimes the symptoms point in the wrong direction, for example in glycogen storage disease type II (Pompe disease). Here, the enzyme defect was shown to affect α-1,4-glucosidase, which was previously not known to be involved in glycogen metabolism.

In other conditions, the clinical symptoms may be so unspecific that no clues as to the metabolic defect exist. Thus, a failure of infants to thrive is associated with many different inborn errors of metabolism affecting various metabolic enzymes.

A small proportion (about 1%) of mentally retarded children resident in institutions have phenylketonuria (PKU). This condition was discovered by Følling in 1934 [86] in two sibs who had a peculiar, mousy odor and excreted large amounts of phenylpyruvic acid in their urine. This discovery provided much hope that many other types of mental retardation could be shown to be caused by various other inborn errors. Many surveys for abnormal urinary metabolites were performed among patients with mental retardation. Unfortunately, the yield was low, and although some other conditions were discovered, such as homocystinuria (see below), most mentally retarded persons were not affected with inborn errors that could be detected by this approach.

While widespread clinical findings involving bone and connective tissue and gross defects are usually not associated with inborn errors of metabolism, there are exceptions, such as homocystinuria, an anomaly in the metabolism of the sulfur-containing amino acid methionine, caused by a deficiency of the liver enzyme cystathionine synthase. The patients suffer from three groups of symptoms: (a) connective tissue and eye anomalies such as osteoporosis, knock knees, spider fingers and toes, and dislocated lenses; (b) anomalies in the function of the central nervous system such as mental retardation in about 50% of the cases; and (c) arterial and venous thrombosis. Some of the findings are similar to those found in Marfan syndrome, a dominant condition that may occur as a new mutation, lead to sporadic nonfamilial cases, and consequently be confused with homocystinuria. Even without this parallel, however, no one familiar with the general symptomats of recessive enzyme defects would have suspected this particular enzyme defect in a disorder with so many different, and principally structural, symptoms. The disorder was discovered in a program for the screening of mentally defective persons.

Clinical Diagnosis of Metabolic Defects. Metabolic defects are quite rare. This means that even busy pediatricians see only a few of them during their careers, and those seen are encountered only once or a very few times. Therefore the complexity of diagnosis and particularly of therapy cannot be expected from every pediatrician. A few departments of pediatrics or medicine in America and Europe are increasingly specializing in the diagnosis (including antenatal diagnosis) and therapy of single or small groups of enzyme defects. This specialization provides the highest possible level of medical care for these patients.

However, every physician, whether general practitioner, pediatrician, or medical geneticist needs to be prepared, to arrange for appropriate diagnosis of these metabolic diseases. Early diagnosis is important not only for disorders in which specific therapy is possible (Sect. 7.2.2.9), but also in cases where births of further affected siblings can be prevented by antenatal diagnosis. Careful diagnosis is therefore important for most inborn errors of metabolism, which often manifest in infants as failure to thrive.

Methods Used for Analysis of Enzyme Defects. The appropriate methods for analysis of enzyme defects are generally those of enzymology. In the elucidation of the genetic basis of enzyme defects in inborn errors one must examine not only quantitative assessments of enzyme activity but also qualitative differences in enzyme characteristics.

Temperature instability has been found, for example, in defective HPRT in some children with Lesch-Nyhan syndrome (308 000). Unusual thermoresistance was found with α-galactosidase in several severe mutations causing Fabry disease, a lysosomal enzyme defect.

Frequently the difference between the normal and abnormal enzyme can be analyzed at the protein level, for example, by altered electrophoretic mobility. Often, the abnormal protein, while losing the catalytic capacity by which it is characterized as an enzyme, may have maintained its immunological properties. It reacts with an antibody that has been produced against the normal enzyme. Such *cross-reacting material* (CRM) was first discovered in bacteria, for example, for trypophan synthetase of *E. coli*. Such CRM proteins are frequent in human enzyme defects.

It is characteristic of enzyme deficiencies in humans that, unlike many enzyme deficiencies in bacteria, qualitatively altered enzymes are more frequently observed than complete or nearly complete loss of an enzyme protein. This finding indicates that most currently known enzyme defects in humans are caused by structural mutations, and not by regulatory mutations, as are often found in bacteria. Wherever analysis of mutants has been possible at the DNA level, this conclusion has been confirmed. These facts are of great significance for the understanding of gene regulation in higher organisms including humans (Chap. 8). Among the many methods used for analysis of enzyme defects, one – described below – has gained special significance.

Examination of Enzyme Defects in Human Fibroblast Cultures. When the genetics of micro-organisms was successfully elucidated in the 1940s and 1950s, many scientists believed that genetic analysis in higher organisms using individual cells would increase the resolving power of analysis by several orders of magnitude. The technical conditions – the growing of cell lines in culture – had been available for several years. However, cell lines capable of growing in cultured medium for an indefinite time had either been derived from malignant tumors – such as the much-used He-La cells – or undergone a change in growth characteristics in vitro, thereby losing their ability for contact inhibition; they were "transformed." These cells were genetically different from normal cells; above all they were almost always aneuploid, with a wide range of chromosome numbers within the same cell line and even in the same culture. Such cells cannot be used for genetic research; methods had to be developed for growing normal euploid cells in culture. Quantitative biochemical work, such as measurement of enzyme activities, is meaningful only when the growth of cells is carefully controlled.

Adequate methods have been developed and are now widely used in human genetics – not only in research but in routine diagnosis as well - especially in prenatal diagnosis of chromosomal and biochemical anomalies.

Some enzyme defects are not expressed in fibroblasts. Occasionally, the use of other tissues such as lymphocytes or red blood cells is successful in these instances. In general, enzyme defects that are not expressed in fibroblasts cannot be assayed in amniotic cells.

7.2.2.2 Typical Group of Enzyme Defects: Erythrocyte Enzymes

A well-examined group of enzyme deficiencies affects the enzymes of the red blood cell [128, 250]. The human erythrocyte has no nucleus and is therefore unable to synthesize mRNA. Protein synthesis in nucleated red cell precursors supplies the erythrocyte with a number of enzyme systems that are active for only a limited time. They gradually lose their activity, and the red cell is removed from the circulation after 120 days. For many of these, enzyme deficiency syndromes are known; some of them cause nonspherocytic hemolytic anemia.

Enzyme Defects in Glycolysis. The most important catabolic pathway for obtaining energy-rich phosphates (ATP) in mature erythrocytes is glycolysis (Fig. 7.3) by the Embden-Meyerhof pathway. This anaerobic pathway leads to the formation of 2 mol lactate from 1 mol glucose. Moreover, 1 mol glucose generates 4 mol ATP, 1 mol of which is needed for phosphorylation when glucose-6-phosphate is transformed to fructose-1,6-diphosphate and also when 1 mol glucose is changed to 1 mol glucose-6-phosphate. Hence, the net gain is 2 mol ATP per 1 mol glucose. ATP is utilized for erythrocyte functions including maintenance of shape as a biconcave disk, for energy of the cation pump, and for the synthesis of such

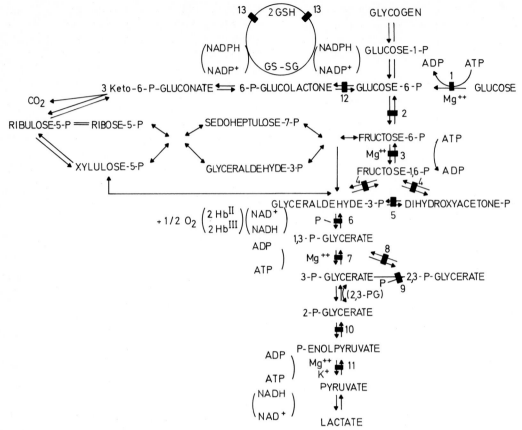

Fig.7.3. The glycolytic pathway and its metabolic blocks in erythrocytes. This pathway is catalyzed by 11 enzymes. The rate-limiting enzyme is hexokinase (1) which converts glucose into glucose-6-phosphate. This is then metabolized in steps (2–6) into 1,3-diphosphoglycerate. This compound may be converted directly by phosphoglycerate kinase into 3-phosphoglycerate and ATP (7). In an alternative pathway (Rapoport-Lübering cycle), however, 1,3-diphosphoglycerate may be converted into 2,3-diphosphoglycerate (8). 2,3-Diphosphoglycerate is cleaved into inorganic phosphate and 3-phosphoglycerate, which is then recycled into the glycolytic pathway. The Rapoport-Lübering cycle produces no ATP. Hence, degradation of glucose may proceed with different net gains of ATP. However, the 2,3-diphosphoglycerate content of erythrocytes is important for normal oxygen dissociation of hemoglobin. Another requirement for functional hemoglobin is the availability of NADH by the glyceraldehyde-P-dehydrogenase reaction. NADH is needed not only for hydrogenation of pyruvate to lactate but also for reduction of methemoglobin. About 5%–10% of glucose-6-phosphate is degraded oxidatively via the hexose-monophosphate cycle (12). In a number of steps, pentose phosphate is converted into fructose phosphate or glyceraldehyde-3-phosphate and recycled into glycolysis. The hexose monophosphate cycle is important as it provides NADPH which is needed for reduction of oxidized glutathione. This reduction is catalyzed by glutathion reductase (13). Glycolysis is controlled via a "multistep control" system in which hexokinase, phosphofructokinase, and the concentration of inorganic phosphates and Mg are important. The metabolic blocks that have been analyzed in humans are also shown by the heavy black bars

metabolites as glutathione (GSH) or AMP. The glycolytic pathway is catalyzed by 13 enzymes.

About 5%–10% of glucose-6-phosphate is degraded oxidatively via the hexose monophosphate cycle or the "shunt" pathway. In a number of steps, pentose phosphate is converted into fructose phosphate or glyceraldehyde-3-phosphate and recycled into glycolysis. The hexose monophosphate cycle is important, as it provides the NADPH needed for reduction of oxidized glutathiones. This reduction is catalyzed by glutathione reductase.

Nonspherocytic Hemolytic Anemias. Dacie et al. (1953) [67] defined a group of hemolytic anemias which they described as nonspherocytic, as contrasted to hereditary spherocytosis. The patients suffer from increased hemolysis and its consequences, including jaundice of varying degrees, slight to moderate splenomegaly, and an increased occurrence of gall stones. Contrary to the findings in hereditary spherocytosis (182900), osmotic fragility of the red cells was not increased, and there was no qualitative hemoglobin abnormality. On the basis of refined hematological criteria the condition was considered heterogeneous, but the findings in the various forms overlapped to some extent. Full analysis had to await development of enzymatic methods.

Enzyme Defects in the Glycolytic Pathway. Between 1961 and 1975 genetic defects were described for 11 of the 13 glycolytic enzymes. In at least eight of these, a causal relationship with hereditary nonspherocytic hemolytic anemia was shown – in some cases with additional disturbances in the central nervous system and muscles. In general, a critical reduction in enzyme activities leads to accumulation of the metabolite prior to and diminution of the metabolite after the block. Secondary effects such as ATP reduction can be expected in some deficiencies. Due to the internal regulatory capacity of the system, however, direct inferences from metabolic, clinical, and hematological findings as to the nature and degree of the enzyme defect are often misleading. In addition, the examinations are usually carried out in an erythrocyte population, which comprises a larger proportion of young red cells. Since enzyme activities are often higher in young than in old cells, enzyme deficiency may be overlooked.
Some observations of a more general significance for human enzyme defects can be made using these examples.

Blood Is Readily Available for Examination. For almost every step of the glycolytic pathway in erythrocytes, enzyme deficiencies are now known. This is in striking contrast to other pathways, for which little if any evidence of such deficiencies is available. One obvious reason is that the affected tissue – blood – is readily available for examination. Repeated venepunctures are not unreasonable impositions on patients, as compared, for example, to skin, let alone brain biopsies. Moreover, the erythrocytes are specialized cells that contain only a portion of the enzyme system found in other cells. This restriction in the enzymatic makeup reduces the number of possibly affected reactions and facilitates analysis. These advantages of blood – especially red blood cells – have been widely exploited, as testified by the work on glucose-6-phosphate dehydrogenase and especially on the hemoglobin molecule, which provides the leading paradigms on molecular aspects of gene-determined proteins (Sect. 7.3) and on natural selection in human populations (Sect. 12.2.1.6).

Analysis at the Enzyme Level Reveals Genetic Heterogeneity. It is shown in Sect. 4.3.5 that analysis of genetic heterogeneity meets with severe limitations at the phenotypic level. If two conditions both have an autosomal-recessive mode of inheritance, and the overlap of phenotypic manifestations is fairly strong, the only clue to heterogeneity would be a mating of two affected homozygotes with only normal unaffected children (Chap. 4.3.5). Once the analysis is carried out at the enzyme level, genetic heterogeneity is obvious under the following conditions:

a) All the enzyme defects in the glycolytic pathway of red blood cells as described in Fig. 7.3 lead to very similar hemolytic anemias. One source of genetic heterogeneity is the fact that mutations of different genes determining the various enzymes of a given pathway may have similar or identical phenotypes. This conclusion could also be made using the example of the glycogen storage diseases.

b) A second source of heterogeneity is provided by the many ways in which an enzyme that is determined by one gene may change its properties due to various mutations. The more methods for examining enzyme properties are applied, the more differences are exposed. Genetic heterogeneity at a given locus is of course expected since the number of mutations causing amino acid substitutions and deletions is very large (Sect. 7.3.5), as shown by analyses at the gene-DNA level.

Residual Activity Is Found Among Homozygotes In Almost All Enzyme Defects. Studies on enzyme activities of homozygotes for glycolytic defects have found residual activity sometimes of considerable magnitude. In some instances, this may have been caused by the activity of another enzyme which catalyzed the same metabolic step. Usually, however, the mutation does not change the protein to such an extent that enzyme activity is completely lost. This maintenance of residual activity has been claimed by Kirkman [147] to be a general property of many or most human enzyme deficiencies, quite contrary to bacteria, in which many mutations lead to complete enzyme blocks. This phenomenon may partially be due to selection for survival. For example, it is easy to imagine that the complete genetic block of an enzyme in a key metabolic sequence would be lethal for the individual. In bacteria, on the other hand, mutants are observed principally when enzyme activity has been entirely or almost entirely lost. A mutant with an incomplete block ("leaky" mutant) often manages to survive on a minimal medium.
The principal difference between human and bacterial mutations is their mode of ascertainment. Bacterial mutations are usually found by the failure of bacterial cultures to grow under certain culture conditions. Most human mutations causing enzyme deficiency are detected in patients with disease. When appropriate techniques are used to screen all types

of bacterial mutations, the expected range of different structural mutations similar to human structural enzyme mutations are found. The difference between residual activity in man and microbes, particularly if one allows for lethality of mutations of key enzymes in man, may therefore be largely spurious.

Clinical Findings Caused by an Enzyme Defect Depend on the Normal Activity of This Enzyme in a Variety of Different Tissues. Multiple forms of a given enzyme may occur within a single organism or even in a single cell. These are called isozymes. Isozymes may be generated by secondary alterations of the enzyme in the tissue and in such cases are of nongenetic origin. Genetically determined isozymes owe their origin to various genetic loci coding structurally distinguishable polypeptide chains that may have a common origin in evolution (Sect. 14.2.3). The term isozyme has also been applied to allelic variants at a single genetic locus that are detectable by electrophoresis. Isozymes are known for many enzymes. They catalyze the same overall reactions but are usually adapted to slightly different conditions under which these reactions occur in various tissues. Since the differences in intracellular milieu are small and largely unexplored, it cannot be predicted on theoretical grounds for which enzymes isozymes may exist.

Enzyme defects are due to mutations affecting single genes. Therefore they usually affect only one in a series of isozymes. If more than one isozyme is affected, these variant isozymes may share a common polypeptide chain, or secondary effects on enzyme structure have occurred.

An enzyme deficiency caused by a mutation of a gene that is active in only one tissue affects the phenotype of its carrier in a different way from an enzyme deficiency affecting many tissues. Gene mutations may show pleiotropy, i.e., a single mutation may have many different consequences in a single individual. An enzyme deficiency affecting more than one tissue is expected to have a pleiotropic effect. This is one mechanism of pleiotropy, but certainly not the only one. Even if an enzyme – or any other protein – is active in only one tissue, its deficiency may affect other tissues by perturbations induced by the primary defect. However, an enzyme defect that is found in all tissues sometimes leads to a phenotypic anomaly in one tissue only – probably because the defect can be compensated more easily in other tissues.

The enzyme deficiencies presented in Fig. 7.3 show examples for all types of pleiotropism. For example, phosphofructokinase (PFK) deficiency (step 3) leads in some cases to relatively mild nonspherocytic hemolytic anemia. Here the pleiotropic pattern consists of mild anemia, slight jaundice, and mild splenome-

galy. All these symptoms can be explained by a shortened life span of the red blood cells. Patients in other families have had very mild hemolytic anemia and severe myopathy associated with glycogen storage disease. Muscle and erythrocyte forms of phosphofructokinase differ from each other, as shown by electrophoresis, chromatography, and immunology. The isozyme pattern seems to be fairly complicated; even within erythrocytes at least two enzyme components have been found [128]. The differences in patterns of pleiotropic action between families can in principle be explained by mutations in genes determining polypeptide chains that may either be present or absent from tissue-specific isozymes.

However, an enzyme deficiency can be present in all examined tissues but the primary phenotypic effects still be confined to a single tissue. Glucose phosphate isomerase deficiency (step 2) is one example. In probands ascertained via the erythrocyte defect the activity of this enzyme is usually reduced similarly in leukocytes, thrombocytes, fibroblasts, muscles, and liver. In all these tissues, the enzyme appears to have the same biochemical properties. In spite of thorough examinations there is no hint of tissue-specific enzymes. Nevertheless the erythrocyte defect dominates the pattern of clinical manifestations. Patients suffer from severe hemolytic anemia which may present at birth as severe jaundice. Many different structural mutations affecting glucose phosphate isomerase have been detected, and compound heterozygotes for these defects are frequent. On the other hand, not all enzyme deficiencies cause any disease at all; they might not be associated with any detectable signs [26].

Pyruvate Kinase Deficiency (266200). Pyruvate kinase deficiency (step 11) is the most frequent defect of a glycolytic enzyme in red blood cells. Homozygotes may show a wide range of hematological symptoms. Some patients have a fully compensated nonspherocytic hemolytic anemia while others suffer from severe and repeated hemolytic episodes. Some features seem to be undisputed:

a) Homozygotes usually have a residual activity of about 5%–20% of the normal enzyme activity. Heterozygotes have values around 50% and are clinically healthy.
b) Examination of qualitative enzyme characteristics such as kinetic properties, nucleotide specificity for ADP and UDP, temperature stability, urea stability, pH optimum, and isoelectric point have exposed the existence of many variants with different properties. It is very difficult from enzyme assays to draw conclusions as to whether affected patients are truly homozygous for the same var-

iant or have two different defective alleles ("compound heterozygotes"; Sect. 4.1.3, Fig. 4.12). Here molecular studies are necessary.

Enzyme Activities and Clinical Symptoms in Heterozygotes. For most of the glycolytic defects of Fig. 7.3 determinations of enzyme activity have been performed in heterozygotes. As a rule, the activities are halfway between those of normal and those of defective homozygotes. This finding exemplifies a rule that applies more generally: In most human enzyme defects analyzed so far heterozygotes show roughly 50% of normal activity. Usually this amount of reduction does not lead to any obvious clinical manifestations; half the enzyme activity is sufficient to maintain function under normal circumstances.

It is of great interest that heterozygotes in fact have only 50% of enzyme activity in most enzyme deficiencies, which in humans have usually been shown to be structural mutants. This finding clearly establishes that the amount of enzyme activity is rigorously specified by the structural gene locus specifying the enzyme activity under its control. Thus normal homozygotes who have two structural genes for the enzyme have 100% enzyme activity while heterozygotes have only 50%. The single normal gene in such heterozygotes is therefore unable to compensate for the mutant structural gene that produces an inactive gene product. This finding is of great importance for a consideration of gene regulation in mammals since it differs from regulatory phenomena in bacteria.

Aerobic Energy Production in the Red Cell: Hexose Monophosphate Pathway [21]. The left side of Fig. 7.3 shows the aerobic pathway via the so-called hexose monophosphate cycle, also called the pentose phosphate or shunt pathway. Its main function is to generate reducing power in the form of NADPH. Glucose-6-phosphate is oxidized through the action of glucose-6-phosphate dehydrogenase (step 12) [53] to 6-phosphogluconate. Through various further steps d-ribose 5-phosphate is formed.

Enzymatic reduction of oxidized glutathione (GSSG) oxidizes NADP. Reduced glutathione (GSH) maintains SH groups in the reduced stage and may also help to protect the cell from damage by compounds such as H_2O_2. An inherited condition has become known in which GSH is completely lacking. The anomaly is caused by glutathione synthetase deficiency, which in some cases also leads to oxiprolinemia [128]. GSH deficiency leads to a nonspherocytic hemolytic anemia which is associated with drug sensitivity to oxidant drugs. Many drugs have an oxidizing effect and stress the reductive capacity of the cell [205].

Deficiency of Glucose-6-Phosphate Dehydrogenase (305900) [25, 84, 169, 171]. A genetic block in the hexose monophosphate pathway (Fig. 7.3) that also manifests as increased drug sensitivity has gained special significance and has become one of the leading paradigms of pharmacogenetics. During the Korean War (1950–1952) United States soldiers received prophylactic treatment with the antimalaria drug primaquine, a quinoline derivative. An intravascular hemolytic reaction was observed in about 10% of black soldiers and in substantially fewer whites >1‰–2‰; usually of Mediterranean origin. Similar hemolytic reactions had been observed earlier in darkly pigmented patients when drugs such as sulfanilamide and pamaquine were administered. Hemolytic reactions have long been known in Mediterranean areas, for example, Sardinia, occurring in some individuals after eating broad beans (*Vicia faba*).

Critical cross-transfusion studies of the life span of "primaquine-sensitive" red cells showed that the defect is intracellular, and that the older red cells are more susceptible to hemolysis in blacks. This explains the short duration of the hemolytic reaction (Fig. 7.4). Once the older cells are destroyed, hemolysis ceases despite continuation of drug treatment.

At first an immune mechanism was suspected. Later the sensitive cells were found to show glutathione instability on incubation with acetylphenylhydrazine. In 1956 Carson et al. [53] demonstrated the specific enzyme defect. The following reactions were examined:

a) GSSG + NADPH
$$+ H^+ \xrightarrow{\text{GSSG reductase}} 2\ GSH + NADP^+$$

b) Glucose-6-phosphate
$$+ NADP^+ \xrightarrow[\text{dehydrogenase}]{\text{glucose-6-phosphate}} \text{6-phosphogluconate}$$
$$+ NADPH + H^+$$

c) 6-Phosphogluconate
$$+ NADP^+ \xrightarrow[\text{dehydrogenase}]{\text{6-phosphogluconate}} \text{pentose phosphate}$$
$$+ CO_2 + NADPH + H^+$$

The critical defect was found to be glucose-6-phosphate dehydrogenase (G6PD) deficiency. Activity of GSSG reductase and 6-phosphogluconate dehydrogenase were normal. These studies clearly established G6PD deficiency as the cause of the hemolytic reactions in primaquine-sensitive men.

It soon became obvious that men were more frequently affected with hemolytic reactions than women. Before the G6PD defect was discovered, the glutathione stability test was used as an indicator of primaquine sensitivity; the quantitative assay using this test measures GSH before and after incubation of

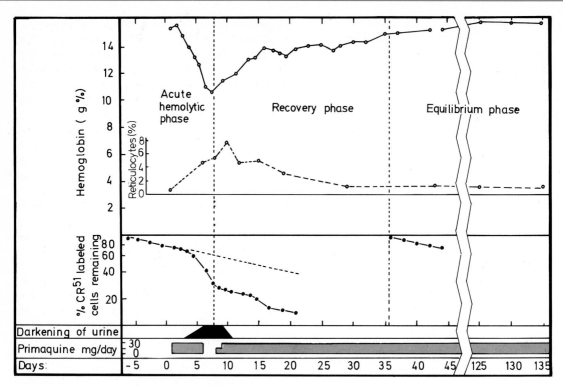

Fig. 7.4. Hemolytic reaction after treatment with primaquine. Within the first days of drug administration many erythrocytes are destroyed by hemolysis. This leads to increased production of new erythrocytes; reticulocytes increase, and hemoglobin rises again. Hemolysis affects only red cells older than about 60 days; during therapy the red cell population consists of younger cells. (From Carson et al. 1956 [53])

red cells with acetylphenylhydrazine. The results showed a clearly bimodal distribution of postincubation glutathione values in 144 African-American males, with an appreciable part of the population having very low values. In 184 African-American females the distribution of glutathione values was shifted to the left, and the number of persons with very low values was much lower than among males. This sex difference suggested an X-linked mode of inheritance, with very low values in male hemizygotes and female homozygotes; female heterozygotes had somewhat intermediate values. The hypothesis of X-linked inheritance was soon confirmed by family studies [55]. Direct G6PD enzyme assays were carried out later in population samples and showed a similar distribution, but the values among female heterozygotes were closer to halfway between normal and abnormal homozygotes (Fig. 7.5), with much overlap with normals.

Difference Between the African and Mediterranean Variants. Within a few years after discovery of the G6PD deficiency differences in severity of the deficiency between African and Mediterranean male carriers became known. In red cells of Africans with G6PD deficiency a residual activity of 10%–20% was regularly found, whereas Mediterranean carriers showed only minimal activity (below 5%). In addition, Africans had a nearly normal activity in their leukocytes, which was moderately or markedly reduced in Mediterraneans.

Electrophoretic methods were developed for examination of affected enzyme proteins. The mobility of the normal wild-type enzyme was designated as B. In blacks with normal enzyme activities, an electrophoretic variant with rapid migration was discovered in 20% of males, which was designated as A. Blacks with enzyme deficiencies always had a G6PD band with strongly reduced activity and the mobility of the A variant. This was called A⁻. The variant of G6PD deficiency in the Mediterranean population (G6PD Med) migrated with mobility similar to normal G6PD and therefore was sometimes referred to as G6PD B⁻. In normal white populations G6PD migrated almost exclusively as the normal B component (Fig. 7.6).

More Detailed Characterization of G6PD Variants. Many additional G6PD variants were discovered in various populations, and standardization of methods for classification became necessary. A proposal by a group of specialists in this field was published in

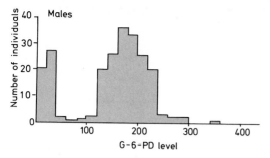

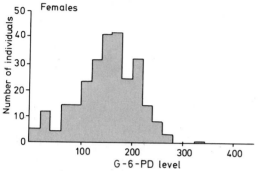

Fig. 7.5. Distribution of G6PD activities in males and females in a black population. Note almost perfect distinction between affected and normals in males. (From Harris 1980 [110])

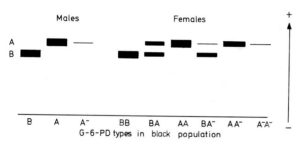

Fig. 7.6. G6PD electrophoretic phenotypes in a black population. Males are hemizygous A, A⁻, or B; females may be homozygous for any of these alleles or heterozygous for all possible combinations. The A⁻ bounds are indicated by *thin lines* since little staining occurs because of enzyme deficiency. Neither BB and BA⁻ nor AA and AA⁻ types can usually be distinguished. (From Harris 1980 [110])

1967 by the World Health Organization [23]. According to this proposal, characterization of a variant should include the following:

Red cell G6PD activity, electrophoretic mobility, substrate specificity (K_m) for glucose-6-phosphate and NAD, use of substrate analogs, thermostability, and optimal pH for enzyme activity. The use of these techniques led to the definition of a large number (over 300) of G6PD variants. In many cases a small difference from normal in many enzymological characteristics led to the conclusion that a new and unique G6PD variant had been found. More recently,

molecular techniques have revolutionized approaches to the study of G6PD. Instead of enzymological and electrophoretic study it is now easier to examine the DNA sequence of the G6PD gene. The resulting nucleotide changes found in G6PD variants are definitive and allow inferences regarding the nature of the amino acid substitutions. Virtually all G6PD mutations have been missense mutations producing amino acid replacements which, depending on the nature of the substitution, may or may not cause enzyme deficiencies. Total enzyme deficiency due to extensive deletions has never been detected. Since G6PD is a "housekeeping enzyme" that is active in cells of all tissues, some residual G6PD activity is probably essential for embryonic survival. Mutations that lead to complete G6PD are therefore presumably lethal. Many mutations that appeared to be unique on biochemical study proved to be identical defects on DNA analysis. The total number of G6PD variants, which had been higher than that observed for any other protein, is therefore smaller than that which had originally been thought.

G6PD Variants Observed in Human Populations. Enzyme variants may be classified as follows:

a) Variants with enhanced enzyme activity. Only two such variants are known: G6PD Hektoen and G6PD Hartford.
b) Variants with almost normal activity. One of these is the variant A mentioned above, which is found in about 20%–25% of the male population of tropical Africa and their American descendants.
c) Variants with moderately reduced activity. These show activities between 10% and 50% among hemizygous males. Sensitivity to hemolysis-provoking drugs may occur; favism is not observed.
d) Variants with severe enzyme deficiency and mild compensated hemolysis. A characteristic member of this group is the Mediterranean variant.
e) Variants with severe enzyme defect and chronic hemolysis even without additional exposure to oxidizing agents. These variants lead to a congenital nonspherocytic hemolytic anemia.

More Incisive Biochemical and Molecular Analysis [25]. An X-linked mode of inheritance has been confirmed for all variants for which family investigations have been carried out. This strongly suggests that all these mutations have indeed occurred within one gene locus, and that no other autosomal loci for red cell G6PD exist. Furthermore, all variants seem to be caused by different mutations within this structural gene.

The active enzyme has a molecular weight of approx. 120 000 and consists of dimers. The subunits are polypeptide chains consisting of about 450 amino acids; their sequence has been analyzed [271]. By 1993 the exact point mutation had been identified in at least 58 different variants, among them the A^+, A^-, and Mediterranean variants.

G6PD activity has been shown to be present in most if not all tissue cells; there appear to be no tissue-specific isozymes, and if there is a G6PD mutation, the altered G6PD can be demonstrated in all tissues. However, the clinical manifestations are limited to the red cells in practically all instances.

Clinical Significance. In addition to drug-induced hemolysis observed in common G6PD variants and chronic nonspherocytic hemolytic anemia with certain rare G6PD variants, several other hemolytic syndromes have been observed in G6PD deficiency. For example, favism or hemolytic anemia on ingestion of fava beans occurs only in subjects with G6PD deficiency. The condition is most severe in hemizygote males and has been widely observed in Mediterranean countries and China where fava beans are grown.

Neonatal jaundice is occasionally seen in G6PD deficient (mostly male) subjects. The reason why only some G6PD-deficient individuals are affected is not entirely clear but may be related to hepatic immaturity, undefined exogenous hemolytic agents, or additional genetic factors predisposing to jaundice.

Lists of various drugs that can cause hemolysis or are safe in G6PD deficiency have been published [24].

Significance of G6PD Variants for Understanding Human Enzyme Deficiencies. The G6PD system has been an excellent model because males with an X-linked structural mutation exhibit only the mutant gene product. Furthermore, the relatively high frequency of some variants in certain populations has permitted a more thorough genetic analysis than is possible with many other human enzyme deficiencies.

Among the more general findings that also apply to other human enzyme defects are:

The phenotypic consequences of the molecular changes caused by mutations form a continuum: there are variants that do not alter the capacity for normal biological function at all and can be detected only by special methods. Defects of other variants are easily compensated for under normal conditions and lead to hemolysis only under special environmental conditions. Yet other variants manifest disease even in the absence of harmful environmental factors. Most variants are benign and do not cause disease. There is a bias in detecting variants associated with illness since patients with hemolytic anemia are more frequently examined for enzyme deficiency than normals.

Most variants are quite rare, some having been identified only in single individuals and their relatives. A few, however, have become frequent in some human populations; the reasons are discussed in Sect. 12.2.1.6.

While there is evidence that most of these conclusions hold true for most if not all human enzyme defects, a further conclusion can be drawn concerning the location of this gene on the X chromosome. In most cells of female heterozygotes either one or the other of the two alleles is functionally active, the other having been inactivated. This inactivation can be used as a tool to examine problems of tumor formation and cell differentiation. It has been shown, for example, that uterine leiomyomata in women heterozygous for two electrophoretic types of G6PD exhibit either one or the other of these G6PD types in the tumor. This finding can be explained by the origin of all the cells of each tumor in a single cell. Similar findings suggesting a so-called monoclonal origin of tumors have been made for most neoplastic processes (see Chap. 10).

Phenocopy of a Genetic Enzyme Defect: Glutathione Reductase Deficiency [81, 82]. An enzymatic reaction related to the shunt pathway is the reduction of GSSG to GSH. The enzyme involved is glutathione reductase (see Fig. 7.3; step 13). The older literature includes a number of observations of families with various hematological disorders and alleged defects of this enzyme. However, the family data do not really fit the usual genetic hypotheses. More recently it has been shown that deficiencies of glutathione reductase are almost always due to a nutritional lack of the coenzyme flavin adenine dinucleotide. The cause is usually riboflavin (vitamin B_2) deficiency; enzyme activity of glutathione reductase is normalized within a few days after administration of riboflavin to deficient individuals.

This condition has been found to be frequent in northern Thailand and is caused by riboflavin deficiency in the customary diet. This example demonstrates that not every common enzyme defect in a population must be genetically determined.

Some cases, however, reflect a genuine, genetically determined glutathione reductase defect [164]. We do know of other hereditary anomalies in humans caused by an abnormally high demand for a certain coenzyme that must be supplied as a therapeutic vitamin. X-linked vitamin D resistant rickets [267] (see also Sect. 4.14) and pyridoxal dependency with epileptic seizures are two examples. It may be that in some families glutathione reductase deficiency is related to genetically determined riboflavin deficiency.

Table 7.1. Etiology and pathogenesis of genetic diseases as elucidated by methods from molecular biology and biochemistry (modified from Scriver et al. [215])

Level of analysis	Type of anomaly	Example
Altered DNA structure	1. Deletion mutations	α Thalassemia, Lepore hemoglobins, hemophilias
	2. Missense mutations	Sickle-cell disease
	3. Splicing mutations	Some β thalassemias
	4. Nonsense mutation	Some β thalassemia variants
	5. Frame-shift mutations	Hemoglobin Wayne
	6. Gene duplications	Hemoglobin Grady; anti-Lepore hemoglobins
	7. Regulatory mutations (For these types of mutation, see Sect. 9.4)	Some β thalassemias
	8. Mutations by amplification of trinucleotides	Myotonic dystrophy
Disturbed protein function: enzymes	1. Absent activity	
	a) Protein detectable immunologically	Some variants of Lesch-Nyhan syndrome
	b) No protein detectable immunologically	Most variants of Lesch-Nyhan syndrome, variants of homocystinuria
	2. Reduced activity	
	a) Decreased affinity for substrates	G-6-PD deficiency, Freiburg variant
	b) Decreased affinity for cofactors	Homocystinuria (pyridoxine-responsive type)
	c) Unstable structures	G-6-PD deficiency, some variants
	3. Enhanced activity	G-6-PD Hektoen variant
	4. Defect of enzyme activator protein	AB variant of G_{M2} gangliosidosis [146]
	5. Reduced availability of cofactors	Pyridoxine (vitamin B_6) dependency
Disturbed protein function: nonenzymic proteins	6. Defective posttranslational modification	α_1-Antitrypsin deficiency, ZZ variant
	7. Enhanced tendency to aggregation	Sickle cell disease
	8. Defective receptor binding	Familial hypercholesterolemia; testicular feminization
Disrupted cell and organ function	1. Altered flux through metabolic pathways	
	a) Accumulation of a toxic precursor (catabolic pathway)	Phenylketonuria; mucopolysaccharidoses and other lysosomal defects
	b) Deficiency of product (anabolic pathway)	Various types of hypothyroidism with goiter
	c) Overproduction of product (anabolic pathway)	Rare form of gout due to altered PRPP synthetase
	2. Disordered feedback regulation of synthetic pathways	
	a) Overproduction of end product due to decreased synthesis or availability of feedback regulator	Acute intermittent porphyria, familial hypercholesterolemia
	3. Disordered membrane function	
	a) Deficient transmembrane transport	Cystinuria, hereditary spherocytosis
	b) Deficient receptor-mediated endocytosis	Familial hypercholesterolemia, receptor-negative and receptor-defective variants
	c) Deficient generation of second messenger	Pseudohypoparathyroidism
	4. Disordered intracellular compartmentation	
	a) Accumulation of unprocessed protein	α_1-Antitrypsin deficiency, ZZ variant
	b) Mislocation of protein	I cell disease
		Familial hypercholesterolemia (internalization-defective varian)
	5. Distorted cellular tissue architecture	
	a) Alteration of cell shape	Sickle cell disease, hereditary spherocytosis
	b) Alteration of organelle structure	Immotile cilia syndrome, especially Kartagener syndrome
	c) Alteration of extracellular matrix	Epidermolysis bullosa, Pasini type, lysyl-hydroxylase deficiency

Mimicry of a genetic defect by an environmental injury is called a *phenocopy* – a term coined by Goldschmidt (1935) [103]. A phenocopy is defined as the simulation of an inherited character by external factors. Goldschmidt, treating wild-type *Drosophila* in various stages of development by heat shocks, succeeded in producing numerous phenotypic abnormalities that were similar to variants usually produced by mutation.

Phenocopy experiments have been carried out in many species and appear to offer insights into the mechanisms of normal embryonic development and the production of malformation. However, their explanatory power has been overrated. Nevertheless, in human metabolic diseases the possibility that a phenocopy may be present should be kept in mind, especially since effective therapy may be available.

From a genetic point of view a common phenocopy of a human metabolic disease is hypothyroidism due to lack of inorganic iodide – a frequent condition in the Alpine countries of Europe and in some other parts of the world. Here, deficiency of an essential inorganic ion leads to the same end result as defective synthesis of the thyroid hormone observed in some families as a consequence of any of a number of enzyme defects in thyroid hormone synthesis (see Fig. 7.21). Nevertheless, while the clinical end results are similar, the pathophysiology of nutritional iodine lack and of genetically defective enzymes in thyroxin biosynthesis are different.

Other Enzyme Defects. Analysis of enzyme deficiencies not present in blood cells poses more difficulties. Many enzymes can be detected in fibroblasts grown, for example, from skin biopsies. In contrast to red cells, the fibroblasts contain nuclei. They are able to divide and to carry out all stages of protein synthesis. Their enzymatic endowment is much more complete than that of red cells. Fibroblasts sometimes lack only those enzymes that are confined to single groups of specialized cells, such as liver cells (i.e., phenylalanine hydroxylase deficient in phenylketonuria). Deficiencies of enzymes for many different metabolic pathways manifest in fibroblasts. This is why enzyme assays from cultured fibroblasts have brought much progress to our knowledge of enzyme defects.

Only a single group of diseases is discussed, which permits some general conclusions regarding properties of human enzyme deficiencies. These are the mucopolysaccharidoses, which are classified with the larger group of conditions due to deficiencies of lysosomal enzymes, other subgroups being the sphingolipidoses and mucolipidoses.

7.2.2.3 Mucopolysaccharidoses

Deficiencies of Lysosomal Enzymes. Enzymes or enzyme systems are usually located in one distinctive cell compartment. For example, the enzymes of electron transport and oxidative phosphorylation as well as other oxidative enzymes for pyruvate, fatty acids, and some amino acids are located in the mitochondria. A number of hydrolytic enzymes are concentrated in organelles called lysosomes. If these enzymes are liberated by destruction of the lysosomal membrane, the entire cell is destroyed by self-digestion. Normally this digestion proceeds within the lysosomes, which degrade not only defective cell particles and material from intercellular connective tissue but also external material taken up by the cell. Among the cell elements degraded by the lysosomes are mucopolysaccharides, mucolipids, and sphingolipids (Fig. 7.7). Deficiencies are now known for many of the enzymatic pathways involved in this degradation [49, 187].

The metabolic pathways causing red cell defects manifesting as hemolytic anemia were already known, and analysis of enzyme deficiencies in these pathways involved testing the known metabolic steps. The situation was different with the mucopolysaccharidoses. Here the genetic diseases were known first, and analysis of the enzyme defects led to elucidation of the enzymatic pathway. Therefore we start with a description of these diseases and proceed stepwise to the biochemical defect, the enzyme defects, and the reconstruction of the normal pathway. This example shows how studying genetic diseases as experiments of nature contributes to the understanding of normal biochemistry and physiology.

Mucopolysaccharidoses: Clinical Picture. Mucopolysaccharidoses are rare diseases with complex and in most cases severe manifestations ranging from abnormalities of the skeletal and vascular systems to impairment and deterioration of mental functions. The clinical symptoms result from excessive and progressive storage of sulfated polysaccharides in various tissues.

These disorders have been classified into seven distinct categories on the basis of clinical, genetic, and biochemical evidence. Table 7.2 presents this classification together with the major clinical findings. With the exception of type II (Hunter), the mode of inheritance is autosomal-recessive. The clinical symptoms range from very severe to relatively mild forms. For type II (Hunter; 309 900) two forms have been described: a severe "juvenile" form in which death occurs before puberty, and a mild "late" form with mild or no mental retardation and a generally better prognosis. Type VI (Maroteaux-Lamy;

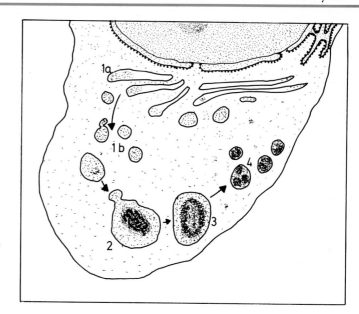

Fig. 7.7. Schematic representation of a lysosome and its functional apparatus. The figure shows the functional cycle of a normal lysosome. *1a*, Golgi apparatus; *1b*, primary lysosome; *2*, phagolysosome; *3*, secondary lysosome; *4*, residual body. (Courtesy of Dr. W. Buselmaier)

Table 7.2. Major clinical findings in the mucopolysaccharidoses (modified from Cantz and Gehler 1976 [49]; Spranger 1983)

Mucopolysaccharidosis		Mental and motor retardation	Growth retardation	Coarse facies	Bone dysplasia	Joint contractures	Hepato-splenomegaly	Corneal opacities	Mode of inheritance
Type[a]	Name								
I H	Hurler (1)	+++	++	+++	+++	++	++	+	Autosomal-recessive
I S	Scheie (1)	−	−	±	+	+	±	+	
I H/S	Hurler-Scheie compound (1)	±	+	++	++	++	+	+	
II A	Hunter, severe (2)	+	+	+	++	+	+	±	X-linked recessive
II B	Hunter, mild (2)	±	+	+	++	+	+		
III A	Sanfilippo A (3 A)	+++	−	+	+	±	++	−	Autosomal-recessive
III B	Sanfilippo B (3 B)	+++	−	+	+	±	++	−	
III C	Sanfilippo C (3 C)	+++	−	+	+	±	++	−	Autosomal-recessive
III D	Sanfilippo D (3 D)	+++	−	+	+	±	++	−	
IV	Morquio (4)	−	+++	+	+++	+	+	+	Autosomal-recessive
V		Vacant							
VI A	Maroteaux-Lamy, classic form (6)	−	++	++	++	+	+	+	Autosomal-recessive
VI B	Maroteaux-Lamy, mild form (6)	−	+	+	+	+	+	+	
VII	Sly (7)	+	±	±	+	−	++	±	Autosomal-recessive

−, Absent; ±, occasionally present; +, mild; ++, less severe; +++, severe. The numbers in parentheses refer to the genetic blocks shown in Figs. 7.10, 7.11.

[a] Classification of McKusick (1972) [163]; supplemented.

253 200) differs from type I (Hurler; 252 800) in that there is normal intelligence and less pronounced facial dysmorphism. Here, two different subtypes have also been observed – one with a fairly rapid course and more severe clinical features, the other with a relatively slow progression. In both types cardiac impairment may lead to death. For type IV (Morquio) two subtypes have also been discovered – IV A with very severe, IV B with milder manifestations.

Table 7.3. The metabolic defects in the mucopolysaccharidoses (from Cantz and Gehler 1976 [49])

Mucopolysaccharidosis[a]		Major storage substance	Enzymatic defect
I H	Hurler	Dermatan sulfate and heparan sulfate	α-L-Iduronidase
I S	Scheie		
I H/S	Hurler-Scheie compound		
II A	Hunter, severe form	Dermatan sulfate and heparan sulfate	Iduronate sulfatase
II B	Hunter, mild form		
III A	Sanfilippo A	Heparan sulfate	Heparan N-sulfatase
III B	Sanfilippo B	Heparan sulfate	α-N-Acetylglucosaminidase
III C	Sanfilippo C	Heparan sulfate	α-Glucosaminidase (?)
III D	Sanfilippo D	Heparan sulfate	N-ac-Glucosamine-6-sulfate sulfatase
IV	Morquio A	Keratan sulfate	N-Acetylgalactosamine 6-sulfatase
VI A	Maroteaux-Lamy, classic form	Dermatan sulfate	N-Acetylgalactosamine 4-sulfatase
VI B	Maroteaux-Lamy, mild form	Dermatan sulfate	(arylsulfatase B)
VII	Sly	Dermatan sulfate and heparan sulfate	β-Glucuronidase

[a] Classification of McKusick (1972); supplemented.

Lysosome Storage and Urinary Excretion. Histochemical studies have shown that these conditions are storage diseases. In many cells – including fibroblasts, liver cells, Kupfer cells, reticulum cells of spleen and lymph nodes, leukocytes, epithelial cells of kidney glomeruli, and nerve cells – enlargement and vacuolization is observed due to large quantities of stored material. The main storage compounds have been identified as sulfated glycosaminoglycans.

Further clarification has come from electron-microscopic studies. The compounds were found to be stored in rounded vacuoles, similar to lysosomes seen in experimental animals that have been injected with nonmetabolizable substances. Therefore it was concluded that these vacuoles are lysosomes engorged with undigested or only partially degraded glycosaminoglycans [244]. This conclusion was confirmed and extended to other tissues. The overloading of the lysosomal system led to classifying these disorders as lysosomal diseases even at a time when the metabolic defects were still unknown. The fact that the most important function of lysosomes lies in the hydrolytic degradation of macromolecules suggested that the storage results from deficiencies of lysosomal hydrolytic enzymes. Other evidence for a disturbed glycosaminoglycan metabolism came from the detection of an excessive excretion of these compounds in the urine. These excretion patterns reflect the basic metabolic lesions (Table 7.3).

Biochemistry of Sulfated Glycosaminoglycans. The sulfated glycosaminoglycans are complex heterosaccharides consisting of long polysaccharide chains covalently linked to a protein core. In dermatan sulfate, heparan sulfate, and chondroitin 4- and 6-sulfates the polysaccharide chains are composed of alternating residues of uronic acid and sulfated hexosamine. Keratan sulfate differs from the other glycosaminoglycans in that the uronic acid residues are replaced by galactose. The polymeric chains may be about 100 residues long and are bound to specific proteins via a distinct linkage region; several sugar chains are attached to the same polypeptide backbone. Such proteoglycans may form even larger complexes through noncovalent bonds. Within the carbohydrate chains there is considerable variation of the constituent sugars, as well as in the degree of sulfation.

For example, the major part of the polysaccharide chain in dermatan sulfate is composed of repeating dimers of L-iduronic acid linked to 4-sulfated N-acetyl-galactosamine (Fig. 7.8). The other glycosaminoglycans have similar structures. They are constituents of connective tissue and the ground substance.

In patients with mucopolysaccharidoses, the glycosaminoglycans of connective tissue consist of the same large proteoglycan entities found in normal in-

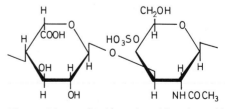

Fig. 7.8. Dimer of L-iduronic acid and 4-sulfated N-acetyl-galactosamine as found in dermatan sulfate

dividuals, indicating that the enzyme deficiency does not affect their synthesis. In the tissues in which abnormal storage occurs, and also in the urine, the molecules are smaller and of varying length. This suggests that the cell has cleaved as many linkages as possible before stopping at a residue for which the specific degradative enzyme is deficient.

Enzyme Deficiencies. The most direct approach to investigating the deficiency in a metabolic disorder is to define the compound whose metabolism is at fault and to measure the activities of the enzymes participating in its turnover. This approach was followed in the analysis of the red cell defects described above. The investigation of the mucopolysaccharidoses was not simple, however, since neither the precise chemical structure of the relevant glycosaminoglycans nor their normal catabolic enzymes were known. A systematic study became possible when fibroblasts cultured from the skin of Hunter or Hurler patients were found to accumulate glycosaminoglycans. The most important step in further analysis was the demonstration that the accumulation can be reduced to normal levels in vitro by a correcting factor from tissue fluids in culture. This was first shown by Neufeld et al. in 1968 [186]. It could be assumed that type I (Hurler) and type II (Hunter) are genetically different since the inheritance of Hurler syndrome was autosomally recessive, while Hunter syndrome was X-linked. The enzyme defect was therefore presumed to occur at different points in the pathway of mucopolysaccharide degradation. If it were possible to fuse the nucleus of a cell from a patient with Hurler syndrome with that of a patient with Hunter syndrome (as has been accomplished by somatic cell geneticists for many different cell types) complementation of each defect by the other should have resulted. During these experiments, however, the problem turned out to be much simpler. Cell fusion was not necessary. Hurler and Hunter cells were found to compensate each other's defects when simply mixed in culture, and even when cells of one genotype were exposed to culture medium preincubated with the other. The accumulation of mucopolysaccharides was measured by $^{35}SO_4$. Figure 7.9 shows these experiments. Compensation of one defect (Hurler syndrome) could be achieved by cells with the other defect (Hunter syndrome) as well as by normal cells [90].

In the following years, such experiments were carried out with the other clinical types, and the correcting factors were characterized by means of biochemical methods. Fibroblasts from patients with Sanfilippo syndrome were found to fall in at least two groups, A and B, each deficient in a respective factor and thus able to cross-correct each other. (Later two addi-

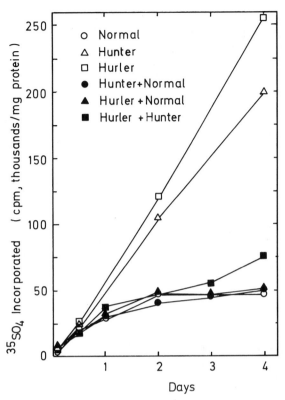

Fig. 7.9. Abnormal incorporation of $^{35}SO_4$ in cells from Hunter and Hurler patients. Mixing of Hunter and Hurler cells with normal cells as well as mixing of Hunter cells with those of a Hurler patient leads to reduction of incorporated $^{35}SO_4$, and the results are similar to those obtained with cells from a normal individual. (From Fratantoni et al. 1968 [90])

tional types were discovered, 3C and 3D). Hence, Sanfilippo disease was genetically heterogeneous. On the other hand, fibroblasts from patients with Scheie syndrome proved deficient in the same factor as Hurler patients – in spite of the wide difference in clinical symptoms.

All corrective factors were soon shown to be specific proteins, and upon more thorough analysis were identified as lysosomal enzymes involved in the degradation of sulfated glycosaminoglycans. In the elucidation of these enzyme defects, recent knowledge on the structure of the compounds was combined with analysis of the function of corrective factors and with direct attempts at identifying the enzymes.

The hypothesis to be tested was that the enzymes that are deficient in these disorders are specific for the different types of bonds occurring in the glycosaminoglycans. The predictions derived from this hypothesis have been confirmed. For example, when Sanfilippo A (252900) corrective factor was incubated in vitro with ^{35}S-labeled heparan sulfate isolated from Sanfilippo fibroblasts, a release of inorganic sulfate was observed. Further investigations revealed that the factor was acting on the *N*-sulfate bond of heparan sulfate. The Hunter corrective factor, when incubated in vitro with ^{35}S-labeled dermatan sulfate or heparan sulfate isolated from Hunter fibroblasts, also catalyzed the release of inorganic sulfate. The Hunter gene is X-linked, and therefore the enzyme defect had to be different from that found in Sanfilippo A disease. Since the two glycosaminoglycans have occasional sulfated iduronyl residues in common, it was hypothesized that the Hunter correction factor is a sulfatase. This was confirmed using an artificial substrate.

A different approach for elucidating the specific type of the enzyme block was to determine the nature of the terminal residues in the polysaccharide chains stored in these diseases. In Hunter disease, for example, the terminal residue of the stored dermatan sulfate was shown to be sulfated iduronic acid. This was expected from the sulfatase defect suggested by the experiments with an artificial substrate. Therefore the experiment confirmed this defect. Moreover, it suggested that the glycosaminoglycans are normally degraded stepwise, and that this degradation is arrested if the enzyme for one step is lacking. As the sequence of the monosaccharides in the chains varies, this stepwise process would explain the varying lengths of the polysaccharide chains stored in these diseases.

The same basic methods were used to analyze the other enzyme defects; determination of the terminal residues invariably showed them to contain the bond for which the enzyme was lacking. The nature of these enzyme defects explained another property of the stored material: a single defect leads to storage of chemically different materials. For example, Hurler patients accumulate dermatan sulfate as well as heparan sulfate. Both contain α-l-iduronic acid residues. Hence the defect of an enzyme specifically directed at this residue causes the accumulation of both types of polysaccharides containing it.

The result of these combined efforts is seen in Fig. 7.10 for chondroitined dermatan sulfate and Fig. 7.11 for heparan and keratan sulfate. The enzymes for which genetic blocks are known are noted. Table 7.3 shows the enzyme defects.

Consequences for Understanding Genetic Heterogeneity. In Sect. 4.3.5, the example of muscular dystrophy was used to show the way in which genetic heterogeneity can be analyzed on the basis of genetic evidence – various modes of inheritance – and evidence from clinical characteristics such as age at onset, pattern of clinical manifestation, and severity of symptoms. In the mucopolysaccharidoses a striking interfamilial variability of all these indicators is

found, whereas within the same family manifestations usually are similar. Therefore a subdivision into various genetic types seemed logical and was in fact carried out before the enzyme defects were analyzed. How does this subdivision on the basis of "indirect" evidence from the phenotypes compare with the "direct" evidence from analysis of enzyme defects?

By and large, the correspondence is satisfactory (Tables 7.2, 7.3). However, there are two exceptions:

1. From the clinical evidence, Sanfilippo disease would have been regarded as a single entity. However, it was shown to consist of four different enzyme defects. This experience has been repeated in the analysis of many metabolic disorders. Different defects within the same pathways may cause the same clinical picture. Other examples are the erythrocyte defects of glycolysis leading to nonspherocytic hemolytic anemia (Sect. 7.2.2.2).
2. On the other hand, the patterns of manifestation of the Hurler and Scheie types are very different, the course being much milder in Scheie disease. However, the enzyme defect is identical. It appears that differences in residual activity of α-l-iduronidase account for the clinical findings. Patients with Scheie syndrome appear to have sufficient enzyme activity to support normal cell function in the central nervous system but not in tissues with high glycosaminoglycane turnovers.

Studies such as those that were successful in eliciting genetic heterogeneity within the gene locus in G6PD deficiency are now being carried out in the mucopolysaccharidoses. Only four point mutations seem to have been identified at the gene-DNA level for types VI and VII; no deletion has been reported

Fig. 7.10, 7.11. In addition to the mucopolysaccharidoses described in the text, some other defects leading to mucolipidoses (Sandhoff disease; M II gangliosidosis) were also indicated. (From Kresse et al. 1981 [145])

Fig. 7.10. **a** Schematic representation of structure and catabolism of chondroitin sulfate. Sequential degradation starts from the nonreducing end *(left)*. Designation of diseases caused by the inactivity of an enzyme is given *in brackets*. *GlcUA*, glucuronic acid; *GalNAc*, *N*-acetylgalactosamine; *S*. *SO₃H*. **b** Schematic representation of structure and catabolism of dermatan sulfate. for details see legend to **a**. *IdUA*, Iduronic acid

Fig. 7.11. **a** Schematic representation of structure and catabolism of heparan sulfate. For details see legend to Fig. 4.11 a. *GlcN*, glucosamine; *GlcNAc*, *N*-acetylglucosamine; *IdUA*, iduronic acid. **b** Schematic representation of structure and catabolism of keratan sulfate. For details see legend to Fig. 7.10 a. *Gal*, galactose; *GlcNAc*, *N*-acetylglucosamine

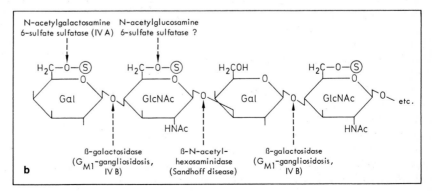

Fig. 7.10 a, b

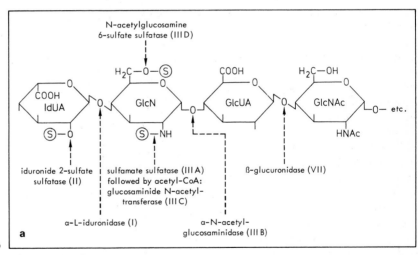

Fig. 7.11 a, b

[63]. The mutations present in double dose in a given patient affected with an autosomal-recessive disease have a common origin if the mutant allele comes from a common ancestor of both parents. This is often the case if the parents are relatives, for example, first cousins, or in genetic isolates (Sect. 13.4.1). In many other instances, however, the mutations found in the affected person have a different origin. Hence they are hardly ever identical. If the term "homozygote" were confined to individuals carrying two absolutely identical mutations, many or even most patients with recessive diseases would not conform to the definition but rather would be considered "compound heterozygotes" (Fig. 4.10). A number of patients have been observed with a clinical phenotype intermediate between the Hurler and Scheie types [228]. Fibroblasts were shown to be devoid of the Hurler correcting factor. Some of these patients may indeed have been compound heterozygotes for both the Hurler and the Scheie alleles. However, in at least four such families consanguinity of parents was observed. Since an increase of consanguinity cannot be expected in families in which two different alleles segregate, occurrence of a third, intermediate allele is likely. The problem will be solved by molecular studies.

Differential Diagnosis and Treatment of Mucopolysaccharidoses. When mucopolysaccharidosis is suspected on clinical grounds, the putative diagnosis should be checked by examining for excessive urinary excretion of sulfated glycosaminoglycans [228]. Definitive diagnosis, however, depends on demonstration of the enzyme defect, which can be studied in cultured fibroblasts, in leukocytes, and, for some enzymes, in the serum [49]. Prenatal diagnosis is possible, the gene being expressed in amniotic cells. However, since the number of amniotic cells that can be cultured in the available time is restricted, the number of enzyme determinations that can be performed is limited. Therefore every attempt should be made to arrive at an enzymatic diagnosis in the affected sib before prenatal testing. Once the defect has been established only the specific enzyme deficiency that has been identified in this sib needs to be looked for in the amniotic cells.

The finding that enzymes can be taken up by deficient cells, thus correcting deficiency, appears encouraging for attempts at enzyme therapy. So far, however, purified enzymes have not been available in sufficient amounts, and infusions of large amounts of plasma or leukocytes have led only to dubious, and in any case slight improvements. The uptake of lysosomal enzymes into a cell is a highly specific process involving the presence of a particular recognition marker on the enzyme protein, which may be differ-

ent for different tissues [186]. Still, the approach appears promising.

Defect of a Recognition Marker for Lysosomal Hydrolases. In 1967 DeMars and Leroy described "remarkable cells" in cultures of skin fibroblasts from a patient thought to have Hurler disease. These cells were filled with acid phosphatase positive inclusions that appeared dense under phase-contrast microscopy. On the basis of the appearance of these inclusions the disease was named I cell disease. Its clinical phenotype, while resembling the Hurler phenotype, is more severe; the mode of inheritance is autosomal-recessive. Fibroblasts from such patients are deficient in β-hexosaminidase, arylsulfatase A, and β-glucuronidase, whereas these enzymes were found in increased concentration in the culture medium. Initially a membrane defect was suspected, but lysosomes from affected patients were shown to take up and retain normal enzyme at a normal rate. However, hydrolases from I cells cannot be endocytosed by normal cells. Therefore, the enzyme molecules themselves are altered. They have been found to lack a recognition marker for endocytosis, i. e., mannose-6-phosphate. Normally mannose-6-phosphate residues are added after synthesis of lysosomal enzymes. The sugar compound acts as a signal allowing lysosomal enzymes to bind to a mannose-6-phosphate receptor that directs lysosomal enzymes into the lysosome, where they are activated. Because this processing enzyme is lacking, most lysosomal enzymes in I cell disease lack mannose-6-phosphate. Instead of being taken up by the cell lysosomes via the specific receptor pathway the enzymes pass through the cell and are secreted into the plasma (Fig. 7.12). The multiple clinical defects in I cell disease can be explained by the single processing defect that fails to add the recognition marker mannose-6-phosphate to these enzymes.

Since the recognition site is common to several enzymes, I cell disease is a condition in which a single gene defect causes more than one enzyme deficiency.

Gaucher Disease (270 800), *a Glycolytic Storage Disease.* Normally there is a relationship between the type of mutation and its effect on the structure of the enzyme protein and the degree of clinical manifestations. Some mutations have few effects on enzyme stability and are therefore expected to have no or mild clinical effects. Other mutations affect critical sites of a protein and are therefore expected to produce more severe effects. This rule, however, has remarkable and so far unexplained exceptions. In Gaucher disease the sphingolipid ceramide accumulates in lysosomes and macrophages due to a deficiency of the enzyme glucocerebrosidase. Inheritance is au-

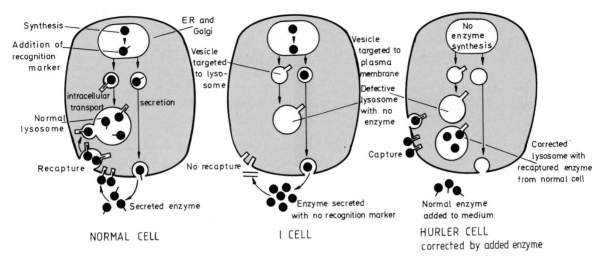

NORMAL CELL   I CELL   HURLER CELL
corrected by added enzyme

Fig. 7.12. Secretion and recapture of lysosomal hydrolytic enzymes by normal and mutant cells grown in culture. Specific receptor proteins located at the plasma membranes of all three cells allow them to take up hydrolytic enzymes into their lysosomes. Hurler cells fail to make α-L-iduronidase, but the defect can be corrected when the enzyme is added to the medium. In I cell disease, all enzymes are present, but a recognition marker for their uptake and transport to lysosomes is lacking. (Modified from Alberts et al. 1983 [3])

tosomal recessive. The clinical symptoms are accumulation of the metabolite in the liver, spleen, and bone. A great number of mutations have been identified. They occur in linkage disequilibrium, with appropriate DNA markers pointing to a single origin for each of them. The condition is usually rare. However, Gaucher disease of adult onset is common in the Ashkenazi Jewish population, where 5% of the population are heterozygotes and one common variant accounts for 70% of all Gaucher mutations in that population (see also [175a]). Surprising differences in age at onset and severity of clinical signs are observed in an appreciable number of such homozygotes – even sometimes among siblings. Gaucher disease is also the principal example of the success for enzyme therapy (see below).

7.2.2.4 Enzyme Defects Involving More than One Enzyme

In examples affecting energy supply of red blood cells or catabolism of glycosaminoglycans, one gene mutation leads to an alteration or deficiency of only one enzyme. These examples are thus in keeping with the one-gene/one-enzyme hypothesis. However, there are cases in which one mutation affects two enzymes. In some of these the activity of one enzyme may be impaired by the deficiency of the other.

Other cases in which both enzymes have a structural mutation, and enzyme activities in heterozygotes are reduced for both enzymes by about one-half cannot be explained in this way. Here, the most likely expla-

Leucine   Iso-leucine   Valine

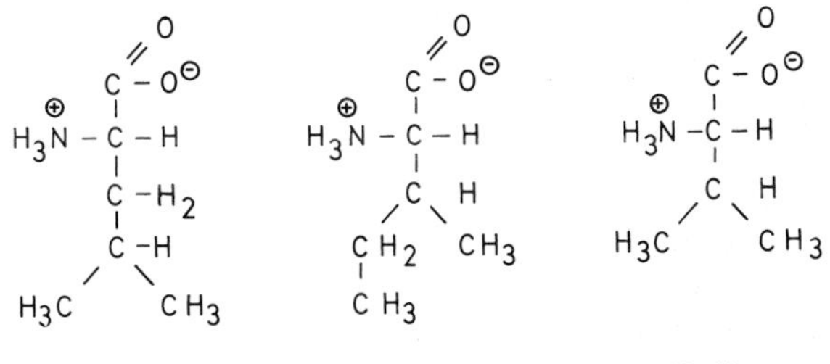

Fig. 7.13. The branched-chain amino acids

nation is that these enzymes are polymers composed of genetically different polypeptide chains and that the enzymes showing the defect have one polypeptide chain in common.

Maple Syrup Urine Disease (Branched-Chain Ketoaciduria; 248 600) [210]. One recessive disorder in which no less than three functionally related enzymes are deficient is maple syrup urine disease. This is a defect in degradation of the branched-chain amino acids leucine, isoleucine, and valine (Fig. 7.13). In the most frequent, classic type of the disease one observes feeding difficulties, vomiting, hypertonicity, and a shrill cry in the first week of life. Loss of tone and apnea may intervene. Later, reflexes are lost; the child frequently has seizures and may die in early infancy. Untreated children who survive suffer from severe mental retardation [214]. Apart from this classic type, "intermittent," "mild," and thiamin responsive types have been described.

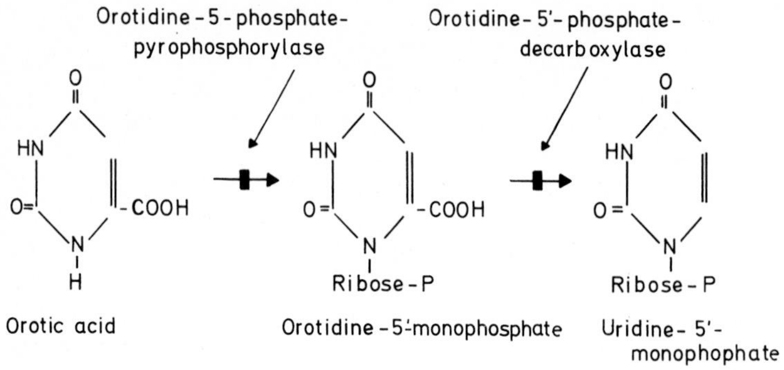

Fig. 7.14. Two genetic blocks in adjacent steps of the synthesis of the pyrimidine base uracil. Both blocks are seen in orotic aciduria

Enzyme analysis shows transaminase activities that transform the three amino acids to the corresponding keto acids to be normal. The anomaly is found in the next step: oxidative decarboxylation. This step is determined for all three amino acids by a single enzyme, a branched chain decarboxylase – a large multienzyme complex located on the outer face of the inner mitochondrial membrane. It consists of four separate proteins; in patients with classical maple sugar urine disease one of these components (E1α) is altered [80]. There have been various reports on similar diseases with alteration of other components [57, 79].

Other Metabolic Defects Involving More Than One Enzyme. Other metabolic defects in which one genetic block involves two enzymes of one metabolic pathway include orotic aciduria (258 900), a deficiency in the formation of uridine, a ribonucleic acid precursor from orotic acid. The two genetic blocks are seen in Fig. 7.14. Enzyme activites in heterozygotes are reduced by about one-half for both enzymes. This rules out a secondary effect of the one enzyme on the activity of the other and argues in favor of one gene being involved in both reactions.

In at least one homozygous case the altered electrophoretic mobility and higher thermolability of the second enzyme, the decarboxylase, show that the enzyme defect must be structural. It turned out that each reaction is performed by a discrete domain within the macromolecule, but all the domains are part of a continuous amino acid sequence [113]. In Sandhoff gangliosidosis, on the other hand, the combined defects of hexosaminidase A and B have been traced back to a mutation affecting their common subunit, the β-chain [208].

A number of patients with simultaneous deficiencies of vitamin K dependent coagulation factors II, VII, IX, and X have been observed. These are caused by a defect in posttranscriptional modification [106].

A Fresh View on the One-Gene/One-Enzyme (or One-Gene/One-Polypeptide) Hypothesis [92]. As noted, single enzyme defects affecting more than one enzyme may sometimes be explained by a common subunit in different enzymes. They are structural mutants.

Another important aspect of enzyme formation is posttranslational processing. Human sucrase-isomaltase, for example, comprises two polypeptide chains, each with a specific enzyme activity. The two chains are derived from proteolytic cleavage of a single precursor macromolecule [220]. The proinsulin molecule must also be processed to form functionally active insulin.

More detailed evidence for the formation of peptide molecules from a longer precursor protein is now available for a number of small peptides in the brain – the enkephalins and endorphins. In these cases, precursor molecules are tailored according to cell type and the stage of cell development, one possible mechanism of differentiation during embryonic development.

Functionally important modification of molecules occurs not only at the posttranslational level but also at levels closer to gene action. As explained in greater detail in Sect. 3.1.3.6, mRNA is formed first from the entire gene, including introns. Intron sequences are then spliced out, and the mature mRNA leaves the nucleus. The splicing process may be modified depending on the state of differentiation and function of the cell, leading to different genes and proteins. This "alternative splicing" is discussed in detail in Sect. 8.1. Even the gene itself may be modified. Such modification was analyzed in the immune globulin genes (Sect. 7.4). These results do not detract from the heuristic value of the one-gene/one-*enzyme* hypothesis for most cases.

A special situation in which several enzymes are functionally disturbed by one mutation occurs when uptake, transport, or binding of cofactors is disturbed.

7.2.2.5 Influence of Cofactors on Enzyme Activity
[214]

Enzyme Cofactors. Many enzymes need nonprotein molecules as cofactors for their activity. These cofactors may be simple ions, for example, Mg^{++} or organic compounds. If the cofactor is a more complex compound, it is known as a coenzyme. Precursors of coenzymes must be supplied by nutrition and are traditionally called vitamins. Often a vitamin acts as a precursor for a coenzyme involved in many enzymatic reactions; nutritional deprivation of vitamins leads to a vitamin deficiency state.

Apart from exogenous deficiency, conditions showing decreased coenzyme function may also be due to genetic defects at various levels of vitamin uptake and utilization (Fig. 7.15). All vitamins must be taken up from the intestine, transported into the cells, and taken up into specific cell organelles. They must be converted into the coenzymes that in turn combine with the apoenzyme to form a holoenzyme. Any of these steps may be impaired by genetic blocks. The exact mechanism of vitamin uptake is known mainly for vitamin B_{12} (cobalamin) and folic acid; for both substances transport and coenzyme synthesis defects have been described.

Folic Acid Dependency (229 030, 249 300, 229 050): *Deficiencies in Transport and Coenzyme Formation.* Folic acid is composed of three residues – a pterin ring, *p*-aminobenzoic acid, and glutamic acid (Fig. 7.16). The amount needed daily is normally present in a large number of foodstuffs. There are five coenzyme forms of folate, all of which are concerned with the transfer of 1-carbon units needed for DNA, RNA, methionine, glutamate, and serine synthesis. The principal steps of absorption and coenzyme formation are presented in Table 7.4.

At least five inherited conditions have been described in which transport mechanisms or mechanisms of conversion into coenzymes are deficient (Table 7.5). Four of these show signs of marked central nervous system dysfunction including mental retardation, and two show megaloblastic anemia. Their most important feature is that they can be treated successfully, provided that the condition has been diagnosed in time. In the intestinal absorption defect, for exam-

Table 7.4. Steps of absorption and coenzyme formation

Step	Enzyme
1. Conversion of polyglutamyl to monoglutamyl form	Conjugase enzyme (intestinal mucosa, stomach, pancreas)
2. Absorption by active transport	Duodenum and jejunum (exact mechanisms?)
3. Transport to the tissues	
4. Conversion of folate to the coenzymes:	
a) Reduction of the pterin ring: formation of tetrahydrofolate	
b) Conversion into five different coenzymes	Five different enzyme reactions

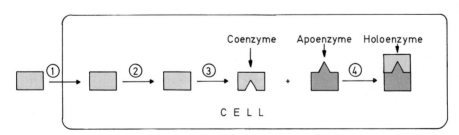

① Defective transport of vitamin into cell   ③ Defective conversion of vitamin to coenzyme
② Defective transport of vitamin within cell   ④ Defective formation of holoenzyme

Fig. 7.15. Mutations may interfere with vitamin-catalyzed reactions at several levels, from vitamin transport into the cell to enzyme formation. (Adapted from Scriver and Rosenberg 1973 [214])

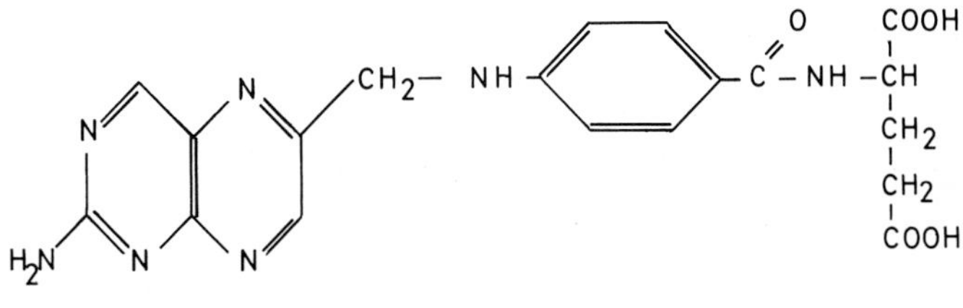

Fig. 7.16. Folic acid. From *left* to *right*, a pterin ring, β-aminobenzoic acid, and glutamic acid

Table 7.5. Inborn errors of folic acid metabolism (From Scriver and Rosenberg 1973 [214])

Phase of metabolism affected	Nature of defect	Manifestation of defect			Folate requirement in vivo
		Serum folate concentration	Megaloblastic anemia	CNS dysfunction[a]	
Intestinal absorption	Undefined	Low	Yes	Yes	Normal
Tissue utilization	Formiminotransferase deficiency	High	No	Yes	Increased
	Cyclohydrolase deficiency	High	No	Yes	NR
	Dihydrofolate reductase deficiency	Normal	Yes	No	Increased
	N^5, N^{10}-methyltetrahydrofolate reductase deficiency	Low–normal	No	Yes	Increased

NR, not reported.

[a] Includes mental retardation, psychotic behavior, seizures, EEG abnormalities, cerebral cortical atrophy.

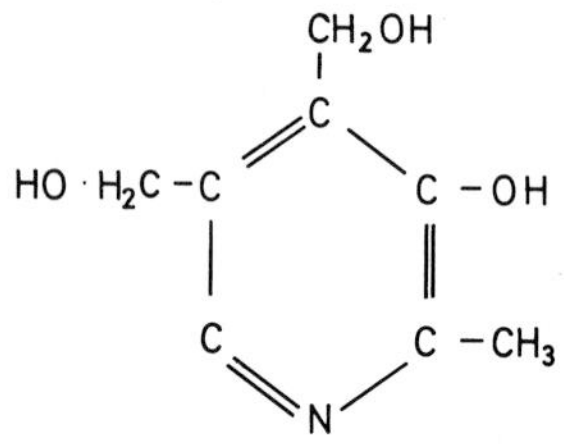

Fig. 7.17. Vitamin B_6 (pyridoxine)

ple, the folate requirement is normal; intramuscular instead of oral administration restores normality. In three of the four enzyme defects increased amounts of folic acid intake improved the metabolic situation, but it is unknown whether the central nervous system improvement could have been prevented had folic acid been supplemented early in infancy.

Deficiencies in coenzyme formation or uptake are expected to affect more than one enzyme at a time – in fact, all enzymes on which this coenzyme acts. On the other hand, deficiencies in the last step, the ability of the protein apoenzyme to form a holoenzyme by combining with the coenzyme, affect only one enzyme. They are more similar to the usual enzyme defects described in the foregoing sections.

Pyridoxine (Vitamin B_6), Dependency (266100). Vitamin B_6 is a substituted pyridine ring that occurs in several natural forms in widely varying foodstuffs (Fig. 7.17). In the cell these precursors are phosphorylated to pyridoxal-5'-phosphate or pyridoxamine-5'-phosphate by a specific kinase. These phosphorylated compounds act as coenzymes for a great many apoenzymes, which regulate the catabo-

lism of amino acids, glycogen, and short-chain fatty acids.

Therefore it is not surprising that numerous defects exist which have pyridoxine dependence in common. All of these are rare, but it is important to diagnose them in time – before seizures have led to irreparable brain damage, since appropriate substitution therapy keeps these children healthy. In newborns suffering from unexplained seizures a therapeutic attempt with vitamin B_6 may be worthwhile even before a final diagnosis has been established.

There are additional case reports in which high doses of a vitamin have improved the clinical and biochemical condition of patients. Further analysis of such diseases will be interesting theoretically, as it promises new insight into the mechanism of coenzyme binding and action. It is also rewarding for medical practice since these conditions will respond most favorably to therapy with high doses of vitamins.

This general concept, which is well-founded for a few very rare inborn errors of metabolism, has been claimed to apply in a variety of more common diseases such as schizophrenia, cancer, and others. The evidence adduced so far, however, is not convincing. However, milder deficiencies of this sort may exist and may contribute to the etiology of common diseases. Further work along these lines may be of interest.

7.2.2.6 X-Linked HPRT Deficiencies (308000)

Enzyme Defects as Tools for Some Basic Questions on Gene Action and Mutation. Some enzyme defects have proven useful as analytical tools for attacking problems of more general significance for our understanding of gene action and mutation. One group of defects has been utilized especially successfully: The HPRT deficiencies, which are classified with the defects of purine metabolism (Fig. 7.19).

Lesch-Nyhan Syndrome. In 1964 Lesch and Nyhan [152] described a peculiar syndrome with athetosis, hyperreflexia, and compulsive self-destructive behavior involving the chewing-off of lips or fingers [141]. All patients show hyperuricemia and excessive excretion of uric acid in the urine, which may lead to uric acid nephrolithiasis with obstructive uropathy. Only boys are affected; the mode of inheritance is X-linked (Sect. 4.1.4). Heterozygotes can be identified but are not affected.

In 1967 Seegmiller et al. [218] discovered almost complete deficiency of one enzyme of purine metabolism, HPRT, in erythrocyte lysates from three patients and

in cultured fibroblasts from another. The enzyme defect was then confirmed in other tissues such as liver, leukocytes, and brain. Various pathways lead to inosine-5-monophosphate in several steps. However, cells can also use preformed purine bases and nucleosides produced during the breakdown of nucleic acids. This "salvage" pathway involves conversion of the free purine bases to their corresponding 5'-mononucleotides. Two enzymes are involved, one specific for hypoxanthine and guanine, the other for adenine (Fig. 7.18). When the first enzyme is deficient, hypoxanthine and guanine are not recycled but are converted into uric acid in large amounts. This leads to hyperuricemia with kidney stone formation; it is not known, however, what mechanism produces the CNS symptoms. This enzyme defect, which can readily be examined in cultured fibroblasts, has been used to investigate a number of problems.

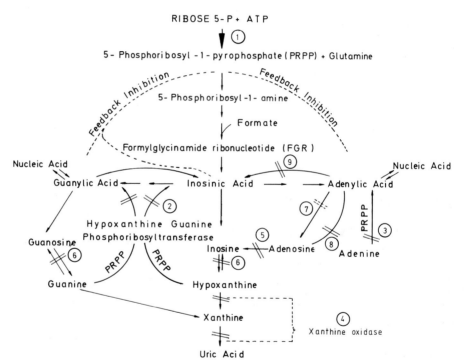

Fig. 7.18. Known enzyme defects in human purine metabolism. *1,* Increased phosphoribosylpyrophosphate synthetase activity in patients with overproduction of uric acid leading to gout. *2,* Gross deficiency of hypoxanthine-guanine phosphoribosyl transferase in children with Lesch-Nyhan disease and partial deficiency of the same enzyme in patients with overproduction of uric acid and gout. *3,* Adenine phosphoribosyl transferase deficiency in patients with kidney stones composed of 2–8 dioxyadenine that are often confused with uric acid stones. *4,* Xanthine oxidase deficiency in patients with xanthinuria who are at increased risk for xanthine urinary calculi and, occasionally, for myalgia caused by xanthine crystals in the muscle. *5,* Adenosine deaminase deficiency associated with a severe combined immunodeficiency state. *6,* Purine nucleoside phosphorylase deficiency associated with isolated defect in T cells. *7,* Purine 5'-nucleosidase activity is low in lymphocytes of patients with agammaglobulinemia that may be secondary to loss of B cells. *8,* Adenosine kinase deficiency has so far been developed only in the human lymphoblast cell lines. Its counterpart in patients is yet to be identified. *9,* Myoadenylate deaminase deficiency is associated in some patients with development of weakness and muscle cramps after vigorous exercise and failure to show a rise in venous blood ammonia in response to muscle exercise. These enzyme defects provide excellent examples for the various, more and often less characteristic phenotypic consequences of different genetic blocks within the same functional network. (From Seegmiller 1983 [217])

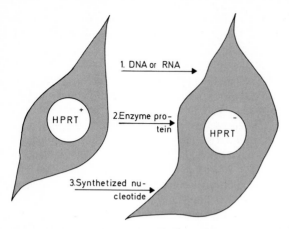

Fig. 7.19. Possibilities for metabolic cooperation between HPRT-active and HPRT-deficient cells in cell cultures from heterozygotes. See text for details

Molecular Heterogeneity. As evidenced by characteristics such as residual activity, Michaelis-Menten constant, thermolability, end-product inhibition by GMP or IMP, and others, the mutations observed in the various families are usually different. Sometimes severe HPRT deficiency has been observed even without the clinical symptoms of Lesch-Nyhan syndrome. Moreover, partial HPRT deficiencies have been observed in some adult patients suffering from gout [140]. The great majority of gout patients, however, have normal HPRT. The minority in which the abnormal enzyme is present show – as do those with the severe defect – an X-linked mode of inheritance. This is additional evidence that the same locus is indeed involved.

Evidence for X-Inactivation. Some of the best evidence supporting the Lyon hypothesis (Sect. 2.2.3.3) comes from analysis of enzyme activity in HPRT heterozygotes at the cell level [144]. These studies have also provided new insight into metabolic interrelationships between the cells of an individual.

Metabolic Cooperation. Heterozygous carriers can be identified by studies on cultured skin fibroblasts. When fibroblasts are cloned, and enzyme activity is measured by autoradiography of the uptake of ³H-labeled hypoxanthine into the cell, about half the clones show full HPRT activity, while the other half is deficient for this enzyme. However, in fibroblast cultures without cloning, the great majority of heterozygote cells show some enzyme activity. HPRT-deficient cells appear to have their metabolic defect corrected when in contact with normal cells [91]. This has been confirmed by artificial mixing of normal and defective cells in culture and is called "metabolic cooperation."

Three mechanisms for this cooperation are possible (Fig. 7.19):

1. Normal cells provide deficient cells with DNA or mRNA, thus enabling them to synthesize a functional enzyme protein.
2. The deficient cells receive preformed enzyme. This possibility is suggested by the observation of correcting factors in the mucopolysaccharidoses (Sect. 7.2.2.3). Incubation of cultured HPRT-deficient fibroblasts with ultrasound-minced material from normal cells shows partial restoration of enzyme function.
3. Normal cells synthesize the nucleotide (the end product), which is transferred to the deficient cell. This mechanism is supported by most of the evidence. After deficient fibroblasts are separated from normal cells, they revert promptly to the mutant phenotype, although normal HPRT is stable for many hours under the experimental conditions. The problems of metabolic cooperation have also been studied in other experimental systems, leading to more detailed knowledge regarding the basic mechanisms.

Other Problems Examined with HPRT Deficiency. HPRT deficiency has proven to be a useful tool for investigations of the mutational process:

a) The possibility of identifying all hemizygotes and heterozygotes by enzyme determinations in fibroblasts makes HPRT deficiency especially suitable for comparison of spontaneous mutation rates in the two sexes (Sect. 9.34). A great number of point mutations (single base substitutions) and quite a few small deletions have been identified up to the end of 1992 (approx. 40 point mutations, approx. 10 deletions [63]).
b) The gene is also expressed in amniotic cells. Therefore HPRT deficiency can be diagnosed by amniocentesis.
c) Using an abnormal substrate, 8-azaguanine, a selective system for point mutations in fibroblast cultures has been developed that permits the examination of problems of spontaneous and induced mutations at the cellular level. Normal cells metabolize 8-azaguanine by HPRT and are killed. HPRT-deficient cells cannot metabolize this compound and survive.

Immune-Deficiency Diseases Associated with Adenosine Deaminase and Nucleoside Phosphorylase Defects. Another defect of an enzyme which is involved in nucleoside metabolism leads to a different phenotype. This enzyme defect involves a rare variant of an enzyme for which a genetic polymorphism is also known. Defects in one or more components of the immune system can result in increased susceptibility to microbial infections. The classic disease in this category is X-linked hypogammaglobulinemia (300 300), which is caused by a maturation defect of B lymphocytes. B lymphocytes are the production sites of humoral antibodies, and their absence causes failure of γ-globulin synthesis. T lymphocytes are involved in cellular immunity and are intact in this disease. With increasingly detailed knowledge of the immune system, (Sect. 7.4), many more genetic immune defects have been described [215]. Enzyme defects related to metabolism of purines, pyrimidines, and nucleic acids have been identified (Fig. 7.18). Combined immune deficiency can be caused by deficiency of adenosine deaminase (ADA; 242 750) or of nucleoside phosphorylase (164 050) [217].

ADA catalyzes the irreversible deamination and hydrolysis of the purine nucleoside adenosine to inosine and ammonia. Nucleoside phosphorylase catalyzes the conversion of the purine nucleoside inosine to hypoxanthine and that of guanosine to guanine. It has little activity in converting adenosine to adenine. These are key enzymes in DNA and RNA synthesis and breakdown.

Adenosine deaminase – located on chromosome 20 (Sect. 5.4) – exists as a polymorphic trait in the population, as demonstrated by starch gel electrophoresis. The common allele is known as ADA[1] and a common variant as ADA[2]. ADA[2] has a gene frequency of about 0.05 in Western populations. Several other rare variants of ADA have also been described. ADA deficiency is an autosomal-recessive trait. Affected children lack any ADA activity in their red cells and other tissues. Their parents usually have intermediate amounts of the enzyme and are clinically normal. Some residual ADA activity can be demonstrated in affected patients. This defect has become well known because it was the first disease in which attempts at somatic gene therapy by introducing normal alleles into lymphocytes were carried out (Sect. 19.2).

Nucleoside phosphorylase is specified by a locus on chromosome 14. No polymorphisms exist at this locus, but several rare variants have been demonstrated. In affected patients, there is complete absence of nucleoside phosphorylase activity while parents have the expected intermediate activity; inheritance is autosomal-recessive.

Patients with ADA deficiency usually have severe B and T cell dysfunction, while in nucleoside phosphorylase deficiency B cell function and therefore immunoglobulin production appears to be intact. T lymphocyte dysfunction as manifest by lymphopenia, failure of lymphocytes to respond to mitogens, and abnormal skin tests to a variety of antigens is striking in both ADA and nucleoside phosphorylase deficiencies.

The exact biochemical mechanism by which these enzyme deficiencies produce immunological abnormalities is not clear. It has been speculated that ADA deficiency produces an accumulation of deoxy-ATP, which inhibits production of pyrimidine deoxyribonucleotides and therefore interferes with DNA synthesis, lymphocyte proliferation, and immune response. The mechanisms for nucleoside phosphorylase deficiency may be similar.

7.2.2.7 Phenylketonuria: Paradigm for Successful Treatment of a Metabolic Disease [214, 215]

Metabolic Oligophrenia. Phenylketonuria (PKU) (261 600) was first described by Følling in 1934 [86] in mentally retarded patients with a peculiar,

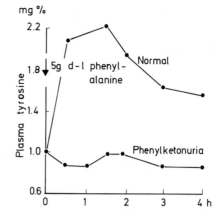

Fig. 7.20. Genetic block in phenylketonuria: ingestion of 5 g *d*-L-phenylalanine leads to an increase of serum tyrosine in the normal but not in the phenylketonuric individual. (Adapted from Harris 1959)

"mousy" odor. The name was coined by Penrose (1935) [197]. It is now one of the best known inborn errors of metabolism in humans; the various aspects of this disease have repeatedly been reviewed [146, 213, 215]. Here only three main aspects are discussed: introduction of a low-phenylalanine diet as the first successful approach to alter the phenotype of a genetic enzyme defect by suitable manipulation of the environment; the genetic heterogeneity of the condition as revealed by mass screening of newborns; and the problem of heterozygote detection and possible phenotypic abnormalities in heterozygotes.

Enzyme Defect in PKU. L-Phenylalanine is an essential amino acid. However, only a proportion of the normal intake of L-phenylalanine can be utilized for protein synthesis; the major part is oxidized primarily to tyrosine and a much smaller part to other metabolites, primarily phenylpyruvic acid. The parahydroxylation of phenylalanine to form tyrosine is a complex reaction. The hydroxylase consists of two protein components, one of which is labile and found only in the liver (and possibly, with lower activity, in the kidneys), the other being stable and found in many other tissues. This stable component contains a pteridine as cofactor.

PKU has been shown to be caused by complete deficiency of hepatic phenylalanine hydroxylase (Fig. 7.20) [127, 243]. The labile component of the enzyme system is affected. More recent enzyme assays from liver biopsies have shown small residual activity (up to about 6% of normal) in about half the cases with classic PKU. Other types of hyperphenylalaninemia with higher enzyme activities are discussed below. The hydroxylase reaction is one step in the metabolic pathways of phenylalanine and tyrosine for which a number of genetic blocks are known (Fig. 7.21). The gene is located on chromosome 12 [270].

Dietary Treatment of PKU. Phenotypic damage due to a genetic block may be caused either by lack of a metabolite behind the block or by accumulation of a metabolite before of the block. Examples of the former include albinism and cretinism with goiter (Fig. 7.21). In PKU it soon became obvious that the abnormalities can hardly be due to tyrosine deficiency, for tyrosine is usually available in sufficient amounts in food. On the other hand, the numerous metabolites found in the urine of PKU patients, along with a high increase in serum phenylalanine levels, indicated the opening of additional overflow pathways. This suggested therapy aimed at reducing phenylalanine intake. Such an attempt was first made by Bickel in 1953 [29]:

On the assumption that the excessive concentration of phenylalanine (or perhaps of some break-down products) is responsible for the mental retardation found in this condition we decided to keep a girl, aged 2 years, with phenylketonuria on a diet low in phenylalanine. She was an idiot and unable to stand, walk, or talk; she showed no interest in her food or surroundings, and spent the time groaning, crying, and banging her head. The diet had to be specially prepared, because a sufficiently low phenylalanine intake could only be attained by restricting practically all the nitrogen intake to a special casein (acid) hydrolysate . . ... This was treated with activated acid-washed charcoal, which removed phenylalanine and tyrosine. Tyrosine, tryptophane, and cystine were then added in suitable amounts . . ..

The child was first treated in a hospital so that careful observations could be made. During a 4-week preliminary period, when no phenylalanine was permitted, no definitive clinical change other than weight loss was observed. The characteristic musty smell disappeared, the levels of phenylalanine in plasma and urine fell to normal, the excretion of phenylpyruvic acid decreased, and the ferric chloride reaction in the urine, indicating phenylpyruvic acid, became negative (Fig. 7.22).

Subsequently, presumably as the result of tissue breakdown, the biochemical abnormalities returned to some extent, along with generalized aminoaciduria. Phenylalanine was therefore added in small amounts in the form of whole milk, a daily intake of 0.3–0.5 g being found sufficient for normal weight gain, with greatly improved biochemical findings. During continued outpatient treatment there was a gradual improvement in the child's mental state over

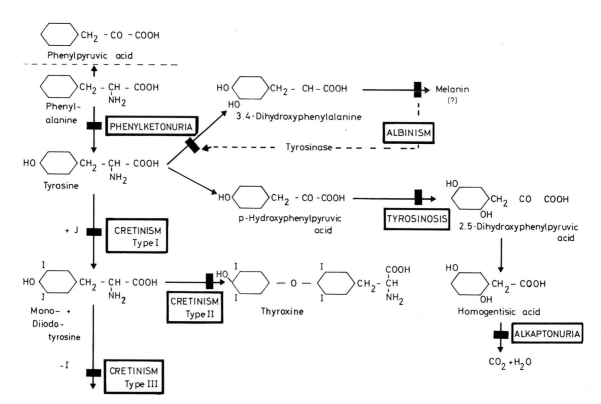

Fig. 7.21. Genetic blocks in metabolic pathways of some aromatic amino acids. The diagram is oversimplified. Genetic block leading to phenylketonuria, albinism, alcaptonuria, tyrosinosis, and three types of hereditary cretinism are included

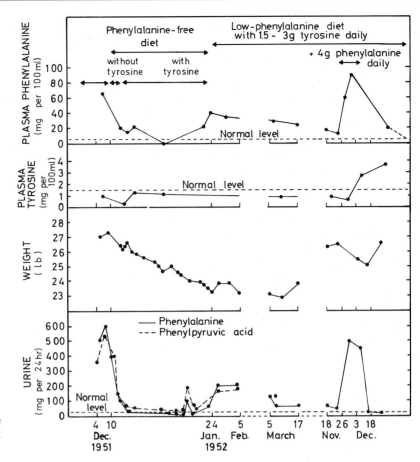

Fig. 7.22. First metabolic studies in a phenylketonuric child treated with a phenylalanine-free diet. Serum and urine phenylalanine were reduced quickly to normal values. Administration of 4 g phenylalanine per day led to a quick increase of phenylalanine level. (From Bickel 1954 [29])

the next few months; she learned to crawl, her eyes became brighter, her hair grew darker, and she no longer banged her head or cried continuously.

To determine whether this improvement was real an experiment was undertaken by adding 4 g phenylalanine per day to the diet. This resulted in a definite deterioration in the child's condition, which was soon reported by the mother. The study was repeated in the hospital under controlled conditions (Fig. 7.22) and again led to the expected biochemical and clinical changes. This case demonstrated the beneficial effects of a low-phenylalanine diet. In the same paper the authors stated that "[F]urther controlled trials are being made, special attention being paid to the very young children, who are likely to benefit most."

The success of dietary treatment was soon confirmed by other groups. The evidence is now quite substantial that the diet does indeed have marked beneficial effect on the development of the PKU patient. There are, however, two qualifications:

1. To prevent brain damage the diet should be started as soon as possible within the first few weeks of life.
2. The metabolic status of the child, especially phenylalanine levels, should be monitored carefully.

Lifelong treatment is not necessary since the adult brain appears resistant to the abnormal metabolite concentration found in PKU.

A number of women with treated PKU have had children. In spite of the fact that these children are only heterozygous, about 90 % of them showed signs of severe mental retardation [151]. Some other birth defects such as congenital heart disease have also been seen frequently. Hence, maternal hyperphenylalaninemia is harmful for the development of the fetus. A carefully controlled low-phenylalanine diet for all PKU patients immediately from the beginning of pregnancy appears mandatory to prevent this complication. This is a serious public health problem since careful search for the rare, previously treated PKU patients becomes necessary to avoid almost certain mental retardation in their offspring. Similar problems may arise with other treatable inborn errors of metabolism.

Genetic Heterogeneity of PKU. The possibility of successful treatment in early infancy before the appearance of clinical symptoms has led to the introduction of screening programs for newborns. In most Western industrialized countries, where PKU has a fre-

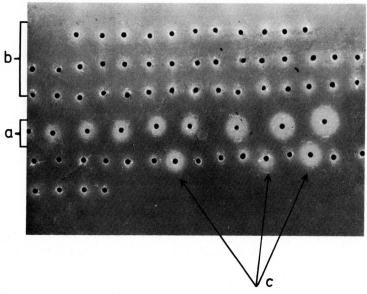

Fig. 7.23. A strain of bacillus subtilis requiring phenylalanine for growth is incubated in an agar plate. The strain is able to grow only if the phenylalanine level of the test blood is increased. This leads to a bacterial growth around the blood spot. The diameter of the growth area is related directly to blood phenylalanine level. *a,* Standard phenylalanine concentrations from left to right; *b,* normal specimens; *c,* abnormal specimens between 6 and 12 mg/% (Courtesy Prof. H. Bickel)

quency of 1:6000–1:20000, practically all newborns are now being screened. Since the screening method should be easy and inexpensive, the so-called Guthrie test is used [108]. This is based on the growth of phenylalanine requiring bacteria on a spot of blood containing phenylalanine above a certein threshold concentration (Fig. 7.23). Thus only blood from infants with high phenylalanine levels supports the growth of these bacteria.

When systematic newborn screening began, it was soon apparent that not every infant in which high blood phenylalanine level is found has PKU. Many have a less severe hyperphenylalaninemia not leading to clinical symptoms. These infants had hyperphenylalanemia which does not lead to mental retardation (non-PKU hyperphenylalanemia). This condition occurs with one-half the frequency of PKU (about 1/20000–1/30000 births). Treatment with a phenylalanine-deficient diet is not required. Several rare hyperphenylalanemias (1 or 2 per 1000000 births) are not associated with phenylalanine hydroxylase deficiency and are designated as malignant hyperphenylalanemias because of their severity and nonresponsivness to the classical therapy. These conditions include dihydropteridine reductase deficiency, guanosine triphosphate cyclase deficiency, and 6-pyruvyl tetrahydropterin synthase deficiency and are disorders of tetrahydrobiopterin homeostasis (261640) causing impaired hydroxylation of phenylalanine and tryptrophone. Not unexpectedly, cloning of the gene for PKU [206] revealed many different mutations, including mis-

sense, deletions [63] and splicing mutations. Based upon haplotype analysis of RFLP variants around the mutational site a multiple origin of even identical mutations was found, providing interesting hints for tracing population history. Non-PKU hyperphenylalaninemia without clinical symptoms is also caused by mutations at the PKU locus. One factor influencing the extent of mutational impairment of the enzyme is residual enzyme activity. Activity of 10% appears sufficient for normal somatic development. Most patients, except offspring of consanguineous matings, carry two different PKU mutations and are compound heterozygotes. Antenatal diagnosis is possible by direct mutational analysis or by indirect linkage diagnosis with RFLP markers. While most couples do not select this option, occasionally parents fear the difficulties of the required dietary treatment and choose selective abortion.

The rare disorders of tetrahydrobioterin homeostasis that cause malignant hyperphenylalanemia do not map to the PKU locus. Antenatal diagnosis is possible. At least for the variety caused by dihydropteridin reductase deficiency.

7.2.2.8 Heterozygote Detection

Heterozygote Detection for PKU and Hyperphenylalaninemia. As in other metabolic diseases, detection of heterozygotes is not only of theoretical interest but can be used in practice for genetic counseling of close relatives, for example, sibs of PKU patients. Such un-affected sibs have a two-thirds risk of being heterozygous. In earlier years heterozygotes were discovered mainly by phenylalanine loading tests. However, results overlap to a certain extent between heterozygotes and normal homozygotes. At present the best method is detection of the mutants present in the affected child by direct examination of the gene.

Health Status of Heterozygotes. At first glance heterozygous parents and sibs of PKU patients appear perfectly healthy. Penrose [196], however, observed in a family of PKU patients a special type of mental illness of a depressive kind with onset at the age of about 50 years in six relatives. He suspected that heterozygotes may run a higher risk of mental disease. In the past 50 years, however, the problem of possible disease liabilities in PKU heterozygotes has received surprisingly little attention, and the few studies devoted to this problem often lack epidemiological sophistication. The problem has been reviewed fairly recently [247].

Recognizable peculiarities in PKU heterozygotes have been reported: deviations in IQ and other psychological tests; a higher risk of mental disease; EEG abnormalities; and disturbances in reproduction. The first three aspects are discussed in greater detail in Sect. 15.2 on behavior genetics. Suffice it to say at this point that the great majority of heterozygotes do *not* suffer from mental disease, but their risk of having a certain type of late-onset schizophrenia may be somewhat elevated. Moreover, a slight reduction in the average IQ – especially its verbal component – has been described, and minor EEG peculiarities appear to be more frequent. According to some disputed data, the risk of abortions and stillbirths may be enhanced. Interestingly, a few reports have asserted an abnormal elevation of blood phenylalanine levels in stress situations, such as influenza with high fever and pregnancy. An interesting ecogenetic suggestion has been made in that aspartame – a high-phenylalanine containing sweetening agent – might injure the fetuses of heterozygous women. (See above for discussion of maternal hyperphenylalaninemia.)

None of the above data for PKU heterozygotes are definite. If confirmed, such results would support the proposition that the reduced enzyme activity makes heterozygotes less able than other persons to cope with a variety of environmental stresses; this subject could constitute a new chapter in ecogenetics (Sect. 7.5.2).

Heterozygote Detection in General. Neel in 1949 [183] was the first to consider the general problem of heterozygote detection in medical genetics and to collect the scattered evidence available at that time. A more comprehensive report was presented by Neel in 1953 [184] and by Franceschetti and Klein in 1954 [89]. Since biochemical genetics has progressed rapidly, heterozygote tests for many diseases have become available. For example, enzyme activities have been found to be reduced to about 50 % in cells in which the gene is active such as in fibroblasts, leukocytes, and erythrocytes. While mean enzyme activity of heterozygotes ranges around 50 %, there is significant variation of enzyme levels in both heterozygotes and normals, leading to overlap of their respective distribution of enzyme values, which, however, is bimodal. Such failure to assign a given individual to a genotypic class definitely applies to many different test systems aimed at heterozygote detection, such as the activity of blood clotting factors (hemophilia). Fortunately, diagnosis of heterozygosity in more and more cases can be based on direct detection of the mutant gene using DNA techniques. Apart from its value for theoretical understanding of enzyme function, diagnosis of heterozygotes is significant practically for two reasons.

First, it helps in genetic counseling of relatives of patients with X-linked or autosomal-recessive diseases. Its practical value is revealed especially in X-linked diseases since female heterozygotes run a risk of 50 % for each son to be affected. In most autosomal-recessive diseases, heterozygote detection is less important provided that the possible heterozygote – in most cases a brother or sister of an affected homozygote – does not plan to marry a relative such as a cousin. There is a risk of homozygous children only if both prospective parents are heterozygous, and most recessive conditions are so rare that the risk of a heterozygote to marry another one is quite low indeed (see the Hardy-Weinberg law, Sect. 4.2).

A second application of methods for heterozygote detection consists in the screening of whole population groups. Such screening has sometimes been introduced in populations in which certain recessive genes are frequent. For example, about 8 % of African-Americans are heterozygous for the sickle cell gene, and about 3 %–4 % of Ashkenazi Jewish populations are heterozygous for the Tay-Sachs gene. However, well-meaning screening programs have repeatedly met with logistical and psychosocial problems (Chap. 18). Such heterozygote screening programs make sense only if introduced for reproductive purposes to allow couples to choose reproductive options such as selective abortion when both partners are found to be carriers. This caveat has sometimes been overlooked (Chap. 18).

The most frequent recessive disease in populations of northern and western European origin is cystic fibrosis (CF) with an incidence of about 1:2000 and a heterozygote frequency of about 4 %–5 %. Here, heterozygote screening for genetic counseling of relatives and particularly for detection of matings involving two heterozygotes (who are not aware of their carrier status) for intrauterine diagnosis of potentially affected offspring is now under discussion. The CF gene has been cloned, and its mutations have been delineated. In populations of northwestern European origins, about two-thirds of the mutations are the Δ 508 deletion. Testing for another five or six mutations common in this population can detect about 85% of CF mutations. The remaining 15% may be due to any one of over 300 different CF mutations. Since it is impractical to test for all these variants, many false-negative test results occur. In populations of southern European or Mediterranean origin the mutational distribution is different. For example, among Ashkenazi Jews five different CF mutations account for 96% of all cases. Mutation testing must consider ethnic origin to select the appropriate mix of mutations for study.

Most informed observers agree that testing for the CF carrier state should be offered to sibs of affected patients. Population testing is more debatable. CF patients in the United States currently have a mean life expectancy of 28 years, and the disease is not as devastating as, for instance, Tay-Sachs disease. The indications for antenatal diagnosis are therefore not as clear in this disease. Since 5% of the white population are carriers, a large-scale population screening program would be logistically complex and expensive, and because of mutational heterogeneity would not detect all mutations. Investigations to assess the receptivity of the public to screening are now under study.

Susceptibility to Common Diseases in Heterozygotes of Recessive Conditions. In Sect. 6.2.4, α_1-antitrypsin deficiency is discussed as an example of a homozygous condition leading in many cases to increased susceptibility to common diseases – principally chronic obstructive pulmonary disease. Many heterozygotes when exposed to environmental agents such as tobacco smoking also may run an increased risk of developing chronic obstructive pulmonary disease. Sickle cell heterozygotes are healthy under normal conditions, but moderate hypoxia, such as an altitude above 2500 m, occasionally causes in vivo sickling and splenic infarcts.

There are many scattered reports [247] for other diseases. For example, heterozygotes for various lipidoses were shown to have a minor decrease in (performance) IQ on average, combined with an increased incidence of personality disturbances. Cystinuria heterozygotes may run a slightly increased risk of kidney stones; galactokinase deficiency heterozygotes may be susceptible to premature cataracts; and in some varieties of Wilson disease, anomalies of kidney tubular function

and minor neurological signs have been described. Heterozygotes for xeroderma pigmentosum, a defect of DNA excision repair (Sect. 10.2) and those for the chromosome instability syndromes often thought to be also caused by repair defects (Sect. 10.1) have been studied with great care for cancer risk. An increased risk of developing cancer at a relatively young age was found in heterozygotes for Bloom syndrome but not in those for Fanconi anemia. Interestingly the incidence of skin cancer is increased only in xeroderma pigmentosum heterozygotes living in the southeastern United States and not in those from other areas of the country [233]. This illustrates again an ecogenetic problem: the high intensity of UV irradiation in the sunlight overloads the reduced capacity for excision repair, which can cope with this problem with lower UV irradiation exposure.

Heterozygotes for ataxia-telangiectasia (208 900) appear to have a higher frequency of a variety of cancers. A prospective study [234] found a 3.5–4 increase for all cancers and a fivefold enhanced risk for breast cancer in women. A significant proportion (more than 8%) of all breast cancers were claimed to be due to heterozygosity for ataxia-telangiectasia. Based on their data, these authors suggest that exposure to ionizing radiation increases the risk of breast cancer in heterozygotes for ataxia-telangiectasia. This important problem needs further study once a specific DNA test for ataxia-telangiectasia has been worked out.

Detection and confirmation of increased risks for common diseases or slight physiological abnormalities requires elaborate statistical designs, such as careful selection of controls and in particular a much larger number of biochemically and genetically well-defined heterozygotes than have so far been included in these studies. One difficulty with studies of this sort is the overlap common in the laboratory values between heterozygotes and normals. In population studies most values from the overlap zone derive from normals and not from heterozygotes since the total number of normals far exceeds that of the heterozygotes. Studies on obligate heterozygotes (parents of homozygous children) are much more promising. Investigation of heterozygotes promises better insight into the genetic conditions underlying common diseases as well as of many other problems. For example, an appreciable proportion of the genetic variability influencing performance and behavior as measured by IQ and personality tests may be caused simply by the high population frequency of heterozygotes for recessive diseases with well-known effects in homozygotes (for discussion, see Sect. 15.2.1). Risk predictions for an increased population load due to mutagenic agents such as ionizing radiation and chemical mutagens would have to be reassessed if slightly increased disease liabilities in heterozygotes were found to be the rule rather than the exception. Therefore studies along these lines are urgently needed, particularly when unequivocal distinction between normals and heterozygotes is possible by a qualitative test system.

Heterozygote Testing in Hemophilia A and B and in Duchenne and Becker Muscular Dystrophies. In these diseases, heterozygote testing can now be done by indirect, and more and more by direct DNA studies (see Chap. 18). The older methods such as the study of creatine phosphokinase (PK) in Duchenne muscular dystrophy and factor VIII or factor IX determinations in the hemophilias are now being replaced by the more accurate DNA methods.

In the entire field of blood coagulation and its disturbances, amazing progress has been made in recent years. The participation of various genetically determined proteins in blood coagulation and clot lysis provides an example for interaction of many genetic factors in a complex physiological process. This network has been disentangled largely with the aid of blood from patients with various genetic coagulation defects. Blood coagulation is not a simple and not even a ramified chain reaction; feedback circles are also involved.

Problems with Heterozygote Detection. As mentioned, there is usually significant overlap between the distributions of heterozygotes and normal homozygotes, so that many normals have enzyme levels of the test substance that are consistent with heterozygosity. The reasons are usually not fully understood but may relate to the presence of undetected "isoalleles" each of which determines a unique range of different activity levels. When using quantitative tests of this type for heterozygote detection it is essential for accurate assessment of the test results that the a priori (or Bayesian) likelihood of heterozygosity is considered.

Table 7.6 shows porphobilinogen deaminase enzyme levels in a population of normals compared with a group of patients with acute intermittent porphyria – an autosomal dominant trait. Note that values below 70 U are found only in porphyrics while values above 129 U strongly suggest the absence of this gene. However, ~30% of the control populations had enzyme values that overlapped with those of the porphyria gene carriers (70–129 U), and only 20% of all those with porphyria had values lower than those of normals (less than 70 U). Table 7.6 also indicates the odds in favor of heterozygosity for a given range of enzyme levels. The actual chance of an individual carrying the porphyria gene strongly depends on the a priori probability that the subject at risk carries this gene.

Table 7.7 provides data on individuals found to have a value of 95 U on testing but different a priori probabilities of carrying the porphyric gene. Note that 16% of porphyria patients and 2.3% of normals have a laboratory value between 90 an 99 U (Table 7.6). The odds of carrying the gene based on the laboratory findings alone is 16:2.3 or 7:1. *These odds alone,*

Table 7.6. Porphobilinogen deaminase level in 217 normals and in 105 acute intermittent porphyria patients (modified from Bonaiti-Pellié et al. 1984 [32])

Units	Porphyria patients and obligate heterozygotes (%) (x)	Normal controls (%) (y)	Laboratory odds (likelihood ratio) in favor of heterozygosity (x) : (y)[a]
< 70[b]	20	0	Very high[b]
70–79	23.8	0.5	48 : 1
80–89	22.9	0.9	25 : 1
90–99	16.2	2.3	7 : 1
100–109	9.5	3.7	2.6 : 1
110–119	5.7	8.3	0.7 : 1
120–129	1.9	14.3	0.13 : 1
> 129[c]	0	70	Unlikely[c]

Note that these values need to be determined for each laboratory separately in a sufficiently large sample size.
[a] Based on laboratory results only
[b] Since 20% of heterozygotes and no normals were found in this range, the odds in favor of heterozygosity are very high.
[c] None of the porphyria patients and 70% of normals had values above 129 U. With increasing enzyme levels the risk of porphyria becomes increasingly unlikely.

however, cannot be used for practical prediction. They must be combined with the a priori probability of the test subject carrying the gene. This a posteriori (Bayesian) probability is definite for a patient's first-degree relatives such as a sib (50% or 1:2). In a population study in which everyone is tested regardless of symptoms, the Bayesian probability of carrying the porphyria gene is the population frequency of the condition (1/10 000). When the disease is suspected clinically, no precise probability can be assigned, but an approximate value, such as 1/10, may be selected for a suggestive clinical impression and a low probability of 1/100 if there is a vague suspicion of the disease being present (Table 7.7).

Note the marked differences for the final practical risk of heterozygosity for the different Bayesian probability values. *These results show how an identical laboratory value (e.g., 95 U) can be of completely different predictive significance depending upon the probability of the disease under study occurring in the test subject.* Most laboratory scientists and physicians often overlook this fact.

The risks of carrying the gene for porphyria with the same laboratory result (i.e. 95 U) would be (a) 1:1500 for a member of the general population with no symptoms who is screened, (b) 7% for a person at 1% a priori risk (i.e., vague clinical suspicion), (c) 44% for someone at 10% a priori risk (i.e., clinical suspicion), and (d) 87% for a first-degree relative at 50% risk. These data show the considerable ambiguity of a diagnosis with such a test result if the a priori expectation of the diagnosis cannot be defined clearly. Note that repeated testing is not necessarily helpful in rendering these final risks more precise. With very low or high enzyme values the interpretation becomes much easier if, as in this example, the exact range of enzyme levels on a large number of normals and heterozygotes is known.

Table 7.7. Different odds with same laboratory value (95 U) with various a priori probabilities for acute intermittent porphyria (based on laboratory values from Bonaiti-Pellie et al. 1984 [32])

Laboratory value (U)	A priori		odds in favor of heterozygosity (See Table 7.6) (Laboratory results only)	Joint odds in favor of heterozygosity	Final risk of heterozygosity[b]
	Odds[a]	Probability			
95	1/9999 (e.g., population screening)	1/10 000	7 : 1	7 : 9999	0.0007 = 1/1500
95	1/99 (e.g., vague clinical suspicion)	1/100	7 : 1	7 : 99	0.07
95	1/9 (e.g., clinical suspicion)	1/10	7 : 1	7 : 9	0.44
95	1 : 1 (e.g., sib or child of definitely diagnosed patient)	1/2	7 : 1	7 : 1	0.87

[a] odds $= p : (1-p)$ where $p =$ probability. Note that these odds are not the final odds to predict whether a patient with a given enzyme level carries the porphyria gene (see Table 7.6).

[b] Calculated by multiplying the a priori odds and laboratory odds for the carrier state and noncarrier state to determine the joint odds for the carrier state as $x/x + z$ where x is the joint odds for the carrier state and z the joint odds for the noncarrier state. Example: a priori odds $1 : 9$; laboratory odds $7 : 1$; joint odds for carrier state to noncarrier state $7 : 9$ derived from $[(1 \times 7) : (9 \times 1)]$. Final risk $7/(7 + 9) = 7/16 = 44\%$.

In contrast to such uncertainties with quantitative testing, a definite diagnosis can be made regardless of the a priori probability if an "all or none" *qualitative* abnormality exists in heterozygotes that can be assessed biochemically or by DNA methods.

7.2.2.9 Treatment of Inherited Metabolic Disease
[70, 193a, 211]

General Principles. In earlier years the conclusion that a certain character is inherited had the connotation that it could not be influenced by environmental manipulation. Hereditary diseases were therefore not considered to be amenable to treatment. These seemingly nihilistic attitudes contributed to the stance of many doctors and behavioral scientists that genetics had little to contribute to their fields. The inborn errors of metabolism offer convincing examples that such beliefs are erroneous. Our ability to influence a disease or a behavioral abnormality often depends upon the depth of insight into mechanisms and not whether the etiology is genetic or nongenetic.

In principle, genetic traits can be influenced at all levels of gene action. The most direct approach would be somatic gene therapy or the introduction of the normal gene into somatic cells of affected patients. Ideally, replacement of the mutation with the normal gene would be most desirable. After appropriate laboratory and animal studies were carried out, investigative trials of gene therapy are now being carried out in diseases such as adenosine deaminase deficiency, familial hypercholesterolemia, and cystic fibrosis using retroviral vectors to carry the genes into the patient's genome. Although early results in ADA deficiency appear promising, no success in the other diseases has been achieved [193a]. Gene therapy is also being tested in various cancers with the aim of preventing proliferation of cancer cells by manipulating immune genes or by introducing genes that produce cytokines to inhibit cancer cell growth. There is general agreement that somatic gene therapy is a sophisticated extension of medical therapy which raises no new ethical issues beyond the usual problems of human experimentation and the assurance that the viral vectors used in gene therapy are safe to the patient and others [7, 93, 167].

Germinal gene therapy or the introduction of genes into germ cells or early zygotes has been carried out in animals. If successful, such manipulation leads to correction of a genetic defect in all body cells. A treated patient therefore could not transmit the mutant gene to his or her descendants, as occurs with somatic gene therapy in which only the affected tissue is treated. Most scientists and others feel that germinal therapy is rarely indicated, too dangerous, and not admissible for ethical reasons particularly since, unlike somatic gene therapy, it raises completely new moral issues. By general consent a voluntary moratorium on germinal gene therapy is therefore being maintained by all investigators in this field. (For further discussion, see Chap.19). Here we discuss therapeutic approaches at the level of gene expression in the phenotype.

Often the metabolic consequences of a genetic block may be influenced by suitable manipulation of the environment. The classic example, treatment of phe-

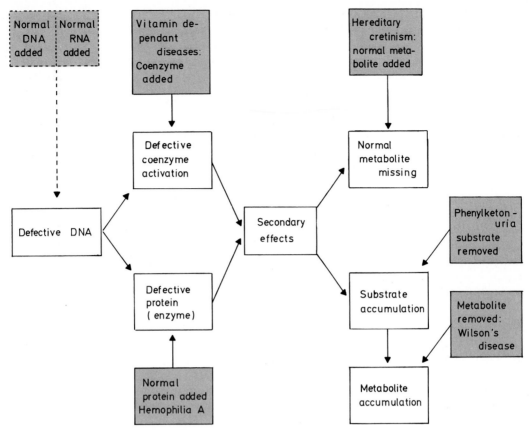

Fig. 7.24. Synopsis of therapeutic approaches to inherited metabolic diseases. Addition of normal DNA or RNA is under study. Normal protein or coenzyme (vitamin) may be added.

The secondary effects of enzyme blocks may be removed by addition of a normal metabolite or removal of excess substrates or metabolites

nylketonuria by a phenylalanine-restricted diet, is discussed in Sect. 7.2.2.9. In other cases the clinical consequences are due not to the accumulation of metabolites ahead of the block but to the lack of a metabolite behind it. In these cases substitution therapy may prove useful.

Finally, a great number of secondary consequences of a genetic disease may be successfully influenced, from the disequilibrium of endocrine regulation caused by a block in hormone synthesis to insufficient oxygen transport in an inherited anemia. Figure 7.24 gives a synopsis of therapeutic possibilities. In the following, some examples are discussed. For a more complete survey see [71].

Substitution (Protein or Enzyme) Therapy. The classic example is hemophilia A. Factor VIII activity at 20%–30% of the normal average controls bleeding. This level of activity can be achieved by factor VIII. Factor VIII concentrates are prepared from human blood or by recombinant DNA techniques [259] and home treatment with control of the bleeding episodes is now possible; hemophilia A patients may lead a

nearly normal life. The high frequency of HIV infections caused by contaminated blood due to use of many hundreds of donors in the early 1980s is no longer a problem due to HIV testing of donor blood and the beginning use of factor VIII produced by genetic engineering techniques.

Another example is substitution of pseudocholinesterase in patients with prolonged apena after succinylcholine administration during surgery (Sect. 7.5.1). Here therapy is facilitated by two favorable conditions:

1. The enzyme deficiency is harmless under normal conditions; substitution is needed only for major surgery that requires a muscle relaxant.
2. After injection of normal plasma the activity decreases to one-half within 12 h. This means that suitable activity can be maintained by one injection for the duration of the operation.

In the majority of patients with pseudocholinesterinase deficiency, enzyme therapy is not necessary because physicians have learned to cope with prolonged apnea by simply prolonging intubation and providing artificial respiration, but in many other diseases substitution of the missing gene product would be very useful.

The endocrine disorders are classical examples. Insulin therapy of diabetes mellitus was introduced as early as the 1920s; in the early decades of this therapy insulin from animals was used. Human insulin produced by recombinant DNA methods is also now available. Another example is pituitary dwarfism due to a genetic defect of growth hormone synthesis (262 400). In earlier years, growth hormone had to be laboriously extracted from many human pituitaries. Therefore it was available only in very small amounts. Recombinant DNA techniques have substantially improved the situation, and now the hormone is readily available and safe.

However, correction of enzyme defects has been more of a problem. Most enzyme defects would require life-long corrective therapy. Here, further difficulties are encountered:

a) The enzyme is eliminated from the body within a relatively short time; continuous supplementation is needed.
b) Enzyme preparations may be recognized by the immune system as foreign proteins, and antibodies may make the injected material biologically ineffective.

The possibilities of overcoming these difficulties vary from one condition to another. Enzyme preparations of human origin are most desirable for therapy; recombinant DNA techniques may be used.

At first glance, the mucopolysaccharidoses (Sect. 7.2.2.3) seemed to be a good candidate for enzyme therapy. So far, however, the results of various therapeutic trials have not been convincing [147]. In Gaucher disease, however, enzyme therapy has been shown to be effective: targeting the missing enzyme – glucocerebrosidase – to mannose-specific receptors of macrophages leads to the uptake of this enzyme, and to clinical improvement [18]. Currently many patients with Gaucher disease receive the missing enzyme with excellent results, and enzyme therapy has become the standard treatment for this disease. Because treatment must be administered frequently, the enzyme, which is laboriously extracted from placental sources, is very expensive. Due to variability in clinical presentation the indications for therapy are not entirely clear. Should a homozygote for Gaucher disease – who has no clinical symptoms even though affected with mild hepatosplenomegaly – be treated with glucocerebrosidase that cost about U.S.$ 300 000 per year? Gene therapy trials are already under way aimed at inserting the missing gene into bone marrow stem cells of patients with Gaucher disease. If this approach is ultimately sucessful, enzyme therapy, which needs to be given repeatedly, would no longer be necessary. Enzyme therapy cannot cure the few patients who have the variety of Gaucher disease that affects the central nervous system; here enzymes are not taken up because of the blood-brain barrier.

Another enzyme that might be used for substitution therapy is α_1-antitrypsin. Its deficiency often leads to chronic obstructive pulmonary disease (Sect. 6.2.4) [66, 258]. The protein can now be produced by recombinant DNA techniques and therapeutic trials have been somewhat successful.

Environmental Manipulation: Removal of a Metabolite Ahead of the Block. The metabolite ahead of the block that acts as substrate for the deficient enzyme may be removed relatively easily if it is not produced in the organism but is taken up as a normal nutrient. One example – phenylketonuria – has been described above. Another example is galactosemia due to deficiency of one of the three enzymes converting galactose into glucose. Here, removal of the substrate is easier, as galactose occurs almost exclusively in milk. The problem becomes more difficult if the metabolite in question cannot be restricted without impairment of normal function. In other cases the phenotypic ill effects of an enzyme block are caused not by accumulation of a metabolite ahead of the block but by lack of a metabolite behind it.

Environmental Manipulation: Substitution of a Metabolite Behind the Block. The best known examples of such substitution therapy are disorders of hormone synthesis mentioned above. Other examples are the glycogen storage diseases types I and III (Fig. 7.2). Here, most of the clinical symptoms are caused not by the glycogen storage itself but by the failure of glycogen to be broken down to glucose, which leads to chronic hypoglycemia. Replacement therapy of blood glucose would meet with insuperable difficulties and, in addition, would lead to still more glycogen storage. Therefore surgical intervention to bypass the liver, with the blood stream coming from the intestine and containing the resorbed glucose, has been tried successfully. A shunt between the portal vein and the inferior vena cava causes most of the blood to bypass the liver and transport glucose directly to the heart muscle and to other organs. Definite improvements have been observed.

Another example is orotic aciduria (Sect. 7.2.2.4). Here, the excess of orotic acid seems to have no major ill effects, but the deficiency of uridine compounds leads to impairment of nucleic acid synthesis and, specifically, to megaloblastic anemia and also to severe growth inhibition. Addition of uridine to the diet provides the missing metabolite and leads to amelioration of the clinical symptoms.

Elimination of the Metabolite Ahead of the Block and Substitution of the Metabolite Behind the Block. In one of the examples mentioned above, the glycogen storage diseases, "internal" substitution of the compound behind the block – glucose – by partial bypass of the liver also helps to reduce accumulation of the metabolite ahead of the block – glycogen. In familial hypercholesterolemia, an autosomal dominant disease (Sect. 7.6), accumulation of cholesterol in coronary arteries leads to coronary heart disease. Reducing cholesterol concentrations by binding to resins (such as cholestyramine or cholestipol) or by inhibiting HMG-CoA reductase (the major rate-limiting step in cholesterol synthesis) with appropriate drugs (such as lovastatin) reduces serum cholesterol levels and prevents coronary heart disease. In other diseases, clinical symptoms are caused by both mechanisms, and therapy should try to influence both. One example is homocystinuria (236 200), which is caused by a defect of the enzyme cystathionine synthetase (Fig. 7.25). Homocysteine is formed from nutritional methionine. Therefore the methionine supply should be reduced. However, since methionine is – as phenylalanine – an essential amino acid, it cannot be eliminated from the diet altogether. Cysteine, on the other hand, is normally formed from methionine via the pathway shown in Fig. 7.25. Many of the numerous symptoms in homocystinuria are due to cysteine depletion; therefore the diet should be enriched by cysteine. A different type of homocystinuria responds to pharmacological doses of vitamin B_6, which acts as coenzyme to cystathionine synthetase.

Treatment by Removing Secondary Effects of the Metabolic Defect. This category is by far the largest group

offering therapeutic possibilities in genetic defects. In contrast to other approaches, specific knowledge of the pathophysiological and genetic mechanisms is not necessary. For example, we know hardly anything about the biochemical basis of developmental conditions, such as polydactyly or cleft lip and palate. Yet this does not prevent successful surgical correction. Very little is known regarding the biochemical basis of mental disease (Sect. 15.2.1.2). Still, drug therapy that was introduced on a purely empirical basis has proven to be fairly successful in managing patients with schizophrenia and affective disorders. In all fields of medicine, most therapeutic measures – including the successful ones – are based on similar empirical evidence, regardless of whether the effects of genetic variability on the disease are large or small. On the whole, our capacities for treating hereditary diseases, and alleviating human suffering are not very impressive at present [65]. This Statement, however, probably also applies to much of medical therapy in general.

Therapeutic intervention that requires specific knowledge of the pathophysiological mechanism is the ultimate aim of most biomedical research. The group of adrenogenital syndromes due to enzymatic blocks in the synthesis of adrenal steroid hormones is one example. Cortisol (17-oxycorticosterone) cannot be formed; therefore the normal feedback for inhibition of ACTH formation in the pituitary cannot work, and 17-ketosteroids are formed in large amounts from 17-oxyprogesterone. These in turn stimulate the development of sexual characteristics and lead to virilization of female patients. Substitution of cortisol restores the feedback circle; ACTH and, consequently, 17-ketosteroid formation is reduced, and virilization is stopped (Fig. 7.26).

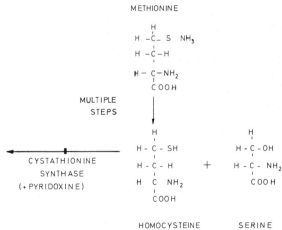

Fig. 7.25. Methionine pathway. Formation of cysteine via this pathway. Cystathionine synthase is inactive in homocystinuria. This leads to an increase of homocysteine and homocystine, on the one hand, and to a deficiency of cysteine, on the other.

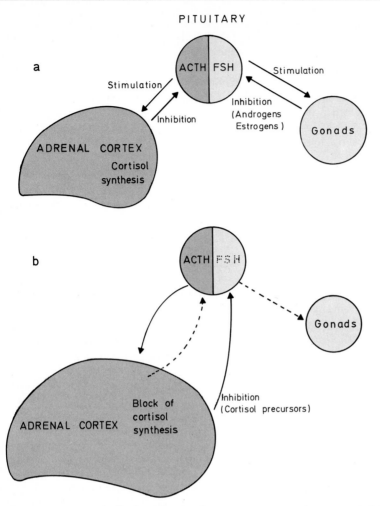

Fig. 7.26. a Negative feedback mechanism between pituitary and adrenal cortex. The adrenal cortex is stimulated by the pituitary hormone ACTH; the end-product of corticosterone synthesis, cortisol, inhibits ACTH formation. At the same time, the gonads are stimulated by FSH produced by the pituitary until the androgens (or estrogens) produced by the gonads inhibit FSH production. **b** In the adrenogenital syndrome, cortisol formation is inhibited by a genetic block. This has two effects on the pituitary; ACTH formation is not inhibited. Abnormally high ACTH formation leads to excessive formation of cortisol precursors, which inhibit FSH production due to their chemical similarity with androgens in the pituitary. Virilization in females results. Cortisol substitution restores the normal feedback cycle

Dietary Treatment of Metabolic Diseases May Be Only the Extreme of a More General "Genetotrophic" Principle. In many metabolic diseases the phenotypic consequences of an enzyme block can be avoided by a suitable change in nutrition. These conditions are considered to be pathological because they are rare. If the great majority of the population had one of these enzyme defects, we would have changed our eating habits accordingly, and such "defects" would be regarded as normal. An example is the intestinal lactose malabsorption found in most Orientals and Africans and in many Europeans. Consumption of *large* amounts of milk and dairy products causes flatulence and bowel irritability in such lactase-deficient persons. Most persons of northwestern Euro-

pean ancestry have no such problems, as they have sufficient intestinal lactase for the breakdown of lactose (Sect. 14.3.1; [83]).

Section 7.2.2.5 describes diseases that are due to abnormal uptake, conversion, and utilization of coenzyme precursors (vitamins). These conditions can usually be treated with high doses of specific vitamins. From an evolutionary standpoint, however, even our normal dependency on vitamins may be regarded as a multiple genetic deficiency, since both *Neurospora crassa* and *E. coli* bacteria are able to synthesize almost all vitamins. L-Ascorbic acid (vitamin C) serves as a potent reducing agent in mammalian metabolism and can be synthesized by all species except humans, higher primates, and guinea pigs. Humans need continuous "substitution therapy," which fortunately is supplied by normal nutrition. In exceptional situations, however, scurvy develops for example, when the food provided to sailors on long voyages in past centuries did not contain enough vitamin C.

Other pathways that have been lost during evolution are those needed for synthesis of the so-called essential amino acids. For some bacteria and fungi, these essential amino acids are not essential at all; they can be synthesized from simple nitrogen sources such as ammonia.

So far we have considered mainly nutritional therapy of rare, exceptional genetic variants with extreme effects. However, even screening for phenylalanine concentration in the serum has identified – apart from the extreme, classic PKU cases – those with mild hyperphenylalaninemia. Such persons do not need a special diet to keep their development within limits commonly regarded as "normal." However, there are some data pointing to a slightly higher vulnerability of heterozygotes whose hydroxylase activity is reduced [247]. If this is confirmed, one can infer that such irritability may depend in part on the amount of phenylalanine intake once the demand for protein synthesis has been satisfied.

Genetic polymorphisms are discussed in Sect. 12.1.2. We note there that one-third of all blood enzymes occur in various molecular forms within the human population. Different molecular forms often show differences in activity. This means that – apart from monozygotic twins – the metabolic pathways are utilized in a slightly different way in each individual, leading to "biochemical individuality," [206]. One aspect of this individuality is that nutritional requirements for optimum development may differ slightly for each individual. This "genetotrophic principle" is a part of the mutual adaptation between the individual, his peculiar genetic makeup, and his environment. This is one aspect of the more comprehensive field of ecogenetics discussed in Sect. 7.5.2.

7.2.2.10 Enzyme Defects That Have Not Been Discovered

How Many Enzymes Are There and What Enzyme Defects Are Known? Some metabolic pathways have not yet been elucidated. Therefore no one knows the exact number of enzymes in humans. Estimates run in the order of magnitude of at least 10 000. For approximately 350 enzymes, or about 3%–4%, enzyme defects are known. What about the other 97%?

First, there are obviously a great number of inherited diseases that may be caused by an enzyme defect but have not been analyzed by appropriate techniques.

Most of the autosomal-recessive conditions listed in McKusick's catalogue may belong to this group [106].

What Enzyme Defects Are Not Known? Well-known enzyme defects occur in:

a) Pathways for obtaining energy from carbohydrates (e. g., glycolytic defects in hereditary hemolytic anemia)
b) Catabolic pathways of some amino acids (e.g., phenylketonuria)
c) Catabolic pathways in lysosomes for degradation of building material of cells and intracellular material (e. g., mucopolysaccharidoses)
d) Catabolic pathways for detoxification and excretion of internal metabolites such as ammonia (e. g., argininemia)
e) A few marginal reactions such as in the salvage pathway of nucleic acid metabolism (e. g., HPRT deficiency)
f) Anabolic pathways for synthesis of biomolecules needed for special regulatory purposes (e. g., defects in thyroid hormone production)
g) Some pathways in transmembrane transport (e.g., cystinuria)
h) Some few DNA repair enzymes (e.g., xeroderma pigmentosum)
i) Some metabolic steps in uptake and utilization of coenzyme precursors (e.g., vitamin D resistant rickets)

Few, if any, enzyme defects are known in:

a) Enzymes concerned with the processes of mitosis and meiosis
b) Enzymes needed for DNA and RNA synthesis, with the exception of a few repair enzymes
c) Enzymes concerned with protein biosynthesis
d) Energy supply sources, especially the cytochrome system
e) Enzymes for synthesis of many of the specialized compounds needed as neurotransmitters in the central and peripheral nervous system
f) Anabolic pathways in synthesis of many amino acids, pentoses, fats, and lipids
g) Anabolic enzymes for synthesis of tissue constituents such as sphingolipids, mucolipids, and mucopolysaccharides
h) The tricarboxylic acid cycle, which serves both catabolic and anabolic functions

In short, our knowledge of enzyme defects in humans is not only incomplete but also very biased. Most enzyme defects affect enzymes related to the "housekeeping" functions of the cell. For most of the central building functions no enzyme defects are known so far. For catabolic pathways and for the biosynthesis of some specialized molecules such as hormones the picture is more complete.

Why Do We Know so Little About Enzyme Defects of Central Building Functions? What are the reasons for this bias? Part of it is methodological. Analysis of enzyme defects in humans depends critically on the availability of organ material. Blood cells are easily available; brain or even liver cells are not. The same problem must be faced in the analysis of genetic polymorphisms. Most polymorphisms detected so far affect blood constituents. If the blood – rather than the brain – were our organ for thinking and feeling, our ignorance in the field of behavior genetics might have been overcome. However, with opportunities to test for genetic alterations at the DNA level directly, it is increasingly possible to search for genetic defects and variants in DNA from white blood cells.

However, it is difficult to believe that methodological difficulties could explain the entire bias. The obvious alternative hypothesis is that deficiencies in these central building metabolic pathways would not be compatible with life; they would be lethal (Sect. 4.1.5). For example, it is difficult to imagine how the nearly complete defect of an essential DNA polymerase, which reduces or even abolishes DNA replication – and hence cell division – could be compatible with life. The same argument applies for the basic steps in the tricarboxylic acid cycle and the synthesis of vital metabolites.

For most enzymes 50% of the normal activity is sufficient to maintain normal function, as evidenced by observations on heterozygotes for enzyme defects. One might surmise therefore that at least heterozygotes for such enzyme defects would be detected. However, such detection would require large-scale population studies of enzyme activities, which have not been carried out. Moreover, many enzymes show marked interindividual variability in activity, which would make identification of heterozygotes difficult. This variability – and especially the observation that in heterozygotes approximately 50% of the "normal" enzyme activity is sufficient for maintenance of function under normal living conditions – indicates an excellent "buffering" of the metabolism against intrinsic genetically determined weaknesses. Many functions are maintained by multiple pathways, and many mutations, even in the homozygous state, may not lead to inborn errors, or do so only under particular environmental conditions such as the presence of a drug (Sect. 7.5).

The conclusion that there may be many recessive lethal mutations affecting essential pathways has far-reaching consequences for population genetics. There is no reason at all to assume that the mutations affecting genes determining these "vital" enzymes are any less frequent than the mutations for which enzyme defects are known. Hence, all these mutations are expected to occur. They may occasionally lead to lethal homozygotes and therefore enhance the proportion of dead zygotes. One would expect this phenomenon to increase the number of miscarriages under conditions favoring segregation of homozygotes in general, i.e., in consanguineous marriages (Chap. 13). This prediction, however, does not seem to be borne out by experience. Most of these zygotes probably die at such an early stage of development that a miscarriage is not realized and therefore not recorded.

7.2.2.11 Some General Conclusions Suggested by Analysis of Human Enzyme Defects

Detection of Enzyme Defects. In our consideration of human enzyme defects there are several points that recur. To be readily detected an enzyme defect should be located in blood cells or should manifest itself in cultured fibroblasts. Moreover, it should lead to clearcut clinical symptoms in affected individuals or should at least lead to alterations that are easily detected by screening techniques, such as excretion of abnormal urinary metabolites. An inborn error with nonspecific symptoms which is not accompanied by currently detectable biochemical disturbances cannot easily be identified. Thus, although several metabolic surveys have been carried out among patients with mental deficiencies, many more inborn errors may exist that have not yet been discovered.

Elucidation of Metabolic Pathways by Use of Enzyme Defects. It is not very difficult to detect enzyme defects if the metabolic pathways are already known. In some cases, however, analysis of enzyme defects may provide a tool for elucidating metabolic pathways that would otherwise be difficult to examine. The mucopolysaccharidoses are a cardinal example.

Characteristics of Mutations Leading to Enzyme Defects in Humans. In many of the enzyme defects analyzed so far, some residual activity of the enzyme has been observed. Moreover, qualitative changes in the enzyme protein are usually discovered, for example, cross-reacting material (CRM), change of kinetic characteristics, and many others. These findings indicate qualitative changes in enzyme proteins due to mutations in structural genes and argue against a major share of regulatory mutations at all possible levels. Such mutations would be expected to cause quantitative changes only in enzyme activity. There is a high degree of genetic heterogeneity within a single gene locus, which adds to the heterogeneity between loci involved in the same pathways.

Mode of Inheritance and Heterozygotes. The mode of inheritance of enzyme defects is usually recessive – either autosomal or, in some cases, X-linked. Healthy heterozygotes almost always have enzyme activities of about half the population average. Therefore the human organism can work perfectly well with a single enzyme at half power. This reveals a remarkable amount of internal regulatory capacity within metabolic pathways. However, if the pathway is loaded with a substance requiring the deficient enzyme for its metabolism, its ability to cope with the metabolite is poorer than normal. Some observations raise the suspicion that this impairment may be less unimportant for the health of heterozygotes than is usually presumed. It may contribute – possibly together with environmental stresses – to their susceptibility to common diseases, somatic or mental. Few systematic, large-scale, controlled investigations have been carried out on the health status of heterozygotes for recessive diseases, especially during middle or advanced age. As with so many other shortcomings of our knowledge in human genetics, this lack of data may have sociological reasons. Work on inborn errors is being carried out principally by pediatricians or medical geneticists with a pediatric background, who are usually not interested in studies of epidemiology and population genetics. Conversely, population geneticists rarely deal with biochemical refinements in field studies.

The observation that virtually all enzyme defects are inherited as recessive traits inevitably raises the question as to the biochemical basis of dominant abnormalities. This problem is discussed below (Sect. 7.6). We first introduce the hemoglobin paradigm. In this special case, many questions raised by the enzyme defects, and even the question as to possible mechanisms of Mendelian dominance, have been answered.

7.3 Human Hemoglobin [42, 119, 149]

The hemoglobin molecule can be studied with greater facility than any other human protein. Blood can easily be drawn from many individuals. Hemoglobin is the principal protein of red blood cells, and its extraction does not require complicated biochemical methods. It is therefore not surprising that we understand more about this protein than about all others. Genetically oriented studies of human hemoglobins have proceeded apace with the elucidation of the amino acid sequence and structure of the molecule. The hemoglobin system is currently a paradigm for the understanding of gene action at the molecular level. Hemoglobin research plays a role in human biochemical genetics similar to that of research on *Drosophila* and phage in basic genetics. Most concepts derived from hemoglobin research apply readily to other proteins. In fact, many conceptual principles of human genetics could be taught by examples from the hemoglobin system.

7.3.1 History of Hemoglobin Research

Sickle Cell Anemia: A "Molecular" Disease. Work on human hemoglobin began with the investigation of a hereditary disease: sickle cell anemia. In 1910 Herrick [114] observed a peculiar sickle-shaped abnormality of red cell structure in an anemic African-American student. It soon became apparent that this condition is fairly common among African-Americans. Affected patients suffer from hemolytic anemia and recurrent episodes of abdominal and musculoskeletal pain. Taliaferro and Huck (1923) [235] recognized that the condition is hereditary. It was shown by Neel (1949) [181] and independently by Beet (1949) [18] that patients with sickle cell anemia are homozygous for a gene that, in the heterozygous state, causes an innocuous condition: sickle cell trait, which is found in about 8% of the African-American population [171].

The decisive step in the biochemical-genetic analysis of this disease was carried out by Pauling et al. (1949) [195]. Pauling, an outstanding chemist, heard about this disease from Castle, a renowned hematologist (and son of one of the pioneers of mammalian genetics) and surmised that a defect of hemoglobin is likely to be the cause: The evidence available at the time that these investigations were begun indicated that the process of sickling might be intimately associated with the state and the nature of the hemoglobin within the erythrocyte.

Therefore the authors examined the hemoglobins of patients with the sickle cell trait and sickle cell anemia, comparing them with the hemoglobin of normal individuals. In accord with the state of methodology for protein analysis at that time, these investigations were performed using Tiselius zone electrophoresis (Fig. 7.27). The peaks in the figure represent the concentration gradients of hemoglobin in a suitable buffer solution; the positions of these peaks depend on the relative number of positive and negative charges in the protein molecule: The results indicate that a significant difference exists between the electrophore-

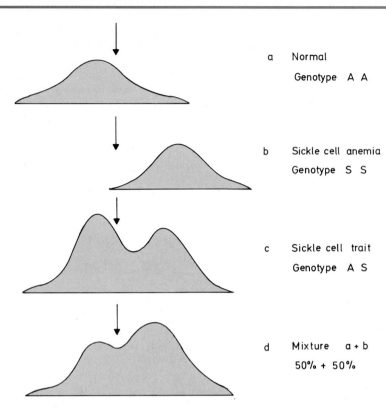

a   Normal
     Genotype A A

b   Sickle cell anemia
     Genotype S S

c   Sickle cell trait
     Genotype A S

d   Mixture   a + b
     50% + 50%

Fig. 7.27 a–d. Zone electrophoresis diagram of hemoglobins at pH = 6.9. **a** Normal homozygote *(AA);* **b** patient with sickle cell anemia *(SS);* **c** sickle cell trait *(AS);* **d** mixture of equal parts of HbA and HbS. *Arrow,* starting point of electrophoresis. (From Pauling et al. 1949 [195])

tic mobilities of hemoglobin derived from erythrocytes of normal individuals and from those of sickle cell anemic individuals.

In the sickle cell trait about 25%–40% of the hemoglobin turned out to be identical with that found in sickle cell anemia, whereas the remainder was indistinguishable from the normal. This result was compatible with the genetic data that sickle cell anemia represents the homozygous state of a gene for which carriers of the sickle cell trait are heterozygous. This investigation reveals, therefore, a clear case of an alteration produced in a protein molecule by an allelic change in a single gene.

Pauling and his collaborators in 1949 [195] used the new finding of a molecular alteration in the protein of a hereditary disease to suggest that sickle cell anemia was the first example of a molecular disease. Hörlein and Weber in 1948 [120] had already demonstrated by a less elegant but imaginative and somewhat difficult method (exchanging heme and globin portions of hemoglobin between normal and affected persons) that the basic abnormality of methemoglobinemia in a family with that condition resides in globin and not in the nonprotein heme portion of hemoglobin. However, unlike Pauling, these investigators did not realize the major and generalizable significance of their finding for the pathogenesis of inherited disease. This illustrates how a discovery that becomes important at a later date (when it fits a novel

paradigm) receives little attention at the time of its discovery unless its general significance is explicated and widely publicized. The work of Pauling was published in *Science,* a general scientific journal with wide circulation and was in agreement with other data on the genetic transmission of sickle cell anemia. The work by Hörlein and Weber, on the other hand, appeared in a German medical journal at a time when hardly any research was carried out in postwar Germany.

Single Amino Acid Substitution. In 1956 Ingram [123] – working in Cambridge in the same laboratory in which Perutz was pursuing his crystallographic work, where Sanger had shown the amino acid sequence of insulin, and where Crick and Watson had demonstrated the DNA model – discovered what precisely distinguishes normal from sickle hemoglobin. Hydrolysis of the globin molecule with the protein-splitting enzyme trypsin yielded about 60 peptides, which were separated on paper in a two-dimensional array by electrophoresis in one direction and paper chromatography in the other. This "fingerprinting" method of protein analysis revealed that sickle cell hemoglobin was identical with the normal molecule in all peptides except one. Further analysis showed that sickle cell hemoglobin differed from normal hemoglobin in only one amino acid: glutamic acid was replaced by valine:

COOH
|
CH$_2$
|
CH$_2$
|
H—C—NH$_2$
|
COOH

CH$_2$
|
H—C—CH$_2$
|
H—C—NH$_2$
|
COOH

Glutamic acid   Valine

Glutamic acid has two COOH groups and one NH$_2$ group, whereas valine has only one COOH group. This charge difference explained the electrophoretic differences between normal and sickle hemoglobin.

Meanwhile, and especially after simpler methods of electrophoresis had replaced the cumbersome Tiselius electrophoresis, an increasing number of other hemoglobin variants were discovered. At present, over 500 of such variants are known [1047]. Further steps of great importance were the establishment and elucidation of the full amino acid sequence of hemoglobin chains by Braunitzer et al. (1961) [1026], and of the three-dimensional structure of hemoglobin [1184, 1280]. Subsequent advances have led to our understanding of structure-function relationships and to detection of various types of mutations, such as deletions and frameshifts. Isolation of the hemoglobin mRNA led to new insights into gene structure and function and opened new paths to the understanding of gene action (Sect. 3.1.3.6).

Molecular work on the hemoglobins has proceeded at a rapid rate. The full DNA sequences of the various hemoglobin genes and their flanking sequences is now known, and the hemoglobin genes are probably better understood than any other mammalian genes. Mutations affecting the hemoglobins, particularly the thalassemias, have been elucidated and are models for the understanding of gene action at the molecular level. The genetics of hemoglobin as currently known is described in the next section.

7.3.2 Genetics of Hemoglobins [42]

Hemoglobin Molecules. Human hemoglobin consists of four globin chains. The general designation of the hemoglobin molecule is $\alpha_2 \beta_2$, signifying that the four globin chains comprise two pairs of identical chains. Most normal human hemoglobins have identical α chains while the non-α chains (β, γ, δ) differ from each other (see below). Each globin chain carries a heme group, a nonprotein molecule attached

at a specific site of the globin molecule (Fig. 7.28). The four globin chains with their respective heme groups constitute the functional hemoglobin molecule that carries oxygen from the lungs to the tissues. A globin chain consists of a string of over 140 amino acids of specified structure (Fig. 7.29). The sequence of the various amino acids in a protein molecule, such as hemoglobin, is known as the primary structure. The spatial relationship between adjacent residues is known as the secondary structure and the three-dimensional arrangement of a protein subunit as the tertiary structure (Fig. 7.28). Quarternary structure refers to the arrangements of the four subunits into a functioning molecule.

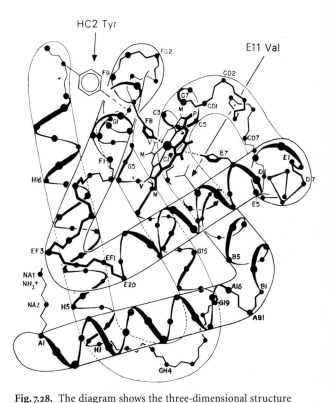

Fig. 7.28. The diagram shows the three-dimensional structure (3° structure) of a typical globin chain, which consists of eight helical and six nonhelical segments. To allow comparisons of different globin chains, the helical segments are labeled *A–H* and nonhelical segments are denoted by two capital letters such as *CD, FG,* etc. *Black wavy line,* the spatial arrangement of the various amino acids (2° structure). Amino acids are numbered from the amino *(N)* terminus starting with *A1. Numbers,* specific amino acids located at the positions, which may differ in various globin chains. Structurally equivalent residues carry the same notation in all hemoglobins regardless of amino acid additions or deletions. Note the insertion of the nonprotein heme chain between *E7* and *F8.* Amino acid residues at *E7* (histidine), *E11* (valine), and *HC2* (tyrosine) are particularly important in the function of mammalian hemoglobins. *M, V, P* in the heme molecule, methyl, vinyl, and propionate side chains, respectively. (From Perutz 1976 [197])

BETA CHAIN

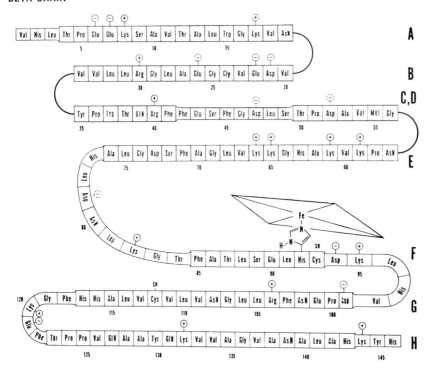

Fig. 7.29. The primary structure of amino acid sequence of the β chain of adult normal human hemoglobin (HbA). Amino acids that are oriented in the characteristic folding of an α-helix are shown as *square boxes*. Nonhelical residues are depicted as *rectangles*. The site for heme attachment is shown. The specific amino acid sequence of the β-globin chain and its various features may be usefully compared with the three-dimensional molecular arrangement shown in Fig. 7.28 [42]

The principal hemoglobin of children and adults is HbA or adult hemoglobin ($\alpha_2 \beta_2$). The characteristic subunit of HbA is the β chain (Fig. 7.29). The α and β chains differ from each other in many amino acids. All adults carry a small amount (2%–3%) of HbA$_2$ ($\alpha_2 \delta_2$). The characteristic δ chains differ in only ten amino acid positions from the β chain. A small amount (<1%) of fetal hemoglobin (HbF: $\alpha_2 \gamma_2$) is also seen postnatally in all individuals (see below). The γ chain differs considerably from both the α and β chains. The α chains of HbA, HbA$_2$, and HbF are identical.

Several hemoglobins characteristic of embryonal and fetal development exist. The ζ chains resemble the α chains in their amino acid composition [134], and ε chains have similarities to the β chains. The ζ chains probably are developmentally the earliest globin chains. The ζ and ε chains disappear after 8–10 weeks of embryonal life (Fig. 7.30) [269]. The principal hemoglobin of fetal development is HbF ($\alpha_2 \gamma_2$) with its characteristic γ chain. There are two types of γ chains with very similar properties: those with alanine at position 136 ($^A\gamma$) and those with glycine at that position ($^G\gamma$). A third type of γ chain with threonine instead of isoleucine at position 75 in the γ chain exists [211]. Its frequency ranges between 0% and 40%, and it does not appear related to any disorder. Only $^A\gamma$ chains carry this variant. Adult hemoglobin can be demonstrated in fetuses as early as at the 6–8 week stage [211, 269].

While γ chain synthesis during fetal life occurs largely in liver and spleen, γ chains can also be produced by marrow erythropoietic cells. Conversely, while β chains in childhood and later are produced in the bone marrow, β chain production can also occur in additional marrow sites [269]. The various normal hemoglobins are listed in Table 7.8.

All of the normal human hemoglobins that have been investigated have an identical three-dimensional structure (Fig. 7.28), which is essential to allow the carrying of oxygen. All globin chains of the various hemoglobins have a common evolutionary origin and originated from each other by genetic duplication (see Sect. 14.2.3). The closer the resemblance between two chains, the more recent in evolutionary terms the duplicatory events occurred. Thus, the $^A\gamma$ and $^G\gamma$ chains with a single difference between them arose more recently while the β and α chain duplication have a more remote origin.

Hemoglobin Genes. The amino acid sequence of each of the globin chains is specified by a unique globin gene. A normal human therefore possesses at least one α, β, γ, δ, ε, and ζ gene in the haploid state or at least two of these genes in the diploid state. The gene for the α chain in most human populations exists in a duplicated state with no known differences

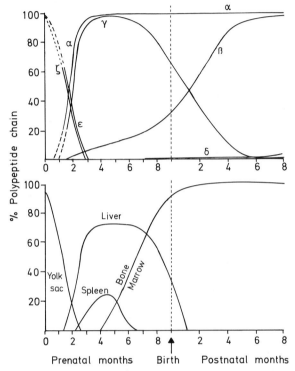

between the two α genes. There are two γ genes – differing in the codon specifying position 136 $^{A}\gamma$ and $^{G}\gamma$. Some $^{A}\gamma$ genes have a variant codon specifying threonine rather than isoleucine at position 75 ($^{T}\gamma$).

Genes that participate in synthesis of the nonprotein heme group by specifying a series of biosynthetic enzymes involved in heme synthesis are not further considered here.

The various globin genes with their respective globin chains and normal hemoglobins are shown in Table 7.8 and Fig. 7.31.

Extensive DNA sequence analysis has been carried out on all hemoglobin genes and their structure has been fully documented [13, 62, 163, 223, 229]. The human hemoglobin genes exist as two separate clusters of related multigene families, a frequent type of organization of mammalian genes (Figs. 7.32, 7.33). The α gene cluster is located on the short arm of chromosome 16 over a 25-kb region. The γ-β-δ family is situated on the short arm of another chromosome, 11, across a 60-kb region. The genetic mechanisms which regulate coordinated gene function on the two different chromosomes to allow equal output of α and non-α gene products (such as β and γ) remain unknown. The structural genes of the Hb α complex [from 5' (upstream) to 3' (downstream)] include: the embryonic ζ gene, a pseudogene for Hb ζ, two pseu-

Fig. 7.30. Ontogeny of human hemoglobin chains before birth and in the first few months after birth. *Above,* characteristic developmental patterns of various globin chains. *Below,* characteristic sites of erythropoiesis during development. There is remarkable similarity in the time sequences of the yolk sac and ε and ζ chain, hepatosplenic and γ chain, and bone marrow and β chain erythropoiesis. (From Motulsky 1970 [170])

Table 7.8. Human hemoglobins

Stage	Hemoglobin	Structure
Embryonic	Gower I	$\zeta_2\varepsilon_2$
	Gower II	$\alpha_2\varepsilon_2$
	Portland	$\zeta_2\gamma_2$
Fetal	F	$\alpha_2{}^{G}\gamma_2$
		$\alpha_2{}^{A}\gamma_2$
Adult	A	$\alpha_2\beta_2$
	A$_2$	$\alpha_2\delta_2$

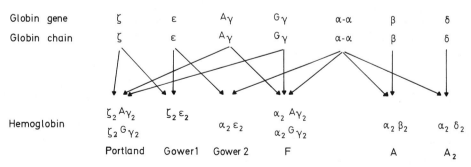

Fig. 7.31. Normal human globin genes. Single genes exist for Hb β, δ, ε and ζ. The genes for Hb α and Hb γ are duplicated. The products of the two hemoglobin γ genes (Hb $^{A}\gamma$ and Hb $^{G}\gamma$) differ from each other by a single amino acid residue, alanine (A) or glycine (G) at position 136. There are no known differences between the two Hb α genes. Tetrameric hemoglobin formation is shown in the lower portion of the figure

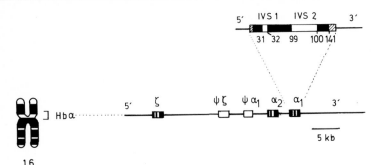

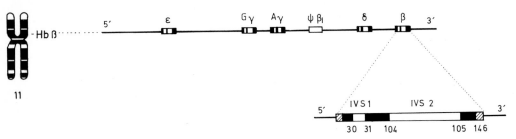

Fig. 7.32. Chromosomal location (16 p) and organization of the human α globin gene cluster. ψ, pseudogene; *IVS*, Introns (intervening sequences, *white boxes*). The numbers underneath the Hb α^1 gene 31, 32, 99, 100 ... refer to the codon numbers of the sequence at which a given intron interrupts the exon sequence. Intron 1 is interspersed between codon 31 and 32. (Only one pseudogene for Hb α shown; newly discovered pseudogene 3' of Hb α_1 is not shown) (Updated Antonarakis et al., 1985 [12])

Fig. 7.33. Chromosomal location (11 p) and organization of the human β globin gene cluster. Symbols and explanation identical as for Fig. 7.32 [12]

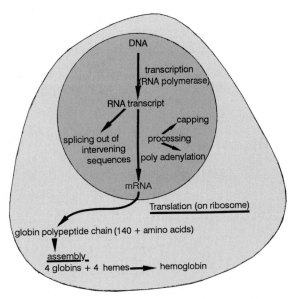

Fig. 7.34. Overview of protein synthesis with hemoglobin as a model. The nucleotides of the DNA hemoglobin gene are transcribed (transcription) by the enzyme RNA polymerase to form the RNA transcript. Intervening sequences not specifying structural information are spliced out. mRNA moves out of the nucleus *(dark gray)* into the cytoplasm *(light gray)* were globin synthesis or translation proceeds on the ribosomes by initiation, elongation, and termination. The globin polypeptide chain is formed and heme is inserted. Four globin chains form the functional hemoglobin molecule

dogenes for Hb α, two identical α genes, and a more recently discovered gene of unknown function (Fig. 7.32). Similarly, the location of the various genes on the β cluster are: the embryonic epsilon gene, two fetal γ genes ($^A\gamma$ and $^G\gamma$), the Hb β pseudogene, and a Hb δ and a Hb β gene (Fig. 7.33). The 5' to 3' arrangement of these genes is in the order of ontogenetic expression during development. Pseudogenes have DNA sequences that resemble those of their homologues. However, various mutational alterations have inactivated transcription, so that there is no functional expression. Pseudogenes are presumably duplication products that arose during evolution and were no longer required for normal function. The Hb δ gene, whose gene product comprises only 2%–3% of total non-α chains, can be conceived of as a gene in transition to becoming a pseudogene.

All the globin genes have many functional similarities in organization. Three exons or coding sequences code for the unique amino acid sequence of each globin chain. Between exons 1 and 2 and between exons 2 and 3 there are unique intervening sequences (IVS) or introns known as IVS-1 and IVS-2, respectively (Figs. 7.33). These introns are transcribed along with exons so that the initial gene transcripts reflect both coding and noncoding DNA sequences of the respective gene. Intervening sequences are excised during nuclear processing so that the terminus of exon 1

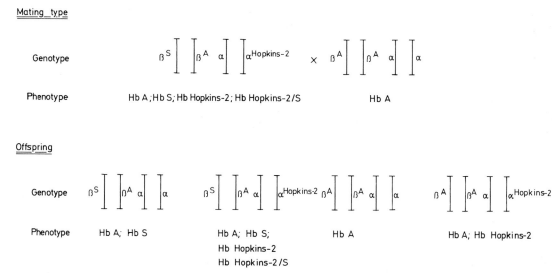

Fig. 7.35. Formal genetics of a mating between a double heterozygote for Hb α ($\alpha^{\text{Hopkins 2}}$) and Hb β (Hb β^{S}) with a normal person. Since the genes for Hb α and Hb β are located on different chromosomes, free assortment of all chromosomes occurs and four classes of offspring in equal proportion were found: normal (HbA); HbA/S: sickle cell trait; HbA/Hb Hopkins-2: Hopkins-2 trait and compound heterozygosity for HbS and Hb Hopkins-2 identical to that of the affected parent. If the genes for Hb α and β were closely linked, the parental phenotypes would not be formed in the offspring except for possible recombination (Sect. 5.1). The closer the linkage, the less the chance of recombination (see Fig. 7.36)

is spliced to exon 2 and the end of exon 2 to exon 3 to form functional mRNA that directs hemoglobin production on the ribosomes. The two intervening sequences of the different genes on the γ-δ-β cluster are largely identical but differ from shorter intervening sequences on the α cluster. Study of certain β thalassemia mutations that interfere with normal excision and splicing (see below) has helped to elucidate the splicing process. All introns start with GT (donor sites) and end with AG (acceptor sites) – these dinucleotides are part of so-called consensus sequences at the splicing sites (see [1, 62, 185] for details). The pathway leading from the gene to the hemoglobin molecule is shown in Fig. 7.34.

The biochemical evidence for nonlinkage of the Hb α and Hb β genes was preceded by genetic evidence that offspring of a mating of the double heterozygote for both an Hb α and Hb β mutation with a normal individual includes four phenotypes: normal, Hb α^{x}, Hb β^{x}, and the double mutation α^{x} and β^{x} (Fig. 7.35). With close linkage of the Hb α and Hb β genes, both Hb α^{x} and Hb β^{x} mutations but not the parental types or normals would have been found among the offspring. Similarly, genetic evidence for close linkage between Hb δ and Hb β came from the failure to find recombinants among children from matings between double heterozygotes for both the Hb δ and Hb β mutations [34] (Fig. 7.36). The existence of Hb Lepore, a δ-β fusion gene, provided biochemical evidence for linkage of Hb δ and Hb β genes on the same chromosome (see below). Linkage between γ and β genes could be inferred by demonstrating that Hb Kenya is a γ-β fusion gene. Thus, correct inferences regarding the genetics of the human hemoglobins were made before DNA could be examined directly.

Regulatory Elements. Three different but similar sequences are located upstream (5′) of any tissue-specific (but not housekeeping) gene and appear to be involved in the regulation of transcription. These are also known as promoter regions [62, 185]. They include the TATA or ATA (Hogness) box 30 base pairs proximal to the initiation site. This sequence serves as a precise site of transcription initiation. Another invariant sequence CAAT (at about minus 80 base pairs) is a recognition site for RNA polymerase. A third distal element is located at 80–100 base pairs and has the characteristic sequence PuCPuCCC (Pu, purine). Mutations of the promoter regions of the globin genes might be expected to reduce hemoglobin production. Thus, β thalassemia has been reported to be caused by mutations both in the PuCPuCCC region and in the TATA box but not yet by CAT box mutations (see below) [137].

Intensive work is being carried out to elucidate the mechanisms of gene regulation by various regulatory elements that are located in the flanking areas of the globin genes [231]. The regulation of globin gene expression is a major model for the study of gene ex-

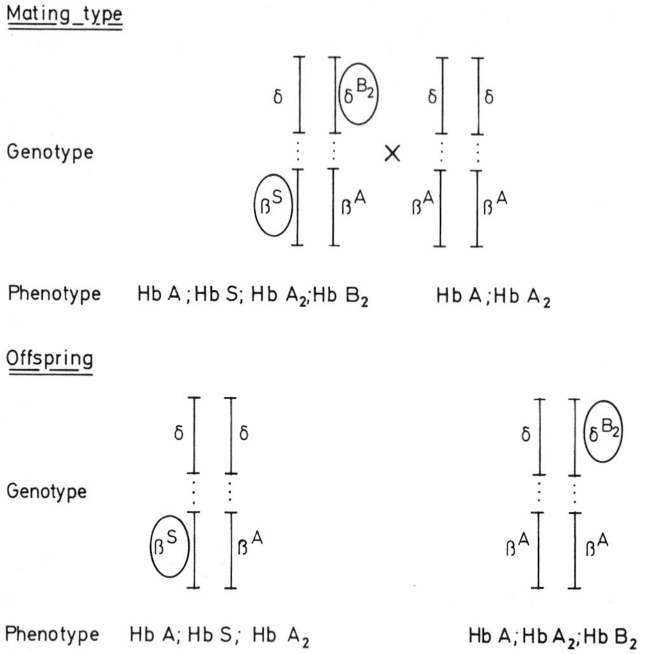

Mating_type

Genotype

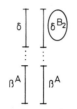

Phenotype   Hb A; Hb S; Hb A$_2$;Hb B$_2$ Hb A; Hb A$_2$

Offspring

Genotype

Phenotype   Hb A; Hb S; Hb A$_2$ Hb A; Hb A$_2$;Hb B$_2$

Fig. 7.36. Formal genetics of a mating between a double heterozygote for Hb β (Hb β^S) and Hb δ (Hb δ^{B2}) who inherited each mutation from a different parent. The genes for Hb β and Hb δ are closely linked. All offspring either inherit the β^S or the δ^{B2} abnormality. No normals or double heterozygotes such as the parental type were seen among the children. These findings are consistent with close linkage of the two genes

pression. In addition to the *cis* acting promoters, *cis* enhancers have been described which may be located several hundred bases upstream or downstream of a given structural gene. Intragenic regulatory elements within both the β globin gene and silencer genes have been reported. Of great interest are the locus control regions (LCR) for the various globin genes which are located 40–100 kb upstream of the globin genes and consist of four major DNase I hypersensitive sites [72, 83]. The globin LCRs are essential for gene expression of the various distal globin gene products and dominate other *cis* active regulatory elements by establishing an active chromatin domain. The globin LCRs confer erythroid cell line specific expression on their linked promoters. Deletion of either the α or the β globin LCR leads to complete inactivation of the downstream globin gene complex producing α thalassemia [204] and $\varepsilon\gamma\delta\beta$ thalassemia [71], respectively (see below).

In addition to these *cis* active regulatory elements, *trans* acting transcriptional regulators have been identified. The most important erythroid factor is the GATA-1 protein, a DNA binding protein required for erythroid differentiation [23]. The gene for GATA-1 is X-linked. The GATA sequence is the obligatory DNA sequence to which this protein binds. Other GATA proteins have been described. There appears to be intricate interaction between the various *trans* acting transcription factors and the different *cis* regulatory sequences of the globin genes. The full elucidation of regulation at $\varepsilon\gamma\delta\beta$ linked loci is likely to provide insight into mechanisms which control the de-

velopmental switch from embryonic (ε gene) to fetal (γ genes) and ultimately to postnatal (β gene) globin production. A favorite hypothesis suggests that globin switching is caused by competition between the different developmental globins genes for access to the LCR [23].

Downstream Sequences. Transcription terminates about 1000 base pairs downstream from exon 3 of the β gene. The highly conserved sequence AAU AAA provides the signal for endonucleolytic cleavage of RNA, which is followed by addition of the polyA tail of 220 residues. This polyA is not coded for by DNA at the globin gene site. The polyA nucleotides are required to stabilize mRNA, which carries the genetic information from the genes of the nucleus to the ribosomes where globin synthesis occurs by joining amino acids in their characteristic sequence (Fig. 7.34).

DNA Polymorphisms at the Globin Genes [12]. Gene mapping by restriction enzyme analysis of the $\gamma\delta\beta$ gene cluster led to the recognition of considerable variation in DNA sequence between different individuals (Fig. 7.37). DNA variants at the Hb β gene complex recognized as single nucleotide substitutions are symbolized as either present (+) or absent (−). Among 17 polymorphic sites at the Hb β cluster 12 are located at flanking DNA, 3 within introns, 1 within a pseudogene, and 1 only within the coding (synonymous) portion of the Hb β gene. This distribution is not unexpected since mutations affecting the cod-

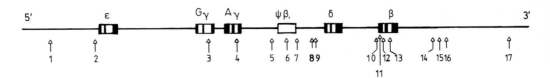

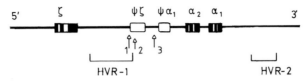

Fig. 7.37. Restriction enzyme polymorphisms at the Hb β *(above)* and Hb α *(below)* genes. *Numbers,* sites at which various restriction enzymes cut the DNA. *HVR,* Hypervariable regions (minisatellites; see also Table 7.9). (From Antonarakis et al. 1985 [12])

Table 7.9. Frequency of DNA polymorphic sites in the β globin gene cluster in different populations (from Antonarakis et al. 1985 [12])

Polymorphisms		Greeks	African-Americans	Southeast-Asians
Taq I	(1)	1.00	0.88	1.00
Hinc II	(2)	0.46	0.10	0.72
Hind III	(3)	0.52	0.41	0.27
Hind III	(4)	0.30	0.16	0.04
Pvu II	(5)	0.27		
Hinc II	(6)	0.17	0.15	0.19
Hinc II	(7)	0.48	0.76	0.27
Rsa I	(8)	0.37	0.50	
Taq I	(9)	0.68	0.53	
Hinf I	(10)	0.97	0.70	0.98
Rsa I	(11)			
Hgi A	(12)	0.80	0.96	0.44
Ava II	(13)	0.80	0.96	0.44
Hpa I	(14)	1.00	0.93	
Hind III	(15)	0.72	0.63	
Bam HI	(16)	0.70	0.90	
Rsa I	(17)	0.37	0.10	

The numbers in parentheses refer to the restriction enzymes in Fig. 7.37.

type. For example, a given array of five polymorphisms may be symbolized as $+ - + - +$ in an upstream (5') to downstream (3') direction.

A remarkable feature of the DNA variants at the β gene cluster is their linkage disequilibrium. (Sect. 5.2.5) If there were free recombination over many generations, one would expect random associations of any two polymorphic sites and a very large number of haplotypes (2^n where n is the number of polymorphisms; with 2^4 there would be 16 expected haplotypes). Instead, only a few haplotypes have been found. For instance, strong linkage disequilibrium for eight polymorphisms upstream from the δ gene exists (1–8 in Fig. 7.37) so that four haplotypes account for 94% of all chromosomes. Similarly, four haplotypes account for 90% of haplotypes for five polymorphisms (12–17 in Fig. 7.37), which are located over an 18-kb region downstream and including the β gene. Surprisingly, complete randomization was found when these upstream und downstream clusters of polymorphisms were compared. The most reasonable interpretation postulates a high recombination rate at a site separating these clusters – a recombinational hot spot; a recombination has already been found in one family.

Hemoglobin Variants. Hemoglobin variants are caused by a variety of mutational events affecting a given hemoglobin gene. The most common hemoglobin variants are amino acid substitutions affecting a single amino acid of a globin chain. Almost 600 such substitutions have been described (Table 7.10). These substitutions are caused by replacements of a single nucleotide in a given codon triplet of the DNA, which changes the mRNA triplet to one that specifies a different amino acid such as GUA (valine) to GAA (glutamic acid) (see third line of Fig. 7.38). If the electric charge of the mutant amino acid is differ-

ing regions would be more likely to cause harmful effects. Presumably, since much of the DNA between coding blocks is not expressed, sequence variation usually has no functional consequences. The various polymorphic sites are of ancient origin since they are found in all racial groups (Table 7.9). Some variants occur as polymorphisms only in blacks and not in other racial groups. Two DNA polymorphisms at the Hb α locus show another frequent type of DNA variation – hypervariable regions.

A specific arrangement of polymorphisms at a gene cluster (or a gene locus) has been termed a haplo-

Table 7.10. Frequency and molecular mechanisms of globin gene mutations (from Bunn 1994 [41]; Carver and Cutler 1994 [54])

Mechanism	Hemoglobin Chain			
	α	β	δ	γ
Missense mutations				
Single nucleotide substitutions	185	313	74	25
Two replacements in subunit	1	14		
Deletions	3	14		
Insertions	3	1		
Deletions/insertions		3		
Extended subunits				
Terminator mutations	4			
Frameshifts	1	3		
Initiator methionine retained	3	1		
Fusion hemoglobins	$\delta\beta$-3			
	$\beta\delta$-4			
	$\delta\beta\delta$-1			
	$\gamma\beta$-1			

The numbers refer to the total numbers of globin gene mutations (exclusive of thalassemias, collated in 1994).

ent, the variant hemoglobin can be recognized by its altered behavior on electrophoresis. Mutations that do not change electrophoretic charge are usually detected only if they affect hemoglobin function deleteriously and cause disease. Most hemoglobin mutations regardless of whether they affect electrophoretic charge have no effect on hemoglobin function and are compatible with normal health. In general, amino acid substitutions of the exterior of the hemoglobin chains cause fewer perturbations of function than those replacing amino acids in the chain interior or close to the insertion of the heme group. Substitutions affecting normal helical turns of the chain often cause hemoglobin instability. Amino acid replacements affecting subunit contacts are often associated with abnormalities in oxygen affinity [230]. Most hemoglobin variants are rare. A few, such as HbS, HbC, and HbE, have reached higher frequencies by natural selection and are further discussed in Sect. 12.2.1.6. Polymorphisms at the nucleotide level in the coding area also exist. The genetic code is degenerate (Table 3.2), i.e., several codons can code for an identical amino acid (see Fig. 7.38). Consideration of two different amino acid substitutions at Hb β^{67} (Fig. 7.38) shows that the original codon for valine at Hb β^{67} from which the mutation occurred must have differed in the two individuals in whom the mutation occur-

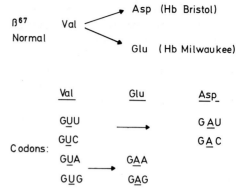

Fig. 7.38. Codon polymorphism. The usual amino acid at position 67 of the Hb β chain is valine. Hb Bristol and Hb Milwaukee are caused by different mutations at that site replacing the original valine with Glu (Hb Bristol) and Asp (Hb Milwaukee). *Below*, the possible codon triplets for valine. A mutation from Val to Asp could have arisen only from GUU or GUC while that from Val to Gln could have originated only from GUA or GUG. Consequently the original persons who underwent the two different mutations must have used different codons for specifying the normal valine at Hb β^{67}

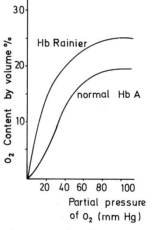

Fig. 7.39. O_2 dissociation curve of a hemoglobin with increased oxygen affinity. Note that the abnormal hemoglobin Rainier does not release oxygen at lower partial pressures of oxygen as readily as normal hemoglobin. Tissue hypoxia results and stimulates erythropoietin formation with resultant erythrocytosis

red and the new hemoglobin arose. Such codon polymorphisms have been shown directly by DNA analysis.

Clinical Effects of Hemoglobin Variants. The results of compromised hemoglobin function can produce different types of disease. There are four principal categories of hemoglobin diseases: (a) hemolytic anemia due to unstable hemoglobins, (b) methemoglobinemia due to more rapid hemoglobin oxidation, (c) erythrocytosis due to abnormal oxygen affinity causing hypoxia with resulting erytropoietin production, and (d) sickle cell disorders due to distortion of the red cell membrane by HbS. In all cases except the sickling disorders, heterozygotes are affected, i.e., the mutations manifest as autosomal-dominants (see also Sect. 7.6).

Unstable Hemoglobins [42, 260]. Over 100 unstable hemoglobins have been described. Approximately three-fourths of these affect the β chain. Many unstable hemoglobins have amino acid substitutions or deletions affecting the heme pocket of the globin chain. Clinical manifestations vary from mild instability which is not clinically apparent to severe instability which causes increased blood destruction. Sulfonamides produced more severe hemolysis in subjects with several unstable hemoglobins. The instability of these hemoglobins is often caused by premature dissociation of the heme from the globin chain. Such heme-depleted globin is precipitated as intracellular material known as Heinz bodies and interferes with cell membrane function. Heinz bodies may be removed ("pitted") by the spleen without destruction of the red cells that carry them. Ultimately, such red cells are removed prematurely by the reticuloendothelial system. In some unstable hemoglobins, splenectomy may ameliorate the severe hemolysis. The diagnosis of unstable hemoglobins, if not associated with electrophoretic mobility alterations, is difficult and may require isolation of the precipitated globin chains for further analysis in specialized laboratories. The unstable hemoglobins contribute to the heterogeneous class of the congenital nonspherocytic hemolytic anemias which are often caused by glycolytic defects of carbohydrate metabolism (Sect. 7.2.2.2). Unstable hemoglobins have been found as fresh mutations and identical hemoglobins (i.e., Hb Köln, Hb Hammersmith) have been found several times as a new mutation in different individuals from different families [42].

Methemoglobinemia Due to HbM [42]. HbM was the very first globin abnormality discovered as a dominant trait; the discovery was made in a family with congenital cyanosis by Hörlein and Weber in 1948 [120] (see above, Sect. 7.3.1). It is interesting that the first discovered human enzyme deficiency was the recessively inherited methemoglobin reductase deficiency, which also produces methemoglobinemia [100]. Methemoglobinemia can therefore be caused either by a dominantly inherited globin abnormality or by a recessively inherited enzyme deficiency. Seven different mutations can produce HbM. (Table 7.11). Methemoglobinemia is caused by the more

Table 7.11. HbM (from Hayashi et al. 1980 [111])

	Location of mutation	Helical residue
α chain		
M Boston	α^{58} His→Tyr	E 7
M Iwate	α^{87} His→Tyr	F 8
β chain		
M Saskatoon[a]	β^{63} His→Tyr	E 7
M Hyde Park	β^{92} His→Tyr	F 8
M Milwaukee 1	β^{67} Val→Glu	E 11
γ chain		
M Osaka	γ^{63} His→Tyr	E 7
M Fort Ripley	γ^{92} His→Tyr	F 8

[a] Hörlein and Weber's classic HbM [126].

rapid oxidation of divalent iron to trivalent iron. Six HbM mutations are caused by tyrosine replacements of the histidine residues that anchor the heme group in its characteristic pocket (Fig. 7.28) of the globin molecule and stabilize the heme iron. The seventh mutation – Hb Milwaukee 1 – cannot yet be fully explained on molecular grounds. Patients with HbM mutations of the α chain are cyanotic from birth. Those with HbM mutation of the β chain do not develop severe cyanosis until 6 months of age, when the γ chains are replaced by β chains. Mild hemolysis is common in patients with HbM. HbM mutations of the γ chain are manifested with cyanosis at birth; this disappears in a few months after β chains have replaced γ chains.

Erythrocytosis Due to Hemoglobins with Abnormal Oxygen Affinity [20, 42]. More than 50 hemoglobins with increased oxygen affinity are known to exist. Substitutions affect $\alpha_1 \beta_1$ contact of the tetramer. Movement of the globin subunits during oxygenation occurs at this interchain contact. Stabilization of the oxy conformation or destabilization of the deoxy conformation by a mutation may result in increased oxygen affinity (Fig. 7.39). Most hemoglobins with high O_2 affinity have substitutions of the COOH terminal of the β chain or at binding sites of diphosphoglycerate (DPG), which are normally involved in maintenance of stability of the deoxy conformation. The increased oxygen affinity reduces oxygen delivery to the tissues with resultant hypoxia (Fig. 7.39). Hypoxia leads to release of the hormone erythropoietin, which stimulates red cell production with resultant erythrocytosis. Patients with erythrocytosis due to abnormal hemoglobins are sometimes erroneously diagnosed as suffering from polycythemia vera. A dominant pattern of inheritance and the absence of splenomegaly, leukocytosis, and thrombocytosis differentiates erythrocytosis due to an abnormal hemoglobin from polycythemia vera. Occasional cases of this type of abnormal hemoglobin have occurred as fresh mutations.

Only a few hemoglobins with *reduced* oxygen affinity have been detected [20]. With increased oxygen delivery to the tissues caused by the reduced affinity for hemoglobin, a lessened production of erythropoietin would be expected. The expected mild anemia usually has been demonstrated.

Sickle Cell Disorders [41, 42, 161, 221]. HbS is caused by the substitution of valine for glutamic acid in the sixth position of the β chain. Unlike all other substitutions, this particular mutation affects the solubility and crystallization of hemoglobin under conditions of hypoxia. Patients with sickle cell anemia inherit the abnormal gene from each of their parents and lack HbA. With a relatively low degree of hypoxia, the HbS of such patients polymerizes into filaments of high molecular weight that associate to form bundles of fibers. These abnormal hemoglobin crystals distort the red cell membrane to its characteristic sickling shape (Fig. 7.40). Some of these cells remain irreversibly sickled and are destroyed prematurely. Sickled cells increase blood viscosity and impede normal circulation in small blood vessels. The resultant hypoxia leads to more sickling with a vicious cycle of more stagnation and characteristic episodic sickle crises with abdominal and musculoskeletal

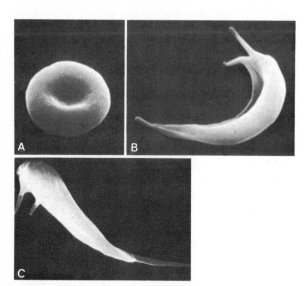

Fig. 7.40. Scanning electron micrographs of oxygenated (**A**) and deoxygenated (**B, C**) red cells from a patient with homozygous sickle cell anemia. Note the normal biconcave shape of the red cell without any HbA and distortion under conditions of hypoxia. The appearance of cells such as depicted in **B** led to the term *sickle* cell since the red cell resembles a sickle – an agricultural tool. (From Bunn et al. 1977 [42])

pain. After several years, necrosis of poorly perfused tissues, such as the spleen, occurs, and this organ atrophies.

Carriers of the sickle cell trait who have one normal (Hb β^A) and one abnormal gene (Hb β^S) have only 25%–40% HbS. These individuals are clinically normal. Their red cells contain both HbA and HbS, and have a normal red cell life span. In vivo sickling occurs only under conditions of severe hypoxia, such as at atmospheric conditions over 3000 m [216].

Certain other hemoglobins when present together with HbS in a red cell decrease the extent of sickling. HbF reduces the gelling and crystallization of HbS so that patients with sickle cell anemia and large amounts of HbF have few or no symptoms of sickle cell anemia. HbF in many of these instances is contributed by a gene for hereditary persistence of fetal hemoglobin (see below). In general, there is an inverse correlation between the amount of HbF and the severity of symptoms in sickle cell anemia. Any manipulation that would increase fetal hemoglobin production would therefore cause clinical improvement in sickle cell anemia [9]. Coexisting α thalassemia in patients with sickle cell anemia is associated with less anemia and improved survival. The clinical manifestations of the thalassemias are covered below.

7.3.3 Other Types of Hemoglobin Mutations

Deletions. The deletion of whole hemoglobin genes has been identified. Deletion of Hb α genes causes α thalassemia, and deletion of both the Hb δ and Hb β genes causes hereditary persistence of fetal hemoglobin (HPFH) or Hb $\delta\beta$ thalassemia (see below).

The deletion of a single nucleotide triplet or codon leads to a deletion of the amino acid specified by that codon. A deletion that removes four codons or 12 nucleotides would cause deletions of four amino acids. Deletions of up to five amino acids corresponding to 15 nucleotides have been seen (see Table 7.12). It is likely that more extensive intraglobin deletions would be incompatible with formation of a viable hemoglobin molecule. Most deletion mutants are either unstable or have increased O_2 affinity or both (Table 7.12). All but three known intragenic deletions affect the Hb β gene. It is not quite clear why so few Hb α deletions have been detected. Possibly, they are more deleterious during embryonic and fetal life where Hb β deletion would be less harmful (see Fig. 7.30).

If a deletion affects a number of nucleotides not divisible by 3, the continued reading of the code in triplets creates new sets of triplets that specify entirely different amino acids ("frameshift" mutants). A re-

Table 7.12. Hemoglobin variants caused by deletions (from Carver and Cutler 1994 [54])

Hb	Site of deletion	Amino acid residue(s) deleted	Properties
Leiden	β 6 or 7	Glu	Unstable, ↑O$_2$ affinity
Lyon	β 17–18	Lys, Val	↑O$_2$ affinity
Freiburg	β 23	Val	↑O$_2$ affinity
Higashitochigi	β 24 or β 35	Gly	Unstable
Korea	β 33 or β 34	Val	Unstable; thalassemia
Bruxelles	β 41 or β 43	Phe	Unstable; ↓O$_2$ affinity
Niteroi	β 42–44 or β 43–45	Phe, Glu, Ser	↓O$_2$ affinity, unstable
Tochigi	β 56–59	Gly, Asn, Pro, Lys	Unstable
Ehime	β 57–59	Asn, Pro	–
St. Antoine	β 74–75	Gly, Leu	Unstable, normal O$_2$ affinity
Vicksburg	β 75	Leu	–
Tours	β 87	Thr	↑O$_2$ affinity, unstable
Gun Hill	β 91–95 or β 92–96 or β 93–97	Leu, His, Cys, Asp, Lys	↑O$_2$ affinity, unstable
McKees Rock	β 145–146	Tyr + His	↑O$_2$ affinity

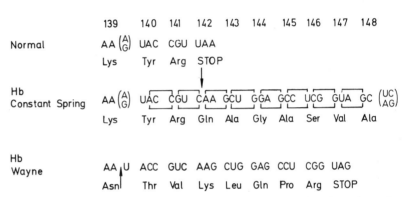

Fig. 7.41. The 3′ end of Hb α gene. Note that in Hb Constant Spring a mutation has changed the stop codon 142 UAA to CAA, allowing translation of flanking nucleotides usually not expressed. The sequence of a few codons among a total of 31 additional codons is shown. Hb Wayne is caused by deletion of the third nucleotide of the 139th codon. The first U nucleotide of the 140th codon is used as a third nucleotide of the 139th codon forming a new codon AAU, which speci- fies Asn. The resultant frameshift in reading the code results in hemoglobin Hb Wayne. The amino acid sequence of Hb Wayne can be predicted from the nucleotides of Hb Constant Spring by reading the shifted codes in multiples of three as shown by the *bracket above and below* the nucleotides of Hb Constant Spring. Hb Wayne has only five additional amino acids since a stop codon UAG is reached after translation of five codons

sultant globin structure can sometimes be identified. Hb Wayne (Fig. 7.41) appears to be caused by a dele- tion of a single nucleotide at the 139th codon near the terminus of the Hb α chain, which consists of 141 amino acids. The nucleotides of the termination codon at position 142 are read out of phase and the shifted reading frame continues until a new termina- tion codon (UAG) is encountered. Thus, a slightly elongated hemoglobin chain with five additional ami- no acids specified by nucleotides of the downstream flanking area of the α gene (see Figs. 7.32, 7.41) results. Since the reading frame is shifted, the sequence of these amino acids differs from the downstream ami- no acid sequence of the terminator mutations of the Hb α gene such as Hb Constant Spring (see Fig. 7.41) which is translated in phase.

It is understandable that the deletion characteristic of Hb Wayne was identified close to the end of the α chain. Any deletions of nucleotides that lead to "read- ing frame" shift errors at positions other than those near the end of the structural globin are unlikely to specify viable globin sequences. The resultant pheno- types would therefore be those of "thalassemia" with- out an identifiable gene product, e.g. β^0 thalassemia (see below).

Deletions are most likely due to mispairing of homologous sequences of nucleotides during either meiosis or mitotic division in germ cell development. Examination of nucleotide sequences around the areas of deletions for the various deletion mutants shows expected homologies that facilitates mispair- ing. Recombination or crossover events following mispairing then may lead to deletions of various sizes.

Fusion genes may be another result of mispairing. The homology of various globin genes may lead to mispairing between similar but not identical genes and nonhomologous crossover may lead to fusion genes that encode the NH_2 terminal portion of one globin and the COOH terminal portion of another. "Hb Lepore" is a Hb δ-β fusion gene (Fig. 7.42), and several kinds of Hb Lepore with differing amounts of δ and β gene material depending upon the site of crossover exist (Fig. 7.42). In the various types of Hb Lepore, portions of both the normal Hb δ and Hb β genes are deleted, and a new Hb δ-β fusion gene replaces them (Fig. 7.44). In Hb Kenya there is misalignment between $Hb^A\gamma$ and Hb β, with crossover and resultant deletion of $Hb^G\gamma$, $Hb^A\gamma$, and Hb δ, and production of a new chromosome consisting of $Hb^G\gamma$ and a $Hb^A\gamma$-β fusion gene (Fig. 7.44).

Duplications. Duplications may affect whole genes, such as the duplications during evolution that led to the various globin chains (α, β, γ, δ, ε, ζ). The existence of two α globin genes and two γ globin genes ($^A\gamma$ and $^G\gamma$) on a single chromosome are examples of more recent evolutionary duplications. Intragenic duplications are known to exist. In Hb α Grady residues 116–118 are duplicated [122].

Duplications of one or two nucleotides may lead to frameshift mutations. Such frameshifts have been discovered near the terminus of the β chain [42]. Hb Tak is caused by duplication of the nucleotide AC following the 146th position, and Hb Cranston owes its origin to duplication of the nucleotide AG following the 144th position of the β chain (Fig. 7.43). Hb Cranston has unique amino acids at positions 145 and 146. Hb Tak has a normal sequence up to and including position 146. The β chain normally has 146 amino acids. The frameshifts by insertion of the two nucleotides in both Hb Cranston and Hb Tak makes for an identical reading frame following positon 146. Both hemoglobins are elongated by the same amino acid sequence at the NH_2 terminal until a new stop codon (UAA) terminates the sequence at position 158. The new elongated sequence reflects the downstream flanking nucleotides of the β gene (Fig. 7.43).

		9	12	22	50	86	87	116	117	124	126
δ - Chain	:	Thr	Asn	Ala	Ser	Ser	Glu	Arg	Asn	Glu	Met
β - Chain	:	Ser	Thr	Glu	Thr	Ala	Thr	His	His	Pro	Val

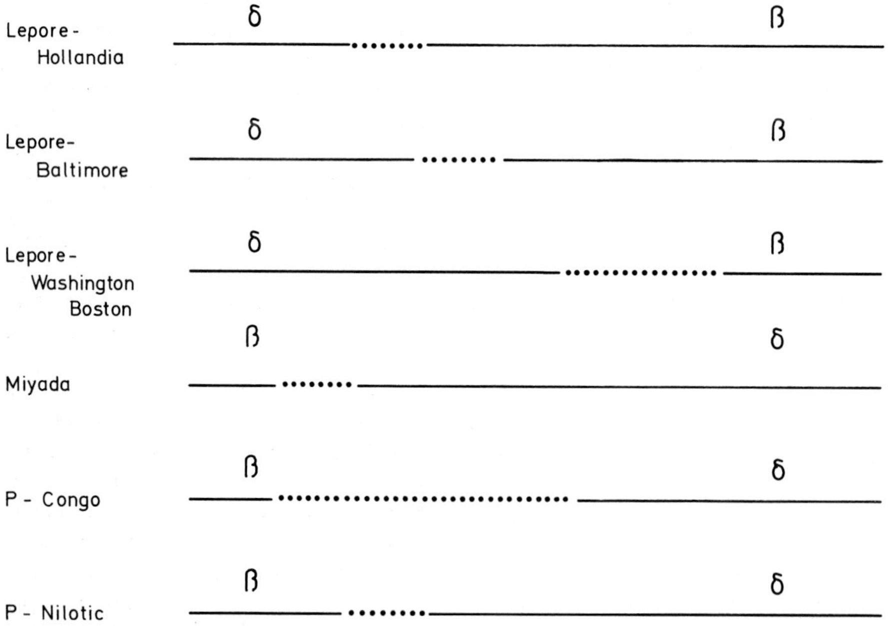

Fig. 7.42. The $\delta\beta$ and $\beta\delta$ fusion genes. *Above,* the ten differences between amino acids in the Hb δ and Hb β. Otherwise Hb δ and Hb β are identical in amino acid structure. Three different types of Hb Lepore have been found. In Hb Lepore Hollandia, the crossover between Hb δ and Hb β occurs between positions 22 and 50. The exact site of crossover is indeterminate since there are no differences between Hb δ and Hb β at these two positions. In Hb Lepore Baltimore the crossover occurs between positions 50 and 86 and in Hb Lepore Washington-Boston between positions 87 and 117. By similar reasoning, the sites of crossover are shown for various Hb β-δ or anti-Lepore hemoglobins. (Adapted from Forget 1978)

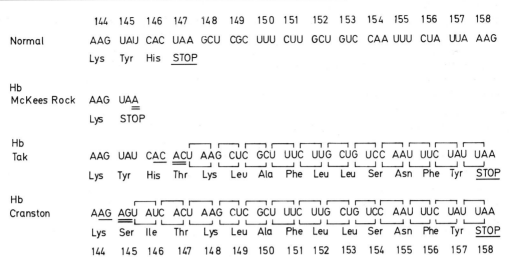

Fig. 7.43. The 3′ end of hemoglobin β gene. The Hb β chain normally has 146 amino acids. Hb McKees Rock has 144 amino acids since a mutation has altered the codon UAU (Tyr) of the 145th position to the stop codon UAA. In Hb Tak and Hb Cranston the last two nucleotides AC at position 146 and the last two nucleotides AG of codon 144, respectively, have been duplicated. The involved nucleotides are *singly* and *doubly* underlined. The resulting coding frameshift by two nucleotides makes for an identical amino acid sequence in Hb Tak and Hb Cranston starting with position 147 until the stop codon UAA at position 158 is reached. *Brackets* define the codon triplets of the normal sequence as shown in the upper part of the figure. The actual amino acid sequences of Hb Tak and Hb Cranston have been determined, and these correspond exactly to those of the actually determined nucleotides of the normal Hb β gene

Duplications of fewer than three nucleotides produce an out-of-phase reading frame and are not likely to give rise to a viable hemoglobin molecule if they occur at portions of the gene other than those specifying the terminus of the hemoglobin chains. Duplication gene products would also be expected from crossover events as the counterpart of fusion genes (Figs. 7.42, 7.44). The resultant (δ, β–δ, β) gene products or Hb anti-Lepore have in fact been identified several times as Hb Miyada, P Congo, and P Nilotic (Fig. 7.42). The expected Hb anti-Kenya product ($^Gγ$, $^Aγ$, δ, β–$^Aγ$, δ, β) (Fig. 7.44) has not yet been found. Duplications presumably have the same origin as deletions and arose from mispairing followed by nonhomologous crossover as shown in Fig. 7.44.

7.3.4 Thalassemias and Related Conditions

[12, 42, 120, 185, 193, 253]

A variety of conditions are characterized by genetically determined diminished or absent synthesis of one or another of the hemoglobin chains. These diseases are known as the thalassemias. This term is derived from *thalassa*, Greek for the Mediterranean Sea, and was originally selected to describe the Mediterranean origin of many gene carriers of these conditions. Although ethnologically and geographically incorrect, the term continues to be used widely. Thalassemias can be subdivided into α, β, δβ, γ, δ and εγδβ forms depending upon which globin chain is absent or reduced. The α and β thalassemias are quite common and clinically most important; δ and γ thalassemias are clinically silent. Some thalassemias are caused by hybrid or fusion genes consisting of fused Hb β and Hb δ or γ and β gene products (see above). Various genetic mechanisms including point mutations and deletions have been demonstrated to cause diminished or absent production of globin chains. The etiology of the thalassemias is therefore highly heterogeneous [253]. The World Health Organization has estimated that as many as 7% of the world's total population are carriers of α or β thalassemia, with the proportion being much higher in the developing world.

Advances in understanding the thalassemias at the molecular level have led to better comprehension of the mutational lesions in human mutations in general [63]. Elucidation of globin gene regulation has been aided significantly by investigation of various thalassemia genes. It has become clear that mutational interference with the various steps involved in globin synthesis can reduce ($α^+$ and $β^+$ thalassemias) or abolish globin production ($α^0$ and $β^0$ thalassemias). Deletion of either the α loci (see below) or the β locus always leads to $α^0$ or $β^0$ thalassemia, respectively [12, 185, 283].

β Thalassemia: Transcription or Promoter Mutations. Thalassemia mutations that affect the noncoding 5′

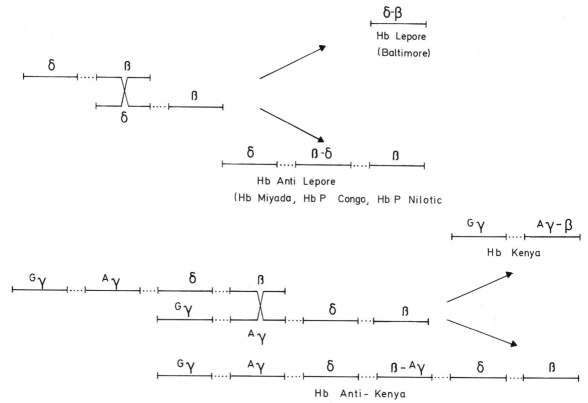

Fig. 7.44. Formation of hemoglobin fusion genes. Mispairing between the Hb δ and Hb β genes followed by recombination within the structural gene leads to a δβ fusion gene (Hb Lepore) with deletion of the normal Hb δ and Hb β gene. The alternate product of such nonhomologous crossovers makes for creation of a β–δ fusion gene preceded by a normal Hbδ gene and followed by a normal Hb β gene. Such Hb anti-Lepore forms have also been discovered (Hb Miyada, Hb P, Hb Congo, see Fig. 7.42). Mispairing between the Hb β and Hb $^A\gamma$ genes followed by recombination has yielded a $^A\gamma$–β fusion gene known as Hb Kenya. The diagram indicates the reason why the normal Hb $^A\gamma$, δ and β genes have been deleted, and the Hb γ gene is preserved in cases of Hb Kenya. A postulated Hb anti-Kenya is pictured but has not yet been discovered

upstream regions of the Hb β gene are regulatory mutations which affect gene transcription. Mutations at the more distal regulatory constant sequence PuC-PuCCC and within the regulatory TATA box have been described (Table 7.13). These mutations diminish hemoglobin synthesis and manifest themselves as relatively mild thalassemias [63, 143]. No mutations have yet been found at the CAAT box.

RNA Cleavage and Polyadenylation Mutations. A β⁺ thalassemia mutation AATAAA → AACAAA in the downstream flanking sequences of the Hb β gene has frequently been discovered among African-Americans, demonstrating that a downstream mutation can affect transcriptional efficiency. The apparent prevalence of the transcription mutation at the TATA box (see above) and this RNA cleavage mutation (Table 7.14) explains the rather mild nature of β thalassemia in African-Americans. Other mutations of this type have also been described [63] (see Table 7.13).

Terminator (Nonsense) and Frameshift Mutations. As explained above, mutations that lead to a terminator signal within an exon coding for hemoglobin would produce a foreshortened nonfunctional globin chain and thus lead to β⁰ thalassemia. Eight such mutations have been found. One of these mutations is common in persons of Mediterranean heritage (β³⁹ C→T; Table 7.14). A restriction enzyme (Mae 1) which recognizes this sequence has been identified and can be used for direct diagnosis of this β³⁹ thalassemia mutation [237].

Deletions or insertions of fewer or more than three bases produce frameshifts with garbled coding, causing effective termination of functional globin synthesis. A total of more than 20 frameshift β⁰ thalassemia mutations of this type have been identified in various populations. (Table 7.13).

RNA Processing Mutations – Splice Mutations. Processing of mRNA transcripts involves excision of inter-

Table 7.13. Molecular pathology of the β thalassemias (from Wheatherall 1994 [253])

Mutation	β^0 or β^+ Thalassemia	Population
1. Nonfunctional mRNA		
Nonsense mutants		
Codon 17 (A → T)	0	Chinese
Codon 39 (C → T)	0	Mediterranean, European
Codon 15 (G → A)	0	East Indian
Codon 121 (A → T)	0	Polish, Swiss
Codon 37 (G → A)	0	Saudi Arabian
Codon 43 (G → T)	0	Chinese
Codon 61 (A → T)	0	African
Codon 35 (C → A)	0	Thai
Frameshift mutants		
− 1 Codon 1 (− G)	0	Mediterranean
− 2 Codon 5 (− CT)	0	Mediterranean
− 1 Codon 6 (− A)	0	Mediterranean
− 2 Codon 8 (− AA)	0	Turkish
+ 1 Codons 8/9 (+ G)	0	East Indian
− 1 Codon 11 (− T)	0	Mexican
+ 1 Codons 14/15 (+ G)	0	Chinese
− 1 Codon 16 (− C)	0	East Indian
+ 1 Codons 27–28 (+ C)	0	Chinese
− 1 Codon 35 (− C)	0	Indonesian
− 1 Codons 36–37 (− T)	0	Iranian
− 1 Codon 37 (− G)	0	Kurdish
− 7 Codons 37–39	0	Turkish
− 4 Codons 41/42 (− CTTT)	0	East Indian, Chinese
− 1 Codon 44 (− C)	0	Kurdish
− 1 Codon 47 (+ A)	0	Surinamese black
− 1 Codon 64 (− G)	0	Swiss
+ 1 Codon 71 (+ T)	0	Chinese
+ 1 Codons 71/72 (+ A)	0	Chinese
− 1 Codon 76 (− C)	0	Italian
− 1 Codons 82/83 (− G)	0	Azarbaijani
+ 2 Codon 94 (+ TG)	0	Italian
+ 1 Codons 106/107 (+ G)	0	African-American
− 1 Codon 109 (− G)	+	Lithuanian
− 2, + 1 Codon 114 (− CT, + G)	+	French
− 1 Codon 126 (− T)	+	Italian
− 4 Codons 128–129	0	
− 11 Codons 132–135	0	Irish
+ 5 Codon 129	0	
Initiator codon mutants		
ATG → AGG	0	Chinese
ATG → ACG	0	Yugoslavian
2. RNA-processing mutants		
Splice junction changes		
IVS-1 position 1 (G → A)	0	Mediterranean
IVS-1 position 1 (G → T)	0	East Indian, Chinese
IVS-2 position 1 (G → A)	0	Mediterranean, Tunisian, African-American
IVS-1 position 2 (T—G)	0	Tunisian
IVS-1 position 2 (T → C)	0	African
IVA-1 3′ end—17 bp	0	Kuwaiti
IVS-1 3′ end—25 bp	0	East Indian
IVS-1 3′ end (G → C)	0	Italian
IVS-2 3′ end (A → G)	0	African-American
IVS-2 3′ end (A → C)	0	African-American
IVS-1 5′ end—44 bp	0	Mediterranean
IVS-1 3′ end (G → A)	0	Egyptian

Table 7.13. *(continued)*

Mutation	β^0 or β^+ Thalassemia	Population
Consensus changes		
IVS-1 position 5 (G → C)	+	East Indian, Chinese, Melanesian
IVS-1 position 5 (G → T)	+	Mediterranean, African
IVS-1 position 5 (G → A)	+	Algerian
IVS-1 position 6 (T → C)	+	Mediterranean
IVS-1 position − 1 (G → C) (codon 30)	?	Tunisian, African
IVS-1 position − 1 (G → A) (codon 30)	?	Bulgarian
IVS-1 position − 3 (C → T) (codon 29)	?	Lebanese
IVS-2 3′ end CAG-AAG	+	Iranian, Egyptian, African
IVS-1 3′ end TAG-GAG	+	Saudi Arabian
IVS-1 3′ end − 8 (T → G)	+	Algerian
Internal IVS changes		
IVS-2 position 110 (G → A)	+	Mediterranean
IVS-1 position 116 (T → G)	0	Mediterranean
IVS-1 position 705 (T → G)	+	Mediterranean
IVS-2 position 745 (C → G)	+	Mediterranean
IVS-2 position 654 (C → T)	0	Chinese
Coding region substitutions affecting processing		
Codon 26 (G → A)	E	Southeast Asian, European
Codon 24 (T → A)	+	African-American
Codon 27 (G → T)	Knossos	Mediterranean
Codon 19 (A → G)	Malay	Malaysian
3. Transcriptional mutants		
− 101 C → T	+	Turkish
− 92 C → T	+	Mediterranean
− 88 C → T	+	African-American, East Indian
− 88 C → A	+	Kurdish
− 87 C → G	+	Mediterranean
− 86 C → G	+	Lebanese
− 31 A → G	+	Japanese
− 30 T → A	+	Turkish
− 30 T → C	+	Chinese
− 29 A → G	+	African-American, Chinese
− 28 A → C	+	Kurdish
− 28 A → G	+	Chinese
4. RNA cleavage + polyadenylation mutants		
AATAAA–rAACAAA	+	African-American
AATAAA–AATAAG	+	Kurdish
AATAA–A)–AATAA)	+	Arab
AATAAA–AATGAA	+	Mediterranean
AATAAA–AATAGA	+	Malaysian
5. CAP site mutants		
+ 1 A–C	+	East Indian

vening sequences with splicing of exons (Sect. 3.1.3.6) to make a functional mRNA molecule (Fig. 7.45). Many different mutations affecting this process have been described. One set of mutations alters the dinucleotide GT at the donor site or AG at the acceptor site of splice junctions. These dinucleotides are part of consensus sequences that include several other nu-

cleotides and are critical for splicing. If altered by single nucleotide alterations at the donor or acceptor site, splicing is markedly compromised and causes β^0 thalassemia. Mutation in the consensus sequence usually cause β^+ thalassemia. So-called cryptic splice sites, which are not used during normal splicing, are sometimes activated by mutations in intervening se-

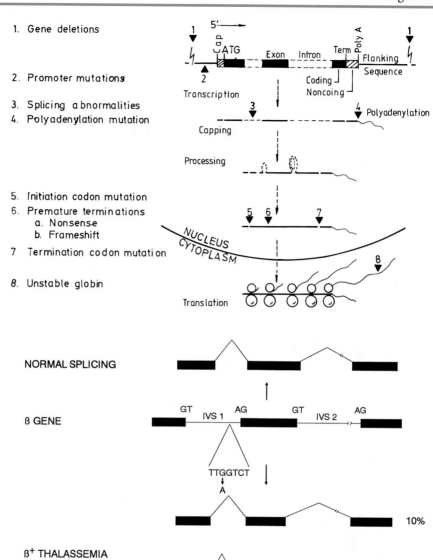

1. Gene deletions

2. Promoter mutations

3. Splicing abnormalities
4. Polyadenylation mutation

5. Initiation codon mutation
6. Premature terminations
   a. Nonsense
   b. Frameshift
7. Termination codon mutation

8. Unstable globin

Fig. 7.45. Transcription and translation of hemoglobin genes. *Numbers,* the sites at which thalassemia mutations occur. (From Kan 1985 [135])

Fig. 7.46. The activation of a splice site in IVS-1 of the β globin gene due to a G → A change at position 110. Since the abnormal splice site is utilized to a greater extent than the normal site, and hence a large amount of abnormal β globin mRNA is generated, this mutation results in a severe β^+ thalassemia phenotype. (From Weatherall 1994 [253])

quences and cause interference with normal mRNA production by creating novel splice sites (Fig. 7.46). Another class of mutations activates existing cryptic sites in coding regions. For example, the common HbE mutation activates such a site and is associated with production of abnormal mRNA causing mild β thalassemia. Two other abnormal hemoglobins (Hb Malay, Hb Knossos) exhibit a similar mechanism. The various splicing mutations are listed in Table 7.13.

Deletion Mutations at the Hb β Globin Gene Cluster and Hereditary Persistence of Fetal Hemoglobin. Unlike the α thalassemias (see below), most β thalasse-mias are not caused by gene deletion. However, a 619-bp deletion extending from within intron 2 to beyond the end of the Hb β gene is the cause of over one-third of β thalassemias among East Indians (Fig. 7.47). Various rare Hb β deletions have also been described in an African-American and in a Dutch person from the Netherlands. Several extensive deletions at the γ-δ-β locus have also been detected. Their location and extent are shown in Fig. 7.47. None of these deletions can be recognized cytogenetically, since they are too small for microscopic detection. Several deletions have led to complete removal of all or almost all of the gene cluster with no synthesis of

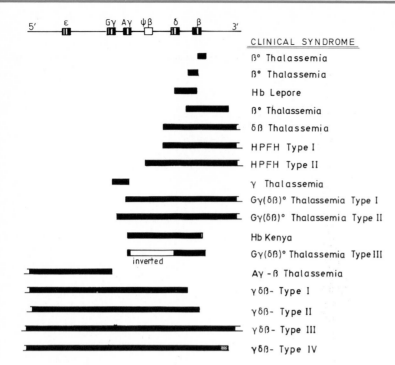

CLINICAL SYNDROME

β° Thalassemia

β° Thalassemia

Hb Lepore

β° Thalassemia

δβ Thalassemia

HPFH Type I

HPFH Type II

γ Thalassemia

Gγ(δβ)° Thalassemia Type I

Gγ(δβ)° Thalassemia Type II

Hb Kenya

Gγ(δβ)° Thalassemia Type III

Aγ -β Thalassemia

γδβ- Type I

γδβ- Type II

γδβ- Type III

γδβ- Type IV

Fig. 7.47. Deletions at the $\gamma\delta\beta$ gene cluster. Most of these deletions are rare HPHF; Hereditary persistence of fetal hemoglobin

γ, δ, and β chains. Deletions far upstream from the δ and β loci which completely abolish ε, γ, δ, and β chain synthesis and produce $\varepsilon\gamma\delta\beta$ thalassemia are of particular interest. They presumably remove the locus control regions (LCR) of the $\varepsilon\gamma\delta\beta$ locus which is required for normal function of these linked genes. Homozygotes for such deletions never have been reported since they presumably would be lethal. Conventionally, a functional distinction is made between deletions that cause thalassemic (i.e., anemic) phenotypes (e.g., δ-β thalassemia) and deletions in which fetal hemoglobin synthesis compensates for the absent δ and β loci (i.e., hereditary persistence of fetal hemoglobin, HPFH). This distinction is not absolute, since full compensation by γ chain production is not achieved in HPFH. The reason why some deletions activate the fetal hemoglobin gene remains unknown and is under active investigation.

HPFH can also be caused by nondeletion mutations. At least 15 different point mutations at the upstream promoter region of either the $^A\gamma$ or $^G\gamma$ locus have been detected [231]. The mechanism by which such mutations upregulate HbF production is not yet known. The codons affected by these point mutations are presumably part of key regulatory regions that are normally involved in transcriptional control of γ chain synthesis. Studies directed at understanding the regulation of the HbF switch have far-reaching implications for treatment of thalassemia and sickle cell anemia, since increased HbF production in these disorders would be of marked therapeutic benefit.

Table 7.14. Frequent β-thalassemias in different ethnic groups (from Antonarakis et al. 1985 [12])

Ethnic group	β-thal mutations	Type	Frequency
African-Americans	TATA box (−29)	β^+	39%
	Poly A site	β^+	26%
Mediter-raneans	Intron 1 (pos 110)	β^+	35%
	β^{39} terminator	β^+	27%
East Indians	Intron 1 (pos 5)	β^+	36%
	Deletion (619 b.p.)	β^0	36%
Chinese	Frameshift (pos 71/72)	β^0	49%
	Intron 2 (pos 654)	β^0	38%

Clinical Implications. The β thalassemias are widespread throughout the tropical and subtropical areas of the world and owe their frequency to a selective advantage vis-à-vis falciparum malaria [173]. β Thalassemia heterozygotes have mild anemia (Table 7.15). HbA$_2$ ($\alpha_2\delta_2$) is slightly increased. The red cells are smaller and less well filled with hemoglobin (MCH and MCV decreased) [42, 254]. Heterozygotes usually do not require medical attention or treatment. The appearance of the red cells is shown in Fig. 7.48.

Severely affected β thalassemia homozygotes have marked anemia requiring blood transfusions. HbA is completely absent in β^0 thalassemia homozygotes

Table 7.15. Clinically important hemoglobinopathies

	Disease	Genetics	Clinical severity
Sickle cell syndromes	Sickle cell anemia	Homozygote for HbS	+ + +
	Sickle β-thal disease	Compound heterozygote for HbS and β-thal	+ + to + + +
	Sickle Hb C disease	Compound heterozygote for HbS and HbC	+ to + +
	Sickle cell trait	Heterozygote for HbS	0
α Thalassemias	Hydrops fetalis	4 Hb α deletions	Lethal
	HbH disease	3 Hb α deletions (or 2 Hb α deletions and heterozygote for Hb CoSp) or point mutation	+ +
	α-thal-1 heterozygote	2 Hb α deletions or point mutation	+
	α-thal-2 heterozygote	1 Hb α deletion or point mutation	0
	Hb Constant Spring (CoSp) heterozygote	α Chain terminus mutant	+
β Thalassemia	β^0 (thalassemia major or Cooley anemia)	Homozygote	+ + + +
	β^+-thal major (Cooley anemia)	Homozygote	+ + + to + + + + [a]
	β^0/β^+ thalassemia	Compound heterozygote	+ + to + + +
	Hb Lepore heterozygote	δ-β fusion	+ (+ + + + for homozygotes)
	β^0, β^+, and $\delta\beta^0$-thal trait	Heterozygous	+
	HbE-β-thal	Compound heterozygotes	+ + + +
Unstable hemoglobin diseases	Congenital nonspherocytic hemolytic anemia of Heinz body type	Heterozygous – dominant (many different varieties)	+ +
Hemoglobins with abnormal oxygen affinity	Familial erythrocytosis (high affinity)	Heterozygote-dominant (many varieties)	+ +
M hemoglobin	Familial cyanosis (methemoglobinemia)	Heterozygote-dominant (5 varieties)	+ +

[a] Milder diseases in β-thal$^+$ homozygotes of African origin.

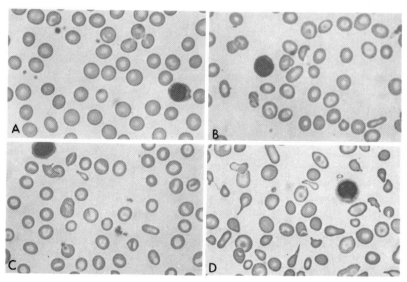

Fig. 7.48. Peripheral blood smears of a normal individual (**A**), of patients with heterozygous β thalassemia (**B**), of heterozygous α-thal-1 (**C**), and of β thalassemia major (**D**). (From Bunn et al. 1977 [42])

and much decreased in β^+ thalassemia homozygotes. Most hemoglobin is of the fetal type (HbF). The disease is associated with growth failure and often leads to death in adolescence or earlier. Homozygosity for β^0 thalassemia and compound heterozygosity for β^+/β^0 thalassemia are severe hemoglobinopathies and are a serious public health problem in countries where these genes are common. The simultaneous presence of α thalassemia (see below) ameliorates the clinical severity of homozygous β thalassemia. HbS/β^+ thalassemia is common in black populations. HbE/β thalassemia is common in Southeast Asia and produces a severe anemia similar to homozygous β^0 thalassemia. This severity is at least partially explained by the fact that the HbE mutation itself also causes mild thalassemia (see above).

Almost 90 different point mutations causing β thalassemia have now been identified (Table 7.13). The remarkable heterogeneity of mutations at the β globin locus explains the frequent finding of compound heterozygotes for β thalassemia, i.e., affected patients are not homozygotes but have inherited a different thalassemia from each parent. The frequency of such compound heterozygotes is somewhat lower in population isolates, where a single thalassemia mutant may account for the majority of thalassemias. For example, while the β^{39} nonsense mutant comprises about 27% of all β thalassemia mutants in general Mediterranean populations (Tables 7.13, 7.14), it accounts for most of the β thalassemia mutations in Sardinia. Since homozygosity for a given thalassemia mutant may range from only mild interference to complete absence of globin synthesis, and compound heterozygosity is frequent, a wide spectrum of thalassemias with different clinical severity will be encountered. Table 7.15 summarizes the clinically important hemoglobinopathies.

α Thalassemia: Deletion α Thalassemia [115, 135]. Most α thalassemias are caused by gene deletions. Gene conversion (see Sect. 3.2) and multiple crossover events occurred during evolution ("concerted evolution") and maintained a high degree of homology in the structural and flanking areas surrounding the two normal Hb α genes. Thus, sequence similarity in that area allows incorrect chromosomal alignment, followed by recombination with subsequent deletions and duplications (Fig. 7.49). The crossover chromosomes bearing either a single Hb α-or a triple α locus ($\alpha\alpha\alpha$) have been observed. Malarial selection has amplified the frequency of the single Hb α gene (-α) among tropical and subtropical populations, and this variant is therefore one of the most frequent thalassemias. The triple α chromosome seems to confer neither an advantage nor deleterious effects on its bearers and is much rarer.

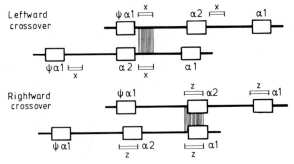

Fig. 7.49. Unique crossover at the X and Z homology boxes in the Hb α gene region. In the *leftward* crossover there is misalignment between the X boxes with crossover and creation of a single Hb α gene (4.2-kb deletion). In the *rightward* crossover there is misalignment between the Z boxes. Crossover inside the α gene causes an α_2-α_1 fusion gene with a 3.7-kb deletion. Note the formation of a triple α gene with these mechanisms in the nondeleted chromosome. (From Kan 1985 [135])

Two principal types of deletion events cause the mild α thalassemia (α-thal-2 or -$\alpha/\alpha\alpha$). Figure 7.49 shows the mechanisms of the deletions. The so-called leftward crossover creates a single Hb α_2 gene by mispairing of a flanking sequence downstream from the pseudo α gene and its homologue at the Hb α_2 locus. Recombination deletes 4.2 kb. Because the area of recombination is further upstream than that observed in the rightward deletion (see below), the term "leftward" is used. The so-called "rightward" crossover derives from misalignment of the Hb α_2 and Hb α_1 genes, with crossover within these genes producing an Hb α_2-α_1 fusion gene and a 3.7-kb deletion. The rightward single α gene is the most common type of α thalassemia in Africa and in the Mediterranean area, while in Asia both leftward and rightward crossovers have been found. The absence of one α gene in either type of mutation causes no hematological alterations in the red cells and is very common all over the world, reaching heterozygote frequencies of up to 33% in parts of Africa and in the Mediterranean. Several deletions that removed the LCR of the Hb α genes were found to silence Hb α gene expression. These deletions illustrate the importance of this regulatory region 60–100 kb upstream of the Hb α loci, which is analogous to the locus control region of the Hb β locus.

Deletions that cause elimination of the Hb α locus are shown in Fig. 7.50. The resultant thalassemias are often designated as α-thal 1. One of these mutations is frequent in Southeast Asia.

All deletions are detectable postnatally and prenatally with DNA techniques. Various phenotypes caused by deletions of one, two, three, or four Hb α genes have been documented (Table 7.16; Fig. 7.51). Absence of a

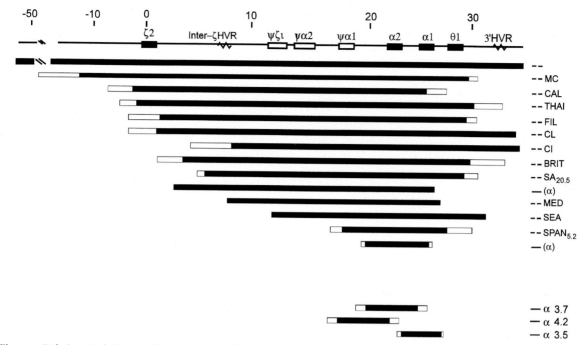

Fig. 7.50. Deletions including the Hb α gene. Most deletions are common (see text). *SEA*, Southeast Asia; *Med*, Mediterranean, ■ ; □ extent of deletion not known. (After Weatherall 1994 [253])

single Hb α gene (-α/$\alpha\alpha$) produces little or no hematological impairment since three genes remain active. DNA analysis is required for detection of mild α thalassemia due to deletion of a single Hb α gene. Deletion of all four Hb α genes (- -/- -) is fatal perinatally and is known as hydrops fetalis, referring to the extensive edema of the stillborn infant. Most of the hemoglobin molecules of such infants consist of four γ chains (Hb γ^4 or Hb Bart's). Survival of the fetus into late pregnancy is thought to be caused by the presence of functional Hb Portland (ζ^2 γ^2). The virtual absence of hydrops fetalis from African infants is related to the nonexistence of the chromosome bearing the two α gene (- -) deletions in that population.

Deletion of two Hb α genes (-α/-α or --/$\alpha\alpha$) produces mild anemia, while deletion of three Hb α genes (-α/- -) causes a more severe anemia characterized by production of HbH – a Hb β^4 tetramer (Table 7.16). HbH is formed because of the deficiency of Hb α chains. Coexisting α thalassemia in patients with sickle cell anemia is associated with less anemia and improved survival.

Hb α Nondeletion Thalassemia [12, 134, 135]. Nondeletion mutations similar to those detected in β-thalassemia would be expected. A variety of such mutations have in fact been found and are listed in Table 7.17. Most of these involve the α_2 gene, where output normally is three times that of the α_1 gene. They are analogous in origin to β thalassemia mutations, but far

Table 7.16. α Thalassemias caused by deletion

Condition	Hb Symbol
Normal	$\alpha\alpha/\alpha\alpha$
Mild α thalassemia (α thal 2)	-α/$\alpha\alpha^a$
Severe α thalassemia (α thal 1)	- -/$\alpha\alpha^a$ or -α/-α^b
Hb H disease (Hb H = β^4)	-α/- -^{b}
Hydrops fetalis with Hb γ^4	- -/- -^{b}

^{a} Transmission from one parent.
^{b} Transmission from both parents.

fewer examples have been detected. No regulatory mutations in the upstream region of the Hb α loci besides an LCR deletion (see above) have been detected. Only one splicing mutation, consisting of a small 5-bp deletion that abolishes an acceptor site in IVS-1, has so far been found. Four different point mutations have altered the UAA termination codon of the α gene to one specifying an amino acid. As a result, the usually untranslated downstream sequence which specifies another 31 amino acids reads through until a new termination signal is reached. These extended abnormal hemoglobin mRNAs are unstable, and only small amounts (5%) of these variants can be detected in the blood. Hb Constant Spring is the most common of these mutants (see Fig. 7.41).

In several mutations the alteration causes extensive instability. A downstream Hb α_2 polyadenylation site

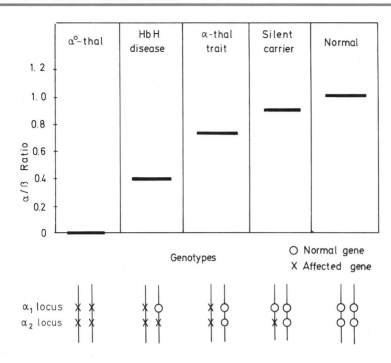

Fig. 7.51. Globin chain synthesis in the α thalassemias. Globin chain synthesis is expressed as the ratio Hb α/Hb β chain synthesis, which is close to 1 in normals, i.e., identical amounts of Hb α and Hb β chains are produced. "Silent" carriers are heterozygotes for α-thal-2 (mild α thalassemia) or -α-$\alpha\alpha$. (From Kazazian et al. 1977)

Table 7.17. Mutations that cause nondeletion forms of α thalassemia (from Weatherall 1994 [253])

Mutation	Population
Non-functional mRNA	
Nonsense mutations	
Codon 116 (G → T)	African
Frameshift mutations	
Codon 30/31 (−4 nt)	African
Initiator codon mutations	
ATG → ACG	Mediterranean
CCCACCATG → CCCCATG	Mediterranean
ATG → GTG	Mediterranean, African
Terminator codon mutations	
α^{CS} Hb Constant Spring	Southeast asian,
(TAA → CAA)	Mediterranean
α^{KD} Koya Dora (TAA → TCA)	Indian
α^{IC} Hb Icaria (TAA → AAA)	Mediterranean
α^{SR} Hb Seal Rock (TAA → GAA)	African
RNA-processing mutants	
Splice junction changes	
IVS-1 donor site	
(GGTGAGGCT → GGCT)	Mediterranean
RNA cleavage; polyadenylation site	
AATAAA-AATAAG	Arab
	Mediterranean
Unstable globins	
$\alpha^{Quong\ Sze}$ (codon 125, Leu → Pro)	Southeast Asian
$\alpha^{Suan\ Dok}$ (codon 109, Leu → Arg)	Southeast Asian
$\alpha^{Petah\ Tikwah}$ (codon 110, Ala → Asp)	Middle East
$\alpha^{Evanston}$ (codon 14, Trp → Arg)	African

mutation that reduces synthesis of the α_2 gene has also been described.

No direct data are available regarding the frequency of nondeletion α thalassemia; however, the nature of the allelic α thalassemia genes in patients with HbH (β^4) who carry at least one double deletion of Hb α (- -/$\alpha\alpha$) on one of their chromosomes has been analyzed. Both in Saudi Arabia and in China as many as 50% of such allelic α thalassemia genes in HbH disease were nondeletion rather than deletion mutants.

α *Thalassemia and Mental Retardation.* α Thalassemia has occasionally been found in patients with mental retardation [98, 262, 263]. In one variety seen among whites, mild mental retardation and various facial and skeletal abnormalities are associated with large deletions of more than 1 million bases in the Hb α region [262, 263]. These cases presumably represent heterozygotes for a contiguous gene syndrome with deletion of the hemoglobin α loci and of other developmental genes on chromosome 16.

In another variety severe mental retardation and a characteristic facial appearance and genital abnormalities are associated with the hematological abnormalities of HbH disease, which are usually caused by three Hb α gene deletions (- -/α-). However, no deletions or structural abnormalities of the hemoglobin α locus were found in these patients. The condition affects only males and has been mapped to the short arm of the X chromosome. The existence of this condition implicates an X chromosomal locus that influences Hb α gene expression. This syndrome is most likely a contiguous gene syndrome with deletion of an X chromosomal transcriptional factor affecting the Hb α locus as well as deletion of nearby yet undefined gene segments essential to normal development. These experiments of nature are of great interest in helping to understand both Hb gene expression and the mechanisms involved in the developmental biology of dysmorphic body structure and of mental retardation.

7.3.5 Population Genetics of Hemoglobin Genes
(see [85]; Sect. 12.2.1.6)

The presence of a relatively large number of DNA polymorphisms at the Hb β locus has made it possible to study the mutational origin and spread of the different hemoglobinopathies (Tables 7.18, 7.19). Taken at face value, the results appear to suggest that the HbS mutation occurred at least four times [85, 179] in different geographic areas (Senegal, Benin, Central African Republic, and Arab/Indian) and then spread by malarial selection (see Chap. 12). HbS in Mediterraneans appears to be of African origin, and both the Benin and Senegalese DNA haplotypes have been found in the Mediterranean area. Three chromosome backgrounds have been observed among HbE chromosomes in Southeast Asia. One of these is derived from crossover at the "hot spot" area (see above). HbC has been observed largely in one haplotype. While the haplotype data suggest that several mutational events were the origin of HbS and HbE, it is more likely that gene conversion events (together with recombination) occurred following a single ancestral mutation and spread the HbS and the HbE

genes into different haplotypes (see Sect. 12.2.16). To cite another example of gene conversion, the same Hb β^{39} thalassemia mutation in Sardinia is associated with several haplotypes that cannot be explained by simple recombination. Unidirectional transfer of sequence information (i.e., conversion) is therefore likely. Since there is strong evidence for gene conversion in globin gene evolution, this mechanism appears to be the best explanation.

The various common β thalassemia mutations usually occurred in a unique haplotype with subsequent

Table 7.19. The common β-thalassemia mutations and associated haplotypes (from Old [190])

Ethnic group/mutations	RFLPs						
	1	2	3	4	5	6	7
Mediterranean							
IVSI-110 (G → A)	+	−	−	−	−	+	+
Nonsense codon 39 (C → A)	−	+	+	−	+	+	+
	+	−	−	−	−	+	+
IVSI-6 (T → C)	−	+	+	−	−	−	+
IVSI-1 (G → A)	+	−	−	−	−	+	−
IVSII-1 (G → A)	−	+	−	+	+	+	−
	−	+	−	+	+	+	−
IVSII-745 (C → G)	+	−	−	−	−	−	+
Frameshift codon 6 (−A)	+	−	−	−	−	+	−
Frameshift codon 8 (−AA)	−	+	−	+	+	−	+
−87 (C → G)	−	+	−	+	−	+	−
East Asian							
619 bp deletion	+	−	−	−	−	−	+
IVSI-5 (G → C)	+	−	−	−	−	−	+
	−	+	+	−	+	−	+
IVSI-1 (G → T)	−	+	−	+	+	+	+
Frameshift codon 41–42 (-TTCT)	−	+	−	+	+	+	−
	+	−	−	−	−	+	+
Frameshift codon 8–9 (+G)	+	−	−	−	−	+	+
Nonsense codon 15 (G → A)	−	+	+	−	+	+	+
Frameshift codon 16 (−C)	+	−	−	−	−	+	−
Chinese							
Frameshift codon 41–42 (-TTCT)	+	−	−	−	−	+	+
	+	−	−	−	−	−	+
Nonsense codon 17 (A → T)	+	−	−	−	−	−	+
Frameshift codon 71–72 (+A)	+	−	−	−	−	−	+
IVSII-654 (C →T)	+	−	−	−	−	+	+
−28 (A → G)	−	+	+	−	+	−	+
−29 (A → G)	+	−	−	−	−	+	−
African							
−29 (A → G)	−	−	−	+	+	+	+
	−	+	−	+	+	+	+
−88 (C → T)	−	−	−	−	+	+	+
codon 24 (T → A)	−	+	−	−	+	+	+

RFLPs used for haplotypes; 1, *Hind*II/ε gene; 2, *Hind*III/Gγ gene; 3, *Hind*III/Aγ gene; 4, *Hind*II/3' of $\psi\beta$ gene; 5, *Hin*II/5' of $\psi\beta$ gene; 6, *Ava*II/β gene; 7, *Bam*HI/5' of β gene.

Table 7.18. β-Thalassemia mutations which affect restriction endonuclease sites (from Old [190])

Mutation	Ethnic group	Site
Nonsense codon 39	Mediterranean	*Mae*I (+)
IVSI-nt6	Mediterranean	*Sfa*NI (+)
IVSI-nt1 (G → A)	Mediterranean	*Bsp*MI (−)
IVSII-nt1	Mediterranean	*Hph*I (−)
IVSII-nt745	Mediterranean	*Rsa*I (+)
Frameshift codon 6	Mediterranean	*Dde*I (−)
−87	Mediterranean	*Avr*II (−)
IVSI-nt5 (G → A)	Mediterranean	*Eco*RV (+)
IVSI-nt1 (G → T)	Asian Indian	*Bsp*MI (−)
Nonsense codon 17	Chinese	*Mae*I (+)
Nonsense codon 43 (G → T)	Chinese	*Hinf*I (−)
−88	Africa	*Fok*I (+)
−29	Africa	*Nla*III (+)

(+) and (−) refer to gain or loss, respectively, of restriction site.

expansion of such chromosomes because of malarial selection. The association of a given mutation with a characteristic haplotype for DNA variants aided greatly in defining the nature of the various β thalassemia mutations. Since different DNA haplotypes were selected for investigation, there was a high chance that the β thalassemia mutation under study was different from those previously investigated, and many different types of thalassemia have been discovered [193].

The various data suggest that practically all thalassemias and hemoglobinopathies first occurred after human racial divergence.

7.3.6 Screening and Prenatal Diagnosis of Hemoglobinopathies [4, 51, 123, 191]

Hemoglobin disorders are often associated with severe and disabling diseases in childhood such as homozygous β thalassemia and sickle cell anemia. These hemoglobinopathies are transmitted by autosomal recessive inheritance. The heterozygote states for these conditions can be detected by a variety of hematological and biochemical methods that are available in hematological laboratories or can be set up relatively easily. These developments and the availability of pregnancy termination of affected fetuses following prenatal diagnosis have led to the establishment of population screening programs for these conditions. Extensive educational campaigns directed at adolescents and young adults have been carried out in some areas of high frequency (see Chap. 18). In such programs, HbS is searched for by electrophoretic techniques. Carriers of β thalassemia and α thalassemia are identified by reduced red cell mean corpuscular volume and mean corpuscular hemoglobin volumes of their red cells as well as by increased levels of HbA_2 in β thalassemia. If testing is first performed during pregnancy, only the woman is tested. If her results are negative, the male partner is not tested. If both partners are heterozygotes, genetic counseling and prenatal diagnosis are offered.

Prenatal diagnosis for detection of hemoglobin abnormalities is often performed on cells obtained by chorionic villus biopsy between the 9th and 10th weeks of pregnancy. Amniocentesis can only be performed later (14–16th weeks of pregnancy) and raises the emotional problems of relatively late pregnancy termination when an affected fetus is found.

The prenatal diagnosis of hemoglobinopathies relies largely on DNA technology. Before proceeding to fetal diagnosis it is essential to define the nature of the specific molecular thalassemia defect to look for in the fetus at risk. Fortunately, a few mutations tend to be common in given ethnic groups, as shown in Table 7.14. The diagnosis of β thalassemia is usually by PCR amplification of Hb β DNA followed by oligonucleotide hybridization with the probe relevant for the mutation(s) to be detected (Fig. 7.52; Sect. 3.1.3.5).

Several β thalassemias and the HbS mutation can be diagnosed directly using one or another restriction enzyme that cuts at a DNA site at which the globin mutation has abolished or created a restriction site producing a characteristic band pattern on Southern blots (Fig. 7.53). Visualization of the various deletions of the Hb β locus (δβ fusion genes, δβ thalassemia, HPFH) and deletions of α° thalassemia also use Southern blotting techniques with radioactive globin probes; but rapid, inexpensive, and nonisotopic PCR-based methods are becoming available for rapid testing.

Fetal blood sampling (using fetoscopy or placental puncture) historically was the first approach to the prenatal diagnosis of β thalassemia. This method allows the assessment of reduced or absent globin chain or mRNA synthesis in fetal erythrocytes but has a relatively high fetal mortality (5%–7%) and is now only rarely required since DNA methods are becoming increasingly available.

Occasionally, a family linkage study using informative RFLP markers closely linked to the β thalassemia gene may be required for fetal diagnosis. In this approach the nature of the β thalassemia defect need *not* be known, but (apart from the fetal cell specimen obtained by chorionic villus sampling or amniocentesis) blood from the heterozygote parents and one or more definitely unaffected family members (ideally grandparents) should be available to allow tracking of the defect by its cosegregation with the RFLP marker.

The use of screening, prenatal diagnosis, and selective abortion (and, less successfully, avoidance of carrier x carrier matings) has been highly successful in reducing the frequency of homozygous β thalassemia in several countries, such as Sardinia, continental Italy, Cyprus, and Greece. Figure 18.2, p. 726 [50] shows the remarkable decline in the birth of infants with β thalassemia major in Sardinia since 1975 when

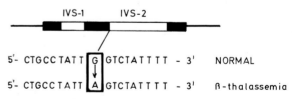

Fig. 7.52. Oligonucleotide probe (19-mer) for normal β globin gene differs from a probe for a β thalassemia gene with a G-A mutation in IVS-2 at one position (G-A). Under appropriate conditions of hybridization the probe for this specific thalassemia mutation recognizes only the thalassemia gene and not the normal gene. Similarly, the "normal" probe does not hybridize with this thalassemia gene

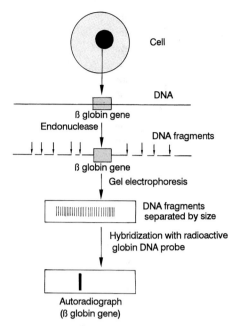

Fig. 7.53. Visualization of globin genes. DNA from any cell nucleus (usually white blood cells) is extracted. This DNA contains the globin genes. Various restriction endonucleases recognize the different sequences of nucleotides and break up the DNA into many fragments. The DNA fragments are separated on gel electrophoresis by size. A specific radioactive DNA probe for globin is prepared and reacted with the DNA fragments. Hybridization of the radioactive β globin probe occurs with a β globin gene and can be visualized following autoradiography. Deletions can also be visualized by this technique. (Adapted from Orkin et al. 1978 [194])

screening was first initiated. These results came from intensive public health oriented screening and provide an excellent example of the almost complete eradication of a devastating genetic disease by voluntary means. However, prenatal methodology has not been utilized in areas where selective abortion is not acceptable for religious and social reasons such as in Islamic populations. While the clinical severity of β thalassemia, with its serious impact on the family and on the health services in high frequency areas, has led to wide acceptance of prenatal diagnosis (even in some Catholic countries); the less severe clinical findings of sickle cell anemia presumably have been the major reason that prenatal approaches in this condition have been used less frequently.

Hemoglobin as a Model System. Hemoglobin is probably the best analyzed genetic system in humans. Therefore experiences with concepts developed in the course of its analysis may help toward a better understanding of phenomena in other fields of human genetics. For example, when hereditary diseases with different phenotypes are found in different fa-

milies, it is generally concluded that they are caused by mutations of different genes. The hemoglobin example shows that this is not necessarily the case. Methemoglobinemia, for example, is quite different phenotypically from erythrocytosis due to abnormal oxygen affinity, yet allelic mutations are involved. The phenotype depends on the type of the molecular anomaly and the ways in which normal function is altered.

Another useful lesson is taught by the way in which the tetrameric structure of the Hb molecule determines its oxygen binding capacity, and how mutations may influence this capacity which may depend on interaction of the products of more than one gene. While the effects of most hemoglobin mutations are innocuous, the phenotypic effects of all pathological structural Hb variants are dominant, with the exception of sickle cell anemia. This finding points to one mechanism for Mendelian dominance: disturbance of interaction among products of alleles (see Sect. 7.6).

Finally, and most importantly, the hemoglobins provide evidence for the action of many possible mechanisms of mutations. Such mutations may occur in the structural gene itself or in adjacent regulatory areas. In most cases they involve exchange of only one base but sometimes base sequences of greater lengths are duplicated or deleted. Other types of mutations, such as frameshifts in reading the genetic code are often encountered. Of great research interest are the various regulatory mutations in the flanking areas of hemoglobin genes, which are increasingly illuminating mechanisms of both gene control and developmental gene switching. As is shown in Chapters 12 and 14, hemoglobin mutations have also contributed significantly to our understanding of the role of mutation in evolution.

7.4 The Defense System

Our ancestors – from the most primitive organisms to primates, as well as human beings of past generations – were constantly exposed to a wide array of infective agents; human existence has been and continues to be challenged by infectious disease. The genome has reacted by developing a complex defense system (Fig. 7.54). It consists of four main components: the humoral defense by antibodies; the cellular defense with T lymphocytes and macrophages; the complement system; and the phagocytic system of the granulocytes. The main purpose of all these systems is defense against foreign germs and parasites, but it also protects the body's integrity against internal enemies such as cancer cells. Its structure can be

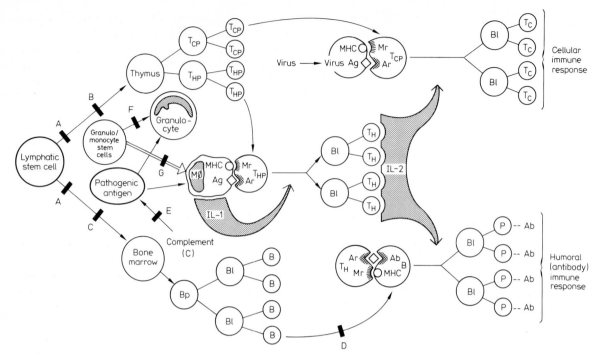

Fig. 7.54. The most important components and sequences of events of the immune system (simplified). *Black bars,* genetic deficiencies may impair the immune defense at various levels. *TCP,* Cytotoxic T cell precursor; *Tc,* cytotoxic T cell; *THP,* helper T cell precursor; *TH,* helper T cell; *Me,* macrophage; *Bp,* B cell precursor; *B,* B cell; *P,* plasma cell; *Bl,* blast; *Ag,* antigen; *Ar,* antigen receptor; *Ab,* antibody (antigen receptor on B cells); *MHC,* major histocompatibility molecules; *Mr,* major histocompatibility molecule receptor; *IL-1,* interleukin 1; *IL-2,* interleukin 2; *A,* developmental abnormalities of lymphocytes; *B,* lymphopenic immune deficiencies and thymus hypoplasia; *C,* various types of B cell defects; *D,* defects of specific immunoglobulins; *E,* defects of complement components; *F,* agranulocytosis; *G,* progressive granulomatosis of childhood

compared with a deep defense system in modern warfare which consists of both fixed and mobile components, allows for counterattacks, and depends on a sophisticated communication network. We do not describe the entire defense mechanism in all details; some aspects are discussed in other chapters of this volume: the major histocompatibility complex (MHC) in Sect. 5.2.5 and natural selection in Sect. 12.2.1.8. Here, some aspects of gene action are described since some of these mechanisms may also apply to other fields such as embryonic development, especially of the brain.

The defense system performs its task in several steps [10]:

1. The intruding antigen is bound to macrophages and presented to T lymphocytes.
2. T cells trigger the T cell cascade which leads to formation and activation of other, specialized T cells.
3, 4. Lymphokines induce B cell proliferation.
5. B cells produce immunoglobulins that fix themselves to the antigen; complement and killer T cells are attracted.

6, 7. The complement system activates mast cells.
8. Inflammation as complex reaction.
9. Cells loaded with immune complexes are killed by macrophages.
10. (Parallel to 9) T cells stimulate killer cells and macrophages.
11. Activation of granulocytes.
12. Infective agents and infected cells are destroyed (killer cells).
13. Debris is removed (macrophages).

Genetic mechanisms range from the simple to the very complex – in some aspects so complex that a special branch of science, immunogenetics, has developed for coping with the scientific problems posed by this complexity. Complement factors may be mentioned as examples of a genetically relatively simple mechanisms. Some of these show conventional genetic polymorphisms, which may be associated with certain aberrations of immune defence (Sect. 6.2.4), as shown by their disease associations. The MHC traits, on the other hand, which are necessary for the presentation of the antigen to the T helper and T killer cells (see Figs. 7.55, 7.56) give a more complex picture.

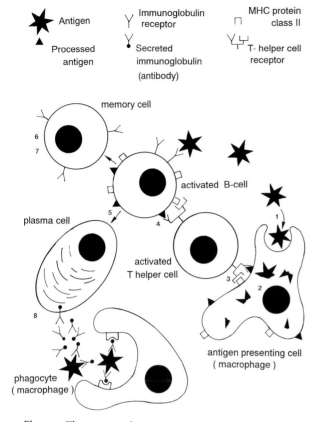

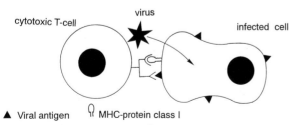

Fig. 7.55. The course of events during an infection (simplified). An infection mobilizes several cooperating populations of immune cells. B cells carrying immunoglobulins as surface receptors recognize and bind circulating antigens. *1*, To be activated the antigen must be taken up by an antigen-presenting cell, often macrophage. *2*, After some modification the antigen appears at its surface. *3*, When it is recognized by a T helper cell, this cell is stimulated. *4*, This cell then activate B cells carrying the same antigen. *5*, These B cells are multiplied and differentiate. *6*, Some become memory cells. *7*, This allows faster immune reactions after a new infection; others become antibody-excreting plasma cell. *8*, Free antibodies bind to the antigen, marking it to be destroyed by other elements of the immune system, for example macrophages. (Modified from Tonegawa 1986 [241])

Fig. 7.56. Virus infection: if a virus enters a cell, it leaves viral proteins in the cell membrane. Cytotoxic T cells (killer cells) recognize such foreign molecules which are presented to them together with MHC class 1 proteins. The infected cell is then destroyed by killer cells. (Modified from Tonegawa 1986 [241])

Here the enormous degree of interindividual genetic variability of the HLA loci is remarkable. At the protein level this leads to interindividual differences in structures necessary for recognition of antigens. Here the "conventional" genetic defense strategy against a variety of antigen-carrying agents shows a high degree of refinement. Linkage disequilibrium is the only slightly unconventional element. Functional differences between allelic proteins lead to varying associations with diseases (Sect. 6.2.4). On the other hand, the genetic basis for the steps that follow – presentation of the antigen to T helper cells and formation of immunoglobulins – is very unconventional. For both the T cell receptors and the immunoglobulins, more generally available genes for cell recognition molecules are used after adaptation to this special function [264]. Figure 7.57 shows a number of these molecules: the immunoglobulin superfamily. Genes of the MHC system belong to the same group; their proteins share some properties with the immunoglobulin and T cell receptor genes. These genes have been adapted to their special function: recognition of a great variety of antigens. In addition to constant segments (see below), the respective molecules also consist of variable parts that are products of multiple, very similar but not identical genes, of which only a few – and always a different combination – are active in a certain cell clone. By this principle an enormous additional diversity and flexibility of defense molecules is created.

In the following, antibody formation is described in somewhat greater detail; followed by a brief outline on T cell receptors.

7.4.1 The Function of B Lymphocytes and the Formation of Antibodies [31, 201, 209]

When a free antigen or its carrier, for example, a bacterium, enters the body, it reacts with B cells that had earlier contact with it (= memory cells) and is taken up by macrophages which present this antigen – together with their own class II antigens – at their surface. This leads to proliferation of this B cell clone: its decendents, the plasma cells, then release a great number of antibody molecules into the blood, where they trap intruding antigen molecules. In the beginning, therefore, the immunoglobulin molecules at the cell surface of memory cells act as receptors for antigens. This function does not differ from that of other surface-bound molecules of the same gene superfamily (Fig. 7.57). When the cells proliferate, however, these molecules are separated from the cell surface and released into the surrounding medium. Some aspects of their structure are very unusual. The antigen-binding portion of the molecule is en-

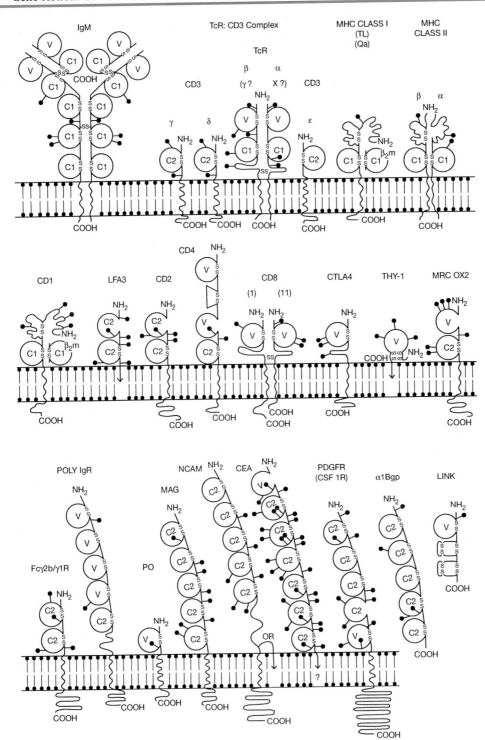

Fig. 7.57. (see text p. 329)

coded by a set of genes (variable part or V genes), and during differentiation of the lymphocyte one of the V genes is randomly combined with a constant part gene (C gene). Thus, each differentiated lymphocyte has the ability to produce only one type of antigen-specific receptor; the lymphocyte population as a whole forms all the various receptors that the organism is able to produce. Exposure to a certain antigen leads to proliferation of those lymphocytes (clones) which possess a receptor fitting the antigen (clonal selection theory [43]). The analysis of these genes is one of the success stories of molecular biolo-

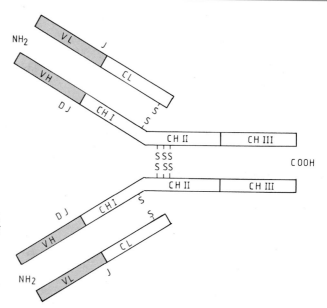

Fig. 7.58. Basic structure of an IgG molecule. It is composed of two identical light chains and two identical heavy chains. Each chain consists of the amino-terminal V region and carboxy-terminal C region. The C region of the light chain has the approximate length of the V region; the C region of the IgG heavy chain is three times as long, consisting of three approximately structure-homologous subsections that were formed during evolution from one common gene

gy. As a first step, the proteins were analyzed; this also permitted direct conclusions regarding the nature of the genes involved [116, 117]. These conclusions were then confirmed and extended by direct analysis at the gene/DNA level.

Myeloma Proteins as Research Tools. The immunoglobulins (antibodies, secreted B cell antigen receptor) of an average human being are a mixture of the gene products of very many different cell clones. At first glance, such heterogeneities seem to be an insuperable obstacle for any chemical analysis of antibodies that requires pure proteins. As in many other situations, however, experiments of nature offer an opportunity to overcome this obstacle. Neoplasias originate from one single cell by somatic mutation (Chap. 10). Plasma cell tumors are therefore expected to produce only a single antibody species in abundance, if differentiation of antibody-forming cells occurs before the beginning of malignant cell growth. Such monoclonal proteins are indeed observed in mice and in humans suffering from myelomatosis – a not uncommon plasma cell tumor. Myeloma proteins can be isolated in

sufficient amounts, purified, and their amino acid sequences determined. In this way the structure of antibodies has been analyzed.

Classes of Immunoglobulins [117]. Several classes of immunoglobulins can be distinguished: IgG, IgM, IgA, IgD and IgE. All these classes consist of several polypeptide chains of different lengths: the smaller light (L) chains and the longer heavy (H) chains. The H chains determine the class to which an immunoglobulin belongs. They may be of γ, μ, α, δ, or ε type. All classes of immunoglobulins utilize one of two L chains: $\varkappa$ or λ; both types of L chains may occur in all classes.

A common immunoglobulin is the IgG molecule; two H chains are connected by S-S bridges with two L chains (Fig. 7.58). The structure of the other classes is more complicated; one IgM molecule, for example, contains five subunits with two H chains each. During normal immunization antibodies of the IgM class are formed first; they are then replaced by those of the IgG class without change of specificity. This switch occurs within the same cells.

◁ **Fig. 7.57.** Phylogenetically related molecules of the immunoglobulin superfamily. One model is shown for each molecular type for one species. In some cases the same model suffices also for the structure named in *parentheses. Circles,* sequences that fold as an Ig domain or are predicted to do so. V and C domains are homologous to those shown in Fig. 7.59. Domain numbers are from the NH₂ terminus of proteins. *TcR, CD3, α and β chains,* T cell receptor and its components; *CD1, CD2, LFA3,* T cell adhesion and related proteins; *CD4, CD8, CTLA4,* T "subset" antigens; *Thy1, MRC, Ox2,* brain/lymphoid antigens; *PolyIgR, Fcγ2b/γ1R,* immunoglobulin re-

ceptors; *NCAM,* neural adhesion molecule; *Po,* myelin protein; *MAG,* myelin-associated protein; *CEA,* carcinoembryonic antigen; *PDGF,* platelet-derived growth factor receptor; *GSF1R,* colony-stimulating factor 1 receptor; *α1Bgp,* non-cell-surface molecules; *LINK,* basement membrane link protein. All molecules except the last two are anchored in the cell membrane; they are linked to carbohydrates (linkage sites: →•). Sulfide bonds within and between chains: ^S_S, ss. (From Williams and Barclay 1988 [264]). (Note that the figure had to be broken up into three parts for technical reasons. *All* molecules are related with each other)

Constant and Variable Parts. It is a common property of all L and H chains that they consist of a *constant* and a *variable* part. The constant (C) part follows the rules that are familiar to us from other proteins; amino acid sequences are always identical for each type of chain, excepting a few positions for which ge-

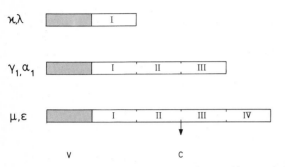

Fig. 7.59. The light chain *(top)* consists of a variable part *(V)* and one constant part, which may be a ϰ or a λ chain. Heavy chains *(center, below)* consist of a variable part *(V)* and a constant part consisting of a threefold or fourfold homologue of the basic antibody gene. The constant parts in γ_1 and α_1 chains, for example, which are components of the IgG and IgA classes of immunoglobulins, consist of threefold homologues of the basic gene. The μ and ε heavy chains consist of four-fold homologues. (From Hilschmann et al. 1976 [117])

netic polymorphisms exist. These polymorphisms are usually recognized indirectly by the ability of some variants to inhibit agglutination of red blood cells by specific antibodies. They are called the GM or KM (Inv) groups and affect the heavy chain and light chains, respectively. Most of their alleles differ by one amino acid substitution only. The variable parts, on the other hand, were shown to be different in their amino acid sequences in all human myeloma proteins analyzed so far. The variable parts of all light and heavy chains have a similar length of 107–120 amino acids. In the light chains the constant part is similar in length to the variable part. In the heavy chains the length of the constant part is an almost exact multiple of that in the light chain (Fig. 7.59). The constant parts of the γ_1 and α_1 chains are three times as long and those of the μ and ε chains four times as long as those of the light chains. Moreover, all segments of the constant parts show some degree of homology with each other, i.e., their amino acid sequences, although different in many details, are much more similar than could be accounted for by chance.

Common Origin of the Genes for All Chains. The most obvious explanation for this similarity is a common

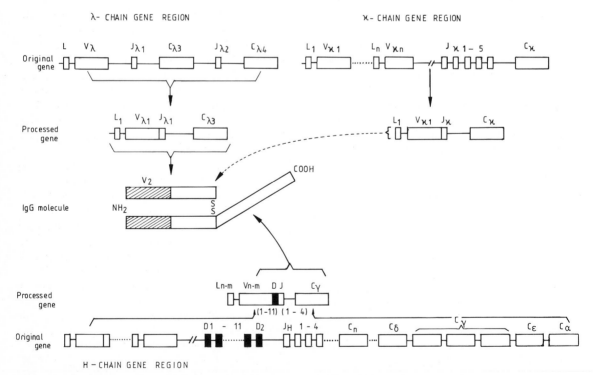

Fig. 7.60. Organization of immunoglobulin gene segments before and after somatic rearrangement. The rearranged state shown here is only one of many possible ones. The genes are rearranged before and during B cell maturation and deter-

mine, in the rearranged state, the IgG molecule (only one half of this molecule is shown). *Dotted line* from the processed ϰ gene to the IgG molecule, either a λ chain or a ϰ chain is contributed to this molecule (See Tonegawa 1983 [240])

origin of all these segments in evolution; at the beginning there was only one gene that determined a polypeptide chain, which had about the length of the constant part of a light chain. During evolution this gene was duplicated repeatedly. Some of these duplications led to longer DNA stretches that determined polypeptide chains in which the same amino acid sequence was repeated three or even four times. These duplicated DNA stretches were then completely structure homologous but no longer position homologous. During the following millions of years, fixation of new mutations led to progressive diversification between these structure-homologous DNA stretches, causing their present differences in amino acid sequence.

The first duplication of a singular gene must have occurred by some chromosome rearrangement. Subsequent duplication can easily be produced by unequal crossing over after mispairing of structure-homologous but not position-homologous closely linked genes (Sect. 5.2.8). This seems to be the obvious mechanism for enhancing the number of homologous stretches in the constant part of different heavy chain genes. Evolution of the various light chain and heavy chain genes requires additional steps of gene duplication and chromosomal rearrangement. The genes for light and heavy chains are not located close to each other on the same chromosome. Genetic polymorphisms of the light chain (the KM or Inv system) and of the heavy chain (the GM system) are not linked.

Genetic Determination of the Variable Chains. So far we have discussed the genetic determination of only the constant parts, which can be explained satisfactorily using classic genetic principles. A simple genetic explanation of the variable parts of the immunoglobulin, however, is not possible. The fact that all their amino acid sequences have so far proven to be different can be explained only by assuming that each person harbors a great many plasma cell clones, each of which produces an immunoglobulin with a different variable part. This postulate immediately suggests that antibody specificity is located in the variable (V) parts. Two questions arise:

1. What genetic mechanisms determine the variable parts?
2. How do they cause antibody specificity?

Somatic Mutation or Selective Activation of Genes? Several hypotheses for the genetic determination of the variable parts have been proposed. Two of these concepts were widely discussed: the "somatic mutation" hypothesis and the "selective gene activation" hypothesis. The somatic mutation hypothesis maintained that there is only one gene that undergoes many random mutations during proliferation of B lymphocytes. Indeed, somatic mutations occasionally occur during proliferation of all cell types. However, the hypothesis required a specific mechanism by which the somatic mutation rate is enhanced specifically for this gene. Such mechanisms are conceivable; for example, the gene loci in question could be inaccessible to repair enzymes.

Somatic mutations are of course random in direction. Hence this hypothesis predicts that the amino acid substitutions within variable chains from various antibodies as assessed by study of myeloma proteins are completely independent of each other. Of course there may be sites at which no mutations are tolerated, and which are therefore identical in all variable parts. Any other regularities, however, would be difficult to reconcile with this hypothesis.

Such regularities, however, were indeed described. The known variable parts could be subdivided into a number of groups that have certain amino acid replacements in common, whereas other substitutions were different even within one group.

This finding suggested an alternative hypothesis for the genetic determination of the V parts: every individual harbors a great number of genes that are arranged in a highly repetitive sequence. However, in every cell only one of these genes could be active. This gene could be connected in some way with the gene for the constant part of the polypeptide chain, permitting continuous mRNA formation. If we assume that this gene sequence was formed by repeated unequal crossing over followed in the course of millenia by random fixation of point mutations, the regularities described above are explained. Mutations that are common to several polypeptide chains have been fixed before the genes for these chains were duplicated; mutations that are unique for one chain are of relatively recent origin.

Both hypotheses – the somatic mutation and the selective gene activation hypothesis – require one element of nonorthodox genetics. For accumulation of so many somatic mutations, cell-specific enhancement of the mutation rate or selection of mutants is necessary. Selective connection of one of the many variable genes with the constant gene required an unusual linking mechanism. The linking could not occur at the protein level and not even at the level of mRNA, as mRNA molecules already comprise the total information. Therefore, it had to occur at the DNA level. On the basis of amino acid sequence data the controversy between the somatic mutation hypothesis and the multiple variable gene hypothesis could not be decided. A solution required direct investigation of the respective genes.

The solution was found when DNA techniques became available. As it turned out, both parties – those who favored the "selective gene" hypothesis and the supporters of the "somatic mutation" hypothesis – were partially right. Figure 7.60 shows the structures of the mouse λ and $\varkappa$ light chain genes and the heavy chain genes. Many of these studies have been performed on mouse genes; these structures are very similar in humans.

All three types of genetic regions are in principle organized identically: they consist of genes for the constant parts, the variable parts, and a "joint region" that connects both. These are referred to below as C, V, and J, respectively, always using roman uppercase letters for the gene products and italic uppercase letters for the gene segments, *C, V*, and *J*. As shown in Fig. 7.61, there are differences in detail. To start with the constant parts, each λ light chain gene sequence has two different *C* genes. Since the protein molecule has only one C region, the gene for this C region must be selected from four *C* genes (two from each of two homologous chromosomes). There are also two *J* genes, each belonging to one *C* gene, and one *V* area. Moreover, on the left side of the *V* segment there is a short segment termed *L* – the signal region where transcription begins. The light chain gene has

one C region but five different *J* regions. There are numerous *V* regions; their number is now estimated at 90–300. In the heavy chain gene sequence eight different *C* regions have been found. The gene product Cμ is present in IgM proteins; the Cδ, γ1, γ2b, and γ2a gene products are parts of IgD, IgG and IgA immunoglobulin. Moreover, there are four different *J* regions. In distinction to the L chains, the H chains contain an additional amino acid sequence that is coded by a *D* region, this *D* region being present in 12 copies. There are also 100–200 L_H–V_H segments.

During differentiation of antibody-forming cells *one L-V segment* is connected with one *J* segment (in H chain producing cell clones with one *D* and one *J* segment) and with one *C* segment. This leads to a great number of possible combinations; if the genome carries 2 Vλ, 3 Jλ, 300 V$\varkappa$, and 4 J$\varkappa$ segments, this gives 1206 different light chains (2×3 plus 300×4). Likewise, if there exist 200 V_H, 12 D, and 4 J_H segments, the maximum number of different V_H regions is $(200 \times 12 \times 4) = 9600$.

Moreover, it was shown by direct examination that the gene region is indeed organized before and during B cell differentiation. Joining of ends is somewhat

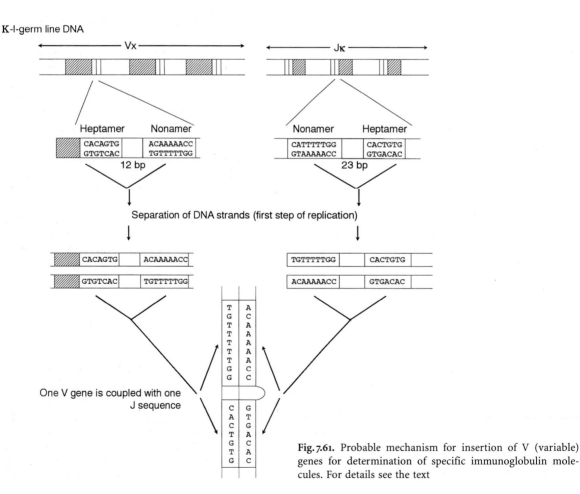

Fig. 7.61. Probable mechanism for insertion of V (variable) genes for determination of specific immunoglobulin molecules. For details see the text

imprecise by some bases; this is an additional source of diversity. In other cases, this lack of precision leads to an out-of-phase reading frame that makes transcription impossible: diversity is achieved at the expense of some waste. Sometimes, one or more nucleotides are inserted at the joints. The molecular mechanisms of this rejoining process have been elucidated: at the 3′ end of each V gene, a 7 bp sequence is found which is identical in all V genes. Twelve random base pairs are then followed by a 12-bp sequence which is also identical. The same heptamer and nonamer sequences are found at the 5′ ends of the J genes; however, here they are oriented in the opposite direction (and separated by 23 random base pairs; Fig. 7.61). Only one stretch with the short spacer (from a V gene) and one with the long spacer (from a J gene) can recombine in one cell. The enzymes necessary for such recombination only occur in B precursor cells; they are controlled by two recombination activating genes [189].

However, the above sources of variation are not quite sufficient for creating antibody diversity. And, indeed, the supporters of the somatic mutation hypothesis were also partially right: there is now conclusive evidence from comparison of homologous sequences of different origin that numerous somatic mutations occur. In almost all instances these are simple base substitutions leading to the replacement of only one amino acid. They have been observed not only in the V but also in the J and D segments.

In conclusion, the somatically generated diversity derives from four sources:

1. Combinatorial; V regions are present in multiple copies, only one of which is connected with the corresponding J, D, and C regions to form the functional gene.
2. and 3. Junctional; Additional diversity occurs in the joint regions and may be named junctional site diversity and junctional insertion diversity: Joining ends are imprecise (2), and nucleotides may be inserted (3).
4. Somatic mutations: Superimposed on these recombinational mechanisms of diversification is somatic mutation in the V, J and D regions.

In this way it is easy to understand how the many thousands of different antibodies can be formed by one individual. In humans the genes for the ϰ chains are located on chromosome 2 (2p12), for the λ chains on chromosome 22 (22q11.12), and chromosome 14 carries all H chain genes (14q32.33). Structural rearrangements such as translocations in which these genes are involved may lead to characteristic malignancies (Chap. 10). For example, in a special type of acute lymphoblastic leukemia H genes on chromosome 14 are brought together with the *myc* oncogene.

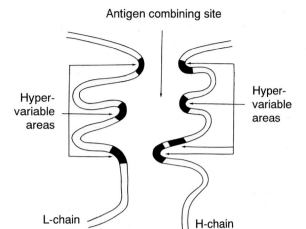

Antigen combining site

Fig. 7.62. The hypervariable areas of immunoglobulin molecules form pockets which act as antigen combining sites

V Parts and the Specificity of Antibodies. As noted, antibody specificity is determined by the variable parts, which differ in amino acid sequences. Even within the variable parts, variability is higher in some regions than in others.

Antigens are bound to the combining site of an antibody molecule. If antibody specificity is indeed determined by differences in amino acid sequences, the combining sites should be found in regions that are especially variable. A method for determining the spatial order of a molecule is X-ray crystallography. Such X-ray data are now available. Three hypervariable regions of the L chains and four such regions of H chains contribute to the combining site with the antigen (Fig. 7.62). The chains form a pocket which differs in shape depending on the amino acid sequences of the seven regions involved in their formation.

Since the early days of immunology, the relationship between antigen and antibody has often been compared to that between a key and a lock. The spatial model in Fig. 7.62 shows that this notion might be more than a metaphor.

7.4.2 T Cell Receptors and Their Genes
[60, 158, 159]

As mentioned above, T cells have an important function in the immune response (Fig. 7.55). T helper cells activate B lymphocytes which produce specific antibodies. T killer cells destroy cells infected by viruses, together with these viruses. In order to be activated, however, these T cells must receive a special signal. This signal comes from cells which have taken up the antigen and present it to these T cells. T cells re-

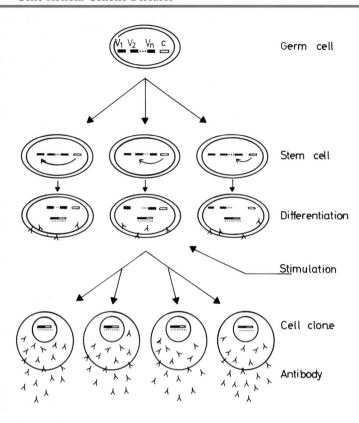

Germ cell

Stem cell

Differentiation

Stimulation

Cell clone

Antibody

Fig. 7.63. During embryonic development and differentiation, several different stem cells for antibody production are formed. In each cell, only one antibody can be formed because the gene for the constant part *(white bar)* has been connected with only one gene for the variable part *(black bar)*. Stimulation by a specific antigen *(arrow from the right)* leads to proliferation of the stem cell clone capable of formation of the appropriate antibody and to increased production of this antibody. (Adapted from Hilschmann et al. 1976 [116])

act to the antigen only if it is presented to them in combination with the MHC pattern of the presenting cells: for presentation to helper cells, HLA class II, and for presentation to T killer cells, HLA class I specifities. In order to accept these signals, T cells need receptors – the T cell receptors. Their function is similar to that of the immunoglobulins in B cells which also have receptor function. The difference is that immunoglobulins recognize an antigen even without MHC specificity and are released from the cell to move freely in the medium, whereas T cells remain bound to the cell surface.

The latter property has made their analysis much more difficult and delayed it for several years. No counterpart to myeloma proteins was available for ready examination. Finally, however, their structure was elucidated. They belong to the same gene superfamily as the immunoglobulins and the MHC proteins (Fig. 7.57); each receptor molecule consists of two polypeptide chains; in most instances there is one α and one β chain, but a small percentage of cells have a receptor consisting of one γ and one δ chain. During maturation of T cells a switch occurs between the production of $\alpha\beta$ and $\gamma\delta$ receptors [232], similar to the switch in hemoglobin synthesis between HbF ($\alpha\gamma$) and HbA ($\alpha\beta$; Sect. 7.3).

As with the immunoglobulins, the T cell receptor molecules consist of constant and variable portions.

They are also determined by a few C genes and many V genes and, in addition, J and D segments. Biological mechanisms for the final, expressed, gene combinations are also the same, except for one difference: there are no somatic mutations in T cell development. The genes for the β and γ chains are located on chromosome 7 (7q35; 7p15-p14) and those for the α and δ chains on chromosome 14 (14q11.2). A good description of the analysis of this system with emphasis on molecular methods is presented in [252].

7.4.3 Genetic Diseases Due to Defects of Genes in the Defense System [68, 215]

The complex and multistep mechanism of the defense system is disturbed if single components are destroyed or impaired by mutations. Some of the genetic blocks are indicated in Fig. 7.54, but there are many more. They may involve B or T cell maturation or both; defects of single immunoglobulin classes; granulocytes [6]; the T cell receptor [2]; or the complement system [267]. In addition to genetic blocks, anomalies in the regulation of the immune respone may lead to disease. For example, if immune tolerance is incomplete, the immune system fails to distinguish between antigens of its own body and foreign ones, and antibodies against components of the individual itself are formed. These autoimmune diseases are often associated with components of the MHC system, as shown by the well-known associations between HLA types and diseases (Sect. 6.2.3). Overactivity in synthesis and action of the IgE immunoglobulins leads to increased reaction to certain foreign proteins and to clinical signs of allergy, especially those of atopic disease: atopic dermatitis, bronchial asthma, and hay fever (Sect. 6.1.2.7). Distribution of this abnormal reaction in the population shows a strong genetic susceptibility, as epitomized by high MZ twin concordance and familial aggregation. However, there is no simple mode of inheritance [167, 248]. The normal function of IgE is defense against parasitic diseases such as intestinal worms. Some data indicate that overreaction of this system, in addition to causing allergies, may afford relative protection against such parasites (Sect. 12.2.1) [107].

7.5 Pharmacogenetics and Ecogenetics

7.5.1 Pharmacogenetics

The development of human biochemical genetics, with its insights into genetically determined enzyme deficiencies, gave rise to the field of pharmacogenetics. Garrod, the founder of human biochemical genetics [96] (Sect. 1.5), and Haldane [109], the great British geneticist, had both suggested that biochemical individuality might explain unusual reactions to drugs and food. In the 1950s several abnormal and adverse drug reactions were shown to be caused by genetically determined variation of enzymes. G6PD deficiency (Sect. 7.2.2.2) explained hemolytic anemias caused by fava bean ingestion and by a variety of drugs in some individuals. Variation in the enzyme pseudocholinesterase (butyrylcholinesterase) [77, 132, 133, 150] was found to underlie prolonged apnea caused by suxamethonium – a drug widely used to relax muscles during surgery. Genetic differences in acetyltransferase activity explained marked interindividual variation in blood levels of isoniazid (INH), a widely used drug in the treatment of tuberculosis [78].

It was therefore suggested by one of the authors (A.G.M) in 1957 that many abnormal responses to drugs may be caused by genetically determined variation, such as enzyme deficiencies [167]. The other author (F.V.) first introduced the term pharmacogenetics [246].

G6PD System (305 900). The G6PD system is discussed in Sect. 7.2. X-linkage of G6PD explains the preponderance of males with hemolytic drug reactions due to G6PD deficiency. G6PD levels intermediate between those of affected males and normals are often seen in heterozygous females but some obligatory heterozygous females have enzyme levels in the deficient range, and many have enzyme levels within the normal range. Heterozygote females with G6PD deficiency possess two red cell populations: normal and mutant cells. The ratio of normal to deficient red cells is generally 1:1 but may range from 1% normal and 99% mutant cells to 1% mutant and 99% normal cells in a few heterozygotes [25]. The frequency of females with G6PD-dependent drug reactions depends upon their population frequency (q^2: homozygotes; $2pq$: heterozygotes, where q is the frequency of affected males [169]), and on the degree of X inactivation as expressed by the ratio of normal and G6PD-deficient cells. Clinically affected females are the few homozygotes and usually only those heterozygotes who have a preponderance of mutant cells with low enzyme levels. Various drugs also differ in their potential for blood destruction.

Several common G6PD variants have been associated with hemolytic reaction. In addition to drugs, hemolysis may be associated with bacterial and viral infection or may be encountered as neonatal jaundice, where the immature liver is unable to clear bilirubin – a metabolic product of hemoglobin released by hemolysis.

Hemolysis is particularly severe in the Mediterranean variant since in this condition G6PD deficiency is associated with both decreased specific activity and molecular instability of the enzyme. In the common type of G6PD deficiency seen in persons of African origin (A⁻), red cells younger than 60 days (red cell life is 120 days) have sufficient amounts of enzymes, and the molecular instability characteristic of the defect affects only the older red cells so that hemolysis is self-limited (Fig. 7.4). G6PD levels in red cells following hemolysis are not as low because only the older cells with deficient enzymes are destroyed. Fatal hemolytic episodes are seldom observed in this milder type of G6PD deficiency. With more severe hemolysis, as occurs in the Mediterranean variety of G6PD deficiency, a fatal outcome may occur. The number of drugs that cause hemolysis is also larger in Mediterranean G6PD deficiency than in the African type. No detailed data on the spectrum of harmful drugs are available yet for most other G6PD variants.

Pseudocholinesterase (Butyrylcholinesterase) Variation. The drug suxamethonium, or succinyldicholine, is commonly used as a muscular relaxant to facilitate surgical operations. Hydrolysis of the drug by the enzyme pseudocholinesterase occurs, and its normal action is brief. In some persons, the enzyme has poor affinity for the drug, and such patients develop prolonged apnea due to depression of respiratory muscles. Under such circumstances many hours of artificial respiration may be necessary. The cause of this drug reaction are various mutations in the homozygous or compound heterozygous state affecting the active site or destabilizing the enzyme, which can no longer effectively hydrolyze its substrate.

Cholinesterase is a tetramer consisting of four identical monomeric subunits. The gene for the enzyme is 80 kb long and is located on chromosome 3q26.1-26.2. It is designated as BCHE (Table 7.20). Several mutations of this gene have been detected and have been characterized at the molecular level. The most common variant is the atypical allele (A) due to a missense mutation and is observed in 3%–4% of the white population in the heterozygote state. This variant is rare in populations of Oriental or African origin. Several mutations lead to complete absence of enzyme activity (silent alleles). Two of these have been characterized as frameshift mutations (Table 7.20). Other variants are also listed. The K variant that reduces cholinesterase activity by one-third is a common polymorphism (heterozygote frequency: 18%), and the rare J mutation is only found on chromosomes carrying the K variant. Similarly, 90% of the common atypical (A) alleles are found in linkage disequilibrium with the K variant.

Cocaine is a substrate that is only slowly hydrolyzed by cholinesterase. Genetic variants prolong the short (1-h) half-life of cocaine but are seldom of clinical significance. However, under conditions of repeated intake, particularly of forms that are absorbed rapidly, such as crack, toxic levels of cocaine may accumulate in blood and tissues [132]. Adverse events, such as cardiac death due to cocaine-mediated vasocon-striction (a known effect of cocaine), might be expected more commonly among homozygotes and compound heterozygotes for the cholinesterase variants, but no data to document this contention are available.

Cholinesterase enzyme status is usually assessed using benzoylcholine as a substrate and inhibiting the activity of cholinesterase by inhibitors such as dibucaine and fluoride. The reduced inhibition of enzyme activity identifies the atypical dibucaine-resistant and the fluoride-resistant enzyme and allows genetic characterization of heterozygotes and homozygotes for these variants. Inhibition status of the enzyme is much better correlated with the clinical phenotype than the level of enzyme activity alone (Fig. 7.64). Heterozygotes for the atypical allele very rarely develop the characteristic prolonged apnea, while compound homozygotes and homozygotes do. In the future PCR-based molecular techniques should make it possible to detect the characteristic mutations without carrying out the enzyme assay and biochemical inhibition tests.

Acetyltransferase Variation [77, 200]. Several drugs are acetylated by the liver enzyme *N*-acetyltransferase. These drugs include isoniazid, hydralazine, procainamide, phenelzine, dapsone, salicylazosulfapyridine, sulfamethazine, and nitrazepam. Human populations can be subdivided into distinct classes (rapid or slow inactivation) based on whether they acetylate a test drug such as isoniazid or sulfamethazine, administered in vivo.

Family studies using blood levels of isoniazid after administration of a standard dose as the test criterion show that slow inactivators are homozygotes for defective acetylation. Heterozygotes usually cannot be

Table 7.20. Characteristics of cholinesterase variants (from Evans 1993 [77]; Kalow and Grant 1995 [132])

Common name	Common abbreviation	Phenotype	Codon alteration	Amino acid alteration	Formal name of genotype	Approximate frequency of homozygotes[b]
Usual	U	Normal	None	None	BCHE	96%
Atypical[a]	A	Dibucaine resistant	70 GAT → GGT	Asp → Gly	BCHE[a] 70 G	1/3500
Silent 1	S₁	No activity	117 GGG → GGAG	Gly → frameshift	BCHE FS 117	⎫
Silent 2	S₂	No activity	6ATT → TT	Ile → frameshift	BCHE FS6	⎬ ~1/100000[c]
Silent 3	S₃	No activity	500 TAT → TAA	Tyr → stop	BCHE[a] 500	⎭
Fluoride 1	F₁	Fluoride resistant	243 ACG → ATG	Thr → Met	BCHE[a] 243 M	~1/150000
Fluoride 2	F₂	Fluoride resistant	390 GGT → GTT	Gly → Val	BCHE[a] 390 V	?
K variant[a]	K	66% activity	539 GCA → ACA	Ala → Thr	BCHE[a] 539 T	1%
J variant[a]	J	33% activity	497 GAA → GTA	Glu → Val	BCHE[a] 497 V	~1/150000

[a] The characteristics of the K variant are also present in all J variants examined and in 90% of atypical variants.

[b] Among whites.

[c] Homozygote frequency refers to *all* silent alleles.

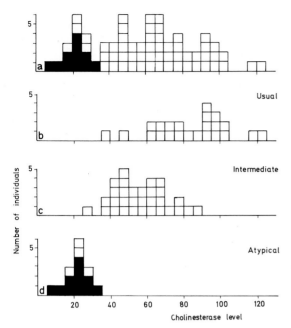

Fig. 7.64 a–d. Distributions of levels of cholinesterase activity in sera from 11 individuals found to be excessively sensitive to suxamethonium and 58 of their relatives. Each *square* represents one individual; *black,* the suxamethonium-sensitive propositi. **a** Distribution of levels of serum cholinesterase activity. The activity levels were determined manometrically with acetylcholine as substrate. **b** Distribution of activity levels of those classified as having the usual phenotype. **c** Distribution of activity levels of those classified as having the intermediate phenotype. **d** Distribution of activity levels in those classified as having the atypical phenotype. Note: distribution **a** represents the sum of distributions **b, c,** and **d.** (From Harris et al. 1960)

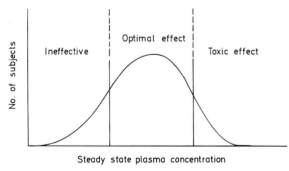

Fig. 7.65. Steady-state plasma concentration of a drug and biological effect

differentiated from wild-type (or normal) homozygotes. In recent years the molecular basis of the acetylation polymorphism has been elucidated (Table 7.21). It has been known for some time that acetylation of some drugs (such as para-aminosalicylic acid and PABA) is monomorphic and does not exhibit the rapid/slow polymorphism demonstrated for isoniazid and some other drugs. Two different acetylation genes were therefore postulated. More recently, two highly homologous (87%) N-acetyltransferase genes (NAT1 and NAT2) were identified on chromosome 8pter. In addition, a third pseudogene (NATP) in this gene cluster with a large number of presumably inactivating mutations was found. It was shown that the NAT1 gene carries out monomorphic acetylation, while variants of the NAT2 gene account for slow inactivation of isoniazid and other polymorphically inactivated drugs. Missense mutations of NAT2 (Table 7.21) render the gene product, i.e., liver N-acetyltransferase less stable and therefore less active. Three principal variants were identified. A markedly lower frequency of slow inactivators among Japanese is caused by the complete absence of the M1 variant in this population. The three variants M_1, M_2 and M_3 of Table 7.21 account for 95% or more of slow inactivators in all populations. Detailed tabulations of the world frequencies are available [77]. Rare variants have been discovered including one mutation that appears to be limited to Africans [154]. The natural substrates of both the NAT_1 and NAT_2 genes remain unknown.

These developments in molecular genetics allow detection of slow acetylation status by PCR techniques from small amounts of blood with clear differentiation of the variant homozygotes, compound heterozygotes (i.e., M_1/M_2) and heterozygotes. Since these methods do not require the cumbersome administration of a test drug followed by collection of blood or urine specimens, more extensive work with the acetylation polymorphisms is now possible.

Clinical consequences of the acetylation polymorphism relate to a higher frequency of vitamin B_6 responsive polyneuropathy among slow inactivators on treatment with isoniazid. A higher frequency of lupuslike side effects are seen when slow inactivators are given hydralazine or procainamide. More hematological side effects in slow inactivators are observed with dapsone and with salicylazosulfapyridine. Rapid inactivators may require higher doses of the drugs to reach satisfactory therapeutic effects.

Many studies have been carried out on the possible association of polymorphic acetylation with a variety of diseases [77]. Slow inactivators appear to be at a 30% higher risk of developing bladder cancer, particularly those living in more industrialized environments (See below, Sect. 7.5.2). Other associations with slow inactivation status such as with non-insulin-dependent, or type 2, diabetes and Gilbert's disease remain to be explained.

Debrisoquine-Sparteine (CYP2D6) Polymorphism [56, 75, 132, 165]. A common polymorphism affecting the

Table 7.21. Polymorphic *N*-acetyltransferase variation (adapted from Evans 1993 [77]; Chinese data from Lee et al. 1994 [148]; African American data from Bell et al. 1993; [19])

Allele	Critical nucleotide change[a]	Amino acid change	Allele frequencies of slow inactivation			
			Whites (n = 372)	African-Americans (n = 238)	Japanese (n = 86)	Chinese (n = 187)
wild type (wt)	–	–	0.25	0.36	0.69	0.51
M1	341 T → C 481 C → T	114 Ile → Thr Nil	0.45	0.30	0	0.075
M2	590 G → A	197 Arg → Gln	0.28	0.22	0.24	0.32
M3	857 G → A	286 Gly → Glu	0.02	0.02	0.07	0.1

[a] Refers to coding exon.

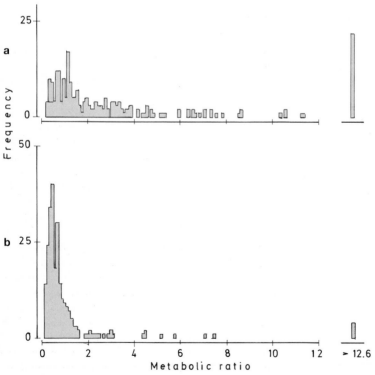

Fig. 7.66. a Oxidation of debrisoquine. **b** Distribution of metabolic ratios $\left(\text{urinary} \dfrac{\text{debrisoquine}}{\text{4-hydroxydebrisoquine}} \right)$ in smoking controls (**a**; above) and lung cancer patients (**b**; below). (Ayesh et al., Nature 312, 169, 1984, modified)

Table 7.22. Drug responses in poor metabolizers with CYP2D6 polymorphisms (from Kalow and Grant 1995 [132]; Eichelbaum and Gross 1992 [72])

Drug reactions	
Captopril	Agranulocytosis
Flecainide	High blood levels with renal impairment – pro arrhythmic effects
Phenacetin	Methemoglobinemia
Phenformin	Lactic acidosis
Tricyclic antidepressants	Overdose
Nortryptyline, desipramine	Sedation and tremors
Beta blockers	Defective metabolism causing a higher frequency of side effects[a]
Propanol, timolol, alprenol	
Lack of efficacy	
Codeine	Reduced analgesia
Encainide	Lack of arrhythmic effect

Many additional drugs are metabolized by CYP2D6 but did not reach the market because of drug reactions, presumably caused by this polymorphism.

[a] Not definitely confirmed.

Table 7.23. Frequency of poor metabolizers in different populations studied by probe drugs (from Kalow and Grant 1995 [132])

Ethnic group	Poor metabolizers
Whites (Europe, North America)	3–10
Chinese	0.7–1.1
Japanese	0–2.3
West Africans	0–8
African-Americans	1.9

Table 7.24. Frequencies of CYP2D6 alleles in different populations (from Masimiriembwa et al. 1993 [161])

	Whites	African-Americans	Zimbabweans	Chinese
D6wt	0.70	0.86	0.94	0.93
D6A	0.02	0.024	0	0
D6B	0.23	0.085	0.018	0.008
D6D	0.05	0.06	0.039	0.056
Total	1.00	1.00	1.00	1.00

wt, wild type

Table 7.25. Number (and frequency) of mutations of CYP2D6 detected by PCR amplification in extensive (EM) and poor metabolizers (PM) of debrisoquine or sparteine (whites; from Meyer et al. 1992 [165])

Genotype by PCR test	EM (n = 98)	PM (n = 46)
wt/wt	52 (0.53)	1 (0.02)
wt/B	43 (0.44)	3 (0.06)
wt/A	3 (0.03)	–
B/B	–	36 (0.78)
A/A	–	1 (0.02)
A/B	–	3 (0.07)
D/D (no amplification)	–	2 (0.04)

wt, wild type

P450 enzyme CYP2D6 causes defective oxidation of many different drugs which require oxidative metabolism (Tables 7.22–7.25). This polymorphism was discovered independently in studies with the antihypertensive drug debrisoquine and the oxytocic and antiarrhythmic agent sparteine. Severe, prolonged hypotension occured occasionally on administration of debrisoquine, and this has been shown to be caused by failure to metabolize the drug by 4-hydroxylation. When sparteine was studied as an antiarrhythmic agent, most patients failed to reach sufficiently high plasma concentrations to attain the desired therapeutic response [67]. Furthermore, severe obstetric complications (such as uterine tetany, abruptio placentae, and unduly rapid labor) occur in 7% of women receiving sparteine to induce labor at term [188] – a complication rate similar to the frequency of poor metabolizers for CYP2D6 (Tables 7.23–7.25). Extensive

studies have shown that the CYP2D6 polymorphism participates in oxidative metabolism of many different drugs, including antiarrhythmic agents, betablockers, neuroleptics, and tricyclic antidepressants. Some unexpected drug responses that can definitely be ascribed to this polymorphism are listed in Table 7.22. The report that poor metabolizers in a large study were found to be more anxiety prone and less successfully socialized than extensive metabolizers is intriguing [157]. The full extent of the potential clinical significance of the CYP2D6 polymorphism for drug therapy remains under discussion [72, 132].

While drug toxicity may often be due to failure to metabolize a given drug, some therapeutic agents require CYP2D6 to activate a precursor to become the actual therapeutic agent. If this reaction is defective, an expected drug effect such as analgesia from codeine or suppression of cardiac arrhythmia by encainide is no longer observed (see Table 7.22), and the drug lacks efficacy with standard dosage.

Family studies have shown that the CYP2D6 polymorphism is a common autosomal recessive trait. With a homozygote frequency of 5%, 35% of the population are heterozygotes. The determination of the CYP2D6 phenotype required in vivo administration

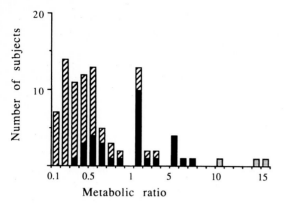

Fig. 7.67. Relationship between debrisoquine metabolic ratio and genotype. Debrisoquine metabolic ratios and CYP2D6 genotypes were analyzed in 93 unrelated volunteers of European origin. ▨, Homozygous extensive metabolizers; ■, heterozygotes; ▦, poor metabolizers. (From Cholerton et al. 1992 [56])

of a small dose of debrisoquine as a drug probe followed by measurement of the ratio of metabolized to nonmetabolized drug in the urine after some hours. More recently, the cough suppressant dextromorphan (an over-the-counter medication) has replaced debrisoquine as the probe drug. The frequency of poor metabolizers using in vivo testing in different populations is shown in Table 7.23. A higher gene frequency of this polymorphism among populations of European origin has been noted.

After purification of the D6 enzyme from livers of organ donors, the CYP2D6 gene was cloned and expressed in mammalian cell culture. It was localized to chromosome 22q11.2-qter. The CYP2D6 gene cluster consists of four genes. Three are nonexpressed pseudogenes, with D6 being the functional gene. Using a combination of Southern blotting and PCR amplification, at least eight alleles have been identified at the D6 locus. In addition to the common wild-type allele (D6wt), a less frequent wild-type allele (allele frequency 0.03) with normal enzyme activity has been detected. Several alleles are associated with absent enzyme function, leading to the poor metabolizer status in the homozygous or compound heterozygous state. The mutations compromising enzyme function (A, B, D) are caused by different deletions and splice site mutations. The most common mutations, D6B, occurs in 23% of Europeans (allele frequency) but is rare in other populations (Table 7.24). Table 7.25 provides comparative data from a study of in vivo phenotyping and in vitro genotyping. Some 8% of poor metabolizers have as yet unidentified mutations. Figure 7.67 shows the relationship between debrisoquine *phenotype* and CYP2D6 *genotype,* indicating considerable overlap between wild-type homozygotes and heterozygotes but demon-

strates that the most active extensive metabolizers are wild-type homozygotes. DNA genotyping for this polymorphism using white blood cells should facilitate work in this area since it can bypass the cumbersome and somewhat invasive phenotyping test that requires drug administration and urine sampling.

Mephenytoin Polymorphism [257]. It has been shown that oxidation of the anticonvulsant mephenytoin is under control of a different gene and is not related to the debrisoquine type of polymorphism. The relatively high frequency of side effects with this drug observed in the past is presumably accounted for by the failure of mephenytoin oxidation in a small percentage (2%–5%) of the population. Phenytoin (dilantin) may be oxidized by the same polymorphic enzyme.

Other Monogenic Pharmacogenetic Traits. A variety of other traits that are inherited as simple Mendelian traits play a role in pharmacogenetics. These are listed in Table 7.26.

Multifactorial Pharmacogenetics. A series of twin studies on the half-life of various drugs have emphasized the importance of genetic factors in drug metabolism. Whenever a drug has been given to identical and nonidentical twins [280], much more similarity in drug metabolism has been noted among the MZ twins. Heritability measurements based on such data have shown that the contribution of heredity to the total variation in drug half-life is high and sometimes reaches 99% (Table 7.27).

When the standard dose of a drug is given to many members of a normal population, considerable variability in blood levels is noted in different persons. While a variety of factors affect such blood levels, differences in drug metabolism as assessed by half-life are usually major determinants. The half-life (or steady level) of a drug is a rather constant parameter in an individual and, as suggested by the twin data, is influenced largely by genetic factors. The underlying biochemical basis of the details of drug metabolism is as yet unknown. Variation in half-life or response for most drugs can be plotted as a bell-shaped Gaussian distribution curve (Fig. 7.65). A certain number of persons at each end of the unimodal distribution curve have either too large or an insufficient amount of drug following administration of an average dose. The result is either toxicity due to excessive blood levels or failure of the drug to exert its effects because of very low blood levels. The demonstration that genetic factors play a role in the metabolism of most drugs transformed pharmacogenetics from a field dealing with a few unusual drug reactions to a discipline of central importance for pharmacology and therapeutics [175].

Table 7.26. Monogenic pharmacogenetic traits

Enzymatic or metabolic abnormalities	Result and/or clinical abnormalities
A. Well-established traits (see text)	
a) *Common traits*	
Some G6PD variants	Hemolysis
N-Acetyltransferase polymorphisms	Reduced acetylation of several drugs (see text)
Poor oxidation (debrisoquine/ sparteine)	Untoward reactions to many drugs (see Tables 7.22–7.25)
b) *Rare traits*	
Pseudocholinesterase variants	Prolonged suxamethonium-induced apnea
Abnormal calcium metabolism	Malignant hyperthermia following inhalation anesthesia
Some unstable hemoglobins	Hemolysis
Various porphyrias	Several drugs precipitate symptoms of disease
Methemoglobin reductase deficiency	Cyanosis with some oxidizing drugs
B. Traits whose clinical significance is less certain	
Paraoxonase polymorphism	Low-activity carriers (~50%) more susceptible to parathione poisoning
Poor mephenytoin oxidation	Severe side effects from mephenytoin
Thiopurine methyl transferase polymorphism (cytosolic)	Lack of effective action of thiopurine drugs (e.g., mercaptopurine)
Catechol-O-methyl transferase polymorphism	Lack of effective action of L-dopa and α-methyl dopa
Epoxide hydrolase deficiency	Phenytoin hepatotoxicity

Pharmacogenetic Variation at the Level of the Target Organ. Genetic variation at the level of the target organ has been discussed for G6PD deficiency of the red cell. The example of alcohol action on the brain is considered in Sect. 15.2.3.5. The side effects of psychotropic drugs on the brain may also have a genetic basis [200]. For example, persons receiving phenothiazine occasionally develop parkinsonism; those having a relative with parkinsonism run a threefold risk [200]. "Tardive dyskinesia" manifesting as abnormal, involuntary movements is not very rare among patients treated with psychopharmacological agents; there is impressive familial aggregation. Since such therapy influences the action of the neurotransmitter dopamine (Sect. 15.2.3.6), it is interesting that an increase in number of dopamine receptors in the nucleus caudatus of rats has been described after administration of neuroleptics; this in-

crease shows marked variation between strains. Schizophrenia – like symptoms can be induced in humans by drugs such as LSD and metaphetamine and even by alcohol abuse (alcohol hallucinosis). The incidence of schizophrenia has been found to be significantly increased among first-degree relatives of patients suffering from these complications.

A rare but dangerous and frequently fatal complication of general anesthesia is malignant hyperthermia, often associated with increased muscular rigidity [132]. An incompletely autosomal-dominant mode of inheritance is often observed. Many patients have minor muscular complaints, such as ptosis, strabismus, cramping, and recurrent dislocations. Abnormal electromyograms and slight histological signs of myopathy have also been described. Slightly increased creatine phosphokinase values are common in these patients.

A test to assess the contractile properties of biopsied muscle on exposure to halothane or caffeine (or both) is often used to detect susceptibility to malignant hyperthermia [132]. Equivocal or false-positive test results, however, are common. A similar condition in swine has been described and is caused by a specific missense mutation of a calcium release channel known as the ryanodine receptor [94]. The gene for malignant hyperthermia in some but not all human families has been localized to the site of the ryanodine receptor locus on chromosome 19 by linkage studies [132]. The specific mutation observed in the porcine disease has been found in only one human family. Malignant hyperthermia, as many other genetic diseases, exhibits both allelic and presumably interallelic heterogeneity.

7.5.2 Ecogenetics [8, 174, 176, 192]

The concept of ecogenetics – first suggested by Brewer in 1971 [36] – evolved historically from pharmacogenetics. Drugs are only a small fraction of environmental chemical agents to which humans are exposed. Various other potentially toxic agents exist in the environment and may damage a fraction of the population who are genetically predisposed. Ecogenetics extends the central concept of genetically determined variable drug responses to other environmental agents. Since twin studies suggest that the metabolism of most drugs is subject to genetic influences, it can be inferred that genetic control of biotransformation and receptor action applies to most chemical agents. The field of human ecogenetics deals with variable responses of humans to environmental agents and attempts to explain why only some of those exposed are injured by harmful agents, and

Table 7.27. Studies in twins on drug elimination rate or under steady-state conditions (from Propping 1978 [199])

Drug	Authors, no. of twin pairs	Measured parameter	Range	r_{MZ}	r_{DZ}	h_2^2
Antipyrine 18 mg/kg p.o. (single dose)	Vesell and Page (1968) 9 MZ, 9 DZ	Plasma half-life (h)	5.1–16.7	0.93	−0.03	0.99
Phenylbutazone 6 mg/kg p.o. (single dose)	Vesell and Page (1968) 7 MZ, 7 DZ	Plasma half-life (days)	1.2–7.3	0.98	0.45	0.99
Dicumarol 4 mg/kg p.o. (single dose)	Vesell and Page (1968) 7 MZ, 7 DZ	Plasma half-life (h)	7.0–74.0	0.99	0.80	0.98
Halothane 3.4 mg i.v. (single dose)	Cascorbi et al. (1971) 5 MZ, 5 DZ	Urinary excretion of sodium trifluoroacetate in 24 h (% of injected dose)	2.7–11.4	0.71	0.54	0.63
Ethanol 0.5 g/kg p.o. (single dose)	Lüth (1939) 10 MZ, 10 DZ	β_{60} (mg/ml · h) EDR (mg/kg · h)	0.051–0.141 50.00–109.63	0.64 0.77	0.16 0.45	0.63 0.67
1 ml/kg p.o. (single dose)	Vesell et al. (1971) 7 MZ, 7 DZ	β_{60} (mg/ml · h)	0.11–0.24	0.96	0.38	0.98
1.2 ml/kg p.o. (single dose)	Kopun and Propping (1977) 19 MZ, 21 DZ	Absorption rate (mg/ml · 30 min)	0.20–1.12	0.56	0.27	0.57
		β_{60} (mg/ml · h) EDR (mg/kg · h)	0.073–0.255 57.6–147.6	0.71 0.76	0.33 0.28	0.46 0.41
Diphenyl-hydantoin 100 mg i.v. (single dose)	Andreasen et al. (1973) 7 MZ, 7 DZ	Serum half-life (h)	7.7–25.5	0.92	0.14	0.85
Lithium 300 mg/12 h p.o. (for 7 days)	Dorus et al. (1975) 5 MZ, 5 DZ	Plasma concentration (mEq/l)	0.16–0.38	0.94	0.61	0.86
		Red blood cell concentration (mEq/l)	0.050–0.102	0.98	0.71	0.83
		RBC/plasma concentration (each after 3 days of treatment)	0.18–0.56	0.84	0.62	0.92
Amobarbital 125 mg i.v. (single dose)	Endrenyi et al. (1976) 7 MZ, 7 DZ	Plasma clearance rate (ml/min)	16.0–67.2	0.87	0.55	0.83
		Weight-adjusted clearance (1/kg · h)	1.76–6.16	0.92	0.60	0.80
		Elimination rate constant (h⁻¹)	2.09–8.17	0.93	0.03	0.91
Nortriptyline 0.6 mg/kg·d p.o. (for 8 days)	Alexanderson et al. (1969) 19 MZ, 20 DZ	Steady-state plasma level (ng/ml)	8–78	−[a]	−[a]	−[a]
Sodium salicylate 40 mg/kg i.v. (single dose)	Furst et al. (1977) 7 MZ, 7 DZ	Slope of serum salicylate decay (mg/dl · h)	0.64–1.02	0.64	0.32	0.86
Aspirin 65 mg/kg·d p.o. (for 3 days)	See Propping	Plateau serum salicylic acid (mg/dl)	11.9–36.4	0.90	0.33	0.98
		Salicylurate excretion rate (plateau) (mg/kg · h)	0.84–1.91	0.94	0.76	0.89

β_{60}, Disappearance rate from blood; EDR, ethanol degradation rate; r_{MZ}, r_{DZ}, intraclass correlation coefficient: in MZ and DZ twins, respectively; h_2^2 (heritability) $= \dfrac{V_W(DZ) - V_W(MZ)}{V_W(DZ)}$; V_W, variance within twin pairs.

[a] Published data do not allow calculation, but MZ twins are much more similar to one another than DZ.

how individuals differ in their adaptation to the environment. The working hypothesis of ecogenetics is the concept that an individual's genetically determined biochemical makeup often determines the response to an environmental agent, particularly in situations in which it is already known that most human beings react differently to the particular agent. Similar to the findings in pharmacogenetics, some ecogenetic reactions are due to the presence of rare mutant genes and cause a grossly abnormal response or idiosyncratic reaction. In other instances the variable response is mediated by a polymorphic system, and a significant proportion (2%–50%) of the population react differently. Most frequently, ecogenetic responses involve several genes and lead to unusual responses in a few individuals whose genetic makeup causes them to fall toward one end of the unimodal distribution curve.

Carcinogens. Recent data (see Sect. 11.2) suggest that most mutagenic substances are also carcinogenic. It is likely that pharmacogenetic principles apply to potentially carcinogenic chemicals. Genetic concepts provide a partial explanation why most persons under equal exposure to a chemical do not develop cancer. Only individuals with variant metabolisms such as slow inactivators or those who transform a substance into a more powerful carcinogen are likely to respond with neoplasia. Genetic variation in repair enzymes (see Chap. 10) or in "immune surveillance" affecting mutant cells may be other sources of cancer. Persons who already carry a germline mutation for a cancer susceptibility gene may be at particularly high risk to develop clinical cancer when exposed to carcinogenic agents which may cause a higher frequency of somatic mutations in the allelic homologous partner of this cancer gene (Chap. 10).

The enzyme system of arylhydrocarbon hydroxylase is involved in the activation of polycyclic hydrocarbons into more potent carcinogenic agents. Arylhydrocarbon hydroxylase levels in humans are under genetic control, as assessed by twin and family studies. The exact mode of inheritance is not clear, but monogenic inheritance has been claimed for humans [138, 139] and for an analogous enzyme system in mice [181]. It is more likely, however, that inheritance is polygenic [181]. In any case, it is conceivable that persons with high arylhydrocarbon hydroxylase activities are at higher risk of cancer induced by polycyclic hydrocarbons, such as lung cancer associated with cigarette smoking [73, 138, 139].

Several other polymorphisms potentially involved in the metabolism of carcinogenic substances have been investigated, such as susceptibility factors for lung cancer associated with smoking. At least ten studies have been carried out with the CYP2D6 (debrisoquine hydroxylase) variant [75, 77, 157, 268]. The data suggest a lower frequency of bronchial carcinoma among poor metabolizers of CYP2D6, who normally comprise about 5% of white populations (Fig. 7.66). The results have not always been statistically significant, and several studies (including one that defined genotypes by molecular techniques [268]) showed no differences in lung cancer frequencies between poor and extensive metabolizers.

Glutathione-S-transferases (GST) are enzymes that conjugate carcinogens such as epoxide and hydroperoxides with glutathione. Several varieties exist. A common polymorphism of the μ type of GST is characterized by a complete deletion of the gene and exists in the homozygote state in about 50% of the population (allele frequency 0.7). The deleted allele can be detected by enzyme measurements with transstilbene oxide as a substrate or by various molecular techniques. Several studies [112, 271] have shown that homozygous individuals lacking the glutathione conjugating enzyme have a higher frequency of smoking-related lung cancer. Other investigators have failed to show statistically significant differences between lung cancer patients and controls. However, in vitro studies show that induction of sister chromatid exchange in lymphocytes is greater among homozygotes for the deleted GST μ allele, providing biological plausibility for the protective effect of the GST μ enzyme against the carcinogenic effects of cigarette smoking [245, 261].

Much more work on the metabolism of carcinogenic substances and their enzymes is necessary to demonstrate the likely role of human variation in environmental carcinogenesis. Variations in repair mechanisms may also play a role as suggested by the higher frequency of cancer in patients affected with genetic lesions of mutational repair (Fanconi anemia, Bloom syndrome, ataxia-telangiectasia, xeroderma pigmentosum; Chap. 10). It is particularly noteworthy that the not infrequent heterozygotes for three of these conditions have a higher frequency of malignancy; however, in heterozygotes for xeroderma pigmentosum malignancy appears only after extensive exposure to sunlight. Since many human cancers are thought to be related to environmental agents to which large portions of the populations are exposed, a genetic approach is likely to provide answers why only some persons develop cancer with similar environmental exposure.

Bladder cancer is more common among industrial workers exposed to amines such as benzidine and naphthylamines. Since these chemicals are polymorphically acetylated by acetyltransferase (see above), it has been suggested that slow inactivators of these substances and others exposed to urban and industrial environments are at higher risk. At least 17 studies (summarized in [77]) have been carried out and indicate a 30% higher risk for bladder cancer among slow inactivators. Heavily exposed industrial workers appear to be at higher risk. The lower frequency of bladder cancer in Japan (6.3/100 000) than in the United States (26/100 000) is consistent with the much lower homozygote frequency of slow inactivators in Japan (11%) than in the United States (~50%).

α₁-Antitrypsin Deficiency (107 400). $α_1$-Antitrypsin deficiency is associated with the Z allele in the homozygous state and predisposes to early chronic obstructive pulmonary disease. Heterozygote smokers for this defect have somewhat impaired pulmonary function. It is possible that with smoking and polluted environments there is an increased frequency of chronic obstructive pulmonary disease among heterozygotes for this defect (see Sect. 6.2.4).

Paraoxonase [184]. Parathione is a widely used insecticide. The compound is metabolized to paraoxone by liver microsomes. Paraoxone is further broken down by the serum enzyme paraoxonase. There is a definite bimodal distribution of widely varying paraoxonase levels in the European population, with 50% having low levels. Family studies have shown that those with low enzyme levels are homozygotes for a low-activity allele (gene frequency 0.7). The low-activity allele is caused by a glutamine to arginine substitution in the enzyme [77]. No epidemiological investigations are yet available regarding the significance of this polymorphism for those exposed to parathione. It can be expected that homozygotes would be at higher risk for poisoning with relatively low exposures. With more massive poisoning the paraoxonase genotype presumably makes no difference for the development of symptoms.

Food. The best example of genetic difference in response to food is represented by adult hypolactasia. All human infants possess the intestinal enzyme lactase necessary for lactose absorption. In most human populations the intestinal lactase disappears after weaning so that most human adults are lactose intolerant. A mutation allows persistence of lactose absorption. This mutation presumably has a selective advantage in agricultural societies where cow milk is available for nutrition. In central and northern European populations, most persons possess this mutation in either single or double dose. Lactose intolerance is caused by the lack of this gene and is inherited as an autosomal-recessive trait. The mutation for persistence of lactose absorption or lactose tolerance is also common in some nomadic pastoralists in Africa and Arabia. Gene frequencies for persistence of lactase activity in some populations are discussed in Sect. 14.3.1. Persons with lactose intolerance tend to develop flatulence, intestinal discomfort, and diarrhea on exposure to milk and other lactose-containing foods [83].

Some, but not all G6PD-deficient persons develop hemolytic anemia on eating fava beans [71]. Decreased urinary excretion of *d*-glutaric acid is a metabolic characteristic of patients with favism and may involve biotransformation to the toxic ingredient of the fava bean. The genetics of glutaric acid excretion is as yet unknown.

One source of genetic variation in susceptibility to environmental influences that should be explored much more intensively is heterozygosity for autosomal-recessive genes, leading in the homozygous state to inherited metabolic disease. The scattered evidence available to date suggests more widespread susceptibilities than have hitherto become known (Sect. 7.2.2.8; [247]). Moreover, phenotypic manifestations in the homozygotes and the nature of the enzyme defects give useful indications as to where disease liabilities of heterozygotes might be found.

The interaction of environmental factors such as high fat intake with the various genetic factors predisposing to coronary heart disease is an unfolding chapter of ecogenetics [175b, 175c]. The clinical expression of the less common monogenic conditions of lipid metabolism such as heterozygous familial hypercholesterolemia is presumably affected by diet, exercise, and smoking since coronary heart disease associated with this condition was less frequent in earlier generations as noted in pedigree studies (Sect. 6.4.2) [264]. However, even more moderate elevations of cholesterol and triglyceride – affecting a large fraction of the population – are likely to be the end result of genetic factors such as the cholesterol-raising apolipoprotein E4 allele interacting with dietary and other factors such as smoking and exercise. Various other lipid-related and other genes (such as those affecting high-density lipoprotein, lipoprotein lipase, lipid absorption, triglyceride, and homocysteine metabolism) as well as variable levels of fat intake create a complex pattern of coronary artery disease risk in any given individual. The relatively large number of lipid-related genes that are likely to affect the response to a given diet as well as likely gene-gene interaction between these genes requires much work in genetic epidemiolgy.

The gene for hemochromatosis is common in various European populations [34]. About 1/500 of the population appear to be homozygotes and absorb iron at an increased rate, but only a fraction of homozygotes have symptoms. Fortification of bread with iron, as practiced in Sweden, has been recommended to prevent iron deficiency, which is a common condition in women and children. This should cause clinically apparent hemochromatosis to occur more frequently and earlier. The common heterozygotes (over 10% of the population) would probably not be harmed.

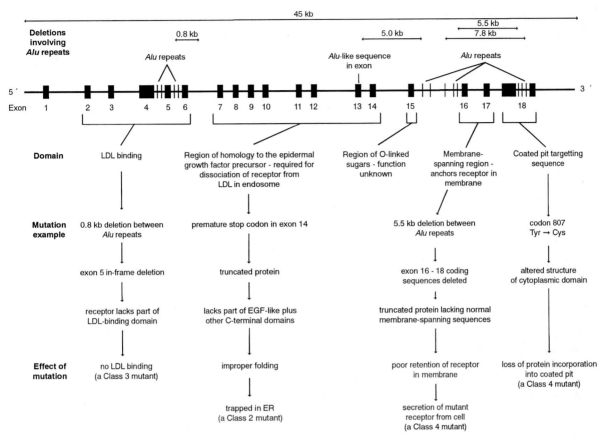

Fig. 7.68. The structure of the LDL receptor gene showing its five domains. Location of selected mutations leading to familial hypercholesterolemia. They are indicated by *horizontal bars above the gene*, sizes of some deletions (*black blocks,* exons). Exons, introns, and Alu repeats are only approximately to scale. (Modified from Goldstein and Brown in Scriver et al. 1989 [215]) (for description see Sect. 7.6.4)

Also, patients with thalassemia major who are already iron loaded would suffer. The practice has therefore been resisted in some countries, although it would undoubtedly reduce iron deficiency.

These considerations show the complex policy problems posed by the existence of genetic heterogeneity. What is helpful to one segment of the population may be harmful to others. Furthermore, the exact scientific details of the extent of benefit and harm cannot always be fully specified. We can expect to face more of these problems as we learn more about genetic individuality.

7.6 Mechanisms of Autosomal Dominance

The basic mechanisms in autosomal-recessive diseases are usually enzyme deficiencies caused by structural mutations of the gene specifying the affected enzyme. The affected enzyme can often be shown to be structurally abnormal or unstable (see

Sect. 7.2.2) [74]. Heterozygotes usually have 50% of normal enzyme activity but are clinically unaffected, indicating that one-half the normal enzyme activity is compatible with normal function. In contrast, in autosomal-dominant inheritance the heterozygote is clinically affected, and the single dose of the mutant gene interferes with normal function.

The mechanisms for autosomal-dominant gene mutations are much more heterogeneous than those demonstrated for autosomal-recessive traits. For a long time the genetic causes and the basic pathogenetic pathways of most dominant diseases defied scientific analysis, and remained unknown. In recent decades, especially since the advent of recombinant DNA techniques, many of these pathways are being elucidated.

If a mutation causes a structural anomaly of the gene-determined protein, this may lead to a functional impairment in a heterozygote and therefore to a dominant disease, if normal function requires some kind of interaction between the products of the two alleles. Recessive inheritance ensues if the products determined by them are functionally more or less independent. This is the case in most genes determining enzyme proteins; therefore enzyme defects are usually recessive. However, if allelic products must cooperate to build a common structure, a defect in 50% of the molecules would make the entire structure defective. Therefore it is not surprising that dominant diseases have often been observed early to affect structural proteins [163], for example, receptors and membranes (Fig. 7.69).

7.6.1 Abnormal Subunit Aggregations

Dysfibrinogenemias (134800) [58]. A person with a dominant disease is a heterozygote. If such heterozygotes carry a protein mutation, there is a mixture of normal and mutant molecules if the protein functions as an aggregate of subunits. Presence of abnormal molecules in a mixture of normal and abnormal ones may interfere with the proper formation of the aggregated proteins (Fig. 7.69). In some dysfibrinogenemias, various mutations involving the fibrinogen molecules may lead to a bleeding tendency. In some mutant forms of fibrinogen the defects appears to be at locations of the molecule that result in interference with fibrin molecule aggregation. In fibrinogen Detroit an amino acid substitution at a site critical for conversion of fibrinogen to fibrin has been observed [32] and is associated with severe bleeding. While the quantitative level of fibrinogen is normal in most abnormal fibrinogens, one fibrinogen defect is associated with lessened amounts of fibrinogen due to shortened molecular survival, presumably caused by molecular instability [160]. Some genetically abnormal fibrinogens are associated with thrombosis. Most fibrinogen variants are not associated with clinical difficulties.

7.6.2 Disturbance of Multimeric Protein Function by Abnormal Subunits

Hemoglobin Diseases. Somewhat analogous mechanisms in subunit formation appear to explain the various types of clinical abnormalities seen in the hemoglobin diseases. Since the functional hemoglobin molecule consists of four subunits produced under specification of two gene loci, heterozygotes form hybrid molecules consisting of the normal and abnormal hemoglobin. (Hybrid molecules such as Hb β^s β^A cannot be demonstrated with the usual methods of hemoglobin separation such as column chromatography and electrophoresis.) Depending upon the characteristics of the respective hemoglobin mutation, various manifestations such as methemoglobinemia, hemolytic anemia, and erythrocytosis may

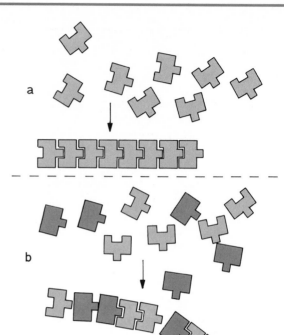

Fig. 7.69 a, b. Diagrammatic structure of a polypeptide chain in normals and heterozygotes. **a** The individual is homozygous and produces only normal polypeptides. **b** The individual is heterozygous. Normal and abnormal polypeptides are formed in equal amounts. The polypeptide chain cannot be properly assembled

occur in heterozygotes (Sect. 7.3). Because of the high degree of cooperation among the four hemoglobin subunits, an abnormality in only one of the four subunits, as encountered in a heterozygote for either an Hb α or Hb β mutation, leads to loss of normal function for the entire molecule. For example, some mutations lead to amino acid substitutions in regions of the α or β chains involved in contact between the four polypeptide chains within the tetramer. Such mutations may impair the heme-heme interaction necessary for oxygen exchange. This leads to a relative oxygen shortage in the tissues, which is compensated by increasing the number of erythrocytes.

In these cases, human mutations are dominant because the resulting protein, to maintain normal function, must undergo a functionally important interaction with a protein formed by the other allele. The amino acid substitution resulting from this mutation is located at a site necessary for this interaction. Functionally analogous amino acid substitutions could easily be responsible for other dominant mutations as well.

7.6.3 Abnormal Feedback Inhibition of Enzymes and Structurally Abnormal Enzymes

Porphyria (176 000): *Decreased Enzyme Activity* [203]. In the various dominant porphyrias (Table 7.28), enzyme deficiencies affecting various steps of heme or porphyrin biosynthesis have been demonstrated. In each case about 50 % of the enzyme, as in the usual heterozygous situation, has been found. The pathophysiology of this group of diseases has been best studied in acute intermittent porphyria. Many patients with the characteristic enzyme deficiency of porphobilinogen deaminase have no symptoms of porphyria such as abdominal pain or neuropathy. Symptoms are usually associated with markedly elevated activity of the enzyme δ-aminolevulinic acid (ALA) synthetase, the first and rate-limiting enzyme of porphyrin synthesis. In fact it was once thought that the primary lesion of acute intermittent porphyria is a regulatory defect causing overproduction of this enzyme. ALA synthetase is induced by many drugs (for example, barbiturates, steroid hormones, and other chemicals) and is normally repressed by feedback inhibition from heme, the end product of the biosynthetic pathway which includes porphobilinogen deaminase. The diminished porphobilinogen deaminase activity in acute intermittent porphyria leads to less heme formation and to derepression of ALA synthetase, with increased ALA formation. Half the normal amount of the enzyme is not sufficient to allow optimal function of the pathway, particularly when the pathway is stimulated by drugs such as barbiturates. In contrast to other enzyme deficiencies, the mutation in acute intermittent porphyria affects a critical rate-limiting enzyme in a tightly regulated biosynthetic pathway.

It is likely that other unidentified environmental chemicals and metabolites also stimulate the pathway. Such a mechanism would explain why only some persons with the enzyme defect develop clinical symptoms. This principle of enzyme repression has been used for therapy. Hematin, as a source of heme, has been given to suppress ALA synthetase activity. A documented decline in ALA and porphobilinogen production has been associated with clinical improvements of acute intermittent porphyria.

Increased Enzyme Activity in Gout. Increased enzyme activity due to a structural lesion of an enzyme with increased specific activity has been demonstrated as a rare cause of gout. Some patients have elevated amounts of phosphoribosylpyrophosphate synthetase activity [17]. Electrophoretic and immunochemical analyses have shown the enzyme in these patients to be structurally abnormal. These findings suggest that the primary defect in this condition affects the enzyme directly.

7.6.4 Receptor Mutations

Receptors. On the surface of cell membranes there are receptors for many hormones, neurotransmitters, and drugs. Many mutations affecting such receptors that are proteins must exist [39], but so far only two groups of receptor mutations have been studied in detail.

One set includes X-linked receptor defects that lead to androgen resistance by failure either to bind dihydrotestosterone to cell surfaces or inability to activate the nuclear binding sites of the hormone. The other set of receptor mutations affects cell binding by low-density lipoprotein (LDL) cholesterol [39, 104]. Cholesterol in the serum is carried mostly by the lipoprotein LDL. A cell surface receptor (Fig. 7.70) exists in specialized structures known as coated pits on fibroblasts and lymphocytes (and by inference in liver cells), which binds LDL to the surface of the cell and transports the LDL cholesterol complex into the cell by endocytosis. The LDL receptor binds only lipoproteins carrying apoprotein B and apoprotein E (B/E receptor). Such receptor-mediated endocytosis is a universal mechanism by which cells take up large molecules, each under the control of a highly specific receptor. Following the movement of LDL cholesterol into the cell, cholesterol accumulation signals the cell to shut off further synthesis of LDL receptors; the LDL protein is degraded by cellular lysosomes. The binding of LDL cholesterol and its transport into the cell further decreases cholesterol synthesis by diminishing the activity of the rate-limiting enzyme HMG CoA reductase. Cholesterol is esterified by the action of acyl-CoA: cholesteryl acyltransferase. The nature of the signals that initiate these various pleiotropic reactions is unknown.

Table 7.28. Enzyme defects in the dominant hereditary porphyrias

Disease	Enzyme defect
Acute intermittent porphyria	Porphobilinogen deaminase
Variegate porphyria	Protoporphyrinogen oxidase
Hereditary coproporphyria	Coproporphyrinogen oxidase
Porphyria cutanea tarda	Uroporphyrinogen decarboxylase
Protoporphyria	Ferrochelatase

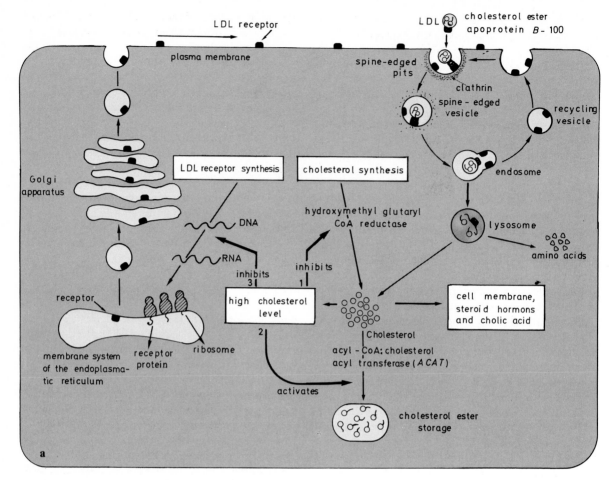

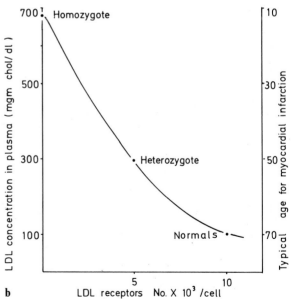

Fig. 7.70 a. Cholesterol metabolism in the cell. LDL carries cholesterol *(above right)*. LDL is bound by a receptor in a spine-edged pit which is then transformed into a spine-edged vesicle. Several such vesicles fuse to form an endosome, where LDL separates from its receptor. The receptor returns to the cell membrane. LDL is taken up by a lysosome, where enzymes degrade apoprotein B-100 into its amino acids and open the ester bonds of cholesterol ester. Free cholesterol is used for production of cell membranes, steroid hormones or cholic acids. The cell regulates its cholesterol level, a high cholesterol level having three different effects: *1*, the enzyme HMG-CoA reductase, the limiting enzyme of cholesterol synthesis, is inhibited; *2*, the enzyme ACAT is activated, which esterifies cholesterol for storage together with fatty acids; *3*, synthesis of new receptors is inhibited by inhibition of transcription of the receptor gene. (Modified from Goldstein and Brown 1989 [105]) **b** Relationship between LDL concentration and typical age at myocardial infarction due to coronary atherosclerosis as a function of the number of LDL receptors on fibroblasts from normal subjects and from persons with heterozygous and homozygous forms of familial hypercholesterolemia. The number of LDL receptors per cell was calculated from experiments in which maximal LDL binding was measured at 4°C in actively growing fibroblasts deprived of LDL for 48 h. (From Goldstein and Brown 1977 [1123])

Familial Hypercholesterolemia. The cause of familial hypercholesterolemia (Fig. 7.68) is one of many different mutations at the same locus (chromosome 19) that affect LDL receptor function [118, 239]. These mutations have been categorized in several classes as (a) no receptor synthesis, (b) defective transport to the cell surface after synthesis, (c) defective binding of LDL, (d) defective internalization, and (e) defective clustering in coated pits. As a result there may be complete absence or defective action of the receptor. About 1/500 persons in the general population are heterozygotes for familial hypercholesterolemia. They have one-half the number of normal LDL receptors and do not remove cholesterol at the normal rate from the circulation (Fig. 7.70). Elevated serum LDL cholesterol levels result with development of atherosclerosis and premature heart attacks. About 50 % of male heterozygotes show overt clinical manifestation of coronary heart disease around the age of 50 years (Sect. 6.4.2). It has become possible to stimulate the heterozygote's normal LDL allele to increase the synthesis of LDL receptors by the oral administration of bile sequestrants (e. g., cholestyramine) which remove bile acids from the intestine [40]. This therapeutic maneuver, together with drugs that are analogs for the substrate of HMG CoA reductase, which blocks cholesterol synthesis, allows normalization of cholesterol levels and prevent coronary heart disease. While bile sequestrants have been used for many years and appear safe, the novel enzyme inhibitors (statins) have now been used for 10 or more years and also appear safe.

Homozygotes for the receptor defect have no functional receptors and, because of their very high LDL levels, develop coronary heart disease and die in childhood or adolescence. Because of mutational heterogeneity many "homozygotes" are really compound heterozygotes for two different LDL receptor mutations. The severity of the clinical disease among true homozygotes and compound heterozygotes depends on the nature of the mutation that they carry. Complete absence of LDL receptors causes more severe disease than lowered receptor activity due to defective LDL binding. Homozygotes do not respond to drug treatment and require other therapeutic approaches. Liver transplantation was performed in one patient to furnish normal LDL receptors, and this markedly reduced LDL cholesterol. Gene therapy has been attempted with indifferent results.

7.6.5 Membrane Defects

In some dominant diseases the mutation seems to affect cellular membranes. An example is hereditary spherocytosis (182900), a common type of hemolytic anemia in which the erythrocytes do not assume the normal disk-like, biconcave shape but show a more spheroid form. There is a decrease in membrane surface area and diminshed membrane lipids and a specific increase in membrane permeability to sodium [126]. Spherocytes are eliminated from circulation by the spleen at an increased rate. The exact membrane defect affects the interaction of spectrin, the main component of the cytoskeleton of red cells, and other proteins to form a normally functioning cytoskeleton [158]. Various mutations appear to be involved.

7.6.6 Deposition of Abnormal Fibrillar Proteins: Hereditary Amyloidoses (104800–105250)

A group of diseases known as the amyloidoses are characterized by the deposition of different abnormal fibrillar proteins with β-pleated structure (for references see [22]). Amyloid fibrils are amorphous and stain with Congo red. Classification of the amyloidoses is based on the specific chemical composition of the characteristic fibrils (Table 7.29). Hereditary amyloidoses are autosomal dominant conditions caused by missense mutations of various proteins such as transthyretin (formerly known as prealbumin), apolipoprotein A1, gelsolin, fibrinogen, and lysozyme. Electron microscopy cannot distinguish between the different biochemical types. Characteristic fibrils appear to carry only the mutant protein although affected patients (who are heterozygotes) synthesize both the normal and abnormal protein.

The most common type, and best studied hereditary amyloidosis, is caused by mutations of the liver protein transthyretin. The onset of clinical symptoms is in middle age or even later. The clinical findings relate to deposits of the mutant transthyretin. Peripher-

Table 7.29. Generalized systematic amyloidoses (from Benson 1995 [22])

	Protein involved
Hereditary amyloidoses	Transthyretin Apolipoprotein A Gelsolin Fibrinogen Lysozyme
Immunoglobulin amyloidosis	Ig light chain ($\varkappa$ and λ)
Reactive amyloidosis (chronic inflammatory disease, familial Mediterranean fever)	Amyloid A
Renal amyloidosis (prolonged uremia or dialysis)	β_2-microglobulin

al neuropathy – often manifesting as carpal tunnel syndrome – is the most common clinical manifestation. Autonomic neuropathy such as gastrointestinal dysfunction, postural hypotension, urinary retention, impotence, and hypohydrosis may occur. Cardiomyopathy and nephropathy are often seen. More than 40 different transthyretin mutations have been described. There is significant resemblance in the clinical pattern of a given mutation in affected members of a family. Homozygotes for some transthyretin mutations do not differ from heterozygotes. For example, the isoleucine 122 mutation found in 2% of African-Americans is associated with cardiomyopathy in both heterozygote and homozygote elderly persons.

The transthyretin molecule has a single chain of 127 amino acids and functions as a tetramer. Since its DNA structure is known, diagnostic tests can be performed with DNA methods. If the transthyretin defect is unknown, direct sequencing (400 nucleotides) is required. If a known molecular defect is defined, PCR or allele-specific oligonucleotide hybridization can be used for diagnosis. Transthyretin variants without clinical symptoms also exist. Since transthyretin binds thyroxin in the plasma, some of these polymorphisms are associated with hyperthyroxinemia without hyperthyroidism.

Gelsolin is a calcium-binding protein. Missense mutation of this gene that produce hereditary amyloidosis have been described. While liver transplantation has been used as successful treatment for transthyretin-associated amyloidosis, this option is not feasible with the gelsolin variety, which is synthesized by muscle.

Hereditary amyloidoses must be differentiated from the more common nonfamilial "immunoglobulin amyloidosis" which is sometimes seen in patients with multiple myeloma. A monoclonal immunoglobin peak is often found in serum and urine. "Reactive amyloidoses" occurs in chronic inflammatory diseases such as tuberculosis, osteomyelitis, and rheumatoid arthritis and is caused by overproduction of the amyloid A protein. This type of amyloidosis is also often observed in familial Mediterranean fever, an autosomal recessive disease.

Dominantly Inherited Alzheimer's Disease [27a, 161a, 201, 243a]. Dementia is a deterioration of mental abilities and has an organic cause. Usually short-term memory disappears first, followed by loss of long-term memory and the ability to understand new information. Other signs are loss of judgment and deterioration of language. The patient becomes disoriented and ultimately loses all cognitive function. The histology of the brain is characterized by a large number of neurofibrillary tangles and so-called senile plaques, which consist of neurites with an amyloid core (Alzheimer 1907 [5]).

Alzheimer's disease with onset in late life (past 65–70 years) shows some familial aggregation. Some observers suggest that the family data can be fitted to a model of autosomal dominant inheritance if one assumes that the age of onset may often be very much delayed (to the late 80s and 90s) so that many individuals would have died of other causes before they could manifest the characteristic dementia. Prevalence increases continually with advancing age: at the age of 90 years at least 5% of the population and about 20% of first-degree relatives of patients are affected.

Definite autosomal dominant inheritance has been demonstrated in kindreds with Alzheimer's disease with relatively early onset (40–65 years of age). At least three different genes have been identified. Linkage studies have shown that rare families carry a gene on chromosome 21. More frequently, the Alzheimer's gene in families is mapped to chromosome 14. In yet another autosomal dominant variety of the disease – among Germans in the Volga region – the gene is located on chromosome 1. The Alzheimer genes on chromosomes 14 and 1 have been cloned and resemble each other.

Another development is the recent discovery that carriers of the apolipoprotein E4 allele have a significantly increased risk for Alzheimer's disease, particularly among families with relatively late age of onset. Their risk of developing Alzheimer's disease is at least three or four times increased but much higher still in E_4 homozygotes. The mechanism by which the apolipoprotein A4 gene causes a higher risk is under investigation.

The pathogenesis of Alzheimer's disease is under active study [27a, 161a]. The localization of the amyloid precursor gene on chromosome 21, the localization of at least one Alzheimer's gene to chromosome 21, and the high frequency of the characteristic neuropathology in trisomy 21 has suggested a mechanistic relationship between amyloid production and Alzheimer's disease. Much further work, however, is necessary.

7.6.7 Heritable Disorders of Connective Tissue

This term was coined by McKusick [164] in a famous monograph for a number of hereditary diseases with predominant involvement of various connective tissues, most but not all of which show an autosomal dominant mode of inheritance. Examples include the various types of osteogenesis imperfecta, Marfan's syndrome, and Ehlers-Danlos syndrome. Studies at the gene-DNA and protein levels have now led to elucidation of their causes and, in part, pathogenetic mechanisms.

In osteogenesis imperfecta, many bone fractures occur; in addition, blue sclerae and sometimes hearing loss are observed. The various types of this disease (120160, 166210, 166220, and some others) have been explained by anomalies in the type I collagen [45–48]. Collagens are the most abundant proteins in the body. More than 20 genes code for them; the complete molecule, which is synthesized in many steps, is a unit of three chains wound around each other in a triple helix. Depending on the way in which various mutations interfere with this structure, milder or more severe phenotypes ensue (Table 7.30) [48]. The mode of inheritance is usually autosomal dominant; this offers a good example of the principle discussed above: the protocollagen chains to be intertwined in this structure come from both alleles; both the normal and the mutated ones. This process is severely disturbed, and the end product is deficient if one-half of these chains are abnormal.

In a few rare cases, a recessive mode of inheritance may be present. However, most instances in which two affected children have been born to healthy parents are due to germ cell mosaicism, which does not appear to be very rare [46, 47] (see Chaps. 9, 10).

Anomalies in other collagens lead to a number of different diseases. The various types of the Ehlers-Danlos syndrome, for example, are caused by anomalies and defects of type III collagen or of enzymes involved in collagen formation (for details see [45]). Mutations affecting type II may cause, for example, the Stickler syndrome and achondrogenesis, while forms of epidermolysis bullosa have been traced to anomalies of type VII.

Marfan syndrome (154700), an autosomal-dominant condition, is conspicuous by long limbs, "spider fingers," laxity of joints, myopia, and skeletal anomalies. Weakness of zonula fibers in the eye often causes dislocation of the lens. A weakness of the elastic tissue in the aortic wall may lead to dilatation and aneurysma formation, with consequent dissection of the aortic wall und early death. This disease is caused by defects of the gene for another component of connective tissue, which has been named fibrillin [212]. The gene has been localized to 15q21.1 [129, 130]; several missense mutations and an occasional deletion have been described [242].

Myosin Genes and Dominant Hypertrophic Cardiomyopathy; Disturbed Interaction of a Mutationally

Table 7.30. Osteogenesis imperfecta phenotypes and their molecular bases (from Byers 1989 [45])

OI Type	Clinical features	Inheritance	Biochemical and genetic abnormalities
I	Normal stature, little or no deformity, blue sclerae, hearing loss in 50%; dentinogenesis imperfecta is rare and may distinguish a subset.	AD	Common: 'nonfunctional' *COL1A1* allele Rare: Substitution for glycine residue in carboxy-terminal telopeptides of α_1(I) Substitution for glycines in the triple helix in proα_1(I) Exon deletion in proα_1(I) triple helix
II	Lethal in the perinatal period, minimal calvarial mineralization, beaded ribs, compressed femurs, marked long bone deformity, platyspondyly (flattened vertebrae).	AD (new)	Common: Substitutions for glycyl residues in the triple-helical domain of the α_1(I) chain and α_2(I) chain Rare: Rearrangement in the *COL1A1* and *COL1A2* genes Exon deletions in triple-helical domain of *COL1A1* and *COL1A2*
		AR (rare)	Small deletion in α_2(I) on the background of null allele
III	Progressively deforming bones, usually with moderate deformity at birth. Sclerae variable in hue, often lighten with age. Dentinogenesis common, hearing loss common. Stature very short.	AD	Point mutations in the *COL1A1* and *COL1A2* gene
		AR (uncommon)	Frameshift (4-bp deletion) in *COL1A2* that prevents incorporation of proα_2(I) chains into molecules
IV	Normal sclerae, mild to moderate bone deformity and variable short stature, dentinogenesis is common and hearing loss occurs in some.	AD	Point mutations in *COL1A1* and *COL1A2* genes Exon-skipping mutations in *COL1A2*

AD, Autosomal dominant; AR, autosomal recessive.

Altered Protein with Another Protein? Myosins are a group of proteins that, together with actins, form the active tissue of the muscles of skeleton, heart, and other organs. They are determined by a gene family which has been localized to various chromosomes and analyzed directly at the molecular level, i.e., not starting with a genetic disease or a variant with a monogenic mode of inheritance – a new principle of genetic analysis that is being utilized in an increasing number of instances (Sect. 3.1.3.10). However, since this gene family has become known, one disease – hypertrophic cardiomyopathy – has turned out to be due to a mutation in a β-myosin gene. This group of diseases affects about 1 in 5000 persons [11]. It is characterized by increased thickening of the heart muscle and often by arrhythmia, which sometimes leads to early death. Some young athletes with this anomaly have suddenly died during physical exertion. It may occur sporadically and for unknown reasons; in the families of many patients, however, an autosomal dominant mode of inheritance is observed. The condition is heterogeneous; linkage studies have revealed mutant genes on chromosomes 1, 11, 14, and 15 [52, 237, 238, 249, 251]. The mutation on chromosome 14 proved especially interesting since a gene for the β chain of the muscular protein myosin was known to be located in the very same area [97, 125]. An increasing number of mutations (single base replacements) have now been identified. Quite a few of these cluster at amino acid position 403, indicating either a mutational hot spot, or a functionally sensitive site of the protein, or both. This site is located in the globular portion of the myosin molecule, which is important for interaction with actin, the other functional protein of muscles in muscular action. Impaired interaction between myosin and actin may well be the primary functional defect; muscular hypertrophy may be secondary as an attempt at compensation [249]. This is another way in which a mutation may have a dominant effect on the phenotype.

Prion Diseases. The prion (protein infectious agent) related protein (176 640) is a sialoglycoprotein that is normally present but appears to be shaped into a different spatial configuration in response to infections with similarly shaped protein particles [61]. It aggregates extracellularly to form structures similar to those of amyloid. The gene has been mapped to 20 p12-pter [164]. Mutations which may fold easily to form this abnormal structure have been described in the autosomal dominant form of Creutzfeld-Jakob disease (123 400), a progressive degenerative disease of the brain with status spongiosus, neuronal degeneration, and corresponding symptoms with onset in adult age; in familial fatal insomnia; and Gerstmann-Straussler disease (137 440). The histology is identical with that found in scrapie, a disease of sheep and cattle; the disease is transmissible between these two species and may even be transmissible between cattle and humans. This group of related diseases is the first example of an infection not by a known type of infective agent – such as a bacterium or virus – but by a specifically folded protein that induces the host protein to assume a specific – and pathogenic – folding structure. In a few individuals, this abnormal folding structure often is formed spontaneously if the protein has undergone a suitable change by a mutation (= autosomal-dominant type); in most instances, infection with the abnormally folded protein from outside (humans or, possibly, animals) is necessary. Another example is Kuru, which was observed in inhabitants of New Guinea who used to eat the brains of their deceased ancestors. As discussed in Sect. 4.1.6, different disease phenotypes have been described, depending on a DNA polymorphism within the prion gene but at a site different from that of the point mutation leading to the disease [102].

Table 7.31. Some mechanisms of dominant disease

Mechanism	Example
Abnormal aggregation of protein subunits	Abnormal fibrinogens
Disturbance of multimeric protein function by abnormal subunits	Unstable hemoglobins
Diminished feedback inhibition by end product due to enzyme deficiency	Porphobilinogen deaminase deficiency in acute intermittent porphyria
Cell receptor defects	LDL cholesterol receptor defects in familial hypercholesterolemia
Cell membrane defects	Hereditary spherocytosis
Deposition of abnormal fibrillar protein in peripheral tissue	Hereditary amyloidoses
Deposition of abnormal fibrillar protein in the brain	Alzheimer disease
Structural deficiencies in ubiquitous and basic proteins	Connective tissue anomalies such as in osteogenesis imperfecta, Ehlers-Danlos syndrome, Marfan syndrome
Functional defect due to impaired interaction with another protein	Hypertrophic cardiomyopathy due to mutation in a myosin gene
Mutation in a homeobox-containing gene	Waardenburg syndrome
Germline mutation plus somatic mutation of allelic homologue	Retinoblastoma (see text)

7.6.8 Dominantly Inherited Tumor Diseases

Mutations of a tumor suppressor gene have often been observed. The classical example is retinoblastoma, the malignant eye tumor of young children. In addition to a dominantly inherited type with approximately 90% penetrance, a nonhereditary type exists. A "tumor suppressor gene" is present in mutant form in all cells of individuals carrying the hereditary type. However, a malignant cell clone is produced only if the normal allele on the homologous chromosome happens to have a somatic mutation in one of its retinal cells. In the nonhereditary type, two somatic mutations – one mutation in each of the two alleles – are necessary to produce a malignant cell clone. Other tumors have been shown to follow the same, sometimes a more complicated, pattern. These problems are discussed in greater detail in Chap. 10.

General Remarks. Table 7.31 lists various mechanisms of dominant disease. Even some enzyme deficiencies such as C1 inhibitor deficiency causing hereditary angioedema (106 100) and antithrombin deficiency (107 300) producing an increased tendency to venous thrombosis are inherited as autosomal-dominant traits. It is not clear why disease develops in these heterozygous enzyme deficiencies. The many dominant anomalies which are caused by disturbances in embryonic development, such as malformations of limbs (for example, brachydactyly; Chap. 4.1.2), are beginning to be elucidated. The examples given indicate that Mendelian dominance may result from a variety of different mechanisms. Their elucidation will help to clarify many aspects of the genetic determination of structure and function. Scientifically, this promises to be much more interesting than the simple one-gene/one-enzyme relationship often found in recessive metabolic diseases. It is of particular interest that several intracellular enzymatic reactions can be disturbed by a single receptor mutation. Such biochemical pleiotropism offers an exciting model for the control of complex biochemical pathways by a single gene. The basic lesions in the common dominant diseases offer exciting scientific challenges.

Conclusions

Genetics attempts to solve the problem of how genes – in cooperation with environmental factors – determine the development and function of the organism. For this, functional defects found in hereditary diseases can be used as analytical tools. Enzyme defects, for example, often lead to recessive disease. Some variants of enzymes and proteins have deleterious phenotypic effects only if carriers are exposed to certain drugs or other environmental agents. Dominant diseases may be caused by a variety of pathogenetic mechanisms. Human hemoglobin variants have been useful as tools in analyzing gene action and its disturbances. The immune system is an especially complex system; here the cooperation and interaction between numerous genes has been studied in detail. In addition to its importance for theoretical understanding, the analysis of gene action offers clues toward the therapy and prevention of genetic diseases.

References

1. Abel T, Maniatis T (1994) Mechanisms of eukaryotic gene regulation. In: Stamatoyannopoulos G, Nienhuis AW, Majerus PW, Varmus H (eds) The molecular basis of blood diseases, 2nd edn. Saunders, Philadelphia, pp 33–70
2. Alargon B, Regueiro AA, Arnaiz-Villena A, Terhorst C (1988) Familial defect in the surface expression of the T-cell receptor – CD3 complex. N Engl J Med 319:1203–1208
3. Alberts B, Bray D, Lewis J, Raff M, Roberts K, Watson JD (1983) Molecular biology of the cell. Garland, New York
4. Alter BP (1985) Antenatal diagnosis of thalassemia. a review. Ann N Y Acad Sci 445:393–407
5. Alzheimer A (1907) Über eine eigenartige Erkrankung der Hirnrinde. Allg Z Psychiatr 64:146–148
6. Anderson DC, Smith CW, Springer TA (1989) Leucocyte adherence deficiency and other disorders of leucocyte motility. In: Scriver CR, Beaudet AL, Sly WS, Valle D (eds) The metabolic basis of inherited disease, 6th edn. McGraw-Hill, New York, pp 2751–2778
7. Anderson WF (1992) Human gene therapy. Science 256:808–813
8. Anonymous (1973) Pharmacogenetics. Report of a WHO scientific group. WHO Tech Rep Ser 524.
9. Anonymous (1978) Fetal haemoglobin in sickle-cell anaemia and thalassaemia – a clue to therapy? (Editorial). Lancet 1:971–972
10. Anonymous (1988) Immunsystem, 2nd edn. Spektrum der Wissenschaft, Heidelberg
11. Anonymous (1993) Diagnosing the heart of the problem (Editorial). Nat Genet 4:211–212
12. Antonarakis SE, Kazazian HH Jr, Orkin SH (1985) DNA polymorphism and molecular pathology of the human globin gene clusters. Hum Genet 69:1–14
13. Baralle FE, Shoulders CC, Proundfoot NJ (1980) The primary structure of the human epsilon-globin gene. Cell 21:621–626
14. Beadle GW (1945) Biochemical genetics. Chem Rev 37:15–96
15. Beadle GW, Ephrussi B (1936) The differentiation of eye pigments in drosophila as studied by transplantations. Genetics 21:225–247
16. Beadle GW, Tatum EL (1941) Genetic control of biochemical reactions in neurospora. Proc Natl Acad Sci USA 27:499–506

17. Becker MA, Kostel PJ, Meyer LJ, Seegmiller JE (1973) Human phosphoribosylpyrophosphate synthetase: increased enzyme specific activity in a family with gout and excessive purine synthesis. Proc Natl Acad Sci USA 70 : 2749

18. Beet EA (1949) The genetics of the sickle cell trait in a Bantu tribe. Ann Eugen 14 : 279

19. Bell DA et al (1993) Genotype/phenotype discordance for human arylamine N-acetyltransferase (NAT2) reveals a new slow-acetylator allele common in African-Americans. Carcinogenesis 14 : 1689–92

20. Bellingham AJ (1976) Haemoglobins with altered oxygen affinity. Br Med Bull 32 : 234–238

21. Benöhr HC, Waller HD (1975) Metabolism in haemolytic states. Clin Haematol 4 : 45–62

22. Benson MD (1995) Amyloidosis. In: Scriver CR, Beaudet AL, Sly WS, Valle D (eds) The metabolic and molecular basis of inherited disease, vol 3, 7th edn. McGraw-Hill, New York, pp 4157–4191

23. Betke K, Beutler E, Brewer GJ, Kirkman HN, Luzzato L, Motulsky AG, Ramot B, Siniscalco M (1967) Standardization of procedures for the study of glucose-6-phosphate dehydrogenase. WHO Tech Rep Ser 366

24. Beutler E (1969) G-6-PD activity of individual erythrocytes and X-chromosomal inactivation. In: Yunis JJ (ed) Biochemical methods in red cell genetics. Academic, New York, pp 95–113

25. Beutler E (1975) Red cell metabolism. A manual of biochemical methods. Grune and Stratton, New York

26. Beutler E (1979) Review: Red cell enzyme defects as non-diseases and as diseases. Blood 54 : 1–7

27. Beutler E (1993) Gaucher disease as a paradigm of current issues regarding single gene mutations of humans. Proc Natl Acad Sci USA 90 : 5384–5390

27a. Beyreuther K, Multhaupt B, Masters CL (1996) Alzheimer Krankheit. Molekulare Pathogenese und deren Implikationen für die Therapieforschung. Akademie-Journal 2/95 : 28–39

28. Bickel H (1953) Influence of phenylalanine intake on phenylketonuria. Lancet 2 : 812

29. Bickel Med Surg 12 : 114–118 (1954)

30. Blackwell TK, Alt FW (1988) Immunoglobulin genes. In: Hames BD, Glover DM (eds) Molecular immunology. IRL, Oxford, pp 1–60

31. Blombäck M, Blombäck B, Mammen EF, Prasad AS (1968) Fibrinogen Detroit – a molecular defect in the N-terminal disulphide knot of human fibrinogen? Nature 218 : 134

32. Bonaiti-Pellié C, Phung L, Nordmann Y (1984) Recurrence risk estimation of acute intermittent porphyria based on analysis of porphobilinogen deaminase activity: a Bayesian approach. Am J Med Genet 19 : 755–762

33. Bothwell TH, Charlton RW, Motulsky AG (1983) Idiopathic hemochromatosis. In: Stanbury JB, Wyngaarden JB, Fredrickson DS, Goldstein JL, Brown MS (eds) The metabolic basis of inherited disease, 5th edn. McGraw-Hill, New York, pp 1269–1298

34. Boyer SH, Rucknagel DL, Weatherall DJ, Watson-Williams EJ (1963) Further evidence for linkage between the β and δ loci governing human hemoglobins and the population dynamics of linked genes. Am J Hum Genet 15 : 438–448

35. Bradley TB, Boyer SH, Allen FH (1961) Hopkins-2-hemoglobin: a revised pedigree with data on blood and serum groups. Bull Johns Hopkins Hosp 108 : 75–79

36. Braunitzer G, Hilschmann N, Rudloff V, Hilse K, Liebold B, Müller R (1961) The haemoglobin particles. Chemical and genetic aspects of their structure. Nature 190 : 480

37. Brewer GJ (1971) Annotation: human ecology, an expanding role for the human geneticist. Am J Hum Genet 23 : 92–94

38. Brown MS, Goldstein JL (1974) Expression of the familial hypercholesterolemia gene in heterozygotes: mechanism for a dominant disorder in man. Science 185 : 61–63

39. Brown MS, Goldstein JL (1976) New directions in human biochemical genetics: understanding the manifestations of receptor deficiency states. Prog Med Genet [New Ser] 1 : 103–119

40. Brown MS, Goldstein JL (1984) How LDL receptors influence cholesterol and atherosclerosis. Sci Am 251 : 58–66

41. Bunn HF (1994) Sickle hemoglobin and other hemoglobin mutants. In: Stamatoyannopoulos G, Nienhuis AW, Marjerus PW, Varmus H (eds) The molecular basis of blood diseases, 2nd edn. Saunders, Philadelphia, pp 207–256

42. Bunn HF, Forget BS, Ranney HM (1977) Human hemoglobins. Saunders, Philadelphia

43. Burnet FM (1959) The clonal selection theory of acquired immunity. Cambridge Universtiy Press, London

44. Butenandt A (1953) Biochemie der Gene und Genwirkungen. Naturwissenschaften 40 : 91–100

45. Byers PH (1989) Disorders of collagen biosynthesis and structure. In: Scriver CR, Beaudet AL, Sly WS, Valle D (eds) The metabolic basis of inherited disease, 6th edn. McGraw-Hill, New York, pp 2805–2842

46. Byers PH (1990) Brittle bones – fragile molecules: disorders of collagen structure and expression. Trends Genet 6 : 293–300

47. Byers PH, Tsipouras P, Bonadio JF et al (1988) Perinatal lethal osteogenesis imperfecta (OI type II): a biochemically heterogeneous disorder usually due to new mutations in the genes for type I collagen. Am J Hum Genet 42 : 237–248

48. Byers PH, Wallis GA, Willing MC (1991) Osteogenesis imperfecta: translation of mutation to phenotype. J Med Genet 28 : 433–442

49. Cantz M, Gehler J (1976) The mucopolysaccharidoses: Inborn errors of glycosaminoglycan catabolism. Hum Genet 32 : 233–255

50. Cao A (1994) 1993 William Allan award address. Am J Hum Genet 54 : 397–402

51. Cao A, Cossu P, Falchi AM, Monni G, Pirastu M, Rosatelli C, Scalas MT, Tuveri T (1985) Antenatal diagnosis of thalassemia major in Sardinia. Ann N Y Acad Sci 445 : 380–392

52. Carrier L, Hengstenberg C, Beckmann JS et al (1993) Mapping of a novel gene for familial hypertrophic cardiomyopathy to chromosome 11. Nat Genet 4 : 311–313

53. Carson PE, Flanagan CL, Ickes CE, Alving AS (1956) Enzymatic deficiency in primaquine-sensitive erythrocytes. Science 124 : 484–485

54. Carver MFH, Cutler A (1994) International hemoglobin information center variant list. Hemoglobin 18 : 77–161

55. Childs B, Zinkham W (1958) A genetic study of a defect in glutathione metabolism of the erythrocyte. Johns Hopkins Med J 102 : 21–37

56. Cholerton S, Daly AK, Idle JR (1992) The role of individual human cytochromes P450 in drug metabolism and clinical response. Trends Pharmacol Sci 13 : 434–439

57. Chuang DT, Fisher CW, Lau KS et al (1991) Maple syrup urine disease; domain structure, mutations and exon skipping in the dihydrolipoyl transacylase (E2) component of the branced-chain alpha keto acid dehydrogenase complex. Mol Biol Med 8 : 49–63

58. Chung D, Ichinose A (1989) Hereditary disorders related to fibrinogen and factor XIII. In: Scriver CR, Beaudet AL, Sly WS, Valle D (eds) The metabolic basis of inherited disease, 6th edn. McGraw-Hill, New York, pp 2135–2153

59. Cleaver JE (1972) Xeroderma pigmentosum: Variants with normal DNA repair and normal sensitivity to ultraviolet light. J Invest Dermatol 58 : 124–128

60. Clevers H, Alarcon B, Wileman T, Terhorst C (1988) The T cell receptor/CD 3 complex: a dynamic protein ensemble. Annu Rev Immunol 6 : 629–662

61. Collinge J, Poulter M, Davis MB et al (1991) Presymptomatic detection or exclusion of prion protein gene defects in families with inherited prion diseases. Am J Hum Genet 49 : 1351–1354

62. Collins FS, Weissman SM (1984) The molecular genetics of human hemoglobins nucleic acid. Res Mol Biol 31 : 315–462

63. Cooper DN, Krawczak M (1993) Human gene mutation. Bioscience Scientific, Oxford

64. Cori GT, Cori CF (1952) Glucose-6-phosphatase of the liver in glycogen storage disease. J Biol Chem 199 : 661

65. Costa T, Scriver CR, Childs B (1985) The effect of Mendelian disease on human health: a measurement. Am J Med Genet 21 : 231–242

66. Cox DW (1995) Alpha-1-antitrypsine deficiency. In: The metabolic and molecular basis of inherited disease, 7th edn. McGraw-Hill, New York, pp 4125–4158

67. Dacie JV, Mollison PL, Richardson N, Selwyn JG, Shapiro L (1953) Atypical congenital haemolytic anemia. Q J Med 22 : 79

68. De Vries DD, de Wijs IJ, Wolff G et al (1993) X-linked myoclonus epilepsy explained as a maternally inherited mitochondrial disorder. Hum Genet 91 : 51–54

69. Desnick RJ (ed) (1980) Enzyme therapy in genetic diseases. Liss, New York

70. Desnick RJ (ed) (1991) Treatment of genetic diseases. Churchill Livingstone, New York

71. Driscoll MC, Dobkin CS, Alter BP (1989) αdβ-thalassemia due to a de novo mutation deleting the 5' β-globin gene activation-region hypersensitive sites. Proc Natl Acad Sci USA 86 : 7470

72. Eichelbaum M, Gross AS (1991) The genetic polymorphism of debrisoquine/sparteine metabolism – clinical aspects. In: Kalow W (ed) Pharmacogenetics of drug metabolism. Elsevier, Amsterdam, pp 625–648

73. Emery AEH, Anand R, Danford N, Duncan W, Paton L (1978) Aryl-hydrocarbon-hydroxylase inducibility in patients with cancer. Lancet 1 : 470–472

74. Epstein CJ (1977) Inferring from modes of inheritance to the mechanisms of genetic disease. In: Rowland LP (ed) Pathogenesis of human muscular dystrophies. Excerpta Medica, Amsterdam, pp 9–22

75. Evans DAP (1993) The debrisoquine/sparteine polymorphism (cytochrome P450 2D6). In: Price Evans DA (ed) Genetic factors in drug therapy. Cambridge University Press, Cambridge, pp 54–88

76. Evans DAP (1993) N-Acetyltransferase. In: Genetic factors in drug therapy. Cambridge University Press, Cambridge, pp 211–302

77. Evans DAP (1993) Genetic factors in drug therapy. Cambridge University Press, Cambridge, pp 137–175

78. Evans DAP, Manley K, McKusick VA (1960) Genetic control of isoniazid metabolism in man. BMJ 2 : 485

79. Fisher CR, Fisher CW, Chuang DT, Cox RP (1991) Occurrence of a tyr393-to-asn (Y393N) mutation in the El-alpha gene of the branced-chain alpha-ketoacid dehydrogenase complex in maple syrup urine disease patients from a Mennonite population. Am J Hum Genet 49 : 429–434

80. Fisher CW, Lau KS, Fisher CR et al (1991) A 17 bp insertion and a phe 215-to-cys missense mutation in the dihydrolipoyl transacylase (E2) mRNA from a thiamine-responsive maple syrup urine disease patient WG-34. Biochem Biophys Res Commun 174 : 804–809

81. Flatz G (1971) Population study of erythrocyte glutathione reductase activity. II. Hematological data of subjects with low enzyme activity and stimulation characteristics in their families. Hum Genet 11 : 278–285

82. Flatz G (1971) Population study of erythrocyte glutathione reductase activity I. Stimulation of the enzyme by flavin adenine dinucleotide and by riboflavin substitution. Hum Genet 11 : 269–277

83. Flatz G (1992) Lactase deficiency: biological and medical aspects of the adult human lactase polymorphism. In: King RA, Rotter JI, Motulsky AG (eds) The genetic basis of common disease. Oxford University Press, New York, pp 305–325

84. Flatz G, Xirotiris N (1976) Glukose-6-phosphat-dehydrogenase. In: Becker PE (ed) Humangenetik, ein kurzes Handbuch, vol 3. Thieme, Stuttgart, pp 494–535

85. Flint J, Harding RM, Clegg JB, Boyce AJ (1993) Why are some genetic disorders common? Distinguishing selection from other processes by molecular analysis of globin gene variants. Hum Genet 91 : 91–117

86. Følling A (1934) Über Ausscheidung von Phenylbrenztraubensäure in den Harn als Stoffwechselanomalie in Verbindung mit Imbezillität. Hoppe Seylers Z Physiol Chem 227 : 169

87. Forbes GB (1953) Glycogen storage disease. J Pediatr 42 : 645

88. Forrester WC, Thompson C, Elder JT, Goudine M (1986) A developmentally stable chromatin structure in the human β-globin cluster. Proc Natl Acad Sci USA 83 : 1359

89. Franceschetti A, Klein D (1954) Le dépistage des hétérozygotes. In: Gedda L (ed) Genetica medica. Orrizonte Medico, Rome

90. Fratantoni JC, Hall CW, Neufeld EF (1968) Hunter and Hurler syndromes. Mutual correction of the defect in cultured fibroblasts. Science 162 : 570–572

91. Friedmann T, Seegmiller JE, Subak-Sharpe JH (1968) Metabolic cooperation between genetically marked human fibroblasts in tissue culture. Nature 220 : 272–274

92. Frézal J, Munnich A, Mitchell G (1983) One gene, several messages. From multifunctional proteins to endogenous opiates. Hum Genet 64 : 311–314

93. Friedmann T (1989) Progress toward human gene therapy. Science 244 : 1275–1282

94. Fujii J, Zorzato F, De Leon S, Khanna VK, Weiler JE, O'Brien PJ, MacLennan DH (1991) Identification of a mutation in porcine ryanodine receptor associated with malignant hyperthermia. Science 253 : 448

95. Garrod AE (1902) The incidence of alcaptonuria: a study in chemical individuality. Lancet 2 : 1616–1620

96. Garrod AE (1923) Inborn errors of metabolism. Frowde, London (Reprint 1963 Oxford University Press, London)

97. Geisterfer-Lowrance AAT, Kass S, Sasazuki T et al (1990) A molecular basis for familiar hypertrophic cardiomyopathy: a β cardiac myosin heavy chain gene missense mutation. Cell 62 : 999–1006

98. Gibbons RJ, Wilkie AOM, Weatherall DJ, Higgs DR (1991) A newly defined x-linked mental retardation syndrome associated with α-thalassemia. J Med Genet 28 : 729

99. Gibson QH (1948) The reduction of methaemoglobin in red blood cells and studies on the cause of idiopathic methaemoglobinaemia. Biochem J 42 : 13–23

100. Gibson QH, Harrison DC (1947) Familial idiopathic methemoglobinemia. Lancet 2 : 941–943

101. Glatt H, Oesch F (1984) Variations in epoxide hydrolast activities in human liver and blood. In: Omenn GS, Gelboin HV (eds) Genetic variability in responses to chemical exposure. Cold Spring Harbor Laboratory, Cold Spring Harbor, pp 189–201(CHS Banbury report 16)

102. Goldfarb LG, Petersen RB, Tabatan M et al (1992) Fatal familial insomnia and familial Creutzfeldt-Jakob disease: disease phenotype determined by a DNA polymorphism. Science 258 : 806–808

103. Goldschmidt RB (1935) Gen und Außeneigenschaft (Untersuchungen an Drosophila) I. und II. Mitt Z Vererbungslehre 10 : 74–98

104. Goldstein JL, Brown MS (1977) The low-density lipoprotein pathway and its relation to atherosclerosis. Annu Rev Biochem 46 : 897–930

105. Goldstein JL, Brown MS (1989) In: Scriver CR, Beaudet AL, Sly WS, Valle D (eds) The metabolic basis of inherited disease, 6th edn. McGraw-Hill, New York

106. Graham JB, Barrow ES, Reisner HM, Edgell CJS (1983) The genetics of blood coagulation. Adv Hum Genet 13 : 1–81

107. Grove DI, Forbes IJ (1975) Increased resistance to helminth infestation in an atopic population. Med J Aust 1 : 336–338

108. Guthrie R, Susi A (1963) A simple phenylalanine method for detecting phenylketonuria in large populations of newborns. Pediatrics 32 : 338

109. Haldane JBS (1954) The biochemistry of genetics. London

110. Harris H (1980) The principles of human biochemical genetics, 4th edn. North-Holland, Amsterdam

111. Hayashi A, Fujita T, Fujimura M, Titani K (1980) A new abnormal fetal hemoglobin, Hb FM Osaka (α₂ α₂ 63 HIS' Tyr). Hemoglobin 4 : 447

112. Heckbert SR, Weiss NS, Hornung SK, Eaton DL, Motulsky AG (1992) Glutathione S-transferase and epoxide hydrolase activity in human leukocytes in relation to the risk of lung and other smoking-related cancers. J Natl Cancer Inst 84 : 414–422

113. Hentze MW (1991) Determinants and regulation of cytoplasmic mRNA stability in eukaryotic cells. Biochim Biophys Acta 1090 : 281–292

114. Herrick JB (1910) Peculiar elongated and sickle-shaped red blood corpuscles in a case of severe anemia. Arch Intern Med 6 : 517

115. Higgs DR, Hill AVS, Micholls R, Goodbourn SEY, Ayyub H, Teal H, Clegg JB, Weatherall DJ (1985) Molecular rearrangements of the human α-gene cluster. Ann N Y Acad Sci 445 : 45–56

116. Hilschmann N, Craig LC (1965) Amino acid sequence studies with Bence-Jones proteins. Proc Natl Acad Sci USA 53 : 1403

117. Hilschmann N, Kratzin H, Altevogt P, Ruban E, Kortt A, Staroscik C, Scholz R, Palm W, Barnikol H-U, Barnikol-Watanabe S, Bertram J, Horn J, Engelhard M, Schneider M, Dreher W (1976) Evolutionary origin of antibody specificity. In: Goodman M, Tashian RE (eds) Molecular anthropology. Plenum, New York, pp 369–386

118. Hobbs HH, Russell DW, Brown GS, Goldstein JL (1990) The LDL receptor locus in familial hypercholesterolemia: mutational analysis of a membrane protein. Annu Rev Genet 24 : 133–170

119. Honig GR, Adams JG III (1986) Human hemoglobin genetics. Springer, Vienna

120. Hörlein H, Weber G (1948) Über chronische familiäre Methämoglobinämie und eine neue Modifikation des Methämoglobins. Dtsch Med Wochenschr 72 : 476

121. Howell DR, Williams JC (1983) The glycogen storage diseases. In: Stanbury JB, Wyngaarden JB, Fredrickson DS, Goldstein JL, Brown MS (eds) The metabolic basis of inherited disease, 5th edn. McGraw-Hill, New York, pp 141–166

122. Huisman THJ, Wilson JB, Gravely M, Hubbard M (1974) Hemoglobin grady: the first example of a variant with elongated chains due to an insertion of residues. Proc Natl Acad Sci USA 71 : 3270–3273

123. Ingram VM (1956) A specific chemical difference between the globins of normal human and sickle cell anaemia haemoglobin. Nature 178 : 792

124. Jackson LG (1985) First-trimester diagnosis of fetal genetic disorders. Hosp Pract 20 : 39–48

125. Jaenicke T, Diederich KW, Haas W et al (1990) The complete sequence of the human β-myosin heavy chain gene and a comparative analysis of its product. Genomics 8 : 194–206

126. Jandl JH, Cooper RA (1978) Hereditary spherocytosis. In: Stanbury JB, Wyngaarden JB, Fredrickson DS (eds) The metabolic basis of inherited disease, 4th edn. McGraw-Hill, New York, pp 1396–1409

127. Jervis GA (1953) Phenylpyruvic oligophrenia: deficiency of phenylalanine oxidizing system. Proc Soc Exp Biol Med 82 : 514–515

128. Kahn A, Kaplan J-C, Dreyfus J-C (1979) Advances in hereditary red cell enzyme anomalies. Hum Genet 50 : 1–27

129. Kainulainen K, Pulkkinen L, Savolainen AC et al (1990) Location on chromosome 15 of the genetic defect causing Marfan syndrome. N Engl J Med 323 : 935–939

130. Kainulainen K, Steinmann B, Collins F et al (1991) Marfan syndrome: no evidence for heterogeneity in different populations, and more precise mapping of the gene. Am J Hum Genet 49 : 662–667

131. Kalckar HM (1957) Biochemical mutations in man and microorganisms. Science 125 : 105–108

132. Kalow W, Grant DM (1995) Pharmacogenetics. In: Scriver CR, Beaudet AL, Sly WS, Valle D (eds) The metabolic and molecular basis of inherited disease, vol 3, 7th edn. McGraw-Hill, New York, pp 293–326

133. Kalow W, Staron N (1957) On distribution and inheritance of human serum cholinesterase, as indicated by dibucaine numbers. Can J Biochem 35 : 1305

134. Kamuzora H, Lehmann H (1975) Human embryonic haemoglobins including a comparison by homology of the human ζ and α chains. Nature 256 : 511–513

135. Kan YW (1985) Molecular pathology of α-thalassemia. Ann N Y Acad Sci 445 : 28–35

136. Kan YW, Chang JC, Poon R (1979) Nucleotide sequences of the untranslated 5' and 3' regions of human α-, β- and γ-globin mRNAs. In: Stamatoyannopoulos G, Nienhuis A (eds) Cellular and molecular regulation of hemoglobin switching. Grune and Stratton, New York, pp 595–606

137. Kazazian HH (1990) The thalassemia syndromes: molecular basis and prenatal diagnosis in 1990. Semin Hematol 27 : 209

138. Kellermann G, Luyten-Kellerman M, Shaw CR (1973) Genetic variation of aryl hydrocarbon hydroxylase in human lymphocytes. Am J Hum Genet 25 : 327–331

139. Kellermann G, Shaw CR, Luyten-Kellerman M (1973) Aryl hydrocarbon hydroxylase inducibility and bronchogenic carcinoma. N Engl J Med 289 : 934

140. Kelley WN, Greene ML, Rosenbloom FM, Henderson JF, Seegmiller JE (1969) Hypoxanthine-guanine phosphoribosyl transferase in gout. Ann Intern Med 70 : 155–206

141. Kelly WJ, Wyngaarden JB (1983) Clinical syndromes associated with HPRT deficiency. In: Stanbury JB, Wyngaarden JB, Fredrickson DS, Goldstein JL, Brown MS (eds) The metabolic basis of inherited disease, 5th edn. McGraw-Hill, New York, pp 1115–1143

142. Kendrew JC, Dickerson RE, Strandberg BE, Hart RG, Davies DR, Phillips DC, Shore VC (1960) Structure of myoglobin – a three-dimensional Fourier synthesis at 2 Å. Resolution. Nature 185 : 422–427

143. Kessel M, Gruss P (1990) Murine developmental control genes. Science 249 : 375–379

144. Kirkman HN (1972) Enzyme defects. Prog Med Genet 8 : 125–168

145. Kresse H, Cantz M, von Figura K, Glössl J, Paschke E (1981) The mucopolysaccharidoses: biochemistry and clinical symptoms. Klin Wochenschr 59 : 867–876

146. Krooth RS, Weinberg AN (1961) Studies on cell lines developed from the tissues of patients with galactosemia. J Exp Med 133 : 1155–1171

147. Kühn A (1961) Grundriß der Vererbungslehre. Quelle and Meyer, Heidelberg

148. Lee EJD, Zhao B, Moochhala SM, Ngoi SS (1994) Frequency of mutant CYPIA1, NAT2 and GSTM1 alleles in a normal Chinese population. Pharmacogenetics 4 : 355–358

149. Lehmann H, Huntsman RG (1974) Man's hemoglobins. North-Holland, Amsterdam

150. Lehmann H, Ryan E (1956) The familial incidence of low pseudocholinesterase level. Lancet 2 : 124

151. Lenke RR, Levy HL (1980) Maternal phenylketonuria and hyperphenylalaninemia: an international survey of the outcome of untreated and treated pregnancies. N Engl J Med 303 : 1202

152. Lesch M, Nyhan WL (1964) A familial disorder of uric acid metabolism and central nervous system function. Am J Med 36 : 561

153. Liebhaber SA, Goossens MJ, Kan YW (1980) Cloning and complete sequence of human 5'-alpha-globin gene. Proc Natl Acad Sci USA 77 : 7054–7058

154. Lin HJ, Han CY, Lin BK, Hardy S (1993) Slow acetylator mutations in the human polymorphic N-acetyltransferase gene in 786 Asians, Blacks, Hispanics and Whites: application to metabolic epidemiology. Am J Hum Genet 52 : 827–834

155. Llerna A, Edman G, Cobaleda J, Ben'itez J, Schalling D, Bertilsson L (1993) Relationship between personality and debrisoquine hydroxylation capacity. Suggestion of an endogenous neuractive substrate of product of the cytochrome P4502D6. Acta Psychiatr Scand 87 : 23–28

156. Loh EY, Covirla SC, Serafini AT et al (1988) Human T-cell-receptor δ-chain: genomic organization, diversity and expression in populations of cells. Proc Natl Acad Sci USA 85 : 9714–9718

157. London SJ, Daly AK, Thomas DC, Caporaso NE, Idle JR (1994) Methodological issues in the interpretation of studies of the CYP2D6 genotype in relation to lung cancer risk. Pharmacogenetics 4 : 107–108

158. Lux SE, Becker PS (1989) Disorders of the red cell membrane skeleton: hereditary pherocytosis and hereditary elliptocytosis. In: Scriver CR, Beaudet AL, Sly WS, Valle D (eds) The metabolic basis inherited disease, 6th edn. McGraw-Hill, New York, pp 2367–2408

159. Marrack P, Kappler J (1987) The T cell receptor. Science 238 : 1073–1079

160. Martinez J, Holburn RR, Shapiro S, Erslev AJ (1974) Fibrinogen Philadelphia: a hereditary hypodysfibrinogenemia characterized by fibrinogen hypercatabolism. J Clin Invest 53 : 600

161. Masirimiriembwa CM, Johannson I, Hasler JA, Ingelman-Sundberg M (1993) Genetic polymorphism of cytochrome P450 CYP2D6 in Zimbabwean population. Pharmacogenetics 3 : 275–280

161a. Masters CL, Beyreuther K, Trillet M, Christen Y (eds) (1994) Amyloid protein precursor in development, aging, and Alzheimer's disease. Springer, Berlin Heidelberg New York

162. May A, Huehns ER (1976) The mechanism and prevention of sickling. Br Med Bull 32 : 223–233

163. McKusick VA (1972) Heritable disorders of connective tissue, 4th edn. Mosby, St Louis

164. McKusick VA (1995) Mendelian inheritance in man, 11th edn. Johns Hopkins University Press, Baltimore

165. Meyer UA, Skoda RC, Zanger UM, Heim M, Broly F (1992) The genetic polymorphism of debrisoquine/sparteine metabolism – molecular mechanisms. In: Pharmacogenetics of drug metabolism. Elsevier, Amsterdam, pp 609–624

166. Meyers DA, Marsh DG (1992) Allergy and asthma. In: King RA, Rotter JI, Motulsky AG (eds) The genetic basis of common disease. Oxford University Press, New York, pp 130–149

167. Miller AD (1992) Human gene therapy comes of age. Nature 357 : 455–460

168. Motulsky AG (1957) Drug reactions, enzymes and biochemical genetics. JAMA 165 : 835–837

169. Motulsky AG (1965) Theoretical and clinical problems of glucose-6-phosphate dehydrogenase deficiency. In: Jon-

xis JHP (ed) Abnormal haemoglobins in Africa. Blackwell, Oxford, pp 143–196 a

170. Motulsky AG (1970) Biochemical genetics of hemoglobins and enzymes as a model for birth defect research. In: Frazer FC, McKusick VA (eds) Congenital malformations. Excerpta Medica, Amsterdam, p 199

171. Motulsky AG (1972) Hemolysis in glucose-6-phosphate dehydrogenase deficiency. Fed Proc 31 : 1286–1292

172. Motulsky AG (1973) Frequency of sickling disorders in US blacks. N Engl J Med 288 : 31–33

173. Motulsky AG (1975) Glucose-6-phosphate dehydrogenase and abnormal hemoglobin polymorphisms – evidence regarding malarial selection. In: Salzano FM (ed) The role of natural selection in human evolution. North-Holland, Amsterdam, pp 271–291

174. Motulsky AG (1977) Ecogenetics: genetic variation in susceptibility to environmental agents. In: Human genetics. Proceedings of the 5th International Congress of Human Genetics, Mexico City, 10–15 Oct 1976. Excerpta Medica, Amsterdam, pp 375–385

175. Motulsky AG (1978) Multifactorial inheritance and heritability in pharmacogenetics. International Titisee Conference, Titisee, 13–15 Oct 1977. Hum Genet 1 [Suppl]:7–12

175 a. Motulsky AG (1995) Jewish diseases and origins. News and views. Nature [Genet] 9 : 99–101

175 b. Motulsky AG (1996) Nutritional ecogenetics: homocysteine-related arterioclerotic vascular disease, neural tube defects, and folic acid (Invited editorial). Am J Hum Genet 58 : 17–20

175 c. Motulsky AG (1996) Human genetic variation in nutrition and chronic disease prevention. 3 rd International Symposium on Infant Nutrition in the Prevention of Chronic Pathology, Sept 18–22, 1995, Alicante. Ergon, Madrid (in press)

176. Motulsky AG, Vogel F, Buselmaier W, Reichert W, Kellermann G, Berg P (eds) (1978) Human genetic variation in response to medical and environmental agents: pharmacogenetics and ecogenetics. International Titisee Conference, Titisee, 13–15 Oct 1977. Hum Genet 1 [Suppl]

177. Mueller RF, Hornung S, Furlong CE, Anderson J, Giblett ER, Motulsky AG (1983) Plasma paraoxonase polymorphism: a new enzyme assay, population, family, biochemical, and linkage studies. Am J Hum Genet 35 : 393–408

178. Mullan M, Crawford F (1993) Genetic and molecular advances in Alzheimer's disease. Trends Neurosci 16 : 398–403

179. Nagel RL, Labie D (1985) The consequences and implications of the multicentric origin of the Hb S gene. In: Stamatoyannopoulos G, Nienhuis A (eds) Experimental approaches for the study of hemoglobin switching. Liss, New York, pp 93–103

180. Nazar-Stewart V, Motulsky AG, Eaton DL, White E, Hornung SK, Leng Z-T, Stapleton P, Weiss NS (1993) The glutathione S-transferase Mu polymorphism as a marker for susceptibility to lung carcinoma. Cancer Res 53 : 2313–2318

181. Nebert DW, Goujon FM, Gielen JE (1972) Aryl hydrocarbon hydroxylase induction by polycyclic hydrocarbons: Simple autosomal dominant trait in the mouse. Nature [New Biol] 236 : 107

182. Neel JV (1949) The inheritance of sickle cell anemia. Science 110 : 64

183. Neel JV (1949) The detection of the genetic carriers of hereditary disease. Am J Hum Genet 1/2 : 19–36

184. Neel JV (1953) The detection of the genetic carriers of inherited disease. In: Sorsby A (ed) Clinical genetics. Mosby, St Louis, p 27

185. Nienhuis AW, Anagnou NP, Ley TJ (1984) Advances in thalassemia research. Blood 63 : 738–758

186. Neufeld EF (1974) The biochemical basis for mucopolysaccharidoses and mucolipidoses. Prog Med Genet 10 : 81–101

187. Neufeld EF, Muenzer J (1989) The mucopolysaccharidoses. In: Scriver CR, Beaudet AL, Sly WS, Valle D (eds) The metabolic basis of inherited disease, 6 th edn. McGraw-Hill, New York, pp 1565–1587

188. Newton BW, Benson RC, McCarriston CC (1966) Sparteine sulphate: a potent carpicious oxytocic. Am J Obstet Gynecol 94 : 234–241

189. Oettinger MA, Schatz DG, Gorka C, Baltimore D (1990) RAG-1 and RAG-2, adjacent genes that synergistically activate V(D)J recombination. Science 248 : 1517–1523

190. Old JM (1992) Haemoglobinopathies. In: Brock DH (ed) Prenatal diagnosis and screening. Churchill, London, pp 425–439

191. Old JM, Weatherall DJ, Wart RHT, Petrou M, Modell B, Rodeck CH, Warren R, Morsman JM (1985) First trimester diagnosis of the hemoblobin disorders. Ann N Y Acad Sci 445 : 349–356

192. Omenn GS, Motulsky AG (1978) "Ecogenetics": genetic variation in susceptibility to environmental agents. In: Cohen BH, Lilienfeld AM, Huang PC (eds) Genetic issues in public health and medicine. Thomas, Springfield, pp 83–111

193. Orkin SH, Kazazian HH Jr (1984) The mutation and polymorphism of the human β- globin gene and its surrounding area. Annu Rev Genet 18 : 131–171

193 a. Orkin SH, Motulsky AG (1995) Report and recommendation of the Panel to Assess the NIH Investment in Research on Gene Therapy. National Institutes of Health, Bethesda

194. Orkin SH, Alter BP, Itay C, Mahoney MJ, Lazarus H, Hobbins JC, Nathan DG (1978) Application of endonuclease mapping to the analysis and prenatal diagnosis of thalassemias caused by globin-gene deletion. N Engl J Med 299 : 166–172

195. Pauling L, Itano HA, Singer SJ, Wells IC (1949) Sickle cell anemia: a molecualr disease. Science 110 : 543

196. Penrose LS (1935) Inheritance of phenylpyruvic amentia (Phenylketonuria) Lancet 2 : 192–194

197. Perutz MF (1976) Structure and mechanism of haemoglobin. Br Med Bull 32 : 195–208

198. Prins HK, Oort M, Loos JA, Zürcher C, Beckers T (1966) Congenital nonspherocytic hemolytic anemia associated with glutathione deficiency of the erythrocytes. Blood 27 : 145

199. Propping P (1978) Pharmacogenetics Rev Physiol Biochem Pharmacol 83 : 124–173

200. Propping P (1984) Genetic aspects of neurotoxicity. In: Blum K, Manzo L (eds) Neurotoxicology. Dekker, New York, pp 203–218

201. Propping P (1989) Psychiatrische Genetik. Springer, Berlin Heidelberg New York

202. Pumphrey RSH (1986) Computer models of the human immunoglobulins. Immunol Today 7 : 206-

203. Romeo G (1977) Analytical review. Enzymatic defects of hereditary porphyrias: an explanation of dominance at the molecular level. Hum Genet 39 : 261–276

204. Romeo L, Osorio-Almeida L, Higgs DR et al (1991) α-globin structural genes. Blood 78 : 1589–1595

205. Rommens JM, Ianuzzi MC, Kerem B-S et al (1989) Identification of the cystic fibrosis gene: chromosome walking and jumping. Science 245 : 1059–1065

206. Rosenthal D, Kety SS (1968) The transmission of schizophrenia. Pergamon, Oxford

207. Sakai LY, Keene DR, Engvall E (1986) Fibrillin, a new 350 kD glycoprotein, is a component of extracellular microfibrils. J Cell Biol 103 : 2499–2509

208. Sandhoff K, Christomanou H (1979) Biochemistry and genetics of gangliosidoses. Hum Genet 50 : 107–143

209. Schimpl A (1991) Antikörper und Antikörpersynthese. In: Gemsa D et al (eds) Immunologie, 3rd edn. Thieme, Stuttgart, pp 16–29

210. Schloot W, Goedde HW (1974) Biochemische Genetik des Menschen. In: Vogel F (ed) Erbgefüge. Springer, Berlin Heidelberg New York, pp 325–494 (Handbuch der allgemeinen Pathologie, vol 9)

211. Schroeder WA, Huisman THJ (1978) Human gamma chains: structural features. In: Stamatoyannopoulos G, Nienhuis A (eds) Cellular and molecular regulation of hemoglobin switching. Grune and Stratton, New York, pp 29–45

212. Scriver CR (1969) Treatment of inherited disease: Realized and potential. Med Clin North Am 53 : 941–963

213. Scriver CR, Clow CL (1980) Phenylketonuria and other phenylalanine hydroxylase mutants in man. Annu Rev Genet 14 : 179–202

214. Scriver CR, Rosenberg LE (1973) Amino acid metabolism and its disorders. Saunders, Philadelphia

215. Scriver CR, Beaudet AL, Sly WS, Valle D (eds) (1989) The metabolic basis of inherited disease, 6th edn. McGraw-Hill, New York

216. Sears DA (1978) The morbidity of sickle cell trait. A review of the literature. Am J Med 64 : 1021–1036

217. Seegmiller JE (1983) Disorders of purine and pyrimidine metabolism. In: Emery AEH, Rimoin DL (eds) Principles and practice of medical genetics. Churchill Livingstone, Edinburgh, pp 1286–1305

218. Seegmiller JE, Rosenbaum FM, Kelly WN (1967) An enzyme defect associated with a sex-linked human neurological disorder and excessive purine synthesis. Science 155 : 1682

219. Seidegård J, Pero RW, Markowitz MM, Roush G, Miller DG, Beattie EJ (1990) Isoenzymes(s) of glutathione transferase (class Mu) as a marker for the susceptibility to lung cancer: a follow-up study. Carcinogenesis 11 : 33–36

220. Semenza G (1981) Intestinal oligo- and disaccharides. In: Randle PJ, Steiner DF, Whelan WJ (eds) Carbohydrate metabolism and its disorders, vol 3. Academic, London, pp 425–479

221. Serjeant GR (1974) The clinincal features of sickle cell disease. Clinical Studies. IV. North-Holland, Amsterdam

222. Sinnott EW, Dunn LC, Dobzhansky T (1958) Principles of genetics, 5th edn. McGraw-Hill, New York

223. Slighton JL, Blechl AE, Smithies O (1980) Human fetal G-gamma- and A-gamma-globin genes: complete nucleotide sequences suggest that DNA can be exchanged in these duplicated genes. Cell 21 : 627–638

224. Sly WS, Achard DT, Kaplan A (1977) Correction of enzyme deficient fibroblasts: evidence for a new type of pinocytosis receptor which mediates uptake of lysosomal enzymes (Abstr). Clin Res 25 : 471A

225. Smithies O (1955) Grouped variations in the occurence of new protein components in normal human serum. Nature 175 : 307–308

226. Smithies O (1955) Zone electrophoresis in starch gels: group variations in the serum proteins of normal human adults. Biochem J 61 : 629–641

227. Spielberg SP, Gordon GB, Blake DA, Goldstein DA, Herlong HF (1981) Predisposition to phenytoin hepatotoxicity assessed in vitro. N Engl J Med 305 : 722–727

228. Spranger J (1972) The systemic mucopolysaccharidoses. Ergeb Inn Med Kinderheilkd 32 : 165

229. Spritz RA, DeRiel JK, Forget BG, Weissman SM (1980) Complete nucleotide sequence of the human delta-globin gene. Cell 21 : 639–646

230. Stamatoyannopoulos G (1972) The molecular basis of hemoglobin disease. Annu Rev Genet 6 : 47

231. Stamatoyannopoulos G, Nienhuis AW (1994) Hemoglobin switching. In: Stamatoyannopoulos G, Nienhuis AW, Majerus PW, Varmus H (eds) The molecular basis of blood diseases, 2nd edn. Saunders, Philadelphia, pp 107–155

232. Strominger J (1989) Developmental biology of T-cell receptors. Science 244 : 943–950

233. Swift M, Chase C (1979) Cancer in families with Xeroderma pigmentosum. J Natl Cancer Inst 62 : 1415–1421

234. Swift M, Reitnauer PJ, Morrell D et al (1987) Breast and other cancers in families with ataxia-teleangiectasia. N Engl J Med 316 : 1289–1294

235. Taliaferro WH, Huck JG (1923) The inheritance of sickle cell anaemia in man. Genetics 8 : 594

236. Thein SL, Wainscoat JS, Lynch JR, Weatherall DJ, Sampietro M, Fiorelli G (1985) Direct detection of β*39 thalassaemic mutation with Mae 1. Lancet 1 : 1095

237. Thierfelder L, McRae C, Watkins H et al (1993) A familial hypertrophic cardiomyopathy locus maps to chromosome 15 q2. Proc Natl Acad Sci USA 90 : 6270–6274

238. Thierfelder L, Watkins H, MacRee C et al (1994) α-tropomyosin and cardiac troponin T mutations cause familial hypertrophic cardiomyopathy: a disease of the sarcomere. Cell 77 : 701–702

239. Tolleshaug H, Goldstein JL, Schneider WJ, Brown MS (1982) Posttranslational processing of the LDL receptor and its genetic disruption in familial hypercholesterolemia. Cell 30 : 715–724

240. Tonegawa S (1983) Somatic generation of antibody diversity. Nature 302 : 575–581

241. Tonegawa S (1986) Somatic generation of antibody diversity. Nature 302 : 575–581

242. Tynan K, Comeau K, Pearson M et al (1993) Mutation screening of complete fibrillin-1 coding sequence: report of five new mutations, including two in 8-cysteine domains. Hum Mol Genet 2 : 1813–1821

243. Udenfriend S, Cooper JR (1952) The enzymatic conversion of phenylalanine to tyrosine. J Biol Chem 194 : 503

243 a. Van Broeckhoven C (1995) Presenilins and Alzheimer's disease. Nature Genet 11 : 230–232

244. Van Hoof F, Hers HG (1964) Ultrastructure of hepatic cells in Hurler's disease (gargoylism). C R Acad Sci [D] 259 : 1281

245. Van Poppel G, de Vogel N, Van Bladeren PJ, Kok FJ (1992) Increased cytogenetic damage in smokers deficient in glutathione S-transferase isozyme Mu. Carcinogenesis 13 : 303–305

246. Vogel F (1959) Moderne Probleme der Humangenetik. Ergeb Inn Med Kinderheilkd 12 : 52–125

247. Vogel F (1984) Relevant deviations in heterozygotes of autosomal-recessive diseases. Clin Genet 25 : 381–415

248. Voigtlaender V (1977) Genetik der Neurodermitis. Z Hautkr 5 [Suppl]:65–71

249. Vosberg H-P (1994) Myosin mutations in hypertrophic cardiomyopathy and functional implications. HERZ (in press)

250. Waller HD, Benöhr A C (1976) Enzymdefekte in Glykolyse und Nucleotidstoffwechsel roter Blutzellen bei nichtspherocytären hämolytischen Anämien. Klin Wochenschr 54 : 803–850

251. Watkins H, MacRae C, Thierfelder L et al (1993) A disease locus for familial hypertrophic cardiomyopathy maps to chromosome 1 q3. Nat Genet 3 : 333–337

252. Watson JD, Gilman M, Witkowski J, Zoller M (1992) Recombinant DNA, 2nd edn. Freeman, New York

253. Weatherall DJ (1994) The thalassemias. In: Stamatoyannopoulos G, Nienhuis AW, Majerus PW, Varmus H (eds) The molecular basis of blood diseases, 2nd edn. Saunders, Philadelphia, pp 157–205

254. Weatherall DJ, Clegg JB (1981) The thalassemia syndromes, 3rd edn. Blackwell, Oxford

255. Wedlund PJ, Aslanian WS, McAllister CB, Wilkinson GR, Branch RA (1984) Mephenytoin hydroxylation deficiency in Caucasians: frequency of a new oxidative drug metabolism polymorphism. Clin Pharmacol Ther 36 : 773–780

256. Weinshilboum RM (1983) Biochemical genetics of catecholamines in humans. Mayo Clin Proc 58 : 319–330

257. Weinshilboum RM, Sladek SL (1980) Mercaptopurine pharmacogenetics: monogenic inheritance of erythrocyte thiopurine methyltransferase activity. Am J Hum Genet 32 : 651–662

258. Wewers M, Casolaro A, Sellers SE et al (1987) Replacement therapy for alpha-1-antitrypsin deficiency associated with emphysema. N Engl J Med 316 : 1055–1062

259. Wiwel NA (1993) Germ-line gene modification and disease prevention: some medical and ethical perspectives. Science 262 : 553–538

260. White JM (1976) The unstable haemoglobins. Br Med Bull 32 : 219–222

261. Wiencke JK, Kelsey KT, Lamela RA, Toscano WA Jr (1990) Human glutathione S-transferase deficiency as a marker of susceptibility to epoxide-induced cytogenetic damage. Cancer Res 50 : 1585–1590

262. Wilkie AOM, Buckle VJ, Harris PC et al (1990) Clinical features and molecular analysis of the αthalassemia/mental retardation syndromes. I. Cases due to deletions involving chromosome band 16 p13.366. Am J Hum Genet 46 : 1112

263. Wilkie AOM, Zietlin HC, Lindenbaum RH et al (1990) Clinical features and molecular analysis of the α thalassemia/mental retardation syndromes. 2 cases without detectable abnormality of the α globin complex. Am J Hum Genet 46 : 1127

264. Williams AF, Barclay AN (1988) The immunoglobulin superfamily-domains for all surface recognition. Annu Rev Immunol 6 : 381–405

265. Williamson R (1976) Direct measurement of the number of globin genes. Br Med Bull 32 : 246–250

266. Winkelstein JA, Colten HR (1989) Genetically determined disorders of the complement system. In: Scriver CR, Beaudet AL, Sly WS, Valle D (eds) The metabolic basis of inherited disease, 6th edn. McGraw-Hill, New York, pp 2711–1738

267. Winters RW, Graham JB, Williams TF, McFalls VC, Burnett CH (1957) A genetic study of familial hypophosphatemia and viatmin D-resistant rickets. Trans Assoc Am Physicians 70 : 234–242

268. Wolf CR, Smith CAD, Bishop T, Forman D, Gough AC, Spur NK (1994) CYP2D6 genotyping and the association with lung cancer susceptibility. Pharmacogenetics 4 : 104–106

269. Woo SLC, Lidsky AS, Guettler F et al (1983) Cloned human phenylalanine hydroxylase gene allows prenatal detection of classical phenylketonuria. Nature 306 : 151–155

270. Woo SLC, Güttler F, Ledley FD, Lidsky AS, Kwok SCM, DiLella AG, Robson KJH (1985) The human poenylalanine hydroxylase gene. In: Berg K (ed) Medical genetics past, present, future. Liss, New York, pp 123–135

271. Yoshida A, Beutler E (1983) G-6-PD variants: another up-date. Ann Hum Genet 47 : 25–38

272. Zhong S, Howie AF, Ketterer B, Taylor J, Hayes JD, Beckett GJ, Wathen G, Wolf CR, Spurr NK (1991) Glutathione-S-transferase Mu locus: use of genotyping and phenotyping assays to assess association with lung cancer susceptibility. Carcinogenesis 12 : 1533–1537

8 Gene Action: Developmental Genetics

Order is heaven's first law.

Alexander Pope, (1688–1744)

8.1 Genetics of Embryonic Development

Biochemical and molecular genetics has taught us much about the structure of genes and about the genetic control of enzymes and functional proteins. The lessons regarding the genetic basis of embryonic development, however, have been much less satisfactory. The genetics of development is only beginning to be charted on the map of our knowledge of molecular genetic mechanisms, but studies with methods from molecular genetics are now starting to elucidate this field.

As in other fields of molecular biology, developmental genetics often uses experimental organisms other than humans because human experimentation is subject to obvious limitations. A textbook of human genetics cannot review the whole field. A broad outline is sketched below, indicating where observations on humans may contribute some additional information. Developmental genetics is historically based on classical developmental mechanics *(Entwicklungsmechanik)* and developmental physiology, which flourished in the first decades of the twentieth century, based on the work of Roux, Driesch, Spemann, Kühn, Waddington, and Hadorn.

The Basic Problem of Developmental Genetics. The fundamental genetic problem during embryonic development is differentiation: how is it possible that groups of cells assume different functions despite the fact that they have identical genomes? Research in molecular biology has highlighted a number of mechanisms for regulation and differentiation of the activity of genes. We know, for example, that methylation of certain bases in the DNA influences not only their rates of spontaneous mutation (Chap. 9) but gene action as well; methylation differences between paternal and maternal genomes lead to variable influences on development of the embryo. Complex models for gene regulation have been proposed theoretically but for a long time, evidence of their applicability has been slender or even nonexistent. The crucial question remains unanswered to this day: what causes differences between cells of the early embryo, enabling them to differentiate? The answer will probably come not from considering merely DNA and its interactions with mRNA, and proteins. The structure of cells – especially cell nuclei and the spatial relationships of chromatin and other components within these nuclei – will have to be considered [12].

We first discuss very briefly some general aspects of gene action and its regulation. We then present an overview of the phases of embryonic development in humans. This leads to the description of discoveries that have improved our understanding of embryonic development and its disturbances, such as alternative splicing, genomic imprinting, homeobox genes mutations, and the use of transgenic animals for research in developmental genetics. Finally, birth defects are discussed. In all of these discussions we return repeatedly to aspects described in the earlier chapters on enzyme defects, hemoglobin variants, pathogenetic mechanisms in dominantly inherited diseases, and immunogenetics.

Gene Action in Eukaryotes, Including Humans. The structure of the genetic material in eukaryotes, including humans, is described in Sect. 3.1 (Fig. 3.17). Genes have an exon-intron structure; the regions in the base sequence where introns are removed from the primary mRNA transcript have specific DNA sequences [45, 87]. Outside the transcribed sequence, at its 5′ end, and separated from it by a number of base pairs, we find the TATA box, a promoter sequence which permits RNA polymerase to start transcription. A special signal helps the polymerase to recognize the coding DNA strand, avoiding transcription of the complementary strand. Farther upstream, DNA sequences are found that can bind to certain proteins, enhance transcription activity, and are therefore termed enhancer regions. This protein binding appears to play an important role for the regulation of gene activities. Some general rules for such regulation have been worked out mainly in bacteria. While these principles are not directly applicable to the regulation of gene activity in higher organisms, they do provide simple models from which further exploration applicable to mammals and humans may start. Two classical examples are presented below, one for negative and the other for positive control.

In bacteria some genes are active only when their specific activity is required. The classical example here is the lactose operon of *E. coli* [39], in which three closely linked structural genes are under the common control of such control genes. They become active when the substrate – in this case, lactose – is available as energy source. In addition to such "negative control," there is also positive control; here transcription requires a special protein in addition to an open operator.

Function of Regulatory Mechanisms. An important function for which bacteria have been selected during evolution is the optimal utilization of varying energy sources for growth. Control systems such as the lactose operon fulfill this function; bacteria use energy resources for producing lactose – cleaving enzymes only when lactose is available. Hence the feedback circle of negative control helps to channel energy into the most useful direction. It can be expected that during evolution from prokaryotes to highly complex mammals such as humans, a stepwise adaptation to the increasingly complex problems of regulation and, especially, differentiation has occurred. It would therefore be logical in research to follow the lead of evolution: to investigate more and more complex biological systems, adding more and more complexity to the interpretive models.

Britten and Davidson [7, 13, 14] have proposed such a model for gene regulation in higher organisms. Based on features of the regulation models in micro-organisms they have introduced logical extensions for the more complex regulation requirements in differentiation. Four kinds of genes are assumed: producer genes; receptor genes that are linked to producer genes and induce transcription under the influence of activator substances produced under the influence of integrator genes; and sensor genes serving as binding sites for agents, inducing specific activity patterns in the genome. During the more than 20 years since this model was proposed, little progress has been made in testing the theoretical framework. A few transcription factors have become known [45], and their binding sites at the DNA as well as binding proteins have been identified [88]. However, no one knows as yet how the entire network fits together.

A number of aspects are discussed below, but we are not yet able to provide an integral view of genetic control of development and its disturbances. This problem remains a key issue problem in biomedical research.

Alternative Splicing [88]. In our discussion of immunogenetics (Sect. 7.4), we have seen how functional diversity between cells can arise at the DNA level by gene rearrangements in which only a few genes from a wide array of possibilities end up in the functional cell. This principle has been identified in immunoglobulins and T cell receptors. It is well suited for cells which have highly specialized functions, but which must also maintain the ability for extensive multiplication [2]. Its shortcoming, however, is that genes are altered irreversibly. Many other specialized and terminally differentiated cells, such as cells in the brain, have lost their ability to divide. Here another mechanism has been established: alternative splicing. The genes themselves, and the pre-mRNA, are identical but the products of mRNA splicing differ in cells with different functions. As in rearrangements at the gene level, alternative splicing increases the coding capabilities "of the genome by expanding the ability of genes to generate protein diversity" [2]. However, the regulation of diversity is shifted to the posttranscriptional level. This mechanism is very widespread in eukaryotes; it is even possible that it developed earlier than the "normal" splicing mechanism.

A early example is the peptide hormone calcitonin (Fig. 8.1) [1]. This hormone is found in the thyroid and the hypothalamus, where it occurs together with a very similar polypeptide, the calcitonin-gene related product (CGRP). This protein is identical to calcitonin over 78 amino acids but has 128 amino acids altogether. The sequence of both proteins is determined by the same gene, whose primary transcript is processed differently: a complete exon is eliminated from the transcript that determines CGRP, and another part is added. This process predominates in the hypothalamus but not in the thyroid. These studies have been performed in rats, but the phenomenon has also been confirmed in humans [2].

There are other instances that follow the same rules, an especially complicated and well-analyzed one being the α-tropomyosin gene [88]. This protein has been found in no less than nine different molecular forms. Another interesting example is the mRNA for myelin basic protein in the mouse (Fig. 8.2). There is also an increasing number of examples in humans [2]. For example, the amino acid sequence difference between the cell wall fixed form and the free form of the immunoglobulin IgM is caused by alternative splicing. Other examples include the genes for the hormone somatotropin, fibrinonectin, γ-fibrinogen, and many more. Looking at the secondary structure of the gene-determined proteins, we can often observe that exons subject to alternative splicing determine insertions in surface loops of the proteins. Here alternative splicing generates "variants with different site affinities and/or functions without disturbing the core structure of the protein. . . . Mutually ex-

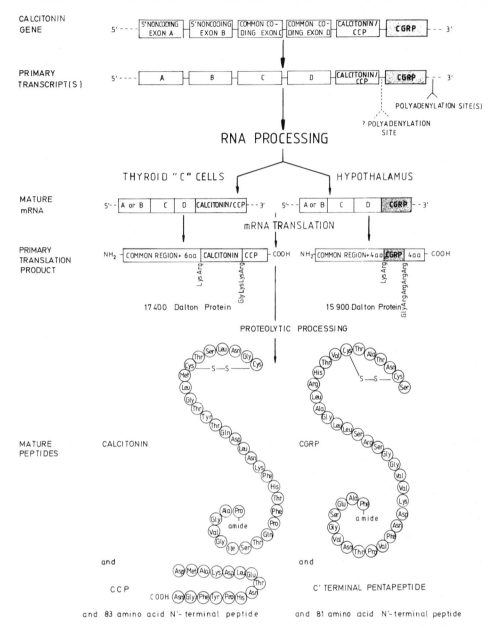

Fig. 8.1. Mechanism of tissue-specific expression of the calcitonin gene by alternative splicing. *Forked arrow,* difference in mRNA processing between thyroid and hypothalamus. Some aspects of this pathway are still hypothetical, but the genes and the end products (calcitonin and CGRP) are established. (From Amara et al. 1982 [1])

clusive exons ... seem to code for protein sequences that interact with other proteins" [2]. Alternative splicing is an effective means for creating differences in cell function.

The structure of the interphase nucleus (Sec. 3.1.3.3) should be considered for a more thorough understanding of differentiation. It has been shown, for example, that localization of chromosome domains may differ in cells with different, specialized functions [60]. This may influence gene action.

Possible mechanisms are now under discussion [12].

The number of diverse mRNA species that are translated in the early embryo is high; the pattern changes with the stage of development. Therefore, a large fraction of the genes seems to be necessary for early development.

In echinoderms the first stages of development until gastrula or even postgastrula development are determined exclusively or predominantly by the maternal

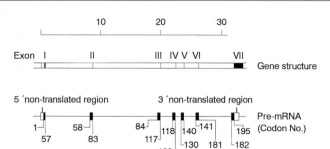

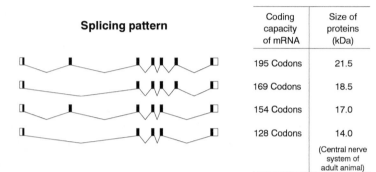

Fig. 8.2. Alternative splicing of myelin basic protein pre-mRNA in the mouse. Exon I comprises, in addition to the sequences of the 5′noncoding area *(open)* the codons for 57 amino acids. Exon VII codes for the 14 terminal codons and the 3′noncoding area *(open)*. *Numbers (pre-mRNA)* codons. The four possible splicing variants are seen at the bottom of the figure. (From Knippers et al. 1990 [45])

genome: the zygote receives a pool of maternal mRNA which directs these early processes. Moreover, tRNA and ribosomes are also of maternal origin. Different parts of the paternal genome are switched on at slightly different times.

Investigations in the mouse give a slightly different picture [17, 59]. Embryonic RNA synthesis starts as soon as the late two-cell stage. However, such analyses do not distinguish between maternal and paternal genomes, and RNA production does not mean that protein biosynthesis occurs as well. This problem can be solved, however, if the father can be distinguished from the mother by genetic markers. In this case the appearance of paternal markers in the embryo indicates the latest possible stage at which embryonic gene products begin to be formed. (It does not exclude the possibility that the maternal genome is turned on earlier.) Such studies have shown that some paternal markers, such as HPRT (see Sect. 7.2.2.6) and the HY antigen, first become visible at the eight-cell stage. Others, such as β-galactosidase and β-glucosidase, can be found as nearly as the four-cell stage, and at least one – the β_2 macroglobulin Fig. 5.15 – turns up even at the two-cell stage. Moreover, it has been shown that for the X-linked HPRT gene both parental genomes are active before X-inactivation (Sect. 2.2.3.3) since enzyme activity in female embryos is twice that in males.

Different Contribution of Maternal and Paternal Genotype to the Child's Phenotype? As noted above, the paternal genotype is switched on early in development. This does not mean, however, that maternal and paternal contributions to the de-

velopment of the young embryo are in fact equal, especially since the young zygote receives a large amount of maternal RNA. There are indeed biological phenomena that suggest a larger contribution from the mother. For example: If in a crossing between horse and donkey the dam is a horse and the sire a donkey, a mule results; if the dam is a donkey, the offspring is a hinny, which looks much more like a donkey than the mule (see also [10]).

Convincing data for this in humans were lacking for a long time. Some findings showed a stronger contribution of maternal than paternal genotypes to ridge patterns of the finger tips, palms, and toes [47, 69]. Such data, however, were set aside as "difficult to interpret." Such data were excluded from the mainstream of thinking and conceptualization in human genetics because they did not conform to Mendelian expectations of equal contributions from the genotypes of both parents (except X-linked genes). The genetic theory founded on Mendel's paradigm is an excellent, strong theory with high explanatory power. However, as potentially the case with any good theory, it seduced scientists to neglect or "explain away" observations that seemed not to fit the paradigm.

8.2 Genomic Imprinting [91]

Differences in impact on the developing embryo between maternal and paternal genomes have been named "genomic imprinting." This is not a fortuitous term, as "imprinting" was introduced by the etholo-gist K. Lorenz [57] for the ability of animals to learn a certain behavior if the appropriate stimulus is pro-vided during a sensitive period of their early life. The term "genomic imprinting" was introduced to refer to quite a different biological phenomenon dur-ing embryological development than the behavioral imprinting studied by ethologists. Hall [34, 35] cites six kinds of observations to suggest its existence: (a) observations on the results of pronuclear trans-plantation experiments in mice, (b) triploid pheno-types in humans, (c) the expression of certain chro-mosomal disomies in mice and humans, (d) the phe-notypic expression of chromosomal deficiencies in mice and humans, (e) the expression of transferred genes in transgenic mice, and (f) the expression of specific genes in mice and humans.

(a) Zygotes in mice with only paternal or only mater-nal genomes as a result of pronucleus transplantation show characteristic anomalies: those with only pater-nal chromosomes exhibit relatively normal develop-ment of membranes and placentas but very poor de-velopment of embryonic structures, whereas zygotes with only maternal chromosomes show much better embryonic development but very poor membranes and placentas. Therefore maternal and paternal ge-nomes are both necessary for normal development; their contributions are different. In humans, paternal disomy may occur, so that zygotes have only paternal chromosomes. The result is a hydatitiform mole – without embryonic tissue [55]. On the other hand, teratomas – which have all three embryonic germ lay-ers but no placental tissue – possess two haploid sets of maternal chromosomes [24]. (b) Human triploids with two paternal and one maternal chromosome sets have a large cystic placenta and a severely growth – retarded embryo with a number of anoma-lies. Embryos with two maternal sets have a small and underdeveloped placenta. These observations on embryonal tumors and on triploidies suggest an especially important role of the paternal genome in the development of placenta and membranes while the maternal contribution may be more important for the embryo proper. (c) Uniparental chromosomal disomies – with absent genetic contribution by one parent – have often been observed in mice. There are definite phenotypic differences, depending on the origin of the duplicated segment; moreover, dis-tortions of sex ratios among offspring indicate a loss of zygotes by early embryonic death. Since identifica-tion of individual chromosomes has become possible with the help of DNA variants, cases with uniparental disomy have repeatedly been found in humans; these may show, among other clinical signs, intrauterine and postnatal growth retardation, mental retardation, and – in mosaics – asymmetric growth of trunk and

limbs [73, 79]. For example, some patients with Pra-der-Willi or Angelman syndrome (see below) were shown to have uniparental disomy. (d) In general, monosomies of entire chromosomes or even of their parts are tolerated very poorly; they often lead to ear-ly death of the zygote [18]. Nevertheless, there are some examples in which small deletions lead to char-acteristic syndromes (see Sect. 2.2.2), and these syn-dromes differ depending on whether the deletion comes from the father or the mother.

Prader-Willi and Angelman Syndromes. Characteris-tics of Prader-Willi syndrome [68] include severe obesity with hyperphagia beginning in early child-hood, hypogonadotropic hypogonadism, small hands and feet, mental retardation, and a characteristic fa-cies (Fig. 8.3; 8.4). In some cases a visible chromoso-mal deletion exists [56], which was shown to come from the father [8, 46], in other instances the deletion is so small that it can be identified only by molecular methods. Patients with Angelman syndrome [3] are

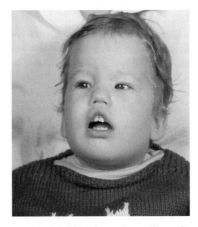

Fig. 8.3. Child with Prader-Willi syndrome

Fig. 8.4. Child with Angelman syndrome. (Both photos, cour-tesy of Dr. G. Tariverdian)

severely mentally retarded and have a characteristic facies (large mouth with drooling, prognathism, protruding tongue). In addition to bizarre atactic movements, they often laugh inappropriately (happy puppet syndrome). The condition is caused by deletions of the very same 15q11-13 segment as in Prader-Willi syndrome except that it comes from the mother. The mechanism in either syndrome may not necessarily be a deletion. In some instances uniparental disomy has been discovered; in others point mutations exist. However, the association between the parental origin of the defect and the respective clinical syndrome is always maintained.

There are a number of other syndromes in which parental origin may be important (see Table 2.7); especially "contiguous gene syndromes" such as Miller-Dieker syndrome (17p-; primarily paternal origin), di George syndrome (22q-; maternal deletion), and cri du chat syndrome (5p-; paternal). As is shown in Chap.10, an increasing number of cancers is being explained by loss of chromosomes and chromosome parts. Here, too, unequal participation of paternal and maternal chromosome losses has been described.

Transgene Expression. Regarding point (e) above (transgene expression), directly transferring genes into mouse embryos and studying their expression has become a powerful tool in developmental genetics. In about 25% of these animals expression of the transferred genes in subsequent generations depends on the sex of the transmitting parent [34]. In such cases nonexpression appears to be associated with methylation of the transgene (see Sect. 9.4).

As regards point (f), imprinting has been shown to explain variation in the age at onset in certain hereditary diseases. The classical examples are Huntington disease (143100) and myotonic dystrophy (160900; see Sect. 4.1.7). Both diseases are caused by a novel and unique mutational mechanism (amplification of a series of repetitive base triplets), and a progressive shift of the age at onset from later to earlier is sometimes observed in successive generations (anticipation). In both diseases the sex of the parent who transmitted the disease may determine the age at onset and the severity of the clinical manifestations. In 5%–10% of families in which the Huntington disease gene is transmitted through the father, a severe, rigid, juvenile form of the disease is observed [70, 71]. In 10%–20% of families in which myotonic dystrophy is transmitted maternally, a severe, hypotonic congenital form of the disease occurs [36]. In several other hereditary diseases such as in cerebellar ataxia, Beckwith-Wiedemann syndrome, familial glomus tumor, and retinoblastoma, imprinting has also been suggested. However, only a few hereditary diseases [62] have been examined systematically. Hall [34] has drawn idealized pedigrees to show how pedigrees in which paternal or maternal imprinting occurs may appear (Fig. 8.5).

Imprinting can be demonstrated more easily and definitely in mice than in humans. It is therefore a good strategy to search for regions of the human genome that are homologous to definitely imprinted sequences in the mouse by consulting the Oxford grid [62; 74], which charts linkage homologies between mice and humans. For example, a distal region of chromosome 2 of the mouse is homologous to the region of human chromosome 15 involved in Prader-Willi and Angelman syndromes in humans. In fact, comparable phenotypic effects have been observed [9] in the two species.

A study was designed to find out how widespread imprinting might be during embryonic development of the mouse [82]. The technique used was two-dimensional (2D) electrophoresis, which allows the separation of a very large number of proteins on a supporting medium such as a large sheet of filter paper [44]. The positions of the protein spots and their staining intensity which reflects the amount of protein can then be studied. The entire spectrum of all gene-determined proteins at a specific phase of development can be investigated.

Livers of mice from an inbred line give identical 2D patterns and approximately 2000 spots can be visualized. 2D analysis of the livers from the F_1 generation resulting from reciprocal crosses, (i.e., crossing males of strain A with females of strain B and vice versa) would be expected to give an identical pattern unless the parental sex influences the outcome of gene expression. Two inbred strains (DBA and C57BL) were crossed in this manner. The strains differ in about 200 protein spots. Differences between the offspring of these reciprocal crosses were observed in 11% of the "spots" and point to imprinting. Imprinting may therefore be fairly common. Most differences were related to staining intensity of the spots and suggested different amounts of gene products depending upon parental sex. All findings were reproducible in repeated experiments. Transmission of the maternal form of the variant was more frequent than transmission of the paternal form.

The basic mechanisms involved are under investigation. It is likely that several mechanisms are involved. Differences in methylation of DNA bases have been advocated most frequently. Methylation of DNA, and especially of bases in critical regions such as CpG islands "upstream" of transcribed genes, appears to prevent such genes from being expressed; demethylation, on the other hand, appears to enhance their mutation rate (Sect. 9.4) [11]. Methylation has been studied extensively [42; 55]. The type of methylation considered most commonly is that from cytosine to 5-methylcytosine (Fig. 9.22). Imprinting probably occurs largely during gametogenesis; differential imprinting inherited from parents must be erased in the germ cell lineage of each individual.

PATERNAL

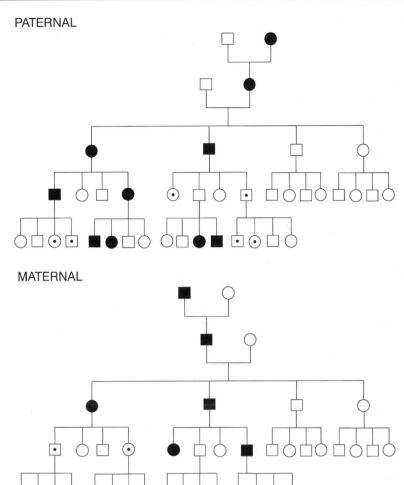

MATERNAL

Fig. 8.5. Idealized pedigrees for maternal and paternal imprinting. An imprinted allele is transmitted according to Mendel's law but its expression is determined by the sex of transmitting parent. The terms "maternal" and "paternal imprinting" are used to imply that there is phenotypic expression of the abnormal allele when transmitted from the mother or the father. Hence, there are a number of nonmanifesting carriers. ⊡, ⊙ gene carriers not manifesting the disease. (From Hall 1990 [34])

What is the biological function of imprinting, if there is any? Here, only speculations are possible; for example, it may play an important role in maintaining sexual reproduction by avoiding the genetically detrimental effects of parthenogenesis [79]. Another possibility is more elasticity during individual development: by using only one copy of particularly important gene, the other allele is kept "in reserve" for critical periods of rapid growth and cell duplication.

8.3 Transgenic Animals and Mouse Teratocarcinomas

8.3.1 Transgenic Animals and Related Methods

In recent years transgenic animals have become one of the most informative systems for the analysis of gene action [27, 28, 40]. The methods of gene transfer are described in Chap. 19. Suffice it to say here that transgenic mice are produced by injecting DNA into the nucleus of the fertilized oocyte before the male and female pronucleus have fused. The gene to be studied is integrated at random and is often expressed. This permits studying aspects of gene action. For example, the role of certain genes during normal embryonic development but also the role of genes in genetic diseases can be studied. An interesting application has been the analysis of specific single genes in transgenic mice as a first approach towards studying their contribution to the pathogenesis of more complex human diseases. Thus, atherosclerosis has been produced in transgenic mice by a variety of genes involved in cholesterol and lipoprotein metabolism. Another example of such research is the influence of certain genes on regulation of normal and elevated blood pressure. The renin-angiotensin system is a key system for electrolyte homeostasis and blood pressure regulation. A mouse renin gene was introduced into rats; it led to hypertension, confirming that renin may under certain conditions cause hypertension. In other studies human angiotensinogen and renin genes were transferred into rats. These animals were then studied regarding the role of these genes in normal and abnormal blood pressure regulation [23, 64].

In many instances, however, it is more interesting to study phenotypic consequences of the loss of certain genes than the effects of additional genes. Here a technique for producing so-called "knock-out" mice has been developed. The defective gene is introduced by a recombinational step which, at the same time, eliminates the normal gene. This cannot be achieved by direct DNA transfer to the fertilized oocyte because spontaneous recombination is too rare. Therefore embryonic stem cells are explanted from the blastocyst. The genes to be studied are transferred, and the recombinant cells are selected by certain selection systems and are reimplanted into the blastocyst. Such an animal model has been constructed, for example, for cystic fibrosis [76]. Mice in which the CFTR gene (Sect. 3.1.3.9) was disrupted by such "gene targeting" showed many features of cystic fibrosis, such as failure to thrive, meconium ileus, and alteration of mucous and serous glands. Such transgenic mice may be reasonable models for studying pathogenic mechanisms and somatic gene therapy. However, in many instances the results with transgenic mice and related techniques have been disappointing since the goal of reproducing a disease identical to one found in humans was not realized.

Results with transgenic animals are discussed in various chapters of this book. However, such animals have been constructed not only for research purposes; they are also starting to be used in the production of therapeutically valuable proteins, for example, in their milk. Sheep producing α_1-antitrypsin in large amounts are one example. This procedure has become popular as "pharming."

8.3.2 Mouse Teratocarcinomas as Research Tools for Investigation of Early Development [38]

Biochemical investigations at the cellular level on early stages of mammalian development are hampered by the scarcity of material. Many critical events take place in very small cell populations, and cells at different stages of differentiation are located close together. Recently it has become possible to circumvent some of these difficulties by the use of mouse teratocarcinoma cells. Testicular or ovarian teratocarcinomata occur spontaneously or can be induced in a number of inbred mouse stains. These contain a wide variety of tissues corresponding to derivatives of the three embryonic germ layers. Teratocarcinoma-derived cells can be obtained in cell culture; here they may grow and differentiate into several tissues. A number of cell lines of early embryolike cells have been established; when injected into mice, they may give rise to tu-

mors containing a variety of differentiated cell types. Such cells show many similarities to normal embryonic cells and are abundant. They can therefore be utilized as model systems for various aspects of differentiation.

Such studies have provided evidence, for example, of the common occurrence of most mRNA species in early embryonic cells and in precursors of blood cells (myeloblasts), while globin mRNA has been found in myeloblasts but not in early embryonic cells. Other experiments have shown both X chromosomes to be genetically active in clonal cultures of undifferentiated female cells; inactivation of one X occurs together with differentiation. The system seems to be especially well suited for investigating the role of cell surface antigens in differentiation. The discussion on the major histocompatibility complex (MHC) in Sect. 5.2.5 mentions that such antigens may have a function in cell differentiation. Early embryonic cells have been shown to be completely devoid of the MHC H-2 antigens (the mouse counterparts of the human HLA specificities). Other antigens, however, are present on these cells.

8.4 Later Phases of Embryonic Development, Phenocopies, Malformations

In later phases of embryonic development the differentiation of organ systems, extremities, head, and brain forms an organism. The occurrence of many inherited abnormalities indicates complex genetic control. These genetic mechanisms may be disturbed by a variety of external influences such as oxygen lack, ionizing radiation, infections with the rubella or cytomegalic inclusion viruses, and drugs such as thalidomide and ethanol. Knowledge of such teratogenic agents is important for preventing damage to the fetus. Details are not given here since, apart from possible differential metabolism in the mothers and variable susceptibility of the fetus to teratogenic agents, the subject of human teratology, although clinically important in the differential diagnosis of birth defects, is beyond the scope of this book.

Indications for Interaction Between Genetic and Nongenetic Factors in Malformation Production. Since St. Hilaire (1832–1836) produced malformations in chicken embryos by inhibiting gas exchange by covering the eggs with lacquer, many investigators have administered a variety of agents in the effort to disturb embryonic development. Most of these experiments have been carried out in the hope of obtaining more information on the mechanisms of normal development. These hopes were fostered by the observation that phenotypes similar to those produced by gene mutation are sometimes obtained (phenocopies; Goldschmidt 1935 [26]). In general these studies have not fulfilled their expectations for an understanding of birth defects and are not discussed further.

One aspect, however, is interesting: Some studies showed that genetic factors may be important even for induction of malformation by exogenous agents [54]. For example, the phenotype "rumplessness" which occurs as a genetic anomaly in certain chicken stocks may be induced in these stocks (but not in others) by a variety of chemicals such as insulin and boric acid. Likewise, cortisone often produces cleft lip and palate in a mouse strain in which this malformation also occurs spontaneously in a certain percentage of animals [54].

8.4.1 The Development of Structure
[15, 25, 43, 61, 66]

After about 1 week of development, the upper and lower body ends of the embryo become visible, and a segmental structure develops; other signs of a more complex pattern appear in due time. Which genes are responsible for these processes, and how do they organize them? This had been an enigma for a long time, but a discovery in *Drosophila* has helped to solve it: the homeobox genes. A homeobox is a region of 183 bp that encodes a domain of 61 amino acids which is able to bind to DNA. It is present in three major classes: bicoid, having a maternal effect; segmentation genes; and homeotic genes which regulate proper development of structures in segments, such as limbs or antennae. Their activity has been analyzed; concentration gradients of mRNA from the anterior to the posterior pole of the larva determine the segments [15, 66] (see also [88]). The homeodomain is a peptide which, similarly to some bacterial repressor, fits into a turn of the DNA double helix (= helix-turn-helix motif). Hybridization experiments on *Drosophila* DNA with the mouse genome had the surprising result that these areas were very similar in base sequence; apparently they had been preserved during evolution, probably over more than 500 millions years. This suggests an important – and very similar – function. Meanwhile, more than 30 homeobox genes have been localized and cloned in the mouse [43]. They are organized in four clusters, each spanning more than 100 kb. In *Drosophila* and probably also in the mouse, Hox sequences within one cluster are ordered in the same sequence in which they are used for gene action during embryonic development. Their pattern of action has sharp anterior expression boundaries (Fig. 8.6, 8.7) [29]; the posterior boundaries are less well delimited. These genes begin to be expressed during early gastrulation; they are active during organogenesis, especially in the neural tube but also in other organ systems such as kidney, lung, intestine, thymus, and germ cells. In germ cells, their precise function is still

unknown. In *Drosophila* alternative models for their mode of protein-DNA interaction have been proposed [37]. As mentioned, all of these gene-determined peptides fit into the DNA double helix, forming a "helix-turn-helix" structure. Binding to the DNA requires an ATTA core sequence as found in promoters (Sect. 3.1.3) The strength of the binding, and therefore the degree of gene activation appears to depend both on the fine structure of the promoter and on the protein, as well.

In addition to the Hox genes, other conserved DNA binding motifs have been discovered; the Pax (= paired box) and the POU Pit-1, octamer binding, Unc 86-proteins genes [33]. Pax genes are also active in various parts of the murine brain. The POU domain occurs in some regulatory proteins. A Pax mutant (undulated) was discovered in the mouse that affects the structure of the axial skeleton. Pax genes are generally much larger than Hox genes, and effects of mutations tend to be dominant; therefore mutant phenotypes can easily be recognized. This property qualifies them especially well for studies on genetic determination of development.

Fig. 8.7 shows an autoradiogram of a mouse embryo, with the expression pattern of the Hox 2.2 gene.

Syndromes Caused by Hox and Pax Genes in Humans. Homeobox genes have also been identified in humans. Table 8.1 present an overview. Hox-7 is deleted (together of course with many other genes) in Wolf-Hirschhorn syndrome (del 4p-). A Pax (paired domain gene; HUP 2) may be involved in Waardenburg syndrome (193500) [5], in which lateral displacement of the inner canthus of the eyes, a white forelock, and heterochromia of irises, together with cochlear deafness are the most prominent signs. Aniridia (106210) is another dominant disease involving development of the iris. A good candidate gene within 11p13 has been cloned; the predicted protein contains a homeobox and a paired box (Pax; HUP) [81]. Moreover, genes for craniosynostosis (147620), Greig craniopolysyndactyly syndrome (175700), and Goldenhar syndrome 141400) have been localized to 7p, close to, or not far away from, the Hox-1 gene. Since they also involve development of the head, involvement of Hox or Hox-related genes is likely [20a].

In conclusion, discovery and analysis of these genes is teaching us much about genetic determination of the gross organization of eukaryotes, including humans. Moreover, it demonstrates a heuristic strategy that has been successful here: a group of genes – and a principle of gene action in morphogenesis – was discovered and analyzed first in an insect, *Drosophila*. The genes were then analyzed in the mouse, and are now being studied successfully in humans. The broad principles of pattern development appear

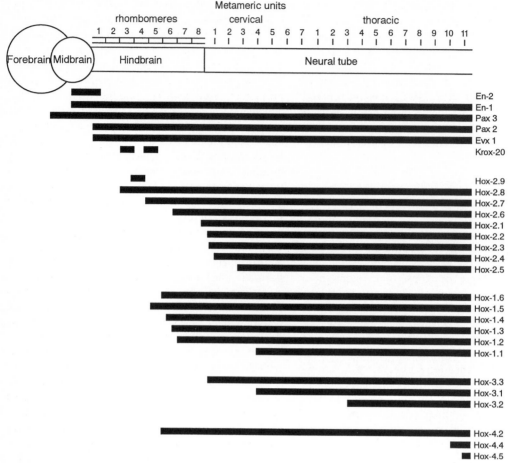

Fig. 8.6. Anterior expression boundaries in the neuroectoderm of the mouse. The indicated anterior borders reflect in principle the boundaries characterizing a gene at its major expression time (usually day 12.5 of gestation). Metameric units are given as the levels of prevertebrae on day 12.5 or as rhombomeres on day 9.5. For explanation see text. (From Kessel and Gruss 1990 [43])

to apply across all species, even if they are as distant as insects and humans.

Timetable of Human Intrauterine Development. To understand prenatal development and its disturbances it is necessary to visualize the normal developmental stages and their timing (Fig. 8.8). The oocyte is ovulated at about the 14th day after day 1 of the menstrual cycle. A short time afterwards it is fertilized (day 1 of pregnancy). One day later the zygote divides; on day three the morula is formed. Days 4 and 5 see the development of the blastocyst, which migrates through the fallopian tube; it begins implantation in the uterine mucosa on day 6. Implantation is completed on day 10; the placenta starts developing. On day 15 the primitive streak becomes visible; during the following days the first structures of the embryo appear, among them the neural fold, which on day 21 develops into the neural groove – the predecessor of brain and spin-

al cord. On day 22 the heart begins to beat, and 1 day later primordia of ears and eyes are present. Three pairs of branchial arches are visible on day 25, and on the following day the upper limb buds appear; the lower ones follow a short time later. On days 36–38 the facial processes unite, and the face is formed. All essential external and internal structures – or at least their beginnings – are present by day 56. This is the end of the 8th week; the embryo is now 30 mm long (crown-rump length), and the fetal period begins. Birth normally occurs after 38 weeks of pregnancy.

8.4.2 Birth Defects in Humans

A small percentage of children are born with birth defects – mild or severe, single or multiple. Occasionally the primary causes are known or can be inferred, for example, when a chromosomal aberration such as in Down syndrome can be found, or when pedigree data suggest a monogenic mode of inheritance. In some cases there is a definitive exogenous factor such as a rubeola infection or maternal alcohol abuse. In many more instances, however, one finds an aggregation of more or less similar cases within families that cannot be explained by a chromosomal anomaly, monogenic mode of inheritance, or exogenous factors. Here, such as in cleft lip/cleft palate the model of multifactorial inheritance in combination

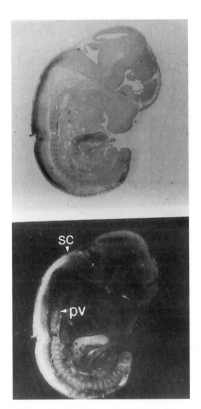

Fig. 8.7. In situ hybridization of the Hox 2.2 gene in a sagittal section of a 2.5-day mouse embryo. *Above,* bright field; *below,* dark ground. *sc,* boundary of expression in the CNS; *pv,* boundaries of expression in the somitic mesoderm, at the seventh prevertebra. (From Graham et al. 1989 [29])

with a threshold often fits the data (Sect. 6.1.2), but despite some observations in animals which have shown the existence of true threshold characters (Sect. 6.1.2.1) [32, 91], this model is too general. Moreover, many birth defects are completely unexplained single events with no evidence of either genetic or environmental causes. In fact, interference with normal development on a completely random or stochastic basis is likely to occur occasionally considering the complex events during organ formation. All these problems have been discussed recently using congenital heart diseases as examples [22]; for other groups of birth defects similar rules apply.

A group of scientists have tried to bring some order to the very heterogeneous group of birth defects by suggesting a number of distinctions, and proposing a unified nomenclature [78] (Fig. 8.9). According to their proposal, a *malformation* is a morphological defect which results from an intrinsically abnormal developmental process. This should be distinguished from a *disruption,* which results from the extrinsic breakdown of, or an interference with, an originally normal developmental process. A *deformation* is an abnormal shape or position of a part of the body caused by mechanical forces; a clubfoot caused by an oligohydramnion is an example. At the opposite end of the range of possible disturbances, *dysplasia* is an abnormal organization of cells into tissues.

To distinguish among various ways in which multiple anomalies in a patient or in several affected family members, the following terms are suggested: a *polytopic field defect* is a pattern of anomalies derived from the disturbance of a single *developmental field.* What is a "developmental field"? The "field" concept is very old but has been rejuvenated by Opitz [67]. Within a field the development of complex structures is controlled and ordered in time and space, and a set of embryonic primordia react identically to different dysmorphogenetic causes. It follows that such primordia must constitute a morphogenetically reactive unit under normal circumstances. On the other hand, the field concept offers

Table 8.1. Human homeobox genes

Designation	Localization	Minimum number of Hox sequences	Homology with mouse chromosome	Remarks
HOX-1	7p21-p14	8	1	90 kb
HOX-2	17	9	11	180 kb
HOX-3	12q12-q13	7	15	160 kb
HOX-4A	2q31-q37	6		70 kb, hematopoiesis
HOX-4B	2q31-2q32			
HOX-7	4p16.1		5	Deleted in Wolf-Hirschhorn syndrome (4p-)
EVX-2	2q		9	100 kb
HOX-10	14q24.2		12	Retina-specific expression

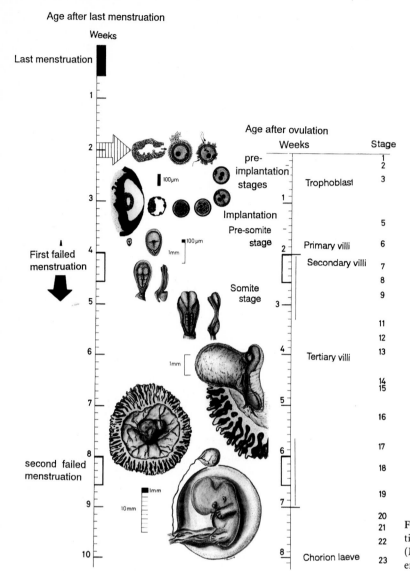

Age after last menstruation

Weeks

Last menstruation

Age after ovulation

Weeks   Stage

pre-
implantation
stages   Trophoblast

Implantation

Pre-somite
stage   Primary villi

Secondary villi

Somite
stage

Tertiary villi

First failed
menstruation

second failed
menstruation

Chorion laeve

Fig. 8.8. Stages of development and timing of the early human zygote. (Modified from Hinrichsen's embryology textbook)

a preliminary explanation for the observation that related malformations may occur as effects of the same cause – as in families with an autosomal dominant mode of inheritance of various, but related types of congenital heart disease. Another example is trisomy 21, where a variety of heart anomalies are observed that can usually be traced to an anomaly of the endocardial cushion.

Another concept is the *"sequence"* (Fig. 8.10): "It is a pattern of multiple anomalies derived from a single ... prior anomaly or mechanical factor." For example, a myelomeningocele may lead to lower limb paralysis, muscle wasting, and clubfoot. A *syndrome*, on the other hand, has been defined as "a pattern of multiple anomalies thought to be pathogenetically related and not known to represent a single sequence or a polytopic field defect." Finally, an *"association"*

is defined as the nonrandom occurrence of multiple anomalies that cannot be explained in any of the above ways. A logical means of diagnostic phenotype analysis based on these concepts has been suggested (Fig. 8.10); this can be combined with available genetic data in a more causally oriented analysis. Rules for genetic counseling can be derived [22]. Unfortunately, practical application of the results of such analysis for genetic counseling is often difficult unless the cause of the birth defect can be established with certainty, such as a chromosomal aberration, a Mendelian defect, or an exogenous cause. In most cases of birth defects the exact etiology remains unknown.

Another shortcoming in our ability to offer rational genetic counseling occasionally arises. Parents often ask about the genetic risk for their future children after

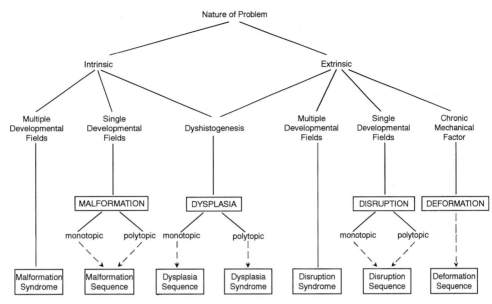

Fig. 8.9. Schema depicting different patterns of morphological defects. (From Spranger et al. 1982 [78])

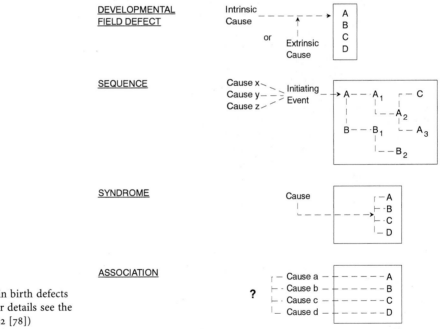

Fig. 8.10. Phenotype analysis in birth defects (errors of morphogenesis). For details see the text. (From Spranger et al. 1982 [78])

having lost one child with a birth defect. Here it is essential to know the specific diagnosis in the deceased child. Sufficient data are often not available; for example, chromosomes were not examined, and in particular a competent postmortem study was not performed. This situation needs improvement. In some countries – for example, Hungary – the autopsy of deceased children is required by law. Everywhere, more expert fetal and neonatal pathologists are needed.

For a better understanding of genetic mechanisms of normal development in humans, naturally occurring abnormalities may be helpful. Here, two genetic situations have proven to be especially revealing: chromosome aberrations and genetic anomalies of sex differentiation.

8.4.3 Genotype-Phenotype Relationships in Human Chromosome Aberrations [18–20]

Most monosomies lead to early embryonic death in the heterozygous state. This is not surprising since a number of gene products are present only once, and genomic imprinting may interfere with their expression. The only monosomy for an entire chromosome which, in a fair proportion of instances, does not lead to embryonic death is X monosomy in Turner syndrome (Sect. 2.2.3.2). Most regions of one X chromosome are inactivated in normal XX women; therefore the functional difference between the XX and the X states is small.

More interesting, particularly for elucidating normal gene action during embryonic development, are the results in trisomy. Clinical signs described in trisomies must not all be present in any one patient; often the findings are variable. Patients with trisomy 21 (Down syndrome), for example, are all mentally retarded (albeit with differences in the degree of retardation), but only about 40% have congenital heart disease. Here the concept of "sequence" (see above) may be helpful for understanding: the basic anomaly is the abnormal endocardial "cushion," which may or may not lead to a morphological defect. The cushion may be caused by an unusual stickiness of certain cells due to abnormal surface structures [53]. Still more important, however, is the fact that trisomy syndromes show a characteristic pattern of clinical signs. Some of these signs are more or less common to all trisomies, such as stunted growth and mental retardation. Others are not always present but are much more common than among chromosomally normal individuals; these include congenital heart disease, malformations in the genitourinary system and kidneys, and anomalies of the face. The latter, however, may be so characteristic that the experienced medical geneticist can diagnose the type of chromosomal aberration at first glance. Down syndrome is the most obvious example, but the same is true for many other aberrations. (For a description of such patterns see [6, 72].) These must be explained, and by their study we may also hope to learn more about normal development.

The most obvious consequence of a trisomy is that a number of genes – those located in the trisomic area – are present not in duplicate but in triplicate. Assuming that there is no further quantitative regulation of expression, approximately 50% more of the gene-determined protein is formed; this expectation has been borne out in a great number of instances [18], especially in Down syndrome (Table 8.2). In this syndrome, however, activities of some other enzymes are changed as well. The question arises: why does excess production of certain proteins or mRNA lead to severe disturbances in development? The answer often is heard that this excess leads to a disturbance in regulatory equilibrium. While probably correct, this answer is nevertheless much too general to provide clues that can be pursued successfully. Specifically, what effects does the increased activity of genes have that are triplicated? Answering this question will provide only partial answers since mechanisms of a higher order are probably also involved. As a first step, however, this analysis may yield results that provide clues for further analysis. So far, studies have been performed mainly on chromosome 21 [20, 48a] (Fig. 8.11).

While it is necessary to study human embryos and fetuses with chromosomal aberrations as carefully and completely as possible, studies on early human embryos are usually not feasible. Therefore animal models are of great help. Such a model does indeed exist: the distal part of the long arm of the mouse chromosome 16 is homologous to human chromosome 21. This homology was confirmed by the presence of some genes that a found in both species and, by comparison, of DNA. Various mouse trisomies can be produced by a breeding scheme introduced by Gropp. Starting with a special mouse with many Robertsonian translocations (tobacco

Table 8.2. Some gene dosage effects in trisomy 21 (modified from Epstein 1989 [19])

Gene	Function	Percentage of normal activity in trisomics
CBS	Cystathionine β-synthase activity in stimulated lymphocytes	1.61
IFNRA	Interferon α-binding to fibroblasts	1.57
PFKL	Phosphofructokinase activity in red cells and fibroblasts	1.46
PRGS	Phosphoribosylglycinamide synthase activity in fibroblasts	1.56
SOD1	Superoxide dismutase-1 activity in red cells, fibroblasts, platelets, lymphocytes, brain, and granulocytes	1.52
CD18	Binding of antibody against LFA-1 β[a] to EBV[b]-transformed β-lymphoblastoid cell lines	1.21

[a] Cell adhesion-associated antigen (116 920).
[b] Epstein-Barr virus.

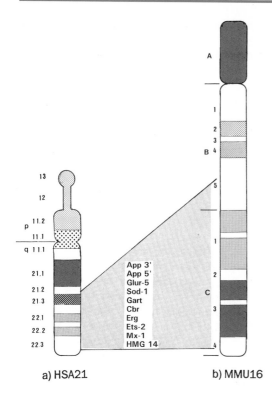

a) HSA21   **b) MMU16**

Fig. 8.11. Comparison of chromosome 16 of the mouse (MMU 16) with human chromosome 21 (HSA 21). Homologous genes are indicated. This mouse chromosome is conventionally subdivided into three Areas (A, B, C). Symbols such as Sod-1 etc. designate genes common to these two chromosomes. The area containing the homologous sequences is slightly bigger in mouse 16 than in human 21. (Courtesy of Dr. W. Buselmaier)

Table 8.3. Studies on growth parameters in trisomy 21 (from Krone and Wolf 1978)

Increased proportion of fibroblasts with intermediate DNA content; suggestion of prolonged S phase
Retarded rate of DNA synthesis in fibroblasts
Significant increase of the population doubling time in fibroblasts; decreased of in vitro life span
Increased population doubling time in fibroblasts
Increased duration of G_2 and possibly of S phase in fibroblasts
Decreased duration of cell cycle time in lymphocytes

mouse; *Mus poschiavinus*), a high fraction of trisomies can be produced. The mouse trisomy 16 shows many similarities with the human trisomy 21; interesting homologies have been shown not only in the external phenotype [18, 19] but also, for example, in anomalies of the heart and the great vessels [4]. Unfortunately, however, trisomy 16 mice die before birth; it is impossible to study later development. Very probably the early death is due to triplication of that portion of

chromosome 16 which is not homologous with the human chromosome 21. To study single triplicated genes in greater detail transgenic mice have been constructed. For example, a mouse containing, in addition, the gene for SOD1 (superoxide dismutase) has been used to study the additional activity of this enzyme [21]. Increased formation of H_2O_2 by this enzyme is toxic and contributes to the aging of cells. Studies using transgenic animals are expected to lead to further elucidation of disturbed pathways in development.

Cellular Studies in Chromosome Aberrations. Many studies have claimed a higher susceptibility of trisomy 21 cells to irradiation or chemical mutagens; some authors have tried to explain the unusually high incidence of leukemia in individuals with Down syndrome in this way. The results, however, are contradictory; no clearcut picture has emerged so far [19].

How are cells with chromosome aberrations different from normal cells? [49, 50, 52] Cultured cells from a variety of chromosome aberrations – mostly from abortions – have been examined for cell cycle, cell morphology, histochemistry, and a number of immunological and biochemical characters. A cell strain with trisomy 7, for example, had among other anomalies a reduced capacity for formation of histotypical structures, low collagen production, poor glycogen content and acid phosphatase activity. In the cell cycle the G_2 period was found to be twice as long as in diploid cells [50]; the S phase was shortened [49]. A different cellular syndrome has been described in a cell strain with trisomy 14. A low growth potential and inability to form histotypical structures have also been found, but the biochemical characteristics are different; for example, acid phosphatase is low, and there is a high concentration of polysaccharides as measured by the PAS reaction [51]. Apart from trisomy 7, the G_2 phase is also prolonged in monosomy 21 and trisomy 21. In trisomy 21, some other growth parameters have been determined, and a number of deviations from normal euploid cells have been reported (Table 8.3). It is especially interesting that the cellular phenotype in triploidy turned out to be almost normal so that no "cellular syndrome" could be established [52]. Apparently the malformations observed in triploidy cannot be related to detectable anomalies of the cells themselves and therefore appear to be produced at a different level of integration: placental insufficiency. (For details see Sect. 8.2.)

Abnormal Phenotypes Due to Chromosome Aberration and Gene Regulation. Regulation of gene activity during embryonic development presupposes a certain equilibrium in the quantity of gene products produced by genes on different chromosomes. These gene products may act as enzymes or structural proteins, or they may have a regulatory function, for example, as repressors of the function of other genes. It is understandable that discrepancies in the amount of genetic material lead to disturbances in the interaction of genes and to regulatory defects in embryonic development. In triploidy the *relative* amounts of chromosome-specific material are undisturbed. Therefore anomalies at the cellular level are relatively minor. In trisomies, on the other hand, some chromosomal material is increased whereas other material is present in normal amounts. If gene regulation requires interaction of gene products from different chromosomes, developmental anomalies at the cellular level are to be expected in trisomies (and monosomies) but not in triploidy.

Examination the mechanisms by which these anomalies are produced requires systematic comparison of all steps in protein biosynthesis and metabolism between normals and those affected with chromosomal aberrations. Studies of the effects of genes located within the trisomic (or otherwise modified) portion of the genome will probably not give the entire answer, but they may provide useful clues. Such studies offer good opportunities to improve our understanding of normal and abnormal embryonic development.

8.5 Sex Differentiation and Its Disturbances

Development of Sexual Dimorphism. The development of sexual dimorphism and sex characteristics is of special interest for the human geneticist.

Four levels of sexual development can be distinguished:

1. Determination of the chromosomal sex (46, XX or 46, XY)
2. Determination of gonadal sex (ovary or testicle)
3. Determination of phenotypic sex (female or male, internal and external sex characters)
4. Determination of psychological sex

The fourth level is discussed in Chap. 15 and the first in Chap. 3. Analysis of both numerical and structural chromosome aberrations involving the *sex chromosomes* has yielded much valuable information not only on chromosomal sex (level 1) but also on determination of gonadal and phenotypic sex (levels 2 and 3).

The gonadal anlage in the early embryo (up to the 5th or 6th week) shows no sex difference and does not contain germ cells. In humans, primordial germ cells become visible during the 3rd week of embryonic development in the ectoderm of the yolk sac. They then migrate under chemotactic influence into the gonads. This migration is independent of sex; in appropriate experimental systems, female germ cells also migrate into male gonads and vice versa. The gonadal anlage may develop into either a testicle or an ovary (level 2). This direction normally depends on the presence of the Y chromosome: a male gonad develops if *one* Y chromosome is present – regardless of the number of X chromosomes.

The mechanisms by which the Y chromsome (Fig. 8.12, 8.13) determines development of male gonads has long been controversial, and our present notions may not be final, but some aspects do appear to be relatively well-founded.

The Role of the SRY Gene. Figure 8.12 shows the Y chromosome and its genes. A sequence of 35 kb, close to the border of the pseudoautosomal region, is essential for sex determination. Here the gene SRY is located [59, 75]. It contains a protein-binding motif and has been conserved through evolution. Its importance for the male differentiation of gonads has been shown in the mouse [31] – especially by studies with transgenic animals: a 14-kb genomic DNA fragment containing this gene was injected into freshly fertilized mouse oocytes. This led to male development of chromosomally female mice – a good example for the usefulness of transgenic animals for solving problems of gene regulation [48]. The SRY gene is present in XX males (Sect. 2.2.3.2) and is not found in XY females (unless they have testicular feminization due to defects of androgen receptors; see below). Some autosomal genes are also involved in sex determination [90]. Other genes formerly suspected of determining the male sex, such as the ZFY gene which codes for a zinc finger protein and the gene determining the HY antigen, may have different – possibly supporting – functions. SRY belongs to a group of protein-binding genes which have a high-mobility group (HMG) box. Another member of this family is the gene for a protein required as cofactor of RNA polymerase-1 for ribosomal DNA transcription [58]. Binding of SRY is sequence specific (consensus sequence AACAAT) and leads to binding of DNA. Since research in this field is proceeding very rapidly, much more will be known soon.

The third level, determination of the secondary sex characters, appears to be more complicated (and liable to error).

Development of Secondary Sexual Characteristics. Somatic sex differentiation follows the differentiation of gonads. The internal genital organs are formed from the müllerian and the wolffian ducts, both descendants of the primordial kidney. In the female, the müllerian duct develops into fallopian tubes and the uterus; the wolffian duct atrophies. In the male a wolffian duct forms the seminal ducts and seminal

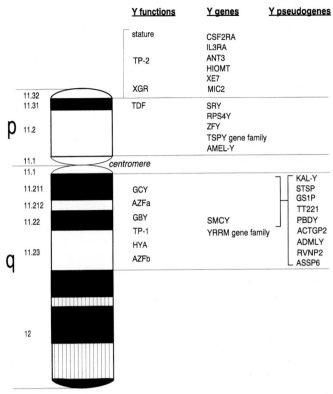

Fig. 8.12. The human Y chromosomes: functions and genes. Y functions: *Stature*, statural determinant; *TP -1, -2*, Turner phenotype prevention; *XGR*, XG blood group; *TDF*, testis determining factor; *GCY*, growth control, Yp influenced; *AZF a, b*, azoospermia factor; *GBY*, gonadoblastoma, Y-influenced; *HYA*, histocompatibility Y antigen. Y-genes: *TSFR2RA*, receptor for the granulocyte-macrophage colony-stimulating factor; *IL3RA*, interleukin 3 receptor, *ANT3*, ADP/ATP translocase; *HIOMT*, hydroxyindole-O-methyltransferase (also known as ASMT: acetyl-serotonin-methyltransferase); *XE7*, gene for a nuclear protein of unknown function; *MIC2*, cell surface antigen involved in cell adhesion processes; *SRY*, sex determining region Y gene; *RPS4Y*, isoform of ribosomal protein S4, Y-linked; *ZFY*, zinc finger protein, Y-linked; *TSPY gene family*, testis specific protein, Y-linked; *AMEL-Y*, Amelogenin; Y-linked; *SMCY*, human homolog to a Y-chromosomal mouse gene; *YRRM gene family*, Y chromosome gene family with RNA-binding protein homology. Y pseudogenes: *KAL-Y*, Kallman syndrome pseudogene; *STSP*, steroid sulfatase pseudogene; *GS1P*, pseudogene of a gene of unknown function expressed from the X chromosome; *TT221*, interrupted Yq homolog of a gene of unknown function expressed from Xq; *PBDY*, a Y-specific duplication of PBDX, a gene spanning the pseudoautosomal boundary; *ACTGP2*, actin, gamma pseudogene 2; *ADMLY*, Y-chromosomal homolog of ADMLX, a gene of unknown function expressed from Xp; *RVNP2*, retroviral sequences NP2; *ASSP6*, arginino-succinate-synthase pseudogene 6. (Courtesy of M. Köhler)

vesicles. Under the influence of maternal human chorionic gonadotropin (HCG), the Leydig cells of the embryonic testicle produce the steroid hormones testosterone and 5-dihydrotestosterone. A hormone called müllerian inhibiting factor (MIF) is produced in Sertoli cells. These hormones meet bipotent anlagen for external and internal sex characters: primarily wolffian ducts, müllerian ducts, and the urogenital sinus. A normal male develops if all these elements act in time and in the right places. In their complete absence, on the other hand, female sexual characters form. Hence, female development does not require any special promoting factors; it is "constitutive." Minor disturbances at several levels of this system lead to incomplete male development despite a male genotype (male pseudohermaphroditism); analysis of such anomalies has taught us much about the normal course of events. As Jost has put it: "Becoming a male is a prolonged, uneasy, and risky venture; it is a kind of struggle against inherent trends toward femaleness." There are four cell lineages in the developing human gonad [58]. They are bipotential, i.e., they may develop into either male or female cells. In addition to the germ cells, they comprise the supporting cell precursors which develop into follicle cells in women and into Sertoli cells in men. Steroid-producing cells develop into theca or interstitial cells in women and Leydig cells in men. Development of male germ cells has been often studied by analysis of genetic defects.

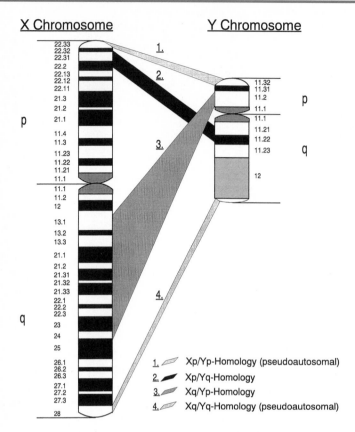

X Chromosome   Y Chromosome

1. ◿ Xp/Yp-Homology (pseudoautosomal)
2. ◢ Xp/Yq-Homology
3. ◿ Xq/Yp-Homology
4. ◿ Xq/Yq-Homology (pseudoautosomal)

Fig. 8.13. Homologies between the human X and X chromosomes. (Courtesy of M. Köhler)

Genetic Control of Spermatogenesis: A Genetic Region on the Long Arm of the Y Chromosome [83–85]. In the euchromatic portion of the long arm of the Y chromosome (Yq11.22/23) a DNA sequence important for the production of mature sperm has been found. The first clue came from infertile patients with arrest of spermatogenesis who had chromosomal deletions in this area [80]. In recent years, this region has been thoroughly analyzed, and many patients with deletions within this critical area were found to be infertile due to arrest of spermatogenesis but did not show any other striking clinical signs. In addition to deletions, other structural anomalies such as a dicentric iso-Yp, ring Y chromosomes, and translocations have been observed in sterile men. Analysis of the breakpoints allowed delineation of the critical area, which was termed the azoospermia factor (AZF) area. It is unknown as yet whether this area comprises one or more such genes. In testicular biopsies either immature spermatogonia and Sertoli cells or only Sertoli cells (Sertoli only syndrome) are found. AZF genes appear to act very early in gametogenesis. At least two AZF genes have been suggested; one appears to be active at the time of spermatogonial development while the other is expressed at the pachytene stage of spermatocyte maturation [86]. An interesting part of the AZF region is a repetitive

sequence (pY6H65) which is present in several copies and appears to comprise a binding motif for proteins with a high molecular weight. Its significance is indicated by its conservation; it is also present in the mouse, and a very similar motif has been found within a fertility gene of *Drosophila* [84]. Again, as in homeobox genes (Sect. 8.4.1), a gene or genes important for development appear to have been conserved for at least 500 million years!

In addition to these regions of the Y chromosome, many genes are involved in the development of male sex characteristics and fertile germ cells. Here, again, the analysis of defects observed in certain patients has led to a better understanding of normal function.

Many different gene defects are known that lead to a disturbance in the differentiation of internal and external male sexual characters. Anomalies occur, for example, in synthesis of androgens, lack of HCG, lack of HCG receptors in Leydig cells, and enzyme defects affecting the enzymes involved in testosterone synthesis. Anomalies may also be caused by lack of sensitivity to testosterone or 5-dihydrotestosterone due to receptor defects in cells of wolffian ducts or the urogenital sinus [30].

The pathways of steroid hormones and potential genetic blocks are shown in Fig. 8.14; the left-hand side

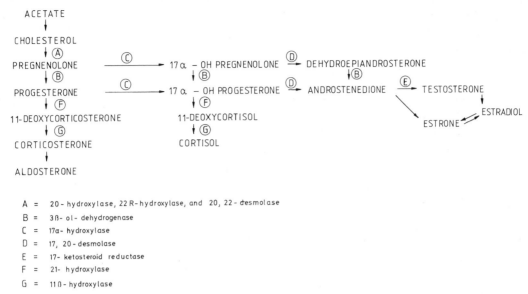

ACETATE

CHOLESTEROL
   (A)
PREGNENOLONE ————(C)————→ 17α – OH PREGNENOLONE —(D)→ DEHYDROEPIANDROSTERONE
   (B)   (B)   (B)
PROGESTERONE ————(C)————→ 17α – OH PROGESTERONE —(D)→ ANDROSTENEDIONE ←(E)→ TESTOSTERONE
   (F)   (F)
11-DEOXYCORTICOSTERONE   11-DEOXYCORTISOL   ESTRADIOL
   (G)   (G)   ESTRONE
CORTICOSTERONE   CORTISOL

ALDOSTERONE

A = 20- hydroxylase, 22 R- hydroxylase, and 20, 22- desmolase
B = 3ß- ol- dehydrogenase
C = 17α- hydroxylase
D = 17, 20- desmolase
E = 17- ketosteroid reductase
F = 21- hydroxylase
G = 11ß- hydroxylase

Fig. 8.14. Pathways of steroid hormone metabolism. Each step (A–G) is a potential site for a genetic block. (From Engel 1982 [16])

of this figure shows the defects (F and G) leading to adrenogenital syndromes (Sect. 7.2) in females but not to pseudohermaphroditism in males. The blocks under A (congenital adrenal lipid hyperplasia) are not very well known; in addition to their largely female external genitals, these males suffer from severe salt wastage. The same holds true for block B. The blocks on the right side lead to male pseudohermaphroditism of various degrees without other manifestations of the adrenogenital syndrome.

An especially interesting enzyme defect not shown in Fig. 8.16 is the defect of 5α-reductase, an autosomal-recessive trait. This enzyme normally reduces testosterone in cells of the urogenital sinus to 5α-dihydrotestosterone. If it is lacking, normal internal male sex organs (seminal vesicles; prostate) develop and the entire body, including muscular development, body hair, etc., is male except for the external genitalia, which upon superficial inspection are female. Hence the descriptive name of pseudovaginal perineoscrotal hypospadia (PPHS; 264600).

These enzyme defects are rare. Other, more common syndromes exist in which androgens are normal, but the target tissue is completely or partially androgen resistant.

Testicular Feminization Syndrome (316800) [30]. At birth, affected individuals appear to be normal females; the anomaly goes undetected during childhood unless a testicle is discovered in an inguinal hernia. Affected persons with this syndrome have a male karyotype and male gonads; the term testicular feminization was coined by Morris (1953) [63]. At puberty, primary amenorrhea and – in the majority of cases – absence or marked deficiency of axillary, pubic, and body hair is striking. In adults, the stature and proportions are normal for females, although the legs are often somewhat longer. Breasts are well developed. The mean body proportions conform more to the present-day ideal of female beauty than to the proportion of the average woman; it is therefore not surprising that affected patients are found repeatedly among models.

The vagina is usually shortened and ends in a blind pouch. Instead of a uterus some remnants of the müllerian ducts are often present; instead of the fallopian tube a fibromuscular streak may be found. Testicles are located in the labia majora, the inguinal canal, or in the abdomen and may show a normal or even increased number of the male hormone producing Leydig cells. Usually there is no spermatogenesis. Occasionally, malignant tumors of the testicle develop. In addition to these complete cases, incomplete ones occur who have reduced or even normal body hair.

The patients secrete normal amounts of androgens, especially testosterone; normal male development would be expected. The most obvious explanation for the disease is an anomaly in end-organ response to the hormone, which had been postulated for many years. An X-linked androgen receptor defect indeed was discovered [89]. The psychological development of testicular feminization patients is entirely female; there is now evidence that interaction between androgens and androgen receptors influences brain development, EEG, and behavior

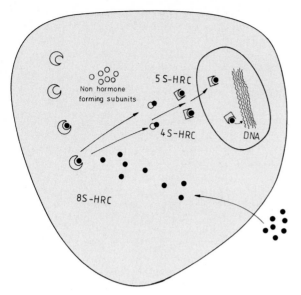

Fig. 8.15. Testosterone and dihydrotestosterone pathways inside the cell. – Three types of receptor molecules can be distinguished (by their sedimentation in the sucrose gradient) with defined ion concentration: 8 S, 5 S and 4 S. After entering, the hormones are bound to the 8 S receptor, forming the 8 S-hormone-receptor complex *(HRC)*. The 8 S-HRC dissociates into a 4 S-HRC and into some – nonhormone forming – subunits. The 4 S complex is transformed into 5-HRC, which is then transported into the nucleus. Testicular feminization may be the result of various – complete and incomplete – blocks within this system. (From Engel 1982 [16])

(Sect. 15.2.3.3). The mode of inheritance is X-linked; the gene has been localized to Xcen-q22.

Interestingly, Patterson and Bonnier had concluded from pedigree information as long ago as in 1937 – at a time when assessment of the genotypic sex was impossible – that the patients are genotypic males, and that the mode of inheritance is either X-linked or sex-limited. Sporadic cases may be caused by new mutations; the selective disadvantage of the gene is strong, the patients being infertile. Therefore a high proportion of new mutants must be expected (Sect. 9.3).

Genetic Heterogeneity. Polypeptide hormones such as insulin are bound to membrane receptors of target cells. Steroid hormones such as testosterone, on the other hand, are bound to cytoplasmic receptors after having entered the cell by diffusion. Figure 8.15 shows the pathway of testosterone and dihydrotestosterone in the cell from its binding to the cytoplasmatic 8 S receptor to the 4 S receptor complex and to the 5 S receptor complex and its movement into the nucleus. Most mutations leading to testicular feminization appear to involve the 8 S receptor, but mutants affecting the 4 S- and 5 S complexes must be expected, and in-

deed cases with testicular feminization and normal 8 S receptor have been observed. In patients with incomplete testicular feminization and intersexual genitals the 8 S receptor has been found to be diminished but not completely lacking.

Conclusions

One of the major riddles of nature is how a fertilized egg develops into an adult organism, with all its multitudinous characteristics. Since all cells have identical DNA content, how is it that genes are turned on and off during development and tissue differentiation? In humans the development from zygote to embryo, fetus, and child has been well described, and many anomalies and birth defects are known. These may result from disturbances due to either exogenous (teratogenic agents) or endogenous anomalies, such as chromosomal aberrations or certain gene mutations. In many instances, however, their causes remain unknown. The basic rules of anatomical and functional differentiation are not yet known, but this process can now be studied by the techniques of molecular and cell biology. For example, the phylogenetically conserved homeobox genes are a useful research tool. Studies on transgenic animals are especially promising. Unequal contributions of maternal and paternal genomes to the developing zygotes (genomic imprinting) are beginning to explain a number of birth defects and, by implication, normal development. Elucidation of phenotypic abnormalities caused by chromosome aberrations has produced some promising results and is expected to open new paths for future research.

References

1. Amara SG, Jonas V, Rosenfeld MG, Ong ES, Evans RM (1982) Alternative RNA processing in calcitonin gene expression generates mRNAs encoding different polypeptide products. Nature 298 : 240–244
2. Andreadis A, Gallego ME, Nadal-Ginard B (1987) Generation of protein isoform diversity by alternative splicing. Annu Rev Cell Biol 3 : 207–242
3. Angelman H (1965) "Puppet" children: a report on three cases. Dev Med Child Neurol 7 : 681–683
4. Bacchus C, Sterz H, Buselmaier W et al (1987) Genesis and systematization of cardiovascular anomalies and analysis of skeletal malformations in murine trisomy 16 and 19. Hum Genet 77 : 12–22
5. Baldwin CT, Hoth CF, Amos JA et al (1992) An exonic mutation in the HuP2 paired domain gene causes Waardenburg's syndrome. Nature 355 : 637–638
6. Borgaonkar DS (1989) Chromosomal variation in man. A

catalogue of chromosomal variants and anomalies, 5th edn. Liss, New York

7. Britten RJ, Davidson EH (1969) Gene regulation for higher cells. A theory. Science 165 : 349–357

8. Butler MG, Meany FJ, Palmer CG (1986) Clinical and cytogenetic survey of 39 individuals with Prader-Labhart-Willi syndrome. Am J Med Genet 23 : 793–809

9. Cattanach BM, Kirk M (1985) Differential activity of maternally and paternally derived chromosome regions in mice. Nature 315 : 496–498

10. Chandley AC (1991) On the parental origin of de novo mutation in man. J Med Genet 28 : 217–223

11. Cooper DN, Krawczak M (1993) Human gene mutation. Bios Scientific, Oxford

12. Cremer T, Kurz A, Zirbel R et al (1994) The role of chromosome territories in the functional compartimentalization of the cell nucleus. Cold Spring Harbor Symp Quant Biol 58 : 777–792

13. Davidson EH, Britten RJ (1973) Organisation, transcription and regulation in the animal genome. Q Rev Biol 48 : 565–613

14. Davidson EH, Britten RJ (1979) Regulation of gene expression: possible role of repetitive sequences. Science 204 : 1052–1059

15. Driever W, Nuesslein-Volhard (1988) A gradient of biocin protein in Drosophila embryos. Cell 54 : 83–93

16. Engel W (1982) Geschlechtsdifferenzierung und ihre Störungen. Verh Dtsch Ges Pathol 66 : 329–343

17. Epstein CJ (1985) Mouse monosomies and trisomies as experimental systems for studying mammalian aneuploidy. Trends Genet 1 : 129–134

18. Epstein CJ (1986) The consequences of chromosome imbalance. Principles, mechanisms and models. Cambridge University Press, New York

19. Epstein CJ (1989) Down's syndrome (trisomy 21) In: Scriver CR, Beaudet AL, Sly WS, Valle D (eds) The metabolic basis of inherited disease, 6th edn. McGraw-Hill, New York, pp 291–341

20. Epstein CJ (1990) The consequences of chromosome imbalance. Am J Med Suppl 7 : 31–37

20 a. Epstein CJ (1995) The new dysmorphology: application of insights from basic developmental biology to the understanding of human birth defects. Proc Natl Acad Sci USA 92(19):8566–8573

21. Epstein CJ, Avraham KB, Lovett M et al (1987) Transgenic mice with increased Cu/Zn-superotide dismutase activity: animal model of dosage effects in Down syndrome. Proc Natl Acad Sci USA 84 : 8044–8048

22. Fuhrmann W (1993) Humangenetische Aspekte der angeborenen Fehlbildingen des Herzens und der großen Gefäße. In: Doerr W, Seifert G (eds) Pathologische Anatomie des Herzens und seiner Hüllen. Springer, Berlin Heidelberg New York, pp 519–548 (Spezielle pathologische Anatomie, vol 22/1)

23. Ganten D, Wagner J, Zeh K et al (1992) Species specificity of renin kinetics in transgenic rats harboring the human renin and angiotensin genes. Proc Natl Acad Sci USA 89 : 7806–7810

24. Garrod AE (1923) Inborn errors of metabolism. Frowde, London (Reprinted 1963 by Oxford University Press, London)

25. Gehring WS (1985) The homeo box: a key to the understanding of development? Cell 40 : 3–5

26. Goldschmidt RB (1935) Gen und Außeneigenschaft (Untersuchungen an Drosophila) I. und II. Mitt Z Vererbungslehre 10 : 74–98

27. Gordon JF, Ruddle FH (1981) Integration and stable germ line transmission of genees injected into mouse pronuclei. Science 214 : 1244–1246

28. Graessmann A (1993) Transgene Tiere. Aus Forschung und Medizin 8 : 49–56

29. Graham A, Papalopulu N, Krumlaut R (1989) The murine and Drosophila homeobox gene complexes have common features of organization and expression. Cell 57 : 367–378

30. Griffin JE, Wilson JD (1989) The androgen resistance syndroms. In: Scriver CR, Beaudet AL, Sly WS, Valle D (eds) The metabolic basis of inherited disease, 6th edn. McGraw-Hill, New York, pp 1919–1944

31. Grubbay J, Collignon J, Koopman P et al (1990) A gene mapping to the sex-determining region of the mouse 4 chromosome is a member of a novel family of embryonically expressed genes. Nature 346 : 245–250

32. Grüneberg H (1952) Genetical studies in the skeleton of the mouse. IV. Quasi-continuous variations. J Genet 51 : 95–114

33. Gruss P, Walther C (1992) Pax in development. Cell 69 : 719–722

34. Hall JG (1990) Genomic imprinting: review and relevance to human diseases. Am J Hum Genet 46 : 857–873

35. Hall JG (1992) Genomic imprinting and its clinical implications (editorial; comment). N Engl J Med 326(12):827–829

36. Harper PS (1975) Congenital myotonic dystrophy in Britain. II. Genetic basis. Arch Dis Child 50 : 514–521

37. Hayashi S, Scott MP (1990) What determines the specificity of action of drosophila homeodomain proteins? Cell 63 : 883–894

38. Jacob F (1978) Mouse teratocarcinoma and mouse embryo. The Leuwenhoek lecture 1977. Proc R Soc Lond [Biol] 201 : 249–270

39. Jacob F, Monod J (1961) On the regulation of gene activity. Cold Spring Harbor Symp Quant Biol 26 : 193–209

40. Jaenisch R (1989) Transgenic animals. Science 240 : 1468–1474

41. Jandl JH, Cooper RA (1978) Hereditary spherocytosis. In: Stanbury JB, Wyngaarden JB, Fredrickson DS (eds) The metabolic basis of inherited disease, 4th edn. McGraw-Hill, New York, pp 1396–1409

42. Jost JP, Saluz HP (eds) (1993) DNA methylation: molecular biology and biological significance. Birkhäuser, Basel

43. Kessel M, Gruss P (1990) Murine developmental control genes. Science 249 : 375–379

44. Klose J (1975) Protein mapping by combined isoelectric focusing and electrophoresis of mouse tissue. Hum Genet 26 : 231–243

45. Knippers R, Philippsen P, Schafer KP, Fanning E (1990) Molekulare Genetik, 5th edn. Thieme, Stuttgart

46. Knoll JHM, Nicholls RD, Magenis RE et al (1989) Angelman and Prader-Willi syndromes share a common chromosome 15 deletion but differ in parental origin of the delition. Am J Med Genet 32 : 285–290

47. Knussmann R (1973) Unterschiede zwischen Mutter-Kind- und Vater-Kind-Korrelationen im Hautleistensystem des Menschen. Hum Genet 19 : 145–154

48. Koopman P, Gubbay J, Vivian N et al (1991) Male development of chromosomally female mice transgenic for Sry. Nature 351 : 117–121

48 a. Korenberg JR, Chen XN, Schipper R, Sun Z, Gonsky R, Gerwehr S, Carpenter N, Daumer C, Dignan P, Disteche C et al (1994) Down syndrome phenotypes: the consequences of chromosomal imbalance. Proc Natl Acad Sci USA 91(11):4997–5001

49. Kukharenko VI, Kuliev AM, Grinberg KN, Terskikh VV (1974) Cell cycles in human diploid and aneuploid strains. Hum Genet 24 : 285–296

50. Kuliev AM, Kukharenko VI, Grinberg KN, Vasileysky SS, Terskikh VV, Stephanova LG (1973) Morphological, autoradiographic, immunochemical and cytochemical investigation of a cell strain with trisomy 7 from a spontaneous abortion. Hum Genet 17 : 285–296

51. Kuliev AM, Kukharenko VI, Grinberg KN, Terskikh VV, Tamarkina AD, Begomazov EA, Vasileysky SS (1974) Investigation of a cell strain with trisomy 14 from a spontaneously aborted human fetus. Hum Genet 21 : 1–12

52. Kuliev AM, Kukharenko VI, Grinberg KN, Mikhailov AT, Tamarkina AD (1975) Human triploid cell strain. Phenotype on cellular level. Hum Genet 30 : 127–134

53. Kurnit DM, Alridge JF, Matsuoka R, Matthyse S (1985) Increase adhesiveness of trysomy 21 cells and atrioventricular canal malformations in Down's syndrome. A stochastic model. Am J Med Genet 20 : 385

54. Landauer W (1957) Phenocopies and genotype with special reference to sporadically occurring developmental variants. Am Nat 91 : 79–90

55. Lawler SD, Povey S, Fisher RA, Pickthal VJ (1982) Genetic studies on hydatidiform moles. II. The origin of complete moles. Ann Hum Genet 46 : 209–222

56. Ledbetter DH, Riccardi VM, Airhard SD et al (1981) Deletion of chromosome 15 as a cause of the Prader-Willi syndrome. N Engl J Med 304 : 325–329

57. Lorenz K (1952) King Salomon's ring. Crowell, New York

58. Lovell-Badge R (1992) Testis determination: soft talk and kinky sex. Curr Opin Genet Dev 2 : 596–601

59. Magnuson T, Epstein CJ (1981) Genetic control of very early mammalian development. Biol Rev 56 : 369–408

60. Manuelidis L, Borden J (1988) Reproducible compartmentalization of individual chromosome domains in human CNS cells revealed by in situ hybridization and three-dimensional reconstruction. Chromosoma 96 : 397–410

61. McGinnies W, Krumlaut R (1992) Homeobox genes and axial patterning. Cell 68 : 283–302

62. McKusick VA (1995) Mendelian inheritance in man, 11th edn. Johns Hopkins University Press, Baltimore

63. Morris JM (1953) The syndrome of testicular feminization in male pseudohermaphrodites. Am J Obstet Gynecol 65 : 1192

64. Mullins JJ, Peters J, Ganten D (1990) Fulminant hypertension in transgenic rats harbouring the mouse Ren-2 gene. Nature 344 : 541–544

65. Noyer-Weidner M, Trautner TA (1993) Methylation of DNA in prokaryotes. In: Jost JP, Saluz HP (eds) DNA methylation: biology and biological significance. Birkhäuser Verlag, Basel, pp 39–108

66. Nuesslein-Volhard C, Frohnhoefer HG, Lehmann R (1987) Determination of anteroposterior polarity in Drosophila. Science 238 : 1675–1681

67. Opitz JM (1985) The developmental field concept. Am J Med Genet 21 : 1–11

68. Prader A, Labhart A, Willi H (1956) Ein Syndrom von Adipositas, Kleinwuchs, Kryptorchismus und Oligophrenie nach myotonieartigem Zustand im Neugeborenenalter. Schweiz Med Wochenschr 86 : 1260–1261

69. Reed T, Young RS (1982) Maternal effects in dermatoglyphics: similarities from twin studies among palmar, plantar and fingertip variables. Am J Hum Genet 34 : 349–352

70. Reik W (1989) Genomic imprinting and genetic disorders in man. Trends Genet 5 : 331–336

71. Ridley RM, Frith CD, Crow TJ, Conneally PM (1988) Anticipation in Huntington's disease is inherited through the male line but may originate in the female. J Med Genet 25 : 589–595

72. Schinzel A (1984) Catalogue of unbalanced chromosome aberrations in man. De Gruyter, Berlin

73. Schinzel A (1993) Genomic imprinting: consequences of uniparental disomy for human disease. Am J Med Genet 46 : 583–684

74. Searle AG, Peters J, Lyon MF et al (1989) Chromosome maps of man and mouse IV. Ann Hum Genet 53 : 89–140

75. Sinclair AH, Berta P, Palmer MS et al (1990) A gene from the human sex-determining region encodes a protein with homology to a conserved DNA-binding motif. Nature 346 : 240–244

76. Snouwaert JN, Brigman KK, Latour AM et al (1992) An animal model for cystic fibrosis made by gene targeting. Science 257 : 1083–1088

77. Solter D (1988) Differential imprinting and expression of maternal and paternal genomes. Annu Rev Genet 22 : 127–146

78. Spranger J, Benirschke K, Hall JG et al (1982) Errors of morphogenesis: concepts and terms. J Pediatr 100 : 160–165

79. Taliaferro WH, Huck JG (1923) The inheritance of sickle cell anaemia in man. Genetics 8 : 594

80. Tiepolo L, Zuffardi O (1976) Localization of factors controlling spermatogenesis in the nonfluorescent portion of the human Y-chromosome long arm. Hum Genet 34 : 119–124

81. Ton CCT, Hirvonen H, Miwa H et al (1991) Positional cloning and characterization of a paired box- and homeobox-containing gene from the aniridia region. Cell:1059–1074

82. Vogel T, Klose J (1992) Two-dimensional electrophoretic protein patterns of reciprocal hybrids of the mouse strains DBA and C57BL. Biochem Genet 30 : 649–662

83. Vogt P (1992) Y chromosome function in spermatogenesis. In: Nieschlag E, Habenicht U-F (eds) Spermatogenesis, fertilization, contraception. Molecular, cellular and endocrine events in male reproduction. Springer, Berlin Heidelberg New York, pp 225–265

84. Vogt P, Keil R, Koehler M et al (1991) Selection of DNA-sequences from interval 6 of the human Y-chromosome with homology to a Y chromosomal fertility gene sequence of Drosophila hydei. Hum Genet 86 : 341–349

85. Vogt P, Chandley AC, Hargreave TB et al (1992) Microdeletions in interval 6 of the Y-chromosome of males with idiopathic sterility point to disruption of AZF, a human spermatogenesis gene. Hum Genet 89 : 491–496

86. Vogt P, Chandley AC, Keil R et al (1993) The AZF-function of the human Y-chromosome during spermatogenesis. Chromosomes Today 11 : 227–239

87. Watson JD, Hopkins NH, Roberts JW et al (1987) Molecular biology of the gene, 4th edn. Benjamin/Cummings, Menlo Park

88. Watson JD, Gilman M, Witkowski J, Zoller M (1992) Recombinant DNA, 2 nd edn. Freeman, New York
89. Wilkins L (1950) The diagnosis and treatment of endocrine disorders in childhood and adolescence. Thomas, Springfield
90. Wilson GN, Hall JG, de la Cruz F (1993) Genomic imprinting: summary of an NICHD conference. Am J Med Genet 46 : 675–680
90. Wolf U (1995) The molecular genetics of human sex determination. J Mol Med 73 : 325–331
91. Wright S (1934) The results of crosses between inbred strains of guinea pigs differing in number of digits. Genetics 19 : 537–551

9 Mutation: Spontaneous Mutation in Germ Cells

"I believe than I am one of the most influential people living today, though I haven't got a scrap of power. Let me explain. In 1932 I was the first person to estimate the rate of mutation of a human gene; and my estimate was not far out. A great many more have been found to mutate at about the same rate since."

J. B. S. Haldane, 1964, in BBC interview broadcast after his death less than one year later

The most important feature of genes is their capacity to be reproduced identically from generation to generation. However, evolution would never have been possible if no change in genetic material had ever occurred. Since there is good evidence that all living beings on our planet have a common origin, genes as the carriers of genetic information must have the capacity for occasional alterations. Such changes are indeed observed; they are called mutations.

Animal and plant breeders since time immemorial had observed that inherited characteristics occasionally change, and that such changes are transmitted to following generations. Darwin was very much interested in these observations; he called them "sports." An example at the time was the so-called Ancon sheep, a chondrodystrophic mutation which was popular since such sheep could not jump over fences.

The term mutation was introduced by de Vries in 1901 (see [7]) for sudden genetic changes in the plant *Oenothera lamarckiana*. In this species de Vries observed for the first time a sudden and genetically stable change in one individual under experimentally controlled conditions. Later, when chromosomes could be examined the relatively frequent mutations in this plant were shown to be due to a particular karyotype: the chromosomes were connected end to end; they formed complicated patterns during pachytene, and centromeres were usually distributed in an orderly fashion during metaphase. This mechanism occasionally failed, producing what de Vries had called "mutations."

9.1 Reappraisal of Genetic Variants That May Occur by New Mutation

In experimental genetics the following types of mutations can be distinguished:

a) *Genome mutations* involve alterations in the number of chromosomes. Whole sets of chromosomes may be multiplied (polyploidy), or the number of copies of a particular chromosome may be increased (trisomy) or decreased (monosomy).

b) *Chromosome mutations* (Sect. 2.2). The structure of chromosomes is changed, allowing microscopic detection. The total number of chromosomes is not altered.

c) *Gene mutations.* Here no changes of chromosomes can be detected microscopically; the mutation can be inferred from a change in the phenotype by genetic analysis or can be detected by DNA studies.

Analysis of human mutants at the protein and DNA levels has taught us much about the molecular nature of gene mutations. With these results and the development of new cytogenetic methods, the distinction of chromosome from gene mutation has become blurred. We know that deletions and insertions are possible at the molecular level, and that unequal crossing-over might alter microstructure [70]. At the same time, new techniques have made it possible to detect previously undetectable chromosomal rearrangements. However, chromosomal alterations visible by classical cytogenetic methods, including banding, involve sizeable alterations that differ by orders of magnitude from changes such as deletions of a structural gene. The categorization into structural chromosomal aberrations and gene mutations is therefore useful for practical purposes.

Cells in Which Mutations May Occur. The *localization* of genetic alterations is of prime importance. Mutations may occur either in germ cells or in somatic cells. Germ cell mutations can be transmitted to individuals of the next generation and are usually found in all cells of the affected offspring. Somatic mutations are found only in the descendents of the mutant cell, making the individual a "mosaic." Phenotypic consequences are observable only if the mutations happen to interfere with the specific functions of the affected cells.

Mutation Rates. One of the most frequently used parameters in mutation research is the *mutation rate*, which is defined in humans as the probability with which a particular mutational event takes place per generation. It has become customary to define mutation rates as the rate at which a new mutation takes place in a fertilized germ cell per generation. It should be kept in mind that an individual is formed by two germ cells. For discussion on somatic cell mutation rates, see Chap. 10.

9.2 Genome and Chromosome Mutations in Humans

9.2.1 Mutation Rates

Methods Used. The direct determination of mutation rates requires assessment of the incidence of those cases with a trait or disease in the population in which parents and other family members are unaffected (sporadic cases). Since the late 1950s, when the first chromosome aberrations in humans were discovered, it has become obvious that the mutational events causing genome mutations must occur much more often than gene mutations leading to hereditary diseases. An exact assessment of chromosomal mutation rates became possible when Court Brown initiated investigations on unselected population samples, such as consecutive series of newborns. The calculation of the mutation rate is simple:

$$\mu = \frac{\text{Number of sporadic cases with a certain anomaly}}{2 \times \text{numbers of individuals examined}}$$

This is the so-called direct method, and it can be applied both to single gene determined traits and to genome and chromosome mutations.

When based on series of newborns, this estimate should be qualified: it is confined to mutations whose carriers survive up to the time of birth. In humans as well as in other mammals, however, the great majority of genome and chromosome mutations are lethal, leading to death of the zygote during embryonic life.

Incidence and Mutation Rates: Genome Mutations [94]. Incidences of sex chromosomal and autosomal abnormalities as estimated from studies on newborns (Tables 9.1, 9.2; see also [95]). Many of these cases, with the exception of Turner syndrome and some mosaics and translocations (Chap. 10) originated by nondisjunction during one of the two meiotic divisions in one of the parents' gonads. These are new mutants. Table 9.3 provides mutation rate estimates.

The lower values for trisomies 13 and 18 than for trisomy 21 are falsely low because many fetuses affected with trisomies 13 and 18 are eliminated during embryonic life; placental mosaicism apparently enhan-

Table 9.1. Incidence of sex chromosome abnormalities in population samples of newborns (from Nielsen and Sillesen 1975 [94])

Karyotype	Total		Rate/1000	
47,XYY	28	} 35	0.81	} 1.02
47,XYY mosaics	7		0.20	
47,XXY	33	} 39	0.96	} 1.13
47,XXY mosaics	6		0.17	
♂ 46,XX	2		0.06	
♂ 45,X/46,XY	1		0.03	
46,X,inv(Y)	9		0.26	
45,X	2		0.10	} 0.39
45,X,mosaics	6		0.29	
47,XXX	20	} 24	0.98	} 1.18
47,XXX mosaics	4		0.20	
Population samples	♀ 20 370, ♂ 34 379, ♀ + ♂ 54 749			

The samples came from Edinburgh (UK), Ontario (Canada), Winnipeg (Canada), Boston (USA), Moscow (Russia), and Århus (Denmark).

Table 9.2. Incidence of autosomal abnormalities (genome and chromosome mutations) in 54 749 newborns (from Nielsen and Sillesen 1975 [95])

Karyotype	Total	Rate/1000
47, + 13	3	0.05
47, + 18	8	0.15
47, + 21	63	1.15
47, + marker chromos.	12	0.22
47, + marker, mosaics	5	0.09
Deletions	5	0.09
Inversions	7	0.13
D/D translocations	43	0.79
D/G translocations	11	0.20
Reciprocal translocations	47	0.85
Unbalanced Y-autosomal translocations	2	0.04

ces their chance of survival [62]. (For a discussion of the problem of chromosome aberration in abortions, see Sect. 2.2.4).

Incidence and Mutation Rates: Chromosome Mutations. The incidence of structural autosomal chromosome abnormalities may be taken from Table 9.2. The surveys on which these figures are based used conventional orcein staining. More recent studies with banding methods have yielded slightly higher frequencies, particularly of inversions but also of balanced reciprocal translocations, but the numbers are relatively small [46]. The cases with deletions may be considered to be due to new mutation, giving a mutation rate estimate of 4.57×10^{-5} for the various types of translocations. Table 9.2 does not show how many occur de novo; therefore no mutation rate estimate can be derived from these data. Jacobs et al. [57, 58] estimated the following mutation rates from more limited data: $1.9 \times 10^{-4} - 2.2 \times 10^{-4}$ for all balanced rearrangements (translocations and inversions) together, 3.24×10^{-4} for unbalanced Robertsonian translocations, and 3.42×10^{-4} for unbalanced non-Robertsonian rearrangements. These mutation rates are defined for those "that survive long enough to give rise to clinically recognized pregnancies," i.e., *including* recognized abortions. These data are of interest since structural chromosome aberrations in newborns are likely candidates for future popula-

tion monitoring programs regarding a potential increase of the mutation rate due to mutagenic agents such as ionizing radiation. It would be interesting to have mutation rates for *single*, specific translocations. As shown by studies using DNA polymorphisms, new mutations leading to chromosomal rearrangements (de novo chromosome mutations), including X-autosome translocations, appear to originate mainly if not exclusively in male germ cells [18, 123, 124]. Zygote loss is still higher with advancing pregnancy, as shown by the higher frequency of Down syndrome at the time of chorionic villus sampling, than that observed of amniocentesis some 5–6 weeks later (Table 9.4).

Mutation rates for both genome and chromosome mutations are higher when calculated from amniocentesis data, indicating that fetal loss between the time of amniocentesis (16th–17th week of gestation) and birth is considerable [54, 127].

Table 9.3. Mutation rates for genome mutations observed in newborns (from Nielsen and Sillesen 1975 [94])

Condition	Calculation	Mutation rate
Sex chromosome trisomies		
XXY including mosaics	$\dfrac{39}{2 \times 34379}$	5.67×10^{-4}
XXX	$\dfrac{24}{2 \times 20370}$	5.89×10^{-4}
XXY and XXX together (X nondisjunction)		5.8×10^{-4}
XYY, including mosaics (Y nondisjunction)	$\dfrac{35}{2 \times 34379}$	5.09×10^{-4}
Autosomal trisomies		
47, + 21	$\dfrac{63}{2 \times 54749}$	5.8×10^{-4}
47, + 18	$\dfrac{8}{2 \times 54749}$	7.3×10^{-5}
47, + 13	$\dfrac{3}{2 \times 54749}$	2.7×10^{-5}

9.2.2 Nondisjunction and the Age of the Mother

Statistical Evidence. In Down syndrome an increased risk with advancing age of the parents has been known for many years. Figure 9.1 shows the relative incidence in different age groups of mothers. The risk changes little up to the age of 29 but rises steeply beginning with the age group 35–39. The population incidence of Down syndrome is therefore expected to vary with the age of the parents. Table 9.4 gives an idea of the absolute risk in different age groups of women, as accepted until recently. Experience from amniocentesis and CVS in women aged above 35 indicates that these estimates for risks of trisomy 21 are too low (Table 9.4). The difference, which is still more marked for trisomies 13 and 18

Table 9.4. Risk of Down syndrome in relation to maternal age at delivery (live birth figures calculated for specific years) at amniocentesis and at chorionic villus sampling

Maternal age	At birth	Following amniocentesis[a]		Following chorionic villus sampling[b]	
	Down syndrome	Down syndrome	All chromosomal aberrations	Down syndrome	All chromosome aberrations
30 years	1/700	NA	NA	NA	NA
35 years	1/450	1/250	1/115	1/240	1/110
37 years	1/250	1/150	1/80	1/133	1/64
39 years	1/150	1/100	1/50	1/75	1/37
41 years	1/80	1/60	1/35	1/44	1/21
43 years	1/50	1/35	1/20	1/23	1/12
All ages	1/650	NA	NA	NA	NA

NA, Not available.

[a] From Harper 1984 [49]; data from Sweden, Australia, and Wales (UK).

[b] From Wilson et al. [148].

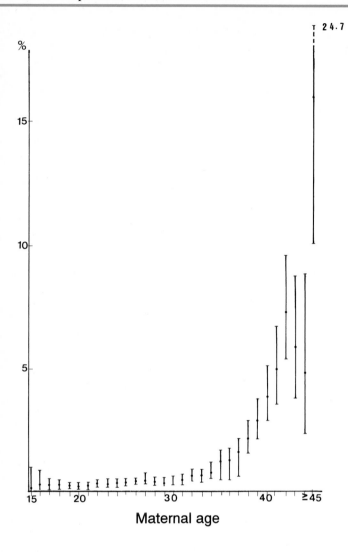

Fig. 9.1. Incidence and confidence intervals (95%) of Down syndrome among newborns. (Data from upstate New York, 1963–1974; Hook and Chamber 1977; 933 cases among 1 729 909 live births)

[54], is explained mainly, if not exclusively, by fetal death after CVS or amniocentesis. Nevertheless recent data indicate a true increase in the risk of having a trisomic child for mothers 35 years old and older.

Some amniocentesis data seem to show a certain leveling off of the increase in risk of trisomies in the oldest age group (mothers aged 46–49). However, a comprehensive study failed to confirm this effect [55].

The ages of mothers and fathers are obviously correlated: older wives tend to be married to older husbands. Therefore it is not immediately obvious whether an increased risk with parental age can be traced to fathers, mothers, or both parents. In Down syndrome, Penrose (1933) [100] showed the increase in risk to be due exclusively to the mother's age. For 150 cases in which both the maternal and the paternal age was known he calculated the following partial correlation coefficients:

a) Between maternal age and incidence of Down syndrome (keeping paternal age constant): $r = +0.221$.

b) Between paternal age and Down syndrome: $r = -0.011$.

The partial correlation techniques permits one to determine the correlations between two variables after the influence of the third had been statistically eliminated.

Cytogenetic data seem to trace about 20% of trisomies 21 to nondisjunction in the paternal germ line. This raised the question of whether, in addition to maternal nondisjunction, there is a (small) increase in risk with paternal age. The once heated discussion came to an end when absence of a paternal age effect was shown even in patients in whom the supernumerary chromosome 21 was identified as of paternal origin based on cytogenetic criteria [53]. A large series (200 trisomy 21 cases) [4, 53] used DNA markers, and demonstrated that 90–95% of the extra chromosomes 21 were of maternal origin. Determination of parental origin by cytogenetically identified chromosome heteromorphisms in the same cases had led to wrong inferences regarding parental origin in several instances. Cytogenetic interpretations are more subjective and may lead to wrong conclusions compared with the definite data using DNA markers (See Sect. 9.2.3). A high frequency of maternal nondisjunction was also shown in trisomy 18 [71].

Higher Risk in Children of Very Young Mothers? It has been claimed that Down syndrome also occurs more frequently in children of very young mothers. The evidence, however, is controversial. In the older literature the most reliable series showed a decreased frequency for age groups under 20 years in comparison to the 20–24 age group. Some series from Canada, Sweden, and Denmark report a somewhat higher incidence in the lowest age group than in the next group [73]. Figure 9.1 shows this increase especially for mothers aged 17–18 years or younger.

Penrose [102] examined the absolute frequencies of trisomic children in relation to maternal ages. Figure 9.2 compares frequencies in different age groups for trisomies 21, 18, 13, XXY, and XXX with the maternal age distributions in four representative populations. Contrary to the distribution in the general population, which is clearly unimodal, the distribution among mothers of trisomy cases suggests bitangentiality. This finding fits the hypothesis of a mixture of an age-dependent and an age-independent group, a conclusion supported by other evidence (Sect. 9.2.4).

Age-Specific Rates in Trisomies. Figure 9.3 shows the effect of maternal age on all clinically recognized pregnancies involving different trisomies (trisomy among liveborn children and spontaneous abortions), assuming a spontaneous abortion rate of 15%. There is a strong maternal age effect, so that for women 42 years and older about one-third of all clinically recognized conceptions are abnormal. Assuming that nondisjunction leads to equal numbers of monosomic and trisomic zygotes, it can be assumed that most oocytes among older women are aneuploid.

Maternal Age Effect in Other Trisomies. The relative incidences at various maternal ages are shown in Fig. 9.4 for trisomy 13 and 18 and in Fig. 9.5 for syndromes due to nondisjunction of the X chromosome (XXY and XXX) [33]. Most trisomies detected in surveys of spontaneous abortions are also more frequent in children of older women (Table 9.5). However, the effect for large chromosomes is small and nonsignifi-

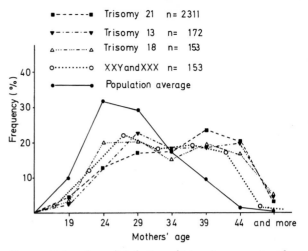

Fig. 9.2. Maternal age distribution of trisomies 21, 13, 18, and a combined sample of XXY, XXX compared with the population average in four representative populations. (Penrose 1957 [102]; Court Brown 1967 [21]; Magenis, see [140])

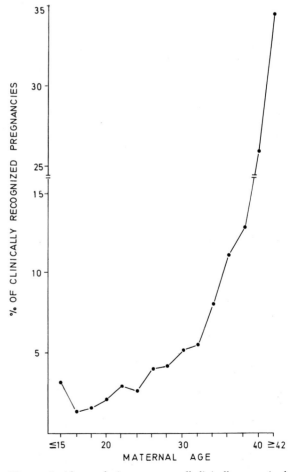

Fig. 9.3. Incidence of trisomy among all clinically recognized pregnancies, including spontaneous abortions, assuming a spontaneous abortion rate of 15%. (From Hassold and Jacobs 1984 [50])

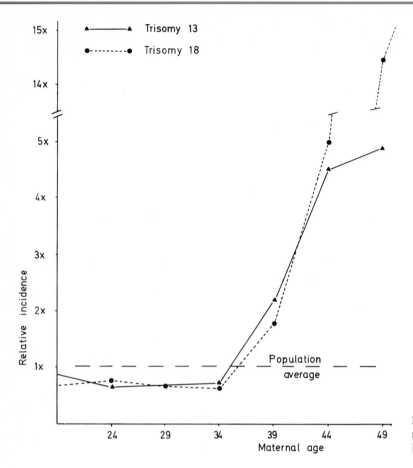

Fig. 9.4. Maternal age effect in trisomies 13 and 18. (Magenis et al. 1968, see [140])

cant (it is slightly larger in some earlier series). Among the smaller chromosomes, the maternal age effect is small – albeit significant – for trisomy 16. No significant age effects have been found for de novo chromosome rearrangements [54].

Little or no paternal age effect was found for the XXY karyotype [15].

9.2.3 In Which Sex and at Which Meiotic Division Does Nondisjunction Occur?

As explained in Sect. 2.2.1, trisomy is caused by meiotic nondisjunction. Two questions arise:

1. Does nondisjunction occur mainly during male or female meiosis?
2. Does it occur mainly in the first or in the second meiotic division?

Since the parental age effect, as described above, was shown to be almost always maternal in origin, it was tempting to conclude that most observed cases of nondisjunction occur in the female germ line. Evidence from X-linked marker studies, however, show that this is not true for the X chromosome.

Evidence for the X Chromosome from X-Linked Marker Studies. The principle of determining the origin of the trisomic germ cell is shown in Fig. 9.6. A color vision defective (deuteranomalous) patient with Klinefelter syndrome has a mother who is heterozygous for deuteranomaly and a father with normal color vision. The trisomic germ cell presumably originates in the mother. Nondisjunction must have occurred during the second meiotic division of oogenesis, in which sister chromatids of the same chromosomes are normally separated.

In principle, the same argument may be used by application of the X-linked blood group Xg. Race and Sanger (1969) [104], reviewing the evidence for nondisjunction in Klinefelter (XXY) syndrome, arrived at an estimate of 40 % occurring in the paternal germ lines, all in the first meiotic division; 50 % occur in the first maternal meiotic division and 10 % in the second. In all four cases of XXXY and XXXXY, the extra chromosomes come from the maternal germ line; it originated during paternal meiosis in the XXYY cases.

Among XO patients whose Xg types contribute information to this analysis, about 74 % posses a maternal

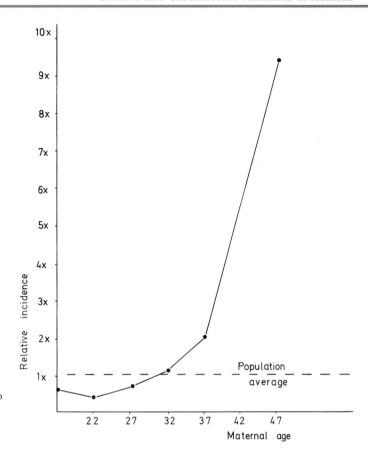

Fig. 9.5. Maternal age effect in syndromes due to nondisjunction of X chromosomes: XXY and XXX combined; 153 cases from the United Kingdom. (Court Brown 1967 [21])

Table 9.5. Maternal age in two series of spontaneous abortions (New York City, 372 abortions; Hawaii, 418 abortions; from Hassold et al. 1981 [51])

Karyotype	Maternal ages	Karyotype	Maternal ages
Normal newborns; XX and XY	NY City: 25.8 years Hawaii: 25.0 years	Abortions with normal karyotypes; XX and XY	26.7 ± 6.2
Trisomies		Trisomies	
47 + 2	28.3 ± 6.2	47 + 13	29.2 ± 5.6
47 + 3	25.5 ± 6.4	47 + 14	25.6 ± 7.4
47 + 4	27.9 ± 6.2	47 + 15	$31.0 \pm 6.3^{***}$
47 + 5	22.0	47 + 16	28.2 ± 5.3
47 + 6	24.0 ± 4.2	47 + 17	32.8 ± 8.6
47 + 7	$30.2 \pm 6.7^{*}$	47 + 18	$33.3 \pm 5.6^{***}$
47 + 8	24.6 ± 4.9	47 + 19	
47 + 9	29.0 ± 6.9	47 + 20	$33.8 \pm 5.5^{***}$
47 + 10	31.8 ± 7.7	47 + 21	$30.0 \pm 7.0^{**}$
47 + 11	25.0	47 + 22	$31.7 \pm 5.3^{***}$
47 + 12	26.6 ± 7.0		

Statistically significant differences between normal and trisomic abortions: ** p 0.01; *** p 0.001.

and about 26% a paternal X. If it is assumed that most XO cases are the result of chromosome loss during the early zygote stages and not of meiotic nondisjunction (Sect. 2.2.1), the findings concur with the results obtained in mice where the paternal X chromosome is particularly vulnerable a short time after fertilization (Sect. 11.1.3).

In principle, the same kind of argument may be used for identification of parental chromosomes by microscopically identifiable chromosome variants.

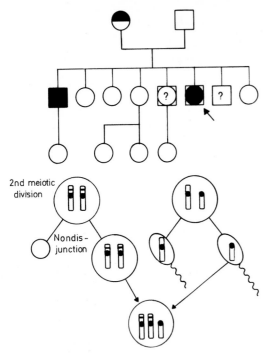

Fig. 9.6. Determination of the origin of trisomic germ cells. A patient with Klinefelter syndrome (*arrow*; XXY) is deuteranomalous. His father has normal color vision. The mother must be heterozygous because she has one deuteranomalous son. If the additional chromosome comes from the father, the Klinefelter son must be heterozygous or homozygous normal. The fact that he is deuteranomalous shows that both his X chromosomes descended from one X of the mother (see Sect. 2.2.3.1). Moreover, nondisjunction must have occurred in the second meiotic division. (Lenz 1964 [76])

Direct Evidence from Chromosome Variants and DNA Polymorphisms. Human chromosomes show individual variants that are constant over many generations and are also known as heteromorphisms. Their overall frequency varies between about 5% and 50%, depending on the methods used (Sect. 2.1.2.3). Variants in the short-arm and satellite regions of acrocentric chromosomes are especially common; they were therefore used for chromosome identification. However, the methods is subjective and error prone, and an appreciable number of cases prove to be noninformative. Therefore, this technique has now been abandoned in favor of DNA polymorphisms. The various types of such DNA polymorphisms are described in Sec. 12.1.2. These can be identified unequivocally; many have a great number of alleles and therefore a large content of information; moreover, there are so many DNA variants that each individual chromosome can be identified if a sufficient panel

of such markers is used. Therefore they are most useful in identifing parental chromosomes in aneuploidy and have replaced chromosome heteromorphisms in this work.

Use of Chromosomal Variants and DNA Markers for Identification of Nondisjunction. Genetic markers can be used to trace a certain chromosome back to father or mother and to the first or second meiotic division

Chromosome 21, as the other acrocentric chromosomes, carries rRNA genes in its short-arm regions; in the interphase nucleus these regions are found close together at the nucleolus (Sect. 2.1.2.3), forming the nucleolus organizer region.

The results of many studies indicate that nondisjunction can occur in either the first or the second meiotic division. Among the cases of maternal origin the great majority result from first division nondisjunction. The 4%–10% of paternal cases may be caused by nondisjunction either in the first or in the second meiotic division, but errors in first meiosis appear to be more common.

In X-chromosomal nondisjunction, the situation turns out to be quite different. Here, the nondisjunction frequency is about the same in both sexes, or it might even be slightly higher in males [59]. The limited data suggest a relatively higher fraction of paternal nondisjunction in fathers of Klinefelter patients but a higher proportion of maternal events in mothers of XXX women. The higher proportion of male nondisjunction in X-chromosomal anomalies may explain the somewhat weaker increase with maternal age as compared with autosomal trisomies (contrast Figs. 9.3 and 9.5).

9.2.4 Nondisjunction, Chromosome Variants, and Satellite Association

Even relatively harsh methods of making metaphase preparations do not destroy completely the intimate relationship of chromosomes to each other.

Satellite Association. In metaphase, chromosomes with satellites (13–15; 21, 22) show a tendency to lie side by side, with their satellite regions facing each other [30]. There is considerable interindividual variability in this phenomenon. It was therefore concluded that human beings with frequent satellite associations would manifest an increased probability of nondisjunction. The definition of a satellite association is shown in Fig. 9.7 [152]. In some studies a higher frequency of such satellite associations has been described in parents – or sometimes only in mothers – of trisomy patients. These data support the hypothesis that an increased tendency to satellite association enhances the risk for nondisjunction.

Thyroid Disease and Antithyroid Antibodies. Altered thyroid function had long been suspected as a risk factor for nondisjunction. Fialkow et al. (1967) [31] found that mothers of children with Down syndrome had a significantly higher frequency of thyroid autoantibodies than the controls (Fig. 9.8). This study was initiated by a clinical observation of one of the authors (A.G.M.): two girls with Turner syndrome due to X isochromosomes also had Hashimoto thyroiditis. It was difficult to believe that this finding was accidental.

In Fialkow's study the frequency of thyroid autoantibodies was about the same in older and younger mothers. Because of the age-dependent increase in positive reactors with age in the control population, a significant difference between mothers of Down syndrome children and controls was found only in the younger age group. Hence there is good evidence that thyroid autoantibodies as a marker for autoimmunity enhance the risk for nondisjunction. Relative to other risk factors, this factor seems to be especially important in younger mothers in whom the age-dependent risk is low.

Do Thyroid Autoantibodies and Autoimmune Disease Also Enhance the Risk for Other Aneuploidies? Nondisjunction of acrocentric chromosomes differs from nondisjunction – and chromosome loss – of X chromosomes. Therefore the results presented above do not necessarily imply that autoimmune thyroid disease enhances the risk for X-chromosomal aneuploidy as well. However, they do provide a hint about where to look for risk factors. Some reports suggest an increased frequency of autoantibodies in patients with gonadal dysgenesis (XO type) and in their parents. An increase in thyroid antibodies is even more pronounced in patients with mosaics and their mothers [32]. Many cases of juvenile diabetes are now considered to be caused by an autoimmune mechanism (Sect. 6.4.1). A surprisingly high incidence of diabetes in close relatives of patients with the XO and XXY karyotypes, especially parents, has been reported [93, 143].

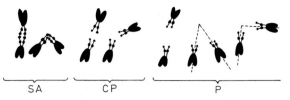

Fig. 9.7. Definition of a satellite association. *SA*, satellite association; *CP*, close proximity; *P*, proximity. (Zellweger et al. 1966 [152])

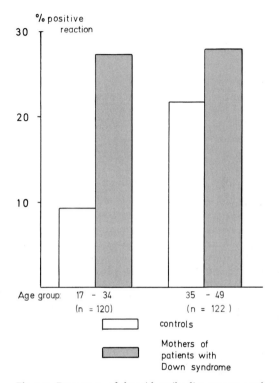

Fig. 9.8. Frequency of thyroid antibodies among mothers of children with Down syndrome compared with age-matched mothers of children not showing Down syndrome. A significant increase was found in young but not in older mothers of Down syndrome children. (Fialkow 1967 [32])

9.3 Gene Mutation: Analysis at the Phenotype Level

Almost all trisomies observed in the human population are caused by new mutations. In these cases the patients are the only ones affected in their families; they are "sporadic" cases. If they have no children, the supernumerary chromosome is not transmitted to the next generation. A person bearing a dominant or X-linked gene mutation usually transmits this mutation to the next generation (Chap. 4). If the mutation does not interfere with health, reproduction is almost normal, and a pedigree with many affected individuals may ensue. Almost every affected individual has affected ancestors and relatives. If the condition prevents many of its bearers from having children, most of the newly mutant genes become extinct soon after they appear, and a relatively large number of all observed cases are sporadic and caused by a new mutation. Large-scale population surveys are required to estimate the mutation rate. Gene mutation rates may be analyzed at the qualitative-phenotypic-Mendelian, at the protein-enzyme, or at the DNA level.

9.3.1 Methods for Estimating Mutation Rates

The following sections describe various methods estimating mutation rates of rare hereditary diseases. In most cases, the methods have been applied to conditions analyzed at the qualitative-phenotypic-Mendelian level, implying that a simple mode of inheritance has been established. In an increasing number of these conditions the genes have been localized, and in some of them genetic heterogeneity has been found.

Direct Method. The principle of the direct method is described (Sect. 9.2.1). Apart from genome mutations and structural chromosome aberrations, this approach can be utilized for dominant gene mutations. It simply requires determination of the *incidence* (frequency at birth) of a certain condition in the population and knowledge of whether the case is indeed sporadic. All sporadic cases are assumed to be new mutations, and the mutation rate can be calculated with ease (see the formula in Sect. 9.2.1). In practice, application of this simple principle meets with difficulties and possibilities of error:

a) The most evident source of error is illegitimacy, a possibility that needs especially to be considered if the selective disadvantage of a trait is not obvious, and if very few sporadic cases are observed among a majority of familial ones. On the other hand, if there is a strong selective disadvantage, and if there are many sporadic cases along with some familial ones, an occasional case of illegitimacy does not substantially disturb the estimate. Since many genetic markers are now available, false paternity can usually be ruled out by appropriate DNA tests.

b) A second possible source of error is the occurrence of phenotypically similar or identical cases that happen to be nonhereditary (phenocopies). The strictest genetic test for this bias is examination of offspring in such sporadic cases. If all such cases are mutants, a 1:1 segregation of their affected and unaffected offspring is expected. However, the test is insensitive if the total proportion of phenocopies is small. A preliminary hint is provided by considerations of genetic equilibrium. Selection against

the condition should be strong enough to account for a fraction of sporadic cases among all cases [130]. This principle leads to the general rule that the proportion of sporadic cases is roughly proportional to the selective disadvantage of the trait (Fig. 9.8). In the extreme cases with no transmission to the next generation because of early death, all cases of autosomal-dominant mutation in the population are sporadic. In intermediate cases a variably large proportion of cases are new mutants.

c) Often different varieties of autosomal-dominant disease exist. For example, mutations of genes at different gene loci may lead to the same phenotype, or different mutations within the same gene may lead to slight (and sometimes not so slight) phenotypic differences. If more than one gene is involved, the mutation rate should be considered as a combined estimate for these genes. If all varieties are lumped together as one disease, the mutation rate may be spuriously elevated. Similarly, in addition to an autosomal-dominant type of disease, an autosomal-recessive variety may exist. Careful clinical and laboratory analysis of the phenotype, age of onset, natural history of the disease, and linkage analysis using, for example, DNA markers, may help in discrimination. Consanguinity, if present, may provide a strong hint of autosomal-recessive inheritance in populations where inbreeding is unusual.

d) Penetrance may be incomplete. If penetrance is not far below 100%, this bias can be corrected.

The most straightforward method is to compare the number of patients carrying new mutations with the total number of children born in the same year. This is equivalent to determining the incidence. This method is feasible mainly for conditions that can be identified in early infancy. Thus, acrocephalosyndactyly (Apert syndrome) (101200) can be identified at birth in all affected infants, and the clinical status of the parents is clearly apparent for this defect. The mutation rate can be calculated on the basis of the total number of births in the population. However, most diseases are not discovered at birth, and prevalence data ("cases at hand") alone are available. For example, the life span of hemophilia patients was one-third that of normals in one mutation study; therefore the prevalence figure had to be multiplied by a factor of 3 to obtain the incidence. Note that in this example the number of hemophiliacs was compared to that of the total population, while in the earlier example the number of cases of acrocephalosyndactyly per year was related to all births of that year alone.

The direct method, in spite of its simplicity, was introduced only after the concept of genetic equilibrium between mutation and selection had been established.

Danforth's Formula. Danforth [23] first formulated the equilibrium concept in 1921 and proposed its use for estimating the mutation rate:

"It may be recalled that there is a considerable number of dominant traits which are slightly unfavorable ... The incidence of these traits is no doubt maintained in part by recurrent mutations. The frequency of such mutations could be estimated if the average number of generations through which they persist were known. In some of these there is evidence that the average duration is for only a few generations ... The rate of mutation must be such as to bring the incidence to its present value and to balance the adverse effect of selection."

He then proceeded to derive a formula to determine the average number of generations during which a mutation persists before elimination as a measure for the mutation rate. On Danforth's proposal, the concept of equilibrium between mutation and selection was clearly formulated. However, it was overlooked only to be rediscovered 15 years later by Haldane [43].

Haldane's Indirect Method for Mutation Rate Calculation. Haldane explained the concept as follows:

"The sex-linked recessive condition haemophilia has been known for over a century. Since only a small minority of haemophiliacs live long enough to breed, and (as will be seen) over one-third of all haemophilia genes in new-born babies are in the X chromosome of males, the condition would rapidly disappear unless haemophilia genes arose by mutation. The only alternatives would be that heterozygous females were more fertile than normal, or that in their meiosis the normal allelomorph ... was preferentially extruded into a polar body. Neither of these alternatives seems likely ...
We now assume, and will later attempt to show, that most large human populations are in approximate equilibrium as regards haemophilia, selection being balanced by mutation ... if x be the proportion of haemophilic males in the population, and f their effective fertility, that is to say their chance, compared with a normal male, of producing offspring, then in a large population of $2N(1-f)xN$ haemophilia genes are effectively wiped out per generation. The same number must be replaced by mutation. But as each of the N females has two X chromosomes per cell, and each of the N males one, the mean mutation rate per X chromosomes is $\frac{1}{3}(1-f)$ x, or if f is small, a little less than $\frac{1}{3}$ x. Hence, we have to determine the frequency of hemophilia in males to arrive at the appropriate mutation rate."

Additionally, Haldane gave a formal treatment that led to the following results. The ratio of heterozygous females to hemophilic males is:

$$1 + \frac{2f\mu + \nu}{2\mu + \nu}$$

Here, μ is the mutation rate in female germ cells, and ν in male. Of all cases of hemophilia, a fraction $(1-f)\mu/(2\mu+\nu)$ should be sons of normal homozygous mothers, i.e., sporadic cases. In the same paper Haldane showed that genetic equilibrium would indeed be reestablished within a very short time after any perturbation.

Haldane's method is practical because information from only one generation is needed. This information, however, can be utilized in different ways. One useful extension concerns a separate assessment of mutations in male and female germ cells (Sect. 9.3.4).

Practical Problems in Applying the Indirect Method. As with the direct method, a number of practical problems arise when the indirect method is actually applied:

a) Illegitimacy is no problem since the estimate is based, in all cases, on conditions in the present-day population and not on previous generations.
b) Phenocopies and genetic heterogeneity raise problems identical to those encountered with the direct method.
c) Incomplete penetrance does not influence estimates of mutation rates provided that those gene carriers who fail to manifest the condition have no selective disadvantage compared with the population average.
d) A problem that is unique to the indirect method is the estimation of f (average fertility of patients relative to the population average). This problem is simple if f = 0, i.e., the affected individuals do not reproduce at all. An example concerns the Duchenne type of muscular dystrophy. The formula for the ratio of patients who are sons of normal mothers is $m = \frac{(1-f)\mu}{2\mu+\nu}$, which becomes $\frac{1}{3}$ if $\mu = \nu$ and f = 0. *This means that one-third of the observed cases are due to new mutations if mutation rates in males and females are identical, and affected individuals are unable to reproduce.*

The problem is much more difficult if the fertility of the trait bearers is only slightly reduced. A useful approximation for f can be reached by comparing reproduction of patients with that of their unaffected sibs, if the patients' fertility is significantly subnormal. Otherwise, biases caused by family planning of sibs may be considerable. The best method is to determine the number of children of a randomly chosen population group in the same age range, with follow-up to the end of their reproductive period in comparison with patients [105, 107]. Even this method may yield distorted data if patients reproduce more frequently than in earlier times because of improved medical therapy. Under such conditions, the population is no longer in equilibrium and mutation rates are underestimated.
These biases make all estimates of f unreliable. Therefore the indirect method can be expected to give a general idea of the correct order of magnitude of the mutation rate only if the fertility of the patient (f) is markedly reduced.

The direct method was used in the great majority of estimates in Tables 9.6 and 9.7; the indirect method was used mainly for X-linked recessive conditions. In hemophilia fertility was markedly reduced at the time that these data were collected. Fertility reaches zero in the Duchenne type of muscular dystrophy and in the two conditions in which hemizygosity is lethal: incontinentia pigmenti and the orofaciodigital (OFD) syndrome. Therefore these estimates can be regarded as fairly reliable.

Mutation Rates Cannot Be Estimated for Autosomal-Recessive Diseases. Obviously the direct method cannot be used in fully recessive conditions because the mutation would most often occur in the germ cell of an individual who is mated with a normal homozy-

Table 9.6. Selected classic mutation rates for human genes (from Vogel and Rathenberg 1975 [138], with additions

Trait	Population examined	Mutation rate	Gene(s) located at	Authors
Autosomal mutations				
Achondroplasia	Denmark	1×10^{-5}	4q	Mørch, corrected by Slatis
	Northern Ireland, U.K.	1.3×10^{-5}		Stevenson
	Four cities	1.4×10^{-5}		Gardner
	Münster, Germany	$6-9 \times 10^{-6}$		Schiemann
Aniridia 1, 2	Denmark	$2.9-5 \times 10^{-6}$	11p13, 2p25	Møllenbach, corrected by Penrose
	Michigan, U.S.A.	2.6×10^{-6}	19q	Shaw et al.
Myotonic dystrophy	Northern Ireland, U.K.	8×10^{-6}		Lynas
	Switzerland	1.1×10^{-5}		Klein, corrected by Todorov et al.
Retinoblastoma	United Kingdom, Michigan, U.S.A.; Switzerland; Germany	$6-7 \times 10^{-6}$	13q14.1–q14.2	Vogel
	Hungary	6×10^{-6}		Czeizel et al.
	The Netherlands	1.23×10^{-5}		Schappert-Kimmijser et al.
	Japan	8×10^{-6}		Matsunaga
	France	5×10^{-6}		Briart-Guillemot et al.
	New Zealand	$9.3-10.9 \times 10^{-6}$		Fitzgerald et al.
Acrocephalosyndactyly (1) (Apert syndrome)	United Kingdom	3×10^{-6}	?	Blank
	Münster, Germany	4×10^{-6}		Tünte and Lenz
Osteogenesis imperfecta, types I, II, IV	Sweden	$0.7-1.3 \times 10^{-5}$	Heterogeneous	Smårs
	Münster, Germany	1.0×10^{-5}		Schröder
Tuberous sclerosis (epiloia)	Oxford, U.K.	1.05×10^{-5}	9q34; 11q23	Nevin and Pearce
	Chinese	6×10^{-6}		Singer
Neurofibromatosis (1)	Michigan, U.S.A.	1×10^{-4}	17q11.2	Crowe et al.
	Moscow, Russia	$4.4-4.9 \times 10^{-5}$		Sergeyev
Polyposis of intestines	Michigan, U.S.A.	1.3×10^{-5}	5q22–q23	Reed and Neel
Marfan syndrome	Northern Ireland, U.K.	$4.2-5.8 \times 10^{-6}$	5q15–5q21.3	Lynas
Polycystic disease of the kidneys	Denmark	$6.5-12 \times 10^{-5}$	16q13.1–16q13.33	Dalgaard
Diaphyseal aclasis (multiple exostoses)	Münster, Germany	$6.3-9.1 \times 10^{-6}$	?	Murken
Sex-linked recessive mutations				
Hemophilia	Denmark	3.2×10^{-5}		Andreassen, corrected by Haldane
	Switzerland	2.2×10^{-5}		Vogel
	Münster, Germany	2.3×10^{-5}		Reith
Hemophilia A	Hamburg, Germany	5.7×10^{-5}	Xp28	Bitter et al.
	Finland	3.2×10^{-5}		Ikkala
Hemophilia B	Hamburg, Germany	3×10^{-6}	Xq27.1–q27.2	Bitter et al.
	Finland	2×10^{-6}		Ikkala
Duchenne type muscular dystrophy	Utah, U.S.A.	9.5×10^{-5}	Xp21.2	Stephens and Tyler
	Northumberland and Durham, U.K.	4.3×10^{-5}		Walton
	Südbaden, Germany	4.8×10^{-5}		Becker and Lenz
	Northern Ireland, U.K.	6.0×10^{-5}		Stevenson
	Leeds, U.K.	4.7×10^{-5}		Blyth and Pugh
	Wisconsin, U.S.A.	9.2×10^{-5}		Morton and Chung
	Bern, Switzerland	7.3×10^{-5}		Moser et al.
	Fukuoko, Japan	6.5×10^{-5}		Kuroiwa and Miyazaki
	Northeast England, U.K.	10.5×10^{-5}		Gardner-Medwin
	Warsaw, Poland	4.6×10^{-5}		Prot
	Venice, Italy	$3.5-6.1 \times 10^{-5}$		Danieli et al.
Incontinentia pigmenti, Mainz type 2 (Bloch-Sulzberger)	Münster, Germany	$0.6-2.0 \times 10^{-5}$	Xq27–Xq28	Essig
Orofaciodigital (OFD) syndrome	Münster, Germany	5×10^{-6}		Majewski

Table 9.7. Comparison of mutation rates

Type of mutation	Example	Mutation rate (at birth)
Trisomy	Trisomy 21 (Down syndrome)	$\sim 5.8 \times 10^{-4}$
Robertsonian translocation	Down syndrome	$\sim 1 \times 10^{-4}$
Reciprocal translocation	Many chromosomal diseases	$\sim 4.3 \times 10^{-4}$
Cytogenetically visible deletions	Cri du chat syndrome (5p-) and others	$\sim 5 \times 10^{-5}$
Deletions invisible by conventional cytogenetic methods but leading to a contiguous gene syndrome	WAGR syndrome (del 15p) and others	$\sim 10^{-5}$–10^{-6}
"Classical" mutation rates estimated from autosomal dominant or X-linked phenotypes	Achondroplasia, Hemophilia A	$\sim 10^{-5}$ ~ 2–5×10^{-5}
Deletion within one gene	Duchenne muscular dystrophy	~ 3–5×10^{-5}
Single base pair substitutions (per base pair)	Hemoglobin β chain, many other examples	$\sim 10^{-8}$–10^{-9} (transitions, especially those involving CpG dinucleotides, are more common than transversions)
DNA polymorphisms outside transcribed genes: RFLP, VNTR, CA repeats		Enormous variation (some VNTR mutations even occur in a small percentage of all germ cells)

Table 9.8. Approximate proportions of patients affected by new mutations in autosomal dominant disorders (modified from Goldstein and Brown 1977)

Disorder	Percentage
Apert syndrome (acrocephalosyndactyly)	>95
Achondroplasia	80
Tuberous sclerosis	80
Neurofibromatosis	40
Marfan syndrome	30
Myotonic dystrophy	25
Huntington disease	1
Adult polycystic kidney	1
Familial hypercholesterolemia	<1

gote, and the mating would therefore produce heterozygous and normal homozygous children. If methods to detect heterozygotes for various conditions were available, this problem might be obviated [107]. Currently this method is not being utilized.

Theoretically the indirect method could be used. With every homozygote who does not reproduce, two mutant genes are eliminated from the population, and the loss of these alleles would need to be compensated by mutation to achieve equilibrium. However, application of the method is subject to two reservations. The selective disadvantage must be confined to the homozygotes, and the heterozygous state must be selectively neutral. According to the Hardy-Weinberg law (Sect. 4.2), the number of heterozygotes is $2pq$, and that of affected homozygotes is q^2. Heterozygotes are therefore found much more frequently than affected homozygotes, especially if the condition is rare. A very small selective disadvantage would therefore require a much higher mutation rate to maintain genetic equilibrium, whereas a small advantage would render mutations unnecessary to explain a genetic equilibrium.

Furthermore, it is demonstrably wrong to assume genetic equilibrium for recessive mutations in present-day human populations. In the past the human population was divided into many isolated groups which showed varying rates of population growth; most of these groups have begun mixing only in relatively recent times. Screening programs for rare inborn errors of metabolism show remarkable differences in incidence of recessive genes even among closely related populations, and the distribution of mutants identified by DNA studies, for example in phenylketonuria and cystic fibrosis, shows a pattern not at all compatible with simple mutation-selection equilibria (Sect. 12.1.3). The almost worldwide decrease in the number of consanguineous marriages has also contributed to the disturbance of any genetic equilibrium that may have existed in the past. In many populations the number of cases of rare recessive disorders is presently below equilibrium value, and the increase to equilibrium is expected to be very slow ([44]; Sect. 13.1.1.2). Depending on arbitrary assumptions, almost any mutation rate estimate can be calculated for a recessive condition using the same data. Therefore such estimates are guessing games of no scientific value.

9.3.2 Results on Mutation Rates

Estimates Based on Population Surveys. Table 9.6 lists mutation rate estimates. Table 9.7 compares these mutation rates with those obtained for visible chromosomal aberrations, single base substitutions, and DNA polymorphisms. The main criterion for inclusion in Tables 9.6 and 9.7 was that the determination of incidence, especially of sporadic cases, be fairly reliable. Some remarks may be useful for some specific conditions.

Achondroplasia (100 800) is a fairly well-defined condition characterized by shortening of the limbs, depressed nasal bridge, characteristic vertebral changes, and occasional internal hydrocephalus. Patients' reproduction is markedly reduced; therefore the majority of observed cases are sporadic and caused by new mutations. At least two superficially similar conditions that lead to death a short time after birth are not included in the estimates: achondrogenesis (200 600, 200 610, 200 700) and thanatophoric dwarfism (273 670). The rapid development of nosology in the skeletal dysplasias with many different subtypes, which may be confused with classic achondroplasia, has made earlier estimates of the mutation rates for this condition suspicious. Genetic heterogeneity is a general problem with most mutation rate estimates (see above) which has been overcome, however, in this instance by the discovery that the classic achondroplasia phenotype, as described above, is caused by a point mutation within the fibroblast growth factor receptor-3 *(FGFR-3)* gene on 4p16.03. The great majority of cases are caused by a $G \rightarrow A$ transition (see below) within the transmembrane domain of this gene [95a, 118a].

In view of the possibility to distinguish the various types of achondroplasia by molecular, clinical and radiographic criteria, a new look at the mutation rates in this condition and in some of its dominant genocopies would be of great interest. Meanwhile, the three currently available estimates for achondroplasia are remarkably similar.

In *aniridia* (Figs. 9.9, 9.10; 106 200), the irises are lacking. Affected persons may suffer additional visual handicaps such as nystagmus, cataracts, or glaucoma. The two estimates, which are based on sound population surveys, gave similar results.

Myotonic dystrophy (160 900) is a muscular disorder described in Sect. 4.1.7 in connection with anticipation.

Retinoblastoma (180 200) is a malignant eye tumor that affects children in the first years of life. In any developed society, every patient is sooner or later seen by a physician, in most cases by an ophthalmologist. Hence, for ascertainment of all cases in a given population, only ophthalmologists and ophthalmological departments need be screened. Series from a number of different populations are available. However, not all cases are due to germ cell mutations. Most unilateral, sporadic cases are nonhereditary and caused by somatic mutations (Chap. 10). Only 10%–12% [132, 135] of such unilateral sporadic cases are hereditary. The bilateral sporadic cases are all hereditary. Even so, the mutation rate estimates in Tables 9.6 and 9.7, which were calculated on this basis, agree very well.

Acrocephalosyndactyly (Apert syndrome; 101 200) is a complex syndrome consisting of skull malformation and complete distal fusion of fingers with a tendency to fusion of the bones (Fig. 9.11). In a number of cases, additional malformations have been reported, and there is an increased risk for early death. While patients rarely have children, transmission has been observed. The conclusion that the condition is due to a dominant mutation is based on these findings and on the very strong paternal age effect (Sect. 9.3.3) [11, 103, 146].

Osteogenesis imperfecta (166 200) includes, apart from the increased fragility of bones, blue sclerae and frequently sensori-

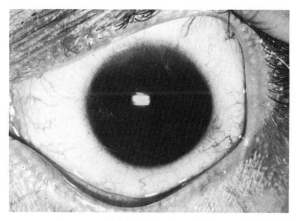

Fig. 9.9. Aniridia. In this case the iris is totally absent. (Courtesy of Dr. W. Jaeger)

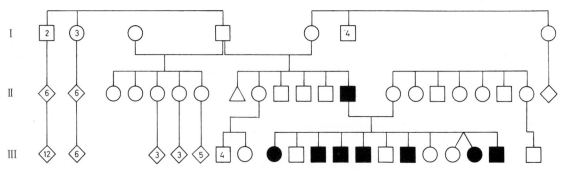

Fig. 9.10. Pedigree with new mutation to aniridia. (From Møllenbach 1947 [86])

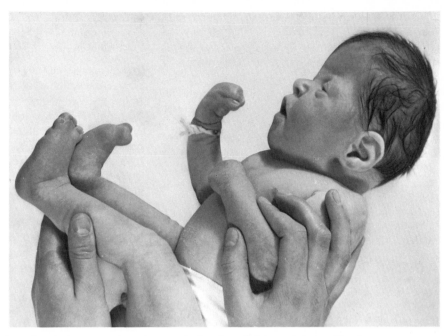

Fig. 9.11. Child with acrocephalosyndactyly (Apert syndrome). Note syndactyly and deformed shape of head

neural hearing loss. Its extremely variable expressivity renders any mutation rate estimate unreliable. Genetic heterogeneity and the existence of recessive types adds to the difficulties.

Tuberous sclerosis (191100) is one of the first conditions for which a mutation rate estimate became available [42]. However, it is not one of the best suited ones, as gene expression is quite variable.

Neurofibromatosis (162200) also shows a quite variable expressivity. The first mutation rate estimate was based on a very careful epidemiological study in Michigan [1441]. The rate was estimated by both the direct and the indirect method. The estimate given in Table 9.6 (1×10^{-4}) is the highest known so far for a human disorder. However, there seems to be no conclusive evidence for genetic heterogeneity apart from families with acoustic neurinoma ("central neurofibromatosis"). A more recent estimate from the former Soviet Union [117] gives a value somewhat more in line with other mutation rate estimates. Here, however, the incidence estimate was based on examination of 16-year-old prospective conscripts. At this age some mild manifestations could still be overlooked.

With *intestinal polyposis* (175100), genetic heterogeneity poses a problem, as there is a least one other syndrome involving multiple colon polyps (Gardner syndrome), but this appears to be intragenic heterogeneity, since in both conditions the same gene is involved.

Marfan syndrome (154700) can be confused with homocystinuria transmitted by autosomal-recessive inheritance.

Polycystic disease of the kidneys (173900) has, after neurofibromatosis, the second highest mutation rates calculated so far.

The estimate for *multiple exostoses* (diaphyseal aclasis) (133700) is based on seven sporadic cases in a relatively small population. Penetrance seems to be influenced by sex-limiting modifying genes (Sect. 4.1.6); the mutation rate estimate might not be exact.

X-Linked Recessive Conditions. For *Hemophilia* (306700, 306900), estimates in various populations agree relatively well. The first estimates (for Denmark and Switzerland) included both hemophilia A and B; later the two conditions were treated separately. The mutation rate for hemophilia A is about one order of magnitude higher than that for hemophilia B.

For the *Duchenne type of muscular dystrophy* (310200) at least 11 mutation rate estimates from various populations are available. As with retinoblastoma, the ascertainment problems can be overcome relatively easily. The diagnosis can be made without difficulty. Application of the indirect method is obviously justified; the patients never have children. Hence selection against this mutation is very strong. All estimates agree amazingly well in order of magnitude.

The *incontinentia pigmenti* (Bloch-Sulzberger syndrome) (308300) estimate is based on the hypothesis (suggested by Lenz [75] and explained in Sect. 4.1.4) that the mode of inheritance of this disorder is X-linked dominant with lethality of the male hemizygotes. This hypothesis has been confirmed by localization of the gene to the X chromosome. This mode of inheritance is bound to lead to strong selection against the mutation.

Are These Mutation Rates Representative of Comparable Mutations in the Human Genome? The mutation rate estimates in Table 9.6 are all of an order of magnitude between 10^{-4} and 10^{-6}. Taken at face value, these data might suggest that they represent the general order of magnitude for human mutation rates that result in more or less detrimental phenotypes with clearcut dominant or X-linked recessive modes of inheritance. This conclusion, however, would be incorrect. The disorders listed in Table 9.6 were selected on the basis of their suitability for a mutation rate estimation. This suitability depends on the ease with which a certain condition may be ascertained and diagnosed and most particularly on its frequency in the population. In all diseases examined so far on an epidemiological scale the surveyed population group has not been larger than about 10 million. It is necessary to use disorders that are relatively frequent to find a sufficient number of cases of a specific disorder to provide the basis for a reasonably acceptable mutation estimate in a population of this size.

This aspect has been examined thoroughly by Stevenson and Kerr [120] for X-linked defects. According to these authors, evidence of frequencies and mutation rates falls into three categories:

1. For a few common disorders, ad hoc studies have been carried out. Here frequency estimates are relatively reliable.
2. With respect to uncommon disorders, the authors undertook to record all X-linked defects known at that time in 875 000 male newborns.

3. For very rare disorders the only guide to frequency is the number of cases and affected families in the world literature.

Stevenson and Kerr [120] analyzed 49 rare conditions. This number excludes polymorphisms whose frequency is evidently not dependent on an equilibrium between mutation and selection (color blindness, Xg blood group, G6PD variants). Figure 9.12 presents the estimated aproximate order of magnitude of the mutation rates for the 49 diseases. The authors do not claim accuracy; they do, however, provide sufficient evidence to make this range of estimates plausible.

There is only one disorder, Duchenne muscular dystrophy, with a mutation rate higher than 5×10^{-5}. In 24 conditions the estimated mutation rate is below 1×10^{-7}, and in another 11 it is estimated to range between 1×10^{-7} and 1×10^{-6}. This distribution makes it extremely difficult to calculate an average, especially since the list is by no means exhaustive. A number of other, mostly very rare X-linked defects, could be added to the list. Obviously the frequency of a specific disorder increases its probability of being known.

The authors' conclusion is acceptable as a first approximation. They arrive at an average mutation rate of 4×10^{-6} per gamete per generation for mutations

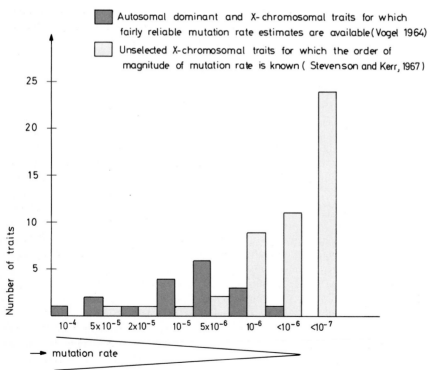

Fig. 9.12. Relatively well-established mutation rates for some autosomal-dominant and X-linked mutations (Table 9.6) in comparison with X-linked mutations for which only an order of magnitude is known (Stevenson and Kerr 1967 [120]) Note much lower mutation rate of the latter group of X-linked diseases

on those loci on the X chromosome that lead to observable phenotypic deviations.

Since the distribution of mutation rates appears to be very skewed toward lower rates, a consideration of the median value in these data may be useful. (The median of a distribution is the value that divides the distribution such that half lie above it and half below.) Cavalli-Sforza and Bodmer [17] calculated a median mutation rate of 1.6×10^{-7} from the same data. This suggests that mutation rates for many traits are very low indeed. Although specific data for autosomal-dominant mutations are not available, a similar rate may be assumed.

Do These Mutation Rates Encompass the Total Mutability of the Gene Loci Concerned? The foregoing does not imply, however, that the total mutability of the various gene loci has been established. The estimates involve only mutations that produce visible changes in the phenotype.

Three other types of mutations defy analysis:

a) Those leading to amino acid substitutions in a specific polypeptide chain that have no noticeable influence on the biological function of the chain. From our experience with known polypeptides, especially those of the hemoglobin molecule, we can conclude that many mutations, perhaps even a majority, belong to this group.

b) Those affecting function to such a degree that they are lethal to the zygote during the course of embryonic development.

c) Most of the mutations affecting nontranscribed DNA segments, unless the required DNA techniques can be feasibly applied at the population level.

The large differences between mutation rates estimated from the phenotypes may have various causes; there may be, for example, a higher degree of genetic heterogeneity for the apparently more frequent diseases. A second possible cause is that numerous mutations within a given gene lead to the same mutant phenotype, whereas for some genes very specific changes are needed to produce a certain phenotype, and most mutations are either lethal or lead to quite different phenotypes. It is of interest that recent work has demonstrated that many different allelic mutations exist for most genetic diseases. In a few instances, such as achondroplasia, a unique single nucleotide substitution is found in all cases, including those caused by fresh mutations. Thirdly, there could even be a genuine difference in mutation rate due to a difference in mutation probability per nucleotide. A good example are the various diseases caused by mutation of the HBβ gene. The most important reason, however, is the huge size difference between genes.

For example, the two genes with the highest mutation rates, the dystrophin gene whose mutations often lead to the muscular dystrophies of Duchenne and Becker, and the neurofibromatosis gene also belong to the largest genes we know (Table 3.4). The problem of total mutability of a gene cannot be solved by analysis at the phenotypic level but needs analysis at the DNA level. Both the gene and the mutational site should be identified. This topic is taken up again in Sect. 9.3.1.

In What Context Should Human Mutation Rates Involving Dominant or X-Linked Phenotypes Be Investigated? Assessment of mutation rates for a genetic disease requires very thorough ad hoc studies. All persons who may possibly be affected by a disorder must be ascertained within a relatively large population as completely as possible, with special attention to "sporadic" cases. Those so ascertained and their families must be examined personally by an experienced investigator to establish the diagnosis and to exclude similar conditions that are genetically different. Such a study would require an unusual effort in time and manpower and would rarely, if ever, be justified simply to establish a mutation rate. Therefore, this effort might conveniently be included in more comprehensive epidemiological studies designed, for example, to classify an ill-defined group of disorders into a number of genetic entities, to examine reproductive behavior of patients under the influence of modern medicine, or to set up a registration system as the basis for genetic counseling or population monitoring. In fact, most of the mutation rates in Table 9.6 were estimated in the context of clinical population genetics studies in the 1940s and 1950s.

The activities of the centers involved in such work are described in Chap. 4. Since about 1960, these institutions have disappeared or have turned to other activities. At present there is no center specializing in this type of genetic epidemiological study, and only very few such studies are being carried out. In our opinion, the most plausible explanation for the abandoning of population studies is supplied by the sociology of science. In the late 1950s and early 1960s the renaissance of biochemical genetics, cytogenetics, somatic cell genetics, and immunogenetics opened completely new prospects for genetic analysis in humans. Methods and concepts of molecular biology became available. Understandably, many research workers were fascinated by these new possibilities and turned away from relatively cumbersome and less satisfying tasks of ascertaining and examining cases in large populations. This trend was reinforced by the development of parts of human population genetics into a highly formalized and somewhat esoteric speciality, which appeared so remote from biology that its significance for a deeper understanding of general biological problems seemed doubtful to many of the biologically and clinically inclined workers.

This development, again, had interesting consequences. Departments and research teams in medical genetics were founded in many countries, but they were no longer organized with epidemiological work in mind. Their emphasis tended to be on cytogenetics and biochemistry and and, later, on molecular genetics. This necessarily funneled the work of new talent in these directions, thus reinforcing the trend. However, problems such as incidence and mutation rate are far from solved. In the face of increasing pollution of our environment by potentially mutagenic chemical and physical agents (Chap. 11), knowledge about human mutations at all levels is needed more urgently than ever before. Although many geneticists speak out publicly about mutations, their pronouncements are based on the same, limited set of old data. Comprehensive knowledge of genetic heterogeneity and phenotypic delineation of disorders is particularly urgent now, since there is an increasing demand for genetic counseling, and many new methods for improving genetic prediction have become available.

What can be done to correct this one-sided development? Obviously the remedy cannot be to abandon or even curb the new methods in favor of old-fashioned population studies. Scientific progress depends on the quality of research workers, and qualified workers cannot be found for research that they regard as uninteresting. Moreover, the old studies had indisputable weaknesses that impaired their usefulness and should deter us from mere repetition. The time is ripe to plan studies that combine the two approaches – analysis at both the molecular or chromosomal level as well as the population level. For the Lesch-Nyhan syndrome, for example, probable mutations with different characteristics can now be observed in single cells in vitro (Sect. 9.4.3). Would it not be interesting to compare the spectrum of these mutations with the spectrum from a comprehensive population study? Similar comparative studies are conceivable in hemophilia and in other diseases of relatively high incidence. They would help not only the population geneticist to find better explanations for the phenomena observed in populations but also molecular and cell geneticists to improve understanding of some findings observed in vitro. Finally, such work could be of great help in management and genetic counseling of patients and their families.

Rates of Mutation of Genes Not Leading to Hereditary Diseases. So far only mutations leading to hereditary diseases have been mentioned. However, mutations have also been observed – and their rates estimated – for normal protein variants as well as for DNA sequences outside of transcribed DNA stretches. The first group of mutations are discussed in the context of radiation-induced mutations because this is the area in which they have mainly been discussed. As for the second group, DNA minisatellites, for example, may have mutation rates higher by several orders of magnitude than those discussed here (see Sect. 12.1.2).

9.3.3 Mutation Rate and Age of the Father

One of Weinberg's Brilliant Ideas. In 1912 Weinberg [147] discussed the genetic basis of achondroplasia. The cases available to him consisted of the pedigrees published by Rischbieth and Barrington (1912) [112]. Weinberg examined the possibility of a simple recessive mode of inheritance and rejected this hypothesis. He found that the data fitted somewhat better with a dihybrid recessive mode of inheritance. He mentioned the opinion of Plate that the condition is dominantly inherited. Analysis of the evidence indicated that later-born siblings are more likely to be affected. Having made some remarks about possible biases, he continued (our translation):

"If a more exact analysis of birth order indeed confirmed a high incidence in the last-born children, this would point to the formation of the "anlage" for dwarfism by mutation."

About 30 years later this prediction was confirmed by Mørch [87], who conducted an epidemiological study of all achondroplastic dwarfs living in Denmark, including some recently deceased at the time of his investigations. He presented convincing evidence that sporadic cases are indeed due to new mutation. He also showed that average maternal as well as paternal age in these sporadic cases is significantly higher than the population average and that the maternal age effect is not due to an influence of birth order. He was not able to determine whether the effect is due to maternal age, or paternal age, or both.

Watson-Crick Model Stimulated New Research on Paternal and Maternal Age Influences. Meanwhile, Watson and Crick [145] had published their model of DNA structure. In addition to explaining replication and information storage, the model suggested a convincing mechanism for spontaneous mutation: the insertion of a wrong base at replication. This mechanism required that the mutation process depend on replication. Moreover, investigations of micro-organisms seemed to confirm that almost all mutations occur in dividing organisms [131]. This concept gave

Table 9.9. Simple models for mutation and their statistical consequences

Model no. Mechanism	1 Mutation depend- ing on time only	2 Mutation depending on cell divisions only	3 Mutation during a certain time before puberty	4 Mutation after ceasing of divi- sions	5 Mutation in mature germ cells
Male germ cells	Linear increase of mutations with age; no sex difference	Increase of mutations with age; higher mu- tation rate in males	No increase with age; no sex dif- ference	No increase with age; lower muta- tion rate in males	No increase with age; maybe somewhat higher mutation rate in males
Female germ cells	Linear increase of mutations with age	No increase with age; lower mutation rate in females	No increase with age	Increase with age; higher mutation rate in females	No increase with age; maybe somewhat lower rate in females

new impetus to statistical analysis of the effect of parental age on human mutation. The argument was set out by Penrose [101] as follows:

"There are very few cell divisions in the female germ cell line but many in the male germ line since the spermatogonia are continuously dividing. Thus the incidence of mutation due to failure to copy a gene at cell division would be unlikely to have any strong relation to maternal age; a marked increase of defects with this origin, however, would be seen at late paternal ages."

The predictions from this mutation mechanism can be compared with those derived from other plausible mechanisms [101, 134]. In Table 9.9, all five possibilities were suggested by one or another result in experimental genetics. The second model (mutations depending on cell division) was the one primarily considered by Penrose. It predicts an increase in mutation rates with age only in males and a higher mutation rate in germ cells of males than in females.

Cell Divisions During Germ Cell Development in Both Human Sexes. To improve predictions of mutation rates beyond the qualitatively correct but very general statement of Penrose, the number of cells and cell divisions in the male and female germ line should be known. The widely scattered evidence in the literature of various fields gives the following picture of early germ cell development, oogenesis, and spermatogenesis [138].

Early Development. The human primordial germ cells emerge from the yolk sac 27 days after fertilization and begin to colonize the gonadal ridges. On day 46 of gestation the indifferent gonad undergoes sex differentiation and becomes an ovary or a testicle.

Oogenesis. Oogenesis (Fig. 9.13) occurs only during fetal life and ceases by the time of birth. After sex differentiation the ovarian stem cells rapidly increase in number by mitotic divisions. From the 2nd month of gestation onward variable numbers of oocytes enter the prophase of meiosis; oogonia

persisting beyond the age of 7 months undergo degeneration. Leptotene and zygotene stages (Sect. 2.1.2.4) are found between the 2nd and 7th months. All stem cells are generally utilized by the time of birth; oogonia have been transformed into oocytes or have degenerated.

The total population of germ cells in the embryo rises from 6×10^5 at 2 months to a maximum of 6.8×10^6 during the 5th month. The population then declines to about 2×10^6 at birth. On the plausible assumption of proliferation by dichotomous divisions, an oocyte has passed through 22 divisions by the time of birth (Fig. 9.13). ($2^{22} \approx 6.8 \times 10^6$) From birth until mature age and fertilization the cell undergoes only two meiotic divisions, regardless of the age at which fertilization occurs.

Spermatogenesis. Cell kinetics differs in spermatogenesis (Fig. 9.14). At the same stage of embryonic life at which the primordial germ cells give rise to oogonia in the female, they become *gonocytes* in the male. From the early embryonic stage until the age of puberty the tubules continue to become populated by so-called Ad spermatogonia (d = dark), and at about age 16 spermatogenesis is fully established. The number of Ad spermatogonia can be estimated in three different ways: from volumetric data; from their average number per tubular cross section and the length of tubules; and from the maximum amount of sperm produced per day. These estimates give values ranging from 4.3×10^8 to 6.4×10^8 per testis. An approximate estimate for both testicles is approx. 1.2×10^9. This value can be reached by about 30 cell divisions.

In contrast to oocytes, however, these Ad spermatogonia undergo a continuous sequence of divisions. Of the two division products one cell prepares for the next division into two Ad cells, whereas the other one divides, giving two Ap (p = pale) cells. These develop into B spermatogonia and spermatocytes, which then undergo meiotic divisions (Fig. 9.14). The timing of these cell divisions is well known, partly from in vivo studies in young men. The division cycle of Ad spermatogonia lasts for about 16 days. This makes it possible to estimate the number of cell divisions according to age (Table 9.10).

If this calculation is approximately correct, the number of cell divisions that a sperm undergoes from early embryonic development until the age of 28 is about *15 times greater than the number of divisions in the life history on an oocyte.* At later ages in males this calculation would give still higher values. Such an ex-

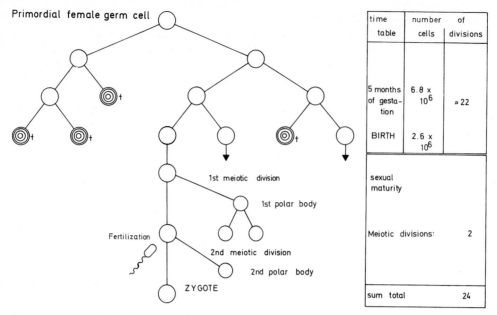

Fig. 9.13. Oogenesis in the human female. All cell divisions – except the two meiotic divisions – are already finished at the time of birth; ◎ cell atrophy

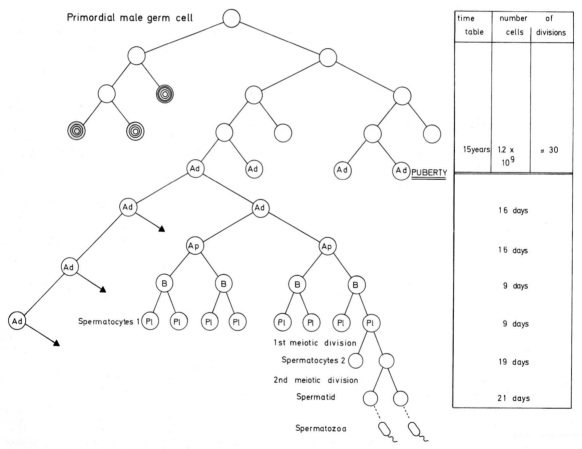

Fig. 9.14. Cell divisions during spermatogenesis. The overall number of cell divisions is much higher than in oogenesis. It increases with advancing age. *Ad (dark),* spermatogonia; *Ap (pale),* spermatogonia; *B,* spermatogonia; *Pl,* spermatocytes; ◎ cell atrophy

Table 9.10. Number of cell divisions in spermatogenesis (from embryonic development to meiosis; from Vogel and Rathenberg 1975 [138])

From embryonic age to puberty	≈30
Ad-type spermatogonia (one division/cycle = 16 days)	≈23/year
Proliferation + maturation	4 + 2 = 6
Total At the age of 28 At the age of 35	≈335 divisions ≈496 divisions

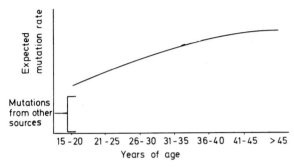

Fig. 9.15. Cumulative distribution of cell divisions in spermatogenesis and expected increase in the mutation rate with paternal age. The distribution curve is derived from Kinsey's data on number of ejaculations at various age groups

trapolation would be dangerous, as the actual examinations on which the estimation is based were carried out in younger people. It is well known, however, that sexual activity as measured by number of ejaculations decreases as early as the 4th decade of life. Some data indicate that sperm counts at that age increase a little, but undoubtedly spermatogenesis decreases with advancing age as shown in Fig. 9.15. Figure 9.15 would also give the cumulative distribution of the number of spermatogonial divisions if slowing down of spermatogenesis were due to prolongation of the division cycles of Ad spermatogonia. Other mechanisms, however, are possible: for example, some Ad spermatogonia could degenerate, while others continue to divide at the same rate. Surprisingly, the changes in human spermatogenesis with advancing age have almost never been examined by histological methods.

Increase in Mutation Rate with Paternal Age. The foregoing digression into histology is necessary to understand the meaning of the increases in mutation rates with paternal age which are actually observed. Figure 9.16 compares the relative mutation rates with the population averages for different age groups of males for a number of conditions; in addition to achondroplasia, these are acrocephalosyndactyly

(Apert syndrome) [11], Marfan syndrome [91], and myositis ossificans (a condition involving progressive ossification of the muscles) [126]. Especially interesting is the increase for maternal grandfathers of sporadic cases of hemophilia A [52].

In all these series there were some problems with adequate control samples from the general population, as many population statistics give the number of newborns in relation to maternal but not to paternal age. However, the paternal age effect is so pronounced that small-scale inconsistencies in the controls do not influence the outcome substantially.

All the curves in Fig. 9.16 have two features in common:

1. The mutation rate in the oldest age group is several times – approximately five times – higher than that in the youngest group.
2. The slope of the increase curve tends to become progressively steeper with advancing age.

The first characteristic is compatible with the assumption that the increase is caused by accumulation of cell divisions. The second one, however, is not compatible with such an assumption; rather, a flattening of the curve would be expected, at least if the division rate of Ad spermatogonia does indeed slow with advancing age. This discrepancy has not been resolved.

Other Dominant Mutations for Which a Paternal Age Effect Is Possible. Conditions known or assumed to be dominant in which there are suggestions for a paternal age effect include: basal cell nevus syndrome, Waardenburg syndrome, Crouzon disease, oculodentodigital syndrome, and Treacher Collins syndrome [61].

Mutations Leading to Unstable Hemoglobins or Hemoglobin M and Paternal Age [115]. As mentioned in Sect. 7.3.2, hemoglobins M (= methemoglobin) and unstable hemoglobins cause clinical syndromes transmitted as autosomal dominants. Stamatoyannopoulos collected worldwide information on pedigrees in which one of these hemoglobins occurred as a new mutation. In all, 50 cases were collected from 14 countries; the individuals were born between 1922 and 1976. The overall average paternal age was 32.7 years, and the average maternal age 28.3 years. To compare the parental ages of the probands with those of the general control population the authors calculated for each year and each country the cumulative frequency distributions of the ages of all parents. The ages of each proband's father and mother were then expressed as percentiles of these distributions. The distribution of paternal age percentiles was shifted towards the upper end of the range; 11 of 50 paternal ages for the probands fell between the 90th and the 100th percentiles (Fig. 9.17). While this result suggests a paternal age effect, the difference from the control distribution was not significant. Adequate paternity testing was impossible in the majority of cases. The series may therefore include an appreciable number of "false mutants" due to

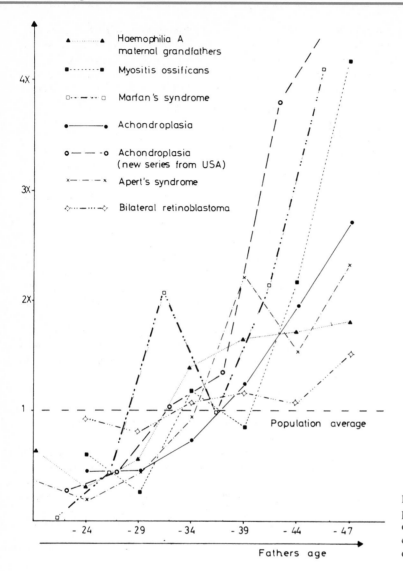

Legend:

▲·······▲ Haemophilia A
maternal grandfathers

■·······■ Myositis ossificans

□··—··□ Marfan's syndrome

●———● Achondroplasia

○———○ Achondroplasia
(new series from USA)

×—·—× Apert's syndrome

◇·—···◇ Bilateral retinoblastoma

Population average

Fathers age

Fig. 9.16. Mutation rates (relative to population averages) in various age groups of males and for a number of dominant conditions; also, maternal grandfathers of hemophilia A patients

wrong assumptions of paternity. A new mutation of a single case of HbM was published; the father of the mutant child was 49 and the mother 37 years old. A new mutation for β thalassemia has been reported and documented unusually carefully [125]. Clinical and biochemical examinations of the child, who was 2 years old at the time of diagnosis, as well as of both parents and three siblings, left no reasonable doubt that the child's thalassemia was indeed caused by a new mutation; illegitimacy was excluded convincingly. The father was 45 and the mother 44 years old at the time of the proband's birth. In the same year (1965) in Switzerland, the average paternal age was 31 and the maternal age 28.2. While a single case report cannot replace a statistical survey, this result points strongly toward a paternal age effect. However, the knowledge that advanced paternal ages are more likely for new mutations would lead to publication bias in such isolated cases.

Some Dominant Mutations Show Only a Small Paternal Age Effect. Penrose [100] noted that not all domi-

nant mutations show a strong paternal age effect. Among those for which the effect is much less are the well-analyzed examples of bilateral retinoblastoma [26, 138] and thanatophoric dwarfism. In the latter condition, according to evidence from a limited number of patients ($n = 41$) the effect appears to be smaller than in achondroplasia but stronger than in osteogenesis imperfecta [99]. Others include tuberous sclerosis, neurofibromatosis, and osteogenesis imperfecta. Figure 9.18 shows the data for these mutation rates. The increase with paternal age is not significant for the three latter conditions, which, however, show a significant birth order effect, suggesting that paternal age probably plays a role. More recently this effect has been confirmed for neurofibromatosis [10]. Comparison of Figs. 9.16 and Fig. 9.18 suggests that not all autosomal dominant mutations have a paternal age effect. There is heterogeneity.

The existence of a paternal age effect for lethal or semilethal mutation of X-linked genes has important consequences. One would expect to find such an effect frequently among the maternal grandfathers of cases of such X-linked diseases on the assumption that the mutation occurs more frequently in the grandparental than in the parental generation. Furthermore, if mutations were replication dependent, such mutations would be often found more in grandfathers than in grandmothers since there are many more cell divisions in spermatogenesis than in oogenesis. This problem was examined in two series of

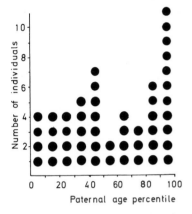

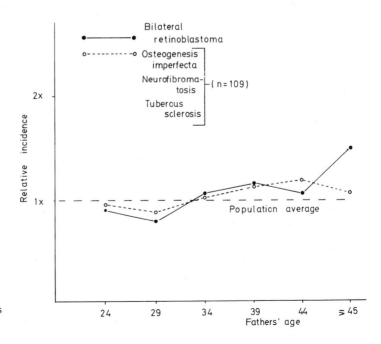

Fig. 9.17. Percentile distributions of ages of probands' fathers for new mutations of unstable hemoglobins or HbM. Expectation on absence of a paternal age effect would be five in each category. Note the high number of cases in the oldest group (≤ 90 % of the population). (From Stamatoyannopoulos et al. (1981) [119])

77 cases with hemophilia A in which a new mutation could be expected [52]. A significant increase in the mutation rate with grandpaternal age was indeed observed (Fig. 9.16 b), and this was confirmed in a later study. This is another argument in favor of a higher mutation rate in male germ cells (see below).

Another X-Linked Disorder: Lesch-Nyhan Syndrome (308000). The X-linked Lesch-Nyhan syndrome is caused by a defect of hypoxanthine-guanine-phosphoribosyltransferase (HPRT). This defect can be diagnosed not only in the male hemizygote but also in cells of the female heterozygote that show lyonization (Sect. 7.2.2.6).

In a survey discussed below [34], five cases were found in which the mother of a patient was heterozygous, but the maternal grandmother was a normal homozygote. The mutation must therefore have originated in the maternal grandfather's germ cell. The mean ages of both grandparents are shown in Table 9.11. They are much higher than those of the average population in the United States. Hence a paternal age effect of the order of magnitude found in the other conditions reviewed above is very likely.

9.3.4 Possible Sex Difference in Mutation Rates

If most mutations depend on cell divisions and DNA replication, one would expect not only an increase in the mutation rate with age in male germ cells but also an absolutely higher mutation rate in males than in females. The male germ cell undergoes a great

Fig. 9.18. Small paternal age effect for some mutations: bilateral retinoblastoma, tuberous sclerosis, neurofibromatosis, and osteogenesis imperfecta

Table 9.11. Ages of maternal grandparents at birth of heterozygous daughters representing a new mutation for the HPRT defect (adapted from Francke et al. 1976 [34]

	Grandfather	Grandmother
5 families	27	24
	35	35
	40	31
	38	32
	40	39
Mean ± SD	36.0 ± 5.43	32.2 ± 5.54
Population average (USA)	28.7 ± 6.8	25.9 ± 6.2

many more cell divisions than does the female germ cell (Figs. 9.13, 9.14).

This problem cannot be examined directly using dominant mutations of humans, as the individual mutation that shows up in the phenotype of a "sporadic" case cannot be located either in the sperm or in the ovum. For X-linked mutations, however, a test is possible. For X-linked conditions – hemophilia A, Lesch-Nyhan syndrome, ornithine transcarbamylase deficiency and Duchenne-type muscular dystrophy – enough case material is available; all four have been analyzed, but with contradictory results.

Sex Difference in the Mutation Rate for Hemophilia A. Haldane [45], analyzing cases from a Danish population survey [2], was the first to postulate a higher mutation rate in male than in female germ cells. According to Haldane [43], of all bearers of an X-linked recessive condition, a fraction (*m*):

$$m = \frac{(1-f)\mu}{2\mu + \nu}$$

(μ = mutation rate in female, ν in male germ cells; f = relative fertility of patients).

must be sons of normal homozygous mothers, i.e., if there is a genetic equilibrium, their disorder must be due to new mutation in the mothers' germ cell. Mutations in mothers' germ cells are expected to lead to sporadic cases, i.e., individuals cases who are the only hemophilics in their sibships. Some isolated cases, however, are to be expected for statistical reasons even if all mothers are heterozygous. Their proportion is increased if, in addition, new mutants occur. This proportion *m* can be estimated from family data. To avoid uncontrollable biases in the ascertainment of sibships with only one, and with more than one patient, however, the proportion of mutant males should be determined, whenever possible, in series of families that have been collected by complete (truncate) selection (Sect. 4.3.4). This means that within a predetermined time period all living hemophilics and their families within a defined population have been ascertained.

In more recent times, the problem has been studied repeatedly using slightly different methods. Vogel (1965) [133] used two different approaches: He tested the excess of sibships with only one affected brother (and no cases among more remote relatives) and compared the quantitative distributions of factor VIII values between mothers of sporadic cases and obligatory carriers. There was no such excess, and distributions of factor VIII values were identical in the two groups of carriers. Both sets of data suggest a very low mutation rate in female germ cells and, by implication, a higher rate in male cells (see also [134]). This result was confirmed by studies using improved techniques. Winter et al. [148] estimated the male/female mutation rate as approx. 10/1. Rosendaal et al. (1990) [113] analyzed the six data sets regarded by them as reliable, including their own data on virtually all hemophilia A families in the Netherlands. Using a refined statistical technique, they concluded that the mutation rate is about 3.1 times higher in male than in female germ cells, with a 95% confidence interval between 1.9 and 4.9. This result is supported by a limited number of observations showing that the mutation originates more often in the germ cell of the maternal grandfather than in that of the sporadic patient's mother.

Hemophilia B. The problem is not yet resolved regarding hemophilia B. While some studies point to a much higher mutation rate in male germ cells, others argue rather for approximately equal rates in both sexes.

Ornithine Transcarbamylase Deficiency (311 250). This disease is characterized by severe hyperammonemia; in addition to the male hemizygotes, some 17% of female carriers are also affected. Bonaiti-Pellie et al. [12] performed a segregation analysis [10]. They failed to find any indication for a deviation of the segregation rate from its expectation if all mothers are heterozygous [10]. This result indicates a higher mutation rate in male germ cells.

Likely Higher Mutation Rate in Male Germ Cells Causing the Lesch-Nyhan Syndrome [34]. The advanced age of maternal grandfathers in this condition has been mentioned. The disorder is so severe in males that they never have children; therefore their relative fertility (*f*) is 0. The formula is simplified to *m* = 1/3 if $\mu = \nu$. With equal mutation rates in the two sexes, one-third of all patients in one generation should have homozygous normal mothers, their disorders being due to a new mutation in their mothers' germ cell.

Epidemiological surveys of all families within a specified population are not currently available. However, in this disease heterozygotes can be diagnosed by laboratory examination. It was therefore worthwhile to determine whether one-third of all known cases do indeed have homozygous normal mothers. A collaborative study [34] analyzed 47 families: 39 from the United States, 3 from the United Kingdom, and 1 each from Canada, Belgium, Germany, Ireland, and Switzerland. In 27 families the only person affected was one male. In all of these cases the mothers were available for heterozygote testing, and only four were normal homozygotes. In the other 23 cases (as well as in all families containing more than one affected), the mothers were heterozygotes. Obviously this proportion is less than the theoretically expected one-third.

There are three hypotheses that would account for the high number of heterozygotes among the mothers:

1. Since families were not located by an epidemiological survey, ascertainment may be biased in favor of families containing more than one case.
2. There could be a reproductive advantage of heterozygotes or segregation distortion.
3. There is a higher mutation rate in male germ cells. Both alternatives 1 and 2 appear unlikely. Hypothesis 3 of a higher mutation rate in the male germ cells is the most plausible explanation for these data. This hypothesis has since been confirmed with additional cases and with an improved statistical method [35].

Fragile X Syndrome. In the Fra X syndrome, the most common X-linked type of mental retardation and one of the most common monogenic anomalies in humans, new mutations exclusively in male germ cells were assumed to be based on statistical evidence [118]. More recently a novel molecular explanation – expanded alleles – has been found [29]. The problem is discussed in detail in Sect. 9.4.2).

No Sex Difference in Mutation Rates in Duchenne Muscular Dystrophy. Another disease for which sufficiently large and carefully examined population samples are available is Duchenne muscular dystrophy (DMD; 310200). Here the relative fertility (*f*) of the patients is, again, 0. They never have children, and if the mutation rates in both sexes were equal, one-third of all patients would be likely to be sons of homozygous normal mothers. This problem has been studied repeatedly, with some contradictory results [90]. Meanwhile, DNA marker profiles have been worked out which permit identification of individual grandpaternal and grandmaternal X chromosomes; information from this source was included in the algorithm for estimating this sex ratio [41, 65]. After calculations regarding the numbers of observations necessary for a precise result if the rate in males were five times that in females [69], a cooperative study was performed by many teams involved in DMD diagnosis. Of 295 sporadic cases, 196 had inherited their X chromosome from the maternal grandmother and 99 from the grandfather. This result leads to an estimate of $m = 1.04$, with a 95% confidence interval between 0.41 and 2.69 [89].

Hence, the overall mutation rate is virtually identical in the two sexes. DMD appears to be a special case: the mutation rate is unusually high, very probably due in part to the unusual length of the dystrophin gene (Table 3.4), and the sex ratio differs from other, carefully analyzed mutations. Again, the explanation has come from molecular analysis of the mutants (see below, Sect. 9.4.2).

Indirect Evidence for a Higher Mutation Rate in Male Germ Cells. The problem of differing mutation rates in male and female germ cells cannot be examined directly in autosomal-dominant conditions. However, inferential evidence may be derived from the increase in the mutation rate with paternal age in, for example, achondroplasia. If the paternal age effect is caused exclusively by the father's age, which has been shown to be very likely [102, 106], the notion of equal mutation rates in the two sexes could not apply, even if all mutations in children of young fathers were due to mutation in female germ cells. If the mutation rates in young parents were equal in both sexes, the surplus of new mutations due to the paternal age effect would bring about a much higher mutation rate in male than in the female germ cells. For details of the argument see Vogel [29].

Sex Difference in Mutation Rates of the Mouse. One would expect this problem to have already been solved in laboratory animals. However, the evidence is suggestive but not entirely conclusive. Only relatively scanty data from mice are available as a by-product of work on mutation induction with the multiple locus test. This method detects recessive new mutations in the F_1 by back-crossing with a test strain homozygous for seven recessive mutations (Sect. 11.1). Table 9.12 shows the data. The sex difference is not very impressive. The seven mutations observed in females include a cluster of six mutants apparently due to a single mutation in the very early development of the ovary. If this cluster is counted only once, the mutation rate in females becomes 1.4×10^{-6}, which is indeed much lower than the mutation rate in males. The hypothesis that the mutation rate in females is lower than that in males is strengthened by the remarkably low yields of mutations obtained by irradiation of females at low dose rates [116].

Statistical Results and Mutation Mechanisms. The various results may be compared with the expectations derived from the five mechanisms listed in Table 9.8. Prepubertal mutations (model 3) and mutations occurring after cell divisions have been completed (model 4) can be excluded for the group of mutations with a strong paternal age effect. Time dependent mutations (model 1) are also very unlikely for this category. For the group with a lower paternal age effect, a linear increase with time remains a possibility, perhaps in combination with mutations in mature germ cells (model 5). Most of the data appear to suggest the cell division dependent model (2). The sex difference and the increase with paternal age are predicted by this model. However, there are two aspects that entail caution in accepting this mechanism entirely:

1. Failure to detect evidence of a strong paternal age effect in some instances.
2. The slope of the age-dependent increase in mutations (in those diseases showing a strong paternal age effect) rises more steeply with advancing age.

As shown, one would have expected a leveling off of the increase in mutations in the higher age groups. The evidence, however, is not decisive, as we know too little about the nature of changes in spermatogenesis with advancing age. Nevertheless, the data suggest that the mutation process is somehow related to DNA replication and cell division. Perhaps the statistical data should not be pushed too hard; more detailed evidence can be expected from analysis at the molecular level.

9.3.5 Germ Cell and Somatic Cell Mosaics for Dominant or X-Linked Mutations

Pedigree Observations. If a mutation occurs during early development of the germ cells, a germinal mosaic can be created, a more or less extended sector of the gonad bearing the mutation. This situation is well-known from mutation work with *Drosophila melanogaster,* and a "cluster" of mutants due to mutation in early stages of oocyte development in the mouse is mentioned in Table 9.12. In humans the chance of finding such clusters is very low; they can be detected only if a fairly large sector of the gonad is affected. The earlier in development the mutation occurs, the larger is the involved gonadal sector. The proportion of affected germ cells is 100% if the first stem cell already bears the mutation; it is $1/2$ if the mutation occurred at the first stem cell division,

Table 9.12. Spontaneous single-locus mutations from wild-type (normal) in the mouse (adapted from Searle 1972 [116]; Russell et al. 1972 [114])

Sex	No. of gametes tested	Mutations	Frequency/ locus
♂	649 227	36	7.9×10^{-6}
♀	202 812	7[a]	6.1×10^{-6}

[a] Includes a cluster of six. Alternative estimate based on one mutation (1.4×10^{-6}).

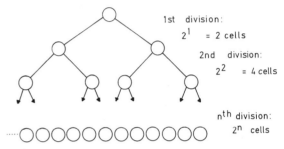

Fig. 9.19. Dependency of mutant germ cells on the stage of development of the germ cell at which the mutation occurred

$(1/2)^2$ if it occurred at the second division; and in general $(\frac{1}{2})^n$ if the mutation occurred at the n'th division (Fig. 9.19). If the total number of cell divisions is of the order of magnitude estimated above (Sect. 9.3.3; Fig. 9.14), and the probability of mutation is equal for all cell divisions, the proportion of new mutations that occur in clusters, revealing germinal mosaicisms, is small but not negligible. If the mutation is dominant, occasional families in which both parents are normal but more than one child shows the mutant phenotype are expected.

Occasional pedigrees have been described in which such a cluster is a possibility, for example, a large pedigree with aniridia [106] and a family with split hands and split feet [80]. It used to be very difficult to exclude the more trivial alternative of incomplete penetrance in one of the parents. Here again, diagnosis at the molecular level has brought new information. Since it became possible to diagnose heterozygous carriers of X-linked mutations such as those causing the muscular dystrophies (DMD, BMD) and hemophilia A, women have been discovered who are germ cell mosaics for these mutations.

Families were earlier often observed in which more than one son was born to a mother who showed no signs of a DMD carrier status. At that time, however, diagnosis of DMD had to be established exclusively on clinical grounds, and the carrier status could be diagnosed in only a certain proportion of carriers. Therefore, two alternative expanations were offered: either the sons suffer from a phenotypically identical autosomal-recessive disease, or they do have DMD and the mother is a carrier. Today the first alternative can be excluded at the protein level if a dystrophin defect is shown in the patient's muscles. As a rule, the second alternative cannot be excluded since the gonads of these women are not accessible. Meanwhile, however, many mothers have been observed who were free of a deletion in the DNA of their white blood cells while their two sons carried the deletion [7]. In one mother the deletion was found in only some of her white cells [84]. In another mother of an affected son and of a carrier daughter, mosaicism was found in muscle but not in blood cells [142]. In the European collaborative study the proportion of mosaics among the mothers of sporadic patients was estimated at about 7% [90]. Therefore it would be wrong to give a mother of a sporadic DMD patient, who had been identified as noncarrier (by examination of blood cells), a risk estimate of nearly 0% for her next son. A precise risk estimate is impossible since the extent of germ cell mosaicism remains unknown.

Mosaicism has also been observed in hemophilia A and appears to be as common as in DMD. For example, a severely affected hemophilia patient as well as his sister had a deletion which was not present in the blood cells of their mother [38]. Occasional observations of mosaicism in several other diseases have been reported [20]. The problem must be kept in mind when parents of sporadic patients with autosomal-dominant and X-linked diseases ask for counseling.

Somatic Mosaicism. Mosaicism caused by mutation occurs not only in germinal but also in somatic tissue. Such mosaicism might not affect only chromosome number, as described in Chap. 10, but applies to gene mutations as well. The pattern of phenotypic manifestation of gene mutations, however, makes detection of such mosaicism very difficult. Still, there is at least one observation. During a population survey on neurofibromatosis [22] four individuals were observed in which the neurofibromata were confined to one sector of the body, such as the extremities, the sacral area, and the back. In these four cases the family history was negative. They produced a total of six children, none of whom was affected. Therefore, these four individuals probably represent somatic mutations affecting relatively early stages of development. Further cases have since been described [83].

Half-Chromatid Mutations? Gartler and Francke (1975) [36] suggested a special mechanism for the production of mosaics for point mutations: half-chromatid mutations (Fig. 9.20). As mentioned, many mutations seem to be due to a copy error in DNA replication. If such a copy error happens to occur in the last DNA replication cycle before germ cell formation, the resulting germ cell contains a mismatching base pair, for example, AG instead of AT (Fig. 9.20). In the first cleavage division A pairs with T, and G pairs with C. Therefore one of the two products of this division has the base pair AT as before; the other one contains the new base pair GC and is mutant.

Incontinentia pigmenti may be an example. The condition is probably caused by an X-linked dominant gene that is lethal in hemizygous males Sect. 4.1.4. A total of 593 cases have been described in females and six in males, who had normal XY karyotypes. The pattern of skin affection was similar in both sexes and resembled the patchy mosaic pattern exhibited by the heterozygotes for some X-linked genes in mice, hamsters, and cats.

Moreover, the male cases were sporadic. The observation that the affected males are phenotypically so similar to the females who are known to be mosaics due to the Lyon effect (Sect. 2.2.3.3) makes it likely that males are also mosaics, and that the mutation occurred in an early stage of embryonic development. Half-chromatid mutation is a good possibility [77].

9.4 Gene Mutation: Analysis at the Molecular Level

DNA analysis offers the opportunity to obtain more specific insight into the mechanisms of mutation. Here, recent studies using analysis at the gene-DNA level have provided new insights permitting us to answer some questions posed by statistical analyses that generally began at the qualitative phenotypic level. Results from both levels should be integrated to obtain a comprehensive model of reality. Here, answers to some intriguing questions have become easier while in other instances the situation is more complicated, and the answer less obvious, than assumed before. A comprehensive survey of gene mutations has been provided by Cooper and Krawczak [20]. These autors are also collecting data on spontaneous mutations in a database that is being updated continuously.

9.4.1 Nucleotide and Codon Mutation Rates

What is the probability that a particular nucleotide or codon mutates in a specific direction, so that one amino acid is replaced by another?

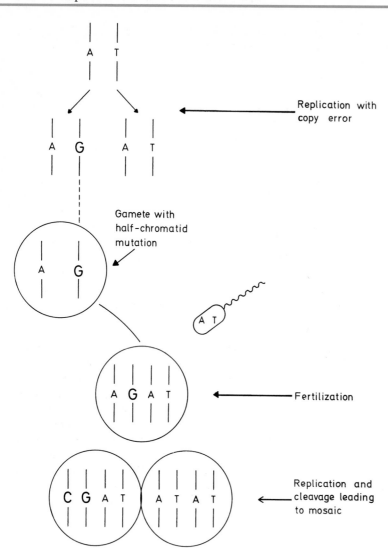

Fig. 9.20. Principle of a half-chromatid mutation: the base replacement in one nucleotide half-strand occurred during the last DNA replication before meiosis. In the first cleavage division the mutant half-strand gives rise to a cell clone having a gene mutation; the normal half-strand produces a normal cell clone. The individual is a 1:1 mosaic

First Examination of This Problem. One of the authors (AGM) noted previously [88]:

Ideally, such calculations could be done most directly by determining the frequency of a given variant in a population followed by testing the parents. In most cases, one of the parents will have transmitted the variant to a child. In families where both parents are normal (keeping the possibility of false paternity in mind), it is assumed that a new mutation has taken place.

At the time the following calculations were made, large-scale electrophoretic population studies were not yet available. Therefore the mutation rate had to be estimated in a more indirect way from the incidence of rare variants in the population (1:2000), the proportion of possible variants that could be detected by electrophoresis, and the few observed mutations for one specific group of variants, methemoglobinemias due to the various types of HbM.

Considering all these factors, the order of magnitude of a mutation rate for a specific nucleotide substitution (μ^n) in hemoglobins was estimated to be:

$$\mu^n = 2.5 \times 10^{-9}$$

meaning the probability for a base to be replaced by a specific other base, leading to replacement of a specific amino acid by another one. The same indirect approach has been followed in a study in which new mutations for HbM and unstable hemoglobins were collected [115] (see also Sect. 9.3.3), and the number of cases observed in one country were compared to the total number of births in that country during one or two generations. The authors were aware that an estimate based on such slender evidence could give only a crude approximation. It presupposes, for example, that in these populations all cases born during a certain period had indeed come to notice. It is not clear, on the other hand, whether all cases included in this study really were new mutants; paternity tests were performed in only 19 of the 55 observations. With all these reservations, the mutation rate for individual nucleotides was estimated at 5.3×10^{-9} for de novo unstable hemoglobins, 10.0×10^{-9} for Hb α^{M} mutants, and 18.9×10^{-9} for Hb β^{M} mutants, giving a nucleotide mutation rate of 7.4×10^{-9} for de novo β-chain mutations, based on unstable Hb and Hb β^{M} mutants together. This estimate can be extrapolated to the entire Hb β gene (excluding, of course, mutations of other molecular types and, especially, those occurring outside the transcribed DNA segment and leading to β thalassemias). The result is: 8.6×10^{-6} per Hb β-chain gene. To repeat: this is a very crude first approximation based on slim evidence with bold extrapolations.

By a quite different – and also indirect – approach, the codon mutation rate was estimated from the rate of evolution of globin pseudogenes, since in the absence of negative selection the codon replacement rate in evolution was shown to be equal to the mutation rate (see our discussions of Kimura's "neutral hypothesis," Sect. 14.2.3). The resulting estimate, 5×10^{-9}, is amazingly similar to the other estimates.

Estimate with More Direct Data. Data from investigations of a huge population sample of several hundred thousand persons are become available from Japan and have been used for a mutation rate estimate [92]. In a total of 539 000 single locus tests, from 36 polypeptides, among which three spontaneous mutations were found, and after correction for the number of nucleotides/gene, a mutation rate of approx. 1×10^{-8} was estimated. Considering the very scanty evidence on which they are based, the various estimates agree fairly well.

In a study designed to utilize blood spots collected during newborn testing for inherited metabolic disease, especially PKU (Sect. 7.2.2.7), a new mutation of a Hb α variant, was discovered and confirmed by paternity testing among 25 000 specimens tested for hemoglobin variants [1]. Ideally this approach should cover the mutations at ten parental gene loci (1 β, 2 α, and 2 γ from each parent), but not all mutations are ascertained by the electrophoretic method.

Analysis at the gene-DNA level permits estimates that are based on more solid ground. For the factor IX gene, for example, a mutation rate of 3.2×10^{-9} per base pair was estimated [65], which was categorized into transitions (27×10^{-10}), transversions (4.1×10^{-10}), and deletions (0.9×10^{-10}). Assuming that the factor IX mutations are representative, extrapolation to the entire human genome would give 8.0 transitions, 1.2 transversions, and 0.27 deletions per genome, 0.19 of which may occur within coding regions. The estimate, however, does not take into account

heterogeneity of mutational sites and, especially, the known high mutation rates in CpG islands (see below).

How Do Nucleotide Mutation Rates Compare with Estimates at the Phenotype Level? Comparison of these estimates at the DNA level with estimates based on specific phenotypes (Tables 9.6, 9.7) show that the latter approach a frequency of 1×10^{-5} within one order of magnitude. However, these rates concern the relatively frequent phenotypes that are well suited for mutation rate calculations. As shown, the average mutation rate for visible pathological phenotypes is probably closer to 1×10^{-6} than to 1×10^{-7} per gamete or lower (Sect. 9.3.2). This value is about 40–400 times higher than the nucleotide mutation rate estimates, but at first glance seems to compare well with the extrapolations for the Hb β gene. For other genes, for example, the retinoblastoma and factor VIII (hemophilia A) gene, the phenotypic mutation rate is much lower than the additive sum of individual nucleotide mutation rates would suggest. This may be principally due to the fact that only a fraction of mutations leading to amino acid replacements cause such a severe impairment of protein function that a disease ensues. Moreover, mutations of the same gene must lead to identical or even similar phenotypes. However, the phenotype produced by a mutation depends on the specific functional alteration of the protein concerned (Sect. 7.3). Mutation within hemoglobin genes may, for example, lead to hemolytic anemias, methemoglobinemia, erythrocytosis, or may not be associated at all with any clinical symptoms.

9.4.2 Various Molecular Types of Mutation

Single Base Pair Substitutions. Analysis at the DNA level has revealed various types of mutations, starting with those causing hemoglobin variants [74]. In recent years mutations of many other genes [20] have been described. The most common type are single base pair substitutions. Depending on the kind of substitution, these are subdivided into transitions and transversions: the replacement of a purine by the other purine or of a pyrimidine by the other pyrimidine is a transition; replacement of a purine by a pyrimidine or vice versa is called a transversion. Four transitions and eight transversions are possible (Fig. 9.21). If the direction of mutations were random and all base replacements occurred with identical probabilities, one-third transitions and two-thirds transversions would be expected.

Missense mutations change the nucleotide sequence so that another amino acid is found in the resulting protein. A classic example is the amino acid substitu-

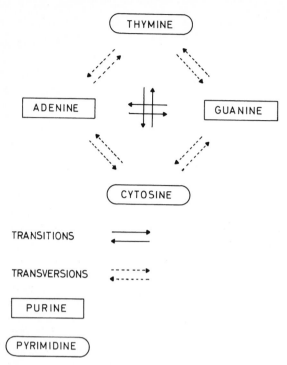

Fig. 9.21. Mechanism of mutations at the nucleotide level. Four transitions ⇆ and eight transversions ⇄ are possible

tion of valine for glutamic acid in the sixth position of the Hb β-chain which leads to sickle cell hemoglobin.

Glutamic acid:			Valine:		
mRNA	DNA coding sequence	Anti-codon	mRNA	DNA coding	Anti-codon
GAA	CTT	GAA	GUA	CAT	GTA
GAG	CTC	GAG	GUG	CAC	GTG
			GUU	CAA	GTT
			GUC	CAG	GTC

A base replacement in the second codon position, T → A, changes the glutamic acid codon into a valine codon. Since T is a pyrimidine and A is a purine, this is a transversion. (It has become customary to describe not the actual DNA codons but their anticodons in such representations. See also the table for the genetic code (Table 3.2).

In other instances, the base sequence may be changed at the third base position of a "degenerate" codon so that a new codon specifies the same amino acid. This is a "same-sense" mutation without phenotypic effects. If the mutation transforms a sense codon into a stop codon, a "nonsense" mutation results.

Single base pair changes may also occur in stop codons, leading to protein chain elongations up to the next stop codon (the classical first example in humans is Hb Constant Spring; see Sect. 4.7.3.2). Fur-

ther, they can occur outside the coding region of the gene, for example, in the promoter regions, in initiation codons, in the mRNA splice junctions – all with various phenotypic effects. For examples see Sect. 7.3 (hemoglobin; see also [20]).

Transitions Are Especially Common. Spontaneous base replacements do not occur at random. Inferring nucleotide alterations in hemoglobin variants from amino acid substitutions using the genetic code, it was shown in 1965 that in hemoglobin genes, transitions are much more frequent than would be expected with random replacement [139]. This result was later generalized for mutations fixed in various genes during evolution [137].

When direct assessment of mutations at the DNA level became possible, the increased occurrence of transitions was confirmed for many other genes. Among 880 directly identified point mutations in various genetic diseases; 275 transversions (31.3%) and 605 transitions (68.7%) were counted, compared to the expected 33.3% for transitions [20]. Some of these transitions could be explained by a defined biological mechanism. In various parts of the genome the DNA sequence — C — G — is found. In most

$$— C — G —$$
$$\quad ||| \quad |||$$
$$— G — C —$$

areas, however, such so-called CpG sequences are much rarer than would be expected if base sequence were random. CpG sequences are found primarily in the CpG "islands," often outside the 5′ ends of coding genes, particularly in "house-keeping" genes. In parts of the genome cytosines are methylated; they are replaced by 5-methylcytosine.

This methylation appears to be important for gene action, but a methylated cytosine – especially if adjacent to a guanine – has an increased risk of being deaminated (Fig. 9.22). Such deamination leads to thymine, which in the next round of DNA replication pairs with adenine, not with guanine. A transition has occurred. This appears to be the reason why CpG base pairs are relatively rare in the genome – except at the above-mentioned CpG islands. In these islands cytosines are not methylated. The mutation rate is therefore not increased, and these islands have been maintained in evolution. They may be "the last remnants of the long tracts of non-methylated DNA" (Bird, cited in [20]).

Obviously the C → T mutation suggested by this mechanism is a transition, and indeed 289 of the 605 transitions cited above have occurred within CpG dinucleotides (outside CpG islands). This mutation therefore explains much, but not all, of the abundance of transitions. There remain 316 transitions and 275 transversions – not one-third but more than one-half transitions. Thus, other mechanisms must

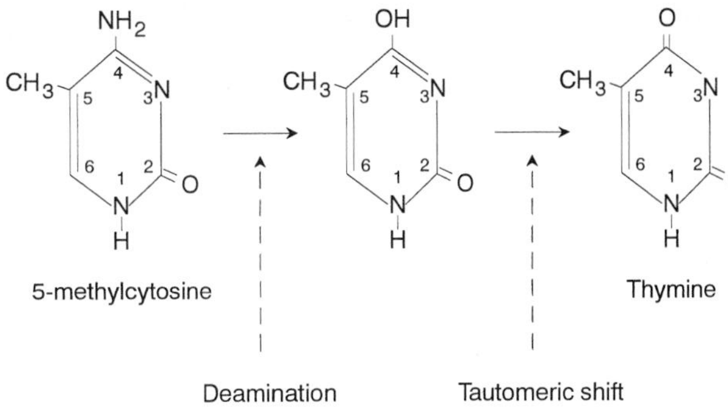

Fig. 9.22. Single base mutation, C → T, by demethylation of 5-methylcytosine; tautomeric shift

be considered. The bases surrounding the mutational site appear to be important. Some observations of independent, repeated, identical mutations suggest that in human genes, just as in the genome of more primitive organisms, mutational hot spots may exist. So far, however, hot spots outside CpG dinucleotides have not been characterized sufficiently well in molecular terms. Among mutations in hemoglobin genes leading to amino acid substitutions, no clear indication for such hot spots has been found.

Mutation Mechanisms, Paternal Age Effect, and Sex Difference. The increase in certain gene mutations with paternal age seemed understandable if mutations were often caused by a copy error during DNA replication. This concept was never perfect since it did not predict an increase in the slope of the curve with increasing age. Nevertheless, the copy error hypothesis was the best explanation not only for the paternal age effect but also for the proportionally greater number of mutations observed in male than in female germ lines. Since spontaneous deamination of methylated cytosine is believed to be a time-dependent and not a replication-dependent process [20], a quasilinear increase in mutation rates with age regardless of sex would be expected. Auxiliary hypotheses are of course possible, such as age – and sex-dependent changes in efficiency of DNA polymerases or repair processes, or an influence of DNA on chromatin configuration in dividing and nondividing cells. Moreover, results in micro-organisms suggest replication dependence of most spontaneous mutations [28]. Probably an appreciable fraction of mutations – especially in male germ cells – are nevertheless replication dependent, since the statistical data are convincing. The precise mechanisms remain unknown.

Deletions [20]. Mutational mechanisms were first elucidated in hemoglobin genes leading to hemoglobinopathies and thalassemias (Sect. 7.3). The great majority of these mutations were single base pair substitutions, but occasionally one or several base pairs were deleted. When data from other genes became available,

similar results were obtained. A relatively small proportion of deletions (compared to single base mutations) were found in most but not in all diseases. In hemophilia A, for example, we know of 77 point mutations and 10 deletions. Since independent but identical mutations are counted only once, and because point mutations often have not been identified with sufficient precision, this figure may be biased in favor of deletions. Antonarakis (1988 [3]) estimated the fraction of deletions for the factor VIII and factor IX genes at about 5% [84]. Even lower frequencies were later shown (1%–2%) [20, 84]. In hemophilia B, deletions also appear to comprise a small percentage of all mutations [65]. For the HPRT deficiency there were 40 single base pair substitutions and 9 deletions.

The situation is quite different for mutations of the dystrophin gene causing Duchenne muscular dystrophy. Here the majority of mutations (approx. 60%–70%) have been identified as deletions [25, 66–68]. Duplications have also been found, although at a lower frequency. Several point mutations have been identified. The distribution of deletions within this gene is decidedly nonrandom.

Deletions and Sex Ratio of Mutation Rates. Mutation rates of the genes for factor VIII, probably factor IX, and the HPRT locus appear to be higher – probably much higher – in male than in female germ cells. Among mutations of the dystrophin gene, on the other hand, no such a sex difference has been found; here the over-all mutation rate may be very similar in the two sexes. This suggests a difference in the underlying mechanisms. While the great majority of mutations are usually single base pair replacements, dystrophin gene mutations are generally deletions. This suggests that single base substitutions may occur more often in the male than in the female germ line and may increase with paternal age. In contrast, deletions occur with equal frequencies in the two sexes, possibly even somewhat more commonly in females. To test this hypothesis Grimm et al. [41] compared the sex ratio of deletions and nondeletions in a comprehensive data set of dystrophin mutants. De-

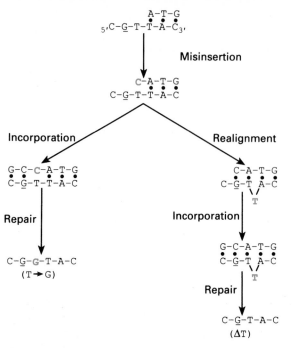

Fig. 9.23. Pathways for single base errors during DNA replication as a result of a misinsertion event (Cooper and Krawczak 1993 [20]; from Kunkel)

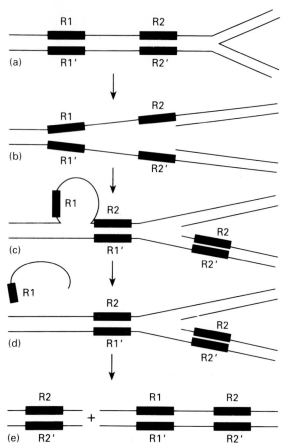

Fig. 9.24. The slipperage mispairing model für the generation of deletions during DNA replication. (*a*) Duplex DNA containing direct repeat sequence. (*b*) Duplex becomes single-stranded at replication fork. (*c*) R2 repeat base pairs with complementary R1′ repeat producing a single-stranded loop. (*d*) Loop excised and rejoined by DNA repair enzymes. (*e*) Daughter duplexes, one of which contains only one of the two repeats and lacks the intervening sequence between R1 and R2. (Efstratiadis et al. 1980; from Cooper and Krawczak 1993 [20])

letions were more common in females; nondeletions, presumably mostly point mutations, were more common in male germ cells. A similar result was reported for factor IX mutations: the mutation rate for point mutations was higher in males, whereas approximately equal mutation rates in the two sexes were found for deletions [64]. Interpretation of such results in molecular terms will become possible by elucidating the molecular mechanisms of deletion (see below).

Molecular Mechanisms of Deletions. Slipping and pairing of structure homologous but not position homologous DNA sequences leads to unequal crossover (Sect. 5.2.8) and has often been discussed as the cause of deletions and duplications. Results of such slipping have been known for a long time: Lepore and anti-Lepore hemoglobins (Sect. 7.3), some rare haptoglobins (Sect. 5.2.8), and the color vision pigment genes (Sect. 15.2.1.5). When analysis at the gene DNA level became possible, such events were seen to be more common than anticipated. Further insights can be expected from analysis of the DNA sequences around the deletion breakpoints. It was shown, for example, that such areas often have Alu sequences (see also Sect. 3.2; Fig. 3.30) [20]. Three possibilities have been discussed: (a) recombination occur between one Alu repeat and a nonrepetitive DNA sequence that has sequence homology with the Alu repeat, (b) crossover takes place between Alu sequences

oriented in opposite directions, (c) recombination occurs between Alu sequences oriented in the same direction (Figs. 9.24–9.28).

However, not all deletions are mediated by Alu sequences; other types of repetitive sequences such as short, direct repeats are also involved. Figure 9.24 shows a possible mechanism ("slipping" between two structure homologous repeats). So far it is not known whether the obviously unequal distribution of deletion breakpoints, such as in the dystrophin gene, can be explained by a corresponding distribution of repetitive sequences. Often the point of recombination is not "clean"; new bases may be introduced, and duplications of the target site are frequent [20]. The length of the deleted segment may vary, but

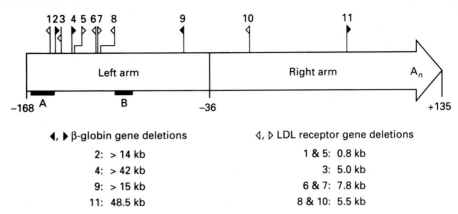

◀, ▶ β-globin gene deletions	◁, ▷ LDL receptor gene deletions
2: > 14 kb	1 & 5: 0.8 kb
4: > 42 kb	3: 5.0 kb
9: > 15 kb	6 & 7: 7.8 kb
11: 48.5 kb	8 & 10: 5.5 kb

Fig. 9.25. Deletion breakpoints within the Alu consenses sequence for mutations in the LDLR *(unshaded arrowheads)* and Hbβ genes *(shadowed arrowheads)*. *Solid bars,* regions of the Alu sequence that correspond to the RNA polymerase III promoters A and B. (Cooper and Krawczak 1993 [20]; after Lehrman)

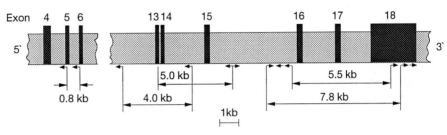

Fig. 9.26. Deletions involving Alu sequences in the human LDLR gene causing hypercholesterolemia. *Vertical black bars,* exons; *horizontal arrows,* extent of the deletions; *arrowheads,* position and orientation of Alu sequences. (Cooper and Krawczak 1993 [20]; data from Lehrman and Horsthemke)

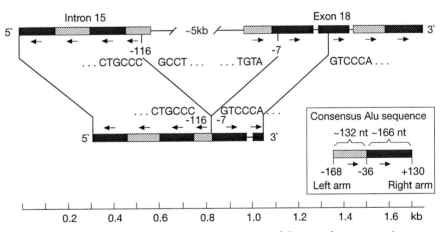

Fig. 9.27. Recombination between Alu sequence elements generating a 5.5-kb deletion of the LDLR gene. *Black, hatched boxes,* right and left arms of Alu consensus sequences, respectively; *arrows,* orientations of he left and right arms of the Alu repeats. Joining of the two Alu sequences is accomplished as a staggered break, and a new T nucleotide is inserted at the deletion. (From Cooper and Krawczak 1993 [20]; data from Lehrman)

short deletions are much more common than longer ones (see Fig. 9.30). The question of whether there are hot spots has been discussed for deletions. The distribution of deletion breakpoints in dystrophin deletions suggests the existence of such hot spots.

Insertions, Duplications, and Inversions [20]. Structural changes other than deletions have also been observed, albeit more rarely. While unequal recombination is one mechanism for forming deletions, duplications are also expected. Theoretically these should

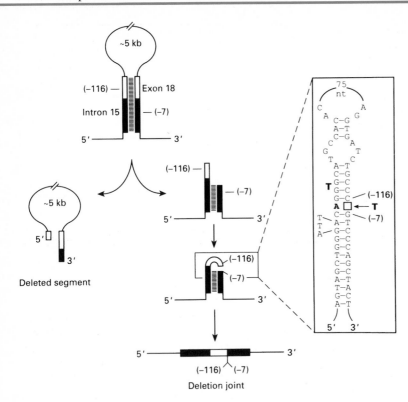

Fig. 9.28. Possible mechanism for the 5.5-kb LDLR gene deletion involving the formation of a loop between inverted repeat sequences located on the same DNA strand. *Shaded, unshaded boxes, right and left arms of Alu consensus sequences, respectively.* (From Cooper and Krawczak 1993 [20])

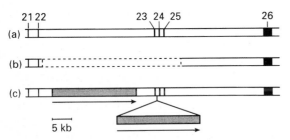

Fig. 9.29. Deletion/duplication lesions in the Factor VIII (F8) gene causing hemophilia A. Only the portion of the F8 gene spanning exons 21–26 is shown. (*a*) Wild-type gene; (*b*) 39-kb deletion of intron 22 and exons 23–25 (*dotted lines,* deleted region). (*c*) Duplication of 23 kb of intron 22 inserted into intron 23 (*duplicated gray region*). (Observation by Gitschier; from Cooper and Krawczak 1993 [20])

be just as common. In fact, however, duplications appear much rarer, although among mutations in the dystrophin gene a small number, about 10 % of the deletions, have been observed. They may lead to disease, and many examples have been documented [56]. Insertions often consist of only a few base pairs. Their location appears to be nonrandom; mechanisms may be similar to those discussed for duplications. Inversions have also been described. For example, a compound γ-δ-β thalassemia observed in India was explained by a complex inversion: two gene seg-

ments were deleted while the intervening segment was inverted [60].

Mutations Leading to Hereditary Diseases by DNA Triplet Expansion [16, 110]. In an increasing number of hereditary diseases a new and unexpected type of mutation has been discovered: the amplification of repeats of a motif consisting of three bases. These diseases include *Huntington disease* [79] (143000; see Sect. 3.1.3.8 for the gene and Sect. 4.1.2 for age at onset), *myotonic dystrophy* [14, 47] (160900; see Sect. 4.1.7 for anticipation), *Fragile X mental retardation* (309550), *fragile XE mental retardation syndrome, spinobulbar muscular atrophy* (SBMA; Kennedy disease; 313200) [1556], and *spinocerebellar ataxia type I.* In two of these diseases, Huntington disease and myotonic dystrophy, an unusually wide range in the age at onset has been observed, and in pre-Mendelian times anticipation was suggested as an explanation for these findings (Sect. 4.1.7). Moreover, in both of these diseases genomic imprinting has been observed (Sect. 4.1.7; 8.2): in Huntington disease early onset is correlated with inheritance of the mutant gene from the father; in myotonic dystrophy onset in the neonatal period (congenital type) may be observed with maternal transmission.

The relatively common Fra X syndrome has a fragile site near the tip of the long arm of the X chromosome (q28), visible in culture media lacking folic acid.

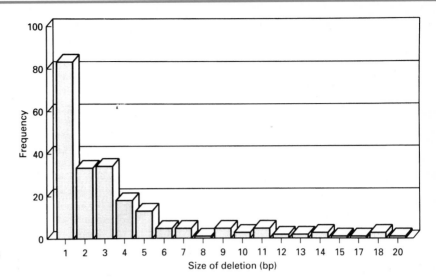

Fig. 9.30. Size distribution of short human gene deletions. (From Cooper and Krawczak 1993 [20])

While the mode of inheritance of the fragile X syndrome is undisputably X-linked, only about 80% of males identified as hemizygotes by pedigree analysis are mentally retarded. About 30% of female heterozygotes are mentally retarded; when their sons inherit the mutant gene, they usually show mental retardation (Chap. 15). This could also occur, however, in sons and daughters of clinically unaffected carrier women.

In distinction to the other syndromes, three phenotypes are observed; persons with the normal number of repeats are normal in every respect. A group of individuals with an increased number of repeats (approx. 60–200) are also normal. When such a somewhat expanded X chromosome is transmitted to a daughter, she is clinically normal, but amplification occurs in her germ cells, and her sons will be affected; daughters may also be mentally retarded. These findings explain the peculiarities of the mode of inheritance, especially among the approx. 20% clinically unaffected male transmitters. These carry the "premutation" with an increased number of repeats, which turns into a full-fledged mutation in the germ cells of their daughters, causing mental retardation among the grandchildren.

The severe mental retardation in this syndrome is associated with a blocked transcription of the *FMR-1* gene. There is excellent correlation between the number of repeats and the extent of mental retardation (see Table 9.13).

Kennedy disease, which is rarer than the other three, usually manifests clinically in adult age, and, as in the other conditions, affects the central nervous system. Here the gene involved specifies an androgen receptor. In the three others, the precise function of the genes has not been elucidated, but in myotonic dystrophy and in Huntington disease the protein sequence is known. In Fra X1 the functional disturbance of the gene appears to be caused by abnormal methylation upstream of its 5' end [97], which may occur early in prenatal development [27]. These results raise the question of how common such triplet repeats are in the normal human genome, and where they occur. Richards and Sutherland [110] reported no less than ten such repeats in various genes, three of which are in coding regions. Repeat length varied between 3 and 11. The mutation mechanism by which such repeats are produced is unknown, and no really convincing model has been proposed so far. However, there is a high degree of variability of repeat length in the normal population (Figs. 9.31; 9.32; 9.33), and differences among population groups of different racial origin have been described [13]. Once a certain length has been achieved, the repeat number per allele becomes unstable. Various repeat lengths have been observed in patients from the same families, despite the fact that all these genes obviously had a common origin. Moreover, clinical parameters such as age at onset and severity of the disease are influenced negatively when the number of repeats increases, although such correlations are not sufficiently striking to be used for clinical prognosis regarding age of onset in Huntington disease. The same changes in repeat number may also occur in somatic tissues [150]. It seemed that most changes would occur in one direction only, i.e., increase in the number of repeats, but this could easily be caused by ascertainment bias since most families are ascertained by severely affected patients in the younger generation. Moreover, a reduction in repeat lengths has occasionally been observed [96].

DNA Triplet Expansion. In each of these diseases, an unusual expansion of a repetitive sequence of DNA triplets has been found which is also present in nor-

Table 9.13. Disorders of trinucleotide repeat expansion (adapted from Mandel 1993 [81])

	Huntington's disease (HD)	Myotonic dystrophy (DM)	Spinobulbar muscular atrophy (SBMA)	FRA X A locus Fragile X (FMR-1)	FRA X E (FMR-1 negative fragile site)	Spinocerebellar ataxia type 1 (SCA1)
Frequency	1/10000	1/1800	<1/50000 ♂	1/2000	~3% all fragile X+	
Inheritance	True autosomal dominant anticipation in ~6% of cases	Autosomal dominant anticipation	X-linked recessive variable severity	X-linked dominant with partial penetrance in ♀	X-linked dominant	AD anticipation
Sex bias for transmission of severe form	♂ (early onset)	♀ (congenital DM)	(♀ due to X-linked inheritance)	♀ (full mutation)	None	♂ (early onset)
Protein/expression	Unknown function, mRNA widely expressed	Putative protein kinase	Androgen receptor	RNA-binding protein? Unknown function, mRNA widely expressed		
Disease causing mechanism	Abnormal protein?; gain of function	Decreased dosage of mRNA/protein? Increased dosage mutant mRNA?	Abnormal protein; gain of function?	Transcription shut down; abnormal DNA methylation	Hypermethylation	Abnormal protein? gain of function?
Repeat/location	CAG Protein coding (Gln)?	CTG (= CAG) 3' (untranscribed region)	CAG Protein coding (Gln)	CGG (interrupted) 5' (untranscribed region)	GCC	CAG Transcribed (Gln?)
Size of repeat	11–34	5–35	11–31	10–50	6–25	19–36
Predisposing normal alleles (frequency)	30–34	20–25	?	38–50		
Linkage disequilibrium	Yes	Yes (absolute)	Not known	Yes		No
Disease alleles	42–100	50–100 (premutation) 100–2000 affected	40–62	52–200 (premutation) 200–2000 (full mutation)	>200	43–81
Effect of increasing size of disease allele	↓ Age of onset	↓ Age of onset ↑ Severity	↓ Age of onset	For ♀ premutation ↑ Risk in offspring: >200 No effect on severity		↓ Age of onset
Parental sex bias for instability	Large expansions paternal	Expansion from both ♂ and ♀ (larger on average from ♀)	Moderate instability predominantly paternal	Large expansion only maternal	Large expansions all maternal	Large expansion paternal

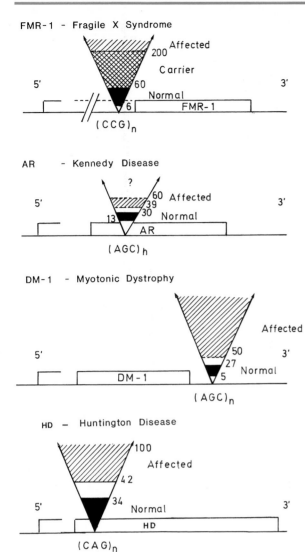

Fig. 9.31. Amplification mutants in four hereditary diseases: Fra X syndrome, Kennedy disease, myotonic dystrophy, and Huntington disease. In the Fra X syndrome three states can be distinguished (affected; carrier, normal). In the others, affected and normal individuals can be distinguished (*black*, normal; *hatched*, affected). The space between *(white)* indicates the unstable region. *Numbers* (right of triangles), range of repeat numbers. (From Richards and Sutherland 1992 [110]; supplemented)

tion occurs, the elongated portion of the protein consists of a polyglutamine sequence. Such polyglutamine sequences are also present in the normal protein gene products, as implied by the repeat numbers of trinucleotides in unaffected individuals. This obviously poses the question regarding their usual function, particularly in the brain. Such glutamine stretches have also been found in transcription factors, for example, in *Drosophila* [81].

In the Fra X syndrome there appears to be no gene product. This finding may be caused by abnormal methylation of the region containing the repeat [97].

Origin of the Mutations. Kennedy disease is very rare; myotonic dystrophy and Huntington disease are moderately common (~1:7000) and the Fra X syndrome is quite common (~1:2000). Therefore the mutation rate leading to the unstable state of the triplet in Kennedy disease is obviously quite low. The same appears to be true for the mutation leading to Huntington disease (see the discussion in Sect. 3.1.3.8). Many patients with this condition come from large families that can be traced back many generations. In a few instances when a new mutation appeared possible, this assumption was refuted by direct study of the gene. Recently a number of new mutants for Huntington disease have been discovered; they appear to occur in paternal germ cells; these men were shown to have an unusually high repeat number in the normal range, and paternal age may be increased [39]. Mutation rate estimates for myotonic dystrophy from Northern Ireland and Switzerland were available before the nature of the defect was known (~1×10^6; see Table 9.6). These data will need to be revised in the light of the new results. On the basis of strong linkage disequilibrium with DNA markers it has been postulated that most families with myotonic dystrophy originated from a single mutation [47].

For Fra X syndrome, Sherman et al. (1984) [118] postulated an unusually high mutation rate exclusively in male germ cells. Vogel [136, 137a] suggested a fairly high mutation rate and a selective advantage due to increased reproduction of clinically unaffected female carriers. Such an increased reproduction was actually shown [141]. In comprehensive studies of apparently unrelated Fra X families from various populations (Australia, France, Spain, North Africa, and others), strong linkage disequilibrium with closely linked DNA markers was found [99, 111] which were transmitted together with the $(CGG)_n$ repeat much more often than in the general population. Linkage disequilibrium may have either of two different causes (see Sect. 5.2.4). Either the mutation originated in a chromosome carrying the marker haplotype,

mal individuals but with a more limited number of repeats. In both Kennedy and Huntington diseases the amplified sequence is located within the coding sequence of the gene; in the other conditions it has been found outside the coding sequence – in myotonic dystrophy, it is localized downstream of the 3′ end (Fig. 9.31). The expansion is $(AGC)_n$ in Kennedy disease and in myotonic dystrophy, $(CCG)_n$ in Fra X, and $(CAG)_n$ in Huntington disease. Interestingly, in Kennedy and Huntington diseases, where transcrip-

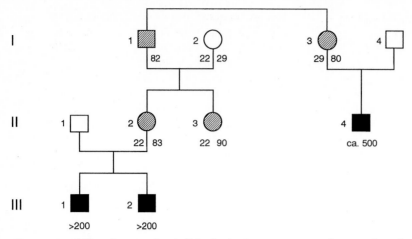

Fig. 9.32. Instability of premutational alleles for the Fra X syndrome. □, ○, Normal (phenotypically and genotypically); ■, ●: affected; ▨, ◌, heterozygotes for a premutation allele. *Numbers,* number of repeats in each of the two alleles. Enlargement occurs during transmission from the mother. (Model pedigree; from Caskey et al. 1991 [16])

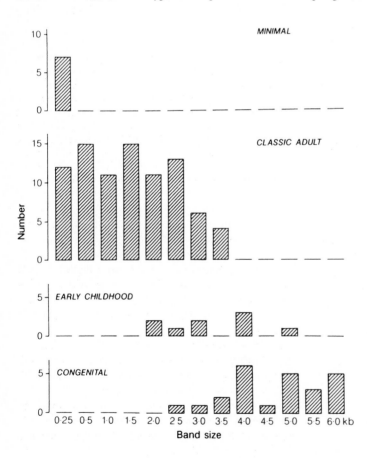

Fig. 9.33. Distribution of sizes of expansion (number of repeats) of the unstable triplet repeat by type of myotonic dystrophy. (From Harley et al. 1992 [47])

and not enough time had elapsed since the time of the mutational event for dilution of this association by crossing over, or the entire haplotype has a selective advantage. In Fra X – and also in myotonic dystrophy – the common origin theory appears more convincing. Scientists who discovered this linkage disequilibrium therefore concluded that the muta-

tions originated in very few individuals [99, 111], suggesting a very low mutation rate. Kennedy disease, being rare, does not really create a problem. The mutation rate must be low, and moreover there is strong selection against this mutation due to its effect on the phenotype of affected males. In Huntington disease there is independent evidence for a very low mutation rate. It has often been discussed whether affected patients may even have had a selective advantage under the living conditions of earlier times. But what about the Fra X syndrome? Affected males do not reproduce. This, together with the reduced reproduction by affected females creates an appreciable selective disadvantage, which must have been compensated someway. It is very doubtful that the advantage of nonaffected carrier women [141] was sufficient to allow a few mutations to spread all over the world, leading to the unusually high incidence of the Fra X syndrome. Alternative explanations include:

1. The entire haplotype, including the Fra X mutation, has a selective advantage compared with other haplotypes containing this mutation.
2. Mutations leading to an increased (and unstable) number of (CGG)$_n$ repeats are more common in the neighborhood of certain DNA sequences, as shown for point mutations [20].
3. The carriers of the premutation, i.e. those carrying a moderately increased number of repeats, had an unknown selective advantage in the past. In some instances, this premutation may be carried through many generations without being converted into the full mutation.

Table 9.13 presents an overview of the most important findings in diseases caused by trinucleotide repeats.

Biological function of trinucleotide repeats. – These trinucleotide repeats raise the question as to their normal function: they are present in small numbers in many normal genes, and their amplification leads to diseases mainly affecting the nervous system. This suggests the hypothesis that a "normal", relatively small repeat number may have something to do with the normal function of these genes within the nervous system. Brahmachari et al. have brought forward the hypothesis – and have given some experimental evidence – that such triplet repeats may act as modulators for "fine-tuning" of transcription [5a].

9.4.3 Mutations in Micro-organisms: Their Contribution to Understanding of Human Mutation

Mutations as Errors of DNA Replication. Data on humans suggest a close relationship between mutation and cell division. Examination of this problem was triggered by the mechanism proposed by Watson and Crick (Fig. 9.34) [145] and by early research in micro-organisms indicating that many spontaneous mutations do indeed occur during DNA replication, and that erroneous introduction of the wrong nucleo-

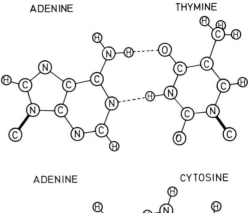

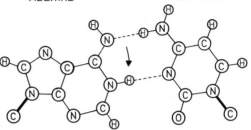

Fig. 9.34. Mechanism of point mutation by base replacement as suggested by the Watson-Crick model. A base may occasionally and for a short time assume a rare tautomeric configuration and pair with another base rather than with its usual partner, for example adenine with cytosine instead of with thymine. By the time of the next replication cycle both bases will have attained their most probable configuration and will pair with their usual counterparts. Hence the two double helices of the next generation are different; a point mutation has occurred. (Watson and Crick 1953 [145]

tide base leads to a different base pair in future cell generations. These results are described in [20, 28].

As expected, amino acid substitutions due to spontaneous point mutation within the tryptophan synthetase A locus of *E. coli* are compatible with the exchange of one base, just as the hemoglobin point mutations [149].

Mutations in dividing and nondividing bacteriophage T4 rII mutants have been examined extensively [28]. The great majority of mutations have arisen by replication-dependent processes, and most are frameshifts. They are especially frequent in two hot spots; when these are excluded from consideration, the ratio of frameshifts to single base substitutions is reduced from 3.3 to 1.6. Evidence in favor of the replication dependence of mutations in bacteria has been presented by Kondo [69].

We may conclude that in micro-organisms many mutations, perhaps even the great majority, are replication dependent. Not all replication-dependent mutations, however, are caused by the exchange of one base: DNA replication also seems to enhance the risk of frameshift mutations.

Mutator Genes. Demerec in 1937 [24] described "unstable" genes in certain stocks of *Drosophila melanogaster*. Since that time numerous examples of genetically determined, unusually high mutation rates have been observed in both eukaryotes and in prokaryotes. Often the enhanced mutability can be traced back to the influence of a "mutator gene." Analysis of the action of such mutator genes has provided valuable information on the interaction of various factors (polymerases, repair processes, etc.) [28, 85] in the mutation process. For human point mutations in germ cells, no evidence for the actual occurrence of such mutator genes seems to exist. Careful search for the extremely rare human families with two mutations would be of interest. For mutations in single cells, however, mutator genes have been identified (see below).

Mutationlike Events Due to Extranuclear Entities Such as Viruses and Transposons. Spontaneous mutation is discussed above with reference to classical concepts, i.e., assuming a genetic change (base replacement; deletion; recombination) within nuclear DNA. However, in discussing results on the structure of chromosomes and DNA in humans (Chap. 3) we also mentioned transposons – or "jumping genes." Their effects cannot be distinguished phenotypically from those of classical mutations. Transposons appear to play a role in the appearance and transmission of germ cell mutations in humans; de novo insertion of a (transposonlike) L1 sequence has been reported in hemophilia A [63].

Latent viruses can be transmitted vertically from generation to generation with no ill effects. They may nonetheless affect the physiology of their host. All King Edward potatoes carry the paracrinkle virus without pathological lesions, but plants freed from the virus look different from the ordinary stocks and give a higher yield. In human in vitro systems certain pathogenic viruses, such as German measles virus, have been shown to induce chromosomal aberrations [8, 48]. Such aberrations could have been induced in germ line chromosomes (first meiotic division) of the male mouse. As long ago as 1963, Taylor [122] discovered that the phage Mu of *E. coli* induces many gene mutations at various sites. Later the Mu phages were identified as transposons. In some phages most "spontaneous" mutations are in fact caused by transposons. More recently it has been discovered that animal viruses such as SV 40 and polyoma virus are able to induce gene mutations in mammalian cells (Chinese hamster and mouse cell lines) [37].

The interaction between mutation and viruses is also considered in Chap. 10, where somatic mutation and cancer are dealt with.

9.5 Examination of Gene Mutations in Single Cells

With the success of genetic analysis in micro-organisms it appeared promising to study human genetics problems in single cells. The development of this approach is described in Sect. 7.2.2.1. Considering the low frequency of spontaneous mutations and the technical obstacles to examining human population samples of a sufficient size to establish even a crude order of magnitude at the level of the individual, such an approach would increase the resolving power of genetic analysis by several orders of magnitude.

First Attempt to Examine Mutations Occurring In Vitro. Atwood and Scheinberg [5] developed a technique that permitted removal by agglutination with an anti-A serum of all blood cells from a blood sample that reacted with this anti-A serum, leaving only those that did not react. In probands of blood group AB some inagglutinable cells were found. These cells were interpreted to have lost the A but not the B antigen, suggesting that nonspecific antigen loss had not occurred. The relative frequency of these cells ranged in various individuals between 0.5 and 10.9/1000 cells. These cells were interpreted as being somatic mutants. However, the very magnitude of the phenomenon made this interpretation unlikely. Moreover, additional immunological observations with this system cast serious doubt on the mutational origin of the inagglutinable cells. This attempt remains interesting in spite of the fact that it was unsuccessful.

The problem of whether cells showing a biochemically or immunologically aberrant phenotype are indeed new mutants or are only the products of some secondary change not affecting the genetic material remains a central issue in mutation research on single cells.

Examination of Mutant Cells In Vitro. The methods for cultivating normal human diploid cells in vitro are described in Sect. 7.2.2.1 [19]. One of the major difficulties involved in the study of mutation in such cells is the low mutation rate. Special selection methods are needed to isolate the very few mutant cells from the great majority of normal ones. The principle of such methods is described in Sect. 7.2.2.6 for the Lesch-Nyhan syndrome (defect of the HPRT enzyme). The cells are offered 8-azaguanine, as a growth substance instead of hypoxanthine; 8-azaguanine when accepted by the normal enzyme, kills the cell. Only cells unable to metabolize this compound because of their enzyme defect in HPRT survive.

Other selective systems are available, for example, in galactosemia, citrullinemia, and orotic aciduria. In another approach, red cells containing hemoglobin variants generated by presumed somatic mutation are being identified by use of specific antibodies against specific variant hemoglobins [108].

As noted a variant isolated in cell culture is not necessarily the product of a true genetic and transmissible alteration. At least two criteria which however are not applicable to all methods are needed to confirm a mutation:

1. Demonstration of stability of the selected phenotype.

Principle of fluctuation test

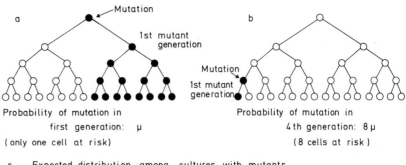

c   Expected distribution among cultures with mutants

Fig. 9.35 a–c. Principle of the fluctuation test of Luria and Delbrück (1943) [78]. If the mode of cell division is dichotomous, there are $2^1 = 2$ cells in the first generation after division has started; $2^2 = 4$ cells in the second; $2^3 = 8$ cells in the third generation, and so on. If the cell cultures to be examined all start with only one cell and if the mutation rates are identical in all cell generations, the relative number of cultures with 0, 1, 2 ... mutant cells can be calculated, depend-

ing on the cell generation in which the mutation has occurred. **a** For example, if the mutation occurred in the first division, half of all cells are mutant. In this case only one cell is at risk. **b** On the other hand, if the mutation occurred in the last division before the culture is examined, only one cell is mutated. **c** For such a mutation, however, 2^{n-1} cells are at risk, where n = number of cell generations. Expected distribution among cultures

2. Positive results with the fluctuation test of Luria and Delbrück [78], which is based on the principle that when a great number of cultures are compared, only very few have numerous mutant cells (mutation early in the development of the culture), whereas most cultures contain very few mutants (mutation late in the development of the culture), and many others do not have even one mutation (Fig. 9.35).

Test methods for determining the spontaneous mutation rate are based on the fluctuation test; in Fig. 9.35 c, for example, every plate would be counted as one mutation if the colonies are clones from one cell each.

Table 9.14 shows a number of estimates of spontaneous mutation rates for the HPRT locus. At first glance these mutation rates seem to be of the same order of magnitude as the germ cell mutation rates calculated at the phenotypic level for hereditary diseases (Tables 9.6, 9.7). These human mutation rates, as explained in Sect. 9.3.2, represent a biased portion of traits with high mutation rates and are the end result of several dozen cell divisions. A comparison with somatic cell mutation rates, which are expressed as the number of mutations for a single cell division only, is therefore not appropriate.

Using the HPRT system, an elevation of the spontaneous mutation rate to $19–23 \times 10^{-6}$ mutants per cell per generation has been described in fibroblasts

Table 9.14. Spontaneous mutation rate in human and Chinese hamster cells: resistance to 8-azaguanine ([138])

Cell line	Ploidy level	Rate
	A. Human	
D98[a]	Aneuploid	4.9×10^{-4}
L54[b]	Diploid	7.0×10^{-5}
Glen[b]	Aneuploid	7.0×10^{-5}
Fibroblasts[c]	Diploid	4.1×10^{-6}
	B. Chinese hamster	
237[b]	Hypodiploid	4.0×10^{-5}
V5[d]	Diploid	2.2×10^{-5}
V25[d]	Tetraploid	4.7×10^{-5}
V68[d]	Octaploid	1.9×10^{-5}
V79[e]		1.5×10^{-8}

[a] Data from Szybalski and Smith
[b] Data from Shapiro et al.
[c] Data from De Mars and Held
[d] Data from Harris
[e] Data from Chu et al.

from two patients with Bloom syndrome compared to a rate of $4.6–4.9 \times 10^{-6}$ in normal fibroblasts [144]. Bloom syndrome is a chromosome instability syndrome; therefore this gene is a human mutator gene. This higher mutation rate has also been demonstrated in vivo.

More recently this increase in mutation rate has been confirmed for lymphocytes in Bloom syndrome and for the other chromosome instability syndromes [128, 129]. The figures are:

Normal control cells:	$2.0–4.4 \times 10^{-4}$
Bloom syndrome (7 patients):	$8.5–24.9 \times 10^{-4}$
Fanconi anemia (2 patients):	$20.0–22.6 \times 10^{-4}$
Ataxia-telangiectasia (1 patient):	8.5×10^{-4}

Conclusions

Once in a while the genes and chromosomes in germ cells change, causing new mutants. Three classes of mutations can be distinguished: numerical and structural chromosome aberrations, and gene mutations. Such mutations are chance events which occur with certain probabilities (mutation rates). Mutation rates for the most common numerical chromosomal aberrations – trisomies – show a steep increase with maternal age. Mutation rates for individual gene mutations causing autosomal-dominant or X-linked hereditary diseases are much lower than those for many numerical chromosome aberrations; some of these show an increase with the age of the father and are usually more common in male than in female germ cells. Newer methods have led to the identification of many mutations at the molecular level. In the majority of hereditary diseases (for example, hemophilia) most mutations are single base substitutions in the DNA. Transitions asre most common. In at least one disease (X-linked muscular dystrophy) about two-thirds of such mutations are submicroscopic deletions. Another recently discovered type of mutation – abnormal amplification of base triplets – may lead to X-linked mental retardation or to dominant diseases with high variability in age at onset and anticipation. Huntington disease and myotonic dystrophy are examples.

References

1. Aaltonen L, Peltomaeki P, Leach FS et al (1993) Clues to the pathogenesis of familial colorectal cancer. Science 260 : 812–816
1a. Altland K, Kaempfer M, Forssbohm M, Werner W (1982) Monitoring for changing mutation rates using blood samples submitted for PKU screening. In: Bonaiti-Pellié C et al (eds) The unfolding genome. Liss, New York, pp 277–287 (Human genetics, part A)
2. Andreassen M (1943) Haemofili i Danmark. Munksgaard, Copenhagen (Opera ex domo biologiae hereditariae humanae universitatis hafniensis, vol 6)
3. Antonarakis SE (1988) The molecular genetics of hemophilia A and B in man. Adv Hum Genet 17 : 27–59
4. Antonarakis SE, the Down Syndrome Collaborative Group (1991) Parental origin of the extra chromosome in trisomy 21 as indicated by analysis of DNA polymorphismus. N Engl J Med 324 : 872–876
5. Atwood KC, Scheinberg SL (1958) Somatic variation in human erythrocyte antigens. J Cell Comp Physiol 52 Suppl 1 : 97–123
5a. Brahmachari SK, Meera G, Sarkar PS et al. (1995) Simple repetitive sequences on the genome: structure and functional significance. Electrophoresis 16 : 1705–1714
6. Bakker E, Veenema H, Den Dumen JT et al (1989) Germinal mosaicism increases the recurrence risk for "new" Duchenne muscular dystrophy mutations. J Med Genet 26 : 553–559
7. Barakat AY, DerKaloustian VM, Mufarrij AA, Birbari A (1986) The kidney in genetic disease. Churchill Livingstone, Edinburgh
8. Bartsch HD (1970) Virus-induced chromosome alterations in mammals and man. In: Vogel F, Röhrborn G (eds) Chemical mutagenesis in mammals and man. Springer, Berlin Heidelberg New York, pp 420–432
9. Batshaw ML, Msall M, Beaudet AL, Trojak J (1986) Risk of serious illness in heterozygotes for ornithine transcarbamylase deficiency. J Pediatr 108 : 236–241
10. Barthelmess A (1952) Vererbungswissenschaft. Alber, Freiburg
11. Blank C (1960) Apert's syndrome (a type of acrocephalosyndactyly). Observations on a British series of 39 cases. Ann Hum Genet 24 : 151–164
12. Bonaiti-Pellié C, Pelet A, Ogie H et al (1990) A probable sex difference in mutation rates in ornithine transcarbamylase deficiency. Hum Genet 84 : 163–166
13. Bora KG, Douglas GR, Nestmann ER (eds) (1982) Chemical mutagenesis, human population monitoring and genetic risk assessment. Prog Mutat Res 3
14. Brook JD, McCurrach ME, Harley HG et al (1992) Molecular basis of myotonic dystrophy: expansion of a trinucleotide (CTG) repeat at the 3'end of a transcript encoding a protein kinase family member. Cell 68 : 799–808
15. Carothers AD, Collyer S, DeMey R, Johnston I (1984) An aetiological study of 290 XXY males, with special reference to the role of paternal age. Hum Genet 68 : 248–253
16. Caskey CT, Pizzuti A, Fu YH et al (1992) Triplet repeat mutations in human disease. Science 256 : 784–789
17. Cavalli-Sforza LL, Bodmer WF (1971) The genetics of human populations. Freeman, San Francisco
18. Chandley AC (1991) On the parental origin of de novo mutation in man. J Med Genet 28 : 217–223
19. Chu EHY, Powell SS (1976) Selective systems in somatic cell genetics. Adv Hum Genet 7 : 189–258
20. Cooper DN, Krawczak M (1993) Human gene mutation. Bios Scientific, Oxford
21. Court Brown WM (1967) Human population cytogenetics. North-Holland, Amsterdam
22. Crowe FW, Schull WJ, Neel JV (1956) A clinical, pathological, and genetic study of multiple neurofibromatosis. Thomas, Springfield
23. Danforth GH (1921) The frequency of mutation and the incidence of hereditary traits in man. Eugenics, genetics and the family. Scientific papers of the 2 nd International Congress of Eugenics, vol 1. New York, pp 120–128
24. Demerec M (1937) Frequency of spontaneous mutations in certain stocks of Drosophila melanogaster. Genetics 22 : 469

25. Den Dumen JT, Grootscholten PM, Bakker E et al (1989) Topography of the DMD gene: FIGE and cDNA analysis of 194 cases reveal 115 deletions and 13 duplications. Am J Hum Genet 45 : 835–847

26. Der Kinderen DJ, Koten JW, Tan KEWP et al (1990) Parental age in sporadic hereditary retinoblastoma. Am J Ophthalmol 110 : 605–609

27. Devys D, Biancalana V, Rousseau F et al (1992) Analysis of full fragile X mutation in fetal tissues in monozygotic twins indicate that abnormal methylation and somatic heterogeneity are established in early development. Am J Med Genet 43 : 208–216

28. Drake JW (1969) The molecular basis of mutation. Holden Day, San Francisco

28a. Easton DF, Bishop DT, Ford D, Crockford GP, Breast Cancer Linkage Consortium (1993) Genetic linkage analysis in familial breast and ovarian cancer: results from 214 families. Am Hum Genet 52 : 678–701

28b. Ellis NA, Groden J, Ye TZ, Straughen J, Lennon DJ, Ciocci S, Proytcheva M, German J (1995) The Bloom's syndrome gene product is homologous to RecQ helicases. Cell 83 : 655–666

29. Emery AEH, Rimoin DL (eds) (1990) Principles and practice of medical genetics, 2nd edn, 2vols. Churchill Livingstone, Edinburgh

30. Ferguson-Smith MA, Handmaker SD (1961) Observations on the satellited human chromosomes. Lancet 1 : 638–640

31. Fialkow PJ (1967) Autoantibodies and chromosomal aberration. Lancet 1 : 1106

32. Fialkow PJ (1967) Thyroid antibodies, Down's syndrome and maternal age. Nature 214 : 1253–1254

32a. Fishel R, Lescoe MK, Rao MRS, Copeland NG, Jenkins NA, Garber J, Kane M, Kolodner R (1993) The human mutator gene homolog MSH2 and its association with hereditary nonpolyposis colon cancer. Cell 75 : 1027–1038

33. Francke U, Benirschke K, Jones OW (1975) Prenatal diagnosis of trisomy 9. Hum Genet 29 : 243–250

34. Francke U, Felsenstein J, Gartler SM, Migeon BR, Dancis J, Seegmiller JE, Bakay F, Nyhan WL (1976) The occurrence of new mutants in the X-linked recessive Lesch-Nyhan disease. Am J Hum Genet 28 : 123–137

35. Francke U, Winter RM, Lin D et al (1981) In: Hook EB, Porter IH (eds) Population and biological aspects of human mutation. Academic, New York, pp 117–130

36. Gartler SM, Francke U (1975) Half chromatid mutations: transmission in humans? Am J Hum Genet 27 : 218–223

37. Geissler E, Theile M (1983) Virus-induced gene mutations of eukaryotic cells. Hum Genet 63 : 1–12

38. Gitschier J (1988) Maternal duplication associated with gene deletion in sporadic hemophilia. Am J Hum Genet 43 : 274–279

39. Goldberg YP, Kremer SE, Andrew SE et al (1993) Molecular analysis of new mutations for Huntington's disease: intermediate alleles and sex of origin effects. Nature [Genet] 5 : 174–179

40. Goldstein JL, Brown MS (1977) The low-density lipoprotein pathway and its relation to atherosclerosis. Annu Rev Biochem 46 : 897–930

41. Grimm T, Mueller B, Mueller CR, Janka M (1990) Theoretical considerations on germ line mosaicism in Duchenne muscular dystrophy. J Med Genet 27 : 683–687

42. Gunther M, Penrose LS (1935) The genetics of epiloia. J Genet 31 : 413–430

43. Haldane JBS (1935) The rate of spontaneous mutation of a human gene. J Genet 31 : 317–326

44. Haldane JBS (1939) The spread of harmful autosomal recessive genes in human populations. Ann Eugen 9 : 232–237

45. Haldane JBS (1947) The mutation rate of the gene for hemophilia, and its segregation ratios in males and females. Ann Eugen 13 : 262–271

46. Hansteen I-L, Varslot K, Steen-Johnsen J, Langard S (1982) Cytogenetic screening of a newborn population. Clin Genet 21 : 309–314

47. Harley HG, Boork JD, Rundle SA et al (1992) Expansion of an unstable DNA region and phenotypic variation in myotonic dystrophy. Nature 355 : 545–551

48. Harnden DG (1974) Viruses, chromosomes, and tumors: the interaction between viruses and chromosomes. In: German J (ed) Chromosomes and cancer. Wiley, New York, pp 151–190

49. Harper PS (1993) Practical genetic counseling, 4th edn. Wright, Bristol

50. Hassold TJ, Jacobs PA (1984) Trisomy in man. Annu Rev Genet 18 : 69–77

51. Hassold TJ, Jacobs P, Kline J, Stein Z, Warburton D (1981) Effect of maternal age on autosomal trisomies. Ann Hum Genet 44 : 29–36

52. Herrmann J (1966) Der Einfluß des Zeugungsalters auf die Mutation zu Hämophilie A. Hum Genet 3 : 1–16

52a. Hogervorst FBL, Cornelis RS, Bout M, van Vliet M, Oosterwijk JC, Olmer R, Bakker B, Klijn JGM, Vasen HFA, Meijers-Heijboer H, Menko FH, Cornelisse CJ, den Dunnen JT, Devilee P, van Ommen G-BJ (1995) Rapid detection of BRCA1 mutations by the protein truncation test. Nature [Genet] 10 : 208.212

53. Hook EB, Regal RR (1984) A search for a paternal age-effect upon cases of 47, + 21 in which the extra chromosome is of paternal origin. Am J Hum Genet 36 : 413–421

54. Hook EB, Schreinemachers DM, Willey AM, Cross PK (1983) Rates of mutant structural chromosome rearrangements in human fetuses: data from prenatal cytogenetic studies and associations with maternal age and parental mutagen exposure. Am J Hum Genet 35 : 96–109

55. Hook EB, Cross PK, Regal RR (1984) The frequency of 47, + 21, 47, + 18 and 47, + 13 at the uppermost extremes of maternal ages: results on 56094 fetuses studied prenatally and comparisons with data on livebirths. Hum Genet 68 : 211–220

56. Hu X, Worton RG (1992) Partial gene duplication as a cause of human disease. Hum Mutat 1 : 3–12

56a. Ionov Y, Peinado MA, Malkhosyan S, Shibata D, Perucho M (1993) Ubiquitous somatic mutations in simple repeated sequences reveal a new mechanism for colonic carcinogenesis. Nature 363 : 558–561

57. Jacobs PA (1981) Mutation rates of structural chromosome rearrangements in man. Am J Hum Genet 33 : 44–45

58. Jacobs PA, Funkhauser J, Matsuura J (1981) In: Hook EB, Porter IH (eds) Population and biological aspects of human mutation. Academic, New York, pp 133–145

59. Jacobs PA, Hassold TJ, Whittington E et al (1988) Klinefelter's syndrome: an analysis of the origin of the additional sex chromosome using molecular probes. Ann Hum Genet 52 : 93–109

60. Jennings MW, Jones RW, Wood WG, Weatherall DJ (1985) Analysis of an inversion within the human b-globulin gene cluster. Nucleic Acids Res 13 : 2897–2906

61. Jones KL, Smith DW, Harvey MAS, Hall BD, Quan L (1975) Older paternal age and fresh gene mutation: data on additional disorders. J Pediatr 86 : 84–88

62. Kalousek DK, Barrett IJ, McGillivray BC (1989) Placental mosaicism and intrauterine survival of trisomies 13 and 18. Am J Hum Genet 44 : 338–343

63. Kazazian HH, Wong C, Youssoufian H et al (1988) Hemophilia A resulting from de novo insertion of L1 sequences represents a novel mechanism for mutation in man. Nature 332 : 164–166

64. Ketterling RH, Vielhaber E, Bottema CDK et al (1993) Germ-line origins of mutations in families with hemophilia B: the sex ratio varies with the type of mutation. Am J Hum Genet 52 : 152–166

65. Koeberl DD, Bottema CDK, Ketterling RP et al (1990) Mutations causing hemophilia B: direct estimate of the underlying rates of spontaneous germ-line transitions, transversions and deletions in a human gene. Am J Hum Genet 47 : 202–217

66. Koenig M, Hoffman EP, Bertelson CJ et al (1987) Complete cloning of the Duchenne muscular dystrophy (DMD) cDNA and preliminary genomic organization of the DMD gene in normal and affected individuals. Cell 50 : 509–517

67. Koenig M, Monaco AP, Kunkel LM (1988) The complete sequence of dystrophin predicts a rod-shaped cytoskeletal protein. Cell 53 : 219–228

68. Koenig M, Beggs AH, Moyer M et al (1989) The molecular basis for Duchenne versus Becker muscular dystrophy: correlation of severity with type of deletion. Am J Hum Genet 45 : 498–506

69. Kondo S (1973) Evidence that mutations are induced by error in repair and replication. Genetics 73 Suppl:109–122

70. Krüger J, Vogel F (1975) Population genetics of unequal crossing over. J Mol Evol 4 : 201–247

71. Kupke KG, Mueller U (1989) Parental origin of the extra chromosome in trisomy 18. Am J Genet 45 : 599–605

72. La Spada AR, Wilson, EM, Lubahn DB et al (1991) Androgen receptor gene mutations in X-linked spinal and bulbar muscular atrophy. Nature 352 : 77–79

73. Lawry RB, Jones DC, Renwick DHG, Trimble BK (1976) Down syndrome in British Columbia, 1972–1973: incidence and mean maternal age. Teratology 14 : 29–34

74. Lehmann H, Huntsman RG (1974) Man's hemoglobins. North-Holland, Amsterdam

75. Lenz W (1961) Zur Genetik der Incontinentia pigmenti. Ann Paediatr 196 : 141

76. Lenz W (1964) Krankheiten des Urogenitalsystems In: Becker PE (ed) Humangenetik, ein kurzes Handbuch in fünf Bänden, vol 3. Thieme, Stuttgart, pp 253–410

77. Lenz W (1975) Half chromatid mutations may explain incontinentia pigmenti in males. Am J Hum Genet 27 : 690–691

78. Luria SE, Delbrück M (1943) Mutations of bacteria from virus sensitivity to virus resistance. Genetics 28 : 491

79. MacDonald ME, Ambrose CM, Duyao MP et al, the Huntington's Disease Collaborative Research Group (1993) A novel gene containing a trinucleotide repeat that is expanded and unstable on Huntington's disease chromosomes. Cell 72 : 971–983

80. Mackenzie HJ, Penrose LS (1951) Two pedigrees of ectrodactyly. Ann Eugen 16 : 88

81. Mandel JL (1993) Questions of expansion. Nature Genet 4 : 8–9

82. McKusick VA (1972) Heritable disorders of connective tissue, 4th edn. Mosby, St Louis

83. McKusick VA (1995) Mendelian inheritance in man, 11th edn. Johns Hopkins University Press, Baltimore

83a. Miki Y, Swensen J, Shattuck-Eidens D, Futreal PA, Harshman K, Tavtigian S, Liu Q, Cochran C, Bennett LM, Ding W, Bell R, Rosenthal J, Hussey C, Tran T, McClure M, Frye C, Hattier T, Phelps R, Haugen-Strano A, Katcher H et al (1994) A strong candidate for the breast and ovarian cancer susceptability gene BRCA1. Science 266 : 66–71

84. Millar DS, Steinbrecher RA, Wieland K et al (1990) The molecular genetic analysis of hemophilia A: characterization of six partial deletions in the factor VIII gene. Hum Genet 86 : 219–227

85. Mohn G, Würgler FE (1972) Mutator genes in different species. Hum Genet 16 : 49–58

86. Møllenbach CJ (1947) Medføbte defekter i ojets indre hinder klinik og arvelighedsforhold. Munksgaard, Copenhagen

87. Mørch ET (1941) Chondrodystrophic dwarfs in Denmark. Munksgaard, Copenhagen (Opera ex domo biologiae hereditariae humanae universitatis hafniensis, vol 3)

88. Motulsky AG (1968) Some evolutionary implications of biochemical variants in man. 8th International Congress of the Anthropology and Ethnology Society, Tokyo

89. Mueller B, Dechant C, Meng G et al (1992) Estimation of the male and female mutation rates in Duchenne muscular dystrophy (DMD) Hum Genet 89 : 204–206

90. Mueller CR, Grimm T (1986) Estimation of the male to female ratio of mutation rates from segregation of X-chromosomal DNA haplotypes in Duchenne muscular dystrophy families. Hum Genet 74 : 181–183

91. Murdoch JL, Walker BA, McKusick VA (1972) Parental age effects on the occurrence of new mutations for the Marfan syndrome. Ann Hum Genet 35 : 331–336

92. Neel JV, Satoh C, Goriki K et al (1986) The rate with which spontaneous mutation alters the electrophoretic mobility of polypeptides. Proc Natl Acad Sci USA 83 : 389–393

93. Nielsen J (1966) Diabetes mellitus in parents of patients with Klinefelters' syndrome. Lancet 1 : 1376

94. Nielsen J, Sillesen I (1975) Incidence of chromosome aberration among 11, 148 newborn children. Hum Genet 30 : 1–12

95. Nielsen J, Wohlert M (1991) Chromosome abnormalities found among 34910 newborn children: result from a 13-year incidence study in Argus, Denmark. Hum Genet 87 : 81–83

95a. Niu DM, Hsiao KJ, Wang NH (et al.) (1996) Chinese achondroplasia is also defined by recurrent G380R mutations of fibroblast growth factor receptor-3 gene. Hum Genet (in the press)

96. O'Hov KL, Tsifidis C, Mahadevan MS et al (1993) Reduction in size of the myotonic dystrophy trinucleotide repeat mutation during transmission. Science 259 : 809–811

97. Oberlé I, Rousseau F, Heitz D et al (1991) Instability of a 550-base pair DNA segment and abnormal methylation in fragile X syndrome. Science 252 : 1097–1102

98. Orioli IM, Castilla EE, Scarano G, Mastroiacovo P (1995) The effect of paternal age in achondroplasia, thanatophoric dysplasia and osteonenesis imperfecta. Am J Med Genet 59 : 209–217

99. Oudet C, Mornet E, Serre JL et al (1992) Linkage disequilibrium between the fragile X mutation and two closely linked CA repeats suggests that fragile X chromosomes are derived from a small number of founder chromosomes. Am J Hum Genet 52 : 297–304

100. Penrose LS (1933) The relative effects of paternal and maternal age in mongolism. J Genet 27 : 219–224

101. Penrose LS (1955) Parental age and mutation. Lancet 2 : 312

102. Penrose LS (1957) Parental age in achondroplasia and mongolism. Am J Hum Genet 9 : 167–169

103. Pfeiffer RA (1964) Dominant erbliche Akrocephalosyndaktylie. Z Kinderheilkd 90 : 301

104. Race RR, Sanger R (1969) Xg and sex chromosome abnormalities. Br Med Bull 25 : 99–103

105. Reed TE (1959) The definition of relative fitness of individuals with specific genetic traits. Am J Hum Genet 11 : 137

106. Reed TE, Falls HF (1955) A pedigree of aniridia with a discussion of germinal mosaicism in man. Am J Hum Genet 7 : 28–38

107. Reed TE, Neel JV (1955) A genetic study of multiple polyposis of the colon (with an appendix deriving a method for estimating relative fitness). Am J Hum Genet 7 : 236–263

108. Reichert W, Buselmaier W, Vogel F (1984) Elimination of X-ray-induced chromosomal aberration in the progeny of female mice. Mutat Res 139 : 87–94

109. Riccardi VM, Dodson II CE, Chatrabarty R, Bontke C (1984) The pathophysiology of neurofibromatosis. IX. Paternal age as a factor in the origin of new mutations. Am J Med Genet 18 : 169–176

110. Richards RI, Sutherland GR (1992) Dynamic mutations: a new class of mutations causing human disease. Cell 70 : 709–712

111. Richards RI, Holman K, Friend K et al (1992) Evidence of founder chromosomes in fragile X syndrome. Nature [Genet] 1 : 257–260

112. Rischbieth H, Barrington A (1912) Dwarfism. University of London, Dulau London, pp 355–573 (Treasury of human inheritance, parts 7 and 8, sect 15A)

113. Rosendaal FR, Broecker-Friends AHJT, van Houwelingen JC et al (1990) Sex ratio of the mutation frequencies in haemophilia A: estimation and meta-analysis. Hum Genet 86 : 139–150

114. Russel WL, Kelly EM, Hunsicker PR et al (1972) Effect of radiation dose-rate on the induction of X-chromosome loss in female mice. In: Report of the United Nations Science Committee on the effect of atomic radiations. United Nations, New York (Ionizing radiation: levels and effects, vol 2: Effects)

114a. Savitsky K, Bar-Shira A, Gilad S, Rotman G, Ziv Y, Vanagaite L, Tagle DA, Smith S, Uziel T, Sfez S, Ashkenazi M, Packer I, Frydman M, Hanik R, Patanjali SR, Simmons A, Clines GA, Sartiel A, Gatti RA, Chessa L, Sanal O, Lavin MF, Jaspers NGJ, Taylor AMR, Arlett CF, Miki T, Weissman S, Lovett M, Collins FS, Shiloh Y (1995) A single ataxia telangiectasia gene with a product similar to PI-3 kinase. Science 268 : 1749–1753

115. Schull WJ, Neel JV, Hashizume A (1968) Some further observations on the sex ratio among infants born to survivors of the atomic bombings of Hiroshima and Nagasaki. Am J Hum Genet 18 : 328–338

116. Searle AG (1972) Spontaneous frequencies of point mutations in mice. Hum Genet 16 : 33–38

117. Sergeyev AS (1975) On mutation rate of neurofibromatosis. Hum Genet 28 : 129–138

118. Sherman SL, Morton NE, Jacobs PA, Turner G (1984) The marker (X) syndrome: a cytogenetic and genetic analysis. Ann Hum Genet 48 : 21–37

118a. Shiang R, Thompson LM, Zhu YZ (et al.) (1994) Mutation in the trans-membrane domain of FGFR-3 causes the most common genetic form of dwarfism, achondroplasia. Cell 78 : 335–342

119. Stamatoyannopoulos G, Nute PE, Miller M (1981) De novo mutations producing instable hemoglobins or hemoglobin MI. Establishment of a depository and use of data for an association of de novo mutation with advanced parental age. Hum Genet 58 : 396–404

120. Stevenson AC, Kerr CB (1967) On the distribution of frequencies of mutation in genes determining harmful traits in man. Mutat Res 4 : 339–352

120a. Struewing JP, Abeliovich D, Peretz T, Avishai N, Kaback MM, Collins FS, Brody LC (1995) The carrier frequency of the BCRA1 185delAG mutation is approximately 1 percent in Ashkenazi Jewish individuals. Nature [Genet] 11 : 198–200

121. Takaesu N, Jacobs PA, Lockwell A et al (1990) Nondisjunction of chromosome 21. Am J Med Genet Suppl 7 : 175–181

122. Taylor AM (1963) Bacteriophage-induced mutation in Escherichia coli. Proc Natl Acad Sci USA 50 : 1043–1051

123. Tommerup N (1993) Mendelian cytogenetics. Chromosome rearrangements associated with mendelian disorders. J Med Genet 30 : 713–727

124. Tommerup N, Schempp W, Meinecke P et al (1993) Assignment of an autosomal sex reversal locus (SRA 1) and campomelic dysplasia (CMPD 1) to 17q24.3-q25.1. Nature [Genet] 4 : 170–174

125. Tönz O, Glatthaar BE, Winterhalter KH, Ritter H (1973) New mutation in a Swiss girl leading to clinical and biochemical b-thalassemia minor. Hum Genet 20 : 321–327

126. Tünte W, Becker PE, von Knorre G (1967) Zur Genetik der Myositis ossificans progressiva. Hum Genet 4 : 320–351

127. Van Dyke DL, Weiss L, Roberson JR, Babu VR (1983) The frequency and mutation rate of balanced autosomal rearrangements in man estimated from prenatal genetic studies for advanced maternal age. Am J Hum Genet 35 : 301–308

128. Vijayalakshmi, Evans HJ, Ray JH, German J (1983) Bloom's syndrome: evidence for an increased mutation frequency in vivo. Science 221 : 851–853

129. Vijayalakshmi, Wunder E, Schroeder TM (1985) Spontaneous 6-thioguanine-resistant lymphocytes in Fanconi anemia patients and their heterozygous parents. Hum Genet 70 : 264–270

130. Vogel F (1954) Über Genetik und Mutationsrate des Retinoblastoms (Glioma retinae). Z Menschl Vererbungs Konstitutionslehre 32 : 308–336

131. Vogel F (1956) Über die Prüfung von Modellvorstellungen zur spontanen Mutabilität an menschlichem Material. Z Menschl Vererbungs Konstitutionslehre 33 : 470–491

132. Vogel F (1957) Neue Untersuchungen zur Genetik des Retinoblastoms (Glioma retinae). Z Menschl Vererbungs Konstitutionslehre 34 : 205–236

133. Vogel F (1965) Sind die Mutationsraten für die X-chromosomal rezessiven Hämophilieformen in Keimzellen von Frauen niedriger als in Keimzellen von Männern? Hum Genet 1 : 253–263

134. Vogel F (1977) A probable sex difference in some mutation rates. Am J Hum Genet 29 : 312–319

135. Vogel F (1979) Genetics of retinoblastoma. Hum Genet 52 : 1–54

136. Vogel F (1984) Mutation and selection in the marker (X) syndrome. Ann Hum Genet 48 : 327–332

137. Vogel F, Kopun M (1977) Higher frequencies of transitions among point mutations. J Mol Evol 9 : 159–180

137a. Vogel F, Krüger J, Brøndum Nielsen K (et al.) (1985) Recurrent mutation pressure does not explain the prevalence of the marker X syndrome. Hum Genet 71 : 1–6

138. Vogel F, Rathenberg R (1975) Spontaneous mutation in man. Adv Hum Genet 5 : 223–318

139. Vogel F, Röhrborn G (1965) Mutationsvorgänge bei der Entstehung von Hämoglobinvarianten. Humangenetik 1 : 635–650

140. Vogel F, Röhrborn G (eds) (1970) Chemical mutagenesis in mammals and man. Springer, Berlin Heidelberg New York

141. Vogel F, Crusio WE, Kovac C et al (1990) Selective advantage of fra (X) heterocygotes. Hum Genet 86 : 25–32

142. Voit T, Neuen-Jacob E, Mahler et al (1992) Somatic mosaicism for a deletion of the dystrophin gene in a carrier of Becker muscular dystrophy. Eur J Pediatr 151 : 112–116

143. Wais S, Salvati E (1966) Klinefelter's syndrome and diabetes mellitus. Lancet 2 : 747–748

144. Warren ST, Schulz RA, Chang CC, Wade MH, Troske JE (1981) Elevated spontaneous mutation rate in Bloom syndrome fibroblasts. Proc Natl Acad Sci USA 78 : 3133–3137

145. Watson JD, Crick FHC (1953) The structure of DNA. Cold Spring Harbor Symp Quant Biol 18 : 123–132

146. Weech AA (1927) Combined acrocephaly and syndactylism occurring in mother and daughter. A case report. Johns Hopkins Med J 40 : 73

147. Weinberg W (1912) Zur Vererbung des Zwergwuches. Arch Rassen Gesellschaftsbiol 9 : 710–718

148. Wilson AC, Carlson SS, White TJ (1977) Biochemical evolution. Annu Rev Biochem 47 : 573–639

149. Winter RM, Tuddenham EGD, Goldman E, Matthews KB (1983) A maximum likelihood estimate of the sex ratio of mutation rates in hemophilia A. Hum Genet 64 : 156–159

149a. Wooster R, Neuhausen SL, Mangion J, Quirk Y, Ford D, Collins N, Nguyen K, Seal S, Tran T, Averill D, Fields P, Marshall G, Narod S, Lenoir GM, Lynch H, Feunteun J, Devilee P, Cornelisse CJ, Menko FH, Daly PA, Ormiston W, McManus R, Pye C, Lewis CM, Cannon-Albright LA, Peto J, Ponder BAJ, Skolnick MH, Easton DF, Goldgar DE, Stratton MR (1994) Localization of a breast cancer susceptibility gene, BRCA2, to chromosome 13 q12–13. Science 265 : 2088–2090

150. Yanofsky C, Ito J, Horn V (1966) Amino acid replacements and the genetic code. Cold Spring Harbor Symp Quant Biol 31 : 151–162

151. Yu S, Mulley J, Loesch D et al (1992) Fragile-X syndrome: unique genetics of the heritable unstable element. Am J Hum Genet 50 : 968–980

152. Zellweger H, Abbo G, Cuany R (1966) Satellite association and translocation mongolism. J Med Genet 3 : 186–189

10 Mutation: Somatic Mutation, Cancer, and Aging

"The essential hypothesis as already formulated by von Hansenmann" (1890) is:
"The cell of the malignant tumor is a cell with a certain abnormal chromatin content."
"The way in which it originates [has] no significance. Each process which brings
about this chromatin constitution would result in the origin of a malignant tumor."

(T. Boveri 1914, translated by U. Wolf 1974 [19])

Mutations may also occur in somatic cells. The effect of a somatic mutation is found in the descendants of the mutant cell, making the individual a mosaic. Mosaics are individuals with a mixed cell population. In the simplest situation a normal cell population and a mutant cell population may coexist in a single individual. Often such different populations exist side by side. In fibroblasts – and possibly in other cell types – there is extensive mixing of genetically different cell groups since very small cell patches may be shown to exhibit different cell markers, such as different G6PD types.

10.1 Formation of Mosaics for Genome Mutations

Mosaics for genome mutations are frequent. In Down syndrome, for example, a ratio of 1 mosaic to 48 standard trisomy patients has been reported. An estimated population incidence of 1:650 for Down syndrome would thus result in a mosaic frequency of 1:31000. Such mosaic cases also show dependence on maternal age, but to a lesser degree than in simple trisomy 21 [59].

Mechanism of Mosaic Formation in Early Cleavage [59]. Analysis of maternal age effects makes it possible to draw some conclusions regarding the origin of mosaics in Down syndrome. A mosaic may be derived from a normal zygote. In such cases nondisjunction would have to occur in an early (but not the first) cleavage division. (Nondisjunction in the first cleavage division would result in a trisomic and a monosomic division product. With loss of the monosomic cell, this could lead to standard trisomy.) The monosomic product of the division is usually lost. Mosaicism can also be derived from a trisomic zygote. One cell strain would lose the extra chromosome by anaphase lagging, or nondisjunction may occur in a somatic cell (secondary nondisjunction; Fig. 10.1). The proportion of mosaics caused by each of these mechanisms can be estimated. With origin in a normal zygote no maternal age increase would be expected. With origin in a trisomic zygote, the increase in maternal age should be similar to that found for Down syndrome in general. The total of mosaics represents a mixture of the two mechanisms; the average maternal age depends on the proportion attributed to each cause. Among 40 mosaics described in the literature 20% were estimated as being derived from normal zygotes. From this calculation one can derive a comparative estimate for the frequency of certain mitotic disturbances in normal and trisomic zygotes. It follows that trisomic zygotes show an almost 40 times greater tendency toward anaphase lagging than normal cells, and nondisjunction is 70 times as frequent. These estimates, however, are applicable only to mosaics that evolve to clinically recognizable Down syndrome. The probability of developing Down syndrome is much higher among zygotes who were originally trisomic than among those who were originally normal.

Mosaics with a small fraction of trisomic cells might occur at a later stage of development. They are often phenotypically normal or show only micromanifestations of Down syndrome, such as abnormal dermatoglyphics. They may have children with Down syndrome if a segment of the ovary or the testicle has the abnormal karyotype. The proportion of such minor mosaics among parents of children with Down syndrome may be significant. The 1% recurrence risk of trisomic Down syndrome could conceivably be related to gonadal trisomy 21 of this type.

10.2 Hereditary Syndromes with Increased Chromosome Instability
[11, 20, 21, 71, 75]

Fanconi Anemia (227650, 227660). Fanconi anemia is a childhood panmyelopathy with bone marrow failure leading to a pancytopenia. Skeletal anomalies, especially of the thumb and radius, and hyperpigmentation are usually found; other malformations

CONSEQUENCES OF MITOTIC NONDISJUNCTION:

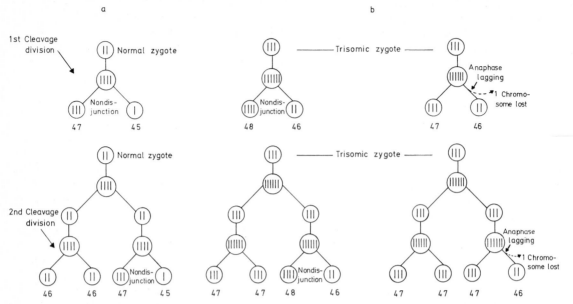

Fig. 10.1a, b. Secondary nondisjunction and anaphase lagging as mechanisms for production of mosaics. **a** Normal zygote, secondary nondisjunction; **b** Trisomic zygote, secondary nondisjunction, anaphase lagging

are frequent. The mode of inheritance is autosomal-recessive. Genetic heterogeneity was suggested by analysis of age at onset [74] and confirmed by mutual correction of chromosome instability after fusion of cells from patients with different clinical types [98]. There is a more common type with onset in the first years of life, and a rarer one (see the description below) with juvenile onset (see also [55]). One gene has been localized at 20q [47]; there is linkage heterogeneity.

Schroeder et al. (1964) [72] described two brothers with this disease, 21 and 18 years old. The parents and a younger brother (7 years) were healthy. The older brother showed, apart from normal karyotypes, metaphases with multiple chromosome aberrations, such as achromatic lesions (gaps), chromatid breaks; isochromatid breaks, acentric fragments, dicentric chromosomes, and chromatid interchanges; 19 of 39 metaphases showed at least one but in some cases multiple anomalies. Endoreduplication was seen in about 10% of all metaphases. The younger, clinically unaffected brother showed a somewhat lower number of mitoses with chromosome aberrations but the same range of anomalies. Six years later he developed clinical symptoms of the disease. He died of multiple hemorrhage at the age of 32. Autopsy revealed clinically unrecognized lung cancer [71].

These were the first published cases of chromosome instability in a hereditary disease. The result was soon confirmed in other cases (Fig. 10.2).

Bloom Syndrome (210900). Bloom syndrome is a condition characterized by a low birth weight, stunt-

ed growth, sun sensitivity of the skin, and a facial butterfly-type lesion with telangiectasia. The mode of inheritance is autosomal-recessive. Most families are of Ashkenazi Jewish origin. German et al. [22], examining metaphases from blood cultures of seven patients, observed in six patients high frequencies (4%–27%) of cells with broken and sometimes rearranged chromosomes. Other cytogenetic anomalies described in Fanconi anemia were present in Bloom syndrome as well. However, the hallmarks of Bloom syndrome are symmetric quadriradial chromatid interchanges that are not seen in Fanconi anemia. These presumably arise from chromatid exchanges between homologous chromosomes. In contrast, in Fanconi anemia asymmetric quadriradials are common, which are caused by random breaks of nonhomologous chromosomes. The frequency of sister chromatid exchanges (Sect. 2.1.2) in Bloom syndrome is ten times higher than in normals or in patients with Fanconi anemia. Although superficially somewhat similar, the basic mechanisms leading to Bloom syndrome and Fanconi anemia are quite different.

The fundamental defect in Bloom's syndrome is due to chain terminating mutations affecting a DNA helicase [28b], an enzyme required to maintain genomic stability in somatic cells. Not unexpectedly, all mutations in Ashkenazi patients were identical, both by direct mutational and by haplotype analysis, indicating a common origin with subsequent expansion of this population.

Ataxia-Telangiectasia (208900) [25]. The two constant clinical features of the ataxia-telangiectasia (Louis-Bar) syndrome (AT) are progressive cerebellar ataxia radiation sensitivity, cell cycle abnormalities, and oculocutaneous telangiectasia. Ataxia is usually recognized at the age of 12–14 months; the patient is confined to a wheelchair before adolescence. Various immune deficiencies have been reported that vary from patient to patient. The most common defect is a low level or complete absence of IgA. The mode of inheritance is autosomal-recessive. Chromosome instability has been reported repeatedly; the number of breaks seems to be lower than in Fanconi anemia and Bloom syndrome [5, 24, 27]. Breaks are apparently random. The level of chromosomal breaks often fluctuates. Pseudodiploid clones are common, and a translocation involving the long arm of chromosome 14 is characteristic. A gene has been located to 11q22–11q23. All mutations in A-T affect a phosphotransferase (phosphatidyl inositol-3' kinase) involved in mitogenic signal transduction, meiotic recombination and cell cycle control [114a].

In these three conditions – Fanconi anemia, Bloom syndrome, and AT – it is reasonable to assume that the clinical symptoms are directly related to chromosome instability. Moreover, chromosomes from patients with all three diseases show an increased sensitivity to various chromosome breaking (= clastogenic) agents – AT, for example, to X-rays.

Chromosome Instability and Cancer. Patients with all three conditions have a strongly increased risk of developing malignant neoplasias. Many patients with Fanconi anemia succumb during childhood and youth from bleeding or infections, but an increasing number of neoplasias are being reported [21]. By 1981, 45 cases had been collected; these included 22 acute leukemias, none of them lymphatic; 16 primary tumors of the liver; the rest carcinomas of other organs. A variety of malignant tumors have been found in ataxia-telangiectasia [21]. Among 108 patients there were 48 with various non-Hodgkin lymphomas; 12 with Hodgkin disease; 26 leukemias, mostly lymphatic; and 22 with other conditions (cancers of stomach, brain, ovary, skin, etc.). Lymphatic neoplasias thus prevail. Of the 99 individuals known to suffer from Bloom syndrome up to 1981, 23 developed at least one neoplasm. Considering the young age of these patients, a 100-fold increase in the risk for neoplasia has been estimated. In distinction to AT a great diversity of type and tissue distribution is observed.

It is reasonable to assume that the increased risk of developing neoplasias in these syndromes is directly related to the increased rate of spontaneous chromosome breakage.

Such chromosome instability leads to many cells with various aneuploidies due to chromosome breakage. Most of these cells die immediately, but some survive a few divisions. In an occasional cell, however, the structural defect provides a selective advantage in that the cell division rate is no longer inhibited.

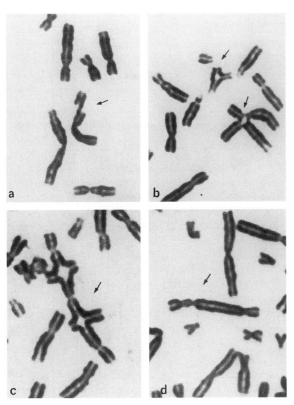

Fig.10.2a–d. Chromosomes from a patient with Fanconi anemia. a Chromatin break. b Two chromatin interchange figures with participation of nonhomologous chromosomes. c Hexagonal interchange figure in which three chromosomes participate. d Tricentric chromosomes. (Courtesy of Dr. T.M.Schroeder-Kurth)

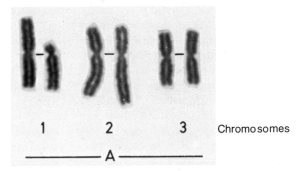

Fig.10.3. Marker chromosome 1p- found in a cell clone from a patient with Fanconi anemia. (Courtesy of Dr. T.M.Schroeder-Kurth)

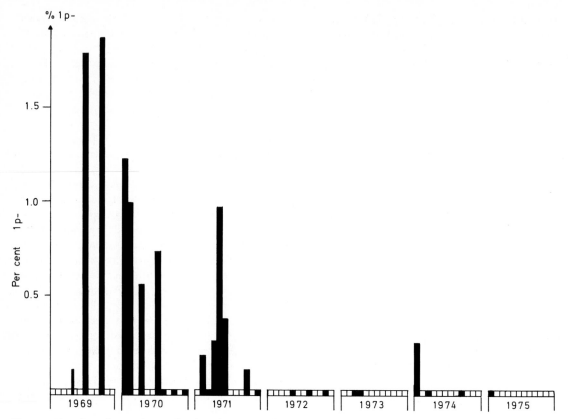

Fig. 10.4. Fraction of metaphases showing the 1p- marker chromosome of Fig. 10.3 in Fanconi anemia. Repeated examinations over many years. (Courtesy of Dr. T. M. Schroeder-Kurth)

Such a cell soon forms a clone of genetically identical cells: the initial cancer cells. Due to their uninhibited growth, the abnormal cell clone gradually replaces the normal cells.

If such a cell clone contains a structurally abnormal chromosome, we should occasionally find specific chromosome aberrations in a certain fraction of cells from patients with one of the three syndromes with chromosome instability. Such cell clones have indeed been observed. Figure 10.3 shows a "marker chromosome," a 1p- that characterizes a clone from a patient with Fanconi anemia that has been observed since 1974 [73]. Figure 10.4 compares the proportion of metaphases with this marker chromosome over several years. This clone probably had a certain selective advantage; however, this is decreasing. Similar clones have been observed in the other two conditions as well. Full development of leukemia by gradual increase of a defined cell clone has been observed in AT [27]. Possible molecular mechanisms of malignant transformation related to chromosome breaks are discussed in Sect. 10.4.2.

10.3 Molecular Mechanisms of Chromosomal Instability and Tumor Formation Due to Somatic Mutation

10.3.1 Xeroderma Pigmentosum (278700–278750) [41]

Chromosome instability and the existence of marker chromosomes in the three syndromes with inherited chromosome instability suggest that repeated chromosome breakage may lead to cell clones that develop into malignancies. This raises the question of the molecular mechanism of chromosome instability. Another hereditary disease, xeroderma pigmentosum (XP) has provided much information.

After exposure to ultraviolet light the skin of patients with XP shows erythrema that is followed by atrophy and telangiectasia (Sect. 4.1.3). Gradually these areas become wartlike, and finally skin cancer develops. It is known from work in micro-organisms that cells have an enzyme system capable of repairing defects of the DNA. Enzymatic repair of defects induced by ultraviolet light has been well analyzed in micro-organisms at the molecular level. XP is characterized by an abnormally high sensitivity to ultraviolet light. Cleaver and Bootsma [10] showed that this disease is caused by a defect in a DNA repair enzyme. A number of different complementation groups (see below) that lead to XP-like phenotypes were later identified (Table 10.1).

Mechanisms of DNA Repair. Three main mechanisms of DNA repair have been analyzed in micro-organisms: photoreactivation, excision repair, and postreplication repair (Fig. 10.5) [89].

1. Photoreactivation. Blue-violet light enhances the chance of survival of bacteria that receive ultraviolet light exposure. The main effect of ultraviolet irradiation is formation of thymine dimers from two neighboring thymines. Photoreactivation is due to an enzyme that restitutes these thymines by cleaving the dimers.
2. Excision repair. The second mechanism, excision repair, does not require light. In a first step, an endonuclease recognizes the dimers and cuts the affected DNA strand close to them, producing free ends. These free ends are recognized by an exonuclease that cuts off nucleotides, beginning at these

free endings. Apart from the UV-induced dimers, up to 100 other nucleotides are removed. A polymerase induces resynthesis of the removed strand using the intact sister strand as template. Finally, the last gap between the resynthesized and the old strand is closed by a ligase.

3. Postreplication repair. If photoreactivation and excision repair are impossible, the damaged DNA strand cannot act as a template during replication since the dimer does not pair with any other bases; a gap remains in the newly synthesized complementary DNA strand. However, the genetic information distorted by dimer formation is available in a newly synthesized DNA strand along the old complementary strand. This information acts as a template for the production of another intact copy that replaces the damaged DNA strand. The exact mechanism of this replacement is still unknown; it seems to be similar to normal recombinational events. One normal DNA strand as a template is necessary for excision repair as well as for postreplication repair. (For numerous details, especially on repair enzymes and the SOS reaction, see [41].)

Enzyme Defects in Xeroderma Pigmentosum-Like Diseases. Cultured fibroblasts of patients with XP were

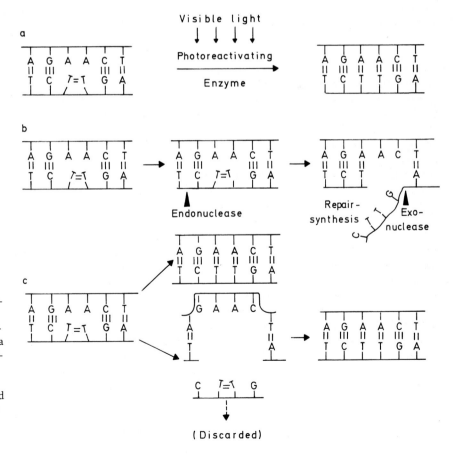

Fig. 10.5 a–c. Three types of DNA repair. **a** Photoreactivation, thymine dimers are reopened and A = T hydrogen bonds are reestablished. **b** Excision repair, half-chromatid sequences containing the thymine dimer are excised, and a new half-chromatid is synthesized. **c** Postreplication (recombination) repair. A half-chromatid sequence is excised and repair occurs after replication, with assistance from the other division product

shown to have a reduced survival time after ultraviolet irradiation. Moreover, the survival of various UV-irradiated viruses grown in XP cells is less than that in normal cells. This shows that the genetic defect of the host cell prevents it from correcting the defect in the virus genome. These results suggested a defect in one of the above repair mechanisms; this was confirmed by direct investigation of these mechanisms in XP cells. Cells from different patients were found to be defective in their ability to perform excision repair. An early stage of excision repair, the ability to excise dimers, is deficient, and this leads to a reduction of the insertion of new bases – repair replication (unscheduled DNA synthesis).

Genetic Heterogeneity [10, 16]. The enzyme system responsible for excision repair comprises a number of enzymes, and the clinical differences between the various XP patients suggest genetic heterogeneity – either by mutations within the genes for different polypeptide chains or at different sites within the same genes. One method for examining this problem is cell hybridization (Sect. 5.1.3) between fibroblasts from different patients. The daughter cell of two fused cells is able to carry out excision repair if their enzyme defects affect different loci. In this case one genome provides one intact enzyme, the other genome produces the other enzyme, and the two defects are mutually compensated. If the enzyme defects are identical, although different mutational sites are affected within the same gene, no such compensation is possible. In this way at least eight complementation groups have been identified (Table 10.1). There are also clinical differences between the complementation groups: only patients of groups A, B, and D have additional neurological findings such as microcephaly, progressive mental deficiency, retarded growth and sexual development, deafness, ataxia, choreoathetosis, and areflexia (de Sanctis and Cacchione 1932). Many XP patients with neurological

manifestations do not show the full spectrum of symptoms. Even within the same complementation group, heterogeneity in the degree of neurological manifestation may be remarkable. In a number of patients diagnosed clinically as XP excision, repair has been found to be entirely normal. They are now classified as XP variant. In these patients postreplication repair was shown to be deficient. Defects in photoreactivation have not yet been observed. Moreover, Table 10.1 shows a very unequal population distribution of these variants; for example, type A and the variant type are especially common in Japan, whereas type C, which is common in populations of European extraction, is rare in Japan.

Malignant Neoplasias in Patients with Xeroderma Pigmentosum. Patients with XP sooner or later develop multiple malignant skin tumors. All cell types that are exposed to UV light may be involved. Basal and squamous cell carcinomas, malignant melanomas, keratoacanthomas, hemangiomas, and sarcomas may develop. Cancer formation can be prevented by minimizing exposure to UV irradiation with sun screens and ointments and by avoidance of sunlight.

Increased Cancer Risk in Heterozygotes [84]. The mode of inheritance of all three chromosome instability syndromes and of XP is autosomal-recessive. Wherever enzyme defects have been identified, enzyme activity in heterozygotes is usually about half that found in normal homozygotes (Sect. 7.2.2.8). Therefore it was reasonable to look for a possible increase in cancer risk among heterozygotes. The best evidence is now available for AT [76, 78]. The study was based on 27 families with 1639 close relatives of AT patients. The data on cancer incidence in this group were compared with expectations calculated from the United States cause-, age-, time-, sex-, and race-specific mortality rates (Table 10.2). There was a definite increase in cancer mortality in the youngest age group of relatives (0–44 years) and in both sexes, in females more than males. In addition, a higher incidence of malignant neoplasms was also found in living relatives. There were many types of malignancies – especially, as in homozygotes, neo-

Table 10.1. Distribution of complementation groups in studies on xeroderma pigmentosum (from Fischer et al. 1982 [19])

Country	Frequency of cases in the respective complementation groups								Number of patients
	A	B	C	D	E	F	G	Variant	
North America[a, b, c]	3	1	5	5	0	0	0	2	16
Europe[a, b, c]	10	0	14	8	2	0	2	5	41
Japan[d]	21	0	1	1	0	3	0	14	40
Egypt[b]	7	0	12	0	0	0	0	5	24
Germany	2	0	7	5	0	0	0	9	23
Number of analyzed cases	43	1	39	19	2	3	2	35	144

[a] Cleaver and Bootsma 1975; [b] Cleaver et al. 1981; [c] Kraemer 1980; [d] Takebe 1979.

plasias of the lymphatic system, but also carcinomas of stomach and ovary.

The incidence of homozygotes for AT in the population was estimated at about 1:40 000, corresponding to a heterozygote frequency of about 1%. In this case it was estimated that "A-T heterozygotes might comprise over 5% of all persons dying from a cancer before age 45 and about 2% of those dying from this cause between ages 45 and 75." In addition to the cancer risk, AT heterozygotes may also have a somewhat increased susceptibility to diabetes, severe scoliosis, and neural tube defects [76]. More recently these investigations were complemented with additional studies by the same group [79, 80]. Another common cancer, that of the female breast, especially with early onset – appears to be much more common. The discovery of the basic defect in A-T [114a] should allow direct heterozygote testing among cancer patients.

Another study on cancer disposition was performed in close relatives of XP patients. There was no *general* increase in cancer mortality. However, an increased incidence of (nonfatal) nonmelanoma skin tumors was found in these heterozygotes [77]. This study was based on 2597 close relatives of XP patients from 31 families in the United States. Interestingly, an increase in skin cancers was found only among those living in the southern United States, where exposure to sunlight is extensive. This, together with the negative result for all other cancers except skin carcinomas, suggests a specific defect in UV repair in epithelial cells manifesting only when the skin is exposed extensively to sunlight, that is to say, an ecogenetic phenomenon (Sect. 7.5.2).

In Fanconi anemia, on the other hand, careful assessment of the available evidence failed to reveal any increased cancer risk of heterozygotes.

10.3.2 Molecular Mechanisms in Syndromes with Enhanced Chromosome Instability

Formation of thymine dimers affects only one of the two DNA sister strands. It therefore does not lead immediately to a chromosome gap or break. If the dimer cannot be excised, however, it cannot act as template in the next replication; the complementary DNA strand will be incomplete, and a break will become visible in the second replication cycle (Fig. 10.6). Hence, if interruption of the DNA double helix is related to microscopically visible chromosome breakage, we would expect a greater number of chromosome breaks after irradiation of XP cells than in normal cells. This increase has indeed been reported. In unirradiated XP cells, on the other hand, no chromosome instability has been observed in contrast to the three syndromes described before (Fanconi anemia, Bloom syndrome, and AT), associated with spontaneously enhanced chromosome instability. Therefore the molecular defects are probably different. However, it is reasonable to assume that some of the mechanisms of DNA replication and repair are involved in these conditions as well. Some observations seem to corroborate this conclusion [42, 64, 65].

Table 10.2. Observed and expected deaths from malignant neoplasms in heterozygous parents and in sibs (from Swift 1982 [76])

Age groups	Ataxia-telangiectasia		Xeroderma pigmentosum	
	Observed	Expected	Observed	Expected
Males				
0–44	6	2.14		
45–74	21	19.02	38	35,4
75 +	4	5.22		
All ages	31	26.38		
Females				
0–44	9	2.97		
45–74	23	18.34	30	33.2
75+	4	5.00		
All ages	36	26.31		
Both sexes and all ages	67	52.7	68	68.6

BREAK FORMATION AFTER DIMER FORMATION

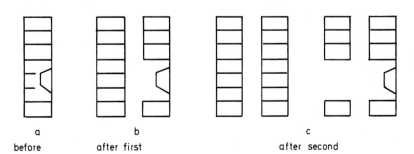

a
before
replication

b
after first
replication

c
after second
replication

Fig. 10.6a–c. Defect of excision repair. If thymine dimers formed by UV irradiation cannot be excised, they cannot act as templates in the next replication cycle. This leads to a chromosome break. **a** Dimer formation. **b** At first replication no complementary bases are assembled opposite to the thymine dimer. **c** Second replication: discontinuity of DNA structure in one division product is the result

The discovery of the different fundamental defects in A-T [114a] and in Bloom syndrome [28b] has demonstrated that these conditions are caused by abnormalities affecting enzymes involved in basic cellular processes, such as a phosphotransferase essential for mitosis as well as cell cycle control (A-T) and proper DNA unwinding (helicase defect) in Bloom syndrome. These findings provide key data for understanding the molecular mechanisms of spontaneous mutations in general and of somatic tumorigenic mutations in particular. For understanding the molecular mechanisms of spontaneous mutations in general and of somatic mutations in their relationship to neoplasia in particular, these findings provide key data.

Chain of Events in the Formation of Malignant Neoplasias by Somatic Mutation [76]. The chain of events that lead to the formation of neoplasias by somatic mutation is depicted in Fig. 10.7. The first step is damage to DNA. This may be caused either by internal factors, such as genetically defective replication and repair mechanisms, or by outside influences, such as ionizing radiation, chemical mutagens, and viruses. The DNA damage may lead to a complete breakdown of the replication mechanism and can therefore be lethal. On the other hand, it might – and, in very many cases, will – be repaired. A second possibility is that a mutation is formed. Here it is of no principal importance which type of mutation occurs. For example, it may be a point mutation caused by a single base exchange, or a visible chromosome anomaly. Most often this mutation may be lethal, leading to death of the affected cell clone due to a selective growth disadvantage in competition with normal cells. Sometimes, there may be normal cell growth, and a marker chromosome may be the only indicator of the mutation. A third rare possibility is that the new cell clone has a selective advantage due to a genetic defect in the normal mechanisms of growth inhibition and regulation and cancer develops. Secondary genetic processes such as the formation of additional aneuploidies may contribute to cell death or may occasionally lead to cell clones with stronger selective advantage. Neoplastic growth therefore occurs unchecked until

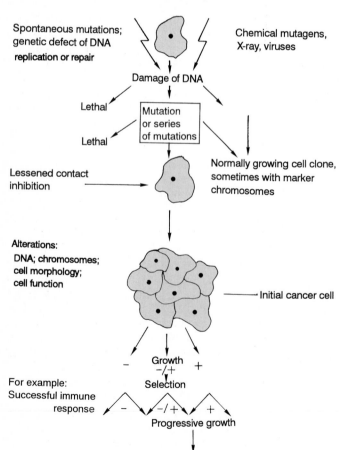

Fig. 10.7. Gradual development of a malignant tumor. A somatic mutation, such a chromosome break, may produce a cell clone with a selective advantage. This clone may develop gradually into a malignant tumor. (Adapted from Schroeder 1972 [70])

the affected individual dies from interference with normal functioning. This general formulation of chromosomal involvement in neoplastic origin was developed in the early twentieth century by Boveri (1914; see [19]) but had to await the development of better cytogenetic and molecular techniques.

10.4 Other Observations Suggesting Somatic Mutation as a Mechanism in Carcinogenesis [37]

History of the Somatic Mutation Hypothesis of Cancer. The hereditary syndromes with enhanced chromosome instability and with deficiencies of DNA replication and repair are remarkable because they offer models for molecular mechanisms of somatic mutation and tumor formation. However, the hypothesis that cancer is due to somatic mutation is much older. Von Hansemann (1890) [87], on the basis of his studies on mitosis, postulated that the cell of the malignant tumor is a cell with an abnormal chromatin content [19]. Boveri (1914) specified this concept further, assuming unequal distribution of the chromosomes of a cell to its daughter cells. He emphasized, however, that the abnormal chromatin constitution as such, and not the mechanism of its production, is important. During the following decades the somatic mutation hypothesis was elaborated by many authors and discussed in many aspects. Burnet (1974) [6] most clearly formulated its most important consequences:

a) Neoplasias must be monoclonal, i.e., they must originate from one single cell.
b) Their incidence can be enhanced by chemical agents or viruses that react with DNA.
c) In a large population of proliferating cancer cells, additional mutations are to be expected in single cells. These produce cells having additional selective advantages; subclones derived from them rapidly overgrow the other tumor cells.
d) The mutation hypothesis also explains the increased incidence of most cancers with age, if somatic mutation can – as a first approximation – be regarded as a time-dependent process. In addition, clones need a number of years to grow before they manifest clinically.

Induction of cancer by chemicals that react with DNA is not discussed further here. Suffice it to mention that many chemical mutagens (Sect. 11.2) have indeed been shown to be carcinogenic as well. Malignant neoplasias usually show a variety of subclones with different karyotypes, indicating multiple anomalies of mitosis during tumor proliferation. Increase in neoplasia with age is a well-known general feature of cancer biology. Evidence from many different sources is now available confirming the monoclonal origin of tumors. It is known, for example, that one B lymphocyte produces only one specific type of γ-globulin light chain – a λ or $\varkappa$ chain. Different B lymphocytes, however, synthesize light chains that differ in the "variable" part of the amino acid sequence (Sect. 7.4). All cells in myelomatosis – a malignant disease – on the other hand, produce light chains with identical variable parts. Uterine muscle cells of women who are heterozygous for a normal and a variant X-linked G6PD allele are mosaics, expressing either the normal or the variant G6PD allele indifferent cells, as expected from random inactivation of X chromosomes. Fibroid tumors of the uterus, on the other hand, always show a single G6PD type in all tumor cells. Similar observations are available for many other tumors [35]: Most tumors are in fact monoclonal in origin. Some hereditary tumors, such as those that appear in neurofibromatosis, have a multiclonal origin, suggesting that the proliferation tendency is inherent in every cell [14].

Virus Etiology Versus Somatic Mutation? There are now many observations, especially in animals, that tumors may be caused by viruses, and it is reasonable to assume that some human tumors have a viral origin as well. This hypothesis does not contradict the somatic mutation hypothesis. Viruses are often site-specific and may induce a mutational event in the chromosomes. The course of tumor formation following viral damage may then be similar to that described for any kind of somatic mutation.

Elucidation of the Origin of Malignant Tumors as a Surplus "Bonus" of Genetic Theory. The following sections describe the use of genetic concepts and data for stepwise elucidation of the causes of tumor formation. Most human tumors are nonhereditary and are often clearly caused by environmental agents. One might therefore expect that elucidation of their causes and mechanisms may not involve genetic concepts and methods. This exceptation, however, turns out to be incorrect. Advances in this field have been almost exclusively the result of progress in genetics – formal genetics and linkage, cytogenetics, and especially molecular genetics. This is an example of the general rule described in Sect. 6.1.1.6: good theories may reward the scientist with an additional bonus, explaining phenomena other than those for which they were originally created. At the same time, this is an example of the way in which genetics has extended its range to currently become the leading science basic to medicine. The following discussion is confined to genetic concepts and to results which have proved helpful for explaining malignant growth in humans.

10.4.1 Neoplasias with Constant Chromosomal Aberrations

The Philadelphia Chromosome. In neoplasias a single specific chromosome aberration limited to the tumor tissue was described some years ago. The classical example is the Philadelphia (Ph[1]) chromosome, which

is almost regularly associated with chronic granulocytic leukemia [57]. A translocation of much of the long arm of chromosome 22 to chromosome 9 is usually found (Fig. 10.8). Such patients are chromosomally normal in all tissues except for the hematopoietic system. The characteristic chromosome abnormality affects all blood cell precursors including megakaryocytic and erythropoietic cells. Biological tagging techniques by G6PD variants has led to similar insights. The clinical and hematological consequences of the translocation, however, affect only the granulocytic elements of the blood, showing that

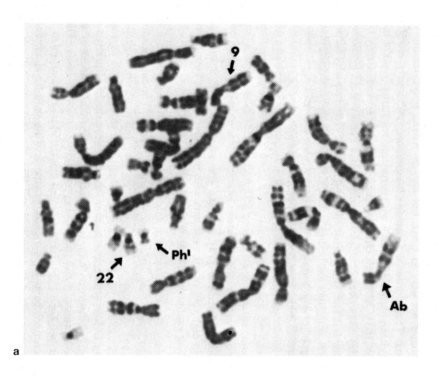

a

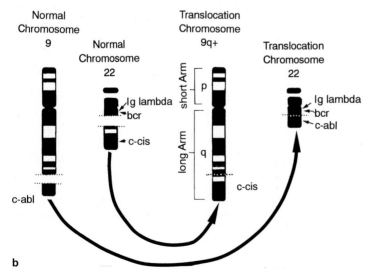

b

Fig. 10.8 a, b. Philadelphia (Ph[1]) chromosome in a patient with chronic granulocytic leukemia. a Translocation between chromosomes 22 and 9; Ab aberrant chromosome. b Formation of the Philadelphia chromosome by reciprocal translocation between chromosomes 9 and 22. One proto-oncogene, *abl*, is moved to no. 22 adjacent to an unknown gene; the break in no. 22 is distal to the Ig λ locus, which is not involved in the translocation. (Modified from Rowley 1986)

a given "mutation," while present in several cell types, may affect the growth pattern of only a single differentiated tissue.

In some exceptional families, several members have died from chronic granulocytic leukemia, and in one family some younger individuals showed the Ph[1] chromosome in their hematopoietic cells without clinical signs of leukemia. In this family, the susceptibility to breakage of chromosome 22 was inherited in an autosomal-dominant manner. In a study of 1129 Ph[1+] patients the 9;22 translocation was identified in 1036 (92%). The others had various translocations, all affecting chromosome no. 9 [63]. During the acute phase of the disease, additional aneuploidies may be observed; a second Ph[1] chromosome, an isochromosome of 17q, or a + 8 are the most common ones. Patients lacking the Ph[1] chromosome were shown not to have myeloid leukemia, but other types of myelodysplasia. Figure 10.9 shows chromosomes of a patient with chronic myeloid leukemia; the drawing includes a number of genes located on the two chromosomes 9 and 22 that may be important for malignant growth.

Chromosomal Patterns in Other Leukemias [60, 61]. Typical chromosomal patterns have been identified in other leukemias (Table 10.3). In acute nonlymphocytic leukemia (ANLL) a cell clone is generally present with a gain of one chromosome no. 8, and a loss of no. 7. t(8;21) is found in acute myeloblastic leukemia (AML; M2), especially in children. For other types of acute nonlymphocytic leukemia (ANLL), see Table 10.3. ANLL leukemias with consistent chromosomal aberrations show a rather similar age distribution which differs from other ANLL leukemias: they are more common in children and younger adults (median age about 40 years or under, compared with 50 years in ANLL patients overall).

A special type of leukemias has been observed after cytotoxic therapy: they often show partial or complete loss of chromosome 5 and/or 7. The region 5q23–5q32 often appears to be critically involved in the development of such mutagen-induced leukemia.

Leukemias and tumors affecting lymphocytes have also been analyzed, with interesting results. The best known example is Burkitt lymphoma, a tumor of viral origin (Epstein-Barr virus) that occurs mainly in the malaria belt of Africa. The tumor cells are consistently shown to have a translocation 8;14. The missing part of chromosome 8 is translocated to the long arm of chromosome 14. In follicular lymphoma with small cleaved cells (FSC) chromosomes 14 and 18 are involved. Rearrangements involving the proximal bands of chromosomes 14 and 7 are characteristic for T cell leukemias. (For T cells and their receptors see Sect. 4.4). Chromosome 14 is also involved in lymphatic leukemias of patients with AT (Sect. 10.2; Fig. 10.9).

In the chromosome instability syndromes – and, more rarely, in blood stem cells of normal individuals – many chromosome breaks, and rearrangements may occur throughout life, usually leading to cell death. However, a few may provide the cell with a selective advantage: the cell and its progeny escape growth regulation, and become a malignant neoplasm. The translocations offer hints regarding the site and the nature of changes necessary for such events.

In lymphomas and lymphatic leukemias, two kinds of chromosome sites are regularly involved; one site is important by carrying genes involved in the function of the lymphatic cell. We have seen that B lymphocytes produce immunoglobulins. B cell leukemias and other neoplasias often involve genes determining parts of immunoglobulin chains. The genes for the heavy-chain complex, for example, are located on 14q32, and the λ light-chain gene has been mapped to 22q11. Chromosomal aberrations, especially translocations, leading to T cell malignancies, are often

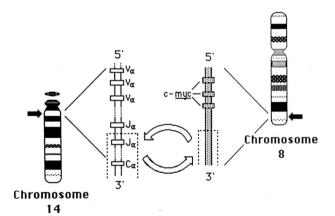

Fig. 10.9. Translocation t(8;14)(q24;q11) in T cell leukemia. *Short dark arrows,* breakpoints in chromosomes 8 and 14. The structure of the *myc* gene on no. 8 and of the *tcr* A gene (v_a; δ_a) on no. 14 and the breakpoints in each are shown in the center (From Rowley 1986)

Table 10.3. Common chromosome changes in leukemia (modified from Rowley 1994 [61])

Type	Gains	Losses	Rearrangements
Myeloid leukemia			
CML (Chronic myeloid leukemia)			
Chronic phase			t(9;22)(q34;q11)
Blast crisis	8, +Ph[1]	Rare; −7	t(9;22), i(17q)
ANLL (Acute nonlymphocytic leukemia)			
AML (M2)	+8	−7; less −5	t(8;21)(q22;q22)
APL (M3)	–	–	t(15;17)(q22;q12–21)
AMMoL (M4) (abn. eosinophils)	+8	−7	inv(16)(p13q22) t(16;16), del(16q)
AMoL (M5)	–	–	t(9;11)(p12;q23), t(11q), del(11q)
M2/M4 (incr. basophils)	–	–	t(6;9)(p23;q34)
M4 (incr. platelets)	–	–	t(3;3)(q21;q26), inv(3)
Lymphoid leukemia			
CLL (Chronic lymphatic leukemia)			
B cell	+12	–	14q + (q32)
T cell	–	–	t(8;14)(q24;q11), inv(14)(q11q32)
ALL (Acute lymphatic leukemia)			
Early B-precursor			t(4;11)(q21;q23)
Common	+21, +6	Rare	t(9;22), del(6q)(q15–q21), near haploid
pre B	–	–	t(1;19)(q23;p13)
B cell			t(8;14)(q24;q32), t(2;8)(q13;q24), t(8;22)(q24;q11)
Early T-precursor			t(9p), del(9p)(p21–22)
T cell			t(11;14)(p13;q11), t(8;14)(q24;q11), inv(14)(q11q32)

near T cell receptor genes; 14q11, for example, is the site of a T cell receptor gene. Two regions of chromosome 7 (p15;q35–q36), where the T cell β and γ genes are located, are also often affected. Hence, one break point is near a gene locus that is involved in the specific function of the cell that develops into a tumor. The other break point is equally interesting: so-called proto-oncogenes have been discovered in its neighborhood.

10.4.2 Oncogenes [11, 86, 90, 91, 94]

Basic Principles. The discovery of oncogenes has enhanced our understanding of molecular mechanisms of cancer formation. Early studies with cell fusion (Sect. 5.1.3) suggested that mutations at certain gene loci might be important for malignant transformation, and treatment of hamster cells with polyoma viral DNA resulted in malignant transformation. The ability of some RNA viruses, the retroviruses, to induce tumors in animals had been known since the work of Ellermann and Bang (1908) and P. Rous (1911). These and later results caused many research workers to look for retroviruses that might be implicated in causing of human tumors. Until the end of the 1970s these studies failed to be convincing. In recent years, however, the introduction of new methods

from molecular biology has led to new and important insights.

The genome of a retrovirus consists of a one-stranded RNA. It comprises the following areas of information (from the 5' to the 3' end): 5'-regulating sequence; genes for proteins necessary for the internal structure; reverse transcriptase gene; genes for surface glycoproteins; 3'-regulating sequence. As soon as the virus particle enters the cell, the reverse transcriptase produces a double-stranded DNA copy of the one-stranded RNA. This DNA is then integrated into the chromosomal DNA of the cell; integration may occur at many sites of the host genome. This DNA induces the cell to produce new viral RNA as well as the proteins necessary for the synthesis of new viral particles.

In addition to this minimal information, genomes of oncogenic retroviruses carry an additional gene which is responsible specifically for malignant transformation of host cells. This gene is called a retroviral oncogene (v-*onc*). DNA hydridization studies (Sect. 3.1.3.3) using DNA probes from v-*onc* genes showed that such genes are homologous with genes occurring at various sites of the host genome. Under normal conditions, however, they do not lead to malignant transformation. These are called proto-oncogenes or cellular oncogenes (c-*onc*). It is now assumed that these genes became integrated into the

Table 10.4. Characteristic rearrangements in solid tumors (modified from Rowley 1994 [61])

	Translocations	Genes
Pleomorphic adenoma	t(3;8)(p21C2)	
Liposarcoma (myxoid)	t(12;16)(p13.3;p11.2)	CHOP(12p)/FUS(16p)
Synovial sarcoma	t(X;18)(p11C1)	OAT1,2(Xp)/?
Rhabdomyosarcoma	t(2;13)(q35–37C4)	PAX3(2q)/?
Clear cell sarcoma of tendons	t(12;22)(q13C2)	ATF1(12)/EWS(22q)
Ewing sarcoma	t(11;22)(p24C2)	FLI1(11q)/EWS(22q)
Peripheral neuroepithelioma	t(11;22)(q24C2)	FLI1(11q)/EWS(22q)

	Deletions-inversions	Genes
Meningioma-acoustic neurinoma	del(22)(q12)	NF2
Papillary thyroid carcinoma	inv(10)(q11q21)	RET/PTC-PKA
Parathyroid adenoma	inv(11)(p15q13)	PTH/CCND1(PRAD1)
Retinoblastoma	del(13)(q14q14)	RB1
Wilms tumor	del(11)(p13p13)	WT1

viral genome at some time during evolution. Three different c-*onc* genes code for three tyrosine-specific protein kinases show homologies in the amino acid sequences of their gene products, and are located on human chromosomes 3, 15, and 20. Such protein kinases phosphorylate proteins and may therefore change their biological activity. This leads to transformation, for example, by changing properties of the cell surface such as by contact inhibition. Proto-oncogenes code for normal growth factors or their receptors. Thus, the *sis* gene codes for one chain of the platelet-derived growth factor and the *erb-b* gene encodes the receptor for the epidermal growth factor [12]. It can readily be visualized how mutant proto-oncogenes (see below) stimulate mitosis and cause cancerous growth. The platelet-derived growth factor is particularly interesting [12]. Together with other growth factors it is involved in stimulating the formation of atheromatous lesions – an event presumably mediated by increased synthesis of the normal growth factor. Its mutant counterpart *sis* is involved in neoplastic sarcoma formation. There are structural differences between v-*onc* and the homologous c-*onc* genes; for example, the c-*onc* genes, as other eukaryotic genes, consist of exons and introns (Sect. 3.1), whereas the corresponding v-*onc* genes have maintained only the exons.

Cellular Transformation. In many cases cellular oncogenes were discovered by direct gene transfer from transformed to normal cells. The first transforming gene characterized in transformed NIH/3T3 mice (c-Ha-*ras-1*) came from the human bladder carcinoma cell line EJ. The genetic lesion leading to activation (i.e., transforming ability) of the oncogene involved in this transformation was found to be a single point mutation, resulting in the replacement of just one amino acid in the gene-determined protein. The codon GGC was replaced by GTC; the transversion G → T led to replacement of glycine by valine in the resulting protein. However, a search for this proto-oncogene in 29 human cancers failed to show additional instances; it appears to be rare in cancers. Interestingly, the viral counterpart of this oncogene in a mouse sarcoma virus showed a point mutation in the same position. Other gene activations by single point mutations have been discovered. Moreover, some oncogenes found in other tumors showed structural similarity with the C-Ha-*ras-1* gene.

A transforming gene with different structure (B-*lym*) was isolated from Burkitt lymphoma cell lines; its gene product is partially homologous with transferrin, the iron-transporting protein. Rearrangement of another gene – C-*myc* – was also found in Burkitt lymphoma cell lines. The fact that two different oncogenes were found in the same tumor type suggests the possibility that mutations at more than one gene locus are sometimes necessary for malignant transformation. The different steps in carcinogenesis are represented by sequential activation of different proto-oncogenes, leading to qualitative and quantitative alterations in gene expression. The mechanisms for such activation are now being investigated with great intensity since they promise insights into molecular events leading to malignant transformation. In addition to point mutations, attachment of c-*onc* genes to strong promoter or enhancer regions of DNA, for example, by insertion of such regions close to c-*onc* genes, proved successful for transformation in vitro.

Table 10.5. Cytogenetic-immunophenotypic correlations in malignant B-lymphoid disease (from Rowley 1994 [61])

Phenotype	Chromosome abnormality	Involved genes
Acute lymphoblastic leukemia		
Pre-B	t(1;19)(q23;p13)	PBX1-TCF3(E2A)
B(Slg+)	t(8;14)(q24;q32)	MYC-IGH
	t(2;8)(p12;q24)	IGK-MYC
	t(8;22)(q24;q11)	MYC-IGL
B or B-myeloid	t(9;22)(q34;q11)	ABL-BCR
	t(4;11)(q21;q23)	AF4-MLL
Other	50–60 chromosomes	
	t(5;14)(q31;q32)	IL3-IGH
	del(9p), t(9p)	
	del(12p), t(12p)	
Non-Hodgkin lymphoma		
Burkitt type	See Slg + ALL	MYC-IGH-IGK-IGL
Follicular	t(14;18)(q32;q21)	IGH-BCL2
Mantle cell	t(11;14)(q13;q32)	CCND1-IGH
Diffuse large cell	t(3;14)(q27;q32)	BCL6-IGH
	t(10;14)(q24;q32)	LYT10-IGH
Chronic lymphocytic leukemia	t(14;14)(q13;q32)	CCND1-IGH
	t(14;19)(q32;q13)	IGH-BCL3
	t(2;14)(p13;q32)	IGH
	t(14q) and/or + 12	
Multiple myeloma	t(11;14)(q13;q32)	CCND1-IGH

However, promoter (or enhancer) insertion close to an oncogene might be only *one* requirement for transformation that might require changes in other proto-oncogenes.

Oncogenes Involved in Carcinogenesis Due to Chromosomal Rearrangements. The observation that the insertion of oncogenes near strong promoters activates neoplastic growth raises the question of whether in chromosomal rearrangements characteristic of certain neoplasias such proto-oncogene rearrangements close to promoter/enhancer regions (and possibly other regulating genes) might be the decisive step. Therefore many groups are now investigating oncogenes and their activities in tumors in relation to their localization in normal and rearranged chromosomes (Fig. 10.10). In Burkitt lymphoma, for example, transcription of the c-*myc* gene may increase 20-fold [95]. The human oncogene c-*abl* is located on the terminal band of the long arm of chromosome 9, the same band is involved in the breakpoint of the 9:22 translocation in chronic myeloid leukemia. This as well as other evidence favors the hypothesis that activation of oncogenes may indeed be involved in carcinogenesis by chromosome rearrangements. Proto-oncogenes have also been found as *amplified* copies in tumor cells. Double minute chromosomes and homogeneous staining regions of chromosomes are cytological counterparts of gene amplification in cancer cells. The *myc* oncogene involved in translocation of Burkitt lymphoma is overexpressed in carcinoma of the lungs, colon carcinoma, and promyelocytic leukemia. The N-*myc* gene is amplified in advanced stages of neuroblastoma and high levels of epidermal growth factor are expressed in squamous cell carcinomas — presumably as a result of amplification of the *erb-β* gene.

Tumor *progression* may be related to activities of particular oncogenes since certain oncogenes are found in the more aggressive forms of the tumor. If characteristic patterns can be identified, specific targets for therapy are available. Oncogene analogues or anti-oncogene antibodies, in contrast to current cancer chemotherapy, would not interfere with normal cell function but might arrest tumor growth.

From the available data, however, one general rule regarding the role of chromosome rearrangements for malignant transformation emerges [60]: such an rearrangement "brings together two genes that are normally far apart under different regulatory control. As a general rule, one of these genes would be related to growth control in a particular cell type and stage of differentiation, and the other gene would produce a protein whose function played a central role in the same cell type in the same stage of differentiation." The types of growth control genes within rearrangements may be quite different (Rowley 1994 [61]). Tyr-

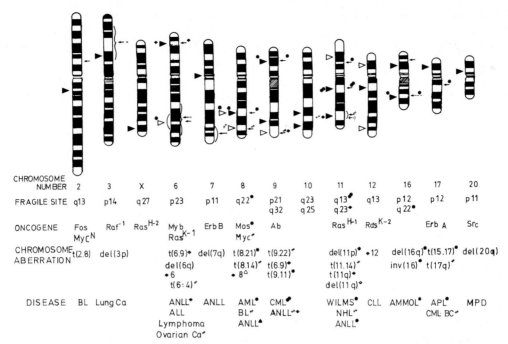

CHROMOSOME NUMBER	2	3	X	6	7	8	9	10	11	12	16	17	20
FRAGILE SITE	q13	p14	q27	p23	p11	q22*	p21 q32	q23 q25	q13* q23*	q13	p12 q22*	p12	p11
ONCOGENE	Fos Myc^N	Raf^{-1}	Ras^{H-2}	Myb Ras^{K-1}	ErbB	Mos* Myc*	Ab		Ras^{H-1}	Ras^{K-2}		Erb A	Src
CHROMOSOME ABERRATION	t(2.8)	del(3p)		t(6.9)* del(6q) •6 t(6:4)*	del(7q)	t(8.21)* t(8.14)* •8$^\Delta$	t(9.22)* t(6.9)* t(9.11)*		del(11p)* t(11.14)* t(11q)* del(11 q)*	•12	del(16q)* inv(16)*	t(15.17)* t(17q)*	del(20q)
DISEASE	BL	Lung Ca		ANLL* ALL Lymphoma Ovarian Ca*	ANLL	AML* BL* ANLL▲	CML* ANLL*•		WILMS* NHL* ANLL*	CLL	AMMOL*	APL* CML BC*	MPD

Fig. 10.10. Chromosomal fragile sites, localizations of oncogenes, and localization of chromosomal breaks in chromosomal aberrations leading to cancer. Diagram of chromosomes containing known fragile sites; the chromosome number, fragile site, oncogene, karyotypic aberrations and associated neoplastic diseases are indicated below the chromosome. *Arrowheads,* to left of each chromosome, bands carrying the fragile site (▶) or celular oncogene (▷); *arrow(s),* to right of a chromosome, specific bands involved in consistent translocations (←) or deletions (---) observed in patients having the disorders listed. Additional symbols (*, #, +) link the aberration to a disease. *BL,* Burkitt lymphoma; *ALL,* acute lymphoblastic leukemia; *ANLL,* acute nonlymphocytic leukemia; *AML,* acute myeloblastic leukemia; *CML,* chronic myelogenous leukemia; *AMMOL,* acute myelomonocytic leukemia; *AMOL,* acute monoblastic leukemia; *BC,* blast crisis; *NHL,* non-Hodgkin lymphoma; *CLL,* chronic lymphocytic leukemia; *APL,* acute promyelocytic leukemia; *MPD,* myeloproliferative disorder. *p,* Short arm; *q,* long arm of the chromosome; *t, del, inv,* translocation, deletion and inversion. (From Yunis and Soreng 1984 [1762])

osine and serine protein kinase genes, cell surface receptor genes, growth factors, genes for mitochondrial membrane proteins, cell-cycle regulator genes, and genes for myosin and for ribosomal proteins may be involved. Such genes may not only have different functions but different structures as well. There are homeobox (Sect. 8.4.1), helix-loop-helix, zinc finger genes, and several other types. This shows that proper regulation of cell growth normally depends on many mechanisms; disturbance of any of them may lead to the loss of control and to malignant transformation.

Oncogenes — at least those known at present — certainly do not provide the *entire* answer to the question on mechanisms of carcinogenesis. In most human tumors, for example, activated oncogenes have *not* been found (which might well be because not all oncogenes are already known). In other tumors, however, another class of genes was found to be important: the "tumor suppressor" genes.

10.4.3 Tumor Suppressor Genes

Retinoblastoma (180 200). Retinoblastoma is an eye cancer of children that occurs as both an inherited or a noninherited type [82, 83]. The inherited type shows an autosomal-dominant mode of inheritance with about 90% penetrance; the penetrance varies somewhat between families [51]. Approximately 68% of all inherited cases are bilateral; the rest are unilateral. In some of the bilateral cases more than one primary tumor in the second eye was observed [36–39].

Many cases of retinoblastoma are sporadic, i.e., they are the first in an otherwise healthy family. Within this group, only some 20%–25% are bilateral. Sporadic cases belong to either of two groups: dominant new mutations with 50% heterozygous and 45% affected offspring and nonhereditary cases. All bilateral sporadic cases are new mutations; the segregation ratio among their offspring is not far below 50% [67]. Among the unilateral sporadic cases about 10%–12% are new mutations, and the rest are nonhereditary

Table 10.6. Recurring structural rearrangements in malignant myeloid disease (from Rowley 1994 [61])

Disease	Chromosome abnormality	Inolved genes
Chronic myeloid leukemia	t(9;22)(q34;q11)	ABL-BCR
Blast phase	t(9;22) with + 8, + Ph + 19, or i(17q)	ABL-BCR
Acute myeloid leukemia		
AML-M2	t(8;21)(q22;q22)	ETO-AML1
APL-M3, M3V	t(15;17)(q22;q12)	PML-RARA
AMMoL-M4Eo	inv(16)(p13q22) or t(16;16)(p13;q22)	MYH11-CBFB
AMMoL-M4/AMoL-M5	t(9;11)(q22;q23)	AF9-MLL
	t(10;11)(p11–p15;q23)	?-MLL
	t(11;17)(q23;q25)	MLL-?
	t(11;19)(q23;p13)	MLL-ENL
	other t(11q23)	MLL
	del(11)(q23)	
AML	t(6;9)(p23;q34)	DEK-CAN
	t(3;3)(q21;q26) or inv(3)(q21q26)	?-EVI1
	+21	
	−7 or del(7q)	
	−5 or del(5q)	
	del(20p)	
	t(12p) or del(12p)	
	−Y	
Therapy-related AML	−7 or del(7q) and/or −5 or del(5q)	IRF1?
	t(11q23)	MLL1
	t(3;21)(q26;q22)	EAP/MDS1/EVI1-AML1
	der(1)t(1;7)(q10;p10)	

cases. The consequences of this situation for genetic counseling have been described [83] (see Appendix 6 for example).

Mutations Necessary to Create a Malignant Cell Clone [35]. Two steps are necessary for creating a malignant cell clone. The first mutation has no untoward effects but makes such a mutant cell vulnerable to tumor formation when a second, independent, mutation occurs. A malignant cell clone may then develop. In the inherited type of retinoblastoma, the first mutation is already present in one of the two germ cells which form the individual – either by transmission from one parent or by new mutation in the parent's germ cell. Therefore such a germinal mutation is present in all cells of the individual. Malignant transformation of such a mutant single cell requires only one additional step. The probability of this additional somatic mutation is low, but not extremely low, and all cells of the embryonic retina are at risk. Such a mutation may occur not at all, thereby explaining incomplete penetrance. If the additional mutation occurs once only, the result is unilateral retinoblastoma. When it happens twice or even more often, bilateral and/or multiocular retinoblastoma results.

In noninherited cases, on the other hand, both mutations must occur independently in the same retinal cell to create the malignant cell clone. The probability of these two events occuring in the same retina is so low that it is likely to occur only once in any individual. Therefore all noninherited retinoblastoma cases are unilateral and unilocular. Knudson (1977) [37] and later references) refined this hypothesis, suggesting that the first mutation may make the cell or (in inherited cases) the body heterozygous for a certain gene. The second mutation affects the allele of the same gene on the homologous chromosome, making the cell homozygous. This hypothesis explained why only one or a few cells become malignant in a multicellular organism, and provides a convincing explanation for the observation that in inherited, but not in noninherited cases, more than one primary tumor may occur. How could this hypothesis be confirmed?

As long ago as 1963, a 13q deletion was described in all cells of a patient with bilateral retinoblastoma and some (not very impressive) additional constitutional abnormalities [44]. In recent years, an increasing number of such patients have been discovered; in many of them only a very small segment of the long arm of chromosome 13 was deleted or involved in a reciprocal translocation (Fig. 10.11). By comparison of many such observations the deleted segment could be identified as 13q14; (or even 13q14.13; Fig. 10.11).

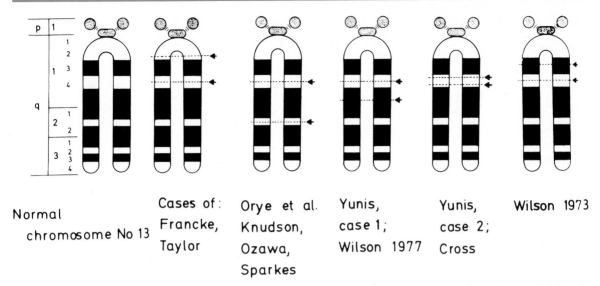

Fig. 10.11. Deletions in patients with retinoblastoma. Segments of chromosome 13 lost in cases with retinoblastoma, reported by Francke and Taylor, Orye et al., Knudson, Ozawa, Sparkes, Yunis, Wilson, Cross, Wilson. The stippled lines in each example refer to the extent of deletions. (From Vogel 1979 [1726])

These observations raised another question: Might the second step – that leading to the development of a malignant cell clone – consist in deletion of a chromosome segment in the region 13q14? In this case all tumor cells should show this deletion. Involvement of the 13q14 region has indeed been confirmed in many cells from patients with the inherited as well as the noninherited variants [2].

The mechanism has been elucidated by studies at the gene – DNA level [3, 8, 23]. The decisive step occurred indeed in the homologous chromosome carrying the normal allele. Sometimes, DNA markers from one parent were lacking in the tumor cells altogether, indicating that the chromosome from this parent had disappeared, been replaced by a second copy of the chromosome carrying the mutant retinoblastoma gene; in other cases recombinational events involving only a part of the normal chromosome led to the same consequence: the hemizygous mutation was made homozygous (Figs. 10.12, 10.13). The significance of this result extends beyond this specific case. It suggests that events such as nondisjunction or recombination could be much more common in somatic tissue than hitherto expected. This might explain various phenotypic peculiarities and even diseases. More specifically, it may offer explanations for other autosomal-dominant tumor diseases.

Once the mutant gene had been localized to region 13q14,11, the way was open for identification of the gene and its function. A suitable DNA probe was found in a DNA library of chromosome 13; the DNA segment identified by this probe was lacking in tumor tissues from retinoblastoma (Rb) patients. The corresponding mRNA is present in normal retinal cells but not in Rb tissue. The Rb gene found using this cDNA is approx. 190 kb long and has 27 exons encoding a mRNA of 4600 base pairs [17, 43, 92]. The gene-determined protein is a nuclear phosphoprotein that limits cell proliferation. It probably is involved in the regulation of the cell cycle and, specifically, in the decision between cell division and differentiation [29]. During G_0–G_1 phases the Rb protein is not – or is very weakly – phosphorylated; phosphorylation is much stronger during S and G_2 phases and, partially, during mitosis. Phosphorylation is performed by cyclin-dependent kinases. For normal function a single dose of this gene is sufficient. Analysis of the function of this protein has already contributed to our knowledge on mechanisms by which adenoviruses turn normal cells into tumor cells: the oncoprotein of an adenovirus transforms infected cells by binding to the protein determined by the normal Rb gene [88].

Investigations on the Rb gene laid the foundation for our knowledge of a new type of genes involved in malignant transformation of cells. These have been named "tumor suppressor genes," and they have also been found later in other tumors. Moreover, these results are now helping in genetic counseling for retinoblastoma, including prenatal diagnosis, since the gene can be identified directly (see Appendix 8).

Genetic Syndromes Associated with Tumors. Osteosarcoma frequently occurs in patients with retinoblastoma. In such cases, as in retinoblastoma, homozygosity of the identical locus at chromosome 13 has also

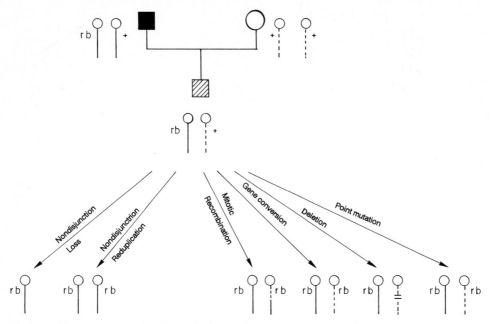

Fig. 10.12. Chromosome mechanisms which could lead to homozygosity of the retinoblastoma (Rb) allele in the somat- ic tissue of a heterozygote for this allele. (From Cavenee et al. 1983)

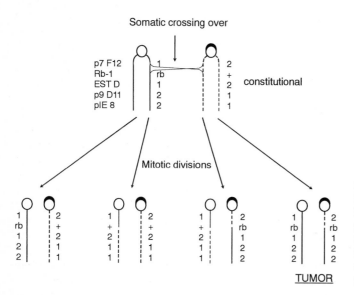

Fig. 10.13. Homozygosity of the Rb allele produced by a mitotic recombinational event (crossing over between two of the four chromatids in the G_2 phase). *ESTD*, Esterase D locus; p7F12, p9D11, pIE8, DNA fragment length polymorphisms. The cap on the chromosome with the wild allele on the Rb locus represents a C-stained heterochromatic marker. (From Cavenee et al. 1983 [1431])

been demonstrated [40]. Homozygosity of an identical site thus can cause either retinoblastoma or osteosarcoma. In Wilms tumor, as predicted by Knudson, a site on chromosome 11, (11p13), is rendered homozygous by similar mechanisms. Here an occasional autosomal-dominant syndrome is found characterized by aniridia, genitourinary abnormalities (gonadoblastoma, ambiguous genitalia) and mental retardation (WAGR). The same locus appears to be involved in Beckwith-Wiedemann syndrome, which is associated with a variety of embryonal tumors (Table 10.8).

The common mechanism of these various tumors also involves development of homozygosity for a site on chromosome 11 [40].

An especially interesting protein has been shown to be lacking (or reduced in function) in another, dominantly inherited family cancer syndrome: the Li-Fraumeni syndrome (151 623). Family members suffer from childhood sarcoma, breast cancer, leukemia, brain tumors, and other tumors. The gene has been mapped to 17p13. Its product is the tumor suppressor p53 protein whose lack or malfunction has been in-

Table 10.7. Functional classification of transforming genes at translocation junctions (from Rowley 1994 [61])

	Location	Translocation	Disease
SRC family (TYR protein kinases)			
ABL	9q34	t(9;22)	CML/ALL
LCK	1p34	t(1;7)	T-ALL
ALK	5q35	t(2;5)	NHL
Serine protein kinase			
BCR	22q11	t(9;22)	CML/ALL
Cell surface receptor			
TAN1	9q34	t(7;9)	T-ALL
Growth factor			
IL2	4q26	t(4;16)	T-NHL
IL3	5q31	t(5;14)	Pre B-ALL
Mitochondrial membrane protein			
BCL2	18q21	t(14;18)	NHL
Cell cycle regulator			
CCND1 (BCL1-PRAD1)	11q13	t(11;4)	CLL/NHL
Myosin family			
MYH11	16p13	inv(16), t(16;16)	AML-M4Eo
Ribosomal protein			
EAP (L22)	3q26	t(3;21)	t-AML/CML BC
Unknown			
DEK	6p23	t(6;9)	AML-M2/M4

CML, Chronic myeloid leukemia; ALL, acute lymphoblastic leukemia; T-ALL, T-cell ALL; NHL, non-Hodgkin lymphoma; Pre B-ALL, pre-B-cell ALL; CLL, chronic lymphocytic leukemia; AML, acute myeloid leukemia; CML BC, CML in blast crisis.

ferred in numerous hereditary and nonhereditary tumors. As with the Rb protein, it is also a phosphoprotein probably involved in normal blocking of the cell cycle at the end of the G_1 phase [45, 85].

A Combination of Mutations in Oncogenes and Tumor Suppressor Genes: Polyposis and Colon Cancer. Two different types of mutations were discussed that may lead to cancer: mutations and structural changes in and around oncogenes turning harmless and useful genes into those that actively promote uninhibited cell growth; and mutations in tumor suppressor genes removing a kind of "brake" of cell growth so that the normal process of blocking of cell division no longer functions. However, turning a normal cell into a cancer cell is never – or only in rare instances – a simple one-step process. Malignant transformation generally occurs in two (or more) stages. These stages are caused by stepwise elimination of various control mechanisms. These mechanisms may be determined either by oncogenes, by tumor suppressor genes, or by both.

An example is familial adenomatous polyposis (FAP; 175100; Fig. 10.14), an autosomal-dominant condition leading to a great number of adenomatous but benign polyps throughout the colon and rectum. Sooner or later one or a few of these polyps turn malignant; metastases occur, and the patient dies from colorectal cancer, often in young adult age. The gene has been localized to 5q22; it is a tumor suppressor gene [32]. Mutations at this locus alone are not sufficient to produce cancer; further steps are necessary, involving, in addition to loss of chromosomes 18 and 17, also a mutation of the *ras* oncogene.

FAP is a rare condition; fewer than 1% of cases with colorectal cancer are caused by this condition. However, in about 15% of patients with colorectal cancer who do not have FAP, familial aggregation is observed; there are no polyps, and no simple monogenic mode of inheritance can be found by pedigree studies. De la Chapelle and Vogelstein discovered a gene on chromosome 2 that appears to have a destabilizing effect on the genome, basically similar to the chromosome instability syndromes discussed above [50]. Tumors of such patients often show characteristic replication errors leading to shifts in the position of di- and trinucleotide DNA markers at electrophoresis. This may indicate a new mechanism of tumorigenesis [1a] caused by a mutation in a DNA replica-

Table 10.8. Some genetic syndromes associated with tumors

Genetic syndrome	Tumor	Chromosome
Beckwith-Wiedemann syndrome	Hepatoblastoma Rhabdomyosarcoma Wilms tumor Adrenal carcinoma	11p15.5
Aniridia, genitourinary abnormalities, retardation, mental (WAGR syndrome)	Wilms tumor	11p13
Neurofibromatosis	Astrocytoma Sarcomatous transformation	17q11.2
Multiple endocrine neoplasia (type II)	Medullary thyroid cancer	10q21.1
Basal cell nevus syndrome	Medulloblastoma Astrocytoma Ovarian cell cancer Hamartoma	
Li-Fraumeni syndrome	Soft tissue sarcoma Osteosarcoma Breast cancer Adenocarcinoma of the lung etc.	13p1 (p53 gene)

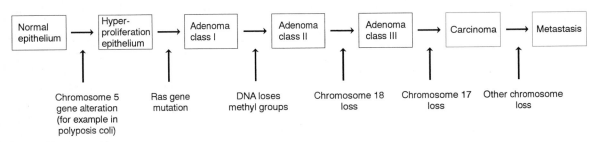

Fig. 10.14. A model for the formation of colorectal tumors. A series of successive genetic alterations is necessary to create a malignant cell, including activation of the *ras* oncogene and (probably) loss of tumor suppressor genes on chromosomes 5, 17, and 18. (Adapted from Vogelstein and Kinzler 1992 [85])

tion factor causing reduced fidelity for replication and repair – a mutator mutation [56a]. The affected gene turned out to be highly homologous to a gene involved in mismatch repair in bacteria and yeast [32a]. A total of 4 different chromosome loci involved in mismatch repair have been identified. Mutations at two of these sites (chr 2p16 and chr 3p21) account for 90% of hereditary nonpolyposis colon carcinomas.

Still other tumors occur as noninherited, solitary events as well as dominantly inherited types with multiple tumors. In skin tumors this is a well-known dichotomy [69]: solitary tumors are usually noninherited, whereas multiple tumors of the same histological type are dominantly inherited. Examples include neurofibromatosis versus single neurofibromas, multiple versus single lipomas, multiple versus sporadic cutaneous leiomyomas, multiple versus sporadic

glomus tumors, or the basal cell nevus syndrome versus single basal cell nevi [4].

Knudson has adduced epidemiological and statistical evidence that similar mechanisms may apply in a variety of other childhood tumors, such as neuroblastoma and pheochromocytoma. In these conditions both sporadic and hereditary cases are postulated. The hereditary cases are more often familial, are frequently bilateral, occur at an earlier age, and involve more than one site (i.e., both adrenals in neuroblastoma) [36, 38, 53].

"Cancer families" are known for other malignant tumors. Carcinoma of the esophagus in many patients of a family with a special type of tylosis palmaris and plantaris has been discussed in Sect. 6.1 (Fig. 6.24). In this as well as in other examples of dominant inheritance of an increased liability for neoplasia, the tumors tend to occur earlier in life within these families than in the more common sporadic variety of the same cancer. There is often an increased tendency to multiple occurrences in the same individual. Again, these observations suggest a genetically determined and dominantly inherited increased probability for malignant transformation.

10.4.4 A Genetic View of Human Cancer

The various data regarding cancer in humans allow some generalizations. Mendelian inheritance in human cancer is rare, but a variety of benign tumors particularly can be inherited as Mendelian traits. Neurofibromatosis, familial adenomatosis polyposis, and the various multiple endocrine neoplasia syndromes are examples [46, 56]. Malignant transformation is common in most if not all of these conditions. The chance of malignant transformation may be related to the turnover time of the affected cell. With more frequent cell divisions a malignant tumor cell is more likely to develop. It may be no coincidence that in the rapidly turning-over epithelium of the large bowel colon cancer invariably appears following benign hereditary polyposis.

Another tumor mechanism may be more generally applicable to some other tumors: homozygosity (as found in retinoblastoma, Wilms tumor, and other embryonal tumors) produced by an inherited germinal mutation and followed by a somatic mutation or somatic rearrangement in a single somatic cell.

Chromosomal abnormalities are often seen in cancer cells but are rarely unique. They are usually considered to reflect abnormalities secondary to disordered growth. In addition to the translocation in chronic myelogenous leukemia, other nonrandom chromosomal changes (Table 10.3) were identified first with increasing frequency in various hematological malignancies where detailed studies were more readily possible. In fact, Yunis has claimed that with high-resolution banding of chromosomes most tumors can be shown to have characteristic chromosomal defects [96]. General rules are emerging such as that suggested by Rowley [60]: in the structural rearrangements, a gene for growth control must come in close apposition to a gene having a specific function such as production of an important protein in a differentiated cell type. Of course, this is not the full explanation, but it does direct our thinking toward such explanations.

The association of tumors with the autosomal-recessive chromosomal breakage syndromes is certain and is particularly interesting in xeroderma pigmentosum, where the basic lesions have been identified as a defect of DNA repair following exposure to UV light. Suggestions that the much more common heterozygotes for these syndromes may also suffer from increased frequency of cancer makes pathophysiological sense but needs further documentation [84]. The possibility that a significant proportion of human malignancy is caused by the carrier state for a variety of breakage syndromes will be of great public health importance for tumor prevention after tests for carrier detection of these defects become readily available.

Various genetic defects affecting chromosome breakage and repair, as well as lack of immune surveillance in the presence of environmental carcinogens are likely to produce neoplasias. Additional genetic factors bearing on environmental carcinogenesis relate to abnormal metabolism of carcinogenic environmental agents. Judging from twin studies in pharmacogenetics (Sect. 7.5.1), it is likely that metabolism of most foreign agents is under genetic control. Both slow biotransformation and the existence of enzyme systems that transform procarcinogens into more potent cancer-causing agents cause a certain proportion of the population to be at higher risk for genetic reasons. Data on arylhydrocarbon hydroxylase levels (an enzyme that transforms polycyclic hydrocarbons into more carcinogenic agents), the debrisoquine-sparteine system and glutathione transferase in lung cancer in humans suggest such mechanisms (Sect. 7.5.2). Many of the chemicals that are mutagenic in bacterial test systems are also carcinogenic in animal systems. Viral causation of human tumors is most certain in Burkitt lymphoma, a tumor that is clearly of monoclonal origin. The fact that chromosome 14 and chromosome 8 are involved in the development of Burkitt lymphoma makes this tumor a model relating viral causation, monoclonal origin, oncogenes, and clonal chromosomal evolution.

Studies on oncogenes helped to bridge the gap between spontaneous and mutagen- and virus-induced somatic mutation by showing that oncogenes can be "activated," i.e., turned from normal and necessary to cancer-promoting genes, either by classical point mutations or by rearrangements effected by passage through a virus. In many laboratories scientists are working on the molecular mechanisms of "activation" of oncogenes or "inactivation" of tumor suppressor genes. Changes and anomalies in DNA methylation, as discussed in Chap. 9 on gene mutation, may be important. For example, genomic imprinting, which is thought to be caused at least in part by methylation differences of maternal and paternal genomes (Sect. 8.2) has been observed in retinoblastoma, and anomalies of methylation have been shown in Wilms tumor [62].

Striking familial aggregation of tumors occurs only in the various hereditary tumor syndromes and occasionally in so-called cancer families, which are characterized by (a) increased occurrence of adenocarcinoma, primarily of colon and endometrium, (b) increased frequency of multiple primary malignant neoplasms, (c) earlier age at onset, and (d) vertical transmission compatible with autosomal-dominant inheritance. With many tumors, such as carcinoma of the breast and stomach, a modest familial aggregation is found.

Such statistical results must be considered with caution. Members of the same families share not only a part of their genes but a part of their environment, so that familial aggregation may be entirely environmental. However, evidence is accumulating that genetic mechanisms such as those discussed here for rarer tumors with predominantly genetic causes may also occur in common types of malignancies. The example of colorectal cancer [1] has been mentioned. Many studies are also being performed on breast cancer. They agree on the conclusion that there is an autosomal dominant gene with approx. 80% penetrance if incidence over lifetime is considered. This gene also predisposes to ovarian cancer. Among cases affected as early as at ages 20–29, about 36% are thought to have this gene; the fraction decreases markedly among patients 30 years or older [9]. Often, both breasts are affected. Linkage studies in families with early onset identified a gene at 17q21 [13], which may also be involved in some families with later onset [48]. The gene has been cloned [83a]. Many different mutations at this locus (BrCA1) have been found [52a]. It is of some interest that 1% of Ashkenazi Jewish women carry a specific mutation at this locus [120a]. A second breast cancer gene (BrCA2) has also been identified [149a]. The fact that these mutant genes can be diagnosed early in life raises intriguing questions regarding appropriate strategies of prevention or early diagnosis.

Chromosome studies in breast tumor tissue have suggested – among various, random anomalies, 3p-deletions in some tumors [58]. Further studies will probably reveal a similar chain of events in this tumor as has been found in colorectal cancers. It is well known that the risk for breast cancer is influenced, in addition to genetic factors, by external factors such as number of pregnancies and possibly by diet. Interaction of these influences with genetic factors still eludes our understanding.

The changing epidemiological pattern of some tumors, such as the increasing frequency of lung cancer and the decreasing frequency of gastric cancer in the past generation, clearly suggests that genetic factors alone cannot explain the predisposition to these cancers. We know that cigarette smoking is related to lung cancer and suspect that disappearance of carcinogens because of better food preservation has re-

duced the frequency of stomach cancer. The fact that only a certain proportion of heavy cigarette smokers develop lung cancer argues against a simple environmental explanation. It is most likely that specific environmental factors, such as the hydrocarbons of cigarette smoke and other irritants, interact with specific genetic factors, such as those affecting hydrocarbon metabolism, DNA repair, genetic variability at sites of proto-oncogenes and tumor suppressor genes, as well as immunological surveillance in determining the establishment of a clinically significant lung cancer. In occasional cases either the environmental factor alone or the genetic factor alone is likely to be exclusively important. In most cases, however, the interaction of heredity and environment is probably operative. Since a large number of environmental agents and numerous genetic mechanisms appear to be involved in various cancers, no simple panacea to prevent all cancers is likely to be found. Hopefully, identification of genetically susceptible subgroups of the population by simple laboratory tests will become possible with better understanding of the various mechanisms underlying genetic susceptibility to various cancers. These groups can then be informed about individual risk factors for specific types of cancer, and regular examinations can be offered at a larger scale than at present, and with better, more individualized surveillance. Warning against certain environmental agents such as smoking may then be more successful.

10.5 Somatic Mutation and Aging

Aging and Death. Humans are the only living beings who know about their inevitable death. The difficulties in accepting this knowledge are epitomized by the enormous importance of death cults in cultural evolution – from the burial rites among Neandertals and highly sophisticated cultures dedicated almost exclusively to taking care of the deceased, such as the Egyptians, to the heaven-and-hell beliefs of some present-day religions. In earlier periods, death generally came suddenly and often at a young age. With the development of modern life styles, and especially modern medicine, we have gained decades of life span. As a result we are now confronted with the slow deterioration of our biological capacities during aging. An increasing proportion of the population, particularly in industrialized societies, consists of the elderly; a large part of medical practice is now devoted to geriatric medicine. This has created many new social, biological, and medical problems.

Twin and family investigations in humans have shown a relatively strong genetic influence on life expectation [31]. Moreover, the onset and natural history of many physical and mental diseases is age dependent.

Studies on Biological Mechanisms of Aging in Single Cells [33]. The notion that aging and death might be reduced to properties of the single cell was first put forward by the zoologist A. Weismann some 100 years ago. He suggested that the origin of "natural death" lies in the limitation of somatic cells' powers of reproduction: "Death takes place because a worn-out tissue cannot forever renew itself, and because a capacity for increase by means of cell division is not everlasting but finite" (Weismann 1891 [93]). Weismann was also the first to suggest that too long a survival of a multicellular organism beyond the period of reproduction and care for offspring would carry a selective disadvantage to the species.

Weismann's hypothesis of somatic cell deterioration seemed to be refuted by the claim of A. Carrell (1912) that chicken somatic cells could be cultured outside the donor indefinitely. For a long time this "potential immortality" hypothesis was generally accepted and was among the most popular biological "results." Failures to reproduce these findings were explained by inappropriate culture conditions and other methodological factors, until the careful experiments of Hayflick in the early 1960s [26] established that Carrell's claim had been wrong. It is now widely accepted that normal diploid, mammalian fibroblasts can undergo only a finite number of cell divisions; for human embryonic fibroblasts this number is about 50 ± 10. On the other hand, heteroploid, *transformed* cells generally appear to multiply indefinitely.

Several studies have explored the link between aging of the organism as a whole and the limited number of cell divisions in somatic cells. If cells stop dividing in culture because they have "exhausted" an intrinsic maximum number of divisions, their in vitro capacity for division should decrease with increasing age of the donor. In fact, this has repeatedly been demonstrated [49, 68]. Another implication of this hypothesis is that cells of short-lived species would be able to undergo fewer divisions than those of long-lived ones. This, too, was confirmed when cells from humans (maximum life span $\approx$ 110 years) were compared with those from mice (3.5 years). Mouse fibroblasts can undergo only 14–28 divisions in vitro. However, when other species are included in the comparisons, the association between species-specific life span and ability of cells to divide in vitro is blurred. Moreover in humans, even cells taken from very old persons maintain the capacity to divide about 20 times.

The significance of such cell studies for the problem of aging appeared to increase when it was discovered [49] that cells affected by Werner syndrome (277700) have a markedly reduced capacity for division in cell culture. The main clinical signs of this autosomal-recessive syndrome are cataracts, subcutaneous calcifications, premature graying, premature arteriosclerosis, skin changes, diabetes mellitus, a higher incidence of malignant tumors, chromosome instability, and a prematurely aged face (Fig. 10.15). Life expectancy is considerably reduced [15, 66]. The cells in Werner syndrome show many different chromosome rearrangements ("variegated translocation mosaicism", see [66]), suggesting an unusual type of chromosome instability. Werner syndrome has sometimes been regarded as a genetic model of premature aging. However, while some of the clinical and pathological findings resemble those found during the normal aging process, many others do not.

Molecular and Chromosomal Mechanisms. Many studies are in progress on molecular and chromosomal mechanisms, but so far no coherent and generally ac-

Fig. 10.15. A Japanese-American woman with Werner syndrome as a teenager and at 48 years of age. She had eight sibs, two of whom were also affected. (From Salk 1982 [66])

cepted theory has emerged. With considerable over-simplification, two groups of hypotheses can be distinguished. One posits that the ceasing of cell divisions is somehow programmed by a biological mechanism of regulation. An important argument in favor of this hypothesis is the observation that transformed cells are not limited in their dividing capacity. The other group of hypotheses postulates that cells lose their dividing capacity by accumulation of "errors" that prevent them from dividing when a certain threshold is exceeded. Such errors may occur somewhere at the translational or posttranslational level; they may also occur at the DNA level in the form of somatic mutations. The latter hypothesis, which was first proposed by Szilard in 1959 [81] and extended by Burnet in 1974 [7], is supported by evidence of an increasing frequency of somatic mutations with age in vitro [18] and in vivo. For instance, the fraction of lymphocytes having a defect of HPRT that makes them resistant to 6-thioguanine increases with age [54]. Accumulation of mitochondrial mutations has been observed in normal cells and may be a further cellular aspect of aging [88] (Sect. 4.1.9). However, the somatic mutation theory does not explain all phenomena of cell aging; accumulation of errors at various posttranslational levels is also likely [28].

How does the transformed cell, however, escape the consequences of such an accumulation of errors, regardless of whether they are produced by somatic mutation or by another error-prone process? So far this problem does not seem to have been solved. Several possible mechanisms [33] are under scrutiny, among them reactivation of normally suppressed correction mechanisms or more rapid selective elimination of cells harboring such errors. In conclusion, it is likely that somatic mutations contribute to the process of normal aging just as they are involved in cancer formation, which is also an age-dependent process. But how much they contribute and how they interact with other cellular processes is largely unknown.

Conclusions

Mutations may occur not only in germ cells but also in somatic cells. Depending on the genes involved, there may be no phenotypic effect at all, or various deviant phenotypes may occur. Probably all malignant tumors and leukemias owe their origins to somatic mutations affecting a single cell which acquires a selective growth advantage over its sister cells. Often several additional somatic mutations are required for the development of malignancy. The Philadelphia chromosome in chronic granulocytic leukemia is a cardinal example of a microscopically visible somatic

mutation. Studies on neoplasias have led to the discovery of two main classes of genes: "oncogenes" and "tumor suppressor genes." The normal function of these genes involves various aspects of the regulation of cell division. Some inherited malignant tumors (such as retinoblastoma) are caused by a transmitted mutation of a tumor suppressor gene via the germ line. A somatic mutation of the allelic partner of this gene initiates malignant growth by homozygous loss of tumor suppression.

Some rare autosomal-recessive DNA repair defects and chromosomal instability syndromes (Fanconi anemia, Bloom syndrome) are associated with a high frequency of various cancers, presumably caused by the characteristic underlying mutations predispoing to malignancies in various somatic tissues.

Somatic mutations probably also cause some of the changes observed in normal aging.

References

1. Aaltonen L, Peltomaeki P, Leach FS et al (1993) Clues to the pathogenesis of familial colorectal cancer. Science 260 : 812–816
2. Balaban G, Gilbert F, Nichols W, Madans AT, Shields J (1982) Abnormalities of chromosome no 13 in retinoblastomas from individuals with normal constitutional karyotype. Cancer Genet Cytogenet 6 : 213–221
3. Benedict WF, Murphree AL, Banerjee A, Spina CA, Sparkes MC, Sparkes RS (1983) Patient with 13 chromosome deletion: evidence that the retinoblastoma gene is a recessive cancer gene. Science 219 : 973–975
4. Berendes U (1974) Multiple tumors of the skin: clinical, histopathological, and genetic features. Hum Genet 22 : 181–210
5. Bochkov NP, Lopukhin YM, Kulsehov NP, Kovalchuk LV (1974) Cytogenetic study of patients with ataxia-teleangiectasia. Hum Genet 24 : 115–128
6. Burnet FM (1974) The biology of cancer. In: German J (ed) Chromosomes and cancer. Wiley, New York
7. Burnet FM (1974) Intrinsic mutagenesis: a genetic approach to ageing. Wiley, New York
8. Cavenee WK, Dryja TP, Phillips RA, Benedict WF, Godbout R, Gallie BL, Murphree AL, Strong LC, White RL (1983) Expression of recessive alleles by chromosomal mechanisms in retinoblastoma. Nature 305 : 779–784
9. Claus EB, Risch NJ, Thompson WD (1991) Genetic analysis of breast cancer in the cancer and steroid hormone study. Am J Hum Genet 48 : 232–242
10. Cleaver JE, Bootsma D (1975) Xeroderma pigmentosum: biochemical and genetic characteristics. Annu Rev Genet 9 : 19–38
11. Cohen MM, Levy HP (1989) Chromosome instability syndromes. Adv Hum Genet 18 : 43–149
12. Deuel TF, Huang JS (1984) Roles of growth factor activities in oncogenesis. Blood 64 : 951–958

13. Easton DF, Bishop DT, Ford D et al (1993) Genetic linkage analysis in familial breast and ovarian cancer: results from 214 families. Am J Hum Genet 52 : 678–701

14. Ehling UH (1982) Risk estimate based on germ cell mutations in mice. In: Sugimura T, Kondo S, Takebe H (eds) Environmental mutagens and carcinogens. Proceedings of the 3rd International Conference on Environmental Mutagens. University of Tokyo Press, Tokyo/Liss, New York, pp 709–719

15. Epstein CJ, Martin GM, Schultz AL, Motulsky AG (1966) Werner's syndrome: a review of its symptomatology, natural history, pathological features, genetics and relationship to the natural aging process. Medicine (Baltimore) 45 : 177–221

16. Fischer E, Thielmann HW, Neundörfer B, Rentsch FJ, Edler L, Jung EG (1982) Xeroderma pigmentosum patients from Germany: clinical symptoms and DNA repair characteristics. Arch Dermatol Res 274 : 229–247

17. Friend SH, Bernards R, Rogelj S et al (1986) A human DNA segment with properties of the gene that predisposes to retinoblastoma and osteosarcoma. Nature 323 : 643–646

18. Fulder SJ, Holliday R (1975) A rapid rise in cell variants during the senescence of populations of human fibroblasts. Cell 6 : 67–73

19. German J (1974) Chromosomes and cancer. Wiley, New York

20. German J (ed) (1983) Chromosome mutation and neoplasia. Liss, New York

21. German J (1983) Neoplasia and chromosome-breakage syndromes. In: German J (ed) Chromosome mutation and neoplasia. Liss, New York, pp 97–134

22. German J, Bloom D, Archibald R (1965) Chromosome breakage in a rare and probably genetically determined syndrome of man. Science 148 : 506–507

23. Gilbert F (1983) Retinoblastoma and recessive alleles in tumorigenesis. Nature 305 : 761–762

24. Gropp A, Flatz G (1967) Chromosome breakage and blastic transformation of lymphocytes in ataxia teleangiectasia. Hum Genet 5 : 77–79

25. Harnden DG (1974) Ataxia teleangiectasia syndrome: cytogenetic and cancer aspects. In: German J (ed) Chromosomes and cancer. Raven, New York, pp 87–104

26. Hayflik L (1965) The limited in vitro lifetime of diploid cell strains. Exp Cell Res 37 : 614–636

27. Hecht F, Kaiser, McCaw B (1977) Chromosome instability syndromes. In: Mulvihill JJ, Miller RW, Fraumeni JF Jr (eds) Genetics of human cancer. Raven, New York, pp 105–123

28. Holliday R, Kirkwood TBL (1981) Predictions of the somatic mutation and mortalisation theories of cellular ageing are contrary to experimental observations. J Theor Biol 93 : 627–642

29. Hollingsworth RE Jr, Hensey CE, Lee W-H (1993) Retinoblastoma protein and the cell cycle. Curr Opin Genet Dev 3 : 55–62

30. Iselius L, Slack J, Littler M, Morton NE (1991) Genetic epidemiology of breast cancer in Britain. Ann Hum Genet 55 : 151–159

31. Kemp T (1940) Altern und Lebensdauer. In: Just G (ed) Handbuch der Erbbiologie des Menschen. Springer, Berlin, pp 408–424

32. Kinzler KW, Nilbert MC, Vogelstein B et al (1991) Identification of a gene located at chromosome 5q21 that is mutated in colorectal cancers. Science 251 : 1366–1370

33. Kirkwood TBL, Cremer T (1982) Cytogerontology since 1881: a reappraisal of August Weisman and a review of modern progress. Hum Genet 60 : 101–121

34. Knippers R, Philippsen P, Schafer KP, Fanning E (1990) Molekulare Genetik, 5th edn. Thieme, Stuttgart

35. Knudson AG (1971) Mutation and cancer: statistical study of retinoblastoma. Proc Natl Acad Sci USA 68 : 820–823

36. Knudson AG (1973) Mutation and human cancer. Adv Canc Res 17 : 317–352

37. Knudson AG (1977) Genetics and etiology of human cancer. Adv Hum Genet 8 : 1–66

38. Knudson AG, Strong LC (1972) Mutation and cancer: neuroblastoma and pheochromocytoma. Am J Hum Genet 24 : 514–532

39. Knudson AG, Hethcote HW, Brown BW (1975) Mutation and childhood cancer. A probabilistic model for the incidence of retinoblastoma. Proc Natl Acad Sci USA 72 : 5116–5120

40. Koufos A, Hansen MF, Copeland NG, Jenkins NA, Lampkin BC, Cavenee WK (1985) Loss of heterozygosity in three embryonal tumours suggests a common pathogenetic mechanism. Nature 316 : 330–334

41. Kraemer KH, Lee MM, Scotto J (1987) Xeroderma pigmentosum: cutaneous, ocular and neurologic abnormalities in 830 published cases. Arch Derm 123 : 241–250

42. Lambert MW, Tsongalis GJ, Lambert WC et al (1992) Defective DNA endonuclease activities in Fanconi's anemia cells, complementation groups A and B. Mutat Res DNA Repair 273 : 57–71

43. Lee WH, Bookstein R, Hong F et al (1987) Human retinoblastoma susceptibility gene: cloning, identification, and sequence. Science 235 : 1394–1399

44. Lele KP, Penrose LS, Stallard HB (1963) Chromosome deletion in a case of retinoblastoma. Ann Hum Genet 27 : 171

45. Li FP, Fraumeni JF (1969) Soft-tissue sarcomas, breast cancer and other neoplasms: a familial syndrome? Ann Intern Med 71 : 747–752

46. Lynch HT (1976) Miscellaneous problems, cancer, and genetics. In: Lynch HT (ed) Cancer genetics. Thomas, Springfield

47. Mann WR, Venkatraj VS, Allen RG et al (1991) Fanconi anemia: evidence for linkage heterogeneity on chromosome 20q. Genomics 9 : 329–337

48. Margaritte P, Bonaiti-Pellié C, King M-C, Clerget-Darpoux F (1992) Linkage of familial breast cancer to chromosome 17q21 may not be restricted to early-onset disease. Am J Hum Genet 50 : 1231–1234

49. Martin CM, Sprague CA, Epstein CJ (1970) Replicative lifespan of cultivated human cells: Effect of donor's age, tissue and genotype. Lab Invest 23 : 86–92

50. Marx J (1993) New colon cancer gene discovered. Science 260 : 751–752

51. Matsunaga E (1976) Hereditary retinoblastoma: penetrance, expressivity and age of onset. Hum Genet 33 : 1–15

52. Matsunaga E (1981) Genetics of Wilm's tumor. Hum Genet 57 : 231–246

53. McKusick VA (1995) Mendelian inheritance in man, 11th edn. Johns Hopkins University Press, Baltimore

54. Morley A, Cox S, Holliday R (1982) Human lymphocytes resistant to 6-thioguanine resistance increase with age. Mech Ageing Dev 19 : 21–26

55. Moustacchi E, Averbeck D, Diatloff-Zito C, Papadopoulo D (1989) Phenotypic and genetic heterogeneity in Fanconi anemia, fate of cross links, and correction of the defect by DNA transfection. In: Schroeder-Kurth, Auerbach AD, Obe G (eds) Fanconi anemia. Springer, Berlin Heidelberg New York, pp 196–210

56. Mulvihill JJ, Miller RW, Fraumeni JF Jr (eds) (1977) Genetics of human cancer. Raven, New York

57. Nowell PC, Hungerford DA (1960) A minute chromosome in human chronic granulocytic leukemia. Science 132 : 1497

58. Pandis N, Yuesheng J, Limon J et al (1993) Interstitial deletion of the short arm of chromosome 3 as a primary chromosome abnormality in carcinomas of the breast. Genes Chromosomes Cancer 6 : 151–155

59. Richards BW (1969) Mosaic mongolism. J Ment Defic Res 13 : 66–83

60. Rowley JD (1987) Chromosome abnormalities and oncogenes in human leukemia and lymphoma. In: Vogel, F, Sperling K (eds) Human genetics. Proceedings of the 7th International Congress, Berlin 1986. Springer, Berlin Heidelberg New York, pp 401–418

61. Rowley JD (1994) 1993 American Society of Human Genetics Presidential Address: Can we meet the challenge? Am J Hum Genet 54 : 403–413

62. Royer-Pokora B, Schneider S (1992) Wilms' tumor-specific methylation pattern in 11p13 detected by PFGE. Genes Chromosomes Cancer 5 : 132–140

63. Russell LB, de Hamer DL, Montgomery CS (1974) Analysis of 30 c-locus lethals by viability of biochemical studies. Biol Div Annu Prog Rep ORNL 4993 : 119–120

64. Sakaguchi K, Harris PV, Ryan C et al (1991) Alteration of a nuclease in Fanconi anemia. Mutat Res DNA Repair 255 : 31–38

65. Sakaguchi K, Zdzienicka M, Harris PV, Boyd JB (1992) Nuclease modification in Chinese hamster cells hypersensitive to DNA cross-linking agent – a model for Fanconi anemia. Mutat Res DNA Repair 274 : 11–18

66. Salk D (1982) Werner's syndrome: A review of recent research with an analysis of connective tissue metabolism, growth control of cultures cells and chromosomal aberrations. Hum Genet 62 : 1–15

67. Schappert-Kimmijser J, Hemmes GD, Nijland R (1966) The heredity of retinoblastoma. Ophthalmologica 151 : 197–213

68. Schneider EL, Mitsui Y (1976) The relationship between in vitro cellular ageing and in vivo human age. Proc Natl Acad Sci USA 73 : 3584–3588

69. Schnyder UW (1966) Tumoren der Haut in genetischer Sicht. Praxis 55 : 1478–1482

70. Schroeder TM (1972) Genetische Faktoren der Krebsentstehung. Fortschr Med 16 : 603–608

71. Schroeder TM (1982) Genetically determined chromosome instability syndromes. Cytogenet Cell Genetics 33 : 119–132

72. Schroeder TM, Anschütz F, Knopp A (1964) Spontane Chromosomenaberrationen bei familiärer Panmyelopathie. Hum Genet 1 : 194–196

73. Schroeder TM, Drings P, Beilner P, Buchinger G (1976) Clinical and cytogenetic observations during a six-year period in an adult with Fanconi's anaemia. Blut 34 : 119–132

74. Schroeder TM, Tilgen D, Krüger J, Vogel F (1976) Formal genetics of Fanconi's anemia. Hum Genet 32 : 257–288

75. Schroeder-Kurth TM, Auerbach AD, Obe G (eds) (1989) Fanconi anemia: clinical, cytogenetic and experimental aspects. Springer, Berlin Heidelberg New York

76. Swift M (1982) Disease predisposition of ataxia-telangiectasia heterozygotes. In: Bridges A, Harnden DG (eds) Ataxia teleangiectasia – a cellular and molecular link between cancer, neuropathology and immune deficiency. Wiley, New York

77. Swift M, Chase C (1979) Cancer in families with Xeroderma pigmentosum. J Natl Cancer Inst 62 : 1415–1421

78. Swift M, Sholman L, Perry M, Chase C (1976) Malignant neoplasms in the families of patients with ataxia-teleangiectasia. Cancer Res 36 : 209–215

79. Swift M, Reitnauer PJ, Morrell D et al (1987) Breast and other cancers in families with ataxia-teleangiectasia. N Engl J Med 316 : 1289–1294

80. Swift M, Morrell D, Massey RB, Chase LL (1991) Incidence of cancer in 161 families affected with ataxia-teleangiectasia. N Engl J Med 325 : 1831–1836

81. Szilard L (1959) On the nature of the ageing process. Proc Natl Acad Sci USA 45 : 30–45

82. Vogel F (1957) Neue Untersuchungen zur Genetik des Retinoblastoms (Glioma retinae). Z Menschl Vererbungs Konstitutionslehre 34 : 205–236

83. Vogel F (1979) Genetics of retinoblastoma. Hum Genet 52 : 1–54

84. Vogel F (1984) Relevant deviations in heterozygotes of autosomal-recessive diseases. Clin Genet 25 : 381–415

85. Vogelstein B, Kinzler KW (1992) p53 function and dysfunction. Cell 70 : 523–526

86. Vogt PK (1983) Onkogene. Verh Ges Dtsch Naturforsch Arzte 235–247

87. Von Hansemann D (1890) Über asymmetrische Zellteilung in Epithelkrebsen und deren biologische Bedeutung. Virchows Arch Pathol Anat 119 : 299–326

88. Wallace DC (1992) Mitochondrial genetics: a paradigm for aging and degerative disease? Science 256 : 628–632

89. Watson JD, Hopkins NH, Roberts JW et al (1987) Molecular biology of the gene, 4th edn. Benjamin/Cummings, Menlo Park

90. Weinberg RA (1983) A molecular basis of cancer. Sci Am 126–142

91. Weinberg RA (1984) Ras oncogenes and the molecular mechanisms of carcinogenesis. Blood 64 : 1143–1145

92. Weinberg RA (1991) Tumor suppressor genes. Science 254 : 1138–1146

93. Weismann A (1891) Essays upon heredity and kindred biological problems, vol 1, 2nd edn. Clarendon, Oxford (1st edn 1889)

94. Willecke K, Schäfer R (1984) Human oncogenes. Hum Genet 66 : 132–142

95. Yunis JJ (1983) The chromosomal basis of human neoplasia. Science 221 : 227–236

96. Yunis JJ, Soreng AL (1984) Constitutive fragile sites and cancer. Science 226 : 1199–1204

97. Zakrzewski S, Sperling K (1980) Genetic heterogeneity of Fanconi's anemia demonstrated by somatic cell hybrids. Hum Genet 56 : 81–84

11 Mutation: Induction by Ionizing Radiation and Chemicals

I regard it as a piece of relatively good news for society that our genetic material does not appear to be as susceptible to the mutagenic effects of ionizing radiation as was at some time feared ... My best guess is that at currently regulated levels of chemical exposures, there is no more of a genetic problem [of chemical mutagenesis] than there is with respect to ... radiation exposure.

(J. V. Neel, *Physician to the Gene Pool*, 1994)

Public Interest in Induced Mutation. The preceding chapter discussed spontaneous mutations. "Spontaneous" here means that these mutations are unpredictable and without known cause, even though we know that some conditions – such as parental age – may enhance the probability of mutation. This probability increases under the influence of certain agents, such as energy-rich radiation and a number of chemicals. Since human beings in their normal environments are exposed to a variety of these agents, research on induced mutation is receiving more and more attention from the general public. Relatively large amounts of money have been allotted to this work, and scientists are expected to advise political authorities as to protective measures. In regard to radiation-induced mutations, an appropriate return from these investments has been forthcoming for a number of years. The World Health Organization, the International Commission on Radiation Protection (ICRP), the United States National Academy of Sciences, and other influential organizations have established expert groups and, with their help, have published estimates of genetic risks. There are still many gaps in our knowledge, particularly regarding the effect of low-level radiation on humans but a fairly coherent picture of the radiation threat is now emerging. Relatively little is known about the possible impact of environmentally induced mutations by chemicals.

There is much confusion within the scientific community concerning the specific nature and extent of the chemical threat and the kinds of information and recommendations that scientists should provide [35, 110]. One reason is that the problem is indeed more complex than that of radiation-induced mutation. Another reason may be that the scientific community rewards success in research within a relatively narrow field, carried out with technically difficult methods. This requires ingenuity in recognizing and precisely formulating problems that by necessity are limited in scope and can be solved by suitable methods. Such an approach does not usually require experience in many different areas of science. Within the natural sciences the barriers between specialties tend to impede free exchange of the information

available at different levels. Hence, scientists who are considered capable of providing expert advice on the basis of their prominence within a given field frequently may not have a balanced view of a complex problem. They tend to see problems mainly from the viewpoint of their own specialty.

The traditional channels of scientific communication – scientific societies, congresses, journals – have so far failed to overcome these difficulties in the field of environmental mutagenesis. International organizations such as WHO have not worked as efficiently in the fields of chemically induced mutation as in the field of radiation. Perhaps the solution will lie in new institutions organizing scientific efforts at all relevant levels. Unfortunately, such institutions have so far been successful in mobilizing broad-scale scientific endeavors only under two conditions: (a) when the goal was clearly defined and purely technological, and (b) when this was supported by a strong and immediate political motivation.

Examples that readily come to mind include the Manhattan Project in World War II, which led to the development of the atomic bomb, and the American and Russian projects to explore outer space.

Below we give our view of the problems. This view may help others to appreciate the complexity of the matter and to participate in the definition of goals.

11.1 Radiation-Induced Mutation

11.1.1 Basic Facts and the Problems Posed by Them
(See also [73–76, 97])

Capacity of Energy-Rich Radiation to Induce Mutation. That energy-rich radiation can induce mutations had been suspected but was first proven by Muller (1927) [54] in *Drosophila melanogaster* and by Stadler (1927–28) in barley after Mavor (1924) [49] had shown before the induction of nondisjunction. A condition for Muller's discovery was the ingenious use of *Drosophila* mutants in developing test systems for counting mutations, especially X-linked lethals. Results of Muller's classic experiment are presented

seen in Table 11.1. Here, dose t_4 is twice as high as dose t_2. Doubling the radiation dose approximately doubled the number of induced mutations. The scope of this experiment was too small, however, to detect spontaneous mutations among the controls.

Within the following two decades, classical radiation genetics developed. Its basic concepts were set out in the books of Lea and Catcheside (1942) [42], Timoféeff-Ressovsky and Zimmer (1947) [90], and Hollaender (1954–1956) [34].

Some Technical Remarks on Radiation. Two types of energy-rich radiation must be considered: electromagnetic waves and corpuscular radiation. The biological activity of electromagnetic waves in relation to their wavelength and energy is seen in Fig. 11.1. For a mutagenic effect the energy should be sufficient to lift an electron from an inner to an outer shell, rendering the atom unstable and more prone to chemical reactions. UV radiation has this property and is therefore mutagenic, provided it reaches the DNA. Its best-known chemical reaction is the linking of two adjacent thymine molecules. This prevents them from pairing with adenine. There-

fore UV photons cause point mutation but few structural defects. For human germ cells, UV radiation is not dangerous, being absorbed in the epidermis. There, however, it may induce somatic mutation – and cancer – of the skin (Chap. 10). Radiation photons of higher energy (X-rays, γ-rays) are able to push electrons out of the external shell so that the atom becomes a positive ion. These electrons may now react with other atoms, converting them into negatively charged ions. Both types of ions – together with free radicals – form the material for secondary chemical reactions. Corpuscular radiation consists not of energy photons but of particles. These may be charged, as are electrons and protons, or uncharged, as neutrons. Their physical effects depend on their kinetic energy. Ionization induced by neutrons is densely concentrated around the track of the particle, whereas electromagnetic waves (X and γ radiation) produce looser ionizations.

The biological effects of all types of radiation depend on the location of the source (inside or outside the body), type of radiation (electromagnetic waves, charged or uncharged particles), their energy, and the properties (density, water content, etc.) of the absorbing material.

Irradiation of any kind has not only direct but also indirect effects. Energy-rich neutrons, for example, may either be accepted into the atom's nuclei or transfer their kinetic energy to, say, hydrogen nuclei (protons). These protons are accelerated and undergo many secondary reactions with other molecules.

The radiation energy dose is usually measured in grays (Gy): 1 Gy is equal to the energy dose produced when 1 J energy is transmitted by ionizing radiation to matter of 1 kg mass under constant and defined conditions. This corresponds to 100 rad (old nomenclature), and, as a rule, to 100 roentgen (R), but the latter is defined with reference to the number of ionizations. Another important measure is the equivalent dose that measures the (negative) biological effect of a certain radiation dose. This is calculated by multiplying the energy dose by a factor that varies from one type of radiation to the other, depending mainly on the energy discharge; it is higher, for example, with densely than with loosely ionizing

Table 11.1. Muller's classic experiment confirming mutation induction by X-rays (from Muller 1927 [54])

Experiment	Number of chromosomes tested	Number of mutations observed		
		Lethals	Semilethal mutations	Visible mutations
Controls	198	0	0	0
X-rays $(t_2)^a$	676	49	4	1
X-rays $(t_4)^a$	772	89	12	3

^{a} The dose t_4 was twice the dose t_2.

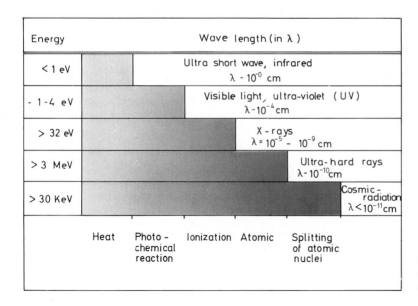

Fig. 11.1. Biological activity of electromagnetic waves in relation to their wavelength and energy

radiation. It is measured in J/kg. An old measure is the rem (1 rem = 0.01 J/kg).

The new measure is sievert (1 Sv = 100 rem). We often cite data from the older literature; we use here the new nomenclature: 1 Gy = 1000 mGy = 100 rad = 100 R. This may not always satisfy the needs of radiation physicists, but for the purposes of this book it should suffice.

Results and Concepts of Classic Radiation Genetics [34, 42, 90]. The most important results and concepts of classic radiation genetics may be summarized as follows:

a) To induce mutations in a certain cell the radiation must reach this cell, for example, a germ cell. The truth of this statement is not quite as self-evident as it may seem. Indirect influences, for example, by induction of a chemical compound that reaches the gonads via the circulation, are not impossible a priori; a limited amount of indirect action has indeed been demonstrated. Nevertheless, the statement is a good approximation for practical purposes. This principle is important for humans because the high absorption rate of some types of radiation (UV, or very low energy X-rays) prevents them from being dangerous to the germ cells. Nevertheless they may still be dangerous to the individual by inducing somatic mutations and cancer.

b) Radiation does not create any new biological phenomena; it only enhances the probabilities of various mutations and cellular events which occur spontaneously from time to time. Phenotypic effects of radiation-induced mutations do not differ basically from spontaneous mutations. This rule has also been confirmed for chemically induced mutations. However, not all types of spontaneous mutation are increased in the same proportion by all mutagenic agents. On the contrary, there are definite differences in the relative frequencies of the various types of both spontaneous and radiation-induced and chemically induced mutations.

The fact that any type of inducible mutation may also occur spontaneously creates a difficult statistical problem in trying to prove that an increase of the mutation rate in a human population is caused by exposure to mutagenic agents. This may be explained by an example from human teratology: thalidomide, which was introduced as a sleep-inducing drug, turned out to be teratogenic when taken during early pregnancy. The pattern of malformations induced by the drug was strikingly characteristic, and the combination of short and malformed limbs (phocomelia) with the frequent occurrence of ear, eye, and internal malformations was unique. It was this specificity that first led physicians to realize that something new

must have happened and then to look systematically for possible teratogenic agents. However, had the drug caused the same number of cases of cleft lip and palate or neural tube defects as it had of phocomelia, it very probably would still be regarded as a useful hypnotic and anti-emetic agent with excellent indications for use in pregnancy.

These considerations suggest that any mutagenic agent inadvertently introduced into our environment and causing the same amount of damage as thalidomide would very probably be overlooked.

c) A third problem that has frequently been discussed is the dose factor in the mutation rate. For mutation requiring only one primary event, the dose-effect curve has been shown to be simply linear:

$$M = \mu + kD$$

(M = number of mutations; μ = spontaneous mutation rate; D = dose; k = mutation rate/dose unit). This linear dose effect holds true as long as only a moderate proportion of all irradiated cells include an induced mutation. For higher mutation rates there is a saturation effect which leads to a flattening of the curve. Here an exponential equation would describe the curve more properly. An example for *Drosophila melanogaster* is shown in Fig. 11.2. The radiation doses used are relatively high; for lower doses the evidence is not very conclusive, as the fitting of a dose-effect curve to data requires very large sample sizes.

These results have been interpreted by the target theory. Every mutational even is produced by one

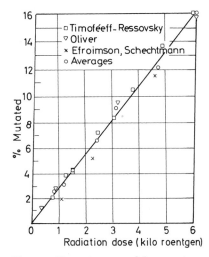

Fig. 11.2. Linear increase of the mutation rate for point mutations and single chromosome breaks in Drosophila melanogaster. Note linearity of effect. Data from various authors. (From Timoféeff-Ressovsky and Zimmer 1947 [90])

hit in a susceptible structure, and the probability of this structure being struck increases linearly with increasing dosage. The saturation effect with very high doses (higher than those used in the experiments depicted in Fig. 11.2) is due to the same structure being struck more than once.

d) The theory predicts a nonlinear dose-effect relation for mutations needing more than one primary event. Many translocations require two breaks within a reasonably short distance and at approximately the same time for the chromosomes to rejoin after they are broken. However, the two breaks need not be produced by two separate hits. They may also be induced by one particle or photon, especially if the density of ionizations is very high, for example, when neutrons are used. In this case a dose-effect curve could be expected which combines a linear (one-hit) and a quadratic (two-hit) component:

$$M = \mu + k_1 D + k_2 D^2$$

Such dose-effect curves have indeed been observed in many experiments. Figure 11.3 shows an example for *Drosophila melanogaster*.

e) For single-hit events the theory makes still another prediction of great practical importance. Obviously the number of hits depends on the dose rather than on the time within which the dose is applied. It should therefore be unimportant whether the same dose is given within a very short time (high "dose rate") or extended over a much longer time (low "dose rate"). The first *Dro-*

sophila experiments seemed to confirm this prediction. More recent results from mice [71], however, have shown that in many cases a certain dose applied at a lower dose rate leads to a lower number of mutations than the same dose applied at a higher dose rate.

Confirmation and Extension of These Results. A great number of studies have helped to confirm and partially modify these results. The developmental stage of the irradiated germ cell has been taken into account; various – for example, biochemical – phenotypes have been utilized, and radiation-induced mutations have been explained in terms of chemical reactions.

Influence of the Chemical Environment, Especially the O_2 Content of Irradiated Tissue. A secondary consequence of irradiation is the formation of highly reactive radicals such as peroxides. Their formation requires oxygen. Therefore it is not surprising that a high oxygen content of irradiated tissue enhances mutation induction. This effect is strong, for example, with X-rays. It is almost or totally absent with densely ionizing radiation, such as α particles.

Molecular Effects of Radiation [92]. Classical radiation genetics defined mutagenic effects at the phenotypic level. However, morphological examination of chromosomes, especially in plants such as *Tradescantia*, was introduced at a relatively early stage, and it was shown that many of the mutational changes

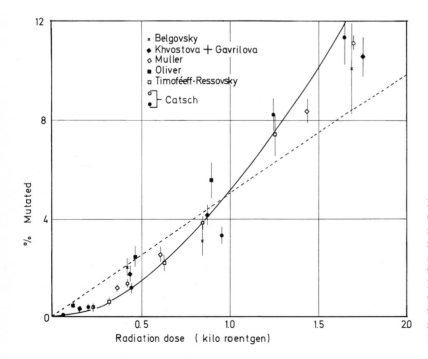

Fig. 11.3. Chromosome aberrations (two-break events) in relation to the radiation dose (in 10 Gy = 1 kilo roentgen). The two-hit curve *(uninterrupted line)* fits the data better than the one-hit curve *(dotted line)*. Data from various authors. The figure gives experimental points and their standard deviations. (From Timoféeff-Ressovsky and Zimmer 1947 [90])

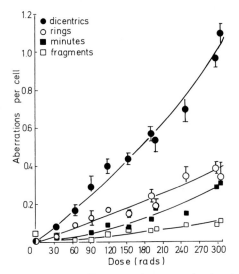

Fig. 11.4. Dose-effect curve from acute in vitro irradiation of human lymphocytes. Dicentrics, ring chromosomes, minutes (fragments containing a centromere), and fragments. (From Bloom 1972 [12])

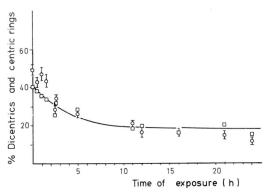

Fig. 11.5. Fraction of metaphases with dicentric chromosomes (□) and ring chromosomes (○) after in vitro irradiation with 2 Gy X-rays. Decrease of the effect with increasing exposure time and decreasing dose rate (= increasing exposure time). (Adapted from Brewen and Luippold 1971 [14])

could be explained by induction of morphologically visible chromosome breaks and their sequelae, such as translocations.

For a long time it was not even known whether ionizing radiation could induce point mutations in genes. Many investigators considered it plausible, but there remained the possibility that all radiation-induced mutations are basically small deletions or chromosome rearrangements. Recently this matter was settled by using the φ X174 phage, which has only one DNA half-strand. Here reverse mutations were induced that could not be explained by any other mechanism than a single point mutation. In bacteria, the induction of gene mutations, among them transitions, has been shown for the tryptophan locus of *E. coli* [92].

The first eukaryote in which radiation-induced gene mutations were found was *Neurospora crassa* [47]. Of the induced mutations 42% were transitions, 37% insertions or deletions of single base pairs, and the rest various origins, among them probably transversions. The occurrence of radiation-induced point mutations has been confirmed by direct DNA analysis in various mammalian systems [84].

Basic Facts of Radiation Genetics Reconfirmed in Human Lymphocyte Chromosomes [11, 12]. Most of work in classical radiation genetics was carried out using organisms such as *Drosophila* which are only remotely related to humans. Irradiation experiments on chromosomes were carried out mainly on plants. Work on humans began soon after the method of chromosome preparations from lymphocytes became available (Sect. 2.1.2.2). Dicentric chromosomes turned out to be an especially good indicator for chromosome breakage. Figure 11.4 shows the dose-effect curve. Its slope is more or less linear but has a tendency to become steeper with higher doses. Such dose-effect curves are produced if some of the primary events are one-hit and others are two-hit events. Formation of a dicentric chromosome requires two chromosome breaks in closely adjacent chromosomes. If our theoretical notions are correct, these breaks may be caused either by a single hit or by two hits. Hence, the dose-effect curve found does not contradict these notions. Figure 11.4 also provides data for ring chromosomes, minutes, and fragments.

Figure 11.5 shows a decrease in the number of dicentric and ring chromosomes with the same dose (2 Gy) but decreasing dose rate, i.e., radiation given over longer periods at lower rates. These data confirm the dose-rate effect that have been established for other organisms, such as the mouse. These few data show that basic phenomena from radiation genetics also apply, in principle, to human chromosomes.

The human population is exposed to ionizing radiation from a number of sources. Geneticists are asked to advise the public on the extent of possible hazards. How can these questions be answered?

11.1.2 Problem of Estimating the Genetic Risk Due to Radiation and Other Environmental Mutagens

The problem of estimating the extent of risk to the human population and from all other mutagenic agents can be specified as follows:

a) In what way, if any, does the agent affect the genetic material?

b) How extensively is the human population exposed to this agent?

c) What is the probable increase in mutations compared to the "spontaneous" mutation rate?

d) What are the long-term consequences of this increase for the population?

These four questions should ideally be answered by the scientist. There is, however, a fifth question for which the scientist can only supply data. The answer should come from society as a whole.

e) How high an increase in mutational damage are we prepared to accept in exchange for the benefits that present-day society enjoys from mutagenic agents – for example, diagnostic and therapeutic X-ray use, nuclear energy, certain drugs, and amenities of all kinds?

Principles of Mutagenicity Testing. The question regarding effects on the genetic material is very complex. It can be subdivided into a number of more specific questions, such as:

1. What kinds of mutations are induced: genome, chromosome, or gene mutations?
2. Are they induced mainly in germ cells or in somatic cells?
3. If they are induced in germ cells,
   a) What stage of germ cell development is primarily affected?
   b) Are the mutations transmitted to the next generation, or are they eliminated, for example, during meiosis?
   c) If they are transmitted, what phenotypic changes may be expected in offspring?
4. If they are induced in somatic cells,
   a) What cells are especially endangered?
   b) What are the consequences for the individual?

All these questions should be investigated by a comprehensive mutagenicity testing program. Which organisms should be tested in such a program? In theory the answer is obvious. We are interested in humans; therefore the best approach would be to examine these problems in humans. This, however, is not possible experimentally for ethical reasons. Only naturally occurring situations can be explored, and here the conditions are usually so complex that a clearcut answer cannot be expected. Therefore an experimental animal must be found for many studies. This animal should fulfill four main conditions:

1. It should be related closely enough to humans for meaningful extrapolation. The inevitable differences between the experimental species and humans should be of such kind and order of magnitude that they can be accounted for either theoretically or experimentally when the comparison is made.
2. The generational turnover should be rapid so that genetic experiments can be carried out within a reasonable time.
3. It should be possible to keep test animals in sufficiently large numbers at a reasonable price.
4. Sufficient knowledge about the genetics of the organism should be available.

The only animal that fulfills all these conditions is the mouse (*Mus musculus*). Therefore radiation genetics of mammals – and mutation genetics in general – is largely that of the mouse. Other species, such as rats, Chinese hamsters, and marmoset monkeys, have been used occasionally. Thus the following sections are devoted mainly to the mutation genetics of the mouse, with inferences to be drawn for human beings. When available, data from humans or from other species are cited for comparison.

In Vivo Test Systems for Mutagenic Agents in Germ Cells of the Mouse. The test systems available for this animal can be subdivided into in vivo and in vitro systems and into systems for examining germ cells and somatic cells.

The available test systems for germ cell mutations are shown in Fig. 11.6. Assume that a mutation has been induced in spermatogonia. If this mutation is a chromosome mutation, it can be initially identified in mitotic spermatogonial divisions. During the first and second meiotic divisions the influence of meiosis on induced aberrations may be observed. The next mitotic divisions occur during early embryonic development of the F_1 generation. At this stage it is possible to flush zygotes out of the fallopian tube and to examine their chromosomes.

When the oocytes are prepared from the ovary, the first meiotic divisions in the female germ cell can be observed. Depending on the time lapse between copulation and killing of the animal, the stage of oocyte and early zygote can be examined. Cleavage stages may even be grown for a number of additional divisions in culture.

Implantation of the zygote in the uterus occurs on the 9th day of pregnancy; the zygote is then in the blastocyst stage. Examination of implanted embryos is possible a few days thereafter. The so-called deciduomata indicate implantation sites of embryos that died a short time after implantation. Also visible are some embryos that died at a later stage, as well as healthy embryos. Deciduomata, together with embryos that died later, represent postimplantation zygote loss. The preimplantation loss may be inferred from the difference between the number of corpora lutea in the ovaries and the total number of implantations.

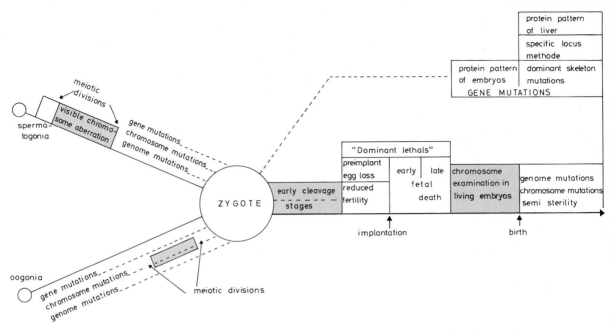

Fig. 11.6. In vivo test systems for examining the effects of a mutagenic agent in the mouse. (For details, see text)

Preimplantation and postimplantation zygote loss, unless shown to be due to nongenetic factors, is usually attributed to "dominant lethals." Despite more modern developments the dominant lethal method remains one of the standard methods for mutagenicity testing [67]. The surviving animals can be examined for chromosomal anomalies during late pregnancy.

Up to this stage of late pregnancy all methods are cytogenetic. Chromosome and genome mutations induced in early germ cell stages can now be followed through all stages of development to the newborn animal. There is no difficulty in examining mitotic chromosomes after birth in somatic tissue or even in following them up into the germ cells of the F_1 generation.

Examination of point mutations in the mouse must await birth. Methods for a group of dominant and a group of recessive mutations then become available. The dominant mutations ascertained are mainly those affecting the skeleton (Sect. 6.1.2.5). [74].

Multiple Recessive Mutation Method. The method for assessing recessive mutations was one of the first to be developed in mouse radiation genetics. It permits detection of mutations at a small number of loci with great precision. Many of the basic results in this field have been achieved using this method. Wild-type animals, for example, males, are irradiated and mated with a test stock are homozygous for seven autosomal-recessive mutations. When no mutations are

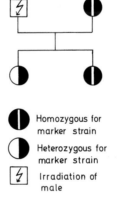

Homozygous for marker strain

Heterozygous for marker strain

Irradiation of male

Fig. 11.7. Principle of the multiple recessive mutation method. A wild-type male *(upper left)* is irradiated and mated with a female of the test stock *(upper right)*. If no mutation is induced, the offspring will be heterozygous for the test loci and therefore phenotypically wild-type *(lower-left)*. If a mutation at one of the seven loci is induced in one of the sperms, one animal in the offspring will be homozygous for this gene and show the mutant phenotype

induced, the F_1 animals are all heterozygous and show the wild-type phenotype (Fig. 11.7). However, if one of the seven mutations has been induced in the paternal germ cell, the F_1 animal becomes homozygous for only this mutation and shows the phenotype. The mutations in the test stock are selected in such a way that each homozygote can easily be visually identified. Figure 11.8 for example, shows a homozygote for one of these genes (piebald spotting).

Fig. 11.8. Mouse with the phenotype of one of the seven test loci *(piebald spotting)* together with the wild-type father, the test stock mother *(white)*, and the heterozygous wild-type sibs. (Courtesy of Dr. U. Ehling)

Both methods used for the screening of gene mutations – the skeleton method and the multiple loci method – permit analysis at the qualitative phenotypic level (Sect. 6.1) of gene action. Hence, the results, important as they are for a quantitative assessment of genetic effects in the mammalian organism, are not satisfactory for analyzing the mechanisms of mutation induction. In addition, they provide information only for a limited number of gene loci. However, results of this work suggested that the induced mutation rates of various genes may differ considerably.

When the methods of protein analysis by electrophoresis became available, a number of attempts were made to introduce such techniques into mutagenicity testing. For example, a "biochemical specific-locus test" has been developed which includes, among other markers, products of the mouse hemoglobin loci; the electric charge of proteins and enzyme activities [24, 74]. The histocompatibility (H) alleles which are homologous to the human HLA system (Sect. 5.2.5) are also being used [74]. Theoretically this approach is impeccable. Induced mutations can be identified at the molecular level, and electrophoretic systems are now available for a great number of loci (Sect. 12.1). The practical difficulty is simply the large number of examinations needed. Inexact as our estimates of nucleotide mutation rates may be, they show an order of magnitude between 10^{-8} and 10^{-9} per nucleotide or amino acid replacement (Sect. 9.4.1). Even if the spontaneous mutation rate per gene for electrophoretically detectable mutations were only about 10^{-6}–10^{-7}, such a rate would require a huge amount of testing to discover very few mutants. This problem is critical in devising ways of screening human populations for mutation rate increases. Is there any way to overcome this difficulty in animal experimentation – apart from setting up huge and expensive testing facilities?

A promising technique is the two-dimensional assay with electrofocusing in one direction and electrophoresis in the other (discussed in Sect. 12.1) [36]. A highly constant pattern of several hundreds of spots can be obtained from protein extracts of inbred mice. Point mutation may lead to the shifting of one "spot," which can easily be recognized (Fig. 11.9).

Test Systems for Somatic Mutations. The simplest test system for somatic mutations is chromosome examination in lymphocyte cultures or in bone marrow. The lymphocyte system has frequently been used with humans. It is in fact the only one that permits examining high-risk groups of humans with a minimum of distress for genetic effects of exposure to mutagenic agents.

A still simpler test system examines chromosomes irradiated in vitro. In most cases human lymphocyte cultures have been used. With this system it was shown that the general rules of mutation induction as worked out in experimental genetics also apply to human chromosomes. Assays for examining point mutations in single cells in vitro have been developed for some biochemical markers (Chap. 9), and the same methods are being utilized for mutagenicity testing. In theoretical terms this approach is elegant. Its resolving power is several orders of magnitude higher than that of in vivo methods for detecting gene mutation since the experimental unit is not the individual but the single cell. Moreover, mutants can be cloned and subjected to further biochemical and molecular investigation. In practice, however, the method still meets with difficulties.

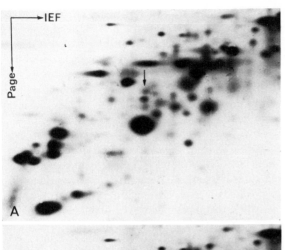

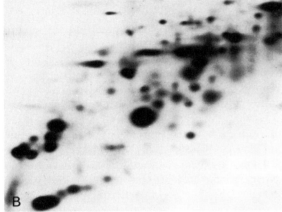

Fig. 11.9 A, B. Two-dimensional, electrophoretic separation of soluble proteins from fetal mouse liver; part of a protein pattern. The protein sample is taken from the supernatant after homogenization and centrifugation of single livers. Separation of proteins is performed on polyacrylamide by isoelectric focusing *(IEF)* in one dimension and by electrophoresis *(PAGE)* in the second. **A** Part of protein pattern of fetal liver after treating the sire with a mutagen (methylnitrosourea). *Arrow,* a new protein spot. **B** Normal protein pattern of a fetal liver from the same inbred mouse strain. (Courtesy of Dr. J. Klose, Berlin)

Methods from Molecular Genetics. In the long run, all the methods discussed use some change in phenotypic characteristics – either of entire individuals or of single cells – as endpoints and may be replaced by direct assessment of changes at DNA level. Various approaches have been suggested, from comparison of the entire genome between parents and offspring by "subtractive hybridization" to DNA sequencing [48, 53]. So far all of these methods, despite their conceptual advances, have remained without practical significance. Nonradioactive in situ hybridization (FISH; Sect 3.1.3.3) allows detection of very small deletions and may offer good prospects for practical application in the near future [19].

11.1.3 Results of Radiation Mutagenicity Testing in Mammals [95]

General Effects of Radiation on Mammalian Germ Cells. The development of germ cells in both human sexes has been described in Chap. 9. This is similar in principle in all mammals; minor differences between species are important for planning of mutation experiments but are not covered here. In the mouse the following general effects of radiation on germ cell development have been observed.

Acute irradiation of the male germ cells with 2–4 Gy kills most spermatogonia, whereas more mature germ cells (spermatocytes and all postmeiotic cell stages) survive. There is therefore little impairment of fertility during the first 6 weeks following irradiation. During this period all germ cells that had already entered the spermatocyte stage become mature sperms. A sterile period of 2–3 months follows, depending on the radiation dose. After this time fertility is resumed. Meanwhile, the testicular tubules become repopulated, starting from a very small population of A spermatogonia.

Female animals, after a short lag phase, become permanently sterile even with low doses. When irradiated with 0.5 Gy X-rays, female mice have three to four litters before they become sterile due to the destruction of oocytes. As oocytes approach ovulation, they become more resistant, and shortly before ovulation their death rate does not differ from that of controls. The high degree of sensitivity of oocytes makes mutation experiments difficult. These differences between the sexes and between the stages of germ cell development regarding radiation sensitivity must be kept in mind when considering the results of mutagenicity experiments.

Chromosome Mutations in Male and Female Germ Cells of Mice [28]. Spermatogonial mitoses show a high frequency of chromosome aberrations (breaks, reunions, dicentric chromosomes) after acute irradiation. Such aberrations can be detected when the first meiotic divisions are examined.

Experimental data from human males are available. The use of human beings for such experiments appears highly questionable. Although the investigators stated that they used prisoner volunteers only under conditions where these men would have no further children [15], such assurances are difficult to adhere to and there is the further possible risk of carcinogenicity.

Testis biopsy material was obtained from nine human volunteers who had received testicular irradiation with high-dose X-rays (0.78, 2, or 6 Gy). The interval between irradiation and sampling varied, depending on the dose. For comparison, adult male individuals of various other animal species were irradiated with similar doses and under comparable conditions (Table 11.2). The doses used in these studies ranged between 1 and 6 Gy. The increase in yield of reciprocal translocation with increasing radiation dose varied in different mammalian species but was very high in humans. In various studies in mice and rabbits the effect was about half that found in humans; guinea

Table 11.2. Comparison of the slope (β) of linear regression for various species of mammals for the induction of reciprocal translocations in spermatogonia (data from various authors [95])

Animal	$10^{-4} \pm$ S.D. 10^{-4}
Rhesus monkey	0.86 ± 0.04
Mouse (various strains)	between
	1.29 ± 0.02
	and 2.90 ± 0.34
Rabbit	1.48 ± 0.13
Guinea pig	0.91 ± 0.18
Marmoset	7.44 ± 0.95
Man	3.40 ± 0.72

The linear regression coefficients (β) were estimated for all species using the data given by the authors on doses giving the peak yield of translocations.

pigs and rhesus monkeys showed still lower values. No Robertsonian translocations seem to have been induced by ionizing radiation [29].

Reciprocal translocations can also be induced by irration of female mice; in this case the details of the outcome vary depending on the experimental conditions (e.g., time interval between irradiation and ovulation).

Direct Evidence of the Outcome of Induced Chromosome Aberrations. There is strong selection against germ cells and zygotes containing chromosome aberrations. After irradiation of female mice with fairly high doses, for example, very few translocations were found among F_1 offspring. After irradiation with as much as 3 Gy, a total of eight translocations were recovered among 1735 progeny (female and male), about half as many as after irradiation of males [95]. This phenomenon has been examined in detail through various stages of embryonic development by direct examination of chromosomes in early embryonic stages after treatment by irradiation of female mice [65]. The decrease in the number of zygotes with chromosomal aberrations after treatment of oocytes during the preovulatory phase was examined through four stages: the second meiotic division, blastocysts, death during embryonic age as evidenced by the dominant lethal test, and proportion of chromosomal aberrations among living embryos in the late embryonic age. About 88% of all meiotic (II) cells showed induced numerical or structural chromosome aberrations. By late embryonic age all cells in which a chromosomal aberration had been identified were eliminated, and the number of embryos carrying an aberration was not higher than in the controls. In 1977 the UNSCEAR report had estimated that approximately 6% of all fetuses with induced chromo-

some aberrations survive until birth. The above study showed that this was a very cautious estimate; the proportion of survivors is probably lower. As noted, the great majority of spontaneous chromosomal aberrations in humans are also eliminated before birth (Sect. 2.2.4). It therefore is reasonable to believe that this occurs in the great majority of induced aberrations.

Radiation-Induced Genome and Chromosome Mutations: Sensitivity of Certain Cell Stages. X-linked genetic markers make it possible in the mouse to distinguish XO from XX animals and to ascertain whether the X chromosome in XO individuals is of paternal or maternal origin. With this method it has been found that XO individuals occur spontaneously at a frequency of 0.1%–1.7%, depending on the mouse strain examined. The single X chromosome is usually maternal. The paternal chromosome apparently is lost between impregnation and the first cleavage division [69, 72, 101].

During this relatively long period – about 4.5 h – the paternal and maternal genomes do not become fused (Sect. 2.1.2.4) but rather form "pronuclei." It is at this time that the paternal X is at risk of being lost. This risk may be enhanced by radiation, especially if the animal is irradiated between impregnation and the first cleavage division, i.e., during the pronucleus phase. Radiation at this phase yields exclusively XO animals and no XXY types. This implicates X chromosome loss and not nondisjunction as the cause of the chromosome defect.

Nondisjunction seems to be enhanced by irradiation of male spermatocytes, mainly in the preleptotene stage, as evidenced by metaphase II examinations.

Loss of the X chromosome a short time after impregnation – and during the first cleavage divisions – is also frequent in humans. In fact this seems to be the most frequent cause of the formation of XO zygotes. Moreover, the X chromosome that is lost is generally the paternal, as shown by marker studies (Sect. 9.2). Thus the similarity to spontaneous X chromosome loss in the mouse is close. It is therefore safe to conclude that human preovulatory oocytes and very early zygotes may also be susceptible to chromosome loss and possibly to induction of nondisjunction and structural chromosome damage [30].

Radiation-Induced Gene Mutation in the Male Germ Line. Methods for chromosome examinations were not available to mammalian radiation genetics in its most fertile period during the 1950s. Therefore emphasis was principally on the induction of gene mutations. The most extensively used method was the multiple loci test for visible recessive mutations. The number of animals tested reaches the millions. As expected, the data show a marked increase in the number of mutations compared with the spontaneous mutation rate in both sexes. The mutation rate was estimated to be about 30×10^{-8}/cGy per locus in the male and around 18.5×10^{-8}/cGy per locus in the female. This induced mutation rate differs widely depending on treatment conditions. What are these conditions?

Dose-Rate Effect. In contrast to the *Drosophila* data discussed above, a dose administered at one time (a high dose rate) has a much stronger effect than the same dose distributed over a longer time (a low dose rate) [71]. Figure 11.10 shows the dose-effect curves for a series of experiments on male mice with high and low dose rates. The dose-rate effect is seen by comparing the upper two curves (acute X-ray radiation) with the lower two, which show the effect of chronic γ irradiation with extremely low dose rates. Despite the huge number of examined animals the 90% confidence intervals for all mutation frequencies were rather high. A dose applied at a high dose rate induced about three times as many mutations as the same dose at a low rate.

Such a dose rate effect was also shown in female germ cells; it is caused by repair processes that are more efficient when the damage is spread over a longer period. The lessened effect for single irradiation with 10 Gy is due to prior elimination of severely damaged cells. Such paradoxic dose-effect curves are not rare in mutation genetics; a mutagenic agent may damage a cell at two levels. It may impair it in a way that prevents it from further division, or it may leave the cell intact, damaging only the genetic material. In analyzing mutagenic activity at the phenotypic level we consider only damage to the genetic material, tacitly assuming that the viability of the cell is not affected. If a relatively low dose affects mainly DNA while a higher dose also damages cell viability, a paradoxic dose-effect curve may result.

Another factor that influences the mutation rate is the developmental stage of the germ cells at which irradiation occurs. In female mice, for example, irradiation administered more than 7 weeks before mating does *not* cause any increase in mutation rate. In males the irradiation of more mature cells (spermatocytes) leads to a higher mutation rate.

Ionizing radiation induces mainly structural chromosome damage and only relatively few point mutations [70, 74]. More thorough examination of many induced mutations at the seven loci of the mouse has shown that structural aberrations are also the most frequent cause of these mutations. Most of these were small deletions.

In addition to recessive mutations, dominant mutations – especially those affecting the skeleton or caus-

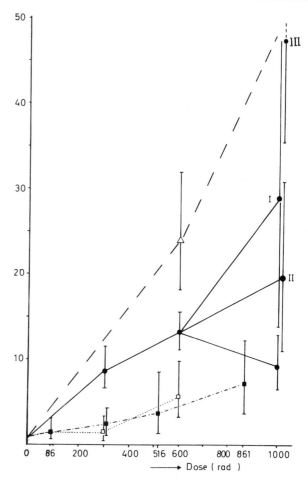

Fig. 11.10. Dose-effect curves and dose-rate effect after irradiation of male mice in the poststerile phase (irradiation of spermatogonia). Number of observed mutants per 100 000 gene loci tested in the single locus test. For all experimental points the 90% confidence intervals are included. –, Irradiation with one acute dose. The three different points for 1000 rad (= 10 Gy) indicate one acute dose (lowest point) and two experiments with fractionated application. △– – –△, Two doses with an interval of 24 h; ■–·–·–■, chronic γ irradiation (0.9 mGy/min); □····□, chronic γ irradiation (0.1 mGy/min). (Adapted from Russell et al. 1972 [72])

ing cataracts of the lenses – have also been studied [95, 100]. The extremely variable expressivity and markedly incomplete penetrance of such mutants is discussed in Sect. 6.1.25.

Doubling Dose. Discussion of genetic radiation hazards for humans frequently consider the so-called doubling dose [46]. The concept of a doubled mutation rate is entirely arbitrary and was selected as a convenient reference to find a dose of radiation which, if delivered to a human population, would double the natural mutation rate. *It is obvious from the foregoing discussion that there cannot be a single doubling dose.* The doubling dose must vary with the type of mutation, the germ cell stage at which irradiation occurs, the specific kind of radiation, and the dose rate. The use of a single doubling dose for all types of exposure of humans is therefore meaningless. Doubling doses for specific situations, such as an acute exposure doubling dose or a chronic exposure doubling dose for radiation, makes more sense. Older data from mouse experiments suggested doubling doses of about 0.18–0.52 Sv for acute and about 1 Sv for chronic irradiation. Doubling doses in humans were thought to be of the same order of magnitude. Recent studies on children of atomic bomb survivors in Hiroshima and Nagasaki have led Neel et al. [61] to suggest a doubling dose estimate in humans which is much higher (i.e. indicating a lower radiation effect). This led to a reassessment of the old estimates for the mouse, with the result that the combined evidence in the mouse may also point to a higher doubling dose [56] (see below).

Population Experiments with Mice and Other Mammals. The experimental results discussed relate only to the first of the five questions of Sect. 11.2: in what way does radiation affect the genetic material? However, some animal experiments were carried out with the fourth question in mind: what are the long-term consequences for the population of a mutation rate increase due to irradiation?

Some experiments with long-term irradiation of animal populations continued over many generations. The radiation doses were of the order of magnitude of several grays per generation; some were administered at high and others at low dose rates. The overall effects were surprisingly minor: litter size was usually reduced, the number of dead implantations (dominant lethals) was increased, and in some experiments there were more sterile animals. Progeny after irradiation of many generations of ancestors in some cases even surpassed the nonirradiated controls in life span [101].

One experiment with rats [62] examined a behavioral characteristic – maze learning. Progeny of irradiated parents were on the average somewhat less "bright" than the controls.

The outcome of these population experiments can be interpreted either optimistically or pessimistically. Optimists may conclude that even long-term irradiation with very high doses causes little genetic damage. Almost all induced mutations are eliminated during meiosis or lead to early zygote or embryonic death. Such effects, however, would not be very important for health effects of living human populations. Pessimists, on the other hand, may object to extrapolation from multiparous animals to humans.

They may argue that embryonic and newborn deaths found in the animal experiments would turn up as a high stillborn rate and as early death of malformed children.

Most of the long-term population studies in mice date from the late 1950s and early 1960s. Little if any work of this kind is continuing because its explanatory power for the human situation is thought to be low.

One study was carried out in a rodent population exposed to a high level of natural irradiation in Kerala (Southern India). The average γ irradiation was 16 mGy per year, 7.5 times higher than in a control area. There was no difference whatever in skeletal abnormalities in comparison with animals from other non-exposed areas [31].

Conclusions from Mouse Radiation Genetics for Genetic Hazards to Humans. In Sect. 11.1.2 a number of questions were formulated, to which radiation genetic research in the mouse and other mammals may help provide answers.

1. What kind of mutations are induced? Ionizing radiation seems to induce mainly chromosome mutations. There is good evidence that genome mutations, especially aneuploidies, are also induced. Many of the induced mutations affect only a single functional gene; some may be gene mutations in the strictest sense. All these findings can probably be extrapolated to humans.
2. Are mutations induced mainly in germ cells or in somatic cells? There is good evidence that mutations are induced in all exposed cells – germ cells as well as somatic cells. It may be assumed that the findings mentioned under headings 1, and 2 hold true for humans as well.
3a. What stage of germ cell development is primarily affected? In the male all germ cell stages may be affected, but meiosis acts as a potent filter, especially for chromosome aberrations. Postmeiotic stages up to and including the time of fertilization seem to be more endangered than premeiotic germ cells. Extrapolation to humans also seems reasonable. In the female mouse, oocytes may be susceptible to mutation induction and to chromosome loss only when irradiated within the last 7 weeks before ovulation. Oocytes during the dictyotene stage that lasts many years without cell divisions are resistant to mutation induction. This result from the multiple-locus experiments in mice has been largely confirmed by cytogenetic evidence. An optimist would conclude that human oocytes would also be radiation resistant during most of their lifetime. A pessimist, on the other hand, would point to possible species-specific differences. We tend to hold the optimistic view. The oocyte from some days before ovulation until several hours after impregnation is especially sus-

ceptible to chromosome loss – especially loss of the X chromosomes – as well as to induction of structural aberrations and possibly meiotic nondisjunction. There is no reason why this should not be true for the human oocyte as well. Careful radiation protection during the weeks around conception is the obvious countermeasure. Are oogonia also sensitive to mutation induction by radiation? Mammalian oogonia are found only during the embryonic age; therefore pregnant dams must be irradiated and their daughters later mated. So far, however, there is only limited experimental evidence that points to induction of translocation at about half the rate of spermatogonia.

3b. Are mutations transmitted to the next generation or are they eliminated? Information is available mainly for chromosome aberrations. Here the meiotic division acts as a good filter. Many aberrations, however, pass this filter, and others are induced during and after meiosis in the female. At least 90% of the induced aberrations are eliminated during embryonic development – more than half of them even before implantation in the uterus and most of the remaining a short time after implantation. A minority, 5% or fewer, survive, giving rise to offspring with translocations or being aneuploid themselves. The high incidence of chromosome anomalies among spontaneous abortions in humans (Sect. 2.2.4) makes it very likely that most induced chromosome aberrations would be eliminated in the same way. More specifically, the experimental results suggest that in humans loss of zygotes before implantation may be at least as high as after implantation. A significant number of aberrations would survive, increasing the number of children with aneuploidies and balanced or nonbalanced chromosome aberrations.

3c. What phenotypic changes are to be expected in the offspring? Zygote loss before implantation in the uterus would probably go largely unnoticed, leading only to a short delay of menstruation. Zygote loss after implantation would show up mainly as spontaneous abortion, the distribution between early and late abortion possibly being similar to the distribution without radiation. Aneuploidies and unbalanced structural anomalies would lead to the well-known chromosomal syndromes. Balanced translocations would be phenotypically unnoticeable mostly in offspring of exposed parents but could lead to unbalanced zygotes in the following generations.

Dominant gene mutations may lead to a certain increase in the well-known dominant phenotypes. The frequency of such traits in the human population is maintained by an equilibrium between mutation and selection (Sect. 9.3.1). However, experience with skeletal mutants of the mouse suggests that other, and possibly much more frequent, dominant mutations may lead to less distinctive and more variable phenotypic changes, such as slight anomalies of the skeleton, connective tissue, or other organ systems. Here our extrapolations are bound to be much less certain than with visible chromosome anomalies.

4. If mutations are induced in somatic cells, which cells are especially endangered, and what are the consequences for the individual? In principle, all types of somatic cells are at risk. However, it is safe to assume that the risk for cells that divide frequently is especially high. These cells may give rise to cell clones with selective advantages comparised to other cells of the same type – and in the end, to malignant neoplasias. Such neoplasias caused by irradiation have actually been demonstrated in human populations (see below).

Altogether, radiation genetics of the mouse has provided results that are useful for estimating the radiation risk to humans. They give us a fairly good idea about induction and transmission of genome and chromosome mutations. As to gene mutations, the evidence is not quite as good, but it is known in principle that such mutations may be induced as well as transmitted.

11.1.4 Human Population Exposure to Ionizing Radiation

How strongly and how widely is to day's human population exposed to ionizing radiation? This is the second question that scientist needs to answer in estimating the extent of potential radiation damage to humans, and it has been addressed frequently. Only a few aspects and a small selection of data are discussed here. First, the natural background radiation and then the increase due to modern civilization – including diagnostic medicine – are considered.

Natural Background Radiation. All humans are exposed continuously to natural sources of irradiation. The average dose per year from cosmic radiation depends on the altitude above sea level and on the latitude. Terrestrial radiation is higher in areas with primordial rock than in alluvial lands. The total mean irradiation received by the gonads of humans in low altitude areas over a 30-year period (one generation) has been estimated 3–4 rem (Table 11.3). More recent estimates put the figure higher; earlier reports did not sufficiently consider the additional radiation load due to radon in the air.

Table 11.3. Average radiation load of human populations[a,b] (adapted from Barthelmess 1973 [10])

Sources	Europe		United States	
	mrem/year	rem/30 years	rem/years	rem/30 years
Natural background radiation	50 (≈ 30–120)			
Cosmic radiation	60			
Mean terrestrial radiation	20			
Incorporation of radioactive elements				
	≈ 130	≈ 3.9	≈ 100	≈ 3.06
Additional radiation due to modern civilization				
Medical diagnostics and therapy				
1958	20	0.6	73	2.2
1971	50	1.5		
Occupational (without nuclear technology)	< 1	< 0.03	0.8	0.024
Nuclear technology	< 1	< 0.03	0.003	0.0001
Fallout (atomic bomb testing)	8	0.24		
Minor sources (television; watches)	< 2	< 0.06	2	0.06
Air traffic	< 1	< 0.03	0.4	0.120
Sum total additional irradiation	≈ 60	≈ 1.8	≈ 80	≈ 2.4
Sum total natural background and additional irradiation	190	5.7	180	5.4

These figures do not include radiation received by the general population from nuclear power plants. Under circumstances of normal operations exposure of the general population is minimal. Workers in the nuclear power industry receive exposure of 0.006–0.008 rem/year. These data are relatively old; however, *there have been no major changes in recent years except a reduction to practically zero exposure in fallout from nuclear weapons testing, and a reduction of exposure to irradiation in medical practice.* The problem of imput from various sources is discussed extensively in the 1988 UNSCEAR report [95].

Additional Irradiation Due to Modern Civilization. Some irradiation estimates for Europe and the United States are listed in Table 11.3. Medical diagnostics and therapy constitute the major source. Within the large group of X-ray diagnostic measures, examination of the abdomen and the pelvis are the chief sources of relevant exposure. Improvement in X-ray technology and strict regulations for supervision have contributed substantially to minimizing this load. However, some X-ray diagnostic measures do increase the probability of mutations; any X-ray exposure should be clearly indicated and be carried out with maximum shielding. The benefit to the individual should be clear and must be weighed against the potential damage to him or her and to future generations. By far the greatest amounts of genetically relevant irradiation is delivered to cancer patients receiving radiation therapy of pelvic organs. Many such patients are of postreproductive age or die of their cancer within a relatively short time. However, if they do desire children, the results of radiation genetics research in mammals may help to minimize the genetic risk by advising them to avoid conception during and some weeks after therapy.

Exposure due to nuclear power plants constitutes a special problem that is receiving substantial attention in all industrialized countries. Depending on the type of reactor and on other conditions, the actual irradiation varies and is said to be much lower than the allowed limit in most cases. This statement is true, however, only in the absence of irregular emissions due to technical failure, accidents or sabotage. The risk of such an incident occurring sooner or later cannot be estimated, but it had been thought to be very low – until the nuclear power plant accident in Chernobyl on 26 April, 1986. In addition to killing a number of persons, this led to the radioactive contamination of a wide area. Many experts maintain that the energy needs of future decades cannot be met without nuclear power, and that the well-being, even the life, of future generations depends heavily on ensuring a sufficient energy supply. The geneticist can only hope that a major effort will be made to develop alternate technologies for clean energy.

Of all the sources contributing to exposure, the medical sector is the only one adding more than a trivial share to the whole load. Some of this is unavoidable. In many cases, however, the radiation load can be kept to a minimum. In activities where heavier exposure cannot be avoided, only persons who are past child-bearing age, or who for other reasons are unlikely to reproduce, should be employed.

It is an open question whether we will have to face increased exposure to radiation in the future. On the one hand, there is the indisputable increase due to professional use of radiation and radioactive substances and due to nuclear power plants. X-ray technology, on the other hand, is being so vastly improved, especially in medicine, that a certain decrease in exposure can be anticipated for age groups that may still have children. As Table 11.3 shows, the average exposure from all sources of modern civilization, excluding nuclear power plants, has not quite reached the average natural background radiation.

11.1.5 How Much of an Increase in the Spontaneous Mutation Rate Must Be Anticipated?

How can the probable increase in spontaneous mutation rate be calculated? This is the third question to be answered in estimating the projected genetic damage due to radiation. We need information on three points:

1. How much irradiation does the average individual receive?
2. How many additional mutations per dose unit are induced?
   When these two questions are answered, we shall know how much the mutation rate increases relative to the spontaneous rate. To turn this relative estimate into an absolute one, another question must be answered.
3. How many mutations would occur "spontaneously," i.e., without additional irradiation due to modern civilization?

Answers the first two questions answers may be derived from Sects. 11.1.3 and 11.1.4, together with some additional, direct observations in human beings. The third question turns out to be the most difficult and will need special discussion. The data in Table 11.3, indicate that the average irradiation per individual within a generation of 30 years increases from 0.03–0.04 Gy ("natural" irradiation) to 0.07 Gy, i.e., that modern civilization approximately doubles the background irradiation. Such a doubling must be carefully distinguished from the dose of irradiation which doubles the mutation rate. Such a dose is much higher than 0.03 Gy. This answers the first question. The response to the second question is somewhat more complex and requires a more elaborate discussion.

How Many Additional Mutations Per Dose Are Induced? The various classes of mutations should be considered separately, and both dose rate and sex should be taken into account (Sect. 11.1.3). Most humans are exposed to very low doses of chronic irradiation administered at a very low dose rate; very few are exposed to high doses at very high dose rates.

In the following we first discuss direct evidence available for irradiated human populations. This evidence is then compared to data from experiments with mice and other mammals as discussed above. Wherever possible, quantitative estimates for various kinds of mutations relative to the spontaneous rate are then calculated.

Phenotypic Characteristics in Irradiated Human Populations. Apart from ethically dubious experiments with human volunteers, genetic radiation effects can be assessed directly only when humans are exposed to radiation either for therapeutic purposes or by accident. In therapy the dose calculations are usually fairly accurate, but the number of individuals is low, and these are selected for various diseases. Radiation is usually administered in a few very high doses during medical therapy or at a low dose rate during professional exposure. In the case of accidents, estimates of the dose may be inaccurate, but dose and dose rate are usually high.

Survivors of Atomic Bombs in Hiroshima and Nagasaki [57]. Following the explosions of the atomic bombs over Hiroshima and Nagasaki genetic studies on the survivors were organized by American and Japanese scientists under the Atomic Bomb Casualty Commission (ABCC) as early as 1946 and have been continuing ever since, since 1975 under the auspices of the Radiation Effects Research Foundation (RERF). This was an early attempt to organize an ongoing "big science" project in human genetics.

1. Are there any differences between children conceived after the parents were exposed to the atomic bomb and children of unexposed parents?
2. If there are differences, how are they to be accounted for?

Data collection was facilitated by the food-rationing system in postwar Japan, which provided additional food for pregnant women. When applying for extra rations, the women were given a questionnaire regarding past pregnancies, exact location when the bomb was dropped, and any symptoms that might indicate radiation illness afterward. When the child was born, relevant data on delivery and health status of the newborn were added, and each child was examined by a physician of the ABCC. About one-third were reexamined 9 months later. In the following years the method of examination and follow-up was modified in various ways. The following parameters were considered:

a) Sex ratio at birth
b) Congenital malformations
c) Stillbirths
d) Body weight at birth
e) Death within the first 9 months of life and death during childhood and youth
f) Anthropometric measurements
g) Autopsy results

The critical reader might ask why the two most important indicators of genetic damage – chromosome aberrations in somatic and germinal cells and dominant or X-linked recessive new mutations – were omitted. The answer is simple: The methods for chromosome examination in humans became available only 10–15 years later, and dominant or X-linked gene mutations are so rare that a sufficient number could not be expected in the examined population.

Germ cell radiation exposure was estimated using parameters such as distance from the hypocenter, type of shelter, and signs of radiation sickness at the time of the bombing. An essential problem in assessing biological effects is calculating the quality and quantity of radiation emitted by the bombs and received by exposed individuals. These physical problems were reevaluated only in 1989 [61]. Neel et al. presented an average conjoint gonad radiation dose (= radiation dose for both parents together) for all parents with increased exposure of 0.4–0.5 Sv. However, the distribution was highly skewed: a small proportion received a much higher dose while the dose received by the majority was much lower. Children of unexposed parents in the same cities were taken as controls. Consanguinity of the parents and maternal age were considered to avoid biases due to these variables.

Shift in Sex Ratio Due to X-Linked Lethals? The results of these studies can be summarized very briefly. Regardless of the extent of irradiation exposure, no significant difference was found in the number of malformations, in stillbirth rate, or in any other of the examined parameters – with one possible exception: the sex ratio.

The first studies [80] showed a small sex ratio shift: reduction in the number of live-born males among the offspring of irradiated females ascribed to X-linked recessive lethals and reduction in the number of females in the offspring of radiation-exposed males because of X-linked dominant lethals. However, these findings were not confirmed later after an increase in sample size [82].

Support for the Sex Ratio Shift by Studies After Exposure to X-Rays. A similar sex ratio shift has been observed among children of parents exposed to X-ray *therapy* for treatment of ankylosing spondylitis, anal and vulval pruritus, and other conditions. Two studies, one from France [43] and the other from the Netherlands [79], together comprised several hundred patients irradiated with high doses, sometimes several grays. The sex ratio among children conceived after irradiation deviated in the expected direction, and the deviations were significant.

Therapeutic irradiation and irradiation by atomic bombs involve high doses and high dose rates. Professionally exposed individuals, on the other hand, usually receive relatively low doses at very low dose rates. Do these doses also cause a shift in sex ratio? A study of radiology technicians in Japan yielded an affirmative answer: not only was the sex ratio altered, but sterility appeared to be more frequent than in the general population [88].

Data from Hiroshima and Nagasaki on the sex ratio may be interpreted in one of three ways: (a) the sex ratio deviations found in the earlier analysis were fortuitous, and the radiation doses received by the atomic bomb survivors did not induce enough X-linked lethals to bring about a sex ratio shift; (b) the sex ratio is actually no indicator for X-linked mutations, the argument being too simplistic; or (c) the effect in the earlier series was real, but in the 10–15 years between radiation exposure and the conception of children of the second series germ cells containing lethal mutations were eliminated. The results showing sex ratio shifts in medically irradiated individuals in Europe [43, 79] favor the latter explanation.

The fact remains, however, that sex ratio is an unsatisfactory measure of genetic effects, being influenced by many other variables, such as age of parents and general living conditions.

A Reassessment Using Additional Data and Methods. A cytogenetic investigation of the children of survivors began in 1968. Aberrations leading to severe malformations and early death could thus no longer be ascertained since the children had died in the meantime. In the mid-1970s studies were also begun using electrophoretic techniques for qualitative and quantitative variation (rare variants) of blood proteins and enzymes. Both studies had negative results: the incidence for both endpoints were even somewhat (no significantly) lower in the irradiated than in the control group (Table 11.4). In the study on electrophoretic variants the possibility of false paternity was examined carefully in all instances in which a new mutation appeared to be possible. The observed mutation rate fits well with mutation rates of genes leading to autosomal dominant or X-linked diseases (Table 11.5).

Table 11.4. Numerical gonosomal aberrations among children of exposed and unexposed parents 1991 (from Neel and Schull 1991 [57])

	No. with gonosomal aberrations					
	No. of children	Males		Females		Total
		n	%	n	%	
Children of exposed parents	8322	12	0.307%	7	0.159%	19
Children of unexposed parents	7976	16	0.435%	8	0.186%	24

Table 11.5. Protein variants in children of survivors of the atomic bombs (data from Neel et al. 1986 [60], Neel and Schull 1991 [57])

Electrophoretically screened gene loci		
	No. of loci screened	New mutants
Proximally exposed cohort	667 404	3
Distally (or not) exposed cohort	466 881	3
Mutation rates for electrophoretically detected protein mutation and mutations leading to reduction of enzyme activity		
	Mutation rate per locus per generation	95% confidence interval
Proximally exposed cohort	6.0×10^{-6}	$2\text{--}15 \times 10^{-6}$
Distally (or not) exposed cohort	6.4×10^{-6}	$1\text{--}19 \times 10^{-6}$

The data regarding stillbirth, congenital malformations, and early death were reevaluated using an improved statistical procedure. None of these endpoints showed a statistically significant difference between irradiated and nonirradiated parents; therefore the evaluation strategy was changed. In the earlier studies exposed individuals had been classified into subgroups according to their distance from the hypocenter at the moment of the detonation, and presence or absence of signs of radiation sickness. The deviation from the null hypothesis (no genetic effect at all) was then tested. In the new studies each exposed individual was assigned an individual dose estimate on the basis of recent information on the kind of radiation produced by the bombs, and taking account of all available data, such as distance from the hypocenter, sheltering, etc. The authors no longer tested the null hypothesis. They argued that since it is known and generally accepted that ionizing radiation induces mutations in all species, any increase in effects that could reasonably be related to mutations, whether statistically significant or not, was probably caused by additional induced mutations. Hence, regressions of an increase of ill effects on exposure were calculated. These regressions, together with certain assumptions on the proportion of new mutations explaining the "spontaneous" occurrence of the events under study, were used to estimate doubling doses. Major congenital malformations, stillbirths and deaths during the first week of life were combined under the heading "untoward pregnancy outcome." Regression analyses were carried out on prevalences of these events and individual radiation exposure of both parents, assuming a linear (one-hit) dose-effect relationship. More than 70 000 pregnancies were analyzed in this way. After controlling for confounding variables (e.g., parental consanguinity) there was an increase of 0.00239 per 1 gonadal Sv [57]. This endpoint therefore showed a small, statistically nonsignificant increase in children of irradiated parents. However, the causes of stillbirth, malformations, and death in the first week of life are seldom genetic. Assessing the extent of increase due to radiation – for which the doubling dose (Sect. 11.1.3) is a convenient measure – therefore required assumptions about the proportion of "untoward pregnancy outcome" with a genetic basis. These were calculated to be $1/200\text{--}1/400$ [83]. In order to err rather on the conservative side, that is, to overestimate rather than underestimate the genetic radiation effects the latter figure was selected. However, they included death up to the age of 19 in their assessment. From these data a doubling dose of 1.5–1.9 Sv was calculated.

This estimate is higher than that for various endpoints in the mouse [46]. The doubling dose estimate conventionally used by UNSCEAR was 1 Sv. However, this – admittedly very crude – estimate referred to chronic radiation with small dose rates, whereas irradiation by the atomic bombs was acute – with high dose rates. Therefore one would have expected a higher effect (= a doubling dose lower than 1 Sv) if the human doubling dose were about equal to that in the mouse. Therefore, Neel and Lewis [56] reconsidered the doubling dose calculations in the literature and the mouse data on which they were based. They arrived at a Figure of 1.35 Sv for acute and approximately 4 Sv for chronic irradiation – in line with their estimates in humans. Discussions on these matters continue. It should be remembered, however, that the human estimate is based on very slender – and largely indirect – evidence.

Another factor to assess germ line mutations was the search for childhood and adolescent tumors among children of atomic bomb survivors [104]. A rationale for the possibility of such an increase could be provided by recent results on the existence of tumor suppressor genes, such as those found in retinoblastoma and other tumors (Sect. 10.4.3). However, no significant increase was found.

Taken together, the studies on children of atomic bomb survivors in Hiroshima and Nagasaki show – contrary to common opinion – that even an atomic holocaust does not lead to a genetic catastrophe among the progeny of the fairly heavily irradiated survivors. No genetic effects were discernible in the first generation of offspring. We shall show that this will probably apply for future generations as well. The horror of such an atomic holocaust lies in the immediate and short-term effects on exposed individuals and, to a lesser extent, in mutations in somatic tissue manifesting in later life, for example, as malignant neoplasias with a latent period of as long as 10–40 years (See below Sect. 11.1.6).

The Chernobyl Accident. On 26 April 1986 a severe accident occurred at the Chernobyl nuclear power station about 100 km northwest of Kiev [13, 95]. Large amounts of radioactive material were released from the plant. In addition to severe contamination of the immediate surroundings, clouds with radionucleotides were distributed over many European countries (Fig. 11.11), especially of iodine (^{131}I) and cesium (^{137}Cs) but also other radionuclides. While the accident was caused by gross neglect of safety regulations, it showed that with worldwide dissemination

of this technology, substantial risks must be faced. It can be concluded from the only precedent – the atomic bombs – that the genetic consequences of this accident for future generations are likely to be minor while a slight but definite increased cancer risk over a long period of time can be expected. Studies on birth defects – possibly related to this accident – are in progress. The best studies are from Hungary, with its well-functioning system of population monitoring. No increase in either "sentinel mutations" (see above), Down syndrome cases, or any other malformation has been observed [21]. However, so-called stochastic effects on somatic tissue were observed. For example, an increased frequency of thyroid cancer in children relatively soon after the accident was noted [39]. If these effects are confirmed as real and not due merely to intensive screening for thyroid malignancies, a further increase in thyroid cancers might be expected in the future [37].

Irradiation of Parents for Medical Reasons and Trisomy 21 in Children. Reports on the frequency of trisomy 21 (Down syndrome) among children of mothers exposed to X-ray diagnostics or therapy are contradictory. Sigler et al. (1965) [86] compared the radiation exposure of 216 mothers of children with Down syndrome with that of 216 other mothers of the same age and origin. Exposure was clearly higher among the mothers of children with trisomy 21. Uchida et al. (1968) [93] compared the frequency of aneuploidy among the children of radiation-exposed and unexposed women. Whereas the other examined parameters did not show differences, there were ten children with aneuploidy among the progeny of the irradiated women – eight cases with Down syndrome and two with trisomy 18. Among the controls only one child with Down syndrome was found; the difference was statistically significant. However, the gonadal radiation doses received by these mothers were very low, between 0.007 and 0.126 Gy. Additional studies suggest a higher incidence of aneuploidy after radiation exposure. Meanwhile, the problem has often been reexamined (for a review, see [95]); most studies, including the reexamination of children of atomic bomb survivors [81], have failed to show a radiation effect. Some of the earliest studies were retrospective, i.e., radiation history was established after the children were born.

This mode of study can introduce a bias if the search for radiation exposure is – understandably enough – more thorough in mothers who have had a child with Down syndrome than among control mothers. The study by Uchida et al. [93], on the other hand, was prospective. Studies yielding negative results have been both retrospective and prospective.

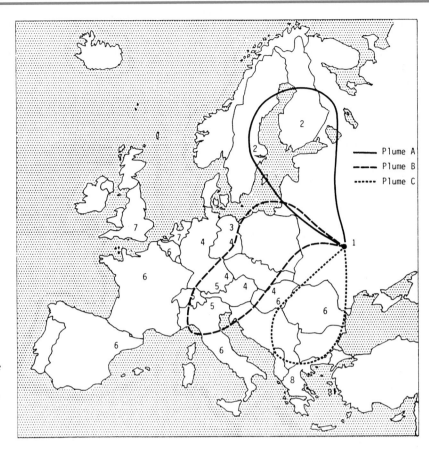

Fig. 11.11. Air mass movements originating from Chernobyl on 26 April *(plume A)*, 27–28 April *(plume B)*, and 29–30 April *(plume C)*. Arrival times: *1*, 26 April; *2*, 27 April, *3*, 28 April; *4*, 29. April; *5*, 30 April; *6*, 1 May; *7*, 2 May; *8*, 3 May. (From UNSCEAR 1988 [95])

From the mouse data discussed in Sect. 11.1.3 one would conclude that maternal irradiation in low doses long before conception should not induce a higher nondisjunction rate in the oocyte. Moreover, women in need of X-ray studies may differ from other women in their health status, which may affect their risk of nondisjunction, and this may very well lead to a spurious correlation. Evidence has been reported, that irradiation of human lymphocytes in vitro with small doses may enhance the frequency of somatic nondisjunction [94].

Higher Incidence of Structural Chromosome Anomalies and Down Syndrome in Human Populations Exposed to High Background Radiation? Some areas in Brazil, southern India (Kerala) and China have a high background radiation, about 10–100 times "normal" levels, due to a high soil content of radioactive elements such as thorium and radium in the monazite sands. In Brazil a chromosome study was carried out in 12 000 chronically exposed inhabitants. There was a significant but marginal increase in chromosome-type aberrations such as deletions, dicentric chromosomes, and ring chromosomes in lymphocyte cultures [7]. A population study of 12 918 individuals living in a high-irradiation area of Kerala found 12 individuals with Down

syndrome, 12 others with severe mental deficiency and additional malformations, and 11 with severe mental deficiency but without additional malformations, compared with 5938 controls where no Down syndrome cases, only one case with severe mental deficiency and malformations, and 2 patients with severe retardation without additional clinical signs were found. Moreover, chromosome counts in blood cultures from exposed individuals showed a slight increase in the number of chromatid and chromosome aberrations [41]. While the increase in chromosome aberrations in somatic cells, which was reported in both studies, is probably real, the higher incidence of Down syndrome in the latter study could easily have been caused by an ascertainment bias or by other differences between the test and control populations, since Down syndrome was unusually rare in controls.

In China, studies on a population of about 80 000 were performed in two high-radiation areas in Yangjiang province. Background radiation here is about three times as high as in the control district, where it corresponds roughly to that found in most other populations of the world. Most families had lived there for six or more generations; there was no major difference in general living conditions between high-radiation and control areas. The studies failed to show

Table 11.6 Prevalence of Down syndrome among children in the irradiated Chinese population and in the control area (from Wei et al. 1987 [102])

Area	Mother's age	No. of children examined	No. of cases with Down's syndrome	Frequency per 10^3 children
High irradiation		25 258	22	0.87
Controls		21 837	4	0.18
High irradiation	> 35	3 076	14	4.61
	< 35	22 222	8	0.36
Controls	> 35	970	3	3.09
	< 35	20 867	1	0.05

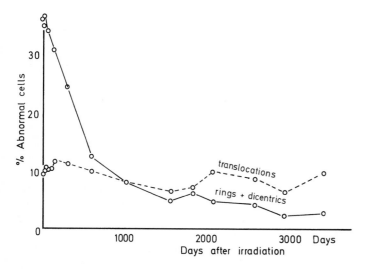

Fig. 11.12. After therapeutic radiation exposure of women with gynecological tumors the fraction of cells showing dicentric chromosomes and ring chromosomes decreased (O–O–O) in the years following exposure. The number of translocations, on the other hand, (O--O--O) remains more or less stable. (Bauchinger 1968 [11])

any increase in cancer mortality or in the proportion of children suffering from specified genetic defects and diseases compared with controls. The proportion of chromosome abnormalities in lymphocytes was slightly but not significantly increased in exposed individuals [17, 20, 26, 33]. An increase in the frequency of Down syndrome in the irradiated area compared with the control area was explained tentatively by a higher frequency of mothers aged over 35 years. This explanation, however, it not entirely satisfactory (Table 11.6) [102, 103]. The frequency of Down syndrome is relatively high also in comparison with other areas in China. There were obvious problems with ascertainment. The problem of an increase in trisomy 21 after very small radiation doses – perhaps only under special conditions – has not been solved so far.

11.1.6 Evidence of Somatic Chromosome Mutations After Exposure to Radiation

Medical Therapy. Tough et al. (1960) [91] were the first to describe structural aberrations in the chromosomes of two patients who had been irradiated for treatment of ankylosing spondylitis. Since then radiation sensitivity of human somatic chromosomes has often been examined [11, 12]. The following results may be noted:

a) Within a certain time after therapeutic irradiation, approximately 25%–35% of cells (in most cases lymphocytes) have structural chromosome aberrations.

b) The number of cells showing dicentric chromosomes, ring chromosomes, and acentric fragments declines with time elapsed between irradiation and examination, mostly within the first 2 years after irradiation. After 10 years, however, such anomalies still were about four times more frequent than among controls. The number of reciprocal translocations, on the other hand, was not much lower after 10 years than immediately after irradiation (Fig. 11.12). Structural chromosome defects in lymphocytes were visible as long as 25 years after combined X-ray and radium therapy of gynecological tumors. Similar aberrations were observed in patients treated with radioisotopes such as ^{131}I or ^{32}P.

Professional Exposure. Chromosome aberrations in individuals exposed professionally to chronic irradiation have frequently been described. Cells with ring or dicentric chromosomes occur very rarely spontaneously (1/2000 cells after 48 h of lymphocyte cultivation, 1/8000 after 72 h). They are therefore good indicators of radiation exposure. However, simple chromosome breaks also increase in number [11]. Exposed groups include workers painting phosphorescent materials on watch dials, nuclear reactor personnel, and persons involved in radiation accidents. It has even become possible to calculate the radiation dose an individual has received from the extent of cytogenetic changes. Such biological dosimetry is a useful test in human radiobiology, especially when the rapid method of CISS hybrization is used (Sect. 3.1.3.3) [19].

Atomic Bomb Survivors. Sasaki and Miyata examined 51 survivors of the atomic bomb about 22 years after exposure, comparing them with 11 untreated controls. No less than 83 506 cells were analyzed. In the irradiated sample, the rate of dicentric chromosomes and rings was 0.0027 per cell (201 in 73 996 cells). Among the controls it was 0.0002 (2 in 9510 cells). The number of cells with stable, symmetric translocations was also increased. There was a clear relationship between the distance from the hypocenter at the time of the bombing and the number of aberrant cells (Fig. 11.13).

Other studies with similar results were carried out by Bloom et al. [12]. They found four heavily exposed individuals who had received radiation doses of between 2.07 and 6.42 Gy to have developed clones with aberrant cytology. As discussed in Chap. 10, clone formation indicates a selective advantage of the cells having this chromosome aberration. Cell clones with this advantage may develop into malignant neoplasias.

Neoplastic Disease in Atomic Bomb Survivors. Follow-up studies among patients who have received extensive radiation and among survivors of the atomic bombs in Hiroshima and Nagasaki have shown an increased frequency of a variety of malignancies. These include leukemia, cancer of the female breast, cancer of the lung, cancer of the thyroid, and cancer of the digestive system. These findings are compatible with the somatic mutation theory of cancer (Chap. 10) and are therefore not entirely unexpected. The "latent" period between radiation exposure and clinical diagnosis of malignancy is usually very long (more than 10 years), particularly for solid tumors (but is shorter (~ 4 years) for chronic granulocytic leukemia). The problem of malignancies after somatic radiation exposure from various sources and of other

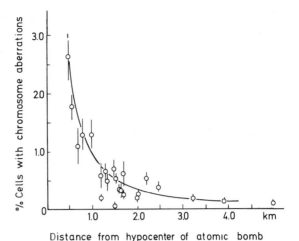

Fig. 11.13. Frequency of cells (lymphocytes) with chromosome aberrations in relation to the distance from the hypocenter at the time of atomic bombing. (From Sasaki and Miyata 1968 [77])

"stochastic" effects has been studied extensively (see the 1986 and 1988 UNSCEAR reports [4]).

There was some concern as to whether children exposed to the atomic bomb in utero would run a higher risk of developing malignancies, but the investigations carried out on this problem failed to produce any evidence of an increase [38].

Patients who have survived childhood cancer have often undergone radiation therapy. Therefore there was some concern about whether their children would have an increased cancer risk. A large and well-controlled cooperative study involving many centers found no such effect; quite a few affected children were found among parents who had been so treated, for example, because of bilateral retinoblastoma, a well-known dominantly inherited disease [55].

As noted in Sect. 10.4.5, some of the changes concomitant with normal aging could be due to somatic mutation. Some animal experiments suggest a shortening of life span after irradiation [95].

11.1.7. Projected Additional Mutations Per Dose

How many additional mutations are induced relative to the spontaneous mutation rate? Here the information from animal genetics, especially mouse genetics, serves as a guideline. In view of possible species differences in radiation susceptibility we would certainly prefer to base all risk estimates on human data, but such data are so scanty that they do not permit any reasonable quantitative basis. However, they can serve as a background and in some cases help to qualify estimates from animal data. We consider in particular the expected relative frequencies at the time of birth but occasionally refer to the expected zygote loss before birth. The various groups of mutations will be considered separately.

Here we compare two recent estimates of prominent committees, the United Nations Committee on Effects of Atomic Radiation (UNSCEAR) in 1988 and the Committee on the Biological Effects on Ionizing Radiation (BEIR) of the United States Academy of Sciences. Each supposed low-level chronic radiation, i.e., radiation with a low dose rate, and each accepted a doubling dose of 1 Sv. These two predictions are considered critically below (Table 11.7) [97].

Autosomal Dominant and X-Linked Recessive Diseases. The two reports are in substantial agreement for Mendelian diseases. It is not difficult to establish a baseline (incidence without additional irradiation) and to calculate its mutational component for autosomal dominant and X-linked diseases with a strong selective disadvantage where population prevalence is maintained by an equilibrium between mutation pressure and selection. Relatively reliable spontaneous mutation rates are available. Autosomal dominant diseases that can be diagnosed early in life have been termed "sentinel" mutations, since they could be used as indicators (or sentinels) to assess mutation rate increases caused by a new agent in the environment. Czeizel [21, 22] listed 15 such sentinel diseases (Table 11.8), differentiating the incidence of these conditions before and after the Chernobyl accident. No difference was noted. However, since sentinel conditions are very rare, a very large population must be screened before statistically significant mutation rate increases could be discerned [87].

Potential sentinel mutations make up only a small fraction of autosomal-dominant anomalies in human populations. The UNSCEAR and BEIR reports (Table 11.7) estimated the incidence of autosomal-dominant (and, in UNSCEAR, X-linked) diseases to be about 1% of newborns. The incidence of potential sentinel mutations was reported to be about 3 per 10000 in Hungary [21]; this is about 30 times lower. Moreover, there are some diseases such as bilateral retinoblastoma and the hemophilias which in the past were maintained by an equilibrium between mutation and selection, an equilibrium which has been upset by successful therapy in recent decades. Assuming constant mutation rates, the incidence of such conditions is bound to increase even with no additional mutagenic agent, unless there are counteracting circumstances, such as artificial selection.

However, it is very unlikely that the most common dominant conditions are maintained by an equilibrium between mutation and selection (Table 11.9). Monogenic familial hypercholesterolemia, for exam-

ple, is among the most common of well-defined dominant conditions (Sect. 6.4.2.2). With today's living conditions in industrialized countries there may be a very small selective disadvantage due to premature coronary heart disease, but the high incidence of this diseases could be explained by a selective advantage in earlier times due to the more austere living conditions. Population genetic problems with diseases caused by abnormally high amplifications of base triplets – principally in myotonic dystrophy, Huntington disease, and Fra X syndrome have been discussed (Sect. 9.4.2). There is certainly no simple mutation-selection equilibrium, and the contribution of mutation and selection to incidence of these conditions in the current population can hardly be assessed. The fraction of new mutants among all cases cannot be determined easily.

It follows that dominant and X-linked diseases cannot be entirely subdivided into those in which the incidence is maintained by an equilibrium between mutation and negative selection, and others in which a selective advantage under certain living conditions has been the decisive factor. The "mutational component" of the entire group has been assumed to be 15%. Intuitively, this figure appears to be rather high. In absence of more precise evidence, however, it is wise to remain cautious and assume a rather high mutational component: the higher the "spontaneous" mutational component, the higher is the radiation risk.

To place any possible mutation rate increase in proper perspective it should be remembered that the "spontaneous" mutation rate is not a natural constant. The best known influence is that of paternal age (Sect. 9.3.3). Model calculations have been performed to determine the degree to which a given shift in the mean paternal age in a population would alter mutation rates [52]. Using published data on achondroplasia, myositis ossificans, and acrocephalosyndactyly as examples, the expected incidence of dominant new mutants was compared to their incidences if all fathers were under 30 years of age at the time of birth of their children (Table 11.10). Ratios vary between 1.22 (Bulgaria, 1980) and 2.67 (Pakistan, 1968). Thus even a relatively small shift in paternal age, especially an increase or decrease of older fathers, could appreciably affect the mutation rate for paternal age dependent mutations. Such an effect would be significantly greater than any conceivable change in exposure to mutagenic agents, such as radiation.

Autosomal-Recessive Diseases. Here the two reports diverge. BEIR calculated the increase in the first generation to be low, remarking in addition: "very small increase." UNSCEAR, on the other hand, gave a fig-

Table 11.7. Risk estimates for genetic disease: increase per 1 Sv (low dose rate) radiation (from UNSCEAR 1988 BEIR 1990)

UNSCEAR

Disease classification	Current incidence per million live births	Effect of 1 Sv per generation		
		First generation	Second generation	Equilibrium
Autosomal dominant and X-linked	10000	1500	1300	10000
Autosomal recessive	2500	5	5	1500
Chromosomal				
Due to structural anomalies	400	240	96	400
Due to numerical anomalies	3400	Probably very small		
Congenital anomalies	60000	Not estimated		
Other multifactorial diseases	600000	Not estimated		
Early acting dominants	Unknown	Not estimated		
Heritable tumors	Unknown	Not estimated		
Totals of estimated risk		1700	1400	12000

BEIR

Type of disorder	Current incidence per million liveborn offspring	Additional cases/10⁶ liveborn offspring/Sv/generation	
		First generation	Equilibrium
Autosomal dominant			
Clinically severe	2500	500–2000	2500
Clinically mild	7500	100–1500	7500
X-linked	400	<100	<500
Recessive	2500	<100	Very slow increase
Chromosomal			
Unbalanced translocations	600	<500	Very little increase
Trisomies	3800	<100	1000–10000
Congenital abnormalities	20000–30000	1000	1000–10000
Other disorders of complex etiology			
Heart disease	600000	Not estimated	Not estimated
Cancer	300000	Not estimated	Not estimated
Selected others	300000	Not estimated	Not estimated

Table 11.8. Follow-up figures for indicator conditions (sentinel anomalies) after the Chernobyl accident (based on number of live births; from Czeizel 1989 [21, 22])

	Baseline (1980–1985) ($n = 807\,939$)		5 May 1986–30 April 1987 ($n = 126\,708$)		1 May 1987–30 April 1988 ($n = 125\,514$)		Total	
	No.	Per 10 000	No.	Per 10 000	No.	Per 10 000	No.	Per 10 000
100 800 Achondroplasia	38	0.47	4	0.32	7	0.56	11	0.44
101 200 Apert syndrome	7	0.09	2	0.16	2	0.16	4	0.16
106 200 Aniridia, bilateral	2	0.02	0	–	1	(0.08)	1	(0.04)
123 500 Crouzon syndrome	5	0.06	1	(0.08)	2	0.16	3	0.12
129 900 EEC syndrome	3	0.04	0	–	0	–	0	–
142 900 Holt-Oram syndrome	12	0.15	2	0.16	2	0.16	4	0.16
154 500 Treacher-Collins syndrome	3	0.04	1	(0.08)	1	(0.08)	2	0.08
166 200 Osteogenesis imperfecta, type I	11	0.14	3	0.24	3	0.24	6	0.24
174 700 Preaxial polysyndactyly, type IV	56	0.69	15	1.18	10	0.80	25	0.99
183 600 Split hand and/or foot, typical	3	0.04	1	0.08	0	–	1	0.04
187 600 Thanatophoric dwarfism	6	0.07	2	0.16	1	(0.08)	3	0.12
308 300 Incontinentia dwarfism	7	0.09	0	–	0	–	0	–
311 200 Gorlin-Psaume syndrome	2	0.02	0	–	0	–	0	–
180 200 Retinoblastoma	26	0.32	2	0.16	1	(0.08)	3	0.12
194 070 Wilms tumor	65	0.80	6	0.47	7	0.56	13	0.52
Total	246	3.04	39[a]	3.08[a]	37[a]	2.95[a]	76	3.01[a]

[a] Preliminary figures

Table 11.9. Estimated incidence at birth per 1000 newborns for some of the most common dominant and X-linked diseases (adapted from Carter 1977 [16])

Incidence/live births (order of magnitude)	Diseases
2.0	Monogenic familial hypercholesterolemia
1.0	Dominant otosclerosis
0.8	Adult polycystic kidney disease, X-linked mental retardation
0.5	Multiple exostoses
0.4	Huntington disease
0.2	Hereditary spherocytosis, neurofibromatosis, Duchenne-type muscular dystrophy
0.1	Hereditary polyposis, dominant form of blindness, dominant form of early childhood onset deafness, dentinogenesis imperfecta, hemophilia A, dominant ichthyosis
0.04	Osteogenesis imperfecta, Marfan syndrome
0.03	Hereditary retinoblastoma, hemophilia B
0.02	Achondroplasia, acute intermittent porphyria, X-linked deafness, ocular albinism, nystagmus
0.01	Tuberous sclerosis, Ehlers-Danlos syndrome, osteopetrosis tarda, variegate porphyria, cleft lip and/or palate with mucous pits of lip, X-linked imperforate anus, X-linked aqueductal stenosis, hypogammaglobulinemia, hypophosphatemic rickets, anhidrotic ectodermal dysplasia, amelogenesis imperfecta

ure for a moderate increase. This was based on calculations regarding induced new mutations and their probabilities of meeting by chance an identical mutation in the population to form a homozygote [85]. This probability depends critically on the percentage of matings between relatives, i.e., the consanguinity rate (Sect. 13.1.1), which cannot be predicted for future centuries or millennia. Moreover, recessive mutants are not in equilibrium at this time. Their incidence in future generations will depend much more on many aspects of population dynamics than on the mutation rate. However, recessive new mutations should not be neglected entirely, especially since heterozygotes may occasionally show small health im-

Table 11.10. Estimated relative mutation rate in European countries (from Modell and Kuliev 1990 [52])

Country and year	Fathers > 35 years (%)	Relative mutation rate (1 = level when all fathers are < 30
Bulgaria 1980	7.3	1.22
East Germany 1980	8.5	1.22
Czechoslovakia 1978	9.6	1.27
Hungary 1980	10.2	1.28
Belgium 1978	11.t	1.33
Scotland 1980	12.0	1.34
Poland 1980	12.1	1.35
Netherlands 1979	12.9	1.38
France 1980	14.9	1.43
England and Wales[a] 1979	15.4	1.44
Finland 1980	16.0	1.45
Denmark 1980	17.6	1.47
Luxembourg 1980	17.8	1.47
Norway 1980	17.6	1.47
Iceland 1980	21.1	1.53
Northern Ireland 1978	19.0	1.53
Switzerland 1979	20.0	1.53
Sweden 1980	23.4	1.53
Malta 1980	19.8	1.54
West Germany 1980	22.2	1.57
Spain 1980	23.6	1.64
Italy 1978	24.1	1.64
Greece 1979	24.3	1.69
Spain 1966	33.5	1.69
Pakistan 1968	46.1	2.67

[a] Correlation for infants born out of wedlock does not alter the figure.

pairments [89]. Moreover, radiation-induced recessive mutations in the mouse are generally deletions that exhibit more severe clinical signs in humans than point mutations, even in the heterozygous state.

Chromosomal Diseases. The risks given for chromosomal diseases are in principle very similar in the two reports. Indeed, fairly reliable figures are available for incidence at birth. Moreover, all trisomies, monosomies, and virtually all nonbalanced structural chromosomal aberrations prevent their bearers from reproduction; most are therefore new mutants. Unbalanced structural aberrations are inherited from one of the parents, and balanced aberrations are relatively rare. Calculating the "mutational" component therefore presents few problems, in distinction to autosomal-dominant diseases. Problems do arise, on the other hand, by the uncertainties regarding actual radiation effects: for monogenic disorders substantial data are available from mouse experiments for calcu-

lating increases in mutation rates with radiation dose, but such data are less abundant for trisomies, a most important category of chromosomal diseases in humans. The uncertainties regarding the possible effect of very small radiation doses on nondisjunction and on Down syndrome (see above) add to the difficulties. Moreover, the baseline (mutation rate without additional, induced mutations) depends critically on the age distribution of mothers in a population.

For numerical chromosomal diseases UNSCEAR gives a figure of 3400/1 000 000; BEIR's estimate is 3800/1 000 000. The two reports agree that the increase with rising mutation dose is probably very small. Adding incidence data in newborns for autosomal and gonosomal numerical anomalies yields an incidence (without radiation) that is closer to the higher figure. The conclusion that the increases with radiation may be small is based on mouse data; it is corroborated by the results of the Hiroshima-Nagasaki studies which failed to show any increase in gonosomal anomalies in children of atomic bomb survivors. At first glance one would expect more X monosomies since in the mouse the time around fertilization has been found to be especially susceptible to X-chromosome loss. However, most X monosomies occurring at this early time do not survive but lead to early abortion. Therefore it is appropriate not to consider them in predictions confined to abnormalities at birth or later.

The situation is simpler regarding structural chromosomal aberrations. Here, not only are good human data available regarding incidence at birth but experimental results in the mouse are supplemented by data from other mammalian species including humans (see above). The two estimates agree very well.

Congenital Anomalies and Multifactorial Diseases. For the incidence of these two categories the figures of UNSCEAR and those of BEIR deviate from one other: those in the former are twice the latter's estimate for congenital anomalies. This discrepancy reflects uncertainties in definition and incidence (see, for example [97], and earlier UNSCEAR reports). Moreover, only a minority of birth defects have a well-defined genetic origin. For many of them genetic differences in liability play a role, but many may be random events during the complex process of differentiation and embryonic development (Sect. 8.4.2). Therefore estimating the "mutational component" becomes practically impossible.

In theory, a system of continuous registration of all birth defects according to carefully defined rules, complemented by ad hoc studies of family data and environmental exposure (radiation, drugs, environmental chemicals) would be the best system for answering questions regarding incidence and causation. The program in Hungary, which has been in operation since the early 1960s, appears to be most informative [23–25].

In multifactorial diseases, the difficulties in arriving at an estimate are still greater than those for birth defects, as should be obvious to readers who have studied the sections on these diseases (Chap. 6). The main problems are the following [40, 97]:

1. The degree to which genetic variability, on the one hand, and variation in the environment, on the other contribute to disease liability differs for different diseases.
2. Even for the same disease (or group of diseases), liability varies over time and with environmental conditions, for example, in Diabetes mellitus and coronary heart disease (Chap. 6).
3. The nature of genetic variation in complex diseases is different, and mostly unexplored. A part of it may not be susceptible to mutation pressure, and may not increase if the mutation rate rises. For example, in some multifactorial diseases, genetic polymorphisms such as the AB0 blood groups and HLA types – as well as other polymorphisms – are involved (Sect. 6.2). Here, very rare mutations may occur. The effect of such mutations on the frequency of multifactorial diseases is presumably negligible. In the mouse paternal irradiation has led to an increase in dominant skeleton mutations with irregular penetrance and expressivity which could serve as models for human malformations.

In conclusion, the estimates for congenital anomalies and multifactorial diseases are extremely unreliable. Predictions in this area would be most interesting from the point of view of public health and radiation protection because these two groups of conditions are much more common than monogenic diseases, where estimates of the correct order of magnitude of the risk are possible.

The only way out of this difficulty is to fall back on empirical data. A prediction of the possible effects of an increased mutation rate on incidence and prevalence of birth defects and multifactorial diseases could be based, for rexample, on the Hiroshima-Nagasaki data set, possibly complemented by data from populations living in high-radiation areas. In any case it is reassuring that the limited evidence available so far – both in humans and in mice (Sect. 11.1.3) – shows that low-dose irradiation has only few, if any ill effects on the frequency of birth defects and multifactorial diseases.

How Many Naturally Occurring "Spontaneous" Mutations Are Caused by Natural Background Radiation? The considerations on doubling doses and number of radiation induced mutations per sievert per locus aid in calculating the proportion of naturally occurring mutations induced by natural background radiation. This radiation is estimated to be 0.03–0.04 Sv over 30 years (Table 11.7). The dose rate is extremely low. Therefore an value of 1 Sv or ever higher [56] for the doubling dose may be reasonable for mutations in spermatogonia. Additional irradiation similar in amount to background radiation would enhance the spontaneous rate by 3%–4% or less, and the same proportion of the spontaneous mutation rate might be caused by natural background radiation.

Improvement of these estimates would require the following: (a) A more thorough knowledge of the genetic factors of multifactorial conditions and of dominants with low penetrance. (b) A much better knowledge of the incidence and prevalence of hereditary diseases. This knowledge can come only from carefully planned, large-scale epidemiological studies in which medicostatistical information is combined with ad hoc studies on single diseases using carefully defined criteria for diagnosis. (c) Better knowledge of the interplay between mutation and natural selection, especially for those diseases in which genetic equilibrium between mutation and very strong selection is not immediately obvious.

In Sect. 11.1.2, four questions were asked:
1. In what way, if any, does the agent affect the genetic material?
2. How extensively is the human population exposed to this agent?
3. What is the probable increase in mutations compared to the "spontaneous" mutation rate?
4. What are the long-term consequences of this increase for the population?

A cumbersome chain of arguments has led us to question 3 and to some extent into question 4. The main thrust of question 4, however, cannot be answered since we do not yet know the proportion of the total sum of mutations that are harmful. The population studies on irradiated mouse populations could lead us to rather optimistic conclusions as to consequences for the human population, but humans are not mice, and extrapolation of these results could be misleading. The discussion of population genetics (Chap. 12) provides some answers regarding the long-term trends.

Under close scrutiny, the problem of the genetic risk to human populations by ionizing irradiation turned out to be unexpectedly complex. However, a trail through the wilderness of problems, although partially covered by weeds and occasionally interrupted by inaccessible ravines, has become visible in gross outline. The situation is quite different for the genetic risk due to chemical mutagens. Here the problems are still more complex, and the scientific community has hardly begun to blaze a useful trail.

11.2 Chemically Induced Mutations

11.2.1 Extent of the Problem

History [9, 10]. Mutation induction by chemical compounds was suspected even in the early days of genetics. In his first publication on radiation-induced mutation Muller (1927) [54] wrote:

"It has been repeatedly reported that germinal changes, presumably mutational, could be induced by X or radium rays, but, as in the case of the similar published claims involving other agents (alcohol, lead, antibodies, etc.), the work has been done in such a way, that the meaning of the data, as analyzed from a modern genetic standpoint, has been highly disputatious at best; moreover, what were apparently the clearest cases have given negative or contrary results on repetition."

Following Muller's publication a one-sided emphasis on radiation genetics developed, and for a long time there was little interest in any other mutagenic agents. In 1941 Muller again commented on the attempts to influence genetic material through chemicals (Cold Spring Harbor Sympium):

"But although many very drastic treatments (killing the majority of the organisms treated) have been tried, none so far

reported has met with marked success and has been tested and had its efficacy confirmed by independent workers ... In view of the high protection ordinarily afforded the genes by the cell which carries them, ... it is not to be expected that chemicals drastically affecting the mutation process while leaving the cell viable will readily be found by our rather hit-and-miss methods. But the search for such agents, as well as the study of the milder, "physiological" influences that may affect the mutation process, must continue, in the expectation that it still has great possibilities before it for the furtherance both of our understanding and of our control over the events within the gene ..."

In 1942 Auerbach and Robson in the United Kingdom achieved unassailably positive results by producing mutations in *Drosophila* with nitrogen mustard (not a "mild" agent). For reasons of military secrecy these results were not published until after the war [6]. The motive of testing this particular substance for mutagenicity was the similarity between the skin lesions induced by nitrogen mustard and those induced by high acute doses of radiation.

Independently, Oehlkers (1943) [63] in Germany had achieved positive results on *Oenothera*, particularly with urethane, a much-used drug which was considered a good hypnotic for children but later proved carcinogenic. Oehlkers was interested exclusively in basic research. His publication gave no indication at all of any particular motive for choosing urethane as a test agent. Rapoport (1946) [64] in Russia described the mutagenic action of carbonyl compounds.

After these first discoveries the new field expanded rapidly, but there was little if any discussion on practical perspectives or concern with mankind's genetic

Mutagen example	Mechanism
Base analogues: (5-bromuracil) incorporation during division	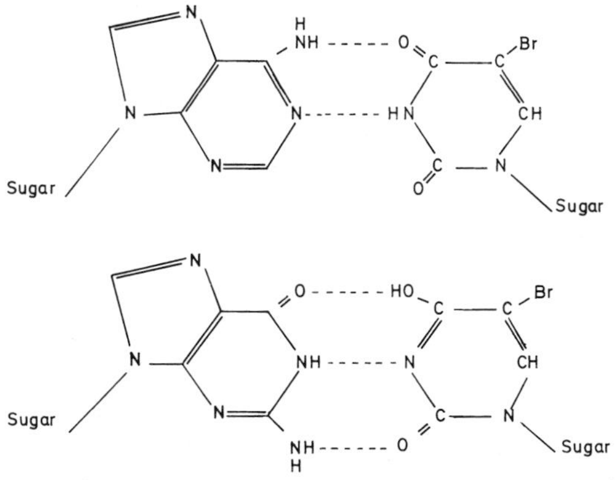 1. BUdR is incorporated during replication instead of thymine 2. BUdR undergoes a tautomeric shift more frequently than thymine 3. In the resulting enol-state, it pairs with Guanine

Fig. 11.44. (Text see p. 485)

Fig. 11.14. *(continued)*

Mutagen example	Mechanism
Nitrous acid: deamination of adenine and cytosine in resting DNA	Adenine to hypoxanthine 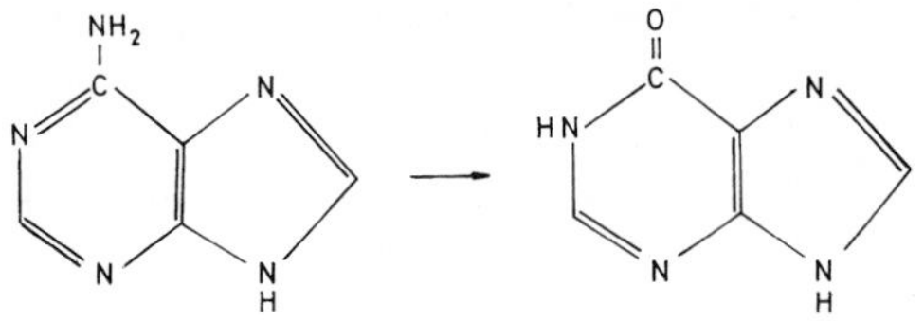 → Base pairing like guanine Cytosine to uracil  → Base pairing like thymine
Alkylating agents: methyl-methane sulfonate (MMS)	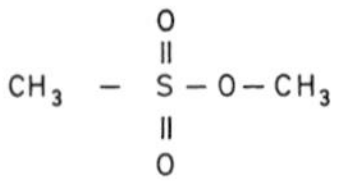 $CH_3 - S(=O)(=O) - O - CH_3$ Alkylation of guanosine at the 7 position with resulting tauto-meric shift (lower right) 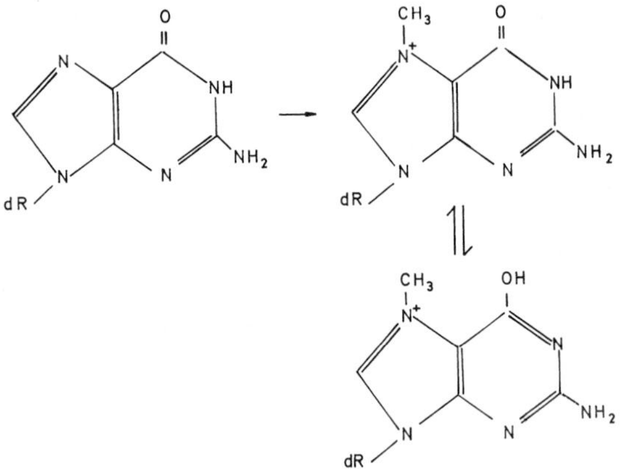
Hydroxylamines	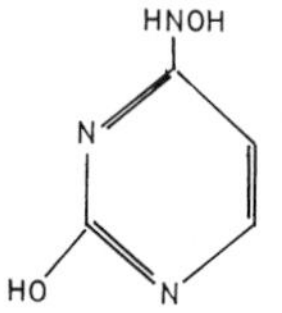 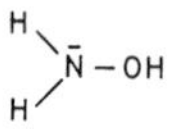 HNOH / $H_2N - OH$ 1. Hydroxylamines react with cytosine, forming derivatives that are N-hydroxylated in 4-, 6-, or both positions 2. The derivatives are in a different tautomeric state than cytosine. They can pair with adenine instead of guanine
Acridine dyes (trypaflavine)	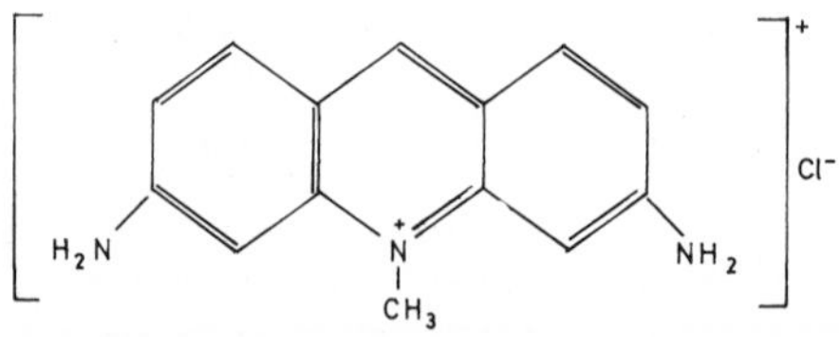 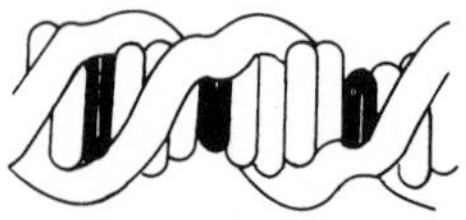 1. The acridine molecule intercalates in the DNA double helix 2. This stretches the transcribing DNA strand leading to a frameshift mutation

health. The first moves in this direction can be found in a lecture by Lüers (1955) [45] and in a survey article by Barthelmess (1956) [8]. In 1961 Conen and Lansky [18] described for the first time chromosome aberrations in the lymphocytes of individuals treated with nitrogen mustard. Since the early 1960s – after a delay of about 20 years since chemical mutagens were first described – the scientific community gradually recognized the possible threat of chemical mutagens to the human population. Röhrborn (1965) [66] was the first to summarize the evidence and to pose clearly the important questions. Since then the field has evolved rapidly, the Environmental Mutagen Society (EMS) was founded, monographs have been published [35, 102], and frequent conferences are held.

Mutagenic Compounds in the Human Environment. Mutagenic effects have been observed for a great many compounds and in a wide variety of test organisms. For most of these effects – such as genome mutations, chromosome breaks and rearrangements, and point mutations – the genetic significance is obvious. Others, such as "stickiness" of chromosomes and chromosome "gaps" (Sect. 2.2.2), are difficult to evaluate. Some substances impair the function of the tubular apparatus necessary for spindle formation during mitosis. Taken as a whole, the effects of chemicals on the genetic material are more varied than those of radiation.

Figure 11.14 presents the mutational mechanisms of selected mutagenic compounds. The strongest mutagens observed so far are alkylating agents, such as nitrogen mustard, ethyleneimine compounds, and methylsulfonic acid esters. Many of these compounds are used to treat malignancies or conditions where an immune reaction or a cell proliferative process is to be inhibited. Other groups of mutagenic compounds used in therapy include antimetabolites of nucleic acids or of folic acid and acridine compounds.

Testing for Mutagenesis by Chemical Compounds Has Now Become a Field of Toxicology. Earlier editions of this book devoted extensive discussion to problems of chemical mutagenesis and mutagenicity testing. Meanwhile, however, this field has been taken over by a different scientific community and is now part of toxicology. This illustrates the way in which concepts and methods of genetics become basic for other sciences and necessary for solving biomedical problems (See Sect. 6.1.1.6). Therefore these problems are not discussed in detail here. Only a few general approaches related to exposure of human populations and to assessment of genetic risks are mentioned.

As with ionizing radiation, risk estimates confront two main questions: (a) What substances are mutagenic? (b) To what extent are humans exposed to them? In comparison with radiation, however, there are substantially more aspects to be considered. With progress in elucidating molecular mechanisms ever more such mechanisms are being analyzed; each group of mutagens may act in a different way. Therefore our first question – whether a chemical is mutagenic or not – is much too general. Important parameters to be examined include: in what cells is it mutagenic – germ cells, somatic cells, or both? If germ cells are endangered – primarily in which sex and in what phase of germ cell development are mutations induced? Here enormous differences have been found by mouse experiments. Acridine dyes, for example, lead mainly to frame shift mutations and cause mutations in all phases of male germ cell development, while trenimon, a cytostatic agent of the ethyleneimine group, induces structural chromosome mutations, the majority of which are eliminated in the course of meiosis. Thus exposure only during the short time of about 8 weeks between meiosis and fertilization leads to an appreciable yield of mutation effects in the offspring.

Another question is: what kinds of mutations are induced? As has been demonstrated extensively above, we must distinguish genome, chromosome, and gene mutations. Even spontaneous mutations of these three types are produced by several quite distinct mechanisms, such as nondisjunction in trisomy and demethylation of CpG dinucleotides in some transitions. Earlier techniques such as in vivo mutagenicity testing in animals tried to consider at least some of these parameters so as to allow conclusions regarding the actual danger for exposed humans. These in vivo tests, however, are expensive and time consuming. Therefore faster and cheaper testing methods were highly desirable. Here the bacterial geneticist Ames [50, 51] provided support. The tests he developed appear helpful in testing not only for mutagenicity but also for carcinogenicity, since it is generally accepted that most if not all tumors are initiated by a somatic mutation (Chap. 10).

Ames Test as a Screening Test for Carcinogens? This test is based on the following principle: most of the enzymes for drug metabolism are found in the microsomal fraction of liver cells. The chemical to be tested is therefore first exposed to microsomal preparations in vitro to mimic in vivo metabolic processes and then tested for mutagenicity in bacterial systems. A few bacterial mutations, especially revertants, such as from histidine dependence to wild type, are regarded as representative for the entire genome. This test has been improved in recent years and has found wide acceptance in toxicology. At the same time, it seemed to solve another urgent problem: how to test substances for their capacity of inducing cancer

◁ **Fig. 11.14.** Molecular mechanism of point mutations induced by chemicals

since cancer is usually due to somatic mutations – (Chap. 10). It is therefore not surprising that many mutagens are, at the same time, carcinogens and vice versa, and that many of the compounds known as carcinogens have been identified as mutagens. At present animal testing for carcinogenicity – as mutagenicity testing of the live animal (in vivo) – is time consuming and expensive. It was therefore suggested to introduce a fast method for mutagenicity testing – and the Ames test is such a fast method – as a screening test for potential carcinogens. In a study of 300 different compounds, carcinogens and noncarcinogens, this test showed 157 of 175 carcinogens to be mutagenic [50, 51]. Only very few noncarcinogens demonstrated mutagenic activity. This result has been discussed extensively in recent years; it is undisputed that some carcinogens may escape observations as they do not act through DNA damage. In general, the problems of extrapolation from the bacterial to the human genome and from the in vitro assay to in vivo conditions are similar to those encountered in testing for germ cell mutagenicity. The method may help to establish priorities in carcinogenicity testing of environmental chemicals as many more compounds can be tested within a short time. However, it cannot replace the in vivo methods.

Meanwhile, a two-step procedure has been established in many agencies interested in mutagenicity testing: the Ames test or similar methods are being used to prescreen new chemicals. This helps to eliminate compounds with strong mutagenic activity. Compounds passing this test – or appearing to be important for medical therapy – are then tested in a second step, generally using the faster in vivo procedures available in mammals, especially the mouse, or in human cells. However, some of the really important questions – such as do pharmacogenetic differences affect mutagen metabolism? – Which type of mutations are mainly induced? – In what phase of germ cell development do they occur? – And what would be their phenotypic consequences in humans? – all these questions remain unanswered. Genome mutations, for example, are very important from a public health point of view since they are so common (Sect. 9.2.1), but in conventional toxicological mutagenicity testing they are not considered.

11.2.2 How Widely Is the Human Population Exposed to the Agent?

An Important but Sometimes Neglected Question. The question of how extensively human populations are actually exposed to the agent is crucial for any estimate as to the genetic threat of chemical mutagens. These considerations have sometimes been neglected in discussions on chemical mutagenesis. Again, as with many other problems, the most plausible explanation can be found by considering the sociology of science. Most research workers concerned with problems of chemical mutagenesis are experimental scientists with experience in mutation research on some test system, for example, the mouse, human chromosomes, or bacteria. Understandably, their principal concern is with the efficiency of test methods. Toxicologists who take up these methods for practical use are usually not initiated along genetic lines.

The problem of population exposure will be considered as follows:

a) How many people are exposed to a specified mutagenic agent?
b) What doses of this agent are used and for how long?
c) How do the age and sex of the exposed populations compare with the age- and sex-specific reproduction rate of the general population?
d) To what extent, if any, do individuals exposed to the agent actually reproduce? For example, do they suffer from diseases that prevent them from having children?
e) Do the offspring show signs of genetic damage?

The first three questions can often be answered approximately on the basis of data available in statistical and medical publications.

Population Exposure to a Frequently Used Drug. Isoniazid (INH) was a drug frequently used in the therapy of tuberculosis, a disease affecting persons of all ages. The mutagenic activity of this drug has repeatedly been shown in bacteria. INH inhibited transcription in an in vitro polynucleotide assay, and it reduced postreplication repair in Chinese hamster cells. On the other hand, a comprehensive research program using many different mammalian in vivo test systems for chromosome aberrations failed to show any definite increase in the spontaneous mutation rate [68]. Induction of point mutations could not be tested. Certainly a drug of this kind should be considered from the standpoint of population exposure. Such a study was carried out for the population of West Germany in 1970. It turned out that more than 35% of the total drug-exposed population were of reproductive age. Assuming that the reproduction of these tuberculosis patients equaled that of their age and sex mates in the normal population, it was estimated that 5600 children per year should be expected from them. This means that one child in every 162 would be at risk as offspring of a marriage in which one parent had been treated for tuberculosis. If the compound were mutagenic, this would represent a small but nonnegligible population exposure [78].

Population Exposure to Highly Mutagenic Drugs. The situation is quite different for cytostatic drugs. Here mutagenicity is undisputed. The analysis, however, cannot be sufficiently exact if only published or available statistical data are used. Life expectancy is usually very short at the beginning of therapy. In addition, the cancer patients treated with these drugs are usually in generally poor health, so that their reproduction cannot be assumed to equal that of the untreated population. An additional step specifically involving the number of children born to patients after the onset of treatment needed to be introduced into the analysis. This, together with all the steps set out earlier for INH, was carried out in another population study in West Germany [99]. The results indicated that only 23 children per year could be expected from cytostatically treated patients after initiation of therapy. Therefore cytostatic treatment would not have enhanced the overall mutation rate of the population to a noticeable extent. However, this conclusion is valid only as long as cytostatic therapy is limited to malignant diseases and to a few other rare conditions.

Similar Studies Are Needed for Other Chemicals. The two studies from West Germany cited above show that available information sometimes makes it possible to estimate the additional load of mutations imposed by specific compounds, provided the compound has been shown to be mutagenic for man. The results for INH, on the one hand, and cytostatic drugs, on the other, were essentially negative but for different reasons. The mutagenicity of INH in the mammalian test system used so far could not be confirmed. As for gene mutations, the matter has not been settled. Although cytostatic drugs are undisputed mutagens, reproduction by the exposed population is so low that there is virtually no genetic risk for the overall population.

Where there is more exposure, and when the compounds involved are confirmed to be mutagenic by in vivo experiments with mammals, a basis for extrapolating the extent of the additional mutational load is available. The time has come to attempt similar estimates for all drugs and environmental chemicals which have been shown to be mutagenic. In this way it would be possible to achieve at least a minimum estimate of the additional genetic load due to chemically induced mutations. This information is urgently needed. In drawing possible conclusions we should always keep in mind that while the exposure to any specific compound may be small, the total exposure to a great variety of possible mutagenic compounds could be appreciable. On the other hand, the human species – as other species – has always been exposed to a great number of naturally occurring chemicals from plants and other sources. Any assessment of an additional burden due to chemical mutagens (and

carcinogens) makes it necessary to consider this "natural" load and its variation in relation to nutrition and other, "naturally" occurring exposures [3].

In Sect. 11.1.4 the exposure of human populations to ionizing radiation is discussed. The data presented there were selected from a huge amount of information collected systematically over a long period of time in many countries and populations. Little work along such lines has been carried out so far to assess human exposure to chemical mutagens.

11.2.3 How High an Increase in the Spontaneous Mutation Rate Must Be Anticipated Due to Chemical Mutagens?

Chemically Induced Versus Radiation-Induced Mutations. In Sect. 11.1.5 no clearcut answer to the problem of increased mutations by ionizing radiation could be given. The expected order of magnitude can be determined with some confidence only for structural chromosome aberrations and, with some qualifications, for dominant and X-linked gene mutations. The uncertainty was due in part to our ignorance of the "baseline," i.e., the overall spontaneous mutation rate in humans. The same element of uncertainty would hamper any attempt to calculate the increase in mutations due to chemical mutagens. Unlike the situation in radiation genetics, however, other and quite essential information is also lacking. We know very little about the actual size of the populations exposed to known mutagens. Our knowledge of the exact pattern of action of many of these compounds is fragmentary. We do not know enough about cell stage specificity, about mutations induced in germ cells and somatic cells, about induction of genome, chromosome, and gene mutations, or about pharmacokinetics. For a great number of compounds to which human beings are exposed it is unknown whether they are mutagenic for mammals. The risk is somewhat higher when mutagenicity has been confirmed in other organisms, such as bacteria, bacteriophages, fungi, or *Drosophila,* but even chromosome breaks in human lymphocyte cultures do not prove mutagenic activity in the living organism, as evidenced in the case of caffeine. Sufficient in vivo studies on mammalian test systems are still lacking for many compounds.

As Grüneberg has put it [31], the radiation risk resembles the risk that we are exposed to in automobile traffic: the enhanced rate can be calculated approximately, and adequate precautions are to a certain extent possible. The risk due to chemical mutagens, on the other hand, resembles the risk involved in a walk through the jungle at night: here a crackling in the underbrush, there an unexplained sound, may signal unknown hidden dangers.

Monitoring of Human Populations for Increased Mutation Rates. In view of these difficulties it is indeed tempting to ask whether we should not try to monitor large populations for new mutations. Any increase would be observed directly, and we could try to relate it to radiation, chemical mutagens, or other possible causes.

There are various possibilities for selecting traits to be sampled in such a monitoring program. For example, it is possible to screen for some "sentinel" mutations – dominant mutations with specific phenotypes for which spontaneous mutation rates are fairly well known (Tables 9.6, 9.7). However, the number of individuals to be surveyed is in the order of several million, and because mutations to a specific phenotype are rare, many of them pose diagnostic difficulties due to genetic heterogeneity and phenocopies that require a high level of medical expertise for diagnosis. While a good estimate of the *order of magnitude* of the mutation rate is feasible, calculating an *increase* with confidence would be difficult. An alternative would be to screen for genome and chromosome mutations. Technically such a task would be much easier as these conditions are more frequent, but obviously this approach does not provide any information about gene mutations.

A clean, but at the same time ambitious, approach is examination of population samples to assess genetically determined protein and enzyme variants from blood and to establish for each variant whether it is transmitted from one of the parents, as is usually the case, or whether it represents a new mutation [58]. We must be sure only to study rare variants and to omit the many polymorphic variants that are frequent.

There are many problems involved in studies of this type. The most critical is statistical. Spontaneous mutation rates for single mutations discoverable at the protein level are very low (Sect. 9.4). Therefore, very large sample sizes are needed to detect an appreciable increase in the mutation rate with certainty [87, 96, 98].

The question may be put as follows: How precisely can a true trend (t)

$$t = \frac{\mu_2}{\mu_1}$$

between the mutation rates in two populations or time periods be estimated from the observed trend (t')

$$t' = \frac{x_2}{x_1}$$

if x_1 and x_2 are the observed numbers of new mutants in the two populations? An approximation can be found for the probability

$p = 0.95$

that t' is found between the following limits depending on the real trend t:

$$t' = t \pm 2t \sqrt{\frac{3.8}{x_1, x_2}}; \quad \frac{|t' - t|}{t} = 2 \sqrt{\frac{3.8}{x_1, x_2}}$$

where $x_{1,2}$ is the sample size of the two samples x_1 and x_2 together. Figure 11.15 shows the confidence limits for the apparent trend t' (ordinate) depending on the sizes of both samples together and the real trend (abscissa). For example, if the real trend t is 1.3, i.e., if the mutation rate in the second sample is 1.3 times the mutation rate in the first sample, 500 new mutants are needed to find an increase, and in this case the apparent trend t' would be found with 95% probability between approx. 1.08 and approx. 1.52. These figures should be compared to the spontaneous human mutation rates of the possible human sentinel mutations (Sect. 9.3, Tables 9.6, 9.7) and of chromosome and genome mutations (Sect. 9.1.2, Table 9.3) to obtain some notion of the sheer size of the problem. Some data are available. Among 133 478 alleles studied in individuals at the Galton Laboratory in London a total of 77 rare biochemical variants were found, and each biochemical variant was found to be transmitted from one parent or the other to the proband. No new mutations were detected. These data allow the estimation of a maximum mutation rate for these biochemical variants, which was 2.24×10^{-5} per gene per generation (See [32]).

Genome and chromosome mutations frequently occur spontaneously so that a monitoring program can remain within reasonable limits. To detect a signifi-

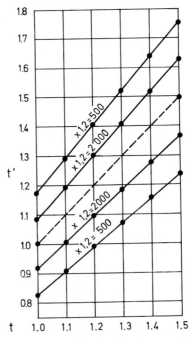

Fig. 11.15. Confidence limits ($p = 0.95$) for an apparent trend between two mutation frequencies (t', *ordinate*) depending on the real trend (t, *abscissa*) and the sizes of the two samples x_1 and x_2 together ($x_{1, 2}$)

cant increase in mutations for the usual dominant genetic diseases, whole populations of large countries would have to be screened thoroughly for decades. For mutations identifiable at the protein level, a large and extremely well-organized program would be successful if the number of genes available for screening were suitably enhanced [58, 59].

Two different approaches have been suggested. The first [59] uses cord blood samples collected from placentas (infant blood) immediately after delivery. At the same time blood samples are also collected from both parents. In these samples as many electrophoretic polymorphisms are examined as possible. Quantitative variation in some enzymes is also assessed. Such a program requires a special organization for assuring the cooperation of parents and collecting of blood. Therefore this method is practicable wherever relatively small populations are exposed to a potentially high risk, and a maximum amount of information must be collected from few individuals. However, such programs are not feasible for continuous monitoring of populations consisting of millions of people. Under such circumstances relevant biological specimens may be available that have been collected for other purposes and could also be used for mutation monitoring. As noted in Chap. 7, newborns are screened for phenylketonuria (PKU) and other inherited metabolic diseases. For PKU, the Guthrie test is normally performed, using blood spots dried on paper. These spots can be tested for hemoglobin variants [98] and a number of other gene products [1]. The methodology is fast and cheap, both technically and logistically. As discussed in Sect. 9.4, one fresh Hb α mutation has been discovered in a pilot study on 25 000 newborns.

Before such a program is introduced into large-scale population monitoring, cooperation within the scientific community must be organized in such a way as to ensure that every variant which may represent a new mutant is carefully verified – including paternity testing – by an experienced laboratory. The paternity problem is particularly vexing. In Western countries, false paternity is common in a small percentage of all family studies. Mutations are much rarer than false paternity. However, with an increasing number of polymorphisms, especially of DNA, paternity exclusion can usually be established. In the long run, human society would have information on mutation frequencies with changing environmental – as well as genetic – conditions. Statistical problems involved in ascertaining an increase in mutations by population monitoring are not trivial but appropriate methods have been suggested. However, even if an increase in mutations were to be demonstrated, the task of finding the causative agent or agents would be indeed daunting.

Present Social Attitudes Toward Mutagenicity Testing. Relatively little effort is presently being invested in any society to estimate risks posed by chemical mutagens. There is, however, growing awareness that chemicals yet to be introduced – for example, as pharmaceuticals or pesticides – should be tested for mutagenicity. The problem is more difficult in the case of those already long in use, but testing at least the most important of these compounds has been initiated. Hence there is agreement *that* chemicals should be tested for mutagenicity; the confusion arises over the question of *how* they should be tested.

Mammalian in vivo test systems are available for all types of germ cell mutations and for most somatic mutations, but these techniques generally require more time and more technical skill than methods using simpler test systems such as bacteria, fruit flies, or lymphocyte cultures. Therefore it is always tempting for policy makers to fall back on the following argument: since geneticists disagree in their recommendations, and since there is a correlation in mutation induction even between the most remotely related species, expensive mammalian in vivo systems need not be applied, and simple systems that are cheaper and quicker are sufficient. As noted, such a policy leaves important questions as to phase specificity, type of mutations, and most pharmacokinetic and pharmacogenetic problems unanswered – even when liver microsomes are used to allow for metabolic variation. Moreover, the differences usually found even among various "simple" systems make it more than likely that many mutagens would be overlooked even if the following two-step procedure were adopted:

1. Test a great number of compounds in a simple system.
2. Select those that are mutagenic and test them in the "relevant" mammalian in vivo systems.

Nevertheless, the sheer size of the problem forces compromises, and a first step involving quick testing of a great number of substances – for example, with microbial test systems involving mammalian liver microsomes – may be inevitable in some cases. However, the possibility of "false negatives" should not be overlooked. Compounds that are suspected because of their chemical constitution and those frequently used by many persons of reproductive age ought to be tested with more relevant systems even when screening systems yield negative results.

Medical and Social Significance of Various Types of Mutations. Complete assessment of the impact of mutations on humans would require an analysis of the types of diseases produced by various kinds of mutations. It is important to differentiate between prenatal and postnatal effects. Mutagenic agents that lead to early abortuses may not even be noticed and may be manifested by a slight delay of menstrual period. Their impact is zero. Abortions during the first trimester are recognized but have relatively little impact. Fetal loss during the last trimester of pregnancy is rare, and stillbirths have a more significant impact on the family but entail less burden than a variety of genetic diseases and birth defects. Here again, those defects associated with relatively early mortality cause fewer mental, social, medical, and family problems than diseases associated with long-term suffering for the patient and his family. Since most mutagenic agents are also carcinogenic, the production of neoplasms is a more direct and immediate effect. All autosomal trisomies apart from trisomy 21 are lethal in infancy, and most are aborted in the first trimester. The largest medical impact therefore concerns Down syndrome, with a current incidence of 1–2/1000. There is profound mental retardation that requires special care in the home or in institutions. The medical and personal impact of spontaneous abortions as the result of chromosome anomalies or lethal genes is much less severe. X-chromosomal aneuploidies are much more common. Their total impact on society comes largely from XXY (Klinefelter) patients, with a current frequency of 1/1000 male births or some what higher. The mean intelligence of such males is somewhat lower than that of the control population. There is some increased failure of social adaptation, and infertility is the rule (Sect. 2.2.3.2). While there is a slightly increased load on medical facilities, the major impact of this condition occurs in the personal and social sphere, which is harder to quantitate. XO females are usually aborted spontaneously. If there is postnatal survival, shortened stature and infertility constitute the main personal and medical impact. The XXX condition has no apparent medical implications but may be associated with mental retardation.

Chromosome mutations, such as deletions and translocations, if unbalanced, usually lead to early fetal loss. The few postnatal survivors with such mutations have serious malformations and may require extensive medical care, but most die early in life.

The total impact of point mutations is more difficult to assess. There are many autosomal-dominant and X-linked diseases clearly maintained in the population by equilibrium between mutation and selection (Sect. 9.3). However, the total frequency of all these diseases is about 1%. Many are almost always transmitted, such as Huntington disease, and few new mutations have been observed. In others, such as achondroplasia, about 80% of cases are new mutations. An increase in the mutation rate would lead to a substantial increase in diseases where most cases now represent fresh mutations. Other common dominant diseases such as familial hypercholesterolemia are unlikely to be maintained by mutational equilibrium. Selective factors have probably been prominent in making for relatively high frequencies. New mutations would therefore have a small impact.

New mutations for autosomal-recessive diseases usually manifest only in the homozygous state. The medical load is minimal since it requires the mating of two heterozygotes, and the disease impact is delayed over many generations. Furthermore, equilibrium for such recessive diseases no longer applies: this has led to a relatively low incidence in recent times.

The multifactorial group of diseases includes many birth defects, most common diseases of middle life, the common psychoses (schizophrenia and affective disorders), and many cases of mental retardation. The total medical and social impact of these diseases is far in excess of the strictly genetic diseases. The estimated impact of mutation depends upon the genetic model applicable to the genetic causation of these disorders. The problems involved in estimating the mutational impact of these diseases have been discussed; no predictions as to the effect of an increased mutation rate are possible.

However, it is possible that serious and large-scale effects of mutations could potentially affect birth defects and diseases that are not usually considered genetic diseases by the public. More genetic work on the specific genetic contribution to these diseases is therefore required.

Conclusions

"Spontaneous" mutations often lead to phenotypic anomalies and disease. It is therefore important to identify factors that enhance the mutation rate. Two such influences have been identified by studies on nonhuman model systems: ionizing radiation and chemical mutagens. Studies on mice and human cells indicate that the human genome is also susceptible to mutation induction by these two types of mutagens. However, a variety of methods used to examine human populations whose parents were exposed to increased radiation, such as the children of survivors of atomic bomb explosions, have failed to provide clearcut, conclusive evidence of an increase in the germ cell mutation rate. Study of the survivors themselves showed an increase in various malignancies due to somatic mutation. No epidemiological evidence is available regarding a possibly increased mutation rate in humans due to chemicals. Careful protection against mutation-inducing agents such as ionizing radiation is required, but there is no reason to anticipate a genetic catastrophe if radiation exposure remains approximately at current levels. Very little information is available to assess the extent of possible mutational damage from chemicals.

References

1. Altland K, Kaempfer M, Forssbohm M, Werner W (1982) Monitoring for changing mutation rates using blood samples submitted for PKU screening. In: Bonaiti-Pellié C, et al (eds) The unfolding genome. Liss, New York, pp 277–287 (Human genetics, part A)

2. Ames BN, McCann J, Yamasaki E (1975) Methods for detecting carcinogens and mutagens with the Salmonella/mammalian microsome mutagenicity test. Mutat Res 31 : 347–364

3. Ames BN, Profet M, Gold LS (1990) Dietary pesticides (99.99% of all natural) and nature's chemicals and synthetic chemicals: comparative toxicology. Proc Natl Acad Sci USA 87 : 7777–7786

4. Anonymous (1982–1992) Ionizing radiation: sources and biological effects. United Nations Scientific Committee on the Effects of Atomic Radiation. United Nations, New York

5. Anonymous (1982) Identifying and estimating the genetic impact of chemical environmental mutagens. National Academy Press, Washington

6. Auerbach C, Robson JM (1946) Chemical production of mutations. Nature 157 : 302

7. Barcinski MA, Abreu MC, de Almeida JCC, Naya JM, Fonseca LG, Castro LE (1975) Cytogenetic investigation in a Brazilian population living in an area of high natural radioactivity. Am J Hum Genet 27 : 802–806

8. Barthelmess A (1956) Mutagene Arzneimittel. Arzneimittelforsch 6 : 157

9. Barthelmess A (1970) Mutagenic substances in the human environment. In: Vogel F, Röhrborn G (eds) Chemical mutagenesis in mammals and man. Springer, Berlin Heidelberg New York, pp 69–147

10. Barthelmess A (1973) Erbgefahren im Zivilisationsmilieu. Goldmann, Munich

11. Bauchinger M (1968) Chromosomenaberrationen und ihre zeitliche Veränderung nach Radium-Röntgentherapie gynäkologischer Tumoren. Strahlentherapie 135 : 553–564

12. Bloom AB (1972) Induced chromosome aberrations in man. Adv Hum Genet 3 : 99–172

13. Bodmer WF, Cavalli-Sforza LL (1976) Genetics, evolution and man. Freeman, San Francisco

14. Brewen JG, Luippold HE (1971) Radiation-induced human chromosome aberrations. In vitro dose-rate studies. Mutat Res 12 : 305–314

15. Brewen JG, Preston RJ (1974) Cytogenetic effects of environmental mutagens in mammalian cells and the extrapolation to man. Mutat Res 26 : 297–305

16. Carter CO (1977) Monogenic disorders. J Med Genet 14 : 316–320

17. Chen Dequing et al (1982) Cytogenetic investigation in a population living in the high background radiation area. Chin J Radiol Med Protect 2 : 61–63

18. Conen and Lansky (1961) Chromosome damage during nitrogen mustard therapy. BMJ 1 : 1055–1057

19. Cremer T, Popp S, Emmerich P et al (1990) Rapid metaphase and interphase detection of radiation-induced chromosome aberrations in human lymphocytes by chromosomal suppression in situ hybridization. Cytometry 11 : 110–118

20. Cui Yanwei (1982) Heredity diseases and congenital malformation survey in high background radiation area. Chin J Radiol Med Protect 2 : 55–57

21. Czeizel A (1989) Hungarian surveillance of germinal mutations. Hum Genet 82 : 359–366

22. Czeizel A (1989) Population surveillance of sentinel anomalies. Mutat Res 212 : 3–9

23. Czeizel A, Sankaranarayanan K (1984) The load of genetic and partially genetic disorders in man. I. Congenital anomalies: estimates of detriment in terms of years of life lost and years of impaired life. Mutat Res 128 : 73–103

24. Czeizel A, Sankaranarayanan K, Losenci A et al (1988) The load of genetic and partially genetic diseases in man. II. Some selected common multifactorial diseases: estimates of population prevalence and of detriment in terms of years of lost and impaired life. Mutat Res 196 : 259–292

25. Czeizel A, Sankaranarayanan K, Szondi M (1990) The load of genetic and partially genetic diseases in man. III. Mental retardation. Mutat Res 232 : 291–303

26. Deng Shaozhuang et al (1982) Birth survey in high background radiation area. Chin J Radiol Med Protection 2 : 60

27. Ehling UH, Charles DJ, Favor J et al (1985) Induction of gene mutations in mice: the multiple end-point approach. Mutat Res 150 : 393–401

28. Ford CE, Searle AG, Evans EP, West BJ (1969) Differential transmission of translocation induced in spermatogonia of mice by X-irradiation. Cytogenetics 8 : 447–470

29. Ford CE, Evans EP, Searle AG (1978) Failure of irradiation to induce Robertsonian translocations in germ cells of male mice. In: Conference on mutations: their origin, nature and potential relevance to genetic risk in man. Boldt, Boppard, pp 102–108 (Jahreskonferenz 1977, Zentrallaboratorium für Mutagenitätsprüfungen)

30. Fuhrmann W, Vogel F (1983) Genetic counseling, 3rd edn. Springer, Berlin Heidelberg New York (Heidelberg science library 10)

31. Grüneberg H, Bains GS, Berry RJ, Riles L, Smith CAB (1966) A search for genetic effects of high natural radioactivity in South India. Mem Med Res Council (London) 307

32. Harris H (1980) The principles of human biochemical genetics, 4th edn. North-Holland, Amsterdam

33. High Background Radiation Research Group, China (1980) Health survey in high background radiation area in China. Science 209 : 877–880

34. Hollaender A (ed) (1954–1956) Radiation biology, 4 vols. McGraw-Hill, New York

35. Hollaender A (ed) (1973) Chemical mutagens. Principles and methods for their detection, vol 3. Plenum, New York

36. Hook EB, Cross PK, Regal RR (1984) The frequency of 47, + 21, 47, + 18 and 47, + 13 at the uppermost extremes of maternal ages: results on 56094 fetuses studied prena-

tally and comparisons with data on livebirths. Hum Genet 68 : 211–220

37. Ilvin LA, Balonov MI, Buldakov LA et al (1990) Radio-contamination patterns and possible health consequences of the accident at the Chernobyl nuclear power station. J Radiol Protect 10 : 3–28

38. Jablon S, Kato H (1970) Childhood cancer in relation to prenatal exposure to A-bomb radiation. Lancet 2 : 1000

39. Kazakov VS, Demidchik EP, Astatchova LN (1992) Thyroid cancer after Tschernobyl. Nature 359 : 21–22

40. King RA, Rotter JI, Motulsky AG (eds) (1992) The genetic basis of common diseases. Oxford University Press, New York

41. Kochupillai N, Verma JC, Grewal MS, Ramalingaswami V (1976) Down's syndrome and related abnormalities in an area of high background radiation in coastal Kerala. Nature 262 : 60–61

42. Lea DE, Catcheside DG (1942) The mechanism of the induction by radiation of chromosome aberrations in Tradescantia. J Genet 44 : 216–245

43. Lejeune J, Turpin R, Rethoré MO (1960) Les enfants nés de parents irradiés (cas particuliers de la sex-ratio). 9 th International Congress on Radiology, July 23–30, 1959, Munich, pp 1089–1096

44. Lu Bingxin et al (1982) Survey of hereditary ophthalmo-pathies and congenital ophthalmic malformations in high background areas. Chin J Radiol Med Protect 2 : 58–59

45. Lüers H (1955) Zur Frage der Erbschädigung durch tumortherapeutische Cytostatica. Z Krebsforsch 60 : 528

46. Lüning KG, Searle AG (1970) Estimates of the genetic risks from ionizing irradiation. Mutat Res 12 : 291–304

47. Malling HV, DeSerres FJ (1973) Genetic alterations at the molecular level in X-ray induced ad-3B mutants of Neurospora crassa. Radiat Res 53 : 77–87

48. Mattei JF, Mattei MG, Ayme S, Siraud F (1979) Origin of the extra chromosome in trisomy 21. Hum Genet 46 : 107–110

49. Mavor JW (1924) The production of nondisjunction by X-rays. J Exp Zool 39 : 381–432

50. McCann J, Ames BN (1976) Detection of carcinogens as mutagens in the Salmonella/microsome test: assay of 300 chemicals: discussion. Proc Natl Acad Sci USA 73 : 950–954

51. McCann J, Choi E, Yamasaki E, Ames BN (1975) Detection of carcinogens as mutagens in the Salmonella/microsome test: assay of 300 chemicals. Proc Natl Acad Sci USA 72 : 5133–5139

52. Modell B, Kuliev A (1990) Changing paternal age distribution and the human mutation rate in Europe. Hum Genet 86 : 198–202

53. Mohrenweiser HW, Branscomb EW (1989) Molecular approaches to the detection of germinal mutations in mammalian organisms including man. In: Jolles G, Cordier A (eds) New trends in genetic risk assessment. Academic, New York, pp 41–56

54. Muller HJ (1927) Artificial transmutation of the gene. Science 66 : 84–87

55. Mulvihill JJ, Connelly RR, Austin DF et al (1987) Cancer in offspring of long-term survivors of childhood and adolescent cancer. Lancet 2 : 813–817

56. Neel JV, Lewis SE (1990) The comparative radiation genetics of humans and mice. Annu Rev Genet 24 : 327–362

57. Neel JV, Schull WJ (1991) The children of atomic bomb survivors. A genetic study. National Academy, Washington

58. Neel JV, Tiffany TO, Anderson NG (1973) Approaches to monitoring human populations for mutation rates and genetic disease. In: Hollaender A (ed) Chemical mutagens, vol 3. Plenum, New York, pp 105–150

59. Neel JV, Mohrenweiser HW, Meisler MH (1980) Rate of spontaneous mutation of human loci encoding protein structure. Proc Natl Acad Sci USA 77 : 6037–6041

60. Neel JV, Satoh C, Goriki K et al (1986) The rate with which spontaneous mutation alters the electrophoretic mobility of polypeptides. Proc Natl Acad Sci USA 83 : 389–393

61. Neel JV, Schull WS, Awa AA (1989) Implications of the Hiroshima and Nagasaki genetic studies for the estimation of the human "doubling dose" of radiation. Genome 31 : 853–859

62. Newcombe HB, McGregor F (1964) Learning ability and physical wellbeing in offspring from rat populations irradiated over many generations. Genetics 50 : 1065–1081

63. Oehlkers F (1943) Die Auslösung von Chromosomenmutationen in der Meiosis durch Einwirkung von Chemikalien. Z Induktiven Abstammungs Vererbungslehre 81 : 313–341

64. Rapoport IA (1946) Carbonyl compounds and the chemical mechanism of mutation. C R Acad Sci USSR 54 : 65

65. Reichert W, Buselmaier W, Vogel F (1984) Elimination of X-ray-induced chromosomal aberration in the progeny of female mice. Mutat Res 139 : 87–94

66. Röhrborn G (1965) Über mögliche mutagene Nebenwirkungen von Arzneimitteln beim Menschen. Hum Genet 1 : 205–231

67. Röhrborn G (1970) The dominant lethals: method and cytogenetic examination of early cleavage stages. In: Vogel F, Röhrborn G (eds) Chemical mutagenesis in mammals and man. Springer, Berlin Heidelberg New York, pp 148–155

68. Röhrborn G et al (1978) A correlated study of the cytogenetic effect of INH on cell systems of mammals and man conducted by thirteen laboratories. Hum Genet 42 : 1–60

69. Russell LB, Saylors CL (1963) The relative sensitivity of various germ cell stages of the mouse to radiation-induced nondisjunction, chromosome losses and deficiency. In: Sobels FH (ed) Repair from genetic damage and differential radiosensitivity in germ cells. Pergamon, Oxford, pp 313–340

70. Russell LB, de Hamer DL, Montgomery CS (1974) Analysis of 30 c-locus lethals by viability of biochemical studies. Biol Div Annu Prog Rep ORNL 4993 : 119–120

71. Russell WL, Russell LB, Kelly EM (1958) Radiation dose rate and mutation frequency. Science 128 : 1546–1550

72. Russel WL, Kelly EM, Hunsicker PR et al (1972) Effect of radiation dose-rate on the induction of X-chromosome loss in female mice. In: Report of the United Nations Science Committee on the effect of atomic radiations. United Nations, New York (Ionizing radiation: levels and effects, vol 2: Effects)

73. Sankaranarayanan K (1991) Ionizing radiation and genetic risks. I. Epidemiological, population genetic, biochemical and molecular aspects of Mendelian diseases. Mutat Res 258 : 349

74. Sankaranarayanan K (1991) Ionizing radiation and genetic risks. II. Nature of radiation-induced mutations in

experimental mammalian in vivo systems. Mutat Res 258 : 51–73

75. Sankaranarayanan K (1991) Ionizing radiation and genetic risks. III. Nature of spontaneous and radiation-induced mutations in mammalian in vitro systems and mechanisms of induction of mutations by radiation. Mutat Res 258 : 75–97

76. Sankaranarayanan K (1991) Ionizing radiation and genetic risks. IV. Current methods, estimates of risk of Medelian disease, human data and lessons from biochemical and molecular studies of mutations. Mutat Res 258 : 99–122

77. Sasaki MS, Miyata H (1968) Biological dosimetry in atomic bomb survivors. Nature 220 : 1189–1193

78. Schmidt H (1973) Wahrscheinliche genetische Belastung der Bevölkerung mit INH (Isonikotinsäure-Hydrazid). Hum Genet 20 : 31–45

79. Scholte PJL, Sobels FH (1964) Sex ratio shift among progeny from patients having received therapeutic X-radiation. Am J Hum Genet 16 : 26–37

80. Schull WJ, Neel JV (1958) Radiation and the sex ratio in man. Science 128 : 343–348

81. Schull WJ, Neel JV (1962) Maternal radiation and mongolism. Lancet 2 : 537–538

82. Schull WJ, Neel JV, Hashizume A (1968) Some further observations on the sex ratio among infants born to survivors of the atomic bombings of Hiroshima and Nagasaki. Am J Hum Genet 18 : 328–338

83. Schull WJ, Otake M, Neel JV (1981) Genetic effects of the atomic bombs: a reappraisal. Science 213 : 1220–1227

84. Searle AG (1972) Spontaneous frequencies of point mutations in mice. Hum Genet 16 : 33–38

85. Searle AG, Edwards JH (1986) The estimation of risks from the induction of recessive mutations after exposure to ionising radiation. J Med Genet 23 : 220–226

86. Sigler AT, Lilienfeld AM, Cohen B-H, Westlake JE (1965) Radiation exposure in parents with mongolism (Down's syndrome). Johns Hopkins. Med J 117 : 374

87. Strobel D, Vogel F (1958) Ein statistischer Gesichtspunkt für das Planen von Untersuchungen über Änderungen der Mutationsrate beim Menschen. Acta Genet Stat Med 8 : 274–286

88. Tanaka K, Ohkura K (1958) Evidence for genetic effects of radiation on offspring of radiologic technicians. Jpn J Hum Genet 3 : 135–145

89. Tang BK, Grant DM, Kalow W (1983) Isolation and identification of 5-acetylamino-6-formylamino-3-methyluracil as a major metabolite of caffeine in man. Drug Metab Dispos 11 : 218–220

90. Timoféeff-Ressovsky NW, Zimmer KG (1947) Das Trefferprinzip in der Biologie. Steinkopf, Leipzig

91. Tough IS, Buckton KE, Baikie AG, Court Brown WM (1960) X-ray induced chromosome damage in man. Lancet 2 : 849–851

92. Traut H (1976) Effects of ionizing radiation on DNA. In: Hüttermann J, Köhnlein W, Téoule R (eds) Molecular biology, biochemistry and biophysics, vol 27. Springer, Berlin Heidelberg New York, pp 335–347

93. Uchida IA, Holunga R, Lawler C (1968) Maternal radiation and chromosomal aberrations. Lancet 2 : 1045–1049

94. Uchida IA, Lee CPV, Byrnes EM (1975) Chromosome aberrations induced in vitro by low doses of radiation: Nondisjunction in lymphocytes of young adults. Am J Hum Genet 27 : 419–429

95. United Nations Scientific Committee on the Effects of Atomic Radiation (UNSCEAR) (1982–1992) Ionizing radiation: sources and biological effects. United Nations, New York (UNSCEAR reports, every four years)

96. Vogel F (1970) Monitoring of human populations. In: Vogel F, Röhrborn G (eds) Chemical mutagenesis in mammals and man. Springer, Berlin Heidelberg New York, pp 445–452

97. Vogel F (1992) Risk calculations for hereditary effects of ionising radiation in humans. Hum Genet 89 : 127–146

98. Vogel F, Altland K (1982) Utilization of material from PKU-screening programs for mutation screening. Prog Mutat Res 3 : 143–157

99. Vogel F, Jäger P (1969) The genetic load of a human population due to cytostatic agents. Humangenetik 7 : 287–304

100. Vogel F, Röhrborn G (eds) (1970) Chemical mutagenesis in mammals and man. Springer, Berlin Heidelberg New York

101. Vogel F, Röhrborn G, Schleiermacher E, Schroeder TM (1969) Strahlengenetik der Säuger. Thieme, Stuttgart (Fortschr Allg Klin Humangen)

102. Wei LX, Zha YR, Tao ZF (1987) Recent advances of health survey in high background radiation areas in Yangjiang, China. In: International symposium on biological effects of low-level radiation. pp 1–17

103. Wei LX, Zha YR, Tao ZF et al (1990) Epidemiological investigation of radiological effects in high background radiation areas in Yangjiang, China. J Radiat Res (Tokyo) 31 : 119–136

104. Yoshimoto Y, Neel JV, Schull WJ et al (1990) Malignant tumors during the first 2 decades of life in the offspring of atomic bomb survivors. Am J Human Genet 46 : 1041–1052

12 Population Genetics: Description and Dynamics

*"The main agent of natural selection
in the human species during the last five thousand years
has been infectious disease."*

Haldane and Jayakar, 1965

Population genetics deals with the consequences of Mendelian laws on the composition of the population with special reference to the effects of mutations, selection, migration, and chance fluctuation of gene frequencies. All these factors together determine the genetic population structure. Knowledge of population genetics is useful for many purposes. For example, it provides a basis for understanding the epidemiology of genetic diseases and helps in the planning of measures for their prevention. Another objective of studies in population genetics is improved understanding of human evolution and the prediction of future trends in the biological evolution of mankind in the face of various environmental changes. Since the human population is much better described than any other species, and much better records are available, there are many advantages to studying the population genetics of man.

The work of Fisher, Haldane, Wright, and their successors has provided an elaborate theoretical framework for population genetics. Empirical data on man and their interpretations, however, have lagged behind mathematical and theoretical considerations. Several excellent expositions of population genetics exist (Li 1955 [77]; Li 1976 [79]; Crow and Kimura 1970 [26]; Cavalli-Sforza and Bodmer 1971 [18]; Jacquard 1974 [63]; Hartl and Clark 1989 [55]; Ewens 1980 [36]; Weiss 1994 [144]). Our treatment of the subject therefore is not exhaustive. Special attention is paid to empirical data collected in human populations and their interpretation.

Work in human population genetics may conveniently be divided into two broad classes: description of populations and their genetic composition and studies designed to understand the causes for changes in the human gene pool. These two approaches are intimately connected. It is impossible to elaborate specific hypotheses or to design studies to test them unless certain underlying facts regarding population structure are known. However, there are so many different human populations and so many known genetic traits that the task of describing the genetic characteristics of all populations is formidable. A selection of the more important problems is necessary. What are the guidelines for such a selection?

In general, similar principles should apply to both the planning of scientific work in human population genetics and to formulating the design of laboratory investigations. Empirical studies conducted without guidance by a specific hypothesis rarely lead to significant insights. The quality of scientific work generally depends on the profundity and specificity of the underlying hypotheses. One cannot expect, however, that all data collection in science will always be guided by a hypothesis. The human genome project is an excellent example. However, usually mere data-collecting activities are scientifically less satisfactory than projects that ask specific questions. Fortunately, descriptive data can often be obtained from a variety of sources:

a) Polymorphisms for DNA variants particularly, and less so for enzymes and other proteins are continually being described. Testing of populations to determine gene frequencies of various genetic markers provides data for assessing population structure.

b) Population testing may be carried out for medical reasons. For example, newborn screening for phenylketonuria and sometimes for other rare recessive disorders is now performed as a routine procedure in many populations. These studies provide valuable information about population differences in gene frequencies.

c) Gene frequency data may be collected while testing a specific hypothesis in a population. Even if the hypothesis is rejected, or if the outcome of the study is equivocal, the descriptive data may be useful.

Categorizing a human population using frequencies of genetic polymorphisms and genetic diseases is only the initial step in understanding the differences in gene frequencies among various human populations. Explanatory hypotheses are required to explain the differences.

Consider, for example, the hypothesis that the high frequency of persistence of intestinal lactase activity in White adults as contrasted with Mongoloids and blacks is caused by a selective advantage in a climate favoring the development of vitamin D deficiency and rickets, since lactose enhances calcium resorption from the intestine and calcium reduces the risk of developing rickets. This hypothesis has a biological basis, is specific, and can easily be refuted by showing that lactose does not increase calcium resorption. Hypotheses of this sort are highly desirable. However, much work in human population genetics remains at the descriptive level.

12.1 Population Description

12.1.1 Hardy-Weinberg Law:
Extended Consideration – Gene Frequencies

Hardy-Weinberg Law for Autosomal Genes [100]. The Hardy-Weinberg Law is discussed above (Sect. 4.2). Let two alleles A_1, A_2 have the gene frequencies $A_1 = p$, $A_2 = q$; $p + q = 1$, let mating be random. The three phenotypes will occur in the following frequencies: $p^2A_1A_1$, $2pqA_1A_2$, $q^2A_2A_2$. This rule may be generalized as follows. If the gene frequencies of n alleles, $A_1, A_2 . . . A_n$, are $p_1, p_2, . . . p_n$ ($\sum p_i = 1$), and if the population breeds at random with respect to the gene locus A, the phenotypes will occur according to chance combinations of these alleles in pairs:

$$(p_1A_1 + p_2A_2 + \ldots + p_nA_n)^2 =$$

$$\sum_{i=1} p_i^2 A_iA_i + \sum_{i<j} 2p_ip_jA_iA_j$$

In the absence of disturbing influences both the gene and the genotype frequencies remain constant from generation to generation. For autosomal genes this "Hardy-Weinberg equilibrium" is reached in the first generation of random mating. This is not so, however, for X-linked genes.

Hardy-Weinberg Law for X-Linked Genes. Let two alleles A_1 and A_2 have frequencies p_M and q_M ($p_M + q_M = 1$) in the male population; the phenotypes A_1 and A_2 are also found with frequencies p_M and q_M. In the female population, on the other hand, the genotypes A_1A_1, A_1A_2, and A_2A_2 occur in frequencies r, $2s$, and t ($r + 2s + t = 1$). Then the frequencies of the alleles A_1 and A_2 in the females are $p_F = r + s$ and $q_F = s + t$, respectively, and these women produce oocytes of the two types A_1 and A_2 also with frequencies p_F and q_F, respectively. Their male offspring are formed in the same proportion. Female offspring, on the other hand, are produced from a combination of the ($p_FA_1 + q_FA_2$) oocytes with the ($p_MA_1 + q_MA_2$) X sperms of the men. The next generation will therefore be composed as follows:

♂♂:

$p_FA_1 + q_FA_2$,

♀♀:

$p_Mp_FA_1A_1 + (p_Mq_F + p_F q_M)A_1A_2 + q_M - q_FA_2A_2$

It follows that the frequencies of A_2 in the males and in the females of the next generation are:

$$q'_M = q_F, \quad q'_F = \tfrac{1}{2}(p_Mq_F + p_Fq_M) + q_Mq_F = \tfrac{1}{2}(q_M + q_F)$$

This means:

a) The gene frequency of A_2 in males of each generation is equal to the gene frequency of A_2 in females of the preceding generation.
b) The frequency of A_2 among females in each generation equals the mean of the frequencies of A_2 in males and females of the foregoing generation.
c) The equation:

$$q'_M - q'_F = -\tfrac{1}{2}(q_M - q_F)$$

holds.

This means that if the gene frequencies in males and females in one generation are not equal their difference is halved in the next generation. In addition, the sign of the difference changes: if q_F is higher than q_M, then q'_M will be higher than q'_F. Both gene frequencies q_M and q_F converge to a common value $\hat{q}$; at the same time, the genotype distributions in the two sexes tend to the equilibrium states:

♂♂:
$(1 - \hat{q}) A_1 + \hat{q}A_2$,

♀♀:
$(1 - \hat{q})^2 A_1A_1 + 2\hat{q}(1- \hat{q})A_1A_2 + \hat{q}^2 A_2A_2$

The mode of approximation to this equilibrium can be seen in Fig. 12.1. This example shows that establishment of a Hardy-Weinberg equilibrium after one generation of random mating is by no means self-evident.

The other limitations of the Hardy-Weinberg Law have been enumerated in Sect. 4.2; a great deal of po-

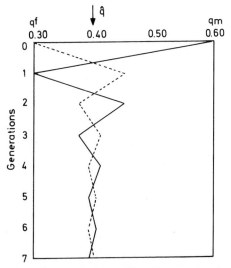

Fig. 12.1. Approximation of gene frequencies in males to equilibrium value, if the mode of inheritance is X-linked $q_M = 0.6$; $q_F = 0.3$; $\hat{q} = 0.4$; —, gene frequencies in males, q_M; ---, gene frequencies in females, q_F; $\hat{q}$, gene frequency at equilibrium

pulation genetics can be regarded as the elaboration of this fundamental rule.

Gene Frequencies. Individuals in a given population can be classified on the basis of their phenotypes. In an increasing number of cases – especially in genetic polymorphisms – these phenotypes provide unequivocal evidence of the genotype. However, description of genetic variability in a population in terms of *genotypes* is usually clumsy. The Hardy-Weinberg Law makes it possible to describe this variability in terms of *gene frequencies.* This notation simplifies the matter; in a system with two alleles, a single number (p or q) comprises all the information needed. The estimation of gene frequencies is described in principle in Sect. 4.2 and in Appendix 1; a detailed outline is given by Race and Sanger [109] and by Mourant et al. [94].

12.1.2 Genetic Polymorphisms

Definition and History. A polymorphism is a Mendelian or monogenic trait that exists in the population in at least two phenotypes (and presumably at least two genotypes), neither of which is rare – that is, neither of which occurs with a frequency of less than 1%–2%. This cutoff point is somewhat arbitrary and has no formal scientific basis. The term polymorphism is generally not used when referring to the relatively frequent heterozygotes for a rare autosomal recessive disease. However, once the heterozygote state becomes detectable in the laboratory, as increasingly happens, it is appropriate to refer to such traits as having polymorphic frequencies. Often we find more than two alleles, and more than two phenotypes, for a single locus. A polymorphism should be contrasted with a *rare* genetic variant. Rare genetic variants are arbitrarily defined as monogenic traits that occur in the population with a frequency of less than 1%–2%, and usually at much lower frequencies. The first human polymorphism was the AB0 blood group discovered by Landsteiner (1900) [76]. Until 1955 polymorphisms were known for only a number of other red cell surface antigens, i.e., the blood groups. In 1955 Smithies [119, 120] described the method of starch gel electrophoresis, which permits separation of proteins on a fixed medium not only by charge but also by molecular size. The new method enabled Smithies to detect polymorphism of a serum protein: the hemoglobin-binding protein haptoglobin. This method proved useful first for detection of serum protein polymorphisms and later – together with methods for specifically identifying enzyme activities – for detection of enzyme polymorphisms.

A great number of polymorphisms for other serum proteins – and later for enzymes of blood plasma, erythrocytes, and leukocytes – have been detected. Many polymorphisms are genetically straightforward, with two alleles determining two variants of the same protein. Others are highly complex, such as the MHC, with multiple, related loci in a complex system on human chromosome 6 (Sect. 5.2.5).

Present Situation. Table 12.1 shows the most important polymorphisms. Some of these exist in only one major racial group. For a few polymorphisms specific hypotheses as to their maintenance in human populations by natural selection have been proposed. These hypotheses are discussed in greater detail in Sects. 12.2.1.6 and 12.2.1.8.

Biochemical Individuality for Polymorphisms. Garrod (1902) [47] ended his paper on alcaptonuria (Sect. 1.5) with the sentence:

If it be, indeed, the case that in alcaptonuria and the other conditions mentioned we are dealing with individualities of metabolism and not with the results of morbid processes the thought naturally presents itself that these are merely extreme examples of variations of chemical behavior which are probably everywhere present in minor degrees and that just as no two individuals of a species are absolutely identical in bodily structure neither are their chemical processes carried out on exactly the same lines.

This "biochemical individuality" is striking when all blood polymorphisms are considered. Let us, for example, select a northwestern European individual having the most frequent alleles of every polymorphism listed in Table 12.1. How many other individuals will have the same phenotype and genotype for all markers? This probability can be calculated by multiplying the relative frequencies of these phenotypes in the white population (Table 12.2), with the result of 9.2×10^{-7}; thus, among one million males fewer than one will have this phenotype, although this particular combination is the most frequent one. All other combinations are rarer still.

Table 12.1 does not even include the phenotypes of the major histocompatibility complex (MHC) (Sect. 5.2.5) or other less well-documented polymorphisms and many enzyme systems for which only rare variants have been observed. Had these been included, it could have been shown that every person on this planet (except for identical twins) is genetically unique. The physiological function for only some of the polymorphic characters listed in Table 12.1 is known. Their possible significance for predictions of health risks under varying environmental conditions is discussed in Sect. 7.5.2.

Table 12.1. Some important human polymorphisms affecting gene products (blood groups, proteins, and enzymes)

Name	Main alleles	Remarks
Erythrocyte surface antigens (blood groups)		
AB0	A_1, A_2, B, 0	For discussions of disease associations see Sect. 6.6.2; for natural selection see Sect. 12.2.1.8.
ABH secretion	Se, se	Interaction with the Lewis system.
Diego	Di^a, Di^b	Allele Di^a present only in Amerindians and Mongoloid populations (Sect. 14.3.1).
Duffy	Fy^a, Fy^b, Fy	Amorphic allele Fy common in Africans; discussion of selection in Sect. 14.3.1.
Kell	K, k	Other, closely linked loci, e. g., Sutter (Js^a).
Kidd	Jk^a, Jk^b	Very few individuals with Jk (a–b–).
Lewis	Le^a, Le^b	Interaction with ABH secretor locus.
Lutheran	Lu^a, Lu^b	
MNSs	MS, Ms, NS, Ns	There are some other, closely linked antigens: Hunter and Henshaw, especially in Africans.
P	P_1, P_2, p	Allele p is very rare.
Rhesus	Gene complexes CDe, cde, cDE, C^wDe, cDe, cdE, CDE, and others in varying combinations	Discussion of maternal-fetal incompatibility; structure of the gene complex and linkage disequilibrium in Sect. 5.2.5; biochemical basis in Sect. 5.2.4.
Xg	Xg^a, Xg	X-linked.
Serum protein groups		
α_1-Antitrypsin (α_1-protease inhibitor)	PI^{M_1}, PI^{M_2}, PI^{M_3}, PI^S, PI^Z	Numerous rare alleles. Discussions of α_1-antitrypsin deficiency, especially in homozygotes of the PIZ allele, in Sect. 6.2.4.
Ceruloplasmin	CP^B, CP^A, CP^C	Most Europeans are homozygous CP^B/CP^B; blacks have a gene frequency of 0.06 for CP^A.
Complement component-3	$C3^S$, $C3^F$	Apart from these two common alleles, there are a number of rare ones.
Group-specific protein	GC^{1F}, GC^{1S}, GC^2	Special variants described, for example GC^{Chip} in Chippewa Indians, GC^{Ab} in Australian aborigines; subdivision of common alleles possible; discussion of natural selection in Sect. 14.3.1.
Haptoglobin	HP^{1S}, HP^{1F}, HP^2	Many rare variants are known; genetic and nongenetic ahaptoglobinemia occurs.
Immunoglobulins IGHG (gm)	$G1m^3$, $G3m^5$, $G1m^1$, $G1m^{1, 2}$	This is a very complicated system with many rare haplotypes and specificities; see genetics of antibody formation, Sect. 7.4.
IGKC (Km)	Km^1, Km^3	More alleles are known, but are usually not easily available.
Properdin factor B (glycine-rich-β-glycoprotein)	BF^S, BF^F	Rare alleles are known.
Transferrin	TF^{C_1}, TF^{C_2}, TF^{C_3}, TF^B, TF^D	Different D and B variants have been described, all of them are rare. D variants occur mainly in Africans.
Red cell enzymes		
Acid phosphatase-1	$ACP1^A$, $ACP1^B$, $ACP1^C$	An additional allele $ACP1^R$ has been observed in Khoisanids.
Adenosine deaminase	ADA^1, ADA^2	Rare alleles ADA^3 and ADA^4 have been described.
Adenylate kinase-1	$AK1^1$, $AK1^2$	Some other, rarer alleles are known.
Esterase D	ESD^1, ESD^2	Rare variants are also known.
Peptidase-A	$PEPA^1$, $PEPA^2$	$PEPA^2$ has a gene frequency of about 0.07 in Africans; whites have almost exclusively $PEPA^1$. Rare variants are known.
Peptidase-D (proline dipeptidase)	$PEPD^1$, $PEPD^2$, $PEPD^3$	$PEPD^3$ observed especially in Africans.

Table 12.1. *(continued)*

Name	Main alleles	Remarks
Phosphoglucomutases		
PGM1	PGM1^{a1}, PGM1^{a2}, PGM1^{a3}, PGM1^{a4}	Rare alleles are known.
PGM2	PGM2^1, PGM2^2	Allele PGM2^2 only in Africans, other alleles very rare.
PGM3	PGM3^1, PGM3^2	Enzymes mostly in leukocytes, placenta and sperms.
Phosphogluconate dehydrogenase	PGD^A, PGD^B	Some other, rare alleles are known.
Other enzyme polymorphisms		
Alcohol dehydrogenase	ADH3^1, ADH3^2	ADH2 active in other organs, ADH3 active in the intestines.
Cholinesterase (serum)-1	CHE1^U, CHE1^D, CHE1^S	Discussed in Sect. 7.4.1.
Glutamic-pyruvic transaminase (alanine aminotransferase)	GPT^1, GPT^2	Activity discussed in Sect. 6.1.1.2.
Liver acetyltransferase	Rapid and slow inactivators	Discussed in Sect. 7.5.1.

Table 12.2. Frequencies of the most common phenotypes of the polymorphisms in Table 12.1 among European populations

Phenotype	Approximate frequency
Blood groups[a]	
A1	35%
CdD. ee	33%
Fy (a + b +)	48%
Jk (a + b +)	50%
K − k +	91%
Lu (a − b +)	92%
MNSs	23%
P$_1$	79%
Se	78%
Xg (a +)	66%
Serum proteins and erythrocyte enzymes[b]	
ADA 1	87%
ACP1 AB	39%
AK1 1	93%
BF S	61%
C3 S	61%
GC 1S	34%
Gm (−1, −2, + 3, + 5)	45%
GPT 2 − 1	49%
HP 2 − 1	47%
km (−1)	87%
PGD A	96%
PGM1 a1	39%
Pi M1	55%
TF C1	57%

Incidence of this most common phenotype: 9.2×10^{-7}.

[a] Frequencies from Race and Sanger (1975) [109].
[b] Frequencies from Hummel (1971, 1977, 1979).

What is the Proportion of Polymorphic Human Gene Loci? For what proportion of their gene loci are humans polymorphic? Are polymorphic loci only a relatively small fraction of all human gene loci, or is the proportion of polymorphic loci appreciable? Blood groups can be detected only if an antibody is found against a certain antigen. However, the antibody could be formed only in an individual not carrying the same antigen. Therefore serological detection of a gene locus usually presupposes genetic variability with respect to this gene locus – either a polymorphism or a rare variant. The detection of genetic variability proceeds differently for enzymes. The search for enzyme variability requires an ability to detect the enzyme readily in an electrophoretic medium. If there is polymorphism, abnormal electrophoretic mobility of the variant gene product often results. Detection of the normal (wild-type) and variant enzyme can be achieved only by localization of the enzyme-specific biochemical reaction on the supporting medium of electrophoresis, for example starch gel. Such localization is performed by developing a test system for enzyme assay that results in a colored end product allowing direct visualization on the gel. For example, red cell acid phosphatase splits phenolphthalein phosphate at pH 6 into phenolphthalein and phosphate; the free phenolphthalein stains the starch gel only at those segments of the gel where the enzyme has migrated. Different patterns of enzyme staining are detected that can be related to major gene differences (Fig. 12.2).

Screening a number of enzymes for variants by analogous methods made possible an unbiased estimate of the proportion of polymorphic enzymes. Such a calculation was carried out by Harris and Hopkinson in 1972 [53]. They collected information from electrophoretic surveys of European populations on 71 different loci determining enzyme structure. The number of different enzymes tested was smaller, as one enzyme may consist of more than one polypeptide chain and hence may be determined by more than one gene. A polymorphism was registered if the gene frequency of the most common alleles did not exceed 0.99. Since the heterozygote frequency ($2pq$) is about twice the genotype frequency (q), about 2% of the population would be heterozygous for such alleles. According to this definition, 20 of the 71 (28.2%) loci showed polymorphisms. This figure

does not comprise the total number of polymorphisms, as some variants cannot be identified by electrophoresis but require different methods of detection (e. g., pseudocholinesterase; Sect. 7.5.1). Since the electrophoretic technique may not have been good enough in some cases to show clearly differences between variants, the figure of 28.2% is almost certainly an underestimate. For one of these enzymes – erythrocyte acid phosphatase with three common alleles – more than 50% of the population are heterozygous. For other loci the degree of heterozygosity is much lower, and for all 71 loci the average proportion of heterozygous individuals was estimated at 6.7% per locus. It can be calculated from the genetic code that only about one-third of all possible base substitutions lead to polar amino acid replacements with an expected change in the electrophoretic mobility [136]. However, there is no a priori reason to assume that mutations leading to nonpolar replacements have a lower probability of occurrence. Hence, the real average heterozygosity per locus may be about three times higher, amounting to approx. 20%. Similar findings regarding the ubiquitous occurrence of genetic polymorphisms have been made in all biological species examined (Fig. 12.3). This finding is therefore not unique to the human species where it was first discovered. Attempts to explain this remarkable genetic heterogeneity, which is not surprising in the human species with its highly differentiated individualization of physiognomic features, has elicited considerable controversy and led to theories of non-Darwinian evolution in which it has been claimed that most genes are selectively neutral (See Sect. 14.2.3).

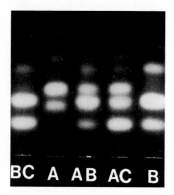

Fig. 12.2. Erythrocyte acid phosphatase polymorphism. Separation in agarose-thin layer electrophoresis. Treatment with 4-methylumbelliferyl phosphate produces of bright fluorescence detectable by UV light. (Courtesy of U. Barth-Witte)

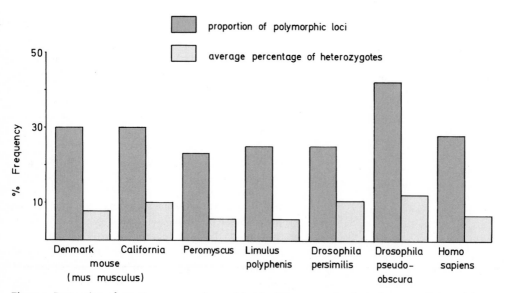

Fig. 12.3. Proportion of genetic enzyme polymorphism in different species, including man. (Adapted from Harris and Hopkinson 1972 [53])

Rare Variants. The limiting frequency of the most common allele, and thus for a genetic polymorphism, was set at 0.99 in the study by Harris and Hopkinson [53]. However, rare variants have become known for many enzymes. Often such rare variants are present in addition to polymorphisms; for many of the other loci only rare and no common variants are known. The probability of detecting rare variants depends on the sample size, and population samples of at least several hundred individuals need to be studied for an assessment. Such samples were available for 43 enzyme loci. Rare variants were defined as having a gene frequency of 0.005 or lower.

At 22 of the 43 loci – 7 of 13 "polymorphic" loci and 15 of 30 "nonpolymorphic" ones – rare alleles were found. There were 56 rare alleles in all, and of these the majority were extremely rare, 45 having gene frequencies of less than 0.001. The influence of sample size on the probability of variant detection was demonstrated by the fact that the mean sample size for loci at which rare alleles were found was 4023, while it was only 1300 for loci at which no rare alleles were detected. For some loci where this study detected no rare variants, such rare mutants have been found in other studies. It is therefore reasonable to assume that they may occur for all enzymes.

Average heterozygosity per locus was calculated for rare alleles in a similar way as for polymorphic variants:

$$\frac{\text{Number of heterozygotes for rare alleles at all loci (204)}}{\text{Sum of individuals screened for each locus (115\,755)}}$$

= 1.76 per 1000 individuals per locus.

Similar results have been obtained for other populations [126, 139]. Chapter 11 describes studies of such polymorphisms on children of atomic bomb survivors in Hiroshima and Nagasaki. The mutation rate found was in the order of magnitude of 10^{-6} per locus, the estimate being based on three new mutants in half a million gene loci among the normal control children. Therefore this mutation rate has the same order of magnitude as those for genes leading to hereditary diseases (Table 9.6).

Genetic Polymorphisms of Other (e.g., Structural) Proteins. On the basis of data showing a high frequency of genetic polymorphism in enzymes it was initially concluded that most genes are highly polymorphic. Some studies have shown, however, that this is not the case. There are suggestions that there are fewer polymorphisms in structural than in soluble proteins [73, 141, 146, 147].

DNA Polymorphisms [23]. The large extent of polymorphisms in expressed gene products such as blood groups, tissue types, and blood proteins is vastly exceeded by genetic variation at the DNA level. Since a large proportion of the genome does not appear to be involved in the direct regulation or specification of gene products, mutations in these nonregulatory and noncoding regions of DNA appear to have no phenotypic effect and might be selectively neutral. Determination of DNA sequences in different individuals has revealed tremendous variability at the level of DNA, usually outside coding genes. Family studies have shown the different DNA variants to be transmitted by Mendelian inheritance. An entirely new set of markers is therefore now available.

Types of DNA Polymorphism. While the proportion of interindividual heterozygosity per base pair in single-copy DNA was estimated to be 0.004, i.e., every 200th or 300th base pair turned out to be polymorphic, estimates for DNA inside of coding genes, and identified at the protein level, were about ten times lower [24, 95]. Therefore the noncoding DNA offers an abundance of polymorphisms which can be used, as markers for linkage studies (Chap. 5) or for studies on population history and evolution. The first DNA polymorphism was discovered by Kan and Dozy in 1978 [66]. It was closely linked to the β-globin gene and permitted prenatal diagnosis of sickle cell anemia. A short time later Solomon and Bodmer [122] and Botstein et al. [16] showed that a roster of a few hundred such polymorphisms, distributed over the entire genome, would allow the localization of genes on all chromosomes provided that a sufficient number of informative families were available. In the following years this prediction has come true on a large scale (Chap. 5). Table 12.3 shows the types of polymorphisms. There are now three types:

1. Restriction length fragment polymorphisms (RFLPs). These polymorphisms were discovered first. They are usually caused by relatively common, inherited single base pair changes in noncoding DNA sequences that are inherited by Mendelian transmission. Such alterations lead either to removal or introduction of a recognition site for one or another of the many different restriction enzymes, causing increases or decreases in the length of restriction fragments. These DNA variants are therefore referred to as restriction fragment-length polymorphisms. They are caused by a difference in the number of cleavage sites that are cut by a certain restriction endonuclease (Sect. 3.1.3.2) in different areas of the genome. They are demonstrated as follows: DNA is usually prepared from white blood cells and cut into fragments by the restriction enzyme. The resultant fragments are separated by electrophoresis; the smaller fragments migrate faster in the electric field than in larger ones. The double-stranded

Table 12.3. Classes of DNA polymorphisms

Marker:	What causes the poly-morphism?	How is the polymorphism demonstrated?	Information content (PIC)
Restriction fragment length polymorphism (RFLP)	Nucleotide differences in restriction sites recognized by restriction enzymes	Restriction with suitable enzyme; electrophoresis; Southern blot; hybridization with DNA probe	0–0.5
Minisatellites, VNTRs	Variable number of consecutive repetitive sequence blocks, size of one repeat 9–60 bp	Restriction with suitable enzymes; electrophoresis; Southern blot; hybridization; sometimes PCR	0.3–0.9
Microsatellites, STRs	Variable number of consecutive, repetitive sequence blocks; size of one repeat unit about 2–4 kp	PCR technique; sometimes with radioactive primers	0.3–0.9

PIC, Probability of information content (1 = 100 %).

DNA is now denatured to become single-stranded by heating and is transferred to a nylon filter (Southern blotting). As a next step, the segment of interest must be identified within the total DNA of the individual. This is accomplished by a radioactive (or otherwise marked) DNA probe which hybridizes with the relevant DNA segment which contains the complementary sequence.

2. Minisatellites (variable number of tandem repeats, VNTR). These were discovered by Jeffreys et al. in 1985 [64] and are found very frequently in noncoding, repetitive areas of the genome. Sequences of about 9–60 base pairs are repeated in tandem order and may be present in a varying number of repeats per chromosome. One repeat number is represented by one band in electrophoresis (Fig. 12.4); as a rule, one individual is heterozygous and has therefore two bands. However, there may be many alleles, up to approximately 100, in a population. Therefore they are likely to be informative as markers (high PIC values). For their identification the Southern blot technique described above for RFLPs or a PCR technique is used. Often several minisatellites can be studied in the same assay. One property, however, diminishes their utility for some applications: each of these polymorphisms has a relatively high mutation rate – a small percentage per generation. This is not entirely surprising. The similarity of base composition between multiple sequences creates ideal conditions for meiotic pairing of structure homologous but not position-homologous DNA segments, leading to unequal crossing over. (Sect. 5.2.8).

3. Microsatellites (short tandem repeats, STRs). These were discovered by Weber and May in 1989 [142]. For example, pairs of two bases $(CA)_n$ may be repeated from a few to very many times. There may be up to approx. 30 different alleles in a population for one such polymorphism. The best method for

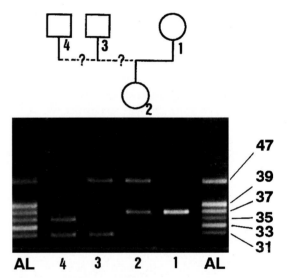

Fig. 12.4. VNTR polymorphism of a region 3′ to the apolipoprotein B gene on chromosome 2pter-24, detected by PCR analysis. Genomic DNA was amplified with primers outside the polymorphic region. The PCR products were separated by gel electrophoresis and visualized by ethidium bromide staining. The results were compared with an allelic ladder (AL), consisting of a mixture of the most common alleles (numbered). VNTR markers of this sort are often used in forensic and paternity testing. The figure shows a mother/child pair (1, 2) and two alleged fathers (3, 4). One (4) can be excluded from paternity. (Courtesy U. Barth-Witte)

their study is the PCR reaction (Sect. 3.1.3.5). Two primers attached to the DNA on both sides of the polymorphism are required; this means that short base sequences outside the polymorphism must be known. Information on primers can now be gathered from a genome data base (Appendix 3). In addition to the two base pair STRs, those comprising variable repeats of three to five base pairs have been described [33, 113]. The common STR

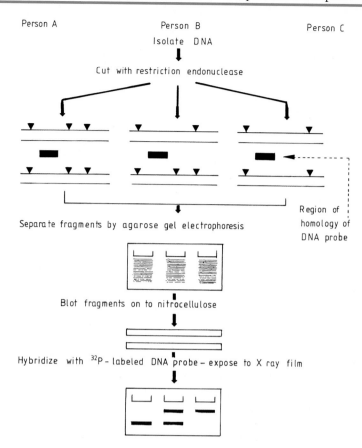

Fig. 12.5. Three individuals, two of them (*A, C*) homozygous for different restriction fragment length polymorphism *(RFLP)* haplotypes, the third (*B*) heterozygous. Principle of analysis and results of agar gel electrophoresis and Southern blotting

markers have become the DNA variants of choice in the study of human and mammalian linkage. Many such markers can be assessed by automated typing. If a sufficient number of families or sib pairs are available, any Mendelian gene can nowadays be localized by a study of a large number of STRs spread over the entire genome.

A radioactive DNA probe complementary to the gene or DNA sequences under study is applied to the filter. The probe hybridizes with the DNA to be tested, which can be recognized by autoradiography. The "needle in the haystack" or the tiny amount of DNA among the 6–7 billion DNA base pairs of an individual can therefore be identified [96].

Applications of DNA Marker Studies. The use of DNA markers extended the theoretical and practical applications of genetic linkage work considerably. For example, the high degree of individuality of DNA patterns together with the fact that DNA can be extracted from all nucleated cells, and even minute amounts can be amplified the PCR reaction, makes DNA polymorphisms excellent tools for identifying individuals even if very little material is available. Thus forensic applications for the identification of blood and sperm residues have come into common use. While there is

no controversy about the conceptual basis of this DNA technique, much discussion has been devoted to statistical issues that arise in calculating the probabilities that a suspect's DNA pattern come from the same person [98a]. It is almost certain that using multiple markers appropriately will make it possible to demonstrate a unique DNA pattern for every person except for identical twins.

Applications to paternity testing are often discussed. More conventional polymorphisms such as the MHC system are already producing satisfactory results in almost all cases. However, DNA studies will probably replace them because they are cheaper.

Mitochondrial DNA Polymorphisms. Mitochondria are transmitted only from the mother to all her children; there is no diploidy, no meiosis, and no recombination. Polymorphisms of mitochondrial DNA are especially useful in population genetics, mainly for the analysis of relationships between population groups [60] and population history: they do not appear to be subject to selection pressures. Therefore comparison of maternally inherited mtDNA restriction patterns between population groups gives an unbiased picture of the population's genetic history (see also Sect. 12.2.4).

12.1.3 Hereditary Diseases

Dominant and X-Linked Recessive Diseases. Two subgroups of dominant and X-linked hereditary diseases may conveniently be distinguished from the standpoint of population genetics:

1. Those in which reproduction is impaired considerably – either because affected persons are so severely sick that they die early, or because they are so severely disabled that they have little chance to marry and have children.
2. Those in which reproduction is not impaired – either because the abnormality is trivial or because it manifests only after reproduction has been completed.

In the first subgroup, that of disabling genetic disease, the incidence and prevalence are determined mainly by the mutation rate: most mutations disappear after one or a few generations. This problem discussed in Chap. 9; data on mutation rates for such conditions (Sect. 9.3; Tables 9.6, 9.7) indicate similar mutation rates in all populations for which evidence is available. The sources of selection against these mutations are disease specific and have been more or less identical in all populations; therefore the incidence is also similar. With different applications of successful treatments in different populations, variable incidences would be expected. For example, successful treatment of hemophilia would lead to an increase of affected patients.

The situation is quite different for conditions that do not impair reproduction. Here the incidence may definitely be unequal between populations, depending on such factors as population size, history, and breeding structure. These problems are discussed in Sect. 13.3.1.

Autosomal-Recessive Diseases. The various influences that have shaped the current frequencies of autosomal-recessive diseases in human populations are largely unknown. How these frequencies will change under the influence of modern civilization therefore cannot be determined with certainty. The incidence of many autosomal-recessive diseases is unknown. Estimates for the incidence of several inborn errors of metabolism have recently been collected as a side effect of the widespread newborn screening programs designed to secure early diagnoses of treatable metabolic defects.

Screening centers that participated in a study to estimate the frequency of phenylketonuria were invited by the organizer to cooperate on the basis of acceptability of their techniques and the number of newborns tested. They were requested to report only when at least 70 000 newborns had been screened. Centers that were aware of regional differences

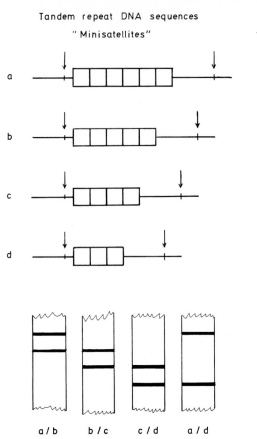

Tandem repeat DNA sequences
"Minisatellites"

a / b   b / c   c / d   a / d

Fig. 12.6. A DNA variant may be caused by different length of tandem repeat sequences. *a, b, c,* and *d* differ each by one repeat segment so that *(a)* has six repeats and *(d)* has three. *Arrows,* the site of a restriction enzyme cut. The cut sites are not altered. However, the size of the DNA fragments varies depending upon the number of repeats. The smallest *(d)* variant moves faster, and the various DNA sizes can be distinguished on the Southern blot by mobility differences. Heterozygotes for various combinations are shown diagrammatically (*a/b; b/c;* etc.). (From Emery 1984 [35])

within the population under study were asked to give separate figures for each region or population group. The data are given in Table 12.4.

Phenylketonuria (2 616 000) and Hyperphenylalaninemia

a) Within Europe there is a higher incidence of PKU in the east than in the west or south. The difference between eastern and western Austria fits into this pattern.
b) The Scandinavian populations, especially the Finns, show an exceptionally low PKU frequency. It is interesting that the Finnish population differs also in regard to other genetic aspects from the rest of the Europeans (Sect. 13.3.1); for example, a unique pattern of other inherited diseases unknown in other populations has been observed.
c) High frequencies of PKU are found in the Republic of Ireland; differences within the United Kingdom, such as a high frequency in the Manchester area, may reflect migration from Ireland.

Table 12.4. Frequency of PKU and HPA in some populations (from Thalhammer 1975 [127])

Region	PKU	HPA
Warsaw, Poland	1: 7782	1: 16885
Prague, Czech Republic	1: 6618	1: 6303
Eastern Germany	1: 9329	1: 52135
Eastern Austria	1: 8659	1: 21982
Western Austria	1: 18809	1: 18809
Switzerland	1: 16644	1: 24106
Evian, France	1: 13715	1: 13143
Hamburg	1: 9081	1: 61297
Münster, Germany	1: 10934	1: 7997
Heidelberg, Germany	1: 6178	1: 14580
Denmark	1: 11897	1: 40790
Stockholm, Sweden	1: 43226	1: 22140
Finland	1: 71111	1: 71111
London, United Kingdom	1: 18292	1: 50304
Liverpool, United Kingdom	1: 10215	1: 112362
Manchester, United Kingdom	1: 7707	1: 80925
Western Ireland (Eire)	1: 7924	1: 68670
Eastern Ireland (Eire)	1: 5343	1: 32594
Boston, Mass., U.S.A.	1: 13914	1: 17006
Portland, Oregon, U.S.A.	1: 11620	1: 33700
Montreal, Canada	1: 69442	?
Auckland, New Zealand	1: 18168	1: 95384
Sydney, Australia	1: 9818	1: 22091
Japan	1: 210851	1: 70284
Ashkenazi (Israel)	1: 180000	1: 15000
Non-Ashkenazi (Israel)	1: 8649	1: 7111

d) In the United States, Boston and Portland, Oregon, frequencies are quite comparable to those in Europe. In Montreal, in the French-speaking part of Canada, there is a much lower frequency of PKU than in most European countries and the two centers in the United States, and a significantly lower rate than that found in France from where this population originated.

e) In Japan, the only Far East country screened for inborn errors, the frequency of PKU is especially low, comparable only to the frequencies in Finland and among Ashkenazi Jews in Israel.

f) The frequencies of hyperphenylalaninemia (HPA), a condition not associated with mental retardation, vary greatly and independently of those for PKU. Interestingly, PKU is more common in certain eastern parts of Germany than in the west. A study on the origin of families of PKU children in northwestern Germany showed that a higher proportion of these families than would be expected by chance were refugees from the east [42]. There appears to be a decline from the east to the (north-)west of continental Europe.

Other Conditions. The study also included galactosemia due to transferase deficiency (230400) (Sect. 7.2), histidinemia (235800), maple syrup urine disease (248600; Sect. 7.2.2.4), and homocystinuria (236200; Sect. 7.2).

For galactosemia, most centers – Hamburg, Vienna, Auckland, Prague, Stockholm, Zurich – report frequencies be-

tween about 1 : 30000 and 1 : 65000. Hence the incidence in most populations of European origin seems to be similar. There is a significant difference between eastern and western Austria.

In the centers with the best screening conditions histidinemia has an incidence between 1 : 13000 and 1 : 19000. Homocystinuria, leucinosis, and – with the exception of Montreal – tyrosinosis (276700) are very rare, showing incidences between about 1 : 100000 and 1 : 600000; in Montreal tyrosinosis reached a frequency of about 1 : 13000, which could be traced to a specific French-Canadian isolate.

High Frequencies of Recessive Diseases in Special Populations. A number of recessive diseases attain high frequencies in special population groups. Tyrosinosis in the French-Canadian population is mentioned above; others include Tay-Sachs (272800), Niemann-Pick (257200), and Gaucher diseases (230800) in the Ashkenazi Jewish population. In the Finnish population a number of such diseases show high frequencies as a consequence of a distinct population structure and history (Sect. 13.3.1). Some diseases have been found only in smaller population groups (isolates) where they became frequent while being more or less unknown in any other group. Others – such as thalassemia and sickle cell anemia – are frequent in certain geographic areas and in some racial groups. In general, recessive diseases unique to certain small populations owe their origin to genes which in the heterozygote state have no particular advantage, while genes causing more common and widespread diseases such as thalassemia and sickle cell anemia have had a heterozygote selective advantage in the past. However, the occurrence of metabolically related gene mutations in Ashkenazi Jews (Tay-Sachs, Gaucher, Niemann-Pick diseases) may suggest selective factors in these instances. A full discussion is devoted to these problems below (Sect. 13.3).

DNA Haplotypes and Recessive Mutations. With the discovery and ubiquity of DNA polymorphisms it has become possible increasingly to characterize mutations not only by identifying the mutational site itself within the gene, but also to determine the unique pattern of DNA variants around a mutant gene. DNA variants, which are found mostly in nontranscribed DNA sites, include nucleotide variation that causes RFLPs, VNTRs, and STRs (see above). The unique pattern of contiguous polymorphic DNA variants is designated as a haplotype, a term initially used for closely linked genes at the MHC locus. (Sect. 5.2.5). The use of this method for characterizing disease origin and the history of genes is presented below using two examples: the phenylalanine hydroxylase (PHA) gene, whose mutations often lead to phenylketonuria (261600), and the cystic fibrosis (219700) (transmembrane conductance regulator, CFTR) gene.

Population Genetics of PHA gene [78]. PKU occurs in various populations with different frequencies (Table 12.4). The disease is caused by a variety of mutations within the PHA gene. The mutations have been identified, and the haplotypes within which they occur have been described. Fig. 12.7 shows many of the alleles identified so far. In some instances hyperphenylalaninemia (HPA) is found without the classical symptoms. Table 12.5 shows the most important haplotypes compared to the same haplotypes in the general population. Since the sample sizes are generally small, some of the differences may not be biologically real. However, there are some striking findings. In eastern Asia, for example, haplotype 4 has been found almost exclusively, but haplotype 4 is common also in the normal population. Therefore no conclusions as to the origin of the mutation are possible. PKU is much rarer in Japan than in China, but the same haplotypes are found. The mutation was probably introduced from China to the Japanese island. In Europe, on the other hand, haplotype 2 appears to be much more common than in the general population, especially in Slovakia, Hungary, and Italy. The

high frequency of haplotype 6 in Italy and Turkey [80] is remarkable since this pattern appears to be infrequent in the general European population.

If a specific haplotype is associated with a certain mutation more frequently than in the general population, such linkage disequilibrium between the marker genes and the disease gene is usually explained as follows: It is assumed that the mutation causing the disease has occurred once within a chromosome carrying the particular haplotype, and that insufficient time has elapsed since the original mutational event to permit randomization or linkage equilibrium by repeated crossing over. With this assumption, a certain haplotype would be expected almost always to be associated with a specific mutation. This appears to be true for mutant PHA alleles in German families and for haplotypes 1, 2, and 3. Haplotype 4, however, has been found with four different mutations.

This "classical" interpretation seldom leads to contradictions for the PHA gene. If a mutant occurs occasionally with different haplotypes, this may be explained either by crossing over within the haplotype after the mutation occurred or by repeated mutations

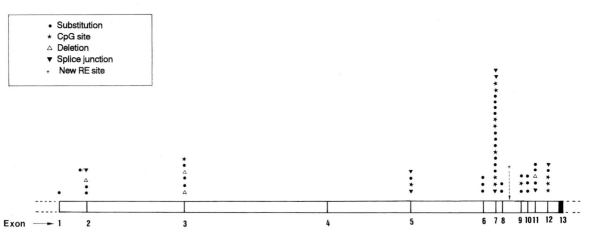

Fig. 12.7. Types of mutations in the phenylalanine hydroxylase (PHA) gene. (Modified from Romano 1983)

Table 12.5. Most important haplotypes in which the PHA gene occurs

Haplotype	*Bgl*II	*Pvu*II(a)	*Pvu*II(b)	*Eco*RI	*Msp*I	*Xmn*I	*Hind*III	*Eco*RV
1	+	–	–	+	+	–	+	+
2	+	–	–	+	+	–	+	–
3	–	–	+	–	+	–	–	+
4	–	–	–	+	+	+	–	+
5	+	–	–	+	+	+	–	+
6	+	+	+	–	+	–	+	–
7	–	–	+	–	+	–	+	–
8	+	+	–	+	–	–	+	–
9	–	–	–	+	–	–	+	–

+, Presence; –, absence.

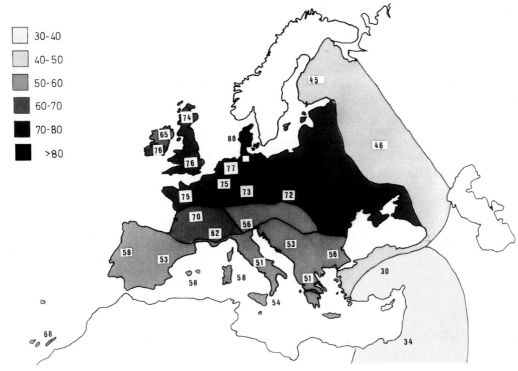

Fig. 12.8. Frequency of the Δ508 deletion of the cystic fibrosis gene in various European populations. *Numbers,* percentage frequencies of the Δ508 mutation. (Modified from Romeo and Devoto 1990 [111]

at the same site, since it has been shown that point mutations (single base replacements) do not occur at random (Chap. 9), and mutational hot spots are well known. Moreover, the distribution of known mutants across the PHA gene appears definitely nonrandom (Fig. 12.7).

Population Genetics of the CFTR Gene. Cystic Fibrosis is the most common autosomal recessive disease in populations of northwestern European origin, with an incidence of approximately 1:2000 at birth (1:1700 Northern Ireland; 1:7700 northern Sweden [14, 111]. This means that heterozygotes for this gene are found with a frequency of 1:23. In other populations, for example African-Americans, this condition is much rarer. Analysis of mutations has given a picture that is quite different from that of the PHA gene: almost two-thirds of all mutants in northwestern Europe have been found to carry one specific mutation: a deletion of three base pairs at code position 508 (Δ508). Figure 12.8 shows the distribution of the Δ508 mutation in the European population. Many other mutations, mainly single base replacements, have also been described; a total of over 300 different mutations are known. Four haplotypes have been defined using two unique DNA sequences located in the 5′ extragenic region. These are defined as follows:

Table 12.6. Haplotypes in 23 European populations

	General population		Δ508 mutation		Other mutants	
	n	%	n	%	n	%
A	1221	34.5	63	1.8	275	18.8
B	581	16.4	3156	92.5	713	48.8
C	1297	36.6	23	0.7	314	21.5
D	440	12.4	167	4.9	158	10.8

– Haplotype A: ZU2C absent, KM19 absent
– Haplotype B: ZU2C absent, KM19 present
– Haplotype C: ZU2C present, KM19 absent
– Haplotype D: ZU2C present, KM19 present

The distribution of these haplotypes in the common Δ508 mutants in the general population differs substantially; the great majority of mutants are found in haplotype B (Table 12.6). Haplotype B is found in more than 90% of cystic fibrosis mutations while only 16.4% have this haplotype in the general population. This strongly suggests a single origin of the mutation: the small minority of Δ508 mutants in other haplotypes can be explained by crossing over – especially since both restriction sites of this haplotype are located at least 200 kb away from the gene [128]. A further noteworthy result is that numerous other mutants are also more common in the B haplotype than in the general population (48.8% vs. 16.4%). Such a massive increase in a single mutation such as Δ508 in a relatively large area points strongly to a selective advantage under the living conditions in northwestern Europe in earlier times. Thus, since other cystic fibrosis mutations are also found frequently at the identical haplotype, it is most probable that the *common haplotype* somehow causes this advantage. Homozygotes, however, can never have had such a selective advantage; they are chronically ill and until recently usually died before the age of reproduction. Therefore the advantage must be confined largely to heterozygotes of a population subgroup carrying haplotype B. To understand this requires an excursion into theoretical population genetics and a discussion of natural selection.

12.2 Systematic Changes in Gene Frequencies: Mutation and Selection

Gene frequencies remain unchanged in populations only in the absence of disturbing influences (see Hardy-Weinberg equilibrium, Sect. 4.2) The most important of these influences are:

a) New mutation
b) Natural selection
c) Migration
d) Chance fluctuations.

Spontaneous and induced mutations in humans are dealt with in Chaps. 9 and 11; these aspects are extended below in the context of other influences. Selection, however, is dealt with in detail. Another source of deviation from the Hardy-Weinberg equilibrium is assortative mating. A discussion of consanguinity and chance fluctuation of gene frequencies leads to a more general consideration of the breeding structure of human populations and sets the stage for a better understanding of human evolution (Chap. 13).

12.2.1 Natural Selection

Natural selection was recognized by Darwin as the main driving force of evolution. His evolutionary theory, which is based largely on the understanding of selection and its consequences for the origin of species, became the leading paradigm of biology in the nineteenth century. The concept of biological "fitness" is central for understanding this paradigm. In the early twentieth century, selection theory was given a sound mathematical foundation, and a number of empirical examples were analyzed.

12.2.1.1 Mathematical Selection Models: Darwinian Fitness

Scope of Mathematical Models in Selection Theory and Their Limitations [77]. In discussing selection we shall use mathematical models on a fairly large scale. These models make a number of assumptions for some parameters, for example, gene frequencies and selective advantages or disadvantages of special genotypes. The consequences of these assumptions for the direction and extent of changes in gene frequencies over time are examined below. Such models help to understand the consequences of certain changes in these parameters by creating some order in the vast and initially unintelligible complexity of genetic differences among human populations.

Such order may be artificial: certain aspects are singled out, while others are deliberately neglected. While the calculations become tractable, major errors may result. The most important oversimplifications are the following:

a) The population is considered to be infinitely large in size. Gene frequency thus remains constant in the absence of other factors. No real population is infinitely large; on the contrary, many human breeding populations were very small throughout the long time periods most important for human evolution. Therefore all results derived from selection models must be scrutinized critically in the light of the theory of small populations (Sect. 13.3) and our knowledge of human population size and mating structure. Unfortunately, this knowledge is relatively scanty, especially for earlier times. Conclusions in concrete cases are thus difficult if not impossible.

b) As a rule, selective advantages or disadvantages are assumed to be constant over long evolutionary periods. Closer scrutiny of their biological mechanisms, however, often show that they may have changed even within relatively short periods.

Deterministic and Stochastic Models: Use of Computers. The various limitations apply mainly to models that assume a functional relationship between parameters. For example, the change in gene frequency through generations is assumed to depend upon a certain mode of selection: The model is "deterministic." In reality, however, all parameters – gene frequencies, selection pressures, mutation rates – show chance fluctuations because population size is not infinite. The available of computers made it possible to include chance fluctuation in the calculations, thereby creating stochastic models. The change in gene frequency over the generations can now be simulated assuming a certain population size. The curve showing, for example, the change in gene frequencies over time does not give the "ideal" outcome but only one of many possible outcomes; it is not even known whether a certain curve is a very likely one. Therefore a single calculation for a certain set of parameters is not sufficient to obtain an unbiased impression of the consequences of certain assumptions; the same calculation must be repeated several times. Such a method provides better information than a deterministic model; in addition to the main trend, possible deviations from this trend caused by chance fluctuations can be demonstrated. The following section makes use principally of deterministic models, but stochastic models are occasionally also considered.

How Should Models Be Used in Practice? Oversimplifications are necessary in all models, deterministic or stochastic, for recognition of the general laws of natural selection. In each concrete case, however, the fact that they are oversimplifications should be kept in mind. While the consequences inferred from such models are *formally* correct, the possibility that they derive from aspects of the model that have no counterpart in the real world is often not considered. Much of human population genetics suffers from too uncritical interpretations of formal results deduced from oversimplified models.

The ideal sequence of events would be that:

a) A certain genetic situation requires theoretical explanation for its existence.
b) A hypothesis is formulated, and a relatively simple model is constructed that includes the main parameter(s) of the hypothesis.
c) The consequences of this model for changes in gene and genotype frequencies over time or for differences among present populations are explored.
d) The result of this exploration is compared with the empirical evidence.
e) Discrepancies between the theoretically expected and the empirically observed results are noted. Critical interpretation may lead to rejection of the hypothesis, change in assumptions regarding important parameters, or refinement of the model.

The recognition that a concrete problem requires closer scrutiny may accompany the acquisition of further knowledge, as problems are not isolated but are parts of the whole context of a special field. This context may change; a problem formerly regarded as holding a key position may become less important,

and a model designed to solve this problem, while not formally shown to be inadequate, may turn out to have trivial or unrealistic consequences. Additional biological data often throw new light on a problem, making new theoretical assessments necessary. We recognize that the ideal sequence of events in the study of problems in population genetics is often impossible because of technical and logistic limitations. We would like to stress, however, that we consider hypothesis-oriented efforts in population genetics to be ultimately of greater explanatory power for understanding population genetics than descriptive studies even if embellished by the most advanced and elegant molecular and statistical methods.

Concept of Darwinian Fitness. The central concept of selection theory is Darwinian fitness. Under given environmental conditions, not all individuals in the population perform equally well. These differences are caused partially by different genetic endowment of the individuals. This obviously has many medical, social, and ethical aspects. In its effect on natural selection, however, only one aspect is relevant: different reproduction rates of individuals with different genotypes. Only reproductive differences can lead to a shift in gene frequencies over time, if population size is regarded as unlimited so that chance deviations can be neglected. This reproductive performance of a certain genotype in comparison with a norm is often called its Darwinian, or biological fitness. This notion of fitness can also be defined for a single allele if it influences the reproduction of its carrier. Fitness of a certain genotype may be reduced or enhanced in two different ways:

1. The genetic constitution reduces the chance of a genotype to survive into reproductive age. This reduction in viability is frequent in hereditary diseases.
2. The genetic constitution reduces the chance of a genotype to produce offspring, i.e., fertility is diminished.

From the point of view of population genetics, there is no difference between a gene that causes spontaneous miscarriage and one that causes sterility in an otherwise healthy person. Medically and socially, of course, there are considerable differences between the impact of these two genes.

12.2.1.2 Selection Leading to Changes in Gene Frequencies in One Direction

Symbols Used. Fitness of a genotype is defined as its efficiency in producing offspring. It is measured not in absolute but in relative units, the fitness of the optimum genotype being taken as unity (1). Deviations from unity are denoted as *s*. For example, for a genotype having 80% fitness of the optimum genotype, $s = 0.2$; this fitness is: $1 - s = 1 - 0.2 = 0.8$. To avoid confusion with the signs, it is sometimes desirable to have a direct measure for the fitness $1 - s = w$. In the literature *s* is sometimes defined with respect to the average, not to the optimum fitness of the population, $\bar{w}$. This convention has disadvantages since the fitness of a genotype varies with the genotype distribution in the population. We therefore define fitness with respect to an "optimum" genotype.

Elimination of Heterozygous "Dominant" Phenotypes. This case is frequent and simple: a mutation changes the phenotype of its carrier so thoroughly as to make reproduction impossible. All numerical and most structural chromosome aberrations and many single gene mutations (Sect. 9.3) have this effect. Whereas chromosome aberrations can be diagnosed directly, dominant new mutations can be definitely recognized only by transmission of the condition to the offspring. Hence, nonreproductive dominant mutations are in principle unrecognizable. Still, extrapolation from a very few individuals who manage to reproduce, or a paternal age effect suggestive of a dominant mutation may make such mutation plausible even for anomalies preventing their carriers from reproduction. Studies at the gene – DNA level may confirm this assumption. Some diseases and birth defects may in fact be caused by such dominant mutations that have not been recognized so far.

Partial Elimination of Autosomal Dominants. Most dominant diseases reduce the average reproduction of their bearers. In the absence of counteracting forces, such as mutation, such gene loss would lead to reduction in gene frequencies in every generation by a fraction which depends on the selective disadvantage of the probands (Fig. 12.9). A genetic equilibrium between new mutation and selection, either at present or in the past, can be assumed in almost every case. Let *p* be the frequency of the dominant allele A; *q* the frequency of the recessive normal allele a. Selection may be appreciable, and the mutation rate, μ, may be low. Then, affected homozygotes AA can be neglected, as p^2 is very small. The genotype frequencies are seen in Table 12.7.:

The loss of Aa individuals per generation is approximately $2ps$. Since only half of the genes are A, the loss of A genes is $\frac{1}{2}(2ps) = ps$. A genetic equilibrium exists if this loss is compensated for by mutation:

$$ps = \mu(1 - p) \approx \mu$$

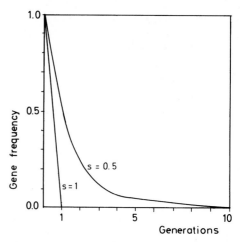

Fig. 12.9. Decrease in the gene frequency of a dominant gene in absence of new mutations with $s = 1$ and $s = 0.5$; *s*, selection coefficient

Table 12.7. Selection and dominance

Genotype	Fitness	Before selection	After selection
aa	$1(= w_{22})$	$q^2 \approx 1 - 2p$	$\dfrac{1 - 2p}{1 - 2ps}$
Aa	$1 - s(= w_{21})$	$2pq \approx 2p$	$\dfrac{(1 - s)2p}{1 - 2ps}$

This gives the equilibrium frequency of *p*:

$$\hat{p} = \mu/s \qquad (12.1)$$

Two limiting cases are interesting:

1. No reproduction of Aa heterozygotes or AA homozygotes occurs, i.e., $s = 1$, $w_{21} = 0$; in concordance with our conclusion from the last section, the frequency of the dominant allele (*p*) equals the mutation rate (μ).
2. No selective disadvantage of the gene applies (i.e., $s = 0$, $w_{21} = 1$). Under such conditions there is no equilibrium, and the gene frequency of the dominant allele increases slowly and monotonically, as new mutations occur: $\mu/s \to \infty$.
3. Finally, an intermediary case with some disadvantage of the dominant gene, e.g., $s = 1/3$, $w_{21} = 2/3$; the final gene frequency of the dominant gene ($\hat{p}$) is three times the mutation rate.

Most pathological conditions have an appreciable selective disadvantage. Their population frequency is maintained by an equilibrium between mutation and selection. Here the prime mover is of course the mutation rate. The frequency of the condition increases until there are so many affected individuals that their selective disadvantage balances the mutation rate increase and a genetic equilibrium is reached. This equilibrium is stable: If the number of affected individuals rises above the equilibrium value, the number of losses of the disadvantageous allele by selection exceeds the number of its production by mutation, and its frequency is reduced in the next generation. On the other hand, if the number of affected individuals falls below the equilibrium value, more disadvantageous genes are produced by mutation than lost by selection, and the gene frequency increases until equilibrium is reached again.

Selection Relaxation. Equation (12.1) can be used to calculate the consequence of selection relaxation: Assume that medical therapy, by eliminating some of the phenotypic consequences of a dominant mutation, succeeds in reducing the selective disadvantage of the allele from:

$$s_1 = 1/2 \quad \text{to} \quad s_2 = 1/4.$$

Let $\hat{p}_2$ be the equilibrium value for s_2. Then Eq. (12.1) gives:

$$\hat{p}_1 = \frac{\mu}{1/2} = 2\mu; \quad \hat{p}_2 = \frac{\mu}{1/4} = 4\mu; \quad \hat{p}_2 = 2\hat{p}_1$$

Therefore the new frequency is twice the earlier frequency. Moreover, the new equilibrium is reached within only a few generations.

Selection Relaxation in Retinoblastoma. Retinoblastoma is a malignant eye tumor of young children. The great majority of cases are sporadic. However, familial cases are frequent and show autosomal-dominant inheritance with about 90% penetrance. All bilateral but only about 10%–12% of unilateral sporadic cases are new mutations (See Sect. 10.4.3).

In earlier times retinoblastoma was almost always fatal; the patients died during childhood, and none could ever reproduce. In 1865 A. v. Graefe introduced the enucleation of the diseased eye; more recently, additional therapeutic methods such as X-ray irradiation and light coagulation have become available, and currently about 90% of unilateral and 80% of bilateral cases can be cured, and the patients can have children [132]. Of all new mutations in hereditary cases 68% are bilateral and 32% are unilateral, as shown by population studies. The changes in frequency of the condition can be predicted as follows:

Let X_0 be the frequency of individuals in the population with hereditary retinoblastoma. Since the condition is very rare, the gene frequency of the normal allele may be assumed to be $q \approx 1$; therefore, X_0 is nearly equal to the frequency of heterozygotes ($2pq \cong 2p$). Let $s_1 = 1$ before selection relaxation. After introduction of efficient therapy the selection against unilateral cases becomes $s_U = 0.1$ (only 10% of unilateral cases die in childhood; 90% survive to reproduce normally); selection against bilateral cases becomes $s_B = 0.2$ (80% survive to reproduce normally). Subscripts u and b refer to unilateral and bilateral cases respectively. This leads to the following overall estimate for s_2 (= remaining selection against the retinoblastoma allele after selection relaxation):

$$\begin{aligned} s_2 &= s_U \times 0.32 + s_B \times 0.68 \\ &= 0.1 \times 0.32 + 0.2 \times 0.68 \\ &= 0.168 \end{aligned}$$

The new selection coefficient after selection relaxation is 16.8% of the coefficient prior to selection relaxation when s is 1. The frequency of heterozygotes in the nth generation, X_n, can be related to that in the $(n + 1)$th generation, X_{n+1}, by the approximate formula:

$$x_{n+1} = X_n (1 - s_2) + 2\mu = X_n (1 - s_2) + X_0$$

From this recurrence formula, a general formula for X_n can be derived:

$$X_n = X_0[1 + (1 - s_2) + (1 - s_2)^2 + \ldots + (1 - s_2)^n]$$

Thus, the new equilibrium value for the heterozygote frequency, X, is the sum of a geometric series with the initial term $X_0 = 2\mu$ and the factor $1 - s_2$:

$$\hat{X} = \frac{X_0}{1 - (1 - s_2)} = \frac{X_0}{s_2} = \frac{X_0}{0.16s} = 5.95 \times X_0$$

This result can also be obtained from Eq. 12.1. It follows that inherited retinoblastoma will be about six times more common several generations after successful medical therapy is introduced. Introducing actual estimates from present populations (see Tables 9.6, 9.7), $X_0 = 2\mu = 1.2 \times 10^{-5}$:

$$\hat{X} = \frac{1.2 \times 10^{-5}}{0.16s} = 7.14 \times 10^{-5}$$

From this value the overall frequency of *all* retinoblastoma cases – including the nonhereditary ones – can be estimated when the fraction of hereditary cases among all retinoblastoma cases in the population before selection relaxation is known. The overall incidence of retinoblastoma is $\approx 4 \times 10^{-5}$; it follows that $\approx 2.8 \times 10^{-5}$ are nonhereditary cases. From this an incidence estimate at equilibrium after selection relaxation can be derived:

$$7.14 \times 10^{-5} + 2.8 \times 10^{-5} = 9.94 \times 10^{-5}$$

This means that incidence increases from $\approx 1 : 25\,000$ to $\approx 1 : 10\,000$, i.e., by $\approx 150\%$. Moreover, while before selection relaxation $\approx 30\%$ of all cases are hereditary, $\approx 72\%$ are hereditary in the new equilibrium state. This equilibrium is established relatively quickly (Fig. 12.10): After nine generations the incidence of hereditary retinoblastoma is more than four times the value before selection relaxation (equilibrium: $\hat{X} = 5.95X_0$). Figure 12.10 also gives calculations for two alternative assumptions: $s_2 = 0.4$; and $s_2 = 0$, i.e., no selection. In the latter case incidence increases in a linear fashion; there is no equilibrium. There is some hope that the prediction given here does not come true; artificial selection by genetic counseling followed by voluntary birth control on the part of gene carriers will partially replace natural selection.

Selection by Complete Elimination of Homozygotes. In many autosomal-recessive diseases the homozygotes usually do not reproduce. Again, two alleles A, a with gene frequencies p, q are considered. This time, however, the homozygotes aa have a selective disadvantage:

$$s = 1; \quad w_{22} = 0$$

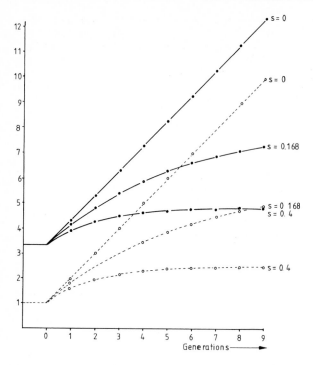

Fig. 12.10. Predicted increase in retinoblastoma in the population due to selection relaxation. *Ordinate: 1,* incidence of hereditary retinoblastoma (●—●; 2 × mutation rate); *2, 3, 4, 5 . . . etc.,* incidence of all retinoblastoma cases, hereditary and non-hereditary, in relation to hereditary retinoblastoma (○----○). 100 % mortality before beginning of selection relaxation is assumed. *Abscissa:* number of generations. Three assumptions regarding the selection coefficient *s:* no selection at all (*s* = 0), weak selection (*s* = 0.168), and strong selection (*s* = 0.4). (From Vogel 1979 [132])

The frequencies of zygotes are seen in Table 12.8. In general, the gene frequencies in two successive generations n and $n + 1$ are related by:

$$q_{n+1} = \frac{q_n}{1 + q_n} \qquad (12.2)$$

This formula reflects the recurrence relation between successive partial sums (= sum of the first n terms) of a harmonic series.[1] It leads to the general formula for the gene frequency q_n after n generations:

$$q_n = \frac{q_0}{1 + nq_0} \qquad (12.3)$$

[1] A harmonic series is one whose successive partial sums are the reciprocals of those of an arithmetic series:

$$u_o = \frac{c}{b}, \ u_1 = \frac{c}{b+k}, \ u_2 = \frac{c}{b+2k}, \text{ etc.}$$

In this special case, $b = 1$, $c = k = u_o$.

The change of gene frequency per generation is:

$$\Delta q_n = \frac{q_n}{1 + q_n} - q_n = -\frac{q_n^2}{1 + q_n} \qquad (12.4)$$

It follows from Eq. (12.3) that $q_n = q_0/2$, if $nq_0 = 1$. Hence, the gene frequency is halved within $n = {}^1/q_0$ generations.

An important practical application is considered: if all homozygotes of the most common recessive disease – cystic fibrosis – refrain from reproduction, how would this affect the gene frequency?

Table 12.8. Selection s = 1 against homozygotes

	AA	Aa	aa	Gene frequency of a
Before selection	p^2	$2pq$	q^2	q
After selection	$\dfrac{p^2}{p^2 + 2pq}$	$\dfrac{2pq^2}{p^2 + 2pq}$	0	$\dfrac{q}{1+q}$

The gene frequency is $q_0 = 0.02$, corresponding to a homozygote frequency in a random mating population of $q_0^2 = 0.0004$. Of 10 000 individuals 4 are affected.

After $n = 1/0.02 = 50$ generations, q_0 is reduced from 0.02 to 0.01. Assuming a generation time of 30 years, this halving of the gene frequency requires 1500 years. With rarer diseases such as galactosemia, $q_0 = 0.005$, $q^2 = 1 : 40\,000$; halving would require 200 generations, or 6000 years.

Hence, the attempt to reduce the number of recessive genes by homozygotes refraining from reproduction is an extremely inefficient process. Moreover, new mutations are not even considered in this calculation.

Partial Elimination of Homozygotes. In some recessive diseases the homozygotes are not completely unable to reproduce, but there is significantly reduced biological fitness. Let their reproduction be reduced by s ($1>s>0$). Gene frequencies before and after selection may be taken from Table 12.9. This gives the following frequency of the recessive allele in the next generation:

$$q_1 = \frac{pq + q^2(1-s)}{1-sq^2} = \frac{q(1-sq)}{1-sq^2} \tag{12.5}$$

The formula for the relationship between the frequencies of the recessive allele in two successive generations is:

$$q_{n+1} = \frac{q_n(1-sq_n)}{1-sq_n^2}$$

This recurrence formula seems to have no general solution. The change per generation is:

$$\Delta q_n = q_{n+1} - q_n = -\frac{sq_n^2(1-q_n)}{1-sq_n^2}$$

thus Δq_n depends on the values of both q_n and $p_n = 1 - q_n$. It is small if one of these terms is small. For example, the following values can be calculated for $s = 0.2$:

q:	0.99	0.50	0.01
Δq:	−0.00244	−0.0263	−0.00000198

With very small q, Δq approximated − sq^2.

A Bit of Calculus. To determine the change in q over a greater number of generations, Δq can be replaced by:

$$\frac{dq}{dt} = -sq^2(1-q); \quad \frac{dq}{q^2(1-q)} = -sdt$$

Integrating both sides over n generations, we obtain:

$$\int_{q_0}^{q_m} \frac{dp}{q^2(1-q)} = -\int_0^n dt = -sn$$

$$sn = \left[\frac{1}{q} + \log_e \frac{1-q}{q}\right]_{q_0}^{q_n} = \frac{1}{q_n} - \frac{1}{q_0} + \log_e \frac{1-q_n}{q_n}$$

$$- \log_e \frac{1-q_0}{q_0} = \frac{q_0 - q_n}{q_0 q_n} + \log_e \frac{q_0(1-q_n)}{q_0(1-q_n)}$$

Calculations for number of generations (n) needed to produce a certain change in q ($s = 0.01$ against the homozygotes) shows the following:

Decrease in q	n generations
0.9999–0.9990	230
0.9990–0.9900	232
0.9900–0.5000	559
0.5000–0.0200	5,189
0.0200–0.0100	5,070
0.0100–0.0010	90231
0.0010–0.0001	900230

Table 12.9. Partial elimination of homozygotes

	AA	Aa	aa	Total
Before selection	p^2	$2pq$	q^2	1
Fitness	1	1	$1-s$	
After selection	p^2	$2pq$	$q^2(1-s)$	$1-sq^2$

Artificial selection against homozygotes of an autosomal recessive gene would therefore need an enormous time span to have a measurable effect. In the meantime, other events may have occurred to disturb the genetic composition of the population much more.

Gametic Selection. In the above, selection is assumed to act on the zygote. However, the gametes may already be affected. Because of their genetic make-up some gametes may have a lower chance for fertilization than others.

A mutation of any kind which affects the probability of fertilization results in a distortion of the segregation ratio for that mutation. For human conditions conformity with Mendelian segregation ratios is usually taken for granted. Since large samples would be required to show small deviations from Mendelian proportions, such a shift in the segregation ratio could easily be overlooked. However, gross deviations are unlikely.

For gametes containing balanced and unbalanced translocations, segregation distortion is beyond any doubt, although the exact mechanism remains to be elucidated (Sect. 2.2.2). If the relative contributions of gametes A and a to the following generation are 1 and $(1-s)$, and their frequencies are p and q, their contribution to the next generation will be p and $q(1-s)$, respectively. Hence, the change per generation is:

$$\Delta q = \frac{q(1-s)}{1-sq} - q = \frac{-sq(1-q)}{1-sp}$$

Selection against gametes is nearly identical formally with selection against homozygotes and intermediary heterozygotes (Table 12.10).

Table 12.10. Selection $2s$ against homozygotes and s against heterozygotes

	AA	Aa	aa	Total
Before selection	p^2	$2pq$	q^2	1
Fitness	1	$1-s$	$1-2s$	
After selection	p^2	$2pq(1-s)$	$q^2(1-2s)$	$1-2sq$

12.2.1.3 Selection Leading to a Genetic Equilibrium

Above, only those models of selection are discussed that lead to an increase in the frequency of one allele at the expense of the other. An equilibrium, and hence a steady state of gene frequencies over generations, could be achieved only by introduction of an external force – mutation. However, there are modes of selection that in themselves lead to an equilibrium. A steady state with no systematic changes in gene frequencies may be created if selection favors heterozygotes.

Selection in Favor of Heterozygotes with Selective Disadvantage of Both Homozygotes. The biological basis of this model – heterosis – has been known in experimental genetics for a long time. It has led both to theoretical discussions and to many practical applications in plant breeding, most notably of corn [25, 32, 116, 117]. Heterosis, or hybrid vigor, means the superiority of heterozygous genotypes with respect to one or more characters in comparison with the corresponding homozygotes.

Heterozygote Advantage: Formal Consequences. Table 12.11 shows the genotypes before and after selection

Table 12.11 Selection in favor of heterozygotes

	AA	Aa	aa	Total
Before selection	p^2	$2pq$	q^2	1
Fitness	$1 - s_1$	1	$1 - s_2$	
After selection	$p^2(1 - s_1)$	$2pq$	$q^2(1 - s_2)$	$1 - s_1p^2 - s_2q^2$

(s_1, s_2: selection against the phenotypes AA and aa). Using the relationship $pq + q^2 = q$, we obtain the change from one generation to the next:

$$\Delta q = \frac{q - s_2q^2}{1 - s_1p^2 - s_2q^2} - q = \frac{pq(s_1p - s_2q)}{1 - s_1p^2 - s_2q^2}$$

Figure 12.11 shows the relation for various values of q and Δq and for $s_1 = 0.15$, $s_2 = 0.35$; Dq may be positive or negative, depending on whether s_1p is larger than or smaller than s_2q. If $s_1p = s_2q$, D$q = 0$. Solving for p or q gives the following equilibrium values:

$$\hat{p} = \frac{s_2}{s_1+s_2}; \quad \hat{q} = \frac{s_1}{s_1+s_2} \tag{12.6}$$

These equilibrium values depend only on s_1 and s_2; they are independent of the gene frequencies at the onset of selection. Moreover, the equilibrium is stable; Δq is positive if $q<\hat{q}$; it is negative if $q>\hat{q}$. If disturbed by accidental influences, the equilibrium tends to reestablish itself.

The leading paradigm for heterozygote advantage in humans is a selective mechanism that has caused the high frequency of the sickle cell gene in some human populations. This is described in more detail in Sect. 12.2.1.6.

12.2.1.4 Selection Leading to an Unstable Equilibrium

Selection Against Heterozygotes. A stable equilibrium may be established in the population if selection favors heterozygotes at the expense of the homozygotes. However, selection may also favor the homozygotes at the expense of the heterozygotes.

Let 1, 1-s, 1 be the fitness of AA, Aa, and aa, respectively. The change in gene frequencies is seen in Table 12.12.

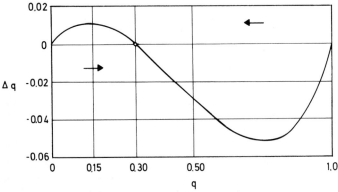

Fig. 12.11. The change in gene frequency Δq for various values of q, and for arbitrary selective disadvantages. $s_1 = 0.15$ and $s_2 = 0.35$. s_1 and s_2 denote the selective disadvantages of the two homozygotes in comparison with the heterozygote. (Li 1955 [100]) Let us assume a pair of alleles a_1, a_2; the gene frequency of a_1 may be p and that of a_2, q. If $q = 0.1$, this value tends to increase in the following generations (Δq has a positive value). If $q = 0.4$, however, this value decreases (Δq is negative). If q = 0.3, there is neither increase nor decrease ($\Delta q = 0$; equilibrium point)

Table 12.12. Selection against heterozygotes

	AA	Aa	aa	Total
Before selection	p^2	$2pq$	q^2	1
Fitness	1	$1-s$	1	
After selection	p^2	$2pq(1-s)$	q^2	$1-2spq$

This leads to:

$$q' = \frac{pq(1-s)+q^2}{1-2spq} = \frac{q-spq}{1-2spq} \quad \text{and}$$

$$\Delta q = \frac{spq(2q-1)}{1-2spq} \approx 2spq\left(q - \frac{1}{2}\right), \tag{12.7}$$

if s is small. Figure 12.11 shows Δq in dependence of q.

If $q = 1/2$, $\Delta q = 0$. This means that there is a genetic equilibrium, but that this equilibrium is unstable. If q becomes higher than 1/2, Δq is positive. The gene frequency tends toward 1. If q is shifted below 1/2, on the other hand, Δq becomes negative: q tends toward 0. An unstable equilibrium alone cannot maintain a polymorphism in the population; even in the artificial case of two populations homozygous for two different alleles mixing in equal proportions, small shifts in gene frequencies by chance fluctuation would soon destroy the equilibrium state, causing the gene frequency to move either toward 1 or toward 0.

Still, such an unlikely event appears to have happened. In humans, two situations are known in which the only selective factor that can be analyzed so far is selection against heterozygotes. In one of these situations, the rhesus system (111700), allelic frequencies conform to our definition of polymorphism. In pericentric inversions, heterozygotes have also been found in polymorphic frequencies, and homozygotes have occasionally been observed.

Pericentric Inversions. In a heterozygote for a pericentric inversion, pairing of homologous chromosomes during meiosis may not proceed properly. The resulting unbalanced gametes may either be eliminated before fertilization or form lethal zygotes. On the other hand, no meiotic disturbance is expected in homozygotes; here, pairing of homologs proceeds properly. A high incidence of reproductive wastage has indeed been observed in the progeny of cases with some pericentric inversions [65]; adequate population genetic studies are lacking. They are needed particularly, as pericentric inversions have played a major role in human evolution, apparently providing a powerful mechanism for reproductive isolation. Pericentric inversions do not affect the health status of their heterozygous carriers but impair their reproductive capacity. This is a classic – and possibly the best – example

Table 12.13. Mating types and classes of progeny against which selection due to Rh incompatibility works *(white areas)*

Fathers→ ↓Mothers	DD p^2	Dd $2pq$	dd q^2
DD p^2	p^4 DD	p^3q DD p^3q Dd	p^2q^2 Dd
Dd $2pq$	p^3q DD p^3q Dd	p^2q^2 DD $2p^2q^2$ Dd p^2q^2 dd	pq^3 Dd pq^3 dd
dd q^2	p^2q^2 Dd	pq^3 Dd pq^3 dd	q^4 dd

of fertility influencing fitness, without any viability component. This may be the reason for its significance in evolution, especially speciation (Sect. 14.2.1).

Selection Against Rh Heterozygotes. The unstable equilibrium in selection against heterozygotes was first discovered by Haldane (1942) [50] for the special case of serological mother-child incompatibility in the Rh factor. This situation is slightly more complicated than that considered for pericentric inversions. The danger of erythroblastosis and therefore of selection against heterozygous children occurs if an Rh-negative mother bears an Rh positive child. The Rh loci have a complex structure with two closely linked genes (Sect. 5.2.4). To understand the principle of selection, however, we need only consider the genes D (positive) and d (negative); this reduces the problem to that of a simple diallelic system. Erythroblastosis children occur in matings dd♀ × DD ♂ or dd♀ × Dd♂. Table 12.13 lists the mating types; the endangered children, against whom selection works, are indicated by shading. Combining all mating types leads to the formula for the change in the frequency of D:

$$\Delta q = \frac{p - \frac{1}{2}q^2(s_1p^2 - s_2pq)}{1 - q^2(s_1p^2 + s_2pq)} - p =$$

$$\frac{p(p - \frac{1}{2})q^2(s_1p - s_2q)}{1 - pq^2(s_1p + s_2q)}. \tag{12.8}$$

Here, s_1 equals selection against children with dd mothers and DD fathers; s_2 equals selection against heterozygous children of dd mothers and Dd fathers. As the risk of immunization increases with the number of Dd children, it is lower when the father is Dd, because on the average only every second child will cause maternal immunization, whereas *every* child of a DD father is heterozygous and may immunize its mother; hence $s_2 < s_1$.

It can easily be shown that $\Delta p = p' - p = 0$ if and only if $p = 1/2$. This means that Eq. (12.8) has the same equilibrium point as Eq. (12.7). Again, the equilibrium is unstable.

In current western European populations d has a gene frequency $q = 0.35$. It follows that its frequency will decrease, unless other selective mechanisms counteract this tendency. How rapidly will this decrease proceed? Figure 12.11 shows Δp for several generations with two assumptions on the initial gene frequencies p for D and q for d, and selection coefficients s_1 and s_2 in the range actually observed in humans. The number of affected children varied with the average number of pregnancies before prophylactic therapy of women at risk with anti-D antiserum was introduced [1803]. Erythroblastosis in about 5% of all Dd children of dd mothers is a reasonable estimate. Earlier, when women had more pregnancies, this figure must have been somewhat higher.

Figure 12.13 shows that the change in gene frequencies under these conditions proceeds very slowly, the reason being that not all heterozygotes are subject to selection (Table 12.13). This may explain why the Rh polymorphism still exists. Other, still unknown modes of selection, even if only very minor, may have upset selection against heterozygotes. Random mating may not have been proceeding long enough for selection against heterozygotes to work, since the time when chance fluctuations in human populations consisting of small and relatively isolated subgroups created huge between-group differences in gene frequencies (Sect. 13.3.1). A full explanation of the population genetics of the Rh polymorphism remains an enigma in view of the high frequency of Rh system alleles D or d in so many populations. Conceivably, other not yet understood functions of this system were involved in selection.

AB0 Blood Group System. Serological mother-child incompatibility also occurs in the AB0 blood group system. On average, erythroblastosis is milder than with Rh incompatibility. However, AB0 incompatibility may also lead to increase in the number of spontaneous miscarriages, although the evidence is controversial [131]. Selection seems to work mainly against A0 and B0 children of 0 mothers. Recurrent equations can be derived in the same way that was shown above for the two-allele case of the Rh system – with the difference that selection is identical against heterozygous children of either homozygous or heterozygous fathers, since, unlike in Rh immunization, the first incompatible child may already be damaged [109]. The following formula describes the change in gene frequency for allele A:

$$\Delta p = \frac{p - \frac{s}{2} p r^2}{1 - s r^2 (1 - r)} - p = \frac{s p r^2 (\frac{1}{2} - r)}{1 - s r^2 (1 - r)} \quad (12.9)$$

An analogous formula can be derived for Δq, the change in frequency for allele B. The formula for Δr, the change in frequency of allele 0, is slightly different:

$$\Delta r = \frac{r - \frac{s}{2} r^2 (1 - r)}{1 - s r^2 (1 - r)} - r = \frac{s r^2 (1 - r)(r - \frac{1}{2})}{1 - s r^2 (1 - r)} \quad (12.10)$$

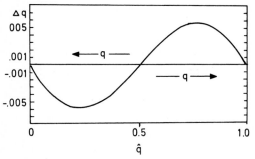

Fig. 12.12. Change in gene frequency Δq. Selection $s = 0.50$ against the heterozygotes. Δq is negative if $q < \hat{q}$ and positive if $q > \hat{q}$. (From Li 1955 [77])

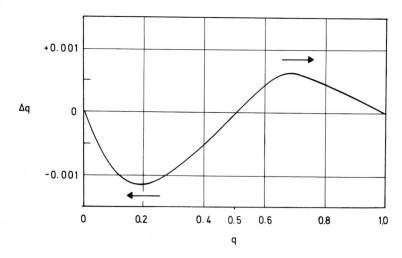

Fig. 12.13. Unstable equilibrium at $p = q = 0.5$ in the special case of mother-child incompatibility (Rhesus factor); selection $s_1 = 0.05$ for heterozygous children of homozygous DD fathers and selection $s_2 = 1/2s_1$, for heterozygous children of heterozygous Dd fathers

If $r = 0.5$, $\Delta p = \Delta q = \Delta r = 0$, independently of the ratio between p and q. This equilibrium is unstable in respect to r, and neutral in respect to p and q. Figure 12.14 shows the speed at which r approaches 1, and p and q approach 0. No other selective mechanism is at work, and initial gene frequencies corresponding to those found in western Europe are assumed (A: $p = 0.3$; B: $q = 0.1$; o: $r = 0.6$). The selection coefficients are those suggested by the empirical evidence. Changes in gene frequencies are much more rapid than those found in the Rh case (Fig. 12.14).

12.2.1.5 Other Modes of Selection

Frequency-Dependent Selection [21, 63]. The discussion above treats selection values as constants. However, these values may be functions of the genotype frequencies as well as of the population density. This type of selection is known as frequency- or density-dependent selection. More specifically, the correlation between the selective value of a genotype and its frequency may be negative, a genotype becoming more advantageous when its frequency declines. Such cases have been observed in nature when selection against a species is influenced by predators that can be fooled if a few individuals of this species change their phenotypes. In humans, one case may be interaction of genetic components of the immune system such as MHC types with infective agents.

Let us assume the following simple model of frequency-dependent selection in a dominant gene:

Phenotype	Frequency	Fitness
A	$1 - q^2$	$w_1 = 1 + s_1(1 - q^2)$
a	q^2	$w_2 = 1 + s_2 q^2$

Then we have:

$$\Delta q = q' - q = \frac{pqw_1 + q^2 w_2}{\bar{w}} - q = \frac{pq^2(w_2 - w_1)}{\bar{w}}$$

$$\bar{w} = (1 - q^2)w_1 + q^2 w_2 = 1 + s_1(1 - q^2)^2 + s_2 q^4$$

Therefore, the equilibrium condition is:

$$w_1 = w_2$$

or:

$$q^2(s_1 + s_2) - s_1 = 0$$

This equation is solvable for the gene frequency, q, if and only if the selection coefficients s_1 and s_2 are both positive or both negative. The solution is:

$$\hat{q} = \sqrt{\frac{s_1}{s_1 + s_2}}$$

For the equilibrium state, the fitness value is:

$$\hat{w}_1 = \hat{w}_2 = \hat{\bar{w}} = 1 + \frac{s_1 s_2}{s_1 + s_2}$$

The equilibrium can be shown to be stable if s_1 and s_2 are both negative (and not less than −1). This means that a stable equilibrium exists when A has an advantage if it is rare, such as in Batesian mimicry (Sect. 5.2.6).

At equilibrium all the genotypes have the same fitness. If there is no dominance, and the fitnesses of all three genotypes are different, calculation becomes more cumbersome [63]. However, it can be shown that in this case frequency-dependent selection can also yield stable polymorphisms when there is no heterozygote advantage. Polymorphisms can even be maintained by such a mechanism despite a selective disadvantage of heterozygotes. Furthermore, in a particular situation, there may be more than one stable equilibrium. In humans, frequency-dependent selection is a plausible mechanism for mutual adaptation

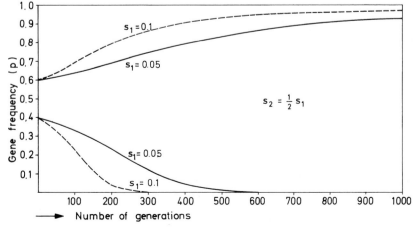

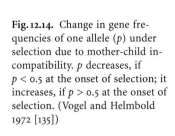

Fig. 12.14. Change in gene frequencies of one allele (p) under selection due to mother-child incompatibility. p decreases, if $p < 0.5$ at the onset of selection; it increases, if $p > 0.5$ at the onset of selection. (Vogel and Helmbold 1972 [135])

of a parasite to the human host and vice versa. The parasite – for example, a bacterium or virus – may become adapted to the commonest biochemical or immunological variety of the host, and rarer varieties may gain a selective advantage. The parasite mimics the antigens of its host, either by acquiring the genetic ability to produce its antigens or by directly utilizing the host's membrane material for synthesis of its own membrane. In either case, the immune defense mechanism of the host is deceived, and the parasite is more successful than if it had no antigen in common with the host. Selection depends on the frequency as the virus adapts mainly to the most frequent genotype; the rarer ones have an advantage. Examples are provided below.

One mechanism by which the host can defend itself against this "strategy" of the parasite is to create a highly polymorphic system with very many different antigenic patterns at its cell surface. This prevents the parasite from gaining selective advantage by adapting to one specific pattern. Such a highly polymorphic system has developed for the MHC, which comprises the HLA loci together with a number of other gene loci involved in the immune response (Sect. 5.2.5).

Frequency-Dependent Selection in Combination with Linkage Disequilibrium. The second significant feature of the MHC complex – apart from the high degree of polymorphism – is the nonrandom distribution of allele combinations for loci forming this complex, especially the HLA loci. This *linkage disequilibrium* is described in Sects. 5.2.4 and 5.2.5. Not one allele at one locus but a certain combination of alleles at the various loci of a gene complex confers a selective advantage to its carriers. Such groups of closely linked loci with related functions may have been created during evolution when certain combinations of genes had a selective advantage. This advantage is usually destroyed in the next generation by free recombination of the involved gene loci; it is maintained if these loci are closely linked. Therefore selection is expected to favor chromosome rearrangements leading to closer linkage between such loci. This is not the only possible cause of linkage disequilibrium. It may also – and possibly more often – exist as the result of a mutation that happened to occur in a certain chromosome carrying an individual combination of variants in the DNA structure outside of transcribed genes. (The PHA and CFTR genes are discussed above as examples.) But the mechanism discussed here – advantage of a certain array of alleles of various genes) may occur much more often than is assumed by many scientists, and its analysis may open up new paths for understanding interaction and cooperation of genes – and human evolution.

Density-Dependent Selection [21]. Selection may vary not only with the relative frequencies of genotypes within a population but also with the absolute population size/habitat, i.e., the population density. It has been shown that density-dependent selection allows the maintenance of a balanced polymorphism under a wide range of conditions. Under some circumstances a polymorphic population can support a larger number of individuals than a monomorphic one. Still more importantly, changes in population density may bring about *genetic revolutions* that mimic the effects of random genetic drift (Sect. 13.3). Changes in population density have been important in human evolution, for example, when agricultural techniques were learned during the Neolithic Era. It is therefore plausible to assume that density-dependent selection may have played an important role in human evolution.

Kin Selection. In recent years another kind of selection has been discussed increasingly by evolutionary theorists: kin selection. Animals generally interact with each other and with the environment in social groups – families, hordes, gangs. Within such groups, behaviors are often observed that seem to contradict the expectation that each individual competes with all other members of the species for survival and reproduction, an expectation that is implicit in the earlier selection models. The biological meaning, for example, of *altruistic* behavior – sometimes even at the expense of the individual's life – is obvious intuitively when we observe a mother defending her offspring against a predator. Sometimes, however, behaviors are observed in which the biological basis is less clear; for example, when a lion sire takes over a harem and kills the young offspring of his predecessor. Theoretical interpretation of such social behaviors, altruistic and otherwise, remained an enigma for evolutionary theorists from Darwin onwards until in 1964 Hamilton [52] developed a theory of kin selection. The full theory is intricate and is only outlined here. The basic idea, however, is simple and obvious: The individual fights for survival and transmission of his or her *genes*. Often the interest of my genes coincide with my interest as an individual, but if I have, for example, two children, sacrificing my life for their survival is equivalent to saving my own life at their expense, since each of them carries half my genes. If I have three or more children, such altruistic behavior is even superior, saving *more* of my genes than if I were to survive myself. The same reasoning applies for sibs and also or for more remote relatives such as grandparents, uncles and nephews, and others. The lion, on the other hand, will probably own the harem for only a limited – and often short – time. During this time he is "interested" in reproducing as extensively as possible. Lionesses who are still nursing cubs from a different father do not ovulate and cannot become pregnant. Therefore it is in the interest of the lion's genes to kill these cubs.

An increasingly important branch of science – sociobiology (Sect. 14.2.4) – uses the theoretical concepts of Hamilton and his successors. The consequences have been popularized in discussions of the "selfish gene" and have led to heated discussions between biologists and social scientists and philoso-

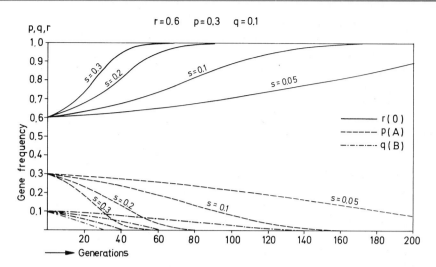

Fig. 12.15. Change in gene frequencies in the AB0 blood group system due to selection s against A0 children and B0 children of 0 mothers. p, Gene frequency of A; q, gene frequency of B; r, gene frequency of 0; r tends to 1; p, q tend to 0. (Vogel and Helmbold 1972 [135])

phers. Understanding of this requires some knowledge of the conceptual background. Its mathematical derivation, which is not simple, cannot be given here. (For a relatively simple treatment see [20].) The following result is important:
For the selective advantage of altruistic behavior, improvement in the chance of survival by genes of the individual practicing this behavior is decisive. This depends on the relative positions of "donor" and "recipient" in a pedigree; the closer their "coefficient of relationship" (Hamilton), the higher the gain from altruistic behavior. This coefficient is identical to the proportion of genes which two individuals have in common by descent: in the absence of additional consanguinity such behavior does not lead to the loss of my genes if I sacrifice my life for at least two children or full sibs, or four half sibs, or eight first cousins. If the donor does not necessarily sacrifice his life but incurs an increased risk of doing so, this risk must be weighted against the increased chances for relatives. For example, at a 20% risk of being killed in a given situation to protect ones sibs, from 100 individuals only 80 will survive. If their sacrifice leads to the survival of 200 instead of 160 sibs, i.e., 40 more, the same number of "donor" genes are transmitted to the next generation. With parents and children the situation may be slightly more complicated because of the age difference.

Selection for Continuously Distributed, Multifactorially Determined Characters. The discussion above examines selection for only single genes. However, many normal and abnormal traits and diseases show a continuous distribution in the population and are determined by an unspecified number of different genes. In principle, the laws of selection theory apply for these characters as well. However, since genetic analysis has not yet penetrated to the Mendelian-phenotypic level (Sect. 6.1), biometric methods must be used [37, 39, 77]. These methods have gained importance for animal and plant breeding. These problems include the following:

a) Obviously the change in a quantitative characteristic in a population under the influence of natural

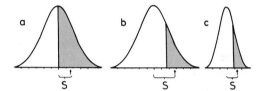

Fig. 12.16 a–c. Selection in continuously distributed multifactorial characters. **a** Selection by elimination of 50% of the population (those below the population mean, shown in white) from reproduction. In the F_1 generation, the mean is shifted by S. It can be shown that this shift must be $0.8\times$ the standard deviation (SD). **b** Selection by elimination of 80% of the population from reproduction. This leads to a shift of $1.4\times$SD of the population mean in the next generation. **c** This shows a population with lower genetic variability. If 80% of the population are eliminated from reproduction as in b, the shift of the mean relative to the standard deviation is the same but absolutely it is much lower (in this case 1/2). (Falconer 1960 [37])

selection is proportional to heritability (for the heritability concept, see Sect. 6.1.1.4).
b) The response of the character to selection depends on the strength of selection (Fig. 12.15 a, b).
c) It also depends on the degree of genetic variability in the population (Fig. 12.16). In the absence of genetic variability, selection is ineffective.

Figure 12.17 shows the effect of artificial selection over generations. The mean shifts toward the direction of selection but genetic variability decreases from generation to generation, until selection ceases to be effective.
Mutations causing slight, barely discernible shifts in multifactorial systems have presumably been of great importance in evolution. Under the influence of selection these mutations have led to a slow and gradual shift in quantitative characteristics.

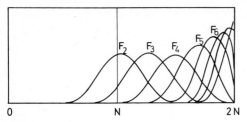

Fig. 12.17. Effect of selection over many generations on a continuously distributed character with multifactorial inheritance. With a shift in the mean the variability gradually decreases. The same is true for the effect of selection in one generation. Ultimately the population becomes genetically homogeneous, and no further selection is possible. (From Falconer 1960 [37]

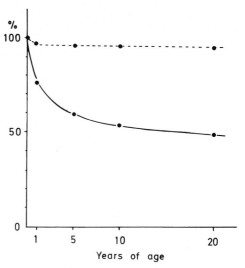

Fig. 12.18. Survival rate of newborn children up to the age of 20 in Prussia (central Europe) in the middle of the eighteenth century. (Based on data from Süssmilch 1786 [125]). *Dotted line*, survival rate for Berlin (1955) for comparison

One genetic property that shows such a continuous distribution, and is therefore thought by many to be due to the interaction of many genes, is the liability to common "multifactorial" diseases such as congenital malformations, common diseases, and mental disorders. Altering selective conditions – for example, by successful treatment of congenital heart disease or schizophrenia – could lead to a shift in the distribution curve of disease liability and thus to a higher incidence of these conditions. It is, however, difficult to calculate the extent of the shift. We do not know sufficiently well whether or in what cases the multifactorial genetic model adequately describes reality (Chap. 6). With the identification of specific genes responsible for aspects of such genetic susceptibilities, population genetic analysis will also become easier, and more reliable predictions will become possible.

12.2.1.6 Selection Due to Infectious Diseases
[91, 131]

The foregoing sections describe the most important mathematical selection models and indicate some practical applications to situations in human populations. As mentioned, probably the most important source of natural selection has been selection due to genetic differences in susceptibility to infective agents.

Selection Due to Infectious Diseases in Historical Populations. Natural selection is especially effective when acting through differential mortality before the age of reproduction, i.e., during childhood and youth. The first reliable statistics on childhood mortality are available for Europe in the eighteenth century [125]. The survival rate in Prussia up to the age of 20 is shown in Fig. 12.18. More than half the population died before reaching the age of 20, and about one-quarter died within the first year of life. What were the causes of these early deaths? Statistics answer this

question only partially, since not all diagnoses reported at that time can be identified as presently known diseases. However, the data leave little doubt that most children died from viral and microbial diseases such as intestinal infections, smallpox, tuberculosis, and measles. Therefore infectious diseases are good a priori candidates for possible selective agents. Endemic infections that affect every generation would be more effective selective agents than epidemic diseases which occur only episodically, such as plague.

History of Some Infectious Diseases. What are the infectious diseases that may have influenced gene frequencies of whole populations in the past? Four groups come to mind:

1. Acute infections that repeatedly invaded whole countries, sweeping away large parts of the populations. Examples include plague, cholera, and smallpox.
2. Chronic infections such as tuberculosis, leprosy, and syphilis.
3. The heterogeneous group of intestinal infections that occurred in all age groups but were fatal mainly in infancy and early childhood.
4. Tropical diseases such as malaria.

The available information on the history of these infections is fragmentary and often misleading, as no bacteriological evidence for historical epidemics is available, and the descriptions often do not permit a diagnosis. Some conclusions seem to be warranted, however [59]:

a) Plague can be traced back to the end of the second or the beginning of the third century B.C. Good descriptions exist from Alexandria and Libya around the time of Christ's birth. During the Middle Ages a number of plagues swept through Europe, killing a large proportion of the population. In more recent times epidemics have come from the Near East (Turkey) and southern or southeastern Asia. Centers of smallpox were in Africa and in Asia, especially India and China. In both countries smallpox has been known for thousands of years; India even has a smallpox goddess, Sitala.

b) Tuberculosis was formerly endemic in large parts of the Old World and was introduced from there wherever white men went. Leprosy requires special living conditions to become endemic. This disease was frequent in Europe during the Middle Ages but disappeared with improved living conditions by the eighteenth century. At present many millions are still affected, especially in India, southeastern Asia, Africa, and South America. Syphilis, according to one hypothesis, was a disease of the New World – Central and South America – and was introduced to Europe immediately after the discovery of America. Following epidemics in the late fifteenth and the sixteenth centuries, syphilis became an endemic disease [12]. This hypothesis is far from being universally accepted, although it is in reasonably accord with many historical facts. Historical hypotheses of this kind are difficult to refute but still more difficult to confirm.

c) It is known that infant mortality due to intestinal infections leading to infant diarrhea was very high in Europe even up to 1900, and in large parts of Asia, Africa, and South America until much more recent times. It is reasonable to assume that such high infant mortality has always been experienced in human history. Beyond this statement, more specific hypotheses are impossible.

d) For tropical countries such mortality statistics as those presented above for Prussia (Fig. 12.18) have become available only recently; however, there is ample evidence that even after World War II many of these countries still had a very high rate of childhood mortality. This mortality was caused by a variety of conditions; infectious diseases such as malaria and intestinal infections prevailed. If there were genetically conditioned susceptibilities to these infections, those individuals who are susceptible would die more often, whereas others who enjoy a higher level of genetic resistance would have a better chance to survive and to transmit their genes to their children. Therefore the genetic composition of present human populations will have been strongly influenced by such differences in resistance to infections. Recent decades have seen much progress in the analysis of mechanisms for such resistance and in studies of the genetic composition of human populations. Below two principal examples are be analyzed:

1. Selection due to malaria in relation to the frequency of hemoglobin genes in populations of tropical countries.
2. Selection in relation to the AB0 blood groups.

This leads below to a comparison of "attack" and "defense" strategies between host and infective agent.

Distribution of Sickle Cell Gene and Other Abnormal Hemoglobin Genes in the World Population. The leading paradigm for heterosis in humans is the selective mechanism causing the high frequency of the sickle cell gene in some human populations. The molecular basis, genetic determination, and genotype-phenotype relationship of sickle cell hemoglobin has been described in Sect. 7.3.

Sickle cell hemoglobin (HbS) is produced by a single nucleotide substitution of the β-hemoglobin gene.

Affected homozygotes suffer from a hemolytic anemia, and episodic attacks of joint and abdominal pain whereas heterozygotes are clinically healthy under normal conditions.

The very unequal distribution of the sickle cell gene in the populations of the world is the most striking feature. From the standpoint of population genetics. Such a pattern is not, however, confined to this gene; a number of other hemoglobin variants such as C, D, E, and the thalassemias show a similarly unequal distribution and are also polymorphic. Within a broad periequatorial belt from Zaire to Tanzania, frequencies for heterozygotes of HbS range from 15% to as high as 40%. The frequency decreases slightly toward the western part of Africa. In northern and southern Africa it is much lower; in both these areas HbS is found only sporadically in various populations. In the Mediterranean HbS occurs especially in Sicily, Calabria, and some parts of Greece. On the Chalcidice peninsula the heterozygote frequency may reach 30% (Fig. 12.19). The gene is relatively common in southern Indian populations and has been found in Arabs. It is absent from Native American populations and practically absent from all northern and northwestern European populations. Three explanations are possible in principle for such an unequal distribution.

1. The mutation rate is different, either for exogenous reasons, for example, by differential exposure to a mutagenic influence (Chap. 11), or for endogenous reasons, for example, unequal distribution of mutator genes (Sect. 9.4).
2. Selection works differently under different environmental conditions.
3. Chance fluctuations of gene frequencies (genetic drift) occur, especially if the effectively breeding populations are small (Sect. 13.3.1).

To begin with the latter possibility, it is very unlikely that differences among large population groups such as those observed for the frequency of HbS have come about by chance. Moreover, why should all known polymorphic hemoglobin variants (S, C, D, E and the thalassemias) be found almost exclusively in the tropical-subtropical belt? The drift hypothesis has therefore never been considered seriously to explain the distribution of these polymorphisms.

Differential Mutation Rates to HbS? Shortly after the discovery of this unusual distribution the hypothesis of differential mutation rates was seriously considered [98]. Application of Haldane's indirect method (Sect. 9.1) led to unrealistically high mutation rate estimates confined to only a few population groups. Moreover, the hypothesis was refuted by direct exam-

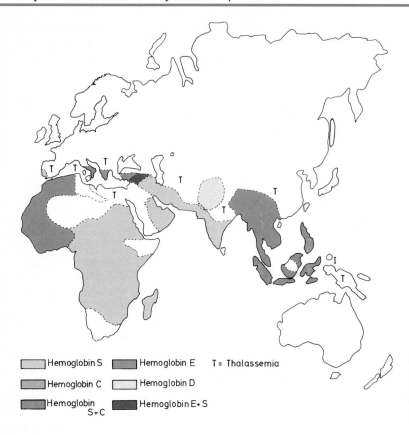

Hemoglobin S

Hemoglobin C

Hemoglobin S+C

Hemoglobin E

Hemoglobin D

Hemoglobin E+S

T = Thalassemia

Fig. 12.19. Areas in which the sickle cell gene and genes for other abnormal hemoglobins are common. (See also Hill and Wainscoat 1986 [57]; Livingstone 1985 [85])

ination of children with sickle cell anemia and their parents.

All evidence points to very low mutation rates for single base substitutions in the DNA (Sect. 9.4), ranging possibly between 10^{-8} and 10^{-9}. This, together with the small overall population sizes in earlier centuries, suggested that ultimately all sickle cell genes could be traced back to very few single mutations or only one. Linkage studies with DNA polymorphisms suggest four different haplotypes comprising sickle cell mutants ([7a]; Sect. 4.3) which are separated from each other by no less than two crossover events. These have been observed in population groups from neighboring regions in central Africa. Some observers have suggested that this may point to four independent mutational events. However, we find the idea attractive that ultimately all HbS alleles derive from only *one* mutation, and that the present location of this allele in four RFLP haplotypes is explained by very rare recombination or by gene conversion (see Sect. 3.2; Sect. 7.3). On the other hand, independent mutations are all the more unlikely, since the $T = A \rightarrow A = T$ mutation necessary for the glutamic acid $\rightarrow$ valine exchange is a transversion, belonging to the rarer type of base replacements (Sect. 9.4.2). Why did these genes become so frequent? There must have been a selective advantage.

Malaria Hypothesis. Geographic distribution and genetic analysis offer two hints:

1. The selective advantage seems to be confined to tropical and subtropical areas.
2. Homozygotes suffer from a severe hemolytic anemia. Their reproductive fitness is estimated to be approximately 20%–25% of normal and under primitive living conditions approaches 0. The homozygous state for the sickle cell gene was unknown in central Africa until the past three decades because affected infants died very early. The high frequency of HbS could be reached only with a selective advantage of heterozygotes who are, at the same time, much more frequent than homozygotes.

The two criteria led to the hypothesis that heterozygotes are less susceptible to falciparum malaria than are normal homozygotes:

Beet (1946, 1947) [8, 9] observed in the Balovale district and in other parts of Zambia that during the dry season when malaria infectivity is in generally lower, heterozygous AS children showed malaria parasites in their blood smears less frequently than normal homozygotes. The difference was not statistically significant but splenomegaly was also less pronounced in heterozygotes. Beet appears to have been the first to suggest malaria as the principal selective agent. In

1951 the Lambotte-Legrands [75] in Zaire had the impression that cerebral malaria was rarer among heterozygotes for sickle cell anemia.

Haldane (1949) [51], noted a similar geographic distribution of thalassemia and malaria and suggested that thalassemia could be maintained in the population by a selective advantage of heterozygotes in the presence of malaria. For the sickle cell gene this hypothesis was tested by Allison [1–4, 6], who formulated it as follows:

1. The homozygous sickle-condition is virtually lethal in Africa.... The rate of elimination of the gene could not be compensated by recurrent mutations.
2. Balanced polymorphism has resulted because the sickle cell heterozygote is at an advantage, mainly as a consequence of protection against Falciparum malaria.
3. Malaria exerts its selective effect mainly through differential viability of subjects with and without the sickle cell gene between birth and reproductive age, and to a much lesser extent through differential fertility.
4. High frequencies of the sickle cell gene are found only in regions where Falciparum malaria is, or was until recently, endemic.
5. In most New World black populations frequencies of the sickle cell gene are lower than would be expected from dilution of the African gene pool by racial admixture. This is probably the result of elimination of sickle cell genes without counterbalancing heterozygote advantage.
6. In regions where two genes for abnormal hemoglobins coexist, and interact in such a way that individuals possessing both genes are at a disadvantage ... these genes will tend to be mutually exclusive in populations.

The following sections take up the evidence for the first four arguments of the "malaria hypothesis" in its specific form. The evidence for points 5 and 6 is deferred to later sections (Sect. 12.2.1.7).

Evidence for the Malaria Hypothesis. Several lines of evidence are discussed in detail below because this example illustrates the methodology applicable to humans for examining hypotheses of natural selection. The evidence may be divided into two parts: (a) results from testing the proposed mechanism of selective advantage and (b) results from examining the proposed consequences for reproduction and population frequency.

1. The hypothesis predicts that young children (during the first 5 years of life) *without* the sickle cell trait will be infected more massively, will become more severely sick, and will die more often from falciparum malaria than will sickle cell heterozygotes. The hypothesis does not predict that school children or adults will be more frequently or severely infected. In hyperendemic areas immunization by frequent infection with malaria has developed early in life so that no differential mortality can be expected between heterozygotes and normal homozygotes in older children or adults.
Table 12.14 shows the incidence of *Plasmodium falciparum* infections in sicklers and nonsicklers. For statistical evaluation, the method of Woolf [144a] was used; the weighted mean relative incidence

Table 12.14. Incidence of severe P. falciparum infections in African children (from Allison 1964 [1773])

Authors	Classification of infection	Sickle cell		Non-sickle cell		Relative incidence[a] (Woolf)	χ^2	Probability
		Severe infections	Total	Severe infections	Total			
Allison (1954)	Group 2 or 3	4	43	70	247	3.86	6.16	0.02 > p > 0.01
Foy et al. (1955)	Heavy	21	241	38	241	1.96	5.44	0.02 > p > 0.01
Raper (1955)	> 1000 µl	35	191	374	1009	2.63	23.74	p < 0.001
Colbourne and Edington (1956)	> 1000/µl	3	173	57	842	4.11	5.59	0.02 > p > 0.01
Colbourne and Edington (1956)	> 1000/µl	5	15	75	177	1.47	0.46	p > 0 .50
Garlick (1960)	> 1000/µl	25	91	147	342	1.99	7.06	0.01 > p > 0.001
Allison and Clyde (1961)	> 1000/µl	36	136	152	407	1.66	5.27	0.05 > p > 0.02
Thompson (1962, 1963)	> 5630/µl	3	123	42	593	3.05	3.38	0.10 > p > 0.05

[a] Incidence of heavy P. falciparum infections in non-sickle cell trait groups relative to unity in corresponding sickle cell trait groups.
Weighted mean relative incidence = 2.17. Difference from unity χ^2 = 51.4 for 1 df p<0.001.
Heterogeneity between groups χ^2 = 5.7 for 7 df, p>0.5.

shows that children without the sickle cell trait (normal homozygotes) incur a risk 2.17 times that of sickle cell heterozygotes of having a heavy falciparum infection. Table 12.14 shows only cases with heavy infections, defined as showing more than 1000 parasites per microliter of blood. However, malaria can lead to a difference in fitness only if a higher percentage of affected nonsicklers die from malaria – or if the disease impairs their reproduction. Table 12.15 presents the number of fatal outcomes among normal homozygous and heterozygous children in several areas of Africa. With one exception, only normal homozygotes died of malaria. In view of the incidence of the sickle cell trait in these populations (8%–29%), this finding cannot be a chance result. Therefore a higher malarial

susceptibility during early childhood and, consequently, a higher death rate were demonstrated for normal homozygotes than for heterozygotes. The results of in vivo studies in which volunteers with sickling trait were infected with malarial plasmodia are less clearcut. One study found [2] sickle cell heterozygotes to have a lower incidence of parasitemia than controls after experimental malaria inoculation; other studies have failed to confirm this result.

2. Another set of investigations examined the consequences of increased disease susceptibility for the population were: (a) a higher nonsickler mortality during childhood should result in a greater frequency of sicklers among adults compared with children of the same population group, as shown in Table 12.16; (b) a higher mortality of nonsicklers should also lead to a greater number of surviving children from marriages in which sickle cell heterozygotes segregate. The evidence from one large study [5, 56] is shown in Table 12.17. Childhood mortality was highest in matings between two heterozygotes. This is not surprising, as one-fourth of them are expected to be homozygotes for the sickle cell allele and hence to suffer from sickle cell anemia. However, relatively fertile matings which, at the same time, show the lowest number of dead children were those between heterozygotes and normal homozygotes (AS × AA). This is to be expected if heterozygotes run a lower risk of death during childhood.

From the differences in frequencies among the genotypes AA, AS, and SS in the adult population of the Musoma district and Hardy-Weinberg expectations Allison [5] calculated the relative fitness (w) of the different genotypes as compared with the population average:

Table 12.15. Malarial mortality in HbS trait (AS; from Motulsky 1964) [92]

	AS (population frequency, %)	No. dead of malaria	Observed no. AS dead of malaria	Expected no. AS dead of malaria
Kinshasa, Zaire	26	23	0	6
Kananga, Zaire	29	21	1	6.1
Ibadan, Nigeria	24	27	0	6.5
Accra, Ghana	8	13	0	1
Kampala, Uganda	19	16	0	3
Total		100	1[a]	22.6

[a] $\chi^2 = 26.77$ $p < 0.001$.

$$w_{AA} = 0.7961; \quad w_{AS} = 1.000; \quad \text{and} \quad w_{SS} = 0.1698$$

Table 12.16. Comparison of prevalence of sickle cell heterozygotes among children and adults (from Allison 1956 [5])

Population	No. of children examined	Sickle cell heterozygotes (%)	No. of adults examined	Sickle cell heterozygotes (%)
Dar es Salaam, Tanzania	753	17.9	283	23.3
Zaire; Baluba	147	16.3	775	23.5
Zaire; Pygmies	119	22.7	327	28.1
Dakar, Senegal	1350	6.2	952	15.5
Rwanda, Burundi	516	14.2	928	13.2
Musoma, Tanzania	287	31.8	654	38.1
Mandingo, Gambia	211	9.0	713	11.5
Jola, Gambia	103	14.5	312	17.0
Fula, Gambia	69	17.3	127	18.9
Jolloff, Gambia	48	18.8	104	17.3

Table 12.17. Fertility and child mortality in Africans living in the Musoma district, Tanzania (from Allison 1956 [5])

Mating type	Number of matings	All living children	Mean number of living children/ mating	Deceased children	Mean number of deceased children/ mating	Deceased children (%)	Total number of children (living or dead)	Total number of children/ mating
AS × AS	18	44	2.44	35	1.94	44.3	79	4.39
AS × AA	84	221	2.63	121	1.44	35.3	342	4.07
AA × AA	74	172	2.32	115	1.55	40.1	287	3.88
All mating types	176	437	2.48	271	1.54	38.2	708	4.02

corresponding to:

s_1 (selection against AA) = 0.2039

s_2 (selection against SS) = 0.8302

Is the ratio of the selection coefficients sufficient to maintain a genetic equilibrium for the actually observed frequency of the sickle cell gene? To test this we return to Eq. 12.6 for equilibrium conditions:

$$\hat{q} = \frac{s_1}{s_1 + s_2} = \frac{0.2039}{0.2039 + 0.8302} = 0.1972$$

A gene frequency of this order of magnitude, corresponding to a heterozygote frequency of 31.7%, has indeed been found in some populations. Therefore the data agree roughly with the predictions from the hypothesis of a balanced polymorphism.

In how many generations would such an equilibrium be established if this selection started anew? Results of some model calculations [104, 118] are in Fig. 12.20. Forty generations correspond to about 1000 years if a generation time of 25 years is taken. Since a millennium is not an unduly long time, the model is realistic from this standpoint as well. Some other results point to an additional loss of fertility due to placental malaria in infected homozygous normal women [38]. The evidence, however, is not entirely convincing.

Some Other Aspects of the Malaria Hypothesis [92, 93]. Two further aspects are:

a) It has been estimated [82] that malaria contributed significantly to approximately 15% of childhood deaths in endemic areas with ubiquitous infection. Considering that childhood mortality approached or even exceeded 50%, this mortality level is enough to provide a high selective advantage for any gene offering at least partial protection.

b) Malaria is an ancient disease; it has probably existed for at least 2000 years (80 generations) in

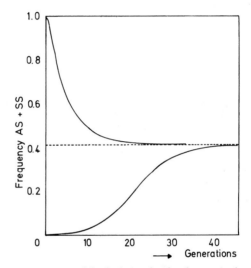

Fig. 12.20. Model calculation for the changes in the combined frequency of sickle cell homozygotes and heterozygotes for the sickle cell gene and the assumption about s_1 and s_2 set out in the text. Fitness of the heterozygote is assumed to be 1.26 times that of the normal homozygote. Fitness of the sickle cell homozygote is one-fourth that of the heterozygote. With initial gene frequencies as different as 0 or 1, an identical equilibrium frequency will be reached. (From Smith 1954 [118])

Mediterranean areas. In some parts of Africa malaria is probably less ancient and was presumably introduced by slash and burn agriculture [85], a practice that facilitated mosquito breeding in water ponds warmed by sunshine. Malaria – and therefore selection in favor of the sickle cell gene – depends on ecological conditions.

What Will Happen If the Advantage of Heterozygotes Disappears? The advantage of sickle cell heterozygotes is present only in an environment in which many children die from falciparum malaria; it vanishes as soon as malaria is eradicated. Malaria has been reduced in many countries very much below its level of 20–30 years ago, although the condition is increasing again in certain areas. This means that little if any heterozygote advantage remains at present. The consequences for the gene frequency of the sickle cell gene are obvious. Selection against the sickle cell gene would lead to its gradual diminution provided only that selection against abnormal homozygotes (patients with sickle cell anemia) continues. The speed of this decline can be calculated from the formulas used in the discussion of complete selection against homozygotes.

Since most New World black populations have been living for several generations in a malaria-free environment frequencies of the sickle cell gene are lower than would be expected from admixture of white genes (Sect. 12.3.4).

The expected reduction in the sickle cell gene was observed in blacks of malaria-free Curaçao, whereas blacks in Surinam had higher HbS frequencies. Both groups reported a similar origin in Africa. Only the Surinamese population was exposed to malaria, which persists to the present time. Results pointing in a similar direction were obtained in African-American blacks in Georgia, when white admixture estimates were calculated for a number of genes. The HbS trait turned out to give much higher admixture values than other markers. Selection against HbS homozygotes in the absence of malaria would explain the results [13].

Population Genetics of G6PD Variants and Falciparum Malaria [67, 92, 93]. Some other genetic traits of red blood cells are also common in tropical and subtropical areas (Fig. 12.19). Examples include HbC in western Africa, HbD in parts of India, HbO in Arabia, HbE in south-eastern Asia, the various α thalassemia and β thalassemia variants that are observed all over the malaria-infested area, and some variants of the enzyme glucose-6-phosphate dehydrogenase (G6PD). It is plausible to conclude that malaria was the major selective factor for these traits as well. The actual evidence, however, is not nearly as good as for the sickle cell gene. Only the G6PD variants, thalassemia, and HbE are considered below.

Soon after the discovery of G6PD deficiency it became apparent that this trait is found mainly in populations originating from tropical and subtropical areas of the globe. The geographic distribution was similar to that of falciparum malaria and suggested that G6PD deficiency, as the sickling trait, owes its distribution to selection by this malarial organism. Malaria requires nonprotein glutathione (GSH) for growth. Since GSH is reduced in G6PD deficiency, proliferation of falciparum malaria might be curtailed in the enzyme deficiency. Lessened parasitization is associated with lower mortality. Microgeographic mapping of the G6PD frequencies in areas of the world where both high and low endemicity of

falciparum malaria exists demonstrates that high gene frequencies are found where malaria is frequent and low gene frequencies are seen in areas with little or no malaria. The results from Sardinia are particularly impressive and show high G6PD frequencies in the plains where malaria was endemic and low G6PD frequencies in the hills where malaria was absent. Hill and plains populations were similar in terms of other genetic markers.

The finding that various genetic types of G6PD deficiency of different mutational origin (such as the A⁻ type, the Mediterranean type, and various Asian types) reach high frequencies in different parts of the globe is strong presumptive evidence for selection. A good correlation of the frequencies of the sickling gene and that of the A– variety of G6PD deficiency exists in African countries and between frequencies of β thalassemia and the Mediterranean type of G6PD deficiency in Sardinia. In other words, both traits in a given area correlated with malarial prevalence in earlier generations. Since G6PD deficiency is X-linked, and HbS and β thalassemia are autosomal traits, no other explanation besides selection can adequately explain these correlations.

The dynamics of selection of an X-linked trait are complex since males have two (normal and mutant) and females have three genetic classes (normal, heterozygotes, and homozygotes). Heterozygote females have two red cell populations: a G6PD-deficient and a normal G6PD population (Sect. 2.2.3.3). The G6PD-deficient red cell population has been demonstrated to show fewer malaria parasites than the normal population, thus providing direct evidence for protection of G6PD-deficient cells vis-a-vis falciparum malaria. Recent data on children in Africa (Gambia and Kenya) [112a] have shown that *both* male hemizygotes and female heterozygotes for G6PD deficiency have a reduced relative risk of about one half to develop *severe* falciparum malaria when compared with those without the trait. Presumably, the selective disadvantage caused by hemolytic episodes in G6PD deficient hemizygote males (but much less frequently in heterozygote females) has prevented the gene for G6PD deficiency to become the predominant allele in the endemic areas. Lisker and Motulsky (1967) [81a] and Ruwende et al. (1995) [112a] have modeled evolutionary trends of G6PD deficiency under various assumptions of biological fitness.

In Vitro Studies of Malarial Growth in Red Cells [46, 84, 88, 90]. The development of a method for growing malarial organisms in red cells in tissue culture provided a direct technique to study the ability of genetically abnormal cells to support malarial growth. However, owing to technical difficulties, work with this test system has not always provided definite answers. Red cells from sickle cell trait heterozygotes were found to be a poor medium for falciparum malaria proliferation under conditions of hypoxia. Invasion of such cells was was somewhat reduced. Sickling itself was not required [45, 102]. G6PD-deficient cells from males with either the African or the Mediterranean variety do not support falciparum growth as well as normal cells [11], although some groups have been unable to find any differences (see [100]). However, all groups agree that there is lessened proliferation of malarial organisms in G6PD heterozygote females, depending upon the number of deficient cells. It had already been shown with staining studies that falciparum parasites are less frequent in the deficient cells of G6PD heterozygotes whose red cells are a mixture of deficient and normal cells [89]. Somewhat low glutathione (GSH) levels are not further diminished in G6PD-deficient cells in culture, a somewhat surprising finding since GSH depletion was thought to be an important cause of the growth differential.

Heterozygotes for HbC and HbE appear to support malarial growth as well as normal cells, while the data from homozygotes for HbEE and HbCC are more equivocal. Most studies have failed to show any proliferation differentials for malaria parasites between normal and β thalassemia cells except with additional oxidative stress [112]. Among the α thalassemias only those with HbH disease, i.e., deletion of three α genes: α-/--, showed definitive decreased growth of parasites [62]. The data in those with two abnormal α genes were less consistent, and normal growth was seen in those with a single Hbα deletion (α-thal 2). HbF appears to inhibit proliferation of falciparum malaria [101]. However, the various conditions characterized by hereditary persistence of fetal hemoglobin with high fetal hemoglobin levels in heterozygotes have not become as frequent as might be expected from the in vitro findings. Ovalocytes resist entry of falciparum organisms in vitro, and ovalocytosis is common in certain areas of Papua New Guinea [70]. However, cells with a rare and widespread blood group En (a-) resist invasion of falciparum malarial organisms in vitro [103], but this blood group has not reached polymorphic frequencies in any population.

The various in vitro data suggest that a direct demonstration of diminished falciparum growth may be easy when the growth differential is relatively great, as with HbS heterozygotes. The occasionally observed proliferative differences in the somewhat rarer homozygotes for HbC and HbE and in HbH disease are immaterial for the population spread of these genes but show that only fairly large genetic differences can be demonstrated by such laboratory studies. With more subtle differences, as in HbE and HbC heterozygotes, the methods may not be sensitive enough to demonstrate small differences in growth. In any case, results of the in vitro methods are still too inconsistent to provide the decisive data that could prove or disprove the malarial hypothesis.

Ascertaining and Measuring of Selection in Humans. This section summarizes some research strategies for ascertaining and measuring selection in humans. Most of these strategies have been used for testing the malaria hypothesis for the sickle cell gene [93]:

a) There are obvious geographic correlations between endemicity of the selecting disease and the protective gene.

b) Affliction with severe forms of the selecting disease and thus mortality is less in heterozygotes of the protective gene than in normal homozygotes.

c) There could also be greater fertility in A/S heterozygotes than in normal homozygotes [92, 93].

d) Mortality differences are expected to lead to age stratification in the population. If the selecting disease selectively killed small children, there should be a relative increase in the frequency of the protective gene with age.

The most obvious way to measure selection in infancy is to compare gene frequencies in infants and in the adult population; this may be supplemented by comparing effective fertility in families (Tables 12.15–12.17). If selection intensity is as strong as with the sickle cell trait, this approach may be successful. However, in human genetics the expected selection intensities are generally much lower. For a recessive disease with 100% selection against the affected homozygotes and a selective disadvantage of 3% of normal homozygotes, the equilibrium value (Eq. 12.6) does not depend on gene frequency and may be calculated as follows:

$$\hat{q} = \frac{0.03}{0.03 + 1} = 0.0291; \quad q^2 = 0.00085$$

This implies a homozygote frequency of a little less than 1 : 1000 – a frequency somewhat higher than that of cystic fibrosis in the western European population. For rarer recessive genes the disadvantage of the normal homozygote (s_1) compared to the heterozygote must be much lower to maintain a balanced system (Table 12.18). An enormous sample size is required to verify a selective disadvantage of this order of magnitude (0.5%–3.0%). If the rarer allele is moderately frequent, i.e., if a genetic polymorphism exists, the selective disadvantage of the normal homozygote is expected to be higher, and the required sample sizes are of more reasonable size. Most studies on selection have been carried out with polymorphisms, such as blood groups. However, the results have proven ambiguous [131]. This is not surprising

Table 12.18. s_1 (selective disadvantage of the normal homozygote) needed for maintaining a balanced system if $s_2 = 1$ (complete selection against the abnormal homozygote) under genetic equilibrium (s_1 depends on the relative frequencies of affected, q^2, and normal homozygotes, p^2; see Sect. 12.2.1.3)

q	q^2	s_1	Example
0.0291	0.000847	0.03	Cystic fibrosis
0.0109	0.000118	0.011	Phenylketonuria
0.00498	0.0000025	0.005	Galactosemia

since present-day gene frequencies reflect selection processes of the past or may not be caused by selection at all. Infant mortality in general has been dramatically reduced. If the studies described in Tables 12.15–12.17 were repeated today in malaria-free areas, it is unlikely that childhood mortality differentials and difference in the number of surviving children would be observed. Therefore, investigations of fertility and mortality differentials in human populations that are conducted to assess selection may be impossible for practical reasons. It is probably misleading to use present-day results for conclusions regarding probable selection in the past when entirely different environmental conditions prevailed.

e) Difficulties of this type are inherent in most present studies on selection and can be partially circumvented by direct examination of a putative selective mechanism. However, using this approach requires a specific hypothesis regarding such a mechanism. For the malaria hypothesis this was not too difficult. The geographic distribution of the sickle cell gene showed striking similarity with the distribution of falciparum malaria, and the *Plasmodium* organism was known specifically to attack red blood cells.

In general, formulation of such a causal hypothesis requires knowledge of the physiological function of the gene concerned. Once a reasonable hypothesis for a mechanism is available, the task of testing for selection is simpler. It has even been asserted that no case of balanced polymorphism in any species has ever been discovered without knowledge, or at least a plausible hypothesis, of the biological mechanisms by which selection works (B. Clarke, personal communication).

12.2.1.7 Natural Selection and Population History: HbE and β Thalassemia

(This discussion is not absolutely necessary for understanding further sections.)

It has often been discussed how genetic data, for example, gene frequencies of genetic polymorphisms, can be used to derive conclusions as to population history and population affinities. The following, well-analyzed example demonstrates how various techniques of population genetics and data from history and linguistics may be combined to answer such questions. This example is also used to examine the problem of two different alleles under selection and to demonstrate the value of computer simulation of population processes for population genetic analysis.

Interaction of Two Abnormal Hemoglobin Genes in a Population. In one of his predictions Allison concluded that in regions where two genes for abnormal hemoglobins coexist and interact in such a way that individuals possessing both genes are at a disadvantage, these genes will tend to be mutually exclusive in populations [5]. This problem is illustrated by the interaction of HbE and thalassemia genes in southeastern Asia (Flatz [40]). The homozygous state of thalassemia has been described in Sect. 7.3. The anemia in HbE disease is much milder than that in HbS homozygotes. β Thalassemia major (Cooley anemia) is associated with severely reduced hemoglobin synthesis. Most compound heterozygotes for both β thalassemia and HbE (thalassemia-HbE disease) suffer from a marked chronic anemia approaching the severity of β thalassemia major. The genes for the Hbβ chain variants such as HbE and the β thalassemias are so closely linked (Sect. 7.3.4) that when they occur *trans*-position, they can be treated as alleles.

Distribution of HbE and Thalassemia. The distribution of HbE in southeastern Asia has a center of maximal frequency in the Khmer-speaking population of northern Cambodia and the adjacent areas of northeastern Thailand; here the gene frequency may reach 0.3, corresponding to a heterozygote frequency of 42%, one of the highest hemoglobinopathy frequencies ever attained. In other parts of Thailand, on the Malaysian peninsula, and in Indonesia, the frequency is much lower. HbE has also been found in China, Assam, and in Bengal (Fig. 12.21). The total number of carriers of this gene may be around 20 million. β thalassemia alleles occur in the same areas; they are, however, much more widespread.

HbE and Malaria. Once the relationship between HbS and falciparum malaria had been worked out, a similar mechanism for maintenance of other hemoglobin polymorphisms was considered plausible. Attempts to test this hypothesis directly met with difficulties such as the problems of field study in populations not benefiting from public health and medical supervision, as well as the presence of other protective genetic mechanisms such as G6PD deficiency and thalassemia in the same populations. A protective effect of the HbE allele in heterozygotes and homozygotes was, however, strongly suggested. Comparisons between the geographical distributions of HbE and malaria had to take into account that in mainland southeastern Asia the main vector is *Anopheles minimus,* a forest mosquito that is abundant in hilly and mountainous areas. This fact causes the distribution of malaria to be opposite that of Mediterranean countries, where swamp and brackish water mosquitos are most important. In southeastern Asia malaria is a disease of the hills and forests. Indeed, it is in these areas that the frequency of HbE tends to be highest [41].

Fitness of the Genotypes Involving HbE and Thalassemia: Problem of a Genetic Equilibrium. What are the conditions for change in gene frequencies and for genetic equilibria in such a system of three alleles ($Hb\beta A$, $Hb\beta E$, $Hb\beta T$; T is thalassemia)? To answer this question the fitness values (i.e., the selective advantages or disadvantages) of the various genotypes must be calculated. On the basis of gene frequencies in the nuclear Khmer group and the clinical manifestations in HbE homozygotes, the following values were derived:

$Hb\beta E/Hb\beta E$: $w_{EE} = 0.7$ to 0.8
$Hb\beta E/Hb\beta A$: $w_{AE} = 1.05$ to 1.2
$Hb\beta A/Hb\beta A$: $w_{AA} = 0.9$ to 0.95

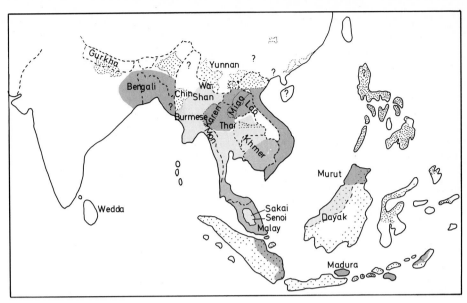

Fig. 12.21. Populations in which HbβE has been observed. *Shaded areas,* high frequency; *lightly shaded areas,* moderate frequency; *stippled areas,* occasional observations. (From Flatz 1967 [40])

HbβT/HbβT: $w_{TT} = 0$
HbβT/HbβA: $w_{AT} = 1.05$ to 1.2
HbβT/HbβE: $w_{ET} = 0.2$ to 0.5

where w is the fitness of a certain genotype compared with average fitness in the population.

Is a stable genetic equilibrium under these conditions possible? Contrary to the two-allele system described in Eq. (12.6), selective advantage of heterozygotes in a triallelic system does not necessarily lead to a stable genetic equilibrium. Such equilibrium can be established only if certain conditions are fulfilled [104]. In this example, HbE and β thalassemia in Southeastern Asia, they are only partially met; the selective disadvantage of the compound heterozygote, Hbβ T/HbβE, is too severe to permit a stable equilibrium.

What do we expect of the distribution of HbβE and HbβT gene frequencies, q_E and q_T, in various population subgroups when there is a stable or semistable equilibrium, as compared with an unstable equilibrium or no equilibrium? A stable equilibrium would result in a clustering of the distribution points, which represents the HbE and β-thalassemia (p_E, q_T) in a two dimensional coordinate system around a certain equilibrium point. If the equilibrium is only semistable, the clustering effect is lessened: After a disturbance of the equilibrium the distribution points do not necessarily return to the same equilibrium point as before but – as Penrose et al. (1956) [104] have shown – to some point lying on the straight line connecting the "unopposed" equilibrium points of HbβE and HbβT (where only one of the alleles exists).

The distribution actually found [40] suggests that no stable or semistable equilibrium obtains. This means that the two alleles HbβE and HbβT tend to reduce one another below equilibrium frequency. This reciprocal effect is caused by the strong selective disadvantage of the compound heterozygote.

Population Dynamics of HbβE and HbβT. If the population is not at equilibrium, how quickly and in which direction are the gene frequencies expected to change? Or – if we look at the problem from the point of view of population history – how was the present distribution of gene frequencies attained?

In the sections above, formulas for change in gene frequency from one generation to the next (Dq) were derived for special cases. In a similar manner it is possible to derive relevant equations to deal with the change in gene frequencies between generations for a three-allele situation. In this way the speed of change in gene frequencies under different selection pressures can be examined.

In Fig. 12.22, for example, gene E is introduced into a population with high T frequency. Both A/E and A/T heterozygotes have a high selective advantage; gene E replaces gene T. If the selective advantage of the A/E heterozygotes is lower, there exists a critical value of this advantage below which gene E can no longer replace gene T. Even if gene E were able to replace T, the speed of elimination would depend strongly on the initial frequency of E. There are also situations, however, in which E cannot replace T, or might even be replaced by T, if T is introduced anew in the population. When both genes are introduced into a population for the first time, they both rise about identically, but after a certain time, E will probably rise to equilibrium frequencies while T drops out.

Selection Relaxation. Malaria may be eradicated in southeastern Asia in the future. The selective advantage of A/T and A/E heterozygotes would then no longer exist, but the disadvantages of E/E and T/T homozygotes as well as of E/T compound heterozygotes would still prevail. What would be the consequences for the frequencies of the E and T genes? Figure 12.23 shows two situations, the first with very high E and very low T frequencies (as in northeastern Thailand), the second with more similar E and T frequencies (as in central Thailand). In both cases the decline of each gene will be fairly rapid, especially at the beginning.

Gene frequency

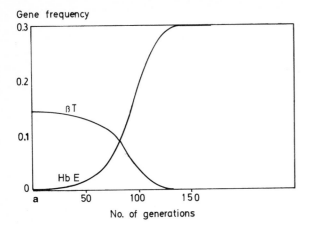

a   No. of generations

Gene frequency

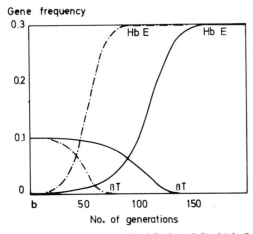

b   No. of generations

Fig. 12.22. a Elimination of HbβT by HbβE; high fitness of HbβA/E and HbβA/T heterozygotes; simulation of conditions with high selective pressure. Assumed fitness values: A/A 1.0; A/E 1.225; A/T 1.2; E/E 0.7; E/T 0.25; T/-T 0. Note that HbE replaces thalassemia. **b** Elimination of HbβT by HbβE; HbβA/T fitness lower than **a**. Fitness values: A/A 1.0; A/E 1.15; A/T 1.125; E/E 0.8; E/T 0.25; T/-T 0.0 –·–·–· A/A 1.0; A/E 1.225; A/T 1.125; E/E 0.7; E/T 0.25; T/T 0.0. (From Flatz 1967 [40])

Implications of These Results for the Population History of Southeastern Asia. Today's gene frequencies are the result not only of selection pressures but also of the demographic history of populations. More specifically, if an ecological situation has led to different means of adapting in two different populations – for example, adapting to malaria in one population by HbβE and in the other by thalassemia – and if the two mechanisms are to a certain degree mutually exclusive, a comparison of gene frequencies with the known facts of population history may give some clues as to the genetic relationship of these populations.

In southeastern Asia there is little evidence of a Neolithic culture comparable to that found in Europe. Three stages of social development and ecological situations can be discerned:

1. Hunter-gatherer groups. Judging from similar groups that exist today, the habitat of these people was in forested areas in the hills and mountains. If it is assumed that the distri-

bution of malaria was similar to current conditions, these hunter-gatherer populations must have been exposed to intense malaria pressure. Nevertheless, conditions for diffusion of a gene carrying heterozygote advantage were unfavorable because of the small size of the breeding population and the few possibilities for gene diffusion among them; the ultimate fate of even favorable mutations would in most cases be extinction.

2. At about 1000 b.c., with the introduction of rice cultivation in irrigated fields, a social organization at the village and district level appeared. Most of the known settlements of this area were located at the margin of valleys. In such a society the conditions for diffusion of a gene maintained by protection against malaria were most favorable. The time available for the HbβT and HbβE genes to become frequent (about 3000 years or 120 generations, assuming a generation time of about 25 years) was sufficient to attain the present-day frequencies as evidenced by Fig. 12.24.

3. At present the majority of the population in most southeastern Asian countries reside in the great river basins and deltas, which were generally uninhabitable in prehistoric times; social and political development since has permitted organized cultivation in the lowlands and led to the continual migration of persons into the plains. Due to the special ecological requirements of the vector, *Anopheles minimus,* malaria is rare in the plains. Therefore the move to the relatively malaria-free plains is believed to have caused a considerable relaxation of selection against the HbβA homozygote and a diminishing advantage of the HbβE and HbβT heterozygotes. The two genes HbβT and HbβE are indeed less frequent in the plains than in the adjacent hill areas.

Comparison with HbβS in Western Africa. In western Africa the principal malaria vectors are mosquitoes requiring open spaces and stagnant water for their propagation. The sickle cell gene was probably introduced to western Africa in the Neolithic Period concomitantly with improved agricultural methods (slash and burn farming). This development opened wide spaces for the malaria mosquitoes, leading to high endemicity, which in turn set the stage for the spread of the sickle cell gene and establishment of its polymorphism.

Similar trends with the introduction of agriculture in two different populations in the presence of different ecological requirements of the mosquitoes led to the establishment of the HbβS polymorphism in the plains in Africa and of the HbβE polymorphism in the hills of southeastern Asia.

Hemoglobin βE in the Austroasiatic (Mon-Khmer) Language Group. The Austroasiatic language group now comprises the Khmer (Cambodia), tribal languages in Vietnam, Mon in lower Burma, western and northern Thailand, tribal languages in Thailand, Burma, and southern China, and several languages in Assam and Bengal. Historical and linguistic evidence suggests that the entire area of mainland southeastern Asia, with the exception of southern Malaysia and parts of Vietnam, was inhabited by Austroasiatic people until the fifth or sixth century AD when large-scale migration began.

Figure 12.25 compares the areas of Austroasiatic languages, past and present, with the areas in which the HbβE gene is polymorphic. The congruence is evident; the most likely explanation is a concomitant diffusion process: HbβE may have emerged in an original Austroasiatic group, and both HbβE and Austroasiatic language and culture gradually dif-

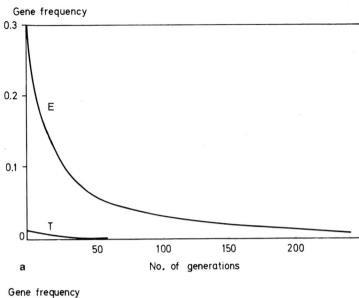

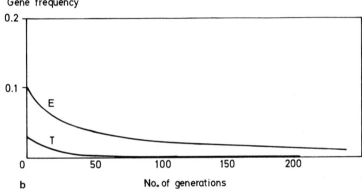

Fig. 12.23. a Simulation of conditions in northeastern Thailand with assumed selection relaxation. A/A = A/E = A/ T = 1.0; E/E 0.7; E/T 0.25; T/T 0. **b** Selection relaxation under conditions found in river basin areas (e.g., central Thailand). A/A 1.0; A/E 1.0; A/ T = 1.0; E/E 0.7; E/T 0.25; T/T 0

fused throughout mainland southeastern Asia. The dynamic models (Fig. 12.22) indicate that this diffusion could have occurred in a population with preexisting high thalassemia frequency, as the HbβE gene replaces thalassemia genes under many selective conditions.

As noted, other austroasiatic groups have migrated into other parts of South Asia. In such groups, HbE is expected to occur, – especially if their present habitat was Malaria-infested. In the Khasi of Assam, and Austroasiatic group, this prediction was confirmed. But a Mongoloid group, the Ahom, also showed high HbE frequency; this group had immigrated to Assam from Thailand – and apparently had picked up the gene HbβE there.

Some General Conclusions from the Studies on HbE and Thalassemia. The studies on HbE and thalassemia in southeastern Asia lead to more general conclusions for the interpretation of population differences in gene frequencies. They show how these differences in gene frequencies may be determined either by population history or natural section.

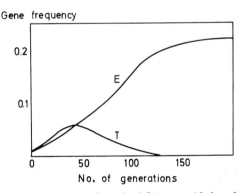

Fig. 12.24. Both HbβE and HbβT start with low frequency; rise of HbβE to equilibrium frequency, initial rise and subsequent elimination of HbβT. A/A 1.0; A/E 1.0; A/T 1.125; E/E 0.75; E/T 0.25; T/T 0. (From Flatz 1967 [40])

Comparing populations exposed to the same relevant ecological agent – in this case malaria – we find a definite genetic difference caused by different population histories. In one population adaptation to the agent was achieved by the HbβE gene and in the oth-

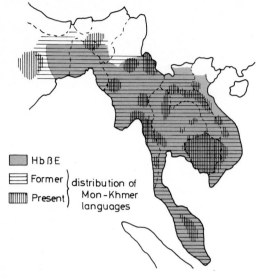

Fig. 12.25. Polymorphic distribution of HbβE in mainland southeastern Asia ▨; former ⊟ and present ⫿⫿⫿ distribution of Austroasiatic (Mon-Khmer) languages. (From Flatz 1967 [40])

ers by β thalassemia genes. Both adaptations were shown to be mutually exclusive up to a certain degree. Since Hb E/E homozygotes are less severely affected than the Hb T/T homozygotes, adaptation through the HbβE gene offers antimalaria protection at a lower price and tends to replace adaptation afforded by the β thalassemia gene. These results will be obtained in the long run in spite of a counteracting force, the partial mutual exclusiveness of the two genes that is caused by strong selection against the compound heterozygote.

Comparing populations that have been exposed to malaria in different degrees, we also find definite genetic differences. In populations of the hilly areas HbβE and thalassemia tend to be more frequent than in the plains, where there has been less malaria. This is expected but provides little information about the ethnic relationship between populations of hills and plains. Resolution of this ambiguity required knowledge of the selective agent – in this case, malaria.

Refinement of Analysis by Study of DNA Polymorphisms [7, 61]. Studies of DNA polymorphisms in the area of the Hbβ gene that carries the mutations for both HbE β thalassemia have led to a refinement of these conclusions, but posing at the same time a new problem regarding the origin of the HbβE mutant. DNA polymorphisms may occur not only in the neighborhood of a gene but also within a gene itself, for example, in introns or even in the coding region. Identical mutations may occur within different haplo-

types. The HbβE mutant has been observed in two different haplotypes.

Haplotypes were identified unambiguously in most subjects either because they were homozygous in all sites, or they were heterozygous in one site only, or by appropriate family studies. The following conclusions are based on a combination of all available data [7, 61, 97]:

1. HbβE is observed in Southeast Asia in two within-gene haplotypes: no. 2, and no. 3 Asian.
2. These within-gene haplotypes occur in combination with gene restriction sites outside forming no less than 11 different haplotypes.
3. Distribution of these haplotypes differs strongly between individuals with the HbβE gene and normal homozygotes: HbβA occurs in more than 80 % in one haplotype, whereas HbβE is combined preferentially with another.
4. Within-gene haplotype no. 2 is observed in HbβE subjects from northern and northeastern Thailand, whereas type 3 Asian occurs exclusively among the Khmer-speaking population group of Cambodia.

Occurrence of various rarer haplotypes in addition to the above common ones can be explained easily by simple or, in some instances, repeated crossing over. However, the occurrence of HbβE mutants in two different within-gene haplotypes found in different population groups cannot be explained easily. Again, as in the sickle cell case, where independent HbβS mutations in no less than four adjacent populations were postulated (see above), the obvious – and most conventional – explanation seemed to be that the mutation had occurred independently in the two groups; after observation of HbβE in Europeans even a third, independent origin has been tentatively postulated [68]. In distinction to the HbβS mutation, which is a rare transversion, the HbβE mutation (position 26; Glu → Lys) is a C   T transition and thus is one of

$$\frac{|||}{G} \rightarrow \frac{||}{A}$$

the more common types of mutations (Sect. 9.4). However, to assume that such an unlikely coincidence – two or more independent, identical mutations not only occurring but also establishing themselves in adjacent population groups and, in southern Asia, even within the same haplotype, and that this happened twice, in Africa and Asia – is too unlikely a coincidence. Alternative explanations must be sought. Repeated, conventional crossing over appears to be very unlikely [7]. In our opinion, it is again much more likely that all HbβE mutants derive from only one mutational event, but that, in addition, gene conversion between different haplotypes within the gene occurred. This alternative had also been envisaged by the original investigators [61, 68].

Studies on Sickle Cell Polymorphism in Africa: A Stochastic Model for Replacement of One Allele by Another. Similar studies have analyzed the population history of western Africa together with exposure to malaria and the frequencies of the genes for HbβS and HbβC. The situation is similar to that encountered for HbβE and thalassemia in southeastern Asia: there are two alleles, HbβS and HbβC, offering protection against malaria – with different fitness values w_i of homozygotes and heterozygotes and with strong selection against the double heterozygote. Figure 12.26 a shows changes in gene frequencies concomitant with replacement of allele HbβC by HbβS, the decisive factor here being the higher selective advantage of the heterozygote for HbβS compared with the heterozygote for HbβC. The selection model is deterministic, as the model used for HbβE and thalassemia. Population size is assumed to be infinite. In Fig. 12.26 b, on the other hand, the size of the effective breeding population (Sect. 13.3.1) is assumed to be 1000, and the resulting chance fluctuations are allowed for. This model is stochastic. The general tendency is the same as in Fig. 12.26 a; however, chance fluctuations are obvious.

α⁺ Thalassemia and Malaria in Melanesia. It is very likely that selection due to malaria has also been the decisive factor causing the high incidence of various types of thalassemia in many tropical and subtropical areas. A recent study on α⁺ thalassemia in Melanesia has provided new and, in our opinion, convincing evidence [43, 44]. As a rule, this type of thalassemia results from the loss of a single α-globin gene, for example, by unequal crossing over (Sects. 7.3; 5.2.8). Quite a few such deletions can be found; even identical mutants may occur within different haplotypes. In Melanesia and, especially, in New Guinea, the population is divided into many small, genetically isolated subgroups. There is a strong correlation with former infection rates by malaria; villages at mountain sites, where malaria infections were low according to statistics from preeradication times, show low frequencies of mutant genes, whereas severely infected areas in the plains display high – sometimes very high – gene frequencies. When islands in Melanesia are compared, the same correlations are found. Some islands in which, according to reliable evidence, malaria had never been present, show a certain number of mutant genes. However, since the population history is fairly well known, this could easily be explained by migration. In addition to the conventional explanation for identical mutants found in different haplo-

types – independent mutation – the authors also discuss gene conversion [44].

In western Africa, as in southeastern Asia, research on abnormal hemoglobins has contributed to our knowledge of population history. There are other areas in the world in which such studies could help in a similar way – in which population history is complicated and further studies of G6PD and hemoglobin variants are required. One such area is India, especially in the south and east.

Several types of G6PD deficiency coexist in polymorphic frequencies in the Philippines and in Thailand, unlike the presence of only two principal G6PD variants (A^- and A^+) in African populations. Presumably, population mixture of groups who originally carried only a single G6PD mutation brought about this situation. More studies on the relevant variants and on population history in these parts of Asia are required for a detailed analysis of this problem.

12.2.1.8 Selection in the ABO Blood Group System and in Other Polymorphisms

ABO Blood Groups and Disease. No other human example could be analyzed as thoroughly as the interactin between hemoglobin and G6PD variants and malaria. However, it may be useful to discuss some difficulties in arriving at a clear picture in another, much more complicated and controversial example: the ABO blood groups. As noted above, one aspect of selection in this system is generally accepted, although there is no agreement as to the extent of selection: serological mother-child incompatibility. However, such incompatibility leads to an unstable equilibrium and to slow changes in gene frequencies (Sect. 12.2.1.4). In the absence of other modes of selection the polymorphism would slowly disappear. Contrary to this prediction, the ABO polymorphism is present in almost all human populations. This finding suggests other selective factors. Do we have positive evidence for such selection?

ABO Blood Groups and Infectious Disease. Widespread disease associations have been reported for the ABO blood groups (Sect. 6.6.2). For example, carriers of type A are more susceptible to a number of malignant tumors and some other diseases, whereas those with type 0 show a higher susceptibility to gastric and duodenal ulcers. Moreover, rheumatic fever for which immune mechanisms are undisputed [131] is also associated with blood group; the risk of being affected is lower for group 0 than for A, B, or AB. While possibly leading to a higher average chance of group 0 carriers to survive to a more advanced age, these associations presumably had little if any influence on natural selection, as most of them affect individuals of middle and older age, i.e., after reproduction. However, they show a fundamental influence of the ABO anti-

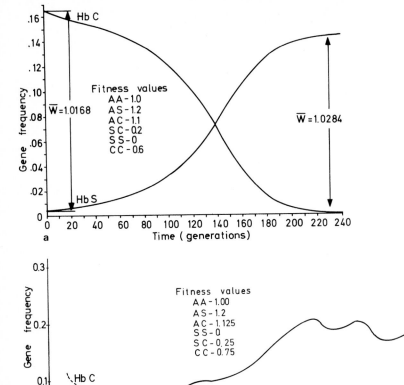

Fig. 12.26. a Replacement of HbβC by HbβS due to higher selective advantage of the heterozygote HbβA/S as compared with the heterozygote HbβA/C. **b** Computer simulation of the model shown in Fig. 12.23a, but assuming an effective breeding population of $N = 1000$ and allowing for chance fluctuation of gene frequencies. (Note slight differences in fitness of AC, SC and CC as compared with 12.23a). (Adapted from Livingstone 1983 [73])

gens on body physiology. More specifically, the data on rheumatic fever suggest that this influence may have something to do with the immune response. Even the associations of A with cancer and of 0 with peptic ulcers may be due to differences in immunological response.

Infectious diseases specifically challenge the immune response of the organism. If this response is influenced by the AB0 blood group, selection by differential susceptibility to infections could cause differential mortality in childhood and youth.

Distribution of AB0 Alleles in the World Population. Figure 12.27 shows the distribution of the alleles A, B, and 0 [94]. This distribution suggests an influence of natural selection. Had it been caused by chance fluctuation of gene frequencies, all possible combinations of gene frequencies of the three alleles should have appeared. This, however, is not the case. Only a limited number of the possible combinations are observed [17].

Certain clues regarding the kind of selection can be derived from the distribution of allele 0. This allele

is usually frequent in populations that have lived for a long time in relative isolation, such as aborigines of Australia and Polynesia, the Arctic, and northern Siberia. Also within Europe certain isolated population groups usually have high frequencies, for example e.g., Irish, Basques, Icelanders, Corsicans, Sardinians, and those in the Valais district of Switzerland. An especially high frequency of allele 0 is found among the Indians of Central and South America, setting them apart from other populations. The differences in other polymorphisms, for example Rh, argue against the hypothesis that all these areas were once inhabited by a homogeneous population with high 0 frequency. The data are suggestive of natural selection. What kind of selection might have led to an increase in gene frequencies in relatively isolated areas – or, conversely, to a decrease in areas in the mainstream of world "traffic"? Plausible candidates include infectious diseases, especially the great epidemics of the past.

In Sect. 12.2.1.6 the following groups of infectious diseases have been mentioned as possibly important for natural selection:

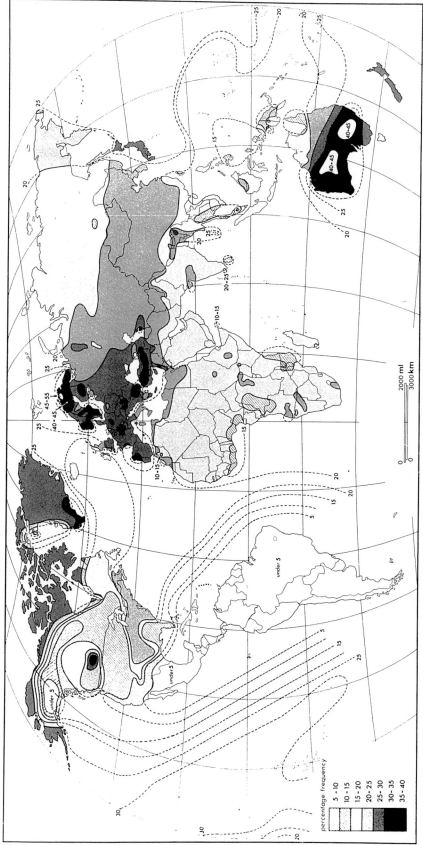

Fig. 12.27. a Frequency distribution of allele A in the aboriginal populations of the world. (From Mourant et al. 1976 [94])

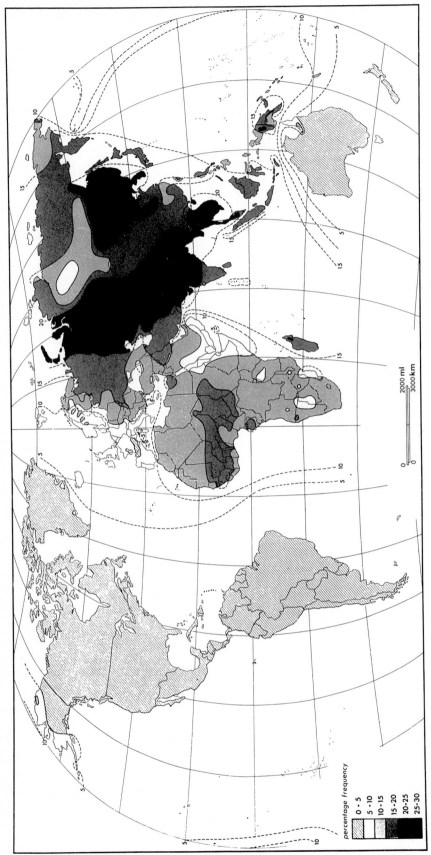

Fig. 12.27. b Frequency distribution of allele B in the aboriginal populations of the world. (From Mourant et al. 1976 [94])

percentage frequency

0 - 5
5 - 10
10 - 15
15 - 20
20 - 25
25 - 30

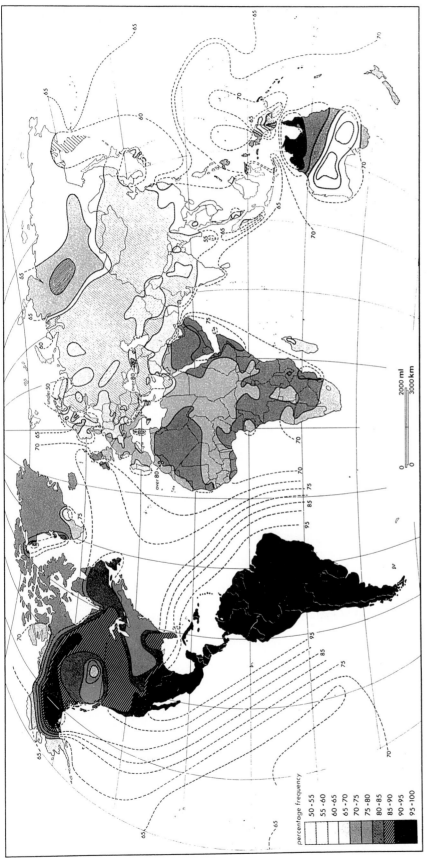

Fig. 12.27.c Frequency distribution of allele 0 in the aboriginal populations of the world. (From Mourant et al. 1976 [94])

a) Recurrent epidemics such as plague, cholera, and smallpox.
b) Chronic infections, for example, tuberculosis and syphilis.
c) Intestinal infections mainly in children.
d) Tropical diseases of children and young adults.

Unlike the polymorphic hemoglobin variants, which are confined to tropical countries, the AB0 polymorphism is found all over the world. Therefore tropical diseases are unlikely to play a major role in selection. Three aspects could be used for constructing a testable hypothesis:

1. The population of Central and South America was almost completely isolated before Columbus' arrival and may have had a special group of infections not shared by the rest of the world population, e. g., syphilis and, possibly, related treponemal diseases. Blood group 0 is extremely frequent in these populations. Is there any evidence for an advantage of group 0 toward infection with *Treponema pallidum*, the microorganism causing syphilis?
2. Plague repeatedly devastated Europe, mainly affecting densely populated areas. Marginal and partially isolated populations that may have been less affected usually show high group 0 frequencies. Is there any evidence for a disadvantage of blood group 0 in coping with the plague bacillus?
3. Smallpox was not eradicated until the mid 1970s. Up to that time it had been frequent in many countries. Therefore modern statistics as to frequency and death rate are available, especially for Africa and the Indian subcontinent. AB0 distributions are fairly well known for these areas. The blood group with the higher susceptibility should be rarer in areas with high smallpox rates. Is there any evidence for this prediction?

Syphilis and Blood Group 0. We are faced with the problem of judging the influence of an infection on differential biological fitness in a population before 1492. However, just as in the example of sickle cell protection vis-à-vis malaria (Sect. 12.2.1.6), the most convincing indirect evidence would be that possession of group 0 conveys an advantage for coping with this infection. Such a hypothesis can no longer be tested since syphilis is treated with penicillin so successfully that individual differences in outcome due to different immune responses no longer apply. In the 1920s, however, penicillin therapy was not available, and at that time comprehensive data on blood groups and syphilis were collected and analyzed, with the following results [131]:

a) There was no association between the risk of new syphilis infection and AB0 blood groups.

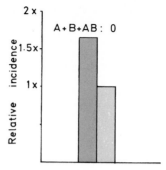

Fig. 12.28. Relative incidence of tertiary syphilis in relation to the AB0 blood groups. (From Vogel and Helmbold 1972 [135])

b) However, after the customary therapy at that time – neosalvarsan – individuals with group 0 had a much better chance of becoming seronegative than did those with the other blood groups.
c) Tertiary syphilis, such as general paralysis, was less frequent in blood group 0 than in the other AB0 blood groups (Fig. 12.28).

Hence, the combined data suggest an advantage of group 0 in the immune response to syphilis. Thus, the prediction derived from the hypothesis can be verified. If syphilis influenced reproduction, this effect resulted mainly through infection of the fetus by a syphilitic mother. Such an infection is known frequently to lead to late fetal death. The evidence thus allows the tentative hypothesis that the high frequency of group 0 in Central and South American Indians is due to selection by syphilis and related treponema infections *if* these infections did come from America, *if* they were widespread in earlier times, *if* they did influence survival of children or fertility of their mothers, and *if* no other selective agents were responsible.

Cholera and Blood Group 0 [48]. A clear association of AB0 type and a lethal endemic infectious disease has come from recent large-scale studies dealing with cholera in Bangladesh. While patients with diarrhea due to rotavirus, *Shigella*, toxicogenic *E. coli,* or nontoxicogenic cholera had blood group 0 frequencies similar to controls (about 30%), patients infected with toxicogenic vibrio cholerae had a blood group 0 frequency of 57%. This difference was statistically highly significant. Among family members infected with the toxicogenic cholera strain there was a statistically significant tendency for the frequency of group 0 to increase with increasing severity of diarrhea. Severe diarrheal epidemics had been described in the past in this area, and were most probably caused by cholera. The low frequency of blood group 0 in this region may have been induced by a higher susceptibility to cholera, with resulting death of blood group 0 carriers. The mechanism of the interaction remains obscure.

Plague and Blood Group 0. Was plague related to the distribution of Group 0 in Europe? In distinction to syphilis and cholera, no blood group data on plague are available. The disease is very rare now; it occurs principally in areas inaccessible to research workers. Therefore indirect evidence must be examined.

In Sect. 12.2.1.5 on frequency-dependent selection it is mentioned that parasites may adapt to their hosts by producing surface antigens common to those of the host, thereby deceiving the host's immune response. Evidence of this has been collected for vertebrates and their parasites [21, 27, 28]. It has been known since the late 1950s that humans share ABH-like antigens with a great number of bacteria, especially those of the *E. coli* group. Even the "normal" anti-A and anti-B isoantibodies are thought to be immune antibodies against ubiquitous infections with intestinal germs. It was therefore of interest to investigate whether *Pasteurella pestis* had ABH-like antigens [105, 106]. H antigen, which is much more abundant in the human group 0 than in other blood groups was indeed detected. This finding suggests a poorer immune response against *P. pestis* in group 0 individuals and agrees with the assumption that 0 individuals may have had a selective disadvantage due to a higher death rate from plague. Needless to say, this finding does not prove the hypothesis.

Does a Common Blood Group Antigen of the Microorganism Impair the Immune Reaction of the Host? E. coli and Infectious Diarrheas. The next step in testing this hypothesis would be to examine whether communality of a blood group antigen impairs the immune response of the human host. As noted, such a study cannot be carried out with plague since no plague patients are available for examination. However, this approach has become possible for another group of diseases that are caused by bacteria whose ability to synthesize ABH antigens is well known – the *E. coli* group. In the 1950s and early 1960s central Europe was swept by a succession of waves of infective infant diarrhea. The causative organisms were identified as *E. coli,* which by serological examination of their antigen profiles could be subdivided into various substrains. The outcome, unlike in earlier times, was rarely fatal because of therapy with antibiotics, plasma and fluid infusions.

In the early 1960s the Austrian pediatrician Kircher [71, 72], observed a more severe course of infant diarrhea in patients of group A than in those of other blood groups. This topic was restudied, drawing on comprehensive data that had been collected over many years and could be compared with suitable controls [138]. Definite heterogeneity was found; in some years A patients were frequently affected and in others 0 carriers were common (Fig. 12.29). From the case histories a number of clinical criteria for severity of the disease were elaborated, for example, infusions given to the more severely afflicted infants (Table 12.19). In the years with higher incidence for group A there was a more severe course of the disease. In the years in which group 0 had a higher inci-

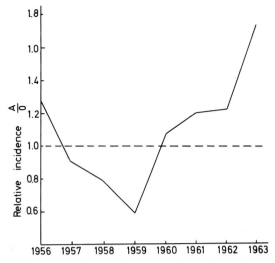

Fig. 12.29. Relative incidence of blood group A as compared with 0 among 1200 children with infant diarrhea in Heidelberg, Germany 1956–1963 (From Vogel and Helmbold 1972 [135]

Table 12.19. Course of infant diarrhea due to *E. coli* in 396 patients (from Vogel et al. 1964 [138])

Variable compared	1956 and 1960–1963		1957–1959	
	A	0	A	0
Patients with plasma infusions (%)	56.6	39.4	41.9	52.3
Loss of body weight (g)	161.0	137.1	154.0	158.6
Average frequency of stools	6.10	5.42	5.95	6.0
Highest body temperature (°C)	38.40	38.18	38.51	38.60
Time in hospital (days)	26.54	26.13	23.67	28.10
Gain of body weight (g)	577.8	549.0	506.3	585.6

dence infants of this blood group were somewhat more severely affected. This tendency was especially pronounced in that part of the data in which specific *E. coli* strains were identified.

Studies on various serologically identifiable *E. coli* strains during these years suggested that the observable differences were probably related to corresponding variations in strains of E. coli.

Blood group associations with infant diarrhea have been shown in other series as well [110, 121], and the antibody titers against *E. coli* 086 were found to be higher in persons of groups A, B, and AB than in 0, indicating that these individuals had more serious infections [34] (Fig. 12.30). The involved *E. coli* strain is known to have B and also A antigen. In view of the specificity of these associations, it is not surprising that some association studies of *E. coli* diarrhea and AB0 blood types also had negative results [129].

These results make it likely that the proposed mechanism – antigen communality between parasite and host – may indeed lead to a more severe infection if the host is human and the antigen is part of the ABH system. By analogy, the H antigen of the plague bacillus could have led to more severe disease among carriers of group 0. Thus, selection against this allele would occur.

One of the AB0 blood group associations that was discovered as early as the 1950s and has been reconfirmed abundantly in the following years was that between group 0 and peptic ulcer (Sect. 6.2.2.1). However, the pathogenetic mechanism eluded all attempts at understanding. Much more recently, however, it has been discovered that a micro-organism, *Helicobacter pylori*, is involved in pathogenesis of gastric and duodenal ulcer; this suggested that this organism is responsible for the blood group association. This hypothesis has been confirmed [15]. The bacterium attaches to Lewis receptors at the surface of mucosa cells; these receptors are diminished in blood groups A as B compared with group 0.

AB0 Blood Groups and Smallpox. The question of a possible blood group association with smallpox is still more controversial than the associations discussed above but is cited here because the experimental rationale may be used as a model for future research on interactions between a virus and a human host.

After the hypothesis was elaborated that the human AB0 distribution is related to great epidemics, and that antigen communalities are the decisive variable, vaccinia virus was examined for ABH activity. For technical reasons these investigations were carried out not with the smallpox (variola) but with the closely related vaccinia virus. Strong A activity was found [105, 137]. This suggested an obvious immunological mechanism for a disease association. During viremia with smallpox a virus having A antigen is partially inactivated by the anti-A antibodies present only in those of group B or 0 but not in A and AB individuals who lack anti-A. Therefore a more severe couse of smallpox is expected in patients with groups A and AB. As smallpox affects children and is often fatal, such a blood group difference would have a strong impact on selection.

The result on A-like antigen in the virus has been challenged [54]. The A antigen was said to be derived not from the virus but from the medium on which the virus was grown. At the time of these studies the possibility that viruses could take up material of the host into their own capsids was still unknown. Evidence for such a mechanism has now been put forward to explain the different clinical reactions to infec-

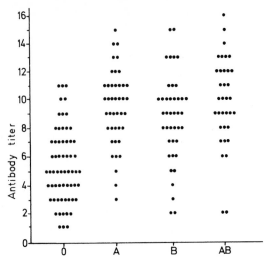

Fig. 12.30. Antibody titers in the AB0 system against *E. coli* 086 in individuals with various blood groups. (From Eichner et al. 1963 [34])

tions with the hepatitis B virus. This virus appears to take up serum proteins – especially γ-globulins – and to transfer them to the new host, whose immune reaction may depend at least in part on the similarity of these components with his own genetically determined proteins.

Association Studies on Smallpox Patients Yield Contradictory Results. The prediction that individuals of groups A and AB would be more frequently and severely afflicted with smallpox has been repeatedly tested, with conflicting results. One study of 986 fresh and former smallpox cases showed the relative incidence was shown to be much higher in persons of groups A or AB than in those with B or 0 [34]. The same tendency was apparent when severity of clinical symptoms and mortality were considered [134] (Fig. 12.31). Furthermore, individuals who had survived earlier smallpox epidemics showed a slight excess of groups B and 0 – an indication of the high death rates of A and AB. Among the survivors severe scarring was, again, more frequent in groups A and AB. This study was carried out in Indian villages during a smallpox epidemic; the sibs of the affected probands who remained healthy despite similar exposure to the infection were used as controls. Almost none of the individuals, affected or healthy, had ever been vaccinated. This research design maximizes differences in blood group distribution between patients and controls, while reducing possible errors due to population stratification as much as possible.

No study with such a design has ever again been carried out. However, two studies on hospital populations in Indian cities [31, 124] and one study from Brazil with a milder variety of smallpox [74] failed to confirm this association. There is one likely explanation of these discrepancies: The study on smallpox in Indian villages [134] was performed mainly in children. No information on the age distribution of the patients was published for the two Indian hospital studies, but background information suggests that most patients may have been adults. As noted (Sect. 12.2.1.6), the selective advantage of HbS heterozygotes is present only in young children; surviving adults are highly immunized irrespective of their HbS

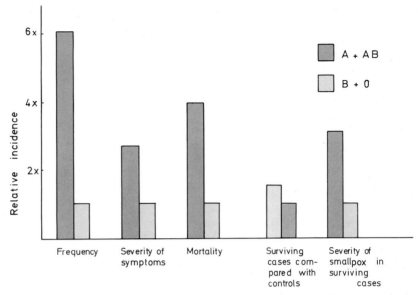

Fig. 12.31. Relative incidence of smallpox in blood group A + AB and B + 0 patients. From left to right: incidence (437 fresh cases vs 428 unaffected control siblings); 300 cases with severe vs 137 cases with mild symptoms; 225 patients who died from smallpox vs 212 survivors; blood group distribution among 428 surviving patients in comparison with 324 controls; severe vs mild smallpox scars among 548 smallpox survivors (From Vogel and Helmbold 1972 [135])

types. The same might easily have been true in smallpox: until the time of its systematic eradication, this infection was almost endemic in large parts of India. High titers of hemagglutination-inhibiting variola antibodies were indeed found in many individuals who lived in these areas but had no history of clinically discernible affliction with smallpox and had never been vaccinated [19]. However, the evidence remains contradictory, with little chance to settle the matter by direct examination since smallpox has been eradicated.

Blood Group A and Smallpox in the World Population. As noted, smallpox was frequent in many populations up to the early 1970s, and comprehensive population statistics are available. If smallpox has been an important selective agent against allele A, a negative correlation between A frequency and smallpox frequency or mortality must be expected. This negative correlation was found for the Indian subcontinent [10] and for Africa [130], when the frequency of type A in various subpopulations was compared with the occurrence of smallpox (Table 12.20).

The correlations are in the direction expected if selection by smallpox were in fact important, but of course they do not prove it.

Distribution of AB0 Blood Group Genes in the World Population and Selection by Infectious Diseases. What aspects of the world distribution of AB0 alleles might be explained by these selective mechanisms, and which ones elude explanation? The following highly tentative conclusions can be drawn:

a) The extremely high 0 frequency in Central and South America could be due to advantage of 0 in the presence of syphilis.

Table 12.20. Negative correlation between incidence of smallpox and frequency of blood group A

	Number of population groups	No. of individuals included in the calculations	Spearman-Rank correlation
India and Pakistan (mortality)	18	87 153	$\varrho = -0.634$, $p < 0.01$
Africa (morbidity)	27	195 313	$\varrho = -0.499$, $p < 0.01$

b) The higher frequency of 0 in marginal European populations could be caused by lower selection against 0 carriers by plague and cholera.

c) The relatively low A frequencies in central and southern Asia could be due to selection by smallpox. In the same areas, gene 0 is also less frequent, the gap being filled by allele B. This advantage of group B could be due to long-standing selection against type A by smallpox as well as against type 0 by plague and cholera.

In view of the strong and long-standing selection against gene A it is justifiable to ask why this blood group is still present in these populations. It is conceivable that another, still unknown, selective advantage of A exists?

One aspect of the world AB0 distribution cannot be explained satisfactorily. Why is allele B so frequent in central and southern Asia and so rare in most other areas? Long-standing selection against types A and 0 may be part of the story. Interaction of the human host with intestinal germs – and possibly foods containing ABH-like antigens – is one of the major unknown factors in ABH selection [99].

These studies were performed at a time when little was known about the physiological function of AB0 specificities – or those of other blood groups such as the Lewis and P systems and the genetically determined ability of secreting ABH substances. When the main steps of the immune response were elucidated, (Sect. 7.4), protein components necessary for cell recognition and interaction of cells – far beyond the requirements of the immune response – became known in great detail; but the glycoproteins and glycolipids of the cell surface went largely unnoticed. Meanwhile this has changed, and their role in cell recognition and, especially, host defense is being recognized [11, 143]. For example, the group P and ABH substances have been shown to act as receptors for attaching E. coli to the cell surface; such receptor binding is necessary, for example, for urogenital infections [81, 86, 87]. These substances belong to a much larger group of compounds, the lectins [115], which were first detected in plants but are present in many other living beings as well. They play an important role not only in the process of infection but also in cell differentiation, organ formation, lymphocyte migration, and metastasis of malignancies. Research on AB0-disease associations began in the late 1950s and 1960s with much enthusiasm but were discontinued in the late 1970s and later; one reason may have been a certain disappointment since it seemed impossible to formulate specific hypotheses as to the mechanisms involved. Here studies on lectins are opening up a new field for statistical studies guided by explanatory hypotheses. It is very well possible, for example, that similarities in surface antigens between hosts and infective germs, as discussed above for E. coli caused diarrheas, may modify germ-host interactions, leading to blood group specific differences in susceptibility to many infections. In view of the widespread occurrence of lectins in plants, it may even be possible to find connections to human polymorphisms of AB0-specific and/or other cell surface antigens and certain aspects of nutrition – a new and promising field of ecogenetics (Sect. 7.5.2).

The experiences with Hbβ E in Thailand (Sect. 12.2.1.7) suggest that such frequency clines in populations are difficult to interpret. They may reflect population history and gene diffusion, but they may also indicate clines in selection intensities.

Lesson of Studies on AB0 Blood Group Selection for Research on Natural Selection in Human Populations. In spite of their shortcomings, studies on selection and the AB0 blood groups have been described extensively for their possible significance:

a) It is an oversimplification to treat selection as constant over long time periods in the same population. For the main selective factor of the hemoglobin variants, *Plasmodium falciparum,* this oversimplification might hold true for many centuries or even millennia since malaria remained endemic as long as the ecological conditions for the mosquito vector did not change. Many other infections, on the other hand, come and go as epidemics. Here, selection may change even over short time periods. Sometimes cataclysmic events may result that will long be remembered in history, such as the plague epidemics of the Middle Ages. In other cases some infections may not be recorded, such as infantile diarrheas. The variety of selective agents and their change over time is an almost all-pervading element.

b) In such a situation genetic variability in itself may be an advantage for the species. If one epidemic kills almost all individuals carrying one genetic variant only, many of those survive who have other variants and are therefore less susceptible. The next epidemic may wipe out this variant but favor the first. This may lead to a dynamic situation in which gene frequencies oscillate over time, depending on the prevalent selective agents. Data on blood group determination of bones from the fifteenth to the seventeenth centuries have been interpreted as evidence of such oscillations [69]. In view of the difficulties of accurate AB0 blood group measurements in ancient bones and mummies due to bacterial cross reactions, these results must be considered with caution. If genetic variability in itself is an advantage, no single optimum genotype exists.

c) The fact that an unstable polymorphic situation conveys advantages for survival of the species does not account for maintenance of such a situation. Considering the small population sizes in isolated groups of earlier human history, one would expect many more populations to have become monomorphic in the course of time. To maintain a polymorphism, a stabilizing element is required. It has been suggested that this stabilizing element may have been frequency-dependent selection (Sect. 12.6.15).

Genetic Susceptibilities and Infectious Disease. The foregoing sections have discussed examples of genetic selection vis-à-vis infectious disease. Future work in this field will most likely be successful when concerned with diseases causing maximum selection. Endemic diseases, because they act at all times, are more effective agents than epidemic diseases, which act episodically. Diseases that affect a large portion of the population are more effective as selective agents than those limited to restricted segments of the population. Diseases that kill children are more effective selective agents than those compatible with survival or those affecting principally adults, particularly after their reproductive period.

Natural Selection by Infectious Agents Is Likely for the MHC Polymorphism. Apart from the AB0 blood groups, associations with infectious diseases have also been claimed for a number of other genetic polymorphisms. The best a priori candidates for such associations might be the polymorphisms of the major histocompatibility complex (MHC) and especially the HLA genes. At present information on associations of HLA types with major infectious diseases is scarce. Most scientists interested in this genetic system are working in countries where the major epidemics have been eradicated or have lost much of their importance to public health. However, studies of HLA and disease have shown the MHC to be a major component in the genetic variability of immune response in humans (Sect. 5.2.5). Therefore associations with epidemic diseases and a strong influence of natural selection on gene frequencies can safely be predicted. In fact, HLA associations with leprosy [29, 140], typhoid, and malaria [108] as well as immune responsiveness to streptococcal and tetanus [114] antigens have been suggested (Table 12.21).

Malaria has been a major selective agent in many tropical and subtropical countries. Therefore it is surprising how little the problem has been studied of whether certain HLA alleles or haplotypes common in those populations may afford protection against this infection. A well-designed case-control study on several hundred children in Gambia [58] showed that both a class I allele, Bw53, and a class II haplotype, DRB1$^+$1302–DQBl$^+$0501, afford relative protection against falciparum malaria. Both were decreased significantly among children up to the age of 10 years when those suffering from severe malaria were compared with mild cases, and when patients as an overall group were compared with adults from the same area. Both HLA types are unusually common in that population. *It is likely that AB0 selection will be regarded in the future as a minor factor compared to selection due to components of the MHC locus.*

Does Genetic Liability to Atopic Diseases Lead to an Increased Resistance to Helminth Infestation [49]? One of the main

Table 12.21. Associations of HLA alleles with response to immunizations and diseases

	MHC genes
Immunizations	
Tetanus	B5 (low response)
Vaccinia	Cw3
Influenza	Bw16
Hepatitis B	B8, DR3
Diseases	
Meningococcal meningitis	B27
Leprosy	A1, B40
Leprosy, tuberculoid	DR2
Measles	Aw32
Tuberculosis	B5, DR5

health risks for children in tropical countries is the almost ubiquitous infestation with intestinal worms, mainly tapeworms, ascaris, and hookworms. Hookworms can cause severe anemias, which in combination with other infections may contribute to premature death. Characteristic clinical signs of worm infestation are an increased level of eosinophilic granulocytes and an increased IgE blood level. Such findings are also observed in atopic diseases such as asthma, hay fever, and atopic dermatitis. There is good evidence that atopic diseases have a multifactorial genetic basis, and that a gene or genes influencing IgE levels are involved (Sect. 6.1.2.7). Atopic diseases are common in present-day populations even though some manifestations of such diseases may impair health significantly. It is therefore conceivable that genotypes associated with atopic manifestations had a selective advantage in earlier times. Studies were therefore carried out to assess whether atopic genotypes in Papua New Guinea [49] have a selective advantage in relation to helminthic infestation. 500 villagers were tested for immediate hypersensitivity responses against a number of allergens. Based on these tests, 10% of these villagers were identified as atopic. Moreover, a clinical diagnosis of asthma was made on the basis of clinical evidence and pulmonary function tests on all inpatients of a local district hospital; asthma was diagnosed in 24 patients. Stool samples from these patients and from 50 nonasthmatic villagers diagnosed as atopic and 139 nonatopic villager controls were examined for hookworm egg counts. The result is seen in Fig. 12.32. Average egg counts were found to be lowest in asthmatic patients, higher in the nonasthmatic atopic individuals, and highest among the controls. Possible biases, such as different distribution of atopic and nonatopic individuals among different villages, were carefully excluded. The study confirms the hypothesis proposed by the authors: The concomitants of atopy, such as IgE elevation, apparently afford relative protection against hookworm infestation. There is also some limited clinical evidence for a relative protective effect of IgE levels mitigating the severity of worm infestation [107].

Interaction Between the Human host and Infective Agents. The examples above refer to selective advantages and disadvantages of human beings having certain genotypes after exposure to infective agents such

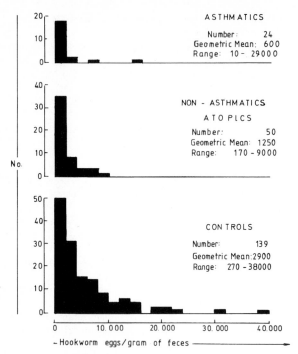

Fig. 12.32. Distribution of the numbers of hookworm eggs/g feces in three samples of a heavily infested population in Papua New Guinea. Note the enormous differences in infestation range between asthmatics, non-asthmatic atopics, and controls. (From Grove and Forbes 1975 [49])

as intestinal helminths, malaria parasites, *E. coli* bacteria causing diarrhea, smallpox viruses, and others. We saw at the beginning that natural selection due to infectious diseases has played a major role in all periods of human evolution, except the past few decades. Among the few theoretical models in which the consequences of certain assumptions regarding modes of selection have been examined, there is one in which the human host and the infective agent were considered together. The consequences turned out to be complex, defying sweeping generalizations. Section 7.4 describes genetic determination of the human immune response in crude outline. We now compare these "defense strategies" of the human host with "strategies of attack" used by various infective agents. Some of these strategies turn out to be surprisingly similar, making use of largely the same genetic principles [133]:

1. Germ cell mutations leading to genetic polymorphisms, with slight differences in interaction with infective agents, are well known in humans; the AB0 and MHC polymorphisms are obvious examples, but there are many more. Many bacteria use mutations toward resistance to antibiotics to cope with antibiotic therapy. Resistant mutants are almost always present in bacterial populations, but they become common under the influence of antibiotics. Other examples abound.

2. Somatic mutations leading to different cell clones within the same individual constitute one of the mechanisms by which variability of immunoglobulins is enhanced; their mutation rate is unusually high. An increased intraclonal mutation rate by a certain lack of precision in forming DNA replicas of the RNA genome is observed, for example, in the HIV virus leading to AIDS. It deceives and diminishes the humoral immune response of the host.

3. Integration of foreign genomes or gene products to help in mimicry against host defense. This mechanism, which is used, for example, by *Schistosoma* and by certain viruses, may be compared in its protective effect with the symbiosis between humans and certain benign bacteria, for example, in the colon, the vagina, and other organs, and formation of "normal" antibodies such as the anti-AB0 isoagglutinins.

4. Multiplication of certain genes to permit rearrangements for creating a great variety of defense cells by a switch mechanism is a common principle in the formation of immunoglobulins and T cell receptors. A similar mechanism, antigen drift, exists in *Trypanosoma* (sleeping sickness in humans and Nagana disease in cattle [30]). These germs protect themselves from attacks of the host defense system by continuously changing their surface antigens. This is made possible by hundreds and possibly even thousands of genes for its surface glycoproteins, the expression of which is controlled by a switch mechanism and varies in space and time.

Only a few examples of interaction between human host and infective agent have been analyzed, but there are many more. Such mechanisms have been active even in the twentieth century and in industrialized countries as shown, for example, by the demonstration of genetic factors in twin studies on tuberculosis and other chronic infections (Sect. 6.3) Another study [123] compared adoptees with their biological and adoptive parents in Denmark: there were similarities in the frequency of death in early adult age – often from infections – not only with adoptive parents (due to obvious, common environmental factors) but also between adoptees and their biological parents from whom they had been separated for many decades.
The change in living conditions which we are experiencing at present is probably the greatest alteration which the human species has ever gone through. It can be compared only with the "Neolithic Revolution" a few thousand years ago. This demographic revolution leads, among other changes, to a break-up of isolates (see below) – with strong genetic effects. Its

influence on the pattern of infectious diseases will probably be much more extensive. Some infections, such as smallpox, have been completely eliminated; others will probably follow. This may lead to relaxation of selection against small weaknesses of the immune system. The almost worldwide epidemic of AIDS – caused by the HIV virus and probably existing in a limited area of Africa for a long time – has taught us that infective agents have an almost unlimited ability for adaptation to changing ecological conditions. Other infections will come, with unpredictable abilities for adaptation to changing living conditions and behaviors of the human host. Theoretical models, of which this chapter presents a small sample, are useful for understanding some of the simpler situations, such as the advantage of sickle cell heterozygotes in the presence of falciparum malaria. However, genetic changes which we are now facing are complex and unpredictable. We can safely predict that the composition of the human gene pool will change, but neither the extent nor the direction of these changes can be foreseen. Careful observation of facts, inclusion of the genetic point of view into studies on the epidemiology of infectious diseases, assessment of biological mechanisms by observation and experiment, and application of appropriate mathematical models are necessary.

Conclusions

There is much individual variability in gene frequencies between human populations. This variability has been studied at the phenotypic-biochemical and at the gene-DNA levels. In addition to mutation, natural selection is important in shaping the genetic composition of populations. Simple mathematical models have been described for studying the genetic consequences of various types of selection in populations. An important selective force that has shaped the genetic composition of human populations, at least during the past several thousand years, has been exposure to various infectious organisms. Examples include genetic adaptation of hemoglobin genes – and of other red blood cell genes – to malaria, and associations of AB0 blood groups and HLA types with various infectious diseases.

References

1. Allison AC (1954) Protection afforded by sickle-cell trait against subtertin malarial infection. BMJ 1 : 290
2. Allison AC (1954) The distribution of the sickle-cell trait in East Africa and elsewhere, and its apparent relationship to the incidence of subtertian malaria. Trans R Soc Trop Med Hyg 48 : 312
3. Allison AC (1954) Notes on sickle-cell polymorphism. Ann Hum Genet 19 : 39–51
4. Allison AC (1955) Aspects of polymorphism in man. Cold Spring Harbor Symp Quant Biol 20 : 239
5. Allison AC (1956) The sickle and hemoglobin C-genes in some African populations. Ann Hum Genet 21 : 67–89
6. Allison AC (1964) Polymorphism and natural selection in human populations. Cold Spring Harbor Symp Quant Biol 24 : 137–149
7. Antonarakis SE, Orkin SH, Kazazian HH et al (1982) Evidence for multiple origins of the β^E-globin gene in Southeast Asia. Proc Natl Acad Sci USA 79 : 6608–6611
7a. Antonarakis SE, Boehm CD, Serjeant GR, Theisen CE, Dover GJ, Kazazian HH (1984) Origin of the β^s-globin gene in blacks: the contribution of recurrent mutation or gene conversion or both. Proc Natl Acad Sci USA 81 : 853–856
8. Beet EA (1946) Sickle-cell disease in the Balovale district of North Rhodesia. East Afr Med J 23 : 75
9. Beet EA (1947) Sickle-cell disease in Northern Rhodesia. East Afr Med J 24 : 212–222
10. Bernhard W (1966) Über die Beziehung zwischen AB0-Blutgruppen und Pockensterblichkeit in Indien und Pakistan. Homo 17 : 111
11. Blackwell CC (1989) The role of AB0 blood groups and secretor status in host defences. FEMS Microbiol Immunol 47 : 341–350
12. Bloch J (1901) Der Ursprung der Syphilis. Fischer, Jena
13. Blumberg BS, Hesser JE (1971) Loci differently affected by selection in two American black populations. Proc Natl Acad Sci USA 68 : 2554
14. Boat TF, Welsh MJ, Beaudet AL (1989) Cystic fibrosis. In: Scriver CL, Beaudet AL, Sly S, Valle D (eds) Metabolic basis of inherited disease. McGraw-Hill, New York, pp 2649–2680
15. Borén T, Falk P, Roth KA et al (1993) Attachment of helicobacter pylori to human gastric epithelium mediated by blood group antigens. Science 262 : 1892–1895
16. Botstein D, White RL, Skolnick M, Davis RW (1980) Construction of a genetic linkage map in man using restriction fragment length polymorphisms. Am J Hum Genet 32 : 314–331
17. Brues AM (1954) Selection and polymorphism in the AB0 blood groups. Am J Phys Anthropol 12 : 559–598
18. Cavalli-Sforza LL, Bodmer WF (1971) The genetics of human populations. Freeman, San Francisco
19. Chakravartti MR, Vogel F (1971) Haemaglutination-inhibiting variola antibodies in blood serum of former smallpox-patients, their healthy siblings and unvaccinated controls in other areas. Humangenetik 11 : 336–338
20. Charnov E (1977) An elementary treatment of kin selection. J Theor Biol 66 : 541–550
21. Clarke B (1975) Frequency-dependent and density-dependent natural selection. In: Salzano F (ed) The role of natural selection in human evolution. North-Holland, Amsterdam; American Elsevier, New York, pp 187–200
22. Clarke LA (1974) Rh haemolytic disease. Original papers commentaries. Medical and Technical Publishing, Newcastle
23. Cooper DN, Schmidtke J (1984) DNA restriction fragment length polymorphism and heterozygosity in the human genome. Hum Genet 66 : 1–16

24. Cooper DN, Smith BA, Cooke HJ, Niemann S, Schmidtke J (1985) An estimate of unique DNA sequence heterozygosity in the human genome. Hum Genet 69 : 201–205

25. Crabb AR (1947) The hybrid-corn makers. New Brunswick

26. Crow JF, Kimura M (19709 An introduction to population genetics theory. Harper and Row, New York

27. Damian RT (1964) Molecular mimicry: antigen sharing by parasite and host and its consequences. Am Nat 98 : 129–150

28. De Vay JE, Charudattan R, Winnalajeewa DLS et al (1972) Common antigenic determinants as possible regulators of host-pathogen compatibility. Am Nat 948 : 185–193

29. De Vries RRP, Fat RFMLA, Nijenhuis LE, van Rood JJ (1976) HLA-linked genetic control of host response to mycobacterium leprae. Lancet 2 : 1328–1330

30. Donelson JE, Turner MJ (1988) Wie Trypanosomen das Immunsystem täuschen. In: Van Burnet M, Nossel GJ, Porter RR et al (eds) Immunsystem, 2nd edn. Spektrum der Wissenschaft, Heidelberg, pp 174–183

31. Downie HW, Meiklejohn G, Vincent LS, Rao AR, Sundara Babu BV, Kempe CH (1966) Smallpox frequency and severity in relation to A, B and 0 blood groups. Bull WHO 33 : 623

32. East EM, Jones DF (1919) Inbreeding and outbreeding. Lippincott, London

33. Edwards A, Civitello A, Hammond HA, Caskey CT (1991) DNA typing and genetic mapping with trimeric and tetrameric tandem repeats. Am J Hum Genet 49 : 746–756

34. Eichner ER, Finn R, Krevans JR (1963) Relationship between serum antibody levels and the AB0 blood group polymorphism. Nature 198 : 164

35. Emery AEH (1984) An introduction to recombinant DNA. Wiley, Chichester

36. Ewens WJ (1980) Mathematical population genetics. Springer, Berlin Heidelberg, New York

37. Falconer DS (1981) Introduction to quantitative genetics, 2nd edn. Oliver and Boyd, Edinburgh

38. Firschein IL (1961) Population dynamics of the sickle cell trait in the Black Caribs of British Honduras. Am J Hum Genet 13 : 233

39. Fisher RA (1930) The genetical theory of natural selection. Oxford University Press, Oxford

40. Flatz G (1967) Haemoglobin E: distribution and population dynamics. Hum Genet 3 : 189–234

41. Flatz G, Pik C, Sundharayati B (1964) Malaria and haemoglobin E in Thailand. Lancet 2 : 385

42. Flatz G, Oelbe M, Herrmann H (1983) Ethnic distribution of phenylketonuria in the North German population. Hum Genet 65 : 396–399

43. Flint J, Hill AVS, Bowden DK et al (1986) High frequencies of α-thalassemia are the result of natural selection by malaria. Nature 321 : 744–750

44. Flint J, Harding RM, Clegg JB, Boyce AJ (1993) Why are some genetic disorders common? Distinguishing selection from other processes by molecular analysis of globin gene variants. Hum Genet 91 : 91–117

45. Friedman MJ (1978) Erythrocytic mechanism of sickle cell resistance to malaria. Proc Natl Acad Sci USA 75 : 1994–1997

46. Friedman MJ, Trager W (1981) The biochemistry of resistance to malaria. Sci Am 244 : 154–164

47. Garrod AE (1902) The incidence of alcaptonuria: a study in chemical individuality. Lancet 2 : 1616–1620

48. Glass RI, Holmgren J, Haley CE, Khan MR, Svennerholm A-M, Stoll BJ, Hossain KMB, Black RE, Yunus M, Barua D (1985) Predisposition for cholera of individuals with 0 blood group. Am J Epidemiol 121 : 791–796

49. Grove DI, Forbes IJ (1975) Increased resistance to helminth infestation in an atopic population. Med J Aust 1 : 336–338

50. Haldane JBS (1942) Selection against heterozygotes in man. Ann Eugen 11 : 333

51. Haldane JBS (1949) The rate of mutations of human genes. Hereditas 35 Suppl:267

52. Hamilton WD (1964) The genetical evolution of social behavior. I. J Theor Biol 7 : 1–16

53. Harris H, Hopkinson DA (1972) Average heterozygosity per locus in man: an estimate based on the incidence of enzyme polymorphisms. Ann Hum Genet 36 : 9–20

54. Harris R, Harrison GA, Rondle CJM (1963) Vaccinia virus and human blood group A substance. Acta Genet 13 : 44

55. Hartl DL, Clark AG (1989) Principles of population genetics, 2nd edn. Sinauer, Sunderland

56. Hiernaux J (1952) La génétique de la sicklémie et l'interet anthropologique de sa fréquence en Afrique noir. Ann Mus Congo Belge [Sci Homme Anthropol] 2 : 42

57. Hill AVS, Wainscoat JS (1986) The evolution of the a- and b-globin gene clusters in human populations. Hum Genet 74 : 16–23

58. Hill AVS, Allsopp CEM, Kwiatkowski D et al (1991) Common West African HLA antigenes are associated with protection from severe malaria. Nature 352 : 595–600

59. Hirsch A (1881) Handbuch der historisch-geographischen Pathologie, part 1: Infektionskrankheiten, 2nd edn. Enke, Stuttgart

60. Horai S, Gojobori T, Matsunaga E (1984) Mitochondrial DNA polymorphism in Japanese. Hum Genet 68 : 324–332

61. Hundrieser T, Sanguansermsri T, Papp T, Laig M, Flatz G (1988) β-globin gene linked DNA haplotypes and frameworks in three south-east Asian populations. Hum Genet 80 : 90–94

62. Ifediba TC, Stern A, Ibrahim A, Rieder RF (1985) Plasmodium falciparum in vitro: diminished growth in hemoglobin H disease erythrocytes. Blood 65 : 452–455

63. Jacquard A (1974) The genetic structure of populations. Springer, Berlin Heidelberg New York

64. Jeffreys AJ, Wilson V, Thein SL (1985) Hypervariable "minisatellite" regions in human DNA. Nature 314 : 67–73

65. Kaiser P (1984) Pericentric inversions. Problems and significance for clinical genetics. Hum Genet 68 : 1–47

66. Kan YW, Dozy AM (1978) Polymorphism of DNA sequence adjacent to the human b globin structural gene. Its relation to the sickle mutation. Proc Natl Acad Sci USA 75 : 5631–5635

67. Kay AC, Kuhl W, Prchal J, Beutler E (1992) The origin of glucose-6-phosphate-dehydrogenase (G6PD) polymorphisms in African-Americans. Am J Hum Genet 50 : 394–398

68. Kazazazian HH, Waber PG, Boehm CD et al (1984) Hemoglobin E in Europeans: further evidence for multiple origins of the β^E globin gene. Am J Hum Genet 36 : 212–217

69. Kellermann G (1972) Further studies on the AB0 typing of ancient bones. Hum Genet 14 : 232–236

70. Kidson C, Lamont G, Saul A, Nurse GT (1981) Ovalocytic erythrocytes from Melanesians are resistant to invasion

by malaria parasites in culture. Proc Natl Acad Sci USA 78 : 5829–5832

71. Kircher W (1961) Untersuchungen über den Zusammenhang von Dyspepsieverlauf und AB0-Blutgruppenzugehörigkeit. Monatsschr Kinderheilkd 109 : 369

72. Kircher W (1964) Weitere Untersuchungen über den Zusammenhang zwischen Verlauf und Häufigkeit der Säuglingsenteritis und AB0-Blutgruppenzugehörigkeit. Monatsschr Kinderheilkd 112 : 415

73. Klose J, Willers I, Singh S, Goedde HW (1983) Two-dimensional electrophoresis of soluble and structure-bound proteins from cultured human fibroblasts and hair root cells: qualitative and quantitative variation. Hum Genet 63 : 262–267

74. Krieger H, Vicente AT (1969) Smallpox and the AB0 system in Southern Brazil. Hum Hered 19 : 654

75. Lambotte-Legrand 1951, p. 404

76. Landsteiner K (1900) Zur Kenntnis der antifermentativen, lytischen und agglutinierenden Wirkungen des Blutserums und der Lymphe. Zentralbl Bakteriol 27 : 357–362

77. Li CC (1955) Population genetics. University of Chicago Press, Chicago

78. Li CC (1963) The way the load works. Am J Hum Genet 15 : 316–321

79. Li CC (1976) First course in population genetics. Boxwood, Pacific Grove

80. Lichter-Konecki U, Schlotter M, Yaylak C et al (1989) DNA haplotype analysis at the phenylalanine hydroxylase locus in the Turkish population. Hum Genet 81 : 373–376

81. Linstedt R, Larson G, Falk P et al (1991) The receptor repertoire defines the host range for attaching Escherichia coli strains that recognize globo-A. Infect Immun 59 : 1086–1092

81 a. Lisker R, Motulsky AG (1967) Computer simulation of evolutionary trends in an X linked trait, application to glucose-6-phosphate dehydrogenase deficiency in man Acta Genet (Basel) 17 : 465–474

82. Livingstone FB (1971) Malaria and human polymorphisms. Annu Rev Genet 5 : 33–64

83. Livingstone FB (1973) Data on the abnormal hemoglobins and glucose-6-phosphate dehydrogenase deficiency in human populations, 1967–1973. University of Michigan, Ann Arbor

84. Livingstone FB (1983) The malaria hypothesis. In: Bowman JE (ed) Distribution and evolution of hemoglobin and globin loci. Elsevier, New York, pp 15–44

85. Livingstone FB (1985) Frequencies of hemoglobin variants. Thalassemia, the glucose-6-phosphate dehydrogenase deficiency, G6PD variants, and ovalocytosis in human populations. Oxford University Press, Oxford

86. Lomberg H, Eden CS (1989) Influence of P blood group phenotype on susceptibility to urinary tract infection. FEMS, Microbiol Immunol 47 : 363–370

87. Lomberg H, Cedergren B, Leffler H et al (1986) Influence of blood groups on the availability of receptors for attachment of uropathogenic Escherichia coli. Infect Immun 51 : 919–926

88. Luzzatto L (1979) Genetics of red cells and susceptibility to malaria. Blood 54 : 961

89. Luzzatto L, Usanga EA, Reddy S (1969) Glucose-6-phosphate dehydrogenase deficient red cells resistance to infection by malarial parasites. Science 164 : 839

90. Luzzatto L, Sodeinde O, Martini G (1983) Genetic variation in the host and adaptive phenomena in Plasmodium falciparum infection. In: Evered D, Whelan J (eds) Malaria and the red cell. Pitman, London, pp 159–173

91. Motulsky AG (1960) Metabolic polymorphism and the role of infectious disease in human evolution. Hum Biol 32 : 28–62

92. Motulsky AG (1964) Hereditary red cell traits and malaria. Am J Trop Med Hyg 13 (1/2):2147–158

93. Motulsky AG (1975) Glucose-6-phosphate dehydrogenase and abnormal hemoglobin polymorphisms – evidence regarding malarial selection. In: Salzano F (ed) The role of natural selection in human evolution. North-Holland, Amsterdam, pp 271–291

94. Mourant AE, Kopec AC, Domaniewska-Sobczak K (1976) The distribution of the human blood groups and other polymorphisms. Oxford University Press, London

95. Murray JC, Mills KA, Demopulos CM, Hornung S, Motulsky AG (1984) Linkage disequilibrium and evolutionary relationships of DNA variants (RFLPs) at the serum albumin locus. Proc Natl Acad Sci USA 81 : 3486–3490

96. Nakamura Y, Leppert M, O'Connell P et al (1987) Variable number of tandem repeat (VNTR) markers for human gene mapping. Science 235 : 1616–1622

97. Nakatsuji T, Kutlar A, Kutlar F, Huisman THJ (1986) Haplotypes among Vietnamese hemoglobin E homozygotes including one with a g-globin gene triplication. Am J Hum Genet 38 : 981–983

98 a. National Research Council (1996) DNA forensic science: an update. National Academy of Sciences, Washington

98. Neel JV (1951) The population genetics of two inherited blood dyscrasias in man. Cold Spring Harbor Symp Quant Biol 15 : 141

99. Otten CM (1967) On pestilence, diet, natural selection, and the distribution of microbial and human blood group antigens and antibodies. Curr Anthropol 8 : 209

100. Pasvol G (1982) The interaction of malaria parasites with red blood cells. Br Med Bull 38 : 133–140

101. Pasvol G, Weatherall DJ, Wilson RJM (1977) Effects of foetal haemoglobin on susceptibility of red cells to plasmodium falciparum. Nature 270 : 171–173

102. Pasvol G, Weatherall DJ, Wilson JM (1978) Cellular mechanism for the protective effect of haemoglobin S against falciparum malaria. Nature 274 : 701–703

103. Pasvol G, Wainscoat JS, Weatherall DJ (1982) Erythrocytes deficient in glycophorin resist invasion by the malarial parasite plasmodium falciparum. Nature 297 : 64–66

104. Penrose LS, Smith SM, Sprott DA (1956) On the stability of allelic systems, with special reference to haemoglobins A, S, and C. Ann Hum Genet 21 : 90–93

105. Pettenkofer JH, Bickerich R (1960) Über Antigen-Gemeinschaften zwischen den menschlichen Blutgruppen und gemeingefährlichen Krankheiten. Zentralbl Bakteriol [Orig A] 179

106. Pettenkofer HJ, Stöss B, Helmbold W, Vogel F (1962) Alleged causes of the present-day world distribution of the human ABo blood groups. Nature 193 : 444

107. Phills JA, Harrold J, Whiteman GV, Perelmutter L (1972) Pulmonary infiltrates. asthma and eosinophilia due to Ascaris suum infestation in man. N Engl J Med 286 : 965–970

108. Piazza A et al (1973) In: Dausset J, Colombani J (eds) Histocompatibility testing 1972. Munksgaard, Copenhagen, pp 73–84

109. Race RR, Sanger R (1975) Blood groups in man, 6th edn. Blackwell, Oxford

110. Robinson MG, Tolchin D, Halpern C (1971) Enteric bacterial agents and the AB0 blood groups. Am J Hum Genet 23 : 135

111. Romeo G, Devoto M (1990) Population analysis of the major mutation in cystic fibrosis. Hum Genet 85 : 391–445

112. Roth E Jr, Raventos-Suarez C, Gilbert H, Stump D, Tanowitz H, Rowin KS, Nagel RL (1984) Oxidative stress and falciparum malaria: a critical review of the evidence. In: Eaton JW, Brewer GJ (eds) Malaria and the red cell. Liss, New York, pp 35–43

112 a. Ruwende C, Khoo SC, Snow RW, Yates SNR, Kwiatkowski D, Gupta S, Warn P, Allsopp CEM, Gilbert SC, Peschu N, Newbold CI, Greenwood BM, Marsh K, Hill AVS (1995) Natural selection of hemi- and heterozygotes for G6PD deficiency in Africa by resistance to severe malaria. Nature 376 : 246–249

113. Santos FR, Pena SDJ, Epplen JT (1993) Genetic and population study of a Y-linked tetranucleotide repeat DNA polymorphism with a simple non-isotopic technique. Hum Genet 90 : 655–656

114. Sasazuki T, Kohno Y, Iwamoto I, Tanimura M, Naito S (1978) Association between an HLA haplotype and low responsiveness to tetanus toxoid in man. Nature 272 : 359–361

115. Sharon N, Lis H (1989) Lectins as cell recognition molecules. Science 246 : 227–234

116. Shull GH (1908) Composition of a field of maize. Rep Am Breeders Assoc 4 : 296–301

117. Shull GH (1911) Experiments with maize. Bot Gaz 52 : 480

118. Smith SM (1954) Notes on sickle-cell polymorphism. Ann Hum Genet 19 : 51

119. Smithies O (1955) Grouped variations in the occurence of new protein components in normal human serum. Nature 175 : 307–308

120. Smithies O (1955) Zone electrophoresis in starch gels: group variations in the serum proteins of normal human adults. Biochem J 61 : 629–641

121. Socha W, Bilinska M, Kaczera Z, Pajdak E, Stankiewicz D (1969) Escherichia coli and AB0 blood groups. Folia Biol (Krakow) 17(4)

122. Solomon E, Bodmer WF (1979) Evolution of sickle variant gene. Lancet 1 : 923

123. Sorensen TIA, Nielsen GG, Andersen PK, Teasdale TW (1988) Genetic and environmental influences on premature death in adult adoptees. N Engl J Med 318 : 727–733

124. Sukamaran PK, Master HR, Undesia JV, Balakrishnan B, Sanghvi LD (1966) ABo blood groups in active cases of smallpox. Indian J Med Sci 20 : 119

125. Süssmilch (1786) Die göttliche Ordnung, part 2, 9th edn. Berlin

126. Tanis RJ, Neel JV, Dovey H, Morrow M (1973) The genetic structure of a tribal population, the Yamomama Indians. IX. Gene frequencies for 17 serum protein and erythrocyte enzyme systems in the Yamomama and five neighboring tribes; nine new variants. Am J Hum Genet 25 : 655–676

127. Thalhammer O (1975) Frequency of inborn errors of metabolism, especially PKU, in some representative newborn screening centers around the world. A collaborative study. Hum Genet 30 : 273–286

128. Tsui L-C (1990) Editorial. Hum Genet 85 : 391–392

129. Van Loon FP, Clemens JD, Sack DA et al (1991) AB0 blood groups and the risk of diarrhea due to enterotoxigenic Escherichia coli. J Infect Dis 163 : 1243–1246

130. Vogel F (1970) Anthropological implications of the relationship between AB0 blood groups and infections. Proceedings of the 8th International Congress of Anthropologic and Ethnologic Sciences, Tokyo 1968, vol 1, p 365

131. Vogel F (ed) (1974) Erbgefüge. Springer, Berlin Heidelberg New York (Handbuch der allgemeinen Pathologie, vol 9)

132. Vogel F (1979) Genetics of retinoblastoma. Hum Genet 52 : 1–54

133. Vogel F (1992) Break-up of isolates. In: Roberts DF, Fujiki N, Toruzuka K (eds) Isolation, migration and health. Cambridge University Press, Cambridge, pp 41–55

134. Vogel F, Chakravartti MR (1966) AB0 blood groups and smallpox in a rural population of West Bengal and Bihar (India). Hum Genet 3 : 166–180

135. Vogel F, Helmbold W (1972) Blutgruppen – Populationsgenetik und Statistik. In: Becker PE (ed) Humangenetik, ein kurzes Handbuch, vol 1/4. Thieme, Stuttgart, pp 129–557

136. Vogel F, Kopun M (1977) Higher frequencies of transitions among point mutations. J Mol Evol 9 : 159–180

137. Vogel F, Pettenkofer HJ, Helmbold W (1960) Über die Populationsgenetik der ABo-Blutgruppen. II. Genhäufigkeit und epidemische Erkrankungen. Acta Genet 10 : 267–294

138. Vogel F, Dehnert J, Helmbold W (1964) Über Beziehungen zwischen AB0-Blutgruppen und der Säuglingsdyspepsie. Hum Genet 1 : 31–57

139. Wade Cohen PT, Omenn GS, Motulsky AG, Chen S-H, Giblett ER (1973) Restricted variation in the glycolytic enzymes of human brain and erythrocytes. Nature 241 : 229

140. Wallace DC (1989) Report of the Committee on Human Mitochondrial DNA. Cytogenet Cell Genet 51 : 612–621

141. Walton KE, Steyer D, Gruenstein EI (1979) Genetic polymorphism in normal human fibroblasts as analyzed by two-dimensional polyacrylamide gel electrophoresis. J Biol Chem 254 : 7951–7960

142. Weber JL, May PE (1989) Abundant class of human DNA polymorphisms which can be typed using the polymerase chain reaction. Am J Genet 44 : 388–396

143. Weir DM (1989) Carbohydrates as recognition molecules in infection and immunity. FEMS Microbiol Immunol 47 : 331–340

144. Weiss KM (1994) Genetic variation and human disease. Principles and evolutionary approaches. Cambridge University Press, Cambridge

144 a. Wilson AF, Elston RC, Tran LD, Siervogel RM (1991) Use of the robust sub-pair method to screen for single-locus, multiple-locus, and pleiotropic effects: application to traits related to hypertension. Am J Hum Get 48 : 862–872

145. Woolf B (1955) On estimating the relation between blood group and disease. Ann Hum Genet 19 : 251–253

146. Zuckerkandl E (1965) The evolution of hemoglobin. Sci Am 212(5):110–118

147. Zuckerkandl E (1976) Evolutionary processes and evolutionary noise at the molecular level. II. A selectionist model for random fixations in proteins. J Mol Evol 7 : 269–311

13 Population Genetics: Consanguinity, Genetic Drift

Humanity is just a work in progress.

Tennessee Williams, Camino Real,
Block 12

13.1 Deviations from Random Mating

Considerations in the preceding chapters presume random mating, and Hardy-Weinberg proportions are assumed to hold true. However, such assumptions are an abstraction. In modern outbreeding populations mating may approximate randomness for some genetic traits, such as blood groups and enzyme types, but is certainly nonrandom for some traits and some hereditary conditions, such as congenital deafness. Because of their need for special schools and professional training, deaf individuals form social groups with intense in-group contacts but remain partially isolated from the outside world. Naturally, so-called "assortative mating" frequently occurs between deaf partners. If both partners carry the same type of recessive gene for deafness, all their children will be deaf. Assortative matings are less conspicuous but much more common with regard to psychologically or socially significant aspects of life, such as social status, income, range of interest, education, or intelligence (Fig. 13.1). Human populations, far from mating at random, comprise a complex and ever-changing system of more or less isolated subgroups. These subgroups may be called "isolates" if they are well delimited, and mating is confined more or less to members of the group. They have been called "demes" if they represent only groups within which the probability of mating is enhanced compared to matings with outside individuals [39]. There is no sharp demarcation between isolates and demes.

One type of assortative mating is mating among relatives. Since relatives share some of their genes by common descent, consanguineous matings influence the incidence of some inherited diseases. Comparison of the progeny of consanguineous with those of nonconsanguineous marriages uncover the manifestation of recessive genes as an increased frequency of specific diseases and provides evidence regarding the role of recessive genes in morbidity and mortality of diseases in which such genes may play a subsidiary role. These studies also provide data to assess the concept of genetic load (see Sect. 12.3.2).

Another aspect is the widespread tendency to prefer marriages within the same subgroup, which in the long run leads to genetic differences between such subgroups. Measures of "population distance" have been developed to assess such differences. Population structure and genetic composition of the population are also influenced by migration of individuals between subpopulations. Migration counteracts the effects of isolation; it is currently of increasing importance.

13.1.1 Consanguineous Matings

13.1.1.1 Inbreeding Coefficient [39]

All Human Beings Are Relatives. Relatives are defined as individuals who have a certain portion of their genes in common by descent. If we take this definition literally, all human beings are relatives. We have common ancestors. The progenitors might even have been one single couple (see Sect. 14.2.1). Why then are our genes so different? For the simple reason that our common descent dates back thousands of generations. During this long period many intervening mutations have caused genetic variability. Obviously it would be meaningless operationally to treat all of mankind as relatives, since no conclusions could be drawn from this – albeit formally correct – assumption. On the contrary, a main point of interest motivating us to measure consanguinity concerns these intervening mutations and the effect of consanguinity on their phenotypic manifestation. However, in measuring consanguinity we should always keep in mind that it is merely a matter of practical convenience as to how many generations we go back.

Degrees of Relationship Normally Considered. Lines of descent are frequently only studied for three generations. This convention was initially established for reasons of convenience. Catholics require a special dispensation for a marriage between second cousins or closer relatives, and the church registers for these dispensations are an easy source of information on frequencies of consanguineous marriages in Catholic

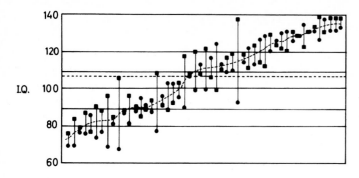

Fig. 13.1. Assortative mating as to intelligence quotient (IQ) in a sample of married couples in the United States. ■, Husband; ●, Wife; *stippled line,* mean of the couples. (From Outhit (1933); Schwidetzky *Das Menschenbild der Biologie* (1959))

Symbol	Description	Inbreeding coefficient
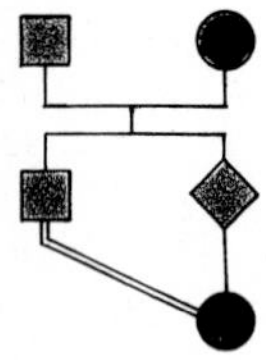	Uncle-niece marriage	1/8
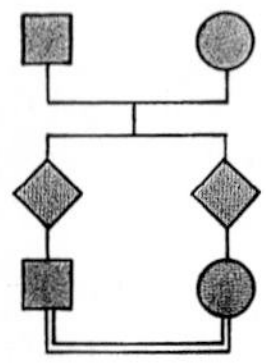	First cousins	1/16
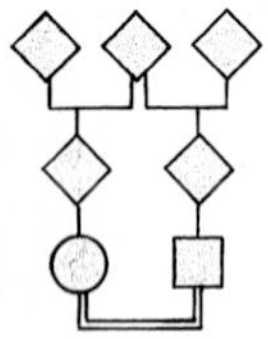	First degree step cousins	1/32
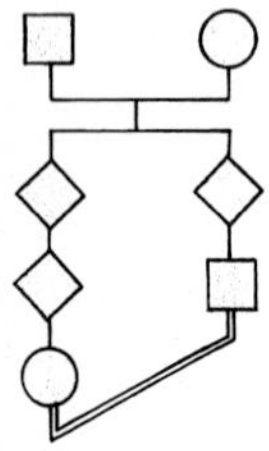	First cousins once removed	1/32
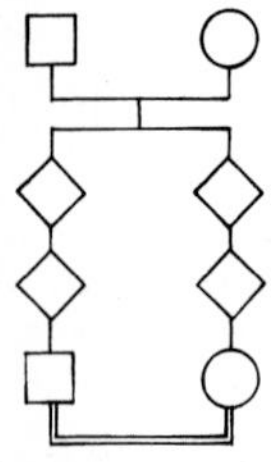	Second cousins	1/64

Fig. 13.2. The most important types of consanguineous marriages

populations. This restriction means that parents, grandparents, and great-grandparents but not more distant ancestors are taken into account. The most remote relatives considered in describing the degree of consanguinity between two individuals are therefore second cousins. Types of consanguineous marriages usually found within this range of relatives are shown in Fig. 13.2.

The convention to limit assessment of consanguinity to more narrow relationships is also theoretically reasonable. Beyond this range the coefficient of inbreeding of an individual increases only very slowly with the number of additional consanguineous marriages in his or her ancestry.

Two Useful Measures: Coefficient of Kinship and Inbreeding Coefficient [48, 52]. Within a population various types of consanguineous matings may occur: those between second or first cousins, between uncle and niece, occasionally even between brothers and sisters or fathers and daughters. It is of course possible to describe the frequencies of all such matings, and such data might offer interesting insights from a sociological point of view. Only one aspect is interesting to the geneticist, however: How closely are the parents of a child related – what is their share of common genes? Whether we intend to compare individuals by the degree to which they are inbred or to describe population groups by the average degree of inbreeding of their members, we need a measure for this proportion. If possible, a single number would simplify our task in the same way as the notion of "gene frequency" simplifies a description of a population in terms of genotypes. Several measures for the degree of inbreeding have been proposed; the choice among them is largely arbitrary. The "inbreeding coefficient" (Wright 1922 [79]) has proven to be the most useful. This is closely related to the "coefficient of kinsip" (Malécot 1948 [52]). These coefficients are defined as follows:

a) The coefficient of kinship, Φ_{AB}, of two individuals, A and B, is the probability that a gene taken at random from A is identical by common descent with a gene taken at random at the same locus from B.

b) The inbreeding coefficient, F, of an individual is equal to the coefficient of kinship, Φ, of his father and mother.

The distinction between the two coefficients is that a coefficient of kinship applies to two individuals who may have common ancestors. An inbreeding coefficient applies to one individual and measures the degree of relationship between his parents and hence the resemblance between the two genes received from his parents at each locus. In fact, *the inbreeding coefficient is equal to the probability that the two genes which the individual has at a given locus are identical by descent.*

Coefficient of Inbreeding and the Hardy-Weinberg Law. Let us consider an autosomal gene pair A, a (gene frequencies p, q). In a randomly mating population the three genotypes occur in the proportions $p^2 : 2pq : q^2$. If the genotype contains N such gene pairs with gene frequencies p_i, q_i ($i = 1, 2, \ldots, N$), the degree of heterozygosity under random mating is:

$$2\sum_{i=1}^{N} \frac{p_i q_i}{N}$$

and the degree of homozygosity:

$$\sum_{i=1}^{N} \frac{p_i^2 + q_i^2}{N}$$

the sum of the two being 1. This degree of heterozygosity indicates the proportion of autosomal genes with two alleles, for which an individual is, on average, heterozygous. For a single gene it indicates the probability of an individual's being heterozygous.
In a consanguineous mating (Fig. 13.2) a pair of alternate alleles A and a are considered. An oocyte may contain the gene a. If the mating is random, the probability of this oocyte being fertilized by a sperm with a is p and by a sperm with A is q. If the parents are related, they have a certain proportion of genes in common by descent; accordingly, p is increased to $(p + Fq)$ and q is reduced to $(q - Fq)$, and correspondingly for oocytes containing allele A. Here the value that corresponds to F in the former delineation is called F'. If the mode of inheritance is autosomal, the two parents show the same distribution of genes A and a. Therefore it must hold that $pq(1 - F) = qp(1 - F')$ and hence $F = F'$.
It can be shown that F is equal to the inbreeding coefficient as defined above. This means that the genotypes of children having an inbreeding coefficient F do not occur in Hardy-Weinberg proportions but rather in the proportions:

AA	:	Aa	:	aa
$(p^2 + Fpq)$	:	$2(1 - F)pq$	:	$(q^2 + Fpq)$

The child's own degree of heterozygosity is diminished on average by a factor F. To put it differently, F is the probability that the two homologous chromosomes carry two genes that are derived from the same ancestral gene at a randomly chosen gene locus. (Fig. 13.3)

Calculation of the Inbreeding Coefficient F. The actual calculation of Φ or F is not necessary for most practically occurring situations in human genetics as the coefficients for the degrees of consanguinity occurring in human populations are known, as shown in Fig. 13.2. An occasional pedigree may require individual calculation. This is quite different from the situation in animal breeding, where very complicated relationships between mates may be encountered. For their assessment Wright proposed the method of path coefficients [46, 80]. The pedigrees of the two mates are drawn, and all their common ancestors are marked. Then one of the least remote common ancestors is selected, and the two mates are connected by all possible pathways, which:

a) Lead to this common ancestor
b) Consist of "steps" (one step being defined as the connection between an individual and one of his parents)
c) Do not lead to one person more than once

The other common ancestors are treated in the same way. The number of steps in each path is counted. For one ancestor, x paths with $m_1 \ldots m_x$ steps may exist, giving for t common ancestors $\left(\sum_{i=1}^{t} x_i \right) = r$ paths. Then:

$$F = \tfrac{1}{2}(2^{-m_1} + 2^{-m_2} + \ldots + 2^{-m_r}) = \tfrac{1}{2}\sum_{i=1}^{r} 2^{-m_i} \qquad (13.1)$$

(If some of the common ancestors come from consanguineous matings, the terms must be corrected considering their inbreeding coefficients.)
The following simple consideration may help to understand this formula. A child shares with each of his parents 1/2 of its genes, with a grandparent 1/4, with a great-grandparent 1/8, etc. In a path of a steps from the mother to an ancestor she has a fraction of $(1/2)^a = 2^{-a}$ genes in common with that ancestor; in a path of b steps from the father to this ancestor he shares 2^{-b} genes with this ancestor. This means that father and mother have $2^{-a} \times 2^{-b} = 2^{-m}$ genes ($m = a + b$) in common. This number divided by 2 gives the probability that a random gene of the mother is identical by descent with a random gene of the father. (For a more rigorous derivation and for other methods of calculation, see Li (1955) [46], Jacquard (1974) [39], and Kempthorne (1957) [41], who gives a useful matrix method.)

Examples. Figure 13.4 shows a first-cousin marriage. The path over the common grandfather of the couple has four steps, and the same is true for the path over the common grandmother. Inserting in Eq. (13.1) gives:

$$F = \tfrac{1}{2}(2^{-4} + 2^{-4}) = \tfrac{1}{2}\left(\tfrac{1}{16} + \tfrac{1}{16}\right) = \tfrac{1}{16}$$

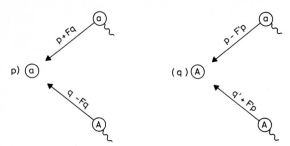

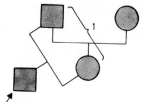

Fig. 13.5. Father-daughter incest. There is only one path

Fig. 13.3. An oocyte may contain the allele a. With random mating, probabilities for this oocyte to be fertilized by a sperm with allele a = p, with allele A = q. In a consanguineous mating, these probabilities are $(p + Fq)$ or $(q - Fq)$. *Left*, fertilization of oocyte with allele *a*; *right*, fertilization of oocyte with allele A. (From Ludwig 1944 [48])

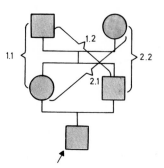

Fig. 13.6. Brother-sister incest. There are two paths with two steps each

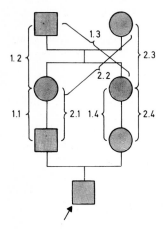

Fig. 13.4. First-cousin marriage. Calculation of *F* by the method of path coefficients. Four steps (1.1–1.4) connect the proband's father and his mother via the common grandfather of the parents. Four other steps (2.1–2.4) connect the father and the mother via the common grandmother. $F = \frac{1}{16}$

$$K = \sum F_i M_i$$

Summation extends over the various types of consanguineous marriages, with F_i and M_i being the inbreeding coefficient and the relative frequency of the *i*-th type of consanguineous marriage. *K* is frequently referred to simply as the mean of *F* or as *F* of the population. In comparing different populations for this parameter one should bear in mind that the convention of including three ancestral generations has often not been followed. In addition, almost all population studies calculate the coefficients of kinship (according to the above definition) of all couples and not the inbreeding coefficients of all individuals. This kinship coefficient gives an unbiased estimate of the mean inbreeding coefficient of the individuals only if inbreeding does not influence reproduction.

Figure 13.5 shows a case of father-daughter incest. There is only one path and it consists of one step:

$$F = \frac{1}{2}2^{-1} = \frac{1}{2} \cdot \frac{1}{2} = \frac{1}{4}$$

The third example is of brother-sister incest (Fig. 13.6):

$$F = \frac{1}{2}(2^{-2} + 2^{-2}) = \frac{1}{2}(\frac{1}{4} + \frac{1}{4}) = \frac{1}{4}$$

As mentioned above, for humans a useful convention limits these calculations to three generations back. The convention is sound quantitatively. For example, if a common ancestor is found five generations back, the corresponding path has ten steps; therefore, it contributes to *F* only $1/2 \times 2^{-10} = 1/2048$.

Inbreeding Coefficient of a Population. Frequently we are interested in an index that measures mean consanguinity in a population, considering all types of consanguineous marriages together. If the calculation is confined to the last three generations, the resulting coefficient *K* of the "apparent consanguinity" is given by:

13.1.1.2 Inbreeding and Inherited Disease

Frequency of Children with Recessive and Multifactorial Diseases in Consanguineous Matings Compared with Nonconsanguineous Matings. Let the allele which in the homozygous state leads to a recessive disease have a gene frequency *q*. The phenotype frequency in the random mating population is then q^2; in the population of individuals with the inbreeding coefficient *F* it is $q^2 + Fpq$. With decreasing *q*, the ratio Fpq/q^2 increases: *The lower the gene (and genotype) frequency, the higher the frequency of consanguineous marriages is among parents of the affected homozygotes.* In other words, the chance that an identical rare allele possessed by one mating partner is also carried by the spouse is very small unless the spouse is related and has derived the rare allele from an ancestor shared by both partners. This does not mean, however, that an autosomal-recessive disease which is observed in a child from a consanguinous marriage necessarily carries two identical alleles from a single, common ancestor. The probability of this event increases with decreasing gene frequency (Fig. 13.7).

This holds true not only for recessive diseases but also for multifactorial characters (Sect. 6.1). Among individuals with inbreeding coefficient F, the variance of a normally distributed liability with heritability $h^2 = 1$ is

$$V_F = V_0(1 + F)$$

V_0 being the variance in a noninbred population. However, with increase in variance the relative number of individuals beyond the threshold also increases. Children from consanguineous matings therefore have a slightly higher risk of being affected with a multifactorial threshold character than children from nonrelated parents (see Fig. 6.1.2).

For recessive genes the argument can be reversed theoretically. If more consanguineous marriages are found than are expected on the basis of the population incidence of a recessive condition, not one but several recessive genes with correspondingly lower gene frequencies may be involved, i.e., genetic heterogeneity may exist. Under the assumption that these genes have equal frequencies, even their number has been estimated. In practice, this approach is almost always futile, for the following reasons:

a) Due to the decrease in inbreeding among modern populations the number of homozygotes for recessive diseases has declined steeply (see below).
b) The decrease in inbreeding has taken place mainly in big cities and densely populated areas. Many of the recessive diseases come from remote rural areas where consanguineous marriages are in general more frequent.
c) Therefore a high ratio of consanguineous to nonconsanguineous matings may be encountered simply due to population heterogeneity, even if there is only one recessive gene.

Deaf-mutism, for example, is often an autosomal recessive condition, and the number of recessive genes has occasionally been estimated from the high consanguinity rates in their families (see [74]). However, deaf-mutes go to school together; both they and their normal family members share many social activities; the social group made up of the deaf and their families has many properties of a social isolate. This contributes to an unpredictable degree to the high proportion of consanguineous matings.

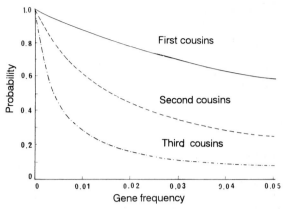

Fig. 13.7. Probabilities of homozygotes of an autosomal recessive gene for having inherited two identical copies of this gene from the same, common ancestor, depending on degree of consanguinity of the parents and frequency of the mutant gene in the population. As the gene frequency increases, the probability of transmission from the same ancestor decreases, particularly with second- and third-cousin matings

Inbreeding Coefficient F in Various Population Groups. Table 13.1 collects the frequencies of consanguineous matings in various populations, listing first-cousin marriages (1-C) and the F values calculated on the basis of the available data. (For Israel only frequencies of first-cousin, uncle-niece, and aunt-nephew marriages are given, with the result that the inbreeding coefficient may be underestimated.) The data in Table 13.2 were collected by the genealogical method, i.e., pedigrees of couples of consanguineous marriages were assessed. Depending on the method of ascertainment, F may be more or less underestimated because:

a) The data from Catholic countries are usually based on dispensation registers for consanguineous matings. However, probably not all Catholics who plan to marry a close relative actually ask for dispensation, and especially in cities the priests may not know the couples well enough to know that they are related.
b) In some studies the investigators relied on the statements of the families, who in many cases concealed consanguinity.

For example, on the Japanese island of Hosojima the seven members of the island council knew of 19 of 45 marriages in which the spouses were related. A check of the Koseki, a Japanese population register, brought this number to 25, which was then increased to 29 by careful analysis of pedigrees.

In almost all European countries and in the United States, the inbreeding coefficients are very low; high coefficients are usually found in some small communities and in religious, geographic, and ethnic iso-

Table 13.1. Frequency of consanguineous marriages and inbreeding coefficients F (×10⁻⁵) in various countries (from von Fumetti 1976 [77] unless otherwise noted)

Country/region/diocese	Method of ascertainment[a]	Time	Population (N)	% First cousin marriages	% Consang. marriages (All types together)	F (×10⁵)	Time	Population	% First cousin marriages	% Consang. marriages	F (×10⁵)
Europe											
Belgium	DA	1918–1959	2404027	0.49	1.47	50	1955–1959	300592	0.22	0.97	29
Czech diocese of Brno	DA	1930–1966	230988	?	0.93	28	1960–1966	?	?	0.20	8
Germany											
Bavaria, Württemberg	DA + CR	1848–1922	16182	0.50	1.18	44	–	–	–	–	–
Five localities near Tübingen	CR + MR	About 1920	453	4.91	20.00	472	–	–	–	–	–
Archdiocese of Cologne	DA	1898–1943	192980	0.37	0.93	35	–	–	–	–	–
Dioceses of Münster, Osnabrück	DA	–	–	–	–	–	1946–1951	119899	0.18	0.59	19
France	DA	1926–1958	6061000	0.52	1.36	49	1956–1958	530000	0.22	0.67	23
Ireland	DA	–	–	–	–	–	1959–1968	190547	0.13	0.53	16
Italy	DA	1911–1960	13687897	1.33	3.00	118	1956–1960	1646612	0.77	1.90	70
Austrian archdiocese of Vienna	DA	1901/1902 1914/1914 1929/1930	117294	0.67	1.28	60	–	–	–	–	–
Switzerland 4 mountain villages	A + CR	About 1920	538	2.79	32.71	509	–	–	–	–	–
Spain											
Total	DA	1930	17000	2.00	5.34	203	1960–1964	2069	2.10	10.63	275
Diocese Ciudad Rodrigo	MR	1940–1964	11394	2.18	9.41	254	1967–1981	893941	0.23	0.66	21.6
North America											
Canada											
Catholics	DA	1885–1995	–	–	–	–	1959	51729	0.37	1.51	45
French speaking population	DA	1915–1925 1945–1965	149992	1.03	4.17	180	1955–1965	50128	0.37	2.10	90
United States Catholics	DA	–	–	–	–	–	About 1958	133228	0.08	0.11	8
Mormons											
Total	FB	1930–1950	132524	0.04	?	?	–	–	–	–	–
Nine rural parishes	FB	–	–	–	–	–	1950	625	1.44	9.92	189
Central and South America											
Argentina 12 dioceses[b]	DA	–	–	–	–	–	1956/1957	51391	0.75	1.12	58
Brazil											
72 dioceses	DA	–	–	–	–	–	1956/1957	212090	2.63	4.82	225
95 dioceses	DA	–	–	–	–	–	1965/1967	198088	2.14	4.00	176
Bolivia 5 dioceses	DA	–	–	–	–	–	1956/1957	4130	0.32	0.63	28

Chile											
8 dioceses	DA	—	—	—	—	—	1956/1957	28 596	0.80	1.31	74
Chile and province of Valparaiso	DA	1917–1966	195 721	0.60	0.98	50	1957–1966	51 828	0.41	0.68	35
Costa Rica (1 dioceses)	DA	—	—	—	—	—	1954	3 833	0.94	3.39	114
Ecuador (3 dioceses)	DA	—	—	—	—	—	1956/1957	3 954	2.17	6.27	229
El Salvador (2 dioceses)	DA	—	—	—	—	—	1956/1957	2 494	1.04	4.85	142
Honduras (3 dioceses)	DA	—	—	—	—	—	1956/1957	3 759	0.56	3.43	110
Columbia (13 dioceses)	DA	—	—	—	—	—	1956/1957	34 470	1.25	2.95	119
Cuba (3 dioceses)	DA	—	—	—	—	—	1956/1957	2 277	0.53	0.83	54
Mexico (10 dioceses)	DA	—	—	—	—	—	1956/1957	28 292	0.17	1.27	31
Panama (1 dioceses)	DA	—	—	—	—	—	1956/1957	350	0	0	—
Peru (3 dioceses)	DA	—	—	—	—	—	1956/1957	565	2.12	4.07	279
Uruguay (3 dioceses)	DA	—	—	—	—	—	1956/1957	8 822	0.85	1.43	65
Venezuela (4 dioceses)	DA	—	—	—	—	—	1956/1957	2 931	1.60	4.46	191
Asia											
Japan											
Total	I + FB	1880–1964	10 403	5.58	14.72	—	About 1950	213 148	5.39	8.16	400
Hirado Island[b]	I + FB	—	—	—	—	461	1960–1964	853	1.76	10.79	200
India											
Bombay[c]	I	—	—	—	—	—	About 1955	3 520	9.29	11.39	612
Southern India	I	—	—	—	—	—	After 1950	26 042	24.58	39.37	2835
Turkey, Muslims [31]	I	—	—	—	—	—	1987	2 604	15.3	35.2	—
Kuwait, women [2]	I	—	—	—	—	—	1983	5 007	30.2	54.3	2190
Pakistan [73]											
Lahore	I	—	—	—	—	—	1980–1983	681	51	61	2690
Mianyammm	I	—	—	—	—	—	1980–1983	70	73	73	2360
Murudke	I	—	—	—	—	—	1980–1983	114	81	83	2400
Sheikynpura	I	—	—	—	—	—	1980–1983	480	80	86	2710
Gujrat	I	—	—	—	—	—	1980–1983	425	80	83	2570
Iheluin	I	—	—	—	—	—	1980–1983	778	51	56	2620
Rawalpindi	I	—	—	—	—	—	1980–1983	781	55	59	2860
Israel											
Total	I	—	—	—	—	—	1955–1957	11 424	5.22	9.68	387
Arabs [38]	I	—	—	—	—	—	1988–1989	610	27	38.70	3360
Africa											
Egypt, nubic populations	I	—	—	—	—	—	1967/1968	1 782	45.29	75.76	3335
Guinea, Fulbe	I	—	—	—	—	—	About 1960	1 280	15.21	29.91	819

[a] DA, Dispensation archives; A, ancestry tables; I, interview; MR, marriage registers; FB, family books; CR, church registers.

[b] The figures for Hirado relate to the time of marriage for those marriages of which at least one spouse was still alive at the time of examination.

[c] Inbreeding coefficient was calculated only from uncle-niece and aunt-nephew matings; therefore, it might be too low by 1/4–1/3.

Table 13.2. Results of various studies on effect of inbreeding on death during infancy, childhood, and young adulthood among Japanese (from Schull and Neel 1972 [67]) (Conceptuses lost before the 7th month of pregnancy not included)

Investigator and locale	A	B	B/A	Size of in-bred sample	Ascertainment
Watanabe					
Fukushima prefecture	0.0881	0.5157	5.8	4594	Through child surviving to high school
Tanaka and Kishimoto					
Shizuoka	0.1253	0.7191	5.7	2205	Through child surviving to elementary school
Schull et al.					
Nagasaki prefecture (Kuroshima)	0.0927	1.4074	15.2	223	Koseki and Catholic church records, followed (average) 15 years, deaths before age 20
Schull and Neel					
Hiroshima prefecture	0.0875	0.5317	6.1	1697	Pregnancy registration at 5th month followed
Nagasaki prefecture	0.0986	0.1060	1.1	2608	(average) to 8 years
Schull and Neel					
Kure	0.0929	0.0405	0.4	564	Pregnancy registration at 5th month, followed (average) to 15 years
Yanase					
Fukuoka prefecture					
Hs	0.0962	1.2535	13.0	277	Household survey, deaths before 6 years
Hi	0.1292	0.3308	2.6	304	
Ta-Ko	0.0916	0.9884	10.8	301	
Fujiki et al.					
Yamaguchi prefecture					
Mis	0.1222	0.3287	2.7	497	Koseki records plus household interviews,
Nuw	0.1985	−0.8107	−4.1	234	followed (average) to midchildhood
Kur	0.1936	−0.9608	−5.0	79	
Nagano					
Fukuoka prefecture (Fukuoka City)	0.0873	0.6765	7.8	5953	Through elementary and junior high school followed through age 12
Schull et al.					
Nagasaki prefecture (Hirado)	0.1157	0.7703	6.7	6626	Household survey, nonaccidental deaths largely through age 20
Freire-Maia et al.					
Japanese immigrants in Bauru state of São Paulo, Brazil	0.1378	0.6995	5.1	105	Household survey, subjects followed through age 21
Average	0.1036	0.6700	6.7	–	–

For a definition of A and B see Sect. 13.1.2.1.

lates. In South America, which has been very well examined, the mean inbreeding coefficient appears to be about twice or three times that in Europe [26, 27]. High values of F have also been found in Japan. The highest values are described for parts of southern India, especially the state of Andhra Pradesh, among the Nubic tribes in Egypt, and among the Fulbe in Guinea.

Decline in Consanguinity in Industrial Countries. In the industrialized countries of the West a decline in consanguineous marriages has been observed since the beginning of this century. The trend began in the highly industrialized areas and big cities and is now spreading into the more remote rural provinces. France has been especially carefully examined; Figures 13.8 and 13.9 show the frequencies of apparently consanguineous marriages in 1926–1930 and in 1956–1958 [39]. In the former period the mean coefficient of inbreeding (more exactly, the coefficient of apparent consanguinity; see above) was 86.1×10^{-5}; by the latter period it had decreased to 23×10^{-5}. The decline is usually explained by the higher mobility of the population in an industrialized society and the wider choice among individuals of the other sex. This explanation is corroborated by studies in which the dis-

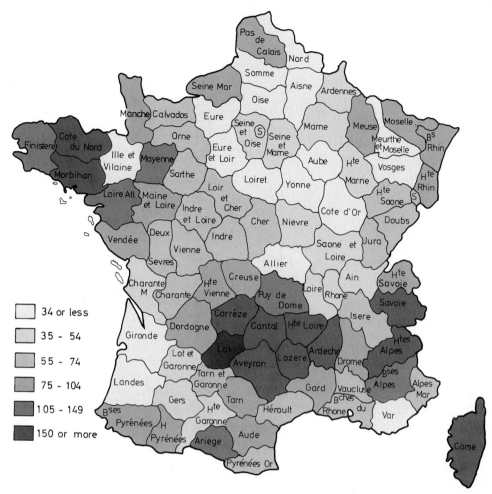

Fig. 13.8. Coefficient of apparent consanguinity in France 1926–1930. The *numbers* $\times 10^{-5}$ refer to *F*. (From Jacquard 1974 [39])

Legend:
- 34 or less
- 35 – 54
- 55 – 74
- 75 – 104
- 105 – 149
- 150 or more

tances between the birth places of spouses were shown to grow over time. This phenomenon is usually referred to as the "breaking up" of isolates. More recently the tendency toward a lower consanguinity rate has been strengthened by the decline in the number of children per marriage, which reduces the number of eligible cousins.

These and other considerations raise the question of the degree to which consanguineous marriages represent an otherwise unbiased sample of all marriages. The question becomes important when consanguineous marriages are used to estimate the "genetic load" (Sect. 13.3.2) due to genes that are lethal or detrimental in the homozygous state.

Social and Psychological Influences on the Frequency of Consanguineous Marriages. Some biases are obvious from inspection of Figs. 13.8. and 13.9; the quality of obstetric and pediatric care is usually better in the developed areas of France with low consangui-

nity rates. Even within these provinces the rates are lower in towns than in villages. Selective migration of healthier persons from rural to urban areas may further add a spurious trend toward higher infant and perinatal mortality in consanguineous marriages.

However, the bias may be much more subtle. In a German study, for example, individuals married to a close relative were shown to differ psychologically from the average population [81]. For example, males living in consanguineous marriages suffered more difficulties in establishing interpersonal contacts than other men and therefore selected a relative rather than a nonrelative as a spouse. On the other hand, completely different sociopsychological conditions for consanguineous marriages may exist; in populations of southern India with very high consanguinity rates, a marriage between daughter and mother's brother is the most preferred marriage type socially.

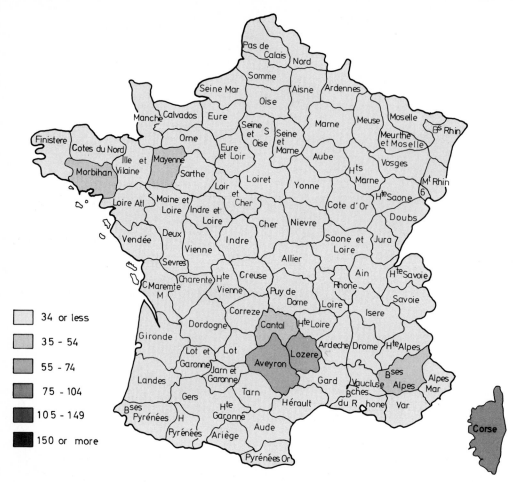

Fig. 13.9. Coefficient of apparent consanguinity in France 1956–1958. *Numbers* × 10⁻⁵ refer to *F*. (From Jacquard 1974 [39])

In both India and in Japan until very recently marriages were commonly arranged by the families of the two spouses. Economic reasons and the advantage of knowing the spouse very well are the most prominent factors in such arrangements. Personality factors such as those described in the above German study seem to be of minor importance. For this and other reasons it is reasonable to conclude that in Japan consanguineous marriages are a less biased sample of all marriages than in European countries. Still, some bias may remain. In a study of consanguinity on the Japanese island Hirado [67–69], for example, there was a greater tendency for older brothers to marry a first cousin. Moreover, it was the older brothers who inherited the land owned by the families and hence tended to stay in their villages, while many younger brothers emigrated to other areas. Consanguineous marriages are still common in certain Hindu population groups of southern India as well as in many Muslim populations (Table 13.1). Within these populations socioeconomic and other differences between consangui-

neous and other couples have been found. A comprehensive study from southern India showed that women in close consanguineous unions marry and begin giving birth to children earlier in life; therefore they tend to have more children [7]. Such biases must be considered, when the influence of parental consanguinity on the health status of their progeny is assessed.

Influence of the Decline of Consanguinity on the Incidence of Recessive Diseases. Let us assume a population with average inbreeding coefficient F in which an equilibrium between mutation rate μ and the selection coefficient s has been established at a gene frequency q. Now, the degree of inbreeding is reduced within a short time period, say, one generation, from F_1 to F_2. Consequently the number of homozygotes drops from $q^2 + F_1 pq$ to $q^2 + F_2 pq$. This change disturbs the genetic equilibrium, selection now being insufficient to eliminate the number of genes produced by new mutations since there are fewer homozygotes. For example, F_1 for European populations once rang-

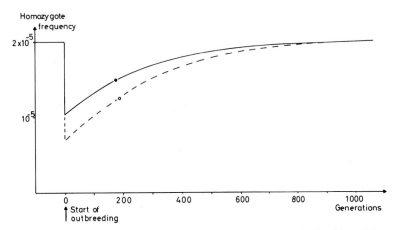

Fig. 13.10. Decrease in the frequency of recessive homozygotes in a population with long-lasting inbreeding by complete cessation of inbreeding and very gradual increase to the old value as the result of an excess of freshly produced compared with selectively eliminated mutants. Mutation rate $\mu = 10^{-5}$; selection coefficients for the recessive homozygotes $s = 0.5$; inbreeding coefficients $F = 0.003$ (—) and $F = 0.005$ (---). At the points ● and ○, respectively, the gene frequency q reaches one-half of the new equilibrium frequency.

The values of the gene frequency c were calculated by solving the equation:

$$q^2 + Fq(1 - q) = \mu/s$$

for q, and then stepwise for each new generation by the recurrence formula:

$$q' = (q^* - sq^{*2})/(1 - sq^{*2})$$

where $q^* = q + \mu p$

ed between about 0.003 and 0.005. Let us also assume a mutation rate of 10^{-5} and selection coefficient (s) of 0.5 against the homozygotes. The equilibrium frequency is $\hat{q} = 2.6 - 3.2 \times 10^{-3}$. In this case complete cessation of inbreeding leads to a drop of the homozygote frequency, as shown in Fig. 13.10. Moreover, 175–185 generations (about 4500 years) are required for the gene to reach one-half the way to the new equilibrium frequency.

The decline in inbreeding is one of the reasons why "modern" societies are now enjoying – and will enjoy for a long time to come – an unusually low incidence of recessive disorders. The other, related reason is that, due to stochastic processes in small populations (Sect. 13.3), genes for recessive diseases have attained unequal frequencies in various population groups. With increasing intermarriage between these groups gene frequencies will level out, and populations with high gene frequencies and hence high homozygote frequencies will disappear.

Physicians and geneticists impressed by somewhat negative effects of consanguinity on health may attempt to dissuade prospective marriage partners from marrying a relative; they even may initiate public health measures to this end. However, consanguineous matings are an important component of the social structure in these populations, for example, in southern India. Attempts to break up marriage customs with long traditions therefore should not be initiated unless all possible social consequences have been carefully considered.

13.1.2 Concept of Genetic Load

13.1.2.1 Theory

Estimation of the Overall Number of Recessive Genes in the Human Population [73]. Homozygotes in general, especially those for rare diseases, are more frequent among children from consanguineous marriages than in the general population. This fact can be used to estimate the number of such recessive genes in each individual of the population.

For example, a gene taken at random from an individual has a probability of 1/2 of being identical by descent to one of the two genes at the same locus of the individual's sib. If one of the sibs in a brother-sister mating is a carrier of a gene leading to a recessive disease in the homozygous state, the other sib also carries the gene with a probability of 1/2, and every child of this mating has a risk of 1/4 of being affected with the disease. Therefore the probability that such a mating produces at least one affected child is $1 - (3/4)^s$, s being the number of children per mating. When an otherwise unselected series of brother-sister matings is examined with respect to the incidence of recessive diseases among their progeny, the average number of individuals carrying such a recessive gene in the general population can thus be calculated. The same argument holds true, for example, for father-daughter matings. These matings are proscribed by law and custom and are rare. Moreover, no one would regard individuals involved in such a mating as an unbiased population sample.

Intuitive Background: Our Load of Mutations. H. J. Muller, the famous geneticist, had been concerned since his teens with the idea that the human species may be in danger of deteriorating biologically, that the evolutionary system could collapse sooner or later, and that our species might finally be submerged in an ocean of suffering from disease, mental defects, miscarriage, and other catastrophes.

We saw at the beginning of this book (Sect. 1.8) that this concern was shared by many scientists at the beginning of the twentieth century and was in fact the motive behind the work of F. Galton and the eugenics movement.

Muller's arguments were set out comprehensively in his paper *Our Load of Mutations* in 1950 [86]. His most important theses may be formulated as follows:

a) A large share of all human zygotes are killed or prevented from reproduction by mutations.

b) The overall mutation rate per individual, i.e., the total number of new mutations contained in both germ cells that form this individual, is one mutation per two to ten germ cells.

c) Every individual is heterozygous for several genes that would kill him if homozygous. These genes are usually deleterious even in the heterozygous state.

d) Natural selection has relaxed; therefore the number of deleterious genes in the human population is increasing dangerously. They may reach a critical threshold above which the whole genetic system may break down, leading to disappearance of the human species.

e) The danger becomes more acute by increased exposure to ionizing radiation.

f) We should try to curb this dangerous development by regulating human reproduction artificially.

Since Muller proposed these theses, our knowledge of human genetics has improved, and some of his questions can now be answered fairly precisely [76]. One aspect is singled out here: the assertion that every human being is heterozygous for several genes that would kill him if homozygous but are deleterious even in the heterozygous state.

Effect of Variation on Fitness. A more formalized and more balanced concept was formulated by Haldane in several papers, especially that titled *The Effect of Variation on Fitness* [33, 34]. One of these conclusions was that, due to recurrent mutations, some genotypes always have a reduced fitness; if fitness is 0, this genotype does not reach the next generation. If, however,

fitness is reduced by 1/1000 one of 1000 of these genotypes is eliminated. In any case reduction in fitness depends on the mutation rate. Therefore these ideas are in principle only generalizations of his indirect method for mutation rate estimation (Sect. 9.3.1).

On the basis of Muller's and Haldane's contributions the concept of genetic load was used to test human populations in another famous paper, *An Estimate of the Mutational Damage in Man from Data on Consanguineous Marriages* by Morton et al. (1956) [54].

Definition of the Genetic Load. Morton et al. [54] distinguish between the total damage due to disadvantageous mutations present in the human genome and the expressed damage. Both are described as lethal equivalents. A lethal equivalent is a group of mutations which, distributed among various individuals, causes an average of one death, for genetic reasons. Such mutations may be, for example, a lethal mutation leading to death in all cases or two mutations each leading to death in 50% of the cases. The total damage per gamete was defined as the average number of lethal equivalents in the zygote when the zygote is formed by doubling all chromosomes of the gamete. The expressed damage per gamete is the average number of lethal equivalents of this gamete that would be manifested if combined in the zygote with another gamete according to the mating system actually prevailing in the population.

The total genetic damage may be calculated as follows. Let us consider one gene locus. The probability for a given zygote to survive deleterious effects of mutations at this locus is given by:

$1 - qFs$	$-q^2(1-F)s$	$-2q(1-q)(1-F)sh$
Probability of death due to homozygosity from consanguinity	Probability of death due to homozygosity not from consanguinity	Probability of death in a heterozygote (13.2)

Here, s is the probability of death for a zygote homozygous for the mutation; h is a measure for the dominance of this mutation ($h = 0$ if the gene is completely recessive, $h = 1$ if the gene leads to death in the heterozygote as often as in the homozygote); F is the inbreeding coefficient.

Another assumption is that genetic and environmental causes of death act independently. With this condition the proportion of survivors may be expressed as follows:

$$S = \prod_{i,j}(1-x_i)[1 - q_jFs_j - q_j^2(1-F)s_j$$
$$- 2q_j(1-q_j)(1-F)s_jh_j] \qquad (13.3)$$

Here x_i is the probability of a certain environmental cause of death. The product comprises all x_i and all q_j, the gene frequencies of deleterious mutations. Both the number of these mutations and the number of environmental factors x_i can be assumed to be large; the single probabilities, however, are small. Therefore this expression is approximated by the following:

$$S = 1 - \Sigma x - \Sigma Fqs - (1 - F)\Sigma q^2 s - 2(1 - F)\Sigma q(1 - q)sh$$

This, in turn, may be approximated by:

$$S = e^{-(A + BF)} \text{ or } -log_e S = A + BF \qquad (13.4)$$

where

$$A = \Sigma x + \Sigma q^2 s + 2\Sigma q(1 - q) \, sh$$

$$B = \Sigma qs - \Sigma q^2 s - 2\Sigma q(1 - q) \, sh$$

The summation includes all environmental factors and all loci with mutant alleles.

In a randomly mating population ($F = 0$) the expressed genetic damage – together with the environmental damage – is represented by A. B, on the other hand, is a measure of the hidden genetic damage that could manifest itself only with complete homozygosity ($F = 1$). The total genetic damage is expressed by Σqs, which is the sum of B and the genetic component of A and therefore lies between B and $B + A$.

A and B can be calculated using weighted regression coefficients of $log_e S$ (S = fraction of survivors) on F. Considering the low degree of inbreeding normally found in human populations and the low death rate among offspring of unrelated couples, the following simplified formula gives a satisfactory approximation:

$$S = 1 - A - BF \qquad (13.5)$$

The actual calculation may proceed as follows:

$$S_1 = 1 - A, \quad S_2 = 1 - A - FB, \quad S_1 - S_2 = FB;$$

$$A = 1 - S_1; \quad B = \frac{S_1 - S_2}{F},$$

S_1 = number of survivors in nonconsanguineous matings, and S_2 = number of survivors in consanguineous matings.

The number of lethal equivalents is derived from the difference between consanguineous and nonconsanguineous marriages in the number of children stillborn and deceased before reproductive age.

An Example. Investigators used data from France for a preliminary calculation [25]. For stillbirth and death during childhood and youth (before reproduction) together, a value of B between 1.5 and 2.5 was calculated; $A + B$ was not very much higher. The ratio B/A, which was to play a major role in later discussions, varied between 15.06 and 24.41. This would mean that the average gamete carries a number of deleterious genes which, if distributed in single individuals and made homozygous, would kill 1.5–2.5 individuals before reproductive age. The total genetic damage is 1.5–2.5 lethal equivalents per gamete; 3–5 lethal equivalents per zygote. This calculation does not include miscarriages and deaths at a later age (for example, during the reproductive period); this approach therefore underestimates the real damage. Every human appears to be heterozygous for several genes that would be deleterious in the homozygous state.

The authors cautiously admit that the differences between consanguineous and nonconsanguineous marriages may in part be nonbiological. Only the pregnancy outcome of consanguineous marriages was determined by direct interview.

Socioeconomic differences between rural and urban populations might add another bias toward higher mortality among children from consanguineous marriages. As we shall see below this caution was only too well justified.

Estimate of the Expressed Genetic Damage. As the next point, the authors concluded that the same genes may have a certain disadvantage even in the heterozygous state, i.e., that their "dominance" h is greater than 0. According to Eq. 13.5, the total probability of a particular mutant being eliminated under the naturally occurring breeding structure is approximately $z \times s$, where $z = F + q + h$ (notations as above). The number of expressed lethal equivalents per gamete can be shown to be equal to the total number of lethal equivalents multiplied by the harmonic mean of the values of z for the particular mutants. Data on human beings were not available to determine h, and data from *Drosophila* were therefore used, which gave values of h for 16 autosomal lethals with a mean of about 0.04. Considering the fact that the more deleterious mutants might be rarer in natural populations, and assuming that most of the adverse effects are produced in heterozygotes (due to their higher frequency), the harmonic mean of z for all deleterious genes was estimated to be 0.02. With 1.5–2.5 as the total number of lethal equivalents per gamete, this corresponds to 3%–5% of expressed lethality per gamete or 6%–10% per zygote.

Estimate of the Overall Mutation Rate of Detrimental Mutations. As noted, Haldane [32] had postulated as early as 1935 a genetic equilibrium between selection and mutation. Over a sufficiently long time the number of new mutations would be equal to the number of detrimental alleles per generation that are lost due to lethality. Therefore the mutation rate was also thought to be approx. 0.03–0.05 per gamete per generation. The authors assumed that one-half to two-thirds of the real genetic damage could not be discovered by analysis of stillbirths and infant mortality; for example, early embryonic death could not be detected. Taking this into account, a total mutation rate of 0.06–0.15 per gamete was calculated [54], a value in line with Muller's estimate in his paper *Our Load of Mutations* [56]. The reader should keep in mind, however, that this figure rests on two assumptions:

1. That the higher incidence of stillbirth and neonatal death in offspring from consanguineous than from nonconsanguineous matings – as analyzed in their paper and leading to a high B/A ratio – is a true biological consanguinity effect.
2. That lethal and detrimental genes reduce the fitness of the heterozygotes as well.

Much of the criticism of conclusions from the theory of genetic load centers around these two assumptions.

Impact of the Genetic Load Concept on Human Population Genetics. The overall picture presented by this concept is rather gloomy. Everyone is heterozygous for a number of genes that lead to genetically determined death not only in the homozygous state but in heterozygotes as well. There is a constant influx of new mutations at a high rate and with deleterious effects. Due to the adverse effects of these mutations virtually everyone is less healthy and carries more defects than if he were free from these mutations.

This concept had a strong impact on theoretical thinking and the planning of research in human population genetics. This effect may have been due in part to its intrinsic appeal, as research along these lines promised an overall view of problems crucial for the future of our species. The scientific reputation of the team that proposed this concept may have contributed strongly to its success: Muller, the Nobel laureate who out of deep concern for the future of our species had left his fruit flies to help save mankind; Crow, the population geneticist of high reputation who guaranteed the solidity of the approach; and Morton, the brilliant young man whose imagination opened the way to a bright scientific future.

Discussions and Controversies Concerning the Load Concept. The concept of genetic load was discussed extensively by population geneticists [47]. On the one hand, it was asserted that investigation of the outcome of consanguineous as compared with nonconsanguineous matings could contribute to the problem of whether detrimental mutations ("mutational load") or balanced polymorphisms due to heterozygote advantage ("segregational load") contribute more to the genetic load of the human species [15–17]. On the other hand, it was shown that the genetic load concept might in some cases lead to absurd consequences [47]. Many geneticists seem to share the feeling that the definition of genetic advantages and disadvantages in the basic mathematic model is too static, and that the concept should therefore be applied with great caution. (For a recent, somewhat more realistic variant of the concept see [18].)

13.1.2.2 Practical Applications of the Theory

Many attempts have been made to estimate the actual genetic load in human populations. Most older studies were based on the theory of genetic load and may be considered as practical applications of this theory. Some of the more recent studies, however, are based on more direct medical evidence. These are discussed in Sect. 13.1.2.4.

Attempts to Assess the Genetic Load by Consanguinity Studies. The effect of parental consanguinity on stillbirth frequency and childhood mortality has been examined in many studies. The most comprehensive, and for many reasons most reliable, set of data is from Japan [58, 59, 65, 66].

Table 13.2 gives an overview of these studies up to 1972. The sizes of the noninbred control samples are not given; they are usually larger than the inbred

samples. The B/A ratio shows variations between $+15.2$ and -5. A negative value means that childhood mortality was even lower in consanguineous than in nonconsanguineous matings. Formally, such a result would establish a "negative genetic load," which biologically is nonsense.

Most consanguinity studies on genetic load yielded B/A values between 5.7 and 7.8; a simple unweighted average gives a B/A ratio of 6.7. This result is probably exaggerated by a lower socioeconomic status of the consanguineous couples. Although consanguineous marriages are socially much more accepted – and also more frequent – in Japan than in other, especially Christian, countries, socioeconomic biases are all-pervading and, moreover, variable in direction [67].

This effect was found to be still stronger in studies carried out in South America, France, the United States, India, and Africa (Table 13.3). For easy comparison the calculation of B/A ratios was restricted here to first-cousin marriages; hence, the data are not strictly comparable with the Japanese data of Table 13.2. However, the variability seems to be still larger than that among the various Japanese series. This is not surprising, as in European countries and the United States consanguineous marriages have become very rare (Table 13.1). In countries with a strong Christian tradition there is social pressure against such unions, and consanguineous couples differ in social and even psychological aspects from the average population, as explained above.

One way to eliminate at least a part of these biases is to use the children of brothers and sisters of consanguineously married persons as controls. A study carried out along these lines in the Vosges mountains of France [29] compared 189 consanguineous marriages with 646 control marriages. The difference in perinatal death rate was small and nonsignificant for first-cousin marriages; it was negligible for the more remote degrees of consanguinity (Table 13.4). The difference in the number of sterile couples is significant between consanguineous and nonconsanguineous marriages. This result may, but need not necessarily, point to a higher intrauterine death rate. Moreover, we should always keep in mind that the A term in the B/A ratio contains not only the genetic component of mortality in a random mating population but all environmental components. (Apart from the biases, the B/A ratio is also sensitive to details of statistical evaluation [26].

Altogether, the data in Tables 13.2 and 13.3 are rather disappointing. Considering the wide spread of B/A ratios, we cannot even conclude with confidence that inbreeding as such enhances the risk of stillbirth and child mortality, although this conclusion is plau-

Table 13.3. Mortality effects from inbreeding studies. (From Fraser and Mayo 1974 [26])

Reference	Population studied		B/A from marriages of first cousins compared with unrelated couples
	Racial classification	Location	
Neel (1963)[a]	Blacks	Brazil	6.9
Neel (1963)[a]	Blacks	Brazil	7.6
Freire-Maia (1963)	Blacks	Brazil	9
Freire-Maia and Azevedo (1971)	Blacks	Brazil	3
Neel (1963)	Africans	Tanganyika	−1.0
Neel (1963)	Whites	Brazil	−0.0
Neel (1963)	Whites	Brazil	−0.6
Freire-Maia (1963)	Whites	Brazil	1.0
Freire-Maia and Azevedo (1971)	Whites	Brazil	3
Freire-Maia et al. (1963)	Mixed	Brasil	16.5
Neel (1963)	Whites	U.S.A.	7.2
Neel (1963)	Whites	Chicago, U.S.A.	6.6
Neel (1963)	Whites	Morbihan France[c]	20.2
Neel (1963)	Whites	Loir-et-Cher, France[c]	13.1
Neel (1963)	Whites	N. Sweden	−3.0
Kumar et al. (1967)		Kerala	20
Roberts (1969)[b]	Indians	Kerala	14.8

[a] Neel (1963) summarizes the work of many authors.
[b] Roberts modified the results of Kumar et al. (1967) [43a] considering consanguinity other than first cousin relationships.
[c] These studies were used in Morton et al. 1956 [54].

Table 13.4. The effects of inbreeding in the Vosges region of France, 1968 (from Jaquard 1974 [39])

	First-cousin marriages		Second-cousin marriages		All couples	
	Consanguine-ous couples	Control couples	Consanguine-ous couples	Control couples	Consanguine-ous couples	Control couples
Mean number of children per family	4.2	5.1	4.8	4.8	4.5	4.8
Sterile couples (%)	–	–	–	–	6.9	4.6
Perinatal deaths (%)	11.1	9.0	8.0	7.9	8.9	8.5

sible. One conclusion, is obvious, however. The very high B/A ratios detected in the data analyzed by Morton et al. [54] were not found in any other studies (see footnote c, Table 13.3). They were most likely caused by ascertainment biases or by socioeconomic differences between consanguineous and nonconsanguineous matings.

Recessive Diseases and Congenital Malformations in the Offspring of Consanguineous Marriages. The discussion above is very abstract. If children from consanguineous matings are more often stillborn or tend to die during infancy and childhood, the question immediately arises: Why do they die? Do they suffer from known recessive diseases or from multifactorial threshold conditions such as malformations?

Here the most comprehensive sets of data come again from Japan [65]; Table 13.5 shows the incidence of major malformation in three different cities. There is a significant difference between consanguineous and nonconsanguineous matings. In the same large cohort of newborn Japanese infants, the overall frequency of death together with major congenital defects was 4.3% in the control children and 6.2% in the offspring of first cousins. Most of the congenital anomalies in the Japanese data were malformations – sometimes complex in nature – for which a recessive mode of inheritance had never been established. The approximate number of identifiable diseases with a confirmed autosomal-recessive mode of inheritance cannot be established even from the most detailed survey [66]. This means that the huge amount of work invested in studies on consanguinity

Table 13.5. Distribution of children with severe congenital anomalies (SCA) by city of origin and parents' degree of consanguinity (from Schull 1958 [65])

City	First-cousin marriages	First-cousin once removed marriages	Marriage of 2nd degree cousins	Parents unrelated	Total
Hiroshima					
Number of children	936	313	384	26012	27645
Number with SCA	17	2	4	293	316
Percentage	0.0182	0.0064	0.014	0.0113	0.0114
Kure					
Number of children	318	113	140	7544	8115
Number with SCA	4	2	1	58	565
Percentage	0.0126	0.0177	0.0071	0.0077	0.0080
Nagasaki					
Number of children	1592	412	637	30240	32881
Number with SCA	27	4	8	300	339
Percentage	0.0170	0.0097	0.0126	0.0099	0.0103
Total					
Number of children	2846	838	1161	63796	68641
Number with SCA	48	8	13	651	720
Percentage	0.0169	0.0095	0.0112	0.0102	0.0105

Analysis[a]			
	χ^2	df	p
Cities	7.269	2	$0.02 < p < 0.05$
Consanguineous marriages *vs* normal marriages	11.775	3	$0.001 < p < 0.01$
Interaction	2.535	6	$0.75 < p < 0.90$

[a] Roy and Kastenbaum's method (1956) is used; see Schull (1958) [65]. There is a significant difference between consanguineous and normal marriages in regard to frequency of major malformations.

effects has provided no information that would enable us to apportion at least part of the inbreeding effects to a clearly defined group of genes – those leading to recessively inherited diseases in the homozygous state.

This does not mean, of course, that all children with autosomal-recessive diseases who are born in a consanguineous marriage have inherited their two mutant genes from a common ancestor. A patient with cystic fibrosis with consanguineous parents, for example, had two different CFTR gene mutations [75]. The proportion of homozygosity due to the consanguinity of parents decreases with increasing gene frequency.

Data on parental consanguinity have been registered systematically since 1967 in Norway, which has a low rate of consanguinity. The stillbirth rate in first cousin marriages was increased to 23.6/1000, compared to 13.4/1000 in controls; neonatal deaths were increased from 14.9 to 34.9/1000, and there were 4.6% malformations detected immediately after birth, compared to 2.2% among controls. The average birth weight of children from consanguineous marriages was also reduced [51]. However, the proportion of clearcut autosomal-recessive phenotypes could not be determined.

Other Parameters Showing an Inbreeding Effect: Cognitive Abilities. In the course of several studies, additional parameters were examined for consanguinity effects such as anthropometric measurements, dental characteristics, blood pressure, coordination, visual and hearing acuity, intelligence, and school performance [28, 59, 66, 67]. On the whole, and neglecting some inconsistencies among the various sets of data, there was usually a slight lowering of performance with inbreeding. This was especially interesting – and only partially accounted for by socioeconomic differences – in the case of intelligence and school performance. After controlling for socioeconomic factors by appropriate statistical techniques, there remained a decrease of about 6 points in the IQ value per 10% of F (= 1.6 times the inbreeding coefficient in first-cousin matings) on the verbal and performance part of a standard intelligence test (Wechsler Intelligence Scale for Children; for comparison: a child from a first-cousin marriage has $F = 1/16 = 6.25\%$). School performance was comparably lower [66].

The conclusion that inbreeding reduces average cognitive performance was corroborated by a study on 3203 Arab school children in Israel. In this Arab population, the rate of first-cousin marriages is about 34%; about 4% are double first-cousin marriages. Socioeconomic conditions were carefully considered; these were practically identical in the inbred group and the noninbred controls. The average performance on three different intelligence tests was significantly lower in the children from first-cousin and especially from double first-cousin matings than in the control group. Average school performance in four major subjects showed the same difference [4].

Overall Estimate of Zygote Loss Due to Parental Consanguinity. Consanguinity data on intelligence or anthropometric measurements are interesting from many points of view. However, they do not contribute to our knowledge of the influence of parental consanguinity on mortality of the zygotes before reproductive age. This is the only important parameter for the problem of "lethal equivalents." Tables 13.2 and 13.3 present data confined to conceptuses who died around the time of birth and during early childhood. The data do not include loss of conceptuses due to early miscarriages those that survived from about the age of 8 up to adulthood. The latter period may safely be neglected, as mortality within this group is generally very low in modern society. Only very few data are available regarding spontaneous abortion. Schull and Neel (1972) [67] attempted an estimate, shown in Fig. 13.11.

13.1.2.3 Critical Evaluation

Theoretical Interpretation. Taken at face value, all these data could be interpreted in terms of lethal equivalents, as proposed by Morton et al. [54], and their number could be estimated. However, the very fact of the variability of B/A values among the various studies (Tables 13.2, 13.3) should be a cause of skepticism. The genetic model underlying the analysis of these data assumes that the effects of lethal genes are independent of each other. Such an assumption is certainly an oversimplification that soon leads to conceptual difficulties, which might be overcome by adoption of other genetic models such as assuming a certain threshold for the tolerable number of homozygous genes compatible with survival. However, even this adjustment does not remove the main difficulty in interpreting such data: the lack of specificity

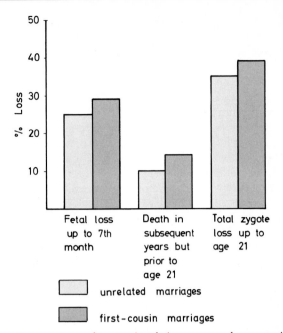

Fig. 13.11. Loss of zygotes in relation to parental consanguinity by miscarriage and by early death (8th month of pregnancy to the age of 21). (From Schull and Neel 1972 [67])

of the phenotypic differences between offspring of consanguineous and nonconsanguineous marriages.

Medical Evidence. Let us look at stillbirth frequency and neonatal death – the most frequently used parameters. Both are known from medical information to be caused sometimes by genetic factors. In the majority of cases, however, no hereditary factors can be implicated. The placenta may detach itself prematurely, or the child is strangled by the umbilical cord, or its position is such that delivery cannot proceed properly, or delivery lasts so long that the child is suffocated, and so on. Death in infancy or early childhood may be due to infection, malnutrition, or many other reasons. In some fatalities a genetic component is often plausible, in others doubtful, and in still others very unlikely. Moreover, pregnancy supervision, improved obstetrical techniques – for example, frequent cesarean sections – and better medical care for the newborn have succeeded in reducing the perinatal death rate to 1%–3% or less in all developed countries. This frequency is less than 10% of the perinatal mortality rate around 1900. One might argue that many children suffer from a gene mutation that would be lethal under more primitive living conditions. However, such children do not appear to suffer any ill effects following the critical perinatal period.

One example that has been analyzed fairly well is pyloric stenosis, which is more frequent in male infants. The recurrence risk is about 2%–6% for first-degree relatives of male probands and 10%–20% for first-degree relatives of female probands. All data point to multifactorial inheritance in combination with a threshold effect (Sect. 6.1); a certain increase with parental consanguinity should therefore be expected. Patients have a hypertrophic pyloric muscle that prevents the stomach from emptying its contents into the duodenum, and in earlier times such patients often died as infants. Today the surgeon cuts the pyloric muscle, which allows normal gastric emptying. This simple operation removes all ill effects, and the children grow up to become normal adults. It has been suggested that they have well-developed muscles and score above-average success in athletics (C. O. Carter, personal communication).

If the effects of many other lethal equivalents causing early death resemble this example, then "so what"? Let us live happily ever after with our lethal equivalents. Other causes of early death, however, may not be quite as trivial. For example, the studies in Japan suggest that one reason for the increased death rate of infants and children from consanguineous marriages is a higher susceptibility to infections [66]. The reader is invited to consult textbooks on obstetrics and pediatrics for causes of stillbirth and neonatal and infant death. Some are genetic and many can now be defined exactly. Other genetic conditions will certainly be discovered by future research in medical genetics; large-scale consanguinity studies have so far contributed little to identifying of genetic causes of neonatal deaths.

13.1.2.4 More Direct Approaches for Calculating the Number of Deleterious Recessive Genes per Individual

More direct approaches have been used in recent years for calculating the average number of deleterious recessive genes per individual. These studies were limited to genes leading to abnormal phenotypes in the homozygote. This means that mainly that part of the overall "genetic load" was considered which is relatively well defined from the genetic and medical viewpoint. One approach is the study of children from incestuous matings between close relatives. Here the highest inbreeding coefficients of all human matings are observed ($F = 1/4$ in father-daughter and brother-sister matings). Hence, if human beings are indeed often heterozygous for deleterious recessive genes, a high proportion of homozygous and therefore severely handicapped children should be detected among the offspring of such matings.

Studies on Children from Incestuous Matings [1, 10, 70]. Four studies have been published on children from matings between first-degree relatives (Table 13.6). All investigations show that a high proportion of children from such matings are severely defective. Taken at face value, this result would suggest that the number of recessive genes carried in the heterozygous state by the individuals involved in these matings is very high indeed. Moreover, diseases with a known autosomal-recessive mode of inheritance such as Sanfilippo disease, a mucopolysaccharidosis (Sect. 7.2.2.3), homocystinuria (Sect. 7.2), cystic fibrosis, and deaf mutism were found among these children.

If all severely defective children in these series are taken to be homozygotes for recessive genes, the number of such genes can be calculated [63]. In the four series of Table 13.6, 85 of 190 surviving children (44%) are affected with a disease that could be caused by homozygosity of a recessive gene. If a man is heterozygous for one recessive gene (Aa), his daughter (or sister) has a chance of 1/2 of being heterozygous as well, and there is a $1/2 \times 1/4 = 1/8$ risk that the child will be homozygous for the anomaly.

Table 13.6. Children from matings between first-degree relatives (father-daughter and brother-sister matings)

Authors	Country	No. of children	No. of severely defectives	Remarks
Adams and Neel (1967) [1]	U.S.A.	18	6	In a control series of the same size that was matched for age, race, weight, stature, and social status, only one defective child was found
Carter (1967) [10]	U.K.	13	8	
Seemanova (1971) [70]	C.S.S.R.	161 (138 surviving)	60 of 138 survivors; (40 severely mentally defective)	In a control series of 95 children from the mothers involved in the incestuous matings with nonconsanguineous spouses, only five severe defects were found; no child was severely mentally retarded
Baird and McGillivray (1982) [3]	Canada	21	9	

If the man is heterozygous for n recessive genes, his child from a father-daughter or brother-sister mating has a chance of $(7/8)^n$ of not being homozygous and therefore of being unaffected. This opens the possibility to calculate n, the number of recessive genes for which an individual is heterozygous. In the data of Table 13.6, $(7/8)^n = 1 - 0.44 = 0.56$; $(7/8)^4 = 0.586$; $(7/8)^5 = 0.513$. Therefore the number of recessive genes per individual would be between 4 and 5, well in line with Muller's estimate.

Most of the defective children, however, suffered from less characteristic abnormalities such as "uncomplicated idiocy." Moreover, the study comprising the most comprehensive data [40] showed clearly that many of the adults involved in these incestuous matings were mentally subnormal. Therefore it is extremely difficult to disentangle the contribution of autosomal-recessive defects that were uncovered by close consanguinity, on the one hand, and the contribution of parental genotypes and adverse environmental conditions, on the other. How many children might have suffered, for example, from the fetal alcohol syndrome? Still, it is reasonable to assume that homozygosity contributes a significant proportion of these defects. This conclusion would mean that, in the parent population, many individuals may be heterozygous for such genes.

Consanguinity in Parents of Severely Mentally Retarded Children. If heterozygosity for genes leading to severe mental retardation is relatively common, homozygotes for such genes should be relatively frequent among severely mentally retarded individuals, and one would expect an increase in parental consanguinity in this group. Moreover, sibs of such patients should fall clearly into two groups, normal and severely mentally retarded. This prediction is confirmed by studies from Israel [14]. A study of 904 families of mental retardates reported 18% individuals regarded as homozygotes among severe retardates with nonconsanguineous parents and no affected sibs, and approximately 75% among severe retardates with affected sibs or first-cousin parents. Most of these probands had no metabolic diseases. Another study [55] on 703 probands and their families in Hawaii observed a certain increase in parental consanguinity, especially in a group called "biological" and comprising mainly severely retarded patients. This, again, suggests a certain proportion of cases with simple recessive inheritance.

Alternative Approach for Calculating the Average Number of Deleterious Recessive Genes in Humans. With the increasing use of screening programs for inherited metabolic disease it is now feasible to calculate the proportion of heterozygotes for such diseases ($2pq$) directly from the homozygote frequency (q^2). Harris (1975) [36] based such an estimate on data from a comprehensive screening program in Massachusetts [45]. From the data in Table 13.7, it can be concluded that an average of 11% of the parent population are heterozygous for one of the 14 conditions screened in this population.

It is tempting to expand this consideration. Let us assume that there are genes for another 100 recessive diseases in the same population, and that each of

Table 13.7. Incidence of certain metabolic disorders among newborn infants in Massachusetts (from Levy 1973 [45])

Disorder	Total screened	No. detected	Incidence	Estimated frequency of heterozygotes (per 1000)
Phenylketonuria	1 012 017	66	1 : 15 000	16
Cystinuria	350 176	23	1 : 15 000	16
Hartnup disease	350 176	22	1 : 16 000	16
Histidinemia	350 176	20	1 : 17 500	15
Argininosuccinic acidemia	350 176	5	1 : 70 000	8
Galactosemia	588 827	5	1 : 118 000	6
Cystathioninemia	350 176	3	1 : 117 000	6
Maple syrup urine disease	872 660	5	1 : 175 000	5
Homocystinuria	480 271	3	1 : 160 000	5
Hyperglycinemia (nonketotic)	350 176	2	1 : 175 000	5
Propionic acidemia (ketotic hyperglycinemia)	350 176	1	<1 : 350 000	3
Hyperlysinemia	350 176	1	<1 : 350 000	3
Vitamin D dependent rickets (with hyperaminoaciduria)	350 176	1	<1 : 350 000	3
Fanconi syndrome	350 176	1	<1 : 350 000	3
				110 = 11%

them has a homozygote frequency of 1:1 000 000 (gene frequency $q = 1/1000$; heterozygote frequency $2pq \approx 1/500$). Neglecting multiple heterozygotes, this would add another 20 % to the number of heterozygotes, giving an overall frequency of 11 % + 20 % = 31 % heterozygosity for any one of 114 (14 + 100) recessive genes in the general population. Since there are more than 1600 recessive diseases, and the frequency of some of these is higher than 1 in 1 million, the figure of 31 % is a minimum estimate of heterozygosity. As noted, there appear to be a number of recessive genes that lead to unspecified severe mental retardation. All these genes together might even be fairly common. No quantitative estimates for the exact frequency of such genes are possible at present.

Consanguinity Effects and the Level of Genetic Analysis. Analyses in population genetics will prove more satisfying the closer the objects of analysis are to gene action. One of the reasons for the satisfying results of analysis of natural selection on the hemoglobin variants is that these variants were analyzed at the level of genes and gene action (Sect. 7.3). This permitted an incisive scrutiny of the mechanism of selection.

The possible genetic damage by lethal or detrimental mutations as revealed by offspring of consanguineous as compared to nonconsanguineous marriages has been analyzed so far principally at the biometric-phenotypic level: genetic variability seems to be present, but simple modes of inheritance or specific genes have not been identified. Data have consequently been collected and possible consanguinity effects have been revealed; interpretations, however, in terms of genetic mechanisms are extremely difficult and, for many questions, controversial. To overcome these difficulties elaborate statistical techniques have been applied, but mainly with the result that socioeconomic variability was shown to obscure most of the biological effects, making specific conclusions regarding genetic mechanisms hazardous if not impossible. As discussed below (Chaps. 15, 16), essentially the same is true for many aspects of human behavior genetics; analysis at the phenotypic-biometric level leads to ambiguous results.

More satisfactory results can be expected from studies on the frequency of heterozygosity for well-defined recessive diseases, such as the metabolic defects included in population screening programs. These diseases have been analyzed at the gene product biochemical and DNA level and often at the level of enzyme activity. Therefore calculating average heterozygosity per individual may lead to fairly clearcut results.

In studies on incestuous matings many of the observed phenotypes are not well-defined, but it is reasonable to assume that for some a single recessive mode of inheritance may also apply. Studies on the outcome of consanguineous marriages show, again, that a global, comprehensive approach with broadly defined phenotypes leads to less satisfactory results than an analysis of specific, well-defined traits or diseases.

Despite these difficulties it is possible to ask a number of specific questions in consanguinity studies.

Effect of Long-Standing Inbreeding. What is the consequence of long-lasting inbreeding – as in Southern Indian populations, where uncle-niece unions with $F = 1/8$ are socially preferred – for the frequency of recessive genes in the population? To what degree are these genes being continuously eliminated? Any hidden deleterious genes can be exposed by inbreeding only if the ancestors of individuals who undertake a consanguineous marriage had mated at random for many generations. A past history of many generations' inbreeding would have "cleaned" the gene pool from deleterious mutations a long time ago. Assuming a constant overall mutation rate, the overall selection against deleterious genes could adapt itself to this mutation rate, but the gene frequencies at equilibrium between mutation and selection would be much lower than in a population with a long history of random mating. Hence, comparison of progeny from consanguineous and nonconsanguineous matings in a population with a history of inbreeding should lead to a smaller B/A value than in a population with a past history of more or less random mating.

This expectation has recently been confirmed in studies on fetal development, incidence of malformations, and reproductive wastage (abortions, stillbirths, death during first year of life) among the offspring of more than 20 000 women in southern India [61, 62]. In the sample examined for the study on reproductive wastage, almost 47 % of women in rural areas and 29 % of those in towns were married to a close relative, in at least 80 % of cases to a maternal uncle ($F = 1/8$). Among more than 70 000 pregnancies there was a marked difference between women living in rural and urban areas, rural women having suffered a much higher incidence of fetal loss and infant death during the first year of life. However, there was only a marginal difference between consanguineously and nonconsanguineously married women in both rural and urban subsamples, and no consistent increase with degree of consanguinity. More than 14 000 pregnancies were followed in the study on fetal development and malformation, which was performed using a prospective research design in the same sample of mothers. No increased incidence of congenital malformations in comparison with controls was observed in children of consanguineous parents. There was no effect of consanguinity, on gestational age, birth weight, or body length.

The entire body of data appears to confirm the expectation that long-standing, high inbreeding depletes the gene pool of deleterious genes. However, even this conclusion cannot be generalized; another study from a different southern Indian population [23] found an extremely high inbreeding coefficient ($F = 0.0414$), but genetic diseases among children, especially those with an autosomal-recessive mode of inheritance, were clearly increased in the inbred group. Inbreeding is also very high in many Muslim populations, where marriages between first cousins are very frequent. Still, an increase in untoward pregnancy outcome has been observed repeatedly [31, 38, 71]. Thus in view of the difficulties discussed above in disentangling biological from socioeconomic effects, the conclusion that long-standing, intensive inbreeding depletes the genome of genes that are detrimental in homozygotes should be regarded with caution, although it is plausible.

This notion explains why, in theory, a comparison of two populations, one having a long history of close inbreeding and the other a history of nearly random mating, would be interesting. However, the two populations should live under otherwise similar conditions and should have a similar anthropological background. There exist, for example, other southern Indian populations in which inbreeding has been so much lower that they could serve as suitable controls. Again, however, the problem for such a research project would be the suitable selection of gene-defined phenotypic characteristics. The aspects of consanguinity for genetic counseling are discussed in Chap. 18.

13.2 Differentiation Between Population Subgroups

13.2.1 Genetic Distance

Real Mating Structure of Human Populations. The usual assumption that in human populations random matings prevail is an abstraction. Even beyond the circle of immediate relatives, the choice of mates is not random at all. It depends on distance of birth places, limitations due to language, race, social class, religion, and other factors. At present this nonrandomness is tending in many societies to decrease strongly. Throughout history, however, nonrandom matings have been much more frequent than today. The choice of mates was especially restricted during the thousands of generations in prehistory during which our ancestors lived in small groups as hunters-gatherers. Studies of present-day populations still living under comparable conditions are especially well suited to give an impression of the population structure at that time.

Population groups living in relative isolation from each other gradually become different genetically – either under the influence of differing selection pressure or due to different adaptations to the same selective agent, or simply due to chance fluctuations of gene frequencies (Sect. 13.3). The ultimate goal of population genetics is a causal analysis of such population differences. In most cases, however, this is impossible. In the absence of more specific evidence it is often useful to begin with the assumption that genetic similarities

among populations are caused by common descent. Therefore the more similar two populations are, the closer is their genetic relationship. This means that their separation is assumed to date back only a short time if they are similar, and a longer time if they are less similar. Hence we draw conclusions from the present-day genetic composition of population groups as to their history. This concept is in principle the method underlying any classification of human beings into subtypes such as races.

Section 12.2.1.7 presents a well-analyzed example of this kind: the spread of HbE in the Austro-Asiatic language group of southeastern Asia. Here the mutation to the HbβE allele became frequent under the influence of a selective advantage due to malaria. It "migrated" with subpopulations of this group into various regions, and a high frequency of HbβE is now an indicator not only of continuing selection due to malaria but also of descent from a certain population group. Thus the HbβE gene is an indicator of population history.

Population History or Selection? At the same time, the southeastern Asian example shows how ambiguous the distribution of such a genetic trait may turn out to be. The HbβE gene is usually rarer in populations of the great southeastern Asian river plains than in the more malarious hills. This differential distribution, however, has nothing to do with a separation in remote history but is simply a consequence of less intense selection pressure due to malaria in relatively recent times.

Genetic similarities between populations are ambiguous. They may, but need not necessarily, reflect a common history. They may result from a parallel development under the influence of similar selection pressures. The southeastern Asian example used only a single gene as an indicator of similarity between populations. One could argue that the greater the number of different genes included in such considerations, the less likely it is that all of them were subject to similar selection pressure. Therefore similarity between populations in many different genes argues very much in favor of their relationship by common descent. Research work describing populations in terms of their overall similarities and differences has usually tried to include as many inherited characters as possible and has used methods of multivariate statistics. Nevertheless, conclusions from such studies should always be considered with the reservation that nothing is usually known of the selection pressures that have been involved. A very strong argument in favor of a common origin of two populations – or at least in favor of substantial gene flow between them some time during their history – is the occurrence in both of them of identical mutations, especially if these mutations do not belong to the more common types, such as $G \rightarrow A$ transitions, and if they are found within identical DNA haplotypes.

Methods for Determining Genetic Distances. A number of excellent reviews on methods for assessing genetic distances between populations are available [11, 39, 78]. For continuously distributed characters such as anthropometric measurements the generalized distance D^2 of Mahalanobis is often used; a simplified version is the index C_H^2 of Penrose. For alternatively distributed traits such as genetic polymorphisms with simple modes of inheritance the arc and chord measure of Cavalli-Sforza and Edwards (1964) [12] has become popular.

13.2.2 Gene Flow

Apart from selection (discussed above) and chance fluctuations of gene frequencies (to be examined below), the composition of a population is also influenced by gene flow. The term "migration" is often used to denote gene flow from one population to another.

Effects of Migration on Gene Frequencies [46]. The effect of migration on gene frequency is examined here using a somewhat oversimplified model. A large population may be subdivided into many subgroups, with an average gene frequency $\bar{q}$; each subgroup exchanges a fraction m of its genes with a random sample of the whole population every generation; q the gene frequency in the first generation in the subgroup to be considered. The gene frequency in this subgroup in the next generation is:

$$q' = (1 - m)q + m\bar{q} = \bar{q} - m(q - \bar{q})$$

$$\Delta q = q' - q = -m(q - \bar{q})$$

where Δq is proportional to the deviation of the subgroup gene frequency q from the overall average $(\bar{q})$, as well as to m. In the long run, and in absence of other factors such as differential selection between subgroups, differences level out, and all subgroups will have a common gene frequency $\bar{q}$.

This model is not realistic, as immigrants often come mainly from neighboring subgroups. If neighbors tend to deviate from the average population mean in the same direction as the "receptor" subgroup, the speed of the leveling out between subgroups is reduced. For actual calculations it is more realistic to regard $\bar{q}$ not as the overall population average but as the average of actual immigrants.

Migration and Selection. If the population subgroups are subject to different forces of selection, the process of leveling-out may be counteracted. Three different situations may be distinguished (for a mathematical treatment of these problems see Li [46] and, at a more sophisticated formal level, Jacquard [39]):

1. If rate of migration and intensity of selection are of the same order of magnitude, the gene frequencies of subgroups may remain very different from each other.
2. If selection intensity is much larger than immigration rate, the subgroup gene frequency is determined largely by selection, with only a weak, diluting effect due to migration.
3. On the other hand, if the proportion of immigrants is much larger than selection intensity, the effects of migration override that of selection.

In any case a stable genetic equilibrium between selection, on the one hand, and migration, on the other, may ensue. This situation is somewhat similar to that of the equilibrium between selection and mutation (Sect. 9.1.3).

Measuring the Admixture of Genes to a Population Subgroup. The proportion of genes that one population has received from another by gene flow is often investigated.

Let q_a be the frequency of a gene in the "pure" ancestor population and q_n that of the same gene in the present-day hybrid population for which admixture of foreign genes is assumed. The frequency of this gene in the "donor" population is q_c. The fraction m of genes in the present-day hybrid population that originated in the donor population can the be calculated as follows:

$$q_n = mq_c + (1 - m)q_a$$

and therefore:

$$m = \frac{q_n - q_a}{q_c - q_a}$$

The (large-sample) variance of m can be calculated as follows:

$$V_m = \frac{1}{(q_c - q_a)^2}[V_{qn} + m^2 V_{qc} + (1 - m)^2 V_{qa}]$$

Rationale for Measuring the Admixture of Genes to a Population Subgroup. In recent years the problem of how many genes from whites (and other groups) are present in the African-American populations has received much attention. Arriving at an answer, while simple in principle, depends on a number of conditions that are difficult to meet [64]:

a) The exact ethnic composition of the ancestral populations and the gene frequencies of the genes used for these estimates should be known.
b) There should be no systematic change in gene frequency between ancient and modern generations within either of the two populations for the gene or genes to be included in such a study. Such systematic changes may be caused by natural selection. In African-Americans, for example, the sickle cell gene is relatively common. This gene has attained its present frequency in Africa despite selection against homozygotes by positive selection due to falciparum malaria (Chap. 12), which does not occur in North America. Therefore the sickle cell gene must have been subject to negative selection due to segregation of homozygotes with sickle cell anemia in the United States; an estimate based on this gene would overrate white admixture.

However, this argument can be reversed: If an estimate of admixture is based on one or ideally many genes that fulfill the preconditions, the difference between this admixture estimate and the estimate from the gene under selection may be used to indicate selection and to measure selection intensity.

Estimates of the Admixture of Genes from Whites to African-Americans. African-Americans are descended from slaves imported from western Africa (e.g., Nigeria, Senegal, Gambia, Ivory Coast, Liberia). Gene frequencies of most genetic markers show definite variation in these ancestral populations. The same is true for the American white population, which originated from immigration of various European groups. Moreover, those genes that entered the gene pool of African-Americans may not be an unbiased and random sample of the genes of all North American whites. It is possible that certain segments of the white population were more extensively engaged in outbreeding than others. Still, careful assessment of possible biases may help to estimate the correct order of magnitude of admixture [64].

Calculations of admixture based on blood group systems and serum factors thought to be under little or no differential influence by selection (Rhesus; Duffy; Gm serum groups) range between $m = 0.04$ and $m = 0.30$ for various African-American subpopulations. Estimates from southern rural areas give lower values than those from northern metropolitan cities, which usually range above 0.2.

An example can be used to show the method of calculation. The allele Fy^a of the Duffy blood group system has a gene frequency $q_c = 0.43$ in American whites. In the population of western Africa its present frequency q_a is less than 0.03; in most subgroups Fy^a is completely absent. It may be assumed that its frequency was very low at the time when slaves were taken. Among today's African-American population of Oakland, Calif. ($n = 3146$) the Fy^a gene was found with a gene frequency of $q_n = 0.0941 \pm 0.0038$; the corresponding value in the white population ($n = 5046$) was $q_c = 0.4286 \pm 0.0058$; q_a (gene frequency in the African population) may be assumed to be 0.0. The formula given above leads to the following estimate of admixture:

$$m = \frac{q_n}{q_c} = \frac{0.0941}{0.4286} = 0.2195$$

If q_a is taken to be 0.02, this estimate would be 0.181. Therefore the admixture of genes from whites as estimated from the Duffy blood groups accounts for 18%–22% of the gene pool of African-Americans in Oakland. An estimate of admixture for the same population from AB0 blood groups leads to a similar result ($m = 0.20$) [1857].

Evidence of Selection. As noted, admixture estimates for m calculated for genes that had been subject to selection in Africa can be used to examine whether and in which direction a change of the selection pattern has occurred in their new habitat. A number of studies on African-Americans have consistently indicated higher estimates for three genetic markers: the sickle cell gene ($Hb\beta$ S); the allele for the African variant of G6PD (Gd^{A-}), and the haptoglobin allele Hp^1. As discussed in Sect. 12.2, $Hb\beta$ S and Gd^{A-} in Africa are subject to selection by falciparum malaria; haptoglobin is a hemoglobin transport protein. The values for admixture were consistently higher than those derived from Duffy or AB0 blood groups; they ranged from about 0.49 (Gd^{A-} in Seattle in the northwest) to about 0.17 (Gd^{A-} in Memphis in the south). These results point to selection against these genes in the United States, a country without malaria. Because the necessary data are lacking, the exact extent of this selection cannot be determined. It has been pointed out that because of the many biases in such studies other approaches should be used to confirm selective factors for genes that do not conform to the migration estimates by this method.

13.3 Chance Fluctuations of Gene Frequencies

13.3.1 Genetic Drift

Deterministic and Stochastic Models. All the discussions above are based on Mendelian segregation ratios and the Hardy-Weinberg Law; parameters such as mutation rates, selective values, and inbreeding coefficients are treated as constants that have certain relationships with each other. The models were deterministic. In real life, however, these parameters are statistical variables that are subject to chance fluctuations. The processes examined in population genetics, such as changes in gene or genotype frequencies under the influence of these parameters, are not strictly deterministic; they are stochastic or random processes.

For infinitely large populations chance fluctuations become negligible, and the operation of stochastic processes has no effect on the results of deterministic models. The use of deterministic models is justified if one is interested in learning how in principle a certain parameter changes the genetic constitution of a population.

One example, the dynamics of the alleles $Hb\beta S$ and $Hb\beta C$ in western Africa (Sect. 12.2.1.7; Fig. 12.26), explicitly considered fluctuations of gene frequencies under selection pressure in a finite population, but even if we compare the observed population frequencies of a blood group system with their Hardy-Weinberg expectations statistically, we implicitly consider chance fluctuations as well. Even if all available individuals could be blood-typed, there still would be chance deviations.

During most of human evolution the population size was relatively small, and the human species was fragmented into several small breeding groups. Such isolates remained frequent until recent times, and some continue to exist today. Some of the theoretical consequences of this need to be considered.

Island Model. Let us look at an extreme example. On each of 160 solitary islands in the Pacific one married couple settles. All 320 individuals have blood group MN. Every couple has a son and a daughter who, an incestuous mating, become the ancestors of each island population. No MN genotype has any selective advantage or disadvantage. After several hundred years we examine the MN blood groups of the island populations. What do we find?

The founder population was identical genetically with respect to this characteristic. Our first assumptions is therefore that the same holds true for their progeny, as there was no selection and – let us also assume this – no new mutation. Upon somewhat closer scrutiny, however, this assumption turns out to be wrong. Since in all cases both parents are MN, their children are expected in a proportion of:

1/4 MM + 1/2 MN + 1/4 NN

This leads to the following probabilities for the genotypes of the sib pairs that are the ancestors of the island populations:

$\text{MM} \times \text{MM} = \frac{1}{4} \times \frac{1}{4} = \frac{1}{16}$

$\text{MN} \times \text{MN} = \frac{1}{2} \times \frac{1}{2} = \frac{1}{4}$

$\text{NN} \times \text{NN} = \frac{1}{4} \times \frac{1}{4} = \frac{1}{16}$

$\text{MM} \times \text{MN} = 2 \times \frac{1}{4} \times \frac{1}{2} = \frac{1}{4}$

$\text{MM} \times \text{NN} = 2 \times \frac{1}{4} \times \frac{1}{4} = \frac{1}{8}$

$\text{NN} \times \text{MN} = 2 \times \frac{1}{4} \times \frac{1}{2} = \frac{1}{4}$

It follows that in the first generation an average of ten islands contain only MM individuals, and another ten have only NN individuals. On 40 islands both children are MN. The other 100 islands contain two genotypes each: 40 MM and MN, 40 MN and NN, and 20 MM and NN.

The genotype frequencies of subsequent generations obviously depend on this distribution. This consequence is most obvious for the ten islands with the MM × MM pair and the ten islands with an NN × NN pair. Their populations must be purely MM and NN, respectively. The other allele has been lost by chance – without any selective disadvantage. We may call these processes random fixation and random extinction, respectively.

M and N are common alleles. However, one of the few individuals founding such an island population may be heterozygous for a rare allele. This allele has a good chance to become frequent in subsequent generations, unless eliminated by random extinction (see below). Such "founder effects" are common in human populations. For example, the dominant condition porphyria variegata (176 200) is common in the Afrikaans-speakers of South Africa and has been traced back to an early immigrant, one of the founders of this population group.

A More General Case. Returning to our island example, random extinction and, correspondingly, random fixation of an allele may occur not only in the first generation – as described above – but in subsequent generations as well. Its probability increases with decreasing size, N, of the breeding population. (Here the capital letter N is used, as n comprises the total population including the aged, the children, and the nonfertile. Moreover, random extinction and fixation are only extreme cases; in less extreme cases both alleles are maintained, but their frequencies fluctuate at random.)

This fluctuation is frequently called "genetic drift." Below, genetic drift is examined in a slightly more formal way [46].

Let us assume a breeding population of N diploid individuals. This population may be visualized as the result of a random sample of $2N$ gametes from the preceding generation. In this generation, let q be the gene frequency of the allele a, and $p = 1 - q$ that of the allele A. The number of genes a in the present generation of N individuals follow a binomial distribution: $(p + q)^{2N}$. This means: the $2N + 1$ possible values of the frequency q of a in this generation are:

$$0, \frac{1}{2N}, \frac{2}{2N}, \frac{3}{2N}, \dots, \frac{i}{2N}, \dots, \frac{2N-1}{2N} \qquad (13.6)$$

and the probability for q to assume a certain value $q_j = j/2N$ is:

$$\binom{2N}{j} p^{2N-j} q^{j} = \binom{2N}{2Nq_j} p^{2Npj} q^{2Nqj} \qquad (13.7)$$

Now let $\delta q = q_i - q$ be the chance deviation of q from one generation to the next. (In contrast, Δq denotes the systematic deviation as caused, for example, by natural selection.) It follows from the above distribution that:

$$\sigma^2_{\delta q} = \frac{q(1-q)}{2N} \qquad (13.8)$$

Hence this variance – which describes the extent of the fluctuation in q from one generation to the next – is inversely related to N, the population size. For example:

$N = 50; \quad q = 0.5$

We obtain:

$$\sigma_{\delta q} = \sqrt{\frac{0.5 \times 0.5}{2 \times 50}} = 0.05$$

The probabilities with which the various values of q occur are given in Table 13.8.

This distribution may be viewed from yet another angle. Let us assume a large number of loci, all with a gene frequency of $q = 0.5$ in the parent generation. Some of them assume a higher and others a lower frequency in the next generation, according to the distribution set out in Table 13.8.

Table 13.8. Probability distribution of gene frequency q in $N = 50$ children if $q = 0.5$ in the parent's generation (from Li 1955 [46])

$q =$	<0.35	0.35 –0.40	0.40 –0.45	0.45 –0.50	0.50 –0.55	0.55 –0.60	0.60 –0.65	>0.65	Total
Probability	0.002	0.021	0.136	0.314	0.341	0.136	0.021	0.002	1000

Decay of Variability [46]. The island example above shows that a given allele may disappear by chance from the population; in this case, the partner allele is fixed. As seen from Eq. 13.7 and Table 13.8, this process has a certain – albeit usually small – probability in a finite population. Once fixation has occurred, however, it cannot be reversed. The probability of fixation (i.e., of q becoming 0 or 1) tends to 1 with increasing number of generations. In the long run therefore a population group will sooner or later become completely homozygous even in the absence of selection if neither mutation nor migration disturb the process. This phenomenon is called "decay of variability."

Let K be the rate of fixation or extinction of alleles per generation, $K/2$ being either the fixation or the extinction rate. It can be shown [53] that $K = 1/(2N)$. Hence in one generation $1/(4N)$ of all such alleles become extinct, and $1/(4N)$ become fixed.

We return to this discussion below and extend it, in the context of molecular evolution (Sect. 7.2.3).

13.3.2 Genetic Drift in Co-operation with Mutation and Selection

Mutation. Let us imagine a large population that is composed of a great number of small or moderately sized subpopulations. The distribution of gene frequencies q in these subpopulations depends on the size of the breeding population (N), the mutation rate μ (A → a), and the back mutation rate ν (a → A). If the mutation rate is constant and identical in both directions (A → a; a → A), the distributions in Fig. 13.12 are possible, depending on N. The mean of q is identical in all cases (in our example: $q = 0.5$). The variance, however, increases with decreasing N. With small N, the marginal classes close to $q = 0$ and $q = 1$ predominate, indicating a high rate of random extinction and random fixation. With high N, on the other hand, the distribution clusters closely around the mean.

Fate of a New Mutation. The preceding section considered mutations occurring at a constant, recurrent rate. What is the fate of a single new mutation in a population? The mutation is present in one sperm or

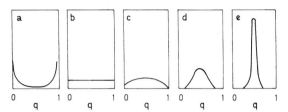

Fig. 13.12 a–e. Distribution of gene frequency q in small populations in relation to the breeding population size N. Mutation rates μ and back-mutation rates ν are assumed to be identical. **a** $N\mu$ very small. **b** $4N\mu = 4N\nu = 1$. **c** $4N\mu = 4N\nu = 1.5$. **d** $4N\mu = 4N\nu = 10$. **e** $4N\mu = 4N\nu = 20$. With small population size (**a**) many populations are homozygous for one of the two alleles ($q = 0$ or $q = 1$); with very large population size (**e**) q clusters around 0.5. (From Li 1955 [46])

one oocyte. The zygote formed from these two germ cells is heterozygous. Each child of the first carrier has a chance of 1/2 to carry this mutation. If this carrier has two children, there is a risk of $(1/2)^2 = 1/4$ for the mutation to be lost by chance after one generation. On the other hand, the chance that the number of new alleles will be two in the next generation is also 1/4. Fisher (1930) [25] calculated the risk of extinction of such a new allele assuming the number of children in sibships to follow a Poisson distribution with mean of 2 (Fig. 13.13). If the mutation is neutral, i.e., if it has neither a selective advantage nor disadvantage, the risk of its disappearing from a population sooner or later is overwhelmingly high. This remains true even if the mutation has a small selective advantage. Such a risk holds, however, only for an infinitely large population; chance fluctuations in a small population may still lead to fixation of this mutation.

Selection. Co-operation between genetic drift and selection may be more complicated, as there are many different modes of selection (Sect. 12.2.1). In Fig. 13.14, the homozygotes AA are assumed to have fitness 1, the heterozygotes Aa, $1 - s$, the homozygotes aa, $1 - 2s$. In an infinitely large population and in the absence of mutations, the gene frequency q approaches 0. For a great number of small subpopulations q becomes 0. In some cases, however, it takes on a higher value, and in a few cases fixations occurs at

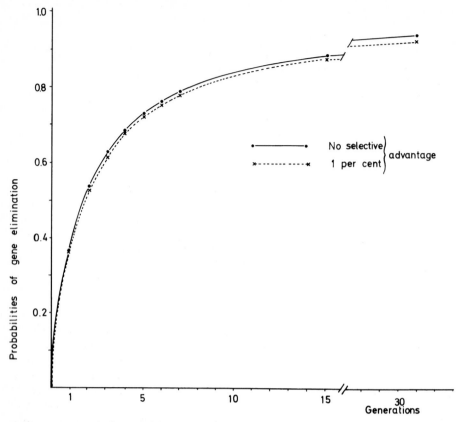

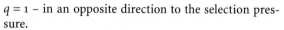

Fig. 13.13. Increase in the probability that a new mutation will disappear from the population. With 1% selective advantage, this probability is almost identical to that in the presence of neither a selective advantage nor disadvantage. Sibship size is assumed to follow a Poisson distribution with a mean of 2. (From Fisher 1930 [25])

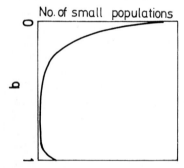

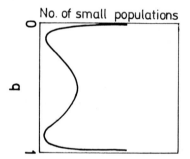

Fig. 13.14. Distribution of gene frequency q for gene a in small populations with $4Ns = 5$ (s, selective disadvantage of heterozygotes Aa; $2s$, selective disadvantage of homozygotes aa). In most populations the gene a is not present at all, in a few it attains a moderate or high gene frequency, and in some it even replaces the other allele. See text for other details. (From Li 1955 [46])

Fig. 13.15. Distribution of gene frequency q for gene a in small populations with $4Ns = 10$ and selective disadvantage s of both homozygotes compared with the heterozygotes. The distribution shows three maxima: in many populations gene frequencies cluster around 0.5; however, there are other populations in which either the allele a or the allele A are completely lacking. (From Li 1955 [46])

$q = 1 -$ in an opposite direction to the selection pressure.

Quite a different distribution of q among subpopulations is found if heterozygotes have a selective advantage (Fig. 13.15). If the disadvantage of the two homozygotes is identical, the values of q cluster around $q = 0.5$, but fixation or extinction still has a small probability of occurring, again, depending on the breeding population size, N. Table 13.9 contrasts the patterns of spread of new genes by drift and selection.

Table 13.9. Patterns of spread of new genes by drift and selection (from Luzzatto 1979 [50])

Drift	Selection
Spread depends on migration of people more than on environment.	Spread depends on environment more than on migration of people.
Increase in gene frequency can be very fast if population is small.	Increase in gene frequency is essentially independent of population size.
Gene spreading may be advantageous or neutral.	Gene spreads only if advantageous in a particular environment.
Mutant gene is "identical by descent" wherever it has spread.	Different mutant genes at the same locus or at different loci that give similar phenotypes may be selected for (evolutionary convergence).

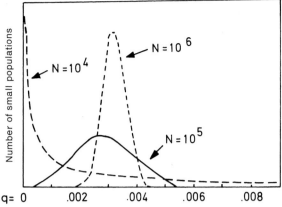

Fig. 13.16. Distributions of gene frequencies q for gene a in relation on the size N of the breeding population with selection $s = 1$ against the homozygotes and a mutation rate $\mu = 10^{-5}$. The distribution of gene frequencies q depends critically on the size of the breeding population N. (From Li 1955 [46])

Mutation and Selection Together. Let us now turn to the distribution in subpopulations of gene frequencies of a completely recessive gene with a mutation rate μ, a selective disadvantage s of homozygotes aa, and, again, a breeding population of size N. Let us also assume that q is small so that back mutations a → A can be neglected. The distribution of q for the extreme case $s = 1$ (complete selection against the homozygotes aa) and for different values of N is given in Fig. 13.16. The mutation rate is taken as $\mu = 10^{-5}$, in accordance with the order of magnitude of the human mutation rates for some visible inherited anomalies (Sect. 9.3). It turns out that, even with a moderate population size of $N = 10^4$, the majority of subpopulations do not harbor the recessive gene at all. In a few populations, however, the gene is much more frequent than in the general population. The affected homozygotes still shows much greater frequency differences since their number corresponds to the square of the gene frequency.

In principle, the same applies for dominant mutations, i.e., if the heterozygote Aa has a selective disadvantage. However, because of this selection against heterozygotes an increase in gene frequency by chance in spite of negative selection pressure becomes less likely. Dominant disorders therefore become frequent in this way only if they cause very little selective disadvantage. With this mode of inheritance the increase in individuals with aberrant phenotypes corresponds only to the gene frequency, not to the frequency of homozygotes.

Rare Inherited Diseases in Human Isolates. These considerations explain why hereditary diseases – especially recessives – occasionally become frequent in small populations that have been living in relative isolation for a long time. Either the allele was introduced by a founder (founder effect) or it happened to be produced by a new mutation. Determining which is the case is generally impossible in concrete instances. In any case, such a gene had a chance to become frequent even in spite of selection against it.

This is one reason why isolates provide much information on rare recessive diseases. Furthermore, consanguineous matings have become less frequent in most modern nonisolate populations. The incidence of homozygotes for rare recessive diseases is therefore reduced below the equilibrium value (Sect. 12.3.1). In relatively isolated – and in many cases rural – populations the traditional ways of choosing a husband or wife, and the old level of inbreeding, tend to be maintained. Consequently there is no overall decline in homozygote frequencies, and the average frequency of recessive homozygotes tends to be higher in isolates than in the general population. This fact, together with the unequal distribution of gene frequencies, makes isolates well suited for the discovery of hitherto unknown recessive conditions. Another possible factor may be selective migration. For a relatively long time the generally fitter and more active individuals have migrated from the isolates into cities and industrial centers. It has been suggested, on the other hand, that heterozygotes for some severe recessive abnormalities might in some cases suffer from minor symptoms (Sect. 7.2.2.8) and may be underrepresented in this migration group.

Investigation of isolates has brought a certain one-sidedness into our knowledge of inherited diseases.

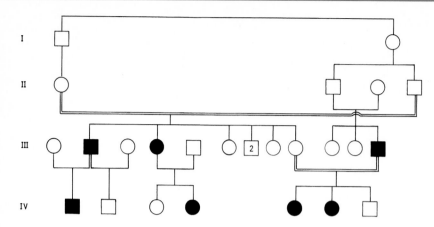

Fig. 13.17. Pedigree from the island Mljet with consanguineous matings and cases with Mal de Meleda. This recessive disease attained a high incidence in the population of this island but is almost entirely unknown elsewhere. (Courtesy of Dr. U. W. Schnyder)

The method can be compared with the microscope: it enables us to examine special parts of a structure very exactly, whereas other parts escape observation completely. There is only one difference: we can direct the microscope to the areas in which we are interested, but in isolate studies chance directs the microscope.

Example: Mal de Meleda. Mal de Meleda is an autosomal-recessive condition categorized in the group of palmoplantar hyperkeratoses; however, unlike the most common dominant type (Unna-Thost), the keratotic changes are not confined to the palms and soles but may also affect other parts of hands and feet. The anomaly was discovered about 150 years ago on the small island of Mljet (Meleda) off the Croatian coast. The population numbers only a few thousand individuals; in 1930 no fewer than 14 of 93 marriages were between first cousins [9]. In the 1960s many persons carrying the anomaly were still alive. Figure 13.17 presents a typical pedigree. Due to the high frequency of heterozygotes and the great number of consanguineous matings, marriages between affected homozygotes and heterozygotes are not rare, and the phenomenon of "pseudodominance" (Sect. 4.1.3) can be observed.

Other Examples. There are many other examples of recessive diseases that have been discovered or described in detail in isolates [24].

One of the first was classical myoclonus epilepsy described by Lundborg [49] (254800) as early as 1913 in the Swedish province of Blekinge. Other examples include the juvenile type of amaurotic idiocy (204200), which was described by Sjögren [72] in a Swedish isolate, Friedreich's ataxia (229300) and a special type of dwarfism investigated by Hanhart in isolated valleys of the Swiss Alps [35]; Werner syndrome (277700; Sect. 10.4.5) in Sardinia [13]; Ellis–van Crefeld syndrome (chondroectodermal dysplasia; 225500) in the Amish, a religious sect living in Pennsylvania [53a]; and tyrosinemia (276700) among French Canadians [5]; sensorimotor polyneuropathy (218000) [21], Tay-Sachs disease [20], and a form of vitamin D dependent rickets (264700) [19]. Structural chromosome variants may become

common in the same way, as shown, for example, by the high incidence of as inverted Y chromosome polymorphism among Indian Muslims from Gujarat in South Africa [6]. This is important for understanding chromosome evolution (Sect. 14.2.1). Diseases that are fairly common might show additional clustering due to genetic drift and founder effects in isolated population groups. Examples include the high frequency of cystic fibrosis in a small area of Brittany in France (1:377 births; about six to eight times as common than in other northwestern European populations) [8] and the same disease (incidence 1:569) in the Ohio Amish [42].

When haplotype and molecular analysis are possible, mutants going back to one mutational event can be identified. In isolates the predominance of one specific mutant and one haplotype is likely. Examples include the phenylalanine hydroxylase mutants in Yemenite Jews and in French Canadians [43], and a Tay-Sachs (272800) mutation in French Canadians that was identified as a deletion not observed in the Ashkenazi Jewish population, among whom Tay-Sachs disease is common [20].

A special problem is posed by the high incidence of three lipid storage diseases due to defects of different lysosomal hydrolases (catabolic enzymes) among the Ashkenazi Jews of eastern Europe [30, 57]. The conditions are: the infantile form of Tay-Sachs disease (gangliosidosis G_{M2}; 272800), Niemann-Pick disease (sphingomyelin lipidosis; 257200), and the adult form (type I) of Gaucher disease (glucosylceramide disease; 230800). A study of different mutations of the Gaucher gene in Israel revealed a quite different distribution in Jewish and non-Jewish patients, and a prevalence of the N370S mutant (an $A \rightarrow G$ transition) in 70% in the Jewish compared with 23% in the non-Jewish subsample [37]. Some conditions in the history of this population seem to favor the hypothesis of genetic drift. During long periods of its history, this population lived as a religious minority in relative isolation, and some claim that the ancestral population of current Ashkenazi Jews was very small, comprising no more than a few thousand individuals (See Motulsky [55a]). However, this evidence is not entirely certain (Neel, in [30]). The population was subdivided into many, sometimes widely scattered, isolates; the population size was very large for drift to be effective, at least in certain periods, and there was some dilution of the gene pool by admixture. Even if all these arguments could be explained away, the fact would remain that no

Table 13.10. Some Ashkenazi Jewish diseases (Modified from Motulsky 1995 [55a])

Disease (McKusick) number	Biochemical defect	Gene structure known	Mode of inheritance	Chromosomal locus	Overall heterozygote frequency in Ashkenazim	Frequency of most common mutation in Ashkenazim	Biological fitness of most common homozygotes
Tay Sachs disease (272 800)	Hexosaminidase A deficiency	Yes	AR	15q23-q24	3%-4%	80% (d)	Lethal
Gaucher disease (230 800)	Glucocerebrosidase deficiency	Yes	AR	1q21	4%-6%	93.5% (d) (in population screening studies) ~70% (d) (among clinically affected patients) See text	At least 1/2 of homozygotes for the common mutation have mild or no clinical illness
Canavan disease (271 900)	Aspartoaclyase deficiency	Yes	AR	17pter-p13	1.7%-2%	83% (d)	Almost lethal
Niemann-Pick disease (258 200)	Sphingomyelinase deficiency	Yes	AR	11p15.4	1%-2%	3 equally frequent mutations (d)	Lethal
Mucolipidosis IV (257 200)	?	No	AR	?	~1%*	?	Lethal but milder variants may exist
Bloom syndrome (210 900)	?	Yes	AR	15q26.1	~1%	97% (i)	Very low
Idiopathic torsion dystonia (128 100)	?	No	AD	9q34	0.1%-0.3%	>90% (i)	Normal? (heterozygotes)
Familial dysautonomia (223 900)	?	No	AR	9q31-q33	3%	75% (i)	Moderately impaired
PTA (factor XI deficiency) (264 900)	PTA deficiency (clotting factor)	Yes	AR	4q35	8.1%	2 equally frequent mutations (d)	Almost normal
Pentosuria (260 800)	Xylitol dehydrogenase deficiency	No	AR	?	2.5%-3%	?	Normal

AR Autosomal recessive; (d), direct estimate; *, uncertain estimate; AD, autosomal dominant; (i), indirect estimate.

less than *three* pathogenetically and biochemically similar genes have spread in the same population. In our opinion, this is likely to be explained by some probably very specific selective advantage of heterozygotes under the special living conditions of this population in the past. Tuberculosis has been suggested as a selective agent, but the proportion of the Ashkenazi Jewish population that appears to be resistant to tuberculosis is much larger than the 4% who are carriers for Tay-Sachs disease. Neel further objects that other urban populations were also infected with tuberculosis, and that these genes have nonetheless not become common. The problem of cooperation between selection and drift is taken up again in Sect. 14.2.3. A number of other diseases have also become common among Ashkenazi jews, reflecting the unique history and demography of this population [55a] (Table 13.10).

"Rare Flora in Rare Soil": Hereditary Diseases in Finland [60]. With the growth of the world population during recent centuries, some populations that began as relatively small and marginal isolates have developed into nations comprising several millions of persons. When, due to favorable geographic and political conditions, this population growth occurred with little disturbance due to migration and admixture by other populations, the random sample of recessive genes present by chance in the founders is represented in the current population. Such a nation displays a unique collection of recessive diseases. Genes that are relatively common elsewhere are lacking, and other genes are found that are unknown in other areas. Since population migration and gene flow in historical times have been intensive, especially in populations of European origin for which most information is available, examples of undisturbed growth of a relatively isolated population are rare. The best is the Finnish-speaking population of Finland (Finland also has a Swedish-speaking minority.). Several favorable features are: an undisturbed, "traditional" population structure, an advanced level of medicine permitting reliable diagnosis of rare diseases, and excellent church records which serve as a reliable population register for roughly the past ten generations.

Population History of Finland. Most ancestors of the current Finnish population immigrated over many centuries during the first millennium from the Baltic region. They were, as the Estonians, descendants of a common basic population, the Baltic Finns, who are also members of a special language group. It is possible that their immigration came to an end long before historical times, i.e., 1000 AD. They settled in the southwest of the country; the scarcity of burial grounds and other remnants indicates that the number of early settlers was probably very limited. In the following centuries settlement spread slowly to the north and east. The total population is estimated to have been 400 000 during the seventeenth century; it reached 1.6 million in 1850, and 4.6 million in 1970. Until recent times the population was largely rural;

this has changed now, as with industrialization many families have been moving into the big cities. However, there still is little migration among rural areas.

The slow immigration of a limited number of settlers and the relatively independent growth of subpopulations with little gene flow between them afford the best conditions for studying founder effects and subsequent shifts in gene frequencies within these subpopulations by genetic drift.

First-cousin marriages were prohibited by law prior to 1872, but more remote, multiple consanguinity between spouses has been, and is still, very common. We might expect to find the following:

a) A relatively high incidence in certain subpopulations of recessive disorders that are otherwise rare or unknown, along with lower frequencies due to relatively recent migrations in adjacent Finnish subpopulations.
b) Very low frequencies of these disorders in the big cities, which are melting pots for the overall Finnish gene pool.
c) Very low incidence or absence of some recessive diseases known in other, non-Finnish populations.

This is exactly what has been found.

Recessive Disorders in Finland. Table 13.11 shows a number of recessive diseases that are found fairly frequently in Finland but only rarely or not at all outside this country. Other disorders listed in Table 13.11 have been observed both in Finland and in other countries. Table 13.11 also shows disorders that occur more commonly in other populations but are rare or nonexistent in Finland. There can be no doubt that Finland presents a number of otherwise very rare or even unknown disorders, whereas some diseases prevalent elsewhere are missing. Most conspicuous among the latter is phenylketonuria, which in a comprehensive newborn-screening program, has been searched for very carefully (Sect. 12.1.3, Table 12.4).

Attempts to locate the origin of specifically "Finnish" disorders led to interesting results. They turned out to have their centers in limited geographic areas, with some scattering outside these areas, and to be virtually absent in the native population of the big cities. Figure 13.18 shows the birthplaces of ancestors (grandparents or great-grandparents) of patients with three of the recessive diseases listed in Table 13.10. As noted, dominant anomalies may also show founder and drift effects, provided there is no strong selection against these genes. A Finnish example is amyloidosis with corneal lattice-like dystrophy and cranial neuropathy (see also Sect. 7.6). The corneal lattice-like dystrophy manifests itself at the age of 20–35 years but affects visual acuity only slowly and to a moderate degree. Symptoms of amyloidosis become visible only later in life. Therefore the disease does not impair reproduction very much. Figure 13.19 indicates the birthplace of the affected parents of 207 patients with this condition.

Table 13.11. Recessive diseases in Finland (from Norio et al. 1973 [60]; for an updated list with some additional diseases, see De la Chapelle 1993 [22])

Rare recessive diseases that are relatively common in Finland	Recessive diseases that are about as common in Finland as elsewhere	Recessive diseases otherwise relatively common that are rare in Finland
Congenital nephrotic syndrome (Finnish type)	Fructose intolerance,	Galactosemia
Aspartyl glycosaminuria (AGU; 208 400)	Tyrosinemia	Hepatic glycogenoses
Early infantile ceroid lipofuscinosis	Mucopolysaccharidosis I (Hurler)	Cystinosis
Progressive dementia with lipomem-	Mucolipidosis II (I cell disease)	Maple syrup urine disease
branous polycystic osteodysplasia	Adrenogenital syndromes (for ex-	Phenylketonuria
Dystrophia retinae pigmentosa-	ample, 21-hydroxylase deficiency)	Homocystinuria
dysacusis syndrome	Polycystic disease of kidney	Cystic fibrosis of the pancreas
Cornea plana congenita	(perinatal form)	Tay-Sachs disease
Åland eye disease (X-linked)	Profound childhood deafness	Gaucher disease
Lysinuric protein intolerance	Xeroderma pigmentosum	
Mulbrey dwarfism		
Cartilage-hair hypoplasia		
Diastrophic dwarfism		
Congenital chloride diarrhea		
Selective malabsorption of vitamin B_{12}		

Fig. 13.18. Origin of patients with three recessive diseases in Finland. *Left,* Congenital chloride diarrhea; greatgrandparents of 11 evident and 3 probable sibships. *Middle,* congenital nephrotic syndrome of Finnish type; 60 grandparents of 57 sibships. *Right,* cornea plana congenita recessiva; grandparents of 32 sibships. *Stippled areas,* to pre-World War II borders. (From Norio et al. 1973 [60])

Conclusions from the Experience in Finland for Research in Population Genetics of Rare Disorders. A population history and breeding structure such as that found in Finland still exists in many populations of the Old World. However, most of them lack the other conditions that are so convenient in Finland: the exact reports on family histories, the excellent medical facilities, and, last but not least, research workers dedicated to exploiting these possibilities. Most other populations lacking extensive medical facilities and research personnel have been subject to large-scale outbreeding by population mixing. Therefore the opportunities for discovering new recessive conditions are unfavorable. In Chap. 7 it was observed that not all of the defects leading to hereditary disease in humans are as yet known and that many more could be discovered in countries that still have a traditional population structure. The considerations of this section have shown why this is the case, and the example of Finland demonstrates that this is not merely a theoretical speculation.

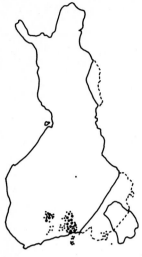

Fig. 13.19. A dominant gene in Finland: familial amyloidosis with corneal lattice dystrophy and cranial neuropathy, 207 patients are indicated by the birth place of the affected parents. *Big dot,* ten parents; *small dot,* one parent. (Norio et al. 1973 [60])

Conclusions

Genotype frequencies in many human populations are the result of random mating. The numbers found are those expected from the Hardy-Weinberg Law. Important deviations from random matings are marriages between close relatives (consanguinity). While the consanguinity rate has decreased in many European and North American populations, it remains quite high in certain other population groups. For the offspring of such matings there is an increased probability of homozygosity, which may lead to a higher incidence of autosomal-recessive diseases. The concept of "genetic load" attempts to quantify these effects and to use them for estimating the number of deleterious mutants in human populations. In addition to changes of gene frequencies due to mutation and selection, random fluctuations are observed especially in small and isolated population groups. "Founder effects" in such groups have led, for example, to unusually high frequencies of otherwise very rare genetic diseases. Traits found among the French-Canadian, the Ashkenazi Jewish, and the Finnish populations provide examples.

References

1. Adams MS, Neel JV (1967) Children of incest. Pediatrics 40 : 55–62
2. Al-Awadi SA, Moussa MA, Naguib KK et al (1985) Consanguinity among the Kuwaiti population. Clin Genet 27 : 483–486
3. Baird PA, McGillivray B (1982) Children of incest. J Pediatr 101 : 854–857
4. Bashi J (1977) Effects of inbreeding on cognitive performance. Nature 266 : 440–442
5. Bergeron P, Laberge C, Grenier A (1974) Hereditary tyrosinemia in the province of Quebec: prevalence at birth and geographic distribution. Clin Genet 5 : 157–162
6. Bernstein R, Wadee A, Rosendorff F et al (1986) Inverted Y chromosome polymorphism in the Gujerati Muslim Indian population of South Africa. Hum Genet 74 : 223–229
7. Bittles AH, Mason WM, Greene J, Appaji Rao N (1991) Reproductive behavior and health in consanguineous marriages. Science 252 : 789–794
8. Bois E, Feingold J, Demenais F, Runavot Y, Jehanne M, Toidic L (1978) Cluster of cystic fibrosis cases in a limited area of Brittany (France). Clin Genet 14 : 73–76
9. Bosnjakovic S (1938) Vererbungsverhältnisse bei der sogenannten Krankheit von Mljet. Acta Derm Venereol (Stockh) 19 : 88
10. Carter CO (1967) Risk of offspring of incest. Lancet 1 : 436
11. Cavalli-Sforza LL, Bodmer WF (1971) The genetics of human populations. Freeman, San Francisco
12. Cavalli-Sforza LL, Edwards AWF (1967) Phylogenetic analysis: models and estimation procedures. Evolution 21 : 550–570
13. Cerimele D, Cottoni F, Scappaticci S, Rabbiosi G, Sann E, Zei G, Fraccaro M (1982) High prevalence of Werner's syndrome in Sardinia. Description of six patients and estimate of the gene frequency. Hum Genet 62 : 25–30
14. Costeff H, Cohen BE, Weller L, Rahman D (1977) Consanguinity analysis in Israeli mental retardates. Am J Hum Genet 29 : 339–349
15. Crow JF (1958) Some possibilities for measuring selection intensities in man. Hum Biol 30 : 1–13
16. Crow JF (1963) The concept of genetic load: a reply. Am J Hum Genet 15 : 310–315
17. Crow JF (1970) Genetic loads and the cost of natural selection. In: Kojima K (ed) Mathematical topics in population genetics. Springer, Berlin Heidelberg New York, pp 128–177
18. Crow JF, Denniston C (1981) The mutation component of genetic damage. Science 212 : 888–893
19. De Braekeleer M (1991) Hereditary disorders in Saguenav-Lac-St. Jean (Quebec, Canada). Hum Hered 41 : 141–146
20. De Brakeleer M, Hechtman P, Andermann E, Kaplan F (1992) The French Canadian Tay-Sachs disease deletion mutation: identification of probable founders. Hum Genet 89 : 83–87
21. De Braekeleer M, Dallaire A, Mathieu J (1993) Genetic epidemiology of sensorimotor polyneuropathy with or without agenesis of the corpus callosum in northeastern Quebec. Hum Genet 91 : 223–227
22. De la Chapelle A (1993) Disease gene mapping in isolated human populations: the example of Finland. J Med Genet 30 : 857–865
23. Devi ARR, Appaji Rao N, Bittles AH (1987) Inbreeding and the incidence of childhood genetic disorders in Karnataka, South India. J Med Genet 24 : 362–365
24. Diamond JM, Rotter JI (1987) Observing the founder effect in human evolution. Nature 329 : 105–106
25. Fisher RA (1930) The distribution of gene ratios for rare mutations. Proc R Soc Edinb 50 : 205–220

26. Fraser GR, Mayo O (1974) Genetic load in man (Review). Hum Genet 23 : 83–110

27. Freire-Maia N, Azevedô JBC (1971) The inbreeding load in Brazilian White and Negro populations as estimated with sib and cousin controls. Am J Hum Genet 23 : 1–7

28. Freire-Maia N, Chautard-Freire-Maia EA, de Aguiar-Wolter IP et al (1983) Inbreeding studies in Brasilian schoolchildren. Am J Med Genet 16 : 331–355

29. Georges A, Jacquard A (1968) Effets de la consanguinité sur la mortalité infantile. Résultats d'une observation dans le département des Vosges. Population 23 : 1055–1064

30. Goodman RM, Motulsky AG (eds) (1979) Genetic diseases among Ashkenazi Jews. Raven, New York

31. Guez K, Dedeoglu N, Lueleci G (1989) The frequency and medical effects of consanguineous marriages in Antalya, Turkey. Hereditas 111 : 79–83

32. Haldane JBS (1935) The rate of spontaneous mutation of a human gene. J Genet 31 : 317–326

33. Haldane JBS (1937) The effect of variation on fitness. Am Nat 71 : 337–338

34. Haldane JBS (1957) The cost of natural selection. J Genet 55 : 511–524

35. Hanhart E (1955) Zur mendelistischen Auswertung einer 33 Jahre langen Erforschung von Isolaten. Novant 'Anni dell Leggi Mendeliane. Orrizante Medico, Roma, pp 397–415

36. Harris H (1980) The principles of human biochemical genetics, 4th edn. North-Holland, Amsterdam

37. Horowitz M, Tzuri G, Eyal N et al (1993) Prevalence of nine mutations among Jewish and non-Jewish Gaucher disease patients. Am J Hum Genet 53 : 921–930

38. Jaber L, Merlob P, Bu X-D et al (1992) Marked parental consanguinity as a cause for increased major malformations in an Israeli Arab community. Am J Med Genet 44 : 1–6

39. Jacquard A (1974) The genetic structure of populations. Springer, Berlin Heidelberg New York

40. Kay AC, Kuhl W, Prchal J, Beutler E (1992) The origin of glucose-6-phosphate-dehydrogenase (G6PD) polymorphisms in African-Americans. Am J Hum Genet 50 : 394–398

41. Kempthorne O (1957) An introduction to genetic statistics. Wiley, New York

42. Klinger KW (1983) Cystic fibrosis in the Ohio Amish: gene frequency and founder effect. Hum Genet 65 : 94–98

43. Konecki DS, Lichter-Konecki U (1991) The phenylketonuria locus: current knowledge about alleles and mutations of the phenylalanine hydroxylase gene in various populations. Hum Genet 87 : 377–388

43 a. Kumar S, Pau RA, Swaminathan MS (1967) Consanguineous marriages and the genetic load due to lethal genes in Kerala. Ann Hum Genet 31 : 141–145

44. Kumar-Singh R, Farrar GJ, Mansberg F et al (1993) Exclusion of the involvement of all known retinitis pigmentosa loci in the disease present in a family of Irish origin provides evidence for a sixth autosomal dominant locus (RP8). Hum Mol Genet 2 : 875–878

45. Levy HL (1973) Genetic screening. Adv Hum Genet 4 : 1–104

46. Li CC (1955) Population genetics. University of Chicago Press, Chicago

47. Li CC (1963) The way the load works. Am J Hum Genet 15 : 316–321

48. Ludwig W (1944) Über Inzucht und Verwandtschaft. Z Menschl Vererbungs Konstitutionslehre 28 : 278–312

49. Lundborg H (1913) Medizinisch-biologische Familienforschungen innerhalb eines 2232-köpfigen Bauerngeschlechtes in Schweden (Provinz Blekinge), 2 vols. Fischer, Jena

50. Luzzatto L (1979) Genetics of red cells and susceptibility to malaria. Blood 54 : 961

51. Magnus P, Berg K, Bjerkdal T (1985) Association of parental consanguinity with decreased birth weight and increased rate of early death and congenital malformation. Clin Genet 28 : 335–342

52. Malécot G (1948) Les mathématiques de l'hérédité. Masson, Paris

53. Mayr E (1982) The growth of biological thought. Harvard University Press, Cambridge

53 a. McKusick VA, Egeland JA, Eldridge R, Krusen DR (1964) Dwarfism in the Amish. I. The Ellis-van Creveld syndrome. Bull Johns Hopkins Hosp 115 : 306

54. Morton NE, Crow JF, Muller HJ (1956) An estimate of the mutational damage in man from data on consanguineous marriages. Proc Natl Acad Sci USA 42 : 855–863

55. Morton NE, Matsuura J, Bart R, Lew R (1978) Genetic epidemiology of an institutionalized cohort of mental retardates. Clin Genet 13 : 449–461

55 a. Motulsky AG (1995) Jewish diseases and origins. News and views. Nature [Genet] 9 : 99–101

56. Muller HJ (1950) Our load of mutation. Am J Hum Genet 2 : 111–176

57. Neel JV (1979) History and the Tay-Sachs allele. In: Goodman RM, Motulsky AG (eds) Genetic diseases among Ashkenazi Jews. Raven, New York, pp 285–299

58. Neel JV, Schull WJ (1962) The effect of inbreeding on mortality and morbidity in two Japanese cities. Proc Natl Acad Sci USA 48 : 573

59. Neel JV, Schull WJ, Kimura T, Yanijawa Y, Yamamoto M, Nakajima A (1970) The effect of parental consanguinity and inbreeding in Hirado, Japan. III. Vision and hearing. Hum Hered 20 : 129–155

60. Norio R, Nevalinna HR, Perheentupa J (1973) Hereditary diseases in Finland: rare flora in rare soil. Ann Clin Res 5 : 109–141

61. Rao PSS, Inboraj SG (1979) Trends in human reproductive wastage in relation to long-term practice of inbreeding. Ann Hum Genet 42 : 401–413

62. Rao PSS, Inboraj SG (1980) Inbreeding effects on fetal growth and development. J Med Genet 17 : 27–33

63. Reed SC (1954) A test for heterozygous deleterious recessives. J Hered 45 : 17–18

64. Reed TE (1969) Caucasian genes in American negroes. Science 165 : 762–768

65. Schull WJ (1958) Empirical risks in consanguineous marriages: sex ratio, malformation, and viability. Am J Hum Genet 10 : 294–343

66. Schull WJ, Neel JV (1965) The effect of inbreeding on Japanese children. Harper and Row, New York

67. Schull WJ, Neel JV (1972) The effect of parental consanguinity and inbreeding in Hirado, Japan. V. Summary and interpretation. Am J Hum Genet 24 : 425–453

68. Schull WJ, Nagano H, Yamamoto M, Komatsu I (1970) The effect of parental consanguinity and inbreeding in Hirado, Japan. I. Stillbirth and prereproductive mortality. Am J Hum Genet 22 : 239–262

69. Schull WJ, Furusho T, Yamamoto M et al (1970) The effects of parental consanguinity and inbreeding in Hirado, Japan. IV. Fertility and reproductive compensation. Hum Genet 9 : 294–315

70. Seemanová E (1971) A study of children of incestuous matings. Hum Hered 21 : 108–128

71. Shami SA, Schmitt LH, Bittler AH (1989) Consanguinity related prenatal and postnatal mortality of the populations of seven Pakistani Punjab cities. J Med Genet 26 : 267–271

72. Sjögren T (1931) Die juvenile amaurotische Idiotie. Hereditas 14 : 197–425

73. Slatis HM (1954) A method of estimating the frequency of abnormal autosomal recessive genes in man. Am J Hum Genet 6 : 412–418

74. Stevenson AC, Cheeseman EA (1955) Hereditary deaf mutism with particular reference to Northern Ireland. Ann Hum Genet 20 : 177–231

75. Ten Kate LP, Scheffer H, Cornel MC, van Lockeren Campagne JG (1991) Consanguinity sans reproche. Hum Genet 86 : 295–296

76. Vogel F (1979) "Our load of mutation": reappraisal of an old problem. Proc R Soc Lond [Biol] 205 : 77–90

77. Von Fumetti C (1976) Inzuchtkoeffizienten und Häufigkeiten konsanguiner Ehen. Thesis, University of Heidelberg

78. Weiner JS, Huizinga J (eds) (1972) The assessment of population affinities in man. Clarendon, Oxford

79. Wright S (1922) Coefficients of inbreeding and relationship. Am Nat 56 : 330–338

80. Wright S (1931) Evolution in Mendelian populations. Genetics 16 : 97–159

81. Zerbin-Rüdin E (1960) Vorläufiger Bericht über den Gesundheitszustand von Kindern aus nahen Blutsverwandtenehen. Z Menschl Vererbungs Konstitutionslehre 35 : 233–302

14 Human Evolution

Time present and time past
Are both perhaps present in time future,
And time future contained in time past.

T. S. Eliot, Four Quartets, "Burnt Norton," I

14.1 Paleoanthropological Evidence

Population Genetics Helps To Understand Evolution. The concepts developed in population genetics and the examples from human genetics help in understanding human evolution and the resulting genetic differences between humans and other mammals, especially our closest relatives, the great apes. These concepts also improve our understanding of genetic variability within and between present-day human populations. We discuss the evidence in three parts: In the very short introductory section we sketch the evidence on human evolution from paleoanthropology. This survey leads to a more elaborate discussion of the genetic mechanisms of human evolution. Finally, we discuss the genetic variability among various present human population groups in the light of the genetic evidence.

Evidence from Paleoanthropology [52, 64]. Evolution of the higher primates including *Homo* is now fairly well known in outline. We know the most important anatomic features of the main ancestral forms and to some extent their geographic distribution, and we can reconstruct their way of living. There are many differences in opinion among paleoanthropologists about details of human evolution and the position of individual specimens and populations within this context, but most specialists agreed until recently on the following conclusions: *Homo,* as well as the great apes, are descended from a common ancestral population, that of the Dryopithecinae that lived in Africa about 15–20 million years ago. For a long period thereafter, however, little precise information is available; in any case, the branch from which both the great apes and *Homo* were to develop separated from the ancestral population of Old World monkeys. Between 5 and 10 million years ago the branches of the great apes – orangutan, gorilla, and two chimpanzee species – separated from that which developed into humans. Somewhat more than 2 million years ago this branch produced a variety of more or less similar forms, whose characteristics place them between apes and modern humans; with the exception of our own species, *Homo sapiens,* all of these later became extinct. The rich development of our species appears to have been promoted by certain geographic, climatic, and ecological conditions in eastern and southern Africa. The Great Rift Valley began to open 20 million years ago, creating a great variety of landscapes and climatic regions in which many different species could develop, a process thought to have been facilitated by climatic change – a decrease in the average general temperatures. In this area there developed a number of hominoid species (Fig. 14.1), the exact relationships among which is still a matter of controversy among anthropologists. Well-known varieties include the australopithecines: *Australopithecus afarensis, A. africanus, A. robustus*). In addition, one variety with larger brains has been found, which used some primitive stone tools: *Homo habilis.* Figure 14.1 shows the time scale of these developments. The oldest *Homo habilis* remains are thought to be 2–2.5 million years old. Bipedalism, i.e., walking on two legs, appears to have been the first evolutionary step toward humanity; much later this was followed by an increase of brain volume and by some other changes. The selecting forces behind the development of increased brain size and intelligence were complex; earlier anthropologists stressed tool making but more recently, emphasis is placed more on evolving social skills within groups [36–38]. About 1.5–2 million years ago the first representatives of *Homo erectus* appeared – a species which about 1 million years ago expanded beyond the limits of Africa; specimens from this time have been found both in Asian countries such as Indonesia and China and in Europe. Their superior intelligence allowed them to live in the colder climate of the north; here they may also have escaped the constant challenge of tropical parasites [25]. A later species, found principally in Europe during the Ice Ages, are known as Neandertals. These disappeared around 40 000 years ago and were replaced by our own species, *Homo sapiens.* Most experts agree that *Homo sapiens* had only one origin, and that this was in Africa [1984, 2973]. Figure 14.2 compares the brain size of humans to those of other mammals, and Fig. 14.3 compares the size of man's brain to that of his evolutionary relatives.

Biological evolution has been supplemented more and more by cultural evolution until cultural conditions have today become the major driving force for the biological changes in our species that can expect in future. Table 14.1 contrasts biological and cultural evolution in humans, Table 14.2 provides an overview of human evolution.

14.2 Genetic Mechanisms of Evolution of the Human Species

Genetic mechanisms involved in the development of the human species can be analyzed mainly by comparing present-day humans with their closest phylogenetic relatives, the great apes. Our goal is twofold:

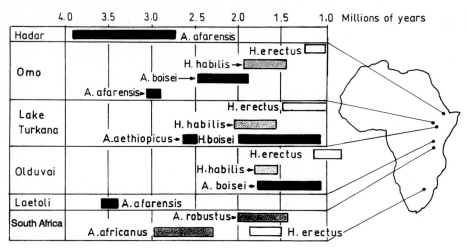

Fig. 14.1. The earliest hominid discoveries in Africa (South Africa, Tanzania, Ethiopia). The sequence of shadings from black to white indicates development from ape-like forms to early humans. (From Leakey and Lewin 1992 [64])

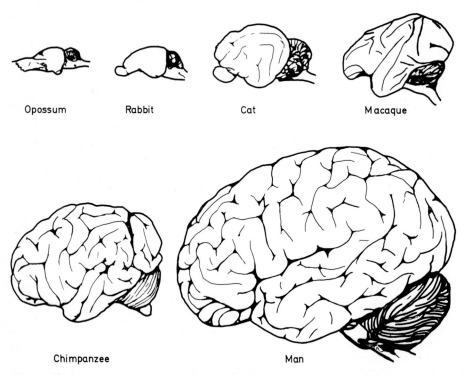

Fig. 14.2. The brain in various species of mammals. Note the developmental differences in size and structure of the cerebral cortex

1. Establishing the degree of relationship among species and constructing a phylogenetic tree that shows the order in which these species have developed from common ancestral populations.
2. Analyzing the genetic mechanisms of evolution and speciation.

The construction of phylogenetic trees requires evidence from skeletal fragments of presumed ancestors as well as evidence from comparative anatomy and comparative genetics. For analyses of the mechanisms of evolution these skeletal remains are of no value, and we must rely principally on comparing genetic differences among present species. (For examination of ancient DNA see Sect. 14.2.4). Here, intraspecies variability may help to improve our understanding of interspecies differences.

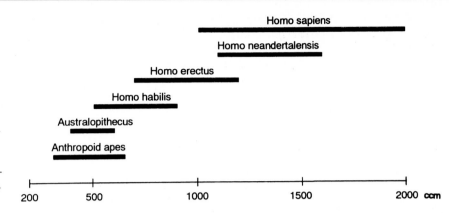

Fig. 14.3. Estimated interindividual variability of skull capacity in populations at different stages of human phylogeny. (Adapted from Heberer, in [3])

Table 14.1. Comparison of biological and cultural evolution (from Omenn and Motulsky 1972 [86])

	Biological evolution	Cultural evolution
Mediated by	Genes	Ideas
Rate of change	Slow	Rapid and exponential
Agents of change	Random variation (mutations) and selection	Usually purposeful; directional variation and selection
Nature of new variant	Rarely beneficial	Often beneficial
Transmission	Parents to offspring	Wide dissemination by many means
Nature of transmission	Relatively simple	May be highly complex
Distribution in nature	All forms of life	Most important in humans
Interaction	Human biology interacts with human culture	Human culture required biological evolution to achieve the human brain
Complexity achieved by	Rare formation of new genes by chromosomal duplication	Frequent formation and cumulative growth of new ideas and technologies

Note that the evolutionary destiny of all species was dependent upon interaction of genetic constitution and environment beyond its control. Only humans have the means to control both their environment and, to some extent, their genetic constitution.

The available evidence is discussed below at four levels:

1. Chromosomes
2. DNA
3. Amino acid sequences of specific proteins
4. Behavior.

14.2.1 Chromosome Evolution and Speciation

Chromosome Number of Humans and Closely Related Nonhuman Primates [26]. The following species are specifically described: chimpanzee *(Pan troglodytes)*, pygmy chimpanzee *(Pan paniscus)*, gorilla *(Gorilla gorilla)*, and orangutan *(Pongo pygmaeus)*; their chromosomes are then compared to those of the gibbon *(Hylobates)*. The chromosome number in all four species was soon established as 48; the main dif-

ference between *Homo* and both chimpanzee species is a fourth pair of acrocentric chromosomes of the D group (Fig. 14.4). In the other two species, gorilla and orangutan, more acrocentric chromosomes are found. There is a strong overall similarity between the chromosomes of *Homo* and *Pan* that confirm the evidence from morphological and biochemical research (Sect. 14.2.3) that *Pan* is our closest living relative.

Comparison of Chromosome Structure with Banding Methods. Comparison of karyotypes between two species should help in reconstructing the number and kind of chromosome rearrangements that have occurred since these species separated in evolution. This reconstruction became possible when banding methods were introduced in 1970 (Chap. 2). Pericentric inversions were soon established as the mechanisms responsible for most of the species differences

Table 14.2. Evolution of humans (Modified from Omenn and Motulsky 1972 [86])

Mean brain volume (cc)	Time scale		Tool use	Life-style	Arts and language
	Years ago	Generations ago			
400–550	1.7 million	85 000	Simplest stone & bone	Hunting & gathering	
900	600 000	30 000	More refined stone tools	Similar	
1300	50 000	2 500	Stone axes	Still hunters	Cave painting Early languages
	30 000	1 500			
	10 000	500	Metal tools	Agriculture	Hieroglyphic, iconic written languages
	8 000	400			
	6 000	300	More complex tools & vehicles for transportation	Cities & agriculture	Alphabetized languages
	3 500	175			
	300	15	Complex machinery	Industrialized centers	Printing
	30	1	Nuclear energy use	Atomic age	Radio, TV
	30		Computers	Postindustrial age	Cyberspace

Evolutionary Events

Bipedalism   Stone tools   Cave painting   Language   Agriculture   Organized society   Industrialization   Atomic & computer age

Time scale

> 2.5 million years   50 000 years   300 years

between humans and apes. The difference in chromosome number could be explained by joining of two different acrocentric chromosomes about the length of D chromosomes to form one large submetacentric chromosome – the human no. 2. Such joining of different chromosomes is well known from the present human population: the prevalent mechanism is centric fusion, which implies loss of the short arms of a chromosome. It is therefore not surprising that centric fusion was thought to be the mechanism for these species differences as well. However, a more detailed analysis of banding patterns has shown that short-arm material is indeed present. The human no. 2 was produced by a telomeric fusion [26].

Such a fusion chromosome has two centromeres; this should lead to mitotic complications known from dicentric chromosomes produced by interchanges, such as after radiation-induced chromosome breakage (Chap. 11). Such problems are avoided by only one of the centromeres carrying on its normal mitotic function. This suppression of one centromere has occasionally been observed in present-day chromosome aberrations as well.

The goal of a detailed analysis of all discernible rearrangements by which the species differ from each other and from humans using all available banding techniques was largely achieved by Dutrillaux in 1975 [26].

Example. The homologues of the human no. 2 are seen in Fig. 14.5. *Pongo* and *Gorilla* differ by an inversion in 2 q, *Gorilla* and *Pan* by another inversion in 2 p. A telomeric fusion must have occurred between *Pan* and *Homo*. In some preparations the human no. 2 shows a secondary constriction at the fusion point (2 qh). Very rarely an endoreduplication of the segments corresponding to the former 2 q chromosome is observed and indicates some independence between the fusion partners. On the basis of these results the evolution of no. 2 can be reconstructed.

Comparison of Overall Karyotypes of the Five Species. Species differences can be used to reconstruct the evolution of all single chromosomes in the same way as shown above for no. 2. Apart from the one telomer-

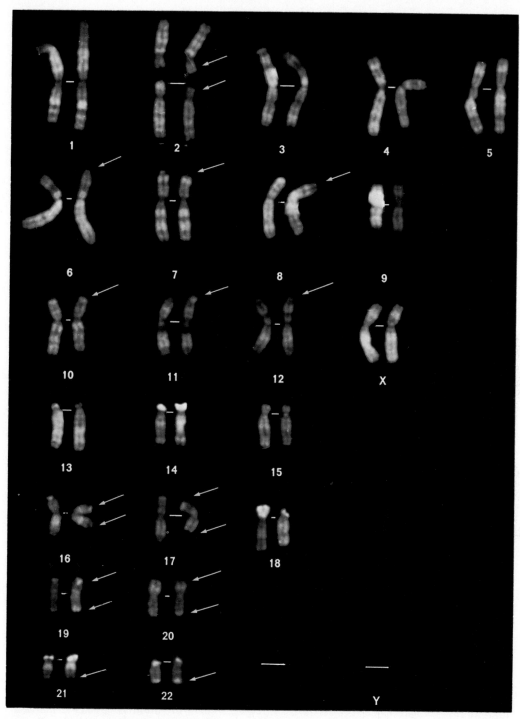

Fig. 14.4. Karyotype of a chimpanzee *(Pan troglodytes),* Q banding. A counterpart of the human chromosome 2 is obviously lacking. Instead, two additional pairs of acrocentric chromosomes are present. *Arrows,* terminal Q bands not present in humans (or orangutans). Numbers correspond to human chromosome numbers. (Courtesy of Dr. B. Dutrillaux, Paris).

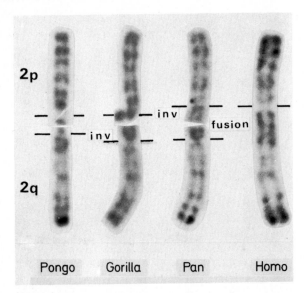

Fig. 14.5. Phylogenetic development of human chromosome 2 through some inversions and one telomeric fusion. G banding. *Pongo* and *Gorilla* differ by an inversion *(inv)* in 2 q; *Gorilla* and *Pan* differ by an inversion in 2 p. *Homo* differs from all the three other species by a telomeric fusion of two chromosomes. (Courtesy of Dr. B. Dutrillaux, Paris)

ic fusion and the pericentric inversions, a number of paracentric inversions have been observed. As expected, the two species of chimpanzees show the highest degree of similarity; they are separated by only one, doubtful, paracentric inversion. *Homo* is most closely related to chimpanzees; there are six pericentric inversions between these species. On the other hand, chimpanzees and gorillas are separated by six pericentric and two paracentric inversions. The most remote relative of *Homo* is *Pongo*; this could be expected from morphological evidence.

There have been some inconsistencies regarding chromosome inversions. *Pan troglodytes* and *Gorilla* have three inversions in common (5, 12, 17), suggesting a common ancestor not shared by *Homo*, whereas *Pan* and *Homo* have two inversions in common that are not found in *Gorilla*. This shows that phylogenetic relationships may not be as simple as suggested by our phylogenetic tree models [26].

Presence and Absence of Certain Segments. In addition to the rearrangements described above, some of these primate species regularly show certain chromosome segments that are lacking in others:

a) Terminal Q bands: After quinacrine staining small Q bands are observed at the ends of many chromosomes in *Pan* and *Gorilla*; in most cases these are located in the same chromosome arms in both species. They are lacking in both *Pongo*

and in *Homo*. There are two possibilities: either they appeared as a fresh mutation in a common ancestor of *Pan* and *Gorilla,* or they existed in a primitive ancestor and disappeared during the evolution of *Pongo* and *Homo*. Both events are difficult to understand on the basis of classic concepts: duplication and deletion of single chromosome segments are isolated and random events.

b) Heterochromatic regions are seen at the short arm of some acrocentric chromosomes. Their number diminishes in the following order *Pongo*→*Gorilla*→*Homo*→*Pan*. Such heterochromatic material is presumably formed from time to time as a new mutation in immediate proximity to the centromeres of acrocentric chromosomes and is then distributed by random chromosome rearrangement to other chromosome parts. The secondary constriction of the chromosome 9 of *Homo* (Sect. 2.1.2) might be such material. The no. 9 of humans and of *Pan* also comprises a heterochromatic block close to the centromere.

c) Variations also exist in the occurrence of T bands. Repeated de novo synthesis of some T band material at the ends of the chromosome and, in some cases, secondary distribution to other chromosome parts by rearrangements is one possible explanation.

d) *Gorilla* and *Homo* show additional Q bands close to the centromeres of chromosomes 3 and 4; only *Gorilla* shows a similar band at no. 9, *Homo* on no. 13, and *Pan* only on no. 3. They are lacking completely in *Pongo* – and in the gibbon.

All three observations – terminal Q and T bands, heterochromatic regions, and juxtacentromeric Q bands – show that the karyotype differences between the five closely related species do not consist only in rearrangements of genetic material that can be explained using classical principles. An additional mechanism seems to be de novo synthesis as well as loss of chromosome material. Studies at the DNA level have helped to understand the nature of such newly synthesized material somewhat better.

Methods of nonradioactive in situ hybridization (Sect. 3.1.3.3) are now being used to study chromosome evolution in greater detail and to solve problems that could not be dealt with by classical cytogenetic techniques. For example, the composition of gibbon (Hylobates) chromosomes and their relationship to those of the great apes and humans seemed to pose insuperable technical problems, but these have now been solved by CISS hybridization [51, 107] (Figs. 14.6, 14.7). In the gibbon karyotype the 22 human chromosomes appear subdivided into 51 blocks, which have been combined anew to form the 21 autosomes of three gibbon species (which have identical karyotypes except for one chromosome). For example, parts of the human chromosome 1 can be found in gibbon 5 q, 7 q, 9 p, and 19. X and Y chromosome, however, are identical in humans, the great ape species, and the gibbons (Figs. 14.6, 14.7).

Another old world monkey, on the other hand, *Macaca fuscata,* has been shown by the same method to have a much more similar chromosome set to *Homo* than gibbon although the ancestors of *Macaca* and *Homo* probably separated much earlier [107]. This may be due to the fact that macaques – as with other, similar species such as baboons – live in large, terrestrial groups whereas gibbons are monogamous, and the pairs form separate units. The latter way of life may have favored matings between close relatives, and fixation of structural aberrations in populations (Fig. 14.8).

A tentative phylogenetic tree of the primates from the prosimians to man has been established by study of the karyotypes of more than 60 species of primates using almost all available banding techniques [27]. The entire euchromatic material, i.e., the nonvariable R and Q bands (Sect. 2.1.2.1) appears to be identical in all species of *monkeys, apes,* and in *humans.* Quantitative and qualitative variations all involve heterochromatin. The types of chromosome rearrangements reconstructed from species differences in chromosome structure vary from one subgroup to the next; for instance, Robertsonian translocations (centric fusions) prevail among the Lemuridae; chromosome fissions are frequent among the Cercopithecinae but are not found elsewhere; and pericentric inversions are common in the evolution of *Homo sapiens* including Pongidae.

Chromosome Rearrangements in Evolution and in the Current Population. There is one major difference between breakage points and chromosome rearrangements in evolution and in the current population. The most frequent chromosome rearrangements currently observed are centric fusions, i.e., stable connection between acrocentric chromosomes with the loss of short-arm material (Sect. 2.2.2). Surprisingly, not even one of these centric fusions has been fixed during evolution of the five species of pongidae, including homo sapiens. A possible explanation is that of selective disadvantage due to formation of aneuploidies, for example, zygotes with trisomy for the long arm of 21 q, which would produce Down syndrome (Sect. 2.2.2), or those leading to miscarriage. To the best of our knowledge, however, not all centric fusions lead to a selective disadvantage. (For a more detailed discussion, see Sect. 2.2.2.2.)

Could the high frequency of centric fusions and the resulting zygote loss be a relatively recent genetic adaptation to the special conditions of child rearing in humans?

Selective Advantage of High Rate of Spontaneous Miscarriage in Humans [50]? About 5%–7% of all recognized conceptions in humans are chromosomally abnormal (Sect. 2.2.4); most of these are lethal. They lead to miscarriage or – in exceptional cases – to delivery of a severely malformed child who would have no chance of survival under primitive living conditions. In the surviving infants – largely those with X-chromosomal aneuploidies – fertility is considerably reduced (Sect. 2.2.3.2). Loss of additional zygotes before implantation goes unnoticed in most cases. At first glance this considerable loss due to chromosomal aberrations seems to indicate considerable impairment in the reproductive fitness of our species. Upon closer scrutiny, however, the problem takes on quite a different light and has led to the following hypothesis: because human offspring require a high amount of parental care for an extended time period, there must be an optimum interval between live births to maximize the probability that a large proportion of the offspring survive to reproduction. Any mechanism that reduces the number of live births from the maximal to the optimal and the birth interval from the shortest to the best, without placing the mother's life into jeopardy, may have been of a selective advantage. Early miscarriage due to chromosomal aberrations may have fitted these requirements well under the primitive living conditions of our ancestors since mothers were thus exposed to the risks of childbirth less frequently. Furthermore, this extended the average breast-feeding period per child and may therefore have protected infants from the hazards of malnutrition and intestinal infections. The overall number of infants and young children per mother is clearly reduced by a high frequency of spontaneous abortions.

The higher incidence of centric fusions may also be caused by selection. This mechanism may be related to some function of the nucleolar region since the chromosomal areas involved in centric fusion are concentrated in the nucleolar region. This hypothesis would predict that centric fusions are rarer in other higher primates than in humans. Because few nonhuman primates have been examined so far, no incidence data for centric fusions – or for chromosome aberrations in general – are available. Trisomies, on the other hand – the most potent source of reproductive wastage due to chromosome anomalies – do occur in nonhuman primates, as shown by the occurrence of trisomy 21 in chimpanzees [72]. These few observations do not permit any conclusions as to incidence; considering the relatively small number of chimpanzees under surveillance, trisomy 21 may not be much rarer in chimpanzees than in humans. Structural rearrangements such as inversions that have occurred frequently during primate evolution are also common in today's human population [15].

Homologies of Chromosomes and Chromosomal Segments Between Humans and Other, More Remotely Related Species. Homologies in chromosome structure and the order of genes can be found not only among primates – including *Homo sapiens* – but also among more remotely related species. The mouse, the most commonly used mammalian model for many genetic problems, has been well studied in this respect, and the linkage maps of humans and mice are relatively well known. By 1992, 665 homologous gene loci had been assigned to chromosomes in both species; many of these are known to be associated with human hereditary diseases. By

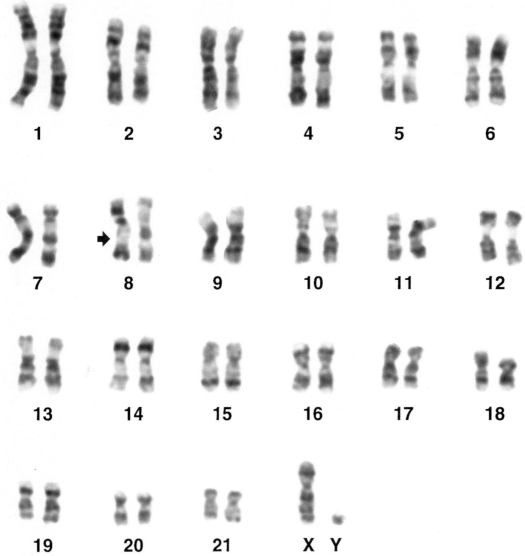

Fig. 14.6. G-banded chromosomes of Gibbons. There is only one difference (→) between Gibbon species. **14.7. a** A plasmid DNA library from flow-sorted human chromosome 1 was hybridized to metaphase chromosomes of *Hylobatis lar.* and detected by FITC *(yellow fluorescence).* The colored gibbon chromosomes were arranged pairwise below the metaphase spread. Chromosome 19 is entirely colored, chromosomes 5, 9, and 7 only partially. (Courtesy of A. Jauch; see Jauch et al. 1992 [51]) **b** Multicolor fluorescence in situ hybridization with human chromosome specific DNA libraries to metaphase chromosome of *Hylobatis syndactylus.* The images were produced by a CCD camera and pseudocolored. *Red,* probe sequences derived from human chromosome 5; *green,* chromosome 16; *violet,* chromosome 17; *yellow,* chromosome 22. Gibbon chromosome 8 reveals signals with probes for chromosomes 5, 16, 17, and 22. Gibbon chromosome 6 is entirely labeled by human chromosome library 5 and 16. Gibbon chromosome 18 reveals signals with human chromosome 5 library and gibbon chromosomes 13 and 16 are partially colored with human chromosome 17 library. In gibbon chromosomes 6 and 16 the hybridization patterns of the re-

spective libraries show two hybridization sites, suggesting a ▷ pericentric inversion. (Courtesy of T. Ried) **c** The hybridization of an X chromosome specific cosmid clone for a subregion of the human dystrophin gene, hybridized to chromosomes of a female *Gorilla gorilla (red fluorescence).* In humans the probe maps to Xp21 *(arrowhead).* The cohybridization with Alu-PCR products generate a R banding *(green fluorescence).* The banding allows the assignment of the probe with respect to cytogenetic bands. (From Ried et al. 1993) **d** FISH with DNA library specific for human chromosome 13 to metaphase chromosomes of the cat *Felis catus (FCA).* Only one coloring signal is detectable and labels the distal region on cat chromosome A1. (Courtesy of J. Wienberg) **e** *Left,* FISH with human chromosome 17 specific library DNA probe to a partial mouse metaphase spread after actinomycin-D/DAPI banding. *Right,* a colored syntenic segment on chromosome 11 at band B–E *(arrows).* (From Scherthan et al. 1994 [93]) **f** FISH with a human telomere specific YAC clone (chromosome 7 q) to metaphase chromosomes of Pan troglodytes. The telomeric map position is maintained on the homologue chimpanzee chromosome (Coursey of T. Cremer and H. Riethman)

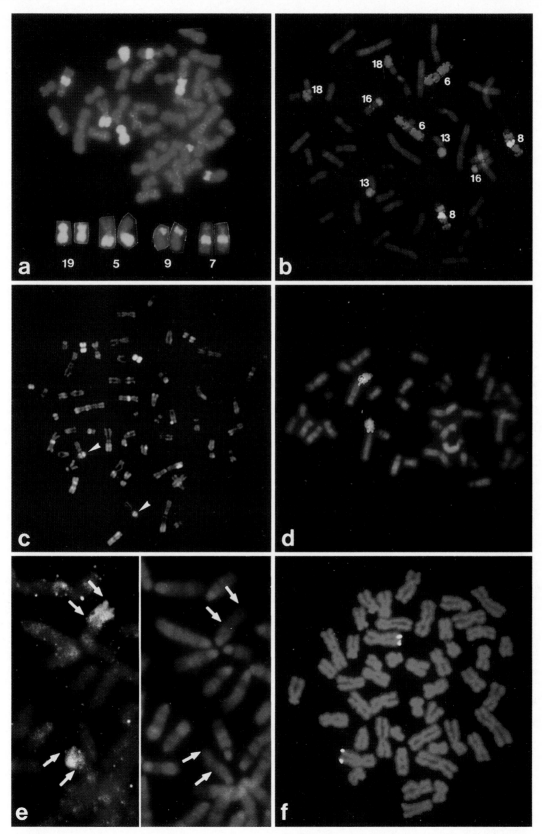

Fig. 14.7 a–f

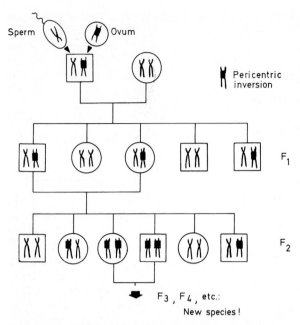

Fig. 14.8. Possible establishment of a new species by close inbreeding. A pericentric inversion is formed in a germ cell of one individual, giving rise to one heterozygous carrier in the next generation. This carrier may have a number of heterozygous offspring who may mate with each other and produce homozygous progeny

1989 at least 50 conserved blocks with two or more loci were known [73, 95]. Location of genes for certain diseases in the mouse may offer hints for the localization of homologous human genes (see also [20, 83]). Many of the homologies of gene clusters are so strong as to suggest that various linkage groups have been kept together in different species by natural selection. This conclusion implies a functional significance of sequence and ordering of the genetic material beyond the level of single genes. The X chromosome has been conserved almost entirely through mammalian evolution [84]; many homologous X-linked loci have been found on the X chromosomes of mouse and man [22, 45]. Ohno [84] proposed that this is caused by phenomena such as X inactivation and gene dosage compensation.

The so-called Oxford grid, a graphic representation of the relationships, permits easy comparison of the two genomes (Fig. 14.9) [73]. Homologous segments have also been found between humans and other, more distantly related mammals [93].

How Can a Chromosome Rearrangement Become Fixed in a Population? As shown in Sect. 13.2.2, the overwhelming majority of all new mutations in a population disappear. Such loss applies not only to selectively neutral mutations but even to those with a small selective advantage. Many chromosome rearrangements, such as pericentric inversions, have a selective disadvantage due to meiotic difficulties. Kimura has shown [54] that the rate of fixation of a near-neutral mutation depends only on its mutation

rate. Mutation rates for pericentric inversions are unknown. In the absence of this knowledge and that of reliable information on the extent of the selective disadvantage – which may vary from one rearrangement to the other – no predictions can be made as to probability of fixation.

Nevertheless, we must bear in mind one special feature of selection against inversions. Such selection works only against heterozygotes. Inversion homozygotes have normal fertility, regardless of the site of the inversion, since pairing of homologous chromosomes at meiosis becomes normal. Do we know a genetic situation in which the "dangerous" state of heterozygosity is overcome quickly – even within two generations? Such a situation may occur when a brother and a sister inherit the same rearrangement from one parent and produce homozygous offspring in a brother-sister mating. Within this homozygous group fertility would now again be normal, whereas matings within the general population would produce only heterozygous offspring with reduced fertility. This mechanism would therefore build an effective reproductive barrier, providing the best conditions for gradual establishment of a new species (Fig. 14.8).

Present-day primates often live in small groups and similar conditions may also have obtained for our prehuman ancestors, thus facilitating the conditions for brother-sister matings or matings between other close relatives. If the "incest taboo" that prevents mother-son or brother-sister matings was already observed in these ancestral groups, one or a few more generations of heterozygotes may have occurred before the two homozygotes that were to form the ancestral couple would be produced. Moreover, homozygosity of a pericentric inversion has been observed in the present human population in a child from a father-daughter mating [6]. Are new primate species founded by one couple? More specifically: Do all human beings have in common one ancestral couple? The myth of Adam and Eve as the ancestral couple of mankind thus may even have a scientific basis.

A comparative study on 1511 species, representing 225 genera of vertebrates, has shown a strong correlation between speed of chromosome evolution and speciation, both processes being very fast, for example, in primates [11]. The authors suggested that population subdivision into small demes has probably been the decisive factor. This conclusion is in full accord with the hypothesis discussed above.

Development of Chromosome Bands [48]. Human chromosomes, as those from other mammals, are organized into bands (Sect. 2.1.2.3). G light bands contain many "housekeeping" genes and short interspersed repeats such as Alu sequences, whereas G

Mouse chromosome

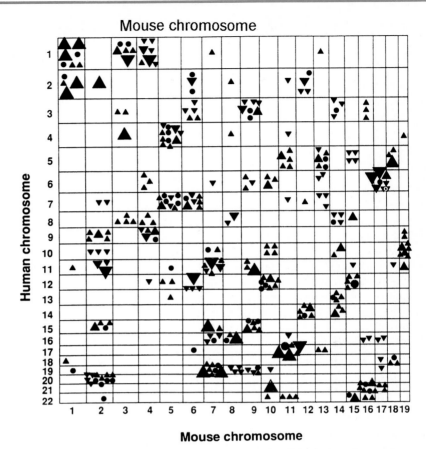

Fig. 14.9. The "Oxford grid" showing synteny between human and mouse chromosomes. *Horizontal length of each rectangle,* proportional length of a mouse chromosome; *vertical length,* proportional length of the given human chromosome. *Triangles and circles* refer to chromosomal location of 665 pairs of homologous loci in mouse and in humans; *triangles point up or down* depending on whether the gene locus is on the long arm or the short arm of the human chromosome; *circles,* arm location is unknown. *Large triangles* refer to 10 loci; *medium-sized triangles* to 5 loci; *small triangles, circles* to one locus. For every human chromosome at least three known loci are also syntenic in the mouse. (According to Searle et al. 1989 [95]; from McKusick 1995 [73])

dark bands have many genes for specialized functions in differentiated cells. This pattern has been maintained in evolution for about 100 million years; in addition to mammals, it is found, for example, in fish and reptiles. This phylogenetic history suggests an important function, such as differences in codon use and chromatin stability.

One group of repetitive DNA has been named satellite DNA; this has been studied extensively from a phylogenetic point of view (see below).

Direct DNA Studies in Human and Animal Fossils. DNA "survives" and is maintained in analyzable, albeit sometimes chemically and physically altered form in very old remains from extinct animals and ancient plants and also from human remains such as mummies or even bones. The polymerase chain reaction (PCR; Sect. 3.1.3.5) makes it possible to amplify and examine even minute amounts of DNA. It is

now being applied to human remains in the hope of obtaining direct information on the genetic constitution of early humans. For example, short segments of human mitochondrial DNA have been amplified [87–88]. This is opening up a field of direct cooperation between molecular biologists and paleoanthropologists [13].

14.2.2 Comparison of Satellite DNA in Higher Primates

Human Satellite DNA. The establishment of a pericentric inversion in a population does not necessarily require mating between two pericentric inversion heterozygotes. A slow increase in the frequency of such an inversion in a limited population by chance or by drift (Sect. 13.3.2) – even in the face of small selective disadvantage – would be another possibility.

However, such an explanation would meet with more difficulties in explaining another, recently discovered phenomenon: species differences in satellite DNA.

Human satellite DNA is described in Sect. 3.1.1.1, where it is explained that the term satellite refers to observations of DNA centrifugation in a cesium chloride density gradient, which in addition to the main DNA peak show smaller peaks that are characteristic for each species. Satellite DNA consists of relatively short, highly repetitive DNA sequences; their biological function is unknown, but they may influence crossing over during meiosis. In humans four satellite DNA fractions, SAT I–IV, have been distinguished, isolated, and characterized; these make up about 4% of the human DNA, or one-sixth to one-fifth of the entire highly repetitive DNA. These four satellite fractions were transcribed to yield radioactive complementary cRNAs, which were hybridized in situ to metaphase chromosomes of humans and great apes to assess their evolutionary origin. *Homo* turned out to share SAT I, III, and IV, but not II, with *Pan*. All four fractions, however, are shared with *Gorilla*, and at least I, II, and III with *Pongo* [44].

It now appears "that the fundamental sequences of all four satellites were present in the common ancestor of the species, ... but possibly only in one or a few copies per chromosome. The subsequent amplification of these sequences may have occurred after speciation, and, although the majority of amplification occurred at homologous sites in different species, sufficient differences exist to provide further evidence for the independent nature of this amplification event or events" (Fig. 14.10) [44]. Such amplifications of short, repetitive DNA sequences have also been discovered as causes of a few hereditary diseases (Sect. 9.4.2).

Comparison with Chromosome Evolution (Sect. 14.1.2). Differences between the karyotypes of *Homo* and the great apes are found in heterochromatin. They partially involve the centromeric regions. Additional telomeric regions show species differences in Q and T bands that do not contain any of the satellite fractions identified so far, but still unidentified satellite fractions are likely to exist. As noted above, euchromatic chromosome bands, which are thought to contain most structural genes (Chap. 3), appear to be identical in the primate species examined so far (Sect. 14.2.1). Variation is found in the satellite DNA and heterochromatin fractions.

In experiments with nonradioactive (CISS) hybridization [51] certain regions of ape chromosomes remain unlabeled with human DNA. These include telomeric heterochromatin in *Pan* and *Gorilla*, Y chromatin in *Gorilla* and *Pongo*, and an interstitial hetero-

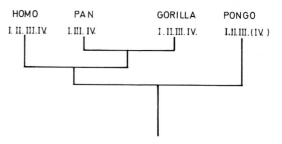

Fig. 14.10. Phylogeny of the Hominoidea showing contemporary distribution of sequences homologous to the four human satellite DNA species I–IV. (For explanation see text)

chromatic band in chromosome 14 of *Pan*. It is not impossible that these – at first glance inconspicuous – differences between chromosomes of *Homo* and the great apes may be associated in some way with the evolution of specific human traits. Reassessment of these problems is needed by a variety of modern techniques (Chap. 3) that allow an incisive analysis.

Species-specific satellite DNA fractions are known notably in the higher primates but in other species as well. Their significance for evolution and gene function remains obscure.

14.2.3 Protein Evolution [109]

Protein Sequences [23, 41]. One of the major achievements of biochemistry has been the determination of amino acid sequences within proteins. When the first sequences became known in the late 1950s and the early 1960s, homologies of sequences between homologous proteins of different species soon became apparent. Similarly, within the same species homologies between different but functionally related proteins were found. The sequences were usually identical at some positions and showed differences in others. At that time it was already known from study of some human hemoglobin variants that point mutations usually lead to the replacement of a single amino acid in a polypeptide chain. When the genetic code was deciphered, such replacements were shown to be caused by substitution of only one base in the transcribed DNA strand. Determination of biological relationships among species then followed by comparing the number of differences in amino acid sequences of their homologous proteins. Phylogenetic trees were thus constructed that could be compared with those derived from classic paleontological and morphological evidence. The methods of tree construction are discussed critically elsewhere [30].

Phylogenetic Tree for Hemoglobin Genes [113]. Figure 14.11 shows a phylogenetic tree for a number of

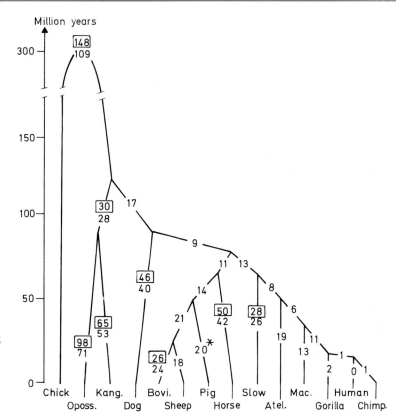

Fig. 14.11. Phylogenetic tree based on myoglobin and the Hb α and β genes. *Ordinate,* the approximate time of the splits between two phyla, on the basis of paleontological evidence; *numerals,* the number of nucleotide replacements between species; *numerals in boxes,* substitutions corrected for multiple-step mutations. The myoglobin sequence of the pig was incomplete at the time of tree construction, and the number of replacements is therefore slightly underestimated. (From Goodman and Tashian 1976 [41])

vertebrate species, including man, chimpanzee, and gorilla using amino acid sequences of myoglobin and the Hb α and Hb β genes. The time scale is derived from paleontological evidence. This tree shows that there is only one base substitution for these genes between man and chimpanzee and three between humans and gorilla.

More detailed evidence has come from comparative DNA sequence analyses of noncoding sequences within the HBB gene cluster [43]. These have confirmed the sequence of relationships; *Homo – Pan – Gorilla – orangutan – rhesus; Pan* showed fewer differences from *Homo* than from *Gorilla.* Occasionally, however, exceptions have been noted, the involucrin gene encoding an epidermal protein, for example, is more similar between *Pan* and *Gorilla* than between *Pan* and *Homo* [24].

Similar trees can be constructed for other proteins and – by combining the evidence – for all known proteins together. They can be extended beyond the vertebrates by including primitive animals and – for ubiquitous proteins such as histones or cytochrome c – even plants, fungi, and micro-organisms.

Rates of Evolution for Different Proteins. The number of necessary mutations for a given number of steps of speciation can be compared for different proteins. Some proteins have turned out to evolve at a much higher rate than others (Table 14.3). The histones, for example, have been amazingly stable, whereas the fibrinopeptides have evolved at a much faster rate. Remembering that only a minute fraction of all mutations are fixed during evolution – on average about 1 in 3.5 million [55] – we can discount differential mutation rates as a possible explanation; rather, the restraints imposed upon the amino acid sequence by protein function seem to be important. Fibrinopeptides, for example, are released in the process of fibrin formation from fibrinogen; their function is not very specific (covering of the fibrin surface, thereby preventing premature formation of the fibrin network). This might explain the high rate of evolution. It is understandable, on the other hand, that histones are very restricted in their conformation. They occur in the chromatin in intimate spatial relationship with DNA that could be impaired by even small molecular changes.

Gene Duplications. As noted in Sect. 7.3, the molecule of hemoglobin A consists of two α- and two β-chains; HbF has γ-chains and HbA$_2$ has δ-chains instead of the β-chains. All four types of chains have many homologous amino acids in common. The most obvious explanation is that all these genes – together with the gene for the myoglobin chain – are derived from one ancestral hemoglobin chain. Stepwise,

Table 14.3. Rates of mutation acceptance in evolution (from Dayhoff 1972 [23])

Proteins	PAMs[a] per 100 million years
Fibrinopeptides	90
Growth hormone	37
Pancreatic ribonuclease	33
Immunoglobulins	32
$\varkappa$ chain C region	39
$\varkappa$ chain V regions	33
γ chain C regions	31
λ chain C region	27
Lactalbumin	25
Hemoglobin chains	14
Myoglobin	13
Pancreatic secretory trypsin inhibitor	11
Animal lysozyme	10
Gastrin	8
Melanotropin β	7
Myelin membrane encephalitogenic protein	7
Trypsinogen	5
Insulin	4
Cytochrome c	3
Glyceraldehyde 3-PO$_4$ dehydrogenase	2
Histone IV	0.06

[a] 1 PAM = 1 accepted point mutation/100 amino acid residues in 100 million years. An accepted point mutation in a protein is one that leads to a replacement of one amino acid by another.

functional differentiation required duplication of these genes so that one copy could maintain the original function while the duplicated one was free to acquire a new function. Figure 14.12 shows the duplication steps for the hemoglobin genes, together with the evolutionary level at which they occurred and the approximate time scale. Such genes with a common origin and, very often, related functions constitute a "gene family" (Sect.3.1.3.10). The largest gene family known so far is that comprising immunoglobulins and cell surface molecules necessary for cell contacts (Sect.7.4).

Duplications of genetic material – either of individual genes or short chromosomal parts or of the whole genome (polyploidization) – have been of major importance in evolution. Apparently no polyploidizations have occurred in the evolution of mammals [85]; minor duplications and deletions have been frequent.

The evolution of multigene families has been studied by Maeda and Smithies [71] using the human haptoglobin system as main example. Haptoglobin is a hemoglobin binding protein of the blood; it was one of the first human examples in which unequal crossing over was studied (Sect.5.2.8). Analysis at both the protein and the DNA levels led to the conclusion that recombinational events, for example unequal crossing over or gene conversion, "often produce much more drastic changes than do simple point mutations. New alleles and new genes are constantly being formed by recombinational events between existing genes." Therefore evolutionary trees such as those shown above present an oversimplified picture of the real events during evolution; they should be interpreted with caution. This also applies to conclusions regarding single or multiple origin of mutations that are based on the DNA haplotypes in which they are observed (see Chap.12).

Evolution of Genes for Protein Domains. Above we considered only changes in amino acid sequences. However, proteins have a characteristic three-dimensional configuration, which is usually formed by a succession of two or more "domains," i.e., sequences, within which there are many contacts, whereas substantially fewer contacts are found *between* domains of one protein. Comparison of domains from different proteins has shown conformational similarities to be more widespread than anticipated from amino acid sequences. Protein domains may be very similar in conformation in the absence of similarity in amino acid sequence. In the course of evolution mutations were accepted only when the resultant amino acid replacement did not disturb conformation [94]. As suggested by model calculations, only about 200–500 such domains may have been the basic units from which the great number of different proteins that occur in nature have been found. As discussed in Sect.3.1.3, genes of higher organisms consist of several exons (= transcribed DNA sequences) separated by introns (= nontranscribed sequences). A single exon may comprise the DNA sequences for determination of such a protein domain.

The factor VIII gene involved in blood coagulation, for example (Sect.3.1.3.7) can be traced back to three domains A, B, and C. The A domain consists of 330 amino acids, the B domain 980, and the C domain 150. These domains are arranged in the order A1–A2–B–A3–C1–C2. The A domains show an unexpected but significant homology to ceruloplasmin, a copper-binding serum protein which also consists of three A domains, but lacks the B and C domains. To mention only one other example, the MHC genes which belong to a very widespread gene family (Sects.5.2.5; 7.4) may have been assembled 400 million years ago from at least three structural components [59].

Advantageous or Neutral Mutations? Why are certain amino acids in a sequence replaced by others over

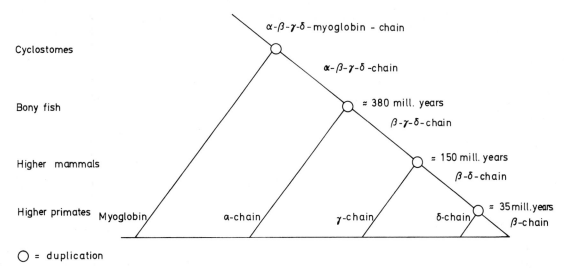

Cyclostomes

Bony fish

Higher mammals

Higher primates   Myoglobin

α-β-γ-δ-myoglobin - chain

α-β-γ-δ-chain

≈ 380 mill. years

β-γ-δ-chain

≈ 150 mill. years

β-δ-chain

≈ 35 mill. years

β-chain

α-chain   γ-chain   δ-chain

○ = duplication

Fig. 14.12. Steps of duplication for the hemoglobin genes and evolutionary stage at which they occurred. Additional duplications have led to the hemoglobin chains of early embryonic age, the ς and ε chains. Moreover, there are, in humans, two γ and two α chains (See Sect. 7.3). Information on the exact time of gene duplication for these chains is not available

the long term? There are two possibilities: either they improve the functional condition of the molecule, which makes for a selective advantage, or they are replaced – although they are selectively neutral or even very slightly deleterious – by genetic drift or random fixation (Sect. 13.3). The latter possibility has been stressed by Kimura [55–57]. In the introduction to his monograph [55], Kimura writes:

> The neutral theory asserts that the great majority of evolutionary changes at the molecular level, as revealed by comparative studies of protein and DNA sequences, are caused not by Darwinian selection but by random drift or selectively neutral or nearly neutral mutants. The theory does not deny the role of natural selection in determining the course of adaptive evolution, but it assumes that only a minute fraction of DNA changes in evolution are adaptive in nature, while the great majority of phenotypically silent molecular substitutions exert no significant influence on survival and reproduction. ...
> The neutral theory also asserts that most of the intraspecific variability at the molecular level, such as is manifested by protein polymorphism, is essentially neutral so that most polymorphic alleles are maintained in the species by mutational input and random extinction. ... [It] regards protein and DNA polymorphisms as a transient phase of molecular evolution and rejects the notion that the majority of such polymorphisms are adaptive and maintained in the species by some form of balancing selection.

On the other hand, Kimura clearly differentiates between positive (adaptive) selection, which he thinks is very rare *at the molecular level,* and negative selection, by which disadvantageous mutations are being eliminated in great numbers. He even estimates that only 10% of all new mutations are neutral (or even slightly deleterious), while about 90% are outright deleterious and have no chance of being fixed.

This "neutral" hypothesis, if applicable to a significant amount of genetic change in time, would have important consequences for our understanding of genetic differences between and also within species that must have originated during evolution. Therefore it is not surprising that it has led to heated discussions among population geneticists. These discussions have been beset with misunderstandings, many of which were clarified when Kimura summarized his views [55].

To understand this hypothesis, two scientific developments should be recalled:

1. In the 1960s an enormous amount of genetic variability became apparent at the protein and therefore at the DNA level. Methods for determining amino acid sequences showed differences between homologous proteins in different species as well as between related proteins in the same species. Moreover, analysis of the genetic code opened up new sources of variation for analysis. Among other aspects, the enormous amount of DNA available in the eukaryotic cell (Sect. 3.1.1.1) raised the question of its function. The large amount of DNA and its considerable variability induced some scientists to question whether natural selection as assumed by the neo-Darwinian theory of evolution accounted for these findings, or whether at the molecular level random processes were more important. If selection were the critical factor, as maintained by the conventional synthetic theory of evolution, the number of DNA sites subject to selection would be enormous.

2. At the same time, Kimura developed mathematical diffusion models that provide answers to such questions as: "What is the probability that a single mutant which appears in a finite-sized population will eventually spread throughout the whole population?" (i.e., the probability of gene fixation, Sect. 13.3).

In a classic paper, he formulated the problem and its solution as follows:

> Consider a population of size N. . . . If we look sufficiently long into the future, the population of genes at a particular locus will all be descended from a single allele in the present generation. . . . This is the result of the inexorable process of random genetic drift.[1] If, in the present generation, an allele A_1 exists in frequency p, the probability is . . . p that the lucky allele from which the whole population of genes is descended is A_1 rather than some other allele.
>
> Now, if mutation occurs at a rate μ per gene per generation, then the number of new mutants at this locus in the present generation is $2N\mu$.[2] Furthermore, the probability that a particular gene will eventually be fixed in the population is $1/2N$. So the probability of a mutant gene rising in this generation and eventually being incorporated in the population is
>
> $$2\,N\mu \times \frac{1}{2N} = \mu$$
>
> The rate of neutral gene substitution is identical with the mutation rate – irrespective of population size.

[1] This statement follows from the rules of random fixation; Sect. 13.3.
[2] Where N = size of populations.

Kimura [55] later amended his reasoning somewhat by noting that a mutant that is selectively disadvantageous in a large population may be neutral in a small one, so that the rate of substitution would in fact be more rapid in small populations.

In his model calculations Kimura treated random mutation as a time-dependent process [55]. He believed that random fixation of mutations, depending only on the mutation rate, would predict accumulation of genetic differences at a linear time scale, regardless of species, generation time, and other parameters. It would be limited only by the constraints imposed by the functional requirements of the genes and their products, the proteins: "negative selection" eliminated base substitutions and amino acid replacements *not* compatible with normal function. Differences in the speed of evolution between different proteins (Table 14.3) can be explained by this negative selection, in agreement with conventional theory. Only the different sequences *maintained* in the course of evolution cannot, in his opinion, be explained by corresponding differences in *positive* selection.

The value of the "neutral hypothesis" for an explanation of aspects of evolution can be tested by examining whether its predictions are verified or refuted by actual data. Such predictions have been made at two levels: amino acid sequences and replacements in proteins; base sequences and substitutions in the DNA.

As observed above, one prediction is the linear time dependence of the amino acid replacement rate – the *evolutionary clock*. Its establishment required measurements of the time when two branches in a phylogenetic tree split, using an independent time scale based upon paleontological evidence. Discussions in the literature on this prediction have often dealt with this evidence. For example, some species of fish have been living in the deep ocean from time immemorial, and the ecological conditions must have been very similar if not identical throughout, but protein "evolution" nevertheless proceeded at a steady state. The α- and β-hemoglobin chains in nonhuman mammals have diverged since their separation by the same rate as those in humans and fish. On the other hand, more detailed examinations of some parts of the "phylogenetic tree" reveal deviations: the protein evolution of primates, for example, has been slower than predicted by the "clock." This has been confirmed by studies of DNA [8, 43, 66].

Arguments Against General Applicability of the Neutral Hypothesis. Much of the discussion about the neutral hypothesis involved arguments from theoretical population genetics. Suffice it to say that, depending on specific assumptions on the parameters of the genetic models used, conclusions other than those derived by Kimura are possible (for example, see [29]). Ideally the theory and the propositions derived from it should be tested by empirical evidence.

The following points are undisputed:

1. Many forms of mutation lead to genetic defects. They have a strong selective disadvantage, however, and are soon eliminated from the population.
2. Other mutations are subject to special modes of selection that maintain genetic polymorphisms, either by heterosis (Sect. 12.2.1.3) or by frequency-dependent selection (Sect. 12.2.1.5).
3. Many amino acids found in proteins would have no measurable selective advantage under present-day conditions compared to the amino acids that they replaced during evolution, probably many millions of years ago.

Some amino acid replacements change biological properties of proteins, such as protein conformation, less than others. They have a higher chance of being selectively neutral and of being fixed by random processes.

It is conceivable that an undetermined proportion of mutations had a slight selective advantage at the time of fixation. Some recent evidence points to such influences:

1. Once the function of a certain gene product is established, selection tends mainly to preserve its functional characteristics or even adapt the molecule somewhat better to its function. A substitution of one amino acid for another with similar conformational and biochemical properties therefore tends to be maintained [14, 21]. There is, indeed a strong correlation between biochemical similarity of amino acids and probability of substitution.

2. Once a mutant is formed, for example, by duplication, selection is expected to adapt it to its new function; the frequency of amino acid replacement presumably increases. This was asserted for the hemoglobin genes, which show increased amino acid replacement after exhibiting duplications [42]. This result has been disputed, however, mainly by questioning the paleontological evidence on which it was based and the method of tree construction [55].

3. The neutrality hypothesis postulates that many, if not most, present-day polymorphisms in the human population are not maintained by selection but are neutral alleles on the way to fixation by drift.

At first glance the distribution of rare and common electrophoretic variants seemed to contradict this postulate (see Sect. 12.1). The observed distribution is strongly bimodal; there is a group with relatively high (intermediary) gene frequencies, which are thought to be maintained by heterozygote advantage or frequency-dependent selection, and a group with low gene frequencies that could comprise genes maintained without selective advantage or disadvantage by genetic drift. The neutral hypothesis would predict a relatively high frequency of different rare and very common (gene frequency ≥ 0.9) variants, and a lower frequency of variants with intermediary gene frequencies; this is apparently at variance with the distribution actually found. However, it is very difficult to distinguish whether such a distribution has been caused mainly by genetic drift of neutral alleles, by a mixture of various modes of selection, or by both.

4. The hypothesis can also be discussed at the level of DNA. For example, base substitutions not leading to amino acid replacements – especially those affecting the third base of a codon – were found to be more common than those causing such replacements; and DNA sequences outside transcribed re-gions turned out to be especially variable. This also applies to variation *within* the human species; there are many DNA polymorphisms of various types (Sect. 12.1). An average heterozygosity/codon might be about ten times as high in noncoding DNA sequences of the human genome, including also introns, as in coding sequences [19]. Moreover, the base substitution rate in functionally inert pseudogenes, for example, an Hb α pseudogene of the mouse (Sect. 7.3), appears to be higher than in their active counterparts. On the other hand, a comparison of the mRNA of the hemoglobin-β-chains for humans, mouse, and rabbit *failed* to show the randomness expected from the neutral hypothesis; on the contrary, the pattern of base substitutions was decidedly nonrandom [33].

As shown by these and other arguments, much of the evidence adduced either for or against the "neutral hypothesis" is ambiguous. Until these questions are settled, some plausible inferences can be attempted.

"Genetic Sufficiency" [113a, 114, 115]. Let us assume that environmental conditions change in such a way that functional adaptation of a certain polypeptide becomes less efficient. Then, if a mutation occurs that meets the new demand in a more efficient manner, its carriers enjoy a selective advantage. The new mutation does not necessarily improve the polypeptide to its optimal conceivable state; there is only some improvement. Moreover, a number of different mutations may bring about this improvement; nature has a number of options by which a given demand can be met, not always in an optimum fashion but often to an adequate degree. The actual substitution that is selected depends on the availability of mutants in the population at the time when the demand arises. Availability, in turn, depends on the mutation rate (in addition to genetic drift). Indeed there is evidence that base transitions are more frequent among mutations that were fixed during evolution than are transversions; transitions, especially those in CpG dinucleotides, are more frequent among fresh mutations (Chap. 9) [18, 104]. There is thus an element of randomness within the boundaries imposed by functional demands and selection.

One example familiar to human geneticists may help to demonstrate the meaning of "genetic sufficiency." When malaria became widespread in populations of tropical countries, an improved resistance against this disease became useful to the population. Genetic adaptations soon occurred in all exposed populations. The precise mode of adaptation, however, differed (Sect. 12.2.1.6). In Africa HbS and HbC were selected, while in the Austroasiatic population it was HbE, and in various populations different thalassemias and various G6PD deficiencies became frequent. The adaptive value of these mutations was by no means identical; HbE, for example, offered protection from malaria at a much lower price than most β thalassemias, as HbE homozygotes have a much milder form of anemia than β thalassemia homozygotes (Sect. 12.2). Nevertheless, both adaptations were sufficient, since the population survived. Obviously the adaptations depended on the kind of mutation that happened to be present and could therefore be favored by selection.

This concept of "evolutionary sufficiency" was proposed by the biochemically oriented biologist Zuckerkandl. However, it is very similar to the position of the population geneticist Ewens [29] that various, not necessarily "optimal," combinations of genes may meet a special demand from the environment. Kimura's ideas [55] are not far from this concept. The main difference seems to be that authors such as Ewens and Zuckerkandl attribute greater importance than does Kimura to *positive selection*, i.e., selection in favor of slightly advantageous substitutions *in addition to random processes*. It appears to be agreed, however, that in a constant environment most selection is of a negative kind, i.e., tends to *preserve* a certain function by eliminating deleterious mutants.

Limitations of Present Knowledge of Natural Selection and Neutral Substitutions in the Evolution of Proteins. Most research workers agree that natural selection has been responsible for substitution of some amino acids in proteins and for some of the genetic protein polymorphisms found in the human population. On the other hand, a portion of the observed variation among species and within the human population is probably caused by random drift; here selective advantages or disadvantages may be trivially small or even completely lacking. However, current evidence is insufficient to decide what proportion of genetic variability is caused by selection and what proportion by random processes. In this context the amount of genetic polymorphism in the human population should be remembered: The human genome may comprise about 50 000–100 000 structural genes [58] that encode for proteins. Among the known proteins up to 30% may be polymorphic.

Most expressed polymorphisms affecting proteins and enzymes have been detected in the blood. As explained elsewhere (Sect. 12.1.2), the amount of expressed polymorphisms in other, less accessible tissues may be much lower, but there are probably hundreds or even thousands of polymorphic loci, only a minute fraction of which are known so far. Moreover, although the physiological function of many polymorphic enzymes is completely unknown, inferences regarding natural selection are much more likely to be correct if founded on knowledge about the physiological function of the polymorphism under study.

In most proteins a characteristic function depends crucially on a few amino acid positions. Functional restrictions can be of such a general kind that they may be met with by many different amino acids; for example, maintenance of three-dimensional conformation. Here genetic drift may shift bases freely, and this may lead to polymorphisms at the protein level. These polymorphisms may even bring about small functional differences that do not or only trivially influence fitness (Sect. 12.1.2) of their carriers. When the ecological conditions change, they might provide a reservoir for fast adaptation. On the other hand, the fact that selective influences are still unknown for most polymorphisms does not mean that selection has been absent. On the contrary, it is difficult to detect selection – especially in human populations, where modern civilization has changed living conditions considerably within a very few centuries, having eliminated potentially important selective agents such as infectious diseases and – for large parts of the world population – malnutrition. The discovery of selective mechanisms requires specific, functionally founded hypotheses. This does not mean, that all functional differences found among polymorphic variants must at some time have influenced fitness. But it would be difficult without such assumption to explain why the rare variants of polymorphic enzymes generally show lower activities (Sect. 12.1). The undisputed fact that selective influences are still unknown for most expressed human polymorphisms does not mean that selection has been absent; rather, it testifies to our inability to propose and test well-founded hypotheses on selective mechanisms. The neutral hypothesis, when applied to the study of human polymorphisms, might even have a counterproductive effect if it discourages the search for sources of natural selection.

Evolutionary Clock and Mutation. The evolutionary clock can be explained if mutations are time dependent irrespective of species, and if they are fixed at random. As discussed in Sect. 9.3, mutation rates in humans are sometimes higher in male than in female germ cells, some mutation rates increase with the age of the father, and many mutations seem to be related to DNA replication. Even the fact of the vastly different generation times found in various animals makes it highly unlikely that mutation is simply time dependent.

For a more realistic approximation to the mutation rate it would be desirable to construct a timetable based on our knowledge of mutation mechanisms

[18], for example, generation time, average ages of reproduction, the number of DNA replication cycles/generation, the position of the mutated site within the base sequence, and many others. Moreover, repair processes must be considered [103]. If a clocklike regularity of base substitutions were indeed shown (which is, in our opinion, dubious) it could certainly not be used as argument in favor of random fixation of mutations. According to the rule derived by Kimura (see above), the rate of fixation depends only on the mutation rate, which cannot be assumed to be solely time dependent in all species.

A possible way out of these difficulties may be to assume that mutations which are slightly disadvantageous in large populations are effectively neutral in small ones and therefore have a higher probability of being fixed. Species with large body size (such as elephants) have a longer generation time on average (and probably also a lower number of DNA replication cycles/*time* unit) but a smaller population size than species with small body size (such as mice). One could also argue that the mutation rate/time unit does *not* depend only on the various molecular mechanisms, such as the number of replication cycles/time unit, but has been adapted to an optimum rate by natural selection. But why, if most mutations are neutral anyway?

In conclusion, in proposing the "neutral theory" Kimura has certainly made an important point by stressing that evolution at the molecular level has some aspects not revealed when studied at the level of phenotypes. There is little doubt that random (or near-random) processes at the molecular level are much more important than most biologists have previously thought, especially in nontranscribed DNA sequences. It is often observed that the explanatory power of new theoretical concepts is overrated by their originators. As Popper says, however, science can proceed only by boldly advancing hypotheses and then submitting them to stringent testing. On the other hand, there may be other functional constraints leading to compartimentalization (so-called "isochores") of both coding and noncoding DNA that introduce an unexpected source of natural selection [5].

Special Problems Posed by Highly Variable DNA Polymorphisms. Many DNA polymorphisms outside coding regions show an unusually high degree of interindividual variation and are often unusually unstable in transmission between parents and children. Is this merely a sloppiness of replication mechanisms that was not eliminated during evolution because it had no influence on survival and reproduction, or does it have a special biological function, and therefore a selective advantage?

Evolution by Reshuffling of Exons. Discovery of the exon-intron structure of genes (Chap.3 and Sect.7.3) opened up a new path for our understanding of protein evolution: exons may be separated from one other and rearranged in a new order, or some exons of only one gene may be transcribed in one species whereas the entire set is used in another. As noted above in the discussion of alternative splicing (Sect.7.2.2.4), such differences in gene usage have been observed even between different tissues of the same individuals; this appears to be one mechanism of differentiation.

Some sets of exons are used to construct different proteins. For example, the low-density lipoprotein receptor has homology with eight exons that code for a precursor of epidermal growth factor. This and other work suggests that functional proteins are mosaics of simpler structures that are shuffled together (see [40]). The complexity of proteins appears to be derived from the combinatorial assembly of a relatively reduced number of smaller genes that specify exon structure.

Comparison of the Protein Data with Data from Chromosome Evolution and Satellite DNA. The data on evolution of proteins show amazingly few differences between *Homo* and the other higher primates, chimpanzee and gorilla. For practical purposes these proteins can be regarded as identical. In the hemoglobin molecule, for example, these species differences are functionally less important than rare variants in human populations that are fully compatible with life but may lead to mild hemolytic anemia. This extremely slow evolution can be explained if the function of these proteins remained largely identical. However, even looking at karyotypes we find only a few chromosomal rearrangements, mainly pericentric inversions. Similar rearrangements are not so rare in the present human population, and they do not influence the phenotype at all. They might explain reproductive barriers that were once an important condition for speciation, but they do not explain the genetic mechanisms that created the specific human phenotype. Little is known about the functions of additional R and T bands and species differences in various DNA fractions. However, centromeric heterochromatin shows much variability within current human populations. An effect of these heteromorphisms on human phenotypes such as behavior has been claimed but is not generally accepted [28].

Thus we are left with the conclusion that the genes important for human evolution during the phase of human brain development are completely unknown. Since most human DNA does not code for proteins and might either be "junk" or of great importance in regulating gene activity, especially during embryonic development, relevant changes may have occurred within this nonstructural DNA [115]. Such alterations might have taken place within the nontranscribed parts of those sections of DNA separating structural genes that are postulated to have regulatory functions. It is conceivable that DNA sequences not important in structural gene function may somehow be required for development, and hence that changes within those DNA species are especially efficient in bringing about improvements in brain function. This concept, however, is very speculative and general. Formulation of more specific hypotheses requires more to be known regarding genetic determination of embryonic development and regarding genes affecting intraspecies variation in human behavior (Chap. 15). Even disregarding any phenotypic effects and only considering the analyzed genetic phenomena such as chromosome rearrangements, addition or reduction of chromosome material, satellite DNA and amino acid sequences of proteins, many aspects are still poorly understood. For example, how were chromosome rearrangements fixed in populations? Are the mechanisms identical to those applying to amino acid replacements? What single events formed satellite DNA and other repetitive DNA fractions? Do such events have specific significance for speciation or for regulation of gene functions?

14.2.4 DNA Polymorphisms and Evolution

As discussed in Sect. 14.2.3, neutrality of base substitutions is more plausible for polymorphisms situated outside coding DNA sequences than for those leading to changes in proteins. Comparisons have therefore been carried out based on nuclear and mitochondrial DNA, between humans and higher apes and between human populations.

A Phylogenic Tree of Mitochondrial DNA. Mitochondrial DNA (mtDNA) is particularly well suited to evolutionary studies. The mitochondrial genome is completely known (Sect. 3.4); with somewhat more than 16 000 base pairs, it is fairly short. It is transmitted by the mother to all children regardless of sex, which much simplifies population genetic analysis. Moreover, changes within mtDNA are fixed in a population much faster than in the nuclear DNA – probably as a consequence of the simple maternal transmission rather than of a higher mutation rate. Wilson's group has published a series of papers comparing mtDNA sequences among humans from different populations (see, for example [108, 110]). Examining 370 restriction sites covering variation in 1550 base pairs – almost 10 % of the mitochondrial genome – in 241 indi-

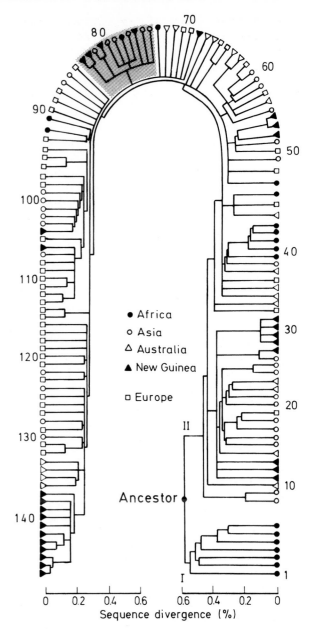

Fig. 14.13. Phylogenetic tree relating 147 types of human mtDNA. An average of 370 restriction sites was used to map the mtDNA of 177 individuals. The tree was constructed on the assumption that it starts with one type of mtDNA (= one individual?), and that the minimum of nucleotides possible be required to reach present-day diversity. *Shaded segments,* control region. (From Wilson et al. 1987 [110])

viduals of various ethnic origins, they described 182 types of mitochondrial DNA. From these types a phylogenetic tree was constructed (Fig. 14.13). These studies were later supplemented by DNA sequence analysis of the control region (shaded in Fig. 14.13) and compared with chimpanzees [102]. The main conclusions from these studies were:

1. The variation of mtDNA types is much greater among Africans than in the rest of mankind. There are two branches of mtDNA in Africans, only one of which is shared with the rest of the human species.

2. The "evolutionary clock" worked relatively regularly in mtDNA evolution but much faster than in the evolution of nuclear genes.

3. All mtDNA lineages found in the various population groups today can be traced back to one woman who lived in Africa about 200 000 years ago.

4. All human populations living today can be traced back to one (or a few) closely related humans who emigrated from Africa. Their "migration" history can be reconstructed in crude outline by comparing population groups within the non-African branch.

This attractive hypothesis has been challenged mainly on statistical grounds [39, 60]. Construction of the "evolutionary tree," for examples has been criticized as the methods which it requires raise difficult statistical problems [38]. Furthermore, the tree's "rootedness," i. e., the fact that it leads back to a single origin, is in fact a condition for its construction – and therefore cannot be represented as a conclusion [60]. Moreover, a great number of different trees may be constructed by appropriate shifting of plausible assumptions and techniques [39]. Trees are often set up using the principle of parsimony, i. e., they are based on the lowest number of mutational steps necessary. However, there are of course many other possibilities, including those positing a non-African origin. Moreover, even if a common origin in Africa is assumed, the time at which the woman was supposed to have lived who had the original mtDNA from which all the others were supposed to originate (200 000 years ago) is probably too close to the present if realistic assumptions on population size and its increase are made. Therefore the nice, romantic picture of "mitochondrical Eve" living some 10 000 generations ago in Africa is probably incorrect. (Here we can mention only some aspects of the criticism; for a more complete assessment see analyses described by Gee [39].)

Two conclusions, on the other hand, probably are correct:

1. Variability of mtDNA appears to be larger in Africans and their descendents than in other racial groups. Together with the fossil record [7] this tends to support the hypothesis that present-day Mongoloid and European racial groups may have originated by immigration from a limited African group in relatively recent times. Still, this concept is challenged by a few other paleoanthropologists who prefer a multifocal origin of *Homo sapiens*

[101]. If the hypothesis is correct, however, which we regard as likely, it raises the question of what happened to the much older *Homo erectus* populations found in Asia and Europe – a question that cannot be answered at present.

2. It is also plausible that all human mtDNA may ultimately stem from one female; however, she most probably lived considerably more than 200 000 years ago [60]. Moreover, the history of the mitochondrial genome does not differ greatly from the history of our chromosomes, as explained in Sect. 14.2.1. We have good reason to believe that each of our chromosomes (or chromosome segments) can be traced back to one individual (in this case, it may also be a "father"). For example, the telomeric fusion leading to chromosome 2, which all human beings have in common, might well be due to one mutational event in one individual. This is also plausible for the pericentric inversions in which humans differ from chimpanzees. As to single base replacements leading to amino acid sequence differences between species in proteins, many may also go back to unique mutational events, whereas others – especially those involving mutations with relatively high probabilities such as transitions in CpG dinucleotides (Sect. 9.4) may well have been fixed twice or more times.

The analysis of mtDNA evolution shows how scientific knowledge often grows. First, there is a theoretical concept that appears to be supported by an elegant set of data. Thus both the result and the conclusions convince many scientists. If the topic is sufficiently interesting, especially if an attractive slogan is found for popularization, such as the "African Eve," the hypothesis may attain wide popularity and be accepted widely but in a way often contrary to the intentions of its inventors. This course of events induces other scientists to assess it critically and results in many questions regarding its validity. Nevertheless, some progress is achieved. This instance demonstrates the need for critical evaluation of hypotheses in order to determine the degree of truth which they do in fact contain. In the present case, for example, we have learned the greater variation in mtDNA among Africans than in other racial groups – a finding which still requires explanation. More importantly, however, is the gain in critical evaluation of such hypotheses generally; this may lead to improved assessment also of other results in related fields, for example, the construction of phylogenetic trees in general [65]. Such critical assessments do not detract from the merits of those proposing the hypotheses nor those trying to test it in a new way.

It is a common, and often commendable, attitude among scientists upon hearing something really novel, to say: "Here, I would be very, very cautious." But this caution alone does not lead to any progress. The studies of Wilson's group on mtDNA – despite all their weaknesses – have improved our understanding of human evolution.

14.2.5 Behavior

Man the Toolmaker. In the face of our deficient knowledge of the genetic determinants of human behavior, it seems premature to speculate regarding the nature of the genetic changes that have brought about the last steps in the development of our species. It is, however, possible to develop hypotheses on the nature of selective pressures under which these changes have occurred, since the living conditions of our early relatives living in Africa can be reconstructed [54, 64, 95]. As mentioned, there were a number of related species such as *Australopithecus afarensis, A. africanus, A. robustus,* and *Homo habilis* (the first specimens of *Homo erectus* have been dated somewhat later). The varied ecological conditions of the Great Rift Valley – with its high mountains and deep valleys, its lakes and rivers – together with the variation of landscapes, after a cooler climate had led to a reduction in tropical forests – were propitious for radiation and speciation within populations of highly developed primates. The food conditions became more difficult than in the tropical forest; no doubt, the various groups (species?) adapted themselves to these conditions in various ways. *A. robustus,* for example, fed mainly upon rough plant material, as shown by his teeth. *Homo habilis* added more meat to his diet; this had the advantage of a higher energy content but required foraging in larger areas, and distribution of labor between the sexes: this led to the use and, later, production of stone tools. It also required a more sophisticated social organization.

It is difficult to reconstruct the living conditions and social organization since we must rely upon only indirect evidence (none of us was present). In addition to careful assessment of the paleoanthropological and archeologic findings, observations of other primates, especially the great apes, and of present-day hunter-gatherer groups must be used. This reconstruction inevitably contains an element of subjectivity; a scientist's prejudices may enter. Feminists, for example, tend to see the roles of males as more similar to that of females, they may criticize the following description as too male oriented. We are impressed, however, by a variety of male/female behavioral differences in many animal species including nonhuman primates which appear to be of biological origin. Human male/female differences in behavior therefore sometimes may also have a biological basis.

Social Structure of Early Prehuman and Human Groups. The context within which all activities of these early prehuman forms must be considered is the structure of the social group. Hunting with primitive tools requires the cooperation of a fair number of individuals. Such cooperation must be carefully planned and requires the exchange of information over a certain distance. At a time when the males are hunting, the females must keep the children safe from predatory animals, while foraging near the campsite for fruits and, probably, small animals. All these activities require a fair group size of between 20 and 50 individuals.

Functioning of such a group could be improved by two conditions. The first is leadership. It is easily conceivable that hunting was planned by one male who convinced other group members to follow his advice. The second conditon is language. The task of a game-hunting group must have been much easier once its members were able to inform and direct one another by acoustic signals.

These simple considerations give an idea of the selection pressures favoring evolution of the two most prominent features that distinguish humans from all other animals: intelligence which enables them to develop abstract concepts and to plan for the future and, closely connected with this, language.

The selective value of a genotype depends on the reproductive rate of individuals with this genotype relative to others. It is, however, easily imaginable that dominant males in such a group – those who planned hunting and influenced the decision as to where the group would move to find favorable living conditions – also had the easiest access to women and would father more children than other males. Given the special requirements of this way of life, these males would, at the same time, also be the most intelligent and, more specifically, those most proficient in mastering language. In a very primitive present-day South American Indian group, the Xavantes, for example, one dominant male has indeed been shown to be the father of many of the children [86]. The same tendency for headmen to father more children than nonheadmen has found in another tribe, the Yanomama [80].

Precursors of Language and Cultural Tradition in Apes and Monkeys. In an effort to better understand evolution of social structure and cultural traditions, such as toolmaking and language, many studies have been performed on higher primates, including chimpanzees, gorillas, baboons, and rhesus monkeys. The social structures of primate groups show vast differences – from baboon species that live in groups of hundreds to the male orangutan who lives alone in his territory and mates with females that happen to inhabit an overlapping territory. The group may be either open or closed; one male may be dominant, or more subgroups consisting of one male and a varying number of females may be present, or there may even be sexual promiscuity. The groups tend to be larger in the open savanna and half-desert than in the tropical rain forest, but even this rule has exceptions. These observations do not permit direct conclusions as to the structure of early human groups. At least three kinds of observations might, however, lend themselves to extrapolation:

1. Most groups show social stratification. Some animals hold a dominant position in the hierarchy, and others are of lower rank. Such stratification is found – in one or another form – in present-day primitive human populations and may well have been one feature of early prehuman groups.
2. Monkeys show cultural transmission of behavior. New behaviors, such as bathing in the sea or dipping a potato in seawater to improve its taste by salt, are invented by single individuals and eventually taken over by the group. Whether a certain behavior is accepted depends on conditions that appear very human to us. A new behavior has a higher chance to be accepted by the group if the inventor is high-ranking than if he is low-ranking. Younger group members take over new behaviors much easier than older ones.
3. While so far no primate has ever learned to vocalize sounds resembling a human language, some chimpanzees can be trained to convey detailed information by giving hand signals using a symbolic sign language similar to that used by the deaf or using computer consoles with symbols.

Chimpanzees are able to figure out ways of producing tools from parts given to them separately, to understand and use concepts such as shape vs. color, and to recognize themselves in a mirror and produce a picture of themselves by combining parts of a puzzle. When challenged sufficiently by the experimental design, apes display amazing intellectual capacities. The question of why they apparently do not use more of these capacities in their daily life in the wild state has puzzled many research workers. Probably they use such capacities not for developing techniques or solving puzzles but to achieve success within the social structure of their group, thereby improving their chance for reproduction [52, 103]. Achieving headman status in primitive human societies is comparable. The abilities developed under the selection pressure of social life in a primate group probably served as preadaptation for human cultural development.

Behavioral Characteristics of Humans in Common with Other Species. The above discussion considers behavioral characteristics that differ between humans and other species. In a great many characteristics, however, humans and other animals are similar. For example, we are frightened, as are animals, if our life is threatened; we are sexually aroused if appropriate stimuli are offered; and an assailant may so enrage us that we react aggressively.

The question of whether aggression is an innate human characteristic has led to many discussions. According to Lorenz (1966) [69], "there is an internal urge to attack. . . . if actual attack has not been possible for some time, this urge to fight builds up until the individual actively seeks the opportunity to indulge in fighting." Many other scientists, however, believe that fighting derives principally from the situation, but they do not deny that the emotional status of the animal is important as well. In Sect. 15.1.2, we see that the ease with which aggressive behavior in the mouse is elicited varies among inbred strains, and that this variation is correlated with the amount of the neurotransmitter epinephrine.

According to Lorenz, humans, unlike other animals engaging in intraspecies fighting, tend to carry on until the enemy is destroyed; the inhibition to killing other members of the species is overruled mainly by development of weapons, especially those acting at greater distance so that eye-to-eye contact is avoided. According to this concept, in a society that does not permit occasional outlets for aggression the aggressive potential builds up, and the society may even be destroyed in the ultimate outbreak.

Most social scientists, and many biologists, feel that this view oversimplifies the situation. To avoid destructive intraspecific aggression, it seems important to look for different means of education apart from providing harmless outlets such as athletics, as proposed by Lorenz. Study of aggressive behavior in animals and possibly in nonhuman primates might help to understand possible biological aspects of the human situation somewhat better.

Some other reactions seem to be inborn in human beings as well. For example, even children who are born blind and deaf smile when they feel that someone is caressing them. The habit of acknowledging the recognition of another person by raising the eyebrows has been observed in all human cultures and may well be an inborn behavior pattern.

Intensified exchange of ideas and concepts between human behavior genetics and ethology may help in the future to understand better the genetic determination of human behavior. These and similar observations and speculations have gained popularity in the United States as "sociobiology" and in Europe as "human ethology."

Human Sociobiology [70]. Sociobiology is the biological and evolutionary study of all forms of social behavior. The term was coined by Wilson in 1975 [111] in his description of existing knowledge of this area. However, the field was not entirely new. Ethology and behavioral ecology were areas of study investigating social behavior in animals. Investigators and theorists in sociobiology are distinguished by invoking genetic and evolutionary principles to understand and predict such behavior. The central theorem of sociobiology states that each individual can be expected to behave so as to maximize his or her inclusive biological fitness [1, 111]. For example, the Darwinian theory of natural selection had difficulties explaining how altruistic behavior could survive in evolution since it diminished the fitness of the "altruist." Such altruistic behavior, however, could be explained if the principle of kin selection is invoked (Sect. 12.2.1.5). Under such selection altruism can evolve if the advantages to survival of the group or the kinship outweigh the negative effects of selection acting on individuals. As Haldane put it succinctly many years ago (quoted in [96]), he was prepared to lay down his life for two brothers and eight cousins since he shared one-half of his genes with his brothers and one-eighth of his genes with his cousins! In general, however, kin selection is a weak selective agent compared to individual selection and can be invoked only under certain extreme conditions [97].

Sociobiologists compare the behavior of hundreds of species in the light of evolutionary principles and hope to shed light on some new aspects of behavior that could not be fully understood before. Using such methods, sociobiology has provided novel explanations for certain genetically fixed behavior patterns in many animals. The initial insights were achieved without attempts to understand the underlying neurobiological mechanisms involved in such behaviors. While this has since changed [70], the particular genes involved in social behavior remain largely hypothetical. Nevertheless, the existence of inherited genetic behavior patterns, such as those determining navigation in migratory birds, cannot be contested.

The possible implications of sociobiological theory for human behavior have elicited considerable interest. By extending their reasoning to the human species sociobiologists are attempting to interpret human emotions, human sexuality, aggression, and social status by evolutionary principles [98, 111]. It has been suggested that a human biogram exists, a pattern of potentials and constraints built into the species. Genes set limits within which cultures can develop. Facial expressions conveying various emotions appear very similar across all human cultures. Sexuality is considered the device of natural selection to ensure pair bondings. Polygyny (mating of one male with many females) is claimed to have a physical basis by conferring a natural advantage to the species. The forms of polygyny – polygamy, mistresses, multiple marriages, etc. – may vary in different cultures. Wilson [112] postulates a physical basis for inherited mythopoetic tendencies and therefore limits to "scientific enlightenment" as a basis for social cohesion. Sociobiologists feel that many more of the recognized human constants are physically rather than socially determined. These include male dominance, sexual division of labor, prolonged maternal care, and extended socialization of the young.

Some ethologists became especially interested in the biological laws influencing individual development during infancy, childhood, and youth. Here, the phenomenon of "imprint-ing," discovered by Lorenz in the wild goose as long ago as 1935 [68] and analyzed later in many other animal species, has influenced thinking on cognitive and emotional learning. The young gosling invariably follows the animal first seen after hatching, even if this is a human being; most often it is the mother. In animals there are many other behavior patterns that can be learned only in a specific phase of individual development, and they must be learned then. It is a matter of controversy whether imprinting occurs in humans at all, but there can be little doubt that human interaction is required, for example, for learning to speak and for social and emotional learning [46].

Wilson admits that the preprogramming of the human brain is much less specific than in other species [112] and therefore allows for much more plasticity of human behavior. Nevertheless, he interprets the evidence of little variation among human cultures to indicate that traits such as incest taboos, use of body ornaments, and elaborate kinship rules are biologically derived.

Sociobiologists see their field as an antithesis to the environmentalism of the social sciences, such as social anthropology and sociology, which often assume human social and cultural life to be entirely culturally determined and constrained only by the most rudimentary biological drives. *The major substantive criticism of sociobiology is its lack of direct evidence for the operation of genetic factors influencing most human behaviors claimed to be under genetic control.* Despite this current lack of direct evidence, it appears highly likely that certain aspects of human behavior have been programmed genetically by natural selection. *It is improbable that the human species is entirely autonomous in its behavior, and that genetic determinants of the central nervous system and their influence on social behavior are entirely overridden by cultural factors* (Chap. 15). The human species and its brain are part of an evolutionary a continuum. The complete independence from biological constraints of traits mediated by the central nervous system is therefore unlikely.

Sociobiology has been vehemently condemned by scientists and others who deny that human biology places any relevant constraints social processes [98]. These critics see sociobiology as another manifestation of social Darwinism used by privileged members of the ruling classes to justify the current status quo of Western societies by "biologizing" the raison d'être of fundamentally unjust and sexist behavior. Critics of sociobiology are intensely conscious of the misuse of past pseudogenetic theories that were used to justify discrimination and social injustice (Chap. 1).

These matters will not be resolved by further polemics nor by the kind of evidence likely to be produced by the current school of sociobiologists. Genetically oriented experimental designs in families that attempt to dissect human behavioral patterns into their biological subcomponents and their interaction with the environment are required to determine the extent of biological programming of social behavior in the human species.

Most conclusions and concepts of sociobiology have been derived from comparisons between humans and other species, or are based on behavioral *similarities* between various human populations in spite of differences in cultural patterns. In contrast to this, classical genetic analysis uses *differences* between individual members of the same populations as analytical tools for elucidating basic mechanisms. This difference in approach between ethology and genetics should be kept in mind in all discussions on genetic determination and evolution of behavior patterns. In this way, misunderstandings in discussions between ethologists and geneticists can be avoided.

Similarities and Differences Between Humans and Animals: The Problem of Emergence. The preceding sections compare humans with other mammals such as chimpanzees, gorillas, and even mice. In terms of chromosome structure, DNA, and amino acid sequences we observe that the overall similarities are overwhelming – particularly with the great apes. Even in some aspects of behavior there are striking similarities between humans and animals. The decisive difference between our species and all others is the superiority of our brain for abstract thinking. But when we compare the human brain with the brains of animals, we also find similarities; the differences are not of a qualitative nature, and there is no entirely new component. Rather, there are quantitative differences. The human neocortex is much larger in relation to other parts of the brain. Chimpanzees, if motivated, are able to perform amazingly well; they appear capable of symbolic thinking and communication, as well as developing simple theoretical concepts. They cannot learn to speak, however; this limitation is caused by differences in the respective brains and to a much lesser extent by differences in speech organs such as mouth and larynx [52].

But these quantitative changes do not explain the qualitative differences between humans and even the highest animals. Why do humans alone create culture? Why can human beings reflect about themselves, their past, present, and future, and the world around them? Why do they pursue science, and why do we find traces of art even in early periods of human history? Only humans know that they have to die. This consciousness has been one of the motives for the creation of rites for burying or burning the dead and for a belief in an afterlife.

Here a notion may be helpful that certainly does not explain human evolution but does show some parallels with other natural phenomena. This is the concept of emergence [9, 10]. With increasing complexity new properties emerge in a system that cannot be predicted from the properties of its parts. Ideally science should attempt to understand a system at its appropriate level. It would be wrong to carry out a scientific analysis only at the most elementary levels with the argument that a complex system is nothing but a set of its component parts. This would be inappropriate reductionism. However, an attempt to connect various findings with each other makes sense, and is often necessary for an understanding of the phenomena under study. The accusation of inappropriate reductionism cannot be made against such an approach. Chemical reactions, for example, can be explained in principle by properties of atoms as revealed by physical studies using quantum mechanics, but these factors do not make chemists abandon the concepts and methods of chemistry. Living organisms are "nothing but" conglomerates of atoms and chemical compounds. Still, biology has its own concepts and methods which require more than physicochemical approaches. A study of biological and genetic mechanisms is best carried out first at the molecular (i.e., physicochemical) level, followed by variety of methods attempting to understand the interaction of genes and products with each other and with the environment. A full understanding of embryonic development, gene regulation, morphogenesis, and central nervous system activity ultimately is likely to require more than mere description of molecular processes.

The most difficult property of humans to explain is the emergence of consciousness. The relationship of mind and body (the "mind-body problem") has long been discussed by philosophers. In our opinion, the mind is a complex system of neuronal processes within the brain. Such a formulation, however, does not mean that the mind is "nothing but" a set of neuronal processes. It is rather an emergent property of these processes that are ordered in a specific way to produce the various properties of mind, including consciousness.

The difference between humans and animals may be understood by the same principle: the structural and functional elements of the human body – from genes to metabolic processes, and even the organizational principles of the brain – are very similar to those found in animals. During evolution, however, the human central nervous system became progressively complex, and this complexity led to the emergence of novel properties. Refinement of mental and emotional processes that evolved in response to everyday challenges such as finding food and shelter, feeding children, and protecting oneself against hardships and natural predators led to a point in evolution at which such abilities came to be used for other purposes, such as art, religion, and, much later, science. Even such uniquely human activities initially served practical purposes, such as improved success in hunting by witchcraft or gaining the help of gods for overcoming dangers in daily life. Humans later became concerned with the human condition itself: an emergent result of increasing complexity of their central nervous system. It would be inappropriate to explain religious phenomena as "nothing but" results of psychological or economic conditions. Nevertheless, such conditions do contribute to a better understanding of religion. Just as there appears to be an innate language capacity, we think it likely that the human brain may have some built-in mythopoetic tendencies which under appropriate external conditions led to the development of religion.

The most challenging problem is the marked similarity between our closest evolutionary relative, the chimpanzee, and humans. While there are significant morphological differences in external features, there are very few differences in DNA and protein sequences between these two species. A major problem confronting human geneticists is to explain the marked qualitative difference between human and chimpanzee behavior. Humans have developed spoken language and the entire superstructure of human culture together with an ability to consider the past and the future. It is considered unlikely that this major difference in central nervous system output is based on entirely novel human genes. It is more likely that a yet undiscovered mechanism in the human brain led to some kind of improved use of the output of existing neural structures that are very similar in the two species – with the result that achievement of human culture and consciousness became possible. A relatively small change in the brain my have led to the emergence of what can be considered a qualitative leap in humans' use of central nervous activity. Thus a small difference at the biological level presumably led to a major difference in how these very closely related species use their central nervous system. More importantly, an entire novel and qualitatively different culture evolved in humans which is unique and is not found in any species elsewhere on this planet.

In conclusion, elements of the "human condition" can be found in one or another form in animals, as well, primarily, but not exclusively in our closest relatives, the great apes. From quantitative changes of such elements a new specifically human quality has emerged. It is unlikely that molecular approaches alone will elucidate the unique quality that makes us humans. Molecular techniques are eminently suitable to resolving many current problems and are likely to provide much insight into the nervous system. However, we feel that more than molecular biology – possibly even an entirely new scientific paradigm – will ultimately be required to understand the "achievements" of the human central nervous system.

14.2.6 Investigation of Current "Primitive" Populations

Most approaches to the study of human evolution are based on indirect evidence. Conclusions are derived from skeletal findings, from chromosomes, proteins, and DNA or from comparative observations of differ-

ent species. A slightly more direct approach exists. Despite the worldwide "progress" of modern civilization, a few human populations are still living as hunters and gatherers, i. e., under conditions differing little from those in which our remote ancestors existed during human evolution. In recent years such populations have increasingly been studied by human geneticists and anthropologists for information on the environmental forces that have shaped our genetic makeup.

Problems for Which Primitive Populations Could Provide Evidence. There are a number of questions for which study of primitive populations may suggest answers [79]:

a) Size of population groups and isolation: As explained in Sect. 13.3, size and isolation of breeding populations are the major factors underlying chance fluctuations of gene frequencies, formation of subgroups such as races, and finally speciation. On the other hand, these parameters are especially liable to differences in environmental conditions.

b) Population control: There is evidence that the overall size of the human populations was fairly constant over long periods, and that an equilibrium existed between population size and the ecological conditions, especially food supply. Modern civilization has deeply disturbed this equilibrium, making today's "population explosion" one of the most dangerous threats to the future of the human species. While study of primitive populations cannot provide meaningful clues for the survival of modern civilization, it is useful to observe how primitive populations manage to adjust population size to their ecological conditions.

c) Natural selection due to differential fertility: As explained in Sect. 12.2.1, natural selection implies that various genes within the gene pool of a population have different chances of reaching the gene pool of the next generation. These chances depend on the mortality and/or fertility of their bearers. Differential mortality cannot be readily studied in primitive populations, and observations of differential fertility are therefore of special interest.

d) Disease patterns: Historical evidence has shown conclusively that during recent centuries diseases – especially those caused by infectious agents – have played a major role in mortality during infancy and childhood. Therefore some infectious diseases have presumably played a major role in natural selection. This raises the question of whether and to what extent genetic selection by resistance to infectious disease applies to primitive hunters and gatherers as well.

e) Selection relaxation: Natural selection has undoubtedly relaxed for many traits that were harmful under primitive conditions. Are there genetic differences between primitive and civilized populations today that suggest selection relaxation?

Populations in Which These Problems Have Been Studied. Investigations on South American Indians living in the jungles of Brazil and Venezuela – the Xavantes, Yanomama, and Makiritare – have proven informative. At the time of these studies – in the 1960s and early 1970s – the tribes were among the least acculturated in South America. Still, they departed in many ways from the strict hunter-gatherer way of life prevalent during much of human evolution. Unfortunately, the remaining true hunter-gatheres are either all greatly disturbed by modern life-styles or are so reduced in numbers and withdrawn to inaccessible areas that appropriate study of a sufficient number of individuals appears impossible. However, the tribes studied are much closer to hunter-gatherers than to civilized humans in their general way of life and breeding structure. They live in primitive villages that serve as bases for their hunting and gathering expeditions. These villages are usually abandoned after a number of years. Primitive agriculture (manioc, squash, sweet potatoes, cooking bananas, and maize) provides a smaller share of the food among the Xavantes and a much larger share among the Yanomama [12].

A few of the more important results of these studies are discussed below.

Size of Population Groups and Isolation. The village is the most important unit; the population size of a village ranges between about 40–50 and about 150–200 [12]. There is a strong tendency to marriage within the village. If the population becomes too large, social rules become increasingly endangered, and part of the community, consisting of several families, may split off. Such small population size – together with the inherent tendency to isolation – obviously favors the creation of many subpopulations with different gene pools and therefore rapid evolution.

Population Control. The maximum human birth rate is much higher than that required for maintenance of a constant population size. In civilized human populations of past centuries the rule has been a high mortality in infancy and youth and during reproductive age. The most important causes of death are infectious diseases, malnutrition, and death of women related to pregnancy and delivery.

Among the primitive Indians these causes of death turn out to be less important; the number of children per woman is limited, and an effective livebirth rate of approximately one child every 4 or 5 years is maintained by a variety of measures, such as intercourse taboos, prolonged lactation (the children are usually weaned at the age of about 3 years), abortion, and infanticide.

Infanticide is practiced especially when a child is grossly defective or when several births follow each other closely [79]. Female newborns are killed more often than males. The health of the surviving children seems to be excellent and remains well up to the age of about 40. The death rate of the population below 40 years of age appears to be lower than that of present-day civilized populations of developing countries such as India – or, presumably, of the western European population of 200 or more years ago. On the other hand, individuals over the age of 40 are only rarely observed. Their causes of death are not evident. Death by warfare, by common intratribal man-to-man fights, or by pneumonia are plausible factors. Among the younger age groups, however, these tribes maintain a health standard lost by our not so distant ancestors as a tribute to continuous settlement and agriculture and regained – and partially surpassed – only recently as a result of modern hygiene and medicine.

Natural Selection Due to Differential Fertility. As explained in Sect. 14.1, the most important aspect of human evolution has been the improvement in innate mental capacities. Changes in brain size have presumably been accompanied by alteration in structure and function of the human brain. Such improvement in mental activities requires a reproductive advantage in favor of individuals bearing genes for such abilities. While our knowledge of the genetic basis of such behavioral variability is limited (Chap.15), it is reasonable to assume that genes for such capacities are found in a higher proportion among individuals holding a leading position in the social hierarchy of their village since they are able to plan hunting trips, provide for food supply, and to settle controversies between members of the community.

Indeed, village headmen have several wifes and disproportionately more children [80]. The effect of their polygamy is all the more striking, as newborn girls are much more endangered by infanticide than boys, creating in the Yanomama a sex ratio of $128\male/100\female$ for the age group 0–14. Together with the polygamy of high-ranking men this can only mean that some males are barred absolutely from reproduction.

Thus in a Xavante group 16 of 37 married men were polygamous; 65 of the 89 surviving children came from these polygamous marriages. The headman had married no fewer than five times – more than any other member of the group. These five unions resulted in 23 surviving children, approximately one-fourth of all children of the group [81].

If the high reproduction of socially high-ranking males has been a general feature of primitive human populations, and if mental ability leading to high so-cial rank has at least a partial genetic basis, a plausi-ble mechanism for relatively rapid evolution of this specifically human characteristic exists.

Balance by Disease [79, 82]. The population up to the age of 40 was usually in excellent health. At the same time, serum gammaglobulin levels were about twice as high as in civilized populations. Hence newborns are expected to have a high level of transplacentally acquired antibodies. To quote Neel:

From the first months of their lifes, these infants are in an in-timate contact with their environment that would horrify a modern mother – or physician. They nurse at sticky breasts, at which the young mammalian pets of the village have also suckled, and soon are crawling on the feces-contaminated soil and chewing on an unbelievable variety of objects. Our thesis is that the high level of maternally derived antibody, early exposure to pathogens, the prolonged period of lacta-tion and the generally excellent nutritional status of the child make it possible for him to achieve a *relatively* smooth tran-sition from passive to active immunity to many of the agents of disease to which he is exposed.
On the other hand, an epidemic from outside may have cata-strophic consequences – not so much because the individuals cannot overcome it but because village life comes to a stand-still, when almost everyone is sick, as was observed when a measles epidemic was introduced to a Yanomama commu-nity.

To mention only one example from history: during the century after the European discovery of America (1492) the indigenous population of Mexico was re-duced from about 20 million to around 1.6 million [25], presumably by newly introduced infectious dis-eases such as smallpox.

Can These Observations on a Few Indian Tribes Be Generalized? The numerous more or less "primitive" populations which are the main objects of ethnological research show an enormous range of variation in most aspects of social life and culture. Unfortunately, stu-dies on the medical genetics and even simpler demo-graphy of such populations are much scantier. The rea-son can probably be found in the sociology of science: most ethnologists and social anthropologists engaged in field work with such populations are usually trained in linguistics, sociology, and other social sciences and are not oriented toward biology and medicine.
In view of the scanty information it may be argued that some of the biological features cited above are specific for the populations examined, and that extra-polations to our primitive ancestors are unwarranted. Interdisciplinary research on the few remaining pri-mitive populations of the world is sorely needed to determine which aspects of their biology are unique to certain cultures and natural environments, and

which can be generalized. Such interdisciplinary re-search is all the more urgent since primitive popula-tions are now rapidly disappearing as a consequence of improved transportation and an increasing open-ing-up of remote areas. The studies on South Ameri-can Indians cited above are paradigmatic for this in-terdisciplinary approach.

Relaxation of Selection. We would expect compari-sons between primitive and civilized populations to give some evidence for the loss or reduction of adap-tive traits in the latter as compared to the former. In-deed, some changes have been suggested based on scattered evidence. For example, color vision defects appear to be more common in populations with a long history of agriculture than in hunter-gatherer populations; for visual and hearing acuity and some other traits, similar differences have been proposed [89], but the actual evidence particularly for color vi-sion which we understand well on a molecular level (Sect. 15.2.1.5) is not impressive. Carefully designed studies are urgently needed.

14.3 Genetics of Group Differences Within the Human Species

14.3.1 Races

Race Classification. All humans living at present be-long to one species: all observed matings between hu-mans of quite different populations have been shown to have fertile offspring. It is impossible to state with certainty whether any of the ancient human types, such as Neandertal man, were members of the species *Homo sapiens.*
The species *Homo sapiens* is divided into populations commonly called races. *A race is a large population of individuals who have a significant proportion of their genes in common and can be distinguished from other races by their common gene pool.* In earlier times members of a race often lived together under similar sociocultural conditions. The connotation of "race" as a broad population group merges without sharp limits into that of smaller units such as "demes." Race classification and race history were one of the major fields of research in classic anthropology in the nineteenth and especially the early twentieth cen-tury. The classifications were based on visual impres-sion and on statistical distributions of measurable traits. With the development of human genetics in re-cent decades such categorizations have been supple-mented by evidence based on frequencies of genetic polymorphisms. Classifications by various authors differ somewhat in detail [49]; subdivision into the

three main races, Negroids, Mongoloids, and Caucasoids, is accepted by practically all observers. Two smaller groups, the Khoisanids or Capoids (the San-speaking aborigines of Southern Africa) and Australoids (Australian aborigines and Negritos) are often added.

Genetic Differences Between Races. The definition of race used here is a genetic one, and it would be desirable to base racial classification on genetically well-defined characteristics that have been analyzed at the gene level. Several groups of such characters can be distinguished.

Many genes are common to all human beings, possibly with small differences in gene frequencies. For example, everyone has the genes that determine the enzymes needed for various basic metabolic processes. Exceptional individuals with rare mutations affecting these genes suffer from inborn errors of metabolism. Many of these genes are more or less identical in humans and other living beings.

Other genetically determined characteristics are the common heritage of all or almost all members of a single race and are lacking in those of other races. The number of such characteristics seems to be fairly small; genetically they are not well defined. An example is the eyelid fold of Mongoloids.

Another group of genetic characters occurs – apart from a few exceptions – in only one of the three main races but is absent from the other two. This group comprises a number of well defined genetic polymorphisms (Table 14.4). One example is the Diego blood type [61–63]. This was discovered in 1953 in Venezuela in four generations of a family and was absent from most whites. Investigations of American Indian populations show phenotype frequencies of type Di[a] between 0.025 and 0.48. White and black populations do not show this allele at all; those of Mongoloid extraction such as Japanese and Chinese, on the other hand, often have it, albeit with a lower average frequency than South American Indians. This finding is consistent with the assertion of classical anthropology that American Indians are part of the main Mongoloid race.

Yet another class of genetic characteristics is more frequent in some populations than in others. It comprises genetic characteristics and alleles that are observed in all human races but with different frequencies. Examples are alleles in most genetic polymorphisms and genes determining continuously varying characteristics, such as stature, body proportions, and physiological functions. Genetic polymorphisms are used increasingly for the genetic characterization of different populations [77]. Tracing anthropological affinities and the possible role of selection in determining some gene frequencies has become possible. The overall results clearly show overlaps in gene frequencies between populations and illustrate the diffi-

Table 14.4. Genetic differences between main racial groups in some expressed genetic polymorphisms (data from Morton et al. 1967 [77])

System	Gene	Approximate range of gene frequencies:		
		Negroids	Mongoloids	Caucasoids
Blood groups				
Diego	DI*A	~0	~0–0.5	0
Duffy	Fy	~0.98–1.00	~0	~0
Kell (Sutter)	JsA*	~0.02–0.20	~0	~0
Serum proteins and blood isoenzymes				
Ceruloplasmin	CP*A	~0.04–0.10		~0.01
Group-specific	GC*CHIP	~0	Cippewa Indians ~0.10	0
component	GC*AB	~0	Australians, New Guineans ~0.02–0.2	0
IGHG (Gm)	IGHG haplo-types	*1,5,13,14,17;1,5,14,17;1,5,6,17;1,5,14,17*	*1,17,21;1,2,17,21; 1,13,17;1,3,5,13,14*	*1,17,21;1,2,17,21;3,5,13,14*
Peptidase A	PEP*E2	~0.07–0.1	~0.003–0.04	0
PGM2	PGM2 (Atkinson)		0	0

Italicized haplotypes are specific for these racial groups.

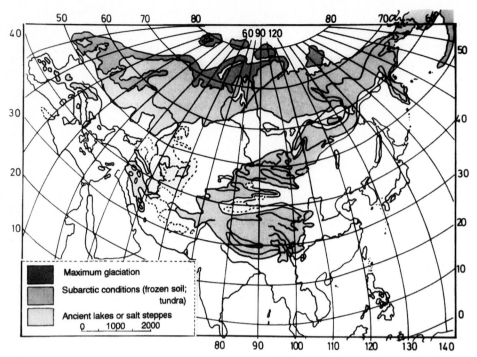

Fig. 14.14. The Eurasian continent about 100 000 years ago. Three habitats are almost completely separated from each other by the Himalayan and Altai Mountains together with their glacier areas. (From von Eickstedt 1934 [104])

culties of fixed racial classifications. While differences, for instance, between a Swedish and Korean population are very clear-cut using all markers, no such differences are often found between neighboring populations. Differences between members within each racial group often exceed those between each of the principal races (Mongoloid, blacks, and whites). Microscopically visible chromosomal heteromorphisms (Sect. 2.1.2.3) also show racial differences [90].

How Did the Genetic Race Differences Evolve? The main factor in evolution of phenotypes – and in the formation of races specifically – *appears to be natural selection in adaptation to different environmental conditions.* For selection to be effective in producing genetic differences, such as those existing between the main races, considerable reproductive isolation between populations is required. Is there a period in early history during which the human species was subdivided into three more or less isolated populations?

During the most recent glacial period about 100 000 years ago much of the earth's expanse was covered by ice (Fig. 14.14). The Himalaya and Altai Mountains together with their glacier areas separated the Eurasian continent into three areas, providing the conditions for separate evolution of whites in the west, Mongoloids in the east, and blacks in the south. Wherever the present living areas of the three main races do not conform with the areas in which they evolved, migrations can explain the discrepancies [105].

Genetic Differences That Can Be Explained by Specific Selective Mechanisms: Skin Pigmentation and Ultraviolet Light. The most conspicuous difference between the main races is the difference in skin pigmentation. Most nonhuman primates today are darkly pigmented, and it is reasonable to assume that in ancient human populations dark pigmentation was also prominent, particularly since the earliest humans most likely lived in Africa. Why then are whites and Mongoloids so lightly pigmented?

According to a plausible hypothesis, there has been an adaptation to low ultraviolet (UV) irradiation in the habitats of these two races. UV light is necessary for conversion of provitamin D to vitamin D in the human skin. Vitamin D is needed for calcification of bones, and its deficiency leads to rickets. One of the most dangerous features of rickets is pelvic deformation, which impairs normal childbirth and under primitive living conditions often leads to the death of mother and child. This effect obviously makes for strong selection pressures. Figure 14.15 shows the degree of skin pigmentation and intensity of UV light irradiation in various areas of the world [106].

The hypothesis implies that UV radiation can penetrate more easily into lightly pigmented skin and therefore that identical

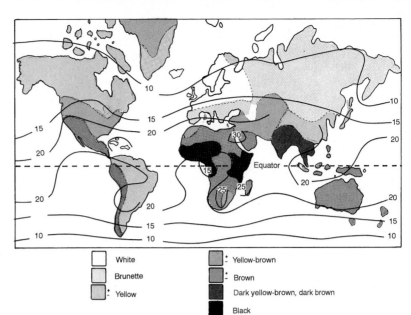

Fig. 14.15. Intensity of ultraviolet light and average skin pigmentation in various areas and indigenous populations of the world. *Numbers,* mean intensities of global solar irradiation on a horizontal plane at the surface of the Earth (mW cm⁻² averaged over 24 h, for the whole year). (Adapted from Walter 1970 [106]; Mourant 1976 [76])

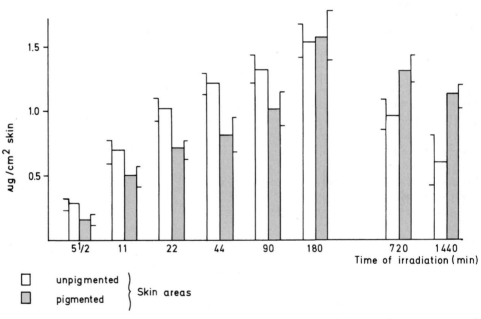

Fig. 14.16. Formation of vitamin D (µg/cm² skin, *ordinate*) in skin samples of saddle pigs after irradiation with UV light (S 300; 50 cm distance); *abscissa,* time of irradiation. After irradiation up to 90 min, vitamin D formation is lower in pigmented than in unpigmented skin. Synthesis in pigmented skin is greatly delayed and higher than in unpigmented skin only after 180 min. *Brackets,* standard deviations of the means. (From Bekemeier 1969 [4])

UV doses lead to more vitamin D formation in lightly pigmented skin. Experiments with skin specimens from saddle pigs have proven informative. These animals are darkly pigmented in the middle of their bodies, while the cranial and caudal regions show little if any pigmentation. Vitamin D formation after UV irradiation in vitro is higher in unpigmented than in pigmented skin of the same animals [4] (Fig. 14.16). Two human populations are apparent exceptions to the geographic correlation with skin pigmentation: Eskimos and African Pygmies. Both, especially the latter, are darkly pigmented despite the fact that UV irradiation in arctic regions and on the ground in tropical rain forests is scanty. Eskimos seem to obtain abundant vitamin D from fish and seal liver, the pygmies possibly from the insect larvae that form part of their nutrition [32].

Frequency of the Fy⁻ Allele in Blacks. The Fy⁻ allele of the Duffy blood group system is frequent among blacks but is very rare or does not occur at all in Mongoloids and whites. Blacks with this allele have a complete resistance to the infective agent of tertiary malaria, *Plasmodium vivax* [74]. Tertiary malaria alone is hardly if ever fatal, and a selective advantage is therefore not immediately obvious. However, under primitive living conditions and in a population exposed to multiple infective agents and parasites, malaria infection can be a severe health hazard indeed.

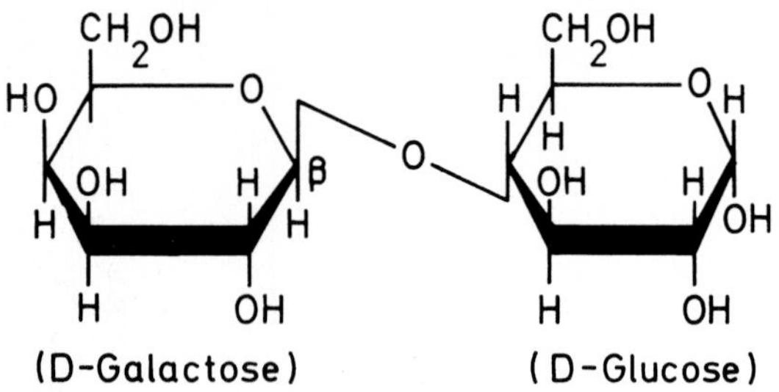

The discovery that the Duffy blood group is involved in receptor activity for *P. vivax* is of great paradigmatic importance because it illustrates the biological significance of a previously discovered red cell polymorphism whose function was unknown, as is the case with most polymorphisms. Since practically all Africans are Duffy negative, this example also demonstrates how an allele that is usually observed in polymorphic frequencies spreads throughout the entire population because of its selective advantage.

An alternate hypothesis has recently been elaborated [67]. It has been suggested that the preexisting high frequencies of the Duffy-negative allele prevents vivax malaria from becoming endemic in western Africa.

Lactose Restriction and Persistence [34, 35]. Lactose is the only nutritionally important carbohydrate in milk (Fig. 14.17). To be absorbed by the small intestine lactose must first be hydrolyzed by a specific enzyme, lactase, which is located in the brush border of the intestinal epithelial cells. The milk of almost all mammals contains lactose; lactase activity is high during the newborn and suckling period in all mammalian species and in all human populations, but declines at the time of weaning. Afterward lactase activities are maintained at low levels, usually at less than 10% of the activity in the newborn.

Until a few years ago humans were thought to represent an exception: they "normally" seemed to maintain their lactase activity into adult age. Such persons with lactase persistence can tolerate large amounts of lactose; after a lactose load they show a considerable rise in blood concentration of glucose and galactose, the sugars that constitute lactose.

Lactose Restriction and Malabsorption [35]. In persons with low lactase activity there is little or no increase in blood glucose after milk ingestion. Many develop clinical symptoms of intolerance after consuming 25–50 g lactose (1 l cow's milk contains up to 50 g lactose). These symptoms are watery, acid diarrhea, colicky abdominal pain, and flatulence. Smaller amounts of milk and milk products from which some lactose has been removed by fermentation, such as yoghurt or whey, are tolerated without untoward effects. Comparative studies of lactose tolerance in American blacks and whites showed that lactose intolerance due to lactose malabsorption

Fig. 14.17. The disaccharide lactose

is much more frequent in blacks [2]. Since that time, studies from many populations have become available [35] (Fig. 14.18). The most reliable method for determining whether intestinal lactase is present is enzyme measurement in intestinal biopsies. Such a method is not suitable for population or family studies. Therefore standard lactose tolerance tests have been designed that measure the rise in blood glucose or H_2 in the exhaled air after ingesting of a standard oral dose of lactose. Family investigations have shown lactase restriction (a better term than lactose intolerance) to be inherited as an autosomal-recessive trait [92]. The lactase gene, alleles of which apparently cause the L.persistence and L.restriction phenotypes, has been localized on chromosome 2; alleles have been named LAC$^+$P and LAC$^+$R (223 100). The allele for persistence (LAC$^+$P) is dominant over the restriction allele (LAC$^+$R), but not all homozygotes for LAC$^+$R suffer from clinical signs of lactose malabsorption; the level of daily milk intake leading to such symptoms depends on many factors, among others, on dietary habits. There may be multiple alleles causing the LAC$^+$C variant: in Finland the switch from high to low enzyme activity occurs in the second decade of life [92], whereas in other populations it takes place as early as after the age of three years.

When lactase restriction leading to lactase malabsorption was first discovered, some observers believed it to be caused by a lack of enzyme induction due to low lactose intake. However, this hypothesis was ruled out experimentally. However, not all patients suffering from lactose malabsorption are homozygotes of the LAC$^+$R gene; some intestinal diseases may have the same effect [35]. So far it is not known whether the lactase formed by malabsorbers differs in its protein structure from that formed by absorbers. The switch-over from high to low lactase activity at the time of weaning in malabsorbers is somewhat similar to that from hemoglobin γ-chain to β-chain production and, hence, from HbF to HbA formation in hemoglobin synthesis; persistence of lactase activity is a similar trait to the persistence of fetal hemoglobin (Sect. 7.3).

Persistence of lactase activity in older children and adults or lactose tolerance is very rare or absent among most Mongoloid populations, including American Indians and Eskimos. A similarly low incidence of lactose tolerance is observed in most Arab and Jewish populations and in the populations of tropical Africa, Australian aborigines, and Melanesians. A consistently high prevalence of persistent lactase activity (over 75%) is found only among persons

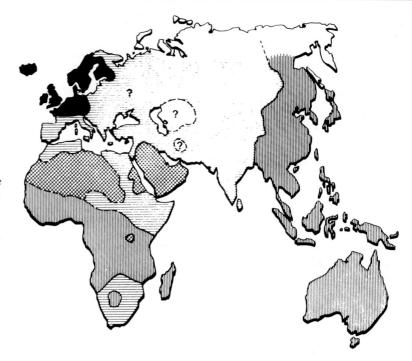

Fig. 14.18. Frequency of Lactose restrictional malabsorption vs. Lactase persistence in populations of the world. Vertical hatching, nonmilking populations with frequencies of lactase restriction of >90%; horizontal hatching, milk-using populations with variable distribution of the lactase phenotypes (frequency of lactase restriction, 30%–90%); cross hatching, predominance of the lactase resistance gene in Afroarabian nomads; solid areas, predominance of the lactase persistence gene in Central and Northwestern Europe. Question marks indicate the dearth of data in Eastern Europe and in Asian nomadic groups (Flatz [35])

from northern and central Europe and in their descendants in other continents. In addition, high frequencies of lactose tolerance have been reported in a few groups of nomadic pastoralists in Africa. Intermediate frequencies (30%–90%) have been observed in Spain, Italy, and Greece. Southern Asia shows a high variability; here the trait may have been introduced by migration. The African-American population shows somewhat higher values than African blacks.

What Is Normal? What Is Abnormal? In conclusion, most human populations show a decrease in lactase activity after weaning; this is the common pattern in two of the three major races; a high prevalance of persistent activity is found only in whites, and even here not in all populations. Therefore the loss of this unique activity following weaning together with its consequence, lactose malabsorption, is the "normal" condition in humans as it is in other mammals.
It is an interesting cultural phenomenon that scientists considered persistence of lactase activity to be normal since this trait was most common in populations of northern European origin where this work was carried out. This conceptual bias even had economic consequences. In an effort to improve the protein intake of children in African and Asian countries, powdered milk was distributed in large amounts on the not unreasonable hypothesis that what is good for European children must also be good for children in developing countries. These programs had to be reconsidered in the light of our

knowledge of population distributions. However, moderate amounts of milk and of milk products are tolerated; suitably organized milk and dairy programs may therefore still be useful.

Natural Selection. The predominance of lactose malabsorption in adults of most human populations together with its regular occurrence in other mammals suggests that the gene responsible for persistence of lactase activity occurred by mutation sometime in human evolution, and that a selective advantage has caused the high frequencies in some populations. What is the nature of this advantage? Two main hypotheses have been proposed:

1. A culture historical hypothesis
2. A hypothesis that advocates improved calcium absorption by lactose.

According to the cultural hypothesis, domestication of milk animals in the Neolithic Period (about 9000 years ago) resulted in a selective advantage of individuals who could satisfy a higher proportion of their nutritional requirements, mainly protein, by milk consumption. In the cattle-breeding tribes discussed above this hypothesis could apply. Its general validity, however, is not certain. For example, there is no parallel between the milking habit and the prevalence of lactose absorption. Large populations in Africa and Asia are milk consumers but have a very low incidence of lactose absorption. Nevertheless, there are always some individual absorbers; hence the gene is present and could have been favored by selection. The highest gene frequency of the gene in Europe is found in southern Scandinavia (0.7–0.75), where breeding of milking animals was introduced more recently because of unfavorable conditions in the postglacial period. Moreover, at a time when artificial refrigeration of milk or production of dried milk had not been invented, milk soured within a short time, and malabsorbers probably discovered easily that sour milk causes fewer digestive symptoms. A specific advantage of milk under the environmental conditions of northern Europe is therefore an alternative worth considering.

In connection with the selective advantage of lightly pigmented skin, vitamin D deficiency in northern areas due to diminished UV irradiation and severe selective disadvantage of rickets by pelvic deformation of childbearing women has also been mentioned. It has been suggested that lactose can replace vitamin D by improving the uptake of calcium. Late rickets and osteomalacia occur at an age at which lactose malabsorption has already developed. The crucial problem of this hypothesis is the mechanism of a possible protective effect of high lactose absorption against rickets. Is there specific enhancement of calcium absorption with hydrolysis of lactose? Animal experiments can give only ambiguous results since adult animals are lactose malabsorbers. However, studies in human subjects have shown that calcium absorption is indeed enhanced by lactose absorption [16].

Regardless of the ultimate fate of the calcium hypothesis, the problem as posed is the type of heuristic hypothesis needed in human population genetics. The hypothesis is specific, provides a mechanism by which selection can work, and suggests experiments by which it can be tested.

There is preliminary evidence for an association of lactase restriction with osteoporosis of elderly women. Other disease associations have also been claimed with lactase persistence, such as higher incidence of coronary heart disease, hyperlipidemia, and senile cataracts; however, definite proof is lacking [35]. Genetic variation at the lactase locus is also an excellect example of ecogenetics (Sect. 7.5.2).

Vitamin D and GC Serum Groups. A genetic polymorphism of an immunologically defined fraction of a human β_2 serum protein has been known since 1959 [47]. Many alleles are known, but most populations are polymorphic for only two of them, GC^1 and GC^2; Australian aborigines have a third allele, GC^{Abo}, and the Chippewa Indians have a fourth, GC^{Chip}. When the first data on gene frequencies became available, the allele GC^2 turned out to have low frequencies in areas of high aridity. This finding was explained when the function of the GC proteins was

discovered, as they are carrier proteins of vitamin D [21].

A subsequent survey suggested a relationship between sunshine and the GC polymorphism; in the aboriginal habitats of the world population high frequencies of GC^2 were found in most populations that had been living for a long time in areas with low sunlight intensity [76].

This geographic distribution suggests a selective advantage due to a more efficient transport of vitamin D, especially if the supply is limited, and therefore a lower incidence of rickets in individuals either heterozygous or homozygous for the GC^2 allele. The exact mechanism remains to be elucidated.

Possible Selective Mechanisms for Other Racial Characteristics. Apart from the examples mentioned above and those in the section on population genetics, little is known about selective advantages or disadvantages of racial characters. It is reasonable to assume that the small and heavyset stature of Eskimos and their relatively thick subcutaneous fat has advantages in a cold climate and the broad and deep thorax of South American Indians living in the high Andes relates to respiratory adaptation to life at high altitudes.

Members of the different racial groups show differences in susceptibility to common diseases under the living conditions of the United States and northwestern Europe and other affluent countries. African-American, for example, are more susceptible to high blood pressure than whites. It has been suggested that genes facilitating sodium absorption may be present at higher frequencies among populations such as African-Americans whose ancestors were exposed to the selective conditions of hot climates with profuse sweating. The presence of such genes under modern conditions of high salt intake would lead to the high frequency of hypertension known in this population group. Some groups of Asian Indians, for example, in Trinidad, were shown to have higher frequencies of diabetes mellitus than other population groups. Better understanding of such racial differences will undoubtedly be forthcoming as soon as medical investigators with intimate knowledge of disease pathophysiology and biochemistry become interested in such studies.

Hypotheses such as the "thrifty genotype" concept [78] and the rapid mobilization of lipids have been intertwined to explain diabetes and atherosclerosis. It is thought that under conditions of starvation the diabetic genotype allows more efficient mobilization of carbohydrates, while atherosclerosis-favoring genes have been explained as permitting more rapid mobilization of fat. This hypothesis is supported by evidence from both humans and animals. For example, type II diabetes (as well as obesity) is very common (and increasing with westernization) among Pima Indians and inhabitants of Pacific islands; mice heterozygous for a diabetes-obesity gene survive fasting much better than normal mice [17, 91].

Such selective mechanisms acting in the past when starvation was common over many generations have also been suggested to explain the high frequency of atherosclerosis (Sects. 6.4.2.2; 7.6.4) in current times. With increasing knowledge of genetics and pathophysiology it will become possible so suggest – and test – more specific hypotheses.

14.3.2 Future of Human Races: Racial-Crossing

Will Races Disappear? The one important condition for the formation of races was isolation. Within their old habitats in Asia, Africa, and Europe this isolation still largely exists. In more recently settled areas such as North and South America huge "melting pots" have formed that comprise elements of all three race groups. In spite of social customs, which keep inter-marriage at a low level, there is little doubt that race mixture will increase, leading sooner or later to a hybrid population.

How long the main races will be considered as separate depends on political developments and cannot be predicted. It is possible that race-specific differences will disappear in the long run by interbreeding.

Interracial Crosses in Hawaii [75]. What are the genetic consequences of racial mixture? At a time when biologists did not think in terms of variabilities within populations but rather of human variability in terms of racial "types," racial mixture was often regarded as disruptive and leading to disharmonious phenotypes. It was agreed that by long selection certain combinations of genes were "coadapted". If such coadaptation were broken up by race mixture, disharmonious phenotypes would result. It therefore came as a surprise that hybrid populations, such as those between Nama (Hottentots) and whites in southwestern Africa, turned out to be fully viable and quite normal in health [31]. Beyond this statement, however, the numerous older studies could not answer the question of possible genetic effects in interracial crosses. Another consideration predicted beneficial effects of racial crossing. Since "hybrid vigor" had been shown in lower forms of life, it was suggested that racial hybrids might be particularly healthy.

Some answers came from a carefully planned study in Hawaii [75]. The population of Hawaii is composed mainly of Hawaiian (Polynesian), white, Chinese, and Japanese elements, with smaller additions of Koreans, Filippinos, Puerto Ricans, and others. Population statistics and medical facilities are reliable, and the environmental conditions under which the various racial groups live differ little and in any case overlap strongly. Intermarriage between racial groups is rather common. The trend toward inter-marriage is relatively recent so that the results in the generation following the initial hybridization can be observed.

Scope of the Study and Data. The scope of the study was defined as follows:

a) What are the genetic effects of out-crossing in humans on the first generation of hybrids?

b) Do human populations represent coadapted genetic combinations that are disrupted after the first generation of out-crossing?

The major part of the data consisted of 172 448 live-birth certificates and 6879 still-birth certificates for all births registered between 1948 and 1958. These data include entries on the racial affiliation of the parents. This affiliation does not mean that the parents were members of "pure races"; they contained varying admixtures from other racial groups that were estimated using data from genetic polymorphisms. The population of parents was 62.7% Pacific (Hawaiian and Mongoloid); the rest were mainly white. Additional information came from records on maternal stature and weight in one clinic.

The analysis was based on a stepwise regression considering environmental factors such as socioeconomic differences and medical care. Such an analysis can examine the question of how differences between certain categories can be explained by concomitant variables, and how much must be attributed to interracial cross can be examined.

Results and Interpretation. The main result of this study can be summarized very briefly. *There were no obvious ill effects of outbreeding either on early or late fetal deaths or on postnatal infant deaths.* Birth weight and maternal weight and stature were not significantly related to maternal hybridity. A maternal effect on birth complications might be expected when the mothers were Japanese or Chinese and the fathers whites because Japanese and Chinese women are smaller, but no such effect was found. As noted in Sect. 6.3.3, the rate of dizygotic twinning is much lower in Mongoloids than in whites and is probably caused by a difference in frequency of polyovulation, i.e., by a maternal factor. This conclusion was corroborated by study of interracial crosses; the DZ twinning rate depended only on the race of the mother, irrespective of the father's race and the hybridity of the children. Mothers who were hybrids between whites and Pacific races had a low dizygotic twin frequency, closely resembling that of Pacific mothers. If the genetic disposition to polyovulation were a threshold character determined by additive gene action, one would expect these women to show an intermediate DZ frequency. As it stands, the result suggests the participation of recessive genes in polyovulation. The data on congenital malformation were not quite as reliable as the mortality data. Taken at face value, there was no difference in the overall incidence of major malformations between racial groups, but frequencies of certain categories of malformations varied between these groups. Spina bifida, for example, was more frequent in the Japanese of Hawaii than in those on the Japanese mainland but still remained significantly less frequent than in whites. Both genetic and environmental factors may be important. The occurrence of severe clubfoot appeared to be high in

Hawaiians and low in other Pacific groups. The incidence of congenital malformations was not influenced by outbreeding.

If anything, there was a small, nonsignificant overall advantage of hybrid children which may be due to a lower risk of recessive detrimental genes to become homozygous.

Questions Not Answered by the Hawaii Study. – The two questions posed by the Hawaii study can be answered as follows:

1. The genetic effects of out-crossing in humans on first generation hybrids do not lead to any ill effects, manifested by perinatal and infant death or major congenital malformations as assessed by vital statistics.
2. There is no evidence for coadapted genetic combinations that are disrupted after the first generation of out-crossing when newborns are assessed indirectly by scrutiny of birth and death certificates.

These conclusions, however, do not answer all possible questions. On the basis of inadequate evidence the older anthropological literature claimed disharmonic combinations between jaws and teeth, an assertion that was severely criticized. Racial crossing in rabbits between races of extreme size showed no such disharmonies [99]. Some of the older authors specifically attributed emotional difficulties of race hybrids to such disharmonic gene combinations; such speculations were usually based on a biologically incorrect understanding of race, where races were considered as specific types rather than as population groups with varying gene frequencies. The role of the environment was generally neglected.

There is no evidence to suggest that racial mixture has any deleterious genetic consequences. Less homozygosity with beneficial consequences for the incidence of many recessive diseases is certain. Whether hybrid vigor and greater physical and mental health are results cannot be answered with certainty. A scientific assessment of the possible genetic effects of racial crossing would require additional studies of physical and mental health of hybrids compared to their ancestral populations under carefully controlled environmental conditions.

Conclusions

The human species evolved from nonhuman primate populations, most likely in Africa. The most important step in this process was the development of a complex brain. Among present-day animals our closest relatives are the great apes, especially chimpanzees, as seen from similarities in DNA and chromosome structure. The "classical" view portrays natural selection as the principal driving force of evolution, but the "neutral hypothesis" (Kimura) stresses the significance of random processes. It is likely that both selection and random drift have interacted in shaping the genetic make-up of present human populations. The human species is divided into subpopulations, often called races, but genetic variation within races tends to be greater than that between races. Race crossing which is now observed at an increasing rate may be advantageous for the health of future generations.

References

1. Barash DP (1977) Sociobiology and behavior. Elsevier, New York
2. Bayless TM, Rosenzweig NS (1966) A racial difference in incidence of lactase deficiency. A survey of milk intolerance and lactase deficiency in healthy adult males. JAMA 197 : 968–972
3. Becker PE (ed) (1964–1976) Humangenetik, ein kurzes Handbuch in fünf Bänden. Thieme, Stuttgart
4. Bekemeier H (1969) Evolution der Hautfarbe und kutane Vitamin D-Photosynthese. Dtsch Med Wochenschr 94 : 185–189
5. Bernardi G, Bernardi G (1986) Compositional constraints and genome evolution. J Mol Evol 24 : 1–11
6. Betz A, Turleau L, de Grouchy J (1974) Hétérozygotie et homozygotie pour une inversion péricentrique du 3 humains. Ann Genet 17 : 77
7. Bräuer G (1984) Präsapiens-Hypothese oder Afroeuropäische Sapiens-Hypothese? Z Morphol Anthropol 75 : 1–25
8. Britten RJ (1986) Rates of DNA sequence evolution differ between taxonomic groups. Science 231 : 1393–1398
9. Bunge M (1980) The mind-body problem. Pergamon, Oxford
10. Bunge M, Ardila R (1987) Philosophy of psychology. Springer, Berlin Heidelberg New York
11. Bush GL, Case SM, Wilson AC, Patton JL (1977) Rapid speciation and chromosomal evolution in mammals. Proc Natl Acad Sci USA 74 : 3942–3946
12. Chagnon NA (1968) Yanomamö, the fierce people. Holt, Rinehart and Winston, New York
13. Cherfas J (1991) Ancient DNA: still busy after death. Science 253 : 1354–1356
14. Clarke B (1970) Selective constraints on amino-acid substitutions during the evolution of proteins. Nature 228 : 159–160
15. Clemente IC, Ponsà M, Garcia M, Egozcue J (1990) Evolution of the simiiformes and the phylogeny of human chromosomes. Hum Genet 84 : 493–506
16. Cochet B, Jung A, Griessen M, Bartholdi P, Schaller P, Donath A (1983) Effects of lactose on intestinal calcium absorption in normal and lactase-deficient subjects. Gastroenterology 84 : 935–940
17. Coleman DL (1979) Obesity genes: beneficial effects in heterozygous mice. Science 203 : 663–664

18. Cooper DN, Krawczak M (1993) Human gene mutation. Bios Scientific, Oxford

19. Cooper DN, Schmidtke J (1984) DNA restriction fragment length polymorphism and heterozygosity in the human genome. Hum Genet 66 : 1–16

20. Copeland NG, Jenkins NA, Gilbert DJ et al (1993) A genetic linkage map of the mouse: current applications and future prospects. Science 262 : 57–65

21. Daiger EI, Schonfield MS, Cavalli-Sforza LL (1975) Group-specific component (Gc) proteins bind vitamin D and 25-hydroxyvitamin D. Proc Natl Acad Sci USA 72 : 2076–2080

22. Dalton DP, Edwards JH, Evans EP, Lyon MF, Parkinson SP, Peters J, Searle AG (1981) chromosome maps of man and mouse. Clin Genet 20 : 407–415

23. Dayhoff MO (1972) Atlas of protein sequence and structure, vol 5 + Suppl. 3 (1978). National Biomedical Research Foundation, Washington

24. Djian P, Green H (1989) Vectorial expansion of the involucrin gene and the relatedness of the hominoids. Proc Natl Acad Sci USA 86 : 8447–8451

25. Dobson A (1993) People and disease. In: Jones S, Martin R, Pilbeam D (eds) Human evolution. Cambridge University Press, Cambridge, pp 411–420

26. Dutrillaux B (1975) Sur la nature et l'origine des chromosomes humains. L'expansion Scientifique, Paris

27. Dutrillaux B (1979) Chromosomal evolution in primates. tentative phylogeny from Microcebus Murinus (Prosimian) to man. Hum Genet 48 : 251–314

28. Erdtmann B (1982) Aspects of evaluation, significance and evolution of human C-band heteromorphism. Hum Genet 61 : 281–294

29. Ewens WJ (1980) Mathematical population genetics. Springer, Berlin Heidelberg, New York

30. Felsenstein J (1988) Phylogenies from molecular sequences: inference and reliability. Annu Rev Genet 22 : 521–565

31. Fischer E (1913) Die Rehobother Bastards und Das Bastardierungsproblem beim Menschen. Fischer, Jena

32. Fischer E (1961) Über das Fehlen von Rachitis bei Twiden (Bambuti) im Kongourwald. Z Morphol Anthropol 51 : 119–136

33. Fitch WM (1980) Estimating the total number of nucleotide substitutions since the common ancestor of a pair of homologous genes: comparison of several methods and three beta hemoglobin messenger RNA's. J Mol Evol 16 : 153–209

34. Flatz G (1987) Genetics of lactose digestion in humans. Adv Hum Genet 16 : 1–77

35. Flatz G (1992) Lactase deficiency: biological and medical aspects of the adult human lactase polymorphism. In: King RA, Rotter JI, Motulsky AG (eds) The genetic basis of common disease. Oxford University Press, New York, pp 305–325

36. Foley R (ed) (1984) Hominid evolution and community ecology. Academic, London

37. Foley R (1987) Another unique species: patterns in human evolutionary ecology. Longman, Harlow

38. Foley R, Lee P (1989) Finite social space, evolutionary pathways and reconstructing hominid behaviour. Science 243 : 901–906

39. Gee H (1992) Statistical cloud over African Eve. Nature 355 : 583

40. Gilbert W (1985) Genes-in-pieces revisited. Science 228 : 823–824

41. Goodman M, Tashian RE (eds) (1976) Molecular anthropology. Plenum, New York

42. Goodman M, Moore GW, Matsuda G (1975) Darwinian evolution in the genealogy of haemoglobin. Nature 253 : 603–608

43. Goodman M, Koop BF, Tagle DA, Slighton JL (1989) Molecular phylogeny of the family of apes and humans. Genome 31 : 316–335

44. Gosden JR, Mitchell AR, Seuanez HN, Gosden CM (1977) The distribution of sequences complementary to human satellite DNAs I, II and IV in the chromosomes of chimpanzee (Pan troglodytes), Gorilla (Gorilla gorilla) and Orang Utan (Pongo pygmaeus). Chromosoma 63 : 253–271

45. Happle R, Phillips RJS, Roessner A, Junemann G (1983) Homologous genes for X-linked achondroplasia punctata in man and mouse. Hum Genet 63 : 24–27

46. Hassenstein B (1973) Verhaltensbiologie des Kindes. Piper, Munich

47. Hirschfeld J (1959) Immuno-electrophoretic demonstration of qualitative differences in human sera and their relation to the haptoglobins. Acta Pathol Microbiol Scand 47 : 160–168

48. Holmquist GP (1989) Evolution of chromosome bands: molecular ecology of noncoding DNA. J Mol Evol 28 : 469–486

49. Howells WW (1993) The dispersion of modern humans. In: Jones S, Martin, R Pilbeam D (eds) Human evolution. Cambridge University Press, Cambridge, pp 389–401

50. Jacobs PA (1975) The load due to chromosome abnormalities in man. In: Salzano F (ed) The role of natural selection in human evolution. North-Holland, Amsterdam, pp 337–352

51. Jauch A, Wienberg J, Stanyon R, Arnold N, Tofanelli S, Ischida T, Cremer T (1992) Reconstruction of genomic rearrangements in great apes and gibbons by chromosome painting. Proc Natl Acad Sci USA 89 : 8611–8615

52. Jones S, Martin R, Pilbeam D, Bunney S (eds) (1992) The Cambridge encyclopedia of human evolution. Cambridge University Press, Cambridge.

53. Leakey R, Lewin R (1992) Origins reconsidered. Little, Brown, London

54. Kimura M (1968) Evolutionary rate at the molecular level. Nature 217 : 624–626

55. Kimura M (1983) The neutral theory of molecular evolution. Cambridge University Press, Cambridge

56. Kimura M (1982) The neutral theory as a basis for understanding the mechanism of evolution and variation at the molecular level. In: Kimura M (ed) Molecular evolution, protein polymorphism and the neutral theory. Japanese Scientific Society, Tokyo/Springer, Berlin Heidelberg New York, pp 3–56

57. Kimura M, Takahata T (eds) (1991) New aspects of the genetics of molecular evolution. Japanese Scientific Society, Tokyo/Springer, Berlin Heidelberg New York

58. King JL, Jukes TH (1969) Non-Darwinian evolution. Science 164 : 788–798

59. Klein J, O'Huigin C (1993) Composite origin of major histocompatibility complex genes. Curr Opin Genet Dev 3 : 923–930

60. Krüger J, Vogel F (1989) The problem of our common mitochondrial mother. Hum Genet 82 : 308–312

61. Layrisse M (1958) Anthropological considerations of the Diego (Di^a) antigen. Am J Phys Anthropol 16 : 173–186

62. Layrisse M, Arends T (1957) The Diego system – steps in the investigation of a new blood group system. Further studies. Blood 12 : 115–122

63. Layrisse M, Arends T, Dominguez Sisco R (1955) Nuevo gropo sanguineo encontrado en descendientes de Indios. Acta Med Venez 3 : 132–138

64. Leakey R, Lewin R (1992) Origins reconsidered. Little, Brown, London

65. Lewin R (1990) Molecular clocks run out of time. New Sci:38–41

66. Li W-H, Tanimura M (1987) The molecular clock runs more slowly in man than in apes and monkeys. Nature 326 : 93–96

67. Livingstone FB (1984) The Duffy blood groups, vivax malaria, and malaria selection in human populations: a review. Hum Biol 56 : 413–425

68. Lorenz K (1935) Der Kumpan in der Umwelt des Vogels. J Ornithol 83 : 137–213, 289–413

69. Lorenz K (1966) On aggression. Methuen, London

70. Lumbsden CJ, Wilson EO (1981) Genes, mind, and culture. The coevolutionary process. Harvard University Press, Cambridge

71. Maeda N, Smithies O (1986) The evolution of multigene families: haptoglobin genes. Annu Rev Genet 20 : 81–108

72. McClure H, Belden KH, Pieper WA (1969) Autosomal trisomy in a chimpanzee. Resemblance to Down's syndrome. Science 165 : 1010–1011

73. McKusick VA (1995) Mendelian inheritance in man, 11th edn. Johns Hopkins University Press, Baltimore

74. Miller LH, Mason SJ, Clyde OF, McGinniss MH (1976) The resistance factor to Plasmodium vivax in blacks. N Engl J Med 295 : 302

75. Morton NE, Chung CS, Mi MP (1967) Genetics of interracial crosses in Hawaii. Monogr Hum Genet 3

76. Mourant AE, Tills D, Domaniewska-Sobczak K (1976) Sunshine and the geographical distribution of the alleles of the Gc system of plasma proteins. Hum Genet 33 : 307–314

77. Mourant AE, Kopec AC, Domaniewska-Sobczak K (1978) Blood groups and diseases. Oxford University Press, London

78. Neel JV (1962) Diabetes mellitus: a "thrifty" genotype rendered detrimental by "progress"? Am J Hum Genet 14 : 353–362

79. Neel JV (1970) Lessons from a "primitive" people. Science 170 : 815–822

80. Neel JV (1980) On being headman. Perspect Biol Med 23 : 277–294

81. Neel JV, Salzano FM, Junqueira PC, Keiter F, Maybury-Lewis D (1964) Studies on the Xavante Indians of the Brazilian Mato Grosso. Am J Hum Genet 16 : 52–140

82. Neel JV, Centerwall WR, Chagnon NA, Casey HL (1970) Notes on the effect of measles and measles vaccine in a virgin-soil population of South American Indians. Am J Epidemiol 91 : 418–429

83. O'Brien SJ, Womack JE, Lyons LA et al (1993) Anchored reference loci for comparative genome mapping in mammals. Nature [Genet] 3 : 103–112

84. Ohno S (1967) Sex chromosomes and sex-linked genes. Springer, Berlin Heidelberg New York

85. Ohno S (1970) Evolution by gene duplication. Springer, Berlin Heidelberg New York

86. Omenn GS, Motulsky AG (1972) Biochemical genetics and the evolution of human behavior. In: Ehrmann L, Omenn GS, Caspari E (eds) Genetics, environment and behavior. Academic, New York, pp 129–179

87. Pääbo S (1989) Ancient DNA: extraction, characterization, molecular cloning, and enzyme amplification. Proc Natl Acad Sci USA 86 : 1939–1943

88. Pääbo S, Higuchi R, Wilson AC (1989) Ancient DNA and polymerase chain reactions. J Biol Chem 264: 9709–9712

89. Post RH (1971) Possible cases of relaxed selection in civilized populations. Hum Genet 13 : 253–284

90. Rodnell 1992, p. 481

91. Rotter JI, Vadheim CM, Rimoin DL (1992) Diabetes mellitus. In: King RA, Rotter JD, Motulsky AG (eds) The genetic basis of common diseases. Oxford University Press, New York, pp 413–481

92. Sahi T (1974) The inheritance of selective adult-type lactose malabsorption. Scand J Gastroenterolog 9 Suppl 30 : 1–73

93. Scherthan H, Cremer T, Arnason U et al (1994) Comparative chromosome painting discloses homologous segments in distantly related mammals. Nature [Genet] 6 : 342–347

94. Schulz GE (1981) Protein-Differenzierung: Entwicklung neuartiger Proteine im Laufe der Evolution. Angew Chem 93 : 143–151

95. Searle AG, Peters J, Lyon MF et al (1989) Chromosome maps of man and mouse. Ann Hum Genet 53 : 89–140

96. Smith JM (1975) The theory of evolution, 3rd edn. Penguin, London, p 180

97. Smith M (1976) Commentary: group selection. Q Rev Biol 51 : 277–283

98. Sociobiology Study Group of Science for the People (1976) Sociobiology: another biological determinism. Bioscience 26 : 182–190

99. Stengel H (1958) Gibt es eine "getrennte Vererbung von Zahn und Kiefer" bei der Kreuzung extrem großer Kaninchenrassen? Ein experimenteller Beitrag zum sogenannten "Disharmonieproblem". Z Tierzucht Zuchtungsbiol 72 : 255–286

100. Stringer C (1990) The emergence of modern humans. Sci Am (December):68–74

101. Thorne AG, Wolpoff MH (1992) The multiregional evolution of humans. Sci Am 266 : 28–33

102. Vigilant L, Stoneking M, Harpendig H et al (1991) African populations and the evolution of human mitochondrial DNA. Science 253 : 1503–1507

103. Vogel C (1983) Personelle Identität und kognitiv-intellektuelle Leistungsfähigkeit im sozialen Feld nicht-menschlicher Primaten. Veröff Joachim Jungius Ges Wiss Hamburg 50 : 23–39

104. Vogel F, Kopun M, Rathenberg R (1976) Mutation and molecular evolution. In: Goodman M, Tashian RE (eds) Molecular anthropology. Plenum, New York, pp 13–33

105. Von Eickstedt E (1934) Rassenkunde und Rassengeschichte der Menschheit. Enke, Stuttgart

106. Walter H (1970) Grundriß der Anthropologie. BLV, Munich

107. Wienberg J, Stanyon R, Jauch A, Cremer T (1992) Homologies in human and Macaca fuscata chromosomes revealed by in situ suppression hybridization with human

chromosome specific DNA libraries. Chromosoma 101 : 265–270

108. Wilson AC, Cann RL (1992) The recent African genesis of humans. Sci Am 266 : 22–27

109. Wilson AC, Carlson SS, White TJ (1977) Biochemical evolution. Annu Rev Biochem 46 : 573–639

110. Wilson AC, Stoneking M, Cann RL et al (1987) Mitochondrial clans and the age of our common mother. In: Vogel F, Sperling K (eds) Human genetics. Springer, Berlin Heidelberg New York, pp 158–164

111. Wilson EO (1975) Sociobiology: the new synthesis. Belknap Press of Harvard University, Cambridge

112. Wilson EO (1978) On human nature. Harvard University Press, Boston

113. Zuckerkandl E (1965) The evolution of hemoglobin. Sci Am 212(5): 110–118

113 a. Zuckerkandl E (1976) Evolutionary processes and evolutionary noise at the molecular level. J Mol Evol 7 : 167–183

114. Zuckerkandl E (1976) Evolutionary processes and evolutionary noise at the molecular level. II. A selectionist model for random fixations in proteins. J Mol Evol 7 : 269–311

115. Zuckerkandl E (1978) Molecular evolution as a pathway to man. Z Morphol Anthropol 69 : 117–142

15 Behavioral Genetics: Research Strategies and Examples

My father gave my stature tall
And rule of life decorous;
Mother my nature genial
And joy in making stories;
Full well my grandsire loved the fair,
A tendency that lingers;

My grandam gold and gems so rare,
An itch still in the fingers.
If no part from this complex all
Can now be separated,
What can you name original
That is in me created?

J. W. v. Goethe, Zahme Xenien (Translation: H. and O. Bosanquet, Zoar, Oxford, 1920)

Scope and Conceptual Difficulties of Human Behavior Genetics. A survey of the genetic aspects of evolution shows great similarity between human beings and higher nonhuman primates in terms of chromosomes, DNA, proteins, and many genetically determined characteristics. The human species differs substantially in only one characteristic: language and abstract thinking. No other species can look into the past nor into the future! Analysis of the essential differences between humans and other species must therefore be directed to the brain – the organ of thought and language. These characteristics have enabled our species to supplement biological evolution with "cultural evolution," with all its consequences for building human civilizations and altering lifestyles (Table 14.1). The uniqueness of the human brain that allowed these developments is part of our genetic heritage. Experiments have attempted to subject nonhuman primates to childrearing practices similar to those for human children and even in a context in which they are raised together with human children. In every such experiment the animals – chimpanzees, our closest relatives – fail to develop spoken language. Although the cognitive functions of chimpanzees have been found to be more advanced than previously assumed, even these animals never reach the level of conceptualizing that older children do.

On the other hand, development of typically human behavior in children requires interaction with other humans and with the environment at large, including sensory stimuli and opportunities for motor behavior. Sensory and motor deprivations – especially if occurring at critical periods in childhood – lead to deficiencies that may be detectable even by changes in the histology of nerve cells and their interconnections. Human genetics is concerned principally with the analysis of genetic mechanisms leading to phenotypic differences between members of our species. The fact that behavioral characteristics can develop only in intimate and continuous interaction with the environment makes genetic analysis conceptually very difficult. Fingers, for example, are formed during a brief period of embryonic development. The trait of brachydactyly or short fingers persists all through later life regardless of environmental changes. The usual dominant mode of inheritance of this condition can easily be traced over many generations (Sect. 4.1.2). Most severe enzyme deficiencies also lead to an abnormal phenotype under all conditions, but specific manipulation of the environment may help to alleviate the symptoms. We are learning now that many simple genetic traits require certain environmental factors for their phenotypic expression such as in pharmacogenetics (see Sect. 7.5.1). Most behavioral characteristics, on the other hand – with the exception of severe mental retardation – need to be considered in the context of a certain environment. From birth on, the individual is not only shaped and modified by the environment – he or she in turn actively manipulates the environment and creates a complex interaction of many components. Analysis of genetic variability in behavior between individuals is difficult since the genetic aspect is only one of many parts in this complicated system.

Practical Difficulties and Possible Resolution. In addition to these conceptual problems, there are also practical obstacles that impede scientific progress in the field. Studies of genetic variability usually require examination of sizable groups of individuals. Such examinations are feasible if the tissue to be studied can be readily obtained, such as blood or even a skin biopsy. This is the main reason why genetic variability of erythrocyte enzymes (Sect. 7.2.2.2) and hemoglobin variants is so well understood (Sect. 7.3). Investigations of human *brain* function, on the other hand, must use more indirect methods since human brain is rarely available except from autopsies. This limitation can in part be overcome today since all genes including those involved in behavioral traits and diseases – even if they are brain specific – can be studied in DNA taken from any tissue. For example, DNA from genes expressed in the brain can be obtained from white blood cells and is widely used to diagnose and study brain disorders such as Huntington disease (Sect. 3.1.3.8). This approach is a powerful method to improve our rudimentary knowledge of genes involved in behavioral variation.

Importance of the Field. We feel that human behavior genetics promises to be the most interesting and important branch of human genetics, with possibly far-ranging consequences. Elucidation of genetic variability will allow better understanding of various human behaviors and emotions. It is probably no mere concidence that work in human genetics began with a problem of behavior genetics – the work of F. Galton on the frequency of high performance among relatives of outstanding men (Sect.1.3).

Thereafter the mainstream of research moved away from behavior genetics, as other fields became more accessible to genetic research methods and concepts. However, some research in human behavioral genetics has always been carried out. This work was usually influenced by concepts of quantitative genetics developed with quite different problems in mind, such as animal breeding (Sect.6.1). The development of the twin method seemed especially well suited for analysis of genotype-environmental interaction and also encouraged such research (Sect.6.3). However, conscious or subconscious prejudice often biased the results of these studies.

Paradigms of Mendel and Galton in Behavioral Genetics. We believe that concepts and methods in human genetics have reached a phase in which they can be applied more frequently to problems of behavior genetics. Research that will be most fruitful and least controversial is likely to be guided by the paradigm of Mendel – the gene concept and its subsequent extensions to the level of molecular biology. Most research in human behavior genetics, on the other hand, has been guided by the paradigm of Galton, using biometric methods in the quantitative analysis of relationships among behavioral phenotypes. This approach, too, has gradually been extended by various methods and has reached a considerable degree of sophistication. However, the passionate polemics that often accompany the subject of the genetics of intelligence suggest that the Galtonian paradigm may have reached the limits of its explanatory power. Such biometric approaches can never provide an explanation of *genetic mechanisms.* For practical reasons the clean experimental designs needed to obtain unequivocal answers to these problems by the Galtonian approach simply cannot be achieved in humans.

We predict that progress in our understanding of the genetic variability underlying interindividual differences in behavior will depend on our ability to apply concepts and methods of the central paradigm of genetics, i.e., analysis of gene action, to research on these problems. This does not mean, however, that the methods of quantitative genetics now used in human behavior genetics will soon become obsolete;

they will retain an important function in data analysis. Decisive progress, however, is likely to come from Mendelian analysis of biological parameters of gene action in the central nervous system.

15.1 Animal Models

Even if the "leap" from the mental capacity of our closest evolutionary relatives seems to be enormous, many basic principles of the action of brain and nervous system are identical in different species, although the degree of complexity is higher in humans. A frequent research strategy in genetics (Sect.7.1) analyzes simpler systems to gradually develop concepts and methods that will eventually provide us the means to tackle more complex situations. This strategy is now being pursued in investigation of the nervous system, especially in relation to behavior (see also [7]). It is not our intention to survey modern brain research, although some animals, including the ocean snail *Aplysia* [240] and the roundworm *Caenorhabditis elegans* [240], have proven excellect models for studying neuronal mechanisms. We confine ourselves here to a few examples that – by analogy or by contrast – may help to elucidate the situation in humans.

15.1.1 Research in Insects

Dialects in the Language of Bees [57, 62]. Insects, especially the colony-forming species, may exhibit complex and meaningful sequences of movement. Honeybees, after having found food in a near-by field with suitable flowers, inform other worker bees of their hive about their find by a circular dance. However, if the field of interest is farther away, they perform another type of dance that indicates the direction in which food can be found. These behavior patterns are genetically closely determined but provide no clues as to genetic mechanisms. Experiments have shown that bees are able to improve their capacity of spatial orientation by learning, and that the learning ability itself is a genetic characteristic in which genetic stocks may differ [118, 119]. (The application of the terms language and dialect to the dancing patterns of bees is semantically inappropriate since such behavioral patterns are quite different from human language and dialects. Their neurophysiology is therefore likely to be very different. We use these terms, following von Frisch [57] in the absence of a simple descriptive term for this behavior.)

Comprehensive experiments on the genetic mechanisms of genetically determined behavior characteristics were performed in another insect more familiar to the geneticist: Drosophila melanogaster.

"Genetic Dissection of Behavior" in Drosophila [10]. A great many mutants are known. Most are charac-

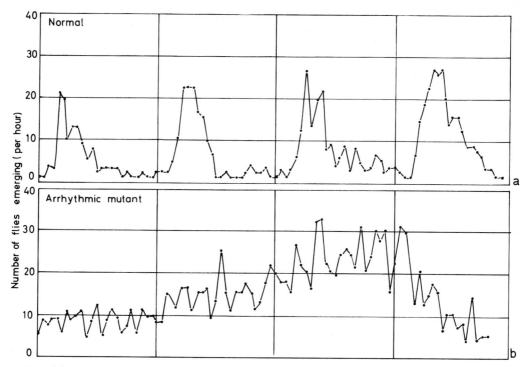

Fig. 15.1 a, b. Number of flies eclosing from the pupae over 4 days. *Ordinate,* number of flies eclosing; *abscissa,* the time of the day. **a** The normal wild-type usually ecloses early in the morning. **b** A mutant in which the normal day-night rhythm has been lost; the flies eclose at about identical rates throughout the day. (Adapted from Benzer 1972 [10])

terized by morphological and color criteria. Sturtevant showed as early as 1915 [229] that the phenotype of the X-linked mutation white-eyed influences choice of mates for pairing. Other mutations specifically influence courtship and mating behavior or motor activity. One mutant does not fly in spite of anatomically normal wings, and others lack the day-night rhythm. Figure 15.1 shows the number of flies eclosing from the pupae over 4 days. Normal flies generally eclose at dawn when the air is moist and cool, and the fly has time to unfold its wings and harden its cuticle while there is still little risk of desiccation or danger from predators. This behavior, as with other characteristics of the fly, is controlled by an "internal clock" independent of environmental influences; for example, this clock works even in complete darkness. A certain mutant, however, has lost this internal clock, and the flies eclose at all times of the day. The morning maximum completely disappears. Genetic analysis has shown his mutation to be X-linked.

Using an ingenious genetic technique, flies have been engineered that are mosaics for two externally visible X-linked traits (white eyes and yellow cuticles) linked to the arrhythmic eclosure mutation. This permitted the behavioral phenotypes of a great number of mosaics with different distributions of the two types of cells to be compared. Such analysis is possible in this insect by a convenient property of embryonic de-

velopment: there is little migration and mixing of cells. Cells maintain their relative position within the body. Using this technique, the internal clock was found to be most closely associated with the head of the insect. Among flies with a mosaic head some show the normal rhythm and others the abnormal one. A few flies, however, show a unique rhythm, suggesting that each side of the brain produces its rhythm independently, and that the fly responds to each of them.

Identifying single cells within the nervous system of the fly by tagging with biochemical markers has made it possible to pinpoint small groups of neurons as the origin of certain behaviors.

Most behavior patterns in *Drosophila* analyzed in this way are genetically predetermined in all details [83]. While simplifying genetic analysis, this high degree of genetic determination makes comparison with humans difficult, as our most important behavioral characteristic is the ability to learn from experience. However, a limited ability to learn has also been shown in *Drosophila*. This property may open up new paths for genetic analysis of learning capacity.

Mouse Mutants Affecting Embryonic Development of the Brain. Deviating phenotypic patterns in mouse mutants were analyzed to elucidate mechanisms of normal embryonic development in the central nervous system. One example is the reeler [2365], a mutation leading to severe difficulties in maintaining body balance. The reeler brain develops normally except for the cerebellar cortex and hippocampal formation, which show disorganization of cell alignment and of intracortical synaptic connections. These regions are the only parts in the mouse brain in which cells destined to establish synaptic contacts with each other become initially transposed. This and other evidence points to a failure in a specific recognition mechanism that normally allows patterned cell alignment in the developing brain. Other mouse mutants have provided material for a continued analysis of brain development in the mouse. The use of such experiments of nature for an analysis of normal biological function is turning out to be a useful investigative tool. It is likely that similar mechanisms apply to brain development of man.

What Can We Learn from Drosophila Experiments for a Genetic Analysis of Human Behavior? These insect experiments, interesting as they may be, provide little help in evaluating the role of genetic factors in human behavioral variability. Their design is based mostly on peculiarities of *Drosophila* not shared by humans. For example, human mosaics can neither be constructed at will nor be tagged by suitable mutations that alter the phenotype; embryonic development is too complicated to make this approach feasible. Most importantly, variations in human behavior are never fixed as irrevocably as in insects. Looked at from this viewpoint, the questions posed by *Drosophila* experiments are less genetic than embryological and – in part – neuroanatomical and neurophysiological. These studies can be compared, for example, with experiments in cats in which certain parts of the brain or of the peripheral nerves are destroyed, and the functional consequences are analyzed to discover how these parts cooperate in normal function. The study of neurological symptoms carried out by neurologists to arrive at brain localization of a given lesion is a human counterpart of this work in *Drosophila*. The *Drosophila* mutants used in these experiments have been produced artificially by chemical mutagens. While they may also occur spontaneously, their fitness is generally low. For example, the mutant with loss of the internal clock hatches all through the day; therefore most animals soon dry out or are devoured by predators. These mutants are analogous to rare hereditary diseases whose mechanism is unknown and shed less light on genes influencing behavioral variability within the "normal" range. However, deleterious mutants, even those affecting the function of certain nerve cell groups, are known in humans. Various methods for analyzing of deviant gene actions are in use, and this research approach has little to learn from work with Drosophila.

Still, a more general lesson can be learned. As other phenotypic variation, behavioral variability is likely to be at least partially determined by genes. To analyze the possible genetic mechanisms affecting behavioral variability, we should analyze the pathways in which gene differences influence behavior. In contrast to *Drosophila,* a straightforward one-to-one relationship between gene and phenotype cannot be expected, however, and a more indirect approach is necessary. Do we have any chance of finding such genes? This topic is taken up in Sect. 15.2.3.4. First, we ask whether experiments with mammals can give us additional clues. The animal that has been examined most extensively is the mouse.

15.1.2 Behavioral Genetic Experiments in the Mouse

Three main approaches have been used in the mouse:

1. Known mutants have been examined for special behavioral characteristics [79]. As in the *Drosophila* mutants described above, mouse mutants have low fitness and are therefore rare in natural populations. They have little or no bearing on the naturally occurring genetic variability of behavior and can be compared to severe hereditary diseases in humans.

2. Various inbred strains have been compared for behavioral characteristics, such as temperature preferences, geotactic reaction as measured by the behavior in climbing slopes of different angles, explorative behavior, general motor activity, ability to find a way through a maze, and emotionality as measured by the frequency of defecation. Inbred strains frequently show between-strain differences in such behavioral characteristics. These characteristics can then be analyzed genetically by between-strain crossings.

3. Artificial selection can be used in randomly mating populations to demonstrate genetic variability for quantitatively varying characteristics. As predicted (Sect. 12.2.1.5), populations usually respond to selection by shifting of the mean in the direction desired by the experimenter and by diminution of the variance of the trait under selection. After a variable number of generations the mean reaches a plateau. The remaining variance at that point is exclusively environmental, and continuing response to selection would be possible only if new genetic variability were created by mutation.

4. Such studies are now being supplemented by examination of mice in which certain genes, for example, those necessary for metabolism of the neurotransmitter monoamine oxidase have been inactivated (knock-out mice, Sect. 8.3.1).

These approaches show that there is genetic variability for measured behavioral characteristics. However, we remain ignorant regarding the genes involved. More specific conclusions may be possible if animals performing differently in the "tests" are studied utilizing physiological variables that could possibly be involved. Selection methods may be especially useful here. If selection for a behavioral trait results in differences in other behavioral or in anatomical, physiological, or biochemical characteristics, such differences may suggest a working hypothesis for a causal relationship [132].

An Example of a Single-Gene Abnormality: The Obese Mouse [2170]. *Inferences to Human Obesity.* – Mice homozygous for the obese gene ob/ob, if reared under standard conditions, eat heavily, become obese, and are relatively inactive. Increased food intake by these animals appears to be caused by the failure of a satiation mechanism.

Recently the genetic mechanism underlying obesity in mice carrying this mutation has been discovered [277]. The body's adipose cells normally produce a substance known as leptin, which is the gene product of the ob gene. Leptin is bound by brain receptors (as well as soluble receptors) with resulting appetite suppression. Mutations of the ob gene in the homozygous state (ob/ob) lead to biologically inactive leptin with failure to inhibit appetite, thereby causing the phenotype of the obese mouse. Administration of leptin corrects the obesity. Homozygous mutations of the leptin *receptor* (db/db phenotype) [119a] also cause obesity in mice.

What is the relevance of these studies for human obesity? Observations in identical and non-identical twins, adoption studies and familial aggregation of obesity point to genetic factors. However, a simple mode of inheritance has not been found [18]. Furthermore, familial eating habits also play an important role. As in other complex human traits, most cases of human obesity are likely to be caused by a combination of several genetic factors (including the db gene and ob gene, which is 85% homologous to its human counterpart) interacting with excessive food intake. Occasional cases of severe and morbid obesity may turn out to be monogenic, and a significant proportion of cases may be largely environmental in origin, caused by excessive food intake. A genetic analysis of human obesity of unselected families without any clues as to mechanism, therefore, starts with a "bouillabaisse" of obesity variants. Biometric analysis of such heterogeneous data leads to the discovery of heritability for human obesity but gives no insight into the causes in individual families.

Genetic Differences in Alcohol Uptake. Another behavior extensively studied in mice and rats is the tendency to prefer alcohol and the degree of susceptibility to the narcotic activity of this substance. Here no single gene has been identified so far, but after the pioneering research of Williams et al. (1949) [271] comparison of inbred strains and selection experiments have shown much of the variability underlying alcohol preference to be genetic in origin (Fig.15.2).

Strain differences have also been found for susceptibility to the effects of alcohol. The sleeping time after intraperitoneal injection of an anesthetic dose of alcohol also shows strain differences. Still more interesting are the effects of alcohol on spontaneous locomotor activity. The activity of C57Bl mice was reduced significantly; that of BALB/c mice was unchanged, and the activity of the C3H strain was enhanced after alcohol exposure [133].

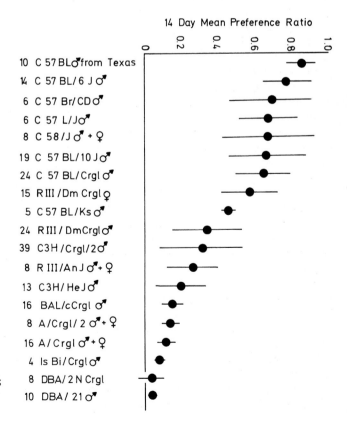

Fig.15.2. Differences in alcohol preference in many inbred mouse strains. *Preference ratio,* alcohol/water; ●, *mean; vertical lines,* ±1 SD. (From Rogers et al. 1963 [67])

Therefore the strain differences in response of the brain to alcohol are not merely quantitative, but the effect of the drug may go in either direction. The data regarding consumption of alcohol and of alcoholism (Sect. 15.2.3.5) [184] show analogous findings in humans.

Quantitative differences in the parameters tested, for example, sleeping time, could be due either to a different rate in metabolism of alcohol or to a difference in the susceptibility of the brain. Blood and brain tissue alcohol levels at various times after injection and the rates of elimination were found to be practically identical between the strains showing the greatest difference in sleeping time – C57Bl and BALB/c. Therefore the differences must lie in the susceptibility of the brain to alcohol. Here, again, similar data in humans are of interest [182]. No data are available regarding the biochemical mechanisms of this strain difference. However, alcohol preference also depends on hepatic metabolism, and here strain differences in alcohol dehydrogenase and aldehyde dehydrogenase have been detected [211]. The low preference in the DBA strain is thought to be due to a higher aldehyde accumulation, as aldehyde dehydrogenase is less active [61].

To extrapolate these results to human problems of alcoholism, correlations of alcohol preference with other behavioral characteristics should be investigated. Alcoholism in humans is most likely the result of complex interaction between postulated genetic differences affecting liver, brain, or other tissues and behavior patterns influenced by the social environment. Aspects of this problem are discussed in Sect. 15.2.3.5.

Learning Ability. One of the most important abilities of humans is the ability to learn from experience. The psychology of learning is a well-developed speciality of modern psychology. Three examples show attempts at analysis of the problem of interaction of genetic and environmental variability.

1. *Simple Mode of Inheritance for Conditioned Avoidance Learning.* A special aspect of learning ability is avoidance learning. Animals are put into a cage with a floor through which an electric current may be applied. This electric shock causes slight pain which can be avoided by jumping into another compartment of the cage. When the animal has learned this reaction, the current is announced by a flash of light, and the animal learns quickly to jump into the other compartment as soon as it sees the flash. The number of flashes necessary for learning this conditioned avoidance reaction is measured.

Examination of several inbred mouse strains show definite strain differences, and extensive breeding experiments using the strains with the most extreme reactions are interpreted to show this difference to be caused by a simple, monogenic mode of inheritance. Exchange of the newborns of the two strains immediately after birth did not change strain-specific learning ability. This behavior pattern could not even be altered by implantation of fertilized oocytes into the uteri of animals from the other strain. Both results – the simple mode of inheritance and the failure of early maternal environment to change the strain-specific behavior pattern – shows that the strain difference in avoidance learning must be genetically determined [28]. For a morphological correlate, see below.

This finding does not mean, however, that no environmental influences could ever modify the genetically determined learning ability. Such modifications are possible and have been analyzed in studies of maze learning.

2. *Heredity and Environment in Maze Learning.* To reach a food source rats are made to pass through a maze. According to the speed with which they achieve this goal, and by counting the number of mistakes made, "bright" and "dull" rats can be distinguished from each other. In a genetically heterogeneous population selective breeding leads after about seven generations to two populations with almost no overlap. This result points to appreciable genetic variability of maze-learning ability in the base population.

One of the many modifications of this experiment has proven to be of special importance for formulating hypotheses on learning ability in humans [41]. Following establishment of the bright and dull rat colonies, three groups were formed from the young of the two stocks. One group was reared as usual. The second group was reared from birth with restricted opportunities for cognitive and explorative experiences. The third group was reared in a specially enriched environment: their cage wall was illustrated with modernistic designs, and the cages contained ramps, mirrors, balls, and tunnels. Figure 15.3 shows the results. The difference in maze running between the two stocks was seen only in the groups that came from the "normal" or usual environment. All deprived rats from both strains showed a high number of errors, and all rats from both strains under the "enriched" environmental conditions performed almost equally well.

It would, of course, be premature to conclude that in humans a suitable environment can level out all inherited differences in learning ability. Moreover, maze learning in animals differs from learning ability in humans, as many experiments have shown.

However, the experiment shows that deprivation at a young age in the rat makes for disadvantage in later performance. On the other hand, an environment rich in opportunities for varied experiences may improve learning ability. This conclusion corroborates similar hypotheses derived from experiences with humans [197].

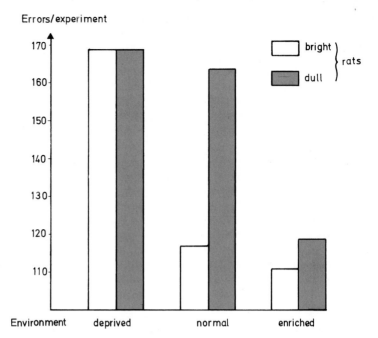

Fig. 15.3. Result of a maze-running experiment with bright and dull rats, in deprived, normal, and enriched environments. There is a definite difference in performance between the groups of rats in normal environment; the difference disappears almost entirely under deprived and enriched conditions. (Data from Cooper and Zubek 1958 [41])

3. *Psychosexual Behavior Must Also Be Learned.* Two inbred stocks of guinea pigs were compared regarding a sequence of activities involved in sexual behavior. Both inbred stocks showed lower sexual acitivty than the heterozygous stock; there were definite interstock differences. One stock, for example, showed more courtship behavior, whereas the other stock proceeded to copulating behavior with few preliminaries. To examine the influence of learning on sexual behavior, males that had been kept isolated from the age of 25 days were compared with others who grew up with a group of females. At the age of 77 days test animals were placed together with a female in heat. In one of the stocks only 6% of the isolated males achieved ejaculation compared with 84% of the controls. Similar results were obtained with the other stock. This experiment shows the importance of environment for the development of normal, adult sexual behavior. The genetic component, on the other hand, is obvious from the strain difference [247, 248].

Attempts at Elucidating the Biological Mechanisms of Behavior Differences. Animal experiments can be tools for elucidating the biological mechanisms of interindividual differences in behavior [36, 200].

A difference in two-way avoidance learning in the mouse, transmitted by a simple Mendelian mode of inheritance, is noted above. The mouse strains used in these experiments underwent morphological study of their brains, with particular emphasis on the hippocampal areas, since earlier work had suggested a relationship of these areas with certain as-pects of learning. The extent of the mossy fiber terminal field, reflecting the number of synapses of the dentohippocampal circuitry in the basal parts of the hippocampal pyramid cell layer, was found to be strongly correlated with poor acquisition of two-way avoidance. Animals with extended mossy fibers and therefore many synapses showed poorer performance than those in which this special field was less well developed. A causal relationship was strongly suggested by two additional experiments: the same correlation was shown by rats selectively bred for differential avoidance performance and by genetically heterogeneous mice tested individually for avoidance performance and subsequently killed for brain study [206, 207].

These studies convincingly verify the association between an aspect of brain structure and behavior. But why is the correlation negative? One would rather expect the opposite: the more synapses there are, the better the ability to learn. Upon closer scrutiny, however, this result makes much more sense. It is necessary to delineate more precisely what "learning" actually means in terms of brain mechanisms. In one experiment the mice had to find the exit (a water-covered platform) from a water-filled maze. Here, too, the mouse strain with more synapses took more time. This experiment permitted observation of the different "strategies" of the two strains: animals with few mossy fiber synapses were simply more active; they used a strategy of trial and error rather than developing a quasi-intelligent concept. This led to a quite different result in another experiment: the length of time needed to learn a task requiring discrimination in a dry maze, when the correct choice was indicated by an additional stimulus. For a certain time a wall was inserted in the maze to provide them "time to think." Here the problem could not be solved by trial and error, and here the result was the opposite: good performance was associated with numerous mossy fiber endings.

To understand these apparently contradictory results the learning tasks must be analyzed in greater detail. In avoidance learning the animal must leave a compartment. Its instinctive behavior would be to stay motionless in one corner; in order to run it must overcome this tendency. In the water maze its instinctive impulse is to escape. The more actively if follows this impulse, the sooner it is successful. In the dry maze, however, there is a chance to form a quasilogical concept before starting to act. With these and other experiments geneticists are now struggling for theoretical explanations. The results could be explained, for example, by "perseverance," on the one hand, vs. "development of concepts," on the other [208, 250]. The kind of concepts to be formed has been studied in still another experiment: here the mice had to find their way in a radial maze. A correlation of mossy fiber density with "spatial working memory" but not with "nonspatial reference memory" was found [44, 208].

Extrapolations from animal experiments to humans – especially in the field of neurobiology and behavior – should be made with caution. However, such experiments may ultimately help to elucidate aspects of learning in humans.

Other attempts have been made, to relate biochemical variables to behavior. Differences have been discovered, for example, in the endocrine system of "emotional" and "nonemotional" rats. Emotional males had larger adrenals and thyroid glands. Endocrine function in humans is one of the most powerful influences on the brain.

Attempts at relating behavioral traits to biochemical features of brain function are of great interest. Here neurotransmitters have recently received much interest. Neurotransmitters are compounds that transmit impulses from one neuron to the other through *synapses* (Sect. 15.2.3.6). Their variation in conditions with normal and psychotic human behavior has attracted much attention in recent years. Early studies had shown cholinesterase levels to be higher in some brain areas of "bright" than in "dull" rats [114]. In more recent experiments [36, 37] two related sublines of one inbred mouse stock, BALB/cJ and BALB/cN, differed in their fighting behavior. Males of the BALB/cJ subline immediately attacked other males that were put into their cage, whereas BALB/cN males remained peaceful. This behavior difference was not caused by a higher general activity of the

first subline, as the sublines did not differ in their general motor activity, but was specific for "fighting." Cross-rearing experiments showed that this type of aggressiveness was not caused by maternal environmental influence. A number of enzymes involved in the metabolism of an important group of neurotransmitters, the catecholamines (Sect. 15.2.3.6), were studied in the adrenals and the brain of these animals. The levels of three enzmyes, tyrosine hydroxylase, dopamine-β-hydroxylase, and phenylethanolamine N-methyltransferase were twofold higher in adrenals of the BALB/cJ, the fighting subline (Table 15.1). The difference in the brain was in the same direction but smaller and not significant.

Breeding experiments between the two sublines were interpreted to show a difference in a single gene pair; nonfighting was dominant over fighting. Enzyme levels of all three enzymes segregated with the behavior. The data point to a lessened degradation of these enzymes as the most likely cause for their elevated enzmye activity.

Not only neurotransmitters may promote aggressive behavior but hormones as well. For example, exposure to a single dose of testosterone on the fourth day of life induced aggressive behavior in female BALB-cBy mice but not in C57BL/6By females. Therefore the effect appears to depend on the genotype [133]. Another environmental factor that enhances male-male aggression in genetically predisposed mice is isolation of the animals for about 14 days [87].

Behavioral differences between inbred mouse strains such as duration of catalepsy were also detected after administration of haloperidol, a compound that binds to receptors for the neurotransmitter dopamine. The expected differences in drug binding were in fact demonstrated; they suggest genetic differences in regulation of the number of dopamine receptors in the stem ganglia [209].

Possible Significance of Experiments with Mice and Other Mammals for Behavioral Genetic Analysis in

Table 15.1. Levels of biosynthetic enzymes involved in catecholamine metabolism in adrenal glands of BALB/c sublines (from Ciaranello et al. 1974 [37]

Subline	Behavioral phenotype	n	Tyrosine hydroxylase	Dopamine β-hydroxylase	Phenylethanol-amine-N-methyltransferase
BALB/cJ	Fighter (F)	9	8.87 ± 1.19	30.23 ± 2.38	0.371 ± 0.014
BALB/cN	Nonfighter (NF)	9	4.51 ± 0.46	17.33 ± 1.18	0.193 ± 0.011
F_1 (F × NF)		12	6.05 ± 0.40	23.82 ± 1.47	0.276 ± 0.016

Activity of each enzyme is expressed in nanomoles of product formed per hour per adrenal pair. In each case, differences between high and low parents and between F_1 and either parent were statistically significant.

Humans. Animal experiments, especially with mammals, are suitable for formulating hypotheses that can be examined in humans. On the other hand, hypotheses suggested by certain observations in humans can be put to a more stringent test in animal experiments. The above examples indicate two different approaches to the problem of genetic determination of behavior.

1. Comparisons between inbred strains and selection experiments designed primarily are to demonstrate that genetic variability exists. The experiments usually permit quantification of the extent of genetic variability. This approach frequently also allows analysis of the interaction of genetic with environmental variability.

2. Very soon, however, this kind of analysis becomes unsatisfactory, as the answers are not specific. To gain insight into the mechanisms of genetically determined behavioral variations, morphological, physiological (usually neurophysiological) or biochemical characters must be investigated. A satisfactory state of this analysis is reached when such variation can be traced to single gene mutations. However, more than one single gene mutation is often involved. It may nevertheless be possible to localize the specific gene or genes by linkage techniques (see Chap. 5). This knowledge allows the bypassing of laborious studies on phenotypes with a variety of different approaches by first elucidating the DNA structure of the gene or genes involved. This information often provides strong clues as to function; it allows a more logical selection of the kind of phenotypic approach needed to understand the mechanisms of behavioral variation.

15.2 Behavioral Genetics in Humans

Normal and Abnormal Behavior. Normal behavior is studied largely by psychologists and abnormal behavior mostly by psychiatrists, while research into the biological causes of major psychoses is often done by neuroscientists. Defining of the normal range of behavior is difficult and depends strongly upon a given society's definition of "normal." However, transcultural studies of recent years have clearly indicated that the major psychoses, such as schizophrenia, occur in all societies and are not produced by artificial labeling, as claimed by some social psychiatrists.

Results of genetic analyses of many traits have shown that abnormal function is often caused by simple genetic defects, including single gene mutations. Variability within the normal range has rarely been traced

to differences in a single allelic series and presumably owes its origin to contributions from several genes intracting with environmental factors. By analogy, gross abnormal behavior is therefore more likely to be caused by single gene defects while normal behavior has a more complex genetic and environmental determination. As a consequence elucidating the genetics of abnormal behavior is a simpler task than genetic understanding of behavior in the normal range.

Our treatment of human behavior genetics approaches the subject similarly to our analysis of gene action (Chaps. 7, 8). More descriptive phenomenological studies are covered first, with a few glimpses at a more causal analysis at a deeper genetic level.

Observation and Measurement of Human Behavior. Human behavior can be observed by outsiders or reported following introspection, or both methods may be applied. The outside observer can note, for example, how quickly an individual moves, whether (s)he prefers to be motorically active or passive, and – or whether (s)he likes to engage in social contacts. On a larger scale, we may be interested in how an individual passes through the stages of life – how childhood, adolescence, and young adulthood are experienced, how and why occupations are selected, and whether a person leads a reasonably happy and successful life. In all these aspects we can compare one individual with others. Furthermore, to simplify the task of measurement and comparison, we may wish to standardize our procedures of observation. For example, certain abilities or interests can be examined and compared as comprehensively as possible within a reasonable time by specifically constructed test systems. Such *psychological tests* should have a high *reliability* i.e., repeated application of the same or analogous tests to the same individual should lead to similar results. The reliability of a test can be assessed relatively easily. Its validity, i.e., whether it measures anything relevant to real life, is much more difficult to determine. Intelligence tests, for example, should be able to predict a person's performance in a situation in which, according to the conventions of our society, intelligence is needed; for example, success in school or university or performance in an occupation.

Psychologists interested in test construction have been successful primarily in measuring sensorimotor skills and the set of abilities usually referred to as called "intelligence." So-called intelligence tests have been applied on a very large scale, and these studies – especially when interpreted in genetic terms – have raised a considerable uproar.

However, such observations and measurements are only one side of the matter. Data obtained from introspection may be just as important. We wish to know how a person feels. Does (s)he feel relaxed or excited, elated or depressed – in brief: What is his or her feeling tone? Does the person feel the urge to take an active part in life or to let things just happen? What attitudes does a person have toward other people – toward family, colleagues, occupational life, cultural activities? All these matters are often lumped together under the label "personality."

Strictly speaking, the psychologist has inside knowledge of only one person – himself. However, human beings communicate with each other – by nonverbal means such as physiognomic expression, movements of their bodies, especially their hands, and above all by that natural instrument unique to our species, language. The investigator also uses these means of communication to gather more indirect information about the feelings and thoughts of other people. The simplest method is the interview. This interview may be structured if the investigator has prepared in advance a number of topics to be explored. Other, more refined methods are *personality questionnaires* with queries on feelings, attitudes, and opinions. Most questionnaires are constructed so that the proband has several choices. The number of statements in one or the other direction on various topics are then compared with those of standard populations. Widely used questionnaires include the Minnesota Multiphasic Personality Inventory (MMPI) and the 16 Personality Factors (16PF) test.

All the above aspects of human behavior show variations between individuals which can be examined by genetic approaches.

15.2.1 Investigations with Classical Phenomenological Methods

15.2.1.1 Reappraisal of Classical Methods (See also Chap. 6)

Family Investigations. The most straighforward – and most unsophisticated – approach for assessing a genetic contribution to the variability of a certain trait is to compare the frequency of this trait between biologically related individuals. If an all-or-none, or alternative characteristic is genetically determined, it usually manifests more frequently among relatives. With increasing biological relationship, the frequency in relatives climbs. Such alternative individual characteristics include various diseases as contrasted to general health, and severe mental retardation in contrast to normal intelligence. If the characteristic is defined on a graded or quantitative scale, the similarity between relatives is generally expected to rise with increasing biological relationship.

One goal of biometric research is to establish the probability among relatives of probands to develop the same conditions. Calculation of such *empirical risk figures* is explained in Sect. 4.3.6.

In quantitative traits the similarities among relatives are usually expressed and measured as correlation coefficients. Such analyses are strictly empirical and require no genetic concepts. The conceptual difficulties begin if we try to interpret empirical risk figures or correlations among relatives in terms of genetic variability. Certain correlation coefficients can theoretically be derived from assumptions regarding the degree of relationship between two persons and the degree of dominance of these genes. The theoretical expectations can then be compared with those empirically observed, and from this comparison, the "heritability," i.e., the proportion of variability contributed by genes, can be estimated. Heritability and methods for calculation – and the assumptions normally inherent in such calculations – are described in Chap. 6.

Heritability calculations can theoretically also be carried out for alternatively varying characteristics, such as diseases. These calculations require information on the occurrence of the test trait in relatives and in the general population. Here, too, a number of assumptions regarding the biological meaning of such a calculation need to be made. Worse still, few of these assumptions can generally be tested with most empirical data. The problem is discussed in Chap. 6.

One objection to such heritability calculations comes to mind immediately. In most cases relatives not only share some of their genes but also live under similar environmental conditions. Heritability estimates, however, assume that genetic and environmental variance components are independent of each other. Attempts at correcting for this bias by introducing estimates of environmental parameters regarded as relevant for the trait in question are of dubious value. The classic approach for overcoming this difficulty is the comparison of monozygotic and dizygotic twins.

Twin Method (see Sect. 6.3 [15]). Monozygotic (MZ) twins are genetically identical; differences found between them must therefore be nongenetic. Dizygotic (DZ) twins are no more similar genetically than normal sibs. As with MZ twins, however, they are usually reared together; therefore, the influence of environmental factors can be taken as essentially similar in each twin. Hence, if MZ twins are more similar than DZ twins in a quantitatively varying trait, or if they are more frequently concordant in an alternately varying characteristic, genetic components in the variability of this characteristic are assumed. This assumption is subject to a number of qualifications that grow out of the observation that twins – especially MZ twins – are in many aspects not a random population sample. These problems are explained in detail in Sect. 6.3.4. For planning of research in behavior genetics, it must be remembered that the special twin situation, especially in childhood and youth, leads to certain deviations from average development. More generally, the environmental differences to which twins – and to a slightly greater degree other sibs as well – are exposed can hardly be regarded as representative for such differences in the general population. Usually such differences are much smaller in relatives.

MZ Twins Reared Apart; Studies on Adopted and Foster Children [127, 159]. To overcome these difficulties two approaches were chosen especially for problems in behavior genetics: first, the comparison of MZ twins who were separated in infancy or early childhood and reared apart. This design theoretically avoids in a very elegant way any biases due to a common environment and interaction between the twins. Secondly, in a comparison of adopted children (or of children living in foster homes) with their *biological* parents, common environmental factors usually acting on parents and their children no longer apply, and the parent-child relationship is reduced to its biological components. Influences of a common environment are excluded.

Unfortunately, matters are somewhat different in reality. Placing of separated MZ twins or of adopted children never occurs at random; selective placement based on socioeconomic status and behavioral characteristics is commonly carried out by social agencies; couples adopting children are a biased sample of all couples; foster homes may offer aberrant living conditions. The ideal research design of a randomized environment cannot be achieved for humans.

15.2.1.2 Mental Retardation and Deficiency

Definition. There are many definitions of mental retardation and deficiency. A useful definition that does not depend on tests or school results is that given in the British Wood Report (1929). A mentally deficient individual was defined as "one who by reason of incomplete mental development is incapable of independent social adaptation." Mental ability is often measured by intelligence tests, which are also used to classify mentally retarded persons in to three categories: mild (high-grade), medium, and severe (low-grade). The "intelligence quotient" (IQ) as devised by Binet and Simon (1907) and later modified was designed primarily as an aid to teachers in Paris in allocating pupils to their correct grades in school [175]. The children were classified into years and months of mental age. By comparing mental and chronological age it was immediately apparent whether a child was advanced or retarded. For example, if a child was 12 years old and his performance was equivalent to the mean of all 9-year-olds, the IQ was 9/12 = 75. For individuals above the age of 14–16 years this formulations is meaningless. Therefore their IQ is defined by comparison with a standardized sample of their age peers. Even this formulation is only a crude approximation, as the mental capacity of a subnormal individual often is qualitatively different from that of a normal individual.

Incidence of Mental Subnormality. Incidence figures of mental subnormality vary widely, depending on the definition. If an IQ of 69 or a mental age of 7–10 is taken as the lower limit of normalcy, incidences formerly clustered at around 2%–3% of the population; more recent studies give lower values [186]. The great majority of mentally subnormal persons can be classified into the mild (high-grade) group; only about 0.25% of the overall population are categorized as severely retarded (IQ < 50). Among severe cases boys are overrepresented.

The frequency of detectable mental retardation increases after the age of 6–7 years due to school attendance, where mental retardation is more likely recognized because of learning difficulties. The detected population frequency decreases again after school age since many individuals unable to succeed in school can achieve satisfactory social adaptation. These data emphasize the importance of school attendance in defining mental retardation.

Two Biological Groups. Mentally subnormal individuals can be divided into two classes, for which the terms "pathological" and "subcultural" became most popular [123]. The pathological group comprises a mixture of cases with various genetic and nongenetic causes. It was once believed that many cases falling into this category were environmentally caused, as parents were usually normal. We know today that this is not entirely true. While this group comprises exogenous cases, for example, those due to brain injury or infectious disease, such as meningitis and encephalitis, many cases are due to genetic causes. Down syndrome has long been the most frequent single diagnosis in this pathological group (Sect. 2.2.2). Most inherited and predominantly autosomal-recessive inborn errors of metabolism that cause mental deficiency are also included in this class which also comprises some autosomal-dominant and, especially, X-linked recessive conditions. Studies of children from consanguineous marriages provide hints that further monogenic autosomal-recessive conditions may be hidden among cases currently diagnosed as nonspecific mental retardation.

X-Linked Mental Retardation (XLMR). The severely affected group contains many more males than females. Moreover, brothers of such males are affected more often than their sisters. Luxenburger postulated as early as 1931 the involvement of X-linked genes [130]. Later, however, some authors explained this sex difference by ascertainment biases [50]. The matter was clarified when a number of X-linked conditions were discovered [168]. A population study in British Columbia, Canada, estimated the prevalence of all types of X-linked mental retardation together as up to 1.8 per 1000 males [245]; studies in populations of mental retardates confirm this order of magnitude. X-linked mental retardation may be as common in the male population as Down syndrome. At present about 80 types are known. These have been subdivided into syndromal and nonsyndromal subtypes (International Workshops on X-Linked Mental Retardation). Scientists have tried to create a certain order mainly by founding a nomenclature committee to which all newly discovered types can be submitted for help in classification and naming [149].

The best-known and most common single type is the Martin-Bell syndrome (approx. 40% of cases with XLMR). Among males with this syndrome a specific anomaly of the X chromosome is found: in approx. 2%–35% of their X chromosomes a "fragile site" is seen at the end of the long arm (Xq 28) in lymphocyte cultures. The same "marker X" can also be found in many female carriers; it "can readily be seen in ... carrier females under the age of about 25, but to demonstrate it in older women is difficult unless they are intellectually impaired" [245]. Even among younger carriers, mar(X)-positive ones are much more common among mental retardates than among normals.

Many male patients show characteristic physical features, including large testicles, large or lop ears, and prominent forehead and jaw (Fig. 15.4, Table 15.2). At birth some have a large head and slightly increased birthweight. In children the diagnosis is more difficult, but experienced clinical geneticists have de-

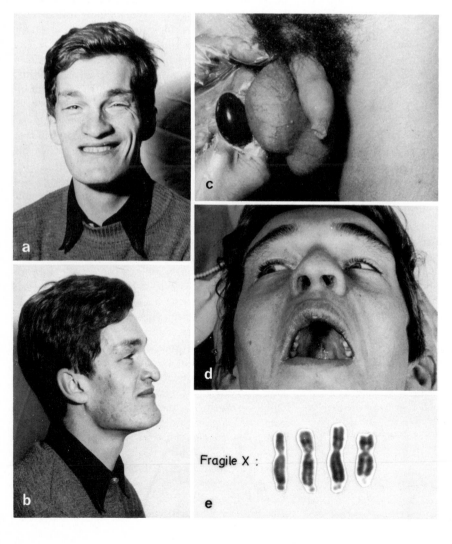

Fragile X:

Fig. 15.4 a–e. Patient with X-linked mental retardation (Martin-Bell syndrome): macro-orchidism, typical face, and high-arched palate. d The size of the testicle is compared with a model indicating average size. The marker X was demonstrated in 35% of the cells. However, clinical features are not always as characteristic as in this case. e Marker (X) chromosome. (From Tariverdian and Weck 1982 [236])

Table 15.2. Clinical features of Fra X mental retardation (Martin-Bell syndrome; modified from Turner and Jacobs 1984 [245])

Intelligence	IQ range 30–65, sometimes borderline normal or even normal. Occasional hyperactivity or autism in childhood; generally friendly, shy, nonaggressive as teenagers; speech anomaly
Growth	Birth weight normal; usually heavier and taller than normal sibs; head circumference above 50th, sometimes above 97th percentile
Facies	Prominent forehead and jaws, long face and big ears
Testicles	May be 3–4 cm^3 in childhood (normal 2 cm^3); postpuberal boys 30–60 cm^3 (normal < 25 cm^3)
Occasional features	Epilepsy; increased reflexes in lower extremities; gynecomastia, striae, fine skin; thickening of scrotal sac

scribed a small, long, and centrally bloated face, narrow eyes, prominent and dysplastic ears, velvety skin, and pudgy hands and feet. These features are rather subtle and may be overlooked; therefore many patients remain undiagnosed during childhood. The IQ may be as low as 30, but in most instances is in the 50–60 range. Speech has been reported as "repetitive," and stuttering appears to be common. Children tend to be hyperactive; later, autistic behavior is common, and the patients have difficulties adjusting socially [195]. Among probands diagnosed as autistic children, Fra X individuals are not especially common. However, there is wide variability within and especially between families in terms of both clinical signs and the frequency with which Fra X can be detected. In an increasing number of pedigrees the mutant gene must have been transmitted by a clinically and cytogenetically unaffected male, and segregation analysis [212] has revealed approximately 80% penetrance of mental retardation in hemizygous males. In many instances the physical features are not nearly as characteristic as in the patient shown in Fig. 15.4. Phenotypic variability among female heterozygotes is still wider. According to one study [212], "mental impairment" among such females was seen in about 30%. There has been some discussion on the mental status of the 70% without outright mental impairment. In one study [171] most subjects had a below-average IQ. Another study confirmed some intellectual impairment, in clinically nonaffected carriers especially a deficiency in nonverbal skills [126]. Carriers were shown to exhibit "a wide spectrum of cognitive deficits ... ranging from subtle to quite ob-

vious" [73] as well as dysfunction in social interaction and affective regulation. In four men with Fra X, magnetic resonance imaging of the brain revealed an increased size of the fourth ventricle and anomalies of the cerebellum [195].

The genetics of the anomaly raised a number of intriguing questions: about 20% of male transmitters fail to show signs of mental retardation; their daughters are found to be phenotypically normal, but the sons of these daughters are often affected. This "Sherman paradox," named after the first author of an important publication [157], has been solved by elucidating the molecular background of this disease (Sect. 9.4.2; Fig. 9.5). The basic mutational event is amplification of a $(CGG)_n$ repeat (or $(CCG)_n$, if the complementary DNA strand is considered) at the fragile site, in normal X chromosomes, n is about 25. In clinically nonaffected male hemizygotes the number of repeats is moderately increased; this "premutation" leads to further amplifications in the germ cells of their (clinically unaffected) daughters. Their sons and daughters often have many more repeats and are affected clinically.

This type of mutation has also been found in some other diseases, such as Huntington disease, myotonic dystrophy, and Kennedy disease (Sect. 9.4.2). The Fra X syndrome also poses a problem in terms of population genetics: why is it so common despite the strong selective disadvantage since affected males almost never have children? Two factors appear to be at play: the mutation rate seems fairly high, and unaffected female carriers may be especially fertile (Sect. 9.1) [126, 263].

The proportion of mitoses showing a fragile X chromosome can be enhanced by folic acid deprivation in cell culture. This has led to speculations on possible pathogenetic mechanisms, but a study on folic acid pathways in Fra X cells failed to reveal a defect [268]. Therapeutic attempts with folic acid were reported to have no success [231].

Heritable fragile sites have been observed in other chromosomes. Most of these are also folic acid sensitive. Heterozygosity of such sites appears to be more common in mental retardates than in the general population of newborns [232]. More recently, another phenotypically similar type of X-linked mental retardation has been discovered which has been termed FRAXE (in distinction to FRAXA, the type described above) [113]. Here the fragile site is also localized at Xq27–Xq28 but is distal to that found in FRAXA. Here a GCC repeat causes the anomaly; in normal persons 6–25 repeats may be present, while in affected individuals about 200 copies are observed. Mosaicism appears to be common; mental retardation may be less severe than in FRAXA patients.

In addition to the Martin-Bell (Fra X) syndrome, many other types of X-linked mental retardation exist. The genes for an increasing number of the approximately. 80 different types have been localized on the X chromosome.

X-Linked Behavioral Disturbance. A large pedigree of 14 males supposedly affected with slightly subnormal intelligence (IQ ~ 85) and an X-linked mode of inheritance has been observed in the Netherlands [21, 22]. There are no prominent physical characteristics; this type therefore belongs to the "nonsyndromic" group. In addition to reduced intelligence, affected males were reported to show periods of severe aggressiveness, such as stabbing, trying to run over a teacher with a car, attempted rape of a sister, and arson. These outbreaks tended to occur as outraged reactions to commonplace provocations. The gene has been localized to the locus for a neurotransmitter enzyme, monoamine oxidase type A in Xp11. Monoamine oxidase is one of the enzymes involved in catecholamine metabolism. A marked disturbance of monoamine metabolism was shown by biochemical studies of urine. A missense mutation has been demonstration in this gene. However, this report has been critisized because of insufficient psychiatric diagnosis [79a]. This observation is reminiscent of the relationship between certain aspects of catecholamine metabolism and aggressiveness in mice. In fact, adult transgenic mice in which transgene integration caused deletion of the monoamine oxidase A gene exhibited a distinct behavioral syndrome, including enhanced aggression and fighting behavior in males [30a]. These results in mice suggest that the behavioral findings in human monoamine oxidase A deficiency might indeed be related to this biochemical defect.

Rett Syndrome [75, 198]. Unlike other syndromes, Rett syndrome is observed exclusively in girls; its in-

cidence is estimated at 1:15000–1:20000. The children are born normal, but the first signs of neurological deterioration appear before 2 years of age. A slow neurodegenerative process leads to severe mental retardation, to continuous, stereotypic hand movements, apraxia, and autistic behavior (Fig. 15.5). Almost all cases are sporadic; an X-linked dominant mode of inheritance with lethality of male hemizygotes (Sect. 4.1.4) has been suspected [39] but cannot be proven since the affected girls do not reproduce. However, both the concordance of MZ twins (Fig. 15.5) [235, 236] and a few reports of affected family members which could be due to germ cell mosaicism strongly suggest a genetic origin. So far the genetic mechanism has not been elucidated [26].

Mild or High-Grade Mental Retardation. A pattern quite different from the severe types of mental deficiency is observed in the mild or subcultural group. There are generally far fewer cases of dramatic exogenous origin although environmental causes do sometimes exist. Few neurological abnormalities or other clinical findings are noted. Instead, familial recurrence is observed. Whereas patients with severe mental retardation usually have normal parents and only occasionally affected sibs, the proportion of affected relatives among those with *mild* mental retardation is high (Fig. 15.6).

Empirical Risk Figures [276]. Severe mental retardation is a typical example of a mixed etiological category. Genetic analysis at the phenotypic Mendelian level has become possible for an increasing number of cases within this class, and many mutations at individual gene loci have been detected. Other cases have been shown to be caused by chromosomal anomalies. For the unclassified rest, empirical risk figures of recurrence have been calculated that must

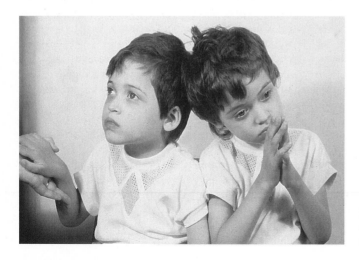

Fig. 15.5. Monozygotic twins with Rett syndrome at the age of 9 years. (Courtesy Dr. G. Tariverdian; see also [237])

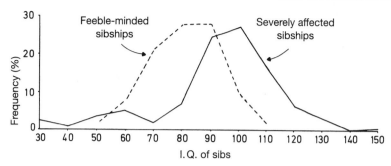

Fig. 15.6. Distribution of IQ in sibships with mildly affected (high-grade mental retardation) and severely affected probands (low-grade mentally retarded probands). High-grade mentally retarded sibships show a nearly normal distribution, which is shifted toward slightly reduced IQ values, suggesting a multifactorial basis. The sibships of severely mentally retarded probands show a bimodal distribution. There is a minority of mentally retarded sibs, but the IQ distribution of most siblings equals that of the average population (the mean is close to IQ = 100). These data point to discrete, major causes, genetic or nongenetic, for severe mental retardation. (From Fraser Roberts 1952)

Table 15.3. Prevalence of mental retardation among parents and sibs of mentally retarded probands (from Penrose 1962 [175])

Degree of mental retardation of proband	n	Above average	Borderline and mildly retarded	Severely retarded	Total retarded
		Parents			
Borderline or mildly retarded (627 probands)	1254	0.32%	27.59%	0.24%	27.83%
Severely retarded (653 probands)	1306	0.53%	15.00%	0.08%	15.08%
All degrees	2560	0.43%	21.17%	0.16%	21.33%
		Sibs			
Borderline or mildly retarded	2321	1.21%	19.52%	2.50%	22.02%
Severely retarded	2549	1.57%	12.24%	4.28%	16.52%
All degrees	4870	1.40%	15.71%	3.43%	19.14%

be used cautiously. Genetic counseling requires careful analysis of the case under study and should include a consideration of exogenous brain damage. A relatively large unclassified core remains, where genetic counseling is difficult.

The calculation of empirical risk figures is more useful for mild mental retardation, especially when the minority of cases with known exogenous causes can be excluded. Table 15.3 shows empirical risk figures derived from Penrose's famous Colchester Survey [174]. There is a notable difference in familial occurrence of mental retardation between severely and mildly affected probands, the latter groups showing higher familial aggregation. Other studies usually give similar figures [85, 186, 273]. Comparison of data from various authors is difficult as the criteria of diagnosis vary. The data on familial aggregation and the absence of clear segregation of normals and mentally retarded in the many families studied are compatible with multifactorial inheritance. If a scale such as IQ is used with continuously varying test results, no threshold exists. However, it may be useful to establish artificial thresholds to delineate the proportion of the population that is unable to benefit from regular schools.

Twin Studies. One way to examine the heritability of a characteristic is to compare MZ and DZ twins. Figure 15.7 describes three twin studies. Considering that discordance in two pairs of MZ twins in Smith's series could be related to exogenous causes, concordance of mental subnormality for which no exogenous causes (early brain damage or infection) can be found, approaches 100%, and hence heritability does indeed seem to be 100%.

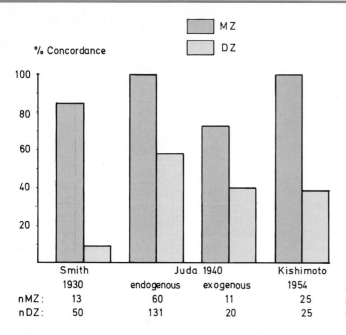

	Smith 1930	Juda 1940 endogenous	exogenous	Kishimoto 1954
nMZ:	13	60	11	25
nDZ:	50	131	20	25

Fig. 15.7. Concordance in MZ and DZ twins in twin studies on mild (high-grade) mental retardation. (From Zerbin-Rüdin 1967 [276])

Somewhat closer scrutiny of the data casts some doubt on this interpretation. First of all, the twins included in these studies lived with their families; even if the parents were not outright subnormal, they were in most cases within the lower range of the intelligence distribution. A certain deprivation from the stimuli necessary for normal intellectual development may safely be assumed to be responsible for part of the mental retardation in these MZ twins. This assumption is corroborated by another interesting observation in the families of Juda's twin series: one-fourth of the fathers of retarded twins and more than one-third of the mothers were recorded as retarded. The significance of the mother for the mental development of the infant and child, especially in the first 3 years of life, is now an undisputed finding of developmental psychology. Hence, the data suggest a maternal influence in addition to X-linked genes.

Juda [103, 104] collected twin data by contacting the families of all children in special schools for retarded children in southern Germany. Among 18 183 students she ascertained 488 twin children, i.e., one twin per 37.3 pupils. The frequency of twin births in Germany at that time was about 1 twin birth in 84 deliveries, i.e., 1 twin individual in 42 mothers. Due to the higher infant mortality of twin children, the probability of finding a twin was calculated to be about 1 twin per 60 children in the population. Therefore twins were overrepresented in these data. This result agrees with the general experience that development of twins is slower [92, 94] and mental subnormality is more frequent than in the general population. The influence of this bias on concordance rate cannot be predicted. In any case, these considerations show that interpretation of empirical risk figures and twin concordances as well as heritability estimates derived from such data may be ambiguous for two reasons: genetic and environmental influences are correlated, and the development of twins differs from that of singletons. As mentioned repeatedly, the genetic model of multifactorial inheritance furnishes only a preliminary description of the genetic situation (Sect. 6.1). A more incisive analysis will be possible in the future for this mild, subcultural group as well, once specific genetic and environmental factors have been identified.

15.2.1.3 Intelligence and Performance in the Normal and Superior Ranges

Superior Achievement. The discussion above concentrates on the lower end of the variation, i.e., those who because of their low intellectual endowment have difficulties in adapting to the demands of society. Galton first reported on individuals whose achievement is regarded as superior by the standards of their society in his classical paper on "Hereditary Talent and Character" (1865) [64] and in the subsequent monograph on *Hereditary Genius* (1869). This work established the Galtonian paradigm of human genetics. He showed that men regarded as "eminent" in British society had many times more male close relatives in the eminent group than would be expected if the distribution of high achievement were random.

Since that time repeated attempts have been made to document the inheritance of genius or of special talents. For example, the pedigrees of famous artists and scientists, such as Bach, Darwin, Galton, and Bernoulli have been reported as evidence for hereditary talent, and comprehensive statistics have been published [105]. These accounts confound heredity and environment, and no specific statements regarding the role of genetic factors can be made.

Variability in the Normal Range: Nature of Intelligence. Many studies have been performed to determine the relative contributions of heredity and environment to behavior within the normal range. Success in life and contribution to human society clearly depend on a variety of factors that can roughly be classified as intelligence and personality. Individual differences in intelligence have long been a principal field of research in psychology. More recently, however, these studies have been criticized fiercely. To understand this controversy somewhat better, we should consider the history of intelligence testing and some modern concepts of intelligence [197].

Intelligence and Intelligence Testing. Mental subnormality and eminent performance have been defined using social criteria: subnormality as making an individual incapable of independent social adaptation and eminence as being recognized by a group of contemporary professional colleagues as one of the leading figures in a field [105]. The "Intelligence quotient", or mental age, had originally been introduced as a device for school assignment and for classifying degrees of mental subnormality. Thus, IQ represented a criterion for delimiting the normal range from mild subnormality in a more clearly defined but somewhat more arbitrary way than independent social adaptation. Indeed, Binet introduced these measurements for the purpose of such classifications. It is said that Binet himself was not happy with the direction which IQ research subsequently took [240]. Emphasis shifted to classifying normal individuals according to their IQ score. An important event that precipitated this development was World War I, which afforded American psychologists the opportunity to test large numbers of army recruits.

Such extensive testing required selection of tasks and questions that would force respondents to give scorable right or wrong answers. Because testers were not communicating directly with individual testees, the reasons for the wrong answers could not be analyzed. Qualitative differences would have to be ignored; scores became all important [240].

Interest in quantitative measurement continued after World War I, and the tests became more sophisticated. The modern intelligence test comprises items that examine ability to handle words and verbal concepts, abstractions, mathematical tasks, spatial visualization, and memory. Such tests have proven useful in practice; despite many criticisms about their meaning, operationally they give fairly reliable predictions about performance in school and university.

Success in a number of professional occupations such as architecture, engineering, science, and medicine requires an above-average IQ, while other occupations make less of an intellectual demand on their practitioners. Failure to reach the IQ result necessary for success in the professions has predictive value in establishing that such persons are not likely to succeed in professional schools and universities preparing for these professions. However, cognitive ability as assessed by IQ tests is certainly not the only criterion for professional success.

The older theoretical work centered mainly around the problem of whether intelligence is one basic ability that influences all single tasks, different as they may be, or whether different tasks demand different abilities. One method for examining such problems is factor analysis. The correlations between single score results were examined for correlation clusters that were believed to indicate such basic abilities. The results were consistent in showing relatively high correlations between single items, pointing to a common factor of "general intelligence" (g) that influences all test scores (Spearman). There is some agreement that in addition to this g factor, specific abilities are necessary for performing verbal and mathematical tasks and for space perception.

Growing Unease over Intelligence Testing Among Psychologists. In spite of its undisputed success in predicting school achievement, the method of measuring intelligence and the IQ concept has met with growing unease among psychologists. It is asserted, for example, that the test measures the ability for "puzzle solving," i.e., for solving problems that are uninteresting by themselves but require certain specific formal skills. These skills are stressed in all school systems [155]. They are also needed in many professional fields, for example, in engineering. The highest puzzle-solving skills are necessary for solution of problems in the physical and some biological and social sciences. Kuhn [115] has even described most science ("normal science") as consisting of puzzle solving (see "Introduction").

Many other problems in daily life, however, require "intelligent performance in natural situations," which may be defined as "responding appropriately in terms of one's long-range and short-range goals, given the actual facts of the situation as one discovers them" [197]. Such abilities are tapped less well by current IQ tests. African natives who have never experienced contacts to Western culture would thus react inappropriately to many test items [67].

New Approaches for a Better Understanding of Human Intelligence [197]. There are a number of approaches to explain how intelligent behavior develops and how individual differences come about. For example, what are the basic cognitive processes that enable us to cope with our environment in an intelligent way? How important are basic processes such as short- and long-term memory, or complex phenomena such as language? Can we learn about these processes by constructing computer programs for solving problems? Can we obtain more relevant information by observing individual behavior in natural environments [34]? Many investigators are now emphasizing the interactions of internal and external factors in the development of the individual; here Piaget's observations on children's gradual aquisition of logical concepts are influencing many psychologists. These attempts do not challenge the practical value of intelligence testing for predicting educational success; they do challenge its relevance for the comparison of groups.

Family and Twin Studies for Assessing the Genetic Contribution to Normal Variability of Intelligence. When methods for measuring intelligence became available, pioneers in this field attempted to determine whether and to what degree normal variability was influenced by heredity or environment. The *Zeitgeist* before and after World War I was very much influenced by the eugenics movement (Sect. 1.8). Galton had impressed on many scientists the idea that intelligence could be measured, and that statistical comparison between close relatives could help to solve the age-old problem of how nature and nurture cooperate in creating intelligence. Siemens [215, 216] showed in 1924 that MZ twins can be readily distinguished from DZ twins (Sect. 6.3); the twin method was quickly adopted and became the most cherished tool for this kind of research.

Success in Schools. The most readily available research materials were school grades; from the point of view of intelligence testing they are a good indicator since they usually show a fairly high correlation with IQ test results. Moreover, school grades may have even higher validity since teachers observe their students for a long time [107]. Many studies show similarities in school grades between students and their parents and sibs and are generally consistent with a model of multifactorial inheritance. However, the data could also be explained in a different way. Parents who were themselves successful at school provide more help for their children – either directly by advising them in their homework and offering rewards for success in school, or indirectly by providing more opportunities for exercising cognitive abilities.

Twin studies may help to distinguish environmental from genetic influences, but they suffer from the dif-

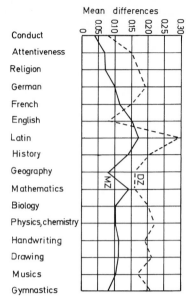

Fig. 15.8. Average differences in school grades among 60 MZ (32♂, 28♀) and 41 DZ (20♂, 21♀) DZ twins from Germany. (From Frischeisen-Köhler 1930 [60])

ficulties discussed in Sect. 6.3. Figure 15.8 shows the differences in school grades in 60 MZ and 41 DZ twin pairs from Germany [58]. The differences in MZ are only half as large as in the DZ twins. These results seem to indicate an appreciable genetic component in school performance. However, even at this relatively low level of sophistication, other results cast some doubt on this explanation. A Finnish study, for example, found a substantial difference in school performance between DZ and MZ twins only among males, not among females [107]. Did the teachers treat girls differently from boys?

Data such as school grades are not sufficiently critical. At the time when these studies were carried out, twins were usually kept together in the same class. Deceiving the teacher about their identity is a popular trick among MZ twins. Shall we really believe that their grades were always assigned independently? More objective methods are necessary.

Intelligence Tests in Families and Twins. Genetic studies with intelligence tests have been carried out on a very large scale. Bouchard and McGue (1981) [14] collected 111 investigations comprising 526 samples of individuals with different degrees of relationship, among them 47 series with parent-child comparisons, 71 comparisons between sibs, 41 series with DZ twins, and 37 series with MZ twins. These results are not discussed in any detail here, as critical analyses of recent years have uncovered errors in reporting, and many of the data were biased, mainly in favor of a

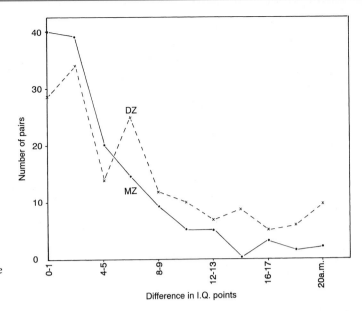

Fig. 15.9. Distribution of differences in IQ points between MZ and DZ twins from three series. Note that some MZ pairs show very large differences. (From Stocks 1930 [224])

high genetic component of the described variability [110, 238].

The lowest correlations were obtained in unrelated individuals and the highest in identical twins. The rank order (from lowest to highest) of these data was as follows: unrelated individuals < parents–adopted children < parents–children = sibs < DZ twins < MZ twins. This result is, of course, compatible with the hypothesis that the variability measured by the IQ test scores has a strong genetic component. The question is whether other hypotheses can be convincingly excluded. Before discussing various kinds of evidence we should examine the actual distribution of differences usually found in such studies between MZ and DZ twins. Figure 15.9 shows the pooled differences in IQ points from three studies. Twin pairs showing no difference are certainly more frequent among MZ than among DZ pairs. However, the number of MZ pairs that are quite different is by no means negligible. This finding shows that genetic factors cannot be exclusively responsible for the variability found in the population.

Heritability Estimates. The concept of heritability was introduced in Sect. 6.1 using another quantitative and measurable trait – stature – as an example. There the following observations are made:

a) Given certain assumptions, the theoretical correlations between relatives can be derived from Mendel's laws and the Hardy-Weinberg Law. Heritability can be analyzed by comparing these theoretical correlations with those empirically observed. Thus the heritability concept has a foundation in Mendelian genetics.

b) Heritability assessment is genetic analysis at the phenotypic-statistical level. The genes determining the measurable trait cannot be identified, and it is not possible to make any specific inferences regarding their number, mode of inheritance, or mode of action.

c) Even an exact quantitative assessment of heritability requires a number of assumptions that can seldom be tested using human data, such as random mating with respect to the trait examined and the absence of correlation or interaction between genetic and environmental influences. The methods may be refined by considering assortative mating or by including environmental – for example, socioeconomic – factors in the calculation. Including such factors, however, creates new problems. Appendix 5 deals with the problem of analyzing heritability from twin data.

d) An unbiased estimate of heritability from twin data is not possible theoretically. Three different means of analysis have been proposed: (i) comparing MZ with DZ twins (h_1^2), which may overestimate or underestimate heritability depending on some assumptions; (ii) comparing MZ twins with control pairs taken from the twin sample and matched by age and sex (h_2^2), which overestimates heritability since the correlation of environmental influences on MZ twins is neglected; and (iii) calculating h_3^2 from the intraclass correlation coefficient. The latter method is in most cases the least adequate. It may overestimate or underestimate heritability depending on assumptions regarding common environmental influences on MZ and DZ twins, biases due to the age distribution in the twin sample, differences in socioeconomic background of MZ and DZ twins, and others. This least adequate method has been almost exclusively the one which has been used for heritability estimates from twin studies.

c) Another possible shortcoming in the twin method is that it treats twins as an otherwise unbiased sample of the population for which conclusions are to be drawn. However, twin pregnancies as well as the twin situation during childhood and youth create special conditions that might bias the results. With these limitations, what conclusions can be drawn from existing data?

Twin Study on Swedish Conscripts. Husén (1960) [93] examined all Swedish male twins born between 1928 and 1933 who were inducted for military service between 1948 and 1952; there were 215 MZ pairs and 416 DZ pairs. A number of tasks common in standard intelligence tests were tested (such as finding synonyms, distinguishing concepts, and complementing of matrices; similar to the Raven test). For whole series the intraclass correlation coefficients for the tests used were 0.90 in MZ and 0.70 in DZ pairs. The test-retest reliability values were 0.92–0.93; This means that the difference between MZ twins in hardly larger than that between two successive examinations of the same individuals. Husén's paper does not provide heritability estimates, which we therefore supply (Table 15.4). At first glance the values seem very high; however, the correlations between DZ twins are also relatively high. This finding results in a large discrepancy between h_1^2, on the one hand, and h_2^2 and h_3^2, on the other, as defined in Sect. 6.3 and Appendix 5. Table 15.4 also contains recalculated heritability estimates based on data from series of twins reared apart (see below). Differences between the various heritability estimates are indeed amazing even considering the small sample sizes in these series. This indicates that the statistical assumptions (Appendix 5) on which such estimates are based may strongly influence the outcome. This is often neglected when such results are discussed. Most correlations between same-sexed sibs that have been reported in the literature are much lower. Another Swedish twin series by Husén [94] comprised 268 MZ and 360 DZ pairs of school age and used a series of school performance tests. The study found correlations between MZ and DZ of the same order of magnitude as those in the conscript study.

Twin Performance on IQ Test Is Lower Than That of Singletons. Both Swedish studies showed still another interesting result. The average test scores of twins were significantly lower than the scores of nontwins; among the conscripts this difference amounted to about one-fourth of one standard deviation; in the school children who were about 12–13 years old, the difference tended to be even larger. In Sect. 15.2.1.2 we noted that twins were found in unexpectedly high frequency among children attending special schools for backward children. This higher frequency is explained by a lower mean IQ level in twins. This is caused in part by biological influences, especially the intrauterine "transfusion

syndrome" of MZ twins (Sect. 6.3.4) since it can also be found when the co-twin had died [151]. However, in Husén's series, DZ twins still performed on average more poorly than MZ twins, contrary to expectations if the reason were predominantly biological in origin. A possible socioeconomic alternative is that the incidence of DZ births but not that of MZ births increases with the age of the mother and with parity. It is probable that not only somewhat older mothers but also mothers with higher fecundability have a higher probability of giving birth to DZ twins (Sect. 6.3.3). It is well known that at the time when Husén's twins were born, women from higher socioeconomic strata had fewer children than those from lower strata. The latter apparently practiced less birth control and relied more on their "natural fecundability," therefore, they had a higher probability of giving birth to DZ twins. As a result DZ twins tended to come from a somewhat lower socioeconomic group than MZ twins and the non-twin population. Whatever the reason, individuals from lower social strata tend to perform less highly on IQ and cognitive tests.

Is there a possible environmental explanation for the twin correlations? DZ twins, being of the same age, normally spend more time together and tend to be exposed to more similar environmental influences than sibs of different ages. Therefore, if they are more similar in their test results, these environmental influences should be of importance. However, we know from a number of studies [265] that MZ twins tend to have much closer relationships than DZ twins: "they are often eager to do the same things, to help each other, not to compete with each other, and to be similar in most aspects." If MZ twins really behave so differently from DZ twins, and if differences in behavior between DZ twins and the genetically equivalent sibs of unequal ages lead to remarkable differences in their test score correlations, should smaller differences in test scores of MZ compared with DZ twins not also be related to their peculiar situation? This question could be answered by finding an experimental design that separates the genetic influence from the special effect of the twin situation.

MZ Twins Reared Apart. In theory, the ideal probands for such studies are MZ twins who were separated immediately after birth, and who have been reared in different environments. The first observations were published in 1922 by Popenoe [180] and in 1925 by Muller [147]. In spite of their different environments, one twin pair, Jessie and Bessie, were very similar in intelligence, both achieving above-average scores. Their feeling tones and temperaments, on the other hand, differed, and this could plausibly be explained by their respective biographies.

Table 15.4. Heritability calculations and intraclass correlations in MZ and DZ twins reared together and reared apart

Reference	Psychological test variable (p)	No. of pairs (n)	SD of $2n$ individuals (σ)	Intrapair correlation (r_p)	Intrapair variance (V_P^w)	Heritability h_1^2	h_2^2	h_3^2	$\hat{r}_E^{(1)}$	$\hat{r}_E^{(2)}$
Husén (1960) [99]	IQ (see text)	MZT 215	30.48	0.894	98.11	–	0.675	0.382	–	–
		DZT 416	31.90	0.703	302.32	–	± 0.039	± 0.057	–	–
Newman, et al. (1937) [159]	Binet mental age	MZS 19	23.55	0.637	201.1	0.637	0.679	0.182	0.785	0.662
		MZT 50	29.5	0.922	67.9[a]	± 0.136	± 0.099	± 0.097	± 0.100	± 0.171
		DZT 50	35.4	0.831	211.8[a]					
	Binet IQ	MZS 19	13.00	0.670	55.84	0.670	0.697	0.540	0.727	0.518
		MZT 50	17.3	0.910	26.9[a]	± 0.126	± 0.093	± 0.174	± 0.128	± 0.244
		DZT 50	15.7	0.640	88.7[a]					
	Otis score	MZT 50	20.7	0.947	22.7[a]	–	0.750	0.294	–	–
		DZT 50	21.3	0.800	90.7[a]	–	± 0.077	± 0.106	–	–
	Otis IQ	MZS 19	13.58	0.727	50.42	0.727	0.789	0.602	0.714	0.603
		MZT 50	16.0	0.922	20.00[a]	± 0.108	± 0.065	± 0.179	± 0.137	± 0.201
		DZT 50	15.8	0.621	94.6[a]					
	Stanford educat. age	MZS 19	23.47	0.502	274.1	0.502	0.657	0.144	0.910	0.847
		MZT 50	30.5	0.955	41.9[a]	± 0.172	± 0.106	± 0.067	± 0.040	± 0.077
		DZT 50	32.3	0.883	122.1[a]					
Shields (1962) [213]	Dominoes intell. test	MZS 37	9.02	0.758	19.68	0.758	–	–	−0.095	0.065
		MZT 34	8.33	0.735	18.40	± 0.070	–	–	± 0.454	± 0.352
	Mill Hill vocabul. scale	MZS 38	5.75	0.741	8.566	0.741	–	–	0.004	0.522
		MZT 36	3.98	0.742	4.097	± 0.073	–	–	± 0.403	± 0.176

MZS = Monozygotic twins separated;

MZT = Monozygotic twins ⎱ reared up together.
DZT = Dizygotic twins ⎰

[a] Recalculated by the formula $V_P^w = \sigma^2(1 - r_p)$.
[b] See App. 5.

$h_1^2 = r_{P,\text{MZS}}$ (unbiased estimate of h^2 if $r_{E,\text{MZS}} = 0$).

$h_2^2 = 1 - V_P^w(\text{MZT})/V_P^w(\text{DZT})$,   SE (h_2^2) calculated by Eq. A 5.14[b].

$h_3^2 = 2(r_{P,\text{MZT}} - r_{P,\text{DZT}})$,   SE (h_3^2) calculated by Eq. A 5.16[b].

$\hat{r}_E^{(1)} = (r_{\text{MZT}} - r_{\text{MZS}})/(1 - r_{\text{MZS}})$ (estimate of $r_{E,\text{MZT}}$ if $r_{E,\text{MZS}} = 0$; see Eq. 5.2[b]).

$$\text{S.E.}(r_E^{(1)}) = \frac{1 - r_{P,\text{MZT}}}{1 - r_{P,\text{MZS}}} \sqrt{\frac{(1 + r_{P,\text{MZT}})^2}{n_{\text{MZT}}} + \frac{(1 + r_{P,\text{MZS}})^2}{n_{\text{MZS}}}} \text{ (from Newman et al. (1937) [159]).}$$

$\hat{r}_E^{(2)} = 1 - V_P^w(\text{MZT})/V_P^w(\text{MZS})$ (estimate of $r_{E,\text{MZT}}$ if $r_{E,\text{MZS}} = 0$; see Eq. 5.9[b]).

SE $(\hat{r}_E^{(2)})$ calculated by Eq. 5.14[b].

It is obviously very difficult to find such twin pairs, but four series of cases are available in which an attempt to ascertain relatively unbiased twin series was successful: Newman et al. (1937) [159] in the United States, Shields (1962) [213] in the United Kingdom, Juel-Nielsen (1965) [106] in Denmark, and Bouchard et al. (1990) [16] in the United States. The older literature usually cites an additional series, that of Burt. There is now evidence that these data appear to have been fabricated by the author.

Newman et al. compared 19 MZ pairs who had often been separated during infancy, but never after the age of 6, with 50 MZ and 50 DZ pairs reared together. The ages at the time of examination was 11–59 years. Every pair was given a careful biological and psychological examination. Table 15.4 compares heritability estimates with those of two other studies, they were recalculated from the originals. The heritabilities in each of the series are high, although not as high as for MZ twins reared together. What are the reasons for the differences? Is it possible that they are correlated with known environmental factors?

In Newman's study, the differences between the conditions in which the twins had been reared were estimated by five observers and classified by separate scoring scales with a maximum of 50 points each for educational, socioeconomic, and physical (i. e., health) advantages. The twin who stayed in school longer and generally had a better educational score also tended to have better test scores. The correlation between education and IQ scores (0.79) was significant. A lower correlation (0.51) was found with social background and a still lower one (0.3) with health status.

One case report illustrates possible environmental differences. Alice and Olive were born in London and were separated at the age of 18 months. Alice's foster parents lived in London; Olive was adopted by relatives who lived in a small Canadian town. The separation lasted to the age of 18; examination took place about 1 year later. Alice's foster parents were classified as lower-middle class; they had four daughters of their own who were much older than Alice. She went to school until the age of 14, took an 18-month business training course, and started working in business offices. Her parents were not able to provide her with much care; due to World War I the quality of her school education was rather poor. Olive, on the other hand, grew up as the only child in a family that was better off than Alice's. She was spoiled by her parents; she attended grade school, took a 2-year commercial course equivalent to high school, and worked as an office clerk, as her sister. There was a significant IQ difference, Alice having a value of 84.9, whereas Olive scored 96.9. The sister who had lower educational opportunities thus had a definitely lower score. Alice and Olive were very similar in temperament, although Olive was more active and domineering. In a separate chapter of their book, Newman et al. explain part of the IQ difference by the fact that the test was designed for American students and was therefore "not quite fair for an English girl."

The second study (Shields 1962 [213]) used only two short tests, one verbal and one nonverbal synonyms section (set A) of the Mill Hill vocabulary scale and the (nonverbal) Dominoes test. The calculation of heritability (Appendix 5) was impaired by lack of control DZ pairs; the other data have been recalculated.

The 12 twin pairs observed by Juel-Nielsen (1965) [106] were between 22 and 77 years of age at the time of investigation. Their ages at separation ranged between 1 day and $5^3/_4$ years. They were tested extensively using the Wechsler-Bellevue (W-B) intelligence scale and Ravens' progressive matrices. On both tests a marked similarity was found between the test scores achieved by the twin partners. The correlations coefficients between the co-twins' scores for W-B values were as follows: total IQ, 0.62; verbal IQ, 0.78; performance IQ, 0.49; Raven raw scores, 0.79. As explained in Appendix 5, these intraclass correlation coefficients are heritability estimates and, more specifically, estimates of h_3^2. Intelligence testing was repeated on nine pairs; the time between the two examinations varied, the average being 12 months and minimum of 6 months. The correlation between the first and second test scores, i.e., the test-retest reliability, was smaller than the difference between co-twins. This means that the study revealed real differences in test intelligence between twins of MZ pairs.

Due to an unfortunate shortage of information on DZ controls – such controls are scanty in the Shields study and absent from the Juel-Nielsen study – heritability was estimated mainly from the intraclass correlation coefficients (h_3^2). As explained in Appendix 5, this may introduce a bias in favor of a spuriously high h^2 estimate if the MZ sample comprises pairs of different ages, and if the characteristic to be tested is age dependent. In theory, intelligence tests are standardized for age, and test scores should be age independent. In practice, however, this is not quite true; in addition, the age differences in each sample of MZ twins reared apart were unusually high.

Each of these studies carried out intelligence testing only as part of a more comprehensive assessment of personality development in relation to environmental differences. In general, striking similarities despite these differences are stressed by the authors; the observed differences could often plausibly be accounted for by corresponding differences in educational opportunities and environmental influences in general.

A recent (and as yet incompletely published) study examined 56 pairs of MZ and 30 DZ pairs who had been separated early in life (Bouchard et al. 1990 [16]). At the time of first examination the average age of MZ pairs was 41 years (range 19–68). In addition to interviews and a medical assessment, the twins were examined using numerous psychological methods, including intelligence tests and personality questionnaires. So far the overall impression is that of surprising similarity. On IQ, for example, a correlation coefficient of 0.7 has been reported, consistent with correlations found in twins reared together. Similarities in other scores such as those measuring information processing and aspects of personality were found to be of the same order of magnitude as in MZ twins reared together. Hence, the results of this very comprehensive and careful study are in line with those of previous investigations.

Overall Results of Studies on MZ Pairs Reared Apart. Altogether, MZ twins reared apart show a remarkable degree of similarity in intellectual performance, not only during childhood and youth but in mature life as well. It is greater than that normally observed between sibs and even between DZ twins reared together. Differences in social status during childhood, educational opportunities, and later experiences influence intellectual abilities to a certain degree and in some cases lead to appreciable differences. On the whole, however, the impression of similarity prevails.

In view of some recent criticism of twin studies assessing intellectual abilities, it should be remembered that the authors of each of the above investigations were keenly aware of the limitations of their studies. Moreover, in their papers we find no preoccupation whatsoever with heritability measurements or with a numerical estimate of the relative contributions of heredity and environment on the variance of test scores. The Figures in Table 15.4 are ours, not theirs. The study of Bouchard et al. [16] examined the correlations between scores of MZ twins reared apart and of those reared together.

Another study [178] studied 223 same-sex twin pairs in Sweden at an average age of 64 years and again 3 years later (34 MZ and 78 DZ reared apart, 48 MZ and 63 DZ reared together). The study was performed between 1986 and 1991. Cognitive abilities were assessed in great detail; heritabilities were approx. 0.8, and there were no differences between twins reared apart and together. Comparing this result with heritability estimates from other twin studies, the authors concluded that the similarity in cognitive abilities tends to increase with age. This conclusion, however, should be regarded very cautiously. The Swedish population has enjoyed peaceful, stable, and comfortable living conditions for many decades; under these conditions major differences in living conditions might have been rare even for twins reared apart.

In our opinion, *the results of these studies support the view that in present white populations and under the environmental conditions of Western society in the twentieth century, genetic variability is responsible for an appreciable part of the variance in test performance on the usual intelligence tests. We make these statements realizing a variety of criticisms* (see also [55, 238]). These include:

a) The sampling process by which the twin pairs of these studies [16, 159, 178, 213] were ascertained may have created a bias in favor of similarity between the twins and may, on the other hand, have induced them to exaggerate the degree of separation in the information provided to the investigators.

b) Some of the twin pairs included in these studies were separated so late in life that common influences during infancy may have helped to shape their attitudes.

c) The homes in which they grew up were generally more similar than random homes.

d) Some of the twins may have met during childhood, for example, at school, so that the known mechanisms of mutual identification may occasionally have come into play.

In view of these criticisms, some observers still deny any heritability of IQ [110]. When one considers only single studies of this sort, such skepticism may be warranted. We are impressed, however, by many different lines of evidence suggesting that some determinants of IQ may have a genetic basis. It is quite clear that the available evidence does not permit exact measurement of the degree to which intellectual abilities in these populations is determined by genetic as opposed to nongenetic influences. One argument against generalizating such results from twin studies is always that twins do not represent a random sample of the general population. Therefore other approaches are needed.

Studies on Adopted and Foster Children. The obvious alternative to avoid the problem of a correlation between genetic and environmental factors is to conduct studies on foster children and adopted children. For children living in foster homes the environmental influences are presumably randomized, leaving only the genetic variability of parents. In adopted children and children living with foster parents the genetic influence of the biological parents can be compared with the environmental influence of the adopting parents [127].

The evidence from studies on adopted and foster children with foster parents was reviewed by Loehlin in 1980 [127]. The earliest studies were performed as long ago as in the 1920s and 1930s. When the data were pooled they gave substantially lower correlations in IQ between adopted children and their adoptive parents than between biological children and the same parents. These studies were criticized for various reasons, and new studies were carried out in the 1970s. Three investigations carried out in the United States included more than 500 adoptive families, with nearly 800 adopted and about 550 biological children. All three yielded substantially the same result: Correlations between adopted children and adoptive parents were lower than those between these parents and their biological children (Table 15.5). In general the correlations were somewhat lower than in earlier studies; this was explained by a more restricted range of IQ values in foster families.

One study compared adopted children with their biological mothers, from whom they had been separated as newborns. The correlation was surprisingly high (0.32), pointing to a substantial genetic contribution.

Studies on adopted and foster children are subject to a number of unavoidable biases. On the one hand, adoption is definitely not carried out at random. Adoption agencies have the understandable tendency to place children in "suitable" homes. Therefore a

Table 15.5. Parent-child and sib-sib correlations in IQ in three recent adoption studies (from Loehlin 1980 [127])

Study	Adoptive children		Biological children		Biological mothers and their children adopted by other parents	Sibs in adoptive families	
	Fathers	Mothers	Fathers	Mothers		Biological-adoptive	Biological-biological
Minnesota I [202]	0.15	0.23	0.39	0.35		0.39[a], 0.30[b]	0.42
Minnesota II [202]	0.16	0.09	0.40	0.41		−0.03	0.35
Texas [127]	0.17	0.19	0.42	0.23	0.32	0.22[a], 0.29[b]	0.35

[a] Adopted-adopted, [b] Correlation between adopted and biological "sibs."

correlation between foster children and foster parents is to be expected. On the other hand, some of these children spent the first year(s) of their lives with their biological mothers, reducing the environmental influence of the foster parents. Moreover, couples willing to adopt children are not an unbiased sample of all parents in the population; their mean IQ is not only higher, but there is also a smaller variance [55]. This may lead to a lower correlation as a statistical artifact [65, 127].

On the whole, the limited evidence derived from all adoption studies may be summarized as follows. *Both genetic and environmental factors influence the intellectual development of the child. Due to unavoidable biases in the examined series, however, the relative contribution of these influences cannot be estimated.*

In view of substantial expenditure in time and money on such studies, this conclusion is certainly disappointing. Current knowledge might well have led one to predict such a conclusion in advance. Although these investigations do not answer more specific questions, they are valuable in providing possibly useful information on social and psychological aspects of adoption.

Two different attempts have been made in recent years to overcome some of the methodological problems and to elucidate the nature-nurture problem in IQ research by applying sophisticated statistical techniques and by including relatives of many degrees and with different degrees of similarity of the environment in the models. The Birmingham school used analysis of variance [102], and the Hawaii school applied the principles of path analysis [141, 142]. Both attempts have been severely criticized [238] because of the many assumptions that had to be made by each group of investigators.

Some Environmental Influences on Intelligence. Another quite different approach examined the influence on IQ, test performance of the position of sibs within their respective sibships

[274]. There was a negative correlation between intelligence (as measured by a standard qualification test) and position in birth order among those aged about 17 years; those born later had lower test scores. This result was not due to the effect of confounding variables such as socioeconomic status. Obviously this result cannot be due to genetic influence. Further support for environmental influences on IQ is the substantial increase in scores over recent decades [225]. In additional to improved educational conditions and environmental stimulation, better nutrition and less malnutrition, especially during childhood, should be considered a possible cause. As noted in Sect. 6.3.10, mean height has increased over the same period, and mean age of menarche has declined – presumably due to improved nutrition.

What Does the Available Evidence Show Regarding Genetic Variability of Intelligence in the Normal Range? The answer is short: very little. With some stretching of the evidence for biases in sampling, statistical evaluation, reporting, and interpretation of ambiguous results, and with some not too implausible assumptions regarding environmental influences of biological and adoptive parents or the way in which MZ twins influence one another, it has even been asserted that genetic variability does not influence IQ performance at all [116], i. e., that heritability is zero or very low [238].

This conclusion does not mean that a zero heritability is the most likely answer. Interpretation of the same data with different assumptions about biases has led other authors to the conclusion that heritability of IQ performance is as high as 0.8 [100, 101]. Most scientists, if asked for an educated guess, would probably settle on values somewhere inbetween, more because they dislike extreme points of view than because of a strong conviction in favor of any positive evidence.

If the same data lend themselves to such diametrically opposite interpretations, the actual evidence must be "soft." As the foregoing discussion has shown, clean experimental designs in humans are in most cases unattainable. Genetic and environmental influences in families are correlated; attempts to separate the two factors by examining adopted children suffer from nonrandomness of adoption. MZ twins living together influence one another in unpredictable ways that change with the cultural setting. MZ twins reared apart are often educated in homes with similar backgrounds. MZ twins tend to show differences in development due to peculiarities of twin pregnancy.

The entire approach of measuring intellectual ability and determining the genetic and environmental contributions to its variance leaves us unsatisfied. Why?

As geneticists, we are interested ultimately in analysis at the level of the individual gene. The absence of such data highlights the contrast between the two paradigms of human genetics: the biometric paradigm of F. Galton (1865, Sect. 1.3) and the gene concept of G. Mendel (1865, Sect. 1.4). These two paradigms are discussed in Sect. 6.1. The genetic analysis of intelligence was long regarded by many research workers as the field in which Galton's paradigm scored its most impressive victories, while analysis on the lines suggested by Mendel seemed to be doomed to failure. The discussion in recent years that started with criticism of Jensen's work on group differences and educability [100] has exposed the weaknesses of the biometric paradigm so mercilessly that one can hardly imagine how it should survive. On the other hand, research on genetic variability in other, more easily accessible fields of human genetics, as well as in population genetics of other species, has revealed an amazing amount of genetic variability in populations (Sect. 12.1). For example, no less than one-third of blood enzymes examined so far show genetic polymorphisms, and the normal variants usually show slight functional differences within the normal range [78]. Studies on the genetic basis of common diseases, as well as the recent progress in pharmaco-and ecogenetics, are now increasingly revealing influences of such normal genetic variability on the health status of the individual under varying environmental conditions. It is our contention that the genetic variability of biological factors influencing intelligence and other aspects of human behavior is likely to be just as extensive. However, its phenotypic expression may be more complex than in somatic traits. The ambiguity of test results in revealing such genetic variability may – at least in part – be caused by the insufficiency of test methods rather than by the weakness of genetic influences on intelligence. But can a Mendelian approach replace current efforts? This problem is deferred to a later section.

Paradigm Clash also in Psychology. The phenomenological and biometric approach associated with Galton has by no means been undisputed among psychologists concerned with problems of genetic variability of human behavior in general and intelligence in particular. On the contrary, its shortcomings have been explained for decades, and repeated attempts have been made to identify specific determinants of intelligence [71]. Such attempts have remained only partially successful (see, for example, attempts at relating intelligence to speed of information uptake [120, 201a]), and considerably more work with attention to such components using a variety of approaches is sorely needed.

Research on IQ and Politics. Adherents of a paradigm frequently form a group that also shares certain convictions that are not immediately connected with the scientific content of this paradigm. Sometimes they even try to separate themselves socially from members of competitive groups. From more general experience in social psychology it is not very surprising that such tendencies lead to identify one's own group as the "good guys" and condemn those of the other group as the "bad guys" – reactionary, politically authoritarian, antisocial, and altogether wicked. This attitude leads to proposals to curb certain kinds of research – and, at a slightly less "academic" level – to threatening certain investigators physically or even destroying their research materials.

Although we agree scientifically with much of the criticism of the biometric approach, we wish to stress that we are fiercely opposed to such politics. The work in theoretical and empirical analysis that has been carried out in recent years in the context of the biometric paradigm [101] has been useful. It has explored the inherent possibilities of this approach and – largely inadvertently – exposed its limitations.

15.2.1.4 Special Cognitive Abilities and Personality

Special Cognitive Abilities. Most of the debate regarding intelligence is centered around the overall outcome of the intelligence test – the IQ. However, most modern tests consist of a number of subtests. Scores on these subtests are positively correlated in the same individuals. However, this correlation is far from complete; the subtests measure partially independent abilities. For example, the Wechsler test consists of verbal and performance subtests. One special ability that shows a relatively low correlation with the other scores is that of visualizing spatial relationships among objects (Sect. 15.2.2). It is not surprising that twin and family studies have also been carried out using such subtests [45]. So far, however, no consistent differences in heritability between such subtests have emerged. The hope that this approach would reveal more basic abilities that are not subject to the problems of heredity-environment interaction has not been fulfilled.

Intelligence Is Not Everything. Almost since the time that the IQ test first became popular it has been obvious that for achievement in school and university, and still more so in professional fields, other qualities are required in addition to intellectual ability. Terman et al. [239, 240] (see also [166]) performed a longitudinal study on individuals selected at school age as especially gifted. Many of them were unusually successful professionally, but some were not; in the unsuccessful group personal instability was much more frequent than in the successful group.

Twin Data for Temperament, Sensory and Motor Functions, and Personality. Abundant literature is available on twin studies regarding temperament, sensory and motor functions, and personality. Most of the methods that psychologists generally use have also been applied to twins in the search for genetic variability in these parameters. Almost all of these studies had the same result: MZ are more similar than DZ twins [13, 62, 131, 249]. A number of attempts have been made to classify the observed variability, the best known of which are those of Cattell and his school [31–33] and those of Eysenck [52–54] using the method of factor analysis. From correlation clusters of a great number of single test scores – such as

questions in a personality questionnaire – Cattell isolated a number of factors that were interpreted as basic dimensions of personality. Eysenck [52], also on the basis of personality questionnaires, distinguished three main dimensions of personality: extraversion, neuroticism, and psychoticism. Such studies were useful for conceptualization in psychology, and Eysenck's studies helped to focus attention on the relationship between behavior and brain function. So far, however, no definite and specific conclusions on genetic mechanisms have emerged.

Longitudinal Study of Twins. The most comprehensive picture of the effect of different environments on genetically identical individuals is the prospective study of the life histories of MZ as compared with DZ twins. Obviously, the investigator who plans such studies must begin early in his or her career and must occupy himself or herself with this project again and again over an entire lifetime. Few investigators are prepared to do this. However, one twin study was carried out for over 40 years by the same investigator, and since his death has been continued by others [71, 72]. (Technical terms used here to describe personality characteristics, such as mental responsiveness, are those indicated to us by the team of psychologists involved in these ongoing studies; F. E. Weinert, personal communication.)

Gottschaldt and coworkers organized two holiday camps for twins on the island of Norderney, Germany, in 1936 and 1937. In these camps they examined 136 twin pairs aged 4–18 years for several weeks. The program consisted of a great number of systematic observations and tests that included not only intelligence tests but also tests of sensory and motor ability and behavior in more complex situations, such as searching for hidden objects. The investigators took great care to avoid the typical "test situation" and to create a socially relaxed climate between twins and examiners. A diary was kept on each participant. This approach comes close to the controlled observation recently suggested by the ethnologist Charlesworth [34] for research on intelligence. It also enabled the investigators to assign scores for qualities such as drive, mental responsiveness, and feeling tone (Fig. 15.10).

Data obtained by comparing the ability to learn by experience were of interest. Pairs of colored figures that occurred in changing orders had to be connected by pencil. After 50 s, the experiment was stopped and the number of connected pairs counted. The experiment was carried out twice a day for 5 weeks. It produced typical learning curves, with an initial increase and a plateau that was reached after different times; this speed in increase of performance differed characteristically between MZ and DZ twins (Fig. 15.11). The test

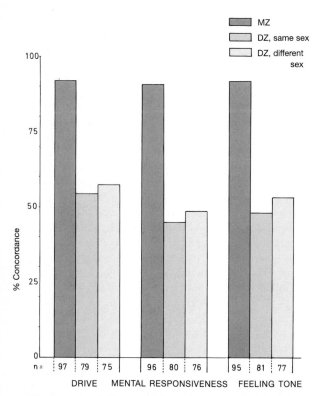

Fig. 15.10. Concordance and discordance in ratings for drive, mental responsiveness, and feeling tone in juvenile MZ pairs, DZ twin pairs of same sex, and DZ pairs of different sex; observations in holiday camps. (From Geyer and Gottschaldt 1939; see [71])

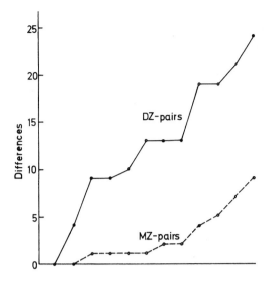

Fig. 15.11. Difference (in minutes) between MZ and DZ twin pairs in learning speed (reaching of a certain learning level in a 50-min experiment). Each point represents one twin pair; see text for further details. (From Wilde, cited in Gottschaldt 1939 [71])

Table 15.6. Results from a longitudinal study of 20 MZ twin pairs (from Gottschaldt 1968 [72])

Mean age	1937 11.7 years	Early 1950s 23.3 years (20.3–31)	1968 41.5 years (34.9–46)
Capacity for information uptake	+ +	+ +	+ +
Abstract thinking	+ +	+ +	+ +
Mental attitudes (range of interest; appraisal of own situation)	+ +	(+)	–
Internal mood	+ +	+	–
Mental responsiveness	+ +	+	–
Control	+ +	+	–
Volitional Mind set	+ +	–	–

+ +, Strong concordance; +, weaker concordance; (+), doubtful concordance; —, no concordance.

results confirmed in all cases the greater similarity of MZ as compared with DZ twins.

The investigators used many tests of their own construction, thus making comparability of their data with other studies difficult. Their approach, however, gives a much more comprehensive, if less well-quantitated, picture of the development of MZ versus DZ twins. Unfortunately, such results do not permit disentanglement of the effects of the special environmental conditions from genetic effects.

However, some of these twins could be re-examined three times: in the 1930s, the early 1950s, and the late 1960s; a fourth round of examination is now in progress. During the course of the study the upheavals of World War II led to thorough changes in life-style, and the ongoing study provided an excellent opportunity to examine the behavior of MZ twins under such conditions.

Table 15.6 gives an overview of the results of three examinations. Two groups of abilities usually considered important as the basis of formal intelligence, i.e., capacity of information uptake and abstract thinking, remained concordant well into middle age. Some pairs became more different, others more similar. A certain concordance up to their 20s is found in their internal mood; here definite differences became apparent in middle age, and these could plausibly be correlated with life histories and life experience. For the other characteristics – volitional control and level of aspirations, mental attitudes, and range of interest – the striking concordances that were seen in childhood tended to weaken in their 20s and to disappear when they reached middle age [72].

The examination of MZ twins in adult age revealed striking differences. These were visible not in the abilities normally described as "intelligence" but in mental attitudes, mind set, volitional control, and internal mood. These discordances were confirmed partially in another longitudinal twin study in Germany that was carried out at about the same time and on twins that were somewhat older when the war ended in 1945 [266]. This study distinguished three classes of MZ twins on the basis of their life experiences: (a) A class of especially gifted and successful twins who were relatively discordant for personality and life experiences; apparently such individuals had a wide range of choices among different lifestyles. (b) The largest group, with about average intelligence and performance, remained relatively concordant for personality and life experiences up to middle age and in some cases until old age. Their choices in life were more restricted by convention. (c) A class of twins with below-average success turned out to be rather discordant for personality and life experiences. In this class the course of their lives seems to have been determined largely by accident.

One could argue that these two studies give an exaggerated picture of the possible differences in psychological development of MZ twins and hence of the plasticity of personality characteristics in the face of differing environmental influences. World War II and its aftermath brought changes in life-style to these study population, including displacement from homes and occupations, interruption of education, and loss of close relatives. Whether and how a normal life could be reestablished was often a matter of luck. Very probably, higher concordances between twins would have been maintained had the living conditions been more normal. The newer studies mentioned above on adult MZ twins reared apart [16] who grew up in the relatively regular and comfortable conditions of the contemporary United States appear to show many more similarities at various psychological levels even in middle age. For example, small idiosyncratic gestures such as frequent voluntary coughing, body postures, body speed, and other gestures were very similar. More recently a longitudinal study on somatic and psychological development of twins has started in Louisville, USA. Here the twins are being examined at relatively short time intervals from birth onward [272]. During the first 2 years of their lives MZ twins, but to a surprisingly high degree also DZ, were found to be highly concordant both for mental development and for the spurts and lags in development.

Possible Consequences for Educational Policy. Educational thinking is traditionally centered around children of school age and young persons. However, twin investigations have shown increasing differences in mental attitudes, volitional control, and mind set between MZ twins in mature life. These findings suggest that there are great potentials for mental and emotional development that our society does not utilize sufficiently. Would it not be a good idea to spend more money and ingenuity on educational opportunities for adults? Could this help them achieve more fulfillment and happiness? Could the society as a whole overcome some of its future problems more efficiently by such approaches? Recent trends in adult education, which started in the United States to occupy the educational establishment in the face of declining enrollments of young persons because of the falling birth rate, will be observed with much interest.

15.2.1.5 Behavior and the Genetics of Human Sensation

Genetics of Scent Differences ("Olfactogenetics") [9]. An interesting aspect of human individuality concerns differences in odor. It has been stated that police dogs are unable to differentiate the scents of identical twins. Presumably the characteristic scent of an individual is under the control of genes that determine the secretion of chemical substances from the skin. These substances give off a distinct scent for each individual. Scents are undoubtedly also influenced by

diet and by the bacteria of the skin. However, it appears likely that the microbial flora of the skin depends largely on the genetically determined biochemical makeup of the various compounds secreted in dermal sweat. It has been shown in mice that the ability to sniff differences in odor depends upon the $H2$ locus – the analog of the HLA complex in man [9]. In these ingenious experiments mice that were genetically identical except for differences at the $H2$ locus were able to detect scent differences between each other. These studies also showed that mice prefer mating partners that differ at the $H2$ locus, which is an interesting evolutionary strategy to promote genetic diversity. It appears that the $H2$ locus (and possibly the human *HLA* locus) is a major genetic determinant that imparts a characteristic smell to each individual. The pathway from genotype to phenotype is unknown. HLA studies in various types of poorly delineated human anosmias (not associated with nervous system diseases) such as anosmia for isobutyric acid (207000) and anosmia for cyanide (304300) would be of interest. It has also been claimed that the supposed presence or absence of the odor of urinary asparagine metabolites is caused by the fact that some persons are unable to smell a urinary metabolite that is always present after eating asparagus [124]. It was previously thought that there was a polymorphism in urinary excretion of a metabolite (108400). Molecular studies on the olfactory system of the rat are begining to elucidate the molecular basis of olfaction [25]. Figure 15.12 shows the olfactory neuroepithelium of a mammal. Sensory neurons have receptors that bind with various "smelling" substances and send specific impulses to the brain. Figure 15.12 B depicts a part of this pathway. The receptors are coded by a great number, probably hundreds, of gene receptors belonging to a large gene family. In any one sensory neuron only a few (approximately 25) of these genes are expressed; the olfactory epithelium of one individual comprises a great number of such neurons with different specificities, which are caused by differences in their amino acid sequences. The arrangement of this system is reminiscent of the immune response (Sect. 7.4). In both instances a specific reaction to a great variety of different impacts from the external world is required; the challenge is met in a similar way: by multiplication and diversification of genes. The olfactory system is one of an increasing number of examples in which a genetic mechanism has been elucidated not by analysis of genetic variants but by direct access to specific functional cells.

Visual Perception. The human eye is among the best studied organs. Many hereditary diseases are known involving either the eye alone or the eye within the context of a more complex pleiotropic pattern. This

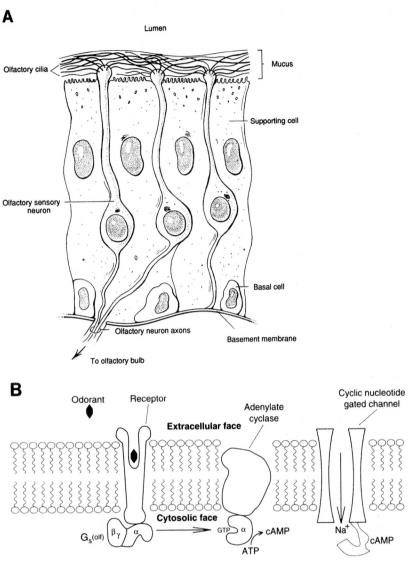

Fig. 15.12 a, b. The olfactory neuroepithelium and a pathway for olfactory signal transduction. **a** The olfactory neuroepithelium. The initial events in odor perception occur in the nasal cavity in a specialized neuroepithelium that is diagrammed here. Odors are believed to interact with specific receptors on the cilia of olfactory sensory neurons. The signals generated by these initial binding events are propagated by olfactory neuron axons to the olfactory bulb. **b** A pathway of olfactory signal transduction. In this scheme, the binding of an odorant molecule to an odor-specific transmembrane receptor leads to the interaction of the receptor with a GTP-binding protein $G_s(olf)$. This interaction in turn leads to release of the GTP-coupled α subunit of the G protein, which then stimulates adenylyl cyclase to produce elevated levels of cAMP. The increase in cAMP opens cyclic nucleotide-gated cation channels, thus causing an alteration in membrane potential. (From Buck and Axel 1991 [25])

reflects the complex embryonic development of the eye, which is determined by many genes and therefore gives rise to many different possibilities for disturbance. The invention of the ophthalmoscope by Helmholtz in the middle of the nineteenth century made possible a thorough study of the inner eye. Some ophthalmologists such as Waardenburg and Franceschetti were early pioneers in describing and classifying hereditary eye diseases. Discovery of color vision defects and their X-linked mode of inheritance also are among the early achievements of human genetics. Hereditary eye diseases, however, are not discussed here. We confine ourselves to genetic and molecular aspects of the primary light receptors, the cones and rods in the retina, and their defects.

Genetics of Color Vision (see also [143, 152, 153]. The cones and rods have different functions in light perception. The rods function principally in dim light when light intensity is low and are found mainly in the periphery of the retina. The cones, on the other hand, are concentrated more in its center. There are three types of cones: those sensitive to short waves, medium waves, and long waves, often referred to as blue, green, and red cones, respectively. These have overlapping spectral ranges, each with a peak of maximum absorption (Fig. 15.13). Each cone type is unique by its conversion of light of a certain wave length into a subjective color experience in the brain. The gene for blue pigment is autosomal on chromosome 7q31; the red and green genes are located as a gene complex near the tip of the long arm of the X chromosome (q28).

The color vision pigment genes and rhodopsin are evolutionarily related. An ancestral gene coded for the precursor gene of several G protein coupled receptors including olfactory, β-adrenergic, serotonin, and muscarinic acetylcholine receptors. Very early in evolution this ancestral gene developed into the rod pigment gene specifying rhodopsin (see below) and a single cone pigment gene (Fig. 15.14). Some 500 million years ago the blue gene and a single middle wave sensitive, undifferentiated (green/red) pigment gene diverged; vision at this stage was dichromatic. About 30 million years ago the red and green pigment genes differentiated by duplication and allowed trichromatic color vision, which is shared by humans and Old World monkeys. These evolutionary events are reflected in the similarities of various visual pigment genes. Green and blue pigment genes share 44% of their amino acids while green and red pigment genes are almost identical (98% homology).

Every human male has a single red gene, but there is a numerical polymorphism for the green genes, and individuals with one, two, three, or more green genes exist [46a]. The red gene is located upstream (5′) from the green gene(s). If there are several green genes, it appears that only the most proximal (5′) one is expressed in the retina [272a]. The close homology of the red and green pigment genes make them prone to illegitimate recombination and causes either deletions of the green gene or leads to the production of green-red or red-green fusion genes (Fig. 15.15). The common color vision defects in the vast majority of cases are therefore caused by such recombinational rearrangements. Point mutations as the cause of color vision defects are quite rare [272b].

There are two principal classes of common color vision abnormalities: severe dichromatic and milder trichromatic defects. In dichromatic color vision the cones for either red color perception (protanopia) or green color perception (deuteranopia) are functionally absent, and color vision is mediated by two rather than by three cone types. In trichromatic defects all three pigments (blue, green, and red) are active, but there is a slight shift in the absorption maximum of

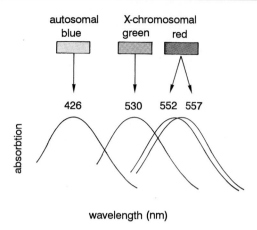

Fig. 15.13. Absorption spectra of visual pigments with maxima of 426 nm for blue, 530 for green, and 552 and 557 nm for the two types of red pigment (Ala/Ser polymorphism). Note overlap between range of absorption spectra for the different pigments. The blue pigment is specified by an autosomal (7q31) gene while the green and red pigments are specified by a X-linked gene (Xq28). (From Passarge 1995 [170])

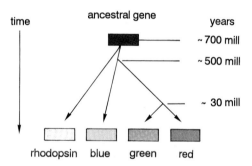

Fig. 15.14. Evolution of visual pigments. Rhodopsin and the blue-green-red pigments share a common ancestor. Differentiation of red and green pigments occurred about 30 million years ago after the division between New World and Old World monkeys. Humans and the higher apes have virtually identical red and green pigment sequences. (From Passarge 1995 [170])

the red visual pigment (protanomaly) or of the green pigment (deuteranomaly). Color perception is most severely compromised in dichromatic protanopes and deuteranopes who cannot discriminate between red and green color. Deuteranomalous and protanomalous individuals exhibit more subtle anomalies with weakened rather than absent color discrimination. Color discrimination by color-defective individuals may be fairly adequate in practical life, even among dichromats who score as grossly abnormal in color vision test systems (see [143]).

Color vision defects are usually detected by various plate tests requiring color discrimination of numbers or shapes

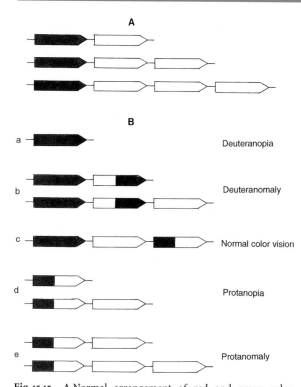

Fig. 15.15. A Normal arrangement of red and green color vision pigment genes. Note one single red pigment gene and one, two, three (or even more) green pigment genes. �▶, Red; ▭▷, green. **B** Typical pigment gene arrangements among males with different types of red-green color vision defects. These defects are caused by unequal crossover between the highly homologous pigment genes to produce deletions or fusion genes (green-red or red-green). ▭▶, Green-red hybrid gene; ▬▷, red-green hybrid gene. **a** Subjects with a single red pigment gene lack green pigment and are therefore dichromats (having "blue" and "red" instead of "blue", "red" and "green" cones). Their more severe color vision defect is known as deuteranopia. **b** Subjects with green-red fusion genes have a milder type of color vision defect known as deuteranomaly. They are trichromats with a slightly shifted absorption maximum of the green pigment. **c** The green-red fusion gene is not expressed when located in a more distal position in the array with resultant normal color vision. **d, e** Subjects with red-green fusion genes either have the more severe dichromatic protanopia (*d*) or the milder trichromatic protanomaly with a slight shift of maximum absorption of the red pigment gene (*e*). The Ser/Ala polymorphism of the red pigment gene (position 180) largely determines whether the color vision defect expresses as protanopia or protanomaly (see Deeb et al. [44a] and text)

(Ishihara or American Optical, Hardy Rand Ritter plates). Definitive assignment of the various kinds of color vision defects requires quantitative anomaloscopy based on standardized color matching.

The genetic basis of dichromatic deuteranopia lies in the absence of the green pigment gene(s) due to dele-

tion, while green-red fusion pigments are associated with deuteranomaly. However, green-red fusion genes located in a more distal position are not expressed and are therefore *not* associated with deutan abnormalities. Among the six exons of the red and green pigment genes, exon 5 is the most important for spectral tuning and for differentiating red from green pigment function since two of the three residues that account for the spectral difference between the red and green genes are located in this exon. Red-green fusion pigments are characteristic of both protanopia and protanomaly. A red-green fusion gene without additional green genes manifests as protanopia [44a]. A serine/alanine polymorphism at position 180 of the red portion of the fusion gene largely determines phenotypic expression of protanopia (if there is serine at position 180) or protanomaly (if there is alanine at this site).

Abnormalities of the blue pigment gene manifest as so-called tritan defects and are caused by amino acid substitutions at the autosomal blue pigment gene. These are much rarer than red/green color vision abnormalities (at most, 1:500).

Deuteranomaly is the most common defect in white populations (4%–5% of males) while the frequency of the other defects (protanopia, protanomaly, deuteranopia) is about 1% each; 8% of males are thus color vision defective. The total frequency of the various red-green color vision defects in non-European populations is significantly lower (around 3%–5%) and is accounted for largely by a lower frequency of deuteranomaly. The common occurrence of color vision defects appears to be due largely to unequal crossover (Sect. 5.2.8) in this highly homologous multigene complex. The high frequency of deuteranomaly among whites is unexplained, and the role of selection remains undefined (see Sect. 14.2.6).

A serine/alanine polymorphism of the red pigment gene is of great interest for normal variation of color vision since males with normal color vision who carry the serine variant of the red pigment gene perceive red color as a deeper red than those who carry alanine at that site [272c]. Serine at position 180 has been found among 62% of whites, 84% of Japanese and 80% of African-Americans. This X-linked polymorphism had been suspected for some years since minor differences in color matching had been observed some time ago [267]. In vitro spectroscopic studies of red pigments with either serine or alanine at position 180 showed a difference of 4–5 nm [135a], consistent with the phenotypic findings detectable in vivo by anomaloscopic color matching (Fig. 15.16). This finding clearly demonstrates the role of a simple polymorphism in explaining variation in sensory perception and can be considered a model for such processes. Each individual's external world is prob-

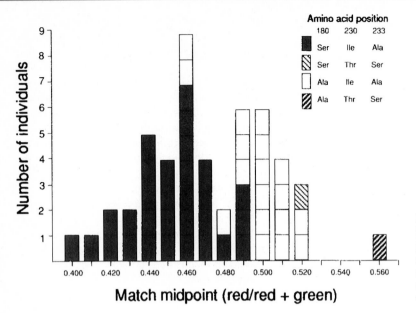

Fig. 15.16. Effects of serine/alanine (position 180) polymorphism of "red" photoreceptor (50 color vision normal male subjects). Note bimodality of color matching midpoints between serine vs. alanine carriers. (From Windericks et al. 1992; [272c])

ably perceived differently depending upon his or her genes!

Female heterozygotes with various red-green color vision defects (about 15% among whites) usually have normal color vision, but minor abnormalities can often be detected by special testing. Skewed X chromosome inactivation causing color vision defects is more common in one of a pair of female MZ twins [102a]. In such instances the X chromosomes carrying the normal color vision allele appear to be completely inactivated. (See Sect. 2.2.3 for mosaic structure of retinal cones in heterozygotes.)

Heterozygotes with a protan defect in one of the X chromosomes and a deutan defect on the other usually have normal color vision since normal red and green cone function is preserved by the normal color vision alleles. However, compound heterozygotes for deuteranopia and deuteranomaly (in *trans* position) manifest phenotypically with a less severe deuteranomaly as expected on molecular grounds. Analogous findings are observed for compound heterozygotes for protanomaly/protanopia.

Blue cone monochromacy is a rare defect with complete absence of both red and green cone function [154]. A critical regulatory area is located 3.4 kb upstream from the transcription start of the red pigment gene [269]. This locus control region is analogous to its counterpart at the hemoglobin locus and normally regulates red and green gene pigment expression [221]. Its deletion in some cases of blue cone monochromacy abolishes both red and green pigment function [154]. Other cases are caused by missense or nonsense mutations in a *single* red pigment gene or in a *single* red-green fusion gene that render their gene products nonfunctional. In such cases a point mutation in a preexisting abnormal gene arrangement inactivates red and green pigment function. Progressive retinal dystrophy may occur in blue cone monochromacy. This is presumably due to unknown mechanisms analogous to the cone degeneration that occurs with rhodopsin mutations in autosomal-dominant retinitis

pigmentosa (see below). In contrast, the various common color vision defects are not associated with any ophthalmological abnormalities.

The Rods; Hereditary Diseases Due to Genetically Abnormal Rhodopsin. Rhodopsin is the photoreceptor protein of rod cells. It is the most abundant protein in the disk membranes of outer segments of rods.

The receptor consists of seven helices that are located in the cellular membrane, establishing a connection between its outside and inside (Fig. 15.17) and a binding pocket for 11-*cis* retinal. Upon photoexcitation rhodopsin changes its conformation. This leads to activation of a G protein, transducin, which then activates cGMP phosphodiesterase, leading to a reduction of cGMP, closure of cation channels in the plasma membrane, and initiation of the neural signal. The phosphorylated and photoactivated rhodopsin is then bound by arrestin; this binding terminates activity of the receptor in the signal transducing process. The amino acid sequence of rhodopsin is known, and the gene has been isolated [152]. Different parts of the surface region of rhodopsin are important for interaction with transducin, the enzyme rhodopsin kinase, arrestin, and 11-*cis* retinal.

After gene and protein were identified, a genetic eye disease was often found to be caused by mutations in rhodopsin: autosomal-dominant retinitis pigmentosa (180380) (See below for description of the disease). The rhodopsin gene (RHO) has been localized to 3q21–q24. Since the first point mutations were shown to cause dominantly inherited retinitis pigmentosa [47, 135], a great number of other mutants have been described that affect various parts of the rhodopsin molecule [48]. Several dozen mutations have been identified. As in other genes, transitions

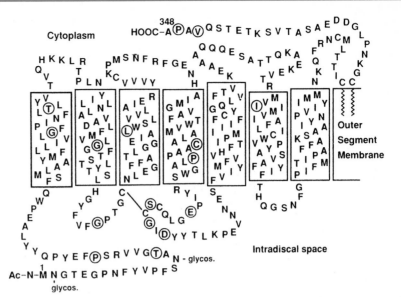

Fig. 15.17. Schematic model of human opsin. *Circled amino acids,* altered by mutations in patients with autosomal dominant retinitis pigmentosa. (From Sokoloff et al. 1992 [220])

(especially those occurring in CpG dinucleotides) are more common than would be expected if point mutations occurred at random (see Sect. 9.3). Study of DNA haplotypes around identical rhodopsin mutations in retinitis pigmentosa shows unique patterns pointing to independent mutational events [48]. Figure 15.18 shows one pedigree.

Meanwhile, mutations in other genes have also been shown to cause autosomal-dominant retinitis pigmentosa, as well. At least five additional such genes may exist, including the peripherin/RDS gene on 6p (170 710) and gene loci on 7p, 7q, 8, and at least one more locus [116]. However, in about one-half of all pedigrees with the autosomal-dominant form of this disease the affected gene has not yet been demonstrated.

Genetic Heterogeneity. The common term retinitis pigmentosa is used for a group of progressive retinal dystrophies with the common characteristics of night blindness, defects of the peripheral visual field, retinal pigment deposits, a characteristic electroretinogram, and in its end stages blindness.

The disease may show autosomal-dominant, X-linked, or autosomal-recessive inheritance. With a population prevalence of 1:4000, retinitis pigmentosa is one of the most common causes of blindness among the middle-aged in industrially developed countries. According to a study conducted in the United States [17], about 84% of cases exhibit autosomal-recessive, 10% autosomal-dominant, and 6% X-linked recessive inheritance. Since at least two genes for the X-linked type have been localized, it can be expected that the specific mutations leading to these varieties of the disease will be known in the future.

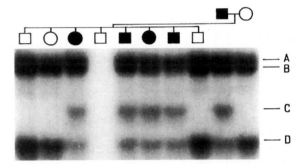

Fig. 15.18. SSCP analysis of family no. 6976 (Pro-171 → Leu). Sequence analysis revealed that the variant band (C) seen in the affected individuals in this family is due to a C → T substitution in codon 171. No DNA was available from the middle, unaffected brother. (From Dryja et al. 1991 [48])

Why Does Autosomal-Dominant Retinitis Pigmentosa Occur? Once a mutation within the gene for a functional protein has been identified, the pathogenic mechanism of the resulting disease can often be explained. It is difficult, however, to explain why rods in retinitis pigmentosa degenerate over a lifetime. A defect present at birth would be easier to explain. Among 18 mutants [48] 3 are located in the cytoplasmic site of the protein [230] (Fig. 15.17), 7 in transmembrane domains, and 8 at the intradiscal side. Many of these amino acid replacements (not all) are expected to alter the tertiary structure. But why should such anomalies lead to degeneration of photoreceptors? One hypothesis is that such structurally altered rhodopsin molecules are less easily transported from their site of synthesis – the rod inner segment – to the pigment epithelium which is the normal site of their degradation. Instead, they are retained in the inner segment, where they cannot be catabolized. This may over time lead to a "clogging" process which kills the rods [48]. More recently genes from this receptor gene family have been shown to be transcribed in olfactory receptor neurons; second messengers are formed upon stimulation with appropriate odorants (see [153]).

Tasting: the Role of Gustducin. Certain substances that stimulate taste receptor cells are organized in groups of 40–60 to form the taste buds in the taste papillae of the tongue. There are different molecular mechanisms for tasting of sweet, sour, bitter, and salty. A recently discovered, taste-cell specific G protein shows sequence similarities with the transducins involved in rhodopsin-mediated light sensitivity [134]. Molecular studies of the sensory receptors are revealing that the heptahelical configuration of rhodopsin and the retinal cone receptors is shared with olfactory and some neurotransmitter receptors, pointing to a common verly early origin in evolution [270].

As noted for color vision polymorphisms, human beings perceive their environment slightly differently depending upon their genetic makeup. The taste of phenylthiocarbamide is a genetic trait; some persons cannot taste it at all. Color vision varies in that, depending upon the specific color vision gene makeup, some persons see colors differently or not at all. Tone deafness exists, and some persons are unable for genetic reasons to recognize different tones, so that they have poor musical ability [109]. The counterpart – high musical ability – seems to aggregate in families. Information processing in the brain varies. Dyslexia is a genetic trait and presumably represents a defective ability to process words by the central nervous system [167].

These examples are presumably only the tip of the iceberg, and many more polymorphic differences in sensory perception (taste, smell, vision, hearing) undoubtedly exist. Each person therefore probably perceives his or her environment somewhat differently; each probably has his and her different external world. Our reactions to this uniquely perceived environment may therefore differ, and variability in behavior may result. The marked similarity of even trivial behavioral characteristics observed in identical twins reared apart [130] may be a partial consequence of their perceiving the environment in an identical manner.

15.2.1.6 "Abnormal" and Socially Deviant Behavior

"Criminality." (We are keenly aware of the fact that "criminality," i.e., conviction by a court for a behavioral deviation regarded as "crime," strongly de-

pends on the value system of a society.) Since Lange (1929) [117] published his monograph on *Crime as Fate*, a number of studies have compared MZ and DZ twins for concordance in "criminal" behavior [62]. The relevant information is collected in Fig. 15.19. The most important conclusions may be summarized as follows:

a) Concordance is higher between MZ than between DZ twins in all series, in many series considerably higher.
b) Absolute concordance between MZ twins varies between series. There is a tendency to higher concordance in older investigations.

The main reason for these differences in concordance rates is the range of criminal offenses covered by these investigations. For example, Lange's original series was selected for severe and repeated convictions, whereas the more recent series, for example those from Denmark, tend to refer to all types of "criminality," including occasional and smaller offenses.

Taken at face value these results suggest that the liability to become a convicted offender depends strongly on the individual's genetic endowment, with particularly striking effects for severe and repeated "criminality." This conclusion, if verified, might elicit either of two responses by society: to isolate law offenders as biologically deviant or to consider them as sick and attempt quasimedical therapy. However, we must be cautious in accepting these MZ concordances as proving genetic influence before social interaction between members of these twin pairs is excluded as an alternative interpretation. On the other hand, could such social interaction be the only important factor? This also seems unlikely. Again, we are left with the ambiguity of results from twin data.

Studies on adoptees are of some interest [43, 95, 205]. One study examined adoptees who were at least 18 years old (the oldest was 47) and had been adopted following the mother's conviction for felony, prostitution, larceny, or other offenses; sufficient information on fathers was not available. With three exceptions they were adopted by persons unrelated to them. A control group of adoptees was matched for sex, race, and age at the time of adoption. The most important results are presented in Table 15.7. There was a significantly higher risk of arrests, incarceration, and psychiatric hospital records among the proband group than among controls. Structured psychiatric interviews, while showing no other difference between the proband and control groups, led to the diagnosis of "antisocial personality" in 6 of 42 cases of the proband group but in only one questionable case of the 42 controls. On the other hand, detailed case histories gave evidence for an influence of environmental fac-

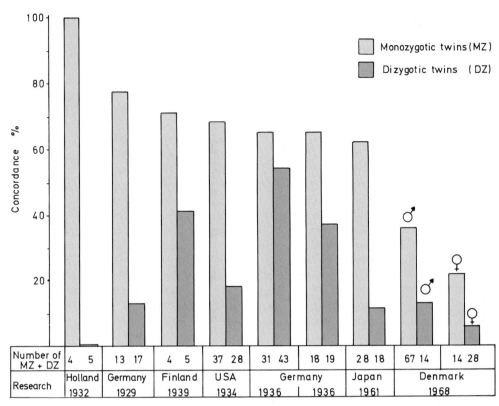

Fig. 15.19. Concordance rates of MZ and DZ twins for "criminality" – as defined by court conviction – in studies from various populations. (From Fuller and Thompson 1978 [63], and some additions)

tors on the manifestation of antisocial behavior. Five of the six antisocial probands spent over 12 months in orphanages and temporary foster homes before adoption and were put up for adoption when already over 1 year of age; most of the other probands had been placed earlier. Socioeconomic status of the adoptive homes was not correlated with the outcome. However, among the six homes in which the antisocial probands were placed, two were broken homes. In the control group adverse environmental influences, such as late adoption, were also found in some cases, but here they did not result in abnormal behavior. On the other hand, in five of the six cases diagnosed as antisocial personality, the alleged father was said to have also had records of offenses.

This study succeeded in a distinguishing genetic and environmental factors fairly clearly. The results point to genetic influences. However, the development of a manifestly deviant personality also required unfavorable environmental conditions. Even the first year of life seems to be important, in accordance with other results from child psychology. On the other hand, the majority of children overcome unfavorable environmental conditions without recognizable ill effects; only a presumably genetically predisposed group reacted by developing a deviant personality. It is com-

Table 15.7. "Criminality" in adoptive children, records follow-up (from Crowe 1974 [43])

	Probands,[a]	Controls,
	$n = 37$	$n = 37$
Arrest records		
Adult arrest	7	2
Adult conviction	7	1 $p = 0.03$[b]
	$n = 42$	$n = 42$
Incarceration		
Juvenile[c]	3	0
Adult	4	0
Either	6	0 $p = 0.01$
	$n = 42$	$n = 42$
Psychiatric hospital records		
Number hospitalized	7	1
Outpatient only	1	1
Total number seen	8	2 $p = 0.04$
Number with both psychiatric and arrest records	6	0 $p = 0.01$

[a] Mothers convicted of felony, prostitution, larceny and other offenses.
[b] Probabilities are Fisher's exact, one-tailed.
[c] One juvenile was actually ascertained through hospital records and was sent to the training school shortly thereafter.

forting that a majority of adoptees developed normally.

In another study [95], psychiatric records of biologic relatives of psychopathic adoptees revealed a higher incidence of psychopathy than among adoptive relatives and controls. (Psychopathy was defined here according to the American definition, which is more restrictive than the European, comprising mainly antisocial behavior.) A third study [205] found biological relatives of "criminal" adoptees to have a higher rate of "criminality" than adoptive relatives and biological or adoptive relatives of control adoptees. However, "criminality" in an adoptee was correlated independently with "criminality" in both biological and adoptive parents. This study thus also gave evidence for genetic as well as environmental influences.

As noted, subnormal intelligence and aggressiveness may be associated with a missense mutation in a monoamine oxidase gene (MAO) [22]. This observation suggests a specifically localized and defined gene associated with "criminal" behavior. Further studies are required to confirm these data, with special reference to better and more detailed definition of the nature and range of behavioral manifestation in affected subjects.

Homosexuality. Kallmann [108] examined concordance of homosexual behavior in 95 male twin pairs, 44 MZ and 51 DZ. He classified the degree of homosexuality using a scale of six grades according to Kinsey. Grades 1–4 indicate low degrees, 5–6 high degrees of homosexual behavior. The probands were taken from court files; hence they were selected for conflict with the law. Concordance between MZ twins was reported to be very high; there was not even one pair in which the other twin was completely free of homosexual tendencies. Of the 51 DZ pairs, on the other hand, 38 proved to be completely discordant in spite of the fact that low-grade homosexual behavior (grades 1–4) is fairly common in the American population (Kinsey).

One obvious argument against the genetic interpretation of the high MZ concordance is the possibility that one twin seduced the other to homosexual activities. This, however, does not seem to be so: all twins denied emphatically any sexual relationship with their co-twin. It did happen occasionally that a twin introduced his co-twin to another sex partner. As a rule, however, homosexual behavior is reported to have developed independently. Concordance between MZ twins was much more specific than indicated by concordance figures. Most pairs played similarly active or passive roles in homosexual relationships. Concordance has also been observed in an MZ pair reared apart [49]. Other authors have subsequently described discordant MZ pairs, and the literature is

now giving the impression that concordance for homosexuality in MZ twins is higher than in DZ twins but by no means complete. Pooled data from recent studies on homosexual males (e.g., [5, 176]) give a concordance rate of 57% for MZ twins, 24% for DZ twins, and 13% for brothers [122]. Female homosexuals – who had been neglected in earlier studies – showed 50% concordance in MZ and 13% in DZ twins (see also [6]). As in other fields of human behavioral genetics, the interpretation of such twin data is difficult [29].

Segregation and linkage studies may help to elucidate biological factors predisposing to homosexuality. A study of 114 families found a tendency to same-sex orientation in brothers and in maternal uncles and cousins, but not in paternal relatives. A linkage study with X-linked DNA markers using the sib pair method (Sect. 5.1.3) indicated significant linkage with markers at Xq28 [77]. This result was confirmed in an additional series.

Neuroses. We do no attempt here a review of the many studies on twins' families and adoptees in which aspects of neurotic or otherwise abnormal behavior have been investigated [76, 96, 98, 169, 203, 218]. Especially interesting are intensive parallel studies on MZ twins with psychological and psychoanalytic methods – in cases in which either one or both twins showed neurotic behavior [60]. Concordance in neurotic symptoms is in generally higher among MZ than among DZ twins. Within these limits, however, development of abnormal behavior, especially the specific symptoms of a neurosis, depends on the familial situation, which might be quite different even for MZ. Here, again, observation of MZ twins over decades has proven especially instructive [145, 146].

Eating Disorders: Anorexia Nervosa and Bulimia [80]. In the affluent Western society two eating disorders have become common in recent decades: anorexia nervosa and bulimia. Both are observed mainly in adolescent girls and young women, but they also occur in older women and, very rarely, in men. Patients with anorexia nervosa eat too little; this leads to extreme leanness and may over time cause death by starvation. Bulimia is the tendency to go on a binge of uncontrolled eating; often the ensuing weight gain is prevented by self-induced vomiting or by the use of laxatives. Limited twin data are available for both conditions [88, 112, 204] and show a higher concordance in MZ than in DZ twins. Family studies on anorexia nervosa show family aggregation [89, 227]. On the other hand, the increased prevalence in Western society suggests an important environmental component – probably related to ideals of slimness, difficulties in accepting the female sex role, and other psychological influences [186]. Sometimes symptoms of both anorexia and bulimia occur in the same patient. Anorectic patients often suffer from depression, obsessive-compulsive neurosis, and occasionally drug addiction and schizophrenia. The results of epidemiological and genetic studies show individual differences in liability that may lead to outright disease if combined with certain environmental conditions. Further studies with reference to normal eating habits and body weight variation may be helpful [80] since even MZ twins reared apart are fairly concordant regarding body weight [228].

Gilles de la Tourette Syndrome. This syndrome was originally described as early as 1885 [66] but it has found much attention in recent years. The major symptoms are repeated, involuntary fast muscular movements ("tics") and inappropriate vocal utterances. These symptoms appear in childhood and often continue throughout the person's life time, although they may also disappear in adult age. Some patients, especially women, develop obsessive-compulsive disorder. Concordance in MZ twins is high (approx. 80% when all manifestations of the syndrome are considered) [181]) family studies reveal an impressive accumulation of secondary cases within these families [40, 173]. The data are compatible with the presence of a dominant, major gene with incomplete penetrance and variable expressivity; the spectrum of manifestation appears to include obsessive-compulsive disorder [40]. However, many nonhereditary cases (phenocopies) appear to exist; only a minority of families of probands with obsessive-compulsive disorder also contains Tourette syndrome patients [173]. Extensive linkage studies have so far failed to localize a gene [172]. While typical cases can easily be diagnosed, delineation of the syndrome and diagnosis of atypical cases is controversial.

15.2.2 Chromosome Aberrations and Psychological Abnormalities

The studies discussed in Sect. 15.2.1 were carried out using classical methods of comparison between twins and other relatives. Such investigations can separate out genetic from environmental influences only with difficulty, and never unambiguously, even with twins reared apart or adoptees. Genetic analysis remains at the phenotypic-biometric level; no individual gene action can be identified, and the biological mechanisms that cause behavioral variation remains unexplored. This approach has been compared with the attempt to understand the way in which a clock works by applying statistics on the movements of its

hands [56]. To solve the riddle of the clock's mechanism one must open it and examine its clockwork. How can this be done?

Human Chromosome Aberrations and Behavior: Possibilities and Limitations. Patients with chromosome aberrations usually show, along with many other findings (Sect. 2.2.2), behavioral abnormalities, which may be mild or severe and may affect all carriers of an aberration or only some of them. For our analysis they offer the unique opportunity to relate the behavioral abnormality to an independently ascertained, undisputable, and clearly defined genetic cause. Unfortunately, however, opportunities for a more incisive analysis are exhausted with this statement at the present state of our knowledge. Chromosome aberrations influence embryonic development in multiple and ill-defined ways. With improved knowledge of the genome – localization of genes and their interactions and regulation during embryonic development – gene action will be analyzed in a stepwise manner (Sect. 8.4.3) [51]. Even at the present state of our knowledge, however, much work can be carried out at intermediate levels of analysis. For example, the influence of brain morphology, endocrine abnormality, and more indirect social and psychological influences on behavior can be examined.

15.2.2.1 Autosomal Aberrations

Down Syndrome. Most unbalanced autosomal aberrations lead to multiple and severe malformations (Sect. 2.2.2) that also affect the brain, causing severe mental deficiency. Down syndrome is the most common of these abnormalities observed after birth. The range among IQ of Down patients is between 20 and 60; the median is 40–50. Many such persons can be educated to read and write, and some can learn to make a marginal social adjustment. The vast majority are unable to function on their own in society.

Deficiencies of psychomotor development are prominent during the first years of life. The children learn to walk only at 3 years of age and remain clumsy [30]. Early motor training improves abilities in other fields, for example, language. Intellectual abilities often are those of 4-year-olds, and abstract thinking is lacking. Most patients learn basic social skills such as dressing, eating without help, and toilet use. Socially they are often friendly and cheerful [175] but temper tantrums may occur especially in boys during puberty. In social interaction there is a lack of sensitivity to the emotional needs of others and of keeping necessary distance in personal space. Behavior of young men toward women may therefore be wrongly interpreted as having sexual intent. As a rule, girls with Down syndrome have fewer problems. Patients kept within families develop better than those institutionalized in early childhood.

One adult patient – a son of teachers – has written an autobiography [91], which is interesting for its revelation of the inner world of such children. Some situations are described fairly vividly and even with some sense of humor, but even the slightest attempts at abstraction are lacking. The father in his two social roles as father and as teacher is not clearly identified as the same person.

Intellectual abilities tend to decrease as early as the third decade of life, and this is often accompanied by gradual loss of social abilities and mood deterioration [128]. These findings are the signs of Alzheimer disease, which develops prematurely in Down syndrome during the third or fourth decade of life and appears identical in its pathological findings (amyloid deposits) to classical Alzheimer disease [129].

Social Problems. Patients with Down syndrome are now surviving longer than formerly when infections could not be controlled. Many reach adulthood, and they can be educated to achieve a certain degree of self-reliance within protected surroundings. When given the chance to select a partner of the other sex, they may even develop stable and satisfying relationships, usually within an institutional framework.

15.2.2.2 Aberrations of the X Chromosome

Numerical and structural aberrations of the X and Y chromosomes generally lead to much milder disturbances in embryonic development than do autosomal aberrations (Sect. 2.2.3). Many somatic abnormalities found in these syndromes are related to abnormal sexual development. The psychological disturbances are less overwhelmingly severe and may sometimes be specific.

Klinefelter Syndrome. The standard karyotype of Klinefelter syndrome is XXY; other karyotypes as well as mosaics occur

(Sect. 2.2.3.1). Adult patients average a few centimeters taller than their normal brothers; especially their legs are longer in relation to overall stature. This growth abnormality can be observed as early as in childhood; at the age of puberty the subnormal sexual development becomes obvious: the testicles are small, and there is aspermy so that these patients are infertile. They are, however, capable of sexual intercourse. Many of their psychological symptoms can be explained by their diminished androgen production, which is normally required for the expression of male-specific psychological development.

The patients show on average slightly reduced intelligence, with special difficulties in learning how to read and write. Their vitality and ability in establishing social contacts is often reduced. Table 15.8 presents data on the IQ of patients with Klinefelter syndrome compared with other anomalies of the X and Y chromosomes. The mean is shifted to below the norm; data from unselected probands range between 88 and 96. However, IQ values well above average are not rare. On the other hand, Klinefelter syndrome has been found more often in series of mildly mentally subnormals. The literature reports do not unanimously support a well-defined, specific defect of mental abilities [278].

School problems are more frequent than expected from intellectual ability and seem to be caused by behavioral problems. Adult patients often hold unskilled jobs; success in higher professional careers has been reported but seems not to be very common.

Psychiatric reports on the personalities of Klinefelter patients show a variety of deviations from the norm [278]. Their behavior has been described as passive-aggressive, withdrawn, self-contented, and mother-dependent, but they have also been characterized as quiet and law-abiding citizens who do not attract attention. The libido is usually reduced and may be lacking completely, but in some patients it is normal. Some patients have no erections and no ejaculations; other have intercourse, but generally only rarely. If

Table 15.8. IQ in patients with abnormal numbers of sex chromosomes (Moor 1967, Pena 1974; from Lenz 1983 [121])

IQ	XXY	XXXY	XXXXY	XYY	XXYY	XXX	XXXX XXXXY
– 19	–	1	4	–	–	1	–
20– 39	–	1	16	–	4	4	1
40– 59	5	5	6	1	12	16	7
60– 79	32	5	3	10	8	14	5
80– 99	23	–	–	8	2	1	2
100–119	12	–	–	1	1	–	1
120–	–	–	–	–	–	–	–
Total	72	12	29	20	27	36	16

sexual activity is present, it usually fades out early – at about the age of 40. On the other hand, a fair number of patients live in stable marriages.

Many of the patients, however, do have difficulties in coping with the normal requirements of life; all too often they find themselves in a kind of outsider position. Therefore it is not surprising that they are more often found among law-offenders than in the general population [160, 243]. Table 15.9 shows their frequencies in two series of law-offenders [273]. Criminal patterns and activities are not specific; however, white-collar crimes are almost completely lacking, probably due to reduced average intelligence. In a recent study of limited but unbiased data, Klinefelter patients turned out to be no more frequent among law-offenders than XY men of the same intelligence and educational achievement [273].

Klinefelter Variants. Variants of Klinefelter syndrome with more than two X chromosomes or with more than one Y are discussed in Sect. 2.2.3.2. Many of the patients described in the literature were ascertained in institutions for the mentally retarded. Table 15.8 presents IQ data; the severity of mental deficiency increases with the number of X chromosomes. Data from psychiatric studies in mosaic Klinefelter patients are scanty.

Therapy and Prevention. Male hormone leads to increased virilization and improvement of libido; this therapy has been recommended to start at about the age of 10–11 [164]. Part of the favorable psychological effect may be due to the virilization, but an additional, direct effect on brain function is possible. Symptoms that can be described as "climacteric," such as moodiness, nervousness, and psychasthenia, seem to disappear. However, in some cases, hormonal therapy seems to enhance restlessness and aggressive tendencies. When a personality has developed over many years under abnormal internal conditions, additional hormone injections cannot normalize the long-established psychological and social behavioral patterns. Psychotherapy is useful in many cases. Psychiatrists and psychotherapists should be aware of the various psychological symptoms in Klinefelter syndrome that have been described and order a chromosome study to rule out the condition. In a few longitudinal studies, the XXY status was diagnosed immediately after birth, or even prenatally, and the children were followed by repeated examinations until the age of about 20 [156, 164, 191, 193, 199]. These studies provide a clear picture of the psychological development of these boys and their special difficulties. Many of them need help especially in tasks involving language capability; the behavioral and educational problems such as lack of endurance, moodiness, and occasionally a tendency toward aggressiveness are not explained sufficiently by the intellectual difficulties.

Table 15.9. Prevalence of the XXY Klinefelter syndrome among inmates of institutions for criminals, juvenile penitentiaries, etc.

Author	Kind of institution	Prevalence of XXY
Tsuboi [243]	Two institutions for criminal psychopaths (Denmark)	$5/480 \approx 1\%$
Murken [150]	Four institutions for criminals and aggressive psychopaths (Germany)	$7/728 \approx 1\%$

Note that the frequency is seven to ten times higher than in the general population (see Table 9.1).

Timely intervention by appropriate educational help is necessary to prevent secondary psychological effects produced by interaction of biologically induced shortcomings with a "normally" reacting environment. Ability to cope with difficulties within the family is poorer than in chromosomally normal boys.

Turner Syndrome. Clinical and chromosomal results are described in Sect. 2.2.3.2; the standard type shows the karyotype XO; however, many mosaics and structural variations are observed.

Earlier investigations reported a significant reduction in mean IQ [278]. While subsequent studies have failed to confirm this [156], a slight overall reduction in IQ does seem to be present in Turner syndrome. This is probably caused by a specific defect. These patients are often successful in school [278]. One study of 126 patients found that 2 achieved a university degree, 10 passed a European school comparable to an American high school, 93 went to elementary school, 21 of them with difficulties, and 21 attended a special school for the mentally backward (see also [161]).

Intelligence Defect in Turner Syndrome. Shaffer [210] observed 20 patients with Turner syndrome (age 5.8–30.9 years, mean 15.9) and found a discrepancy between verbal and performance IQ in the Wechsler Intelligence Scales for Adults and Children, respectively. Verbal abilities were within the normal range while performance was impaired. At a somewhat more anecdotal level, Shaffer pointed to the similarity of these results with those often found in cases with certain types of organic brain damage:

There were several bits of information to suggest that the test deficiencies . . . had their counterparts in every-day behavior. Almost without exception, the subjects reported they had great difficulties in understanding mathematics, especially algebra. One girl had an extremely poor sense of direction and frequently became lost. Another went through an elaborate ritual when putting away kitchen utensils since any departure from this procedure left her thoroughly confused.

This specific cognitive defect has been confirmed and characterized more precisely [138, 139, 210, 278]. The items on which the patients score especially low concern perceptual organization. Dyscalculia may also be present but is usually not quite so severe. The space-form blindness also leads to difficulties with right-left directional discrimination. The patients did very poorly, for example, in a road map test that requires orientation to right and left [1]. Money [138] has suggested a functional defect of the parietal lobe, perhaps involving the nondominant more than the dominant hemisphere. This defect has been confirmed by neuropathology studies [196]. Other alleged anomalies of psychological development such as infantilism and difficulties in social adjustment, may be secondary effects caused by overprotection. Parents and teachers tend to communicate with these patients on the basis of their size, not their age. Here contact groups can be helpful [161]. Among the more intelligent patients a compensatory activity, especially in sports and school, has been described. There seems to be little or no antisocial behavior, in contrast to XXY and XYY patients. The sexual drive is usually underdeveloped; however, patients are often married. By and large, they seem to suffer more from their small stature than from their sexual underdevelopment. Estrogen treatment usually leads to a better development of secondary sexual characteristics, to menstrual spotting, and in some cases to improved sexual responsiveness, but it does not affect the space-form blindness.

Triple-X Syndrome. This syndrome is described in Sect. 2.2.3.1. Many women with the karyotype XXX are normally developed and have children. Of 119 patients described in the literature 12 had epileptic seizures; in a home for epileptic patients 2 of 209 patients had an XXX karyotype [86]. Their intelligence tends to be well below average (Table 15.8) [156]; they are found more frequently in institutions for the mentally retarded than is normal in the general population. Somatic symptoms seldom lead to medical examination, and many of the described patients have been ascertained by screening populations of mental institutions. It cannot be determined definitely how much the XXX karyotype enhances the liability to psychoses, but some authors estimate the rate of schizophrenialike psychoses to be threefold increased [185]. In recent years chromosome studies on newborns have been carried out in many countries, showing a frequency of about 1:1000 female births. Limited prospective studies are in progress [192].

Epilepsy seems to be more frequent in XXX women than in the general population. The same may possibly be true for XXY men.

15.2.2.3 XYY Syndrome

Somatic Symptoms. For a description of the XYY syndrome, see Sect. 2.2.3. The mean stature of these patients is appreciably taller than that in the population from which they come. Many show normal sexual development and are fertile. The distribution of IQ is shifted to a lower range; some of the patients show average intelligence, but the mean IQ is 80–88 (Table 15.8). Other studies have reported higher values [156].

Higher Prevalence Among "Criminals." The XYY syndrome has become widely known since Jacobs et al. (1965) [99] carried out a survey of patients who were mentally subnormal and under surveillance in a special institution because of "dangerous, violent, or criminal propensities." Among 196 probands 12 had an abnormal karyotype; seven with XYY and one with XXYY. This frequency was much higher than expected; however, the authors stated that they could not determine whether these men had been institutionalized mainly because of mental subnormality, aggressive behavior, or some combination of these factors. Their results were soon confirmed in a number of studies from institutions for mentally subnormal men with behavior problems, especially among especially tall inmates. On the basis of such evidence it was concluded that their antisocial behavior was caused by the additional Y chromosome, and that they were genetically predisposed to criminality. The explanation seemed simple. Normal men are more aggressive than normal women; normal men have one Y chromosome, women do not. Hence, if someone has two Y chromosomes, he should be twice as aggressive as normal men; his aggressiveness may fall outside the socially acceptable range, and he may commit acts of violence. He is a "supermale."

Thus the "murderer chromosome" was born. Discussions started on "whether society is justified in restricting an XYY's freedom before he violates the law. The XYY individual is a perpetual threat since at any time he may face a situation in which he will be unable to control his behavior". At the same time, during the trial of a man in Paris who had murdered an elderly prostitute the defense attorney claimed that his client was not legally responsible for his act, as he had an additional Y chromosome. He was given a reduced sentence, presumably because of his anomaly. Similar cases soon followed.

Gradually, however, some pertinent questions were asked: above all, how frequent is the XYY karyotype in the general population of nonconvicts? Studies on the incidence among male newborns showed a frequency of around 1 : 1000, or even higher, similar to that of Klinefelter syndrome (Sect. 9.2). Even in the absence of reliable prevalence studies among the male adult population it was fair to conclude that the prevalence differs little from the incidence at birth, i.e., that there is no preferential mortality. This, however, could only mean that the great majority of XYY men do not come into conflict with the law.

Another question was whether the nature of their crimes revealed a certain pattern and, more specifically, whether acts of violence and – as had been suggested – sexual aggression prevailed. This was not the case: the pattern was quite similar to that of other groups of law offenders with the same degree of intelligence impairment and, more specifically, to that found in patients with Klinefelter syndrome, who were also detected more often among law offenders than in the general population [150, 243]. Offenses against property were the most frequent; so-called white-collar crimes were lacking, presumably due to the lower average intelligence of the probands. These results detracted from the romantic picture of the savage supermale. This image was completely shattered when someone asked how XYY men behave when institutionalized. Are they more aggressive than other men detained in the same institutions? They turned out to be, in fact, more agreeable; on average they had better relationships with supervisory personnel [226]. Many more psychological and psychiatric studies were carried out. While varying in details, their overall picture seldom differed from that of chromosomally normal inmates of the same institutions and having the same range of intelligence.

All these results suggested alternative explanations for the undisputedly higher frequency of XYY probands in institutions for law offenders, especially those specifically for criminals suffering from mental subnormality, psychopathy, and behavior problems.

Intellectual Dysfunction or Simply Stature? Many studies have been carried out on convicted and imprisoned law offenders. Their mean IQ is generally low. Intellectually subnormal persons are more often involved in criminal activities – or they run a higher risk of being apprehended. If they are arrested, they are less able to hire good lawyers and have a lower chance of escaping imprisonment. Is the supposedly higher crime rate of XYY men only a result of their reduced average intelligence?

Support for this contention came from studies that showed convicted XYY offenders to come – often but less frequently than other criminals – from broken families and families of lower socioeconomic status [125, 160]. It even seemed possible that their tall sta-

ture influenced the rate of detention. A huge and strong man may arouse in a judge or jury the feeling that it would be safer to detain him in jail than to let him go free.

Studies on Unbiased Samples. To establish the range of variation in phenotypic manifestation, to determine the intervening variable between the abnormal karyotype and criminal behavior, and to explore the influence of socioeconomic and educational differences, studies on unbiased samples of cases were needed.

a) From chromosome studies in newborns (Sect. 9.2.1) data on a fair number of XYY boys are available. Long-term, careful surveillance of their development should give the most reliable information. Interesting information is now available from such studies [156, 192] (see below).

b) In theory, a sufficiently large, unselected population of adult males should help to answer our questions about both chromosome examination and behavior characteristics. Considering the incidence at birth (1 : 1100), such a study would require an extremely large sample size. However, since XYY males are prominent by their stature, most of them would be found in the upper range of stature, thereby reducing the screening effort.

Such a study has been carried out in Denmark [273]. The population from which the sample was drawn consisted of all male Danish citizens born between 1 January 1944 and 31 December 1947 to women who were residents of Copenhagen. These men were at least 26 years old when the study began; the military draft boards provided information on the stature of 28 884 men who were available for study.

A cutoff point of 184 cm was used in establishing the tall group among whom the search for sex chromosome anomalies was to be conducted; the resulting sample consisted of 4591 men. After preparation of the general public through the mass media, these men were contacted, and if possible a buccal smear for X and Y chromatin determination and blood for karyotyping was taken. The investigators succeeded in securing the cooperation of almost 91%.

Limited but reliable data on the relevant behavioral characteristics were available for the entire study group. Convictions for criminal offenses were documented in police registries. Military draft boards give all draftees a screening test for cognitive abilities, the so-called BPP test, which covers a limited number of cognitive dimensions. However, educational level was used as an additional index for intellectual achievement. In Denmark school examinations are given at the end of the 9th, 10th, and 13th years. The index was constructed simply from the

Table 15.10. Criminality rates and mean values for background variables of XY, XYY, and XXY (from Witkin et al. 1976 [273])

Group	Criminality		Army selection test		Educational index[a]		Parental SES[b]	
	Rate (%)	n	Mean ± SD	n	Mean ± SD	n	Mean ± SD	n
XY	9.3	4096	43.7 ± 11.4	3759	1.55 ± 1.18	4084	3.7 ± 1.7	4058
XYY	41.7**	12	29.7 ± 8.2***	12	0.58 ± 0.86**	12	3.2 ± 1.5	12
XXY	18.8	16	28.4 ± 14.1***	16	0.81 ± 0.88*	16	4.2 ± 1.8	16

Significance level pertains to comparison with the control group (XY) using a two-tailed test. For criminality rate an exact binomial test was used; for all other variables a *t* test was used.

* $p < 0.05$; **$p < 0.01$; ***$p < 0.001$.

[a] Refers to educational achievement by type of school attended successfully (maximum: 3).

[b] Socioeconomic status.

number of examinations a man had passed, ranging from none to three.

Results of the Study. In this population of tall men 12 XYYs and 16 XXYs were identified. A search in the penal registers showed that 5 of the 12 XYYs (41.7 %), 3 of the XXYs (18.8 %), and 9.3 % of all the XY men had been convicted of one or more criminal offenses (Table 15.10). The difference between XYY and XY was significant; the difference between XXY and XY was not. Both groups with abnormal karyotypes (XYY and XXY) showed a marked reduction in the indices for intellectual achievement – BPP and educational index – despite the fact that the socioeconomic status of their parents was identical to that in the control sample.

Three hypotheses on intervening variables between abnormal karyotype and criminality were examined: Is it their aggressiveness that brings these men into conflict with the law, is it their reduced intelligence, or simply their tall stature? A preliminary answer was provided by the nature of the crimes of the five convicted XYY men. Only one of them had committed an act of aggression against other persons. The other offenses included theft, arson, burglary, and procuring for prostitution. With one exception (imprisonment for somewhat less than 1 year), all penalties were mild, indicating small offenses. The one act of aggression was an isolated occurrence in a long criminal career of the only man who could be regarded as a "typical" criminal. The range of offenses of the three XXY men was similar to that of the control population. Thus in both XXY and XYY men the nature of the crimes committed resembled that of the control population.

Together with the other results discussed, these data, despite the small sample size of XYY men, permit the conclusion that the aggression hypothesis is most probably wrong. Is the higher liability of the

probands to break the law caused by their lower intellectual capability?

The data showed, first, that normal XY men with no record of crime had a mean intelligence score (not to be confounded with the IQ) of 44.5, whereas those who had committed one or more such crimes had a score of 35.5. The educational index showed a similar trend for criminality (1.62 for noncriminals, 0.74 for criminals). Can the criminality rate of XYY and XXY men be accounted for completely by their intellectual dysfunction? To examine this question their criminal rates were compared with that of the control group after statistically controlling for background variables such as intellectual functioning, socioeconomic status of the parents, and stature. The analysis consisted of three stages. The first established the probability that an XY male with a particular set of values on the background variables is a criminal. The second step established for each XYY or XXY man the probability that he would be a criminal if he were an XY man with the same background. In the third step the observed frequency of criminals in the proband group was compared with the frequency predicted in the second step. This analysis gave the following results:

a) Background variables, i.e., lower intelligence and socioeconomic status of the parents, account for some of the difference in the criminality between the XYY and XY groups. However, an elevated crime rate among the XYYs remains even after these adjustments are made ($p < 0.05$).

b) XXY men are not significantly different in criminality from the XY control group after controlling for background variables.

Therefore in the XYY state, intellectual functioning seems to be an important mediating variable between the abnormal karyotype and the above-average liability to criminal behavior. However, this behavior is not completely explained in this way. Either the indices used (BPP test and educational index) give only an incomplete account of the cognitive defect, or an additional "personality" factor is involved. We consider the latter hypothesis more likely; psychological examination of unselected XYY probands will help to answer this question.

The aggression hypothesis can now be rejected with confidence; the hypothesis that stature enhances the risk of XYY men being convicted could also not be confirmed. XYY probands without a criminal record even tended to be a little taller than those with such a record.

This investigation together with many other studies led to the following conclusions:

a) There can be little doubt that men with the chromosome constitution XYY run a higher relative risk than normal XY men of showing antisocial behavior and coming into conflict with the law.
b) Part of this risk may be traced to their impaired intellectual function.
c) However, many of them seem to be afflicted with an additional, more specific personality disturbance that may lead to difficulties in adjusting to the social environment. Conviction for criminal acts probably is only the "tip of the iceberg;" social difficulties not leading to conflict with the law may be much more widespread.

The conclusion that XYY men are psychologically different is confirmed by a double-blind study on seven randomly chosen young XYY men in comparison with 28 XY men from the same French population [165]. It was possible to distinguish the XYY men from the XY controls on the basis of psychological results. Examination of 14 XYY males (i.e., including some other cases in addition to the randomly chosen ones) by a number of psychological tests (Rorschach, Thematic Apperception Test, interview) uncovered increased impulsiveness after emotional stimulation, predominant responses to immediate gratification, and lack of emotional control. In some of the probands this lability was checked by rigid self-control. Defense mechanisms against anxiety were poor, and the concept of self was weak, easily fragmented, and often infantile. Such personality characteristics could indeed constitute an increased risk for antisocial behavior. In contrast to most other studies, including the Danish study described above, the average IQ in this series was not subnormal.

EEG studies on eight XYY men (from the Danish study described above) revealed that slowing of the α frequency in the resting (awake) EEG appeared to be a fairly constant finding in this group. The *maximum* α frequency in these subjects was *lower* than the *minimum* frequency in 16 XY controls [264]. This result, together with others, led to the (admittedly very general) hypothesis that "neural factors" contribute to behavioral difficulties in some XYY men [90].

Social and Therapeutic Consequences. The evidence shows that the legal consequences for preventing crimes by XYY men as proposed in the heyday of the aggression hypothesis have no basis at all. Still, problems remain. If the XYY status is discovered in a study on newborns, should parents be informed? Could such information have the effect of a self-fulfilling prophecy in that parents would treat their boy differently, and could this enhance his tendency to deviatant behavior? In our opinion, all information should be provided; however, great care is needed in conveying the facts to the parents in a form that causes as little embarrassment as possible and, above all, no damage. The parents should understand that their boy might possibly need somewhat more special attention during his education than an XY boy, but that given a stable environment and the same amount of parental protection as other boys enjoy, normal social adjustment is the most likely outcome. No effective somatic therapy is known; the blood androgen levels have normal mean values with variance a little higher than normal.

The behavioral problems with XYY individuals (as well as with other persons having deviant sex-chromosomal karyotypes) probably could be alleviated if these conditions were diagnosed at birth, and if the children (as well as their parents) received special care, as shown by the ongoing studies on a series of Danish children [161–163] and by other series of children from the United States and United Kingdom [199]. Children show marked improvement with appropriate care, for example, by training their motoric abilities, not only in psychomotoric but also in intellectual development [189, 190]. In an increasing number of countries, support groups have been founded to help with these problems.

Inevitably, with the widespread use of antenatal diagnostics, XYY as well as XXY and XXX karyotypes are discovered by amniocentesis. Parents usually should be fully informed of the findings and the implications of the anomaly explored. The option of abortion as a possibility needs careful discussion; genetic counseling should be nondirective, and the decision should be left to the parents. On the other hand, the attitude and feelings of the genetic counselor regarding the impact of these states are likely to differ depending upon individual interpretations of the facts cited above. A completely objective and neutral stance is therefore difficult and usually not achievable. Some parents at the current state of knowledge would not hesitate to proceed with the pregnancy while others would prefer an abortion.

What Can We Learn from the XYY Story About the Attitude of Scientists in the Face of a Problem of Great Public Concern? The surprisingly high incidence of XYY men among convicts coincided with a growing public concern about increasing violence. In this situation some scientists reacted almost as precipitously as representatives of public opinion – mainly the mass media. Premature conclusions and sweeping generalizations were made on the basis of results that obviously came from biased data. The principal reason for such misinterpretation may have been that the cytogeneticists who primarily worked on the XYY problem were influenced by their experience in clinical cytogenetics of a fairly straightforward relationship between cause – an abnormality of the karyotype – and effect – a certain phenotype. These workers were not sensitized to the careful consideration of interaction between different influences that shape psychological development and fate of a human being and are usually studied by behavioral scientists, such as psychologists, social anthropologists, and sociologists. This trend played into the hands of the media, which preferred simplistic interpretations to complex explanations.

The discussion of intelligence (Sect. 15.2.1.3) has made it abundantly clear that we do not consider the biometric analysis of complex phenotypic relationships to be the appropriate approach for analyzing genetic *mechanisms*. However, once a genetic mechanism has been established by other, more incisive methods, the combined sophistication of social science, epidemiology, and statistics is needed to assess the phenotypic impact of this genetic mechanism.

Chromosome Aberrations and Behavior: Some General Conclusions. The chromosome aberrations, especially those involving X and Y chromosomes, provide a model to show how genetic variability and environment may interact in producing a psychological phenotype and point to those intervening variables that should be considered: abnormalities in brain physiology or biochemistry and in the endocrine system. The example of sex chromosome aberrations has also indicated a research strategy: first identify a variant genotype, then explore its influence on the phenotype, considering concomitantly the intra- and interindividual differences in the environment. This strategy is the opposite of the usual approach, which starts with the phenotype. However, such strategy has already led to success in the genetic analysis of monogenic diseases (positional cloning, Sect. 3.1.3.9) and of complex diseases (association with genetic markers such as blood groups, HLA types and DNA variants; identification of contributing genes by linkage studies). This problem is taken up below in the discussion of mental diseases. Disturbances in em-

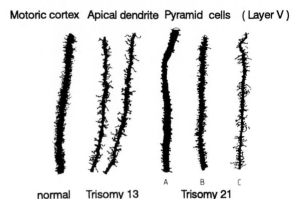

Fig. 15.20. Golgi-stained neuronal dendrites among normals, and in trisomies 13 and 21. In trisomy 13 spines are thinner, unusually long, sometimes dilated, and reduced in number. In trisomy 21 three types of structural anomalies are found: reduction of spines *(A)*, unusually small spines *(B)*, and very long and thin spines *(C)*. (Marin-Padilla; from Propping 1989 [186])

bryonic development and physiology caused by chromosome aberrations are poorly understood. We would be happier if this strategy were feasible not only for chromosome aberrations but also for genetic variability at single gene loci with known physiological mechanisms.

Brain Morphology in Chromosomal Aberrations. (See also Fig. 15.20) Since chromosomal aberrations often lead to reduced intellectual performance and increased behavioral difficulties, and since EEG results in these disorders suggest anomalies in brain development and maturation, we would expect morphological investigations to provide clues regarding the mechanisms by which such aberrations impair brain function. The literature, however, is scanty, although increasing numbers of affected fetuses are available [74] following abortion of affected fetuses diagnosed by prenatal diagnosis. In a series of 274 fetuses and children with trisomies 13, 18, and 21 and some other aberrations, gross neuropathological changes were detected in only two thirds of the cases. The most common findings were holoprosencephaly/arhinencephaly, mainly in trisomy 13 but also in one case of trisomy 18; corpus callosum defects, found mainly in trisomy 18 but sometimes also in trisomies 13 and 21; and neurons that had not reached their normal location during embryonic development (= heterotypy and microdysplasias), mainly in trisomy 21 but also in cases of other chromosomal syndromes. It should be noted that such anomalies may occur at lower frequencies in fetuses and infants without chromosomal aberration; for example, in about-one third of cases with holoprosencephaly, *no* chromosomal aberration is found. Moreover, cerebellar nerve cell heterotopies are frequently found in chromosomally normal newborns, mostly prematures, and disappear with increasing age. Disorders of neuronal migration have also been described in many other anomalies [8]. There are some scattered reports on the brain pathology of sex chromosomal aberrations; for example cortical dysplasia, disturbances of cell migration, and delayed cellular maturation in 45,X and similar anomalies [20] or abnormal cortical convolutions of the frontal cerebral cortex [4] in XYY patients. The significance of these findings remains unclear until the frequency of such anomalies in individuals *not* suffering from a chromosomal aberration is firmly determined. In our opinion, the application of refined neuropathological methods to fetuses and newborns with chromosomal aberrations – always keeping in mind variability in the normal range – might open new approaches to the study of biological mechanisms affecting behavior-genetic variation. On the basis of other evidence, such as the results of aphasia studies, the anomaly described above in XO Turner cases has been tentatively localized to the parieto-occipital region of the cerebral cortex.

A Common Morphological Substrate in Various Types of Mental Retardation? Micromorphological analyses in various types of mental deficiency have been claimed to share a common morphological substrate, an anomaly in dendrites of cortical neurons [186]. Normally these neurons show signs of intensive development during the last 3 months of fetal life: numerous spines are formed that establish contacts with other neurons. In various types of mental deficiencies their numbers are diminished, and they often appear altered (Fig. 15.20).

15.2.3 Suggested Novel Approaches to Human Behavior Genetics

Most material covered in this section relates to possible applications of genetic concepts to behavioral phenomena. These genetic concepts are elaborated in the previous chapters, especially Chaps. 4–6 (formal genetics) and Chaps. 7 and 8 (gene action). So far their explanatory power has not been utilized fully for behavior genetic problems; therefore we advocate *increasing attention to these concepts and methods. We fully realize that the application of such concepts will not be as straightforward* in explaining genetic determination of human behavior, development and variability as, for example, structural mutations of the hemoglobin loci explain the various hemoglobinopathies.

We noted in the introduction to this book that the genetic approach to biological phenomena is "reductionistic:" genetic analysis is regarded as successful when an inherited difference can be traced to a specific difference in the structure of a gene. In many cases – for example, when we want to understand an enzyme deficiency – this approach leads to satisfactory results. Its limitations become apparent when we try to analyze genetic variability in embryonic development and its deviations that lead to congenital malformations. Here a complex system of feedback mechanisms apparently regulates the activity of genes in the various cell groups and in different phases of development. The development of human behavior

is more complex since it is never finished and proceeds through our lifetime. The feedback mechanisms involved appear to be more complex than those operative during embryonic somatic development. It is possible that for some mental functions new principles of gene action have appeared during evolution. Thus our current genetic concepts, which have proven so successful in other fields, may have only limited success in behavior genetic analysis. Still, we must start with these concepts.

Here we are encouraged by an analysis of Bunge and Ardila [27]. With regard to psychology, they write: "The correct and badly needed synthesis is the merger of all the branches of psychology on the basis of neuroscience, together with developmental and evolutionary biology, and in tandem with social science. This is the correct synthesis because behavior and mentation happen to be biological processes". While psychological processes are entirely determined by neurobiological mechanisms, the psychological processes that involve an increasing complexity of the interactive system will make for the emergence of new properties that cannot be derived from the separate analysis of its components. As to the required research procedures, the authors distinguish two approaches: "top-down" and "bottom-up." The "top-down" approach starts with complex phenomena and tries to analyze them down to their single components. The "bottom-up" approach attempts an integration of elements into more complex units. The analogy with modern research strategies in human genetics is striking.

The approach advocated here is principally, but not exclusively, the bottom-up approach. This strategy, which has been successful in other fields, represents the optimal way to determine the limits of genetic principles [250]. In addition, new concepts may emerge that can help us overcome current constraints. For example, the space perception defect in the Turner syndrome may have a definite morphological substrate (see above).

15.2.3.1 Genetic Variability That Could Influence Human Behavior

Few actual data exist in these areas. We describe possible approaches in some detail because we feel that this field should be further explored (Fig. 15.21).

General Metabolism. With gross disturbance of metabolism consciousness becomes blurred, and mental processes come to a halt. Nongenetic examples include hepatic failure leading to liver coma and renal failure causing uremia with its attendant effects on the brain. The mechanisms of both conditions are

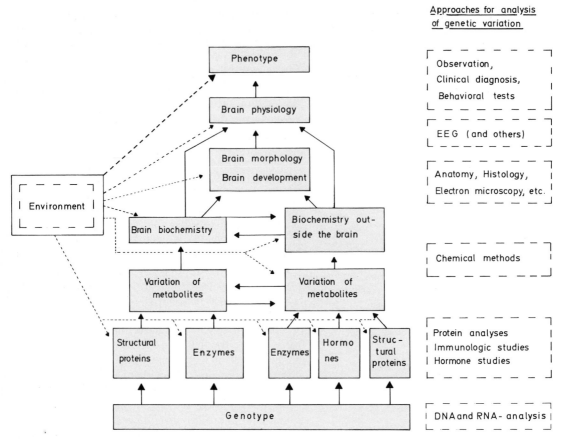

Fig. 15.21. Levels at which genetic variability of brain function can be investigated. ☐ Level at which genetic variability might occur; ⸬ method of investigation

obvious, at least in gross outline. Due to the failure to properly metabolize or to excrete certain compounds their concentration increases, they disturb the normal metabolic processes in the brain, and if the disturbance is severe enough, abolish mental functioning.

Brain function may be altered by intoxication not only when a toxic agent is produced by metabolic processes within the body, but also when it is taken up from the outside. The alteration of consciousness and feeling tone that follows ingestion of naturally occurring toxic foods was observed fairly early in the development of the human species, as witnessed by the fact that some of these effects were known in almost all primitive populations. In modern society drug addiction is one of the most threatening social problems. Modern medicine, on the other hand, has learned to utilize drug action affecting the nervous system in a great variety of ways – from general anesthesia to psychopharmacological agents that help to manipulate the psychic state of patients with mental diseases. Before drugs enter the brain and alter its functions, they must be metabolized, with changes

in chemical composition and modification of their action. These metabolic changes are mediated mainly by various enzymes. Enzymes, in turn, may show genetic variability that influences activity, substrate specificity, and other characteristics (Sect. 7.2). The resulting differences in both drug action and side effects, as well as genetic variation of drug receptors are the object of pharmacogenetics (Sect. 7.5.1). In principle, genetic variability may be found among enzymes that metabolize compounds of our normal food needed for certain metabolic processes. In some cases the genetic alteration of the enzyme molecule may be so profound that its activity is totally or almost totally abolished. Examples involving development and proper function of the brain include the phenylalanine hydroxylase defects in phenylketonuria, which leads to increased brain concentrations of phenylalanine and other abnormal toxic metabolites, and the chronic states of ammonia intoxication that ensue from enzyme defects in the urea cycle. Many inborn errors of metabolism are associated with mental retardation through analogous mechanisms.

Variability of Hormones. Many of the genetic regulatory processes in the human organism are mediated by *hormones* (Chap. 8). We know a great many qualitative and quantitative genetic defects of hormonal function. Most of them also affect human well-being and behavior. Obvious examples include defects of thyroid function that lead to hypothyroidism with severe mental retardation in infants and lack of alertness among adults. Sexual hormones such as testosterone and estrogen also have a profound effect on embryonic and adolescent development and on the mental state and behavior of the adult. In looking for genetic mechanisms that can influence human behavior and could therefore be screened for genetic variability, endocrine glands represent major candidates (Fig. 15.22).

Genetic Variability Within the Brain. Genetic differences in quantitative hormone production and in the structure of hormone molecules are only two possible sources of genetic variability in hormone function. Another source is their effect on the target organs. Here *hormone receptors* are important mediators. Some of the genetic defects that are now described as receptor diseases (Sect. 7.6.4) have implications for brain function. It has been suggested that genetic variability in hormone receptors within the brain affect brain function and thus produce genetic variability in behavior.

To find other possible mechanisms, we must survey all levels at which brain structure and function can be examined – from anatomy and histology to electrophysiology and to the basic biochemical processes involved in excitation and inhibition of nerve cells. A huge amount of information is available in all these fields, but surprisingly little has been undertaken to screen this information for possible genetic variability. This may in part have technical reasons. Assessment of genetic variability normally requires examination of fairly large series of individuals. This is

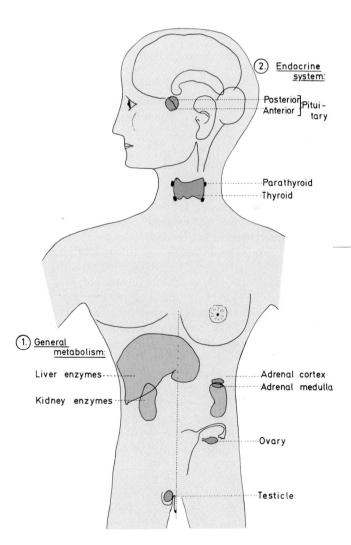

Fig. 15.22. Genetic variability outside the brain that may influence human behavior. In general, metabolic influences of liver and kidney enzymes are especially important. Other influences may come from endocrine glands, such as the anterior or posterior pituitary, the adrenals, gonads, thyroids, and parathyroids

now becoming increasingly possible by studying DNA in white blood cells for genes that are expressed only in the brain. It should be quite clear, however, that even complete understanding of the genes and the genetic polymorphisms affecting the brain is not likely to explain human behavior since behavior is so much more complex than, for example, red cell function. We therefore need to look for alternative strategies.

It is often believed that analysis of the genes themselves can provide a shortcut for such studies, and that genes – even those that are active exclusively in the brain – can also be studied in other tissues, for example, blood cells. This is certainly true, and investigation of these genes, their modes of action, and their interaction in producing certain phenotypes is the final goal of such studies. However, to know which genes can be studied with the best chances for success, it is necessary to assess the problem at various levels. The "bottom-up" strategy – identification of a gene by linkage studies and positional cloning, and analysis of its functions at different levels – is only one possible approach. However, this requires that following a genomic search or search for plausible candidate genes, the responsible genes are successfully mapped and cloned. Convincing success of this approach in complex traits – diseases as well as normal characteristics in which genetic analysis had failed to indicate a monogenic mode of inheritance – has been rare so far so that it would be premature to rely exclusively on this strategy. Alternative levels of study of "intermediate" phenotypes, i. e., phenotypic characteristics in metabolic pathways, between genotype and phenotype should be considered.

In addition to linkage studies, investigating associations with genetic markers has been proposed as an alternative technique. The rationale of such studies is that genetic markers, such as DNA polymorphisms (Sect. 12.1.2) may occur in linkage disequilibrium (Sect. 5.2.4) with genes, various alleles of which may lead to differences of psychological phenotypes. Such linkage disequilibrium may have two reasons: either the time that elapsed to separate by recombination the suspect genes from the tightly linked DNA polymorphisms may not have been sufficient for the disequilibrium to disappear, or the marker allele itself influences the phenotype to be studied. This approach has been pursued mainly in attempts to study various psychiatric diseases, but it is now being used in searching for genes involved in the genetic determination of quantitatively varying normal traits such as cognitive abilities measured by IQ [177].

a) The first possibility is to look for genetic variability in a physiological parameter that can be measured directly by noninvasive techniques. Here the genetic analysis of the normal electroencepha-

logram has so far offered the best opportunities [252, 255]. An obvious disadvantage of this approach is that the relationship between gene action and physiological phenotype is less direct and therefore more subject to distortion by intervening variables than when examining, for example, an enzyme.

b) The genetic variability in its relationship to behavior can be examined in the brains of experimental animals. This approach relies on the well-known homology in many physiological processes between humans and other mammals that has its foundation in their common phylogenetic origin. Animals are in widespread use as model systems for genetic analysis in situations where experimentation in humans is impossible for ethical reasons. For behavioral genetics the most obvious advantage is direct access to the brain and the opportunity to introduce genes of interest (transgenics; knock out mice); the obvious disadvantages are the likely existence of species differences and the unique role of the human brain. This approach can therefore provide us with models for analyzing the role of certain genes in behavioral processes and study of the mechanisms by which variation in brain physiology lead to behavioral differences. These models may provide hints where such differences might also be found in humans. Below we rely heavily on the results of animal experimentation.

Experimental animal populations are often derived from inbred strains and differ radically from human populations in breeding structure. Animals of an inbred strain are genetically identical; they may be compared to human MZ twins. Results achieved in one inbred strain can be compared with those in other inbred strains and with natural "wild" populations.

These approaches are recommended on the assumption that important aspects of genetic variability of brain function that lead to behavioral differences between human beings can be analyzed with the well-established principles of genetics – i. e., the gene concept, gene-determined enzymes and receptors, regulation of gene action and widespread genetic variability at the protein level. This is not to deny that ultimate understanding of complex intelligent behavior probably may require additional principles of information processing and organization.

15.2.3.2 Genetic Variability Outside the Brain That Influences Human Behavior

Enzyme Defects Leading to Mental Deficiency (Fig. 15.22). Many of the known enzyme defects in humans lead, apart from their other widely different phenotypic effects, to mental deficiency (Table 15.11). For our understanding of normal brain function, however, these observations have contributed very little so far. It is plausible, for example, that nerve cells cannot work properly if they are stuffed with insufficiently degraded macromolecules, such as those found in the mucopolysaccharidoses [157] or mucolipidoses; for understanding of normal function, however, this is not very revealing. In general, the brain's development and function appear to be especially sensitive to changes in the biochemical environment. Many different inborn errors tend to increase or decrease a variety of biochemical substances that often cause mental retardation.

In other instances the orderly and well-controlled cooperation of neurons is impaired; mass discharges of large groups of neurons manifest themselves as epileptic fits. Such seizures are observed in a great number of hereditary diseases [12]. Metabolic disturbances may also contribute to susceptibility for mental diseases included in the diagnostic category of "schizophrenia." Examples include the adult type of metachromatic leukodystrophy, various types of porphyria, and even Huntington chorea [185]. The topic is taken up again in Chap. 16.

Self Mutilating Behavior in the Lesch-Nyhan Syndrome: Uric Acid. Lesch-Nyhan syndrome is described in Sect. 7.2.2.6 as caused by a defect of the enzyme hypoxanthine-guanine phosphoribosyltransferase (HPRT) with failure in recycling hypoxanthine into guanine synthesis. Instead, high amounts are converted into uric acid and excreted. Apart from hyperexcitability of the nervous system, this leads to hyperreflexia and almost continuous movements; patients suffer from a compulsive tendency to self-destruction. In spite of the pain, they bite their fingers and lips and mutilate themselves. This destructive tendency has no counterpart in "normal" or "abnormal" human psychology; still, we know of other types of compulsive neuroses. Some patients, for example, feel compelled to wash their hands over and over again. Moreover, schizophrenic psychoses seem to occur among relatives of patients with compulsive neurosis more often than in the general population. Analysis of the specific damage caused by this enzyme defect in the brain could possibly be somewhat instructive regarding neuronal mechanisms of compulsive behavior. Moreover, it would be interesting to explore whether heterozygotes for the HPRT defect show behavioral peculiarities, especially when they reach middle or advanced age.

A positive correlation between elevated blood uric acid level and IQ has repeatedly been asserted and is usually backed by an impressive list of outstanding men in history who suffered from gout [219]. If uric

Table 15.11. Types of selected inherited metabolic diseases leading to mental deficiency

Kind of defect	Metabolic disease (examples)
Amino acid metabolism	Phenylketonuria Maple syrup urine disease Hyperammonemias
Carbohydrate metabolism	Galactosemia
Endocrine disorders	Types of cretinism with goiter
Disorders in binding to cofactors (vitamins)	Pyridoxine dependency
Lysosomal diseases	Mucopolysaccharidoses

acid slightly enhances neuronal excitability, this nonspecific stimulation could conceivably have a positive effect on intelligence and performance. On the other hand, a twin study failed to show a correlation between uric acid level and intelligence (P. Propping, personal communication).

Heterozygotes of Recessive Disorders. Enzyme activity in heterozygotes for inherited metabolic diseases is usually only half the normal levels. Therefore, we would not be surprised if phenotypic abnormalities were detected, at least when the specific metabolic pathways are placed under stress or possibly with advancing age. However, systematic studies on such heterozygotes, especially on their mental status and performance, are remarkably scanty (see also Sect. 7.2.2.8) [254]. The most comprehensive studies have been performed on phenylketonuria. Despite deficiencies in the epidemiological methods used in some of these studies, some conclusions are of interest: (a) On average, PKU heterozygotes show a slight reduction of a few points in IQ compared with controls; the verbal part of the test appears to be reduced more than the performance portion. (b) Some studies suggest a slightly higher risk of psychotic disorder, the psychoses having a late onset and a benign course and involving depressive symptoms. (c) Some heterozygotes have shown increased levels of cortical irritability as evidenced by EEG. (d) These slight anomalies could be caused at least in part by an increased level of intracellular phenylalanine and tyrosine since these amino acids are found to be increased in the (more easily accessible) lymphocytes.

Other autosomal-recessive diseases for which minor phenotypic deviations have been described include several lipidoses, such as the late infantile form of metachromatic leukodystrophy (250 100), globoid cell leukodystrophy (Krabbe) (245 200), and a few heterozygotes for Sandhoff disease (268 800), Niemann-Pick disease (257200), and Wolman disease (278000). Here the slight deficits were found mainly in the performance IQ, especially in subtests for space perception. Moreover, scores on personality questionnaires testing for psychosomatic disorder, depression, and emotional lability were increased. Again, accumulation of abnormal metabolites could be a causative factor since some accumulation of mucolipids is found not only in homozygotes but also in heterozygotes for some lipid storage disorders [45a].

Two studies on autosomal-recessive microcephaly (251200) from different parts of the world, Canada and the former Soviet Union, found about one-third of normocephalic heterozygotes to be mentally subnormal. According to Quazi and Reed [191], who carried out the Canadian study, such heterozygotes may constitute an appreciable proportion of all individuals suffering from mental subnormality "of unknown origin." According to these authors, homozygotes occur at a frequency of about 1 : 40000; heterozygote frequency is about 1 : 100 (Sect. 4.2.1). If one-third of these are mentally subnormal, this would mean that about one third of 1% "of the population at large is mentally retarded because they carry one gene for microcephaly. The prevalence of mental retardation in the United States and United Kingdom using an IQ of 69 as a threshold has been estimated to be about three percent" [175]. Thus about one of nine mentally retarded individuals from the general population could be a heterozygote for the gene for microcephaly. Of course this estimate can be regarded only as a crude approximation; in other populations recessive microcephaly appears to be much rarer, and, moreover, there may be genetic heterogeneity. In principle, however, this argument is plausible. It could even be generalized: when one-third of such heterozygotes are mentally defective, the other two thirds probably have an IQ in the lower normal range. Moreover, about 1 in 50 individuals in our population are heterozygous for PKU (assuming a homozygote frequency of $\approx$ 1 : 10000), and heterozygotes for other recessive diseases for which slight reductions in IQ have been described do also occur. It should be remembered that many types of autosomal or X-linked recessive mental retardation have been only incompletely characterized so far. If the rather frequent types of X-linked mental retardation are added, it could be argued that a significant proportion of the "normal" genetic variability of IQ in the lower range is caused by heterozygosity of autosomal or X-linked recessive diseases. More definitive studies are needed.

15.2.3.3 Hormone Action

How Do Hormones Act? As explained in Chap. 8, hormones usually act on special cells that have receptors by which the hormones are bound. This triggers synthesis of specific proteins within these cells; here cyclic AMP acts as a mediator – as the "second messenger." Some "receptor diseases" have now become known; the best analyzed examples are familial hypercholesterolemia (Sect. 7.6) and testicular feminization (Sect. 8.5). In the latter condition, a deficiency of androgen receptors causes a female phenotype in individuals having an XY karyotype with testicles.

Hormone receptors are also present on cells of the CNS; their development and function can be influenced at least at three levels: by hormone-induced metabolic processes in other tissues that influence brain function indirectly, by quantitatively or qualitatively abnormal hormone supply, and by individual differences in receptors. For instance, the effect of thyroxin on mental and neural activity, which is so impressive when hypothyroid patients respond to thyroid therapy, is probably indirect since thyroxine enhances the basal metabolic rate in all tissues except the brain.

Sex hormones, on the other hand, appear to have a direct influence on brain development [140]. This influence begins as early as in embryonic age, as shown by experiments mainly with rats [186]. Within a certain, sensitive period, the preoptic zone in the hypothalamus is imprinted. Female rats that have been treated with testosterone a few days before and about 10 days after birth behave as males during adult age.

In the human brain, a certain cell group within the preoptic zone is about 2.5 times larger – and contains more cells – in the male than in the female brain [234]. There are also some differences of dendrite structure and in the corpus callosum. The influence of testosterone is shown by the rare enzyme defect of 5 α-reductase deficiency (264600). This enzyme determines conversion of testosterone into dihydrotestosterone; its defect leads to pseudohermaphroditism. Generally the infants are regarded, as girls at birth, and are reared accordingly. During puberty, however, increased testosterone production leads to virilization, and they change their gender identity, physically and psychologically developing into normal, sexually functioning males. Hormonal status thus overrides the sociocultural effect of education as a girl [97].

Therapeutically administered steroid hormones may have a similar effect. For example, some pregnancy complications have been treated with progestins, which usually contain an androgen component. These progestins influence development of the embryonic brains, as witnessed by observations on girls exposed to synthetic progestins.

Tomboyism in Girls Prenatally Exposed to Masculinizing Compounds [140]. Tomboyish girls like to join boys in outdoor sports; their self-assertiveness in competition for positions in childhood groups is strong enough to permit successful competition with boys. They like to dress as boys and do not play with dolls. Often the tomboyish girls reaches the boyfriend stage later than her peers; romance and marriage take second place to achievement and career. There is, however, no tendency to lesbianism.

While such behavior may develop in response to environmental influences, it has been reported that tomboyish behavior can be primed by hormone action during embryonic development. Progestins are steroids related in chemical structure to androgens that can substitute for progesterone in their pregnancy-preserving function. When first introduced in therapy, their masculinizing effect was unknown. In the 1950s some mothers gave birth to otherwise normal daughters but with masculinization of the clitoris. Such children grew up as girls; the masculinizing effect ceased at birth.
Another group of masculinized girls are those with adrenogenital syndromes due mainly to the autosomal-recessive defect of 21-hydroxylase, one of the enzymes needed for synthesis of cortisol (Sect. 7.2). The condition is treated by cortisol substitution, which by reducing the ACTH production of the pituitary curbs synthesis of cortisol precursors with androgenlike effect. Exposure is thus limited to the prenatal period, as encountered in daughters of progestin-treated mothers.

Tomboyism was present in nine of ten girls with the progestin-induced syndrome, in 11 of the 15 girls with adrenogenital syndrome, but in none or only very few Turner and testicular feminization patients. The attitudes of all these girls toward their future roles as married women and mothers were altered accordingly [140].
Interpretation of these results was corroborated in experiments with female rhesus monkeys which had been artificially androgenized during embryonic development, and which during childhood showed behavior that resembled that of their *male* age mates [140]. These data favor the view that sex-specific brain patterns influence the attitudes of children and adolescents in accepting their sex roles.
Repeated studies show boys to be more aggressive than girls, and their aggressiveness is correlated with blood testosterone levels [35]. This led to the hypothesis that boys whose mothers had been treated with progestins during pregnancy would display more aggressive tendencies than their untreated brothers. Eight such boys and 17 girls were asked how they would react in certain provocative situations. The boys (but not the girls) showed much higher aggressiveness than their untreated brothers.

Hence, not only the present testosterone level but also the degree of androgen imprinting of their brains appears to influence their aggressiveness [194].

Testicular Feminization. The testicular feminization syndrome (313700) is caused by androgen insensitivity. If the brain also lacks functional androgen receptors, one would expect in these individuals "typically female" attitudes toward marriage and motherhood. Such tendencies were, indeed, found in most of the ten patients that could be examined [140]. However, the interpretation is more ambiguous than of data in prenatally masculinized girls. Patients with testicular feminization are almost perfect girls externally and grow up as such; therefore, their identification with the female role could also be explained in a straightforward psychological manner.
The problem of androgen influence on the female brain was reexamined repeatedly both in animals and in humans, not always with the same results [186]. However, it is our impression from the bulk of the literature that prenatal androgen exposure in some way influences brain development.

Homosexuality and Hormones. The twin and family studies discussed in Sect. 15.2.1.6 suggest a genetic component in the homosexual behavior of males. It has been suggested that the distortion of the sexual release scheme of male homosexuals may be caused by an abnormally low androgen effect. This problem has been investigated repeatedly by measuring urinary excretion or serum levels of steroid hormones [140]. Different proportions of various selected metabolites of sex hormones were claimed to be detectable in urine; however, the studies were poorly controlled for age, amount of sexual activity, or general health status. Moreover, the degree of variation of these metabolites in the urine of controls is considerable and strongly overlaps with that of homosexuals.

Even males awaiting estrogen therapy and sex-assignment surgery had normal androgen levels [46].

These results show that there is no simple relationship between androgen production and homosexuality. However, they do not exclude the possibility of minor differences in androgen receptors and their relationship to homosexuality. A possible influence of an X-linked gene on male homosexuality is of interest in this connection [77]. However, the X-linked androgen receptor gene was ruled out as a candidate gene to explain these findings. Independent evidence of sex-specific differences in brain function comes from studies with a noninvasive method that enables us to assess the functional state of the central nervous system – the electroencephalogram (EEG).

15.2.3.4 Brain Physiology: Genetics of the EEG

Investigations of genetic variability in brain function in humans are fraught with technical difficulties. Therefore more indirect approaches are necessary. Several have been mentioned: direct examination of brain physiology, investigation of genetic variability outside the brain that might be related to brain function, and animal experimentation. For studies of human brain physiology the EEG has mainly been used [252, 255].

Human EEG. The most important features of the human EEG are described in Sect. 6.1 where criteria for a simple autosomal-dominant mode of inheritance in a continuously varying character are discussed using the low-voltage EEG as an example. The resting EEG develops during childhood and youth from irregular forms with relatively slow waves to the final pattern, which is reached latest by about 19 years. This adult pattern is dominated by α waves (frequencies 8–13/s) with variable admixture of β waves ($\geq$ 14/s) and ϑ waves (4–7/s); α waves are usually most prominent over the occipital region of the brain. Under normal conditions, and in healthy individuals, there is little variation between leads from the same individual taken at different times; variation between individuals, on the other hand, is considerable.

Twin Studies. The EEG records a highly complex trait that varies in many dimensions, for example, distribution of frequencies and amplitudes in one lead, variations between leads from the various parts of the head surface, and form of the waves. To obtain an overall impression of the role of genetic factors in this variability, it was reasonable to begin by comparing MZ and DZ twins. As the EEG shows a characteristic development from infancy to adult age, with varying speed between individuals, twins in their first and second decades of life, together with a few young adults, were the most appropriate probands. In the absence of disturbing factors such as severe fatigue, brain disease such as epilepsy and brain tumors, and severe metabolic anomalies, the brain wave pattern under standard conditions (relaxed, with closed eyes) is determined almost completely genetically. This conclusion also applies to the speed of brain maturation as evidenced by EEG patterns [251].

It could be argued that this concordance was caused by the common environment of the twins. EEG studies carried out on MZ pairs reared apart showed the same degree of EEG concordance [106, 222]. This strong similarity is maintained until old age.
The adult EEG pattern is identical even in twin pairs with quite different emotional histories, for example, when one suffered from a severe neurosis. Organic brain diseases, on the other hand, such as stroke and epilepsy, may cause definitive and permanent differences even between MZ twins [241].

Children regarded by psychologists as "immature" or showing behavioral abnormalities often have irregular EEG patterns. Such data suggest the hypothesis that the pattern of EEG maturation is related to differences in psychological maturation in the normal range, as measured by developmental tests and performance [261]. EEG frequency therefore measures part of the genetic variability influencing individual differences in normal psychological development. This relationship needs to be studied in much greater detail.

Family Studies. The results of twin studies encouraged attempts to find more clearcut evidence of genetic mechanisms. If monogenic inheritance is expected, alternatively distributed characteristics should be chosen for study. The most evident is the so-called low-voltage EEG with little or no occipital α activity. The mode of inheritance turns out to be autosomal-dominant (Sect. 6.1; Fig. 6.9), with little overlap between the two phenotypic classes [3]. A gene for this EEG type has been localized in 20q [223], but there is genetic heterogeneity. In the same segment of 20q, a gene for a subunit of the nicotinic acetylcholine – receptor has been mapped – a good candidate gene for an EEG variant [222a].
Another hereditary EEG type is dominated by monomorphic α waves. The average α rhythm has a maximum over occipital parts of the cerebral cortex and is more irregular and mixed with other waves over frontoprecentral areas. In the monomorphic α wave type very regular α waves of high amplitudes are usually seen over the whole cerebral cortex. Twin and family studies leave no doubt that this type is hereditary. Family data suggest simple dominance [252]. However, delineation from the more average EEG is more difficult than with the low-voltage EEG.
In a certain way this EEG type can be regarded as the "countertype" of the low-voltage EEG; while the low-voltage EEG has a weak α rhythm, this rhythm seems to be especially strong in the monomorphic α EEG.
Some other traits with simple dominant modes of inheritance have also been identified, for example, a variant in which the occipital α waves are replaced by 16–19/s waves that show the general characteristics of α waves, such as blocking after eye opening and other stimuli. In other families, genetic variants of the fast component, the β waves, were discovered. In distinction to the α waves, β waves are in most cases concentrated above the frontal and precentral parts of the brain; in some families these β waves form characteristic spindlelike groups. Two autosomal-dominant types have been distinguished [252].

Sex Difference in EEG Patterns. Most EEGs with β waves show definite familial aggregation but the data do not conform to simple monogenic inheritance. They are most easily explained by multifactorial inheritance in combination with threshold effects (Sect. 6.1.2).
Furthermore, the prevalence of EEGs with (mainly diffuse) β waves increases with age, and there is a definite sex difference. In all age groups – with the exception of childhood – the prevalence of EEGs with β waves is higher in females than in males [59].

How Is the EEG Produced in the Brain? Hereditary variation in EEG patterns suggests differences in physiological function of the human brain. To understand the nature of these differences and their possible influence on behavior it is necessary to know how the EEG is produced [2]. The EEG waves, especially α waves, are formed by the interaction of neurophysiological processes at no fewer than three or four levels (Fig. 15.23). The EEG "battery" is located in the cerebral cortex; here groups of neurons discharge in rhythmic order. However, their activity is coordinated by a "pacemaker" (more precisely, a group of interrelated pacemakers) in the thalamus. The thalamus in turn is influenced in its activity by input from brain structures at lower levels. The ascending reticular activating system (ARAS) is located in the reticular formation, mainly in the pons and medulla oblongata. The ARAS has a leading function, for example, in sleeping and dreaming. In the waking state, it maintains a level of "tonic arousal," which is influenced by input from specific, centripetal pathways involving, for example, sensory stimulation. A high level of arousal causes EEG desynchronization. Other influences on the EEG come from the limbic system, a functional unit comprising the hippocampus, amygdala, mammillary bodies, and connecting structures. The limbic system is involved in emotion, activity, and motivation. Neurophysiological studies have also provided some hints as to the physiological function of α activity [2]: the α rhythm seems to modulate and selectively amplify afferent stimuli.

Influences of Inherited EEG Variations on Personality. The personality and performance of an individual depends on the way in which its brain works spontaneously, reacts to incoming stimuli, and handles information. Individual differences in such neurophysiological parameters therefore result in psychological differences. What influences on "personality" and intellectual performance can be expected in individuals carrying the genetic EEG variants described above, considering the neurophysiological results outlined in the proceding paragraph? Probands with monomorphic α waves might be expected to be "strong modulators and amplifiers," whereas probands with a low-voltage EEG might be weak modulators and amplifiers. The β waves, especially those of the diffuse variant (which seems to be multifactorially inherited) are generally assumed to be the result of a high level of tonic arousal in the ARAS. Therefore disturbances in the modulating function of α activity are expected in probands with this EEG variant.
Indeed, the results of comparative studies on 298 probands with various EEG variants were broadly compatible with these tentative expectations [257, 259–261]:

a) The monomorphic α wave pattern tends to be on average sthenic, stable, and reliable. The probands

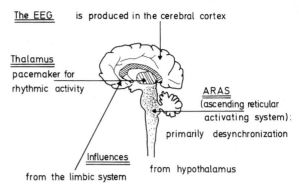

Fig. 15.23. Oversimplified drawing of the human brain showing the structures involved in production of the EEG

are more likely to show signs of high spontaneous activity and toughmindedness; precision of work, especially under stress conditions, and short-term memory are especially good. On the other hand, information processing is not very quick.

b) The low-voltage category shows little spontaneous activity; the probands tend to be group-oriented and extraverted. Spatial orientation is especially good. The expected difference in information processing between groups (a) and (b) have been shown directly by studying averaged evoked EEG potentials [262].

c) The group with diffuse β waves shows a high error rate in spite of low working speed in tests measuring concentration and precision. Stress resistance seems to be low. The disturbance of the α mechanism by high tonic arousal leads to a disturbance in intelligence test performance, especially in tasks measuring the ability for spatial orientation.

An oversimplified picture of the proposed relationship between EEG variation and personality is given in Table 15.12, which also lists some additional EEG variants. For example, probands with fast occipital α variants (16–19 c/s) seem to excel in abstract thinking and in motor skills. They are probably able to process information rapidly. The EEG literature has reports on a possible positive correlation between α frequency and intellectual performance (see [257]).
Another, very rare EEG variant not shown in Table 15.12 differs from superficially similar types by a number of characteristics; for example, the 4–5 cps waves are blocked by eye opening and are immediately replaced by α waves after minute disturbances. The genetic basis is not quite clear; two concordant MZ pairs and a small number of families with more than one affected member point to genetic factors, but most probands are the only affected persons in otherwise normal families. Many probands with this EEG

Table 15.12. Hereditary EEG variants, their genetic bases, and their psychological consequences [257]

EEG type and mode of inheritance	Genetic variation of EEG	Functional consequences	Psychological consequences[a]
Monomorphic α waves (probably autosomal-dominant)		Strong selection and amplification	Sthenic (tough-minded), stable, resistant to stress
Fast α variants (16–19/s) (autosomal-dominant)		Fast information processing	High intelligence and high motor skill
Low-voltage EEG (autosomal-dominant)		Weak amplification	Relaxed, little activity, conformistic
Low-voltage borderline (mixed group)		Weak amplification; disturbed	(Mixed group)
Frontoprecentral β groups (autosomal-dominant)		?	(Inconspicuous; undisturbed)
Diffuse β waves (polygenic)		Disturbed information processing due to high tonic arousal	Tense, disturbed spatial orientation, susceptible to stress
Normal EEG (polygenic)			

[a] The psychological characteristics for the various EEG variants are much oversimplified.

variant show emotional disturbances and abnormalities in the autonomic nervous system; among psychiatric patients the variant is much more common than in the general population [148]. An anomaly in the limbic system may explain these results.

Association Between α Waves and Spatial Ability. The mean performance of women is poorer than that of men in test items requiring visual orientation in space (Sect. 15.2.2.2). It was shown to be especially poor in patients with Turner syndrome. As parieto-occipital parts of the cortex are involved in visual perception, a relationship of occipital α activity to performance in tests for spatial perception is possible. Some studies suggest a decrease in spatial ability with decreasing α and increasing β activity as expected from the sex difference in both measures. However, the evidence is ambiguous. If the above reasoning is correct, we would expect low α and high, diffuse β activity. A study on 62 Danish females with Turner syndrome (age 6–47 years) found precisely this when the probands were compared with an age-matched control sample of non-Turner women [244]. Probands also showed more slow waves in the ϑ and δ ranges as signs of slight functional disturbance, but there were no focal abnormalities.

Averaged Evoked EEG Potentials. The brain reacts to a stimulus by producing a characteristic wave. This reaction, however, is so small that it is lost in the "noise" produced by the resting EEG. This problem can be overcome by repeating stimuli (e.g., light flashes and sounds) many times and adding up the reactions. In this way background noise will level out and the *averaged evoked potential (AEP)* remains. Its characteristic form has been found to be concordant in MZ twins [23]. Some of its properties have been claimed to be correlated with measures of intelligence [84], personality characteristics, and susceptibility to mental disease [24]. However, there are many controversies regarding details of methods, artifacts, epidemiological basis of samples examined, and results. A comprehensive study failed to confirm the correlation with intelligence [262a].

Considerably more work needs to be done on large numbers of individuals to validate psychological and cognitive correlates of EEG patterns.

15.2.3.5 Genetic Aspects of Alcoholism

Alcoholism is defined as a dependence on alcohol which leads to social disability. Alcoholism depends strongly on the environment. In a society in which alcohol-containing drinks are not available no one can become an alcoholic. Even in Western countries where opportunity is abundant and social pressure tends to favor alcohol consumption, individuals who become alcohol-dependent often suffer from personality disturbances for which no psychological or social explanation can be found. There are individual differences in alcohol susceptibility, and the possible genetic basis of such differences can be studied [187].

Animal Models. Experiments with mice and rats suggest that genetic variability influences the susceptibility to alcoholism (Sect. 15.1.2). Definite differences in alcohol preference have been found between inbred strains. These differences were shown to be associated not only with differences in alcohol metabolism but also with quantitative and qualitative differences in response of the brain to alcohol as well. These results suggest that genetic differences between humans should also be sought at two levels – in alcohol metabolism and in influence on brain physiology.

Studies with Classic Methods: Family, Twin, and Adoption Studies. Numerous studies in families of alcoholic probands have been performed. Some years ago Cotton [42] reviewed no less than 27 studies published in English, in which the families of altogether 6251 alcoholics and 4083 nonalcoholics were compared. Despite the fact that these studies are not of equal caliber since controls were not always investigated, some general conclusions are possible: almost one-third of all alcoholics had at least one affected parent, and in the majority of these cases (25%; range of variation studies: 2.5%–50%), the father was an alcoholic. Most studies found relatives of female alcoholics to be affected more often than those of affected males. This could indicate that women become alcoholic more often for endogenous reasons, whereas among men the immediate cause of alcoholism is related more frequently to nonfamilial environmental factors. Among family members of patients with schizophrenia and affective disorders, alcoholism was much rarer than in families of alcoholics. An increased susceptibility to alcoholism is therefore not a general feature of patients with major psychiatric disorders. No specific personality type among alcoholics has been discovered.

Åmark in Sweden studied the relatives of 645 alcoholic probands. In addition to alcoholism, he also found an increased risk of "psychopathy," "psychogenic psychoses," and criminality, but not of mental retardation, epilepsy, or endogenous psychoses. In this connection, it should be remembered that in Scandinavian countries, alcohol consumption in daily life is shunned more strongly by society than in countries such as Italy or France, where alcoholic drinks are a part of daily life.

Of course, the aggregation of alcoholics in certain families and the results of twin studies do not by themselves provide definite evidence of genetic susceptibility. Therefore family investigations were supplemented by twin and adoption studies. Table 15.13 summarizes data from six twin studies on alcohol abuse or alcoholism [187]. Populations from which these samples were drawn, as well as methods of investigation, and to some degree results differed between these studies, but the overall result was that of higher concordance of drinking habits and signs of addiction among MZ than among DZ twins.

A number of adoption studies have been performed [253], in Denmark.

Table 15.14 compares 55 Danish adoptees (age range 20–40) of whom at least one biological parent had been hospitalized for alcoholism with 78 controls [68, 69]. The data come from a pool of 5483 adoption cases originally established for the study of schizophrenia. No difference was found between the adopting families of probands and controls as to drinking habits or socioeconomic background. There was a strong dissimilarity between the two groups as to alcoholism; a difference also emerged in the frequency of divorce, whereas other types of psychopathology were similar. The two groups differed only as to alcoholism, defined as social and job problems, loss of control, withdrawal hallucinations, and psychiatric treatment. Heavy and even problem drinkers without these symptoms were found among the controls in about the same frequency. This study was supplemented by one in which 20 sons of alcoholics who were adopted by other families were compared with 30 of their biological brothers who had not been adopted and grew up in the homes of their biological parents [68]. As expected, the average socioeconomic conditions in these homes were poorer than those in the families of the adoptive parents. The result of the study, however, was unexpected: among the nonadopted brothers, the proportion of alcoholics was about the same as in the adopted group. Had the environment played an important role in causing alcoholism, one would have expected many more alcoholics in

Table 15.13. Twin studies on alcohol abuse or alcoholism (from Propping 1992 [187]), with additions

Reference	No. of pairs	Characteristics studied	Concordance rates (%)		
			MZ	DZ	h^2
Partanen et al.	172 MZ, 557 DZ males	Amount of intake			0.36
		Density of drinking			0.39
		Lack of control			0.14
Loehlin	850 same-sex pairs	"Had a hangover"			0.54
		"Used alcohol excessively"			0.36
		"Drink before breakfast"			0.62
		"Never any drinking"			0.36
Kaij	48 MZ, 126 DZ males	Five degrees of alcohol abuse	55.6–71.4	20.0–32.3	
Hrubec and Omenn	15924 males	Alcoholism, including alcohol psychosis	26.3	11.9	
Gurling et al.	28 MZ, 28 DZ, males and females	Alcohol dependence	21	25	
Kaprio et al.	879 MZ, 1940 DZ males	Frequency of beer drinking	40.8	21.5	0.39
		Frequency of spirits drinking	32.3	13.2	0.38
		Density of drinking	43.6	23.8	0.40
		Quantity of drinking	37.4	19.3	0.36
Pickens et al.	81 MZ, 88 DZ, males and females	Alcohol abuse and dependence	♂ 76.0	60.9	0.35
			♀ 35.5	25.0	0.24

h^2, Heritability.

Table 15.14. Comparison of drinking problems and patterns in two adoptive groups (percentages; from Goodwin 1976 [68], Goodwin et al. 1973 [69])

	Probands[a] (n = 55)	Controls (n = 78)
Hallucinations*	6	0
Lost control**	35	17
Amnesia	53	41
Tremor	24	22
Morning drinking**	29	11
Delirium tremens	6	1
Rum fits	2	0
Social disapproval	6	8
Marital trouble	18	9
Job trouble	7	3
Drunken driving arrests	7	4
Police trouble, other	15	8
Treated for drinking, any*	9	1
Hospitalized for drinking	7	0
Drinking pattern		
Moderate drinker	51	45
Heavy drinker	22	36
Problem drinker	9	14
Alcoholic**	18	5

* $p < 0.02$; ** $p < 0.05$.
[a] Adult male adoptees: At least one biological parent hospitalized for alcoholism.

the nonadopted group. Therefore these data suggest that in a society in which alcoholic drinks are universally available genetic factors largely determine whether an individual becomes an alcoholic.

A similar study compared daughters of alcoholics who gave them up for adoption with their sisters who had stayed with their alcoholic parents. Despite the pronounced difference in familial environment alcoholism was about equally frequent in the two groups, and more common than in the female population [70]. The results of a number of other adoption studies are summarized in [187].

Psychiatric Symptoms Preceding Alcoholism [2328]. Psychological and psychiatric signs and symptoms preceding alcoholism might be useful for planning preventive measures, especially in children of alcoholics. Hyperactive children appear to run a higher risk; the three psychological dimensions of novelty seeking, harm avoidance, and reward dependence at the age of 10–11 years have been found to be correlated with alcohol abuse in young adult age. Moreover, alcoholism is often combined with depression, but the occurrence of alcoholism is not increased in relatives of depressive, nonalcoholic probands. Cloninger [38] tried to differentiate between various personality types in children in terms of risks for alcoholism.

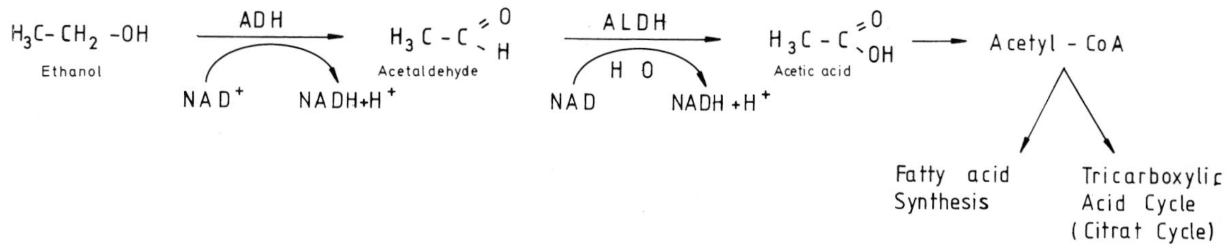

Fig.15.24. Alcohol degradation in two successive steps, the first step controlled by alcohol dehydrogenase *(ADH)*, the second by acetaldehyde dehydrogenase *(ALDH)*

The problem is not simple. Just as there are different reasons (or combinations of reasons) for a person to become an alcohol addict, predisposing factors may also differ. Some of these can be identified.

Genetic Variability of Alcohol Metabolism. If there is genetic variability in susceptibility to alcoholism, what is its mechanism? The answer can be sought at two levels: alcohol metabolism [1a] and alcohol action on the brain.

Twin studies have shown a strong genetic influence on alcohol metabolism [183]. Figure 15.24 shows the two most important steps of ethanol oxidation. The two most important enzymes are alcohol dehydrogenase (ADH) and aldehyde dehydrogenase (ALDH). Both are localized in the liver. ADH is determined by three autosomal gene loci (ADH_1, ADH_2, ADH_3). The genes ADH_1 and ADH_3 are active mainly during fetal life; in adults ADH_2 is responsible for most of the activity in liver and kidney. In 5%–20% of individuals of European extraction, but in 90% of Japanese, an atypical variant is discovered. Since at physiological pH the atypical enzyme shows much more activity than the more common one, it has been suggested that alcohol oxidation procedes more rapidly in carriers of the atypical enzyme. Moreover, it is known that many Japanese show the phenomenon of "flushing" after the intake of relatively small amounts of alcohol; i.e., the face reddens, pulse rate increases, and the individual feels poorly. A similar effect can also be produced in carriers of the more common ADH variant when they take the drug disulfiram (Antabuse) with alcohol. Since this drug is known to enhance the level of acetaldehyde by inhibiting ALDH (Fig.15.24), the elevated acetaldehyde levels are thought to be the cause of "flushing."

The enzyme aldehyde dehydrogenase also shows a genetic polymorphism in the Japanese population, allele frequencies each being around 50%, whereas in those of European descent this enzyme variant is rare. The Japanese variant is associated with decreased ALDH activity and appears to account for the frequent flushing phenomenon. The combination of faster production of aldehyde with diminished breakdown of this compound presumably accounts for the flushing.

Does this genetic difference in alcohol metabolism have something to do with susceptibility to alcoholism? As mentioned above, flushing causes not only an increased pulse rate but is also associated with discomfort. This discomfort may prevent carriers of the ALDH variant from drinking too much and therefore protect them against becoming alcoholics. This variant has indeed been shown to be much less common among Japanese alcoholics than in the general population [275]. Moreover, studies on normal, nonalcoholic Japanese men have shown that "flushers" tend to drink less alcohol than "nonflushers" [233]. The ALDH gene appears to be an anti-alcoholism gene!

The ALDH variant with decreased activity is also common among Chinese. Just as in Japanese patients, this variant is found to be much rarer in Chinese alcoholics than in the normal population of Taiwan. In addition, the atypical ADH variant is also reduced among these alcoholics [241].

Reaction of the Brain to Alcohol as Measured by the EEG. Hints to the genetic determination of alcoholism were found, when the reaction of the brain to alcohol was examined using the EEG [182]. A twin sample yielded measurements of EEG characteristics, such as amplitudes and distribution of frequencies, with high heritabilities; the EEGs of MZ twins had a tendency to become even similar after alcohol. Even more interesting were the differences in reaction when the resting EEG was considered. Individuals with a prominent and stable α rhythm in the resting stage showed relatively little change after alcohol intake (Fig.15.25). On the other hand, persons whose EEG at rest showed less well-developed α waves had a stronger response to alcohol intake: their α waves were much more prominent and stable after alcohol than in the resting EEG (Fig.15.26). These reactions were highly concordant in MZ twins but in some cases discordant in DZ twins.

Some other twins showed qualitatively deviating reactions, for example, β waves. Individuals with low-

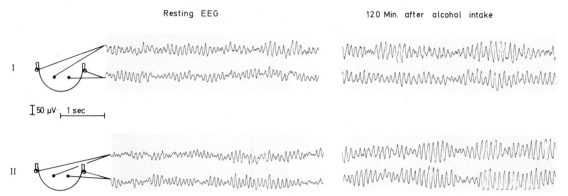

Fig. 15.25. A male adult MZ twin pair with well-developed occipital α rhythm. Loading with 1.2 g/kg ethanol leads to a relatively small increase of α activity 120 min after ethanol intake. (From Propping 1977 [182])

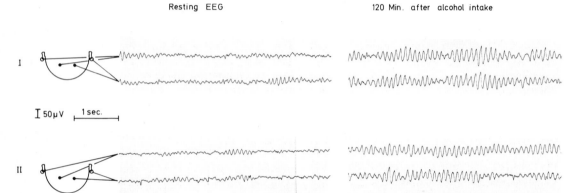

Fig. 15.26. A male adult MZ twin pair with relatively poorly expressed α waves in the resting EEG; 120 min after intake of 1.2 g/kg ethanol, the α rhythm is strikingly enhanced. (From Propping 1977 [186])

voltage EEG, a variant with a single dominant mode of inheritance (Sect. 6.1), did not react to alcohol by increased α wave production. Hence there are strong genetically determined differences between individuals in the reaction of their brains to alcohol. These differences were not caused by a corresponding variability in alcohol metabolism. There was no relationship between alcohol blood level and EEG reactivity. Still more important, the alcohol-induced modification of EEG pattern persisted for a relatively long time after most of the alcohol was metabolized. The response of the brain itself appears to be subject to a genetically determined difference.

Can Our Knowledge of Neurophysiological Mechanisms of the EEG Explain the Differential Reaction to Alcohol? The mechanisms of EEG production, which are explained in Sect. 15.2.3.4, might help to understand brain function as altered by alcohol. The pacemaker in the thalamus is influenced only slightly in individuals with well-developed α rhythm. Its syn-

chronizing function is improved, on the other hand, in subjects in whom the normal spontaneous α rhythm shows a tendency to desynchronization. This tendency may be caused by a stronger desynchronizing input from the reticular system. Is there any evidence that could connect these two pieces of information, suggesting a specific hypothesis for a genetic component of alcoholism?

It has been noted that some alcoholics tend to have a poor α rhythm [182]. More specifically, their α rhythm is often described as being similar so that depicted for twins showing a strong alcohol reaction (Fig. 15.26). It is not quite clear, however, whether this EEG pattern is one of the causes of alcoholism or a consequence of alcohol action on the brain either directly or via liver damage. On the other hand, meditation techniques, such as transcendental meditation, enhance α activity. The subjective result of meditation is that of relaxation and peacefulness. The same result can sometimes be achieved by direct attempts at enhancing α activity by biofeedback. One can feed the EEG into a device that emits a sound as long as α waves are produced. The proband is asked to try to maintain this sound as long as possible. In this way α activity can be enhanced for a limited time. Often the probands describe their feeling tone as relaxed, bordering on happiness; their condition seems to resemble that achieved by meditation [11].

These results suggest a neurophysiological and genetic hypothesis for some cases of alcoholism. Susceptibility to alcoholism is high in persons who normally suffer from a relatively high level of tonic arousal or a weak resistance of the thalamic pacemaker. This high arousal level is attenuated by alcohol, and the person feels better. This positively reinforces alcohol consumption and may result in alcoholism. This hypothesis has been confirmed by EEG studies on nonalcoholic family members of alcoholics [179, 188]. Some of them showed similar EEG patterns; therefore the EEG could not be the effect of chronic alcohol abuse. In one study, however, the reduction in average α activity was found only in female and not in male alcoholics. At the same time, these women were found by independent psychiatric evidence to become alcoholic more often regardless of external circumstances, whereas most of the men had become alcoholics as a result of social pressure and other external factors [188].

This example shows how genetic susceptibility may lead to different phenotypic effects, depending on the sociocultural environment. In Western society with its strong emphasis on social drinking, the genetically susceptible individual is in danger of becoming an alcoholic. In a Buddhist setting the same person would probably become one especially dedicated to meditation.

The arousal-attenuating effect on the reticular system is not the only alcohol effect on the brain. Twin studies suggest genetic variability for other aspects of brain function as well. The example above has been described in detail, since it is one of the first instances in which neurophysiological data and concepts have been used to develop a genetic hypothesis. Neurophysiology is a highly developed field of science into which genetic concepts and methods are beginning to be introduced. Human geneticists have only rarely tried to include concepts of neurophysiology in their thinking. The main reason for this barrier is the compartmentalization of science, such that until recently physiology in general and neurophysiology in particular have been less influenced by genetics than most other fields in biology. The reason may be that physiology is less reductionistic and deals more with interpretation of integrated systems and feedback processes than other areas in biology [137]. Studies on the molecular basis of EEG variants and epilepsies are now bridging this gap. There are very few studies on genetic aspects of non-alcohol substance abuse [74 a].

15.2.3.6 Brain Physiology: Genetic Variability Affecting of Neurotransmitters

Analysis at the Biochemical Level Is Needed: The Synapsis. Genetic analysis at the EEG level, while conceptually more satisfactory than analysis of behavioral phenotypes, in the long run will have limited success. Resolution at the DNA level and at the level of enzymes and proteins will have to be achieved. Where could we find genetic variability of enzymes and proteins that may influence brain function?

The main functional components of the nervous system are neurons. These are single cells with one nucleus, one long branch that is called neurite or axon and functions as the effector organ of the neuron, and a number of elaborately branched dendrites that establish contacts with other nerve cells via the so-called synapses. One neuron may have many thousands of synapses. A reduction in their number has been observed in many etiologically different types of mental retardation (Fig. 15.20). Figure 15.27 shows the main organelles of a synapse. The presynaptic terminal and the postsynaptic membrane are separated by a narrow synaptic cleft. When the nerve impulse reaches the presynaptic terminal, transmission across the synapse is effected by chemical means. The specific transmitter substances are prepacked in quanta of some thousands of molecules in the vesicles of the presynaptic terminals. The arriving impulse causes one or a very few vesicles to liberate their quanta of transmitter into the synaptic cleft. Thus, the transmitter can act on specific receptor sites of the postsynaptic membrane. This reaction causes Na$^+$ ions to diffuse across the membrane and bring about a change in electric potential. There are two types of synapses: excitatory and inhibitory. When a neuron receives a sufficient number of impulses from excitatory synapses, the axon "fires," i.e., it releases an impulse. An inhibitory synapsis, on the other hand, may cause an inhibitory hyperpolarization of the postsynaptic membrane, which prevents depolarization from reaching the critical level above which the neuron fires. In this way an excitatory impulse can be transmitted to an ever-increasing number of excitatory nerve cells; this "chain reaction" is prevented from evolving into an "explosion" by intercalation of inhibitory nerve cells.

This chain of events suggests a number of possibilities for genetic variability. For example, the enzymes for synthesis or breakdown of neurotransmitter molecules may have different activities, the membrane may show structural differences that influence its permeability to neurotransmitter or enzyme molecules, there might be differences in receptors, or outside regulating processes may influence the function of the synapsis at various levels. Some results in mental diseases do indeed suggest abnormalities in neurotransmitter function.

Chemical Types of Neurotransmitters. Several compounds are used in the brain as neurotransmitters; synapses are specialized for one type. The best known examples are norepinephrine (adrenergic synapses) and acetylcholine (cholinergic synapses). The reason that their analysis has so far been most successful is technical: both can be studied in cells of the peripheral nervous system. For example, neurons of the sympathetic nervous system are adrenergic,

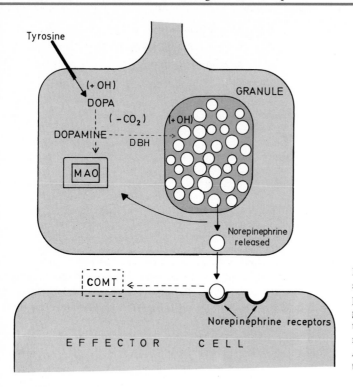

Fig. 15.27. Schematic representation of an adrenergic synapsis with its most important organelles. Norepinephrine is synthesized from tyrosine, stored in granules, released into the synaptic cleft, and bound to receptors of the effector cell on the postsynaptic membrane (see text for further explanation). *MAO,* Monoamine oxidase; *COMT,* catecholamine *O*-methyltransferase; *DBH,* dopamine β-hydroxylase

neurons of the parasympathetic nervous system are cholinergic. In the brain, however, these two types of synapses together constitute only a minority of all neurons; a variety of amino acids (histamine, glutamic acid, aspartic acid, glycine, and possibly others) act as neurotransmitters. Not only synthesis but also inactivation is important for their activity. The main groups of neurotransmitters are:

– Norepinephrine
– Acetylcholine
– Dopamine
– Gamma-aminobutyric acid (GABA)
– Glycine
– L-glutamic acid
– L-aspartic acid
– Histamine
– Serotonin

Two classes of substances have recently attracted much attention as they appear to be altered in both affective disorders and schizophrenic psychoses: the catecholamines norepinephrine and epinephrine and their precursors, and especially dopamine, and the indoleamines, especially 5-hydroxytryptamine (serotonin). Here we confine ourselves mainly to one group, the catecholamines.

Catecholamines. Epinephrine and norepinephrine are formed from tyrosine. The function of an adrenergic synapsis is shown in Fig. 15.27. This example demon-

strates some of the possibilities for genetic variability and, at the same time, for experimental approaches to analyze this variability. In the present context we can give only a simplified picture. Norepinephrine, when not used as a neurotransmitter, or after having performed this function, must be inactivated. Two enzymes have often been examined here: catecholamine-*O*-methyltransferase (COMT) and monoamine oxidase (MAO). The concentration of norepinephrine may be enhanced or lowered – or the synapses may even be depleted of this neurotransmitter – either by altering synthesis or degradation of this substance.

Genetic analysis of variability in these and other enzymes is difficult as the human brain is not directly accessible to our analysis. There are three ways to overcome this difficulty:

1. Experiments with animals.
2. Examination of these enzymes in other, more easily accessible tissues.
3. Study of genes for enzymes and receptors involved in neurotransmitter function at the gene-DNA level. In the long run this is probably the most promising approach.

Animal Experiments on Genetic Variability in Catecholamine Metabolism [36, 37]. The enzymes tyrosine hydroxylase, dopamine β-hydroxylase, and phenylethanolamine-*N*-methyltransferase have been shown to have about twice the activity in the adrenals of the inbred BALB/c as in the BALB/cN mouse line. In F_1, F_2, and back-cross progenies single genes control these enzyme activities, suggesting that either the structural genes for these enzymes are closely linked or that they are under a common genetic regulatory control.

Another manifestation of genetic variability is the level of cAMP, which acts as a second messenger for various hormones and neurotransmitters [270]. cAMP content was found to be differ in the brains of four inbred mouse strains.

These experiments suggest complex regulation of the quantity of norepinephrine in adrenergic synapses of the brain; it may be recalled that the resulting differences in activity are correlated with differences in aggressive behavior (Sect. 15.1.2). In view of these complexities it seems to be a difficult undertaking to examine the same enzymes in humans and to draw conclusions about differences in neurotransmitter in the human brain on the basis of measurements of its activity. However, there are some indications that the approach may lead to the detection of genetic variability that could help to understand the genetics of normal and deviant behavior.

Psychotropic Drugs [186]. Research on mental disease was encouraged by the observation that the symptoms of affective and mental disorders can be influenced by psychopharmacological agents. These drugs were shown to influence synaptic transmitter function, especially the function of norepinephrine. It was observed, for example, that some patients with depression responded better to MAO inhibitors and others to tricyclic antidepressants, such as imipramine. Moreover, relatives who suffered from depressions responded positively to the same drug as the proband. This familial tendency to respond to one class of drugs and not to the other suggests genetic factors. Each drug affects the function of norepinephrine on adrenergic synapses. MAO inhibitors reduce the degradation of epinephrine, thereby enhancing the amount of available at the synapse. Tricyclic antidepressants, such as imipramine, reduce reabsorption of norepinephrine into the neuron from which it was released, thereby enhancing the available amount of norepinephrine for neurotransmission. The interfamilial differences in therapeutic efficiency of those drugs could conceivably point to different genetic anomalies at the synapses in these families. Such specific conclusions, however, are made difficult by the observation that there are genetic differences between humans in the metabolism of these drugs and hence in their steady-state blood levels. The differences between families may thus be caused by genetic differences in drug metabolism rather than in target organs. Such metabolic differences have been demonstrated for the tricyclic antidepressant nortriptyline, which only differs slightly from imipramine [186].

In interpreting possible genetic differences in psychopharmacological reactions both the drug metabolism and the target organ – mainly the brain must be considered. Experiments are needed in which bio-

synthesis and blood level are held constant so that effects at the level of brain can be studied. Such investigations in humans are necessary not only to obtain a deeper understanding of the genetic basis of affective and other mental diseases but to design rational drug therapy as well.

Possible Genetic Variability at the Level of Receptors: Isoreceptors. The action of neurotransmitters depends, in addition to their availability in sufficient quantities, on their binding to specific receptors at the postsynaptic membrane of the effector cell. Analysis of receptors has become an important field of molecular genetic research in medicine since their genetic variation has turned out to cause genetic disease. [See, for example, the role of deficient androgen receptors for testicular feminization (Sect. 8.5) and that of LDL receptors in familial hypercholesterolemia (Sect. 7.6.4). Genetic variation of receptors for neurotransmitters is now being studied intensively. Genes for such receptors have been localized and in part analyzed in the human genome. Study of their genetic variation is therefore possible at the DNA level. There are indications that genetic variation in the "normal" range may exist in receptor genes as found previously for genes determining enzymes and other proteins (Sect. 12.1.2). Just as we refer to "isoenzymes," the concept of "isoreceptors" [81] may become familiar in the future. The dopamine receptor may be one example.

There are five dopamine receptor genes (D_1–D_5), localized on different human chromosomes. Their structure is compared in Fig. 15.28. They are part of a superfamily of heptahelical receptor transmembrane domains that are coupled to their intracellular transduction system by a G protein. The dopamine receptors have been identified in various parts of the brain and react with partially different antagonists [220] (Table 15.15). Interindividual variation of the receptor structure in the normal range is likely and a new field for linkage and association studies is opening up. Studies on the association of a variant of the D2 dopamine receptor gene with various psychiatric conditions including alcoholism have failed to produce consistent results so far.

Other receptor families include the acetylcholine, GABA, glycine, and adrenoreceptors [82, 220]. An increasing number have now been cloned and are available for further study.

Conclusions

While the genetic determination of human behavior may be the most interesting field of human genetics it is also the most controversial. Its study is made dif-

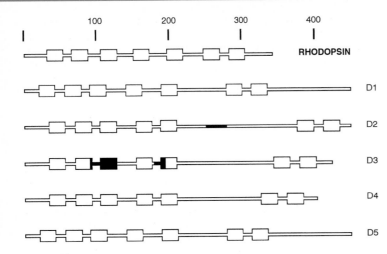

Fig. 15.28. The family of dopamine receptor genes and comparison with rhodopsin. The scale indicates the length of the amino acid sequence. *Blackened areas,* alternative exons. (From Sokoloff et al. 1992 [220]; modified)

Table 15.15. Dopamine receptor subtypes (from Sokoloff et al. 1992 [220])

	D_1	D_2	D_3	D_4	D_5
Chromosome localization	5q31–34	11q22–23	3q13.3	11p	4p
Highest brain density	Neostriatum	Neostriatum	Palaeostriatum, (islands of Calleja; n. accumbans)	Medulla; frontal cortex	Neostriatum; hippocampus
Affinity for dopamine	Micromolar	Micromolar	Nanomolar	Submicromolar	Submicromolar

ficult by technical factors. The following approaches have been used: (a) Studies on animal models, mainly mice, but also insects and more primitive animals; (b) studies using measurable psychological phenotypes such as intelligence in the normal as well as in the subnormal range; (c) examination of individuals showing specified genetic defects, such as chromosomal aberrations; (d) studies of genetic mechanisms influencing variability of sensory functions such as vision, hearing, and taste; (e) examination of the influence of genetically determined, neurobiological traits on behavior, such as EEG or neurotransmitters; (f) investigations of abnormal and borderline phenotypes such as alcoholism, homosexuality, and socially deviant behavior. Analyses in humans carried out at the quantitative-phenotypic levels have often failed to disentangle heredity (nature) from environment (nurture). Human studies often utilized the twin method. Definite understanding of human behavioral variation in health and disease will require the analysis of specific genes affecting neurobiological mechanisms.

References

1. Alexander D, Walker AT, Money J (1964) Studies in direction sense: I. Turner's syndrome. Arch Gen Psychiatr 10 : 337–339

1a. Agarwal DP, Goedde HW (1984) Alkoholmetabolisierende Enzyme: Alkoholunverträglichkeit und Alkoholkrankheit. In: Zang KD (ed) Klinische Genetik des Alkoholismus. Kohlhammer, Suttgart, pp 65–89

2. Andersen P, Andersson SA (1968) Physiological basis of the alpha rhythm. Appleton-Century Crofts, New York

3. Anokhin A, Steinlein O, Fischer C et al (1992) A genetic study of the human low-voltage electroencephalogram. Hum Genet 90 : 99–112

4. Austin GE, Sparkes RS (1980) Abnormal cerebral cortical convolutions in an XYY fetus. Hum Genet 56 : 173–175

5. Bailey JM, Pillard RC (1991) A genetic study of male sexual orientation. Arch Gen Psychiatr 48 : 1089–1096

6. Bailey JM, Pillard RC, Neale MC, Agyei Y (1993) Heritable factors influencing sexual orientation in woman. Arch Gen Psychiatr 50 : 217–223

7. Baringa M (1994) From fruit flies, rats, mice: evidence of genetic influence. Science 264 : 1690–1693

8. Barth PG (1987) Diseases of neuronal migration. Can J Neurol Sci 14 : 1–16

9. Beauchamp GK, Yamazaki K, Boyse EA (1985) The chemosensory recognition of genetic individuality. Sci Am 253 : 86–92

10. Benzer S (1973) Genetic dissection of behavior. Sci Am 222 : 24–37

11. Birbaumer N (1975) Physiologische Psychologie. Springer, Berlin Heidelberg New York

12. Blandfort M, Tsuboi T, Vogel F (1987) Genetic counseling in the epilepsies. I. Genetic risks. Hum Genet 76 : 303–331

13. Bouchard TJ (1994) Genes, environment, and personality. Science 264 : 1700–1701

14. Bouchard TJ, McGue M (1981) Familial studies of intelligence: a review. Science 212 : 1055–1059

15. Bouchard TJ, Propping P (eds) (1993) Twins as a tool in behavioral genetics. Wiley, Chichester

16. Bouchard TJ, Lykken DT, McGue M et al (1990) Sources of human psychological differences: the Minnesota study of twins reared apart. Science 250 : 223–228

17. Boughman JA, Coneally PM, Nance WE (1980) Population genetic studies of retinitis pigmentosa. Am J Hum Genet 32 : 223–235

18. Bray GA (1992) Obesity. In: King RA, Rotter JI, Motulsky AG (eds) The genetic basis of common diseases. Oxford University Press, New York, pp 507–528

19. Bruck I, Phillipart M, Givaldi D, Antoniuk S (1991) Difference in early development of presumed monozygotic twins with Rett syndrome. Am J Med Genet 39 : 415–417

20. Brun A, Gustavson K-H (1982) Letter to the editor. Hum Genet 60 : 298

21. Brunner HG, Nelen MR, van Zandvoort P et al (1993) X-linked borderline mental retardation with prominent behaviorial disturbances: phenotype, genetic localization, and evidence for disturbed monoamine metabolism. Am J Hum Genet 52 : 1032–1039

22. Brunner HG, Nelen MR, Breakefield XO et al (1993) Abnormal behavior associated with a point mutation in the structural gene for monoamine oxidase A. Science 262 : 578–580

23. Buchsbaum MS (1974) Average evoked response and stimulus intensity in identical and fraternal twins. Physiol Psychol 2 (3A):365–370

24. Buchsbaum MS (1975) Average evoked response augmenting/reducing in schizophrenia and affective disorders. In: Freedman DX (ed) The biology of the major psychoses: a comparative analysis. Raven, New York, pp 129–142

25. Buck L, Axel R (1991) A novel multigene family may encode odorant receptors: a molecular basis for odor recognition. Cell 65 : 175–187

26. Bühler EM, Malik NJ, Alkan M (1990) Another model for the inheritance of Rett syndrome. Am J Med Genet 36 : 126–131

27. Bunge M, Ardila R (1987) Philosopy of psychology. Springer, Berlin Heidelberg New York

28. Buselmaier W, Vierling T, Balzereit W, Schwegler H (1981) Genetic analysis of avoidance learning by means of different psychological test systems with inbred mice as model organisms. Psychol Res 43 : 317–333

29. Byne W (1994) The biological evidence challenged. Sci Am 5 : 50–55

30. Carr J (1970) Mental and motor development in young mongol children. J Ment Def Res 14 : 205–220

30 a. Cases O, Seif I, Grimsby J, Gaspar P, Chen K, Pournin S, Muller U, Aguet M, Babinet C, Shih JC, DeMaeyer E (1995) Aggressive behavior and altered amounts of brain serotonin and norepinephrine in mice lacking MAOA. Science 268 : 1763–1766

31. Cattell RB (1955) The inheritance of personality. Am J Hum Genet 7 : 122

32. Cattell RB (1964) Personality and social psychology. Knapp, San Diego

33. Catell RB (1965) Methodological and conceptional advances in evaluating heredity and environmental influences and their interaction. In: Vandenberg SG (ed) Methods and goals in human behavior genetics. Academic, New York, p 95

34. Charlesworth WR (1976) Human intelligence as adaptation: an ethological approach. In: Resnick LB (ed) The nature of intelligence. Erlbaum, Hillsdale, pp 147–168

35. Christiansen K, Knussmann R (1987) Androgen levels and components of aggressive behavior in man. Horm Behav 21 : 170–180

36. Ciaranello RD, Hoffman HF, Shire JGCh, Axelrod J (1974) Genetic regulation of the catecholamine biosynthetic enzymes. J Biol Chem 249 : 4528–4536

37. Ciaranello RD, Lipsky A, Axelrod J (1974) Association between fighting behavior and catecholamine biosynthetic enzyme activity in two inbred mouse sublines. Proc Natl Acad Sci USA 71 : 3006–3008

38. Cloninger CR (1987) Neurogenetic adaptive mechanisms in alcoholism. Science 236 : 410–416

39. Comings DE (1986) The genetics of Rett syndrome: the consequences of a disorder where every case is a new mutation. Am J Med Genet 24 : 383–388

39 a. Zhang Y, Proenca R, Maffei M et al. (1994) Positional cloning of the mouse obese gene and its human homologue. Nature 372 : 425–43283–388

40. Comings DE, Comings BG, Devor EJ, Cloninger CR (1984) Detection of a major gene for Gilles de la Tourette syndrome. Am J Hum Genet 36 : 586–600

41. Cooper RM, Zubek JP (1958) Effects of enriched and restricted early environments on the learning ability of light and dull rats. Can J Psychol 12 : 159–164

42. Cotton NS (1979) The familial incidence of alcoholism. J Stud Alcohol 40 : 89

43. Crowe RR (1974) An adoption study of antisocial personality. Arch Gen Psychiatry 31 : 785–791

44. Crusio WE, Schwegler H, Brust I (1993) Covariations between hippocampal mossy fibres and working and reference memory in spatial and non-spatial radial maze tasks in mice. Eur J Neurosci 5 : 1413–1420

44 a. Deeb SS, Lindsey DT, Hibiya Y, Sanocki E, Winderickx J, Teller DY, Motulsky AG (1992) Genotype-phenotype relationships in human red/green color vision defects: Molecular and psychophysical studies. Am J Hum Genet 51: 687–700

45. DeVries JC, Vandenberg SG, McClearn GE (1976) Genetics of specific cognitive abilities. Annu Rev Genet 10 : 179–207

45 a. Diebold K, Häfner H, Vogel F, Schatt E (1968) Die myoklonischen Varianten der familiären amaurotischen Idiotie. Hum Genet 5: 119–164

46. Dörner G (1976) Hormones and brain differentiation. Elsevier, Amsterdam

46 a. Drummond-Borg M, Deeb S, Motulsky AG (1989) Molecular patterns of X chromosome-linked color vision

genes among 134 men of European ancestry. Proc Natl Acad Sci USA 86 : 983–987

47. Dryja TP, McGee TL, Reichel E et al (1990) A point mutation of the rhodopsin gene in one form of retinitis pigmentosa. Nature 343 : 364–366

48. Dryja TP, Hahn LB, Cowley GS et al (1991) Mutation spectrum of the rhodopsin gene among patients with autosomal dominant retinitis pigmentosa. Proc Natl Acad Sci USA 88 : 9370–9374

49. Eckert ED, Bouchard TJ, Bohlen J, Heston LL (1986) Homosexuality in monozygotic twins reared apart. Br J Psychiatry 148 : 421–425

50. Egeland JA, Gerhard DS, Pauls DL et al (1987) Bipolar affective disorders linked to DNA markers on chromosome 11. Nature 325 : 783–787

51. Epstein CI (1986) The consequences of chromosome imbalance: principles, mechanisms and models. Cambridge University Press, New York

52. Eysenck HJ (1952) The scientific study of personality. Routledge and Kegan Paul, London

53. Eysenck HJ (1980) Intelligenz-Struktur und Messung. Springer, Berlin Heidelberg New York

54. Eysenck HJ (ed) (1982) A model for intelligence. Springer, Berlin Heidelberg New York

55. Farber SL (1981) Identical twins reared apart. Basic Books, New York

56. Feldman MW, Lewontin RC (1975) The heritability hangup. Science 190 : 1163–1168

57. Fischer M (1972) Umweltfaktoren bei der Schizophrenie. Intrapaarvergleiche bei eineiigen Zwillingen. Nervenarzt 43 : 230–238

58. Fischer M, Harvald B, Hauge M (1969) A Danish twin study of schizophrenia. Br J Psychiatry 115 : 981–990

59. Friedl W (1977) Untersuchungen des Ruhe-EEG normaler weiblicher und männlicher junger Erwachsener mit Hilfe der elektronischen EEG-Analyse. MD dissertation, University of Heidelberg

60. Frischeisen-Köhler I (1930) Untersuchungen an Schulzeugnissen von Zwillingen. Z Angew Psychol 37 : 385

61. Fuller JL, Collins RL (1972) Ethanol consumption and preference in mice: a genetic analysis. In: Nature and nurture in alcoholism. Ann NY Acad Sci 197 : 42–48

62. Fuller JL, Thompson WRT (1960) Behavior genetics. Wiley, New York

63. Fuller JL, Thompson WR (1978) Foundations of behavior genetics. Mosby, St Louis

64. Galton F (1865) Hereditary talent and character. Macmillan Mag 12 : 157

65. German J (1974) Chromosomes and cancer. Wiley, New York

66. Gilles de la Tourette G (1885) Étude sur une affection nerveuse characterisée par l'incoordination motrice accompagnée de l'echolalie et de coprolalie. Arch Neurol 9 : 158–200

67. Goodnow JJ (1976) The nature of intelligent behavior: question raised by cross-cultural studies. In: Resnick LB (ed) The nature of intelligence. Erlbaum, Hillsdale, pp 169–188

68. Goodwin D (1976) Is alcoholism hereditary? Oxford University Press, New York

69. Goodwin DW, Schulsinger F, Hermansen L, Gruze SB, Winokur G (1973) Alcohol problems in adoptees raised apart from alcoholic biological parents. Arch Gen Psychiatry 28 : 238–243

70. Goodwin DW, Schulsinger F, Knop J, Mednick S, Gruze SB (1977) Psychopathology in adopted and nonadopted daughters of alcoholics. Arch Gen Psychiatr 34 : 1005

71. Gottschaldt K (1939) Erbpsychologie der Elementarfunktionen der Begabung. In: Just G (ed) Handbuch der Erbbiologie des Menschen, vol V/1. Springer, Berlin Heidelberg New York, pp 445–537

72. Gottschaldt K (1968) Zwillingsuntersuchungen vom zweiten bis zum sechsten Lebensjahrzehnt. In: Herz und Atmungorgane im Alter. Psychologie und Soziologie in der Gerontologie. Steinkopf, Darmstadt (Veröffentlichungen der Deutschen Gesellschaft für Gerontologie, vol 1)

73. Grigsby JP, Kemper MB, Hagerman RJ, Myers CS (1990) Neuropsychological dysfunction among affected heterozygous fragile -X females. Am J Med Genet 35 : 28–35

74. Gulotta F, Rehder H, Gropp A (1981) Descriptive neuropathology of chromosomal disorders in man. Hum Genet 57 : 337–344

74 a. Gynther LM, Carey G, Gottesman I, Vogler GP (1995) A twin study of non-alcohol substance abuse. Psychiatr Res 56 : 213–220

75. Hagberg B, Aicardi J, Dias K et al (1993) A progressive syndrome of autism, dementia, ataxia, and loss of purposeful hand use in girls: Rett's syndrome. Report of 35 cases. Ann Neurol 14 : 471–479

76. Hallgren B (1960) Nocturnal enuresis in twins. Acta Psychiatr Scand 35 : 73–90

77. Hamer DH, Hu S, Magnuson VL et al (1993) A linkage between DNA markers on the X chromosome and male sexual orientation. Science 261 : 321–327

78. Harris H, Hopkinson DA (1972) Average heterozygosity per locus in man: an estimate based on the incidence of enzyme polymorphisms. ann Hum Genet 36 : 9–20

79. Hawkins JD (1970) Single gene substitutions and behavior. In: contributions to behavior genetic analysis. The mouse as a prototype. Appleton-Century-Crofts, New York, pp 139–159

79 a. Hebebrand J, Klug B (1995) Specification of the phenotype required for men with monoamine oxidase type A deficiency. Hum Genet 96 : 372–373

80. Hebebrand J, Remschmidt H (1995) Anorexia nervosa viewed as the extreme left end of the normal distribution of the body mass index: genetic implications. Hum Genet 1 : 1–11

81. Hebebrand J, Friedl W, Propping P (1988) The concept of isoreceptors: application to the nicotinic acetylcholine receptor and the gamma-aminobutyric acid A/benzodiazepine receptor complex. J Neural Transm 71 : 1–9

82. Hebebrand J, Reichelt R, Körner J (1990) Receptor heterogeneity at the molecular level: implications for neuropsychiatric research. Psychiatr Genet 1 : 19–34

83. Heisenberg M, Wolf R (1984) Vision in drosophila: genetics of microbehavior. Springer, Berlin Heidelberg New York

84. Hendrickson AE (1982) The psychophysiology of intelligence. Part I, II. In: Eysenck HJ (ed) A model for intelligence. Springer, Berlin Heidelberg New York, pp 151–230

85. Herbst DS, Baird PA (1982) Sib risk for nonspecific mental retardation in British Columbia. Am J Med Genet 13 : 197–208

86. Hiroshi T, Akio A, Shingi T, Eiji i (1968) Sex chromosomes of Japanese epileptics. Lancet I:478

87. Hoffmann H-J, Schneider R, Crusio WE (1993) Genetic analysis of isolotion-induced aggression. II. Postnatal environmental influences in AB mice. Behavior Genet 23 : 391–394

88. Holland AJ, Hall A, Murray R et al (1984) Anorexia nervosa: a study of 34 twin pairs and one set of triplets. Br J Psychiatry 145 : 414–419

89. Holland AJ, Sicotte N, Treasure J (1988) Anorexia nervosa: evidence for a genetic basis. J Psychosomat Res 32 : 561–571

90. Hook EB (1979) Extra sex chromosomes and human behavior: the nature of the evidence regarding XYY, XXY, XXYY and XXX genotypes. In: Vallet HL, Porter IH (eds) Genetic mechanisms of sexual development. Academic, New York, pp 437–461

90 a. Hu S, Pattatucci AML, Patterson C, Li L, Fulker DW, Cherny SS, Kruglyak L, Hamer DH (1995) Linkage between sexual orientation and chromosome Xq28 in males but not in females. Nature [Genet] 11 : 268–256

91. Hunt N (1966) The world of Nigel Hunt – diary of a mongoloid youth. Darwen Finlay-Son, London

92. Husén T (1953) Twillingstudier. Almquist and Wiksell, Stockholm

93. Husén T (1959) Psychological twin research. Almquist and Wiksell, Stockholm

94. Husén T (1960) Abilities of twins. Scand J Psychol 1 : 125–135

95. Hutchings B, Mednick S (1975) Registered criminality in the adoptive and biological parents of registered male criminal adoptees. In: Fieve RR, Rosenthal D, Brill H (eds) Genetic research in psychiatry. Johns Hopkins University Press, Baltimore, pp 105–116

96. Ihda S (1961) A study of neurosis by twin method. Psychiatr Neurol Jpn 63 : 681–892

97. Imperato-McGinley J, Peterson RE, Gautier T, Sturla E (1979) Androgens and evolution of male-gender identity among male pseudohermaphrodites with 5-reductase deficiency. N Engl J Med 300 : 1233–1237

98. Inoue E (1965) Similar and dissimilar manifestations of obsessive-compulsive neurosis in monozygotic twins. Am J Psychiatry 121 : 1171–1175

99. Jacobs PA, Brunton M, Melville MM, Brittain RP, McClermont WF (1965) Agressive behavior, mental subnormality and the XYY male. Nature 208 : 1351–1352

100. Jensen AR (1969) How much can we boost IQ and scholastic achievement? Harvard Educ Rev 39 : 1–123

101. Jensen AR (1973) Educability and group differences. Methuen, London

102. Jinks JL, Fulker DW (1970) Comparison of the biometrical genetical MAVA, and classical approaches to the analysis of human behavior. Psychol Bull 73 : 311–349

102 a. Jorgensen AL, Philip J, Raskind WH, Matsushita M, Christensen B, Dreyer V, Motulsky AG (1992) Different patterns of X-inactivation in monozygotic twins discordant for red-green color vision deficiency. Am J Hum Genet 51 : 291–298

103. Juda A (1940) Neue psychiatrisch-genealogische Untersuchungen an Hilfsschulzwillingen und ihren Familien. II. Die Kollateralen. Z Ges Neurol Psychiatr 168 : 448–491

104. Juda A (1940) Neue psychiatrisch-genealogische Untersuchungen an Hilfsschulzwillingen und ihren Familien.

III. Aszendenz und Deszendenz. Z Ges Neurol Psychiat 168 : 804–826

105. Juda A (1953) Höchstbegabung, ihre Erbverhältnisse sowie ihre Beziehungen zu psychischen Anomalien. Urban and Schwarzenberg, Munich

106. Juel-Nielsen N (1965) Individual and environment. A psychiatric-psychological investigation of monozygotic twins reared apart. Acta Psychiatr Scand [Suppl] 183

107. Just G (1970) Erbpsychologie der Schulbegabung. In: Just G (ed) Handbuch der Erbbiologie des Menschen, vol V/1. Springer, Berlin Heidelberg New York, pp 538–591

108. Kallmann FJ (1953) Heredity in health and mental disorder. Norton, New York

109. Kalmus H, Fry DB (1980) On tune deafness (dysmelodia): frequency, development, genetics and musical background. Ann Hum Genet 43 : 369–382

110. Kamin LJ (1974) The science and politics of I.Q. Erlbaum, Potomac

111. Kandel ER, Schwartz JH (1981) Principles of neural science. Elsevier, New York

112. Kendler KS, McLean C, Neale M et al (1991) The genetic epidemiology of bulimia nervosa. Am J Psychiatry 148 : 1627–1637

113. Knight SJL, Flannery AV, Hirst M et al (1993) Trinucleotide repeat amplification and hypermethylation of a CpG island in FRAXE mental retardation. Cell 74 : 127–134

114. Krech D, Rosenzweig MR, Bennet EL, Kräckel BA (1954) Enzyme concentrations in the brain and adjustive behavioral patterns. Science 120 : 994–996

115. Kuhn TS (1962) The structure of scientific revolutions. University of Chicago Press, Chicago

116. Kumar-Singh R, Farrar GJ, Mansberg F et al (1993) Exclusion of the involvement of all known retinitis pigmentosa loci in the disease present in a family of Irish origin provides evidence for a sixth autosomal dominant locus (RP8). Hum Mol Genet 2 : 875–878

117. Lange J (1929) Verbrechen als Schicksal. Thieme, Leipzig

118. Lauer J, Lindauer M (1971) Genetisch fixierte Lerndispositionen bei der Honigbiene. Inf Org 1, Akademie der Wissenschaft und Literatur. Mainz

119. Lauer J, Lindauer M (1973) Die Beteiligung von Lernprozessen bei der Orientierung. Fortschr Zool 21 : 349–370

119 a. Lee G-H, Proenca R, Montez JM, Carroll KM, Darvishzadeh JG, Lee JI, Friedman JM (1996) Abnormal splicing of the leptin receptor in diabetic (db) mice. Nature 379 : 632–635

120. Lehrl S, Fischer B (1990) A basic information psychological parameter (BIP) for the reconstruction of concepts of intelligence. Eur J Personal 4 : 259–286

121. Lenz W (1983) Medizinische Genetik, 6th edn. Thieme, Stuttgart

122. Le Vay S, Hamer DH (1994) Evidence for a biological influence in male homosexuality. Sci Amer May:44–49

123. Lewis EO (1933) Types of mental deficiency and their social significance. J Ment Sci 79 : 298

124. Lison M, Blondheim SH, Melmed RN (1980) A polymorphism of the ability to smell urinary metabolites of asparagus. BMJ 281 : 1676–1678

125. Little AJ (1974) Psychological characteristics and patterns of crime among males with an XYY sex chromosome complement in a maximum security hospital. BA

Sp Hon Thesis (Quoted from Borgaonkar and Shah), Sheffield University

126. Loesch DZ, Hay DA (1988) Clinical features and reproductive patterns in fragile X female heterozygotes. J Med Genet 25 : 407–414

127. Loehlin JC (1980) Recent adoption studies of IQ. Hum Genet 55 : 297–302

128. Louis-Krutina C (1988) Untersuchungen zur Persönlichkeitsstruktur des Patientien mit Down-Syndrom. Eine vergleichende Literaturübersicht. Med Diss Homburg/Saar

129. Ludlow CL, Cooper JA (eds) (1983) Genetic aspects of speech and language disorders. Academic, New York

130. Luxenburger H (1932) Endogener Schwachsinn und geschlechts-gebundener Erbgang. Z Neurol Psychiatry 140 : 320–332

131. Manosevitz M, Lindzey G (1969) Thiessen DD. Behavioral genetics: methods and research. Appleton-Century-Crofts, New York

132. McClearn GE (1972) Genetics as a tool in alcohol research. In: Nature and nurture in alcoholism. Ann N Y Acad Sci 197 : 26–31

133. McClearn GE, Rodgers DA (1972) Differences in alcohol preference among inbred strains of mice. Q J Stud Alcohol 20 : 691–695

134. McLaughlin SK, McKinnon PJ, Margolskee RF (1992) Gustducin is a taste-cell-specific G protein closely related to the transducins. Nature 357 : 563–569

135. McWilliam P, Jarrar GJ, Kenna p et al (1989) Autosomal dominant retinitis pigmentosa: localization of an ADRP gene to the long arm of chromosome 3. Genomics 5 : 619–622

135 a. Merbs SL, Nathans J (1992): Absorption spectra of human cone pigments. Nature 356 : 433

136. Michard-Varnhée C (1988) Aggressive behavior induced in female mice by an early single injection of testosterone is genotype dependent. Behavior Genet 18 : 1–12

137. Mohr H (1977) The structure and significance of science. Springer, Berlin New York Heidelberg

138. Money J (1968) Cognitive deficits in Turner's syndrome. In: Vandenberg S (ed) Progress in human behavior genetics. Johns Hopkins Press, Baltimore, pp 27–30

139. Money J, Alexander D (1966) Turner's syndrome: Further demonstration of the presence of specific congitional deficiencies. J Med Genet 3 : 47

140. Money J, Ehrhardt AA (1972) Man and woman, boy and girl. John Hopkins University Press, Baltimore

141. Morton NE (1982) Outline of genetic epidemiology. Karger, Basel

142. Morton NE, Chung CS (eds) (1978) Genetic epidemiology. Academic, New York

143. Motulsky AG, Deeb SS (1994) Color vision and its genetic defects. In: Scriver CR, Beaudet AL, Sly WS, Valle D (eds) The metabolic and molecular bases of inherited disease, 7 th edn. McGraw Hill, New York

144. Motulsky AG, Vogel F, Buselmaier W, Reichert W, Kellermann G, Berg P (eds) (1978) Human genetic variation in response to medical and evirnomental agents: pharmacogenetics and ecogenetics. Human Genet [Suppl 1]

145. Muhs A, Schepank H (1991) Aspekte des Verlaufs und der geschlechts-spezifischen Prävalenz psychogener Erkrankungen bei Kindern und Erwachsenen unter dem Einfluss von Erb- und Umweltfaktoren. Z Psychosomat Med 37 : 194–206

146. Muhs A, Schepank H, Manz R (1990) 20 Jahres-Follow-up-Studie eines Samples von 50 neurotisch-psychosomatisch kranken Zwillingspaaren. Z Psychosomat Med 36 : 1–20

147. Muller HJ (1925) Mental traits and heredity. J Hered 16 : 433–448

148. Müller-Küppers M, Vogel F (1965) Über die Persönlichkeitsstruktur von Trägern einer seltenen erblichen EEG-Variante. Jahrb Psycholog Psychother Med Anthropol 12 : 75–101

149. Mulley JC, Kerr B, Stevenson R, Lubs H (1992) Nomenclature guidelines for X-linked mental retardation. Am J Med Genet 43 : 383–391

150. Murken JD (1973) The XYY-syndrome and Klinefelter's syndrome. Topics human genetics. Thieme, Stuttgart

151. Myrianthopoulos NC, Nichols PL, Broman SH (1976) Intellectual development of twins – comparison with singletons. Acta Genet Med Gemellol (Roma) 25 : 376–380

152. Nathans J, Hogness D (1984) Isolation and nucleotide sequence of the gene encoding human rhodopsin. Proc Natl Acad Sci USA 81 : 4851–4855

153. Nathans J, Merbs SL, Sung C-H et al (1992) Molecular genetics of human vision pigments. Annu Rev Genet 26 : 403–424

154. Nathans J, Maumenee IH, Zrenner E et al (1993) Genetic heterogeneity among blue-cone monochromats. Am J Hum Genet 53 : 987–1000

155. Neisser U (1976) Academic and artificial intelligence. In: Resnick LB (ed) The nature of intelligence. Erlbaum, Hillsdale, pp 135–144

156. Netley CT (1986) Summary overview of behavioral development in individuals with neonatally identified X and Y aneuploidy. Birth Defects 22 : 293–306

157. Neufeld EF (1974) The biochemical basis for mucopolysaccharidoses and mucolipidoses. Prog Med Genet 10 : 81–101

158. Newcombe HB, McGregor F (1964) Learning ability and physical wellbeing in offspring from rat populations irradiated over many generations. Genetics 50 : 1065–1081

159. Newman HH, Freeman FN, Holzinger KJ (1937) Twins: a study of heredity and environment, 4 th edn. (1968) University of Chicago Press, Chicago

160. Nielsen J (1970) Criminality among patients with Klinefelter's syndrome. Br J Psychiatry 117 : 365–369

161. Nielsen J, Stradiot M (1987) Transcultural study of Turner's syndrome. Clin Genet 32 : 260–270

162. Nielsen J, Sillesen I, Sørensen AM, Sørensen AM, Sørensen K (1979) Follow-up until age 4 to 8 of 25 unselected children with sex chromosome abnormalities, compared with sibs and controls. Birth defects: original article series, vol XY, pp 15–73

163. Nielsen J, Sørensen AM, Sørensen K (1981) Mental development of unselected children with sex chromosome abnormalities. Hum Genet 59 : 324–332

164. Nielsen J, Pelsen B, Sorensen K (1988) Follow-up of 30 Klinefelter males treated with testosterone. Clin Genet 33 : 262–269

165. Noel B, Duport JP, Revil D, Dussuyer I, Quack B (1974) The XYY syndrome: reality of myth? Clin Genet 5 : 387–394

166. Oden MH (1968) The fulfillment of promise: 40-year follow-up of the Terman gifted group. Psychol Monogr 77 : 3–93

167. Omenn GS, Weber BA (1978) Dyslexia: search for phenotypic and genetic heterogeneity. Am J Med Genet 1 : 333–354

168. Opitz JM, Sutherland GR (eds) (1984) Conference report: international workshop on the fragile X and X-linked mental retardation. Am J Med Genet 17 : 5–385

169. Parker N (1964) Twins: A psychiatric study of a neurotic group. Med J Aust 51 : 735–742

170. Passarge E (1995) Color atlas of genetics. Thieme, Stuttgart

171. Paul J, Froster-Iskenius U, Moje W, Schwinger E (1984) Heterozygous female carriers of the marker-X-chromosome: IQ estimation and replication status of fra (X) (q). Hum Genet 66 : 344–346

172. Pauls DL (1993) Behavioural disorders: Lessons in linkage. Nature Genet 3 : 4–5

173. Pauls DL, Raymond CL, Stevenson JM, Lackman JF (1991) A family study of Gilles de la Tourette syndrome. Am J Hum Genet 48 : 154–163

174. Penrose LS (1938) (Colchester survey) A clinical and genetic study of 1280 cases of mental defect. Spec Rep Ser Med Res Counc (London) 229 His Maj Stat Off London

175. Penrose LS (1962) The biology of mental defect, 3rd edn. Grune and Stratton, New York

176. Pillard RC, Weinrich JD (1986) Evidence of familial nature of male sexual orientation. Arch Gen Psychiatry 43 : 808–812

177. Plomin R, McClearn GE, Smith DL et al (1994) DNA markers associated with high versus low IQ: the IQ quantitative trait loci (QTL) project. Behav Genet 24 : 107–118

178. Plomin R, Pedersen NL, Lichtenstein P, McClearn GE (1994) Variability and stability in cognitive abilities are largely genetic later in life. Behav Genet 24 : 207–215

179. Pollock VE, Volavka J, Goodwin DW, Mednik SA, Gabrielli WF, Knop J, Schulsinger F (1983) The EEG after alcohol administration in men at risk for alcoholism. Arch Gen Psychiatry 40 : 857–861

180. Popenoe P (1922) Twins reared apart. J Hered 5 : 142–144

181. Price RA, Kidd KK, Cohen DJ et al (1985) A twin study of Tourette syndrome. Arch Gen Psychiatry 42 : 815–820

182. Propping P (1977) Genetic control of ethanol action on the central nervous system. Hum Genet 35 : 309–334

183. Propping P (1978) Alcohol and alcoholism. In: Motulsky AG et al (eds) Human genetic variation in response to medical and environmental agents: Pharmacogenetics and ecogenetics. Hum Genet [Suppl] 1 : 91–99

184. Propping P (1980) Genetic aspects of alcohol action on the electroencephalogram (EEG). In: Biological research in alcoholism. Begleiter H, Kissin (eds) Plenum, New York, pp 589–602

185. Propping P (1983) Genetic disorders presenting as "schizophrenia." Karl Bonhoeffer's early view of the psychoses in the light of medical genetics. Hum Genet 65 : 1–10

186. Propping P (1989) Psychiatrische Genetik. Springer, Berlin Heidelberg New York

187. Propping P (1992) Alcoholism. In: King RA, Rotter JI, Motulsky AG (eds) The genetic basis of common disease. Oxford University Press, New York, pp 837–865

188. Propping P, Krüger J, Mark N (1981) Genetic disposition to alcoholism. An EEG study in alcoholics and their relatives. Hum Genet 59 : 51–59

189. Puck MH (1981) Some considerations bearing on the doctrine of self-fulfilling prophecy in sex chromosome aneuploidy. Am J Med Genet 9 : 129–137

190. Puck MH, Bender BG, Borelli JB, Salbenblatt JA, Robinson A (1983) Parents' adaptation to early diagnosis of sex chromosome anomalies. Am J Med Genet 16 : 71–79

191. Quazi RH, Reed TE (1975) A possible major contribution to mental retardation in the general population by the gene for microcephaly. Clin Genet 7 : 85–90

192. Ratcliffe SG, Paul N (eds) (1986) Prospective studies on children with sex chromosome aneuploidy. Liss, New York (Birth defects original article series 22 [3])

193. Ratcliffe SG, Murray L, Teague P (1986) Edinburgh study of growth and development of children with sex chromosome abnormalities III. Birth Defects 22 : 73–118

194. Reinisch JM (1981) Prenatal exposure to synthetic progestins increases potential for aggression in humans. Science 211 : 1171–1173

195. Reiss AL, Freund L (1990) Fragile X syndrome. Biol Psychiatry 27 : 223–240

196. Reske-Nielsen E, Christensen A-L, Nielsen J (1982) A neuropathological and neuropsychological study on Turner's syndrome. Cortex 18 : 181–190

197. Resnick LB (ed) (1976) The nature of intelligence. Erlbaum, Hillsdale

198. Rett A (1966) Über ein eigenartiges hirnatrophisches Syndrom bei Hyperammonämien im Kindesalter. Wien Med Wochenschr 116 : 723–738

199. Robinson A, Bender BG, Borelli JB et al (1986) Sex chromosomal aneuploidy: prospective and longitudinal studies. Birth Defects 22 : 23–71

200. Rodgers DA (1970) Mechanism-specific behavior: An experimental alternative. In: Contributions to behavior-genetic analysis. The mouse as a prototype. Appleton-Century-Croft, New York, pp 207–218

201. Rogers DA, McClearn GE, Bennett EL, Herbst M (1963) Alcohol preference as a function of its caloric utility in mice. J Comp Phys Psychol 56 : 666–672

201a. Ruchalla E, Schalt E, Vogel F (1985) Relations between mental performance and reaction time: new aspects of an old problem. Intelligence 9 : 189–205

202. Scarr S, Weinberg RA (1976) IQ test performance of black children adopted by white families. Am Psychol 31 : 726–739

203. Schepank H (1974) Erb- und Umwelsfaktoren bei Neurosen. Springer, Berlin Heidelberg New York

204. Schepank H (1983) Anorexia nervosa in twins: is the etiology psychotic or psychogenic? In: Krakowski AJ, Kimball CP (eds) Psychosomatic medicine. Plenum, New York, pp 161–169

205. Schulsinger F (1972) Psychopathy, heredity, and environment. Int J Ment Health 1 : 190–206

206. Schwegler H, Lipp H-P (1983) Hereditary covariations of neuronal circuitry and behavior: Correlations between the proportions of hippocampal synaptic fields in the regio inferior and two-way avoidance in mice and rats. Behav Brain Res 7 : 1–38

207. Schwegler H, Lipp H-P, Van der Loos H, Buselmaier W (1981) Individual hippocampal mossy fiber distribution in mice correlates with two-way avoidance performance. Science 214 : 817–819

208. Schwegler H, Crusio WE, Brust I (1990) Hippocampal mossy fibers and radial-maze learning in the mouse: a correlation with spatial working memory but not with non-spatial reference memory. Neuroscience 34 : 293–298

209. Severson JA, Randall PK, Finch CE (1981) Genotypic influences on striatal dopaminergic regulation in mice. Brain Res 210 : 201–215

210. Shaffer JW (1962) A specific cognitive deficit observed in gonadal aplasia (Turner's syndrome). J Clin Psychol 18 : 403–406

211. Sheppard JR, Albersheim B, McClearn GE (1970) Aldehyde dehydrogenase and ethanol preference in mice. J Biol Chem 245 : 2876–2882

212. Sherman SL, Morton NE, Jacobs PA, Turner G (1984) The marker (X) syndrome: A cytogenetic and genetic analysis. Ann Hum Genet 48 : 21–37

213. Shields J (1962) Monozygotic twins brought up apart and brought up together. Oxford University Press, London

214. Sidman RL, Greene MC (1970) Nervous new mutant mouse with cerebellar disease. In: Les mutants pathologiques chez l'animal. CNRS, Paris, pp 69–79

215. Siemens HW (1924) Die Zwillingspathologie. Springer, Berlin

216. Siemens HW (1924) Die Leistungsfähigkeit der zwillingspathologischen Arbeitsmethode. Z Induktive Abstammungs Vererbungslehre 33 : 348

217. Simpson GG (1951) Zeitmaße und Ablaufformen der Evolution. Musterschmidt, Göttingen

218. Slater E (1964) Genetic factors in neurosis. Br J Psychol 55 : 265–269

219. Sofaer JA, Emery AE (1981) Genes for super-intelligence? J Med Genet 18 : 410–413

220. Sokoloff P, Lannfelt L, Martres MP et al (1992) The D_3 dopamine receptor gene as a candidate gene for genetic linkage studies. In: Mendlewicz J, Hippius H (eds) Genetic research in psychiatry. Springer, Berlin Heidelberg New York, pp 135–149

221. Stamatoyannopoulos G (1991) Human hemoglobin switching. Science 252 : 383

222. Stassen HH, Lykken DT, Propping P, Bomben G (1988) Genetic determination of the human EEG. Survey of recent results on twins reared together and apart. Hum Genet 80 : 165–176

222a. Steinlein O, Anokhin A, Mao Y-P et al (1992) Localization of a gene for the human low-voltage EEG on 20 q and genetic heterogeneity. Genomics 12 : 69–73

223. Steinlein O, Smigrodzki R, Lindstrom J, Anand R, Köhler M, Tocharoentanophol C, Vogel F (1994) Refinement of the localization of the gene for neutonal nicotinic acetylcholine receptor a4 subunit (CHRNA4) to human chromosome 20 q13.2-q13.3. Genomics 22 : 493–495

224. Stocks P (1930) A biometric investigation of twins and their brothers and sisters. Ann Eugen 4 : 49–108

225. Storfer MD (1990) Intelligence and giftedness: the contribution of heredity and early environment. Jossey-Bass, San Francisco

226. Street DRK, Watson RA (1969) Patients with chromosome abnormalities in Rampton Hospital. In: West DJ (ed) Criminological implications of chromosome abnormalities. Cropwood Round Table Conference, Institute of Criminology, University of Cambridge, pp 61–67

227. Strober M, Lampert C, Morrell W et al (1990) A controlled family study of anorexia nervosa: evidence of familial aggregation and lack of shared transmission with affective disorders. Int J Eating Dis 9 : 239–253

228. Stunkard AJ, Harris JR, Pedersen NL, McClearn GE (1990) The body-mass index of twins who have been reared apart. N Engl J Med 322 : 1483–1487

229. Sturtevant AH (1915) Experiments of sex recognition and the problem of sexual selection in Drosophila. J Anim Behav 5 : 351–366

230. Sung C-H, Schneider B, Agarwal N et al (1991) Functional heterogeneity of mutant rhodopsins responsible for autosomal dominant retinitis pigmentosa. Proc Natl Acad Sci USA 88 : 8840–8844

231. Sutherland GR (1983) The fragile X chromosome. Int Rev Cytol 81 : 107–143

232. Sutherland GR, Hecht F (1985) Fragile sites on human chromosomes. Oxford University Press, Oxford

233. Suwaki H, Ohara H (1985) Alcohol-induced facial flushing and drinking behavior in Japanese men. J Stud Alcohol 46 : 196–198

234. Swaab DF, Fliers E (1985) A sexually dimorphic nucleus in the human brain. Science 228 : 1112–1115

235. Tariverdian G (1990) Follow-up of monozygotic twins concordant for the Rett syndrome. Brain Dev 12 : 125–127

236. Tariverdian G, Weck B (1982) Nonspecific X-linked mental retardation. A review. Hum Genet 62 : 95–109

237. Tariverdian G, Kantner G, Vogel F (1987) A monozygotic twin pair with Rett syndrome. Hum Genet 75 : 88–90

238. Taylor HF (1980) The IQ game. A methodological inquiry into the heredity-environment controversy. Harvester, Brighton

239. Terman LM, Merrill MA (1937) Measuring of intelligence. Houghton Mifflin, Boston

240. Terman LM, Oden MH (1959) Genetic studies of genius, vol V, The gifted group at midlife. Stanford University Press, Stanford

241. Thomasson HR, Edenberg HJ, Crabb DW et al (1991) Alcohol and aldehyde dehydrogenase genotypes and alcoholism in Chinese men. Am J Hum Genet 48 : 677–681

222. Thorne AG, Wolpoff MH (1992) The multiregional evolution of humans. Sci Am 266 : 28–33

243. Tsuboi T (1970) Crimino-biologic study of patients with the XYY syndrome and Klinefelter's syndrome. Hum Genet 10 : 68–84

244. Tsuboi T, Nielsen J, Nagayama I (1988) Turner's syndrome: a qualitative and quantitative analysis of EEG background activity. Hum Genet 78 : 206–215

245. Turner G, Jacobs PA (1984) Mental retardation and the fragile X. Adv Hum Genet 13

246. Tyler LE (1976) The intelligence we test. An evolving concept. In: Resnick LB (ed) The nature of intelligence. Erlbaum, Hillsdale, pp 13–26

247. Valenstein ES, Riss W, Young WC (1954) Sex drive in genetically heterogeneous and highly inbred stains of male guinea pigs. J Comp Physiol Psychol 47 : 162–165

248. Valenstein ES, Riss W, Young WC (1955) Experimental and genetic factors in the organization of sexual behavior in male guinea pigs. J Comp Physiol Psychol 48 : 397–403

249. Vandenberg SG (1968) Progress in human behavior genetics. Johns Hopkins University Press, Baltimore

250. Vernon PA (ed) (1993) Biological approaches to the study of human intelligence. Norwood, Ablex

251. Vogel F (1958) Über die Erblichkeit des normalen Elektroenzephalogramms. Thieme, Stuttgart

252. Vogel F (1970) The genetic basis of the normal human electroencephalogram (EEG). Hum Genet 10 : 91–114

253. Vogel F (1981) Humangenetische Aspekte der Sucht. Dtsch Med Wochenschr 106 : 711–714

254. Vogel F (1984) Relevant deviations in heterozygotes of autosomal-recessive diseases. Clin Genet 25 : 381–415

255. Vogel F (1986) Grundlagen und Bedeutung genetisch bedingter Variabilität des normalen menschlichen EEG. Z EEG EMG 17 : 173–188

256. Vogel F (1988) Genetic determination and experience. In: Scheibe E (ed) The role of experience in science. De Gruyter, Berlin, pp 82–104

257. Vogel F, Schalt E (1979) The electroencephalogram (EEG) as a research tool in human behavior genetics: Psychological examinations in healthy males with various inherited EEG variants. III. Interpretation of the results. Hum Genet 47 : 81–111

258. Vogel F, Sperling K (eds) (1987) Human genetics. Human Neurobiology. Springer, Berlin Heidelberg New York

259. Vogel F, Schalt E, Krüger J (1979) The electroencephalogram (EEG) as a research tool in human behavior genetics: Psychological examinations in healthy males with various inherited EEG variants. II. Results. Hum Genet 47 : 47–80

260. Vogel F, Schalt E, Krüger J, Propping P (1979) The electroencephalogram (EEG) as a research tool in human behavior genetics: psychological examinations in healthy males with various inherited EEG variants. I. Rationale of the study; material; methods; heritability of test parameters. Hum Genet 47 : 1–45

261. Vogel F, Schalt E, Krüger J, Klarich G (1982) Relationship between behavioral maturation measured by the "Baum" test and EEG frequency. A pilot study on monozygotic and dizygotic twins. Hum Genet 62 : 60–65

262. Vogel F, Krüger J, Höpp HP, Schalt E, Schnobel R (1986) Visually and auditory evoked EEG potentials in carriers of four hereditary EEG variants. Hum Neurobiol 5 : 49–58

262 a. Vogel F, Krüger J, Schalt E et al (1987) No consistent relationship between oscillations and latencies of visual and auditory EEG potentials and measures of mental performance. Hum Neurobiol 6 : 173–182

263. Vogel F, Crusio WE, Kovac C et al (1990) Selective advantage of fra (X) heterozygotes. Hum Genet 86 : 25–32

264. Volavka Jan, Mednick SA, Rasmussen L, Sergeant J (1977) EEG spectra in XYY and XXY men. Electroencephalogr Clin Neurophysiol 43 : 798–801

265. von Bracken H (1969) Humangenetische Psychologie. In: Becker PE (ed) Humangenetik, ein kurzes Handbuch, vol I/2. Thieme, Stuttgart, pp 409–562

266. von Verschuer O (1954) Wirksame Faktoren im Leben des Menschen. Steiner, Wiesbaden

267. Waaler GHM (1967) Heredity of two types of colour normal vision. Nature 251 : 406

268. Wang JCC, Erbe RW (1984) Folate metabolism in cells from fragile X syndrome patients and carriers. Am J Med Genet 17 : 303–310

269. Wang Y, Macke JP, Merbs SL et al (1992) A locus control region adjacent to the human red and green visual pigments genes. Neuron 9 : 429–440

270. Watson JD, Gilman M, Witkowski J, Zoller M (1992) Recombinant DNA, 2 nd edn. Freeman, New York

271. Williams RJ, Berry LJ, Beerstecher E (1949) Biochemical individuality. III. Genetotrophic factors in the etiology of alcoholism. Arch Biochem 23 : 275–290

272. Wilson RS (1972) Twins: early mental development. Science 175 : 914–917

272 a. Winderickx J, Battisti L, Motulsky AG, Deeb SS (1992) Selective expression of human X chromosome-linked green opsin genes. Proc Natl Acad Sci USA 89 : 9710–9714

272 b. Winderickx J, Sanocki E, Lindsey D, Teller DY, Motulsky AG, Deeb SS (1992) Defective colour vision associated with missense mutation in the human green visual pigment gene. Nature Genet 1 : 251–266

272 c. Winderickx J, Lindsey DT, Sanocki E, Teller DY, Motulsky AG, Deeb SS (1992) A Ser/Ala polymorphism in the red photopigment underlies variation in colour matching among colour-normal individuals. Nature 356 : 431–433

273. Witkin HA, Mednick SA, Schulsinger F, Bakkestrøm E, Christiansen KO, Goodenough DR, Hirschhorn K, Lindsteen C, Owen DR, Philip J, Rubin DB, Stocking M (1976) Criminality in XYY and XXY men. Science 193 : 547–555

274. Zajonc RB (1976) Family configuration and intelligence. Science 192 : 227–236

275. Zang KD (ed) (1984) Klinische Genetik des Alkoholismus. Kohlhammer, Stuttgart

276. Zerbin-Rüdin E (1967) Idiopathischer Schwachsinn. In: Becker PE (ed) Humangenetik, ein kurzes Handbuch, vol V/2. Thieme, Stuttgart, pp 158–205

277. Zhang Y, Proenca R, Maffei M et al. (1994) Positional cloning of the mouse obese gene and its human homologue. Nature 372 : 425–432

278. Züblin W (1969) Chromosomale Aberrationen und Psyche. Karger, Basel

16 Behavioral Genetics: Affective Disorders and Schizophrenia

Though this be madness, yet there is method in't

W. Shakespeare, Hamlet

16.1 Affective Disorders

Genetic Investigations in Affective Disorders and Schizophrenia. Genetic studies on these conditions have a long history. After many case reports in the pre-Mendelian area, the classic work by Rüdin appeared in 1916 [45]. This owed its statistical sophistication to collaboration with Weinberg, of "Hardy-Weinberg" fame, and became a paradigmatic model for the phenotypic-biometric approach to such disorders. Following this paradigm, many family and twin studies were carried out. These studies clearly established that genetic variability plays a major part in causing affective diseases and schizophrenia (Fig. 16.1, 16.2). Moreover, this work contributed much to increasing the sophistication of statistical methodology for establishing empirical risk figures needed in genetic counseling [28]. These empirical results, however, left an increasing number of research workers dissatisfied, and there is a long list of – mostly futile – attempts to move genetic analysis to a level closer to gene action and to discover the biological bases for these disorders.

Affective disorders include manic-depressive or bipolar disease and unipolar depression. Another large group of common psychoses is usually classified as schizophrenia. Affective disorders are characterized mainly by cyclic anomalies of feeling tone – depression or mania – whereas in schizophrenia, anomalies of thought patterns and loss of contact with reality are the main symptoms.

Twin and Family Studies of Affective Disorders [14, 40, 51, 61]. Older studies usually regarded the group of affective disorders as a single entity. Twin and family studies began with unselected series of patients or with affected twins to establish empirical risk figures for the various degrees of family relationship (Sect. 4.3.6). Figure 16.1 presents reported twin series. Concordance in MZ twins is obviously much higher than in DZ twins, suggesting – if the straightforward interpretation is accepted – a genetic contribution. It is particularly important that the concordance rate of 12 MZ pairs reared apart was 67% and thus of the same order of magnitude as that MZ twins reared together [61]. Concordance even between MZ twins is far from complete. This finding confirms the importance of environmental factors. Unfortunately, the concordance rates in Fig. 16.1 were calculated without age correction; it is therefore possible that some discordant pairs sooner or later become concordant.

Figure 16.1 does not contain data from the most comprehensive twin study [4]: this requires a more detailed discussion. It was based on a complete registration of Danish twins; there were 55 MZ and 52 DZ pairs in which at least one partner had been diagnosed with an affective disorder. In addition to a "narrow" diagnostic classification, the authors also used a "broad" one. Table 16.1 presents the results. Moreover, concordance in unipolar female MZ twins was higher than in unipolar male MZ twins. Such a sex difference was not found in bipolar twins. In only a few pairs did one partner show unipolar and the other bipolar disease; the great majority were also concordant regarding disease type. Bipolar disease is therefore almost completely genetically determined, when somewhat atypical and less severe manifestations are included. In unipolar diseases, on the other hand, environmental influence may be stronger. If the environmental pathogenic influence were the affected parents, we would expect a higher proportion of affected children from the affected than from the nonaffected partner of a discordant MZ pair. This, however, was not found [3]. Therefore this factor does not appear to be important. The same strategy has been applied in schizophrenia, with the same result (see below).

In nonpsychotic (neurotic, reactive) depression the concordance rate in MZ twins is lower than in the "psychotic" types discussed above but still higher in MZ than in DZ twins [50]. Delineating this type of depression from the psychotic types may be diagnostically difficult.

Bipolar and Unipolar Types: Empirical Risk Figures. Neither the twin data nor the older family studies separated bipolar patients, i.e., those with both manic and depressive phases, from unipolar cases, i.e., those suffering only from depression. Leonhard [31] first

% concordance rate for affective illness in twins

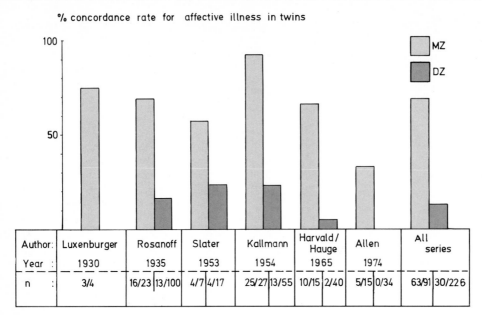

Author:	Luxenburger	Rosanoff		Slater		Kallmann		Harvald/Hauge		Allen		All series	
Year :	1930	1935		1953		1954		1965		1974			
n :	3/4	16/23	13/100	4/7	4/17	25/27	13/55	10/15	2/40	5/15	0/34	63/91	30/226

Fig. 16.1. Concordance and discordance for affective illness in MZ and DZ twins. *n*, Number of concordant pairs/total number of twin pairs

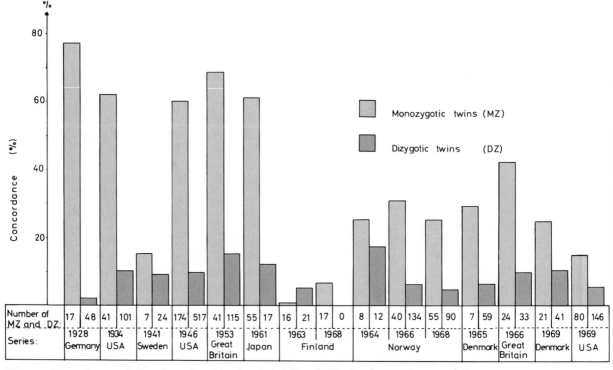

Fig. 16.2. Concordance and discordance for schizophrenia in MZ and DZ twins from various populations

suggested that these two disorders are genetically distinct. Subsequent studies have consistently confirmed that bipolar patients have more bipolar relatives than unipolar patients. However, the prevalence of unipolar depression is much greater among relatives of bi-

polar patients than in the general population. These conclusions were confirmed by a comparative study of biological and adoptive parents of bipolar patients [35]. The degree of psychopathology, especially affective disorders, in the biological parents of these adoptees was found to be similar to that of the parents of nonadopted bipolar patients, whereas the rate of psychiatric disorder in the adoptive parents of manic-depressive patients was similar to that of the adoptive parents of unaffected controls.

In another study [59] which included a broader spectrum of diagnoses, biological relatives also showed somewhat higher frequencies of affective anomalies; their rate of suicides was also increased.

An adoption study on depressive disorders, mainly of the unipolar type, produced a quite different result [56]. The number of registered psychiatric illnesses in *adoptive* fathers was about fivefold that in the controls (adoptive fathers of carefully matched, psychiatrically normal controls). On the other hand, a threefold increase in psychiatric illness was found only in biological mothers of female patients. These data suggest a major contribution of familial, exogenous factors in unipolar affective disease. Moreover, some other psychiatric disturbances, such as a personality type characterized by unusual mood swings, mild to moderate depressions, alcoholism, and acute nonrecurrent psychosis, are more frequent among relatives of affective disorder patients. Another interesting aspect is a sex difference. Female first-degree relatives of bipolar probands are affected 1.5–2 times more frequently than male first-degree relatives. No consistent sex difference has been shown among first-degree relatives of unipolar probands. Table 6.2 shows the most important empirical risk figures.

Simple Modes of Inheritance: Problems with Linkage Studies. Taken at face value, these empirical risk figures are not compatible with a simple mode of inheritance. However, some observations fairly consistently suggest participation of major genes in at least some cases. Some pedigrees point to an autosomal-

Table 16.1. Twin data on concordance in affective disorders (percentages; from Bertelsen et al. 1977 [4])

	Narrow diagnostic classification	Broad disagnostic classification
MZ twins (*n* = 55)		
Bipolar	79	97
Unipolar	54	77
DZ twins (*n* = 52)		
Bipolar	24	38
Unipolar	19	35

Table 16.2. Empirical risk figures for affective disorders (six series; both bipolar and unipolar disease in relatives; from Gershon et al. 1976 [14])

a) Influence of age at onset in proband

Type of disease	Age at onset	Number of first-degree relatives	Affected (%)
Bipolar	<40	561	19.9
	>40	276	11.2
Unipolar	<40	886	16.7
	>40	933	9.5

b) Influence of sex of proband (age corrected[a])

Sex of proband	Sibs				Children			
	n		Affected (%)		*n*		Affected (%)	
	♂	♀	♂	♀	♂	♀	♂	♀
Bipolar probands								
♂	146.9	135.1	12.3	15.5	115.9	122.2	8.6	21.3
♀	136.9	137.3	11.0	19.7	179.0	167.8	13.4	16.7
Unipolar probands								
♂	296.6	307.0	16.2	12.1	305.5	335.8	10.5	11.0
♀	743.9	789.1	7.8	13.6	717.4	755.2	7.8	15.2

[a] Correction for age may lead to decimals in numbers of relatives at risk.

dominant mode of inheritance. Other pedigrees suggest X-linked dominant inheritance; such cases seem to be characterized by relatively early onset and good response to lithium treatment [60]. Both severe and mild, questionable, bipolar and unipolar psychoses have been observed in such families. Delineation of an X-linked dominant from an autosomal-dominant mode of inheritance is difficult (see Sect. 4.1.4), especially in the case of a clinically indistinguishable autosomal-dominant type.

An argument in favor of X-linked inheritance would be linkage with X-linked markers. The hypothesis of an X-linked dominant mode of inheritance was suggested by the fact that unipolar depression is more common in females [60]. Some pedigrees suggested but did not quite prove X-linked inheritance. Definite demonstration of linkage with X-linked traits (color vision, G6PD deficiency, etc.) or X-linked DNA polymorphisms would confirm X linkage. Such positive linkage results have been reported repeatedly but have not been able to be replicated in studies on other pedigrees. The most plausible explanation seemed to be that a mutant X-linked gene causes bipolar affective disorder in a few but not in all such pedigrees, i.e., that there is genetic heterogeneity. Moreover, genetic heterogeneity could also exist within the group of X-linked types; the example of the numerous types of X-linked mental retardation shows that different genetic mechanisms may lead to phenotypically similar anomalies. More recently, however, a critical assessment of pedigrees with affective disorders supposedly pointing to X-linked inheritance has led to the conclusion that this mode of inheritance is unlikely in most pedigrees [22].

These difficulties bring up a more general discussion of linkage studies in psychiatric and other complex diseases. As explained in greater detail in Chap. 5, a key strategy for disentangling the genetic components of complex diseases is detection and localization of genes by linkage with genetic markers. This strategy is based on the plausible hypothesis that in some instances one gene may cause so much of the variation in disease liability that a linkage study with such a "major gene" would produce a positive result. Two conditions are required:

1. A single gene locus must contribute so much to the variation leading to the disease that a linkage study with realistic sample size could show a positive result.
2. The implicated gene locus must be located so close to the genetic marker that linkage can be shown using customary statistical methods.

Two approaches are available in principle: (a) the method using a logarithm of differences (lod score) and (b) the method of affected sibs (or close relatives;

Sect. 5.1). The lod method gives optimal results when there is a clearcut monogenic, preferably dominant, mode of inheritance, and when the affection status of an individual in families can be inferred from phenotypes or can otherwise be identified unequivocally. It may lead to wrong results if the assumption of a monogenic mode of inheritance requires additional hypotheses such as incomplete penetrance. There is a risk that assumptions regarding attribution of individuals to certain genotypes are incorrect. Under these conditions there may even be unconscious manipulation of the data so that lod scores are maximized in favor of linkage. This risk does not exist in the affected sibs (or affected relatives) method because here only the marker profile of two affected individuals is considered, and no assumptions regarding mode of inheritance are necessary. However, this method suffers from another drawback: a much larger sample of sib pairs is needed, and pairs of sibs or relatives are normally derived from many different families. Therefore an accumulation of evidence for linkage is possible only if the same major gene is involved in a substantial proportion of these families. If the sib pairs are a mixture of cases in which different major genes are involved and include instances in which the disease is caused by an unlucky combination of several minor genetic deviations, the sib pair method may not yield positive results.

These considerations suggest the following alternate strategy: select large families with many affected members. Here one has the best chance of identifying a major gene. Localization and analysis of such a gene and its mechanism of action after positional cloning (Sect. 3.1.3.9) may then be successful. A positive linkage result, however, cannot be generalized to all other cases and families with this disease. However, once the pathophysiological mechanisms involved have been suggested by the nature of the detected gene, a search for functionally related genes that may be implicated in other cases can be initiated. Large families with many members showing psychiatric diseases are very rare; families with two or a few affected members are much more common. Here a certain amount of homogeneity can be achieved if families are selected from a population that has been isolated for a long time so that founder effects or genetic drift or both could have led to an increase in certain gene frequencies (See Chap. 14).

One of the most ambitious attempts to localize a major gene for bipolar affective disorder was made in a huge family with many affected members among the Amish, a religious isolate in Pennsylvania. The data were compatible with the assumption of an autosomal-dominant mode of inheritance with 63% penetrance; a linkage study using lod scores seemed to give clearcut evidence of linkage with two markers on the short arm of chromosome 11 – the insulin gene and the gene for the HRAS1 oncogen [8]. This localization seemed plausible since the structural gene for tyrosine hydroxylase is located in the same area; This is the rate-limiting enzyme for the synthesis of catecholamines that may be involved in affective disorders. Unfortunately, however, this linkage was not confirmed by further studies in families from other populations and not even in a restudy with the addition of more cases in the same family [25]. Lack of confirmation within the same family indicates a serious weakness of the lod score method since the shift of assignment in disease status of merely a few patients in the pedigree can substantially alter the lod scores and render positive results no longer compatible with linkage.

Schizoaffective Disorder [40, 55]. Modern psychiatry began with Kraepelin, who postulated two main groups of mental diseases: affective disorders and schizophrenia. Most patients can be assigned to one of these two groups, and family studies confirm this subdivision. Secondary cases in families of schizophrenic probands tend to have schizophrenia; those in families of patients with affective disorders are affected principally with affective disorders. However, there are exceptions. Among first-degree relatives of patients with affective disorders schizophrenia is slightly more common than in the general population [37, 43], and among relatives of schizophrenia probands the proportion of patients with "major depression" is decreased [32]. Moreover, some patients suffer from a "schizoaffective psychosis," i.e., they show clinical signs of both groups of diseases; among their relatives both diseases are increased, in addition to schizoaffective disorders [40, 51]. MZ twin pairs are occasionally observed in which one partner was   diagnosed as schizophrenic and the other as having affective disorder. There may thus be genes creating a nonspecific vulnerability which leads to either of theses diseases, depending on additional conditions. Moreover, in a few families one finds cosegregation of some genes increasing liability for affective disorders and others that for schizophrenia.

There is no evidence whatever for linkage with X-linked markers in families in which only *unipolar psychoses* are found. Furthermore, in this group women are affected more often than men, age at onset is usually later, and in a large subgroup depressions occur in phases of hormonal instability, such as pregnancy, after delivery, and especially during menopause. In conclusion, the family data on affective disorders, in addition to confirming the hypothesis of a strong genetic contribution, suggest genetic heterogeneity and different biological mechanisms.

16.2 Schizophrenia

Diagnosis and Epidemiology [16, 20, 40]. Most psychiatrists would agree on the diagnosis of a "classical" case of schizophrenia, but there is much disagreement regarding less typical cases. This disagreement has been overcome, at least in part, by adoption of strict criteria for clinical diagnosis, generally that of either the International Classification of Diseases (ICD) or the third edition of the Diagnostic and Statistical Manual of Mental Disorders (DSM III). Most of the data discussed in this chapter were not collected using either of these systems; differences in the diagnostic criteria used may therefore exist between different authors. Psychoses classified as schizophrenia have been observed in all populations studied; the lifetime prevalence appears to be amazingly similar (approx. 1%). This makes the claim that schizophrenia was a new disease and originated in the eighteenth century [16] not very likely, despite the lack of convincing historical evidence before this time. While the genetic problems posed by this disease have repeatedly been studied, no clearcut results have yet emerged.

Twin and Family Studies in Schizophrenia [17, 18, 21, 40, 44, 61]. Figure 16.2 summarizes twin studies on schizophrenia. While showing generally higher concordance in MZ than in DZ twins, these data indicate a considerable degree of statistical heterogeneity for concordance in MZ pairs. More recent studies usually show lower concordances than older studies. In part this discrepancy is caused by the mode of ascertainment. Older studies were based on "limited representative" sampling: probands were an unselected series of patients, and the authors ascertained whether these probands had twins, and whether these twins were affected or unaffected. More recent series, on the other hand, are based on "unlimited representative" sampling [11, 30, 48]: all twin pairs in the population are first ascertained, then all pairs in which at least one twin is schizophrenic are selected. In contrast to "limited representative" sampling, this method also includes probands who are less severely affected and therefore have little or no chance of being included in a series based largely on hospital admissions. Lower concordance rates are therefore not surprising. Lower concordance rates for MZ twins with milder clinical symptoms have repeatedly been reported in other studies [57].

In some studies the analysis of discordant MZ pairs has revealed differences in the premorbid personality during childhood and youth [48, 49]. Table 16.3 shows some of the environmental factors studied in a Finnish twin series. Differences in health and in attitudes of parents and sibs, especially the co-twin,

Table 16.3. Environmental factors in 16 discordant MZ twins discordant for schizophrenia (from Tienari 1963 [48])

	Twins with schizophrenia	Non-psychotic twins
First born	10	6
Lower birth weight	8	6
More difficult delivery	5	2
Bigger during childhood	7	9
Dominating during childhood	1	14
More lively during childhood	6	8
"Speaker" during childhood	4	12
More adjusted during childhood	8	–
More sensitive during childhood	5	–
More timid during childhood	8	4
Better in school	3	10
Started working earlier	1	6
Left parental home earlier	–	6
Married earlier	1	10
Unmarried (twin married)	7	–
Lower social status	6	–
More withdrawn	7	–
More often sick during childhood	3	2
Stronger during childhood	7	9

Table 16.4 Empirical risk figures for schizophrenia in relatives of affected patients (from Propping 1989 [40])

	Number of studies	Life-long risk	
		mean %	range
Parents	14	5.6	(5–10)
Sibs (all)	13	10.1	(8–14)
Parents not affected	9	9.6	
One parent affected	5	16.7	
Both parents affected	5	39.2	
Children	7	12.8	(9–16)
Half-sibs	5	4.2	
Nephews/nieces	6	3.0	(1–4)
Grandchildren	5	3.7	(2–8)
(First) cousins	3	2.4	(2–6)
Population (life-long prevalence)		~1.0%	

combined with external stress lead to a severe psychosis in some cases, whereas others are more mildly affected or even healthy in spite of their identical genetic constitution. Table 16.3 includes the most important biographical events that are considered by psychodynamically oriented authors to be responsible for the onset of schizophrenia. The effect of these events has been shown convincingly by descriptions of life histories such as that of the famous Genain quadruplets – monozygotic quandruplets who were observed for many decades. All four sisters suffered from schizophrenia, but the course of the disease was shown to depend on social interactions both with the parents and among themselves, as well as on many other events in their lives [36, 42].

One argument against a genetic interpretation of concordance in MZ twins reared together is thus based on their close social interaction, which would reinforce schizophrenic symptoms in the second twin if one were affected (Sect. 6.3.4). However, there are reports on at least 12 twin pairs reared apart in which at least one twin became schizophrenic; in seven pairs the other twin was also schizophrenic [18], a figure well in line with that reported for MZ twins reared together (Fig. 16.2). Another argument is that a schizophrenic parent would provide especially poor conditions for mental development of the child and would therefore enhance the child's risk of becoming schizophrenic too. This question can be an-

swered by comparative study of children of discordant MZ twins. If the presence or absence of a schizophrenic parent were a significant factor, the children of the clinically healthy twin would be less often affected with schizophrenia than children of the twin with schizophrenia. Such a study in Denmark failed to show this difference [10], thus supporting the genetic interpretation.

Analysis of interaction patterns between parents and twins showed that parents do generally not spontaneously *act* in a more similar way toward MZ than toward DZ twins, but their *reactions* to MZ twins are more similar, since these twins *act* similarly [26]. Therefore, the higher concordance of MZ than of DZ twins cannot be explained plausibly by the presumption of parents treating MZ twins more similarly than DZ twins.

Empirical Risk Figures. Empirical risk figures for schizophrenia suggest the following conclusions (Table 16.4):

a) First-degree relatives of schizophrenics have a 10–20 times increased risk of developing the disease compared with the general population. The actual absolute risk is about 6%–15%.

b) The risk is higher in relatives of probands with catatonic and hebephrenic symptoms than in relatives of probands with paranoid or so-called simple schizophrenia.

c) Within families there is a correlation of clinical subtypes. Relatives of catatonics have a higher risk of becoming catatonic than do relatives of the paranoid type. However, the risk of relatives for catatonics and hebephrenics of becoming paranoid is also enhanced compared with the population average, and vice versa.

d) The risk for sibs of affected probands is lowest (about 10%) when both parents of the proband are unaffected; it is higher with one (actual risk: 17%) and still higher with two affected parents (actual risk, 40%).

e) Among relatives of schizophrenics one often observes personality types who, while not actually abnormal, deviate from the normal and are often classified as schizoid [6].

These figures must be used with care in actual genetic counseling; for example, the higher the number of affected relatives in the pedigree, the higher is the risk for future children to be affected. Hanson and Gottesman [21] gave a table in which the risk is differentiated further according to the number of affected and unaffected relatives – especially sibs – of the counseling client. These figures are based on a biometric model, and although somewhat theoretical, may still be helpful. (For a more complete discussion see [40].)

Familial aggregation of a psychiatric disorder does not necessarily mean that genetic factors are involved. Pathogenic familial environments could lead to a similar result, and social psychiatrists are actively searching for such family factors. While the results from twin studies provide more critical evidence in favor of genetic susceptibility, the usefulness of twin data, especially in behavior genetics, can also be disputed (see Chap. 6.3.4). The best way to separate the relative contribution of genetic and environmental effects is to compare adopted children with their biological and adopted parents.

Adoption Studies in Schizophrenia. The first adoption study on schizophrenia was published by Heston in 1966 [23]. He examined adults who were adopted early in life, and whose mothers suffered from schizophrenia. The probands had been separated from their biological mothers, had no contact with them, and did not live with maternal relatives.

Five of the 47 offspring in this study were affected with schizophrenia. This incidence is similar to that among children who lived with their schizophrenic parents. On the other hand, none of 50 individuals in a matched control group of adopted persons with nonschizophrenic parents became schizophrenic. Half of the children of schizophrenic women who were not affected with overt schizophrenia showed a significant excess of psychosocial disability; the other half were notably successful adults.

More elaborate studies were carried out by a team of American and Danish workers in Denmark, where records of good quality are available on adoption as well as on schizophrenia. It was shown that the incidence of schizophrenia in biological relatives of schizophrenic adoptees was about three times as high as in their adoptive relatives; that adoptive parents of schizophrenic probands did not have schizophrenia more often than adoptive parents of nonschizophrenic children (they more often showed other types of psychological abnormality); and that children of schizophrenic parents had the same risk of schizophrenia regardless of whether they were reared by their parents or were adopted early in life. Intrauterine or perinatal influences from the biological mothers of schizophrenic patients were excluded by examining adoptees who were the paternal half-sibs of schizophrenics and therefore showed only common heredity with the father. In this group schizophrenia was also more frequent than among the controls (adoptees without schizophrenic half-sibs or other relatives).

Biological Hypotheses in Schizophrenia. The yield of biological research in schizophrenia has been low despite all the research efforts that have been carried out over decades. Interpretations are difficult because there is no qualitative or quantitative somatic measure by which schizophrenia can be diagnosed. Diagnosis depends on clinical evidence which may be interpreted differently by different psychiatrists and may vary between countries. The diagnostic difficulties are highlighted by the fact that some studies consider it necessary to use a panel of several psychiatrists in order to reach a diagnosis.

All data from twin, family, and adoption studies on schizophrenia can most easily be explained by the model of multifactorial inheritance in combination with a threshold effect. Some investigators using family data have attempted to set up and to test more sophisticated theoretical models, for example, by introducing two thresholds or by assuming, in addition, major genes [16]. Similar attempts have also been made for affective disorders [51]. No convincing results, however, have yet been achieved. As emphasized throughout the present volume, multifactorial models permit only a general and preliminary description of the genetic findings (Chap. 6). For example, they do not exclude the possibility that in some families major genes may contribute to the genetic disposition since genetic heterogeneity between families is likely. At this preliminary level of genetic analysis the situation is described by a *diathesis-stress hypothesis* [6, 17]: stress situations that are overcome by most individuals and may lead to "neurotic" symptoms in others can trigger a schizophrenic psychosis in genetically predisposed persons. This explanation, however, is not quite satisfactory. The question is: what genetically determined physiological changes enhance the risk of an individual to become schizophrenic?

Again, as in affective disorders, the current hypotheses are centered around anomalies of neurotransmitter metabolism [16]. There is some evidence, for example, that this group of diseases may be caused by hyperactivity of dopamine. For example, substances leading to a release of dopamine, such as amphetamine, may trigger a psychotic phase. Neuroleptics, on the other hand, which are known to mitigate schizophrenic symptoms, block dopamine receptors. Many mechanisms have been discussed that might lead to such dopamine hyperactivity. For example, there may be genetic anomalies of dopamine receptors. Methionine may lead to acute psychotic reactions in chronic schizophrenics. On the other hand, the psychotomimetic drug mescaline is a methylated derivative of dopamine, a norepinephrine precursor. It is possible that some schizophrenics have an abnormally high capacity for dopamine methylation, and that methionine, being a general source of methyl groups, enhances this effect. Other candidates for a possible endogenous psychotoxin are methylated derivatives of serotonin (5-hydroxytryptamine). For example, brain uptake of the serotonin precursor tryptophan could be reduced; this would also be compatible with the methionine effect, as methionine blocks tryptophan uptake competitively. Another possibility is an imbalance between an overactive brain dopamine system and an underactive serotonin system. This would be similar to the situation in Parkinson disease, where there is an imbalance between an overactive acetylcholine and an underactive dopamine system (see the list of neurotransmitters, Sect. 15.2.3.6). The combination of tryptophan loading with a MAO inhibitor – to prevent serotonin degradation – has indeed been claimed to lead to behavioral improvement of some schizophrenics.

These examples show the direction in which concepts and experimentation are currently moving. However, quite different hypotheses are still being discussed, for example, those involving membrane properties and abnormal immune processes. It is possible that in many cases not one major biochemical abnormality but a set of a few or even several minor abnormalities, together with external stress, may push the individual beyond the threshold and into psychosis. In other cases, one major abnormality may be decisive. The history of somatic theories in schizophrenia is not encouraging. In such a situation it may be useful to ask a seemingly simple question: What is schizophrenia? Does it really exist as a disease unit?

"Schizophrenia" in the Light of Human Genetics [40a]. We have noted above that the diagnosis of schizophrenia may

be difficult, and that it often depends on somewhat arbitrary criteria. In Sect. 6.4.2 the contribution of human genetics to a theory of disease is discussed in more general terms. The concept of a disease determined by a single, main cause originated in the last decades of the nineteenth century, when medical bacteriology was the leading biomedical science. This concept showed its explanatory power when the causative organisms of tuberculosis and of syphilis were discovered. The success of specific therapies with chemotherapeutic agents or antibiotics would have been impossible without this concept of disease.

The diagnosis of dementia praecox by Kraepelin (later termed schizophrenia by E. Bleuler) was conceived in the light of such disease models by combining a set of clinical signs with the gradual deterioration observed in longitudinal studies of patients. This diagnostic concept implied one major common cause. Human geneticists are intuitively sympathetic to such an idea because their ideal disease concept is provided by Garrod's "inborn errors of metabolism" [13] or, even more specifically, by the hemoglobinopathies, where *one* specific mutation determines *one* protein anomaly, leading to *one* characteristic disease (Sect. 7.3.2). Thoughtful psychiatrists such as K. Jaspers and E. Bleuler realized early that too direct an application of this model to this group of diseases might lead to wrong conclusions. During the following decades, however, Kraepelin disease concept proved remarkably viable, surviving even the discovery that many cases did *not* show deterioration, and that much of the observed deterioration (*not* all of it) had been an artifact caused by long-term hospitalization [7]. Survival of this diagnostic concept was achieved – at least in part – by an interesting strategy: whenever symptoms characteristic of schizophrenia were observed in association with findings that suggested organic disease, the diagnosis of schizophrenia was withheld; often it was replaced by diagnoses such as "schizophrenic reaction." When all patients with schizophrenic symptoms who also showed signs of a specific organic disease were excluded, a disease group remained for which specific causative factors could not be found.

Schizophrenialike symptoms have been described more often than would be expected by mere chance in a large number of organic brain diseases [39]. Moreover, there are many reports of brain atrophy in chronic schizophrenics. It is true that in many of these cases no genetic disposition for "true" schizophrenia could be shown by family studies, and the natural history of the condition was that of the underlying disease and not that of schizophrenia. It is difficult, on the other hand, to escape the conclusion "that had the organic diagnosis not been reached independently of the psychiatric symptomatology, most of the cases would have been regarded as indubitably schizophrenic". Moreover, more EEG abnormalities were found in sporadic than in familial schizophrenia [29], suggesting an "organic," noninherited subgroup.

There are also a number of well-defined genetic conditions in which schizophrenialike psychoses appear to be more common than would be expected by chance (Table 16.5). Such psychoses have also been described in a number of other conditions in which an increased risk is possible, but evidence is so scanty that no conclusion can be drawn. Examples include: 45,XO and XYY karyotypes, adult types of various lipidoses, congenital adrenal hyperplasia, homocystinuria, Wilson disease, and several others [39]. Some of these conditions, when considered in the light of pathogenetic hy-

Table 16.5. Genetic disorders in which an increased risk of developing a schizophrenialike psychosis is probable (modified from Propping 1983 [39]; for more data see [40])

Condition	Reported findings
XXY (Klinefelter)	The rate of schizophrenialike psychoses is probably increased by a factor of 3.
XXX	The rate of schizophrenialike psychoses is probably increased by a factor of 3.
18q⁻ or r(18)	Moderate mental retardation, poor speed development, psychotic episodes (schizophrenialike or of manic-depressive type) in childhood or adulthood.
Huntington disease	Schizophrenialike psychoses in the initial phases of the disease.
Acute intermittent porphyria	Various psychiatric symptoms including "schizophrenia" are occasionally reported.
Porphyria variegata	One case report of schizophrenialike psychosis; serine loading produced psychotic symptoms.
Metachromatic leukodystrophy, adult type	Numerous reports on psychoses presenting as "schizophrenia"; cases were detected by chance or by systematic screening of psychotics.
Cerebrotendinous xanthomatosis	4 of 35 homozygous patients showed definite signs of a schizophrenialike psychosis.
Familial basal ganglia calcification	Concordant schizophrenialike psychosis in a pair of identical twins, and familial occurrence.

potheses discussed on the basis of other evidence, even suggest plausible biological mechanisms, for example, those leading to a reduced supply of folic acid or influencing metabolism and function of sulfated amino acids such as methionine. Such patients may have an abnormally high capacity for methylation of dopamine; methionine, being a general source for methyl groups, might enhance this effect.

Most of the patients with a diagnosis of schizophrenia cannot be classified as having clearcut genetic or nongenetic organic disease. However, in many cases slight morphological functional deviations have been described, such as anatomical reduction and dysfunction of the prefrontal cortex (as measured, for example, by computed tomography, positron emission tomography, in vivo measurements of cerebral blood flow [58]), reduction in MAO and other enzymes, a "choppy" rhythm together with some reduction in α activity in the EEG, or slight deviations in visually or acoustically evoked EEG potentials. Here studies of MZ twin pairs discordant for schizophrenia provide interesting results [15]. For example, the hippocampus area is usually smaller in the affected twin (as measured by magnetic resonance imaging) [47]. However,

none of these deviations can be found in all schizophrenics, and there is always the question of whether they are related to the cause of the disease. An additional complication is introduced by the fact that most patients observed for the first time have been treated with psychotropic drugs, which may have changed many of the parameters of possible interest.

These small deviations, however, suggest still another question: some of the conditions listed in Table 16.5 have an autosomal-recessive mode of inheritance; in a dominant condition, such as porphyria, clinical signs are seen only under special stress conditions. Is it not possible that heterozygotes – for metachromatic leukodystrophy or homocystinuria, for example – have an increased liability of becoming schizophrenic, especially if this genetic "weakness" combines with other such liabilities, or if somatic or psychological stress factors are added? A possible slight increase of the risk for psychosis in PKU heterozygotes has been mentioned before (Sect. 7.2.2.8). The genetic model of multifactorial inheritance serves only to describe observations at a preliminary level: it poses questions, rather than answering them. In schizophrenia answers will be easier if the monocausal disease concept suggested by Kraepelin and his successors is replaced by a multicausal concept. The human brain is a complex system, in which a great number of structural and biochemical subsystems interact. Its reaction to exogenous and endogenous stress factors depends on the individual, genetically determined variation within these subsystems, on the life history of the individual, and on the kind and localization of stress. Many combinations of extrinsic and intrinsic stress factors may lead to the same, or a similar, end result. Apparently, the brain has only a limited number of ways of reacting to such stress, the reaction depending on the specific subsystems toward which the stress is directed and their intrinsic susceptibilities.

This leads from the question of *causes* to the question of pathogenesis. Which functional *mechanisms* of the brain are altered in schizophrenia, and how are the clinical signs and symptoms produced by these alterations? Models have been discussed – some of them very elaborate [19] (see also the various discussion remarks on this paper) – but so far no coherent theory has won the support of the various experts. However, there is some evidence that the common final pathway in which all etiological factors come together could have something to do with attentional dysfunction or "faulty filtering of information" (Erlenmeyer-Kimling; see also [33]). Once abnormal functioning has begun, there will be an inherent tendency for self-perpetuation. Evolution has provided the human brain with the ability to "learn." This means that patterns of function change the structure of connections between neurons in a way that repetition of this function becomes easier. Under normal conditions this ability has selective advantages; it helps the individual to cope with demands from a wide variety of environments. As with many other biological adaptations, however, it can become disastrous under special conditions, i.e., when the functional pattern offered for "learning" is counterproductive.

A similar mechanism is well established for another brain disease (or group of brain diseases): epilepsy. Every epileptic seizure can be seen as helping to prepare for the next. Apparently the epileptic seizure is another of the very few ways that the brain has to react to many different stimuli. Many of the principles discussed at some length in this section on schizophrenia could also have been derived using the etiology and pathogenesis of epilepsy as an example [1].

Research Strategies for Further Elucidation of the Genetic Basis of Schizophrenia [40a]. As explained above, the disease concept of "schizophrenia" is phenomenological; unlike tuberculosis or phenylketonuria it does not imply a single major cause. This does not necessarily detract from its value for diagnostics as a basis for therapy. As discussed in greater detail in Sect. 6.4, there is no "natural system of diseases." Disease units and diagnostic categories ultimately are meant to serve a practical purpose: to aid the physician in his attempts at helping his patients [54]. Moreover, analysis of familial aggregation, in addition to its value in genetic counseling, provides clues for a better understanding of causes. For further elucidation, the following strategies are suggested:

1. Studies on "multiplex families" (families with several affected members) offer the best chance of discovering major genes and therefore single, major biochemical abnormalities.
2. Long-term prospective studies on high-risk children (e.g., children of schizophrenic parents) promise valuable information. Such studies are now being performed by a number of groups [9, 12, 24, 34].
3. Genetic variability in the normal range of parameters thought to be of possible importance should be studied.
4. In some families an unfortunate combination of slight quantitative variation in biochemical and structural variables may have caused the disease in approximately the same way as described for squinting (Sect. 6.1.2.7).

Studies could be performed at two levels: at the gene product – biochemical and at the gene – DNA levels (Chap. 6). Studies at the biochemical level include neurotransmitters, their enzymes, and receptors. At present, however, much more is being expected from studies at the gene – DNA level despite the fact that such studies have so far failed to produce convincing results. In a disease such as schizophrenia the straightforward lod score method to detect linkage runs into difficulties because the phenotypes do not permit unequivocal assessment of genotypes. The sib pair method [38, 41], on the other hand, is more useful. Elaborate linkage studies can be short circuited if the disease is shown in some families to cosegregate with a visible chromosomal anomaly, for example, a translocation. Such an unbalanced translocation was found in an uncle and his nephew in 5q11–5q13.3; both suffered from schizophrenia [2]. Linkage studies in a large Icelandic family seemed to confirm linkage of a main gene in the same area of chromosome 5 with schizophrenia [46], but this linkage was soon excluded in a large northern Swedish pedigree [27]. This result may be interpreted in two ways: either the positive linkage result was accidental, or the outcome reflects genetic heterogeneity. In Iceland, a different main gene may cause schizophrenia than in northern Sweden and in other areas, where linkage studies have had negative results. These results show the difficulties of such an analysis. They can be overcome only by international cooperation and careful planning strategies.

Results from such an international collaboration have been published recently. Analysis was performed in three steps: in a first step, the genome was searched systematically using the Généthon linkage map [36a], consisting of 413 highly polymorphic microsatellite markers (Sect. 12.1.2) which cover 84% of the human genome. These markers were used to screen large schizophrenia pedigrees from Iceland, a geographic isolate. Twenty-six loci were suggestive of linkage. In a second step, 10 of these loci were followed-up in families from Austria, Canada, Germany, Italy, Scotland, Sweden, Taiwan, and the United States. Potential linkage on chromosomes 6p, 9, and 20 was observed again. A third, independent sample from China utilized fine mapping of the 6p region and showed evidence for linkage or linkage disequilibrium by association studies (See Sect. 6.2). These and other studies provided evidence for a schizophrenia gene located in the 6p area, distal from the MHC complex ($p = 0.00004$). (See Section 5.2.5). At the same time, this study showed that this is not the only gene locus mutation involved in schizophrenia. Comparison of linkage results from large pedigrees in various countries gave evidence for locus heterogeneity, i.e., different loci contributed to disease liability; a few major genes (Sect. 6.1.2.3) might be involved (= oligogenic transmission).

Such studies are cumbersome; they require much time and effort by many scientists. The reward is worth the effort however. Once a gene is definitely localized, it could be identified (See Chap. 3), and its function in health and disease elucidated.

Critical Assessment of the Attempts to Relate Behavioral Variability to Biochemical Differences in Brain Function. In spite of suggestive hypotheses attempts at explaining abnormal or normal behavior in terms of a genetically determined biochemical mechanism of deviant brain function have largely failed so far. Still, this approach appears worthwhile following.

We should keep in mind, however, that norepinephrine, dopamine, and serotonin are only a fraction of all neurotransmitters; little if any genetic research has been carried out on some of the other neurotransmitters listed in Sect. 15.2.3.6. However, neurotransmitters and their receptors may only be one aspect of brain function in which individual differences can be expected. So far, other potential variables such as individual differences in growth of the brain, number of nerve cells, number of connections between nerve cells, and myelinization have not been studied to a necessary degree. In addition, it has been shown that even the development of synaptic connections between nerve cells can proceed properly only when the nerve cells are functioning. For example, the number of apical dendrites in the optical region of mouse brain is reduced if the animals are raised in darkness [52]. The occipital α rhythm of the human EEG is on average less well developed in adults who were born blind [5]. As noted above, interindividual variability of the EEG, especially the development of the α rhythm, is exclusively genetically determined [53] under normal conditions. However, the brain structures that produce the EEG of the optical cortex can develop properly only if they receive adequate sensory input. The genetic program of brain development can be realized only in interaction with the environment.

Comparing our insight into genetic variability in the function of the brain influencing human behavior with our knowledge of the genetic endowment of the red blood cell which is of course much simpler (Sects. 7.2, 7.3), we realize how fragmentary our information is regarding behavior. Whereas in a hemoglobin gene we can follow the influence of a well-defined alteration in the DNA sequence step-by-step to the altered phenotype, in behavioral genetics we are left with measurements and comparisons of phenotypes that are far removed from gene action. We have taken only the first steps in the analysis of intervening variables. Are the mechanisms more complicated? Is their development in the individual less strictly programmed and less autonomous? Will interaction with the environment turn out to be more essential for proper development of brain function and even of brain structure? Such findings would not be surprising since the human brain is the latest and most complex product of evolution. The few available results endorse this view.

At the beginning of Chap. 15 we noted that the field of human behavioral genetics is conceptually the most important area of *human* genetics. At the same time, however, our knowledge in this field is still unsatisfactory and fragmentary, and our theoretical framework is the least elaborate. In such a situation scientific hypotheses find little objective foundation. Scientists are human beings with prejudices and emotions; they are influenced by their personal biases stronger than in fields with a more elaborate theoretical framework and a sounder empirical foundation. In behavioral genetics one specific problem has raised the most bitter controversies: the claimed existence problem of genetically determined intellectual differences between ethnic groups (Chap. 17).

Conclusions

There are two principal groups of mental disease: affective disorders, such as depression and mania, and the schizophrenia group. Studies using classical methods of human genetics, such as the determination of disease risks among close relatives, concordance rates of mono- and dizygotic twins, and comparison of adoptees with their biological and adoptive parents, have led to the conclusion that genetic variability has a strong influence on disease suscept-

ibility. The mechanisms remains unknown. Linkage studies using DNA markers have been performed to identify individual genes that may be involved in causing these diseases. So far, however, these studies have yielded no generally accepted and reproducible results.

References

1. Anderson VE, Hauser WA, Penry JK, Sing CF (eds) (1982) Genetic basis of the epilepsies. Raven, New York
2. Bassett AS (1989) Chromosome 5 and schizophrenia: implications for genetic linkage studies, current and future. Schizophr Bull 15 : 393–402
3. Bertelsen A (1985) Controversies and consistencies in psychiatric genetics. Acta Psychiatr Scand 71Suppl 319 : 61–75
4. Bertelsen A, Harvald B, Hauge M (1977) A Danish twin study of manic-depressive disorders. Br J Psychiatry 130 : 330–351
5. Birbaumer N (1975) Physiologische Psychologie. Springer, Berlin Heidelberg New York
6. Bleuler M (1972) Die schizophrenen Geistesstörungen. Thieme, Stuttgart
7. Bodmer WF, Cavalli-Sforza LL (1976) Genetics, evolution and man. Freeman, San Francisco
8. Egeland JA, Gerhard DS, Pauls DL et al (1987) Bipolar affective disorders linked to DNA markers on chromosome 11. Nature 325 : 783–787
9. Erlenmeyer-Kimling L, Bernblatt B, Fluss J (1979) High-risk research in schizophrenia. Psychiatr Ann 9 : 38–51
10. Fischer M (1971) Psychoses in the offspring of schizophrenic monozygotic twins and their normal co-twins. Br J Psychiatry 118 : 43–52
11. Fischer M, Harvald B, Hauge M (1969) A Danish twin study of schizophrenia. Br J Psychiatry 115 : 981–990
12. Garmezy N (1974) Children at risk: the search for the antecedents of schizophrenia. II. Ongoing research programs, issues and intervention. NIMH Schizophr Bull 9 : 55–125
13. Garrod AE (1923) Inborn errors of metabolism. Frowde, London (Reprinted 1963 by Oxford University Press, London)
14. Gershon ES, Bunny WE, Leckman JF, van Eerdewegh M, de Bauche BA (1976) The inheritance of affective disorders: a review of data and of hypotheses. Behav Genet 6 : 227–261
15. Goldberg TE, Ragland JD, Fuller Torrey E et al (1990) Neuropsychological assessment of monozygotic twins discordant for schizophrenia. Arch Genet Psychiatry 47 : 1066–1072
16. Gottesman I (1991) Schizophrenia genesis. The origins of madness. Freeman, New York
17. Gottesman I, Shields J (1972) Schizophrenia and genetics. Academic, New York
18. Gottesman I, Shields J (1982) Schizophrenia, the epigenetic puzzle. Cambridge University Press, Cambridge
19. Gray JA, Feldon J, Rawlins JN et al (1991) The neuropsychology of schizophrenia. Behav Brain Sci 14 : 1–84

20. Häfner H (1992) Epidemiology of schizophrenia. In: Ferrero EP, Haynal AE, Sartorius N, Libbey J (eds) Schizophrenia and affective psychoses. Nosology in contemporary psychiatry. Libbey, Rome, pp 221–236

21. Hanson DR, Gottesman (1992) Schizophrenia. In: King RA, Rotter JI, Motulsky AG (eds) The genetic basis of common diseases. Oxford University Press, New York

22. Hebebrand J, Henninghausen K (1992) A critical analysis of data presented in eight studies favoring X-linkage of bipolar illness with special emphasis on forward genetic aspects. Hum Genet 90 : 289–293

23. Heston LL (1966) Psychiatric disorders in foster home reared children of schizophrenic mothers. Br J Psychiatry 112 : 819–825

24. Itil TM, Hsu W, Saletu B, Mednik S (1974) Computer EEG and auditory evoked potential investigations in children at high risk for schizophrenia. Am J Psychiatry 131 : 892–900

25. Kelsoe JR, Ginns EL, Egeland JA et al (1989) Re-evaluation of the linkage relationship between chromosome 11p loci and the gene for bipolar affective disorders in the Old Order Amish. Nature 342 : 238–243

26. Kendler KS (1983) Overview: a current perspective on twin studies of schizophrenia. Am J Psychiatry 140 : 1413–1425

27. Kennedy JL, Giuffra LA, Moises HW et al (1988) Evidence against linkage of schizophrenia to markers on chromosome 5 in a northern Swedish pedigree. Nature 336 : 167–170

28. Koller S (1940) Methodik der menschlichen Erbforschung. II. Die Erbstatistik in der Familie. In: Just G, Bauer KH, Hanhart E, Lange J (eds) Methodik, Genetik der Gesamtperson. Springer, Berlin, pp 261–284 (Handbuch der Erbbiologie des Menschen, vol 2)

29. Kendler KS, Hays P (1982) Familial and sporadic schizophrenia: a symptomatic, prognostic, and EEG comparison. Am J Psychiatry 139 : 1557–1562

30. Kringlen E (1967) Heredity and environment in the functional psychoses. Heinemann, London

31. Leonhard K (1957) Aufteilung der endogenen Psychosen. Akademie, Berlin

32. Maier W, Lichterman D, Minges D et al (1993) Continuity and discontinuity of affective disorders and schizophrenia: results of a controlled family study. Arch Gen Psychiatry 50 : 871–883

33. Matthyse S, Spring BJ, Sugorman J (eds) (1978) Attention and information processing in schizophrenia. J Psychiatr Res 14 : 1–331

34. Mednik SA, Mura E, Schulsinger F, Mednik B (1973) Perinatal conditions and infant development in children with schizophrenic parents. Soc Biol 20 : 111–112

35. Mendlewicz J, Rainer JD (1977) Adoption study supporting genetic transmissions in manic-depressive illness. Nature 268 : 327–329

36. Mirsky AF, Quinn OW (1988) The Genain quadruplets. Schizophr Bull 14 : 595–612

36a. Moises HW, Yang L, Kristbjarnarson H et al. (1995) An international two-stage genome-wide search for schizophrenia susceptibility genes. Nature [Genet] 11 : 321–324

37. Odegard O (1972) The multifactorial theory of inheritance in predisposition to schizophrenia. In: Kaplan AJ (ed) Genetic factors in schizophrenia. Thomas, Springfield, pp 256–275

38. Ott J (1990) Cutting a Gordian knot in the linkage analysis of complex human traits. Am J Hum Genet 46 : 219–221

39. Propping P (1983) Genetic disorders presenting as "schizophrenia." Karl Bonhoeffer's early view of the psychoses in the light of medical genetics. Hum Genet 65 : 1–10

40. Propping P (1989) Psychiatrische Genetik. Springer, Berlin Heidelberg New York

40a. Propping P, Nothen MM (1995) Schizophrenia: genetic tools for unraveling the nature of a complex disorder. Proc Natl Acad Sci USA 92 : 7607–7608

41. Risch N (1990) Linkage strategies for genetically complex traits. II. The power of affected relative pairs. Am J Hum Genet 46 : 229–241

42. Rosenthal D (ed) (1963) The Genain quadruplets. Basic, New York

43. Rosenthal D (1970) Genetic theory and abnormal behavior. McGraw-Hill, New York

44. Rosenthal D, Kety SS (1968) The transmission of schizophrenia. Pergamon, Oxford

45. Rüdin E (1916) Studien über Vererbung und Entstehung geistiger Störungen. Springer, Berlin

46. Sherrington H, Brynjolfsson J, Petursson H et al (1988) Localization of a susceptibility locus for schizophrenia on chromosome 5. Nature 336 : 164–170

47. Suddath RL, Christison GW, Fuller Torrey E et al (1990) Anatomical abnormalities in the brains of monozygotic twins discordant for schizophrenia. N Engl J Med 322 : 789–794

48. Tienari P (1963) Psychiatric illnesses in identical twins. Acta Psychiatr Scand Suppl 171

49. Tienari P (1971) Schizophrenia and monozygotic twins. Psychiatr Fenn:97–104

50. Torgersen S (1986) Genetic factors in moderately severe and mild affective disorders. Arch Gen Psychiatry 43 : 222–226

51. Tsuang MT, Faraone SV (1990) The genetics of mood disorders. Johns Hopkins University Press, Baltimore

52. Valverde F (1967) Apical dendritic spines of the visual cortex and light deprivation in the mouse. Exp Brain Res 3 : 337–352

53. Vogel F (1970) The genetic basis of the normal human electroencephalogram (EEG). Hum Genet 10 : 91–114

54. Vogel F (1990) Humangenetik und Konzepte der Krankheit. Springer, Berlin Heidelberg New York, pp 335–353 (Sitzungsberichte der Heidelberger Akademie der Wissenschaften, Math.-Naturwiss. Klasse, 6. Abhandlung)

55. Vogel F (1992) Für und wider die Einheitspsychose aus genetischer Sicht. In: Mundt C, Sass H (eds) Für und wider die Einheitspsychose.Thieme, Stuttgart, pp 110–119

56. Von Knorring AL, Cloninger R, Behman M, Sigvardsson S (1983) An adoption study of depressive disorders and substance abuse. Arch Gen Psychiatr y40 : 943–950

57. Wahl OF (1976) Monozygotic twins discordant for schizophrenia: a review. Psychol Bull 83 : 91–106

58. Weinberg DR (1988) Schizophrenia and the frontal lobe. Trends Neurosci 11 : 367–370

59. Wender PH, Kety SS, Rosenthal D et al (1986) Psychiatric disorders in the biological and adoptive families of adopted individuals with affective disorders. Arch Gen Psychiatry 43 : 923–929

60. Winokur G, Tanna VL (1969) Possible role of X-linked dominant factor in manic-depressive disease. Dis Nerv Syst 30 : 89–94

61. Zerbin-Rüdin E (1967) Endogene Psychosen. In: Becker PE (ed) Humangenetik, ein kurzes Handbuch, vol 5/2, Thieme, Stuttgart, pp 446–573

17 Behavioral Genetics: Differences Between Populations

*The significant problems we face
cannot be solved at the same level
of thinking we were at when we created them.*

Albert Einstein

Differences in IQ and Achievement Between Ethnic Groups

Group Differences in Behavioral Traits? The human population is subdivided into subpopulations; these are called races if they have a certain amount of their genes in common in which they differ from other subpopulations. The term "ethnic group" is often used when historical and cultural aspects are included in the criteria of classification. Genetically related groups also tend to have common cultural traditions and social systems, and the concepts of "race" and "ethnic group" therefore overlap strongly (Chap. 14). The genetic compositions of races differ because they have developed in reproductive isolation, which may have produced chance fluctuations of gene frequencies and especially random fixation of alleles. Moreover, there may have been different selective conditions. Nowadays, migration with its accompanying gene flow tends to diminish group differences. As is seen in Chap. 14, genetic susceptibilities to disease, such as diabetes, and the ability to digest certain foods, such as lactose, show different distributions in the various racial groups. It is therefore conceivable that the various environments in which human groups lived in the past made different demands on their behavior, thereby selecting for different combinations of genes that influence behavior. Chance fluctuation may sometimes also have affected such genes, since human groups early in evolution and for many generations consisted of small numbers of individuals, thereby providing frequent opportunities for genetic drift. The existence of some genetic group differences in behavior would therefore not be surprising.

Beyond this very general statement, however, more specific predictions are hazardous. We know hardly any human genes that influence behavior in the normal range; we know very little about the special abilities that human groups needed to cope with the different microenvironments to which they were exposed in the past; we have even less knowledge about everyday living conditions of our remote ancestors. Did people living in cold areas with long winters need a better ability to plan in advance for food supply? Did hunters and gatherers in the tropical rain forest need more alertness and versatility in coping with sudden dangers? Were people who lived in open savannas and semideserts accustomed to larger social groups, while the rain forest rather favored smaller band size, as appears true of subhuman primates? We simply do not know. Ethnologists have observed enormous, and in some cases extreme, differences in behavior patterns in the same race and between groups living under conditions that appear very similar. These differences may in turn influence the genetic composition of such populations, especially for genes affecting behavioral characters. Therefore inferences from present-day primitive populations as to the behavioral patterns of our ancestors who supposedly lived under similar conditions are somewhat hazardous. Nevertheless, such inferences are often made, but these depend significantly on whether the primitive population selected for study is peaceful or aggressive, sexually restrained or uninhibited, cooperative or selfish. These very differences strongly indicate that simplistic genetic interpretations cannot apply.

To compare behavioral characteristics that could show genetic variability between races or ethnic groups within present-day "civilized" populations, comparisons should be confined to groups living under identical conditions, such as family patterns, education, chances to enter into various occupational careers, and other such functions. In conceptually simpler situations, for example, in animal breeding, we would not dream of drawing conclusions for selective breeding of animal stocks unless we had carefully kept the environment constant. Among humans, on the other hand, comparable conditions hardly if ever exist. This difficulty makes all judgments ambiguous.

Two principal differences have been documented between groups: the higher average intelligence and intellectual achievements of Ashkenazi Jews compared with the gentile European and North American populations among whom they live, and the lower average IQ of African-Americans than the white and Oriental population groups of the United States. The Jewish-gentile difference may have contributed to – but certainly was not the only reason for – the antisemitic movements in many European countries that led to the genocide of most of the central and eastern European Jewish population by the Nazi regime. The black-white difference is now providing racists with pseudoscientific arguments for discrimination against African-Americans.

Intelligence and Achievement of Ashkenazi Jews [12; 13; 13a]. Ashkenazi Jews of Europe lived for many centuries under conditions of severe discrimination. They were confined to restricted quarters within the cities, called ghettos; they were not allowed to own property and were barred from many occupations. The situation changed in the nineteenth and early twentieth centuries when civil rights were achieved in western European countries. Social discrimination of various degrees continued beyond this time. Nevertheless, many occupational careers opened up, with the result that Jews were soon frequently found in many professions requiring high intellectual abilities.

In 1907, for example, about 1% of the German population was Jewish, but 6% of physicians and 15% of lawyers were Jewish [9]. Among university professors the proportion of Jews (including those baptized – the usual requirement for such a position) in 1900–1910 was: 14.2% in law, 12% in arts and sciences, and 16.8% in medicine. During the winter term of 1924/1925 Jews were about six times more frequent among university students than in the general population. Comparable and more recent data are available from the United States. For example, 27% of the Americans who received Nobel prizes from 1901 to 1965 were of Jewish origin, while Jews constitute only about 3% of the American population [5].
Most observers agree that Marx, Freud, and Einstein were among those who profoundly influenced civilization over the past century. All three were Jews of Ashkenazi origin.
In comparing test scores between ethnic groups carried out in the United States and Canada, the mean of Jewish subjects (almost entirely of Ashkenazi origin) is 5–10 IQ points higher than that of non-Jewish whites – especially in the verbal parts of IQ tests [11].

What are the reasons for this unsually high performance of a relatively small ethnic group? Here, cultural explanations could undoubtedly account for most or even all of the difference. Socioeconomically the group lived for centuries under conditions in which only intellectual performance could secure survival. A high emphasis on intellectual ability was characteristic of the cultural climate. High aspirations, encouragement of superior performance, and an intellectually rich environment favor intellectual development. In recent decades the relatively smaller number of children in Jewish families might provide an additional explanation as high performance of

children from small sibships, especially first-borns, has often been observed [21]. The challenge of being the "marginal outsider" to higher performance in most societies may also have contributed to the results.
On the other hand, genetic explanations cannot be entirely dismissed. Selection in favor of intellectual capability within the Jewish communities favored the "scholars," i.e., those who were especially able to interpret traditional texts such as the Talmud. These scholars were maintained by their communities and were given the opportunity to marry the wealthiest girls. Since they lived under more favorable economic conditions than the majority of Jews who were economically rather poorly off in Poland and Russia, the mortality of their infants may have been lower than that of the general Jewish population.
In fact, data from Poland from the middle of the eighteenth century suggest that the poorer Jewish families had 1.2–2.4 *surviving* children per family while the more prominent Jews had 4–9 children who reached adulthood. Another conceivable genetic factor is selective survival under conditions of persecution over the centuries. Possibly the more nimbler and smarter young adults were more readily able to escape violent death and therefore transmitted their genes to their descendants. The effect of such "IQ-dependent" mortality can be significant [12, 13].
In the absence of specific knowledge of the genetic mechanisms that may underlie individual differences in intellectual performance *we have no way to decide whether genetic factors have contributed to the intellectual excellence of Jews.* The means to solve this problem do not exist, and there is no way of tackling the question unambiguously at the present state of knowledge in human behavioral genetics.

Difference in Mean IQ Between Ethnic Groups in the United States, Especially Between African-Americans and Whites. Similar difficulties apply to a problem that has aroused an unusual degree of public controversy: the difference in average IQ between ethnic groups in the United States, especially African-Americans and whites [15, 18]. The reader who has followed our considerations of human behavioral genetics, especially the discussion of the heritability concept and its application to IQ test performance, should by now be able to provide his own answers. There are enormous difficulties in setting up an unassailable experimental design to assess the causation of a variable, such as IQ, that is determined by a complex interaction of a genetic disposition with many different influences from the physical and sociocultural environment. The problem is even more complex than that posed by IQ differences in Jews since the environmental conditions under which

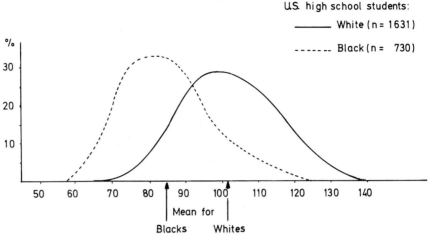

U.S. high school students:

———— White (n = 1631)

------- Black (n = 730)

Fig. 17.1. Distribution of IQ in a series of white (–) and African-American (—) high school students. In other series some African-Americans were observed whose test results equaled those of the best whites. (Data from Roland and Swan 1965; see Walter H, Grundriß der Anthropologie, BLV, Munich, 1970)

African-Americans have been living are much more different than those between Jewish and non-Jewish whites.

However, the problem has attracted much attention. Therefore we describe briefly the most essential facts and some arguments in favor and some against genetic interpretations [7, 8, 11, 15].

Since intelligence testing was first carried out on United States Army conscripts during World War I, African-Americans have fairly consistently shown lower average IQ values than white Americans. Figure 17.1 shows a typical result. Certain findings seem to be undisputed:

a) There is a mean difference of about 15 IQ points in most comparisons between African-Americans and whites.

b) The distributions of African-Americans and whites overlap strongly; differences between individuals of each group may be much larger than the between-group differences of mean values. Some African-Americans reach highest levels of performance; the African-American population as a whole comprises the full range of human talents [7].

c) There is considerable variability of means and distributions between subgroups within the white and African-American populations, depending on whether the samples come from the south or the north, from urban or rural areas, from children or adults, etc. [11].

Explanation: Genetic or Socioeconomic? Two groups of explanations have been offered for this difference: genetic and socioeconomic. The following tabulation lists some of the arguments opposed to one other.

Genetic explanation	Nongenetic explanation
IQ has a high heritability within the white population [7, 8]	The within-population heritability estimates are of highly dubious scientific value [20] (see Sects. 6.1, 6.3; Appendix 5) The data are often based on politically biased scientific evidence and some data are fraudulent [1, 3, 4, 10]
Cautious extrapolation from heritability of within-group differences to that between groups is possible [8]	Extrapolation from heritability of within-group differences to between-group differences requires too many untestable assumptions and is scientifically unsound (see Sect. 6.3.10 on stature)
Sociocultural differences can explain only a minor part of the differences: a) Other, equally discriminated groups such as native American fail to show an IQ difference of similar order b) Differences are especially pronounced in so-called culture-free tests c) Differences in standard intelligence tests are more pronounced in performance than in verbal tests	The sociocultural and educational differences between whites and blacks are fully sufficient to explain existing differences Culture-free tests are not really culture free. Performance depends on: a) Intellectual training. b) Active interest in solving puzzles that are not related to everyday problems.
The IQ differences are not removed when tests using	IQ test are based on the language of the white middle-

Genetic explanation	Nongenetic explanation
the special language of African-Americans are applied.	class population. Tests using the language of African-Americans show improved results.
Contrary to their poor test performance (and their correspondingly poor school success), African-American children often show a remarkable degree of practical intelligence in daily-life situations. This contrast can be explained by two different levels of intellectual performance: level 1, the ability to process information in a simple, straightforward way that is sufficient for most purposes of daily life, and level 2, the ability to process and rearrange information in a more complex way that is necessary for abstract thinking. The deficiency of most African-Americans relates to level 2, not level 1 [8]	The difference between practical and theoretical intelligence in African-Americans can be explained sociopsychologically: a) African-Americans consider abstract thinking as less interesting. b) African-Americans experience school and its values as imposed from the outside, in which they do not actively take part. c) African-American children are not motivated and supported by their parents to strive for intellectual achievements.
Only a small part of the test difference is removed when the tests are given by African-American instructors.	IQ tests have in most cases been given by white instructors with diminished motivation on the part of African-American children.
Even if the average IQ of African-American children can be boosted by adoption (see opposite column) into especially favorable white homes, there remains an IQ difference among these children in the following rank order (from high to low): white children of biological parents/white adopted children/adopted African-American children with a higher proportion of white genes/adopted African-Americans with a lower proportion of white genes [16].	African-American children who have been adopted by white parents show IQ values that are above the average of the white population. This shows the powerful influence of environment, which is favorable in families who adopt children [16]. The IQ difference between adopted African-American children with a higher and those with a lower proportion of white genes can be explained fully by specified environmental differences [16].
The U. S. Army refused a much higher proportion of African-American than of white army conscripts because of poor performance in intelligence tests. The African-American fathers of German children (see opposite) were selected for higher IQ.	There is no IQ difference between German children whose fathers were African-American soldiers and appropriate German controls [11].
The differences in white admixture between the examined children [15, 17] are so small that the expected IQ correlation could only be shown in a much larger sample. The correlation of skin color and African-American marker genes was so small in this study that lack of correlation between IQ and white admixture is difficult to interpret.	There is no correlation between IQ and white admixture in African-American children [15, 17].
Those not considering genetic reasons for the IQ difference between African-Americans and whites are so preoccupied by their liberal ideology that they have lost their ability to think clearly and to face facts.	Those considering genetic reasons for the IQ differences between African-Americans and whites are racist reactionaries who consciously or unconsciously want to discriminate against ethnic minorities to maintain the privileges of their own class or ethnic group. Moreover, they are poor scientists.

These arguments show how little even very sophisticated argumentation based on biometric studies may help in gaining decisive evidence as long as nothing is known regarding the biological mechanisms involved. Given the present state of our knowledge of the biological mechanisms that underlie genetic differences in intelligence within the normal range, attempts at elucidating the possible genetic reasons for group differences in intellectual performance – especially the IQ difference between African-Americans and whites – appear futile.

Is All Research That Has Been Done in This Field Scientifically and Socially Worthless? We do not consider the work that has been carried out so far in this field as scientifically worthless. Studies on heritability have helped to expose the severe inherent limitations of this concept, especially in its application to human populations. Moreover, once well-defined genetic variability at the physiological and biochemical level can be established, many of the methods and considerations will find useful applications. Experiences with numerical chromosome aberrations, for example, the XYY karyotype, have already shown the extent to which biological studies can be improved by statistical, biometric, and epidemiological sophistication.
Moreover, many results are interesting in their own right, even if the principal purpose – to collect evi-

dence for the causal explanation of group differences – cannot be achieved. It is, for example, useful to know how much IQ and school performance of adopted children can be boosted under the favorable conditions of adoption. The concept of two levels of information processing, even if basically incorrect in biological terms, may help to develop teaching strategies better adapted to different requirements of various children than those used by our present-day school systems. In these and other ways some of these attempts that seem to be futile when considered from the standpoint of a genetic analysis may gain new significance.

If Genetic Group Differences Did Indeed Exist, Would They Suggest Any Consequences in Social Policy? Let us assume for the sake of argument that a part of the IQ difference between African-Americans and whites is indeed genetically determined. What conclusions would we have to draw for social policy?

All considerations of this problem must start with the contention that the individual and not the ethnic group should be the target of social policy. It would certainly be the goal of a society to create conditions in which every individual has a fair chance to develop his or her abilities and fit into society in a way that secures the highest possible degree of personal fulfillment – in a manner which, at the same time, best serves the requirements of society as a whole. One way of achieving this goal is to offer adequate facilities for education. This condition includes teaching methods that have an optimum effect on individual learning abilities and motivations, regardless of whether and to what degree any differences in abilities are genetically determined. All experience of modern genetics teaches us that the phenotype is the result of an interaction between genotype and environment and that specific genotypes need appropriate, and in many cases also specific, environmental conditions for optimal development. It is the task of behavioral genetics to define these conditions and to provide advice regarding development of individually oriented educational strategies that best fit the genetically determined strengths and weaknesses of the child. Current educational approaches to dyslexia, which is frequently genetically determined [2, 6, 14] are an example of such an approach.

Whether programs of compensatory education for children who show certain difficulties should be part of such strategies cannot be answered fully and will require future experience. These decisions have nothing whatever to do with the question of whether and to what degree genetic variability influences these deficiencies, not to mention the question of whether the individual belongs to a minority whose population average differs from that of the majority. Should we push a

mentally slightly subnormal Jewish child through all levels of higher education only because such a student belongs to a subpopulation with a higher average level of achievement? This is obviously absurd. Logically it would be just as absurd to deny an African-American child compensatory education that could increase his or her chance of success in life because the child belongs to a minority in which a higher proportion of children are in need of compensatory measures. Nevertheless, racial sensibilities are often aggrieved when compensatory education is suggested.

We realize that emphasis on optimum educational opportunities and consideration of the individual's abilities regardless of ethnic identification, while theoretically based on sound principles, may cause many practical problems at the present time. What teaching methods should a teacher use in a multiracial or multiethnic class with a wide range of abilities? Should the teacher give most attention to the lowest ability group? To the average? We have no ready answers. Is equal representation of all ethnic minorities desirable if some minorities excel in certain areas? Should there be different standards for minority representation in various occupational groups? In our opinion, the ideal society would provide each individual, regardless of race or ethnic origin, with opportunities for maximum development of his or her abilities. It is conceivable that such a scheme may lead to overrepresentation or underrepresentation of some racial or ethnic groups, whether for genetic or cultural reasons.

Sect. 6.1.1.5 describes the increase of stature over the past century. Stature has a high heritability under most environmental conditions. Despite high heritability better nutrition has brought about a very significant increase in the population average. Average IQ has also increased in populations of industrialized countries over the past century [19]. This, too, has occurred much too rapidly to be explained by genetic changes and therefore must have been caused by the environment. This observation is encouraging.

It does not mean, however, that every environmental change has effects; environmental alterations must be appropriate. Compensatory programs may fail not because of high heritability of the characteristics which they are meant to influence but because of their inadequacy to compensate for what is lacking. The geneticist has no reason whatever to discourage any such programs; (s)he should strongly encourage all attempts by social and behavioral scientists to explore the specific conditions that cause some persons to do less well than others and to try to change these conditions.

Intermarriage. Marriage is a matter not of ethnic groups but of individuals. Individual members of two different ethnic groups may be much more similar genetically, i.e., they have many more genes in common, than two random individuals of a single ethnic group. This conclusion also applies to genes that can influence behavioral variability. We interpret all evidence of racial admixture to indicate that no biological ill effects for the children from racial intermarriage have been observed (Sect. 14.3.2). In modern society spouses are selected individually, and marriages are not arranged by families; hence there is strong assortative mating for behavioral phenotypes, such as intelligence (Sect. 13.1; Fig. 13.1). Such assortative mating will remain, regardless of whether intermarriage between groups becomes more frequent or not.

Unlike the old nations of Europe, the society of the United States consists of one ethnic majority and several strong ethnic minorities. The fact that such minority groups exist creates tensions and conflicts. One could suggest that these conflicts might be solved most easily if the minorities were absorbed by the majority. It could also be strongly argued, on the other hand, that the richness of a country's cultural heritage can best be preserved if minority groups maintain their respective biological and cultural identities. Such an advantage may be worth the sacrifice of living with the ensuing tensions and conflicts.

Conclusions

Family, twin, and adoption studies suggest the role of undefined genetic factors together with environmental determinants in affecting cognitive performance as measured by IQ tests. Heritability values range widely between 20% and 80%.

Differences have been observed between human population subgroups in the distribution of such traits. The Asian-American and Ashkenazi Jewish populations, for example, often show higher average IQ than other whites, whereas a lower average IQ is observed among African-Americans. The question whether such differences are caused, at least in part, by genetic factors or are entirely environmental in origin has led to highly emotional discussions among both scientists and the general public. The demonstration that a quantitative trait within one population is heritable does not imply that differences between populations for that trait have a genetic origin. A variety of other arguments regarding a possible genetic origin of such population differences are compared. The present evidence does not permit final conclusions regarding the role of genetic factors to explain population differences.

References

1. Feldman MW, Lewontin RC (1975) The heritability hangup. Science 190 : 1163–1168
2. Finucci JM, Childs B (1983) Dyslexia: family studies. In: Ludlow CL, Cooper JA (eds) Genetic aspects of speech and language disorders. Academic, New York, pp 157–167
3. Goldberger A (1976) Jensen on Burks. Educ Psychol 12 : 64–78
4. Goldberger AS (1977) Models and methods in the IQ debate. I. University of Wisconsin (Social Systems Research Institute workshop series 7710)
5. Gulotta F, Rehder H, Gropp A (1981) Descriptive neuropathology of chromosomal disorders in man. Hum Genet 57 : 337–344
6. Herschel M (1978) Dyslexia revisited. Hum Genet 40 : 115–134
7. Jensen AR (1969) How much can we boost IQ and scholastic achievement? Harvard Educ Rev 39 : 1–123
8. Jensen AR (1973) Educability and group differences. Methuen, London
9. Lehrl S, Fischer B (1990) A basic information psychological parameter (BIP) for the reconstruction of concepts of intelligence. Eur J Personal 4 : 259–286
10. Lewontin RC (1975) Genetic aspects of intelligence. Annu Rev Genet 9 : 387–405
11. Loehlin JC, Lindzey G, Spuhler JN (1975) Race differences in intelligence. Freeman, San Francisco
12. Motulsky AG (1979) Possible selective effects of urbanization on Ashkenazi Jewish populations. In: Goodman RM, Motulsky AG (eds) Genetic diseases among Ashkenazi Jews. Raven, New York, pp 201–212
13. Motulsky AG (1980) Ashkenazi Jewish gene pools: admixture. In: Eriksson AW et al. (ed) Population structure and genetic disorders studies on isolates. Academic Press, London, pp 353–365 (Sigrid Juselius symposium 7)
13 a. Motulsky AG (1995) Jewish diseases and origins. News and Views. Nature (Genet.] 9 : 99–101
14. Propping P (1989) Psychiatrische Genetik. Springer, Berlin Heidelberg New York
15. Scarr S (1981) Race, social class, and individual differences in IQ. Erlbaum, Hillsdale
16. Scarr S, Weinberg RA (1976) IQ test performance of black children adopted by white families. Am Psychol 31 : 726–739
17. Scarr S, Pakstis AJ, Katz SH, Barker UB (1977) Absence of a relationship between degree of white ancestry and intellectual skills within a black population. Hum Genet 39 : 69–86
18. Shuey AM (1966) The testing of negro intelligence, 2 nd edn. Social Science, New York
19. Storfer MD (1990) Intelligence and giftedness: the contribution of heredity and early environment. Jossey-Bass, San Francisco
20. Taylor HF (1980) The IQ game. A methodological inquiry into the heredity-environment controversy. Harvester, Brighton
21. Zajonc RB (1976) Family configuration and intelligence. Science 192 : 227–236

18 Genetic Counseling and Prenatal Diagnosis. Human Genome Project

*An ounce of prevention is worth
a pound of cure.*

Old English proverb

The expanding knowledge in human genetics has led to practical applications at an increasing rate – especially in genetic counseling and genetic screening. These approaches are promoted to avoid unnecessary hardships for families today. However, widespread genetic counseling and genetic screening will also influence the genetic composition of future generations. More recently, molecular biology has provided increasingly efficient techniques for genetic diagnosis and therapy which will be described in Chap. 18. Human geneticists need to consider whether these influences are beneficial or not. What will be the impact of all these new developments on the human species? These problems are considered in Chapter 19.

18.1 Genetic Counseling [5a, 18, 19, 22, 52]

Genetic counseling has become an important area of applied human genetics, and an increasing number of patients are requesting advice or are referred by their physicians for counseling about the diagnosis, impact, and recurrence risks of genetic diseases. As the public media and the medical literature disseminate more news about genetics, public and medical interest in genetic disease is growing further. What is genetic counseling? Genetic counseling refers to the totality of activities that (a) establish the diagnosis, (b) assess the recurrence risk, (c) communicate to the patient and family the chance of recurrence, (d) provide information and sympathetic counsel regarding the many problems raised by the disease and its natural history, including the potential medical, economic, psychological, and social burdens, and (e) provide information regarding the reproductive options to be taken including prenatal diagnosis, and refer the patients to the appropriate specialists.

The range of problems seen during genetic counseling covers a wide area. There is some variation because local expertise for certain diseases differs, but many conditions are usually encountered. Generally only 30%–50% of the patients and families prove to have classical genetic illnesses, such as monogenic diseases or chromosomal aberrations. Many consultations deal with various birth defects, mental retardation, delayed development, dysmorphic looking children, short stature, and similar problems, which may or may not have a genetic cause.

Genetic counseling is usually carried out by specialist physicians, and many physicians all over the world are now specializing in medical genetics. The new profession of "genetic counselors" has also emerged in the United States. Genetic counselors are trained in specialized postbaccalaureate 2-year university programs. They usually work with physicians in medical genetics clinics and carry out much of the information gathering, counseling, and follow-up. Their participation in genetic service activities allows medical manpower to be used more efficiently. Their counterparts in European countries are often trained social workers.

Genetic counseling is a medically oriented rather than a "eugenic" endeavor. Most observers consider it inappropriate to advise couples regarding reproductive decisions that are based on eugenic considerations, although the outcome of a decision to have a child with a genetic disease who later in life will have offspring may worsen the genetic load by adding harmful genes that would otherwise be eliminated. Couples asking for advice are encouraged to make those reproductive decisions which are most appropriate for them regardless of possible deleterious effects on the gene pool of the population. This practice places genetic counseling squarely within the framework of medical practice, where the individual and his family rather than the general population is the focus of advice and treatment. Fortunately, the course of action selected by most couples (i.e., limitation of reproduction when there is a high risk) coincides with a favorable impact on the human gene pool (see Sect. 12.3).

Diagnosis. The accurate diagnosis of a genetic disease using all the modalities of modern medicine is essential. Accurate diagnosis is emphasized since similar phenotypes may sometimes have different modes of inheritance or may not be inherited at all. The family history is often important because a clearcut pattern of inheritance such as in autosomal-dominant traits often provides the basis for counseling when a definitive diagnosis may not be clear. Previous medical and hospital records are often helpful in making a precise diagnosis. Since many genetic diseases are associated with somewhat characteristic facial features, inspecting of photographs of family members may be helpful. Chromosomal examinations are frequently required in the diagnosis of complex birth defects (see Sect. 2.2.2). Since many genetic diseases are rare, even trained medical geneticists and specialists in a given field of medicine may have difficulty in arriving at an accurate diagnosis. They cannot be equally knowledgeable about all genetic diseases in every area of medicine but do need to be aware of recent monographs and computerized expert systems (Appendix 3) to establish the appropriate diagnosis. The *Catalogue of Mendelian Traits in Man* by McKusick [38] and its computerized version OMIM are helpful but often need to be supplemented with other references, especially for non-Mendelian diseases (see Table A 3.1). A good library and the knowledge of how to consult the current literature in clinical genetics is essential. Because of the rapid expansion of knowledge, utilization of the journal literature, as opposed to textbooks and monographs, is more important in clinical genetics than in most fields of medicine. This is facilitated today by computerized searches.

Parents whose infants are stillborn or die in the neonatal period often request genetic counseling regarding recurrence risks. Usually little or no information is available regarding the specific abnormality of the stillbirth since no pathological or other diagnostic study was conducted. It has been recommended that as a minimum a gross autopsy, photography, radiography, and bacterial cultures be performed in all cases of stillbirth or early neonatal death to allow a diagnosis, since this is needed for genetic counseling [19]. Chromosomal studies and histopathology are not likely to provide diagnostic information if the gross autopsy findings are normal. However, such studies are usually indicated if multiple anomalies are found on autopsy.

A definitive diagnosis often cannot be made even by experienced specialists due to the enormous complexity of development and its possible perturbation by frequently unknown genetic, epigenetic, and environmental factors. Fewer diagnostic uncertainties occur with monogenic diseases than with various birth defects. However, even in this area the growth of the McKusick catalogue over the years (i.e., from 866 definitive loci in 1971 to 4967 definitive loci in 1996) attests to the rapid expansion of knowledge in this field.

To keep up with this explosive advance in knowledge several research groups have stored information about clinical findings in genetic diseases and birth defects on computer and furnish programs to allow diagnosis using this information. With the increasing number of possible hereditary diseases and syndromes no medical geneticist can perform his daily work satisfactorily without the assistance of these data bases, which have been developed into fairly sophisticated expert systems (Appendix 3). However, proper use requires knowledge, experience, and a critical mind.

Recurrence Risks. Genetic risks in Mendelian diseases are clearly defined and depend upon the specific mode of inheritance (Table 18.1). The actual clinical risks to the patient, particularly in autosomal-dominant inheritance, depend upon variable penetrance and expression and late onset of many disorders. Patients are more interested in the actual recurrence risk of the clinical symptoms than in the formal genetic risks alone. In diseases with lowered penetrance the actual recurrence risk is lower than the formal risk of genetic transmission. For example, an

Table 18.1. Risks for rare Mendelian disorders in families of affected patients

Mode of inheritance	First-degree relatives at risk	Risk	Other relatives at risk	Risk
Autosomal-dominant	Sibs, parents, children (both sexes)	50%	Uncles, aunts, nephews, nieces, first cousins	25% 12.5%
Autosomal-recessive	Sibs (both sexes) Children	25% Negligible[a]	Uncles, aunts, nephews, nieces, cousins	Negligible
X-linked recessive	Brothers, sisters as carriers	50%[b]	Maternal uncle, maternal aunt as carrier	50%[b]

[a] Risks for children of patients affected with common autosomal-recessive diseases depend upon heterozygote frequency.

[b] The risks for X-linked recessive disease apply only if the disease is familial and not if the mother's carrier status is caused by a new mutation. Risks are negligible when the disease in an affected patient is caused by a new mutation. Except in Duchenne muscular dystrophy the proportion of new mutations may be much less than the expected 33% in X-linked lethal diseases (Sect. 9.3.4). Moreover, some apparent new mutants are in fact products of germ cell mosaicism in their mothers – with an increased risk for their brothers.

offspring's risk of an autosomal-dominant disease with 70% penetrance is 35% rather than 50% ($0.5 \times 0.7 = 0.35$). The risk declines with late onset diseases, as a person remains unaffected beyond the age at which the disease first becomes manifest. McKusick's catalogue is available as a computerized data base on-line (OMIM) and is updated frequently. The most recent information can therefore be readily obtained with appropriate computer access (see Appendix 3). In addition, by introducing key words more comprehensive cross-searches that also allow access to abstracts of articles are now possible.

Molecular Diagnosis. As more genes are being cloned and the molecular nature of mutations causing disease becomes known, *direct* DNA diagnosis of genetic disease is becoming increasingly possible. Unlike indirect diagnosis by linked DNA markers, a family study is not required for study. However, the exact nature of the mutation to be detected must usually be known (Table 18.2). This raises problems if many different mutations are implicated, as is often the case. However, test systems that can detect several mutations at one locus are under development. It is always good practice to isolate and store DNA from patients with genetic diseases for appropriate future study. The resultant information may be of great help in counseling family members in the future. Table 18.6 lists some autosomal dominant disorders in which the genes have been mapped or cloned.

In many cases one can use indirect DNA diagnosis by gene tracking (which requires several family members). If the gene has been isolated, but the mutation remains unknown, an intragenic DNA polymorphism (RFLP, VNTR, CA repeat) can be used to track the mutant gene by cosegregation of the marker without knowing the exact nature of the mutation (Tab. 18.2). The possibility of nonallelic genetic heterogeneity, however, must be kept in mind as this leads to misdiagnosis.

Where neither the gene nor its mutations have been defined, linked DNA polymorphism can still be helpful. However, as with intragenic DNA polymorphisms, the exact "phase" of the DNA marker in relation to the disease gene must be known (*cis* versus *trans*). Often a sufficient number of family members is not available, and it is therefore impossible to obtain the necessary information. Problems of genetic heterogeneity and possible recombination between the linked DNA marker and the disease gene make this approach less than 100% accurate.

The examples of Duchenne and Becker muscular dystrophy illustrate these principles. About two-thirds of all Duchenne and Becker mutations are caused by deletions of the X-linked dystrophin gene. Direct DNA diagnosis for these deletions is relatively simple (for example, by in situ hybridization) (Fig. 3.11) [54] and can be used for prenatal diagnosis and for carrier detection. If a deletion cannot be detected, a search for one or another of the many different missense point mutations that can cause the disease is not always

Table 18.2. DNA testing for monogenic (Mendelian) diseases

	Direct mutational analysis	Indirect linkage analysis
Principle	Search for a known molecular defect	Search for a closely linked DNA marker gene cosegregating with the disease gene.
Source of specimen	Blood (white cells), other tissues, archival specimens	Blood (white cells), other tissues, archival specimens.
Family study	Not necessary, only individual at-risk is tested	Essential: both affected and unaffected family members must be included.
Mutational defect	Specific DNA defect to be detected must be known to allow appropriate molecular diagnosis[a]	The DNA defect can be unknown.
Role of allelic heterogeneity	Specific DNA mutation must be known[a]	No effect; all allelic mutations at a single disease locus are detectable.
Role of nonallelic heterogeneity	Specific DNA mutation must be known	May lead to missed diagnosis if an unlinked defect causes the same phenotypic disease.
Role of recombination	No effect	Wrong diagnosis if DNA marker and disease gene are separated by crossing over. Increasingly unlikely with very tight linkage.

[a] Different nonsense mutations may be detectable by single truncation test

practical. However, indirect DNA diagnosis using DNA markers of the dystrophin gene together with a family study can be attempted. Since the gene is very large, intragenic crossovers are relatively frequent (about 5%). This problem can be overcome by using flanking markers on either side of the disease gene (Appendix 6; Fig. A 6.4). In view of these complexities, measurement of creatine phosphokinase levels that are elevated in Duchenne muscular dystrophy carriers (but less so in Becker muscular dystrophy carriers) may aid further in carrier detection.

Genetic advice concerning multifactorial conditions such as birth defects, common diseases of middle life, and major psychoses, lacks the precision possible in counseling involving Mendelian genes since the number of genes and their relative contributions are usually unknown. Empirical risk figures need to be used, based on the frequency of recurrence of the disease in many affected families. These recurrence risks are usually lower than those in the Mendelian diseases and range between 3% and 5% for many common birth defects, such as the neural tube defects and cleft lip and palate. Risks to first-degree relatives (sibs, parents, children) for the more common diseases of middle life such as hypertension, schizophrenia, and affective disorders are about 10%–15%. Careful search for the rare monogenic variety of a disease that appears multifactorial must always be kept in mind. For example, rare patients with gout may have an X-linked disease due to hypoxanthine-guanine phosphoribosyl-transferase deficiency (308 000), or their gout may be caused by the autosomal-dominant phosphoribosylpyrophosphate synthetase deficiency (138 940). Among male patients with coronary heart disease under the age of 60 years about 5% have familial hypercholesterolemia – an autosomal-dominant trait (144 400; see Sect. 7.6.4).

Transmitted chromosome abnormalities, such as translocations do not segregate by Mendelian ratios, and counseling must be based on empirically derived risk figures.

Citing percentage figures of absolute recurrence risk is more meaningful to a family than relative risks based on the relative likelihood of the disease compared to the general population. A 100-fold increase for a condition that occurs in the population with a frequency of 1 : 100 000 carries an actual risk of only 1 : 1000 – a negligible recurrence risk. For Mendelian conditions the recurrence risks are fixed regardless of whether several or no affected children have preceded. Chance has no memory! In multifactorial diseases such as congenital heart disease or cleft lip and palate, if two or more first-degree relatives are affected in a given family, more disease-producing genes are operative in that family, and the risk for future offspring becomes higher than the usual 3%–5% [19]. However, differentiation from an autosomal-re-

cessive variety of the condition with a recurrence risk of 25% may sometimes be difficult. Detailed discussions of the approaches to genetic counseling and risk data for many different types of diseases can be found in recent books [19, 22].

Communication. The meaning of genetic risks must be conveyed in terms understandable to patients. The probability that 3%–4% of all children of normal parents develop serious birth defects, genetic diseases, or mental retardation should be communicated as a measure against which additional risk can be gauged. There may be problems in communicating the extent of uncertainty. For example, with a sporadic case of an undiagnosable birth defect the risk might be zero if the disease is nongenetic, 2%–3% if there is a multifactorial etiology, and 25% if it is caused by an autosomal-recessive trait. An empirical risk based on the probability of the various possibilities is often given as an empirical risk. Such a risk in this example might be 5% on the assumption that monogenic recessive varieties of this birth defect tend to be rare. However, many counselees prefer to be told about the full extent of uncertainty rather than being offered a single risk figure [5]. The burden of the disease must be clearly explained. Very severe but invariably fatal conditions in early life carry a less severe burden to the family than those associated with chronic crippling disease. Various reproductive options and alternatives must be discussed. Since problems under discussion may be complex and prove to be emotionally difficult to the patient, it may be necessary to have several counseling sessions. In any case, the counselor should provide a written summary in lay language.

Consanguinity. First cousins and more remote relatives who contemplate marriage occasionally ask for advice about the risks of having children with inherited diseases. Marriages between first cousins are illegal in 30 states of the United States. Consanguinity definitely increases the risks of disease caused by homozygosity for recessive genes (Chap. 13), but the absolute risks are relatively low. It has been estimated that the rate of various diseases, birth defects, and mental retardation among offspring of first-cousin matings is at most twice the background rate faced by any given couple; thus the chance that a child from such a mating will be normal is around 93%–95%. These risks are still lower for more remote consanguinity and are difficult to discriminate from the population background rate for such disorders. There are no additional risks for offspring of a normal person married to an unrelated person when one partner has consanguineous parents. On the other hand, the risks are considerable for children of incestuous matings involving first-degree relatives, such as sib-sib and father-daughter matings; there is an almost 50% risk that the child will be affected by a severe abnormality, childhood death, or mental retardation (Sect. 13.1.2.4). It is therefore advisable that children of incestuous matings who are to be placed for adoption be observed for about 6 months before the adoption is finalized. By that time many potential defects and recessive diseases should have become evident.

It is remarkable that generally defects in offspring of consanguineous matings show up as nonspecific congenital malformations, childhood death, and mental retardation rather than as well-defined autosomal-recessive diseases. However, detailed searches for the many different recessive inborn errors of metabolism have rarely been carried out, and it is likely that a significant proportion of childhood deaths involve unrecognized inborn errors.

It is possible that in societies where inbreeding has been practiced for many generations (such as in parts of southern India) the risks for offspring of consanguineous matings are sometimes lower since selection against homozygous gene combination will have removed many such genes over the generations (Sect. 13.1.2.4).

Heterozygote Detection. It is particularly important to detect heterozygotes in sisters of boys affected with X-linked recessive diseases, such as hemophilia (306700) and Duchenne-type muscular dystrophy (310200). Regardless of their partner's genetic constitution, there is a 50% risk that the sons of female heterozygotes will be affected with these diseases. In contrast, autosomal-recessive diseases become evident only when *both* parents are heterozygotes; a heterozygote sib of an affected patient must mate with another heterozygote for the disease to occur. The chance that an unrelated mate of a person who is a carrier of an autosomal-recessive disease carries the same mutation is usually quite low.

Specialized laboratory tests may be somewhat helpful in carrier detection (such as creatine phosphokinase assays for Duchenne-type muscular dystrophy and a combined assay of antihemophilic globulin clotting and antigenic activity for hemophilia A; 306700). However, these are being replaced increasingly by various DNA tests.

Biochemical and functional tests must be carefully standardized on normal subjects and obligate heterozygotes before applying them for individual carrier identification. The detection of heterozygotes is accurate and simple in the hemoglobinopathies. An increasing number of heterozygote states for various autosomal enzyme deficiencies such as the hexosaminidase deficiency of Tay-Sachs disease (230700) can also be recognized by enzyme assays as well as by DNA tests [31].

If there is overlap in laboratory results of tests such as for enzyme activity between normals with a low value and carriers with a high value, the significance of an identical laboratory result in various individuals may differ depending upon the a priori probability of the tested person being a carrier. Tests that are excellent for carrier detection in sisters of males affected with X-linked diseases may give too many false positives in screening studies of extended kindreds or particularly in the general population where the probability that the tested subject is a carrier is slight [21, 45]. For example, 5% of the normal female population would be identified as carriers of hemophilia using the same standards that identify sisters of hemophilic boys as heterozygotes with a high probability. The principle of widely varying significance of the same laboratory test result is discussed in detail for enzyme tests used in the detection of porphyria (Sect. 7.2.2.8).

In some of these situations additional statistical techniques are occasionally helpful for refinement of a genetic prognosis. For example: A woman's brother is affected with an X-linked recessive condition; a maternal uncle is also affected. She has therefore a 50% risk of being heterozygous. Assume that she already has two nonaffected sons, and that a test for heterozygote detection is not available. The information that her two sons are normal reduces her chance of her being a carrier. Alternately, such a woman may have a negative result for a test that detects 90% of heterozygotes. In this case her risk of being a carrier is very low. Appendix 6 and Murphy and Chase's book [47] deal with the statistical principles of calculating the exact recurrence risks in such situations.

The increasing availability of DNA markers is making carrier diagnosis in X-linked diseases more effective; this is discussed above. In any given diagnostic problem the most direct and simple approach should be selected, which increasingly is direct DNA diagnosis. However, use of biochemical tests is often necessary and complements the diagnostic armamentarium. The information from DNA markers can be combined with biochemical tests and pedigree information for more precise diagnosis (see Appendix 6 for an example with detailed calculations). The fragile X mental retardation syndrome is another example in which DNA diagnosis has become very helpful (Table 18.2). Since the number of CGG triplet repeats responsible for the syndrome can be assessed, DNA diagnosis can discriminate between affected males who have a large triplet expansion and normal transmitting males who carry a premutation characterized only by less expansion of CGG triplets (see Sect. 9.4.2). The nonaffected carried daughters of such transmitting males will have a moderate expansion of CGG triplets while carrier sisters of affected males will have a greater expansion, causing mental retardation in about one-third of this type of heterozygotes. There is good correlation between the number of CGG repeats and the extent of mental retardation.

Reproductive Options and Alternatives. If a couple decides that the risks of further reproduction are too high, several options besides contraception should be discussed. Adoption is becoming less practicable because fewer babies are available. Sterilization of either husband or wife may be considered, but it must be emphasized that this usually is an irreversible procedure. Sterilization is therefore undesirable for preventing autosomal-recessive conditions because remarriage to a noncarrier after possible divorce or spouse's death could eliminate the genetic risks almost entirely. Artificial insemination by a donor other than the husband may be acceptable for rare couples to prevent autosomal-recessive disease or autosomal-dominant disease contributed by the husband.

Detection of Genetic Diseases in Relatives. Optimal genetic counseling in some diseases should include the testing of relatives at risk (Table 18.3). For some conditions the detection of latent disease in relatives

may be lifesaving if followed by suitable therapy. A sib of a patient with Wilson disease has a 25% chance of being affected but may be too young to exhibit overt symptoms. Sibs of patients with hereditary polyposis (175100) [62] have a 50% chance of being affected and therefore risk malignant transformation in one of the many polyps in this condition. In general, vigorous attempts should be made to examine relatives when a genetic condition causes serious preventable or treatable diseases. A case can also be made for early detection of diseases such as polycystic kidneys (173900) [12, 23] to allow those affected better reproductive decisions, choice of life-style and appropriate occupations, and better preparation for ultimate renal transplantation or dialysis. Possible carriers of serious X-linked diseases (such as hemophilia and Duchenne-type muscular dystrophy) and of chromosomal carrier status (such as Down syndrome associated with translocation) should be searched for in families to allow antenatal diagnosis of potentially affected offspring (see below).

Directive vs Nondirective Genetic Counseling. After genetic advice that includes an estimate of the recur-

Table 18.3. Treatable and preventable *adult* genetic diseases with autosomal-dominant inheritance for which search in family members of affected patients is mandatory

Disorder	Method of diagnosis	Treatment	Advantages of early diagnosis and treatment
Hemochromatosis (autosomal-recessive)	Transferrin saturation, ferritin levels, liver biopsy most reliable	Venesection	Prevents liver, heart, and pancreatic disease
Hereditary spherocytosis	Incubated osmotic fragility test	Splenectomy	Prevents anemia and gall stones; protects against splenic rupture
Hereditary polyposis	Colonoscopy, DNA diagnosis	Colectomy	Prevents colon cancer
Gardner syndrome	Colonoscopy, Benign cysts, lipomas, fibromas on physical examination, DNA diagnosis	Colectomy	Prevents colon cancer
Familial hyperparathyroidism	Serum calcium, phosphorus, parathyroid hormone	Surgery	Prevents renal damage and other complications of hypercalcemia
Multiple endocrine adenomatosis	Serum calcium, phosphorus, blood sugar, gastrointestinal, and skull X-ray, DNA diagnosis	Surgery	Prevents complications of hyperparathyroidism, hypoglycemia, peptic ulcer
Medullary thyroid carcinoma pheochromocytoma syndromes	Calcitonin, measurement of blood pressure, DNA diagnosis	Surgery	Prevents thyroid carcinoma and complications of hypertension
Familial hypercholesterolemia	Serum cholesterol, LDL receptor	Diet, drugs	Prevents premature coronary heart disease
Malignant hyperthermia	Serum creatine phosphokinase	Avoid general anesthesia	Prevents fatalities during general anesthesia
Acute intermittent porphyria	Porphobilinogen deaminase in red cells, DNA diagnosis	Avoid precipitating drugs	Prevents abdominal and neurological symptoms

Note that DNA diagnosis in relatives of patients with familial breast cancer and familial nonpolyposis colon cancer is increasingly possible (See Sect. 10.4.3). Prevention in those testing as positive will allow more frequent monitoring for early signs of disease.

rence risk (see below) has been given, parents need to decide whether to have further children or not. Many physicians are paternalistically inclined and are accustomed to giving specific directive advice for or against future pregnancies. In the practice of medical genetics, however, a fairly strong tradition of nondirectiveness has developed. Some of this nondirectiveness may have sociological reasons. When genetic counseling began in the United States some 40 years ago, it was usually carried out by nonphysician geneticists, who lacked the medical profession's tradition of dispensing directive advice. However, the nondirectiveness of genetic counseling fits well with recent trends to increasing patient autonomy. Since each family is unique and reactions to risks vary, nondirectiveness fosters mature decision making. However, absolutely neutral advice is rarely possible or even desirable. The person or family requesting advice usually wants and needs more than a computerlike professional who only dispenses facts. The counselor may unconsciously emphasize the more positive or the more frightening aspects of a given disease. These feelings tend to affect the counseling process directly or indirectly, often by nonverbal clues. A cup may be half full or half empty, and the positive or the negative aspect of such facts may be stressed more vigorously. Nor do all couples have the necessary educational background and social and emotional maturity to make fully informed decisions. A hasty assessment of the family in these matters, however, may lead to a more paternalistic stance in advice giving than is desirable. Nevertheless, many couples expect the medical geneticist, who has the required knowledge and experience, to aid them in making their decisions. "What would you do if you were in our position?" is a frequent question from counselees, regardless of background. However, since a couple's economic situation, religious affiliation, and cultural background may differ substantially from that of the counselor, the counselor's choice for his/her own circumstances may not necessarily be appropriate. Reproductive decisions differ between couples even when the genetic facts and the disease burden are identical. Furthermore, cultural traditions vary within different countries.

As predictive testing for late-onset diseases becomes increasingly possible, nondirective advice no longer applies when a disease can be prevented or treated by appropriate measures. Reproductive advice regarding genetic transmission is rarely sought under such circumstances. Instead, the relevant medical recommendations must be given (including all options) for how to prevent and treat the disease. While most medical geneticists and counselors consider nondirective counseling as the appropriate policy regarding reproductive decisions, specific medical advice and guidance (i.e., directive counseling) is the recommended course of action under these circumstances and is expected by patients.

Assessment of Genetic Counseling and Psychosocial Aspects [17, 34]. Genetic counseling is a relatively new field, and its practice has not been standardized. Most professionals engaged in genetic counseling agree that counselees should achieve sufficient understanding of the medical significance and social impact of the disease to allow them to make appropriate reproductive decisions. Some observers have measured the success of genetic counseling by subsequent reproductive behavior. If more couples with a high risk (>10%) were deterred from reproduction than couples with a low risk, genetic counseling is considered to be successful. This result has in fact been noted in several studies [9]. Such a narrow end-point is considered an inadequate objective of genetic counseling. It would be better to know whether full sharing of information and comprehension of the disease and its recurrence risks has been achieved, and whether all needs for information and psychological and social support have been met.

Various studies of genetic counseling agree that many patients after counseling are confused about recurrence risks and do not fully understand the nature of the disease. A large study was carried out by a group of sociologists in the late 1970s in 47 genetic counseling clinics in the United States, involving 205 counselors and over 1000 female counselees [60]. Many different conditions were included, and both counselors and counselees were questioned about their experiences and assessment of the counseling process. The results showed that counselors tended to emphasize recurrence rates during the counseling sessions while counselees were often interested in causation, prognosis, and treatment of the disease – an area which according to the counselees was seldom discussed as fully as desired. While both counselors and counselees were generally more interested in the medical and genetic aspects of the consultation, counselees occasionally had psychosocial concerns which were not addressed by the professionals.

This study found that 54% of counselees who were given a risk and 40% of those given a diagnosis were unable to report these data shortly after counseling. This failure of learning occurred independently of whether MDs, PhDs, or genetic counselors had carried out the counseling and was unrelated to the experience of the counselor. Counselors with many years of experience had no better results than more recent graduates. Several other studies have reported results that are substantially better but by no means perfect [17]. Usually, but not always, education is found to be correlated with a better level of understanding.

Genetic counseling services have been used more extensively by families with good educational backgrounds than by less advantaged population groups. Couples who are motivated to learn about the disease and its recurrence risks are more likely to be affected in their reproductive decisions by the information provided than those who have been referred and are not always certain about the purpose of the genetic consultation. Thus, self-referred patients also tend to have better comprehension of genetic counseling information.

Another study investigated perception of counseling information [36, 37]. Perception of recurrence rates was often not used by the counselees in the probabilistic sense represented by the figures given. Percentage risks were more frequently perceived as binary, i.e., even with lower risks it was believed that the disease either would or would not occur, with all the attendant fears of recurrence. Parents were then overwhelmed by multiple uncertainties, such as how to make reproductive choices, how others would react to their decision, what it would mean to have an affected child, and whether they would be able to fulfill their role as parents. Such perceptions appeared more important for decision making than the actual facts of diagnosis, prognosis, and risk. These data show that there is often a discrepancy between the mental set of the scientifically oriented counselor and the thought processes of the counselees, who find it difficult to deal with probabilistic information. Bridging this gap is a real challenge.

Genetic counseling – as currently practiced in most countries – places less specific emphasis on emotional aspects than "counseling" activities in other areas such as psychological and marriage counseling. Some observers have recommended that more attention be given to the psychodynamic aspects of genetic disease [8, 14, 33]. It is our experience, however, that psychologically oriented genetic counseling, in which a significant amount of time is spent on psychodynamics, is rarely required. If there are deep psychological problems, referral to a psychiatrist or psychotherapist is the most appropriate course of action. An empathic and understanding approach to families with an awareness of the many social and psychological aspects of the disease and support in these matters, however, needs to be encouraged. Genetic counseling is more than mere diagnosis, risk assessment, and "cold" dispensation of information.

There are imperfections in the genetic counseling process as currently practiced. Nevertheless, most educated counselees who receive definite information about the matters troubling them usually appear to be satisfied, and the majority of counselees with low risks are relieved to find that their actual risks are much lower than they had feared.

The interaction of patients and professionals in any encounter has many variables, and scientific study of this process is difficult. Nevertheless, genetic counseling as a new field demands further investigation of the process, its psychosocial effects, and outcomes, so that optimum results can be worked out. Controlled studies comparing patients who received counseling with those with a similar disease who did not would be interesting.

18.2 Prenatal Diagnosis [5a, 15, 26, 46, 55]

The field of prenatal diagnosis has grown rapidly and has altered the practice of genetic counseling. Specific information regarding the possibility of prenatal diagnosis is now usually provided in genetic counseling situations. Prenatal diagnosis substitutes definite information for a probability of recurrence – a much more acceptable outcome to most individuals.

Prenatal diagnosis includes a variety of techniques, among which amniocentesis, chorionic villus sampling, and ultrasonography are most frequently used:

Amniocentesis or chorionic villus sampling
    Chromosomal disorders
    Fetal sexing
    Inborn errors of metabolism
    All disorders detectable with DNA methods
    Open neural tube defects (amniocentesis only)
Ultrasonography
    Neural tube defects
    Structural malformations
Fetoscopy and biopsy
    Epidermolysis bullosa and some skin diseases
    Liver biopsy
Maternal blood screening
    α-Fetoprotein (neural tube defects and Down syndrome)
    β-HCG and estriol (Down syndrome)
    Fetal cell analyses (under study)

Amniocentesis. Amniocentesis (Fig. 18.1 a) is carried out at the beginning of the second trimester of pregnancy (15–17th week of pregnancy) by transabdominal puncture. The procedure has proven safe in the hands of trained obstetricians, but it is not 100% harmless. There is a slight risk of fetal loss ($\sim$ 0.5–1%). Infection and hematomas are much rarer still, and other obstetric complications are even rarer. The procedure is performed in an outpatient setting in conjunction with ultrasonography, a procedure that decreases the failure ratio and the frequency of blood-stained fluid and fetomaternal hemorrhage. A chromosomal study requiring culture of the aspirated amniotic cells of fetal origin is usually carried out, and the results are obtained 2–3 weeks later.

In addition to chromosomal aberrations, many enzyme deficiencies and other biochemical defects can

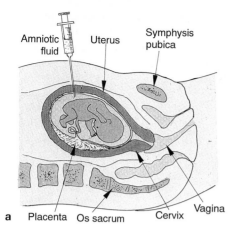

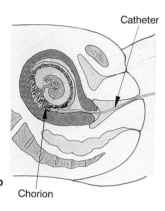

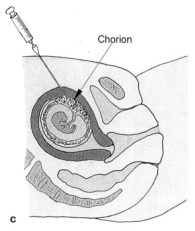

Fig.18.1. a Amniocentesis. Puncture of the amniotic cavity through the abdominal wall. **b** Chorion villus sampling. The uterus is entered through the vagina and portio vaginalis uteri. **c** The chorionic villus sampling by the abdominal route

be detected in amniotic cells by appropriate assays or, increasingly, by DNA diagnosis (Table 18.4). Since individual enzyme deficiencies are rare and the technical problems of assay considerable, specialized reference laboratories ideally carry out the appropriate testing. In view of the rarity of most inborn errors,

enzyme and DNA analyses in prenatal diagnosis (unlike the search for chromosomal errors in older mothers) is never performed routinely but only on specific indications in high-risk pregnancies (e.g., a previously affected child).

Chorionic Villus Sampling [6, 29, 42, 59, 63]. This procedure may be performed by sampling chorionic villi by the cervical (Fig.18.1b) or by the abdominal approach (Fig.18.1c), which is now generally preferred because the risk of infection is lower. Chorionic villus sampling must be performed under ultrasound guidance. Chorionic tissue of fetal trophoblastic origin can be used for cytogenetic, biochemical, or DNA testing. The procedure can be carried out between the 8th and 10th weeks of pregnancy and therefore has psychological advantages compared with amniocentesis, which is performed during the 15–17th weeks of pregnancy. Cytogenetic results are available within a few days. The risk of miscarriage after chorionic villus sampling is slightly higher than after amniocentesis (2%–3% vs. 0.5–1%) even in experienced hands. In most centers amniocentesis is the routine procedure while chorionic villus sampling is reserved for high-risk cases and molecular studies.

Ultrasonography. Wide use of noninvasive ultrasound examination of fetuses allows prenatal diagnosis of a variety of fetal anomalies. Ultrasonography has changed rapidly in recent years, leading to improved diagnostic precision. The diseases that can be detected prenatally by ultrasound include the following [22]:

Obstetric indications
    Accurate gestational dating
    Multiple pregnancy
    Placental localization
CNS disorders
    Anencephaly
    Hydrocephaly
    Encephalocele
    Meningomyelocele
    Spina bifida
    Holoprosencephaly
    Microcephaly
Abdominal/Gastrointestinal
    Gastroschisis
    Omphalocele
    Duodenal atresia
    Esophageal atresia
Various fetal tumors
Skeletal defects
    Severe bone dysplasias
    Congenital types of osteogenesis imperfecta
    Limb defects
Chest
    Diaphragmatic hernia
    Intrathoracic cysts

Table 18.4. Prenatal diagnosis of inherited metabolic disorders (autosomal-recessive inheritance unless specified; from Harper 1993 [22]

Disorder	Usual enzyme deficiency	Comments
Acid phosphatase deficiency (lysosomal)	Acid phosphatase	Needs confirmation.
Adenosine deaminase deficiency (combined immunodeficiency)	Adenosine deaminase	
Adrenogenital syndrome[a]	21-Hydroxylase	Amniotic fluid analysis; also indirectly by HLA linkage; treatable.
Adrenoleucodystrophy[a]	Long-chain fatty acid defect	X-linked; DNA analysis only.
Argininosuccinic aciduria	Argininosuccinase	Argininosuccinic acid also raised in amniotic fluid.
Citrullinemia	Argininosuccinate synthetase	
Cystic fibrosis[a]	CFTR protein	Mutation detection or linked DNA markers.
Cystinosis	Unknown	Accumulation of intracellular ^{35}S-labelled cystine.
Fabry disease[a]	α-Galactosidase	X-linked; variable expression in female.
Farber disease	Ceramidase	
Fucosidosis	α-L-Fucosidase	
Galactosemia (classical)[a]	Galactose 1-phosphate uridyl transferase	Treatment available.
Galactosemia (galactokinase deficiency)	Galactokinase	Relatively benign and treatable disorder.
Gaucher disease[a]	Glucocerebrosidase	Heterogeneous.
Generalized gangliosidosis	β-Galactosidase	
Glucose 6-phosphate dehydrogenase deficiency[a]	G6PD	Often mild; many enzyme variants; X-linked.
Glutaric aciduria	Glutaryl-CoA dehydrogenase	
Glycogenosis type I[a] (von Gierke disease)	Glucose 6-phosphatase	
Glycogenosis type II[a] (Pompe disease)	α-1,4-Glucosidase	Heterogeneous.
Glycogenosis type III	Amylo-1,6-glucosidase	
Glycogenosis type IV (Andersen disease)	Brancher enzyme	
Hemoglobin S disease[a]	β-Chain substitution	Severity variable: DNA analysis or fetal blood.
Hemophilia A[a]	Factor VIII (C)	DNA or fetal blood; X-linked.
Hemophilia B[a]	Factor IX	DNA or fetal blood; X-linked.
Homocystinuria[a]	Cystathionine synthetase	Heterogeneous.
Hyperammonaemia; X-linked[a]	Ornithine transcarbamylase	DNA; X-linked; variable expression in female.
Hypercholesterolemia, familial[a]	Low-density lipoprotein receptors	
Hypophosphatasia[a]	Alkaline phosphatase	Only severe infantile type detectable.
I cell disease (mucolipidosis II)	Lysosomal membrane defect?	Increase in multiple lysosomal enzymes.
Krabbe disease	β-Galactosidase	
Lesch-Nyhan syndrome[a]	Hypoxanthine-guanine phosphoribosyltransferase	X-linked recessive; milder partial deficiencies exist.
Mannosidosis[a]	α-Mannosidase	
Maple syrup urine disease	α-Ketoacid decarboxylase	
Menkes disease[a]	Defective copper metabolism	X-linked; abnormal copper uptake.
Metachromatic leucodystrophy[a]	Arylsulphatase A	
Methylmalonic aciduria	Methylmalonyl-CoA mutase	Methylmalonic acid detectable in amniotic fluid; may be treatable in utero; heterogeneous.

Disorder	Usual enzyme deficiency	Comments
Mucopolysaccharidosis I[a] (Hurler syndrome)	α-L-Iduronidase	MPS IS (Scheie syndrome) has the same enzyme deficit; amniotic fluid MPS levels useful in types I, II, and III.
Mucopolysaccharidosis II[a] (Hunter syndrome)	Iduronate sulphatase	X-linked; enzymatic diagnosis possible from amniotic fluid as well as cells.
Mucopolysaccharidosis III A (Sanfilippo A syndrome)	Heparan sulphate sulphatase	
Mucopolysaccharidosis III B (Sanfilippo B syndrome)	α-N-Acetylhexosaminidase	Carrier detection feasible on serum.
Mucopolysaccharidosis IV (Morquio syndrome)	Chondroitin sulphate sulphatase	Heterogeneous; other forms also detectable.
Mucopolysaccharidosis VI (Maroteaux-Lamy syndrome)	Aryl sulphatase B	
Niemann-Pick disease[a]	Sphingomyelinase	Heterogenous.
Phenylketonuria (classic)[a]	Phenylalanine hydroxylase	Treatable; DNA analysis only.
Phenylketonuria (dihydropteridine reductase type)[a]	Dihydropteridine reductase	Severe and difficult to treat.
Porphyria, acute intermittent[a]	Porphobilinogen deaminase	Autosomal dominant; treatable.
Porphyria, congenital erythropoietic[a]	Uroporphyrinogen cosynthetase	
Propionic acidemia	Propionyl-CoA carboxylase	Also directly detectable from amniotic fluid.
Refsum disease	Phytanic acid oxidase	Possible; not actually confirmed.
Sandhoff disease[a]	β-N-Acetylhexosaminidase (A and B)	
Thalassemia (β)[a]	Defective β-chain synthesis	DNA or fetal blood.
Tay-Sachs disease[a]	β-N-Acetylhexosaminidase A	Carrier detection and high-risk population screening feasible.
Wolman disease	Acid lipase	Heterogeneous.
Xeroderma pigmentosum[a]	DNA repair enzymes	Heterogeneous.

Unless otherwise indicated, the diagnosis is made from cultured amniotic fluid cells or from chorionic villi; molecular analysis is specifically noted only if other methods are not possible but is likely to be feasible for all disorders in which the gene has been isolated

[a] Disorder in which the gene has been isolated; molecular prenatal diagnosis likely to be feasible

   Pulmonary hypoplasia
   Small chest wall (various skeletal syndromes)
Renal/genitourinary
   Renal agenesis
   Polycystic kidney (infantile)
   Severe obstructive uropathy

While all current studies indicate that ultrasound is harmless to the developing fetus, the indiscriminate application of the procedure in all pregnancies is of some concern in the absence of absolute proof of its innocuousness. In some countries, however, repeated ultrasonographic examination has become part of routine pregnancy surveillance, and many instances of previously unsuspected malformations have been diagnosed. Various authoritative bodies (National Institutes of Health, USA; World Health Organization) suggest caution and recommend the use of ultrasonography only when definite maternal or fetal indications exist. The increasing technical perfection of ultrasonography complements other forms of prenatal diagnosis, particularly in the detection of neural tube defects.

Fetoscopy [39, 41, 56]. Fetoscopy with small fiberoptic instruments allows entry into the amniotic cavity and is usually carried out between the 18th and 22nd weeks of pregnancy. Even in experienced hands this procedure carries a 5%–10% abortion rate. Inspection of the fetus to detect defects has limitations because of the restricted field of vision. Sampling of fetal blood under direct vision is possible, and any genetic condition that manifests in fetal blood can be diagnosed. Fetal skin biopsies may be carried out, and even fetal liver biopsies have been performed to diagnose diseases that are expressed only in the liver. However, the increasing availability of DNA diagnosis and the high rate of fetal loss have markedly restricted the use of this technique. Some hereditary skin diseases are diagnosed prenatally by ultrastructural anomalies made visible by appropriate methods in skin biopsies from fetuses at risk [4, 23].

Maternal Blood Sampling [11, 20, 39]. The sampling of maternal blood by venepuncture for α-fetoprotein (AFP) elevations as a screening procedure to detect neural tube defects and some other fetal anomalies has been carried out in many centers. Abnormalities that increase or decrease amniotic fluid levels of AFP include:

Increased AFP
   Neural tube defects
   Spontaneous intrauterine death
   Omphalocele
   Gastroschisis
   Nephrosis (Finnish type)
   Sacrococcygeal teratoma
   Bladder extrophy
   Some skin defects
   Meckel syndrome
Lowered AFP
   Down syndrome

The decreasing incidence of neural tube defects in most countries and their increasing detectability by ultrasonography has made definite recommendations regarding the universal adoption of such screening difficult. AFP screening may also be useful to detect Down syndrome, since fetuses with trisomy 21 have lower AFP levels than normal fetuses [2456, 2477], i.e., the median for cases of Down syndrome approximates 0.7 of the median for normal individuals. Other useful biochemical markers include unconjugated estriol and chorionic gonadotropin. Multiple marker screening for fetal Down syndrome (2nd trimester) using these three markers in addition to maternal age achieves a detection rate of about 60%–70%, with a false-positive rate of approx. 5% [40]. It has been shown that if amniocentesis for chromosomal study were offered to all women whose blood AFP levels were at or below a specified level (i.e., 0.5 of the median level) an additional 20%–40% of Down syndrome cases would be found over those detected with current methods using amniocentesis at specified maternal ages. However, for each case of trisomy 21 so detected, 150–200 additional amniocenteses on normal fetuses would have to be performed.

Other methods, such as the detection of fetal cells in the maternal circulation, are under investigation. The successful application of such a technique would be useful in screening maternal blood for fetal cells with chromosomal and biochemical aberrations. However, although some fetal cells exist in the maternal circulation, many technical difficulties need to be overcome before this procedure can be applied routinely.

Indications for Prenatal Diagnosis. With newer modalities of prenatal diagnosis, more indications exist, and an increasing number of fetal conditions can be diagnosed. Chorionic villus sampling is used for the same indications as amniocentesis and has the advantage of being available significantly earlier in pregnancy.

1. *Maternal Age.* Amniocentesis is most frequently performed to rule out Down syndrome and other chromosomal aberrations in women of "advanced" maternal age. In most countries this age is set somewhat arbitrarily at 35 years, where the risk for Down syndrome at birth is about 1/400, rising to 1/100 by age 40 and to 1/40 by age 44 (Table 9.4). The incidence of Down syndrome and other chromosomal aberrations is significantly higher at amniocentesis than at birth, since many aneuploidies are spontaneously aborted prior to birth (Chaps. 9).

2. *Previous Aneuploidy:* A previous child with Down syndrome or other autosomal trisomy slightly increases the risk of recurrence. The risk for Down syndrome is about 1/250 for those under 35 years of age and is probably about twice the age-specific risk for those aged over 35 years.

3. *Parental Chromosomal Rearrangements* [13]. The carrier status for translocations or pericentric inversion gives an increased risk of unbalanced, abnormal fetuses (e.g., translocation Down syndrome; Sect. 2.2.2). The risks do not correspond to those expected from chromosomal segregation but are based on empirical data, presumably because of selection against unbalanced gametes. The risk for translocation trisomy 21 is about 15% when the mother is the carrier and only 3% if the father is the carrier (t14q21 and t21q22q). In reciprocal translocations the risks of future affected offspring are significantly higher (~20%) if ascertainment occurs via an affected live offspring as opposed to ascertainment by recurrent abortions (5% risk; for details see Sect. 2.2.2.2). The more extensive unbalanced duplications/deletions (3–6 chromosome bands out of 200 total) are associated with lower recurrence risks (9%–16%) than those with duplications/deletions affecting only 1–2 bands (34%). Presumably larger defects are often not viable and abort spontaneously prior to amniocentesis.

4. *Risk for X-Linked Disorders.* When prenatal diagnosis became available, but specific diagnosis of X-linked diseases was not yet possible, diagnosis of the male sex was offered; sons of carrier mothers had a disease risk of 50%. Fortunately, however, diagnosis of such diseases at the DNA level has become possible in recent years, and sex diagnosis has become less and less common for this type of indication. It is unfortunately practiced in some societies where males are valued higher for sociological reasons in order that female fetuses can be aborted.

5. *Fragile X Syndrome.* Prenatal diagnosis of this common syndrome (see Sect. 15.2.1.2) can now be achieved by DNA diagnosis of the expanded CGG triplet (Table 18.5).

6. *Hemoglobinopathies* [41]. Diagnosis of the various thalassemias and of sickle cell anemia is usually carried out by direct DNA diagnosis and rarely by the linked marker approach.

7. *Inborn Errors of Metabolism:* Enzyme assay or DNA diagnosis of fetal cells needs to be done. The list of such conditions is long (Table 18.4). Diagno-

Table 18.5. Fragile-X mental retardation (from Harper 1993 [22])

	Clinical features	Risk to children	Cytogenetic results	DNA analysis
Index cases (male or female)	Mental retardation and typical facies	Rarely reproduce	Fragile site present, frequent	Large DNA expansion
Normal transmitting male	Normal	All daughters carriers; sons normal	Fragile site not normally seen	Small expansion (premutation)
Carrier daughters of normal transmitting male	Normal	Around 75% of affected sons and 1/3 affected daughters retarded	Fragile site absent or occasional	Small to moderate expansion (heterozygous)
Carrier sister of affected male	Variable; 1/3 retarded	All affected sons and half affected daughters retarded	Fragile site usually present	Moderate to large expansion (heterozygous)

Table 18.6. Some autosomal-dominant disorders in which the gene has been mapped or cloned (modified from Harper 1993 [22])

Disease	Chromosomal assignment	Mutation or gene product
Adenomatous polyposis coli	5q21–q22	APC gene
Amyloidosis I, neuropathic, Portuguese, Japanese, Swedish types and others	18q11.2–q12.1	Transthyretin
Cardiomyopathy, familial hypertrophic (one locus)	14q11–q12	Cardiac myosin alpha and/or β heavy-chain lesion
Charcot-Marie-Tooth neuropathy I	17p13.1–p11.1	Peripheral myelin protein
Cystic fibrosis	7q31–q32	Cystic fibrosis transmembrane regulator (CFTR)
Endocrine neoplasia, multiple, type 2	10p11.2–q11.2	*Ret* oncogene
Huntington disease	4pter–p16	CAG triplet repeat
Marfan syndrome	15q	Fibrillin
Myotonic dystrophy	19q13	Myotonin protein kinase (CTG unstable triplet repeat)
Neurofibromatosis 1 (von Recklinghausen disease)	17q11	Neurofibromim (related to GPTase activating protein)
Neurofibromatosis 2 (bilateral acoustic neuroma)	22q11	Specific tumor suppressor gene
Retinitis pigmentosa, autosomal dominant (some families)	3q21–q24	Rhodopsin (RHO)
von Hippel-Lindau disease	3q26–p25	Tumor suppressor gene
Waardenburg syndrome, type I	2q37	PAX3 developmental gene

sis at the gene – DNA level is possible in many conditions, including phenylketonuria.

8. *Various other Genetic Diseases allowing direct or indirect DNA diagnosis.* A list of such diseases is given in Table 18.6. Same of these diseases are discussed in Chap. 5. This list will grow rapidly in the next few years.

9. *Neural Tube Defects.* Amniocentesis (*not* chorionic biopsy) for amniotic fluid AFP is usually carried out in high-risk women such as those with previously affected children or following a confirmed maternal high blood AFP level. Ultrasound is very useful and should always be performed in these circumstances. Other markers in amniotic fluid (for example, acetylcholinesterase level increases) may provide hints to the presence of neural tube defects and other anomalies (See below).

Prenatal diagnosis is widely used in today's industrialized countries. Extensive public information via women's magazines in Denmark has led to the highest acceptance rate, with 80% of eligible women being studied. The availability of prenatal diagnosis often encourages parents to start a pregnancy under circumstances in which fear of an affected infant would previously have deterred childbearing. While abortion for fetal indications has become accepted in many countries, a significant proportion of the population in the United States and elsewhere has strong feelings for religious or other reasons against termination of pregnancy. Some antiabortionists are particularly concerned about the making of value judgments regarding the continuation of fetal life as is done with genetic disorders. They feel that such practices are the beginning of the "slippery slope" that would ultimately lead to rejection of relatively minor defects in the search for the "perfect" baby and lead to a resurgence of eugenic and racist schemes. Fears have also been expressed that society would be less inclined to pay large sums of money to take care of children with genetic diseases when abortion could have prevented the birth of the disabled child. However, largely because most disabilities cannot be diagnosed in utero at present, such trends have not emerged. Furthermore, the fact that society in recent years has generally given better financial and social support for the handicapped argues against the validity of these fears [46].

18.3 Genetic Screening [40, 44, 46a, 49]

With better understanding of various genetic diseases, public health applications have developed. It is reasoned that all members of a population at risk should be screened for a given defect if treatment or preventive measures are possible. Similarly, screening for certain genetic carrier states has been recommended to allow genetic counseling or intrauterine diagnosis before a sick person has been born. These programs are distinct from the usual retrospective genetic counseling, in which patients and families ask the advice of genetic counselors because someone in the family has a genetic disease.

Phenylketonuria Screening: Prevention of Mental Retardation [35, 49]. Phenylketonuria (261 600) is one of the most common inborn errors of metabolism in populations of European origin, with a frequency of about 1/10 000 births. The condition is an autosomal-recessive trait and is caused by a mutation affecting the enzyme phenylalanine hydroxylase, which fails to metabolize phenylalanine. The resultant buildup of metabolites damages the developing brain and leads to profound mental retardation. If diagnosed shortly after birth, an appropriate diet that restricts phenylalanine can prevent mental retardation. Before dietary treatment was widely instituted, about 1% of residents in institutions for the mentally retarded had phenylketonuria. The disease can be diagnosed by simple and inexpensive tests in blood obtained by puncture of the infant's heel before leaving the hospital (Sect. 7.2.2.7). Most developed countries have introduced phenylketonuria screening as a routine test for all newborns. A positive screening test does not necessarily mean that the baby has phenylketonuria because variants of hyperphenylalaninemia exist that may not cause mental retardation. Follow-up by a team of an experienced biochemist and pediatrician is therefore required to assure that treatment is administered only if necessary since a phenylalanine-restricted diet may be injurious.

Furthermore, classical phenylketonuria must be distinguished by appropriate tests from defects that cause malignant hyperphenylalanemia, such as dihydropteridine reductase deficiency and errors in biopterin synthesis [58]. Affected patients with these defects are not clinically influenced by phenylalanine-restricted diets.

The success of the PKU program has raised new problems in that girls with PKU who had effective treatment during childhood and are no longer on diets are now becoming pregnant. The pregnant woman's phenylalanine levels are high, so that the developing fetus (who is an obligatory heterozygote for PKU) is injured by the high phenylalanine levels of the mother. Multiple abortions, microcephaly with mental retardation, cardiac defects, and intrauterine growth retardation have invariably been found. Reinstatement of a phenylalanine-restricted diet before the beginning of pregnancy should prevent these abnormalities. In view of logistic difficulties in identifying all mothers who have had PKU and the occurrence of unplanned pregnancies, it is difficult to ascertain all women who need renewed phenylalanine- restricted diets.

Screening of newborns for congenital hypothyroidism has become well established and is based on assay of blood thyroxine (T_4) followed by measurement of thyroid-stimulating hormone (TSH) if T_4 levels are elevated. Treatment with thyroid hormone is highly effective by preventing mental retardation and other signs and symptoms of hypothyroidism. The frequency of congenital hypothyroidism is about 1/4000 – two to three times more common than phenylketonuria. The etiology of congenital hypothyroidism is usually non-Mendelian and often nongenetic. Currently phenylketonuria and congenital hypothyroidism can be recommended unequivocally for routine screening of all newborns.

Testing for sickle cell anemia is required in most states of the United States in *all* newborns since it has been found difficult logistically to identify African-American infants for testing. The rationale is prevention of lethal infections by antibiotic prophylaxis.

Several other potentially or partially treatable inborn errors of metabolism can be tested for on newborn blood. These conditions – such as maple syrup urine disease (248 600), homocystinuria (236 200) and galactosemia – are much rarer than phenylketonuria (261 600). No *general* policy regarding testing of these conditions has evolved. It is particularly important that prior to newborn screening for such conditions parents give fully informed consent for testing of diseases that only partially or poorly respond to treatment [2].

Similarly, there is no agreement whether newborn screening should be carried out for diseases such as cystic fibrosis and Duchenne muscular dystrophy since there is no clear evidence that knowledge of the diagnosis can alter the clinical course. However, such information may be useful so that parents can know their carrier status for reproductive purposes [22]. More pilot studies are needed to assess the psychosocial, genetic, and medical impact of such programs.

Cost-benefit calculations have sometimes been carried out to justify the cost of screening programs for inborn errors of metabolism. Such calculations contrast the financial costs of caring for a diseased child with the expenses of the screening program. Humanitarian aspects cannot be readily quantified and by necessity must be neglected in cost-benefit analyses. Even economic considerations are often treated simplistically in such analyses. For example, the disappearance of all cases of phenylketonuria would reduce the number of patients in hospitals for the mentally retarded by 1%. It is quite unlikely that such a change would have any but the most trivial effect on the budgets of such institutions. The full cost of all personnel, such as professionals who are paid from other funds but spend much of their time on screening programs, are often not taken into account [28, 50]. It is therefore considered unwise to base the initiation of screening programs on cost-benefit analyses alone.

Screening Mothers at Risk for Chromosomal Malformations. The existence of a maternal age effect for many chromosomal defects makes it desirable to search for abnormalities such as trisomy 21 in older mothers. Currently women over the age of 35 are offered prenatal diagnosis since the risks for trisomy 21 rise steeply with increased maternal age (Sect. 9.2.2). Because of small family size, such as is found in the United States and most European populations, most cases of trisomy 21 are born to younger women [25], and the population impact of prenatal diagnosis for the prevention of Down syndrome would be less than ideal. Together with the use of various maternal blood indicators (i. e., triple screening for α-fetoprotein, human chorionic gonadotropin, and estriol), a larger number of chromosomal abnormalities – particularly Down syndrome – can be identified. With these techniques and selective abortion a significant reduction of Down syndromes could therefore be expected. The promise of identifying chromosomal abnormality in the fetus by maternal blood sampling may make it possible to identify most if not all chromosomal defects during pregnancy. Under such conditions, anomalies due to numeric or structural chromosomal aberrations will become completely avoidable.

Screening for Autosomal-Recessive Traits. Certain heterozygote traits are common in some populations. Screening of these populations for the carrier state would identify carrier x carrier matings who have a 25% chance of affected children.

Following screening, carriers should be counseled regarding the genetic and medical risks of the disease and provided with information regarding the various reproductive alternatives. These would include: (a) avoidance of mating with another carrier, (b) avoidance of childbearing if married to a carrier, and (c) prenatal diagnosis and abortion of an affected fetus if married to another heterozygote. The mating of heterozygotes with persons who are not carriers has no untoward medical or genetic consequences. It is not difficult to see that the option of avoiding a mate who happens to carry the identical genetic trait or preclusion of childbearing are not very popular.

Screening programs have been most successful in relatively well-informed populations for traits where intrauterine diagnosis of the disease is available. The carrier state for Tay-Sachs disease (272 800) occurs at a frequency of about 4% among Ashkenazi Jewish populations. The hexosaminidase deficiency characteristic of the defect can be readily detected in serum, and heterozygotes can be identified. Fetuses with Tay-Sachs disease can be identified by assay of enzyme in fetal cells aspirated by amniocentesis and grown in tissue culture. Programs for Tay-Sachs screening have been carried out in many metropolitan areas of the United States, and many fetuses with Tay-Sachs diseases have been identified and aborted [30, 31]. In families with previous cases of Tay-Sachs disease the program has been life-giving by allowing couples to have healthy children, as assured by intrauterine diagnosis. Without amniocentesis such couples would practice contraception. These programs have required a fairly high level of publicity to attract those at risk for testing. Even with the relatively high heterozygote frequency of 4%, only 1 in 2000 Ashkenazi Jewish children are affected ($0.04 \times 0.04 \times 0.25$) and only as the result of matings in which both parents are of Ashkenazi ancestry. Obstetricians are often not acquainted with the disease and may not test Jewish patients for the trait. One community has refused to initiate screening on the grounds that the frequency of the disease does not justify the potential emotional disturbances in those identified as carriers. In general, however, with adequate education, public response to the programs has been good.

Sickle cell *carrier* screening has been less successful [16, 43, 49]; this must be distinguished from the *newborn* program which is designed to detect infants with sickle cell anemia. Black populations in which the sickle cell trait is common are often less well educated, and the purpose of a genetically oriented screening program has not always been made clear to those to be screened. Particularly, the distinction between the innocuous sickle cell trait and sickle cell anemia has often not been properly explained. Occasional sickle cell trait carriers have been discriminated against in occupations, life insurance, and even in marital choice. These consequences illustrate the importance of extensive public education before screening programs are initiated.

While the sickle cell trait always could easily be detected, prenatal diagnosis for sickle cell anemia has also become available by direct DNA study [32]. However, the disease is less serious than Tay-Sachs disease or β thalassemia major (see below). There is considerable variability in clinical expression, and some patients are not very sick. Intrauterine diagnosis for sickle cell anemia is therefore requested less often than for the more severe genetic diseases.

β Thalassemia screening to identify couples for prenatal diagnosis has been very successful in several Mediterranean areas, such as Cyprus, the Ferrara area of Italy, Sardinia, and Greece, where the frequency of the condition has fallen dramatically since the late 1970s when fetal diagnosis (on blood obtained during fetoscopy) and selective abortion was initiated [976, 2467] (Fig.18.2). The results in Mediterranean countries demonstrate that screening and a prenatal diagnostic program is possible and can have a great public health impact.

Screening for Neural Tube Defects [2453, 2476, 2499]. Neural tube defects have a multifactorial etiology. Genetic factors appear to apply, but their specific nature is unknown. Their frequency ranges from 1/200 in southeastern England to 1/1000–1/1500 in Germany

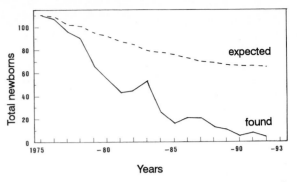

Fig. 18.2. Fall in the birth rate of babies with homozygous β thalassemia in Sardinia. *Y-axis,* number of new borns expected and observed with thalassemia major. Carrier screening began in 1975. (From Cao et al. 1989 [7])

and the United States. Rates have been declining in recent years. Population screening for these defects can be carried out by maternal AFP determinations in serum at 16–18 weeks of pregnancy, often in conjunction with acetylcholinesterase (AChE) assays (see above). Most mothers carrying a fetus with an open neural tube defect have elevated AFP levels. Such elevated levels, however, can be caused by many other fetal diseases and by multiple pregnancy, intrauterine death, or underestimation of gestational age. There is an overlap between normals and affected subjects [58]. The programs detect 80%–90% of open neural tube defects, and still more if AChE is included. A blood AFP test is repeated if AFP levels are elevated, and ultrasound is performed to rule out twins, to confirm gestational age, and to find indications of neural tube defect. If an elevated blood AFP level is confirmed, amniocentesis to detect elevated amniotic fluid AFP is carried out. Such screening programs are complex and somewhat controversial since most women with elevated AFP levels do not have an affected fetus. The lower the frequency of neural tube defects in a given population, the higher is the proportion of false-positive AFP results without neural tube defects. Nevertheless, those who have directed such programs feel that maternal AFP screening is valuable in populations in which the frequency of neural tube defects is not rarer than 1/1000, and if a quality laboratory and trained personnel are available to allow the extensive counseling and ultrasound studies that are required [3]. It has been estimated that 1500 fetuses with neural tube defects are aborted per year in the United States as a result of such programs.

Extensive Future Screening of All Newborns for Many Polymorphisms? Until now screening programs have been introduced only for certain inherited diseases. However, as noted in discussions on the genetic basis of common disease (Chap. 6), on genetic polymorphisms in relation to disease (Chap. 6), and on pharmacogenetics and ecogenetics (Sect. 7.5), some "normal" genes whose products can be recognized may influence susceptibility to common disease, interacting sometimes with specific environmental conditions. This raises the question of whether it could be useful in the future to screen every newborn for many of these polymorphisms to allow an individual prognosis regarding disease risks. Preventive measures, such as avoidance of certain food stuffs, tobacco smoking, alcohol, other drugs, or occupational exposure to specific environmental conditions, such as dust or chemicals, might be recommended depending upon the specific results. It might thus become possible to reduce the risks for certain diseases to which some genotypes are more susceptible.

Such screening would have significant societal impact. How would society react to the knowledge that some of its future members will have a good chance of reaching an advanced age in reasonably good health, whereas others will need lifelong protection from certain ubiquitous environmental influences and still others may have a relatively restricted life expectancy under the best conditions? Who in a society should have access to this information? How could confidentiality be ensured in storage by computers? Will the solidarity within human communities be able to cope with this kind of genetic inequality, or will there be conflict between groups with different genotypes? How will detailed knowledge of one's liabilities influence individual happiness and fulfillment? Because of all these uncertainties, there is current agreement not to test in the newborn period for late onset disease genes or for various genetic susceptibilities unless effective preventive or treatment measures are available [2].

These are only some of the ethical problems raised by the development of human genetics and specifically by the prospects of genetic screening. Mankind does not yet appear to be psychologically prepared for such developments.

18.4 Human Genome Project

A Large-Scale Science Project. The Human Genome Project is the first large-scale science project in biology [51]. Compared to a number of large, coordinated programs in engineering, physics, and the space sciences, biological research has generally been something of a "cottage industry" carried out by single or more recently small groups of investigators. The Human Genome Project aims at full elucidation of the genetic, physical, and nucleic acid sequence map of the human genome and of several model organisms (*E. coli,* yeast, roundworms, *Drosophila,* and mice). This goal involves coordinating the work of many researchers in different locations in various countries. The United States budget for the project in the initial years of 1991–1993 was $ 135 million, $ 166 million, and $ 171 million per year, respectively. Other countries including the members of the European Union and Japan are also spending relatively large amounts of money for this purpose. The project aims to conclude its efforts after 15 years [10].

The project was designed and planned at a time when developments in molecular biology and the "new genetics" could be applied to elucidate the detailed structure of the genetic material. Many scientists were initially skeptical about the need for such an effort. It was argued that since 95 % of DNA may be evolutionary waste or "junk," sequencing of most of the genome would not provide biologically significant data. Attention to cDNA or expressed genes would lead to more interesting biological data at much less cost. Genomic work was often depicted as repetitive, dull, and directed from above rather than driven by hypothesis-testing investigators who ask intelligent and solvable questions. The most exciting questions, such as how genes are turned on and off during development, and how genes work in the nervous system, would not be elucidated by the approach taken by the genome project. Some critics argued that by its very existence the initiative promotes "geneticization" by suggesting that complex traits, diseases, and behavior are fundamentally genetic in origin rather than being caused by environmental factors, with genetics playing a minor role [27].

Advantages. Proponents argued that concerted work would be less expensive than piecemeal efforts by many investigators, such as was required for the cloning of the cystic fibrosis gene. They pointed out that the information provided by the Genome Project would once and forever give a definitive basis and infrastructure for all of biology and biomedical research in the future [51]. Just as one needs to know gross and microscopic anatomy to understand the human organism in health and disease, full knowledge of the genetic map and the DNA sequence is now required to explore, elucidate, understand, and ultimately even manipulate the genetic material.

The Genome project is also designed to improve the research infrastructure of human genetics by launching efforts in the physical, engineering, and computer science aspects of DNA research since most emphasis so far has been directed to genetic and molecular biological approaches. On a more practical basis, knowledge from the Genome Project is likely to contribute to many applications for diagnosis and management not only of the well-defined genetic diseases but also of the many common diseases with genetic determinants. Knowledge gathered by the Genome Project is likely to aid the relatively new biotechnology industry by discovering new genes that will lead to the manufacturing of diagnostic and therapeutic agents. Such an economic argument is well received by legislators who allocate grant funds for research work.

The First Initiative. It is noteworthy that the initiative for the Human Genome Project did not originally come from human or medical geneticists but from policy makers in the Department of Energy (formerly the Atomic Energy Commission) who were seeking a use for their large-scale facilities and experience in physical biology. Later, the National Institutes of Health became committed, largely through the efforts of molecular biologists, particularly James Watson, the codiscoverer of the DNA double helical structure [61] who became the first director of the United States Genome Project.

The Human Genome project owes much to the report of a committee of the National Research Council published in 1988 [48]. This committee brought together both proponents and skeptics who recommended that a major effort in genetic mapping and physical mapping be instituted. Delay in the sequencing phases of the project was advised until more efficient methods became available. These recommendations have generally been followed. New molecular methods (such as PCR) developed outside the Genome Project has helped the project immensely.

Methods. A substantial portion of the funds of the Human Genome Project in the United States are being spent on physical aspects of mapping. In physical mapping, a DNA sequence is localized to a specific physical piece of the chromosome by methods described elsewhere in the present volume, such as chromosome banding (Sect. 2.1.2.3), the various methods for in situ hybridization of chromosomes (Sect. 3.1.3.3), and DNA contig building (Sect. 5.1.3). Chromosomal banding is physical mapping by which a given genetic characteristic is localized to a defined chromosomal band. In situ hybridization documents the localization of a probe for a certain gene to a specific location on a chromosome and is usually carried out by autoradiography with radioactive probes (Sect. 3.1.3.3). More recently, fluorescence in situ hybridization has allowed detection of human gene segments using fluorescent dyes for visualization of the chromosomal sites (Sect. 3.1.3.3). Large-scale physical maps are constructed by the process of contig building. A contig is an organized set of DNA clones that as a group covers a region that is too long to clone a

single piece. The most useful vector to carry large pieces of DNA are yeast artificial chromosomes (YACs) which allow cloning of large pieces of linear DNA segments (up to 1000 kb) in yeast *(S. cerevisia)* [6a]. This methodology has allowed the completion of complete YAC-based maps of human chromosome 21 and Y chromosome and more recently of *all* human chromosomes by a French group. Unfortunately, there are both gaps and YAC rearrangements in such maps which makes them somewhat inaccurate, thus requiring further work. Another useful development for physical mapping have been DNA landmarks designated as sequence-tagged sites (STS) [51]. An STS is a unique DNA sequence that can be amplified by PCR techniques. They can be stored electronically and are readily available to scientists all over the world.

It has became apparent that an understanding of the regulation of gene expression could not be solved by cDNA analysis since regulatory genes may exist upstream and downstream from structural genes as well as in intronic sequences, thus justifying the sequencing of the entire genome. The sequencing of genes in model organisms (*E. coli*, roundworms, yeast) has revealed the existence of a much greater number of genes than had been suspected. A large portion of the open reading frames of the genomic DNA of model organisms appears to reflect the presence of previously unknown genes. Sequencing in several model organisms has already identified homologies and identical motifs across species, including humans. This approach is becoming increasingly useful for identifying critical genes involved in regulation and transport [51]. Although not all criticism has been silenced, the Human Genome project has become fairly widely accepted by the biomedical research community and is thriving. The development of new and faster sequencing technology, however, has lagged. Contributions by the physical, engineering, and computer science community have generally been modest so far.

Research Organizations. Several organizations have sprung up around the Human Genome Project. The Center for Study of Human Polymorphisms (CEPH) in Paris that existed earlier has been particularly useful for coordinating the exchange of DNA specimens from large families between investigators all over the world for linkage studies with genetic markers. The existence and operation of such a repository has been a major factor in hastening the construction of genetic maps. An international organization (HUGO; Human Genome Organization) has been set up to coordinate genome projects in different countries, since most nations of the developed world have established their own human genome projects.

Human Diversity Project. A small offshoot of the human genome initiative has been the Human Diversity Project. Standard large populations such as Japanese, Germans, Russian, etc. are readily available for study so that data on gene distribution will ultimately become available. However, there are many small populations (such as the pygmies, certain native American tribes) all over the world that may vanish in the not so distant future. The task of the Human Diversity Projects is to collect a sufficient number of specimens from members of each of these unique populations in a concerted effort for study of DNA markers to allow characterization of these groups to assess their origins and affinity with other populations. Despite the unexceptional aim of such studies, accusations of scientific imperialism, exploitation, and racism have been raised by some critics who may not have been appropriately informed regarding the scientific aims of such studies.

The question has sometimes been raised whether the genome of a single individual will be chosen as the standard human type. Because of the marked variation between humans, the final gene map will of course be a composite of the DNA sequence of many different individuals. One can hope that sequencing technology will become so advanced in the first quarter of the twenty-first century that the entire DNA sequence of a person (~ 3–3.5×10^9 base pairs of the haploid genome; Sect. 3.1.1.1) can be rapidly obtained for storage on a computer disk. Considerable technical progress will be required to achieve such an aim.

Ethical, Legal, and Social Aspects. Many scientific and technological developments such as those of the Industrial Revolution have had a major impact on human societies, but their effects emerged long after the various innovations were planned and introduced. Many social, ethical, and legal consequences are likely to arise as the role of genes in health and disease becomes better understood. A significant proportion of the budget of the United States Genome Project (about 5%) has therefore been set aside to explore the societal impact of the new insights into human genetics. This is the first time in the history of science and technology that a considerable portion of the budget of a large science project has been devoted to studying its societal and ethical impacts while the scientific work is still being carried out. Many problems that are being discussed in this connection are not entirely new, since the scientific developments in medical genetics over the past two decades have already raised a variety of identical issues. Bioethicists, sociologists, and theologians have pondered these issues, often in governmental or quasi governmental committees and have made recommendations regarding public policies (see [2]). Many symposia to consider these matters have been held.

Genetic risk assessment for individual patients and their families will be carried out increasingly by a variety of DNA and other tests. Generally, genetic testing for differential diagnostic purposes such as in cancer detection raises no ethical problems. What then are some of the problems? The possibility of misuse of genetic knowledge for occupational discrimination has been raised (see [57]). Should a person's higher genetic risk of developing a late-onset disease be permitted as a criterion for job selection or job assignment? What if the disease is a psychiatric condition? Most such scenarios at this time remain theoretical since test systems to make such predictions are rarely available. Most observers feel that the capacity to fulfill the responsibilities of a job *now* rather than *in the future* should be the sole criterion for occupational placement. The problems of genetic susceptibility to developing a disease on exposure raises difficult question in occupational medicine [57]. It has often been discussed whether individuals with a genetic susceptibility to developing an illness from certain chemical agents should be kept away from the offending substance. One would not advise a hemophiliac adolescent to become a butcher! But who should ultimately be responsible for making such decisions? Some would let the individual decide even if damage might occur. Others would let society be the arbiter, particularly if harm to others might occur. For example, we do not allow persons with severe genetic color vision defects to be employed in the transportation industry, where color discrimination is essential.

Unlike in most European countries, health insurance problems in the United States have been a thorny issue. Insurance companies would like to use knowledge of a preexisting genetic disease or its possible future occurrence to exclude coverage. Life insurance raises somewhat different problems. A useful compromise here would be to allow everyone regardless of genetic risk to obtain, say, $ 75 000 life insurance. When higher benefits are involved, life insurance companies might be allowed to use any information (including genetic history or even testing) to estimate the potential longevity of an applicant. Not allowing the use of such predictive information might lead to "adverse selection" since those at high risk would tend to take out insurance policies in high amounts. Life insurance companies already use predictive information when testing for high blood pressure which predicts a higher probability of heart disease and stroke. The detection of hypertension leads to uprating of insurance premiums or to complete rejection of the applicant and is currently generally accepted as a probabilistic indicator of shortened life expectancy.

Apart from occupational and insurance problems there are many psychological, existential, and social issues that arise with genetic risk assessment [2]. There is general agreement that genetic tests which diagnose treatable and preventable medical conditions should be widely offered, followed by treatment of those who require it (including children if the appropriate medical intervention needs to be carried out early in life). Specific and sensitive tests are required before recommending wide population screening. Often, particularly with common diseases of complex inheritance, the tests are probabilistic and not definitive. Commercial pressures often encourage the use of tests before they have been sufficiently assessed in pilot studies. Academic scientists, who counsel caution, may be accused of "dragging their feet" at the expense of the health of those at risk. Government agencies and professional organizations need to cooperate in drafting enforceable criteria as to when a test is to be released to the public, and how quality control is to be carried out [2].

Our rapidly expanding knowledge often allows diagnosis of a genetic disease (or genetic susceptibility) before therapeutic or preventive measures are available. Under such conditions some persons wish to know their status so as to order their lives, while others would rather prefer to remain ignorant. The decision of whether to undergo testing should be left entirely up to the individual. Many observers feel that parents should *not* have the right to order predictive tests of their under-age children who are at risk for a *nontreatable* or *nonpreventable* condition or disease. There should be wide education of the medical profession to avoid the "technological imperative." The mere fact that a test is available does not necessarily mean that it must be used. *A "genetic" test is different from the usual laboratory tests which are generally ordered without discussing them with the patient.* Pregnant women rarely realize the full implications when certain tests such as AFP screening is carried out on maternal blood to detect Down syndrome and neural defects in their fetuses. New ways need to be developed for providing information about such tests and particularly about the consequences if the results are positive.

In population testing for heterozygote traits (such as thalassemia), the screening of pregnant women (followed by further testing of the partner if the woman's result is positive) is often logistically much simpler than large-scale premarital screening of young adults. When testing is not performed before pregnancy, prenatal diagnosis (and possibly selective abortion) is the only alternative to having an affected child if both partners are heterozygote carriers. Premarital screening programs allow several other reproductive options that do not require termination of pregnancy,

such as avoidance of marriage with another carrier, artificial insemination by an artificial donor, contraception, and even sterilization. Nevertheless, reliance on prenatal diagnosis has found wide acceptance in many countries for diseases such as thalassemia major and Tay-Sachs disease and is more practical and less cumbersome to carry out than screening before pregnancy.

Most of the problems under discussion are solvable. It is likely, however, that the societal impact of genetics will assume major dimensions when we obtain increasing knowledge regarding the genetic basis of personality, behavior, and cognition in health and disease. Although monogenic mechanisms will only be rarely found, there is sufficient evidence from developmental, family, adoption, and twin studies (see Chaps. 15; 16) that genes affect a wide variety of human behavioral traits and psychiatric diseases. Considerable scientific work in behavioral genetics needs to be carried out to identify the underlying genes and to elaborate their interaction with various developmental and environmental factors. Since major and serious misuse could occur with premature application of such knowledge, *particular attention must be paid to monitoring the applications of behavioral genetics.*

A pervasive thread in all areas is the problem of privacy and confidentiality of genetic data. Modern computer technology has made it extremely difficult to keep any public or medical data confidential. The expansion of information and advances in information technology will require that we establish better ways of dealing with the confidentiality of genetic information.

Conclusions

Scientific knowledge in human genetics can be used for genetic counseling of families in which an increased risk for hereditary anomalies and diseases in future children is suspected. Genetic counseling refers to the totality of activities that (a) establish the diagnosis of such diseases, (b) assess the recurrence risk, (c) communicate to the client and family the chance of recurrence, and (d) provide information regarding the many problems raised by the disease, including natural history and variability of the disease. Formal methods applied in risk assessment include the use of genetic ratios, statistical considerations, and empirically derived risk factors if the mode of inheritance is unknown. Laboratory techniques involve prenatal diagnosis by ultrasonography, amniocentesis, and chorionic villus sampling and the diagnostic application of techniques from molecular biology. In populations with high risk for certain diseases, screening for such diseases has often been successful. Genetic counseling is not concerned merely with scientific questions. It must provide sympathetic counsel and deal with the many psychological aspects of the process. Genetic counseling is an important part of comprehensive medical care.

References

1. Alter BP (1985) Antenatal diagnosis of thalassemia. a review. Ann N Y Acad Sci 445 : 393–407

2. Andrews LB, Fullarton JE, Holtzman NA, Motulsky AG (eds) (1994) Assessing genetic risks: implications for health and social policy. National Academy, Washington

3. Anonymous (1985) Maternal serum alpha-fetoprotein screening for neural tube defects. Results of a consensus meeting. Prenat Diagn 5 : 77–83

4. Anton-Lamprecht I (1981) Prenatal diagnosis of genetic disorders of the skin by means of electron microscopy. Hum Genet 59 : 392–405

5a. Becker R, Fuhrmann W, Holzgreve W, Sperling K (1995) Pränatale Diagnostik und Therapie. Wissenschaftliche Verlagsanstalt, Stuttgart

5. Bloch EV, DiSalvo M, Hall BD, Epstein CJ (1979) Alternative ways of presenting empiric risks. Birth Defects 15(5C)

6. Brambati B, Simoni G, Danesino C, Oldrini A, Ferrazzi E, Romitti L, Terzoli G, Rossella F, Ferrari M, Fraccaro M (1985) First trimester fetal diagnosis of genetic disorders: clinical evaluation of 250 cases. J Med Genet 22 : 92–99

6a. Burke TD, Carle GF, Olsen MV (1987) Cloning of large segments of exogenous DNA into yeast by means of artificial chromosome vectors. Science 236: 806–812

7. Cao A, Rosatelli C, Galanello R et al (1989) The prevention of thalassemia in Sardinia. Clin Genet 36 : 277–285

8. Capron AM, Lappe M, Murray RF, Powledge TM, Twiss SB, Bergsma D (eds) (1979) Genetic counselling: facts, values, and norms. Birth Defects 15(2)

9. Carter CO, Fraser Roberts JA, Evans KA, Buck AR (1971) Genetic clinic: a follow-up. Lancet 1 : 281

10. Collins F, Galas D (1993) A new five-year plan for the US human genome project. Science 262 : 43–46

11. Crandall BF, Robertson RD, Lebherz TB, King W, Schroth PC (1983) Maternal serum a-fetoprotein screening for the detection of neural tube defects. West J Med 138 : 531–534

12. Dalgaard OZ (1957) Bilateral polycystic disease of the kidneys. A follow-up of two-hundred and eighty-four patients and their families. Acta Med Scand Suppl 328

13. Davis JR, Rogers BB, Hageman RM, Thies CA, Veomett IC (1985) Balanced reciprocal translocations: risk factors for aneuploid segregant viability. Clin Genet 27 : 1–19

14. Epstein CJ, Curry CJR, Packman S, Sherman S, Hall BD (eds) (1979) Risk, communication, and decision making in genetic counseling. Birth Defects 15(5C)

15. Epstein CJ, Cox DR, Schonberg SA, Hogge WA (1983) Recent developments in the prenatal diagnosis of genetic diseases and birth defects. Annu Rev Genet 17 : 49–83

16. Erbe RW (1975) Screening for the hemoglobinopathies. In: Milunsky A (ed) The prevention of genetic disease and mental retardation. Saunders, Philadelphia, pp 204–220

17. Evers-Kiebooms G, van den Berghe H (1979) Impact of genetic counseling: a review of published follow-up studies. Clin Genet 15 : 465–474

18. Fletcher JC, Berg K, Tranøy KE (1985) Ethical aspects of medical genetics. A proposal for guidelines in genetic counseling, prenatal diagnosis and screening. Clin Genet 27 : 199–205

19. Fuhrmann W, Vogel F (1983) Genetic counseling, 3rd edn. Springer, Berlin Heidelberg New York (Heidelberg science library 10)

20. Fuhrmann W, Weitzel HK (1985) Maternal serum alpha-fetoprotein screening for neural tube defects. Report of a combined study in Germany and short overview on screening populations with low birth prevalence of neural tube defects. Hum Genet 69 : 47–61

21. Galen RS, Gambino SR (1975) Beyond normality: the predictive value and efficiency of medical diagnoses. Wiley, New York

22. Harper PS (1993) Practical genetic counseling, 4th edn. Wright, Bristol

23. Hausser I, Anton-Lamprecht I (1990) Prenatal diagnosis of genodermatoses by ultrastructural diagnostic markers in extra-embryonic tissues. Defective hemidesmosomes in amnion epithelium of fetuses affected with apidermolysis bullosa Herlitz type (an alternative prenatal diagnosis in certain cases). Hum Genet 85 : 367–375

24. Hellerman JG, Cone RC, Potts JT, Rich A, Mulligan RC, Kronenberg HM (1984) Secretion of human parathyroid hormone from rat pituitary cells infected with a recombinant retrovirus encoding preproparathyrpoid hormone. Proc Natl Acad Sci USA 81 : 5340–5344

25. Holmes LB (1978) Genetic counseling for the older pregnant woman: new data and questions. N Engl J Med 298 : 1419–1421

26. Holtzman NA, Leonard CO, Farfel MR (1981) Issues in antenatal and neonatal screening and surveillance for hereditary and congenital disorders. Annu Rev Public Health 2 : 219–251

27. Hubbard and Wald (1993) Exploding the gene myth. Beacon, Boston

28. Inman RP (1978) On the benefits and costs of genetic screening. Am J Hum Genet 30 : 219–223

29. Jackson LG (1985) First-trimester diagnosis of fetal genetic disorders. Hosp Pract 20 : 39–48

30. Kaback MM (ed) (1977) Tay-Sachs disease: screening and prevention. Liss, New York

31. Kaback MM, Zeiger RS, Reynolds LW, Sonneborn M (1974) Approaches to the control and prevention of Tay-Sachs disease. Prog Med Genet 10 : 103–134

32. Kazazian HH Jr, Boehm CD, Dowling CE (1985) Prenatal diagnosis of hemoglobinopathies by DNA analysis. Ann N Y Acad Sci 445 : 337–368

33. Kessler S (1980) The psychological paradigm shift in genetic counseling. Soc Biol 27 : 167–185

34. Leonard C, Chase G, Childs B (1972) Genetic counseling: a consumer's view. N Engl J Med 287 : 433

35. Levy HL (1974) Genetic screening. Adv Hum Genet 4 : 1

36. Lippman-Hand A, Fraser FC (1979) Genetic counseling – the postcounseling period. I. Parents' perceptions of uncertainty. Am J Med Genet 4 : 51–71

37. Lippman-Hand A, Fraser FC (1979) Genetic counseling – the postcounseling period. II. Making reproductive choices. Am J Med Genet 4 : 73–87

38. McKusick VA (1995) Mendelian inheritance in man, 11th edn. Johns Hopkins University Press, Baltimore

39. Mibashan RS, Rodeck CH (1984) Haemophilia and other genetic defects of haemostasis. In: Rodeck CH, Nicolaides KH (eds) Prenatal diagnosis. Wiley, Chichester

40. Milunsky A (1975) The prevention of genetic disease and mental retardation. Saunders, Philadelphia

41. Modell B (1984) Haemoglobinopathies – diagnosis by fetal blood sampling. In: Rodeck CH, Nicolaides KH (eds) Prenatal diagnosis. Wiley, Chichester

42. Modell B (1985) Chorionic villus sampling. Evaluation, safety and efficacy. Lancet 1 : 737–740

43. Motulsky AG (1973) Screening for sickle-cell hemoglobinophathy and thalassemia. Isr J Med Sci 9 : 1341–1349

44. Motulsky AG (1975) Problems of screening for genetic disease. In: Went L, Vermeij-Keers C, van der Linden AGJM (eds) Early diagnosis and prevention of genetic disease. University of Leiden Press, Leiden, pp 132–140

45. Motulsky AG (1975) Family detection of genetic disease. In: Went L, Vermeij-Keers C, van der Linden AGJM (eds) Early diagnosis and prevention of genetic disease. University of Leiden Press, Leiden, pp 101–110

46. Motulsky AG, Murray J (1983) Will prenatal diagnosis with selective abortion affect society's attitude toward the handicapped? In: Berg K, Tranoy KE (eds) Research ethics. Liss, New York, pp 277–291

46 a. Motulsky AG (1994) Predictive genetic diagnosis

47. Murphy EA, Chase GA (1975) Principles of genetic counseling. Year Book Medical Publishers, Chicago

48. National Research Council (1988) Mapping and sequencing the human genome. National Academy, Washington

49. National Research Council Committee for the Study of Inborn Errors of Metabolism (1975) Genetic screening: programs, principles, and research. National Academy of Sciences, Washington

50. Nelson WB, Swint JM, Caskey CT (1978) An economic evaluation of a genetic screening program for Tay-Sachs disease. Am J Hum Genet 30 : 160–166

51. Olson MV (1993) The human genome project. Proc Natl Acad Sci USA 90 : 4338–4344

52. President's Commission for the Study of Ethical Problems in Medicine and Biomedical and Behavioral Research (1983) Screening and counseling for genetic conditions. The ethical, social, and legal implications of genetic screening, counseling, and education program. US Government Printing Office, Washington

53. Reeders ST et al (1986) Two genetic markers closely linked to adult polycystic kidney disease on chromosome 16. BMJ 292 : 851–853

54. Ried T, Mahler V, Vogt P et al (1990) Direct carrier detection by in situ suppression hybridization with cosmid clones of the Duchenne/Becker muscular dystrophy locus. Hum Genet 85 : 581–586

55. Robinson A (1985) Prenatal diagnosis by amniocentesis. Annu Rev Med 36 : 13–16

56. Rodeck CH (1984) Obstetric techniques in prenatal diagnosis. In: Rodeck CH, Nicolaides KH (eds) Prenatal diagnosis. Wiley, Chichester

57. Rüdiger HW, Vogel F (1992) Die Bedeutung der genetischen Disposition für Risiken in der Arbeitswelt. Dtsch Arztebl 89 : 1010–1018

58. Scriver CR, Clow CL (1980) Phenylketonuria and other phenylalanine hydroxylase mutants in man. Annu Rev Genet 14 : 179–202

59. Simoni G, Brambati B, Danesino C, Terzoli GL, Romitti L, Rossella F, Fraccaro M (1984) Diagnostic application of first trimester trophoblast sampling – 100 pregnancies. Hum Genet 66 : 252–259

60. Sorenson JR, Swazey JP, Scotch NA (eds) (1981) Reproductive pasts reproductive futures. Genetic counselling and its effectiveness. Birth Defects 17(4)

61. Ton CCT, Hirvonen H, Miwa H et al (1991) Positional cloning and characterization of a paired box- and homeobox-containing gene from the aniridia region. Cell: 1059–1074

62. Veal AM (1965) Intestinal polyposis. Cambridge University Press, London (Eugenics laboratory memoirs 40)

63. Ward RHT (1984) First trimester chorionic villus sampling. In: Rodeck CH, Nicolaides KH (eds) Prenatal diagnosis. Wiley, Chichester

19 Genetic Manipulations and the Biological Future of the Human Species

He who knows where to halt is not in danger.

From a Japanese screen by Nukina Kioku
(1778–1863)

19.1 Genetic Manipulation

Human beings have been manipulating the genes of plants and animals by means of domestication for several thousand years. The development of agriculture is thus itself a form of genetic engineering. The breeding of different types of dogs is another example of manipulating genes affecting behavior [17].

We are constantly and increasingly manipulating the genetic constitution of the human species indirectly, by altering the human environment and by treating diseases that have genetic determinants. Both therapy and public health measures affect the human gene pool by preserving harmful genes that in the absence of these measures would be eliminated. For example, it is very likely that genes involved in the predisposition to various infections will increase in frequency because of the wide use of antibiotics during the past two generations. Previously such genes disappeared with the death of the affected patients. Another example: Since marital partners tend to resemble each other in intelligence due to assortative mating, the distribution of genes affecting intelligence (see Sect. 15.2.1.3 for evidence of the existence of such genes) tends to concentrate a larger share of high-intelligence genes among the offspring of gifted couples.

However, as soon as the topic of genetic manipulation is raised, the public usually imagines something quite different, such as the creation of human beings in the laboratory by genetic specifications and similar bizarre scenarios. Such speculations have been voiced occasionally ever since the foundation of modern genetics. However, they have found a quasigenetic basis only since the Watson-Crick model of DNA inaugurated a new era in genetic research.

In the early 1960s, several symposia became important as prototypes of these discussions. The most notable was the symposium *Man and His Future* in 1963, in which a number of prominent scientists discussed the prospects of genetic manipulation without restraint. The rapid developments in molecular biology over recent years have again led to many discussions regarding genetic engineering. There is a great deal of concern among the public about "mad scientists" and their "tampering" with the human gene pool to modify human characteristics. Since the genetic determination of complex human traits such as personality, intelligence, and stature remains poorly understood and is affected by many genes, such traits can not be manipulated in the foreseeable future.

It may be useful to subdivide the proposed methods of genetic manipulation into two groups: (a) the more conservative steps that make use of well-established biological principles and methods that require only some technical improvements, and (b) the more revolutionary approaches requiring major breakthroughs in molecular biology.

"Conservative" Approach: Germinal Choice and Artificial Insemination [33]. H. J. Muller was the main promoter of germinal choice. He repeatedly encouraged prospective parents not to rely solely on their own germ cells but to choose freely from the germ cells of many individuals, selecting the future phenotypes of their children by knowledge about the personality and achievements of the individuals from whom the germ cells would be taken. According to Muller,

[A] choice is not a real one unless it is a multiple choice, one carried out with maximum foreknowledge of the possibilities entailed and hampered as little as possible by irrational restrictions. . . . Moreover . . ., the final decision regarding the selection to be made should be the prerogative of the couple concerned. These conditions can be fulfilled only after plentiful banks of germinal material have been established, representing those who have proved to be most outstanding in regard to valuable characteristics of mind, heart and body . . . . Catalogued records should be maintained, giving the results of diverse physical and mental tests and observations of all the donors, together with relevant facts about their lives, and about their relatives . . . . The germinal material used should preferably have been preserved for at least twenty years . . . [obviously because the performance of a donor can only be finally assessed after this time]. Such an undertaking by a couple would assume the character of an eminently moral act, a social service that was in itself rewarding . . . .

From a technical point of view this proposal can already be carried out now; storage of human sperm is possible. In fact, artificial insemination with stored sperm is being used on a large scale in cattle breed-ing. In humans, artificial insemination – albeit in most cases with fresh semen obtained from healthy donors – is carried out fairly extensively in women who cannot conceive because of their husbands' sterility. Genetic investigation of the donor is usually not performed by physicians involved in artifical insemination. However, with increasing attention to human and medical genetics, guidelines for genetic screening of artificial insemination donors have been published and are being implemented in some centers. Such genetic screening is based largely on a careful family history questionnaire. Chromosome testing is not usually performed, and heterozygote testing is reserved for high-frequency traits in certain ethnic groups (e. g., Tay-Sachs screening in Ashkenazi Jews; sickle cell testing in African-Americans). While there are few problems with defining rejection crite-ria for a family history of clearcut monogenic dis-eases, it is more difficult to establish such criteria for the more common multifactorial and polygenic diseases.

Sperm banks have been established in some metropo-litan areas of the United States, particularly for de-positing the sperm of men who undergo vasectomy but want to ensure the possibility of having a child in case they change their minds. Such sperm banks are also useful for allowing future offspring to men with neoplastic diseases who undergo treatment with high doses of cytostatic agents or irradiation and will be sterile or genetically damaged following such treatments. A sperm bank in California solicited sperm samples from proven high achievers in the natural sciences (such as Nobel Prize winners and those of similar distinction) for artificial insemina-tion of self-selected women volunteers, and several babies have been born. Based on the heritability of intelligence (see Sect.15.2.1.3) it is hoped that the do-nors' children will have an increased likelihood of in-tellectual distinction – a reasonable premise. It is, of course, highly unlikely that such a child will have the exact genetic and environmental determinants which made the father an exceptionally creative per-son. This sperm bank has rightly received consider-able unfavorable publicity in the media. A commer-cial sperm bank in the United States circulates to physicians a listing of the ethnic backgrounds and social and professional characteristics of the potential donors available for selection by couples. The desired sperm sample is then shipped to the physician of the infertile couple for insemination.

Artificial inovulation is also possible. Oocytes are gathered from human ovaries during laparoscopy and fertilized in vitro. Extensive efforts in recent years have led to successful in vitro fertilization of human eggs [36], many such babies have in the mean-

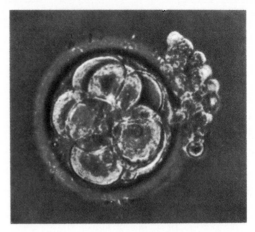

Fig. 19.1. Human oocyte after fertilization in vitro and culti-vation in Ham's F10 medium. (From Edwards and Fowler 1970 [13])

time been born. Multiple eggs are collected by laparo-scopy following medicinal and hormonal stimulation of the ovary and are fertilized with the husband's sperm in vitro. After several cell divisions have occur-red, several eggs are placed in the uterus in the hope that at least one fertilized egg will implant and grow to maturity. Other fertilized eggs are frozen for possi-ble use in the future. A great number of clinics in dif-ferent parts of the world now offer this technique to couples of which the female partner is infertile.

This technique opens up the possibility of in vitro fertilization of sperms and eggs from any human source. Some pregnancies under such a scheme have been carried by "surrogate" mothers rather than by the biological mothers. There has been concern about such practices and their possible abuse as well as about the legal status of frozen fertilized eggs, their use, and their disposal. Some observers have criti-cized the practice of implanting multiple zygotes into one uterus, a practice meant to ensure a reason-ably high rate of sucess since at present each single zygote runs a high risk of early death. This practice has led to a relatively high incidence of multiple births, with additional risks to the children (Sect. 6.3). Human reproduction can now be manipu-lated in many ways at different levels. It is hard to predict how this will influence the attitudes of future generations.

Are Large-Scale Attempts at Breeding Human Beings by Such Methods Inevitable? It was Muller's intention not only to prevent rare hereditary diseases by occa-sional artificial insemination but also to improve hu-man quality by large-scale selective breeding. Are such attempts inevitable in the future? And would they be successful?

We cannot imagine that selective breeding will ever become popular in an open democratic society. It will probably remain restricted to a small minority of the population. It has been suggested that a dictatorship might set out to breed nuclear physicists, for example, by using the sperm of Nobel prize winners in this field. It is likely that many children produced in this way would perform above average in the sciences and some might even be outstanding if provided the appropriate environment. Still, the danger of such a scheme being attempted is relatively small. Even dictatorships usually have more immediate concerns; they will probably not invest their – necessarily limited – resources in such an undertaking that would not yield results for a least 20–30 years.

Molecular Biology and Speculations on Genetic Manipulation. Speculations regarding possible future methods of genetic engineering are based on the following results of molecular biology:

a) The mutagenic activity of certain chemicals (Chap. 11) may be used to induce specific mutations at well-defined gene loci.
b) DNA can be incorporated by extrasexual means (transformation or transduction) – not only in micro-organisms, where this phenomenon was discovered, but in eukaryotes, as well, including human beings.
c) Defective genes may be replaced by a variety of techniques.
d) Artifically synthesized genes may be included into the human genome.

Induction of Specific Mutations. In the majority of gene mutations one base of the DNA sequence is replaced by another base. Such nucleotide substitution may lead to an amino acid replacement in a specific protein that then becomes functionally inadequate. Some chemical mutagens, on the other hand, selectively attack specific bases and induce such point mutations. Only a few years ago any attempt to attack single specific sites by a mutagen seemed doomed to failure by the sheer size of the problem; far too many identical sites are present in the human genome. More recently, however, methods of targeted mutagenesis have become available [47]. Hence the problem of localized mutagenesis is now accessible in principle, even if the practical problems are still formidable.

Gene Transfer and Expression in Eukaryotes. Two principal methods are known for introducing foreign genetic material into a micro-organism. In transformation pure DNA may, under insufficiently defined conditions, enter a microbial cell and be integrated into the genetic material. In transduction a bacteriophage incorporates a particle of the bacterial genome. When the virus is released from the host and infects another bacterium, the bacterial material carried by the virus is transferred to the new host, where it becomes genetically active. Transformation experiments have played an important role in the history of genetics, for example, they established that DNA is the genetically active material (Avery et al. 1944; see [47]).

In eukaryotes, the transformation and transduction of DNA and also expression of transferred genes have repeatedly been reported. Early examples were plants and cultured animal cells. Prokaryotic gene expression in eukaryotic cells has been achieved in an increasing number of instances, with the prokaryotic DNA coming in most cases from viruses but sometimes from bacteria. A famous example was the transfer of the galactose operon of *E. coli* to human fibroblasts in 1971 [27]. In humans galactose is metabolized via the same pathway as in *E. coli*, and deficiency mutants for the three enzymes involved are known. The most common is galactosemia (230 400), a defect of P-gal-uridyltransferase. Incubation of such cells in vitro with λ phages carrying *E. coli* Gal operon led to transferase production in these cells.

Replication of this result in the years thereafter proved difficult, and progress was slow. However, the subsequent introduction of new DNA techniques (Chap. 3) brought much faster progress, and the practical application of gene transfer for somatic gene therapy now appears to be within our reach.

Four methods are available for transferring cloned genes into such cells: (a) viral (RNA or retroviruses and DNA viruses, for example; adenoviruses), (b) chemically mediated, for example, with calcium phosphate; (c) fusion of DNA-loaded liposomes, red blood cell ghosts, or protoplasts to cells; and (d) physical (microinjection or electroporation). Fusion techniques are currently under development. DNA microinjection has been used for many experiments in developmental biology of other vertebrates [42], but the amount of material to be injected is normally very large and is difficult to control. Retrovirus vectors have been used in the first therapeutic trials in humans. Up to 100 % of the target cells can be infected, and the DNA can integrate as a single copy at a single (albeit random) site. Moreover, the structure of the inserted DNA sequence is known. Figure 19.2 shows the principle of this system, and Fig. 19.3 shows the experimental approach in the mouse. A protocol in which the normal gene would not simply be inserted at random but would replace the mutant gene by recombination would be preferable; certain concrete ideas on this problem exist [43]. Some trans-

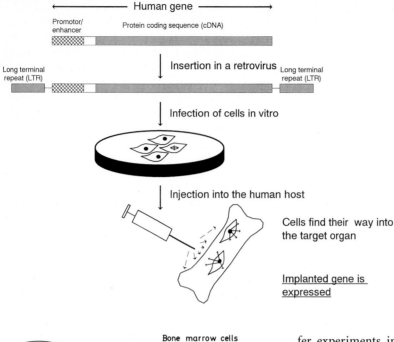

Fig. 19.2. Principle of gene therapy by using a retrovirus vector. The cells are first infected in vitro and are then injected into the human host

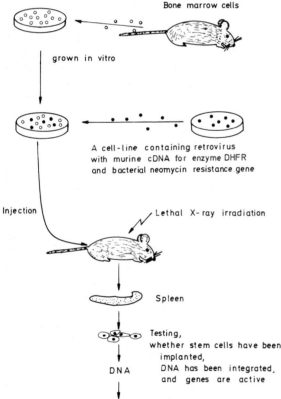

Fig. 19.3. Principle of gene transfer by a retrovirus with inserted cDNA for murine genes into a receptor animal, after its own bone marrow has been destroyed by X-rays. *DHFR*, Dihydrofolate reductase. (Principle from French-Anderson [15])

fer experiments in animals have been successful. A functioning gene for neomycin resistance was introduced into hematopoietic cells of adult mice [48], and the human gene for the enzyme hypoxanthine-guanine phosphoribosyltransferase (HPRT; see Sect. 7.2.2.6.) was transferred and brought to expression in a HPRT-deficient cell line [29].

Safety. Before any attempts in humans can be undertaken, very strict requirements regarding safety need to be met. Human oncogenes (Sect. 10.4.2) are in part structure-homologous with retroviruses. There must be safeguards against the production of malignancies by infection with such a virus that might have been modified, for example, by recombination. Risks are also imaginable for other delivery systems. Experiments at several levels – human bone marrow in vitro, mice and monkeys in vivo – will help to eliminate such risks as far as possible. In very severe and so far untreatable diseases, the requirements for patient safety are less stringent than those in milder diseases or in diseases for which other therapies are available. The situation thus differs little from that encountered in other, more traditional fields of medical therapy [15].

19.2 Human Gene Therapy
[2, 3, 10, 16, 18a, 34a]

Advancing knowledge of molecular genetics has led to the development of techniques allowing the transfer of genes into mammalian cells. Since such approaches theoretically permit the correction of genetic defects, a great deal of experimental work is being done, and an increasing number of human studies are underway. By the end of 1995 over 100 different protocols for human gene therapy were under active study, and over 400 patients had been "treated" [34a]. Since most of these studies investigate the feasibility and safety of this approach, few definitive results of successful treatments in humans have been reported. While the public media have been responsible for exaggerated hopes, the field does have a solid scientific basis and offers much promise for the future.

Somatic Gene Therapy. Two types of gene therapy must be carefully distinguished: somatic and germinal gene therapy. Somatic gene therapy refers to gene transfer into somatic cells, with the attempt to cure or partially improve defective function caused by a mutationally altered gene. Gene therapy is a sophisticated approach to medical treatment that has similarities to organ transplantation. Instead of giving the patient a healthy organ, a small segment of normal DNA is the therapeutic agent. As with all novel therapies, rigorous safeguards are necessary to protect patients against possible deleterious effects of gene therapy (see above). The aim of gene therapies does not differ conceptually from that of other medical therapies and therefore presents no new ethical problems [32]. Gene therapy is directed at somatic tissues and affects only the patient who is being treated. Since germ cell DNA is not affected, successfully treated patients with genetic diseases, although healthy, would continue to transmit mutant genes to their offspring in Mendelian proportions. All current *human* gene therapy involves somatic gene manipulation.

Germinal Gene Therapy [49]. The aim of germinal gene therapy, in contrast, is to modify mutant genes in the germ line by DNA modification of sperm or eggs or their precursors before fertilization or in early embryonic development. If successful, somatic tissues of the treated organism would also be normalized. Germ line therapy by replacing the germ line mutation with a normal gene would prevent transmission of genetic disease to future generations, thereby reducing the frequency of the disease. Although germ line modification has been successful in animals, most observers consider germ line manipulation to be inappropriate in humans since the procedure cannot be deemed safe with current methods. Even if safety could be assured, however, germinal gene therapy is widely condemned on ethical grounds. There is worry that the technique might be used for enhancement, such as altering stature or skin color, influencing intelligence, etc., rather than for treatment of genetic disease. The introduction of germ line manipulation would be a revolutionary step for the human species because it alters genes directly and permanently, thereby affecting future generations and human evolution. Much more needs to be known about human genetics before such a step can be undertaken. Most importantly, before ever considering germ line manipulations, human society must determine how and by whom decisions regarding germinal gene therapy are to be made.

Medical indications for germ line therapy are rare; they include couples who want a normal child when both partners are affected by an identical recessive disease and *all* their offspring would be similarly affected. Much more frequently in recessive conditions both parents are heterozygotes, and only 25% of their offspring inherit the homozygous disease. Such couples can have normal children by prenatal diagnosis followed by selective abortion if the fetus is abnormal. For those who do not accept abortion, preimplantation diagnosis of zygotes following in vitro fertilization of several eggs is increasingly possible. The relevant DNA can be obtained from a few cells of the fertilized zygotes and tested for the mutant gene by PCR techniques. A homozygous affected zygote under such circumstances would not be reimplanted into the uterus. Although already successful, the procedure is much more complex than prenatal diagnosis and is unlikely to be widely available.

Some diseases affecting the brain cannot be treated by somatic gene therapy after birth because the Central Nervous System is already damaged. These include Lesch-Nyhan syndrome, Tay-Sachs disease, and metachromatic leukodystrophy. Germ line therapy for prevention of such disorders has therefore been proposed as a possible future approach that in the opinion of sound observers should not be categorically rejected [49].

Approaches to gene therapy. [6, 34, 34a] Successful gene therapy requires the safe transfer of a gene into the correct target cell, incorporation of the DNA into the cell nucleus, and expression of the gene's product for a sufficiently long time to be effective. The procedure must be safe for the recipient. In some diseases the extent of gene expression would need to be regulated by cotransfer of the gene's normal regulatory signals. This requirement is not necessary for many diseases, such as hemophilia and enzyme deficiencies, where increased production of a missing gene product even in small amounts is sufficient to treat the disease successfully [6] (Sect. 7.2.2.9).

The initial and obvious candidates for gene therapy are monogenic diseases with absent or defective function of a gene product, such as the various enzyme deficiencies. Before applying gene therapy to human diseases, extensive in vitro and in vivo animal experimentation is required. Many gene transfer protocols remove cells from the body, carry out the required manipulation, and return the modified cells into the organism in which they and their cellular descendants are expected to function (ex vivo strategies). Methods that transfer appropriately treated DNA directly into the body (in vivo strategies) are logistically simpler and are being increasingly applied. These approaches requires only a single treatment if a normal gene can be inserted into a stem cell followed by normal functioning of the descendant cells.

Vectors. [1–3, 34] Viruses that carry DNA into cells have been most often used as the vectors for gene transfers; retroviruses have been most frequently employed. They transduce dividing cells only but stably integrate into the host cell genome. These viruses are "disarmed" by removing the wild-type structural genes which are replaced by the therapeutic gene. Such engineered retroviral products are noninfectious. Since they provide a relatively low titer of viral particles, they are poor candidates for direct or in vivo gene transfers. Viral integration into chromosomes is random and theoretically could produce malignancies through insertional mutagenesis of oncogenic or tumor suppressor genes.

Adenoviruses [20] may be more useful for in vivo gene therapy since they infect a large number of nondividing postmitotic cells. Their effect, however, is transient and requires multiple administrations of the vector. Delivery of adenovirus to bronchial and liver cells appears feasible. Other types of viruses such as adeno-associated viruses, herpes simplex, and vaccinia viruses are also under study.

A variety of nonviral systems have also been proposed for gene delivery. These include liposomes, DNA protein conjugates, and DNA protein-defective virus conjugates. Effective delivery of genes into cells remains a major problem with all techniques, and methods which do not use viruses would be ideal to avoid potential problems of viral toxicity. An optimal method might be one based on site-specific homologous recombination where the mutant DNA region is replaced by the normal DNA sequence.

Although somatic gene therapy entails no special ethical problems, its emergence was considered so novel and different, that a special committee – the Recombinant DNA Committee – was established in the United States in the late 1980s. The approval of this committee is required for all human gene therapy under government sponsorship, such as by the National Institutes of Health. In practice almost all human gene therapy studies come before this committee, which scrutinizes proposals for safety to patients and the environment since potentially harmful viruses are frequently used. The underlying science is assessed to ensure that meaningful data are obtained. Problems of informed consent and strict adherence to the principles of ethical human experimentation are carefully considered. Similar committees are active or are being established in other countries. With the increasing number of human gene therapy protocols, only proposals that raise new scientific, ethical, or safety problems are now being considered by this committee in the United States. The Food and Drug Administration, however, requires that all gene therapy proposals come before it for permission to use genes and gene vectors.

Indications for Gene Therapy: Monogenic Diseases. Many genetic diseases with known genetic defects are candidates for gene therapy. The molecular genetics of the hemoglobinopathies was understood before that of other diseases. Gene therapy for these diseases was therefore considered some time ago, but has not yet been able to be carried out because of the need for hemoglobin synthesis to be tightly regulated. Nevertheless, the hemoglobinopathies, which affect many thousands of patients all over the world, remain important targets for gene therapy in the future.

The very rare adenosine deaminase (ADA) deficiency – an autosomal-recessive trait (Sect. 7.2.2.6) – was first selected for human trials (September 1990) [7a]. ADA deficiency is associated with a severe immune defect that leads to childhood death from overwhelming infections. Initially, gene-corrected T lymphocytes were administered periodically. Later, when enzyme therapy for this condition became available, ADA enzyme in polyethylene glycol (PEG) was administered simultaneously to these children, who remain without infections and are developing normally. In interpreting the role of gene therapy it is difficult to disentangle the importance of the exogenous ADA enzyme from that of the genetically engineered cells [7a]. Gene-treated stem cells have been administered to two other newborns with ADA deficiency who also require simultaneous enzyme therapy [2].

Five patients with homozygous familial hypercholesterolemia (see Sect. 7.6.4) have been treated with ex vivo gene therapy. A large portion (20%–35%) of the liver was removed surgically, normal low-density lipoprotein (LDL) receptor genes were transferred into liver cell suspensions, were reinfused into the patient's liver via the portal vein. The results have not been impressive [18a].

Many trials are being carried out in cystic fibrosis [14, 21, 41] (Sect. 3.1.3.9). The normal membrane channel gene that is defective in this disease is transferred into the respiratory tract (i.e., bronchi) by direct installation using adenovirus as the vector. If this yet unsuccessful type of therapy becomes feasible, periodic treatments would be required.

Based on favorable in vitro and animal experiments, enzyme therapy is soon to be initiated for Gaucher disease, Fanconi anemia, and α-antitrypsin deficiency (Sect. 6.2.4). Table 19.1 lists other conditions that are current targets for somatic gene therapy.

Indications for Gene Therapy: Other Diseases. The use of gene therapy has been broadened to include a variety of other diseases. In fact most human gene trials (about 60% in the United States) are currently not carried out on genetic diseases but on various types of cancer [2, 10]. As with gene therapy for genetic diseases, human experimentation for cancer gene therapy is in its early stages, and no successful results have yet been reported. The rationale for tumor cell suppression by gene therapy is wide ranging. Some strategies attempt to block oncogene expression by antisense techniques, while others insert a normal tumor-suppressor gene (p53) into cancer cells with defective p53 function – a common abnormality of human cancers (Chap. 10). Another approach introduces genes that encode cytokines such as various interleukins into immune cells to enhance immune function for tumor suppression. A widely used experimental method places the thymidine kinase gene of the herpes simplex virus into cancer cells which makes such cells sensitive to the antiherpes drug ganciclovir. Another technique protects stem cells from the toxic affects of chemotherapy by introducing genes for multiple drug resistance. Other methods block the mechanisms by which new tumor cells evade immunological destruction or insert toxic genes to destroy tumor cells. Many problems must be overcome before any of these approaches can be used successfully in everyday medical practice.

Gene therapy techniques are also being investigated for application to AIDS therapy [12]. The methods involve interference with HIV replication, induction of immune responses, and use of ribozymes as a RNA species with antisense sequences. Gene therapy is also being studied for the treatment of peripheral vascular disease by transferring angiogenic genes into occluded arteries to stimulate collateral circulation of blood vessels. Other approaches use modified growth factor genes to release anti-arteriosclerotic cytokines to prevent restenosis following angioplasty of occluded coronary arteries. Other applications, such as for neurological diseases [16] and rheumatoid arthritis are also underway. Critical assessment of the status of gene therapy in 1995 was carried out for the National Institutes of Health (United) States [34 a]. While the promise of gene therapy is very great, the current inadequacies of vectors and the difficulties

Table 19.1. Human recessive diseases as candidates for gene replacement therapy (after Beaudet et al. 1995 [6])

Disorder	Alternative treatment	Disease frequency	Requirement for tissue specificity
Adenosine deaminase and nucleoside phosphorylase deficiencies	Transplant, enzyme replacement: fair to good	Very rare	Bone marrow
Gaucher disease	Enzyme therapy: good	1 in 3000 (Ashkenazi Jews)	Liver
Cystic fibrosis	Supportive: fair/poor	1 in 2500 (whites)	Lung
Familial hypercholesterolemia Homozygotes Heterozygotes	Liver and heart · transplants: heroic drugs: good	1 in 1 000 000 1 in 500	Liver Liver
Hemophilia A and B[a]	Replacement: excellent	1 in 10 000 males	? Any organ
Hemoglobinopathies	Transfusion: fair to poor Transplants: good	1 in 600 in ethnic groups	Erythroid bone marrow
Leukocyte adhesion deficiency	Transplant: fair to poor	Very rare	Bone marrow
Urea cycle disorders	Diet, drugs: poor to good	1 in 30 000 (all types)	Liver
Phenylketonuria	Diet: good	1 in 12 000	Liver
α_1-Antitrypsin deficiency	Enzyme therapy: fair	1 in 3500	Liver
Glycogen storage disease Ia	Diet, drug: fair	1 in 100 000	Liver
Duchenne muscular dystrophy[a]	Poor	1 in 3000 males	Muscle
Lysosomal storage diseases	Poor	1 in 1500 for all types	Brain for many
Lesch-Nyhan syndrome[a]	Poor	Rare	Brain

[a] X linked

with gene expression have limited the practical applications of gene therapy.

Impact. These developments indicate how basic scientific concepts initially aiming at the treatment of relatively rare genetic diseases have now penetrated into most fields of medicine. While gene therapy has attracted considerable attention and carries great expectations, much additional work in the experimental laboratory and clinic such as the development of better vectors and improved gene expression will be required before benefits are realized. The biotechnology and pharmaceutical industries are beginning to invest large resources into this field [11]. Industry pays particular attention to common diseases such as cancers and others which command a large market and promise high profits. This illustrates the way in which market forces determine the direction of medical research; if successful, however, many patients affected with common diseases are likely to benefit.

This raises a more general problem of modern medicine. At first glance, somatic gene therapy appears a promising concept with potential significance for a widening field of applications. However, many of the diseases that might be treated in this way are monogenic – and most are rare. On the other hand, several hundred of such diseases are known, and with the progress in human molecular genetics new ones are being discovered constantly. Moreover, even in common diseases of great social impact, rare subtypes are increasingly being delineated, and their molecular mechanisms are analyzed (see Sect. 6.4). This is making medical diagnosis much more difficult than it used to be. Moreover, somatic gene therapy together with other therapeutic approaches (Sect. 7.2.2.9) will increasingly offer therapeutic chances that depend on specific diagnoses. Here, data banks and expert systems will help (see Appendix 3). However, such sophisticated – and therefore expensive – approaches are possible only in affluent societies such as those of western Europe, the United States, and Japan. And even here it is doubtful how much of their social product societies will be willing to spend for treatment of rare diseases. Health problems of populations in the rest of the world, such as in South America, southern and eastern Asia, and especially Africa, are quite different and much more urgent. HIV infectious there are becoming widespread, malaria is coming back due to drug resistance of *Plasmodia,* and deteriorating social conditions open the way for other – possibly novel – infections. In the face of such problems, how long will we be able to develop our highly sophisticated medicine, of which somatic gene therapy is likely to be a part?

Public Reaction to New Achievements and Prospects of Molecular Biology. The achievements and, especially, the prospects of molecular biology have aroused strong reactions in the general public, especially among opinion leaders (theologians, philosophers, journalists). Biologists and medical geneticists have often been shocked by fierce attacks; they often feel that their benevolent intentions are sadly misinterpreted, and that an image has been created of the ambitious and ruthless scientist who will soon start manipulating human populations for sinister purposes unless stopped by an alert public. In this connection, however, we should not forget that scientists were among the first to "ring the alarm" when the possibilities for experimental recombination of DNA by restriction endonucleases became apparent. Still more, these warnings were unduly alarming and could have been avoided had the protagonists acted in a more circumspect way.

The basic scientists who were involved in early research with recombinant DNA became concerned about hypothetical dangers of the new scientific developments that allowed splitting of genes at random, joining genes from different organisms across species, and using the ubiquitous *E. coli* organism for gene transfer. They called a conference to discuss potential dangers of the spread of uncontrollable infections and cancer using these new methods [7]. No microbiologists or epidemiologists experienced in human infectious disease or in human cancer were consulted or invited to the widely publicized conference. The public became greatly alarmed, and laws were rapidly passed in the United States to regulate work with the new techniques. It soon became clear, however, that the envisaged dangers were largely hypothetical. The organisms used were so enfeebled that the kind of epidemic infections feared could never be established. It was also soon realized that cross-species DNA transfers have occurred in nature for millenia. Using the newer experimental laboratory data and the experience of a century in clinical microbiology and epidemiology, many informed and knowledgeable observers regarded the early fears as overblown. The initial turmoil regarding "recombinant DNA research" is an excellent example of failure of communication between molecular biologists who knew little about the realities of cancer and infectious disease and the relevant medical and biological scientific community. Nevertheless, a few eminent scientists continue to be worried about possible dangers of cross-species DNA transfers.

Further Speculations on Gene Manipulation. Starting from the various results and prospects, a few biologists have speculated on much more ambitious goals for gene manipulation: in their opinion, human beings with novel capacities should be created. If, for example, replacing the skin of the head or back by tissue containing chlorophyll would give the individual the capacity for photosynthesis – this could be a possible partial solution to the problem of food shortage in an increasingly overpopulated world.

Cloning of frogs has been achieved by introducing an intestinal frog cell nucleus into an enucleated frog egg. The genetic information of the nucleus was able under appropriate conditions to specify the normal development of a frog. Some scientists and the public have been fascinated by the possibility of cloning a human being by analogous principles. Research work toward this end, however, is not being carried out to our knowledge and is strongly discouraged by the scientific community. A science writer in 1978 claimed in a book that the cloning of a human being had already been achieved [40]. No proof was given, and the media were full of stories about the implications. It has been phantasized that the cloning of human beings would make it possible to duplicate outstanding and creative human beings such as Einstein or Mozart. However, it is obvious that the genetic material of an Einstein alone would not guarantee another Einstein. Others have suggested that dictatorships could clone groups of military scientists or brutal soldiers in the service of the state. If such a feat were ever to become possible, it is unlikely that a country would embark on such an undertaking. Since clones would take a generation to reach their full potential, politicians and statesmen would be more likely to look for more rapid ways of ensuring political and military success.

Other scenarios have implied the creation of subhuman creatures following fusion of cells with chromosomes from human and subhuman primates and subsequent insertion of the hybrid nucleus into an enucleated egg. Such humanoids have been envisaged to carry out dull and repetitive tasks of no interest to normal humans. Again, although scientists are far from being able to carry out such schemes, these scenarios have been strongly condemned.

Along more "conventional" lines Lederberg [23] proposed manipulation of the Central Nervous System by as yet nonexistent chemical or growth factors to improve the efficiency of the human brain. Such schemes of course would not change the genetic material – a process that was termed "euphenics," in contrast to eugenics.

Brainstorming of this sort and discussion of "far-out" scenarios [19] are useful for recognizing novel possibilities for research. Perhaps more importantly they alert us as to potential abuses of science. It is unfortunate and dangerous for the public's understanding of science that the media often leave the impression that these novel reproductive schemes have been seriously planned or are already being carried out by scientists [22].

The Need for a Dialogue on Ethical Issues. Many results of molecular biology are already being applied in various branches of medical genetics, from cytogenetic diagnosis to genetic counseling. Some of the more pretentious goals of some visionaries have been mentioned above, and their list could be extended; it includes the creation of human beings according to genetic specifications, and increasing life span. Compared with our present-day knowledge of the molecular biology of higher organisms, and our ignorance of the genetics of much of the normal variation in humans, many of these proposals are somewhat analogous to the idea that a boy who has just been given his first electronic set for Christmas, could successfully improve on the latest generation of computers. One could also argue that even if the technical conditions for realizing a few of these phantasies do become available, their practical application will probably remain impossible for sociological reasons.

However, the possibility of misuse on a much smaller scale cannot be entirely dismissed. We should therefore be relieved by the fact that these problems are being widely discussed, even if much of the discussion is uninformed regarding the technical facts and concepts. As scientists we should enter the public dialogue wherever we regard this as meaningful and should attempt to enhance its standard of scientific and ethical sophistication. We can learn from the questions that are being asked and become more critical toward our own goals. A continuing public dialogue on ethical issues is necessary. Both authors of this book have been involved repeatedly in such dialogues [4, 32, 37].

19.3 Biological Future of Mankind

Human Evolution Is Not Finished [46]. Evolution of the human species is not confined to the past. The mechanisms that bring about changes in gene frequencies from one generation to the next are still in operation. Knowledge of these mechanisms should help to predict future trends in the genetic composition of human populations. To understand such predictions and to make proper use of them we should keep in mind the limitations that are inherent in all such attempts.

a) We can extrapolate only from trends that are already visible at present; human history, however, has often been shaped by unpredicted events. This is also true for the biological history of our species, which is now inextricably intertwined with cultural, social, and political history as well as with the future development of human biology and medicine with the potential to actively influence the future evolution of mankind.

b) It is well-documented that the visible signs of revolutionary future scientific developments, which will be easily recognized in retrospect, are often not clear to contemporary scientists. It would be presumptuous of the present authors to claim an exception. Future colleagues may cite this book as an example of how important trends have been overlooked.

c) All predictions made here assume that modern civilization will continue to exist, and that modern society's concern for health and medicine will maintain its prominent position. In this context we assume that preventive medicine will continue to gain in importance.

If we assume that attitudes toward these problems will be increasingly rational, there is little doubt that concepts of human and medical genetics will find ever-growing application. One could argue that this assumption is unrealistically optimistic, as it presupposes that human societies will learn to keep their social and technological structures from destruction. It is by no means impossible that a large proportion of the human species will sooner or later be doomed to destruction by an atomic holocaust. We are also intensely aware that many of the concerns expressed here have a much lower priority in developing countries where the problems of overpopulation, undernutrition, and infectious disease are much more immediate problems needing solution [46a].

The following extrapolations are thus subject to the conditions of human survival and continuance of social progress.

Major Forces Determining Evolution. The principal mechanisms that determine evolution are discussed in Chaps. 12 and 13; their effect on the development of present-day human populations is covered in Chap. 14. What can be predicted for the future?

Genetic Drift. Chance fluctuations of gene frequencies may lead to appreciable genetic differences between subpopulations, provided that these populations are isolated from each other, thus keeping gene flow between them very low. This effect becomes stronger with smaller size of a breeding population and even leads to lessened variability due to random loss of alleles at a predictable proportion of loci (Chaps. 12, 13, 14).

In present human populations we are observing a strong overall tendency toward breaking down of isolates and increasing intermarriage between different populations. There is no reason to assume that this trend will reverse in the foreseeable future, and that new isolates of small size will form. Therefore chance fluctuations, in contrast to their indisputable importance for human evolution in the past, will become less significant in the future. If this trend in the human breeding structure persists, new human species will not develop since speciation always requires reproductive isolation of a population subgroup (Sect. 14.2). The creation of a humanlike new species is sometimes discussed by science fiction writers. Currently existing genetic variability coupled with selective breeding of humans could not lead to such a new species. One would have to create novel genetic combinations by presently nonexistent techniques and enforce selective breeding. These possibilities are therefore extremely remote. For future shifts of gene frequencies within the species, two factors causing systematic changes in gene and genotype frequencies remain: mutation and selection.

Mutation. Our current knowledge regarding spontaneous and induced mutations is discussed in Chaps. 9; 11, and the problem of a genetic load due to mutations is reviewed in Sect. 13.1.2. We can safely assume that practically all chromosome aberrations and many gene mutations are unfavorable for both the individual and the population. Most chromosome aberrations kill the zygote during embryonic development; a minority survive until birth or still longer, but affected patients suffer from severe malformations. Gene mutations often lead to inherited diseases with simple modes of inheritance or to defects in multifactorial genetic systems. However, a large proportion of point mutations leads to changes in amino acid sequences of proteins that cause no apparent functional deficiency, as indicated by many hemoglobin variants (Sect. 7.3). When considered in terms of *all* mutations, the proportion that are advantageous constitute at best a small minority.

The combined evidence justifies the conclusion that an enhanced overall mutation rate would be unfavorable. The extent of the impact of various mutations on health is discussed in Chap. 11 and suggests surprising differences depending on mutational mechanisms and phenotypic manifestation.

Trends in Spontaneous Mutation Rates: Chromosome Mutations. The rates of numerical chromosome mutations increase with the age of the mother. Therefore changes in maternal age lead to a corresponding alteration in the overall incidence of such chromosome mutations. In many modern populations there is a trend toward a decrease in the number of children per family and a concentration of childbirths within the age groups with the lowest risk: women in their

20s. It has been calculated that in Western countries and Japan this trend should reduce the number of children with Down syndrome by about 25%–40% [26, 39, 45]. Will this trend continue? Some more recent figures show that the tendency of many modern women to postpone childbearing to a somewhat higher age could easily lead to a reversion of this trend [9].

On the other hand, prenatal diagnosis (Sect. 18.1) is most efficient for the early recognition of chromosome abnormalities. Many countries offer this diagnostic possibility to all women over the age of 35. If all pregnant women aged over 35 would use this procedure, one could expect a significant reduction in Down syndrome, depending on the age distribution of mothers [30].

Gene Mutations. The mutation rate for many gene mutations increases with the age of the father (Sect. 9.3.3). Therefore any trend in the age distribution of fathers influences the mutation rate accordingly [31]. The paternal age effect on dominant and X-linked mutation rates is smaller than the effect of maternal age on incidence of numerical chromosomal anomalies. The total medical impact of the paternal age effect is probably low, and the actual absolute risk of an older father to having a child affected with a dominant mutation is quite small although the relative risk compared with that among younger fathers is substantial (threefold for fathers aged about 40) (See Sect. 9.3.3).

Ionizing Radiation and Chemical Mutagens. It is shown in Sect. 11.1.5 that any conceivable increase in radiation exposure probably enhances the mutation rate by a small percentage. In view of the fluctuation of the "spontaneous" mutation rate, due, for example, to changing age distributions of parents, any increase due to radiation is recognizable only with refined epidemiological techniques. Still, the effect does exist. Therefore one of the major goals of preventive medicine in the future will be to keep irradiation as low as possible. At present the main source of radiation exposure is diagnostic medicine; here the improvement in technology can have a substantial impact. As for occupational and general background irradiation, it can only be hoped that technology for producing the world's long-term energy demands by sources other than nuclear energy will be improved in the future so that nuclear energy can be deemphasized in the long run.

Too little is known regarding the exposure of our population to chemical mutagens (Sect. 11.2) to venture any predictions. However, we shall probably have to live with a certain number of chemically induced mutations, as human society will probably not be prepared to forego the immediate advantages of some chemicals in exchange for long-term avoidance of small and undefined genetic damage. Current, often excessive reactions to possible health dangers of chemicals can, we hope, be replaced by more appropriate responses once we have learned more about chemical mutagenesis and carcinogenesis.

In conclusion, a certain, presumably small increase of the natural mutation rate will have to be faced. This increase will lead to a corresponding increase in numeric and structural chromosome aberrations and dominant or X-linked hereditary disease (Sect. 11.2). Whether this increase will be offset by a decrease due to a shifting parental age distribution cannot be predicted. Since cancer is often caused by somatic mutations (possibly sometimes induced by environmental agents Chaps. 10; 11), increases in neoplastic diseases are possible. The decrease in the frequency of gastric cancer for unknown causes and the anticipated decline in lung cancer with lower levels of cigarette smoking indicate that overall beneficial trends may compensate for certain unfavorable trends.

Selection: Dominant and X-Linked Diseases. It is a widespread contention that natural selection has relaxed due to modern medicine. This statement, however, is only partially true. No therapy has succeeded, for example, in preventing miscarriages caused by chromosome aberrations. Patients with Down or Klinefelter syndrome still do not reproduce. Natural selection has not changed for these conditions.

Selection did relax for some pathological traits with autosomal-dominant or X-linked recessive modes of inheritance. These are genetic diseases that have been maintained so far by genetic equilibrium between mutation and selection. An example is hemophilia A, for which substitution therapy with factor VIII now enables patients to lead an almost normal life. Life expectancy and the chance to have children is vastly improved. For such conditions an appreciable increase in incidence within only a few generations can be predicted quantitatively when the present frequency, the selection coefficients before and after selection relaxation, and the mode of inheritance are known; for retinoblastoma [44], such a calculation is presented in Sect. 12.2.1.2 (Fig. 12.10).

There are many other dominant and X-linked conditions, however, for which no satisfactory therapy is available, and natural selection still acts with full strength. Examples include neurofibromatosis, tuberous sclerosis, and Duchenne type muscular dystrophy.

The future will undoubtedly bring therapeutic progress for these diseases as well, which will lead to selection relaxation. On the other hand, the entire group of dominant diseases is an easy "target" for genetic counseling, and X-linked conditions can often be avoided by prenatal diagnosis. We therefore have reason to hope that a growing proportion of patients affected with such disorders will voluntarily refrain from transmitting such genes to the succeeding generation. Under such circumstances, artificial selection will replace natural selection, and keep population incidence of a given disease close to the mutation rate.

Figure 19.4 illustrates this trend in a more anecdotal way: a man in the grandparents' generation who was blind due to congenital cataract had seven living children, four of whom were affected and blind. All four were married to spouses who were blind for other reasons, demonstrating assortative mating. It is remarkable, however, that three of these couples voluntarily – and without genetic counseling – refrained from having children. The fourth couple had only one son, who asked for genetic counseling even before marriage. The mutation that had multiplied no less than fourfold at a time when contraceptive measures were not yet in general use was apparently wiped out within two more generations by the voluntary decision of its carriers. Every medical geneticist with experience in genetic counseling knows of such examples.

Natural Selection: Recessive Diseases. The most striking success in therapy of genetic diseases has been achieved in recessive enzyme defects (Sect. 7.2.2.7). Treatment allows individuals affected by some of these diseases to grow up healthy and able to have children. Moreover, it can be predicted that if gene therapy is successful in the foreseeable future, it will be so for some of these enzyme defects. However, reproduction of abnormal homozygotes leads only to a very slow increase in gene frequency (Sect. 12.2.1.2), which gives little cause for concern.

Still, an increase in the genes for recessive diseases must be anticipated for another reason: at present most populations are not in equilibrium for recessive genes. A breakdown of isolates and the steep decrease in the number of consanguineous marriages have created a situation in which the overall number of homozygotes is far below the expected equilibrium value. In the absence of other factors such as changes in known selective advantages of heterozygotes this trend should lead to a slow increase in homozygotes over hundreds of future generations (Sect. 12.3.1.2). Since 100 generations correspond to about 2500–3000 years, such an increase need not worry us now. Living conditions are likely to change in an entirely unpredictable manner over such a long time period.

In conclusion, we expect a very slow increase in the incidence of recessive diseases above their present frequency in randomly mating populations of the industrialized countries of Europe and the United States. However, this prediction does not take into account artificial selection by genetic counseling, prenatal diagnosis, and possibly screening programs for heterozygotes. If prenatal diagnosis becomes a more routine procedure, the most frequent recessive diseases, for example, the β thalassemias in Mediterranean and southeastern Asian populations and cystic fibro-

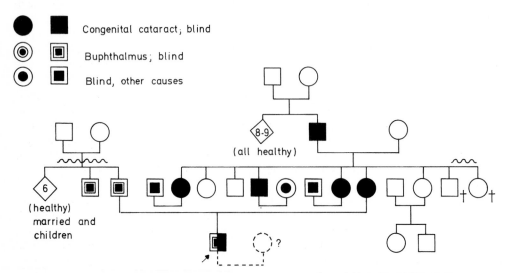

Fig. 19.4. Pedigree with congenital cataract, demonstrating assortative mating for blindness of different causes, and voluntary restriction of reproduction in these marriages, The grandfather of the family was blind due to congenital cataract; nevertheless he had nine children; four were also blind. All are married to blind persons (assortative mating). Only one of these couples had one single son, who being blind, asked for genetic counseling (Observation F. V.)

sis among northwestern European and North American whites, are likely to be included. The rarer defects are unlikely to be screened.

Considering together all the evidence for genetically well-defined abnormalities, such as chromosome aberrations, dominant, X-linked diseases, and autosomal-recessive diseases together, we can predict fairly safely that modern civilization will *not* bring about a large increase in the incidence of these abnormalities. Whether there is a small and very gradual increase, whether the overall incidence stays more or less at its present rate, or whether even a decrease occurs depends largely on many developments in our society. Can we expect success in keeping population exposure to mutagenic agents within reasonable limits, as explained above? How effectively will artificial selection through genetic counseling and prenatal diagnosis reduce genetic damage? And, to mention another, increasingly important point, how long will society be prepared to pay for the relative "luxury" of working out schemes for preventing and treating rare diseases? We cannot make any accurate predictions at this time.

Gradual Loss of Functions That Are Now Being Maintained by Multifactorial Genetic Systems. Apart from the well-defined genetic defects discussed above, there is also genetic variability within the population for functional systems that depend on a complex but ordered interplay of various genes during embryonic development. Examples include the heart, eyes, and immune system. During evolution these systems have developed under constant and intensive selection pressure. As soon as this selection pressure abates, mutations accumulate that lead to small functional infirmities, and over very long evolutionary periods these systems slowly but inevitably "fall apart." Among animals the best known examples are found in species that have lived for many generations in the absolute darkness of caves or at great oceanic depths where an intact visual system brings no selective advantage. Typically, the variability of the eyes first increases, individuals with small defects becoming more frequent. Later most animals have more or less severely defective eyes, and finally an eye-less species results. In a civilized society, minor defects of the visual system fail to carry any selective disadvantage. This trend has already led to a considerable increase in variability in visual acuity. Conditions such as myopia, hyperopia, and astigmatism are probably more frequent than in primitive populations that lived until more recent times under conditions of stronger selection (Sect. 14.2.5). Will this process continue? One could argue that such a trend would be bearable – more persons would simply have to use glasses. Their "defect" provides a living for other professions, such as ophthalmologists, optometrists, and opticians, and supports the spectacle-making industry.

A slow deterioration in the immune system would be more dangerous. As explained in Sect. 12.2.1.6, mortality during infancy and childhood was high until very recent times, with infections being the principal contributory factor. Under the influence of strong selection there has evolved a complex system for recognizing and eliminating infective agents (Sect. 7.4). We know a great number of genetic defects which impair the efficiency of this system. Formerly such defects usually led to the individual's death from infection; today, with antibiotic therapy, many of these patients survive and have children. For extreme defects we can hope for therapies such as bone marrow transplantation, which substitute normally functioning immune cells. However, mutations may change protein molecules in a much more subtle way, only slightly impairing their function. Studies on the hemoglobin molecule have shown that such mutations are probably very frequent. Will relaxation of selection against such variants lead to a slow deterioration of the whole system? Will our descendants gradually become more susceptible to infections of all kinds that need to be counteracted by a more elaborate combination of antibiotics and immune therapy? Possibly still more important, will this deterioration of the immune defense system enhance the frequency of various types of cancer, since potentially malignant cell clones may often be kept from proliferating by so-called immune surveillance? Furthermore, recent experiences especially with HIV infection but also with other "new" infections indicate that the challenge to the human species by infections will continue.

It is also possible that the relaxation of selection against certain multifactorial traits will lead to their higher frequency. Trends in the past, for instance, made for fairly strong selection against cleft lip and palate. Nowadays, because of effective surgical therapy such children survive more readily and have children. The frequency of this trait in the next few generations is therefore likely to increase. The exact extent of such increase is difficult to predict. Similar considerations apply to congenital heart defects.

Increase in the Number of Intellectual Subnormals? As noted above, conscious planning of reproduction is one of the main factors to be considered in every forecast of trends. This factor, however, may have favorable as well as unfavorable consequences. There is one subgroup of the human population that have a reduced ability to plan for their own future – the mildly mentally subnormal. Indeed their mean reproduction exceeds the population average. While environmental factors that may help to bring about mental subnormality are well known; there is evidence to suggest that genetic variability is important especially for the mildly retarded group (Sect. 15.2.1.2). Therefore, a continuing higher reproduction of mildly retarded persons might enhance their population frequency. This conclusion has been challenged by Penrose [35] with the argument that in families with mildly subnormal individuals a certain proportion of the more severely handicapped will segregate who are not able to have children at all. This counterbalancing effect might keep the gene frequency low even in the face of higher reproduction of the married subnormals. Comprehensive studies of Reed and Reed (1965) [38] seem to support this argument: the number of unmarried and childless retarded family members was indeed increased in families of the mentally subnormal. We cannot be sure, however, whether this compensation will prove sufficient in the future; too much depends on unpredictable social conditions: mentally subnormal women can be taught to take birth control pills regularly and may even do so more reliably than women of average or above-average intelligence.

It is possible that genes for mental subnormality will increase; although the number of mentally subnormal individuals may be higher, the mean ability of the rest of the population will not be affected. There is strong assortative mating for intelligence in general and especially for mental subnormality. This trend may create social problems, as modern societies with increasing demands for technical skills will provide progressively fewer jobs for mildly retarded persons.

A Favorable Selective Trend: Abandoning of Genetic Adaptations with Otherwise Unfavorable Effects. Our discussion has focused principally upon unfavorable selective trends, the only benefical trend being artificial selection by genetic counseling. Another favorable trend, discussed below, will probably lead to genetic improvement much more rapidly than most of the unfavorable trends can lead to genetic deterioration.

Some anomalies and diseases of the erythrocytes, most notably hemoglobin variants, attained a high incidence in tropical countries despite the fact that homozygotes suffer from severe blood diseases such as sickle cell anemia or thalassemia (Sect. 7.3). However, heterozygotes have a selective advantage vis-à-vis falciparum malaria, which remains endemic in many of these areas (Sect. 12.2.1.6). Unless malaria reappears in the future, selection against the affected homozygotes will reduce the gene frequencies. Here nature has evolved a compromise: the advantage of increased resistance against malaria came at the price of many cases with inherited disease. As soon as malaria is eradicated, this compromise will no longer be needed and the harmful gene will gradually disappear (Sect. 12.2.1.6).

A similar mechanism may exist for the AB0 blood groups (Sect. 12.2.1.8). It is possible that this genetic polymorphism arose as an adaptation to multiple and varying infective agents and that selection has frequency been dependent. The price has been paid in zygote loss due to serological mother-child incompatibility. If selection due to infections disappeared, and only selection by incompatibility remained, a gradual and slow elimination of the rarer alleles, A and B, and fixation of the most frequent one, 0, would ensue. Similar "compromises" are likely for other genetic systems as well.

The Human Species in the Future. These considerations lead to the following picture of the future genetic composition of the human species: Its overall genetic composition will be similar to that at present. A tendency toward diminution of racial and ethnic differences will probably continue. Genetic defects may be slightly more or less frequent than at present but will be under effective control by genetic counseling and prenatal diagnosis. Living cases with autosomal chromosome aberrations may become rarities. Diseases caused by polygenic factors are likely to increase because of medical and surgical treatments and other cultural factors leading to relaxation of selection. Unfortunately, the extent of this increase cannot be accurately assessed because of our lack of knowledge regarding the specifically genetic contributions to these diseases. Table 19.2 compares the favorable and unfavorable trends.

However, these predictions will hold true only if the general socioeconomic conditions remain stable, their present state extends from the developed countries to the populations in developing countries, and progress of medical sciences is increasingly able to cope with the health challenges [46a]. Two major developments might lead to untoward consequences:

1.) All over the world we are now observing major changes in infective agents of almost all kinds. They are adapting genetically to antibiotics and other chemotherapeutic agents. Return of malarial infections to formerly malaria-free tropical areas, and untreatable pneumonias caused by resistant hospital germs are only two examples. It is possible that a high mortality rate due to some infectious diseases will soon return with full force, requiring new intellectual and genetic adaptations by the human host. We cannot be sure that human ingenuity will, in the long run, win the fight against infections.

2.) It is possible that socioeconomic conditions will change. The living condition of a "golden age" which

Table 19.2. Favorable and unfavorable trends (without genetic engineering) affecting genetic composition of future human populations

Unfavorable trends	Probable significance
Mutation rate increase due to ionizing radiation	Not very significant
Mutation rate increase due to chemical mutagens	Unknown
Higher reproduction of patients with inherited diseases	Probably not very significant
Increase in recessive diseases (new equilibrium)	Insignificant in the next few centuries
Deterioration in normal functions due to "selection relaxation"	May be significant in the long run
Favorable trends	
Elimination of genetic adaptations with otherwise unfavorable effects (infections or malnutrition)	Possibly significant
Decrease of mutation rates for chromosome aberrations and point mutations due to decreasing parental age	Significant
Voluntarily decreased reproduction in families with genetic diseases	Unknown; will become more important in the future
Genetic counseling including antenatal diagnosis	Important, even after a short time

al diseases. Somatic gene therapy is an extension of conventional medical therapy using gene-specific DNA as therapeutic material. Apart from the rigorous safeguards necessary when introducing any new therapy and the special attention given to the safety of the viral vectors that carry DNA, somatic gene therapy raises no novel ethical questions. On the other hand, germinal gene therapy which aims to modify germ cells (or very early zygotes) remains an experimental procedure of animals and is not safe for human applications at this time. Its potential indications in the prevention and treatment of human diseases are restricted and raise the ethical problems of directly altering the gene pool of future generations. There is fear that the procedure may be abused to enhance normal body characteristics in offspring, such as IQ, body height, and similar traits. It is prohibited by law in some countries.

Our knowledge of human genetics permits cautious predictions about the genetic health of future generations. Extrapolation from current trends suggest that mutation rates leading to anomalies and diseases may diminish somewhat, due to a younger age at reproduction, because of fewer offspring in families. As to changes in natural selection due to hereditary defects and weaknesses, positive as well as negative trends may occur; we cannot predict which will prevail. However, we see no indication for a marked deterioration of the genetic quality of future generations.

populations of developed countries are enjoying at present may deteriorate due to political, economic, and mainly ecological turmoils, and medicine may have to abolish highly sophisticated goals such as diagnosis, prevention, and treatment in medical genetics, and concentrate on mere survival. Hence, the scenario outlined for the future in the foregoing chapter may be too optimistic. Improvement of living conditions and – together with them – continued progress in the biomedical sciences is not at all self-evident.

Conclusions

Modification of the genetic material by direct manipulation of DNA is often discussed. More conservative approaches include "germinal choice," i.e., the use of germ cells for artificial fertilization and reproduction. Among molecular approaches, gene therapy of somatic cells has elicited much interest and is being investigated in experimental human studies for sever-

References

1. Anderson WF (1992) Human gene therapy. Science 256 : 808
2. Anderson WF (1994) Gene therapy for cancer. Hum Gene Ther 5 : 1–2
3. Anderson WF (1994) End of the year potpourri – 1994. Hum Gene Ther 5 : 1431–1432
4. Andrews LB, Fullarton JE, Holtzman NA, Motulsky AG (eds) (1994) Assessing genetic risks: implications for health and social policy. National Academy, Washington
5. Anonymous (1995) Gene therapy: when, and for what? Lancet 345 : 739–740
6. Beaudet AL et al (1995) Somatic gene therapy. In: Scriver et al (eds) Genetics, biochemistry and molecular basis of variant phenotypes. The metabolic and molecular basis of inherited disease, vol 1. McGraw-Hill, New York, pp 106–107
7. Berg P, Baltimore D, Boyer HW, Cohen SN, Davis RW, Hogness DS, Nathans D, Roblin RO, Watson JD, Weissman S, Zinder ND (1974) Potential biohazards of recombinant DNA molecules. Science 185 : 303
7a. Blaese RM et al (1995) T lymphocyte-directed gene therapy for ADA-SCID: initial trial results after 4 years. Science 270 : 475–485

8. Brown MS, Goldstein JL, Havel RJ, Stenberg D (1994) Gene therapy for cholesterol. Nature [Genet] 7 : 349–350

9. Cornel MC, Breed ASPM, Beekhuis JR et al (1993) Down syndrome: effects of demographic factors and prenatal diagnosis on the future livebirth prevalence. Hum Genet 92 : 163–168

10. Culver KW, Blaese BM (1994) Gene therapy for cancer. Trends Genet 10 : 174–176

11. Dodet B (1993) Commercial prospects for gene therapy – a company survey. Trends Biotechnol 11 : 182–189

12. Dropulic B, Jeang K-T (1994) Gene therapy for human immunodeficiency virus infection: genetic antiviral strategies and targets for intervention. Hum Gene Ther 5 : 927–939

13. Edwards RG, Fowler RE (1970) The genetics of preimplantation human development. Mod Trends Hum Genet 1 : 181–213

14. Engelhardt JF, Yang Y, Stratford-Perricoudet LD et al (1993) Direct gene transfer of human CFTR into human bronchial epithelia of xenografs with E1-deleted adeno-viruses. Nature [Genet] 4 : 27–33

15. French-Anderson W (1984) Prospects for human gene therapy. Science 226 : 401–409

16. Friedman T (1994) Gene therapy for neurological disorders. Trends Genet 10 : 210–214

17. Fuller JL, Thompson WR (1978) Foundations of behavior genetics. Mosby, St Louis

18. Grossman M, Raper SE, Kozarsky K et al (1994) Successful ex vivo gene therapy directed to liver in a patient with familial hypercholesterolemia. Nature [Genet] 6 : 335–341

18 a. Grossman M (1995) A pilot study of ex vivo gene therapy for homozygous familial hypercholesterolaemia. Nature Med 1 : 1148–1154

19. Haldane JBS (1963) Biological possibilities for the human species in the next ten thousand years. In: Wolstenholme G (ed) Man and his future. Churchill, London, pp 337–361

20. Knowles MR et al (1995) A controlled study of adenoviral-mediated gene transfer in the nasal epithelium of patients with cystic fibrosis. N Engl J Med 333 : 823–831

21. Kohn DB et al (1995) Engraftment of gene-modified umbilical cord blood cells in neonates with adenosine deaminase deficiency. Nature Med 1 : 1017–1023

22. Lawless EW (1977) Technology and social shock. Rutgers University Press, New Brunswick

23. Lederberg J (1963) Biological future of man. In: Wolstenholme G (ed) Man and his future. Churchill, London, pp 263–273

24. LeMeur M, Gerlinger P, Benoist C, Mathis D (1985) Correcting an immunoresponse deficiency by creating $E_a$gene transgenic mice. Nature 316 : 38–42

25. Lever AML, Goodfellow P (eds) (1995) Gene therapy. Br Med Bull 51 : 1–242

26. Matsunaga E (1965) Measures affecting population trends and possible genetic consequences. United Nations World Populations Conference, Belgrad

27. Merrill CR, Geier MR, Petricciani JC (1971) Bacterial virus gene expression in human cells. Nature 233 : 398–400

28. Miller AD (1992) Human gene therapy comes of age. Nature 357 : 455–460

29. Miller AD, Eckner RJ, Jolly JD, Friedmann I, Verma IM (1984) Expression of a retrovirus encoding human HPRT in mice. Science 223 : 630–632

30. Modell B, Kuliev A (1989) Impact of public health on human genetics. Clin Genet 36 : 286–298

31. Modell B, Kuliev A (1990) Changing paternal age distribution and the human mutation rate in Europe. Hum Genet 86 : 198–202

32. Motulsky AG (1983) Impact of genetic manipulation on society and medicine. Science 219 : 135–140

33. Muller HJ (1963) Genetic progress by voluntarily conducted germinal choice. In: Wolstenholme G (ed) Man and his future. Churchill, London, pp 247–262

34. Mulligan RD (1993) The basic science of gene therapy. Science 260 : 926

34 a. Orkin SH, Motulsky AG (cochairpersons) (1995) Report and recommendations of the panel to assess the NIH investment in research on gene therapy. National Institutes of Health, Bethesda

34 b. Passarge E (1995) Color atlas of genetics. Thieme, Stuttgart

35. Penrose LS (1962) The biology of mental defect, 3rd edn. Grune and Stratton, New York

36. Plachot M, Mandelbaum J (1984) La fécondation in vitro: 5 ans, bientôt l'age de raison. Ann Genet 27 : 133

37. President's Commission for the Study of Ethical Problems in Medicine and Biomedical and Behavioral Research (1983) Screening and counseling for genetic conditions. The ethical, social, and legal implications of genetic screening, counseling, and education program. US Government Printing Office, Washington

38. Reed EW, Reed SC (1965) Mental retardation: a family study. Saunders, Philadelphia

39. Richards BW (1967) Mongolism: the effect of trend in age at child birth and chromosomal type. J Ment Subnormality 13 : 3

40. Rorvik DM (1978) In his image: the cloning of a man. Lippincott, Philadelphia

41. Rosenfeld MA, Yoshimira K, Trapnell BC et al (1992) In vivo transfer of the human cystic fibrosis transmembrane conductance regulator gene to the airway epithelium. Cell 68 : 143–155

42. Trendelenburg MF (1983) Progress in visualization of eukaryotic gene transcription. Hum Genet 63 : 197–215

43. Vega MA (1991) Prospects for homologous recombination in human gene therapy. Hum Genet 87 : 245–253

44. Vogel F (1957) Die eugenische Beratung beim Retinoblastom (Glioma retinae). Acta Genet 7 : 565–572

45. Vogel F (1967) Wie stark ist die theoretische Häufigkeit von Trisomie-Syndromen durch Verschiebungen im Altersaufbau der Mütter zurückgegangen? Zool Beitr 13 : 451–462

46. Vogel F (1973) Der Fortschritt als Gefahr und Chance für die genetische Beschaffenheit des Menschen. Klin Wochenschr 51 : 575–585

46 a. Vogel F (1995) Gedanken über die Zukunft der Menschheit aus dem Blickwinkel der Humangenetik. Med Gen 7 : 23–29

47. Watson JD, Gilman M, Witkowski J, Zoller M (1992) Recombinant DNA, 2nd edn. W.H. Freeman, New York

48. Williams DA, Lemischka IR, Nathan DG, Mulligan RC (1984) Introduction of new genetic material into pluripotent haemotapretic stem cells of the mouse. Nature 310 : 476–480

49. Wivel NA, Walters L (1993) Germ line modification and disease prevention: some medical and ethical perspectives. Science 262 : 533–538

Appendix 1

Methods for Estimating Gene Frequencies

In the context of the present work only the principles of gene frequency estimation are demonstrated (Sect. 4.2). For details, see Race and Sanger [68], Mourant et al. [60], and others. We begin with the simplest example.

One Gene Pair: Three Genotypes Can Be Identified by the Phenotype. Here every single allele (*M* or *N*) can be identified, and the gene frequency can be established by direct counting. The *MN* blood types may be used as examples:

$$\hat{p} = \frac{2\overline{M} + \overline{MN}}{2(\overline{M} + \overline{MN} + \overline{N})} = \frac{\overline{M} + \frac{1}{2}\overline{MN}}{\overline{M} + \overline{MN} + \overline{N}}; \quad \hat{q} = 1 - \hat{p} \qquad (A\,1.1)$$

The calculation of variance *V* is also straightforward:

$$V = \frac{\hat{p}\hat{q}}{2(\overline{M} + \overline{MN} + \overline{N})} \qquad (A\,1.2)$$

The gene frequencies $\hat{p}$ and $\hat{q}$ can now be used to test whether the phenotype frequencies agree with their expectations according to the Hardy-Weinberg Law. Using the following formula, the explicit calculation of these expectations can be spared:

$$\chi^2 = \frac{(\overline{M} + \overline{MN} + \overline{N})(\overline{MN}^2 - 4 \times \overline{M} \times \overline{N})^2}{(2\overline{M} + \overline{MN})^2(\overline{MN} + 2\overline{N})^2}$$

In principle the same counting method can be used when there are more than two alleles, and when all genotypes can be distinguished phenotypically, such as in polymorphisms.

One Gene Pair; Only Two Genotypes Can Be Identified by the Phenotypes. The problem becomes slightly more complicated if one of the two alleles is dominant, i.e., if the heterozygote is phenotypically identical to one of the two homozygotes. In this case the homozygote for the recessive gene provides the information necessary for gene frequency calculations. The frequency of the homozygotes is q^2. The Diego blood factor (Sect. 14.3.1.) may serve as example. There are two classes in Native American and Mongoloid populations: those with a positive agglutination reaction with anti-Di^a serum and those without this reaction. Family studies have shown the negative type to be recessive.

q (Frequency of the gene Di^a) $= \sqrt{\dfrac{Di(a\,-)}{Di(a\,+) + Di(a\,-)}}$

$Variance = \dfrac{1 - q^2}{4[Di(a\,+) + Di(a\,-)]}$

There is no degree of freedom left to test for a Hardy-Weinberg equilibrium.

If an anti-Di^b serum is available, the heterozygotes can be identified, and gene frequency can be calculated by the gene counting method as shown above for the MN blood group.

More Than Two Alleles: Not All Genotypes Can Be Distinguished in the Phenotype. A special case, the AB0 blood group, is discussed in Sect. 4.2.2.

Maximum Likelihood Principle of Estimation. We are confronted with the general problem that a parameter is not known a priori but must be estimated from empirical data. According to R. A. Fisher, an estimate should fulfill the following conditions:

a) It should be consistent. This means that the estimate converges stochastically on the parameter with an increasing number of observations.
b) The estimate should be sufficient. It must be impossible to extract additional knowledge about the parameter by calculating other statistics from the data.
c) The estimate should be efficient. It should extract the maximum possible amount of information from the data. The variance should be as small as possible.

The best solution of the estimation problem is generally the maximum likelihood principle established by Fisher. Details on this principle can be found in standard textbooks on statistics. Suffice it to say that the methods recommended here give maximum likelihood estimates.

Calculation of AB0 Gene Frequencies by the Bernstein Method.

Bernstein, in his investigations on the genetic basis of the AB0 system (Sect. 4.2), derived a method for estimating AB0 gene frequencies. He later refined this method by first estimating provisional frequencies p', q', r' and then correcting them to calculate the definite gene frequencies p for gene frequency of A, q for gene frequency of B, r for gene frequency of O:

$p' = 1 - \sqrt{(\overline{B} + \overline{O})/n}$ $p = p'(1 + D/2)$

$q' = 1 - \sqrt{(\overline{A} + \overline{O})/n}$ $q = q'(1 + D/2)$

$r' = \sqrt{\overline{O}/n}$ $r = (r' + D/2)(1 + D/2)$

D is the difference between 1 and the sum $p' + q' + r'$. Estimates using this improved Bernstein method were shown to be practically identical with the maximum likelihood estimate.

Example: Estimation of Gene Frequencies by Gene Counting.

Race and Sanger [66] gave the following phenotype frequencies for individuals from London, Oxford, and Cambridge:

$\overline{M}$	$\overline{MN}$	$\overline{N}$	Sum total
363	634	282	1279

The gene frequencies p of allele M, and q of allele N are therefore, according to Eq. A.1.1:

$$p = \frac{363 + \frac{1}{2} \times 634}{1279} = 0.5317$$

$$q = \frac{282 + \frac{1}{2} \times 634}{1279} = 0.4683$$

It follows that: $p^2 = 0.2827$; $2pq = 0.4980$; $q^2 = 0.2193$.

To calculate the expected numbers of individuals of the three genotypes (E), these figures must to be multiplied by the total number of individuals, 1279:

$$E(\overline{M}) = 361.6$$

$$E(\overline{MN}) = 636.9$$

$$E(\overline{N}) = 280.5$$

These expectations are now compared to the observed numbers of individuals:

$$X_1^2 = 1279 \times \frac{[634^2 - (4 \times 363 \times 282)]^2}{[(2 \times 363) + 634]^2 [634 + (2 \times 282)]^2} = 0.027;$$

$$P \gg 0.05$$

There is no statistically significant difference between observed and expected gene frequencies.

Example: Estimate of AB0 Gene Frequencies. In 21104 individuals from Berlin the blood group distribution was found to be:

$$\overline{A} = 9123$$

$$\overline{B} = 2987$$

$$\overline{O} = 7725$$

$$\overline{A}\,\overline{B} = 1269$$

By the improved Bernstein method, this leads to the following results. (For details see Sect. 4.2.2)

$$p = 0.287685 \pm 0.002411$$

$$q = 0.106555 \pm 0.001545$$

$$r = 0.605760 \pm 0.002601$$

The maximum likelihood method has been shown to lead to exactly the same results [38].

The standard deviations listed above are the square roots of the variances.

Expected genotype frequencies can now be calculated from these gene frequencies in the same way as shown above for the MN blood types and compared with the observed frequencies by a conventional χ^2 test.

Still more complicated problems are encountered for the Rh blood groups and in general for all systems for which many different combinations of antigens are inherited together. Here computer programs have been developed (for the Rh system see [16]). A number of authors have published guidelines for calculating gene and haplotype frequencies for the HLA system [50, 59, 100].

Computer processing cannot compensate for inadequate sampling, however. All methods discussed above are based on the assumption that sampling of individuals is independent, i.e., that the selection of any person does not enhance or diminish the chance of any other person in the population being selected. This rule is violated, when for example, relatives are recruited into the sample. Samples containing relatives are not necessarily useless for gene frequency calculations, but such inclusion must be noted, together with the degree of the relationships, and special statistical methods must be used for analysis [90].

Appendix 2

Testing and Estimating Segregation Ratios; Correction of Ascertainment Biases in Rare Diseases; Multifactorial and Mixed Models; Related Statistical Problems.

As explained in Sect. 4.3, in families in which rare alleles segregate, such as in those having hereditary diseases with monogenic inheritance, ascertainment is usually by sibship with at least one affected sib. This introduces an ascertainment bias, since sibships in which by chance no affected person has appeared escape notice by the investigator. This bias should be corrected when observed ratios are compared with those expected from Mendelian laws or when segregation ratios are estimated. Today computerized programs are available, and these are used in most instances. However, scientists who use this approach should understand the principles of these methods.

As discussed in Sect. 4.3, there are two main modes of ascertainment: single selection ($k = 0$) and complete or truncate selection ($k = 1$). With single selection, each family is ascertained by a single proband. The most common example is a family survey on a disease which starts with a series of hospital patients. With complete selection, each patient is ascertained independently as a proband. In practice this occurs only when a determined effort is made to ascertain each patient carrying a certain disease in a defined population. Computer programs designed to estimate segregation ratios offer a third option: more than one patient in a family but not all of them are ascertained independently as probands (incomplete multiple or proband selection). On the other hand, one patient may be ascertained not once but twice or even several times. Theoretically this can be included in a refined estimate of segregation ratios. However, all these refinements require the probability of ascertainment to be the same for each individual and repeated ascertainments to be independent of each other. Obviously both conditions are almost never fulfilled. It is therefore sufficient to perform the estimates either for single selection ($k = 0$) or for truncate selection ($k = 1$), depending on the method of ascertainment, or to perform both calculations, assuming that the real ratio lies somewhere between these two [85]. In most instances the supposedly refined estimate (incomplete multiple selection) is no more precise than such interpolation.

Earlier editions of this book recommended the method of Finney, as worked out by Kaelin, and the genetic analysis of a comprehensive – and reasonably completely ascertained – data set on deaf-mutism in Northern Ireland [87] was performed step by step in order to demonstrate how such an analysis can be performed in a relatively simple way, and how the results should be assessed critically. Such problems have lost somewhat in interest in recent years; therefore this analysis is not repeated here. Instead we list in table A 2.1 the available computer programs for segregation analysis. Some of these programs are designed for nuclear families, i.e., families consisting of parents and a number of children, while others, such as SAGE, include extended families.

The scientist who collects such family material for genetic analysis should follow a number of rules:

1. The sample to be drawn and the degrees of relationship with the probands to be included in the sample should be defined in advance.
2. Diagnosis should be confirmed as carefully as possible not only in probands but also in their family members, whenever possible by personal examination. As a rule, such comprehensive studies are used to refine clinical criteria such as age at onset, course of the disease, and often additional clinical signs.
3. The probands should be indicated, and the method of their ascertainment should be noted as carefully as possible.
4. The design should be planned, before the study begins in cooperation with the responsible statistician.
5. The results should be assessed critically. Not every result spat out by a computer is biologically meaningful.

More Complex Ascertainment Problems, for Example, in Some Chromosomal Aberrations. Families are often observed that consist of more than one sibship. Moreover, various kinds of probands might exist. For example, families with reciprocal translocations may be ascertained by an unbalanced translocation carrier, in most cases a child with multiple malformations; or the proband may be a balanced translocation carrier; he or she might have been ascertained in the

Table A 2.1. Some computer programs for analysis of family data

Name	Special features of the program	Contact address	Reference
Mendel/Fisher	Mapping, ascertainment correction, fitting of multifactorial models; quantitative traits; paternity problems	K. Lange, Dept. of Biostatistics Univ. of Michigan, Ann Arbort MI. 48109, USA	Lange et al. (1988) [48]
PAP	Pedigrees; all models of segregation; risk calculation	S. J. Hasstedt, Dept. of Medical Physics and Computing, University of Utah, Salt Lake City UT 84943, USA	Hasstedt (1982) [37]
Segran	Nuclear families; program permits testing for sporadic cases	N. E. Morton, CRC Research Group in Genetic Epidemiology, Dept. of Community Medicine, University of Southampton, South Academic Block, Southampton General Hospital, Southampton S 09 4 X, UK	Morton et al. (1983) [58]
POINTER	Nuclear families; large pedigrees must be subdivided; testing for mixed model		Lalouel and Morton (1981) [46]
SAGE (2.1)	General segregation models for quantitative and discrete traits; variable age at onset distribution; marker-trait associations; familial correlation; two-point linkage analysis and sibpair analysis; mapping of a trait to a map of markers (the program is being improved continously)	C. Elston, Dept. of Epidemiology and Biostatistics Case Western Reserve University 2500 Metro Health Drive, Cleveland, OH 44109, USA	S. A. G. E. Handbook (From Cleveland address)

course of chromosome screening carried out, for example, in adult normal populations, in populations of newborns, in mental retardates, in any individuals carrying certain malformations, or in a study of spontaneous abortions. If ascertained by way of abortions, families are normally examined only if at least two abortions have occurred. Furthermore, there are differences depending on whether the analysis is based on single case reports of families or on families collected within the framework of a collaborative study. Such collaborative studies are preferable since there is less danger of a combination of "interesting cases." Schäfer [73] in his study on segregation of translocations discusses these problems and gives suggestions for correcting the most important biases.

Once a collection of cases has been obtained, the type of ascertainment must be determined. The majority of published cases of translocations are usually ascertained via a child with an unbalanced translocation. At first glance, statistical correction according to the single selection model ($k = 0$) in the sibship of this childproband seems appropriate. One could argue that ascertainment and/or publication of the family also depends on the clinical status of other relatives; it is therefore prudent to repeat the calculation with a model of truncate selection; the true segregation ratio may be closer to the single selection result. However, this is true only if analysis is based on families ascertained via a clinically affected individual. In the

future more and more family studies will be based on probands from long-term, complete collections of all malformed newborns in entire populations, as is the case in studies already being carried out in Hungary, for example [17, 18]. In such cases, correction according to the model of truncate selection would be adequate.

As a rule, however, pedigrees consist of more than one sibship. In these cases additional – secondary or even tertiary – probands occur. All individuals whose presence induces the investigator to extend his studies to a further generation or sibship must be regarded as probands. This is explained with reference to a model pedigree (Fig. A 2.1; Table A 2.2; [73]).

The family was discovered through III, 16, a carrier. She would not have been karyotyped if she had not suffered from multiple abortions. Since an indication for karyotyping is normally assumed when at least two abortions have occurred, two abortions are considered as "probands" among her children and are eliminated from the risk calculation for abortions (note that abortions might be "probands"). III, 16 is a "secondary proband." Moreover, if her mother (II, 5) were to have had a normal karyotype, III, 16 would be regarded as de novo translocation, and the sibship to which II, 5 belongs would not have been examined. Therefore II, 5 is another secondary proband and must be eliminated from risk calculations in her sibship. II, 4 was not karyotyped; since she has no children it is unknown whether she is a carrier. Therefore she too must be excluded from risk calculations. II, 11, on the other hand, is a carrier, being a sister of the proband III, 16, and would have been examined in any case, irrespective of her children; therefore she does *not* count as a proband; she must be *included* in the risk calculation for children of translocation carriers. She has three children who would have been examined anyway regardless of their phenotypes. Therefore no correction is necessary; they can all be included in the risk calculation.

The next step of the analysis must be carried out carefully: If generation II would have been examined in any case, then II, 5 is the only proband in this generation, and all other sibs can be used for risk calculation (except of course II, 4). Moreover, the sibships III, 2–III, 9; IV, 1, 2; and IV, 3, 4 were ascertained through an affected parent; therefore no correction is necessary. However, if, for example, sibship IV, 3, 4 was examined only because III, 11 tells the investigator that her cousin also has a malformed child (and if this sibship would not have been examined otherwise), the unbalanced child IV, 4 is a (tertiary) proband and must be eliminated from risk estimation. This example shows how important explicit reporting of the ascertainment process is. The following analysis assumes that the sibships on the left side of Fig. A 2.1 were indeed ascertained through an affected parent. The result can be seen in Table A 2.2. The following risk estimates can be derived:

a) For patients carrying unbalanced translocations: 2/21
b) For abortions: 4/21
c) For balanced carriers: 9/21
d) For normal offspring: 6/21

For these estimates the single cases from all sibships are simply pooled (= precumulation). This procedure could be criticized on the grounds that larger sibships carry too much weight in relation to smaller ones. It is also possible to make risk estimates for all sibships separately and then to pool them (= postcumulation). However, most studies in the literature on translocations were performed with precumulation. This procedure appears justified especially if the sibships come from larger pedigrees, since it can safely be assumed that within such a pedigree the real risks are identical in all sibships. On the other hand, it is necessary to perform such risk calculations separately for families in which at least one patient with an unbalanced translocation is observed as proband, those ascertained through abortions, and those by balanced carriers, since only some of all unbalanced translocations can lead to unbalanced offspring; many unbalanced zygotes cannot survive the early zygote stage. Basically the same rules for risk calculation should be applied to large pedigrees with autosomal-dominant or X-linked diseases. Computer programs are available for inclusion of extended families in such estimates (Table A 2.1), but critical assessment of ascertainment in each pedigree remains the task of the investigator.

Multifactorial Inheritance and Major Genes. The study of segregation ratios is relatively straightforward when the analysis has advanced to the qualitative phenotypic level (Sect. 6.1.1), i.e., when a simple Mendelian mode of inheritance of clearly distinguishable phenotypes can be anticipated. For many human characteristics, however, such an analysis is not yet possible; they must be studied at the quantitative phenotypic-biometric level (Sect. 6.1.1). Normally dis-

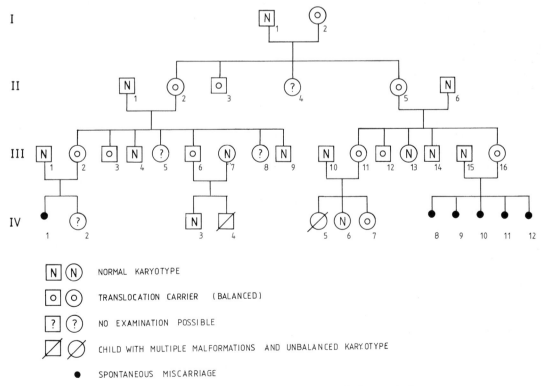

Fig. A 2.1. Model pedigree with translocations. For analysis see Table A 2.2. (From Schäfer 1984 [73])

Table A 2.2. Data for risk calculation for the family in Fig. A 2.1

Sibships at risk	No. of children (excluding probands and unclassified persons)	Probands	Unbalanced cases	Balanced carriers	Abortions	Normals
II, 2–5	2	1	0	2	0	0
III, 2–9	5	0	0	3	0	2
III, 11–16	5	1	0	3	0	2
IV, 1, 2	1	0	0	0	1	0
IV, 3, 4	2	0	1	0	0	1
IV, 5–7	3	0	1	1	0	1
IV, 8–12	3	2	0	0	3	0

tributed traits, such as stature or IQ, and physiological and biochemical characteristics, such as serum cholesterol levels, fall into this category, but so do also most common diseases. Some approaches to the analysis of such traits are described in Chap. 6; the heritability concept is explained, and strategies are suggested for a stepwise dissection of characters and traits subsumed by the model of multifactorial inheritance with and without threshold. Among these strategies we discuss searching for phenotypic subclasses, physiological components, and associations with genetic polymorphisms.

More recently, several authors have proposed statistical methods for more rigorous testing of multifactorial vs. single-gene models and for identifying effects of major genes in the presence of a multifactorial background [13, 20, 55, 57]. In principle these methods consist of two steps. In the first, certain assumptions are made regarding the mode of inheritance of a certain condition, and the consequences of these assumptions are calculated for frequency (of alternatively distributed characters) or distribution (continuously distributed characters) among certain groups of relatives. This creates a tentative "model" of a particular mode of inheritance. Then the goodness of fit of a set of empirical data is tested by a statistical method against the expectations derived from a given model. Therefore this approach to analysis does not differ in principle from that described in Sects. 4.3.3 and 4.3.4 for testing whether family data fit the expectations of simple Mendelian inheritance. Often several alternative models are constructed and are compared with the actual data.

Models cannot be constructed without simplifying assumptions. This is inevitable and does not cause much harm, provided that all assumptions are set out clearly. However, the fact that a data set fits expectations based on a certain model does not prove that this model adequately describes the real situation. Other plausible models must be excluded. Very often such exclusion is simply not possible for models normally encountered in human genetics, for example, multifactorial inheritance vs. autosomal-dominant inheritance with incomplete penetrance. Geneticists who usually work with simple Mendelian models are "spoiled"; there are few situations that convincingly mimic a monogenic mode of inheritance without additional assumptions. As a rule, they are on solid ground, but not necessarily in the morass of multifactorial models.

When we study an alternately distributed trait in the population, such as a disease, and the data obviously do not fit the assumption of monogenic inheritance, two main alternatives are usually considered: a monogenic mode of inheritance with incomplete penetrance or multifactorial inheritance in combination with a threshold. With the second alternative, three models are often considered. In the simplest case a number of genes contribute additively to disease liability; however, in some instances, one gene locus may contribute much more than the others to the genetic variability. Attempts at defining such "major gene" models and distinguishing them from the simple additive/polygenic model have been made, and computer programs for such tests are available (Table A 2.1). A "mixed model" is frequently used in which, in addition to major genes and an additive polygenic component, environmental factors are also considered.

Expectations from such models may then be compared with an empirically collected data set, using standard statistical techniques such as the χ^2 method.

In practice, however, convincing separation of various models is only rarely achieved since there is usually substantial overlap between expectations under the various models. This is true particularly if the trait under study is fairly common. Thus a simple additive model with a threshold, and dominant inheritance with incomplete penetrance may give similar results.

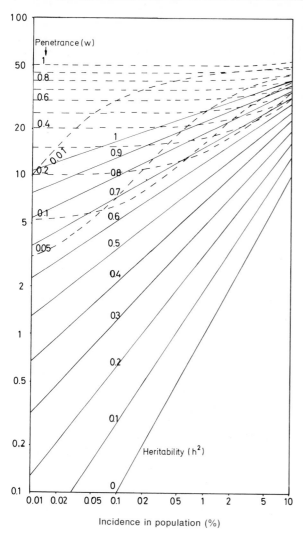

Fig. A 2.1. Frequency of a character among sibs of probands with parents of type affected × unaffected in the diallelic model (- - -) and the multifactorial model (—) [45]

Figure A 2.1 shows only one example – computed expectations for the frequency of affected children if one parent is affected and the other one unaffected. These expectations are computed for the multifactorial model (multiple, additive genes and a threshold) and for the monogenic, diallelic model with incomplete penetrance [45]. Expectations from the two models overlap especially if penetrance in the monogenic, diallelic model is low, and if the condition studied is very common.

Two criteria have emerged that help to distinguish multifactorial from diallelic models: in multifactorial inheritance, and if the trait is not too common, MZ twins are concordant much more often than DZ twins, while this difference is smaller in the diallelic case. Thus if concordance is more than four times higher in MZ than in DZ twins, multifactorial inheri-

tance is more likely [41]. The other criterion comes from a comparison of the number of affected children in terms of whether one or both parents are unaffected. In multifactorial models, the frequency of affected children when both parents are healthy is lower than in monogenic inheritance with low penetrance.

Calculations involving the multifactorial model include an estimate of heritability. The heritability concept is discussed in Chap. 6, and its limitations are set out. In principle, heritability can be estimated in a meaningful way only for quantitatively varying characters. Since a quantitatively varying liability is assumed in the multifactorial model, heritability estimates have a logical basis under such conditions. Such estimates may be based on a comparison between MZ and DZ twins or a comparison of any groups of probands' relatives versus the general population. Computation of heritability estimates in twin studies for continuously varrying traits are discussed in Appendix 5; tables for comparison of first-degree relatives with the general population are available [84], but in our opinion these provide little useful information. Theoretically the expectation of an individual with any degree of relationship being affected can be calculated when heritability estimates are available either from twin or other data. Such expectations are sometimes thought to be useful in genetic counseling, where they could replace empirical risk figures. However, since heritability estimates depend on largely untestable assumptions (Sect. 6.1), such calculated recurrence risks must be used with great caution in practice.

Methods for calculation of such models are not discussed here in detail; the interested reader may find useful information in [45].

Criteria from extended pedigrees could in principle be used. For example, in diallelic inheritance with reduced penetrance or in multifactorial inheritance with major gene involvement we expect the gene contributing to the disease liability to come exclusively, or at least mainly, from one side of the pedigree, whereas in truly polygenic causation more or less equal contributions from paternal and maternal families are expected. This argument was proposed by Slater (1966) [83] many years ago but, to our knowledge, has not been developed further.

In conclusion, when applying these methods, one must be particulary critical toward the results. Especially the demonstration of "major gene effects" within a mixture of (often variable) phenotypes in probands and – even more so – in their relatives could very easily lead to conclusions that are not only premature but misleading. This was demonstrated convincingly by a study by McGuffin and Huckle [51] who applied complex segregation analysis to a family study on medical school attendance in relatives of medical students; they found excellent agreement with the most general transmission model, but the hypothesis of a recessive main gene could not be refuted completely. The authors performed this study of course to demonstrate unavoidable pitfalls in this kind of reasoning. In our opinion, a much better way to disentangle the genetic basis of complex diseases is to analyze pathophysiology – in close association with formal and molecular genetics as well as by linkage studies – at as many levels as ever possible. This is shown in Chap. 6 for coronary heart diseases and for diabetes.

The Choice of a Suitable Program Package for Genetic Analysis. The number and variety of program packages for genetic analysis of data sets is increasing rapidly, and it is becoming more difficult to select and obtain access to the programs most suitable for a given data analysis. Some general rules for access to such packages are given in Appendix 3, but some additional guidelines may be useful for estimating parameters such as segregation ratios or comparison of genetic models.

1. Before deciding in favor of one program, its scientific basis must be considered. Understanding the program to be used helps in assessing critically whether it really provides answers to the questions that are posed.

2. A few programs permit comparison of various genetic hypotheses: for example, a dominant mode of inheritance vs. a polygenic threshold model. The detailed analysis depends on the program used. For example, data are compared first with the mixed model (polygenic inheritance together with a major gene), and their parameters are estimated. Single parameters are then fixed, and the reduced model is computed again. For example, fixing the polygenic component creates a monogenic model. As the next step, calculations for various modes of inheritance are performed. The alternative of no genetic influence at all should also be considered. With results of these computations at hand, one then determines which of the models provides the best fit to the data. This is achieved by comparing pairwise the results of the various models that have been tested by calculating their likelihood ratio, i.e., which genetic hypothesis is the most probable, by maximum likelihood statistics. It is important to emphasize, however, that the most likely model is not necessarily the biologically correct representation of the actual situation. A yet untested model may apply, or the results may have been obtained by chance.

In addition to such model calculations, some simple questions may be asked: for example, what is the ratio of concordance between MZ and DZ twins? Are affected but more remote relatives to be found mainly in one branch of the pedigree, or in both branches? Finally and most importantly, might a better distinction between genetic models be obtained if biochemical and physiological parameters, for example, intervening or intermediate phenotypes (such as blood cholesterol levels in coronary heart disease), are considered?

Some Other Statistical Problems That Might Turn Up in Genetic Analysis: Sporadic Cases and Birth Order. We now return to the estimation of segregation ratios in monogenic diseases. Often this ratio proves lower than expected from Mendel's laws. Admixture of sporadic cases (phenocopies or dominant new mutations) is often the most obvious explanation. In this case the number of sibships with only one affected sibling is increased above expectation *(E)*, which, according to the binomial distribution, is:

$$E = \sum_{s=2}^{\infty} E_s = \sum_{s=2}^{\infty} n_s \cdot \frac{spq^{s-1}}{1-q^s}$$

(complete selection) (A 2.1)

$$E = \sum_{s=2}^{\infty} E_s = \sum_{s=2}^{\infty} n_s \cdot q^{s-1}$$

(single selection) (A 2.2)

where n_s is the number of sibships with sizes *s*. This expectation can be calculated if estimates of the segregation ratio *p* are available; this is provided, for example, by the program SEGRAN. The problem has come up in practice mainly in X-linked recessive diseases in which the only affected son of a healthy mother in an unaffected family could be due to a new mutation in her germ cells Chap 9. Here the investigator should take care that the proportion of sporadic cases is calculated only in families in which no additional affected patients are observed outside the nuclear family. In families in which the mutant has already been shown to segregate it is extremely unlikely that a single affected son is caused by a new mutation in his mother's germ cell.

Another problem that occurs occasionally is the sequence of affected and unaffected sibs within sibships. Mendelian segregation implies that this sequence is random. This randomness can be tested by the theory of runs [24], if necessary. Sometimes the probability of being affected with a certain disease rises with increasing birth order. Useful methods have been proposed [34] to test for such birth order effects.

Appendix 3

Data Bases and Expert Systems

Useful Data Bases and Expert System for Diagnostics in Medical Genetics and for Comparison of DNA and Protein Sequences Data for Diagnosis. With advances in clinical genetics and refinement of diagnostic criteria, the number of distinct hereditary diseases and well-defined syndromes has increased so much that no medical geneticist can remember all diagnostic possibilities when a patient or family with an unusual clinical picture turns up. The traditional way out of this difficulty is consultation of various reference

Table A 3.1. Some useful books and reviews for diagnosis of hereditary diseases and malformation syndromes

Category of anomalies	Title	Authors
General references in medical genetics	Mendelian inheritance in man	McKusick [55]
Common diseases	The genetic basis of common disease	King et al. [44]
References for many hereditary diseases	Principles and practice of medical genetics	Emery and Rimoin [22]
Genetic counseling and prenatal diagnosis	Chromosomal abnormalities and genetic counseling	Gardner et al. [28]
	Practical genetic counseling	Harper [35]
	Genetic disorders and the fetus	Milunsky [53]
	Prenatal diagnosis of congenital anomalies	Romeo [70]
	Congenital malformations, prenatal diagnosis, etc.	Seeds et al. [77]
	Catalogue of prenatally diagnosed conditions	Weaver [97]
	Introduction to risk calculation in genetic counseling	Young [101]
Birth defects	Birth defects encyclopedia	Buyse [12]
	Syndromes of the head and neck	Gorlin et al. [30]
	Smith's recognizable patterns of human malformation	Jones [40]
	Pathology of the human embryo and previable fetus	Kalousek et al. [41]
	Multiple congenital anomalies	Winter et al. [98]
	The malformed fetus and stillbirth	Winter et al. [99]
Metabolic diseases	The metabolic and molecular basis of inherited disease	Scriver et al. [76]
	Inborn metabolic diseases	Fernandes et al. [25]
Neurological and muscle diseases	Genetics and neurology	Bundey [11]
	Epilepsy in children	Aicardi [1]
	Epilepsies, genetic risks	Blandfort et al. [8]
	Genetic strategies in epilepsy research	Anderson et al. [2]
	Neuromuscular disease	Swash et al. [88]
	Duchenne muscular dystrophy	Emery [21]
	Huntington disease	Harper [35]
Skin diseases	Prenatal diagnosis of heritable skin diseases	Gedde-Dahl et al. [29]
	Neurofibromatosis	Riccardi [69]
Eye diseases	Goldberg's genetic and metabolic eye disease	Renie [68]
Connective tissue and internal organs	Connective tissue and its heritable disorders	Royce et al. [71]
	The kidney in genetic disease	Barakat et al. [6]
	The genetics of renal tract disorders	Crawfurd [15]

books. The first part of the bibliography of this text provides many examples. The more recent and important books are mentioned in Table A 3.1. Chromosomal aberrations can be looked up in Schinzel's [154] and Borgaonkar's [22] compendia. Other monographs on various diseases have been mentioned in various chapters of this book.

Frequently these books and the numerous available journals are not sufficient. By relying exclusively on published data, important information may be missed since some data are not published or have not yet appeared in print. Here modern data bases are helpful. Elaborate algorithms and additional picture displays in some of these programs have turned such computer aids into real expert systems. Table A 3.2 gives an overview. The POSSUM system includes data from the London Dysmorphology Data Base and GENDIAG, France. More recent information on the various systems is available from their originators.

Data for DNA and Protein Sequences. Whereas the data given above on clinical signs and symptoms of hereditary diseases and birth defects are of interest mainly for the medical geneticist and for certain groups of physicians, data on DNA and protein sequences are important for many more groups of scientists working in many fields of molecular biology, research on evolution, and in various areas of theoretical medicine such as biochemistry, microbiology,

pharmacology, and many others. Therefore the number and variety of available systems is much larger. Due to the lack of space we can present only a relatively small selection (Table A 3.3).

An overview of the existing databases is given by Kamel [730]; a table in this paper lists no fewer than 102 databases. A "database of databases" has been established in the Los Alamos National Laboratory; information on data bases is available free of charge either by post service or by electronic mail (Table A 3.3). According to Kamel [730], the available databases can be grouped into three classes: the commercially available systems, those supported by international or national agencies, and those in the public domain. These offer sequences (DNA, amino acids, probes, and others), genomic and vector sequences, literature citations, enzymes, macromolecular structure and function, properties of biochemical substances, some organisms and strains, and others. The services are improved and supplemented continuously. All these databases and additional ones are readily available in the World Wide Web.

Some General Suggestions for Using of Databases and Expert Systems. Information systems for human genetics are growing fast, are changing continuously, and are becoming more and more sophisticated. These developments increase the availability of data and methods and provide more advanced informa-

Table A 3.2. Data bases and expert systems for diagnosis in medical genetics

Name	Short description	Available from
POSSUM: Pictures of Standard Syndromes and Undiagnosed Malformations	Based on clinical phenotypes, gives possible syndromes for any combination of clinical signs, supplemented by photos; frequent updates for identification. OSSUM: Specialized for skeletal anomalies	C. P. Export Pty. Ltd., 613 St. Hilda Road Melbourne 3004, Victoria, Australia
OMIM: Online Mendelian Inheritance in Man	Online for actual and updated information, of McKusick: *Mendelian Inheritance in Man*	Johns Hopkins University, School of Medicine, and William H. Welch Medical Library, 1830 East Monument St., Baltimore MD 21205-2100, USA
GDB: Genome Data Base	Online data on the human genome	
LDD: London Dysmorphology LND: London Neurogenetics Database	Based on clinical phenotypes; numerous descriptions of dysmorphic syndromes; many references; CD with photos available; LND with emphasis on neurogenetic disease	Dr. R. M. Winter, Dr. M. Baraitser, Oxford University Press, Electronic Publishing, Walton St., Oxford OX2 6DB, UK
Micro BDIS: Birth Defects Information Service	Birth defect; about 1000 traits; data stored hierarchically	Dr. Mary Lou Buyse, Center for Birth Defects Information Service, Inc., Dover Medical Building, Box 1776, Dover MA 02030, USA
TERIS: Teratogen Information System	Information about data on teratogens	Dr. Janine E. Polifka, TERIS MJ-10, Dept. of Pediatrics, University of Washington, Seattle WA 98195, USA

Table A 3.3. Brief overview of databases for molecular biology and human genetics (See [7, 26, 42])

Name	Short description	Contact
Listing of Molecular Biology Databases (LIMB)	Intended as "database of databases"; provides information on up to 54 items from each of the listed databases, including size, type of data, respondent with telephone number, services available a. s. o.	Theoretical Biology and Biophysics Group, Respondent: C. Burks, T-10, MS K710, Los Alamos Natl. Lab., Los Alamos NM 87545, USA
European Molecular Biology Laboratory (EMBL)	Nucleotide sequence data	ftp. embl-heidelberg. de
EMBOPRO	Protein sequences automatically generated from EMBL	German EMB net mode, GENIUSnet, Abteilung Molekulare Biophysik, DKFZ, Im Neuenheimer Feld 280, 69120 Heidelberg, Germany e-mail: genome @ dkfz-heidelberg. de
GenBank	Reported nucleotide sequences	
PTG	Translations of all protein coding regions in GenBank	

tion. On the one hand, obtaining this information is becoming increasingly simpler when the Internet and World Wide Web systems are used. On the other hand, using these systems adequately has to be learned, and some exercise is needed. Useful reviews have been provided by Bishop [7] and Fischer et al. [26]. Institutions interested primarily in diagnosis need a system for storage and retrieval of patient data. Such systems have been developed: for example, MEGADATS. Programs for calculating genetic risks, such as in complex genetic situations and with use of multiple DNA linkage markers may often be useful and sometimes essential. Often computer-aided drawing of pedigrees [26] can be helpful.

Tables containing the appropriate information may be found, in addition to this chapter, in Appendices 6 and 7. More current data are available through international computer networks and especially through Internet. In addition to programs, data are available on DNA and protein sequences, localization of human genes, and availability of DNA probes from the Genome Data Bank (GDB; see Table A 3.2). Many computer programs can be obtained on file servers, i. e., programs in the public domain that can be taken over on one's own computer free of charge. Data for studies in molecular biology are available, for example, from Integrated Genomic Database, which integrates information from several data bases. Access to these sources usually requires certain passwords, and some identification of the user. As a rule, the easiest access is through university computer centers and similar institutions; their personnel can usually help with technical problems in access and use of such facilities.

Appendix 4

Diagnosis of Zygosity

Diagnosis of Zygosity Using Genetic Markers. MZ twins, being genetically identical, show no differences either in sex or in any genetic markers. DZ twins, on the other hand, are no more similar than normal sibs; about half of them are of different sex, and many differ in genetic polymorphisms. The study of polymorphisms therefore allows the determination of zygosity. If a twin pair differs only in one marker, we can be sure that it is dizygotic (provided that laboratory errors have been excluded). However, dizygotic twins may be identical in all markers studied simply by chance. This source of error must be excluded by statistical calculations. Classical genetic markers such as blood and serum groups were used formerly; the statistical details of their application to zygosity diagnosis were worked out in detail. (See Appendix 5 in the first or second edition of this book, which may still be useful for research workers to whom DNA techniques are not available.) The method was reliable but costly and time consuming. Today DNA markers are generally used, especially microsatellites (Sect. 12.1.2). However, the apparent simplicity of this method should not lure us into sloppiness; well-established rules for statistical reasoning should not be neglected. Here the mathematical principle established by Bayes as early as 1793 is a useful tool.

Bayes' Principle of Conditional Probabilities. Consider one twin pair for which the probability of monozygosity or dizygosity is to be determined. More precisely, our question is: what proportion of all twin pairs with the same combination of genetic markers are expected to be dizygotic? Or, to put it somewhat differently: how often would the twin pair be misclassified, if monozygosity were assumed in every such case? The general formula of Bayes is:

$$P(A_1/B) = \frac{P(A_1) \times P(B/A_1)}{P(A_2) \times P(B/A_2) + P(A_1) \times P(B/A_1)}$$

$$(A\,4.1),$$

where A_1 and B are different events, and A_2 means the event "not-A_1."

In our case $P(A_1/B)$ is the probability of monozygosity among twins with identical blood types. Then, $1 - P(A_1/B)$ is the probability of the twin pair being dizygotic or the probability of error when the twin pair is classified as monozygotic. $P(A_1)$ is the a priori probability for MZ twins among all twins in the population. It is about 30 % in populations of European origin. $P(A_2)$ is the a priori probability for a twin pair to be dizygotic. $P(A_2) = 1 - P(A_1) = 0.7$; Eq. A 4.1 may be simplified as follows:

$$P(A_1/B) = \frac{1}{1 + (Q \times L)} \qquad (A\,4.2)$$

Here Q is the ratio DZ/MZ in the population (if 30 % of all twin pairs are MZ, then $Q = 2.33$). L is the likelihood ratio of conditional probabilities of DZ and MZ twins being identical in a given combination of genetic markers. Its value can be calculated by multiplication from the L_i of the various marker systems used:

$$L = L_1 \times L_2 \times \ldots \times L_n \qquad (A\,4.3)$$

An Example. Tables A 4.1 and A 4.2 present an example. From the list of genetic markers it turns out that some are not informative, parents and children being identical genetically [IGHG (Gm), IGHG (Km), HP,

Table A 4.1. Example of a zygosity diagnosis

	Father	Mother	Both twins
Sex	♂	♀	♂
Blood groups	A₂	0	A₂
	MS/Ms	MS/Ms	MS/Ms
	Kk	kk	Kk
	Fy(a + b +)	Fy(a–b +)	Fy(a–b +)
	R₁ r	R₂ r	rr
Serum proteins	G1m(−1)	G1m(−1,−2)	G1m(−1,−2)
	Km(−1)	Km(−1)	Km(−1)
	HP 2–2	HP 2–2	HP 2–2
	GC 2–2	GC 2–1	GC 2–1
Isoenzymes	ACP B	ACP AB	ACP AB
	PGM1 2–1	PGM1 2–1	PGM1 1–1
	AK1 1–1	AK1 1–1	AK1 1–1

Table A 4.2. Calculations for the data from Table A 4.1

	P_{DZ}	P_{MZ}
A priori probability	0.70	0.30
Conditional probabilities[a]		
Sex	0.50	1.00
AB0	0.50–1.00	1.00
MNSS	0.25	1.00
Kell (K)	0.50	1.00
Duffy (Fy)	0.50	1.00
Rhesus (Rh)	0.25	1.00
GC	0.50	1.00
ACP	0.50	1.00
PGM1	0.25	1.00

[a] Conditional probability of the second twin being phenotypically identical with the first for given phenotype of the first twin.

$$L = \frac{0.7}{0.3} \times 0.5 \times 0.25 \times 0.5 \times 0.5 \times 0.25 \times 0.5 \times 0.5 \times 0.25$$

$$= 0.0011$$

A priori sex MNSs K   Fy Rh GC ACP PGM1
$P(A_1/B) = 0.9989$

Table A 4.3. Random mating table for a system of two alleles (from Maynard Smith and Penrose 1955 [86])

Mating	Frequency	Children		
		AA	Aa	aa
AA × AA	p^4	p^4	–	–
AA × Aa	$4p^3q$	$2p^3q$	$2p^3q$	–
AA × aa	$2p^2q^2$	–	$2p^2q^2$	–
Aa × Aa	$4p^2q^2$	p^2q^2	$2p^2q^2$	p^2q^2
Aa × aa	$4pq^3$	–	$2pq^3$	$2pq^3$
aa × aa	q^4	–	–	q^4
Total	1	p^2	$2pq$	q^2

AK]. For most other markers the mating types and hence the expected segregation ratios among the children are obvious. For example, in the GC system the father is homozygous 2–2, and the mother is heterozygous 2–1. Therefore the expected segregation ratio among the children is 1 : 1, and if twin 1 has type 2–1 and the twins are dizygotic, the probability for twin 2 to be 2–1 as well is 0.50. For the AB0 types, the situation is not as obvious since the father (phenotype A_2) may have the genotypes A_2A_2 or A_20. If he is A_2A_2, both twins must have the phenotype A_2 even if they are dizygous. If he is A_20, the probability of the second twin being A_2 too is 0.50. Sometimes the genotype of the parent may be established, for example, by another child showing type 0. Otherwise the AB0 blood groups may be regarded as uninformative and omitted. For the other systems the calculation proceeds as follows (Eq. A 4.3):

Hence the probability that the twin pair is dizygotic in spite of its concordance in all informative marker systems is extremely low (0.9989). For all practical purposes monozygosity can be assumed. Inclusion of additional marker genes increases the probability that a pair of twins are monozygotic.

Genotype of the Parents May Not Be Known. In the cited example, genetic markers were known not only in the twin pair but also in the parents. In many cases, however, the parents are not available for examination. Under these circumstances, the known gene frequencies of the marker systems in the population can be used for the calculation. The rules have been set out by Maynard Smith and Penrose (1955) [86]. The conditional probability $P_{i,}$ DZ of twin 2 having the same phenotype as twin 1 if this twin has phenotype i is calculated from the frequencies of mating types in the population (Table A 4.3) and from the relative number of children with different genotypes expected from these matings. These tables can be used not only for the classical polymorphisms for which they were designed but also for DNA polymorphisms. Most investigators

Table A 4.4. Sib-sib frequencies for a system of two alleles (from Maynard Smith and Penrose 1955 [86])

Sib	Genotypic				Sib	Phenotypic		
	AA	Aa	aa	Total		$\overline{A}$	$\overline{a}$	Total
AA	$\frac{1}{4}p^2(1+p)^2$	$\frac{1}{2}p^2q(1+p)$	$\frac{1}{4}p^2q^2$	p^2	$\overline{A}$	$p(1+q)-\frac{1}{4}pq^2(3+q)$	$\frac{1}{4}pq^2(3+q)$	$p(1+q)$
Aa	$\frac{1}{2}p^2q(1+p)$	$pq(1+pq)$	$\frac{1}{2}pq^2(1+q)$	$2pq$	$\overline{a}$	$\frac{1}{4}pq^2(3+q)$	$\frac{1}{2}q^2(1+q)^2$	q^2
aa	$\frac{1}{4}p^2q^2$	$\frac{1}{2}pq^2(1+q)$	$\frac{1}{4}q^2(1+q)^2$	q^2				
Total	p^2	$2pq$	q^2	1	Total	$p(1+q)$	q^2	1

who plan a twin study prefer DNA polymorphisms simply because they are much less time consuming and expensive. Moreover, the PCR technique does not require blood since DNA can be studied in cells from buccal smears. However possible technical errors and, especially, the risk of misclassifying DZ twins as MZ because they have identical patterns should be considered carefully. The formula for P(conc)/DZ) for a codominant marker system with n alleles and allele frequencies p_i can be derived in a straightforward way by setting up a random mating table (Table A 4.3 See also Tables 4.4, 4.5) and calculating the frequencies for two concordant children for each parental mating type. Summation over all mating types and some algebraic transformations yields:

$$P(\text{conc/DZ}) = \left(1 + 2\sum_{i=1}^{n} p_i^2 + 2\left(\sum_{i=1}^{n} p_i^2\right)^2 - \sum_{i=1}^{n} p_i^4\right)/4$$

An equivalent formula (which needs more computational effort) has been derived by Selvin. More complex marker systems do not necessarily have lower misclassification probabilities. It depends on whether there are alleles with high frequencies. $P(\text{conc/DZ})$ takes on its minimum value when $p_i = 1/n$. For a given set of markers the expression must be calculated for each and multiplied to obtain the misclassification probability for this special system. Below we present an example.

A set of VNTR markers has been made available (Table 4.6). (We thank Professor J.-J. Cassiman, Leuven, for kindly providing us with this table.) All six have many alleles and fulfill all other conditions for useful systems; for example, the bands can be easily identified by polyacrylamide gel electrophoresis, and the mutation rate is low. Moreover, they have been used extensively for identification of individuals and in cases of disputed paternity in forensic medicine [18 a]. Other markers can of course also be used, the only condition being that MZ twins show 100 % identical patterns, and that the probability of misclassifying DZ for MZ twins is very low. For the marker system presented in Table A 4.6. together with sex the misclassification probability is 2.9×10^{-3}, which is sufficiently low for many purposes. Determining the probability level of statistical error depends on the consequences of misclassification. If, for example, as twin pair is tested for monozygosity because one of them has developed a genetic disease, more markers or, of course, the marker information from the parents may be preferred (see also Fig. A 4.1).

On the other hand, if twins are identical for these markers and for sex, their probability of being monozygotic is at least 98.97 %. This value can be derived approximately by calculating the MZ-probability for twins which are monozygotic and carry the most frequent allele for every used marker. When used in

Table A 4.5. Relative chances in favor of dizygotic twin pairs in a system of two alleles (from Maynard Smith and Penrose 1955 [86]

Genotypic		Phenotypic	
Twin pair both	Relative chance in favor of dizygotic twins	Twin pair both	Relative chance in favor of dizygotic twins
AA	$\frac{1}{4}(1 + p)^2$	$\overline{A}$	$1 - \frac{1}{4}q^2(3 + q)/(1 + q)$
Aa	$\frac{1}{2}(1 + pq)$		
aa	$\frac{1}{4}(1 + q)^2$	$\overline{a}$	$\frac{1}{4}(1 + q)^2$

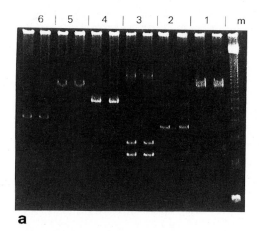

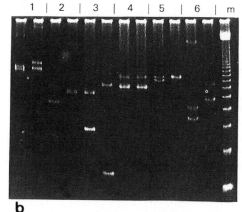

1 = DXYS17
2 = D1S111
3 = D17S5
4 = APOB
5 = D1S118
6 = IGHJ
m = 123 bp ladder (BRL)

Fig. A 4.1 a, b. Diagnosis of zygosity with six VNTR markers (see also the text). **a** MZ pair, **b** DZ pair. Some runs show a third band due to heteroduplexes (Courtesy Professor Cassiman and Dr. Decorte, Leuven)

Table A 4.6. Allele frequencies and their 95% confidence limits for the VNTRs D1S118, APOB, D17S5, D1S111, IGHJ, and DXYS17 in a population of Belgian origin (Courtesy of Prof. Cassiman)

D1S118 (n = 446)			APOB (n = 307)			D17S5 (n = 460)		
Allele	Frequency	Confidence limit	Allele	Frequency	Confidence limit	Allele	Frequency	Confidence limit
11	0.004	0.001–0.017	29	0.003	0.001–0.020	01	0.054	0.036–0.080
12	0.037	0.023–0.060	31	0.094	0.065–0.135	02	0.151	0.120–0.189
13	0.013	0.006–0.030	33	0.064	0.040–0.099	03	0.171	0.138–0.210
14	0.188	0.153–0.229	35	0.237	0.192–0.291	04	0.274	0.232–0.318
15	0.535	0.485–0.581	36	0.003	0.001–0.020	05	0.039	0.024–0.062
16	0.058	0.039–0.085	37	0.370	0.315–0.428	06	0.035	0.021–0.057
17	0.098	0.072–0.130	39	0.044	0.025–0.075	07	0.015	0.007–0.032
18	0.004	0.001–0.017	41	0.015	0.006–0.037	08	0.030	0.018–0.052
19	0.053	0.035–0.079	43	0.005	0.001–0.022	09	0.080	0.058–0.111
20	0.007	0.002–0.020	45	0.008	0.002–0.027	10	0.074	0.052–0.103
21	0.001	0.000–0.011	47	0.046	0.027–0.077	11	0.022	0.011–0.041
22	0.001	0.000–0.011	49	0.086	0.059–0.125	12	0.048	0.031–0.073
31	0.001	0.000–0.011	51	0.020	0.009–0.044	13	0.005	0.002–0.018
			53	0.005	0.001–0.022	16	0.002	0.000–0.013

D1S111 (n = 463)			IGHJ (n = 447)			DXYS17 (n = 420)		
Allele	Frequency	Confidence limit	Allele	Frequency	Confidence limit	Allele	Frequency	Confidence limit
09	0.015	0.007–0.033	07	0.007	0.002–0.021	H 2	0.001	0.000–0.013
10	0.022	0.011–0.041	08	0.173	0.139–0.214	H 1	0.054	0.035–0.082
11	0.008	0.003–0.022	09	0.021	0.011–0.041	G 5	0.001	0.000–0.013
12	0.111	0.084–0.148	10	0.383	0.335–0.430	G 3	0.013	0.005–0.031
13	0.005	0.002–0.019	11	0.002	0.000–0.014	G 2	0.008	0.003–0.024
14	0.006	0.002–0.020	12	0.242	0.202–0.286	G 1	0.219	0.179–0.265
15	0.339	0.293–0.386	14	0.001	0.000–0.012	F	0.180	0.143–0.223
16	0.014	0.006–0.031	16	0.151	0.119–0.189	E 2	0.147	0.114–0.187
17	0.028	0.016–0.050	17	0.011	0.005–0.027	E 1	0.096	0.070–0.132
18	0.290	0.247–0.336	19	0.001	0.000–0.012	E 3	0.008	0.003–0.024
19	0.063	0.043–0.091	20	0.001	0.000–0.012	E 4	0.001	0.000–0.013
20	0.019	0.010–0.038	22	0.002	0.000–0.014	D 2	0.006	0.002–0.021
21	0.043	0.027–0.068	23	0.003	0.001–0.015	D 1	0.002	0.000–0.015
22	0.025	0.013–0.045	24	0.001	0.000–0.012	C	0.024	0.012–0.045
23	0.002	0.000–0.014	30	0.001	0.000–0.012	B 2	0.001	0.000–0.013
24	0.003	0.001–0.015				B	0.010	0.003–0.026
25	0.003	0.001–0.015				A 1	0.217	0.177–0.262
26	0.002	0.000–0.014				A 2	0.010	0.003–0.026
27	0.002	0.000–0.014				A 3	0.002	0.000–0.015

n, Number of individuals

other population groups, allele frequencies of these markers should be checked.

Methods from Classic Anthropology. A fairly reliable method for distinguishing MZ from DZ twins was available even before most genetic polymorphisms were known. This was established by Siemens in 1924 [81, 82] and is based on the comparison of twins in a great number of visible physical characteristics and on anthropological measurements. The following characteristics have proven useful: color, form, and density of hair; shape and proportions of face; detailed structure of facial regions such as the eye, including eyebrows, color and structure of the iris; details of the nose and mouth region; chin, ears, shape of hands and feet; dermatoglyphics; color and structure of the skin, including freckling. Various anthropometric measurements of the body, head, and face

Table A.4.7. Random mating table for a system of n alleles with the probability of identical twins

Mating type	Frequency	Children	Probability of identical twins
AiAi × AiAi	p_i^4	AiAi	1
AiAi × AjAj $(1 \leqslant i < j \leqslant n)$	$2p_i^2 p_j^2$	AiAj	1
AiAi × AjA$_K$ $(1 \leqslant j < k \leqslant n)$	$4p_i^2 p_j p_K$	$\frac{1}{2}$AiAj + $\frac{1}{2}$AiA$_K$	$\frac{1}{2}$
AiAj × AiAj $(1 \leqslant i < j \leqslant n)$	$4p_i^2 4p_j^2$	$\frac{1}{4}$AiAi + $\frac{1}{2}$AiAj + $\frac{1}{4}$AjAj	$\frac{3}{8}$
AiAj × A$_k$A$_e$ $(1 \leqslant i < j \leqslant n,$ $1 \leqslant k < e \leqslant n,$ $(i, j) \neq (k, e))$	$4p_i p_j p_K p_e$	$\frac{1}{4}$AiA$_k$ + $\frac{1}{4}$AiA$_e$ + $\frac{1}{4}$AjA$_k$ + $\frac{1}{4}$AjA$_e$	$\frac{1}{4}$

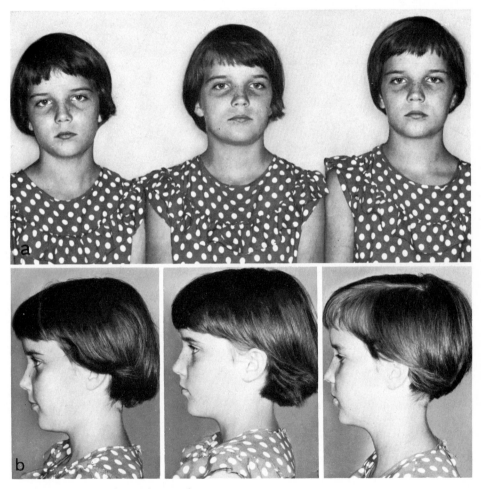

Fig. A 4.2 a, b. MZ triplets at the age of 10

are also helpful. The anthropological literature provides lists of informative characteristics. The experienced investigator bases his diagnosis less on the comparison of single characters as on the whole *gestalt*. Comparison of anthropological and serological methods shows no divergence between the two approaches [74].

This does not mean, however, that zygosity diagnosis based on physical characters is always easy. Due to differing living conditions MZ twins may sometimes look so different that the layman would not even recognize them as sibs, and only painstaking anthropological examination identifies them as monozygotic. On the other hand, DZ twins, as other sibs, may occasionally

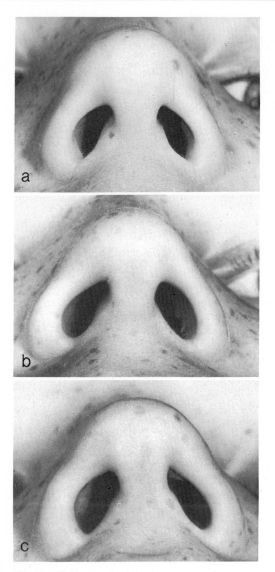

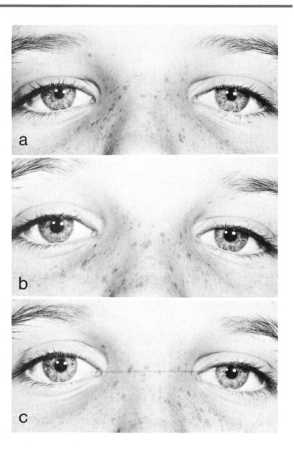

Fig. A 4.3 *(left)*, **A 4.4** *(right)*. Physiognomic details of the MZ triplets shown in Fig. A 4.1

look very similar. Figures A 4.2–A 4.4 give an impression of the degree of similarity found in MZ, as compared with (very similar) DZ twins (Fig. A 4.5).

How Should We Proceed in Practice? From the above discussion it seems as if a comprehensive study of marker systems – DNA or classical – would be the most appropriate and also a sufficient method for reliable zygosity diagnosis. This conclusion, however, needs some qualification. On the one hand, laboratory methods are not immune to errors. An error in only one marker in only one of the two twins will lead to misclassification of an MZ pair as dizygotic. The investigator should therefore trust his eyes. Wherever possible he should add (and docu-

ment) a physiognomic comparison of the twin pair and should insist on repetition of laboratory examinations whenever a twin pair is regarded as monozygotic in spite of discordance in a marker system. Sometimes, especially if a well-equipped DNA laboratory is not available, assistance by an experienced physical anthropologist may be helpful. As a general rule, twins whose physical appearance led to identity mix-ups by teachers and others are almost always MZ in type. Anthropological diagnosis requires experience and judgment; it is therefore more subjective. DNA diagnosis, on the other hand, is objective, but laboratory errors are possible.

Twin examination ideally also includes a registration of placentation and embryonic membranes: DZ twins have two placentas, two amnions, and two chorions, whereas MZ twins may have one chorion or even one amnion and one placenta. As discussed in Sect. 6.3.3, presence of only one chorion may be harmful for intrauterine development of MZ twins. In practice, however, reliable information is only rarely available, and DZ twins may have a single fused placenta which resembles that of MZ twins. The attempt to include such data into zygosity diagnosis is therefore often futile and may be misleading.

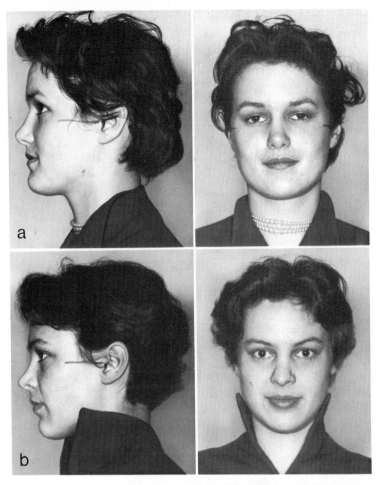

Fig. A 4.5 a, b. DZ twins at the age of 19. Note the conspicuous similarity

Appendix 5

Heritability Estimates from Twin Data

The concept of heritability is discussed in Sect. 6.1, and methods are presented for estimating heritability in threshold characteristics. The criterion is the ratio between the incidence in close relatives of affected probands and that in the general population. In a continuously distributed characteristic, such as stature, heritability is calculated by comparing parents and children.

Twin data can be utilized as an alternative way to get heritability estimates. In Sect. 6.1, heritability is defined as

$$h^2 = \frac{V_A}{V_P}$$

i.e., the ratio of the additive genetic variance (V_A) to the total phenotypic variance (V_P). It was noted that in human genetics h^2 is often referred to as "heritability in the narrow sense," and is contrasted with $H^2 = V_G/V_P$ (heritability in the broad sense; also called the degree of genetic determination), where V_G is the total genetic variance including dominance, epistasis, and interaction variance.

In this appendix the estimation of heritability, h^2, is restricted to twin data. It is not possible to estimate V_D, the dominance variance, from twin data alone. Furthermore, V_D is usually negligible compared to V_A [54]. Thus the error in assuming that the total genetic variance (V_G) is identical to the additive variance (V_A) is likely to be small. We therefore make the assumption of $V_G = V_A$, and therefore:

$$h^2 = \frac{V_G}{V_P} \tag{A 5.1'}$$

As shown in Sect. 6.1:

$$V_P = V_A + V_D + V_E + V_I + V_M + \text{Cov}_{GE}$$

Here, V_E = environmental variance, V_I = variance due to interaction between heredity and environment, and V_M = the variance between repeated measurements of the same characteristic, representing either truly different values – such as blood pressure on different days – or in the case of a constant characteristic, measurement errors. Cov_{GE} is the covariance between the genetic and environmental components of the phenotypic value. Heritability estimates from twin data require that V_I and Cov_{GE} be 0. This assumption is not realistic in most cases, especially in behavioral genetics, but estimation of these parameters, if they deviate from 0, poses almost unsuperable practical difficulties. The measurement variance, V_M, should, however be considered but is often disregarded in many twin studies. Equation A 5.1 becomes:

$$h^2 = \frac{V_G}{V_G + V_E + V_M} \tag{A 5.1'}$$

It may be appropriate to omit the measurement variance:

$$h'^2 = \frac{V_G}{V_G + V_E}$$

If $\text{Cov}_{GE} = 0$, the following relation exists between the correlation coefficient $r_{P_1 P_2}$ of the phenotypic values P_1, P_2 of two relatives, the correlation coefficient of their genotypic values G_1, G_2, and the correlation coefficients of their environmental values E_1, E_2:

$$r_{P_1 P_2} = r_{G_1 G_2} h^2 + r_{E_1 E_2} E^2$$

with

$$E^2 = \frac{V_E}{V_P}$$

In analogy with heritability, E^2 may be called the "environmentability." The correlation coefficients can be defined as intrapair correlations if the two relatives are twins:

$$r_{P, MZ} = h^2 + r_{E, MZ} E^2 \quad \text{for MZ twins} \tag{A 5.2}$$

$$r_{P, DZ} = r_{G, DZ} h^2 + r_{E, DZ} E^2 \quad \text{for DZ twins} \tag{A 5.3}$$

Here we use the theoretical genotypic correlation $r_{G, MZ}$, which is 1 for MZ twins, (Sect. 6.1).

If the environmental correlation between the twins of an MZ pair is taken to be identical to the environmental correlation between the twins of a DZ pair, it follows that:

$$h^2 = \frac{r_{P, MZ} - r_{P, DZ}}{1 - r_{G, DZ}} \tag{A 5.4}$$

This expression is known as the H index (Holzinger). The phenotypic intrapair correlation coefficient is given by:

$$r_{P,\,twins} = \frac{V_P^B}{V_P^B + V_P^W}$$

where V_P^B = phenotypic variance between pairs and V_P^W = phenotypic variance within pairs.

The variance components V_P^B and V_P^W can be estimated from the phenotypic values p_{i1}, p_{i2} ($i = 1, 2, \ldots, n$) observed on n twin pairs:

$$V_P^W = DQ_W,$$

$$V_P^B = (DQ_B - DQ_W)/2 \tag{A5.5}$$

Here DQ_W and DQ_B are the within-pairs and between-pairs mean squares:

$$DQ_W = \frac{1}{2n} \sum_{i=1}^{n} (p_{i1} - p_{i2})^2 \tag{A5.6}$$

$$DQ_B = \frac{2}{n-1} \sum_{i=1}^{n} (\bar{p}_{i.} - \bar{p}_{..})^2 \tag{A5.7}$$

with

$\bar{p}_{i.} = (p_{i1} + p_{i2})/2$ = mean phenotypic value for the i-th pair

$\bar{p}_{..} = \dfrac{1}{n} \displaystyle\sum_{i=1}^{n} \bar{p}_i$ = total mean of all measurements in the twin sample

For the actual calculation, the following formula may be used instead of Eq. A5.7:

$$DQ_B = \frac{1}{2(n-1)} \left[\sum_{i=1}^{n} y_i^2 - \frac{1}{n} \left(\sum_{i=1}^{n} y_i \right)^2 \right] \tag{A5.7'}$$

with $y_i = p_{i1} + p_{i2}$ ($1 = 1, 2, \ldots, n$).

The variance component V_P^W may be decomposed in the following way:

$$V_P^W = V_G(1 - r_{G,\,twins}) + V_E(1 - r_{E,\,twins}) + V_M \tag{A5.8}$$

This equation can be applied to MZ pairs, DZ pairs, or unrelated control pairs from the general population (CP).

$$V_P^W(MZ) = V_E(1 - r_{E,MZ}) + V_M \tag{A5.9}$$

$$V_P^W(DZ) = V_G(1 - r_{G,DZ}) + V_E(1 - r_{E,DZ}) + V_M \tag{A5.10}$$

$$V_P^W(CP) = V_G + V_E + V_P = V_M \tag{A5.11}$$

For further analysis two steps are suggested. First, one should examine whether h^2 deviates significantly from 0; then h^2 should be estimated.

Testing of the Null Hypothesis ($h^2 = 0$). Under the assumption $r_{E,MZ} \doteq r_{E,DZ}$ it follows from Eqs. A5.2 and A5.3 that the hypothesis $h^2 = 0$ is equivalent to the hypothesis:

$$r_{P,MZ} = r_{P,DZ}$$

To test the latter hypothesis the fact is used that under this hypothesis

$$z = 1/2 \log_e \frac{1 + \hat{r}_{P,MZ}}{1 - \hat{r}_{P,MZ}} - 1/2 \log_e \frac{1 + \hat{r}_{P,DZ}}{1 - \hat{r}_{P,DZ}}$$

has approximately a normal distribution with mean 0 and variance

$$\frac{1}{n_{MZ} - {}^3\!/_2} + \frac{1}{n_{DZ} - {}^3\!/_2}$$

Here, $\hat{r}_{P,MZ}$ and $\hat{r}_{P,DZ}$ are estimates for $r_{P,MZ}$ and $r_{P,DZ}$ from n_{MZ} MZ pairs and n_{DZ} DZ pairs.

Estimates for h^2 (and h'^2):

$$h_1^2 = \frac{V_P^W(DZ) - V_P^W(MZ)}{V_P^W(DZ) - V_M} \tag{A5.12}$$

$$h_2^2 = \frac{V_P^W(CP) - V_P^W(MZ)}{V_P^W(CP) - V_M} \tag{A5.13}$$

To estimate h^2 (or h'^2), the intrapair variances are replaced by their estimates from the analysis of variance. An exact formula for the standard error of these two estimates of h^2 does not exist. If V_M is neglected compared to V_P^W the following formula holds approximately:

$$\text{S.E.}^2 = 2F^2 \frac{n_2^2(n_1 - 1)(n_1 + n_2 - 4)}{n_1^2(n_2 - 3)^2(n_2 - 5)} \tag{A5.14}$$

where F means the observed value of the ratio V_P^W (MZ)/V_P^W (DZ) or V_P^W (MZ)/V_P^W (CP), and n_1, n_2 are the number of pairs from which variances in numerator and denominator have been estimated. Apart from sampling errors the estimates by Eqs. A5.12 and A5.13 are biased, as is now explained in detail:

Equation A5.12: If environmental correlations between MZ and DZ are taken to be identical: $r_{E,MZ} = r_{E,DZ}$, then it follows from Eqs. A5.9 and A5.10:

$$h_1^2 = \frac{V_G(1 - r_{G,DZ})}{V_G(1 - r_{G,DZ}) + V_E(1 - r_{E,DZ})}$$

$$= h'^2 \frac{1 - r_{G,DZ}}{1 - r_{G,DZ} + (1 - h'^2)(r_{G,DZ} - r_{E,DZ})}$$

Hence $h_1^2 \lesseqgtr h'^2$, if $r_{E,DZ} \lesseqgtr r_{G,DZ}$.
Therefore h_1^2 overestimates h^2, if $r_{E,DZ} > r_{G,DZ}$, because always $h^2 \leq h'^2$. For other cases no prediction of the bias of h_1 in estimating h^2 is possible.

Equation A5.13: It follows from Eqs. A5.9 and A5.11 that:

$$h_2^2 = \frac{V_G + V_E r_{E,MZ}}{V_G + V_E} = h'^2 + (1 - h'^2)r_{E,MZ} \geqq h'^2 \geqq h^2$$

(Here $r_{E,MZ} \geq 0$ is assumed). This means that h_2^2 usually overestimates h^2.

These two estimates for h^2 use exclusively intrapair variances. Frequently h^2 is also estimated from *intrapair correlation coefficients,* which have been calculated using the variance of the whole sample of MZ or DZ twins, i.e., also the variance between twin pairs. From Eq. A 5.4, the following estimation formula can be derived:

$$h_3^2 = 2(r_{P,MZ} - r_{P,DZ}) \qquad (A\,5.15)$$

This assumes that $r_{G,DZ} = 1/2$, which holds true only if mating is random, and there is neither dominance nor epistasis. In practical terms the condition is at best only approximately fulfilled; moreover, it is unknown whether $r_{G,DZ} > 1/2$, or $r_{G,DZ} < 1/2$. Therefore the bias in estimating h^2 from Eq. A 5.15 cannot be predicted.

A correction is possible if $r_{G,DZ} > 1/2$ due to assortative mating and there is neither dominance nor epistasis. The standard error of the estimate h_3^2 can be calculated only very approximately:

$$\text{S.E.}^2(h_3^2) \approx 4 \left[\frac{(1 - r_{P,MZ}^2)^2}{n_{MZ}} + \frac{(1 - r_{P,DZ}^2)^2}{n_{DZ}} \right] \qquad (A\,5.16)$$

Comments on These Methods of Heritability Estimation. The above considerations show that an unbiased estimate of h^2 from twin data is impossible even if such components as covariance between heredity and environment (Cov_{GE}) and the interaction term V_I are neglected, and if the very unlikely assumption of identity is made between the environmental correlations $r_{E,MZ}$ and $r_{E,DZ}$, i.e., the common environmental influences affecting MZ pairs and DZ pairs. Even with these oversimplifying assumptions there remain systematic errors that can be only partially controlled.

An empirical way to overcome this difficulty partially is to calculate alternative estimates from the same data to determine how well they coincide. The three alternative estimates proposed above can be characterized as follows: h_1^2 is calculated from the classic comparison of MZ and DZ twins. The bias of this estimate includes the genotypic correlation between sibs, $r_{G,DZ}$. This value is $1/2$ with random mating. With regard to many characters for which heritability estimates are used, for example, intelligence quotient or stature, mating is, however, known to be assortative. The direction and the degree of bias depends on the difference in genotypic and environmental correlations between the sibs, which is usually unknown. Therefore the heritability estimate based on control persons (h_2^2) may be a useful second choice, although depending on environmental correlations $r_{E,MZ}$ and $r_{E,DZ}$ means that h^2 is overestimated systematically.

The additional comparison of control pairs was proposed by Vogel and Wendt in 1956 [92] but this has

apparently never been used since then. A similar procedure was suggested by Kamin (1974) [43]. Control pairs from twin samples can easily be matched for age and sex, thus eliminating variance components contained in most twin samples but without significance for the problem to be examined.

This nuisance variance is the main argument against using the estimate from the intraclass correlation coefficients (h_3^2), which contains these additional variance components unless the twin samples are very homogeneous, for example, a single cohort of army conscripts. The problem is taken up in Sect. 15.2.1.3.

Heritability of IQ as an Example. The twin sample consists of 50 German adult male twin pairs (25 MZ and 25 DZ) aged between 23 and 30 years. The twins were military conscripts, and the sample was therefore unbiased regarding socioeconomic status, education, and test intelligence. In this sample the Intelligence Structure Test (IST; Amthauer) [93] was administered. Here only the total intelligence score, cor-

Table A 5.1. Total intelligence scores observed in twin pairs

MZ pairs Pair no.				DZ pairs Pair no.			
(i)	p_{i1}	p_{i2}	$|p_{i1}-p_{i2}|$	(i)	p_{i1}	p_{i2}	$|p_{i1}-p_{i2}|$
1	107	105	2	1	86	98	12
2	88	80	8	2	112	100	12
3	89	102	13	3	89	84	5
4	96	110	14	4	125	128	3
5	84	84	0	5	105	99	6
6	100	89	11	6	90	84	6
7	87	78	9	7	103	98	5
8	79	87	8	8	91	102	11
9	96	97	1	9	94	84	10
10	111	113	2	10	97	107	10
11	114	114	0	11	112	109	3
12	106	111	5	12	106	110	4
13	114	113	1	13	90	85	5
14	120	117	3	14	98	100	2
15	110	107	3	15	116	104	12
16	87	87	0	16	78	79	1
17	92	93	1	17	104	115	11
18	103	101	2	18	95	113	18
19	107	99	8	19	113	115	2
20	83	84	1	20	84	83	1
21	99	105	6	21	110	109	1
22	86	95	9	22	98	93	5
23	107	101	6	23	77	85	8
24	122	117	5	24	76	86	10
25	118	115	3	25	117	117	0
				26	117	110	7

Σy_i	5009	5180
Σy_i^2	1017589	1048268
$\Sigma(p_{i1} - p_{i2})^2$	1005	1628

rected for age, is considered; this score is proportional to IQ. The following procedure was used to construct a set of control pairs with intrapair age differences as small as possible. All twin pairs, regardless of zygosity, were arranged in ascending order of age. In this arrangement the first and second pairs, the third and fourth pairs, and so on were combined as quadruples that were transformed into two new pairs, each by exchanging one co-twin between the original pairs; the co-twins to be exchanged were selected at random. The observed value pairs of total intelligence score (p_{i1}, p_{i2}) and the quantities Σy_i, Σy_i^2, and $\Sigma(p_{i1} - p_{i2})^2$ derived from them (for notations see the foregoing text) are given in Table A 5.1 for the twin pairs and in Table A 5.2 for the control pairs.

From these tables one calculates:

a) *For MZ pairs*   According to Eq.

$$\hat{V}_P^W = 1{,}005/50 = 20.100 \tag{A 5.6}$$

$$DQ_B = (1{,}017{,}589 - 5{,}009^2/25)/48 = 291.370 \tag{A 5.7}$$

$$\hat{V}_P^B = (291.370 - 20.100)/2 = 135.635 \tag{A 5.5}$$

$$\hat{r}_{P,MZ} = \frac{135.635}{135.635 + 20.100} = 0.871$$

Confidence limits (99 %) for $^1/_2 \log_e \dfrac{1 + r_{P,MZ}}{1 - r_{P,MZ}}$:

$$^1/_2 \log_e \frac{1 + 0.871}{1 - 0.871} \pm 2.58 \frac{1}{\sqrt{25 - 1.5}} = 0.805 \text{ and } 1.869$$

to which correspond

$$\frac{e^{2 \times 0.805} - 1}{e^{2 \times 0.805} + 1} = 0.667, \quad \frac{e^{2 \times 1.869} - 1}{e^{2 \times 1.869} + 1} = 0.954$$

as confidence limits for $r_{P,MZ}$.

b) *For DZ pairs:*

$$\hat{V}_P^W = 1682/52 = 31.308$$

$$DQ_B = (1\,048\,268 - 5{,}180^2/26)/50 = 325.052$$

$$\hat{V}_P^B = (325.052 - 31.308)/2 = 146.872$$

$$\hat{r}_{P,DZ} = \frac{146.872}{146.872 + 31.308} = 0.824$$

Confidence limits (99 %) for $^1/_2 \log_e \dfrac{1 + r_{P,DZ}}{1 - r_{P,DZ}}$:

$$^1/_2 \log_e \frac{1 + 0.824}{1 - 0.824} \pm 2.58 \times \frac{1}{\sqrt{26 - 1.5}} = 0.648 \quad \text{and}$$

1.690

with corresponding confidence limits

0.570   and   0.934

for $r_{P,DZ}$.

Table A 5.2. Total intelligence scores observed in control pairs (see text for choice of control pairs)

Pair no.				Pair no.			
(i)	p_{i1}	p_{i2}	$\lvert p_{i1}-p_{i2}\rvert$	(i)	p_{i1}	p_{i2}	$\lvert p_{i1}-p_{i2}\rvert$
1	89	88	1	26	113	78	35
2	102	80	22	27	94	89	5
3	106	114	8	28	84	84	0
4	114	110	4	29	90	100	10
5	96	114	18	30	112	84	28
6	97	113	16	31	110	109	1
7	125	86	39	32	112	107	5
8	128	95	33	33	103	76	27
9	77	83	6	34	101	86	15
10	85	84	1	35	99	117	18
11	78	101	23	36	120	105	15
12	107	79	28	37	96	87	9
13	104	93	11	38	110	87	23
14	92	115	23	39	122	105	17
15	84	99	15	40	117	99	18
16	107	83	24	41	117	113	4
17	103	110	7	42	111	117	6
18	98	109	11	43	98	86	12
19	91	97	6	44	93	98	5
20	102	107	5	45	84	106	22
21	98	115	17	46	84	111	27
22	118	100	18	47	107	104	3
23	95	100	5	48	116	105	11
24	113	89	24	49	90	110	20
25	87	115	28	50	117	85	32

Σy_i 10023
Σy_i^2 2024211
$\Sigma(p_{i1} - p_{i2})^2$ 16659

c) *For the control pairs:*

$$\hat{V}_P^W = 16\,659/100 = 166.590$$

$$DQ_B = (2\,024\,211 - 10\,023^2/50)/98 = 153.066$$

$$\hat{V}_P^B = (153.066 - 166.590)/2 = -6.762$$

$$\hat{r}_{P,CP} = \frac{-6{,}762}{166.590 - 6.762} = -0.042$$

The above confidence limits for $r_{P,MZ}$ and $r_{P,DZ}$ show that the intrapair correlation coefficient of IQ deviates significantly from 0 ($P < 0.01$) in both types of twins. This means that twins, regardless of their zygosity, are more similar in IQ than are two unrelated persons. This result – although expected if IQ has a genetic basis – does not yet exclude the possibility of a purely nongenetic explanation because twins also partially share their environment. To examine this possibility, we test the hypothesis $r_{P,MZ} = r_{P,DZ}$ (null hypothesis).

$$z = {}^1/_2 \log_e \frac{1 + \hat{r}_{P,MZ}}{1 - \hat{r}_{P,MZ}} - {}^1/_2 \log_e \frac{1 + \hat{r}_{P,DZ}}{1 - \hat{r}_{P,DZ}} = 0.168$$

$$\text{var } z = \frac{1}{25 - 1.5} + \frac{1}{26 - 1.5} = 0.0834$$

$$z/\sqrt{\text{var } z} = 0.582$$

Under the null hypothesis the probability of a z value as extreme as or more extreme than the value found is greater than 10 %. This means that from the comparison of the intrapair correlation coefficients in our two twin series we cannot reject the hypothesis $h^2 = 0$, i.e., that there is no genetic contribution to the variation of IQ in the population. Consequently the estimate for h^2 according to Eq. A 5.15:

$$h_3^2 = 2(\hat{r}_{P\text{MZ}} - \hat{r}_{P\text{DZ}}) = 2 \times (0.871 - 0.824) = 0.094,$$

also cannot be regarded as significantly different from 0; this is confirmed by considering the standard error of h_3^2: its approximate value, calculated by Eq. A 5.16 is 0.159. It is possible that the difference in intrapair correlations of IQ in MZ and DZ twin pairs is biased, the between-pairs variance of IQ being, for some unknown reason, smaller in the DZ twins than in the MZ twins. In this case only the two estimates of h^2 based solely on within-pair variances must be used. However, even if there is no such bias – as obvious in our data – the calculation of these other estimates of h^2 (Eqs. A 5.12 and A 5.13) is strongly recommended. In our case:

$$h_1^2 = \frac{V_P^W(\text{DZ}) - V_P^W(\text{MZ})}{V_P^W(\text{DZ})} = \frac{31.308 - 20.100}{31.308} = 0.358$$

$$h_2^2 = \frac{V_P^W(\text{CP}) - V_P^W(\text{MZ})}{V_P^W(\text{CP})} = \frac{166.590 - 20.100}{166.590} = 0.879$$

(The subtrahend V_M in the denominators representing the variance between repeated measurements of IQ is omitted; here the test reliability of IST may be inserted.)

Standard errors for these estimates can be calculated by Eq. A 5.14:

$$\text{SE}(h_1^2) = 0.301, \quad \text{SE}(h_2^2) = 0.045$$

giving (nearly) 95 % confidence intervals

$$- 0.23 \text{ to } 0.95 \text{ for } h_1^2$$

and

$$0.79 \text{ to } 0.97 \text{ for } h_2^2$$

The two intervals overlap partially, but the subinterval common to both is remote from h_3^2 by more than twice the standard error of the latter estimate, the three estimates for h^2 thus appearing incompatible. On the other hand, h_2^2 presumably overestimates h^2; the assumption inherent in this estimation – that environmental intrapair differences are identical for twins and unrelated control pairs – is incorrect. Thus the differences between the three estimates may be explained by this bias combined with sampling errors that, due to the small size of our twin series, are large. In any case these results are hardly compatible with the high values of heritability reported for IQ by some authors [39]. This example illustrates the problems in estimating heritability from human twin data and should make one cautious of accepting such data for definitive scientific conclusions.

For those interested in the use of more refined quantitative-genetic models for the study of quantitative traits in twins the introduction by Eaves [19] is recommended; this also discusses the enormous sample sizes needed for such an analysis.

Appendix 6

Genetic Counseling

A Problem in Estimating Genetic Risks. As noted in Sect. 18.1, the estimation of the genetic risk is based either on segregation ratios in Mendelian diseases or on empirical risk figures when the mode of inheritance is complex. The use of such figures for estimating the specific risk to a certain proband or family is straightforward if no additional information is available. For example, every future child of an affected member of the large Farabee family with brachydactyly (Sect. 4.1.2, Fig. 4.2) suffers a risk of 50% for brachydactyly. However, there are many situations in which additional information should be included in the risk estimate.

Example: Inherited and Sporadic Retinoblastoma. As noted in Chap. 10, retinoblastoma, a malignant eye tumor of young children, may occur either as a dominant disease with about 90% penetrance or as a nonhereditary condition due to somatic mutation. In the latter case both parents – and all other family members – are unaffected, and the genetic risk for children is not higher than the incidence in the general population – about 1 : 15 000–1 : 25 000. Moreover, a somatic mutation always leads to unilateral retinoblastoma. However, about 10% of sporadic unilateral cases are caused by mutation in the germ cell of one parent (who occasionally is also a germ cell mosaic). Parents and other family members are therefore also unaffected; however, the risk for every child of the individual affected with sporadic unilateral retinoblastoma to carry the tumor is now about 45% (90% penetrance with a 50% segregation ratio). Consider the following situation: A sporadic, unilateral proband asks about the genetic risks to his or her children. As a first step we may study the chromosomes with cytogenetic and molecular methods, for example, FISH (Sect. 3.1.3), since the client may have a small deletion in 13q14. If no such deletion is found, and no other information is available, the risk of having children with retinoblastoma is $0.9 \times 0\%$ (for the noninherited proportion) $+ 0.1 \times 45\%$ (for the germ cell mutations) $= 4.5\%$. The situation becomes more complicated, however, if the proband already has one or more healthy children. If the disease had been caused by a dominant mutation, each of these

children would have had a risk of 45% to be affected. The fact that they are not affected therefore increases the chance of having the noninherited form of the disease and hence decreases the risk to future children. How can this risk be calculated?

Probability of Being a Hereditary Case [89]. As noted, the prior probability of our proband being a hereditary case is $P(H) = 0.1$. If he is such a case the conditional probability that his first child would be unaffected (event U), i.e., the probability that it is unaffected despite the fact that the proband carries the gene, is $P(U/H) = 0.55$. On the other hand, the prior probability that the proband is a nonhereditary case is $P(\text{not } H) = 0.9$. In this case, the conditional probability that his child would be unaffected would be $P(U/\text{not } H) = 1$, as there is (almost) no risk. From these considerations we can derive the formula for his posterior probability of being a hereditary case:

$$P(H/U) = \frac{P(H) \times P(U/H)}{P(H) \times P(U/H) + P(\text{not } H) \times P(U/\text{not } H)} \quad \text{(A 6.1)}$$

Inserting the figure from our example leads to:

$$P(H/U) = \frac{0.1 \times 0.55}{0.1 \times 0.55 + 0.9 \times 1} = \frac{0.055}{0.955} = 0.058$$

Hence one unaffected child reduces the probability of the proband being a hereditary case from 0.1 to 0.058. The risk of his next child being affected is now:

$$R_2 = 0.058 \times 0.45 = 0.0261$$

or a reduction from 4% to 2.6%. With two nonaffected children the conditional probability $P(U/H)$ becomes $0.55^2 = 0.3025$. Inserting into Eq. A 6.1 leads to $P(H/U) = 0.0325$, $R_2 = 0.015$. For n children $P(U/H)$ becomes 0.55^n. The principle can easily be grasped by considering Fig. A 6.1.

A Convenient Notation System and a Form of Graphic Representation. Murphy [61] suggested a clear and convenient system of notation that makes the calculation described above more evident, especially for those who have difficulties with abstract mathematical concepts. A table is constructed which visualizes the stepwise calculation. Table A 6.1 shows the above

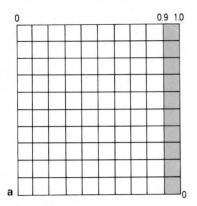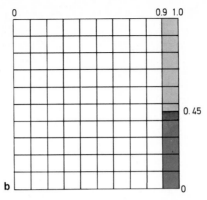

Fig. A 6.1 a, b. Graphic representation of the risk calculation for children of parents with unilateral retinoblastoma. **a** *Lightly shaded area,* the parents with hereditary retinoblastoma (≈ 10 % of all unilateral sporadic cases). **b** After the first child is born, 45 % of these parents (= 4.5 % of *all* parents) are revealed as carriers of the hereditary form. Only $\frac{5.5}{95.5}$ % or 5.8 % of the rest (= first child normal) will be carriers of the genes for the hereditary form

calculation for retinoblastoma. From the a posteriori probability of our client being a hereditary case (0.058) the probability that the first child will be affected can now be calculated as shown in the preceding paragraph: $0.058 \times 0.45 = 0.0261$ or 2 %–3 %.

The principle of calculation can also be shown graphically (Fig. A 6.1); in this square the *white area* represents the nonhereditary group and the *lightly shaded area* the hereditary cases. Once the first child has outlived the risk period (the first few years of life), the parents of affected children, namely 45 % of the hereditary cases (Fig. A 6.1, *dark area*), are eliminated from the group of sporadic unilateral retinoblastoma patients as a whole. These 45 % have been clearly established as hereditary, with the corresponding consequence of a 45 % risk for subsequent children. The risk for clients with healthy children must now be calculated on the basis of the total area *exclusive* of the *dark portion*. The hereditary cases are now no longer represented by 10/100 squares but only by 5.5/95.5 = 5.8 %.

If the client is a proven hereditary case, molecular analysis of the mutation in blood cells should be performed; prenatal diagnosis by studying genetic markers (= indirect approach), and, in an increasing fraction of families, also by direct study of the mutation (= direct approach) is possible.

Example: Huntington Disease. A healthy man at the age of 35 comes for genetic counseling; his father and his grandmother are affected with Huntington disease, and he is concerned about the risk for himself and for his future children. Huntington disease is an autosomal-dominant disease with full penetrance; the age at onset, however, varies between about 20 and 70 years of age (Sect. 4.1.2; Fig. 4.4). If the proband shows signs of the disease, the problem would be simple: each of his children has a risk of 50 %. If he has not yet reached the age of manifestation, the problem is also simple: he has a risk of 50 %, and the risk to his children is 50 % of 50 %, or 25 %. In fact, however, he has already passed through part of the manifestation period without being affected. This increases his chance of being homozygous for the normal allele and of remaining unaffected. How does this situation influence the risk to his children? By the age of 35 about 30 % of all heterozygotes show clinical signs of the disease. This leads to the calculation in Table A 6.2; here we see that the child's risk is reduced from 0.25 to 0.206.

Such calculations can be performed in many other specific situations involving autosomal-dominant and recessive diseases (for a detailed discussion, see [27]).

Since mutations leading to Huntington disease can be identified by direct DNA analysis (Sect. 3.1.3.8), such calculations will often be unnecessary. Many clients, however, wish not to know whether they are or are not carriers of a mutation that will eventually lead to Huntington disease; such clients many be interested in a risk figure that may be less than 50 %.

Heterozygotes of X-Linked Recessive Diseases. This type of calculation has its most important practical application in counseling of women who are at risk of being heterozygous for an X-chromosomal recessive trait and therefore of having affected sons. Consider the pedigree in Fig. A 6.2 a. We are virtually certain that Alma is heterozygous. Her daughter Barbara therefore also has a prior probability of 50 % of being heterozygous. This means a risk of $0.5 \times 0.5 = 0.25$ for any son to manifest the trait. If there is no further information, the above values must form the basis of any counseling.

Table A 6.1. Probability calculation for recurrence risk of retinoblastoma; sporadic, unilateral case (see text)

	Client is a hereditary case	Client is a non-hereditary case
a) The a priori probability (the client's chance of belonging to either group)	0.1	0.9
b) Conditional probability (that the first child will not be affected, given the group to which the parent belongs)	0.55	1.0
c) Combined probability (the chance that both a and b will occur)	$0.1 \times 0.55 = 0.055$	$0.9 \times 1.0 = 0.9$
d) The a posteriori probability (the client has a normal child and is/is not a hereditary case)	$\dfrac{0.055}{0.055 + 0.9} = 0.058$	$\dfrac{0.9}{0.055 + 0.9} = 0.942$

Table A 6.2. Calculation of recurrence risk for Huntington disease: hypothetical case of 35 year old offspring of a patient with HD (see text)

	Client heterozygous	Client homozygous normal
Prior probability	0.5	0.5
Conditional probability of not having clinical symptoms	0.7	1.0
Joint probability	$0.5 \times 0.7 = 0.35$	0.5
Posterior probability	$\dfrac{0.35}{0.5 + 0.35} = 0.412$	$\dfrac{0.5}{0.05 + 0.35} = 0.588$

Risk to child: $0.412 \times 0.5 = 0.206$

The situation is different in the pedigree of Fig. A 6.2 b. In this case Barbara already has a normal son. The conditional probability of having a normal son although she is heterozygous is 0.5. The calculation is presented in Table A 6.3.

The calculation is performed in a similar way if the pedigree is more complicated, for example, if Barbara has a daughter, and this daughter wants to know the risk for her sons, etc. In this case the posterior probability of Barbara's being heterozygous is used to calculate the prior probability for her daughter. For a number of specific examples, see [27].

A basically *new* situation occurs when the carrier of the disease concerned is a sporadic case (Fig. A 6.2 c). In this case he may either be a new mutant, when his mother is homozygous normal, and there is no increased risk to sons of his sisters; or the mother is heterozygous, in which case his sisters also have a prior probability of 0.5 of being heterozygous. As noted in Sect. 9.3, the proportion of new mutants among the bearers of a (rare) X-linked recessive condition is:

$$m = \frac{(1 - f)\mu}{2\mu + \nu}$$

(f = relative fertility of trait bearers in relation to the general population; μ = mutation rate in female germ cells; ν = mutation rate in male germ cells). When mutation rates are equal in the two sexes, and when $f = 0$, the formula is reduced to $m = \frac{1}{3}$. This means that the mother has a prior probability of two-thirds of being heterozygous. This leads to the risk calculation for Barbara's son as presented in Table A 6.4.

This simple calculation holds true only when the two above conditions ($\mu = \nu$; $f = 0$) are fulfilled, and when there is genetic equilibrium between mutation and selection. This is approximately true in Duchenne muscular dystrophy, the most common X-linked recessive disease in most centers. For other mutations, such as hemophilia A and HPRT deficiency, mutation rates appear to be much higher in male than in female germ cells (Sect. 9.3.4). Here the fraction m must be calculated on the basis of empirical evidence, $\nu = 3\mu$ being a useful approximation since the male mutation rate is about $3 \times$ as high as that in females. When no specific data are available, assuming a prior probability of 1 for Alma and $\frac{1}{2}$ for Barbara (Fig. A 6.4), may be a good choice (this overestimates the risk slightly). The genetic counselor should keep in mind – and should explain to the client – that ~ 3–5% of mothers of sporadic cases are gonadal mosaics who may have a second affected son. Such mosaics have been observed mainly in Duchenne disease but also in hemophilia.

The following example is slightly more complicated (for other examples see [27, 61, 36]). Figure A 6.3

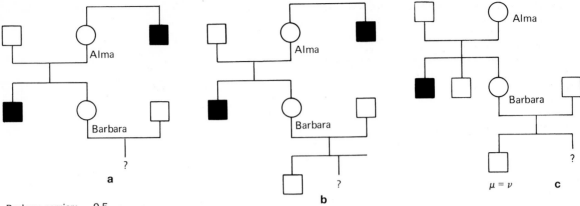

Barbara carrier: 0.5

Risk for son: 0.5 × 0.5 = 0.25

Fig. A 6.2 a–c. X-linked inheritance, Hypothetical pedigrees. **a** Risk to Barbara's son: 0.5 × 0,5 = 0.25. **b** Risk to Barbara's son: 0.333 × 0.5 = 0.167 (see Table A 6.3). **c** Risk to Barbara's son: $^1/_5 × ^1/_2 = ^1/_{10}$

Table A 6.3. Calculation of recurrence risk for X-linked disease: hypothetical case (see Fig. A 6.2 b and text)

	Barbara carrier	Barbara not carrier
Prior probability	0.5	0.5
Conditional probability	0.5	1.0
Joint probability	0.5 × 0.5 = 0.25	0.5 × 1.0 = 0.5
Posterior probability	$\frac{0.25}{0.5 + 0.25} = 0.333$	$\frac{0.05}{0.5 + 0.25} = 0.667$

Risk to son: 0.333 × 0.5 = 0.167

Table A 6.4. Calculation of recurrence risk: for sporadic case of (lethal) X-linked recessive disease (see Fig. A 6.2 c)

	Barbara carrier	Barbara not carrier
Prior probability	$^2/_3 × ^1/_2 = ^1/_3$	$^2/_3$
Conditional probability	$^1/_2$	1
Joint probability	$^1/_3 × ^1/_2 = ^1/_6$	$^2/_3$
Posterior probability	$\frac{^1/_6}{^1/_6 + ^2/_3} = ^1/_5$	$\frac{^2/_3}{^1/_6 + ^2/_3} = ^1/_5$
Risk to son:	$^1/_5 × ^1/_2 = ^1/_{10}$	

shows the pedigree. Barbara has an affected and a normal brother, but she also has a sister, Bettina, who is the mother of two normal sons. Bettina is either a normal homozygote ($^1/_2$ if Alma is heterozygous), in which case normal sons are to be expected, or she is heterozygous, in which case the conditional probability of having two normal sons is $^1/_4$. This aspect ($^1/_2 + ^1/_4 × ^1/_2 = ^5/_8$) is included in the calculation of the conditional probability for Alma, in which her conditional probability of having an unaffected son if she is heterozygous is also considered. The evaluation is presented in Table A 6.5 (again for $\mu = \nu$; $f = 0$).

These various calculations give the risk of the carrier state based on the pedigree information alone. In actual practice, additional information based on biochemical studies and DNA studies can often be included to refine the estimate of risk. In Duchenne

muscular dystrophy (DMD), for example, the mutation of the affected brother can often be identified as a deletion if this brother is available for study, or if the mother can be examined. If there is no deletion, the indirect method of DNA diagnosis using various polymorphisms could be applied. The dystrophin gene is very large (Sect. 3.1.3); therefore several markers are necessary to avoid misclassification due to intragenic crossing over. A practical example follows:

The example demonstrates how a statement about the carrier risk of two clients, derived only from pedigree information, can be drastically altered by additional use of DNA marker information (Fig. A 6.4). Subjects III,1 and III,2 had a deceased brother with DMD and want to know their risk of carrying the mutant gene. Since their maternal grandmother I,2 also had a brother with DMD, she and her daughter II,2 can be assumed to be obligate carriers of the mutation. Thus the prior probability of being a carrier is 50 % for both sisters III,1 and III,2. For III,1 this is the final pedigree risk of being a carrier. For III,2, considering her two normal sons decreases her prior probability of being a carrier to 20 %, as shown by the calculation in Table A 6.6.

When we examine the genotype pattern of the family members for the three DNA markers Dys II, STR 49, and Mz 18/19, we find that the loci of the first and last of these markers are situated at the proximal and distal ends, respectively of the DMD gene, while the locus of the second lies between these. The 2-1-2 haplotype of II,2 must have been transmitted from her deceased father (I,1) since the mother I,2 does not carry this haplotype. Since I2 and her daughter (II,2) have a brother and a son, respectively, with DMD, haplotype 3-3-1 which is shared by mother and daughter carries the DMD mutation. Only one (III,1) of II2's daughters inherited the haplotype with the DMD mutation (3-3-1), the other daughter III,2 did not, unless a double crossing over transferred the mutation to the homologous chromosome, an event which has a very low probability. The noncarrier state of III,2 is ad-

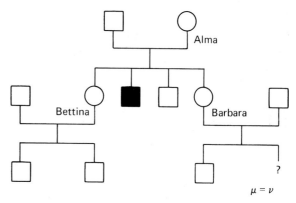

Fig. A 6.3. X-linked inheritance, risk calculation. Risk to Barbara's son: $^5/_{47} \times ^1/_2 = ^5/_{94}$ (see Table A 6.5)

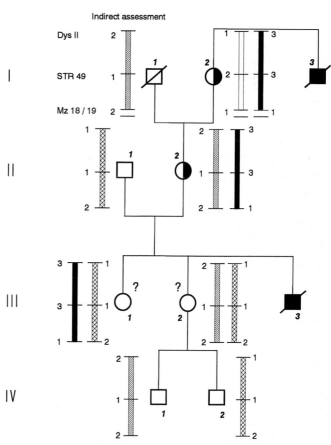

Fig. A 6.4. Linkage analysis in a DMD family with no living affected male for assessment of the carrier risk of the women III-1 and III-2. The following PCR-detectable polymorphic markers were used: Dys II (located at the very 5′ end of the dystrophin gene); Str 49 (located in the intron of exon 49), and Mz 18/19 (located at the 3′ untranslated region). The woman II,1 is a carrier since she has an affected son III,3. The only combination of markers which she has in common with her mother I,2 is 3,3,1; this must have been the marker combination of the mutation-carrying X chromosome of her affect-

ed brother I,3. It follows that the daughter III,1 is most likely a carrier since she does have the marker combination 3,3,1. Her sister III,2, on the other hand, does not have the combination 3,3,1; obviously, she has inherited from her mother the X chromosome with the combination 2,1,2, and from her father the 1,1,2 chromosome. It is therefore not surprising that her two sons are unaffected. (This argument ignores the unlikely possibility of double crossing over). (See text) (Courtesy of Dr. M. Cremer)

Table A 6.5. Calculation of recurrence risk in X-linked recessive inheritance: hypothetical case (see Fig. A 6.3) (For details see the text)

	Alma carrier	Alma not carrier
Prior probability	$2/3$	$1/3$
Conditional probability	$1/2 \times 5/8 = 5/16$	1
Joint probability	$2/3 \times 5/16 = 5/24$	$1/3 \times 1 = 8/24$
Posterior probability	$5/13$	$8/13$

	Barbara carrier	Barbara not carrier
Prior probability	$5/13 \times 1/2 = 5/26$	$21/26$
Conditional probability	$1/2$	1
Joint probability	$5/26 \times 1/2 = 5/52$	$21/26 \times 1 = 42/52$
Posterior probability	$5/47$	$42/47$

| Risk to Barbara's son: | $5/47 \times 1/2 = 5/94$ | |

Table A 6.6. Calculation of recurrence risk in X-linked recessive inheritance for a daughter of a DMD carrier with two normal sons (without DNA marker information) (see Fig. A 6.4)

	III,2 carrier	III,2 not carrier
Prior probability	$1/2$	$1/2$
Conditional probability of two normal sons:	$1/4$	1
Joint probability	$1/2 \times 1/4 = 1/8$	$1/2 \times 1 = 1/2$
Posterior probability	$(1/8)/(1/8 + 1/2) = 1/5$	$(1/2)/(1/8) + 1/2) = 4/5$

ditionally confirmed by the Dys II types of her healthy sons; as the two sons show different types, she could be a carrier only if a further (single) crossing over had occurred in the formation of her gametes, in addition to the double crossing over in the gametogenesis of her mother.

These considerations are corroborated by computation of the carrier probability for the two sisters using the computer program LINKAGE: depending on the actual position of the DMD mutation relative to the marker loci (which is not known) a value between 99.8 % and 99.9 % is obtained for the probability of III,1 being a carrier, while the corresponding value for III,2 varies between 0 and 0.008 %. For this computation the proportions of recombination between the marker loci were taken as 6 % (between Dys II and Str 49) and 4 % (between Str 49 and Mz 18/19).

Hence inclusion of DNA marker information changes the previous uncertainty of the carrier status for both clients to almost complete certainty, in opposite directions.

Similar considerations apply to the diagnosis of hemophilia A and B carriers by identification of mutants at the DNA level, wherever possible, or by indirect diagnosis using DNA markers (Figs. A 6.5, A 6.6). In the hemophilias, only a small proportion of mutants have been identified as deletions; most are base pair substitutions. This makes diagnosis more difficult (Sect. 9.4).

Until recently, biochemical studies of women at risk of being carriers were often performed using, for example, the muscle enzyme creatine phosphokinase in DMD and immunological determination of factor VIII in hemophilia A. Introduction of DNA methods is now making these methods largely superfluous. As shown by this example, risk computation might be complicated. Laboratories involved in counseling of such families are well advised to use a computerized system. Such systems are available.

Recurrence Risks for Children of Unaffected Gene Carriers in Autosomal Dominant Inheritance. Occasionally a clinically unaffected person who has relatives affected with an autosomal-dominant disease of lowered penetrance seeks advice regarding the disease risk to his or her offspring. Regardless of the exact penetrance it has been shown that the clinical risk to children of a person at 50 % genetic risk is never more than 9 % [2366]. The reason for this is that the unaffected parent is unlikely to carry the gene in diseases with high penetrance. Conversely, in diseases with low penetrance, although the parent is a gene carrier, the chance of a child being clinically affected is small.

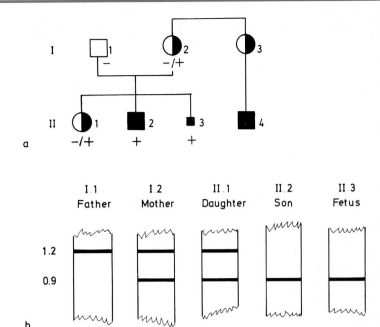

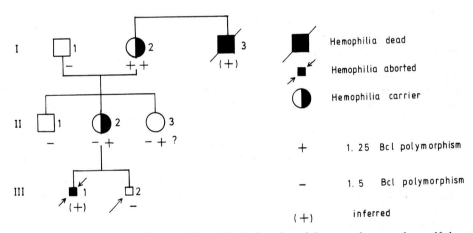

Fig. A 6.5. a, b. Principle of heterozygote identification and prenatal diagnosis in hemophilia A. The mother (I,2) is double heterozygous for the hemophilia allele and the RFLP marker +. The father (I,1) is healthy and has the marker −. Since the affected son II,2 has inherited the hemophilia allele as well as the marker + from his mother, the marker + must be on the same chromosome as the hemophilia allele (= cou-

pling, Sect. 5.1.). Since the daughter II,1 is heterozygous for − / +, she must have inherited the chromosome containing + and the hemophilia gene from her mother; she must be heterozygous for hemophilia. The fetus II,3 is +; he must have inherited the same chromosome, and is affected with hemophilia

Fig. A 6.6. Hemophilia A pedigree. Note that hemophilia A could be ruled out in a fetus (III-2) by prenatal diagnosis using a factor VIII probe and restriction enzyme *Bgl* without actually testing the DNA phenotypes of the two hemophiliacs (III-1 and I-3). The presence of the DNA variant in the fetus

(III-2) identical to that of the normal maternal grandfather (I-1) indicated that the fetus was normal. Note that the mother's sister's (II, 3) carrier state cannot be determined using the DNA variant. (From Din et al. 1985)

Appendix 7

Linkage Calculations: Programs and Examples

Chapter 5 describes methods for examining pedigree material for linkage. The use of lod scores is described as the method of choice, and the derivation of this method from pedigree probabilities is presented for of a monogenic inheritance.

A study of linkage between a trait showing monogenic inheritance and a genetic marker (which has previously been localized) consists of three steps:

1. Linkage must be established. As discussed in Sect. 5.1.2, it has become a convention to regard linkage as confirmed if the combined lod score exceeds + 3, i.e., if the odds in favor of linkage are 1000 : 1 (i. e., the logarithm of 1000). This strict criterion, proposed by Morton [54], has been adopted because the a priori probability of autosomal linkage between two loci chosen at random is very low, and the risk of error is therefore high, especially when multiple markers are tested. If the lod score is lower than − 2, linkage is regarded as excluded.

2. When linkage is established, the most likely recombination fraction Θ must be estimated. There is a sex difference: for most areas of the human genome, Θ is higher in the female than in male meiosis; therefore it should be calculated separately for the two sexes.

3. Very often a linkage study is carried out simultaneoulsy with several pedigrees. Increasingly, identical phenotypes turn out to be caused by mutations at different gene loci, i. e., there is genetic heterogeneity. Hence some pedigrees show linkage while others do not. On the other hand, we always find some degree of chance variation between pedigrees in the number of meioses that do or do not show recombination even if there is no linkage at all. The investigator who is eager to find linkage should not be seduced into selecting only those pedigrees that show a positive lod score. Adding such lod scores may produce spurious evidence for linkage. Special statistical methods have been developed to distinguish true heterogeneity with linkage in some pedigrees from spurious evidence for linkage [62].

Computer programs are available for linkage studies, including heterogeneity testing (Table A 7.1). Only one example is presented below, in which linkage of the gene for a human disease with a "classical" marker (the GC serum polymorphism) was shown in a large pedigree.

Dentinogenesis Imperfecta and the GC Serum Groups. Figure A 7.1 shows a large pedigree in which an autosomal-dominant condition, dentinogenesis imperfecta (125 490), segregates with the GC protein types. There are three alleles, GC1S, GC1F, and GC2. Full evaluation of the pedigree would entail calculation of probable genotypes for individuals who could not be typed for the marker (I, 1, 2; II, 1, 2) on the basis of gene frequencies. Then, the full pedigree can be scored, using the LIPED program (Table A 7.1).

Table A 7.2 gives the lod scores for Θ_M (recombination proportion in males) and Θ_F (recombination proportion in females). The maximum lod score (7.9238) is found at $\Theta_M = 0.05$ for males and $\Theta_F = 0.25$ for females. Table A 7.3 presents a more detailed calculation for the "critical" area of Table A 7.2; the best estimates are $\Theta_M = 0.05$ and $\Theta_F = 0.24$; lod score $z = 7.9277$.

The linkage has been confirmed by subsequent studies; the gene DGI 1 for dentinogenesis imperfecta has been localized to 4 q 13–4 q 21. The GC locus is located at 4 q 12 [52].

This example shows the simplest possible case: study of one big pedigree. In many instances, numerous small pedigrees are available. Here, the problem of linkage heterogeneity often comes up: the same or similar phenotypes are caused by mutations at different gene loci. This linkage heterogeneity can be studied by special statistical methods ([62]).

Such two-point linkage calculations are usually the first step of analysis when the question of mapping a disease locus is adressed. A more powerful method is to use a fixed map of several markers to calculate the likelihood of the location of the presumed disease locus at various points along the known map (= multipoint linkage analysis). The question of locus ordering and simultaneous estimation of several recombination rates requires complex multivariate linkage

Table A 7.1. Computer programs for analysis of linkage

Name	Special features of program	Contact address	Reference
LINKAGE (Includes LIPED)	Two-point and multipoint linkage; special versions for CEPH family data; risk calculation	J. Ott, Rockefeller Univ. New York NY, USA	Ott (1991) [63]
HOMOG	Test for heterogeneity of linkage data		Ott (1983) [62]
SLINK	Simulation program (together with LINKAGE)		
FASTMAP	Fast techniques for constructing a multipoint map from two-point analyses	D. Curtis, Academic Dept. of Psychiatry, St. Mary's Hospital, Praed St., London W 2 7NY, UK	Curtis and Gurling (1993) [16]
MAP	Multiple pairwise linkage for locus ordering; interference possible; quality control for mispairing	V. Andrews, CRC Research Group in Genetic Epidemiology, Dept. of Community Medicine, University of Southamptom, South Academic Block, Southampton General Hospital, Southamptom S 09 4X, UK	Morton and Andrews (1989) [56]
FASTLINK	Faster version of LINKAGE due to improvement of algorithms	A. Schafter, Dept. of Computer Science, Rice University, Houston TX 7725, USA	Cottingham et al. [14]
MENDEL	Same range of functions as LINKAGE	K. Lange, Dept. of Biostatistics, U. of Michigan, Ann Arbor, Mich	Lange et al. (1988) [48]
LINKSYS	Easily usable data bank and generation of data sets for LINKAGE and LIPED	J. Atwood, Genetics and Biometry Dept. UCL Wolfsson House 4, Stephenson Way, London NW1 2HE, UK	Atwood and Bryant (1988) [4]
Mapmaker	Construction of multipoint maps in CEPH-type families	Mapmaker Distribution, Landerlab, Whitehead Institute, for Biomedical Research, Nine Cambridge Center, Cambridge MA 02142, USA	Lander et al. (1987) [47]
CRI-MAP	Construction of multipoint maps	Dr. Phil Green, Dept. Molecular Biotechnology, Univ. of Washington, Seattle WA 98195, USA	Green et al. (1989) [31]

Programs for detection of linkage and mapping are also found in some of the programs of Table A 2.1, for example SAGE. The SAGE program also permits linkage analysis by the sib pair (or close relatives) method. Other programs for sibpair analysis include KIN and PEDSCORE.

analyses [63, 64]. Stepwise strategies have been used to construct a comprehensive genetic map of presently 1266 intervals with an average distance of 2.9 cM. Many problems arise in multipoint linkage analysis, such as what model of interference is to be used, what technical constraints exist, and should sex-specific map distances be estimated. A detailed discussion is provided by Ott [63, 64]. Fig. A. 7.2 gives a recent survey [52] of disease loci in the human genome.

Table A 7.2. Pedigree with dentinogenesis imperfecta and GC blood types

Θ_M	Θ_F									
	0.05	0.10	0.15	0.20	0.25	0.30	0.35	0.40	0.45	0.50
0.05	5.7385	7.0780	7.6460	7.8825	7.9238	7.8286	7.6263	7.3316	6.9506	6.4831
0.10	5.4992	6.8370	7.4048	7.6418	7.6841	7.5906	7.3908	7.0997	6.7240	6.2639
0.15	4.9709	6.3028	6.8690	7.1058	7.1488	7.0569	6.8596	6.5723	6.2019	5.7492
0.20	4.3034	5.6209	6.1826	6.4177	6.4605	6.3697	6.1747	5.8911	5.5262	5.0811
0.25	3.5484	4.8381	5.3897	5.6203	5.6613	5.5704	5.3772	5.0971	4.7376	4.3001
0.30	2.7353	3.9794	4.5123	4.7332	4.7690	4.6761	4.4831	4.2057	3.8512	3.4214
0.35	1.8921	3.0723	3.5733	3.7754	3.7997	3.6999	3.5039	3.2269	2.8764	2.4540
0.40	1.0653	2.1707	2.6243	2.7935	2.7940	2.6774	2.4705	2.1884	1.8390	1.4252
0.45	0.3170	1.3584	1.7647	1.8954	1.8641	1.7217	1.4957	1.2033	0.8581	0.4764
0.50	0.2835	0.7003	1.0796	1.1946	1.1545	1.0094	0.7890	0.5196	0.2383	0.0000

Table A 7.3. Part of Table A.7.2, with finer subdivisions

Θ_M	Θ_F										
	0.20	0.21	0.22	0.23	0.24	0.25	0.26	0.27	0.28	0.29	0.30
0.01	7.1367	7.1579	7.1721	7.1797	7.1813	7.1771	7.1676	7.1529	7.1334	7.1092	7.0807
0.02	7.5746	7.5958	7.6100	7.6177	7.6193	7.6152	7.6057	7.5911	7.5716	7.5475	7.5190
0.03	7.7649	7.7862	7.8005	7.8082	7.8098	7.8058	7.7963	7.7818	7.7624	7.7384	7.7100
0.04	7.8524	7.8737	7.8880	7.8957	7.8974	7.8934	7.8840	7.8695	7.8502	7.8263	7.7979
0.05	7.8825	7.9038	7.9182	7.9260	7.9277	7.9238	7.9144	7.9000	7.8807	7.8569	7.8286
0.06	7.8757	7.8970	7.9114	7.9193	7.9211	7.9171	7.9079	7.8935	7.8743	7.8505	7.8223
0.07	7.8427	7.8642	7.8786	7.8865	7.8883	7.8844	7.8752	7.8609	7.8418	7.8181	7.7900
0.08	7.7902	7.8117	7.8261	7.8341	7.8360	7.8321	7.8230	7.8087	7.7897	7.7660	7.7380

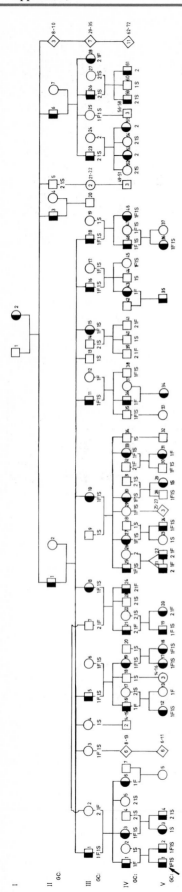

Fig. A7.1. Pedigree in which the autosomal-dominant gene for dentinogenesis imperfecta and the genes for the GC blood types segregate [559]. For explanation and linkage analysis see the text

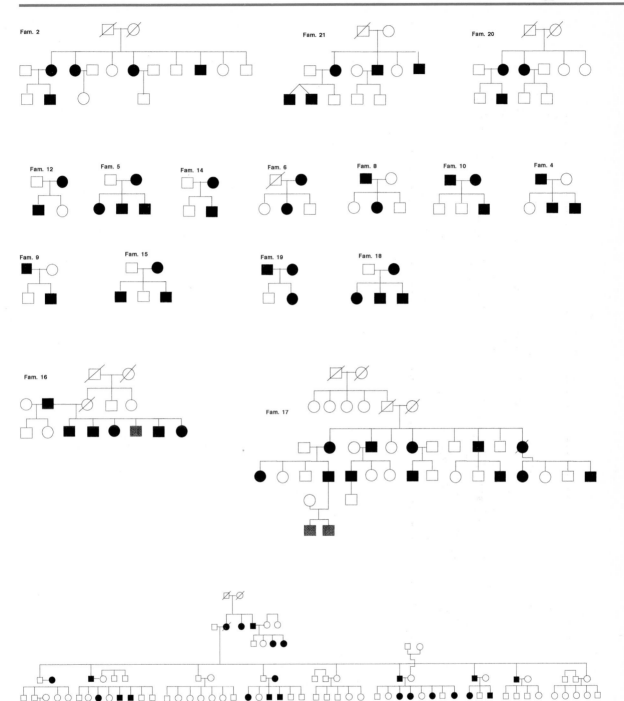

Fig. A 7.2. 17 pedigrees included in a linkage study on the human low-voltage EEG (LVEEG) [3].

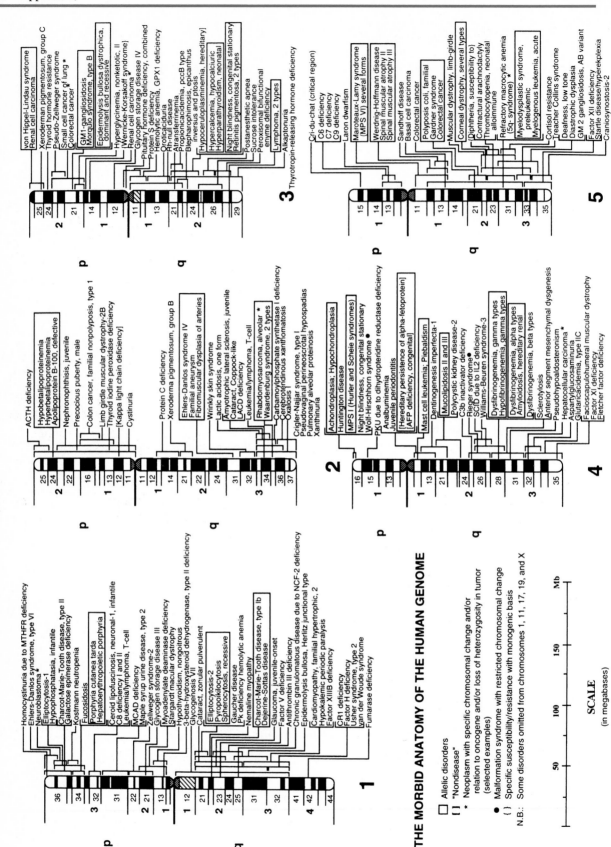

THE MORBID ANATOMY OF THE HUMAN GENOME

☐   Allelic disorders

[ ] "Nondisease"

*   Neoplasm with specific chromosomal change and/or
     relation to oncogene and/or loss of heterozygosity in tumor
     (selected examples)

●   Malformation syndrome with restricted chromosomal change

{ } Specific susceptibility/resistance with monogenic basis

N.B.: Some disorders omitted from chromosomes 1, 11, 17, 19, and X

SCALE
(in megabases)

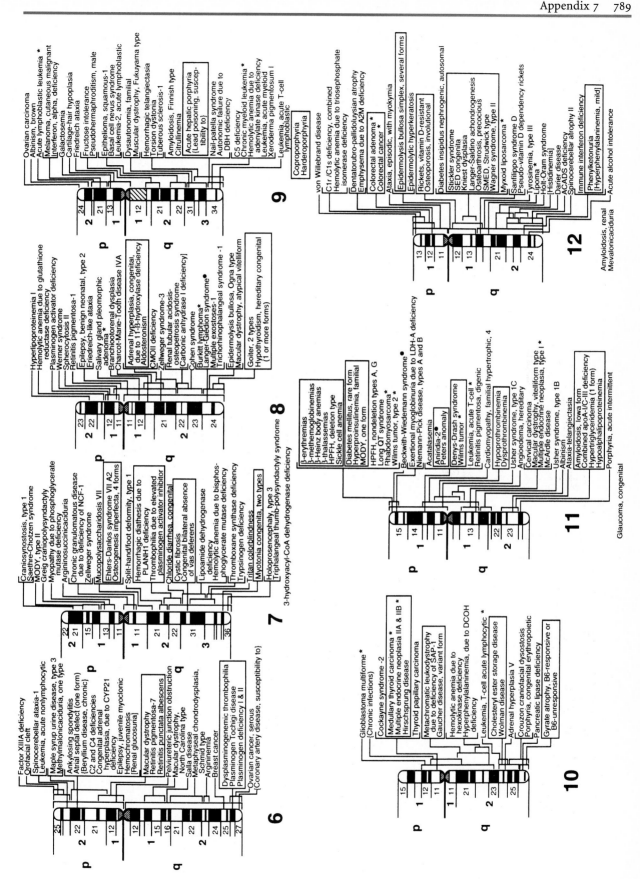

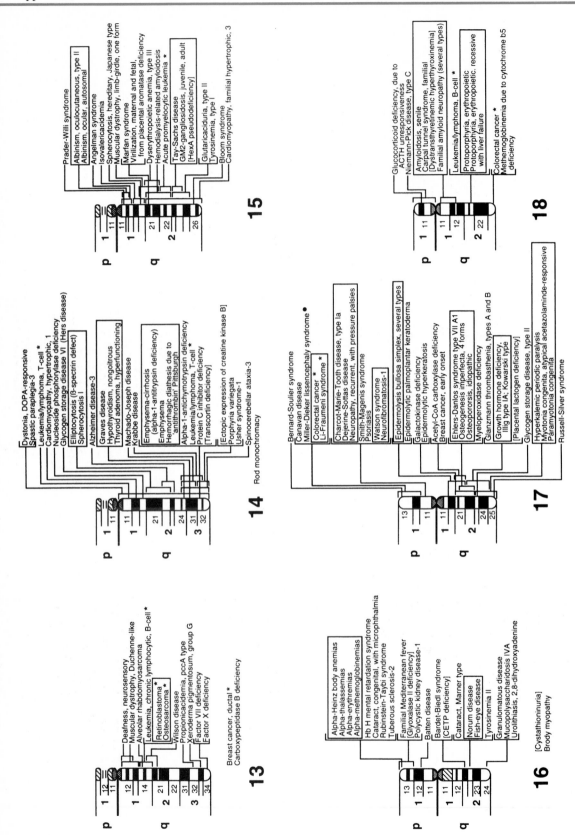

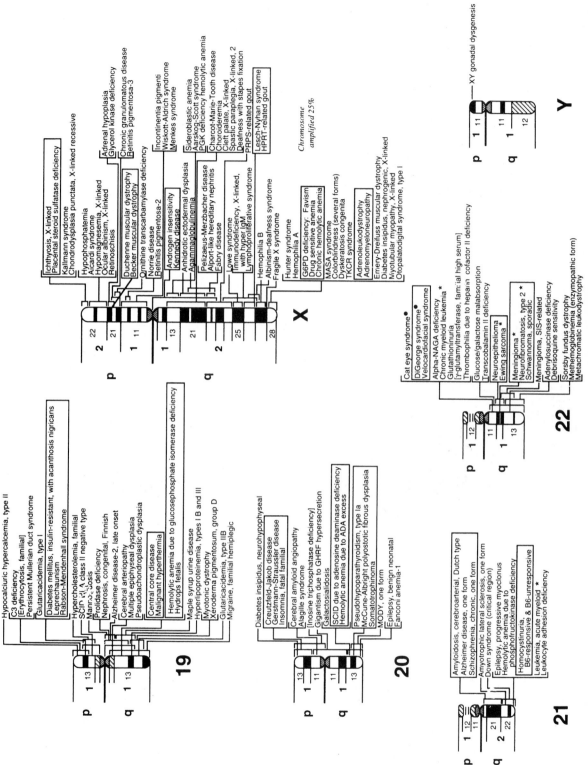

Fig. A7.2. The morbid anatomy of the human genome. Disorders with confirmed or provisional assignments have been included. Because of the large number of disorders assigned to specific regions of the X chromosome, only selected ones are represented here. From McKusick 1995 [52])

Appendix 8

Standardized Nomenclature for Human Genes

The increasing number of identified human genes has made rules for their nomenclature necessary. An international committee in 1979 drafted the following set of rules [79, 80]:

1. Gene loci are designated by uppercase letters or by a combination of uppercase letters and Arabic numerals. Three characters is optimum, and no more than four characters are recommended; these characters are italicized. Examples: *G6PD* (glucose 6-phosphate dehydrogenase), *CBP* (Protan color blindness). The initial character should always be a letter. All characters in a gene symbol should be written on the same line; no superscripts or subscripts may be used. No Roman numerals or Greek letters are admissible; Greek letters are transliterated into their Latin equivalents.

2. Where gene products of similar function are coded by different genes, the corresponding loci are designated by Arabic numerals immediately after the gene symbol. Examples: *ADH1, ADH2, ADH3* (three alcoholic dehydrogenase loci). A final character in the gene symbol may be used to indicate a specific characteristic of the gene. Example: *PKL*, where L refers to the liver form of pyruvate kinase.

3. Allele designations are written on the same line as gene symbols. They are separated from the locus characters by an asterisk which serves to combine gene and allele symbols. Examples: *HLAA*2, HLAA*3* for alleles at the HLAA locus; *HBB*6V* identifies substitution of GLU by VAL at the 6th position of the hemoglobin β chain. The allele symbol should always be limited to four characters, with an optimum of three.

4. In genotypes a horizontal line or slash separates the alleles and indicates chromosomal location. For example, an individual heterozygous at the ADA locus and homozygous at the AMY1 locus is designated as follows:

$$\frac{ADA^*1}{ADA^*2}; \frac{AMY^*A}{AMY^*A}, \text{ or } ADA^*1/ADA^*2; AMY1^*A/AMY1^*A.$$

If the locus is X-linked, females are depicted as shown above for autosomal genes; males are identified as follows: *G6PD*B/Y.*

5. Linkage and phase: loci not located on the same chromosome are separated by a semicolon; as shown above (4). Loci located on the same chromosome when the phase is known are joined by a horizontal line but separated by a space. Loci on the same chromosome, but when the phase is unknown, are separated by a comma. If the order of genes on the chromosome is known, they are listed in order from the end of the short arm to the end of the long arm, and separated by a space. More detailed information is found in the committee reports [860].

An International Committee for Standardizing Human Gene nomenclature has been founded; when new human genes are discovered, and a symbol is proposed, this committee should be contacted. The address is A.J.Cuticchia, Johns Hopkins Hospital Medical School, Baltimore, MD, USA.

A mutation nomenclature is in preparation. Please contact E.Beutler, The Scripps Research Inst., Dept. of Molecular and Experimental Medicine, 10666 North Torrey Pines Rd., La Jolla, CA 92037. Phone (001619) 554-8049; fax (001619) 554-6927

References

1. Aicardi J (1986) Epilepsy in children. Raven, New York
2. Anderson VE, Hauser WA, Leppik IE et al (1991) Genetic strategies in epilepsy research. Elsevier, Amsterdam
3. Anokhin A, Steinlein O, Fischer C et al (1992) A genetic study of the human low-voltage electroencephalogram. Hum Genet 90 : 99–112
4. Atwood J, Bryant S (1988) A computer program to make linkage analysis with LIPED and LINKAGE easier to perform and less prone to imput errors. Ann Hum Genet 52 : 259
5. Ball SP, Cook PJL, Mars H, Buckton KE (1982) Linkage between dentinogenesis imperfacta and GC. Ann Hum Genet 46 : 35–40
6. Barakat AY, DerKaloustian VM, Mufarrij AA, Birbari A (1986) The kidney in genetic disease. Churchill Livingstone, Edinburgh
7. Bishop MJ (1994) Guide to human genome computing. Academic, London
8. Blandfort M, Tsuboi T, Vogel F (1987) Genetic counseling in the epilepsies. I. Genetic risks. Hum Genet 76 : 303–331

9. Borgaonkar DS (1989) Chromosomal variation in man. A catalogue of chromosomal variants and anomalies, 5th edn. Liss, New York

10. Boyd WC (1955) Simple maximum likelihood method for calculating Rh gene frequencies in Pacific populations. Am J Phys Anthropol 13 : 447–453

11. Bundey S (1992) Genetics and neurology, 2nd edn. Churchill Livingstone, Edinburgh

12. Buyse ML (ed) (1990) Birth defects encyclopedia. Center for Birth Defects Information Service, Dover

13. Cantor RM, Rotter JI (1992) Analysis of genetic data: methods and interpretation. In: King A, Rotter JI, Motulsky AG (eds) The genetic basis of common diseases. Oxford University Press, New York, pp 49–70

14. Cottingham RW, Idury RM, Schaeffer (1993) Faster sequential linkage computations. Am J Hum Genet 51:A33

15. Crawfurd dÁM (1988) The genetics of renal tract disorders. Oxford University Press, London

16. Curtis D, Gurling H (1993) A procedure for combining two-point lod scores into a summary multipoint map. Hum Hered. 43 : 173–185

17. Czeizel A (1978) The Hungarian congenital malformation monitoring system. Acta Paediatr Acad Sci Hung 19 : 225–238

18. Czeizel A, Kiss P, Oszotovics M et al (1978) Nationwide investigations of multiple malformations. Acta Paediatr Acad Sci Hung 19 : 275–280

18 a. Decorte R, Cassiman J-J (1993) Forensic medicine and the polymerase chain reaction technique. J Med Genet 30: 625–633

19. Eaves LJ (1982) The utility of twins. In: Anderson VE et al (eds) Genetic basis of the epilepsies. Raven, New York, pp 249–276

20. Elston RC (1981) Segregation analysis. Adv Hum Genet 11 : 63–120

21. Emery AEH (1993) Duchenne muscular dystrophy, 2nd edn. Oxford University Press, Oxford

22. Emery AEH, Rimoin DL (1990) Principles and practice of medical genetics, 2nd edn. Churchill Livingstone, Edinburgh

23. Falconer DS (1981) Introduction to quantitative genetics, 2nd edn. Oliver and Boyd, Edinburgh

24. Feller W (1970/1971) An introduction to probability theory and its applications, 2nd edn. Wiley, New York

25. Fernandes J, Saudubray JM, Tada K (eds) (1990) Inborn metabolic diseases. Springer, Berlin Heidelberg New York

26. Fischer C, Schweigert C, Spreckelsen, Vogel F (1996) Programs, databases and expert systems for human geneticists – a survey. Hum Genet 97 : 129–137

27. Fuhrmann W, Vogel F (1983) Genetic counseling, 3rd edn. Springer, Berlin Heidelberg New York (Heidelberg science library 10)

28. Gardner RJM, Sutherland GR (1989) Chromosomal abnormalities and genetic counseling. Oxford University Press, Oxford

29. Gedde-Dahl T, Wuepper KD (eds) (1987) Prenatal diagnosis of heritable skin disease. Curr Probl Dermatol 16

30. Gorlin RJ, Cohen M, Levin LS (1990) Syndromes of the head and neck, 3rd edn. Oxford University Press, Oxford

31. Green P, Falls K, Crooks S (1989) Documentation for CRI-MAP, Version 2.1 Collaborative Research,

32. Gyapay G, Morisette J, Vignal A et al (1994) The 1993–94 Généthon human genetics linkage map. Nature [Genet] 7 : 246–249

33. Haataja L, Schleutker J, Laine A-P et al (1994) The genetic locus for free sialic acid storage disease maps to the long arm of chromosome 6. Am J Hum Genet 54 : 1042–1049

34. Haldane JBS, Smith CAB (1947) A simple exact test for birth order effect. Ann Eugen 14 : 116–122

35. Harper PS (ed) (1991) Huntington's disease. Saunders, London

36. Harper PS (1993) Practical genetic counseling, 4th edn. Wright, Bristol

37. Hasstedt SJ (1982) A mixed-model likelihood approximation on large pedigrees. Comput Biomed Res 15 : 295–307

38. Helmbold W, Prokop O (1958) Die Bestimmung der ABo-Genfrequenzen mittels der maximum-likelihood-Methode und anderer Verfahren anhand forensischer Blutgruppenbestimmungen in Berlin. Blut 4 : 190–201

39. Jensen AR (1973) Educability and group differences. Methuen, London

40. Jones KL (1988) Smith's recognizable patterns of human malformation, 4th edn. Saunders, Philadelphia

41. Kalousek DK, Fitch N, Paradice B (1990) Pathology of the human embryo and previable fetus. An atlas. Springer, Berlin Heidelberg New York

42. Kamel NN (1992) A profile for molecular biology data bases and information resources. CABIOS 8 : 311–321

43. Kamin LJ (1974) The science and politics of I. Q. Erlbaum, Potomac

44. King RA, Rotter JI, Motulsky AG (eds) (1992) The genetic basis of common diseases. Oxford University Press, New York

45. Krüger J (1973) Zur Unterscheidung zwischen multifaktoriellem Erbgang mit Schwellenwerteffekt und einfachem diallelem Erbgang. Hum Genet 17 : 181–252

46. Lalouel JM, Morton NE (1981) Complex segregation analysis with pointers. Hum Hered 31 : 312–321

47. Lander ES, Green P, Abrahamson I et al (1987) MAPMAKER: an interactive computer package for construction of primary genetic linkage maps of experimental populations. Genomics 1 : 182–189

48. Lange K, Weeks D, Boehnke M (1988) Program for pedigree analysis: MENDEL, FISHER and dGENE. Genet Epidemiol 5 : 471–472

49. Lathrop GM, Lalouel JM, Julier C, Ott J (1984) Strategies for multilocus linkage analysis in humans. Proc Natl Acad Sci USA 81 : 3443–3446

49 a. Leppert M, Anderson VE, Quattlebaum T et al (1989) Benign familial neonatal convulsions linked to genetic markers on chromosome 20. Nature 337: 647–648

50. Matthiuz PL, Ihde D, Piazza A, Ceppellini R, Bodmer WF (1970) New approaches to the population genetics and segregation analysis of the HL-A system. In: Terasaki P (ed) Histocompatibility testing. Munksgaard, Copenhagen, pp 193–205

51. McGuffin P, Huckle P (1990) Simulation of mendelism revisited: the recessive gene for attending medical school. Am J Hum Genet 46 : 994–999

52. McKusick VA (1995) Mendelian inheritance in man, 11th edn. Johns Hopkins University Press, Baltimore

53. Milunsky A (1992) Genetic disorders and the fetus, 3rd edn. Plenum, New York

54. Morton NE (1955) Sequential tests for the detection of linkage. Am J Hum Genet 7 : 277–318

55. Morton NE (1982) Outline of genetic epidemiology. S Karger, Basel

56. Morton NE, Andrews V (1989) MAP, an expert system for multiple pairwise linkage analysis. Ann Hum Genet 53 : 263–269

57. Morton NE, Chung CS (eds) (1978) Genetic epidemiology. Academic, New York

58. Morton NE, Rao DL, Lalouel JM (1983) Methods in genetic epidemiology. Karger, Basel

59. Morton NE, Simpson SP, Lew R, Yee S (1983) Estimation of haplotype frequencies. Tissue Antigens 22 : 257–262

60. Mourant AE, Kopec AC, Domaniewska-Sobczak K (1978) Blood groups and diseases. Oxford University Press, London

61. Murphy EA, Chase GA (1975) Principles of genetic counseling. Year Book Medical Publishers, Chicago

62. Ott J (1983) Linkage analysis and family classification under heterogeneity. Ann Hum Genet 47 : 311–320

63. Ott J (1991) Analysis of human genetics linkage. Johns Hopkins University Press, Baltimore

64. Ott J (1992) Strategies for characterizing highly polymorphic markers in human gene mapping. Am J Hum Genet 51 : 283–290

65. Pauli RM, Motulsky AG (1981) Risk counselling in autosomal dominant disorders with undetermined penetrance. J Med Genet 18 : 340–343

66. Race RR, Sanger R (1975) Blood groups in man, 6th edn. Blackwell, Oxford

67. Reed EW, Reed SC (1965) Mental retardation: a family study. Saunders, Philadelphia

68. Renie WA (ed) (1986) Goldberg's genetic and metabolic eye disease. Lea and Febiger, Philadelphia

69. Riccardi VW (1992) Neurofibromatosis, 2nd edn. Johns Hopkins University Press, Baltimore

70. Romeo R (1987) Prenatal diagnosis of congenital anomalies. Appleton and Lange, East Norwalk

71. Royce PM, Steinmann B (eds) (1993) Connective tissue and its heritable disorders. Wiley-Liss, New York

72. Schaefer EJ (1984) Clinical, biochemical, genetic features in familial disorders of high density lipoprotein deficiency. Arteriosclerosis 4 : 303–322

73. Schäfer MSD (1983) Segregation and Pathologie autosomaler familiärer Translokationen beim Menschen. PhD thesis, University of Kaiserslautern

74. Schiff F, von Verschuer O (1933) Serologische Untersuchungen an Zwillingen. II. Mitt Z Morphol Anthropol 32 : 244–249

75. Schinzel A (1984) Catalogue of unbalanced chromosome aberrations in man. De Gruyter, Berlin

76. Scriver CR, Beaudet AL, Sly WS, Valle D (eds) (1989) The metabolic basis of inherited disease, 6th edn. McGraw-Hill, New York

77. Seeds JW, Azizkhan RG (1990) Congenital malformations. Antenatal diagnosis. Perinatal management and counselling. Aspen, Rockville

78. Selvin S (1977) Genetic systems for diagnosis of twin zygosity. Acta Genet Med Gemellol (Roma) 26 : 81–82

79. Shows TB, Alper CA, Bootsma D et al (1979) International system for human gene nomenclature, ISGN (1979) Cytogenet Cell Genet 25 : 96–116

80. Shows TB, McAlpine PJ et al (1987) Guidelines for human gene nomenclature: an international system for human gene nomenclature (ISGN, HGM 9). Cytogenet Cell Genet 46 : 11–28

81. Siemens HW (1924) Die Zwillingspathologie. Springer, Berlin

82. Siemens HW (1924) Die Leistungsfähigkeit der zwillingspathologischen Arbeitsmethode. Z Induktive Abstammungs Vererbungslehre 33 : 348

83. Slater E (1966) Expectation of abnormality on paternal and maternal sides: a computational model. J Med Genet 3 : 159–161

84. Smith C (1971) Recurrence risks for multifactorial inheritance. Am J Hum Genet 23 : 578–588

85. Smith CAB (1959) A note on the effect of ascertainment on segregation ratios. Ann Hum Genet 23 : 311–323

86. Smith SM, Penrose LS (1955) Monozygotic and dizygotic twin diagnosis. Ann Hum Genet 19 : 273–289

87. Stevenson AC, Cheeseman EA (1955/56) Hereditary deaf mutism with particular reference to Northern Ireland. Ann Hum Genet 20 : 177–231

88. Swash M, Schwartz MS (1988) Neuromuscular disease, 2nd edn. Springer, Berlin Heidelberg New York

89. Vogel F (1957) Die eugenische Beratung beim Retinoblastom (Glioma retinae). Acta Genet 7 : 565–572

90. Vogel F (ed) (1974) Erbgefüge. Springer, Berlin Heidelberg New York (Handbuch der allgemeinen Pathologie, vol 9)

91. Vogel F, Fujiya Y (1969) The incidence of some inherited EEG variants in normal Japanese and German males. Hum Genet 7 : 38–42

92. Vogel F, Wendt GG (1956) Zwillingsuntersuchung über die Erblichkeit einiger anthropologischer Maße und Konstitutionsindices. Z Menschl Vererbungs Konstitutionslehre 33 : 425–446

93. Vogel F, Schalt E, Krüger J, Propping P (1979) The electroencephalogram (EEG) as a research tool in human behavior genetics: psychological examinations in healthy males with various inherited EEG variants. I. Rationale of the study; material; methods; heritability of test parameters. Hum Genet 47 : 1–45

94. Wallace DC (1989) Report of the Committee on Human Mitochondrial DNA. Cytogenet Cell Genet 51 : 612–621

95. Wallace DC (1989) Programs, databases and expert systems for human geneticists – a survey. Hum Genet 97 : 129–137

96. Wallace DC (1994) Mitochondrial DNA sequence variation in human evolution and disease. Proc Natl Acad Sci USA

97. Weaver DD (1992) Catalog of prenatally diagnosed conditions, 2nd edn. Johns Hopkins University Press, Baltimore

97a. Weeks DE, Lange K (1988) The affected-pedigree member method of linkage analysis. Am J Hum Genet 47: A 204

97b. Weitkamp LR, Lewis RA (1989) Pedscore analysis of identical by descent (IBD) marker allele distribution in affected family members. Cytogenet Cell Genet 51: 1105–1106

98. Winter RM, Baraitser M (1991) Multiple congenital anomalies. A diagnostic compendium. Chapman and Hall, London

99. Winter RM, Knowles SAS, Bieber FR, Baraitser M (1988) The malformed fetus and stillbirth. A diagnostic approach. Wiley, Chichester

100. Yasuda N, Tsuji K (1975) A counting method of maximum likelihood for estimating haplotype frequency in the HL-A system. Jpn J Hum Genet 20 : 1

101. Young ID (1991) Introduction to risk calculation in genetic counselling. Oxford University Press, Oxford

Further Reading

Abrahamson S, Bender MA, Conger AD, Wolff S (1973) Uniformity of radiation induced mutation rates among different species. Nature 245 : 460–462

Acampora D, D'Esposito M, Faiella A et al (1989) the human HOX gene family. Nucl Acid Res 17 : 10385–10402

Adam A (1986) Polymorphisms of red-green vision among populations of the tropics. Soc Study Hum Biol Symp 27 : 245

Anderson RM, May RM (1982) Coevolution of hosts and parasites. Parasitology 85 : 411–426

Anderson RM, May RM (1991) Infectious diseases of humans: dynamics and control. Oxford University Press, Oxford

Anonymous (1982) Identifying and estimating the genetic impact of chemical environmental mutagens. National Academy, Washington

Anonymous (1983) International Commission for Protection Against Environmental Mutagens and Carcinogens. Committee 4 report. Mutation Res 115 : 255–291

Anonymous (1984) Human gene therapy – a background paper. US Congress, Office of Technology Assessment, OTA-BP-BA-32, December, Washington

Anonymous (1985) Low maternal serum alphafetoprotein and Down syndrome. Lancet I:259–260

Anonymous (1990) Office of technology assessment. Genetic screening in the workplace OTA-BA-456. US Government Printing Office, Washington DC

Anonymous (1993) Editorial: diagnosing the heart of the problem. Nature Genetics 4 : 211–212

Aoyama T, Tynan K, Dietz H et al (1993) Missense mutations impair intracellular processing of fibrillin and microfibril assembly in Marfan syndrome. Hum Molec Genet 2 : 2135–2140

Applebaum EG, Firestein SK (1983) A genetic counseling casebook. Free, New York

Austin MA (1994) Genetic and environmental influences on LDL subclass phenotypes. Clin Genet 46 : 64–70

Ayesh R, Idle JR, Ritchie JC, Crothers MJ, Hetzel MR (1984) Metabolic oxidation phenotypes as markers for susceptibility to lung cancer. Nature 312 : 169–170

Bailey-Wilson JE, Elson RC (1987) Statistical analysis for genetic epidemiology (SAGE): introduction. Louisiana State Medical Center, New Orleans

Bank A, Mears JG, Ramirez F (1979) Organization of the human globin genes in normal and thalassemia cells. In: Stamatoyannopoulos G, Nienhuis A (eds) Cellular and molecular regulation of hemoglobin switching. Grune and Stratton, New York, pp 521–539

Baraitser M (1982) The genetics of neurological disorders. Oxford University Press, Oxford

Baringa M (1994) From fruit flies, rats, mice: evidence of genetic influence. Science 264 : 1690–1693

Barker D, Schafer M, White R (1984) Restriction sites containing CpG show a higher frequency of polymorphism in human DNA. Cell 36 : 131–138

Beauchamp GK, Yamazaki K, Boyse EA (1985) The chemosensory recognition of genetic individuality. Sci Am 253:86–92

Becker PE (1977) Myotonia congenita and syndromes associated with myotonia. Thieme, Stuttgart

Benjamin R, Parham P (1990) Guilt by association: HLA-B27 and ankylosing spondylitis. Immunol Today 11 : 137–142

Berg K (ed) (1985) Medical genetics; present, past, future. Liss, New York

Berginer VM, Foster NL, Sadowsky M et al (1988) Psychiatric disorders in patients with cerebrotendinous xanthomatosis. Am J Psychiat 145 : 354–357

Bergsma D (1979) Birth defects compendium, 2nd edn. An encyclopedic guide to birth defects. Liss, New York

Bernstein F (1930) Fortgesetzte Untersuchungen aus der Theorie der Blutgruppen. Z. Induktive Abstammungs-Vererbungslehre 56 : 233–237

Bernstein F (1930) Über die Erblichkeit der Blutgruppen. Z. Induktiven Abstammungs-Vererbungslehre 54 : 400

Beutler E (1978) Hemolytic anemia in disorders of red cell metabolism. Plenum, New York

Beutler E (1994) G6PD deficiency. Blood 84 : 3613–3636

Bickel H, Guthrie R, Hammersen G (eds) (1980) Neonatal screening for inborn errors of metabolism. Springer, Berlin Heidelberg New York

Birch J, Chisholm IA, Kinnear P et al (1979) Clinical testing methods. In: Pokorny J, Smith VC, Verriest G, Pincker AJLG (eds) Congenital and acquired color vision defects. Grune and Stratton, New York, pp 83

Birch-Jensen A (1949) Congenital deformities of the upper extremities. Opera ex Domo Biologiae Hereditariae Humanae Universitatis Hafniensis Munksgaard, Kopenhagen 19

Birth defects (1972) Part XV. The cardiovascular system. Williams and Wilkins, Baltimore

Birth defects (1974) Part XVI. Urinary system and others. Williams and Wilkins, Baltimore

Bodmer WF (1981) The William Allan Memorial Award Address: gene clusters, genome organization and complex phenotypes. When the sequence is known, what will it mean? Am J Hum Genet 33 : 664–682

Bodmer WF (ed) (1978) The HLA system. Br Med Bull 34 (3): 213–319

Bodmer WF, Bodmer JG (1978) Evolution and function of the HLA system. Br Med Bull 34 : 390–316

Bond DJ, Chandley AC (1983) Aneuploidy. Oxford University Press, Oxford

Borberg A (1951) Clinical and genetic investigations into tuberous sclerosis and Recklinghausen's neurofibromatosis. Munksgaard, Copenhagen

Bordarier C, Robain O, Rethoré O, Dulac O, Dhellemes C. Inverted neurons in Agyria – a Golgi study of a case with abnormal chromosome 17

Bottema CDK, Ketterling RP, Li S et al (1991) Missense mutations and evolutionary conservation of amino acids: evidence that many of the amino acids in factor IX function as "spacer" elements. Am J Human Genet 49 : 820–838

Bowden DK, Vickers MA, Higgs DR (1992) A PCR-based strategy to detect the common and severe determinants of α-thalassemia. Br J Haematol 81 : 104

Bowmaker JK (1991) The evolution of vertebrate visual pigments and photoreceptors. In: Cronly-Dillon JR, Gregory RL (eds) Vision and visual dysfunction, vol 2. Evolution of eye and visual systems. Macmillan, London

Bowman JE (ed) (1983) Distribution and evolution of hemoglobin and globin loci. Elsevier, New York

Brewer GJ (1971) Annotation: human ecology, an expanding role for the human geneticist. Am J Hum Genet 23 : 92–94

Brown CJ, Ballabio A, Rupert J et al (1991) A gene from the region of the human X inactivation center is expressed exclusively from the inactive X chromosome. Nature 349 : 38–44

Brown MS, Goldstein JL (1984) How LDL receptors influence cholesterol and atherosclerosis. Sci Am 251 : 58–66

Buchsbaum MS, Coursey RD, Murphy DL (1976) The biochemical high-risk paradigm: behavioral and familial correlates of low platelet monoamine oxidase activity. Science 194 : 339–341

Bunn HF, Forget BG (1986) Hemoglobin: molecular, genetic and clinical aspects. Saunders, Philadelphia

Burke BE, Shotton DM (1983) Erythrocyte membrance skeleton abnormalities in hereditary spherocytosis. Br J Haematol 54 : 173–187

Burke W, Hornung S, Copeland BR, Furlong CE, Motulsky AG (1984) Red cell sodium-lithium countertransport in hypertension. In: Villarreal H, Sambhi MP (eds) Topics in pathophysiology of hypertension. Nijhoff Boston, pp 88–99

Burke W, Motulsky AG (1985) Genetics of hypertension. Pract Cardiol 11 : 159–173

Burke W, Motulsky AG (1985) Hypertension – some unanswered questions. JAMA 253 : 2260–2261

Butterworth T, Ladda RL (1981) Clinical genodermatology, vol 1, 2. Praeger, New York

Carney RG Jr (1976) Incontinentia pigmenti: a world statistical analysis. Arch Dermatol 112 : 535–542

Carr DH (1971) Chromosomes and abortion. Adv Hum Genet 2 : 201–257

Carter CO (1969) Genetics of common disorders. Br Med Bull 25 : 52–57

Carter CO (1976) Genetics of common single malformation. Br Med Bull 32 : 21–26

Carter CO, Fairbank TJ (1974) The genetics of locomotor disorders. Oxford University Press, London

Carter CO, Hamerton JL, Polani PE, Gunalp A, Weller SDV (1960) Chromosome translocation as a cause of familial mongolism. Lancet II:678–680

Caskey CT (1985) New aid to human gene mapping. Nature 314 : 19

Ceppellini R, Siniscalco M, Smith CAB (1955/56) The estimation of gene frequencies in a random-mating population. Ann Hum Genet 20 : 97

Chakraborty R, Weiss KM, Majumder PP et al (1984) A method to detect excess risk of disease in structured data: cancer in relatives of retinoblastoma patients. Genet Epidemiol 1 : 229–244

Chakravarti A, Buetow KH, Antonarakis SE, Waber PG, Boehm CD, Kazazian HH (1984) Nonuniform recombination within the human β-globin cluster. Am J Hum Genet 36 : 1239–1258

Chandley AC (1989) Why don't the mule and the hinny look alike? Q J Br Mule Soc 43 : 7–10

Chandley AC, Mitchell AR (1988) Hypervariable minisatellite regions are sites for crossing over at meiosis in man. Cytogenet Cell Genet 48 : 152–155

Chargaff E (1976) On the dangers of genetic meddling. Science 192 : 938

Chaudhuri A, Das B (1991) Efficient sequential search of genetic systems for diagnosis of twin zygosity. Acta Genet Med Gemellol 40 : 159–164

Childs B (1972) Genetic analysis of human behavior. Ann Rev Med 23 : 373–406

Childs B, Finucci JM, Preston MS, Pulver AE (1976) Human behavior genetics. Adv Hum Genet 7 : 57–97

Childs B, Moxon ER, Winkelstein JA (1992) Genetics and infectious diseases. In: King RA, Rotter JI, Motulsky AG (eds) The genetic basis of common diseases. Oxford University Press, New York, pp 71–91

CIBA Found Symp (1963) Man and his future. In: Wolstenholme GEW (ed) Churchill, London

Cloninger CR, Rice J, Reich Th, McGriffin P (1982) Genetic analysis of seizure disorders as multidemensional threshold characters. In: Anderson VE (ed) Genetic basis of the epilepsies. Raven, New York, pp 291–309

Colledge WH (1994) Cystic fibrosis gene therapy. Curr Op Genet Dev 4 : 466–471

Comings DE (1972) The structure and function of chromatin. Adv Hum Genet 3 : 237–431

Coneally PM, Rivas ML (1980) Linkage analysis in man. Adv in Hum Genet 10 : 209–266

Constantini F, Chada K, Magram J et al (1986) Correction of murine β-thalassemia by gene transfer into the germ line. Science 233 : 1192–1194

Cooper GM (1990) Oncogenes. Jones and Bartlett, Boston

Courtney M, Jaliat S, Tessier L-H, Benavente A, Crystal RG, Lecocq J-P (1985) Synthesis in E coli of a₁-antitrypsin variants of therapeutic potential for emphysema and thrombosis. Nature 313 : 149–151

Creutzfeldt W, Köbberling J, Neel JV (eds) (1976) The genetics of diabetes mellitus. Springer, Berlin Heidelberg New York

Crow JF (1983) Chemical mutagen testing: a committee report. Environ Mutagen 5 : 255–261

Crowe RR (1975) Adoption studies in psychiatry. Biol Psychiatry 10 : 353–371

Cruz-Coke R (1982) Nomogram for estimating specific consanguinity risk. J Med Genet 19 : 216–217

Czeizel A, Pazonyi I, Métreki J, Tomka M (1979) The first five years of the Budapest twin register, 1970–1974. Acta Genet Med Gemellol 28 : 73–76

Dalgaard OZ (1957) Bilateral polycystic disease of the kidneys. Opera ex Domo Biologiae Hereditariae Humanae Universitatis Hafniensis Munksgaard, Copenhagen 38

Danieli GA, Mostacciuolo ML, Pilotto G, Angelini C, Bonfante A (1980) Duchenne muscular dystrophy. Data from family studies. Hum Genet 54 : 63–68

Das BM, Deka R (1975) Predominance of the haemoglobin E gene in a Mongoloid population in Assam (India). Hum Genet 30 : 187–191

De Braekeleer M, Larochelle J (1991) Population genetics of vitamin D-dependent rickets in northeastern Quebec. Ann Hum Genet 55 : 283–290

Deisseroth A, Nienhuis A, Lawrence J, Giles R, Turner P, Ruddle FH (1978) Chromosomal localization on human β globin gene on human chromosome 11 in somatic cell hybrids. Proc Natl Acad Sci USA 75 : 1457–1460

Deisseroth A, Nienhuis A, Turner P, Velez R, Anderson WF, Ruddle F, Lawrence J, Creagan R, Kucherlapati R (1977) Localization of the human α-globin structural gene to chromosome 16 in somatic cell hybrids by molecular hybridization assay. Cell 12 : 205–218

Den Dumen JT, Grootscholten PM, Dauwerse JG et al (1992) Reconstruction of the 2.4 Mb human DMD gene by homologous YAK recombination. Hum Molec Genet 1 : 19–28

Der Kaloustian VM, Kurban AK (1979) Genetic diseases of the skin. Springer, Berlin Heidelberg New York

Desnick RJ (1979) Prospects for enzyme therapy in the lysosomal storage diseases of Ashkenazi Jews. In: Goodman RM, Motulsky AG (eds) Genetic diseases among Ashkenazi Jews. Raven, New York, pp 253–270

Dickson G, Love DR, Davies KE et al (1991) Human dystrophin gene transfer: production and expression of a functional recombinant DNA-based gene. Hum Genet 88 : 53–58

Dombrowski BA, Mathias SL, Nanthakumar E et al (1991) Isolation of an active human transposable element. Science 254 : 1805–1808

Dutrillaux B, Viegas-Péguignot E, Aurias A, Mouthuy M, Prieur M (1981) Non random position of metaphasic chromosomes: a study of radiation induced and constitutional chromosome rearrangements. Hum Genet 59 : 208–210

Dworkin RB, Omenn GS (1985) Legal aspects of human genetics. Ann Rev Public Health 6 : 107–130

Edwards JH (1960) The simulation of mendelism. Acta Genet (Basel) 10 : 63–70

Edwards RG, Steptoe PC (1973) Biological aspects of embryo transfer. In: Law and ethics of AID and embryo transfer. Ciba Found Symp (new series)17 : 11–18

Ehrman L, Parsons PA (1976) The genetics of behavior. Sinauer, Sunderland

Ehrman S, Omenn GS, Caspari E (1972) Genetics, environment, and behavior. Academic, New York

Emery AEH (1976) Methodology in medical genetics. Churchill Livingstone, Edinburgh

Emery AEH, Pullen I (1984) Psychologic aspects of genetic counseling. Academic

Epstein CJ, Hassold TJ (eds) (1989) Molecular and cytogenetic studies of non-disjunction. Liss, New York

Falconer DS (1965) The inheritance of liability to certain diseases, estimated from the incidence among relatives. Ann Hum Genet 29 : 51

Falk R, Motulsky AG, Vogel F, Weingart P (1985) Historische und ethische Aspekte der Humangenetik. Ein Interdisziplinares Kolloquium im Wissenschaftskolleg, in Wissenschaftskolleg – Institute for Advanced Study – Berlin Yearbook 1983/84. Siedler, Berlin, pp 75–121

Farber SL (1981) Identical twins reared apart. Basic Books, New York

Fialkow PJ (1977) Clonal origin and stem cell evolution of human tumors. In: Mulvihill JJ, Miller RW, Fraumeni JF Jr (eds) Genetics of human cancer. Raven, New York, pp 439–453

Fields C, Adams MD, White O, Venter JC (1994) How many genes in the human genome? Nature Genet 7 : 345–346

Fincham JRS, Sastry GRK (1974) Controlling elements in maize. Ann Rev Genet 8 : 15–50

Fischer M (1972) Umweltfaktoren bei der Schizophrenie. Intrapaarvergleiche bei eineiigen Zwillingen. Nervenarzt 43 : 230–238

Fisher RA, Race RR (1946) Rh gene frequencies in Britain. Nature 157 : 8

Fitch WM, Langley CH (1976) Evolutionary rates in proteins: neutral mutations and the molecular clock. In: Goodman M, Tashian RE (eds) Molecular anthropology. Plenum, New York, pp 197–219

Flatz G (1995) The genetic polymorphism of intestinal lactase activity in adult humans. In: The metabolic and molecular basis of inherited disease, vol III. McGraw-Hill, New York, pp 4441–4450

Flemming W (1897) Über die Chromosomenzahl beim Menschen. Anat Anz 14 : 171

Ford CE (1969) Mosaics and chimaeras. Br Med Bull 25 : 104–109

Fraccaro M, Kaijser K, Lindsten J (1959) Chromosome complement in gonadal dysgenesis (Turner's syndrome). Lancet I:886

Frankel AD, Kim PS (1991) Modular structure of transcription factors: implications for gene regulation. Cell 65 : 717–719

Fraser GR (1976) The causes of profound deafness in childhood. John Hopkins University Press, Baltimore

Friedmann T (1989) Progress toward human gene therapy. Science 244 : 1275–1282

Fu Y-H, Kuhl DPA, Pizzuti A et al (1991) Variation of the CGG repeat at the fragile X site results in genetic instability: resolution of the Sherman paradox. Cell 67 : 1047–1058

Fuller JL, Thompson WRT (1960) Behavior genetics. Wiley, New York

Galjaard H (1980) Genetic metabolic diseases. Early diagnosis and prenatal analysis. Elsevier/North Holland, Amsterdam

Garmezy N (1974) Children at risk: the search for the antecedents of schizophrenia. I. Conceptual models and research methods. NIMH Schizophrenia Bull 8 : 14–90

Gebhardt E (1974) Antimutagens. Data and problems (review). Hum Genet 24 : 1–32

Gebhardt E (1981) Sister chromatid exchange (SCE) and structural chromosome aberration in mutagenicity testing. Hum Genet 58 : 235–254

Gershon ES, Targum SD, Kewssler LR, Mazure CM, Bunney WE (1978) Genetic studies and biologic strategies in the affective disorders. Prog Med Genet NS vol II, 101–166

Gilbert W (1978) Why genes in pieces? Nature 271 : 501

Gitschier J, Drayna D, Tuddenham EGD, White RL, Lawn RM (1985) Genetic mapping and diagnosis of haemophilia A achieved through a BciI polymorphism in the factor VIII gene. Nature 314 : 738–740

Gitschier J, Kogan S, Diamond C, Levinson B (1991) Genetic basis of hemophilia A. Thromb Haemost 66 : 37–39

Gitschier J, Levinson B, Lehesjoki AE, de la Chapelle A (1989) Mosaicism and sporadic hemophilia: implications for carrier determination. Lancet I:273–274

Goedde HW, Altland K, Scholler KL (1967) Therapie der durch genetisch bedingte Pseudocholinesterase-Varianten verursachten verlängerten Apnoe nach Succinylcholin. Med Klin 62 : 1631–1635

Goldberger AS (1978) Pitfalls in the resolution of I.Q. inheritance. In: Morton NE, Chung CS (eds) Genetic epidemiology. Academic, New York, pp 195–222

Goodman RM, Gorlin RJ (1977) Atlas of the face in genetic disorders, 2nd edn. Mosby, St Louis

Goodnow JJ (1976) The nature of intelligent behavior: question raised by cross-cultural studies. In: Resnick LB (ed) The nature of intelligence. Erlbaum, Hillsdale, pp 169–188

Gorlin RJ, Pindborg JJ, Cohen MM Jr (eds) (1976) Syndromes of the head and neck, 2nd edn. McGraw Hill Book Co, New York

Green PM, Montadon AJ, Bentley DR et al (1990) The incidence and distribution of CpG-TpG transitions in the coagulation in the factor IX gene: a fresh look at CpG mutational hotspots. Nucleic Acids Res 18 : 3227–3231

Grosveld F, Kollias G (eds) (1992) Transgenic animals. Academic, London

Grzeschik K-H (1973) Utilization of somatic cell hybrids for genetic studies in man. Hum Genet 19 : 1–40

Gurdon JB (1992) The generation of diversity and pattern in animal development. Cell 68 : 185–199

Haag EVD (1969) The Jewish mystique. Stein and Day, New York

Hadorn E (1955) Developmental genetics and lethal factors. Wiley, London

Haldane JBS (1939) The spread of harmful autosomal recessive genes in human populations. Ann Eugen 9 : 232–237

Haldane JBS, Moshinsky P (1939) Inbreeding in Mendelian populations with special reference to human cousin marriage. Ann Eugen 9 : 321–340

Hall JM, Lee MK, Newman B et al (1990) Linkage of early-onset familial breast cancer to chromosome 17q21. Science 250 : 1684–1689

Haller M (1963) Eugenics: hereditarian attitudes in American thought. New Brunswick

Hamilton WD (1964) The genetical theory of social behavior, I, II. J Theor Biol 7 : 1–52

Hammer RE, Pursel VG, Rexroad CEJr, Wall RJ, Bolt DJ, Ebert KM, Palmiter RD, Brinster RL (1985) Production of transgenic rabbits, sheep and pigs by microinjection. Nature 315 : 680–683

Harris H (1975) Prenatal diagnosis and selective abortion. Harvard University Press, Cambridge

Harris H, Hopkinson DA (1976) Handbook of enzyme electrophoresis in human genetics. North-Holland, Amsterdam

Hauge M, Harvald B, Fischer M, Gottlieb-Jensen K, Juel-Nielsen N, Raebild I, Shapiro R, Videbech T (1968) The Danish twin register. Acta Genet Med Gemellol (Roma) 18 : 315–332

Hayden MR (1981) Huntington's chorea. Springer, Berlin Heidelberg New York

Heigl-Evers A, Schepank H (eds) (1980, 1982) Ursprünge seelisch bedingter Krankheiten, vols I, II. Vandenhoeck und Rupprecht, Göttingen

Hein DW, Ferguson RJ, Doll MA et al (1994) Molecular genetics of human polymorphic N-acetyltransferase: enzymatic analysis of 15 recombinant wild-type, mutant, and chimeric NAT2 allozymes. Hum Molec Genet 3 : 729–734

Heitz E (1928) Das Heterochromatin der Mouse. I. Pringsheims Jb Wiss Botanik 69 : 762–818

Helmbold W (1959) Über den Zusammenhang zwischen ABo Blutgruppen und Krankheit. Betrachtungen zur Ursache der ABo-Frequenzverschiebung bei Patienten mit Carcinoma ventriculi, carcinoma genitalis und ulcus pepticum. Blut 5 : 7–22

Hess D (1972) Transformationen an höheren Organismen. Naturwissenschaften 59 : 348–355

Heuschert D (1963) EEG-Untersuchungen an eineiigen Zwillingen im höheren Lebensalter. Z Menschl Vererbungs-Konstitutionslehre 37 : 128

Hirszfeld L (1928) Konstitutionsserologie und Blutgruppenforschung. Springer, Berlin Heidelberg New York

Hoch M, Jaeckle H (1993) Wechselspiel der Gene. Aus: Forschung und Medizin, Schering 8 : 27–38

Holmquist GP (1992) Review article: chromosome bands, their chromatin flavor, and their functional features. Am J Hum Genet 51 : 17–37

Hook EB (1981) Unbalanced Robertsonian translocations associated with Down's syndrome or Patau's syndrome: chromosome subtype, proportion inherited, mutation rates and sex ratio. Hum Genet 59 : 235–239

Hook EB, Cross PK, Jackson L et al (1988) Maternal age-specific rates of 47, + 21 and other cytogenetic abnormalities diagnosed in the first trimester of pregnancy in chorionic villus biopsy specimens: comparison with rates expected from observations in amniocentesis. Am J Hum Genet 42 : 797–807

Howell RR, Stevenson RE (1971) The offspring of phenylketonuric women. Soc Biol 18 : 519–529

Hultén M (1974) Chiasma distribution at diakinesis in the normal human male. Hereditas 76 : 55–78

Hultén M, Lindsten J (1973) Cytogenetic aspects of human male meiosis. Adv Hum Genet 4 : 327–387

Husén T (1960) Abilities of twins. Scand J Psychol 1 : 125–135

Ingram VM (1957) Gene mutations in human haemoglobin: the chemical difference between normal and sickle cell hemoglobin. Nature 180 : 325–328

Ionasescu V, Zellweger H (1983) Genetics in neurology. Raven, New York

Iselius L, Lindsten J, Aurias A, Fraccaro M and many other authors (1983) The 11q;22q translocation: a collaborative study of 20 new cases and analysis of 110 families. Hum Genet 64 : 343–355

Jacobs PA, Frackiewitz A, Law P (1972) Incidence and mutation rates of structural rearrangements of the autosomes in man. Ann Hum Genet 35 : 301–319

Jacobs PA, Sulzman AE, Funkhauser J et al (1982) Human triploidy: relationship between parental origin of the additional haploid complement and development of partial hydatidiform mole. Ann Hum Genet 46 : 223–231

Jalbert P, Sele B (1979) Factors predisposing to adjacent −2 and 3 : 1 disjunctions: study of 161 human reciprocal translocations. J Med Genet 16 : 467–478

Jayakar SD (1970) A mathematical model for interaction of gene frequencies in a parasite and its host. Theor Popul Biol 1 : 140–164

Jeffreys AJ, Wilson V, Thein SL (1985) Individual-specific "fingerprints" of human DNA. Nature 316 : 76–79

Jordan B (1993) Traveling around the human genome. Inserm/CNRS, Marseille

Jorde LB (1989) Inbreeding of the Utah Mormons: an evaluation of estimates based on pedigrees, isonymy, and migration matrices. Ann Hum Genet 53 : 339–355

Juel-Nielsen N, Harvard B (1958) The electroencephalogram in monovular twins brought up apart. Acta Genet (Basel) 9 : 57–64

Kalter H, Warkany J (1983) Congenital malformations. N Engl J Med 308 : 491–497

Kan YW, Dozy AM (1978) Antenatal diagnosis of sickle cell anaemia by DNA analysis of amniotic-fluid cells. Lancet II:910–912

Karel ER, te Meerman GJ, ten Kate LP (1986) On the power to detect differences between male and female mutation rates for Duchenne muscular dystrophy, using classical segregation analysis and restruction fragment length polymorphisms. Am J Hum Genet 38 : 827–840

Keats BJB, Sherman SL, Morton NE et al (1991) Guidelines for human linkage maps. An international system for human linkage maps (ISLM) 1990. Ann Hum Genet 55 : 1–6

Kelly P (1977) Dealing with dilemma. A manual for genetic counselors. Springer, Berlin Heidelberg New York

Kessler S (1979) Genetic counseling, psychological dimensions. Academic, New York

Kevles DJ, Hood L (1992) The code of codes. Scientific and social issues in the human genome project. Harvard University Press, Cambridge

Klein D, Franceschetti A (1964) Mißbildungen und Krankheiten des Auges. In: Becker PE (ed) Humangenetik. Ein kurzes Handbuch, vol 4. Thieme, Stuttgart, pp 1–345

Klein J (1975) Biology of the mouse histocompatibility-2 complex. Springer, Berlin Heidelberg New York

Klever M, Grond-Ginsbach C, Scherthan H, Schroeder-Kurth TM (1991) Chromosomal in situ suppression hybridization after Giemsa banding. Hum Genet 86 : 484–486

Knight SJL, Flannery AV, Hirst M et al (1993) Trinucleotide repeat amplification and hypermethylation of a CpG island in FRAXE mental retardation. Cell 74 : 127–134

Koch M, Fuhrmann W (1985) Sibs of probands with neural tube defects – a study in the Federal Republic of Germany. Hum Genet 70 : 74–79

Koller S (1983) Risikofaktoren der Schwangerschaft. Springer, Berlin Heidelberg New York

Konigsmark BW, Gorlin RJ (1976) Genetic and metabolic deafness. Saunders, Philadelphia

Kornberg A (1992) DNA replication, 2 nd edn. Freeman, New York

Kouri RE, McKinney CE, Levine AS, Edwards BK, Vessell ES, Nebert DW, McLemore TL (1984) Variations in arylhydrocarbon hydroxylase activities in mitogen-activated human and non-human primate lymphocytes. In: Toxicol Pathol 12 : 44–48

Kouri RE, McKinney CE, Slomiany DJ, Snodgrass DR, Wray NP, McLemore TL (1982) Positive correlation between arylhydrocarbon hydroxylase activity and primary lung cancer as analysed in cryopreserved lymphocytes. Cancer Res 42 : 5030–5037

Kuhlo W, Heintel H, Vogel F (1969) The 4–5/sec rhythm. EEG Clin Neurophysiol 26 : 613–619

Kunze J, Tolksdorf M, Wiedemann H-R (1975) Cat eye syndrome. Hum Genet 26 : 271–289

Kupfer A, Preisig R (1984) Pharmacogenetics of mephenytoin: a new drug hydroxylation polymorphism in man. Eur J Clin Pharmacol 26 : 753–759

Kyle JW, Birkenmeier EH, Gwynn B et al (1990) Correction of murine mucopolysaccharidosis VII by a human β-glucuronidase transgene. Proc Natl Acad Sci USA 87 : 3914–3918

Latt SA, Schreck RR, Laveday KS, Dougherty CP, Schuler CF (1980) Sister chromatid exchanges. Adv in Hum Genet 10 : 267–331

Law PK, Bertolini TE, Goodwin TG et al (1990) Dystrophin production induced by myoblast transfer therapy in Duchenne muscular dystrophy. Lancet 336 : 114–115

Lawton JR, Martinez FA, Burks C (1989) Overview over the LiMB database. Nucl Acid Res 17 : 5885–5891

LeBeau M, Rowley JD (1984) Heritable fragile sites in cancer. Nature 308 : 607–608

Leder P (1978) Discontinuous genes. N Engl J Med 298 : 1079–1081

Lewontin RC (1967) An estimate of average heterozygosity in man. Am J Hum Genet 19 : 681–685

Lewontin RC (1974) The genetic basis of evolutionary change. Columbia University Press, New York

Lindgren D (1972) The temperature influence on the spontaneous mutation rate. I. Literature review. Hereditas 70 : 165–178

Livingstone FB (1958) Anthropological implications of sickle cell gene distributions in West Africa. Am Anthropologist 60 : 533–562

Livingstone FB (1962) The origin of the sickle cell gene. Conference on African Historical Anthropology. Northwestern University, Chicago

Lockridge O (1992) Genetic variants of human serum butyrylcholinesterase influence the metabolism of the muscle relaxant succinylcholine. In: Pharacogenetics of Drug Metabolism. Pergamon, New York, pp 15–50

Lott JT, Lai F (1982) Dementia in Down's syndrome, observations from a neurology clinic. Appl Res Ment Retard 3 : 233–239

Luzzato L, Mehta (1995) Glucose-6-phosphate dehydrogenase deficiency. In: The metabolic and molecular basis of inherited disease, chap III, vol III. McGraw-Hill, New York, pp 3367–3389

Lynas MA (1956/57) Dystrophia myotonica with special reference to Northern Ireland. Ann Hum Genet 21 : 318–351

Lyon MF, Philipps RJS (1975) Specific locus mutation rates after repeated small radiation doses to mouse oocytes. Mutation Res 30 : 375–382

MacSorley K (1964) An investigation into the fertility rates of mentally ill patients. Ann Hum Genet 27 : 247

Madan K (1983) Balanced structural changes involving the human X: effect on sexual phenotype. Hum Genet 63 : 216–221

Magenis RE, Overton KM, Chamberlin J, Brady T, Lovrien E (1977) Parental origin of the extra chromosome in Down's syndrome. Hum Genet 37 : 7–16

Mann JD, Cahan A, Gelb A, Fisher N, Hamper J, Tipett P, Sanger R, Race RR (1962) A sex-linked blood group. Lancet I:8

Manosevitz M, Lindzey G (1969) Thiessen DD. Behavioral genetics: methods and research. Appleton-Century-Crofts, New York

Marin-Padilla M (1975) Abnormal neuronal differentiation (functional maturation) in mental retardation. Birth Defects 11(7):133–153

Martin GM (1978) The pathobiology of aging. University of Washington Medicine 5 : 3–10

Marx JL (1989) Many gene changes found in cancer. Science 246 : 1386–1388

Marx JL (1985) Making mutant mice by gene transfer. Science 228 : 1516–1517

Mason AJ, Pitts SL, Nicolics K et al (1986) The hypogonadal mouse: reproductive functions restored by gene therapy. Science 234 : 1372–1378

May KK, Jacobs PA, Lee M et al (1990) The parental origin of the extra chromosome in 47, XXX females. Am J Hum Genet 46 : 754–761

Maynard-Smith S, Penrose LS, Smith CAB (1961) Mathematical tables for research workers in human genetics. J and A Churchill, London

Mayr E (1985) The growth of biological thought. Harvard University Press, Cambridge, Mass

McConkey EH (1993) Human genetics. The molecular revolution. Jones and Bartlett, Boston

McLaren A (1985) Prenatal diagnosis before implantation: opportunities and problems. Pren Diag 5 : 85–90

McMichael A, McDewitt H (1977) The association between the HLA system and disease. Prog Med Genet [New Series] 2 : 39–100

Medvedev Z (1977) Soviet genetics: new controversy. Nature 268 : 285–287

Mendelsohn ML (1989) Potential DNA methods for measuring the human heritable mutation rate. Genome 31 : 860–863

Mielke JH, Crawford MH (eds) (1982) Current developments in anthropological genetics, vol 1, 2. Plenum, New York

Mikkelsen M, Stene J (1970) Genetic counseling in Down's syndrome. Hum Hered 20 : 457–464

Miller LH (1994) Impacts of malaria on genetic polymorphism and genetic diseases in Africans and African Americans. Proc Natl Acad Sci USA 91 : 2415–2419

Miller M, Opheim KE, Raisys VA, Motulsky AG (1984) Theophylline metabolism: variation and genetics. Clin Pharmacol Therap 35 : 170–182

Minder EI, Meier PJ, Muller HK, Minder C, Meyer UA (1984) Bufuralol metabolism in human liver: a sensitive probe for the debrisoquine-type polymorphism of drug oxidation. Eur J Clin Invest 14 : 184–189

Mitelman F (1991) Catalog of chromosomal aberrations in cancer. Wiley-Liss, New York

Moll PP, Berry TD, Weidman WH, Ellefson R, Gordson H, Kottke BA (1984) Detection of genetic heterogeneity among pedigrees through complex segregation analysis: an application to hypercholesterolemia. Am J Hum Genet 36 : 197–211

Mollon JD, Jordan G (1988/89) Eine evolutionäre Interpretation des menschlichen Farbensehens. Die Farbe 35/36 : 139

Monnat RJ, Hackmann AFM, Chiaverotti TA (1992) Nucleotide sequence analysis of human hypoxanthine phosphoribosyltransferase (HPRT) gene deletions. Genomics 13 : 777–787

Morgan RA, Anderson WF (1993) Human gene therapy. Annu Rev Biochem 62 : 191

Morton NE (1983) Outline of genetic epidemiology. Karger, Basel

Morton NE, Keats BJ, Jacobs PA et al (1990) A centromer map of the X chromosome from trisomies of maternal origin. Ann Hum Genet 54 : 39–47

Morton NE, Lindsten J (1976) Surveillance of Down's syndrome as a paradigm of population monitoring. Hum Hered 26 : 360–371

Motulsky AG (1964) Current concepts of the genetics of the thalassemias. Cold Spring Harbor Symp Quant Biol 29 : 399–413

Motulsky AG (1968) Human genetics, society, and medicine. J Hered 59 : 329–336

Motulsky AG (1974) Brave new world? Current approaches to prevention, treatment, and research of genetic diseases raise ethical issues. Science 185 : 683–663

Motulsky AG (1976) The genetic hyperlipidemias. N Engl J Med 294 : 823–827

Motulsky AG (1977) The George M Kober lecture. A genetical view of modern medicine. Trans Assoc Am Physicians 40 : 76–90

Motulsky AG (1979) Possible selective effects of urbanization on Ashkenazi Jewish populations. In: Goodman RM, Motulsky AG (eds) Genetic diseases among Ashkenazi Jews. Raven, New York, pp 201–212

Motulsky AG (1980) Approaches to the genetics of common diseases. In: Rotter JI, Samloff IM, Rimoin DL (eds) The genetics and heterogeneity of common gastrointestinal disorders. Academic, New York, pp 3–10

Motulsky AG (1980) Ashkenazi Jewish gene pools: admixture, drift and selection. In: Eriksson AW, Forsius H, Nevanlinna HR, Workman PL, Norio RK (eds) Population structure and genetic disorders. Academic, London, pp 353–365

Motulsky AG (1981) Some research approaches in psychiatric genetics. In: Gershon EL, Matthysse S, Breakefield XO, Ciranello RD (eds) Genetic strategies in psychobiology and psychiatry. Boxwood, Pacific Grove, CA, pp 423–428

Motulsky AG (1982) Genetic counseling. In: Wyngaarden JB, Smith LH Jr (eds) Cecil textbook of medicine, 16th edn, Saunders, Philadelphia, pp 23–26

Motulsky AG (1982) Interspecies and human genetic variation, problems of risk assessment in chemical mutagenesis and carcinogenesis. In: Bora KC, Douglas GR, Nestmann ER (eds) Chemical mutagenesis, human population monitoring and genetic risk assessment. (Progress in Mutation Research, vol 3), Elsevier Biomedical, Amsterdam, pp 75–83

Motulsky AG (1984) Editorial: genetic epidemiology. Gen Epidemiol 1 : 143–144

Motulsky AG (1984) Environmental mutagenesis and disease in human populations. In: Chu EHY, Generoso WM (eds) Mutation, cancer, and malformation. Plenum, New York, pp 1–11

Motulsky AG (1984) Genetic engineering, medicine and medical genetics. Biomedicine and Pharmacotherapy 38 : 185–186

Motulsky AG (1984) Genetic research in coronary heart disease. In: Rao DC, Elston RC, Kuller LH, Feinleib M, Carter C, Havlik R (eds) Genetic epidemiology of coronary heart disease: past, present, and future. Liss, New York, pp 541–548

Motulsky AG (1984) Medical genetics. JAMA 252 : 2205–2208

Motulsky AG (1984) The "new genetics" in blood and cardiovascular research: applications to prevention and treatment. Circulation 70[Suppl III]:26–30

Motulsky AG (1985) Hereditary syndromes involving multiple organ systems. In: Wyngaarden JB, Smith LH (eds) Cecil textbook of medicine. Saunders, Philadelphia, pp 1172–1173

Motulsky AG, Boman H (1975) Genetics and atherosclerosis. In. Schettler G, Weizel A (eds) Atherosclerosis, vol III. Springer, Berlin Heidelberg New York, pp 438–444

Motulsky AG, Boman H (1975) Screening for the hyperlipidemias. In. Milunsky A (ed) The prevention of genetic disease and mental retardation. Saunders, Philadelphia, pp 306–316

Motulsky AG, Fraser GR (1980) Effects of antenatal diagnosis and selective abortion on frequencies of genetic disorders. Clin Obstet Gynecol 7 : 121–134

Motulsky AG, Murray JC (1983) Conference summary: current concepts of hemoglobin genetics. In: Bowman JE (ed) Distribution and evolution of hemoglobin and globin loci. Elsevier, New York, pp 345–355

Muller HJ (1941) Induced mutations in drosophila. In: Genes and chromosomes, structure and organization. Cold Spring Harbor Symp on Quant Biol vol IX:151–167

Mulley JC, Kerr B, Stevenson R, Lubs H (1992) Nomenclature guidelines for X-linked mental retardation. Am J Med Genet 43 : 383–391

Mulley JC, Yu S, Gedeon AK et al (1992) Experience with direct molecular diagnosis of fragile X. J Med Genet 29 : 368–374

Murphy EA (1981) Only authorized persons admitted: the quantitative genetics of health and disease. Johns Hopkins Med J 148 : 114–122

Murphy EA (1982) Muddling, meddling, and modeling. In. Anderson VE et al (eds) Genetic basis of the epilepsies. Raven, New York, pp 333–348

Nance WE, McConnell FE (1973) Status and prospects of research in hereditary deafness. Adv Hum Genet 4 : 173–250

Nasse CF (1820) Von einer erblichen Neigung zu tödlichen Blutungen. Horns Archiv, p 385

Neel JV (1966) Between two worlds. Am J Hum Genet 18 : 3–20

Neel JV (1981) Genetic effects of atomic bombs. Science 213 : 1206

Neel JV (1982) The wonder of our presence here: a commentary on the evolution and maintenance of human diversity. Perspec Biol Med 25 : 518–558

Neel JV, Schull WJ (1954) Human heredity. University of Chicago Press, Chicago

Neel JV, Schull WJ et al (1956) The effect of exposure to the atomic bombs on pregnancy termination in Hiroshima and Nagasaki. Natl Acad Sci Natl Res Counc Publ, Washington DC, 461

Nei M (1975) Molecular population genetics and evolution. North-Holland, Amsterdam

Nei M, Li WH (1979) Mathematical model for studying genetic variation in terms of restriction endonucleases. Proc Natl Acad Sci USA 76 : 5269–5273

New SI, Levine LS (1973) Congenital adrenal hypoplasia. Adv Hum Genet 4 : 251–326

Newton CR, Graham A (1994) PCR. Bios Scientific, Oxford

Niebuhr E (1974) Triploidy in man. Hum Genet 21 : 103–125

Nürnberg P, Roewer L, Neitzel H et al (1989) DNA fingerprinting with the oligonucleotide probe (CAC)₅/(GTG)₅: somatic stability and germline mutations. Hum Genet 84 : 75–78

Nyhan WL, Sakati NA (1987) Diagnostic recognition of genetic disease. Lea and Febiger, Philadelphia

Omenn GS, Gelboin HV (eds) (1984) Genetic Variability in responses to chemical exposure. Cold Spring Harbor Laboratory

Orr HT, Chung M-Y, Banfi S et al (1993) Expansion of an instable trinucleotide CAG repeat in spinocerebellar ataxia type 1. Nature Genetic 4 : 221–226

Padmos MA, Roberts GT, Sackey K et al (1991) Two different forms of homozygous sickle cell disease occur in Saudi Arabia. Br J Hematol 79 : 93–98

Partridge TA, Morgan JE, Coulton GR et al (1989) Conversion of mdx myofibres from dystrophin-negative to positive by injection of normal myoblasts. Nature 337 : 176–179

Pearson M, Rowley JD (1985) The relation of oncogenesis and cytogenetics in leukemia and lymphoma. Ann Rev Med 36 : 471–483

Pembrey M, Fennell SJ, van den Berghe J et al (1989) The association of Angelman's syndrome with deletions within 15 q 1–13. J Med Genet 26 : 73–77

Penrose LS (1953) The genetical background of common diseases. Acta Genet (Basel) 4 : 257–265

Penrose LS, Quastel JH (1937) Metabolic studies in phenylketonuria. Biochem J 31 : 266–271

Powars D, Hiti A (1993) Sickle cell anemia: βˢgene cluster haplotypes as genetic markers for severe disease expression. AJDC 147 : 1197–1202

Price RA, Kidd KK, Cohen DJ et al (1985) A twin study of Tourette syndrome. Arch Gen Psychiatry 42 : 815–820

Raming K, Krieger J, Strotmann J et al (1993) Cloning and expression of odorant suppressors. Nature 361 : 353–356

Rao DC, Morton NE, Elston RC, Yee S (1977) Causal analysis of academic performance. Behav Genet 7 : 147–159

Rao PN, Johnson RT, Sperling K (1982) Premature chromosome condensation. Application in basic, clinical and mutation research. Academic, New York

Reed TE, Kalant H, Gibbins RJ, Kapur BM, Rankin JG (1976) Alcohol and aldehyde metabolism in Caucasians, Chinese and Amerinds. Can Med Assoc J 115 : 851–855

Reik W (1988) Genomic imprinting: a possible mechanism for the parental origin effect in Huntington's chorea. J Med Genet 25 : 805–808

Reilly P (1975) Genetic screening legislation. Adv Hum Genet 5 : 319–376

Reinisch JM (1981) Prenatal exposure to synthetic progestins increases potential for aggression in humans. Science 211 : 1171–1173

Reiss AL, Freund L (1990) Fragile X syndrome. Biol Psychiatry 27 : 223–240

Renwick JH (1956/57) Nail-patella syndrome: evidence for modification by alleles at the main locus. Ann Hum Genet 21 : 159–169

Rhoads GG, Jackson LG, Schlesselman SE et al (1989) The safety and efficacy of chorionic villus sampling for early prenatal diagnosis of cytogenetic abnormalities. N Engl J Med 320 : 609–617

Rieger R, Michaelis A, Green MM (1968) A glossary of genetics and cytogenetics. Springer, Berlin Heidelberg New York

Rimoin DL (1975) The chondrodystrophies. Adv Hum Genet 5 : 1–118

Rimoin DL, Schimke RN (1971) Genetic disorders of the endocrine glands. Mosby, St Louis

Risch N (1991) A note on multiple testing procedures in linkage analysis. Am J. Hum Genet 46 : 1058–1064

Risch N, Baron M (1982) X-linked and genetic heterogeneity in bipolar-related major affective illness: reanalysis of linkage data. Ann Hum Genet 46 : 153–166

Roberts RG, Bobrow M, Bentley DR (1992) Point mutations in the dystrophin gene. Proc Natl Acad Sci USA 89 : 2331–2335

Rodewald A (1992) Chromosomale Polymorphismen: geographische und ethnische Variabilität. Homo 43 : 72–96

Roots I, Drakoulis N, Brockmöller J (1992) Polymorphic enzymes and cancer risk: Concepts, methodology and data review. In: Pharmacogenetics of drug metabolism. Pergamon, New York, pp 815–841

Rosenberg SA, Aebersold P, Cornetta K et al (1990) Gene transfer into humans-immunotherapy of patients with advanced melanoma, using tumor-infiltrating lymphocytes modified by retroviral gene transduction. N Engl J Med 323 : 570–578

Roth EF Jr, Raventos-Suarez C, Rinaldi A, Nagel RL (1983) Glucose-6-phosphate dehydrogenase deficiency inhibits in vitro growth of plasmodium falciparum. Proc Natl Acad Sci USA 80 : 298

Rothschild H (ed) (1981) Biocultural aspects of disease. Academic, New York

Rowley JD, Testa JR (1983) Chromosome abnormalities in malignant hematologic diseases. Adv Cancer Res 36 : 103–148

Rüdiger HW, Dreyer M (1983) Pathogenetic mechanisms of hereditary diabetes mellitus. Hum Genet 63 : 100–106

Russell WL (1965) Effect of the interval between irradiation and conception on mutation frequency in female mice. Proc Natl Acad Sci USA 54 : 1552–1557

Rutter WJ (1984) Molecular genetics and individuality. In: Fox SW (ed) Individuality and determinism. Chemical and biological bases. Plenum, New York, pp 61–71

Salzano FM (ed) (1975) The role of natural selection in human evolution. North-Holland/American Elsevier, Amsterdam New York

Sandberg A (ed) (1983) Cytogenetics of the mammalian X chromosome, vols 1,2. Liss, New York

Sanghvi LD, Balakrishnan V (1972) Comparison of different measures of genetic distance between human populations. In: Weiner JS, Huizinga J (eds) The assessment of population affinities in man. Clarendon, Oxford, pp 25–36

Sankaranarayanan K (1982) Genetic effects of ionizing radiation in multicellular eukaryotes and the assessment of genetic radiation hazards in man. Elsevier, Amsterdam

Scarr S (1981) Race, social class, and individual differences in I.Q. Erlbaum, Hillsdale

Schmidtke J, Cooper DN (1983) A list of cloned DNA sequences. Hum Genet 65 : 19–26

Schmidtke J, Cooper DN (1984) A list of cloned human DNA sequences – Supplement. Hum Genet 67 : 111–114

Schnedl W (1974) Banding patterns in human chromosomes visualized by Giemsa staining after various pretreatments. In: Schwarzacher HG, Wolf U (eds) Methods in human cytogenetics. Springer, New York Heidelberg Berlin, pp 95–116

Schneider EL, Epstein CJ (1972) replication rate and life span of cultured fibroblasts in Down's syndrome. Proc Soc Exp Biol Med 141 : 1092–1094

Schroeder TM, Kurth R (1971) Analytical review. Spontaneous chromosomal breakage and high incidence of leukemia in inherited disease. Blood 37 : 96

Schroeder T-M, Tilgen D, Krüger J, Vogel F (1976) Formal genetics of Fanconi's anemia. Hum Genet 32 : 257–288

Scriver CR (1980) Predictive medicine: a goal for genetic screening. In: Bickel H, Guthrie R, Hammersen G (eds) Neonatal screening for inborn errors of metabolism. Springer, Berlin Heidelberg New York

Sedano HO, Sauk JJ, Gorlin RJ (1977) Oral manifestations of inherited disorders. Butterworths, Boston

Setlow JK, Hollaender A (eds) (1979–1983) Genetic engineering, vols 1–5. Plenum, New York

Shoffner IV JM, Wallace DC (1990) Oxidative phosphorylation diseases: disorders of two genomes. Adv in Hum Genet 19 : 267–330

Shoffner JM, Brown MD, Torroni A et al (1993) Mitochondrial DNA variants observed in Alzheimer disease and Parkinson disease patients. Genomics 17 : 171–184

Shows TB, Sakaguchi AY, Naylor SL (1982) Mapping the human genome, cloned genes, DNA polymorphisms and inherited disease. Adv Hum Genet 12 : 341–452

Sibley DR, Monsma FJ Jr (1992) Molecular biology of dopamine receptors. Trends in Pharmacol Sci 13 : 61–69

Simpson GG (1951) Zeitmaße und Ablaufformen der Evolution. Musterschmidt, Göttingen

Sing CF, Zerba KE, Reilly SL (1994) Traversing the biological complexity in the hierarchy between genome and CAD endpoints in the population at large. Clinical Genetics 46 : 6–14

Sinsheimer R (1977) An evolutionary perspective for genetic engineering. New Scientist 20 : 150

Sinsheimer RL (1977) Recombinant DNA. Ann Rev Biochem 46 : 415–438

Smith CAB (1956–1957) Counting methods in genetical statistics. Ann Hum Genet 21 : 254–276

Smith DW (1982) recognizable patterns of human malformation, 3rd edn. Saunders, Philadelphia

Smith DW, Patau K, Therman E, Inhorn SL (1960) A new autosomal trisomy syndrome: multiple congenital anomalies caused by an extra chromosome. J Pediatr 57 : 338–345

Smith HO (1979) Nucleotide sequence specificity of restriction endonucleases. Science 205 : 455–462

Smithies O (1964) Chromosomal rearrangements and protein structure. Cold Spring Harbor Symp Quant Biol 29 : 309

Sofaer JA, Emery AE (1981) Genes for super-intelligence? J Med Genet 18 : 410–413

Solari AJ (1980) Synaptonemal complexes and associated structures in microspread spermatocytes. Chromosome 81 : 315–337

Solomon SD, Geisterfer-Lorance AAT, Vosberg HP et al (1990) A locus for familial hypertrophic cardiomyopathy is closely linked to the cardiac myosin heavy chain genes, CRI-L436 and CRI-L329 on chromosome 14 at q11-q12. Am J Hum Genet 47 : 389–394

Sommer SS (1992) Assessing the underlying pattern of human germline mutations: lessons from the factor VIII gene. FASEB J 6 : 2767–2774

Stamatoyannopoulos G (1979) Possibilities for demonstrating point mutations in somatic cells, as illustrated by studies of mutant hemoglobins. In: Berg K (ed) Genetic damage in man caused by environmental agents, Academic, New York, pp 49–62

Stamatoyannopoulos G (1991) Human hemoglobin switching. Science 252 : 383

Stamatoyannopoulos G, Nienhuis AW (eds) (1985) Experimental approaches for the study of hemoglobin switching. Liss, New York

Stamatoyannopoulos G, Nienhuis AW, Majerus PW, Varmus H (1990) The molecular basis of blood diseases, 2nd edn. WB Saunders, Philadelphia

Starlinger P, Saedler H (1972) Insertion mutations in microorganisms. Biochemie 53 : 177–185

Steinberg AG, Bearn AG (eds) (1961–1974) Progress in medical genetics, vols I-X. Grune and Stratton, New York

Steinberg AG, Bearn AG, Motulsky AG, Childs B (eds) (1985-) Progress in medical genetics (new series). Saunders, Philadelphia

Steptoe PC, Edwards RG (1978) Birth after the reimplantation of a human embryo. Lancet II:366

Steriade M, Gloor P, Llinás RR et al (1990) Basic mechanisms of cerebral rhythmic activities. EEG Clin Neurophysiol 76 : 481–508

Stern C (1936) Somatic crossing-over and segregation in Drosophila melanogaster. Genetics 21 : 625–730

Stewart RE, Prescott GH (1976) Oral facial genetics. Mosby, St Louis

Strömgren E (1967) Neurosen und Psychopathien. In: Becker PE (ed) Humangenetik, ein kurzes Handbuch, vol V/2. Thieme, Stuttgart, pp 578–598

Sutherland GR (1982) Heritable fragile sites on human chromosomes. VIII. Preliminary population cytogenetic data on the folic acid sensitive fragile sites. Am J Hum Genet 34 : 452–458

Szybalska EH, Szybalski W (1962) Genetics of human cell lines. IV. DNA-mediated heritable transformation of a biochemical trait. Proc Natl Acad Sci USA 48 : 2026–2034

Tattersall R (1976) The inheritance of maturity-onset type diabetes in young people. In: Creutzfeldt E, Köbberling J, Neel JV (eds) The genetics of diabetes mellitus. Springer, Berlin Heidelberg New York, pp 88–95

ten Kate LP, Boman H, Daiger SP, Motulsky AG (1982) Familial aggregation of coronary heart disease and its relation to known genetic risk factors. Am J Cardiol 50 : 945–953

Terwilliger JD, Ott J (1994) Handbook of human genetic linkage. Johns Hopkins University Press, Baltimore

Thalhammer O, Havelec L, Knoll E, Wehle E (1977) Intellectual level (IQ) in heterozygotes for phenylketonuria (PKU). Hum Genet 38 : 285–288

Therman E, Meyer-Kuhn E (1981) Mitotic crossing-over and segregation in man. Hum Genet 59 : 93–100

Therman E, Susman B (1990) The similarity of phenotypic effects caused by Xpand Xq deletions in the human female: a hypothesis. Hum Genet 85 : 175–183

Thompson MW, McInnes RR, Willard HF (1991) Genetics in medicine, 5th edn. Saunders, Philadelphia

Tinbergen N (1968) On war and peace in animals and men. Science 1960 : 1411–1418

Usdin E, Mandell AJ (eds) (1978) Biochemistry of mental disorders. Dekker, New York

Utermann G (1990) Coronary heart disease. In: Emery AEH, Rimoin DL (eds) Principles and practice of medical genetics, 2nd edn. Churchill Livingstone, Edinburgh, pp1239–1262

Van Tol HHM, Wu CM, Guan H-C et al (1992) Multiple dopamine D4 receptor variants in the human population. Nature 358 : 149–152

Vesell ES (1992) Pharmacogenetic perspectives gained from twin and family studies. In: Pharmacogenetics of drug metabolism. Pergamon, New York, pp 843–864

Viskochil D, White R, Cawthorn R (1993) The neurofibromatosis type 1 gene. Ann Rev Neurosciences 16 : 183–205

Vogel F (1958) Gedanken über den Mechanismus einiger spontaner Mutationen beim Menschen. Z Menschl Vererbungs-Konstitutionslehre 34 : 389–399

Vogel F (1958) Über die Erblichkeit des normalen Elektroencephalogramms. Thieme, Stuttgart

Vogel F (1975) Mutations in man. Approaches to an evaluation of the genetic load due to mutagenic agents in the human population. Mutation Res 29 : 263–269

Vogel F (1990) Mutation in Man. In: Emery AEH, Rimoin DL (eds) Principles and practice of medical genetics, 2nd ed. Churchill Livingstone, Edinburgh

Wald NJ (ed) (1984) Antenatal and neonatal screening. Oxford University Press, Oxford

Wald NJ, Cuckle HS (1984) Open neural tube defects. In: Wald NJ (ed) Antenatal and neonatal screening. London

Wallace B (1981) Basic population genetics. Columbia University Press, New York

Walter H (1976) Körperbauform und Klima. Kritische Überlegungen zur Übertragbarkeit der Bergmann'schen Regel auf den Menschen. Z Morphol Anthropol 67 : 241–263

Warkany J, Lemire RJ, Cohen MM (1981) Mental retardation and congenital malformations of the central nervous system. Year Book, Chicago

Watkins Winifred M (1966) Blood-group substances. Science 152 : 172–181

Weatherall DJ, Wood WG, Jones RW, Clegg JB (1985) The developmental genetics of human hemoglobin. In: Stamatoyannopoulos G, Nienhuis AW (eds). Experimental approaches for the study of hemoglobin switching. Liss, New York, pp 3–25

Weinberg RA (1989) The Rb gene and the negative regulation of cell growth. Blood 74 : 529–532

Weinberger DR (1988) Schizophrenia and the frontal lobe. TINS 11 : 367–370

Weiner AM, Deininger PL, Efstradiatis A (1986) Non-viral retroposons: genes, pseudogenes, and transposable elements generated by the reverse flow of genetic information. Ann Rev Biochem 55 : 631–661

Weiner W, Lewis HBM, Moores P, Sanger R, Race RR (1957) A gene, y, modifying the blood group antigen A. Vox Sang 2 : 25–37

Weitz CJ, Miyake Y, Shinzato K et al (1992) Human tritanopia associated with two amino acids substitutions in the blue-sensitive opsin. Am J Hum Genet 50 : 496

Went LN, Pronk N (1985) The genetics of tritan disturbances. Hum Genet 69 : 255

Wertz DC, Fletcher J (eds) (1989) Ethics and human genetics. A cross-cultural perspective. Springer, Berlin Heidelberg New York

White GC, McMillan CW, Kingdon HS, Shoemaker CB (1989) Use of recombinant antihemophilic factor in the treatment of two patients with classic hemophilia. New Engl J Med 320 : 166–170

Wiedemann H-R, Grosse F-R, Dibborn H (1982) Das charakteristische Syndrom. Schattauer, Stuttgart

Wienberg J (1990) Molecular cytotaxonomy of primates by chromosomal in situ suppression hybridization. Genomics 8 : 47–59

Wiener AS (1941) Hemolytic reactions following transfusion of blood of the homologous group II. Arch Pathol 32 : 227–250

Williams RR, Hunt SC, Hopkins PN, Wu LL, Schumacher MC, Stults BM, Ball L, Ware J, Hasstedt SJ, Lalouel JM (1992) Evidence for gene-environmental interactions in Utah families with hypertension, dyslipidemia and early coronary heart disease. Clin Exp Pharmacol Physiol Suppl 20 : 1–6

Williams RJ (1956) Biochemical individuality. Wiley, New York

Williams RR, Hunt SC, Hopkins PN et al (1994) Evidence for single gene contributions to hypertension and lipid disturbances: definition, genetics, and clinical significance. Clinical Genetics 46 : 80–87

Wilson AF, Elston RC, Tran LD, Siervogel RM (1991) Use of the robust sib-pair method to screen for single-locus, multiple-locus, and pleiotropic effects: application to traits related to hypertension. Am J Hum Genet 48 : 862–872

Winkler U (1972) Spontaneous mutations in bacteria and phages. Hum Genet 16 : 19–26

Winter RM, Pembrey ME (1982) Does unequal crossing over contribute to the mutation rate in Duchenne muscular dystrophy? Am J Med Genet 12 : 437–441

Witkowski R, Prokop O, Ullrich E (1995) Lexikon der Syndrome und Fehlbildungen. Springer, Berlin Heidelberg New York

Wivel NA (1994) Gene therapy: molecular medicine of the 1990's. Int J Technol Assess Health Care 10 : 655–663

Wolf U, Reinwein H, Porsch R, Schröter R, Baitsch H (1965) Defizienz an den kurzen Armen eines Chromosoms Nr 4. Hum Genet 1 : 397–413

Wolfe KH (1991) Mammalian DNA replication: mutation biases and the mutation rate. J Theoret Biol 149 : 441–451

Wolff JA, Malone RW, Williams P et al (1990) Direct gene transfer into mouse muscle in vivo. Science 247 : 1465–1468

Wood L, Trounson A (1984) Clinical in vitro fertilization. Springer, Berlin Heidelberg New York

Wood WG, Clegg JB, Weatherall DJ (1977) Developmental biology of human hemoglobins. In: Brown EB (ed) Progress in hematology, vol X. Grune and Stratton, New York, pp 43–90

Wood WI, Capon DJ, Simonsen CC, Eaton DL, Gitschier J, Keyt B, Seeburg PH, Smith DH, Hollingshead P, Wion KL, Delwart E, Tuddenham EGD, Vehar GA, Lawn RM (1984) Expression of active human factor VIII from recombinant DNA clones. Nature 312 : 330–337

Work TS, Burden RH (eds) (1983) Laboratory techniques in biochemistry and molecular biology. Elsevier, Amsterdam

World Health Organization Working Group (1983) Community control of hereditary anaemias. Bull WHO 61 : 63–80 (1983). Also published in French, Bull WHO 61 : 277–297

Wyngaarden JB, Smith LH (eds) (1985) Cecil textbook of medicine. Saunders, Philadelphia

Yoshida A (1982) Molecular basis of difference in alcohol metabolism between Orientals and Caucasians. Jpn J Hum Genet 27 : 55–70

Young W, Goy RW, Phoenix CH (1964) Hormones and sexual behavior. Science 143 : 212–218

Zacharov AF, Benusch VA, Kuleshov NP, Baranowskaya LI (1982) Human chromosomes (atlas) (in Russian). Meditsina, Moscow

Zaleski MB, Dubiski S, Niles EG, Cunningham RK (1983) Immunogenetics. Pitman, Boston

Zankl H, Zang KD (1972) Cytological and cytogenetical studies on brain tumors. IV. Identification of the missing G chromosome in human meningiomas as no 22 by fluorescence technique. Hum Genet 14 : 167–169

Zerbin-Rüdin E (1967) Idiopathischer Schwachsinn. In: Bekker PE (ed) Humangenetik, ein kurzes Handbuch, vol V/2. Thieme, Stuttgart, pp 158–205

Subject Index